New Frontiers in Operational Oceanography

Edited by
Eric P. Chassignet, Ananda Pascual, Joaquin Tintoré, and Jacques Verron

GODAE OceanView
Tallahassee, Florida

Copyright © 2018 by GODAE OceanView.

All rights reserved. No part of this publication may be reproduced, distributed or transmitted in any form or by any means, including photocopying, recording, or other electronic or mechanical methods, without the prior written permission of the publisher, except in the case of brief quotations embodied in critical reviews and certain other noncommercial uses permitted by copyright law. For permission requests, write to the publisher, addressed "Attention: Permissions Coordinator," at the address below.

GODAE OceanView
c/o Eric P. Chassignet
Center for Ocean-Atmospheric Prediction Studies
Florida State University
2000 Levy Avenue
Building A, Suite 292
Tallahassee, FL 32306-2741
www.godae-oceanview.org

New Frontiers in Operational Oceanography / [Edited by] Eric P. Chassignet, Ananda Pascual, Joaquin Tintoré, & Jacques Verron. —1st ed.
ISBN-13: 978-1720549970
ISBN-10: 1720549974
doi:10.17125/gov2018

Cover photo credit: Surface Reflectance and Ocean Temperature, Jacques Descloitres, MODIS Land Rapid Response Team, NASA/GSFC, visibleearth.nasa.gov/view.php?id=55878.

Contents

Preface v
Eric P. Chassignet, Ananda Pascual, Joaquin Tintoré, Jacques Verron

List of Lecturers and Students vii

Introduction

1. An Overview of Operational Oceanography 1
Andreas Schiller, Baptiste Mourre, Yann Drillet, Gary Brassington

2. Notions for the Motions of the Oceans 27
Baylor Fox-Kemper

Observations

3. The Global Ocean Observing System 75
Bernadette M. Sloyan, Moninya Roughan, Katherine Hill

4. Shelf and Coastal Ocean Observing and Modelling Systems: A New Frontier in Operational Oceanography 91
Moninya Roughan, Colette Kerry, Peter McComb

5. Multi-Platform Observations and Numerical Simulations to Understand Meso and Submesoscale Processes 117
Simón Ruiz, Amala Mahadevan, Ananda Pascual, Mariona Claret, Joaquín Tintoré, Evan Mason

6. Biogeochemical In Situ Observations – Motivation, Status and New Frontiers 131
Maciej Telszewski, Artur Palacz, Albert Fischer

7. Satellites and Operational Oceanography 161
Pierre-Yves Le Traon

8. Fine-scale Altimetry and the Future SWOT Mission 191
Rosemary Morrow, Denis Blumstein, Gerald Dibarboure

9. An Operational Interpolated Ocean Colour Product in the Mediterranean Sea 227
Gianluca Volpe, Bruno Buongiorno Nardelli, Simone Colella, Andrea Pisano, Rosalia Santoleri

10. Turbulent Heat Fluxes and Wind Remote Sensing 245
Mark A. Bourassa, Paul J. Hughes

11. Satellite SST and SSS Observations and Their Role to Constrain Ocean Models 271
Tong Lee, Chelle Gentemann

Modeling

12. Ocean Circulation Modelling for Operational Oceanography: Current Status and Future Challenges   289
Julien Le Sommer, Eric P. Chassignet, Alan J. Wallcraft

13. A Primer on Global Internal Tide and Internal Gravity Wave Continuum Modeling in HYCOM and MITgcm   307
Brian K. Arbic, Matthew H. Alford, Joseph K. Ansong, Maarten C. Buijsman, Robert B. Ciotti, J. Thomas Farrar, Robert. W. Hallberg, Christopher E. Henze, Christopher N. Hill, Conrad A. Luecke, Dimitris Menemenlis, E. Joseph Metzger, Malte Müeller, Arin D. Nelson, Bron C. Nelson, Hans E. Ngodock, Rui M. Ponte, James G. Richman, Anna C. Savage, Robert B. Scott, Jay F. Shriver, Harper L. Simmons, Innocent Souopgui, Patrick G. Timko, Alan J. Wallcraft, Luis Zamudio, Zhongxiang Zhao

14. Wind Waves   393
Fabrice Ardhuin, Alejandro Orfila

15. Sea Ice Modelling and Forecasting   423
Sylvain Bouillon, Pierre Rampal, Einar Olason

16. Coupled Atmosphere-Ocean Modelling   445
Chris Harris

Data Assimilation

17. Data Assimilation in Oceanography: Status and New Directions   465
Ibrahim Hoteit, Xiaodong Luo, Marc Bocquet, Armin Köhl, Boujemaa Ait-El-Fquih

18. Operational Ocean Data Assimilation   513
Gregg A. Jacobs, Charlie N. Barron, Cheryl A. Blain, Matthew J. Carrier, Joseph M. D'Addezio, Robert W. Helber, Jackie C. May, Hans E. Ngodock, John J. Osborne, Mark D. Orzech, Clark D. Rowley, Innocent Souopgui, Scott R. Smith, Jay Veeramony, Max Yaremchuk

19. Ocean Reanalysis   545
Keith Haines

Ocean Prediction Systems and Applications

20. The Mercator Ocean Global High Resolution Monitoring and Forecasting System   563
Jean-Michel Lellouche, Eric Greiner, Olivier Le Galloudec, Charly Régnier, Mounir Benkiran, Charles-Emmanuel Testut, Romain Bourdallé-Badie, Marie Drévillon, Gilles Garric, Yann Drillet

21. A Coastal Ocean Forecast System for U.S. Mid-Atlantic Bight and Gulf of Maine   593
John Wilkin, Julia Levin, Alexander Lopez, Elias Hunter, Javier Zavala-Garay, Hernan Arango

22. Marine Biogeochemical Modelling and Data Assimilation for Operational   625
Forecasting, Reanalysis and Climate Research
David Ford, Susan Kay, Robert McEwan, Ian Totterdell, Marion Gehlen

23. Understanding and Predicting El Niño and Southern Oscillation   653
Michael J. McPhaden

24. Assessment of High-Resolution Regional Ocean Prediction Systems Using   663
Multi-Platform Observations: Illustrations in the Western Mediterranean Sea
Baptiste Mourre, Eva Aguiar, Mélanie Juza, Jaime Hernandez-Lasheras, Emma Reyes, Emma Heslop, Romain Escudier, Eugenio Cutolo, Simon Ruiz, Evan Mason, Ananda Pascual, Joaquin Tintoré

25. Learning about Copernicus Marine Environment Monitoring Service   695
"CMEMS": A Practical Introduction to the Use of the European Operational
Oceanography Service
Marie Drévillon, Pierre Bahurel, David Bazin, Mounir Benkiran, Jonathan Beuvier, Laurence Crosnier, Yann Drillet, Edmée Durand, Michèle Fabardines, Isabel Garcia Hermosa, Cédric Giordan, Elodie Gutknecht, Fabrice Hernandez, Stéphane Law Chune, Pierre-Yves Le Traon, Jean-Michel Lellouche, Bruno Levier, Angelique Melet, Dominique Obaton, Julien Paul, Mathieu Peltier, Diane Peyrot, Elizabeth Rémy, Karina von Schuckmann, Cécile Thomas-Courcoux

26. Operational Oceanography and the Management of Pelagic Resources: The   713
Mediterranean Sea as a Case-Study
Patricia Reglero, Diego Alvarez-Berastegui, Francisco Javier Alemany, Vincent Rossi, Asvin P. Torres, Rosa Balbin, Manuel Hidalgo

27. Operational Oceanography at the Service of the Ports   729
Enrique Alvarez Fanjul, Marcos García Sotillo, Begoña Pérez Gómez, José María García Valdecasas, Susana Pérez Rubio, Pablo Lorente, Álvaro Rodríguez Dapena, Isabel Martínez Marco, Yolanda Luna, Elena Padorno, Inés Santos Atienza, Gabriel Díaz Hernandez, Javier López Lara, Raúl Medina, Manel Grifoll, Manuel Espino, Marc Mestres, Pablo Cerralbo, Agustín Sánchez Arcilla

28. Diagnosis, Prognosis and Management of Jellyfish Swarms   737
Laura Prieto

29. Measuring Performances, Skill and Accuracy in Operational Oceanography: 759
New Challenges and Approaches
Fabrice Hernandez, Greg Smith, Katrijn Baetens, Gianpiero Cossarini, Isabel Garcia-Hermosa, Marie Drévillon, Jan Maksymczuk, Angélique Melet, Charly Régnier, Karina von Schuckmann

Index   797

Preface

The implementation of operational oceanography in the past 15 years has provided many societal benefits and has led to many countries adopting a formal roadmap for providing ocean forecasts. Continuing the tradition of two very successful international summer schools held in France in 2004 (Chassignet and Verron, 2006) and in Australia in 2010 (Schiller and Brassington, 2011), a third international school that focused on frontier research in operational oceanography was held in Majorca in 2017. In the coming years, graduate students and young scientists will be challenged by many new observations (SWOT, Sentinel, AUVs, floats, etc.), complex high-resolution numerical models and data assimilation (high resolution, predictability, uncertainty, changing computing platforms, etc.), and the need to work on many scales (open ocean-shelf interactions, coupled ocean-ice-atmosphere, biogeochemistry, etc.). The latter school brought together senior experts and young researchers (pre- and post-doctorate) from across the world and exposed them to the latest research in oceanography, specifically how it will impact operational oceanography. This book is a compilation of the lectures presented at the school and presents a summary of the current state-of-the-art in operational oceanography research.

Acknowledgements

The authors and editors of this book would like to express their gratitude to Meredith Field and Tracy Ippolito for tirelessly working on the book and editing the chapters.

Eric P. Chassignet, COAPS, Florida State University, Tallahassee, USA
Ananda Pascual, IMEDEA (CSIC-UIB), Esporles, Mallorca, Spain
Joaquin Tintoré, SOCIB & IMEDEA (CSIC-UIB), Palma, Mallorca, Spain
Jacques Verron, IGE, Grenoble, France

August 2018

Note: Access to eBook and Color Figures

For access to the electronic, full-color version of this book, please visit:
https://www.godae-oceanview.org/outreach/education-training/gov-summer-school-2017/publication/

List of Lecturers and Students

Lecturers

First Name	Last Name	Country
Enrique	Alvarez Fanjul	Spain
Brian	Arbic	USA
Fabrice	Ardhuin	France
Mounir	Benkiran	France
Jerome	Benveniste	Italy
Jonathan	Beuvier	France
Sylvain	Bouillon	Norway
Mark	Bourassa	USA
Eric	Chassignet	USA
Marie	Drévillon	France
David	Ford	UK
Baylor	Fox-Kemper	USA
Keith	Haines	UK
Chris	Harris	UK
Fabrice	Hernandez	France
Ibrahim	Hoteit	Saudi Arabia
Gregg	Jacobs	USA
Johannes	Karstensen	Germany
Tong	Lee	USA
Jean-Michel	Lellouche	France
Julien	Le Sommer	France
Pierre-Yves	Le Traon	France
Benjamin	Loveday	UK
Nicolai	Maximenko	USA
Michael	McPhaden	USA
Rosemary	Morrow	France
Baptiste	Mourre	Spain
Ananda	Pascual	Spain
Laura	Prieto	Spain
Alejandro	Orfila	Spain
Patricia	Reglero	Spain
Moninya	Roughan	Australia
Simón	Ruiz	Spain
Andreas	Schiller	Australia
Maciej	Telszewski	Poland
Joaquin	Tintoré	Spain
Jacques	Verron	France
Martin	Visbeck	Germany
Gianluca	Volpe	Italy
John	Wilkin	USA

Students

First Name	Last Name	Country
Eva	Aguiar	Spain
Adekunle	Ajayi	France
Alfatih	Ali	Norway
Jessica	Anderson	USA
Theo	Baracchini	Switzerland
Barbara	Barcelo-Llull	Spain
Filipe	Bitencourt Costa	Brazil
Frederic	Bonou	Benin
Sourav	Chatterjee	India
Nurul Rabitah	Daud	Malaysia
Josephine Dianne	Deauna	Philippines
Matias	Dinapoli	Argentina
Romain	Escudier	USA
Nicholas	Foukal	USA
Beatriz Ixetl	Garcia Gomez	France
Laura	Gomez Navarro	France
Johannes	Hahn	Germany
Jaime	Hernandez Lasheras	Spain
Nariaki	Hirose	Japan
Rodrigue Anicet	Imbol Koungue	South Africa
Md Jakiul	Islam	Bangladesh
Svetlana	Karimova	Belgium
Ahon Jean-Baptiste	Kassi	Cote d'Ivoire
Hanna	Kauko	Norway
Christina Eunjin	Kong	Korea
Georgios	Krokos	Saudi Arabia
Lilian	Krug	Portugal
Oliver	Kruger	Germany
Valerie	Le Guennec	UK
Eva	Le Merle	France
Stephanie	Leroux	France
Leonardo	Lima	Brazil
Georgina	Long	UK
Manuel	Lopez Radcenco	France
Guy Rodier	Mabiala Boutoto	Congo
Anna Katharina	Miesner	Denmark
Diego	Moreira	Argentina
Arnab	Mukherjee	India
Wendy Paola	Navarro Ariza	Columbia
Arielle Stela	Nkwinkwa Njouodo	South Africa
Nektaria	Ntaganou	USA
Erick	Olvera	USA
Jenna	Palmer	USA
Joaquim	Pereira Bento	Chile
Beatriz	Perez Diaz	Spain
Olav Safo	Piñeiro Rodriguez	Spain
Sisi	Qin	China
Marcel	Ricker	Germany

First Name	Last Name	Country
Shaun	Rigby	UK
Alejandra	Rodriguez	Spain
Pablo	Rodriguez Ros	Spain
Aurpita	Saha	Germany
Hao	Sai	China
Rafael	Santana	Brasil
Rene	Schubert	Germany
Timothy	Smith	USA
Natalia	Tilinina	Germany
Virginie	van Dongen-Vogels	Australia
Dakui	Wang	China
Clifford	Watkins	USA
Olmo	Zavala-Romero	Mexico
Peng	Zhan	Saudi Arabia
Xueming	Zhu	China
Mihhail	Zujev	Estonia

CHAPTER 1

An Overview of Operational Oceanography

Andreas Schiller[1], Baptiste Mourre[2], Yann Drillet[3], and Gary Brassington[4]

[1]*CSIRO Oceans and Atmosphere, Hobart, Tasmania, Australia;* [2]*SOCIB, Palma de Mallorca, Spain;* [3]*Mercator Océan, Parc Technologique du Canal, Ramonville-Saint-Agne, France;* [4]*Bureau of Meteorology, Sydney, Australia*

Operational oceanography is like weather forecasting for the ocean, it provides estimates of ocean variables (temperature, currents, surface height, etc.) for the past, present, and future. There is a systematic focus on sustained operational ocean observing systems, estimates of the current state, short-range predictions and ocean reanalyses. Operational oceanography systems provide routine and fully supported production and delivery of oceanographic information at pre-determined and agreed upon service levels. Nowadays, many operational oceanography systems cover global-to-coastal marine environments, and physical and biogeochemical properties, with active research underway to eventually include ecosystems. Operational oceanography involves and benefits marine industries, service providers, government agencies, and research and development (R&D) providers.

Introduction

This chapter provides a high-level overview of the key elements of state-of-the-art operational oceanography systems, including observing systems, modelling, data assimilation, service delivery, applications and benefits derived from operational oceanography. The efforts of the international GODAE OceanView team are being described, which targets consolidation and improvement of global and regional ocean analysis and forecasting systems as well as assessment of contributions to the global ocean observing systems and scientific guidance for its improvement. The chapter concludes with a discussion of current and future research challenges in operational oceanography that aim to increase forecast skill and support Earth system modelling efforts.

What Is Operational Oceanography?

There is no widely accepted, unambiguous definition of "operational oceanography." The European component of the Global Ocean Observing System provides the following working definition: *Operational oceanography* can be defined as the activity of systematic and long-term routine measurements of the seas and oceans and atmosphere, and their rapid interpretation and dissemination. Ocean forecasting as part of operational oceanography is based on the near real-time

---

Schiller, A., et al., 2018: An overview of operational oceanography. In "*New Frontiers in Operational Oceanography*", E. Chassignet, A. Pascual, J. Tintoré, and J. Verron, Eds., GODAE OceanView, 1-26, doi:10.17125/gov2018.ch01.

collection of ocean observations and proceeds by the rapid transmission of observational data to data processing centres, which conduct quality controls and provide the data to forecasting centres. There, powerful computers using numerical ocean forecasting models assimilate the observations into the models to improve and create initial conditions for ocean forecasts, with forecasts typically up to 10 days in advance. Data analysis and ocean forecasting centres are operating at a routine, fully supported production and delivery level of services to user-defined levels. At present, ocean forecasting centres cover global-to-coastal marine environments, and physical and biogeochemical properties. Ecosystems analysis and forecasting are active areas of research.

Outputs from the models are used to generate data products, often through intermediary, value-adding service providers. Examples of final products include optimised shipping routes, coastal flood warnings, and forecasts of harmful algae blooms. Because ocean conditions are constantly changing over time, the final forecasts and products must be distributed rapidly to marine industry, governments, regulatory authorities and the public.

A Brief History of Operational Oceanography

This subsection provides a brief introduction to the history of operational oceanography with a focus on elements that laid the scientific foundation of ocean forecasting, i.e., ocean observations, ocean general circulation models, and data assimilation tools. It then describes the scientific achievements of the first phase of internationally coordinated efforts in the development of global- and basin-scale operational ocean forecasting systems during the Global Ocean Data Assimilation Experiment (GODAE) from 1997–2008.

Three key developments took place in the 1980s and 1990s that had a profound impact on the development of operational oceanography. First, there was the launch of Earth observation satellites designed to observe the oceans and its mesoscale structures. Second, "supercomputers" surpassed a threshold of availability and capability in providing simulations of the ocean general circulation at basin and global scales. Third, adopting advances being made in meteorology enabled oceanographers to leapfrog developments in ocean data assimilation. All three of these (observations, modelling, and data assimilation) are key elements of today's ocean forecasting systems.

Before satellites became more commonly available in the 1980s, oceanographers were "data poor." However, since then, significant technological and scientific advances in satellite remote sensing have made it possible to obtain near real-time measurements of sea surface height anomalies, sea surface temperature (SST), and ocean colour. These key observations have, for the first time, enabled ocean forecasting applications (Fu and Cazenave, 2001). The realisation of a network of 3,000 Argo profiling floats freely reporting temperature and salinity profiles to 2,000 m depth in a timely fashion has transformed the in situ ocean measurement network in the new millennium. This enables oceanographers to continuously monitor the temperature, salinity, and velocity of the upper ocean, with all data being relayed and made publicly available within hours after collection.

The advent of "supercomputers" in the 1970s and 1980s provided researchers with a new capability to describe the ocean within a mathematical framework and allowed the development of the first basin- and global-scale numerical ocean circulation models. Since that time, there has been increased emphasis on the application of computers in oceanography to allow numerical simulations and predictions of the state of the ocean. Based on significant advances in supercomputing technologies, the 1990s saw the emergence of the first large-scale, eddy-resolving models (Semtner and Chervin, 1992) and the first ocean-atmosphere coupled climate change projections (see e.g., IPCC First Assessment Report, 1990).

Over the last 20 years, the global ocean observing system (in situ and remote sensing) has been progressively implemented and led to a revolution in the amount of data available for research and forecasting applications. The ocean observing system, primarily designed to serve climate research, is the backbone for most operational oceanography applications. Although significant progress has been made, sustaining the global ocean observing system remains a challenging task (Clark et al., 2009). This recent progress in the global ocean observing system was complemented by advances in supercomputing technology, allowing for the development and operational implementation of eddy-resolving (~10 km), basin-scale ocean circulation models.

GODAE was established in 1997 with the two primary goals: (i) to demonstrate the feasibility and utility of global ocean monitoring and forecasting on the daily-to-weekly timescale and on eddy-resolving spatial scales, and (ii) to assist in building the infrastructure for global operational oceanography (Smith and Lefebvre, 1997; GODAE Strategic Plan, 2000; Bell et al., 2009). GODAE has had a major impact on the development of global operational oceanography capability. Global modelling and data assimilation systems have been progressively developed, implemented, and inter-compared (Dombrowsky et al., 2009; Cummings et al., 2009; Hernandez et al., 2009). There has been increased attention to the development of products and services and the demonstration of their utility for applications such as marine environment monitoring, weather forecasting, seasonal and climate prediction, ocean research, maritime safety and pollution forecasting, coastal and shelf-sea forecasting, national security, as well as forecasting for the oil and gas industry and fisheries management (Davidson et al., 2009; Hackett et al., 2009; Jacobs et al., 2009). GODAE as an experiment ended in 2008 having achieved most of its goals. It has been demonstrated that global ocean data assimilation is feasible and GODAE made important contributions to the establishment of an effective and efficient infrastructure for global operational oceanography that includes the required observing systems, data assembly and processing centres, modelling and data assimilation centres, and data and product servers.

Components of Operational Oceanography

Overview of Components of Operational Oceanography Systems

Operational Oceanography is comprised of five key components. Fig. 1.1 captures the main functional components required by any state-of-the-art ocean forecasting system. It also illustrates many of the interactions required to ensure or enhance the quality of the ocean forecasting systems and their outputs. These are: the observation networks, data management and monitoring, prediction and assessment, service delivery and dissemination, and uptake of products by end users/clients (and their feedback about fit-for-purpose products to the operational centres). The rest of this chapter is structured in line with these components.

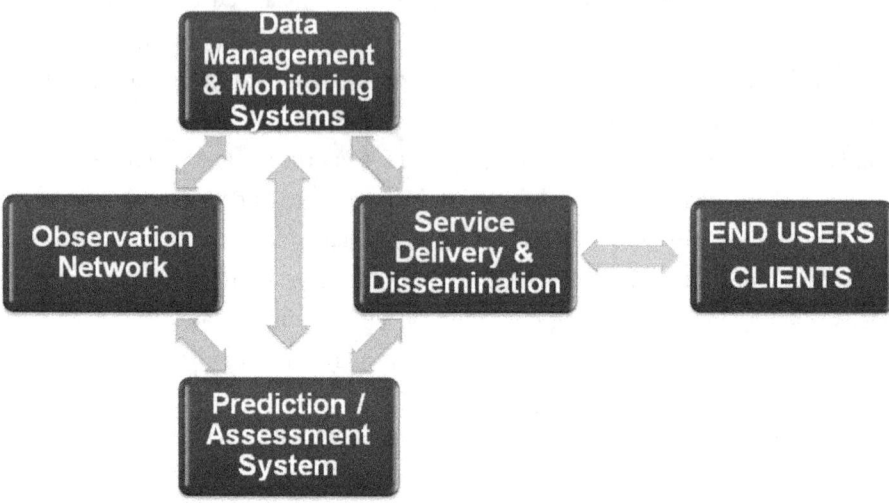

Figure 1.1. Operational oceanography components.

Our ability to monitor and predict the evolution of energetic motions in the ocean mesoscale (such as the meanders in the western boundary currents and the rings that break off from them) at mid-latitudes is based on five key technologies:

1) satellite altimeters, which measure the ocean's sea level at the mesoscale;
2) surface forcing fields such as heat and freshwater fluxes plus remotely sensed and in situ SST;
3) the Argo system of profiling floats, which measures temperature and salinity profiles within the ocean (but does not resolve the ocean mesoscale); and
4) high-resolution ocean models that resolve the ocean's mesoscale motions and data assimilation capabilities that combine the measurements from the above three technologies plus other ocean monitoring platforms with model predictions to provide accurate initial conditions for future predictions.

Observations

Altimetry

Satellite altimetry provides global, real-time, all-weather measurements of sea surface heights (SSH) (sea level) at high space and time resolution. At mid-latitudes, the sea level is, to a good approximation, an integral measure of the density variations in the upper ocean interior. Consequently, sea level provides a strong constraint for inferring the four-dimensional (4-D) ocean circulation through data assimilation. This explains the unique and fundamental role of satellite altimetry in data assimilation and operational oceanography. The quality of mesoscale predictions in ocean analysis and forecasting models is heavily reliant on the availability of altimeter data from multiple satellites. It is commonly agreed that real-time analysis and forecasting of mesoscale circulation requires at least three or preferably four altimeters to be flown simultaneously in appropriately selected orbital configurations (Le Traon, 2013). A very precise long-term altimeter system (e.g., Jason series) is also needed as a reference for the other missions (particularly for climate monitoring). Jason-2, CryoSat-2, HY-2A, and SARAL/AltiKa altimetry missions are the result of a period of intense cooperation between the space agencies (NASA, NOAA, ESA, CNES, EUMETSAT, CNSA, and ISRO).

For the next decade, future missions include Sentinel 3A&B, Jason-3, and Jason CS. Sentinel 3A&B and Jason CS will use an improved along-track SAR mode (higher along-track resolution, lower noise level). Although, at the time of this writing, there are uncertainties about the launch dates and funding of some of these missions, it is hoped that the altimeter constellation will be satisfactory (although not ideal) in the coming years. For example, the Sentinel 3A mission was launched in 2016 and sea level anomaly observations are already available and assimilated in several operational forecasting systems. It is also hoped that by 2020 the new Surface Water Ocean Topography (SWOT) concept will have been demonstrated, providing new capabilities for very high-resolution observations of the ocean mesoscale over a swath. SWOT should be seen as an essential contribution to operational oceanography, helping observation capabilities to keep pace with our steadily increasing model resolutions.

Argo

A major challenge at the end of the 1990s was to set up a real-time global in situ observing system to complement satellite observations. Based on technological progress made during the World Ocean Circulation Experiment (WOCE, 1990-2002), proposals were made to develop a global array of profiling floats taking temperature and salinity profile measurements down to 2000 m every 10 days throughout the deep global oceans. The resulting Argo international programme (Roemmich & the Argo Science Team, 1999) was initially developed as a joint venture between GODAE and CLIVAR (Climate and Ocean: Variability, Predictability and Change). Argo has been an outstanding achievement, and in November 2007 it reached its initial target of 3,000 profiling floats. Argo delivers data both in real-time for operational users and after careful scientific quality control for climate change research and monitoring. Argo has brought remarkable advances in ocean and climate change research (Freeland et al., 2009) and ocean forecasting capability (Oke et al., 2009; Dombrowsky et al., 2009). There are strong and unique complementarities between Argo and satellite altimetry, and Argo data are now systematically used together with altimeter data for ocean analysis and forecasting. The main ocean forecasting requirements for Argo are to maintain Argo global coverage and sampling for the long term. As reanalysis is as important as real-time prediction, reprocessing of past data with improved quality must be conducted in addition to the delivery of products in real time. Finally, the need for an improved vertical sampling has been identified for both regional prediction systems and coupled ones; near surface sampling needs to be improved to better reproduce interactions with the atmosphere. Operational oceanography should also benefit greatly from extending Argo capabilities towards deeper observations (below 2000 m) and towards the observation of biogeochemical variables. This could be quantified through Observing System Simulation Experiments (OSSEs) and tested as part of Argo extension pilot projects. The main challenges for Argo over the next decade will be first to maintain the global array and ensure its long-term sustainability, and second to prepare the next phase of Argo with an extension towards biogeochemistry, the polar oceans, the marginal seas, and the deep ocean. Given the prominent role of Argo in constraining ocean models, meeting such challenges is essential for the long-term sustainability and evolution of global operational oceanography.

Measurements of SST and other variables

GODAE recognized the inadequacy of the suite of SST products in the late 1990s and launched a pilot project that would not only meet the requirements of GODAE, but also those of weather prediction and climate centres. The Group for High Resolution Sea Surface Temperature (GHRSST) (Donlon et al., 2009) was initiated in 2001 and has overseen the introduction of standardised and verified SST products for multiple uses, including ocean analysis and prediction. The GODAE-sponsored project has resulted in a coordinated network of centres disseminating SST data in real time in a common format to agreed standards from a wide range of microwave and infrared instruments on polar orbiting and geostationary satellites. The basis for this success is the complementary polar orbiting and geostationary satellite radiometers, which yield high-resolution infrared and microwave estimates of SST. Complementary data from a wide variety of in situ sources are also used. Products are widely available from a number of sources and have both high-resolution and accuracy; documented bias and error characteristics; meet timeliness and temporal resolution demands, and are dependable and fit-for-purpose (Donlon et al., 2010). There is also a wide usage of information on sea-ice concentration from satellites by groups participating in GODAE OceanView (GOV) and information on sea-ice thickness, sea-ice drift, and sea-ice temperature should be widely used in future. A number of other systems contribute measurements of temperature, salinity, sea level, and currents, among other variables. Argo is now the workhorse for broad-scale temperature and salinity measurements. The tropical moored buoy arrays (McPhaden et al., 2010) provide high-frequency sampling in the rapidly changing tropical waters (temperature, salinity, velocity); the Expendable Bathy-Thermograph (XBT) networks (Goni et al., 2010) provide high-resolution sections to complement the broad-scale Argo network (particularly important for ocean transport calculation); L-band microwave satellite-borne radiometers (Lagerloef et al., 2012; Font et al., 2013) are now, for the first time, providing measurements of surface salinity with of the order of 0.5 psu accuracy for 10-day averages in 50 km squares and are expected to have a positive impact on ocean forecasting systems (Le Traon et al., 2015). Although the current accuracy of remotely sensed salinity limits its impact on the skill of ocean forecasting systems, it is expected to increase in the coming years with the launch of new sensors. The Global Drifter Program (Dohan et al., 2010) provides important surface drift (and temperature) data for validation. Further information about future plans regarding SST remote sensing can be found in Le Traon et al. (2015); future developments of all components of the global observing system are described elsewhere in this book.

Data Management and Monitoring Systems

Measurement network and data assembly and processing centres provide the main inputs to assimilation centres. This includes the in situ and satellite components of the current global observing system.

Near real-time quality control of observational data and the joint use of in situ and satellite data are key elements of any operational ocean forecasting system, as indicated by the product servers. Product servers also provide the underpinning concepts and technologies that enable the observed data to be discovered, visualised, downloaded, intercompared, and analysed all over the world.

Prediction and Assessment Systems

Models

Despite the increasing wealth of observations from satellites, floats, moorings, and ships, observational coverage is still sparse for vast parts of the deeper ocean, especially on the ubiquitous mesoscale with horizontal scales of the order of 100 km. With the growing exploitation of the seas, mesoscale prediction of ocean currents and, more recently, their biophysical variability have been key goals of the ocean forecasting community. Using numerical ocean models, national forecasting centres simulate the global and regional ocean circulation from the surface down to the abyss. Ocean models evolve according to physical and dynamical constraints. They have the ability to produce forecasts using information on the atmospheric surface forcing, the ocean's bathymetry, and the recent state of the ocean obtained from ocean observations and introduced into the model through data assimilation. Ocean modelling is an active field and new knowledge stemming from observations and theoretical studies produces a continuous stream of improvements to ocean models, which leads to more accurate ocean analyses and forecasts.

The first operational models applied to short-term ocean forecasting in the late 1990s and early 2000s typically had horizontal resolutions between 1° and ¼° in global configurations and higher resolutions in regional applications (Dombrowsky et al., 2009). This range of horizontal resolution is insufficient outside the equatorial domain to predict the mesoscale variability accurately. Such problems have been solved with the current generation of operational global ocean forecasting systems at model resolutions of about 1/10° or higher (and up to 80 vertical levels or layers). Work is progressing at some centres towards operational implementation of global models with at least 1/25° grids in the next few years.

The practical impact of weather, ocean, and climate prediction on the world's population and economy drives the use of high performance computing (HPC), which includes supercomputing and data management, for earth system modelling. The computational and operational requirements for ocean simulations of appropriate spatial and temporal scales are immense and require HPC to provide forecasts and services in practical timeframes. Supercomputers have enabled the weather, ocean and climate research and operational communities to produce results in the shortest amount of time possible while investigating and predicting increasingly complex and detailed phenomena, such as eddies. Nowadays, the world's most powerful supercomputers (many of them are being used in earth system modelling and forecasting) have peak performances of tens of petaflops (10^{15} calculations per second) [http:// www.top500.org/lists/].

In general, nearly all basin- and global-scale ocean models use forms of the primitive equations, whereas the first systems developed in the 1990s were based on quasi-geostrophic approach. These equations relate the variables of velocity, temperature, and salinity, and their evolution over space and time. The primitive equations are a set of nonlinear differential equations that are used to approximate the ocean circulation (and atmospheric flow in atmospheric models). They consist of three main sets of balance equations:

- A *continuity equation*: Representing the conservation of mass.
- *Conservation of momentum*: Consisting of a form of the Navier–Stokes equations that describe hydrodynamical flow on the surface of a sphere under the assumption that vertical acceleration is much smaller than horizontal acceleration (hydrostatic) and that the fluid layer depth is small compared to the radius of the sphere.
- A *thermal energy equation*: Relating the overall temperature of the system to heat sources and sinks. As similar equation applies to salt in the ocean (absolute salinity).

Further details about ocean circulation modelling can be found in, e.g., Haidvogel and Beckmann (1999) and Griffies (2004).

Data assimilation

GODAE was predicated on the expectation that profiles of temperature and salinity data, as well as altimeter data would provide complementary information and that their assimilation would control the evolution of ocean models; the altimeter data controlling the ocean mesoscale and the profile data controlling the vertical water mass structure on larger horizontal scales. Initial evidence that the altimeter data could be effectively assimilated into models came mainly from idealised experiments using simple models (Hurlburt, 1986), statistics on vertical structure (De Mey & Robinson, 1987), and ideas on conservation of water mass properties (Cooper & Haines, 1996). The groups within GODAE chose to adopt widely differing approaches to data assimilation (Cummings et al., 2009; Martin et al., 2015). For example, the core algorithms for estimating the covariances of the errors in the model fields differ a great deal, including the Ensemble Kalman filter approach, the SEEK filter (Pham et al., 1998), the static error covariance estimated from an ensemble of integrations, and the use of 3-D variational assimilation schemes with geographically varying covariances calculated from observation minus model (and other) statistics using balance operators (Dobricic and Pinardi, 2008). Four-dimensional variational methods have also been explored (Stammer et al., 2003; Sugiura et al., 2008). There have also been large differences between the centres in terms of prioritizing assimilation of different observation types, the pre-processing and quality control of the observations, the specification of the time window for the observations, the methods for adding increments into the models, and the methods for assimilation into models with significant biases (whose drifts can be exacerbated by data assimilation particularly near the equator). While each of these systems has had some success in constraining their ocean models, a better understanding of their relative effectiveness is very desirable as it would assist in the improvement of the systems.

Examples of prediction and assessment of forecasting systems

Forecasting centres comprise modelling and assimilation components and cover coastal and/or basin-scale ocean forecasting. Most centres now operate systems with 1/10° or finer horizontal grid spacing in global models; make use of community ocean models like HYCOM (Bleck, 2002, Chassignet et al., 2003); MOM4 (Griffies et al., 2004) or NEMO (Madec, 2008) and assimilate in situ profile data, altimeter data, some form of surface temperature data, and sea ice observations. Product assessments and interactions with research users have been key activities since the inception of operational ocean forecasting systems. An important component of product assessment is procedures developed to intercompare forecasts produced by different centres. The intercomparisons of results from the systems are being used to assess the consistency and uncertainty of the state estimates.

Today, there are numerous ocean forecasting consortia covering the coastal-to-basin/global scales. Many of these initiatives provide forecasts based on multiply nested models, with finer resolution in shelf-scale and coastal models. Here, we describe two initiatives: one with a regional focus (SOCIB) and one with a quasi-global to regional focus (Bluelink).

SOCIB, the Balearic Islands Coastal Observing and Forecasting Initiative (Tintoré et al., 2013), started its activities in 2009 in Mallorca Island, Spain, in order to implement and operate a multi-platform observing and forecasting research infrastructure providing streams of oceanographic data and modelling services that could support operational oceanography at a national, European, and international level. As part of its mission, driven by science, technology, and society needs, SOCIB acquires, processes, analyzes, and disseminates multi-disciplinary information about the sea in a systematic way, participates in the development and testing of new technologies, and provides science-based tools and products for society and coastal management.

SOCIB operates a variety of observing platforms, including a coastal research vessel, two high-frequency radar antennas, gliders, Argo floats, surface drifters, fixed moorings, weather stations, and beach monitoring systems. Additionally, SOCIB's modelling and forecasting facility develops and implements operational ocean prediction systems to complement and integrate these observations in line with the initiative's objectives. These modelling systems aim to represent: (i) the ocean circulation in the Western Mediterranean Sea (WMOP system), (ii) hazardous sea level oscillations in Ciutadella (locally known as "rissaga" phenomenon, Menorca, Spain; BRIFS system), and (iii) waves around the Balearic Islands (system operated in collaboration with Puertos del Estado). The first two prediction systems are briefly described next.

The 2 km-resolution ocean circulation model WMOP (Western Mediterranean OPerational System, Juza et al., 2016) is used to investigate processes and to provide short-term numerical predictions, which are then integrated in SOCIB products and services. WMOP is a regional configuration of the ROMS over the Western Mediterranean Sea, extending from the Strait of Gibraltar to the Sardinia Channel. It is initialized from and nested in the larger scale Mediterranean Model distributed through the European Copernicus Marine Environment Monitoring Service

(CMEMS-MED). High-resolution atmospheric model outputs are used to force the system (Spanish National Meteorological Agency HIRLAM model), with a 0.05° spatial resolution and a 1h temporal resolution. Due to the lack of data assimilation in the present operational WMOP system and to avoid a drift from realistic conditions, the model restarts every week from the outputs of a three-week spin-up simulation initialized from CMEMS-MED model fields, which include assimilation of sea level anomalies and temperature and salinity profiles. Data assimilation algorithms based on a local Ensemble Optimal Interpolation approach have recently been developed and tested in the model. They will soon be implemented in the operational chain, allowing models to ingest observations from satellites and in situ platforms, including altimeter sea level anomaly (gridded or along-track), SST, temperature and salinity profiles from Argo, CTDs and gliders, as well as high-frequency (HF) radar sea surface currents.

The operational system is run once a day, starting at 07:30 local time and delivering model outputs (72-hour forecasts) in the morning. In recent years, WMOP model outputs have been used to support efforts to deal with emergencies at sea related to oil spills, search-and-rescue, or for forensic purposes. Moreover, WMOP hindcast simulations over the period 2009-2015 have also been used to investigate ocean variability, regional connectivity, larval and plastics drifts, and to simulate future SWOT altimeter measurements and virtual observations to study innovative high-resolution multi-tracer analysis methods.

SOCIB has developed a second ocean forecasting system (BRIFS; Renault et al., 2013, Licer et al. 2017) aimed at representing the occurrence and magnitude of meteotsunamis (Monserrat et al. 2006; locally also known as "rissagues") in Ciutadella harbour, Menorca, Spain. A meteotsunami or meteorological tsunami is a tsunami-like wave of meteorological origin. Meteotsunamis are generated when rapid changes in barometric pressure cause the displacement of a body of water, which is amplified under specific resonant conditions. The representation of such phenomena requires high-resolution ocean simulations forced by very high temporal resolution atmospheric forcing. As a consequence, BRIFS is an atmosphere-ocean modelling system using a very fine horizontal grid resolution (10 m) around the Menorcan harbour of Ciutadella, and a two-minute temporal resolution of atmospheric pressure forcing. WMOP and BRIFS outputs are illustrated in Fig. 1.2.

The atmospheric part is based on the WRF model with a 4 km resolution in the inner grid. The atmospheric pressure outputs of the WRF model are used to force two nested configurations of the ROMS model.

The Australian ocean forecasting initiative Bluelink operates two systems: a global system called OceanMAPS and a relocatable regional system called ROAM. OceanMAPSv3 represents a major upgrade in operational ocean forecasting for Australia over the previous version, with near-global (75°S-75°N) 0.1° horizontal resolution and improved vertical resolution. OceanMAPSv3 was declared operational in April 2016. The model is an implementation of the Modular Ocean Model version 4p1 (MOM4; Griffies et al., 2004). The data assimilation system is based on an Ensemble Optimal Interpolation (EnOI) implementation of an Ensemble Kalman Filter (EnKF-C; Sakov, 2014). The analysis assimilates all available altimetry, satellite SST, and in situ profiles of temperature and salinity. The forecast cycle is a three-day cycle with an analysis window for all observations of three days. A behind real-time analysis is performed -6 days behind real-time with a near real-time analysis performed at -3 days behind real-time. The surface is forced by the Bureau of Meteorology's global numerical weather prediction system (ACCESS-G) based on the UMv8.2 (UKMetOffice). Boundary conditions are based on GAMSSA and NCEP 1/12° sea ice analysis. The ocean model is forced by the atmospheric fluxes for wind stress and total heat flux (solar radiation, longwave radiation, sensible heat and latent heat). A climatological river discharge is applied based on Dai and Trenberth (2002).

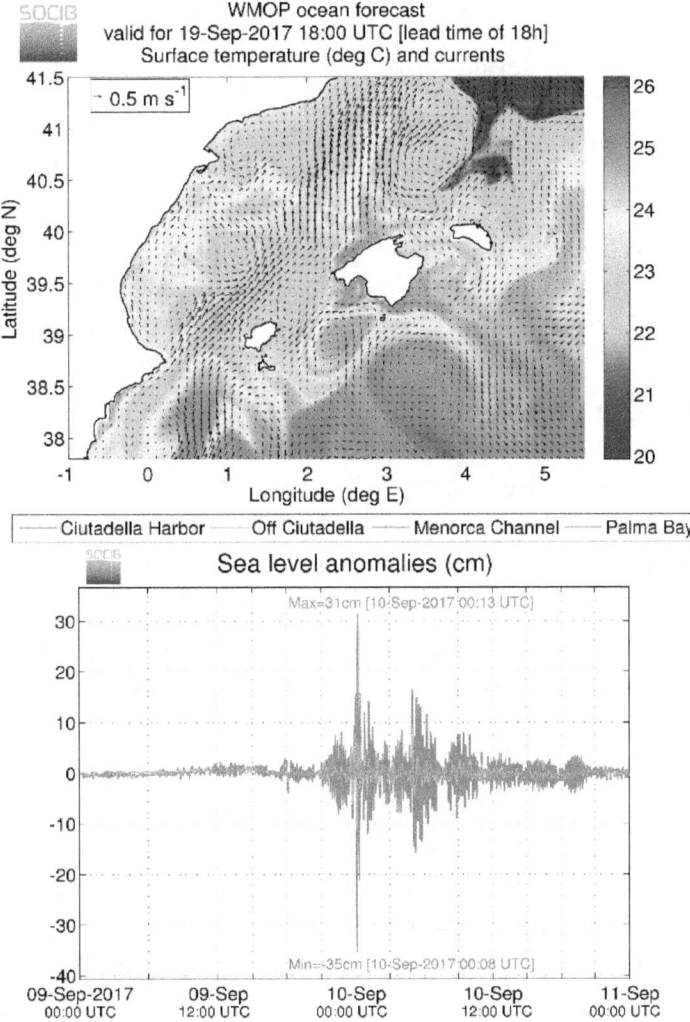

Figure 1.2. Top: WMOP prediction of SST and currents around the Balearic Islands, valid for 19 September 2017. Bottom: BRIFS prediction of sea level anomalies associated with a meteotsunami event in Ciutadella on 9-10 September 2017.

The Relocatable Ocean Atmosphere Model (ROAM) can be configured to have a high grid resolution (variable, resolution as high as one km horizontally, resolving tides and other high-frequency events) and output to file multiple times per day. This high spatial and temporal resolution allows in the ocean component for complex topography, tidal dynamics, and short-lived sub-mesoscale features to be modelled. The ROAM system is presently nested within OceanMPAS (data-assimilating) near-global models run on a 0.1° grid (≈ 10 km). ROAM-SHOC is initialized from fields derived from OceanMAPS and is subsequently forced with boundary conditions from the same model. OceanMAPS does not have tides, it has a coarse bathymetry due to the near-global grid resolution, and it is primarily designed for use in the deep ocean. ROAM is primarily designed for use over the continental shelf and slope, including areas with complex bathymetries or if a high temporal resolution is needed.

Service Delivery and Dissemination

Ocean forecasting systems deliver a broad range of products and services. Here we provide an overview of one of the most advanced systems, the European Copernicus Marine Environment Monitoring Service (CMEMS) (http://marine.copernicus.eu). CMEMS provides regular and systematic reference information on the physical state, variability, and dynamics of the ocean and marine ecosystems for the global ocean and the European regional seas. Some variables extracted from the global high-resolution forecasting system are shown as an example in Fig. 1.3. Other products, such as wave parameters and biogeochemistry variables, are also available at a global scale and for the European seas based on dedicated regional forecasting systems. Reanalyses, which are long time series produced based on homogeneous analysis systems, are available from the 1990s when altimetry observations first became available.

The CMEMS service is based on a network of production centres (Le Traon et al., 2017) with a cross-cutting coordination and strong links to scientific research to ensure a continuous evolution of the service (Fig. 1.3). The observations and forecasts produced by the service support all marine applications (Fig. 1.4). For instance, the provision of data on currents, waves, winds, and sea ice helps to improve ship routing services, offshore operations, or search and rescue operations, thus contributing to marine safety.

The products delivered by the CMEMS are provided free of charge to registered users through an interactive catalogue available on a web portal and with a strong support for all user needs through a dedicated service desk. These products encompass a description of the current situation (analysis), the variability at different spatial and temporal scales, the prediction of the situation a few days ahead (forecast), and the provision of consistent retrospective data records for recent years (reanalysis). The web portal allows for global and regional searches and includes over 40 variables, including physical variables (e.g., temperature, sea ice, waves) and biological variables (e.g., plankton, nutrients, turbidity).

CMEMS also contributes to the protection and the sustainable management of living marine resources, especially aquaculture, fishery research, and regional fishery organisations. Physical and marine biogeochemical components are useful for water quality monitoring and pollution control. Sea level rise helps to assess coastal erosion. SST is one of the primary physical impacts of climate change and has direct consequences on marine ecosystems. As a result, the service supports a wide range of coastal and marine environment applications. Many of the data delivered by the service (e.g., temperature, salinity, sea level, currents, wind, and sea ice) also play a crucial role in the domain of weather, climate, and seasonal forecasting.

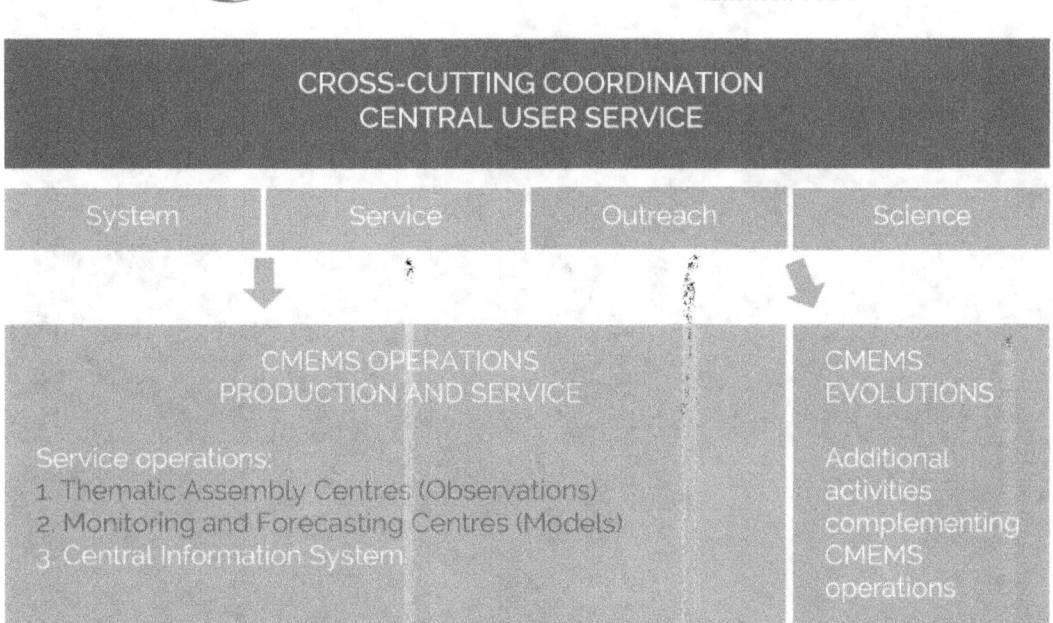

Figure 1.3. Organization of the CMEMS service with the four Thematic Assembly Centres (TACs for the Sea Level, In Situ, Ocean Colour, Ocean and Sea Ice observations) in charge of the observations and the seven Monitoring and Forecasting Centres (MFCs for the global ocean, the Arctic Ocean, the Baltic Sea, the North West Shelf area, the Iberian Biscay and Irish Seas, the Mediterranean Sea and the Black Sea) in charge of the models reanalysis, real-time analysis and forecast.

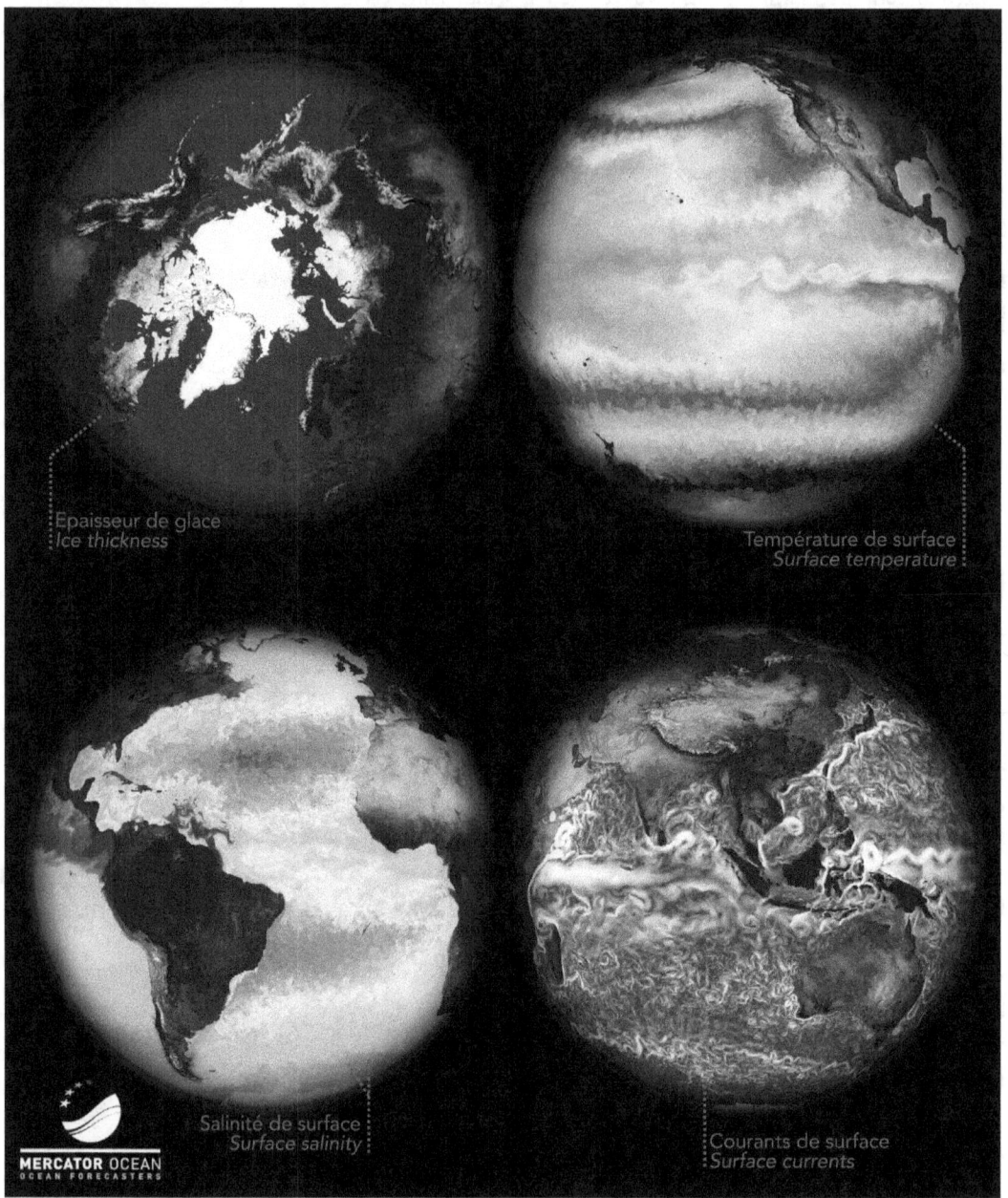

Figure 1.4. Example of variables from the global high-resolution forecasting system available in CMEMS. Ice thickness (top left), temperature (top right), salinity (bottom left), and current (bottom right) are available at 1/12° of resolution from the surface to the bottom, in the past (from 2007) and in the future (10-day forecast are provided daily).

Client Applications and Benefits

Broadly speaking, the primary sectors supported by ocean forecasting and reanalysis products are marine operations and activities with a focus to make these safer and more efficient. Exploitation of products is also heavily dependent on their fitness for purpose, the information that is provided with them on their expected accuracy, and the robustness and reliability of the service. Data policies for service and access conditions to the products are also believed to have a significant impact on their uptake in downstream services by small-to-medium enterprises and by the research sector. Open access to freely available products, free of charge at point of use, with minimal restrictions on their use, has been promoted strongly by the GEO (Group on Earth Observation) and by the European Copernicus programme. From its inception, GODAE also strongly supported open exchange of information.

Data access and visualisation
State-of-the-art information technology and visualisation techniques are being developed and updated by forecasting centres and intermediary service providers to facilitate uptake of products by end users.

For example, the 3-hourly WMOP forecasts (temperature, salinity, currents, and sea level) produced by the SOCIB system are made available through the SOCIB Thredds server. The SOCIB website (www.socib.es) also provides access to plots and animations of oceanic surface conditions, as well as interactive visualizations allowing users to superpose model results with other datasets such as the present platform deployments. Operational model validation based on observations from satellite SST, surface geostrophic currents, Argo, moorings, glider, and HF radar is also performed operationally and provided on the SOCIB website. Regarding the local meteotsunami phenomenon, BRIFS predictions are available on the SOCIB website as pictures of sea level time series in four points of interest, including Ciutadella harbour. Moreover, figures and videos are provided, illustrating the evolution of atmospheric pressure over the whole domain of the WRF model so as to visualize the occurrence of small-scale instabilities responsible for the generation of meteotsunamis. BRIFS outputs are used to complement the official warning system from the Spanish Meteorological Agency based on the analysis of synoptic atmospheric conditions.

The European Commission has launched an initiative to develop Copernicus Data and Information Access Services (DIAS) that facilitate access to Copernicus data and information from the Copernicus services (http://copernicus.eu/news/upcoming-copernicus-data-and-information-access-services-dias). By providing data and information access alongside processing resources, tools, and other relevant data, this initiative is expected to boost user uptake and stimulate innovation based on Earth observation data and information.

Marine transport, search and rescue, oil and gas industry
All of these industries rely on predictions of near-surface conditions, in particular the surface currents. Predictions of transit times and advice on route choice support the efficiency of the shipping industry. Information on ocean surface temperatures can also be valuable as they affect

the efficiency of the ship engines and the durability of some cargoes. Most search and rescue operations are confined to coastal waters, but some operations occur in deeper waters and in these cases information on currents and temperatures is required very rapidly to optimize the search radius.

Responses to spills of oil, chemicals, hazardous substances, and other ship cargo also require rapid initial predictions of the expected drift and dispersion and more detailed, longer period simulations later in the response cycle. Some nations also use surface currents to trace spills resulting from illegal discharges back to individual ships. Again, the surface temperatures are relevant and, in the case of sub-surface releases (e.g., from well heads), the sub-surface temperatures and currents are critical.

In the oil and gas sector, there are a number of key players who require products from ocean forecasting centres (such as government regulators and agencies, major oil companies, commercial operators, and environmental service suppliers). Generally speaking, environmental support is required throughout the life cycle of oil and gas production including exploration, production, and decommissioning. In addition to its impact on the environment, the oil and gas industry require information to predict the impact of the environment on their operations so they can adapt them to optimize their safety and efficiency. For example, subsurface current shears can generate severe stresses on risers (the pipes conveying oil from the well to the surface), which affects operations.

Naval defence operations

Naval defence operations have a diverse range of requirements for information (Jacobs et al., 2009) the most well-known being that the temperature variations in ocean mesoscale phenomena have a major impact on sonar propagation. Ocean prediction systems are now used to generate sonar predictions for the shelf seas as well as the open ocean. They are also used to produce hindcasts and re-analyses for the last 20–50 years, as well as real-time predictions.

Environmental protection and fisheries

The reanalyses for shelf seas has great potential to support important, widely held aspirations for a more holistic ecosystem-based approach to the management of the marine environment (European Parliament, 2008) and fisheries. Adapting the ocean prediction systems to meet these demands and to exploit the potential of biogeochemical models provides a number of important challenges for the future (Berx et al., 2011). Real-time ocean prediction systems can also help to improve the efficiency of fisheries by giving guidance on the best areas for fisheries (e.g., regions of frontal upwelling).

Short-range weather and seasonal prediction

Weather and seasonal prediction are closely related, the difference between them being the timescale. Short-range weather prediction, which covers forecasts a few days ahead, requires good surface temperature fields in part because the development of mid-latitude cyclones is strongly influenced by them. Coupled atmosphere-ocean models have been developed for seasonal predictions of many years. There is growing evidence that relatively high ocean model resolution is needed to improve the skill of the seasonal forecasts (Scaife et al., 2011) and that short-range weather forecasts can also be improved using atmosphere models coupled to sea-ice, wave, and ocean models (Goni et al., 2009).

Some sectors, such as the oil and gas industry, have extremely demanding requirements for the accuracy of currents at small spatial scales, which continues to challenge the ocean predictions. The nature of the decisions based on weather forecasts has changed dramatically in the last ten years as the forecasts and confidence in them have improved. Systems to provide ocean environmental services such as the Integrated Marine Observing System (IMOS) in Australia, the Integrated Ocean Observing System (IOOS) in the United States, and the Copernicus marine service in Europe are now being developed to address these challenges.

International Coordination

International coordination facilitates progress in all functional components of operational oceanography systems shown in Fig. 1.1 (observations, data management/provision, prediction systems, and service delivery/products). GOV currently coordinates multiagency efforts to optimally support the research, development, and operational implementation of physical and biogeochemical ocean forecasting systems through its science team (www.godae-oceanview.org; Bell et al., 2015). For example, GOV coordinates international activities in support of ecosystem assessments (coral reef and other habitats), forecasts (harmful algal blooms, spills), and the development of associated prediction applications (climate impacts, living marine resource management). GOV continues the legacy of GODAE[2] with collaborators from more than 50 academic and national agencies worldwide. The research focus is on improving short- to medium-range operational ocean forecasting systems, and on enhancing and sustaining their development and routine operations. A formal expert review of GOV in 2013 (www.godae-oceanview.org/files/download.php?m=documents&f=150107120408-GOVStrategicPlan20152020.pdf) recognized the enormous benefits reaped by this globally coordinated operational oceanography effort over the last decade. The review identified areas of research where improvement of interinstitutional scientific coordination could deliver greater societal benefits. The recommendations have informed the GODAE OceanView Strategic Plan 2015–2020 (GODAE OceanView Science Team, 2014), which guides internationally coordinated research in short-term ocean prediction, data assimilation, application development, and service delivery for years to come. The development of an operational support for end-to-end capabilities (i.e., from research through to service delivery) is important to GOV and its sponsoring agencies, and includes routine and sustained ocean observing, data management, and the prediction system, as well as operational production and dissemination. The core objectives of the GOV Science Team (GODAE OceanView Science Team, 2014) are to:

---

[2] GODAE—Global Ocean Data Assimilation Experiment, predecessor of GOV from 1997 to 2008 (www.godae.org)

- assess forecast system and component performance combined with component improvements;
- establish initiatives aimed at exploiting the forecasting systems for greater societal benefit; and
- evaluate the dependence of the forecasting systems and societal benefits on the components of the observation system.

These overarching objectives are aligned with those of the World Weather Research Program (WWRP), the World Meteorological Organization (WMO), the Intergovernmental Oceanographic Commission (IOC) Joint Technical Commission for Oceanography and Marine Meteorology (JCOMM), the Committee on Earth Observation Satellites (CEOS), and the Blue Planet initiative of the intergovernmental Group on Earth Observations (GEO). In this context, GOV contributes to the prioritization, advocacy, implementation, and exploitation of the Global Ocean Observing System (GOOS) and the Global Climate Observing System (GCOS).

Trends and Future Developments

There has been significant progress in ocean forecasting in recent years, which can be summarized as follows (Bell et al., 2015; Schiller et al., 2015):

- Improvements of forecasting systems included increased resolution (horizontal and vertical) tides, sea ice drift and thickness, ecosystem approaches, improvement to mixing biases, and extending regional mode areas (e.g., polar regions and progress of coupled modelling [wave coupling, sea ice, hurricane models, etc.]).
- Data assimilation schemes vary among ocean forecasting groups, ranging from Ensemble Optimum Interpolation (EnOI) and Ensemble Kalman Filter (EnKF) to three- and four-dimensional variational methods (3DVar and 4DVar). Observations assimilated in ocean forecast systems now include ocean colour, surface velocities, sea ice, and data from gliders. Many systems now employ multi-model approaches or ensemble modelling techniques.
- All major numerical weather prediction systems have now transitioned to ensemble forecast systems. Taking advantage of an ensemble in air-sea forcing within an equivalent ocean ensemble forecast system will provide crucial uncertainty information in the upper ocean properties.
- As the demand on forecasting products from ocean predictions is growing, the communication and dissemination of information to downstream users has been improved. Today, dissemination of outputs from forecasting systems is akin to the approaches taken in numerical weather prediction.
- Most ocean forecasting systems are now investing in verification and validation efforts to be able to show the value of their products to their users.

- Research in high-latitude (Arctic and Antarctica) operational ice and ocean prediction has increased significantly, driven by increased demand for accurate forecasts by industry and in support of sovereignty.
- Ocean reanalyses (three-dimensional analyses of the past) as well as the present ocean state at global-to-coastal scales based on the same modelling and assimilation infrastructure used for ocean forecasting are increasingly being used in, e.g., climate research and industry applications (extreme events).
- Boundary and forcing conditions from global ocean-atmosphere climate projections are being used in physical-biogeochemical-ecological modelling to downscale projections at regional and local scales, based on ocean forecasting infrastructure.

With the maturing of oceanographic forecast systems and research, the core ocean forecasting disciplines of ocean modelling, data assimilation, forecast verification, and observing system evaluation are now enabling new research and operational areas to flourish. Some of these focus areas for research and development are described in the subsequent subsections.

Model Development

Model grid resolution

Horizontal ocean model grid resolution has been steadily increasing over the last two decades, accompanied by increases in forecast skill (Tonani et al., 2015). By 2020, typical horizontal grid resolutions will be of the order of 5–10 km for global ocean prediction systems and will approach 1 km or less for systems that resolve sub-mesoscale processes by 2025 (with local/regional implementation much earlier). This will depend on continued growth in supercomputer power and evolution of ocean modelling techniques to make best use of computing power. For instance, the use of unstructured grids or grid nesting will allow models to increase their grid resolution specifically where needed by applications. These are particularly attractive options for coastal and regional forecasting models where there is a growing demand to provide accurate information to decision-makers looking after increasingly populated and urbanized coastal areas (including coastal river plumes from sediments and nutrients).

Because of the computational expense of resolving the highly energetic ocean mesoscale, most of the ocean forecasting community has been slower to implement state-of-the-art ensemble prediction systems than their numerical weather prediction and seasonal atmospheric counterparts. However, it is now computationally feasible to develop global and regional ocean ensemble prediction systems that will provide uncertainty and event predictability estimates.

Coupled prediction

Short-to-medium-term (three days to two weeks) coupled ice–ocean–wave–atmosphere prediction to improve weather forecasts is a key research focus of the international forecasting research community. International groups active in coupled prediction research are pursuing a wide range of applications, including global weather forecasting systems and predictions of tropical cyclones, hurricanes and typhoons, extratropical storms, high-latitude weather, and sea ice, as well as coastal upwelling, sea breezes, and sea fog. In many cases, progress is being accelerated through the developments already made in the seasonal prediction and climate projection communities. Research has moved beyond case studies and sensitivity studies to controlled experiments to obtain statistically significant measures of impact. Some first systems are already run in prototype coupled prediction mode (Brassington et al., 2015). The modelling systems being employed include regional and global coupled models of atmosphere–wave, atmosphere–ocean, atmosphere–wave–ocean, and atmosphere–sea ice–ocean. Despite relatively unsophisticated configurations, the results obtained thus far are generally positive and have encouraged more research and development in this area, including coupled initialization and error propagation. Another related area of increasing interest is that of interactions across the dynamic land–sea interface, including coupled watershed and hydrodynamic modelling efforts for heavily populated coastal zones impacted by both natural as well as anthropogenic phenomena.

Biogeochemical ocean forecasting

Biogeochemical, biological, and ecological forecasting, is another research focus in ocean forecasting, noting that the maturity and reliability, at this stage, of physical ocean forecasts are greater than those of biologically-related forecasts. However, there is a growing capability to accurately simulate and predict key components of the marine biogeochemical cycle, including carbon and nutrient cycles. Combined physical–biogeochemical systems will increasingly resemble "environmental prediction systems" for end use in stock assessment, fisheries and habitat management, marine pollution, carbon cycle monitoring, and functional ecosystem understanding. These developments are happening because of growing demand by users for multidisciplinary information, supported by progress in relevant science areas including new observations (satellite and in situ) of environmental properties. Simultaneously, user demand for interoperable prediction systems accessing a large variety of observational and modelling products to produce their own "scenarios" of the marine environment is increasing. We can expect this area to grow significantly as new communication technologies will open new opportunities for society to use ocean information in a much more accessible and interactive way than experienced previously. Despite some prototype biogeochemical ocean forecasting systems currently operating as part of integrated biogeophysical systems, unresolved challenges to increasing skill remain. This includes appropriate representation of ecosystem complexity and limited observations compared to physical ocean models and observations (Gehlen et al., 2015).

Data Assimilation and Observing System Requirements

Ensemble reanalysis and prediction also offers an opportunity for multi-model ensembles from participating centers similar to the approach used in climate projections by the Intergovernmental Panel on Climate Change (IPCC), and in short-term operational weather prediction. However, this approach has yet to be explored by the ocean forecasting community in terms of efficiency and possible gains in forecast accuracy.

As the ocean forecast models progress in the future, it will become increasingly important to end users that ocean forecasting centres define and project what type of events will be predictable by their systems with useful accuracy and confidence intervals. With the implementation of ensemble forecasts at high resolution, determining how to deliver and present the forecast and accuracy information to the end user/decision-maker is an important challenge.

The ensemble methods allow research on prediction controllability, which includes predictability, observability, and the ability of observations to constrain initial conditions of ocean models. Regular increases in computing power have enabled the development of higher resolution and ensemble models. However, while computing power versus cost ratio increases rapidly every year, the observing network capacity versus cost ratio is relatively fixed, particularly for in situ data. An important question to ask is what level of observation density is needed in future ensemble prediction systems to accurately initialize features like fronts with typical scales of 1–10 km, of currents across the shelf break, and of errors propagated in the ocean through air–sea fluxes? Sub-mesoscale filaments and coastal eddies with horizontal scales of less than 10 km, tidal fronts, and freshwater plumes generated by river run-off are generally well-simulated by coastal ocean forecasting systems with horizontal grids of 1 km or less. A challenge is how to use the fine-scale but spatially limited coastal observations, such as HF radar observations, as part of high-resolution, shelf- and basin-scale ocean forecasting systems. The advent and deployment of new observing systems (e.g., HF radars, gliders, and low-cost buoys) will provide the necessary in situ observations density, at least on a regional scale. The SWOT wide-swath altimeter mission scheduled to launch in 2021 is expected to provide high-resolution sea surface height (Fu and Ferrari, 2008). This dataset will resolve the sub-mesoscale and improve parameterizations in and forecasts with global-, basin-, and shelf-scale models. SWOT should also help us better understand and monitor estuaries, and link properties and fluxes from continent to coastal ocean (and vice versa). The impact of these planned future satellite missions (such as SWOT) on ocean forecasting systems can be tested beforehand through Observing System Simulation Experiments (OSSEs), and is currently the subject of research. For ocean biology and biogeochemistry, significant benefits will be realized by incorporating a constellation of geostationary ocean colour radiometry missions into predictions systems.

Operational ocean prediction systems transform data from satellite and sparse in situ measurement systems into value-added comprehensive and vetted oceanographic data and information products with "uniform-gridded" coverage (e.g., mitigating cloud cover and other data dropout issues). They also enable Lagrangian applications, e.g., oil spill forecasting and search and rescue activities. The increased international focus on developing shelf-scale analysis and prediction capabilities brings with it the additional challenge of developing cost-effective in situ coastal observing systems that enhance prediction system performance. Work currently undertaken by GOV scientists and collaborators is paving the way for fully automated multi-model ensemble Observing System Evaluations assessing all components of the Global Ocean Observing System (GOOS) from global-to-shelf scales. By issuing associated Observation Impact Statements, ocean forecasting centres will contribute to coherent, effective, and scientifically robust advocacy for the

GOOS. This effort will allow observing system agencies to assess the impact of past, present, and future observations on forecast and (re-)analysis skills. Consequently, this will enable future observation strategy and scenario evaluation at a fraction of the cost of implementing a new observing system. Furthermore, this activity maximizes the return on investment for the GOOS, a crucial need given limited funds.

New Observing System Requirements

To further increase observation network capability, encouraging end users of prediction information to collect and contribute ocean observations from their marine (e.g., fishing vessels, sailboats) as well as shore-based (e.g., piers, docks) platforms on a best-effort but consistent and quality-controlled basis will complement prediction systems in two ways. First, these "citizen science" observations would provide a prediction validation mechanism at the end-user location of interest and, second, they would enhance ocean forecast initial conditions. These efforts will be facilitated by the rapid growth in wireless communication capabilities and mobile computing platforms, such as smartphones.

A key point is the evolution of low-cost, efficient observing systems with minimal operating costs similar to the already operating "ship of opportunity" network. As with all volunteer observing systems (e.g., commercial ships deploying observing instruments), it will be imperative for operational centres to develop feedback to these observation-contributing end users. This is achieved by producing standard observation-based validation metrics (Ryan et al., 2015) from all available prediction system output at the end user's observation location and time. Furthermore, this approach also addresses the need for intercomparisons of different forecast systems and their respective forecast skill to allow for steady improvements to the systems.

Conclusion

The above advances require an increasingly multidisciplinary effort in physics, chemistry, biology, geomorphology (especially in the littoral zone), IT/visualization, and exploitation of "big data" expertise, which furthers the science, engineering, and infrastructure leading to sustained and integrated applications. GOV has already embarked on this route through specific task teams for biogeochemical/ecological and coupled ocean–atmosphere–wave analysis and forecasting (GODAE OceanView Science Team, 2014). Associated data assimilation tools are being extended to other branches of marine environmental prediction but require new approaches (e.g., ensemble and parameter estimation techniques, coupled initialization) to capitalize on an increasingly diverse ocean observing system. There are ample opportunities for GOV scientists and collaborators to advance the science of ocean forecasting and to improve its skill. Although speculative at this point in time, scientific developments and prioritization will evolve over time, which might eventually lead to reorganization of the forecasting community itself to better respond to new challenges and societal needs. This could involve increased collaboration with international and intergovernmental organizations, providing recommendations and advice on questions related to the GOOS, and expanding operational ocean forecasting capabilities and systems in developing countries (through summer schools, training, and communication).

Apart from the scientific challenges, there are a wide range of additional factors that will influence the progress of ocean forecasting. For operational oceanography to excel, there needs to be mutualistic benefit across all its components (observations, data management, prediction system, production/service delivery, and the clients). For example, prediction systems become stronger with better observing systems and vice versa. More specifically, ocean forecasting systems are critically dependent on both the satellite and in situ components of the physical GOOS, and the sustainability and expansion of the biological and biogeochemical GOOS (Legler et al., 2015). New opportunities arise in regional seas with the advent of "intelligent" new in situ sensors, sensor networks/webs, and new and improved remote sensing technologies.

Based on the GODAE OceanView Strategic Plan 2015–2020 (GODAE OceanView Science Team, 2014), the analysis and forecasting systems developed by GOV and partners are open to further input from the research community, and contribute back to the research community by providing vetted ocean information products of past, present, and near-future states of the ocean. The facilitation of cooperation between research teams, operational groups, and the wider science and user community will remain a key characteristic of the future GOV Science Team. These collective activities will result in improved research, applications, and services for both developed and developing regions, and significant socioeconomic benefits for a world that increasingly depends on and cares for the health of its oceans.

Acknowledgements

We thank the students of the GODAE OceanView Summer School Beatriz Perez Diaz, Safo Piñeiro Rodríguez, and Marcel Ricker for their review of the manuscript. The comments of the students and an anonymous reviewer helped the authors to significantly improve the manuscript.

References

Bell M.J., M. Lefebvre, P.-Y. le Traon, N. Smith N, and K. Wilmer-Becker, 2009: GODAE: The Global Ocean Data Assimilation Experiment. *Oceanography*, 22: 14–21.

Bell, M.J., A. Schiller, P.-Y. LeTraon, N. R. Smith, E. Dombrowsky and K. Wilmer- Becker, 2015: An introduction to GODAE OceanView. *J. Oper. Oceanogr.*, 8, s2–s11, doi:10.1080/1755876X.2015.1022041.

Berx, B., and Coauthors, 2011: Does Operational Oceanography Address the Needs of Fisheries and Applied Environmental Scientists? *Oceanography* 24 (1): 166–171.

Bleck, R., 2002: An oceanic general circulation model framed in hybrid isopycnicCartesian coordinates, *Ocean Modelling*, 37: 55-88.

Brassington, G., and Coauthors, 2015: Progress and challenges in short- to medium-range coupled prediction. *J. Oper. Oceanogr.*, 8, s239–s258, doi:10.1080/175587 6X.2015.1049875.

Chassignet, E.P., L.T. Smith, H.T. Halliwell and R. Bleck, 2003: North Atlantic simulations with the Hybrid Coordinate Ocean Model (HYCOM): Impact of the vertical coordinate choice, reference Pressure, and thermobaricity, *J. Phys. Oceanogr.*, 33: 2504-2526.

Clark C., In Situ Observing System Authors, S. Wilson, and Satellite Observing System Authors, 2009: An overview of global observing systems relevant to GODAE. *Oceanography*, 22(3): 22–33. Special issue on the Revolution of Global Ocean Forecasting—GODAE: 10 years of achievement.

Cooper M, and K. Haines K., 1996: Altimetric assimilation with water property conservation. *J. Geophy Res.*, **101**, C1: 1059–1077.

Cummings J., L. Bertino, P. Brasseur, I. Fukumori, M. Kamachi, M.J. Martin, K. Mogensen, P. Oke, C.E. Testut, J. Verron, A. Weaver, 2009: Ocean data assimilation systems for GODAE. *Oceanography*, **22**: 96–109.

Dai, A., and K. E. Trenberth, 2002: Estimates of freshwater discharge from continents: Latitudinal and seasonal variations. *J. Hydrometeorology*, **3**, 660-687.

Davidson F.J.M., A. Allen, G. Brassington, Ø. Breivik, P. Daniel, M. Kamachi, S. Sato, B. King, F. Lefevre, M. Sutton, H. Kaneko, 2009: Applications of GODAE ocean current forecasts to search and rescue and ship routing. *Oceanography*, **22**: 176–181.

De Mey P., and A.R. Robinson, 1987: Assimilation of altimeter eddy fields in a limited-area quasi-geostrophic model. *J. Phys. Oceanogr.*, **17**: 2280–2293.

Dobricic S., and N. Pinardi, 2008: An oceanographic three dimensional variational data assimilation scheme. *Ocean Modelling*, **22**: 89–105.

Dohan K., and co-authors, 2010: Measuring the Global Ocean Surface Circulation with Satellite and In Situ Observations. In Proceedings of OceanObs'09: Sustained Ocean Observations and Information for Society (Vol. 2), Venice, Italy, 21-25 September 2009, Hall J, Harrison DE, Stammer D, Eds., ESA Publication WPP-306, doi:10.5270/ OceanObs09.cwp.23.

Dombrowsky E., L. Bertino, G.B. Brassington, E.P. Chassignet, F. Davidson, H.E. Hurlburt, M. Kamachi, T. Lee, M.J. Martin, S. Mei, M. Tonani, 2009. GODAE systems in operation. *Oceanography*, **22**: 80–95.

Donlon C.J., and Coauthors, 2010: Successes and challenges for the modern sea surface temperature observing system: The Group for High Resolution Sea Surface Temperature (GHRSST) Development and Implementation Plan (GDIP). Community white paper for OceanObs09, Venice, Italy, 21-25 September, 2009.

Donlon C.J., and Coauthors, 2009: The GODAE high-resolution sea surface temperature pilot project. *Oceanography*, **22**: 34–45.

European Parliament, 2008: Directive 2008/56/EC and of the Council of 17 June 2008 establishing a framework for community action in the field of marine environmental policy (Marine Strategy Framework Directive). [http://eur-lex.europa.eu/legal-content/EN/TXT/?uri=CELEX:32008L0056].

Font J., and Coauthors, 2013: SMOS first data analysis for sea surface salinity determination. *Int. J. Remote Sens.*, **34** (9–10): 3654–3670.

Freeland H. J., and Coauthors 2009: Argo - A Decade of Progress, Proceedings of OceanObs'09: Sustained Ocean Observations and Information for Society (Vol. 2), Venice, Italy, 21-25 September 2009, Hall J, Harrison DE, Stammer D (Eds.,), ESA Publication WPP-306, 2010.

Fu, L.-L., and R. Ferrari, 2008: Observing oceanic submesoscale processes from space. *Eos*, **89**, 488–489, doi:10.1029/2008EO480003.

Gehlen, M., and Coauthors, 2015: Building the capacity for forecasting marine biogeochemistry and ecosystems: Recent advances and future developments. *J. Oper. Oceanogr.*, **8**, s168–s187, doi:10.1080/175587 6X.2015.1022350.

GODAE OceanView Science Team 2014: *GODAE OceanView Strategic Plan 2015–2020*. UK Met Office, 48 pp.

Goni G., and co-authors, 2010: The Ship of Opportunity Program in Proceedings of OceanObs'09: Sustained Ocean Observations and Information for Society (Vol. 2), Venice, Italy, 21-25 September 2009, Hall J, Harrison DE, Stammer D. Eds., ESA Publication WPP-306, doi:10.5270/OceanObs09.cwp.35.

Griffies, S.M., 2004: Fundamentals of Ocean Circulation Models. Princeton University Press, Princeton and Oxford. ISBN: 9780691118925. 528 pp.

Griffies, S.M., M.J. Harrison, R.C. Pacanowski and A. Rosati, 2004: A Technical Guide to MOM4. GFDL Ocean Group Technical Report No. 5. Princeton, NJ, NOAA/Geophysical Fluid Dynamics Laboratory.

Hackett B., E. Comerma, P. Daniel, and H. Ichikawa, 2009: Marine oil pollution prediction. *Oceanography*, **22**: 168–175.

Haidvogel, D.B. and A. Beckmann, 1999: Numerical Ocean Circulation Modeling. Imperial College Press. ISBN: 1-86094-114-1. 344 pp.

Hurlburt H.E. 1984: The potential for ocean prediction and the role of altimeter data. *Mar. Geod.*, **8**: 17–66.

Hurlburt H.E., 1986: Dynamic transfer of simulated altimeter data into subsurface information by a numerical ocean model. *J. Geophys Res.*, **91**(C2): 2372–2400.

International GODAE Steering Team (IGST), 2000: The Global Ocean Data Assimilation Experiment Strategic Plan. Bureau of Meteorology, Melbourne 26 pp.

IPCC First Assessment Report (1990) Scientific Assessment of Climate Change—Report of Working Group I, 1, Houghton JT, Jenkins GJ, Ephraums JJ (eds), Cambridge University Press, UK, p. 365.

Jacobs G.A., R. Woodham, D. Jourdain, J. Braithwaite, 2009: GODAE applications useful to Navies throughout the world. *Oceanography*, **22**: 182–189.

Juza M., B. Mourre, L. Renault, S. Gómara, K. Sebastián, S. Lora, J.P. Beltran, B. Frontera, B. Garau, C. Troupin, M. Torner, E. Heslop, B. Casas, R. Escudier, G. Vizoso,, and J. Tintoré, 2016: SOCIB operational ocean forecasting system and multi-platform validation in the western Mediterranean Sea, *J. Oper. Oceanogr.*, **9**, s155-s166.

Kourafolou V.K., and Coauthors, 2015: Coastal Ocean Forecasting: Science drivers and user benefits. J. Oper. Oceanogr. **8**, s147-s167. http://dx.doi.org/10.1080/1755876X.2015.1022336.

Lagerloef G., F. Wentz, S. Yueh, H-Y. Kao, G.C. Johnson, and J.M. Lyman: 2012: Aquarius satellite mission provides new, detailed view of sea surface salinity. State of the Climate in 2011. *Bull. Am. Met. Soc.*, **93**(7): S70–71.

Legler, D., and Coauthors, 2015: The current status of the real-time in situ global ocean observing system for operational oceanography. *J. Oper. Oceanogr.*, **8**, s189–s200, doi:10.1080/1755876X.2015.1049883.

Le Traon P-Y, and Coauthors, 2015: Use of satellite observations for operational oceanography: recent achievements and future prospects. J. Oper. Oceanogr. **8**, s12-s27, DOI: 10.1080/1755876X.2015.1022050.

Le Traon P-Y., 2013: From satellite altimetry to Argo and operational oceanography: Three revolutions in oceanography. *Ocean Sci.*, **9**(5): 901–915. http://dx.doi.org/10.5194/os-9-901-2013.

Le Traon P-Y, and Coauthors, 2017: The Copernicus Marine Environmental Monitoring Service: Main scientific achievements and future prospects. Mercator Ocean Journal, Special Issue #56, September 2017. In press.

Licer M., B. Mourre, C. Troupin, A. Krietemeyer, A. Jansá and J. Tintore, 2017 : Numerical study of Balearic meteotsunami generation and propagation under synthetic gravity wave forcing. *Ocean Modelling*, **111**, 38-45.

Madec, G. 2008: NEMO ocean engine, Note du Pôle de modélisation, Institut Pierre-Simon Laplace (IPSL), France, No 27, ISSN No 1288-1619.

Martin M.J., and Coauthors, 2015: Status and future of data assimilation in operational oceanography. *J. Oper. Oceanogr.*, **8**, s28-s48, DOI: 10.1080/1755876X.2015.1022055.

Monserrat S., I. Vilibic and A.B. Rabinovich, 2006: Meteotsunamis: atmospherically induced destructive ocean waves in the tsunami frequency band. *Nat. Hazards Earth Syst. Sci.*, **6**, 1035-1051.

Oke P., and Coauthors: 2015. Assessing the impact of observations on ocean forecasts and reanalyses: Part 1: Global studies. J. Oper. Oceanogr., **8**, s49-s62, DOI: 10.1080/1755876X.2015.1022067.

Oke P.R., M.A. Balmaseda, M. Benkiran , J. A. Cummings, Dombrowsky E, Fujii Y, S. Guinehut, G. Larnicol, P-Y. Le Traon, M.J. Martin, 2009: Observing system evaluations using GODAE systems. *Oceanography*, **22**: 144–153.

Pham D.T., J. Verron, M.C. Roubaud, 1998: A singular evolutive extended Kalman filter for data assimilation in oceanography. *J. Mar. Syst.*, **16**: 323–340.

Renault L., G. Vizoso, A. Jansá, J. Wilkin, and J. Tintore, 2011: Toward the predictability of meteotsunamis in the Balearic Sea using regional nested atmosphere and ocean models, *Geophys. Res. Lett.*, **38**, 3.

Roemmich D., and the Argo Science Team, 1999: On the design and implementation of Argo: An initial plan for a global array of profiling floats. International CLIVAR Project Office Report, 21, GODAE Report 5, GODAE International Project Office, Melbourne, Australia, 32 pp.

Ryan, A. G., and Coauthors, 2015: GODAE OceanView class 4 forecast verification framework: Global ocean inter-comparions. *J. Oper. Oceanogr.*, **8**, s98– s112, doi:10.1080/1755876X.2015.1022330.

Sakov, P., 2014: EnKF-C user guide. (http://github.com/sakov/enkf-c).

Scaife A.A., D. Copsey, C. Gordon, C. Harris, T.J. Hinton, S.P.E. Keeley, A. O'Neill, M.J. Roberts, and K.D. Williams, 2011: Improved Atlantic Winter Blocking in a Climate Model. *Geophys Res Lett.*, **38**. doi:10.1029/2011GL049573.

Schiller, A., and Coauthors, 2015: Synthesis of new scientific challenges for GODAE OceanView. *J. Oper. Oceanogr.*, **8**, s259–s271, doi:10.1080/175587 6X.2015.1049901.

Semtner A.J. and RM Chervin. 1992: Ocean general circulation from a global eddy resolving model. *J. Geophys. Res.*, **97**:5493–5550

Smith, N., and M. Lefebvre, 1997: The Global Ocean Data Assimilation Experiment (GODAE). *Proc.Monitoring the Oceans in the 2000s: An Integrated Approach, International Symposium*, Biarritz, France.

Stammer D., C. Wunsch, R. Giering, C. Eckert, P. Heimbach, J. Marotzke, A. Adcroft, C.N. Hill, and J. Marshall, 2003: Volume, heat and freshwater transports of the global ocean circulation 1993–2000, estimated from a general circulation model constrained by World Ocean Circulation Experiment (WOCE) data. *J Geophys Res.*, **108**(C1): 3007, doi:10.1029/2001JC001115.

Sugiura N., T. Awaji T, S. Masuda, T. Mochizuki, T. Toyoda, T. Miyama, H. Igarashi, and Y. Ishikawa, 2008: Development of a four dimensional variational coupled data assimilation system for enhanced analysis and prediction of seasonal to interannual climate variations. *J. Geophys. Res.*, **113**: C10017. doi:10.1029/2008JC004741.

Tintoré J., et al., 2013: SOCIB: The Balearic Islands coastal ocean observing and forecasting system responding to science, technology and society needs. *Mar. Technol. Soc. J.*, **47**, 101–117.

Tonani, M., and Coauthors, 2015: Status and future of global and regional ocean prediction systems. *J. Oper. Oceanogr.*, **8**, s201–s220, doi:10.1080/175587 6X.2015.1049892.

CHAPTER 2

Notions for the Motions of the Oceans

Baylor Fox-Kemper

Dept. of Earth, Environmental, and Planetary Sciences, Brown University, Providence, RI, USA

Operational oceanography combines models and observations to assess and predict the ocean state. While the mechanisms of forcing and equations of fluid motion for the ocean are known, present computers cannot solve all of the relevant scales of motion at the same time. Thus, choices of scales to emphasize must be made in operational systems. This chapter explores some of the issues in choosing a scale, including evaluating approximations to the equations of motion, typical kinds of variability at different scales, and parameterizations of unresolved processes.

Introduction

Operational oceanography aspires to combine observations with models to infer and predict the state of the oceans. Which aspects of the ocean state are of interest vary by application, but almost always it is important to know the basic physical properties: where we expect the water to be and where it is moving. Biological and chemical predictions rely on these physical basics and sometimes can affect the physics (e.g., through limiting solar illumination at depth), but the focus of this chapter will be on the physical aspects: our notions for the motions of the oceans. Ocean motions result from the overall forcing of the climate system (first section), but are filtered through a variety of processes on different scales before arriving at the motions of interest for a particular application (third and fourth sections). For example, an oceanic mesoscale eddy has the same area as about 1-10 Rhode Islands (0.85 to 8.5 Mallorcas), but the dissipation processes that ultimately remove the energy that energizes the 1-2 year eddy lifetime occur on the scale of millimeters. Operational oceanography follows an even grander vision–not only to infer the motions but to make them amenable to applications. They are the building blocks needed for practical questions such as: "What is the best way to get where I'm going on my sailboat?", "Where are the fish today?", "What marine ecosystems are most vulnerable to climate change?", "When is high tide here?", "Are there seamounts or oil on the seafloor here?", and "Are submarines hiding near my naval fleet?"

An ocean model is a combination of the state variables that describe the ocean state, their interrelationships (i.e., the equations of motion), key physical parameters (e.g., the gravitational acceleration near the Earth's surface), and external forcing at the simulation boundaries (e.g., winds, rivers, runoff, air-sea energy and mass exchange, and precipitation – all of which are inferred from observations, other models, or climatology). As the relationships tend to be complex mathematically, we normally formulate and "solve" a model using computers. A model "solution"

Fox-Kemper, B., 2018: Notions for the motions of the oceans. In "*New Frontiers in Operational Oceanography*", E. Chassignet, A. Pascual, J. Tintoré, and J. Verron, Eds., GODAE OceanView, 27-74, doi:10.17125/gov2018.ch02.

or "simulation" in the absence of observations is just an integration of the initial and boundary condition problem over a window of time. A data-assimilating model "solution" is similar, except that ocean observations are used to realign the simulation toward the observed ocean state, which may require multiple iterations of running and readjusting the model initial and boundary conditions, a series of backwards and forward integrations between each time of observation, or just steps in the integration where the equations of motion are violated or "nudged" to adjust the model state toward the observations. Observations may be used just for forcing and checking the state prediction, which can be called *vetting a forward model*, or observations may be used with an inverse model designed to infer which ocean states are consistent with the constraints of observations collected. In a hybrid approach, data is *assimilated* into a model through the steps outlined above and detailed in other chapters of this book.

The fundamental constraints on the physical and physical-chemistry aspects of seawater have been well understood for more than a century, and they are encapsulated in the conservation equations describing the motion and evolution of seawater. However, these equations are difficult to solve exactly, and even their approximated, discretized form (e.g., on a model *grid*) requires significant computational effort.

Thus, either the scope in terms of domain or duration of the ocean state predicted by the model or the degree of detail captured in its discrete form (the grid *resolution*) is limited. Likewise, observations are limited in scope, accuracy, and precision because getting out on and into the water is expensive and hazardous. Management of the limitations of computation, observations, and understanding to provide the best ocean state prediction is the central challenge of operational oceanography.

This chapter presents the framework and many known aspects of oceanic motions over the scales from global to microscopic that delineate the management of these limitations. The unifying theme will be examining the equations of motion on a variety of scales, together with a few key descriptors of fluid motions at important scales, intended to convey what approximations to the equations are useful in modeling practice and where the near future may take us.

What Drives the Motions of the Oceans?

Early after the Earth formed, its surface was still hot from the gravitational potential energy released during the accretion of Earth-stuff out of the early solar system. The late impact that created the Moon warmed things up again, and the nuclear decay by fission of radioactive elements and continuing release of latent heat as the liquid core solidifies contribute extra sources of energy. A little less than 0.1 W/m^2 of this geothermal energy escapes the seafloor on average. The Sun, on the other hand, uses nuclear fusion as well as fission. Fusion provides the sun with a mechanism and fuel to continue to radiate energy brightly for billions of years.

Kelvin estimated the age of these celestial bodies, based on the rate of cooling and their initial energy source, but his estimate was about ten times younger than geologists estimated Earth's sediments to be! Puzzling through the age of the Earth and Sun helped push forward geology and physics. Known as Kelvin's age of the Earth paradox, this discrepancy was one of earliest applications of the conservation of total (thermal and mechanical) energy in astrophysics and geophysics (Richter, 1986; Stacey, 2000).

Fusion keeps the solar system warm, so presently the vast majority of the heating of the Earth and thus the motions of the atmosphere and oceans comes from the Sun's energy (Fig. 2.1a). At the top of the atmosphere, 341 W/m^2 of solar power arrives as primarily shortwave radiation, and about 102 W/m^2 of this shortwave radiation is reflected back to space. Infrared radiation, emitted by varying sources throughout the earth system, sends about 239 W/m^2 back to space. The brightness temperature, or the temperature that would produce this amount of outgoing infrared radiation by the Stephan-Boltzmann law, must be 255 K for the Earth as a whole to achieve 239 W/m^2. However, the satellite observed brightness temperature varies from around 190 K to about 320 K, reflecting variation in the latitude, altitude, and components of the earth system (clouds, atmosphere, upper ocean, land, ice) that are responsible for the infrared emissions. Radiation from lower altitude emitters has to penetrate the shield of greenhouse gasses of the atmosphere (mainly water vapor, ozone, carbon dioxide, methane, and nitrous oxide). These species of gasses are identified as part of the "greenhouse" effect because they are special in their tendency to absorb infrared radiation, blanketing the earth with an insulating layer that warms the surface to a comfortable average of 288 K. Global warming, or the effects of human-induced, or *anthropogenic*, emissions of additional greenhouse gasses is a primary reason why the average temperatures in 2016 and 2017 were 0.94 K and 0.84 K above the 20th century mean of 287.0 K (NOAA, 2018). Overall, the integrated energy budget of the Earth is very nearly in balance at the top of the atmosphere, at least when averaged over a few decades of variability and aside from the systematic small imbalance due to anthropogenic greenhouse gasses (Trenberth and Fasullo, 2010; Hansen et al., 2011). The small imbalance is most readily detected by ocean warming and sea level rise–key targets for operational oceanography (Von Schuckmann et al., 2016).

But latitude-by-latitude, the energy budget is not in balance (Fig. 2.1b). The equator and tropics receive an excess of incoming radiation over outgoing, while the extratropics and polar latitudes emit more energy back to space than arrives from the Sun. Fig. 2.1b represents the annual average as estimated from satellites and atmospheric data by Trenberth and Caron (2001). As the seasons change or as day turns to night, the balance between incoming and outgoing radiation changes latitude-by-latitude and longitude-by-longitude (or timezone-by-timezone).

Consider the amount of power associated with these variations, roughly 100 W/m^2 over the seasons and 500 W/m^2 day versus night. The specific heat capacity of dry air is 1 J/gK and of seawater is 4 J/gK. The hydrostatic pressure at 10 m ocean depth is about 2 atmospheres–1 from the weight of the atmosphere and one from the weight of the ocean. Thus, ten meters of seawater weighs the same as the whole atmosphere above it (10^4 kg/m^2). Using this weight to estimate their relative masses, the atmosphere has a total heat capacity near 10^7 J/m^{2}K–equal to about 3 meters of seawater. More precisely, a dry atmosphere has about the same heat capacity as 2.5 m of seawater, but if water vapor (specific heat capacity of 2 J/gK) is included a typically moist atmosphere has the same heat capacity as about 3.4 m of seawater–so let's just remember 3 m. If 100 W/m^2 heating is applied to either the atmosphere or 3 m of the ocean, they would warm at 10^{-5} K/s, or a bit more than a degree Kelvin per day. In Providence, day-to-night temperature swings are about 5 K, which is consistent with removing the solar input of energy with this atmospheric heat capacity. Winter-to-summer warming is about 20 K over 6 months–a much slower rate due to the more subtle changes in sunlight over the seasons, but more importantly due to the amelioration of warming and cooling by the heat capacity of the roughly 50 m ocean mixed layer–which has an order of magnitude more heat capacity than the atmosphere. Melting ice and evaporating water takes 84 and 560 times as much energy as warming the same mass of water by 1 K, thus even though there are only a few meters worth of evaporation, precipitation, and formation of sea ice these terms contribute at leading order to the overall energy budget. However, this power analysis is still only providing a vertical picture of the energy exchanges, when understanding the forcing of the climate system requires appreciating the meridional gradients as well.

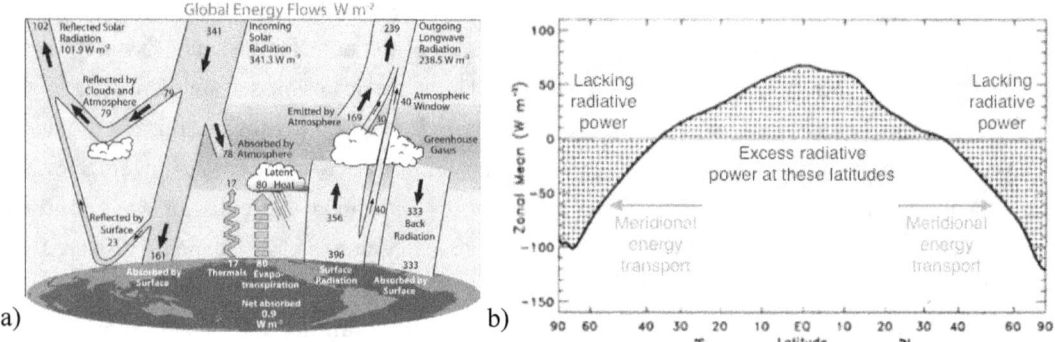

Figure 2.1. a) Estimated flows of energy through the earth system, based on observations from March 2000 to May 2004 (Trenberth and Fasullo, 2009). b) Annual mean top of atmosphere net radiative power (i.e., incoming minus outgoing) by latitude, schematized to indicate meridional energy transport (adapted from Trenberth and Caron, 2001).

The large heat capacity of the ocean makes its temperature changes less sensitive to brief variations in forcing and also makes it the dominant reservoir of thermal energy in the earth system. Thus, the global warming signal is less noisy when detected in ocean warming even though ocean measurements (by ship, buoy, etc.) are practically challenging (Von Schuckmann et al., 2016), and it is the reason why atmospheric variability is "reddened", or reduced at high frequencies versus low frequencies, by the ocean in the coupled system (Frankignoul and Hasselmann, 1977). Thus, on large scales, the variability of the atmospheric weather is dampened and less detectable in ocean temperature changes. On small scales, ocean eddies–the dynamical equivalent of atmospheric weather–produce variability that can drive the atmosphere above (Frenger et al., 2013).

As Fig. 2.1b shows, different latitudes receive an excess of radiative power while others receive a deficit. If the whole system is to be in balance, there must be an exchange of energy between the excessive regions toward the deficient ones. A maximum of about 5.7 PW of poleward energy transport in each hemisphere is needed near 40° where the zero-crossings in Fig. 2.1b are. There are three primary mechanisms by which energy is transferred. Only a negligible amount of potential or kinetic energy is transported (Peixoto and Oort, 1992), so the three primary mechanisms are sensible heat transfer in the atmosphere, sensible heat transfer in the ocean, and latent heat transfer in the coupled system.

First, the atmospheric sensible heat transport is carried by warm air moving poleward on average and colder air moving equatorward, which carries a maximum of about 3 PW. The Hadley Cells ascend near the equator to the tropopause and move poleward to about 30° where they descend and return back toward the equator. The upper branch carries warm air poleward and the lower branch carries cold air southward, thus a net poleward exchange of sensible energy. Because the air in the Hadley cell is heated at low altitudes/high pressure and cooled at high altitudes/low pressure, there is a conversion of thermal energy to mechanical energy that drives the Hadley cell, much like a Carnot cycle (Pauluis, 2011). The Polar Cells, which circulate poleward of about 60° exhibit similar thermally direct overturning. Importantly, these cells do not move directly from equator to pole, but are veered by the Coriolis force, so they have dominantly zonal winds at the surface. The equatorial easterlies and midlatitude westerlies are a consequence of such veering.

Second, the oceanic sensible heat transport is carried by warm seawater moving poleward on average and colder seawater moving equatorward, and it moves a maximum of about 2 PW poleward when zonally integrated over all ocean basins (Trenberth and Caron, 2001). In old-fashioned thinking this *thermohaline circulation* was thought to result from the formation of North Atlantic Deep Water (NADW) or Antarctic Bottom Water (AABW) by cooling & increasing salinity at polar latitudes and heading equatorward at depth. These waters are resupplied by warmer water at the surface, thus there is a transport of sensible heat poleward. However, modern analyses have revealed that this picture is too simple for a number of reasons. Firstly, the oceans are heated and cooled at the surface which means that unlike the atmosphere they cannot extract mechanical energy for circulation as a heat engine or Carnot cycle, because heating and cooling occur at the same pressure.[1] Thus, the energy for the circulation cannot result from the thermal forcing–instead it must come from other sources, e.g., winds and tides. Secondly, the direction of the ocean heat transport is not consistent with this thermally-driven picture. The oceanic meridional heat transport is poleward when zonally-averaged over the whole Earth, but the Atlantic Ocean heat transport is northward at all latitudes—even in the Southern Hemisphere--and the Indian Ocean heat transport is southward at all latitudes (Trenberth and Caron, 2001). Thus, the ocean circulation is not everywhere consistent with poleward heat transport, and regional transports reflect important variations in currents. Thirdly, the old notion of a *wind-driven* circulation and a separate thermohaline circulation is not sensible as these two mechanisms are not distinct. The more modern concepts of a *Meridional Overturning Circulation* (MOC), which can be driven by any force but overturns (after zonal integration) in the meridional-vertical plane can be cleanly distinguished from a gyre, or *barotropic circulation*, that circulates in the zonal-meridional plane after vertical integration. The MOC and barotropic circulations are distinct in terms of volume transport. However, *both* of these circulations contribute to oceanic meridional heat transport. The NADW &

---

[1] Solar radiation that penetrates about 100 m depth and geothermal heating at the seafloor versus sensible at latent cooling right at the ocean surface, as well as meridional variations in atmospheric pressures due to the circulation, are modest exceptions to heating and cooling at equal pressure.

AABW are good examples of branches of the MOC, but so too are the Subtropical Cells (STC)--wind-driven upper ocean cells importantly related to equatorial Ekman upwelling and zonal winds.

Third, the atmosphere/ocean latent heat transport is moist air moving poleward and dry air and liquid water in oceans and rivers moving equatorward, and it moves a maximum of about 1 PW poleward (Peixoto and Oort, 1992). In this exchange of water, the latent heat of vaporization is transported from the location where evaporation occurs to where the precipitation occurs. Pauluis (2011) notes that this "steam cycle" transport generates mechanical energy much like a steam engine. As already noted, it takes 560 times as much energy to evaporate water as to warm it by one degree, so even though the hydrological cycle transports only a small mass of water (about 1 Sv) compared to ocean currents (about 40 Sv in global MOC) or atmospheric winds (about 60 Sv in the Hadley Cells), it carries a lot of heat.

It is a fascinating aspect of the earth system that the transport of heat and moisture by the three mechanisms above may differ significantly between the mean mass transport and the mean heat and moisture transport. At latitudes between 30° and 60°, transient storms carry enough warmth and moisture intermittently to reverse the direction of the air mass transported by the Ferrel Cells, which are the average circulations at these latitudes (Peixoto and Oort, 1992; Held and Schneider, 1999). Similarly, Southern Ocean eddies reverse the heat transport to oppose the mass transport in the Deacon Cell (Karsten and Marshall, 2002).

Celestial forcing, including seasonal and diurnal as well as longer timescale orbital variations, change these energy budgets temporarily. For example, peak Northern Hemisphere winter atmospheric sensible plus latent transport exceeds 8 PW. Climate variability can also alter these patterns. For example, a large El Nino can change the meridional energy exchanges significantly and flush out excess energy in the upper ocean (Sun and Liu, 1996; Deser et al., 2006).

Overall, the ocean receives most of its energy near the global scale, due to the global scale of solar input (and thus winds) and the global scale of tidal forcing. In total, the amount of kinetic energy forced to occur in the oceans is about (24 TW=2.4×10^{13} kg m^2/s^3 from Wunsch and Ferrari, 2004) – only a small fraction of the imbalance in solar forcing.

The climate system has periodic forcing and resonances with a given period, but also broad-banded turbulent variability. The timescales of forcing are precise for the celestial forcing: tidal, diurnal, and seasonal. Atmospheric forcing timescales have to do with weather, which is a chaotic and turbulent process with broad-banded variability rather than discrete periods, and the ocean's greater heat capacity means that atmospheric-forced ocean variability has a preference for lower frequencies. On smaller length scales and longer timescales ocean eddies may drive the atmosphere, again in a chaotic and turbulent process without countable frequencies. Another source of useful timescales for the ocean is the time required for waves to cross basins of a particular dimension. These crossing timescales govern the *resonance* of ocean basins to tidal and seasonal forcing and also set the adjustment timescale of a basin to perturbations. Thus, although tidal forcing is global, the strongest responses (e.g., Bay of Fundy, English Channel) are in basins where the wave crossing timescale matches the period of a tidal forcing. Coupled phenomena, such as El Nino with its 2 to 7 year repeat timescale, are informed by ocean wave crossing timescales but also retain relatively broad-banded variability due to the chaotic nature of atmospheric forcing and response. The next section will clarify some of the notions of spatial and temporal scale for the ocean's motions.

Scales of named oceanic motions

The largest length and longest timescales on the Earth are the circumference of the Earth (4.0×10^7 m) and Earth's age (4.54×10^9 years=1.43×10^{17} s), which set the top and right bounds of Fig. 2.2, but most ocean variability occurs on smaller scales. The ocean is subdivided into basins which change slowly by plate tectonics, but usually these changes are neglected because they are too slow to resonate with an oceanic response. Fast tectonic changes (i.e., earthquakes), do lead to tsunamis, which are among the fastest propagating oceanic signals. The speed of sound is a bit faster, which sets the bottom limit of Fig. 2.2. The slowest ocean and climate variability is normally taken to be paleoclimatic responses (e.g., ice ages) to the variability in Earth's orbit (Milankovitch, 1930). These largest scales define the upper limits in Fig. 2.2.

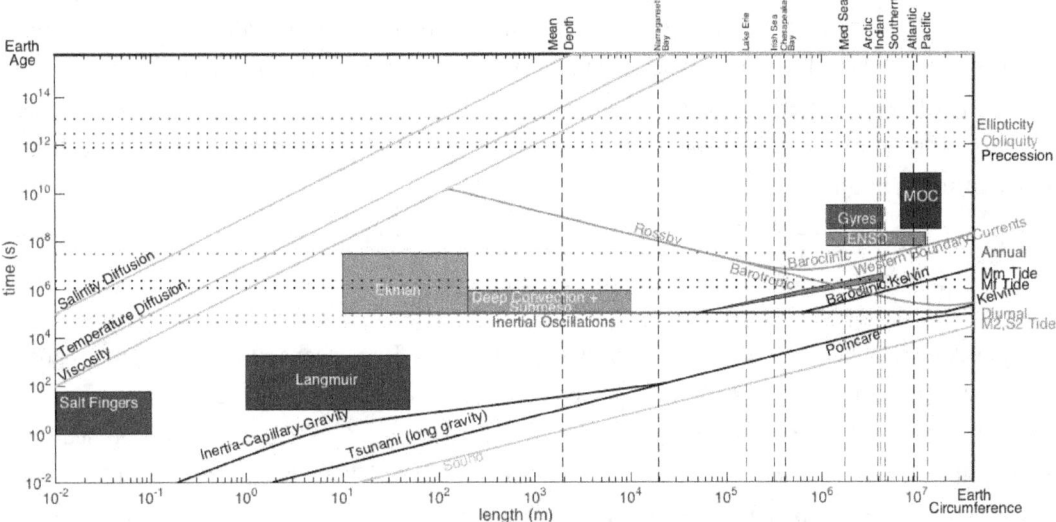

Figure 2.2. Taken from Haidvogel et al. (2017): Approximate length and timescales of oceanic motion, showing many wave dispersion relations–that is, the relationship between timescale and lengthscale of each type of waves–the scaling of viscous and diffusive motions, and the length and timescales of basins and common forcing mechanisms. A latitude of 25° is used to estimate Coriolis parameters.

The smallest scales of *fluid* motion, which might be taken to bound Fig. 2.2 on the left, are the scale of separation of water molecules or the mean free path of a moving molecule of water before it interacts with another, both of which are a few Å or $O(10^{-10}$ m$)$. In a numerical model, however, we normally accept that once the *dissipation scales* of fluid motion, which are the smallest scales of turbulent variability, are resolved then the simulation is accurate and is called a *Direct Numerical Simulation* (DNS). Kolmogorov (1941) theorizes these smallest scales of dissipative motion to be $\nu^{3/4} \epsilon^{-1/4} \approx 0.3$ cm for length and $\nu^{1/2} \epsilon^{-1/2} \approx 10$ s for time. In this estimate the kinematic viscosity of seawater is taken as $\nu \approx 1 \times 10^{-6}$ m^2/s. The kinetic energy dissipation rate per unit mass, E, varies by about 4 orders of magnitude throughout the world (Waterhouse et al., 2014; Pearson and Fox-Kemper, 2018), but it can be roughly yet directly estimated to be $O(10^{-8}$ m^2/s$^3 = 10^{-8}$ W kg$^{-1})$

using the total energy input to the ocean (24 TW=2.4 10^{13} kg m^2/s^3 from Wunsch and Ferrari, 2004) divided by the ocean mass of 1.4 10^{21} kg, under the possible oversimplification that the fraction lost to potential energy by mixing and the fraction lost by waves lifting beach sediments, etc., do not change the order of magnitude of the kinetic energy. The scales of temperature and salinity diffusion are a bit slower or smaller than those of kinetic energy dissipation (Obukhov, 1949; Corrsin, 1951, Fig. 2.2). Finally, as waves tend to oscillate faster than turbulence overturns, waves lead to considerably faster timescales than the turbulent estimate of 10 s. For example, the sound waves in a typical blue whale song are a hundred times faster, with a fundamental frequency of 10 to 40 Hz. Generally, larger scale motions are slower than smaller scale ones, with the intriguing exception of barotropic Rossby waves (Fig. 2.2) where graver modes are able to sense larger variations in the curvature of the Earth and are speedier.

Fig. 2.2 also indicates rough ranges for other types of oceanic motions of interest. A range of overturning scales for branches of the Meridional Overturning Circulation are indicated, as is the El Nino/Southern Oscillation, the circulation timescale and wave crossing timescales of gyres (at the intersections of waves with basin dimensions), and the depth and timescale of the Ekman (1905) layer, etc. Note that all of these phenomena lie in the region of scales between viscosity, sound waves, Earth's circumference, and Milankovitch (1930) forcing of ellipticity. This is the phase space domain of oceanography on Earth, from which an ocean simulation spanning a particular range of scales might be chosen.

Reduced Models

Much is known about the equations governing the motions of the oceans. Practical approximations to the equations are known for all of the named ocean phenomena shown in Fig. 2.2. Improvements are still possible, however, which seek to better incorporate the effects of unresolved scales of motion through *parameterizations*. The small scales in Fig. 2.2 have been directly studied in the laboratory, so their behavior is well-known and may be directly simulated in small domains. On these scales, the equations of motion have been long known (Éuler, 1757; Laplace et al., 1829), and effective phenomenological laws can be formulated to capture the bulk behavior of molecular motions assuming *local* thermodynamic equilibrium (Fourier, 1822; Fick, 1855; Navier, 1822; Stokes, 1845; Onsager, 1931a,b). First, let's examine the limits of computation, then return to approximations that make the equations of motion less expensive to solve.

Reduced to fit computational limits

Present computing capabilities can be quantified in units of a *tera-grid*, which is the computational equivalent of 10^3 discrete steps in each of the 3 spatial directions and time, for 10^3 x 10^3 x 10^3 x 10^3 = 10^{12} unique spatio-temporal locations, or a tera(10^{12})-grid. A typical graduate student computational project might be a few teragrids in computational cost, while present grand challenge simulations are about a hundred teragrids. As the ocean is relatively shallow on average (3.7 x 10^3 m) compared to its horizontal dimension (4.0 x 10^7 m), often many more degrees of freedom are used in the horizontal rather than vertical direction. Resolving the stratification of water properties does require somewhat finer vertical resolution than this purely geometric scaling. Paleoclimate applications can require long duration simulations to rival the Milankovitch (1930) timescales, which necessitates fewer spatial grid points in the horizontal and vertical.

In Fox-Kemper et al. (2014), it is shown that the empirical exponential increase in computing power (Moore, 1965) has resulted in an empirical exponential refinement of ocean model grid scale for climate modeling applications. Moore's observed increase in computation predicts a doubling of model resolution– halving all grid spacings and time step–every 6 years. In practice, models become increasingly complex as well as refined, so the observed resolution doubling rate is closer to 6.9 years for basic atmosphere-ocean models and 10.2 years for complex earth system models. At the basic rate, we will leave behind the now typical ocean tera-grid for a peta-grid in about 18 years. Thus, an undergraduate can expect to exceed today's most computationally-costly simulation on a regular basis as a senior scientist. Looking to the past, the first weather simulation used a deca-kilo-grid (Charney et al., 1950), and from this datum the basic rate of increase predicts reaching a tera-grid in 61 years. However, the first weather run was a grand challenge taking the whole of the fastest computer available for 24 hours, including a daunting output of 100,000 punch cards of data! The output of a tera-grid calculation today is more manageable–saving every 20th time step for 5 ocean state variables (u, v, w, S, T) on a tera-grid at single precision requires only 1 terabyte, although this quantity of data was a considerable challenge about a decade ago (e.g., Maltrud and McClean, 2005). It is important to note that computing power, data storage, and internet connectivity do not accelerate at the same exponential rates, so mismatches are likely to limit future ocean models. The capability to simulate a basin filled with mesoscale features in models and satellite observations shifted the focus of oceanography in the 90s, as resolving basins full of submesoscale phenomena is now. The historical pattern is that resolution increases sufficient to qualitatively change ocean simulations occur slowly but surely, averaging a handful of times in a scientist's career.

The problems studied in operational oceanography vary in scope, but present forward global models at 2 km horizontal resolution have been run for a few years (Rocha et al., 2016a,b). This massive calculation has $O(10^{10})$ wet grid points, so a tera-grid is reached in only 100 time steps. The additional costs of data assimilation mean that operational regional coastal ocean forecast systems have similar horizontal grid scales (Kourafalou et al., 2015), while regional and idealized forward models may have much finer resolutions, from hundreds of meters to meters (Hamlington et al., 2014; Haney et al., 2015; Stamper and Taylor, 2017; Barkan et al., 2017), although in domains of limited extent. The slow diffusion of salinity limits the simulation domain for DNS to considerably less than a cubic meter of seawater (e.g., Penney and Stastna, 2016).

Alongside the costs of extra grid points to increase resolution, refining the grid also requires a change in the time step of the model. In forward, explicit models the maximum stable time step is governed by the Courant-Friedrichs-Levy (CFL) condition. In words, this condition states that the time step must be shorter than the time it takes for the fastest propagating signal in the model to cross one grid cell. Unfortunately, in common numerical implementations the smallest grid cell and/or the fastest propagating signal set the limit for the whole domain. Implicit scheme models can take larger steps stably, but are generally more expensive per time step and accuracy decreases with step size.

To read the CFL and tera-grid requirements off of Fig. 2.2, choose a length scale for the grid and then follow downward to the fastest allowed timescale (e.g., the sound wave dispersion curve). Count upward the number of orders of magnitude needed to reach the timescale of interest. Subtract that count from 12, and that will be the power of ten available to distribute as grid points at that scale that can delineate the domain.[2] The challenge is quickly apparent. For example, by this method only about 100 spatial grid points are available to study ellipticity in the Pacific if sound waves must be resolved!

Thus, on the larger scales of oceanographic interest, approximations to the equations and parameterizations are used to make a tera-grid more valuable. For example, Charney et al. (1950) approximated their equations so that the fastest propagating signal was the barotropic Rossby wave (shown in red in Fig. 2.2), increasing the time step by many orders of magnitude–particularly so in the oceanic mesoscale between 1×10^4 m and 1×10^5 m. At smaller scales, this *quasigeostrophic* theory (i.e., the motions are nearly horizontal, hydrostatic, and geostrophic) breaks down as ageostrophic effects become important at the submesoscale (Boccaletti et al., 2007; Thomas et al., 2008) and vertical flows become important for deep convection (Julien et al., 2006). The next sections in this chapter describe some of these approximations and the assumptions on which they rely. The final section provides some example parameterizations of smaller, unresolved, scales of motion.

Reduced to fit known equations: Direct Numerical Simulations

Directly resolving all scales of fluid motion is called *Direct Numerical Simulation* (DNS). A useful notion in assessing when the fluid scales are resolved is the idea that small parcels of fluid are in local thermodynamic equilibrium so they can be considered to have uniform thermodynamic properties (p, T, S). These properties would remain unchanged if each parcel were isolated, an idea consistent with molecular diffusion acting to homogenize thermodynamic properties on small scales. The dissipation scales (Kolmogorov, 1941; Obukhov, 1949; Corrsin, 1951) are effective estimates of parcel size. It is not the case that the whole ocean is in thermodynamic equilibrium, which would require it to have uniform temperature and salinity everywhere (Fermi, 1956). For seawater, the scale of a thermodynamic equilibrium parcel tends to be smaller than the Kolmogorov viscous dissipation scale over which energy or velocity is homogeneous.

The state variables that describe each fluid parcel, along with square brackets to denote their SI units, are mass density ($[\rho]$ = kg/m^3), three-dimensional velocity ($[\mathbf{u}]$ = m/s), *in situ* temperature ($[T]$ = K), pressure ($[p]$ = Pa = N/m^2 = kg/m s^2), and mass fraction salinity (mass fractions are

---

[2] Don't forget you may need a 3D grid!

dimensionless, so [S] = 1).³ The partial enthalpies of salt and freshwater (h_s and h_w) are thermodynamic variables that differ by fluid and solute. The strain rate is $D_{ij} = 1\ (\partial_i u_j + \partial_j u_i)$.⁴

Following Müller (2006), a convenient form for the fundamental equations of motion for seawater in a rotating frame of reference are:

$$\frac{D\rho}{Dt} = -\rho \underbrace{\nabla \cdot \mathbf{u}}_{\text{divergence}}, \tag{1}$$

$$\rho \frac{D\mathbf{u}}{Dt} = \underbrace{\nabla \cdot \left(-p\mathbf{I} + \sigma^{mol}\right)}_{\text{stress tensor}} - \underbrace{2\rho \mathbf{\Omega} \times \mathbf{u}}_{\text{Coriolis}} - \underbrace{\rho \nabla(\phi)}_{\text{geopotential}}, \tag{2}$$

$$\frac{DT}{Dt} = \underbrace{\left(\frac{\partial T}{\partial p}\right)_{\eta,S} \frac{Dp}{Dt}}_{\text{compress}} + \frac{1}{\rho c_p}\left[-\underbrace{\nabla \cdot \left(\mathbf{q}^{mol} + \mathbf{q}^{rad}\right)}_{\text{heat flux}} + \underbrace{D_{ij}\sigma_{ij}^{mol}}_{\text{friction heat}} - \underbrace{\mathbf{I}_S^{mol} \cdot \nabla(h_s - h_w)}_{\text{mixing heat}}\right], \tag{3}$$

$$= \underbrace{\left(\frac{\partial T}{\partial p}\right)_{\eta,S} \frac{Dp}{Dt}}_{\text{compress}} - \frac{\nabla \cdot q^{all}}{\rho c_p} + \frac{\dot{q}}{\rho c_p},$$

$$\frac{DS}{Dt} = -\frac{1}{\rho}\underbrace{\nabla \cdot \mathbf{I}_S^{mol}}_{\text{salt flux}}, \tag{4}$$

$$\rho = \underbrace{\rho(p, T, S)}_{\text{eqtn. of state}}. \tag{5}$$

The molecular stress, flux, and heating terms indicate the form for the phenomenological flux law approximations which will be discussed more fully below. Standard values of the coefficients—planetary angular rotation rate ($\Omega \approx 7.29211576/\text{s}\,\hat{\mathbf{k}}$, which is 2π per sidereal day oriented along a vector pointing toward the North Pole and perpendicular to the plane of the celestial equator), specific heat at constant pressure ($c_p \approx 3850$ J/kg K), and thermal expansion coefficient ($\alpha \approx 2.5 \cdot 10^{-4}$/K)—are useful guidelines for simulations and the magnitudes of these terms. Oceanic density varies about 4% with temperature, salinity, and pressure (i.e., depth), but the mean density is 1025 kg/m³ at the surface. Frequently, constant values of these parameters are used, but if precise estimation of these properties are needed, they can be determined from practical relations to the ocean thermodynamic state (p, T, S). Likewise, the equation of state (5) must be provided to have a closed set with as many equations as unknowns in (1)-(5).

The *geopotential* ($[\phi] = $ m²/s²) contains the gravitational potential (approximately that of a spherical Earth: $\phi_g \approx g_0 r_e^2/r$, $r_e = 6.370 \times 10^6$ m, $g_0 = 9.81$ m/s²) plus tidal perturbations to the potential plus the centrifugal potential ($\phi_c = (\mathbf{\Omega} \times \mathbf{r})(\mathbf{\Omega} \times \mathbf{r})/2$). The position relative to the center of the Earth is $\mathbf{r}$, and $r = |\mathbf{r}|$. A typical value of the geopotential gradient is $\nabla \phi \approx g_0$, and its orientation

---

[3] For convenience, salinity is often written in grams per kilogram or in practical salinity units (psu). The psu are a measure of conductivity that is empirically tuned to be nearly the same numerically as grams per kilogram.

[4] Note that Cartesian indices are used on some tensors where needed for clarity. Einstein summation of repeated indices is assumed.

defines the direction we call "vertical" with unit vector **k**, which is not quite the same direction as **r** because the centrifugal force is oriented outward from the Earth's rotation axis as opposed to gravity which is oriented toward the point at Earth's center of mass. In a local Cartesian coordinate system tangent to the ocean surface (which is perpendicular to **k**, not **r**), we use z as the vertical coordinate and x, y to be eastward and northward horizontal coordinates.

Other tracers of interest such as nutrients or dissolved gasses, with τ representing mass fraction, can be evolved using an equation similar to that for salinity,

$$\frac{D\tau}{Dt} = C_\tau^{\text{react}} + D_\tau^{\text{mol}}. \tag{6}$$

Here we admit a source term, C_τ^{react}, in addition to diffusion, so that chemical reactions producing ($C_\tau^{\text{react}} > 0$) or reducing ($C_\tau^{\text{react}} < 0$) the tracer in question or radioactive decay ($C_\tau^{\text{react}} \propto -\tau$) might be represented. The salts that constitute salinity are generally assumed not to react or precipitate, although evaporites, manganese nodules, and other precipitates are found in small quantities.

Forcing at the boundaries is also required. The boundary conditions are, to be evaluated at the position of the surface η,

$$\text{Normal stress: } \underbrace{\mathbf{n} \cdot \left(-p\mathbf{I} + \sigma^{\text{mol}}\right) \cdot \mathbf{n}}_{\text{air, ice, seafloor}} = \underbrace{\mathbf{n} \cdot \left(-p\mathbf{I} + \sigma^{\text{mol}}\right) \cdot \mathbf{n}}_{\text{ocean}} + \underbrace{\gamma \nabla \cdot \mathbf{n}}_{\text{surf. tension}}, \tag{7}$$

$$\text{Tangential stress: } \underbrace{\mathbf{t} \cdot \sigma^{\text{mol}} \cdot \mathbf{n}}_{\text{air, ice, seafloor}} = \underbrace{\mathbf{t} \cdot \sigma^{\text{mol}} \cdot \mathbf{n}}_{\text{ocean}} + \underbrace{\mathbf{t} \cdot \nabla \gamma}_{\text{Marangoni}}, \tag{8}$$

$$\text{Heat: } \underbrace{\mathbf{n} \cdot \mathbf{q}^{all}}_{\text{air, ice, seafloor}} = -\mathbf{n} \cdot \underbrace{\left(\mathbf{q}^{mol} + \mathbf{q}^{rad}\right)}_{\text{ocean}}, \tag{9}$$

$$\text{Kinematic/Freshwater: } \frac{D\eta(x,y)}{Dt} = \mathbf{u} \cdot \mathbf{k} + P - E, \quad \mathbf{I}_S^{mol} \cdot \mathbf{n} = 0. \tag{10}$$

where **n** and **t** being normal outward from the ocean and arbitrary tangential unit vectors at the surface. The surface tension ($\gamma \approx 0.0728$ N/m) participates in both the normal and tangential stress balances, but the latter depends only on the gradient of γ, which depends on S, T, and surfactants (Marangoni, 1865). The location of the sea boundary $\eta(x, y)$ can be altered by vertical motions, precipitation (P), or evaporation (E) at the upper surface. At the lower surface, this boundary condition is equivalent to no-normal flow through the boundary ($\mathbf{u} \cdot \mathbf{k} = \mathbf{u} \cdot \nabla \eta$).

Another way to consider the boundary conditions (7)-(10) is as a flow of a conserved property toward the boundary in one continuum (ocean, atmosphere, or seafloor), and then a flow away from the boundary in the other. If the boundary itself is not capable of storing the conserved property, then the flow toward the boundary in each fluid must be equal and opposite. On the other hand, capillary waves are an example of how energy can be stored in the boundary, because their presence increases potential energy stored in surface tension. For example, any downward momentum flux in the atmosphere must cause a downward momentum flux in the ocean–minus any momentum absorbed by the boundary–which is the essence of (7). Energy, momentum, and mass of freshwater and salt all obey a similar rule. If the boundary does not resist the flow through it, then chemical potentials must be constant across the boundary as well for equilibrium to occur, which sets the humidity and partial pressures of gasses. This thought process is a bit less scale- oriented than (7)-(10), and it can be adapted to account for mushy layers including bubbles, spray, etc., more easily.

Heat and freshwater fluxes also arrive at the ocean surface. In practice, the simulation of evaporation and precipitation fluxes, including bubbles, spray, and droplets, is rather complicated in DNS (Schlottke and Weigand, 2008). The heat exchange equations above, while somewhat idealized, are sufficient for our purposes. The heat fluxes through the ocean surface may be in the form of sensible atmospheric heating (≈ 7 W/m^2), latent heat of evaporation (≈ 100 W/m^2), energy that arrives through condensation or rainfall, and electromagnetic radiation at a variety of wavelengths. Positive numbers indicate ocean cooling by convention here. The shortwave radiation that arrives at the ocean surface is primarily solar (≈ -170 W/m^2), while longwave (infrared) radiation is exchanged back and forth between the ocean and atmosphere, with almost no net exchange but regional and temporal variability of order 10 W/m^2. Shortwave radiation penetrates seawater but decays exponentially with depth through absorption of its energy into seawater. Typical e-folding depths for visible light in seawater are $1 - 100$ m depending on wavelength (blues & greens penetrate more deeply) and water clarity. Longwave radiation is absorbed and emitted by only the surface ≈ 1 cm. Daily maximum fluxes can be much larger, with solar and latent heat fluxes near 1000 W/m^2 in magnitude. Useful conversions are that 1 mm of evaporation takes an hour with -635 W/m^2 of ocean heating, while freezing 1 mm of water per hour releases 95 W/m^2. The flux values here are taken from the global means estimated by Grist and Josey (2003) and variability is from Bates et al. (2012).

Note that the details of the air-sea interface, including bubbles, spray, and conversions between molecular and turbulent sensible fluxes complicate these exchanges! Reflection of light at the ocean surface, evaporation, and conversions between different forms of energy make it exceptionally hard to be sure about which fluxes balance which. For this reason, often ocean models are forced with a net energy flux at the surface, or a simple breakdown into net surface fluxes and net penetrative (i.e., radiation) fluxes.

Onsager relations
Lars Onsager, while a junior faculty member at Brown, discovered some amazing relationships that help to better understand what thermodynamical equilibrium and diffusion mean in fluid systems (Onsager, 1931a,b). Unfortunately, this work did not overcome his difficulties as a teacher, and he was asked to leave the faculty soon after publication.[5]

The Onsager relations provide functions for the molecular fluxes in (1)-(5) in terms of the gradients (or "forces" in Onsager's description) of the local thermodynamic properties. What makes the work interesting is that it relates the assumption of local thermodynamic equilibrium to what will eventually occur under global thermodynamic equilibrium if the fluid is left alone. If the fluid is isotropic and locally homogeneous (i.e., in local thermodynamic and dynamic balance), then the relations are

---

[5] He did win the 1968 Nobel Prize in Chemistry for this work as a consolation.

$$\mathbf{q}^{mol} = a\nabla\left(\frac{1}{T}\right) - b\frac{1}{T}\nabla(\mu_s - \mu_w)_T, \tag{11}$$

$$\mathbf{I}_s^{mol} = b\nabla\left(\frac{1}{T}\right) - c\frac{1}{T}\nabla(\mu_s - \mu_w)_T, \tag{12}$$

$$\sigma_{ij}^{mol} = 2\mu D_{ij} - \frac{2}{3}\upsilon D_{kk}\delta_{ij}. \tag{13}$$

The coefficients a, b, c, μ, υ need to be determined for the material under consideration, e.g., the subscripts s and w on μ indicate the Gibbs energy per unit mass of salt or pure water at a given temperature and pressure, respectively. Thus, their difference indicates the change in energy as mass is exchanged between salt and freshwater. Combining these relationships with (1)-(5), it becomes clear that a fluid will not be in equilibrium until: 1) *in situ* temperature is constant, 2) $\mu_s = \mu_w$, and 3) the fluid is strain free, or $D_{ij} = 0$, which is resting or in solid-body motion. For seawater, dynamic viscosities range from $\mu = \upsilon \approx 1.1 \times 10^{-3}$ Pa s for ocean surface conditions and 1.9×10^{-3} Pa s at near freezing temperatures. Taking advantage of the derivatives of the thermodynamic properties and the fact that these diffusivities depend on molecules being present and therefore must be proportional to density, we can express the heat and salt flux in more familiar terms.

$$\mathbf{q}^{mol} = -\rho\left(\kappa_T \nabla T - \kappa_{TS}[\nabla S - \gamma\nabla p]\right), \tag{14}$$

$$\mathbf{I}_s^{mol} = -\rho\left(\kappa_S[\nabla S - \gamma\nabla p] + \kappa_{ST}\nabla T\right). \tag{15}$$

$$\kappa_T \equiv \frac{a}{\rho T^2}, \quad \kappa_S \equiv \frac{c}{\rho T}\frac{\partial(\mu_s - \mu_w)_T}{\partial S},$$

$$\kappa_{TS} \equiv \frac{b}{\rho T}\frac{\partial(\mu_s - \mu_w)_T}{\partial S}, \quad \kappa_{ST} \equiv \frac{b}{\rho T^2}, \quad \gamma \equiv \frac{\partial(\mu_s - \mu_w)_T/\partial p}{\partial(\mu_s - \mu_w)_T/\partial S}.$$

Thus, the molecular fluxes of heat and salt are proportional to the gradients of temperature, salinity, and pressure (albeit with diffusivities that depend on these same properties). Typical values are $\kappa_T = 1 \times 10^{-7}$ m²/s and $\kappa_S = 1 \times 10^{-9}$ m²/s. The mixed κ_{TS} and κ_{ST} are less familiar, but are to be expected because the chemical potentials depend on temperature, pressure, and salinity. They are generally negligible in the oceanic case, i.e., if salinity and temperature both vary in such a way as to have comparable magnitude of effect on seawater density (Caldwell and Eide, 1981).

The implications of constant *in situ* temperature and salinity are likewise rarely considered in the ocean, perhaps because of how long it would take for this system to come into equilibrium. The kinematic viscosity of water is about $\mu/\rho_0 = 1 \times 10^{-3}$ Pa s/1025 kg/m³ $= 1 \times 10^{-6}$ m²/s. For a 4 km deep ocean, it would take roughly 500,000 years of spin-down after forcing for this viscosity alone to silence the flows in the ocean. The thermal diffusivity is 7 to 14 times slower than viscosity, and salinity and carbon dioxide diffusivity are about 100 times slower than temperature, so diffusive ocean equilibration is a 5-500 million year process. Since we know that the forcing of the ocean varies dramatically on diurnal, synoptic, tidal, seasonal, Milankovitch, and tectonic timescales–all of which are faster than these equilibration timescales: the ocean has never been in global thermodynamic equilibrium.

Heat and temperature

Equations (1)-(5) together with appropriate boundary conditions (7)-(10) and molecular bulk formulae (14)-(15) are a closed set of 7 equations in 7 unknowns (ρ, p, T, S and three components of **u**). However, they are rarely numerically integrated in this form.

Common oceanographic simplifications remove the compression term on the right side of (3) by choosing a different temperature-like variable instead of *in situ* temperature. One might call (3) the "temperature equation" since it has DT/Dt on the left side, but note that it also has the compression term proportional to Dp/Dt on the right hand side. This equation is actually a recast form of the first law of thermodynamics (work done equals energy change) for a parcel of fluid, which relates changes in energy to external energy fluxes and adiabatic expansion (work done).

The traditional recasting of this equation is in terms of the potential temperature θ, which is defined as the temperature that a fluid parcel would take if relocated to a reference pressure adiabatically, isentropically, and without changing salinity. As we just noted, the molecular viscosity and diffusivity of seawater are quite slow, so it is reasonable to consider such reversible displacements.

A more accurate temperature-like variable is the conservative temperature, Θ, which is part of the TEOS-10 (McDougall and Barker, 2011) framework. This framework is a considerable improvement over the potential temperature and conductivity empirical relations, because it is based on a thermodynamic potential approach using the Gibb's function which provides consistently accurate relationships between all thermodynamically important quantities. The Gibb's function is $\mu_w(1 - S) + \mu_s S$, that is, the salinity-weighted sum of the chemical potential of salt in seawater and freshwater in seawater. These chemical potentials are not constants, but are functions of pressure, temperature, and salinity.

The conservative temperature is proportional (with coefficient C_p^0 for convenience) to the potential specific enthalpy. Akin to the potential temperature, the potential specific enthalpy is the specific enthalpy a parcel would have if relocated to a reference pressure (usually the surface). Propagating these definitions through (3) yields (McDougall, 2003),

$$\frac{D\Theta}{Dt} = \frac{\alpha}{\tilde{\alpha}} \frac{\dot{q} - \nabla \cdot \mathbf{q}^{all}}{\rho C_p^0} - \left(\frac{\alpha}{\tilde{\alpha}} [\mu_s(p) - \mu_w(p)] - [\mu_s(p_r) - \mu_w(p_r)]\right) \frac{1}{\rho C_p^0} \nabla \cdot \mathbf{I}_S^{mol}, \quad (16)$$

$$\Theta(p, \eta, S; p_0) = h(p_0, \eta, S)/C_p^0, \qquad C_p^0 \equiv 3989.24495292815 J/kgK.$$

While very similar to the potential temperature formulation, in practice (16) is considerably more accurate. It is much easier to implement than (3), because the material derivative for pressure is eliminated so that pressure becomes a diagnostic variable. The coefficients ($\alpha/\tilde{\alpha}$, C_p^0) vary extremely little, as C_p^0 is a constant by definition and $\alpha/\tilde{\alpha} \approx 1$ to within 0.2%. For the same reason, when energy exchanges at the surface are used to alter the conservative temperature, assuming $\alpha/\tilde{\alpha} \approx 1$, the resulting heat uptake is accurate. Finally, specific enthalpy $h = e_i + p/\rho$ is closely related to the specific internal energy e_i, but unlike energy, entropy, or potential temperature, it is particularly easy to model because it is mixed simply by mass proportion (like salinity or mass) when two water parcels are combined at uniform pressure (although mixing also changes the entropy, which affects this simple rule). While potential specific enthalpy (and therefore conservative temperature) is only approximately mixed by mass proportion, it is much closer to being so than temperature, potential temperature, entropy, etc.

Reduced by approximate dynamics: Sound waves and the Boussinesq approximation

Our detailed equations of motion for seawater, including the local thermodynamic properties can be written as

$$\frac{D\rho}{Dt} = -\rho \nabla \cdot \mathbf{u}, \tag{17}$$

$$\frac{D\mathbf{u}}{Dt} = \frac{1}{\rho}\nabla \cdot \left(-p\mathbf{I} + \sigma^{\text{mol}}\right) - 2\mathbf{\Omega} \times \mathbf{u} - g\mathbf{k}, \tag{18}$$

$$\frac{D\Theta}{Dt} \approx \frac{\dot{q} - \nabla \cdot \mathbf{q}^{all}}{\rho C_p^0} \tag{19}$$

$$\frac{DS}{Dt} = -\frac{1}{\rho}\nabla \cdot \mathbf{I}_S^{mol}, \tag{20}$$

$$\rho = \rho(p, \Theta, S). \tag{21}$$

The molecular effects are given by (11)-(13) and the boundary conditions are given by (7)-(10). However, these equations are slow to compute solutions for, as the sound speed limits the possible time step. In this section, we'll explore how sound waves arise and what can be done to ameliorate this issue.

For reversible motions (i.e., those where entropy and salinity are constant), we can convert between the rate of change of density and pressure.

$$\left(\frac{Dp}{Dt}\right)_{S,\eta} = \left(\frac{\partial \rho}{\partial p}\right)_{S,\eta} \left(\frac{D\rho}{Dt}\right)_{S,\eta} = c^2 \left(\frac{D\rho}{Dt}\right)_{S,\eta}. \tag{22}$$

The proportionality constant is the sound speed ($c \approx 1484$m/s) squared, which is a function of the ocean thermodynamic state (p, T, S). If we allow some irreversible processes to occur, they can be absorbed into a diffusion of pressure in a rewritten form of (1).

$$\frac{Dp}{Dt} = -\rho c^2 \nabla \cdot \mathbf{u} + D_p^{\text{mol}}. \tag{23}$$

If this equation and (18) are linearized into small perturbations (indicated by primes, neglected when multiplied together) about a homogenous, stationary background state at rest (indicated by bars, with density and sound speed taken as a constant), they can be combined to eliminate ρ and $\mathbf{u}$.

$$p = p' + \bar{p}, \rho = \rho' + \bar{\rho}, \mathbf{u} = \overset{0}{\bar{\mathbf{u}}} + \mathbf{u}', \rho\mathbf{u} = \bar{\rho}\mathbf{u}' + \rho'\mathbf{u}' \approx \bar{\rho}\mathbf{u}', \tag{24}$$

$$\frac{\partial p'}{\partial t} = -\bar{\rho}c^2 \nabla \cdot \mathbf{u}' + D_p^{\text{mol}}, \qquad -\nabla \bar{p} = \nabla \phi, \tag{25}$$

$$\bar{\rho}\frac{\partial \nabla \cdot \mathbf{u}'}{\partial t} = -\nabla^2 p' - 2\bar{\rho}\nabla \cdot \mathbf{\Omega} \times \mathbf{u}' + \nabla \cdot \sigma^{\text{mol}'}, \tag{26}$$

$$\frac{\partial p'}{\partial t} + c^2 \nabla^2 p' = -2c^2 \bar{\rho} \mathbf{\Omega} \cdot \nabla \times \mathbf{u}' - c^2 \nabla \cdot \sigma^{mol'} + D_p^{mol}. \tag{27}$$

The last equation is the sound wave equation. This equation presents difficulties for numerical modeling, because of the CFL limit for such fast waves (time steps of a few milliseconds for typical grids!). Implicit numerical schemes can be used to remove these waves, but a more common approach is to use a set of reduced equations that emphasize phenomena that travel slower than the speed of sound. This can be accomplished through a number of reduced systems (e.g., the anelastic equations), but the most common oceanographic set relies on the Boussinesq approximation.

Equation (23) connects the change in pressure and density that comes with compression of seawater. But, as seawater is a liquid and not a gas, we know that it is highly resistant to compression. Indeed, even over the immense pressure differences between the surface and the abyss, seawater density changes by only a few percent. So, let us consider a background value of density, a horizontally-averaged density contribution that changes only in the vertical to represent stratification, and a density variation in space and time. For convenience, we write these in terms of buoyancy, which is just minus the density perturbations rescaled to have units of gravitational acceleration.

$$\rho = \rho_0 \left(1 + \bar{b}(z)/g + b(x, y, z, t)/g\right). \tag{28}$$

If we assume that density is nearly its background value, i.e., $b(x, y, z, t)$ and $\bar{b}(z)$ are much smaller than g, and neglect all higher order contributions, then our equations become

$$0 \approx \nabla \cdot \mathbf{u}, \tag{29}$$

$$\frac{D\mathbf{u}}{Dt} = \frac{1}{\rho_0} \nabla \cdot \left(-\phi \mathbf{I} + \sigma^{mol}\right) - 2\mathbf{\Omega} \times \mathbf{u} + b\mathbf{k}, \tag{30}$$

$$\frac{D\Theta}{Dt} \approx \frac{1}{\rho_0 C_p^0} \left[-\nabla \cdot \left(\mathbf{q}^{mol} + \mathbf{q}^{rad}\right) + D_{ij} \sigma_{ij}^{mol} - \mathbf{I}_S^{mol} \cdot \nabla(h_s - h_w) \right] \tag{31}$$

$$\frac{DS}{Dt} = -\frac{1}{\rho_0} \nabla \cdot \mathbf{I}_S^{mol}, \tag{32}$$

$$b = b(z, \Theta, S). \tag{33}$$

Buoyancy appears only in the gravitational term, and only the spatially variable buoyancy at that, because only it can contribute to horizontal gradients of dynamic pressure ϕ, unlike the ρ_0 and $\bar{p}$ which contribute only to the background hydrostatic pressure which is constant in the horizontal. These background fields are related to one another by

$$\frac{\partial}{\partial z}(p - \phi) = -\rho_0(g + \bar{b}), \qquad N^2 = \frac{\partial \bar{b}}{\partial z}. \tag{34}$$

The equation of state (33) depends on depth, not pressure, unlike (21). There are a few reasons for this choice. First, calculating the total thermodynamic pressure is a bit awkward in the Boussinesq system. Second, in a nearly incompressible fluid depth and background pressure are monotonically related and thus interchangeable. Finally, Vallis (2006) notes that this choice makes defining energy in the Boussinesq equations more consistent.

In general, care is needed with energetics as Boussinesq models have no conversion of internal energy to mechanical energy, but do allow conversion between potential and kinetic energy via sinking of dense water or rising of light water via vertical velocity times buoyancy: wb (Young, 2010). In a stably-stratified ocean, wb keeps water parcels near a fixed location. The restoring force for internal waves (where $wb < 0$) is a good example. A second important role for wb is as the source of energy for baroclinic instabilities, which convert mean potential energy to eddy energy by correlating eddy motions with water buoyancy (thus $wb > 0$ on average). Likewise, unstable density profiles convect with $wb > 0$. Also, the care required to handle the compressible aspects resulting in conservative temperature are not lost! Equation (31) is significantly different than (3) in terms of the stratification it predicts in (34), which in turn sets the level of stratification felt by the flow.

These equations behave differently from the compressible set. The sound waves above required a three-dimensional divergence in velocity. In these Boussinesq (1897) approximation equations, the velocity field does not diverge, so sound waves are eliminated as desired. Furthermore, the subtle distinctions between properties per unit mass and properties per unit volume are lost–those conversions now all occur with the background density ρ_0. Finally, the equation that plays the role of conservation of mass (29) doesn't even feature the units of mass! Now, the volume of fluid is conserved. Because the density is still allowed to vary somewhat (through the buoyancy), actual mass is no longer conserved. This fact means that the calculation of subtle mass budgets, e.g., those required to calculate sea level rise, require careful thought (Griffies and Greatbatch, 2012).

Dimensionless equations and scales

The Boussinesq equations (29)-(33) are an efficient alternative to the compressible equations (17)-(21) when we are not interested in sound waves. Perhaps there are other simplifications we can arrive at by neglecting other terms in these equations. The general technique for exploring the size of different terms in the equations is scale analysis, which is most natural in a dimensionless form of the equations of motion. In dimensionless equations, the expected size of each term is made explicit so small terms that may be neglected are highlighted. Taking the Boussinesq equations (29)-(33) in a rotating frame, and simplifying friction terms, we arrive at a set of equations convenient for dimensional analysis (following Lilly, 1983; McWilliams, 1985; Bachman et al., 2017a). One simplifying assumption in addition to the preceding equations is that we now assume that the rotation axis, and hence **f**, is aligned with the local vertical **k**, which is sometimes called the *traditional approximation*.

Next, we choose the approximate scales of the dimensional variables. The 3D velocity in the Boussinesq equations (29)-(33) (**u**) is partitioned into the horizontal velocity (v_h with typical scale V_*) and vertical velocity (w) based on gridscale parameters of vertical grid scale (Δz) and horizontal grid scale (Δs). A locally-linear equation of state[6] will be assumed, so that $\bar{b}(z) + b(x, y, z, t) = \bar{b} + b = g(\alpha[\bar{\theta} + \theta] - \beta[\bar{S} + S])$. Scalings for perturbation pressure ($\phi \sim \phi_0 = \max(V_* f_0 \Delta s, V_*^2)$),

---

[6] That is, α, β may vary but their gradients will be neglected in favor of the larger S, θ variations here.

stratification ($\partial_z \bar{b} \sim N_*^2$, $\partial_z \bar{S} \sim N_*^2/g\alpha$, $\partial_z \bar{\Theta} \sim N_*^2/g\beta$), perturbation active tracers ($b \sim \phi_0/\Delta z$, $\theta \sim \phi_0/\Delta z g\alpha$, $S \sim \phi_0/\Delta z g\beta$), and vertical velocity ($w \sim \phi_0 V_*/N^2\Delta z\Delta s$) are used.

Taking these dimensional scalings for all of the variables, we can construct a set of equations for all of the dimensionless variables. A dimensionless variable is just the dimensional one divided by its scale, e.g., the dimensionless horizontal velocity is just the dimensional horizontal velocity divided by V_*. After substituting in all of these scale factors into (29)-(33), they are collected into dimensionless factors which will be named and then interpreted below. The equations for the dimensionless variables are:

$$\text{Ro}_* [\partial_t \mathbf{v}_h + \mathbf{v}_h \cdot \nabla \mathbf{v}_h + \epsilon w \partial_z \mathbf{v}_h] + \underbrace{\left(1 + \frac{y\text{Pl}_*}{\Delta y}\right) \mathbf{z} \times \mathbf{v}_h + M_{R_*} \nabla_h \phi}_{\text{geostrophic}} = \frac{\text{Ro}_*}{\text{Re}_*} \nabla_i \sigma_{ih}, \tag{35}$$

$$\text{Fr}_*^2 \frac{\Delta z^2}{\Delta s^2} [\partial_t w + \mathbf{v}_h \cdot \nabla w + \epsilon w \partial_z \mathbf{v}_h] + \underbrace{\partial_z \phi - b}_{\text{hydrostatic}} = \frac{\text{Fr}_*^2 \Delta z^2}{\text{Re}_* \Delta s^2} \nabla_i \sigma_{iz}, \tag{36}$$

$$\partial_t S + \mathbf{v}_h \cdot \nabla S + \epsilon w \partial_z S + w \partial_z \bar{S} = \frac{1}{\text{Pe}_*} \nabla \cdot \mathbf{I}_S^{all}, \tag{37}$$

$$\partial_t \Theta + \mathbf{v}_h \cdot \nabla \Theta + \epsilon w \partial_z \Theta + w \partial_z \bar{\Theta} = \frac{1}{\text{Pe}_*} \nabla \cdot \mathbf{I}_\theta^{all}, \tag{38}$$

$$\partial_t b + \mathbf{v}_h \cdot \nabla b + \epsilon w \partial_z b + w \partial_z \bar{b} = \frac{1}{\text{Pe}_*} \nabla \cdot \left(\alpha \mathbf{I}_\theta^{all} - \beta \mathbf{I}_S^{all}\right), \tag{39}$$

$$\nabla \cdot \mathbf{v}_h + \epsilon \partial_z w = 0, \tag{40}$$

$$M_{R_*} \equiv \max(1, \text{Ro}_*), \quad \epsilon \equiv \frac{\text{Fr}_*^2}{\text{Ro}_*} M_{R_*} = \begin{cases} \text{Fr}_*^2 & \text{Ro}_* \geq 1, \\ \text{Ro}_* \text{Bu}_*^{-1} & \text{Ro}_* < 1 \end{cases} \tag{41}$$

When h or z are used as a subscript, only the horizontal components or vertical component of that variable are relevant, respectively. Repeated indices are summed over all three directions. Note that σ_{ih}, σ_{iz} represent projections of an overall (symmetric) stress tensor into horizontal and vertical directions. Only the leading order term involving Pl_* is retained, given its small size on the grid scale of operational models.

Working from the perspective of model development, it is most useful to consider the scaling behavior with lengthscales based on the model grid scale as this set has. Following Fox-Kemper and Menemenlis (2008), an asterisk denotes resolved model fields, as opposed to the abstract field variables. The dynamics that should occur at the grid scale are precisely the largest features that the model cannot resolve, and thus the phenomena whose effects that are likely most important to parameterize within our chosen model. These convenient dimensionless parameters are formed using the discretization scales of the grid (horizontal: $\Delta s = \sqrt{\Delta x \Delta y}$, vertical: Δz). We begin by identifying the grid Rossby number ($\text{Ro}_* = U_*/f_0 \Delta s$), baroclinic Froude number (see below), and planetary number ($\text{Pl}_* = \Delta y \frac{\partial f}{\partial y}/f_0$) as important. M_{R_*}/Ro_* is the Euler number, and can be taken to be the maximum of 1 and Ro_*^{-1}. Also, the grid aspect ratio ($\delta = \Delta z/\Delta s$), Reynolds number ($\text{Re}_* = U_* \Delta s/\nu_*$), and Péclet number ($\text{Pe}_* = U_* \Delta s/\kappa_*$) are needed to establish how much damping is required. Note that ν_* denotes an "eddy viscosity" which is to be used in a numerical model rather than measured experimentally. Numerical stability requires Re_* and Pe_* to be $O(1)$. Parameterization examples determining viscosity are given in the last section before the conclusions.

The first baroclinic deformation radius L_d is approximated in variable stratification by Chelton et al. (1998) as

$$L_d = \frac{1}{|f|\pi} \int_{-H}^{0} N_*(z)\,dz. \tag{42}$$

If the Froude number is defined similarly (LeBlond and Mysak, 1978), then

$$\text{Fr}_* = \frac{V_*}{\int_{-H}^{0} N_*(z)\,dz}. \tag{43}$$

With these definitions the gridscale Burger number (Bu$_*$) relates the deformation radius to grid scale: $\pi L_d/\Delta s = \text{Ro}_*/\text{Fr}_* = (\text{Bu}_*)^{1/2}$, so a large Burger number implies a well-resolved deformation radius. In order for this to be consistent with the values of N_*^2 used above, the vertical average of N_* should be used as the scale factor.

With either small Froude number (Fr$_*$) or a grid with shallow aspect ratio ($\Delta x \gg \Delta z$), or both, (36) reduces to the hydrostatic balance. A large Froude number, on the other hand, implies very weak stratification which is likely to allow 3D overturning and large vertical velocities. However, in this case, the overturning plumes may be horizontally small, so a convective parameterization or square or tall aspect ratio grid is needed to capture these effects. Hydrostatic models have a simplified vertical momentum balance, which means that the pressure is simply the weight per unit area of fluid above and the vertical accelerations are small compared to gravity. Hydrostatic models are typically used for most oceanographic applications, because ocean models tend to have shallow aspect ratios and small Froude number. Hydrostatic models are appropriate to study the submesoscale and larger scales. The hydrostatic, Boussinesq equations are sometimes called the *primitive* equations of oceanography.

Similarly, small Rossby number (Ro$_*$) at the gridscale leads to a dominantly geostrophic balance in (35). The Rossby number compares the Coriolis force to advection and acceleration. Fluids with small Ro$_*$ can be thought of as *rapidly rotating*–that is, the rotation of Earth is fast compared with the rotation and acceleration of the fluid relative to Earth.

The quasigeostrophic equations–the system that results from hydrostatic (small Froude number and shallow aspect ratio) and nearly geostrophic motions (small Rossby number)–remain a popular tool for studying mesoscale dynamics and building subgrid schemes for mesoscale-permitting models (Jansen and Held, 2014; Straub and Nadiga, 2014; Shevchenko and Berloff, 2015; Chen et al., 2016). One reason for their popularity is that inertial, internal gravity, and Kelvin waves are filtered out of explicit treatment in the quasigeostrophic equations, similarly to how sound waves are filtered out by the Boussinesq approximation. As a result, the quasigeostrophic system can take time steps that are orders of magnitude larger than the hydrostatic, Boussinesq equations (Fig. 2.2). Indeed, it was the quasigeostrophic system which Charney discovered during his Ph.D. work and that Charney et al. (1950) used for the first successful weather models. The quasigeostrophic equations have only non-divergent, horizontal, geostrophic motions at leading order, and while this removes the gravity waves formally they cannot be applied globally because of the way the Coriolis parameter is scaled and their inability to handle the equator.

Extremely small Rossby number can make a different approximation appropriate: the planetary geostrophic equations. These equations neglect acceleration and advection of momentum altogether. The only D/Dt terms that are retained are those in the salinity, temperature, and thus buoyancy equations. The planetary geostrophic equations can be used globally, and they are very efficient for low-resolution, long-duration simulations. These equations are often used for theory (Luyten et al., 1983; Huang and Flierl, 1987; Fox-Kemper and Ferrari, 2009), but only rarely (Samelson and Vallis, 1997; Olbers and Eden, 2003) are equations in this regime used for numerical modeling.

Characteristic Motions by Scale

It is a start to have an idea of what reduced, closed sets of equations might result at different scales, but there are behaviors that aren't easily described this way. For one, we will want to distinguish motions that are relatively steady (i.e., currents) from waves and turbulence. Also, not every behavior noted in Fig. 2.2 has its own equation set, but still they are all different enough to be named distinctly! Without being too exhaustive, here we will examine some of the most commonly discussed scales and their dominant behaviors.

Currents, waves, and turbulence

Currents
A current is a persistent pattern of velocity that can transport seawater, along with other properties, from place to place. On large scales away from the equator, currents tend to be in geostrophic balance. On small scales currents can be topographically constrained (e.g., river plumes, tidal channels, island wakes, etc.).

Waves
Unlike a current a wave transports information by the propagation of the wave, which may not require moving the seawater along with the wave. Often we think of a wave as a repeating sinusoid or as a crest approaching shore and breaking, and the sinusoidal wave is helpful in understanding the dispersion curves in Fig. 2.2. Mathematical descriptions of waves can be linear or nonlinear, but the waves that are solutions to (35)-(40) neglecting all nonlinear terms are useful to consider. As a simple example, we can neglect the effects of advection altogether by simply linearizing perturbations about a state of rest in (29)-(33). Furthermore, let's simplify by neglecting radiation, friction heating, and mixing heating, and removing the background stratification in Θ and S that is already in balance with diffusion, and note that small deviations away from this background (with primes) will not distinguish between Θ and T, then

$$0 = \nabla \cdot \mathbf{u}', \qquad (44)$$

$$\frac{\partial \mathbf{u}'}{\partial t} = \frac{1}{\rho_0} \nabla \cdot \left(-\pi' \mathbf{I} + \sigma^{\mathrm{mol}}\right) - 2\mathbf{\Omega} \times \mathbf{u}' + \left(\frac{\partial b}{\partial \Theta}\Theta' + \frac{\partial b}{\partial S}S'\right) \mathbf{k}, \qquad (45)$$

$$\frac{\partial \Theta'}{\partial t} + w'\frac{\partial \overline{\Theta}}{\partial z} \approx \frac{1}{C_p^0}\nabla \cdot \left(\kappa_T \nabla \Theta' - \kappa_{TS}\nabla S'\right) \tag{46}$$

$$\frac{\partial S'}{\partial t} + w'\frac{\partial \overline{S}}{\partial z} = \nabla \cdot \left(\kappa_S \nabla S' + \kappa_{ST}\nabla \Theta'\right). \tag{47}$$

This set of linear equations is closed, and every term in all of the equations depends linearly on a perturbation variable. One obvious solution to this set is thus $\mathbf{u}' = 0$, $S' = 0$, $\Theta' = 0$, $b' = 0$.

However, waves are the nonzero free modes of this system, which can occur only if the set of equations is singular. Again, for simplicity consider the case of plane sinusoidal waves, where every partial derivative of a primed quantity is replaced by multiplication by a frequency or a wavenumber (2π divided by the wavelength, which is 2π times the number of wave crests per unit distance). This set of equations then becomes a matrix set like,

$$\mathcal{L}(\omega, k, l, m)\begin{bmatrix} u' \\ v' \\ w' \\ \Theta' \\ S' \\ \pi' \end{bmatrix} = 0. \tag{48}$$

If the matrix $\mathcal{L}$ can be inverted, then the solution is just zero. If it can't be inverted, then the matrix is singular. The characteristic equation of that matrix will be a sixth-order polynomial, which will depend on the background stratification, frequency, and wavenumber. Each root of that equation will be the dispersion relation of a plane wave under the assumptions just described.

Linear, steady plane waves are reversible in time (you can check this by reversing time in the equations and see if you can rewrite them to be identical to the original set), thus they may wobble back and forth or up and down everywhere, but they do not permanently alter the state of the ocean after averaging over a wave period. Thus, currents and linear, steady plane waves do not interact meaningfully. While these waves are beautiful and interesting objects on their own, if you are interested in mapping the currents in a region—averaging over many wave periods—you don't need to consider them. Furthermore, a matrix system like (48) has a family of solutions for each wavenumber and frequency. Linear systems that result from the fluid equations tend to be *orthogonal* and thus the different waves do not interact with one another. Similarly, a linear plane wave with a given frequency and wavenumber tends to preserve those properties unless it enters a region of different stratification or shear (then it is no longer a plane wave!). However, if you reintroduce the nonlinear or damping terms, this is no longer true. Damped or strongly nonlinear (often called *breaking*) waves can and do affect the mean state and currents of the ocean, as well as each other. Energy gets transferred from wave to wave and frequency to frequency. Understanding these wave-wave interactions is an active topic of research (Nazarenko, 2011).

Of course, plane waves are not the whole story, because the ocean has a surface and a bottom, and the stratification in Θ and S may vary in the vertical as well. Shear and stratification also vary horizontally from region to region as well. The set of waves in such a complex domain is an even larger set of phenomena, as the kinematic and pressure boundary conditions must now be brought to bear on variations in sea surface height. These new degrees of freedom bring surface gravity waves and other free modes of the system.

When waves depend on location such as these surface-bound examples, another important effect occurs: the Stokes drift. The Stokes drift is the difference between the average velocity of all waves and currents averaged sitting at one location (Eulerian velocity) and the average velocity following the trajectory of a fluid parcel in those waves and currents (Lagrangian velocity). The Stokes drift allows waves that vary in space to participate in the processes that permanently transport material and water parcels. Thus, waves with Stokes drift can move flotsam, form windrows, and drive turbulence.

Turbulence

Waves are easier to handle when they are linear or weakly nonlinear. Turbulence is the opposite extreme, intermittent flows that are most easily understood in the strongly nonlinear limit. In oceanography usage, "turbulence" covers a wide range of phenomena. What they all have in common is strong nonlinear effects through the advection of momentum, energy, or tracers. Turbulence research began with the study of isotropic, 3D, incompressible, constant density turbulence (Reynolds, 1895; Kolmogorov, 1941). Oceanographers are very concerned as well with temperature and salinity effects and stratification and the effects of the planet's rotation. Thus, the traditional approach needed, and in many important ways still needs, adaptation for relevance to oceanography.

The Onsager relations, and the beautiful form of the resulting diffusivities and stress tensor, relies heavily on the idea that the molecular motion is isotropic and locally homogeneous. Oceanographic turbulence, particularly submesoscale and mesoscale turbulence, is not isotropic and often not homogeneous. If turbulence behaved just as molecules do–resisting the development of gradients in temperature and salinity through isotropic mixing that leads to fluxes oriented down the gradient of these properties, then Onsager's theory could be modestly adapted. Müller (2006) works through this analysis, and provides "eddy" viscosity and diffusivity closures.

However, the vertical direction is special in that it orients the stratification, and the orientation of the rotation axis is also an important broken symmetry. The Taylor-Proudman theorem states that rapidly rotating, unstratified fluids will "stiffen" along the rotation axis and resist shear in the direction of the rotation axis. In contrast, the presence of stratification "lubricates" the flow leading to larger velocities in the horizontal than the vertical. Furthermore, a major ingredient of the molecular theory is that there is a scale separation between the paths of the molecules and the gradients they act on. Turbulence tends to mimic the properties and locations of its energy source, so that eddies shed from the Gulf Stream are strongest near the stream, nearly the same size, flow with similar velocity, and are comprised of similar water types. Thus, no scale separation exists. On the other hand, turbulence within a boundary layer is limited to the scale of the boundary layer, which might be significantly smaller than a large-scale ocean model, which means in this case there is a scale separation.

To manage the modeling when there is sometimes a scale gap and sometimes not, two methods of analysis and parameterization have arisen. The first, Reynolds-averaged theory (Reynolds, 1895), relies on a scale separation between the turbulence and the resolved flow. The key mathematical aspect of this approach is that once averaging of the turbulence up to the gridscale of the model has occurred, averaging again gives the same result. The second approach is Large Eddy Simulation (Smagorinsky, 1963; Deardorff, 1970). This approach expects that the simulation takes place within the middle of a range of turbulent motions. The largest are resolved and the smallest are parameterized. The model grid lies in between. Here the key concept is the idea of simulation of a filtered version of reality. The filter reduces the small scales and doesn't affect the large scales. However, if the filter is applied twice, the effect is different than applying it only once.

One excellent way to begin to quantify turbulence and waves is using a *power spectrum*. The power spectrum breaks up the energy, or temperature variance, or salinity variance, etc. into contributions from different frequencies and/or wavenumbers. Waves will have their energy concentrated along their dispersion relations, which may be peaked near forcing frequencies (e.g., tides or seasons) with the corresponding wavenumbers to match. Turbulence tends instead to have power spectra that obey a power-law distribution. For example, the kinetic energy power spectrum of 3D, isotropic, homogeneous turbulence is nearly proportional to wavenumber to the -5/3 power. This means that energy is not concentrated in only some particular modes of turbulence, but distributed over all of the modes in a special way as a function of scale. This is one of the most powerful notions for the motions of the oceans.

Basin scales

The largest scales of ocean motions reflect the large scale forcing patterns of winds, tides, and thermal forcing. Two key concepts are the Ekman (1905) and Sverdrup (1947) balances. These balances of Coriolis force and winds form constraints on the currents that are consistent with large-scale steady wind patterns.

The Ekman transport is the mass transported directly by the wind, and this transport can converge and diverge. Ekman effects are particularly profound near coasts and the equator. These convergences and divergences set up patterns in vorticity and pressure that are the key drivers of the gyres and the Antarctic Circumpolar Current (ACC), so these phenomena are indirectly driven by the wind. The Ekman transport also participates importantly in the MOC–forcing the surface branches of the subtropical cells and the Southern Ocean overturning. The Ekman transport is a balance involving the Coriolis force, so it is only achieved after Earth has rotated a few times–it is a subinertial motion. Transient winds that are more rapid than the inertial period tend to drive near-inertial oscillations and gravity waves. Overall, atmospheric forcing on large scales drive transient ocean responses that tend to be more responsive as the frequency of the forcing gets slower (Hasselmann, 1976).

Above, a contrast was made between the gyres–with their horizontal circulations and vertical vorticity– and the cells of the MOC–with their overturning circulations and zonal axis of vorticity. Outside of the regions where they concentrate into boundary currents, the gyres are a result of Ekman and Sverdrup balances (or similar theories that may include bottom slope as well, Wunsch and Roemmich, 1985). The planetary geostrophic equations can be used together with the Ekman transport to simulate the gyres outside of boundary currents. These scales have mostly horizontal motions with only small divergences due to deepening layers and the latitudinal variations of the Coriolis parameter. Thus, the vertical vorticity exceeds the divergence. Depending on basin, the gyres have different mass transport (here expressed with vertically-integrated volume streamfunction Ψ, which is related to the vertical vorticity by a Laplacian derivative), basin width L_b, and depth H, but a rough estimate can be made for the vertical relative vorticity nonetheless:

$$\zeta_z = \nabla^2 \frac{\Psi_z}{H} \approx \frac{\Psi}{L_b^2 H} = \frac{50\text{Sv}}{L_b^2 H} \approx 5 \times 10^{-6}\,\text{f}. \qquad (49)$$

The vorticity is compared to f, so that the coefficient can be interpreted as a (very small) Rossby number, Ro$_* \approx 5 \times 10^{-6}$. The typical Froude number of a gyre interior is also quite small ($O(2 \times 10^{-4})$), given the velocity scales from Ekman and Sverdrup balances of only O (1 cm s^{-1}).

The MOC has important forcing from wind and tidal mixing (Munk and Wunsch, 1998), but increasingly has become understood as a blend of wind-driven overturning, cooling, and mixing (Gnanadesikan, 1999). A key method by which the MOC has been measured is through tracking watermasses as they interleave between one another (Talley, 2008). One interesting question is whether the spread of these watermasses is best understood as a current or as a diffusive or other kind of turbulent dispersion (Lozier, 2010). It is important to recognize that while the MOC and gyres can be distinguished in effect using different vorticity components, their forcing mechanisms are deeply and intricately linked (Yeager, 2015). The horizontal vorticity of the MOC is larger than the relative vorticity of the gyres,

$$\zeta_x = \nabla^2 \frac{\Psi_x}{L_b} \approx \frac{\Psi}{L_b H^2} = \frac{50\text{Sv}}{L_b H^2} \approx 0.05\,\text{f}. \qquad (50)$$

However, most of this overturning is very shallow, and the deeper circulations of NADW and AABW are very slow as they spread through the deep oceans, taking hundreds to thousands of years to refresh the abyss (Gebbie and Huybers, 2012). These slow currents lead to a small Froude number, despite the relatively low stratification below the pycnocline.

Another basin scale feature is the ACC. It is the ocean current most like an atmospheric circulation, as it is not bounded by continents and so it flows around the Earth relatively unimpeded. Dynamically, the winds and surface Ekman transport away from Antartica provide this current with potential energy and momentum. The many smaller eddies and fronts that form within the current play an important role in balancing these budgets (Johnson and Bryden, 1989; Radko and Marshall, 2006). The Ferrel cell is the atmospheric circulation most like the Deacon cell, which is the MOC associated with the ACC, because the mean flow in both cells work to strengthen thermal gradients, but eddies and storms work to overturn in the other direction and dominate the heat transport overall.

There are a number of basin scale modes of variability. Perhaps the most famous is El Nino, which is a coupled oscillation of the ocean and atmosphere. The Pacific Decadal Oscillation resembles El Nino but is slower, and variability of the Atlantic and Indian Oceans' equatorial regions have variability that is similar dynamically to El Nino as well. However, there are other basin-scale oscillating modes of the ocean that are not coupled between the atmosphere and ocean: Rossby basin modes. These large-scale modes are basin-filling Rossby-Kelvin waves that may be excited by long tides, seasons, or other low-frequency climate modes. While these waves are basin-wide, as Rossby waves are fastest when they are large, these modes can be surprisingly fast, especially for the barotropic modes (basin transit time on O(weeks)). Decadal variability in sea surface temperature, sea surface height, and regional climate generally may have predictability due to baroclinic Rossby modes and related phenomena (Meehl et al., 2014).

Based on the idea of a critical Reynolds number, beyond which all motions will be turbulent, and the enormous Reynolds number of basin scale motions (Re $\approx$ 1 x 10^{11}), one might expect that the motions on the basin scale might directly stimulate a cascade of turbulence directly, filling in a power law from the largest scales all the way down to the Kolmogorov scale. However, the effects of rotation and stratification prevent this from occurring. With small Ro_*, Fr_* and ϵ the nonlinear terms in the momentum equations critical to turbulence are emasculated by the hulking Coriolis and stratification terms. The huge potential energy on these large scales vastly exceeds their kinetic energy, as the geostrophic flows enclose stores of potential energy and there are few instabilities or turbulent features capable of accessing it. Only on the mesoscales will turbulence arise.

Mesoscales

At the mesoscale a few different effects combine to make this range of scales the largest scale where ocean turbulence (of the rotating, stratified kind) is found. Because these eddies are the largest and most energetic turbulent eddies in the ocean (Fig. 2.3), they tend to have the largest horizontal turbulent transport of properties. Thus, from the modeling perspective, if these scales are not resolved they must be parameterized, and if they are resolved care is needed to handle them (ensembles, numerics, and scale-aware parameterizations schemes to avoid double-counting the resolved features).

The equations governing the gyres and MOC are actually not capable of closing these circulations at only the basin scale. Western boundary currents (Stommel, 1948; Munk, 1950; Charney, 1955) are *required* to close the Ekman and Sverdrup circulations without a basin-filling inertial circulation (Fofonoff, 1954; Veronis, 1966; Fox-Kemper and Pedlosky, 2004). The MOC, too, has deep western boundary currents for similar dynamical reasons (Stommel and Arons, 1960). Wherever Ekman transport is directed away from the coast, upwelling (or eastern[7]) boundary currents form. These boundary currents focus the large scale kinetic and potential energy into mesoscale-sized intense currents.

So, just how big is the mesoscale? It is deformation radius-sized. The deformation radius compares the scale of the Coriolis parameter to the stratification and width to depth. When the width is just right (the deformation radius), the Coriolis effect is the same size as the stratification effect. Another way to understand the deformation radius is that it is the scale where kinetic and potential energy in a geostrophic current are the same magnitude. Very loosely, the mesoscale can be considered to be in the 10 to 300 km range, but it is considerably more accurate to relate the

---

[7] Although they need not necessarily be on the eastern side of oceans.

mesoscale to motions near the first baroclinic deformation radius which varies globally and with stratification.

We can follow the same rough vorticity scaling to estimate the vorticity in a boundary current of width L_f, and hence its Rossby number,

$$\zeta_z = \nabla^2 \frac{\Psi_z}{H} \approx \frac{\Psi}{L_f^2 H} = \frac{50 \text{Sv}}{L_f^2 H} \approx 0.05 \, \text{f}. \quad (51)$$

So, Ro ≈ 0.05. Similarly, we can estimate the Froude number for such a current (≈ 1 m s^{-1}) to be Fr$_*$ ≈ 0.1 with a typical value of N ≈ 0.01 cycles/s. Now, with the same scales, the deformation radius is (L_d ≈ 30 km) and ϵ ≈ 0.2. So, all of our small parameters are still small, in comparison to geostrophic and hydrostatic balances. Why is there turbulence at the mesoscale?

The key has come from studying the instabilities of the quasigeostrophic equations. These equations do have a form of turbulence, quasigeostrophic turbulence (Rhines, 1979), which is very similar to two-dimensional turbulence (Kraichnan, 1967; Charney, 1971). However, unlike Kolmogorov (1941) turbulence has a *forward* energy cascade of large-scale energy toward small-scale energy, two-dimensional and quasigeostrophic turbulence has an *inverse* cascade from a central range of instability scales of energy toward *larger* scales and a forward cascade of (potential) enstrophy toward smaller scales, which is (potential) vorticity squared.

The character of turbulence and the instabilities that lead to it depend on which of the different dimensionless parameters are small, but also critically that they are equally small. Mesoscale turbulence is what occurs when the Burger number is near 1–that is, motions that are near the Rossby deformation radius. However, the mesoscale also is typified by small Rossby and Froude numbers. The mesoscale is frequently populated by the baroclinic and barotropic instabilities. The baroclinic instabilities are just a bit larger than the deformation radius in size, and the barotropic instabilities are a shear instability that forms from the horizontal shear in the boundary currents— which are also part of the mesoscale based on typical size, rotation, and stratification. The effect of latitudinal variation in the Coriolis parameter (beta-effect) is also important on the mesoscale, leading to jets in the ACC, beta-plumes, and limits on the size of the largest mesoscale eddies, but it is not required to have mesoscale turbulence. The *Rhines* scale $L_R = (V_*/\beta)^{1/2}$ is the largest scale that the beta-effect allows to be turbulent–larger scale perturbations tend to be wave-like. Not coincidentally, Charney (1955) finds that this same scale is the governing one to determine the width of inertial western boundary currents (as opposed to cruder metrics based on frictional or other parameterizations of mesoscale eddies, Stommel, 1948; Munk, 1950; Grooms et al., 2011). The quasigeostrophic equations are horizontally nondivergent at leading order, and so too are mesoscale eddies and currents, with vertical vorticity dominating their divergences.

The mesoscale is also home to many waves–Rossby waves that are subinertial like eddies, but also Kelvin, equatorial, and internal gravity. Indeed, in Fig. 2.2, the mesoscale is best identified as the region where all of the subinertial wave dispersion curves meet. Rossby wave-like features have been observed (Chelton and Schlax, 1996), but often their amplitude is so large that distinguishing waves from nonlinear waves from eddies is difficult (Chelton et al., 2007). The faster Kelvin and internal waves are more commonly observed at the mesoscale.

Until recently, global models resolving the mesoscale were rare. Through process studies and basin-scale simulations, the dynamics of the mesoscale became well-understood enough to recognize the primary behaviors of mesoscale instabilities and their effects on larger scales. In models that do not resolve the mesoscale, these behaviors are approximated with parameterizations of their effects on diffusion of tracers (Redi, 1982) and advection of buoyancy (Gent and McWilliams, 1990), as well as setting a larger horizontal viscosity to ensure numerical stability by $Re_* \sim O(1)$. The key aspects of mesoscale eddies that these parameterizations capture is that mesoscale eddies are much wider (100 km) than they are tall (4 km) so the flux more easily in the horizontal than the vertical, and their energy density is low compared to the stratification. Thus, they prefer to flow along density surfaces–or more precisely surfaces that are energetically neutral–rather than mixing together water masses of different densities which costs energy. These remedies are fairly well-understood now, and they can be made to switch off as latitudes and regional stratification vary and the mesoscale becomes resolved (Hallberg, 2013).

Submesoscales

The Rossby and Froude numbers remain small at the mesoscale while the Burger number is $O(1)$, but what happens when Rossby and Froude are not small? This is the regime known as the submesoscale. There have been many recent papers surveying the dynamics in this regime (Thomas et al., 2008; McWilliams, 2016), so only a brief summary will be included here. In Fig. 2.2 the submesoscale is indicated to be a range of scales from near the inertial period up to weeks in time, and roughly 100 m to 10 km in horizontal scale. However, it is much more accurate to define the submesoscale as the range of scales where the Rossby number and Froude number are $O(1)$. Away from the upper and bottom boundary layers of the ocean, the stratification is large on all scales and the Froude number remains small. Thus, the conditions for the *existence* of a submesoscale may only be satisfied in some layers or regions of the ocean. Given the relationship between the Burger, Rossby, and Froude numbers, this directive can be taken to indicate that the submesoscale occupies primarily the boundary layers of the ocean at scales near the mixed layer deformation radius, i.e., the deformation radius based on the stratification and width of only the mixed layer of the ocean. As the mixed layer varies seasonally, so too does the activity and scale of the submesoscale. High-resolution models and observations (Fig. 2.3) indicate that there is an active submesoscale in most surface and bottom boundary layers in most of the world at least seasonally (Mensa et al., 2013; Brannigan et al., 2015; Callies et al., 2015).

When the Rossby number is not small (58) indicates that geostrophy is no longer a good approximation. The quasigeostrophic equations are therefore no longer valid, although a number of examples of behaviors resembling mesoscale phenomena have been found to occur at the submesoscale. The submesoscale has eddies that are not geostrophic, but still are governed by a balance similar to geostrophy involving potential vorticity (Boccaletti et al., 2007). There are also fairly persistent fronts that approach geostrophic balance for the along-front velocity, even though the cross-front direction does not (Hoskins, 1975; Capet et al., 2008b). However, a critical difference between such fronts at the mesoscale and submesoscale is the strength of the cross-frontal or cross-filamentary circulation and thus the degree of surface convergence. At the mesoscale, such convergences are small, while at the submesoscale surface convergences can be as large as the vertical vorticity (D'Asaro et al., 2018). The magnitude of such convergences stems from the $O(1)$ Rossby number effects in (35)-(40) and from an intense interaction of boundary layer mixing and frontal or filamentary convergences (McWilliams et al., 2015) or straining by submesoscale eddies and a Stokes drift effect explained below (Suzuki et al., 2016). The concentration of energy into these fronts, which then may break up into submesoscale instabilities or be dissipated by boundary layer mixing, is an important way that large-scale energy is transferred to smaller scales (Capet et al., 2008c; Callies et al., 2016), as mesoscale eddies tend to have an energy cascade toward larger scales.

Another aspect of the submesoscale, closely related to the surface convergences, is the rapid restratification due to submesoscale features (Boccaletti et al., 2007; Fox-Kemper et al., 2008; Capet et al., 2008a). While mesoscale features, particularly fronts and baroclinic eddies, do increase the stratification of the ocean by tilting horizontal density gradients into the vertical, this effect is modest and the background stratification of the ocean interior can be considered roughly independent of this effect. Submesoscale fronts and eddies, however, overturn horizontal density gradient much faster–rivaling the rate of overturning associated with geostrophic adjustment after mixing (Tandon and Garrett, 1995). An important dimensionless parameter governing restratification is the Richardson number Ri, which is closely related to the Froude number, but not identical as it depends on velocity shear not velocity magnitude:

$$\text{Ri} \equiv \frac{N^2}{\frac{\partial u}{\partial z}} \sim \frac{N_*^2 H_*^2}{V_*^2} \sim \frac{V_*^2}{\left(\int_{-H}^{0} N_*(z)\,dz\right)^2} \equiv \frac{1}{\text{Fr}_*^2}$$

The Richardson number indicates whether there is enough vertical shear to drive mixing and overturning of the stratification. Geostrophic adjustment can increase stratification, but it also affects the vertical shear, settling down into inertial oscillations that vary about Ri = 1. Thus, geostrophic adjustment cannot extract all of the available potential energy by overturning, just as on the planetary scale geostrophic currents balance the tendency for fronts to overturn. Symmetric instabilities are a small, fast category of submesoscale instabilities, which can also rapidly restratify toward Ri = 1 (Thomas et al., 2013; Bachman et al., 2017b). To continue restratifying beyond Ri = 1, only baroclinic instabilities (sometimes called mixed layer eddies for the submesoscale surface variety) remain. The combined effect of mixing and submesoscales is an active topic of research (Taylor and Ferrari, 2010; Hamlington et al., 2014; Whitt and Taylor, 2017; Callies and Ferrari, 2018; Sullivan and McWilliams, 2018).

Unlike the mesoscale, submesoscale dynamics allow fewer approximations in (35)-(40). Neither Rossby nor Froude numbers are small, but the aspect ratio of the submesoscale still tends to be small, so the hydrostatic approximation is usually valid. However, the quasigeostrophic equations are not valid in the submesoscale, as they require small Rossby and Froude numbers. Recall that a key advantage of the quasi-geostrophic equations was that inertial oscillations and internal gravity waves were filtered out, just as sound waves were filtered out by the Boussinesq approximation. In the submesoscale, it is not as easy to distinguish these waves from submesoscale motions exhibiting balanced dynamics–the balanced fronts and eddies exhibit convergence like waves, and their timescales are much faster and approach the timescale of these waves. For this reason, great care is needed to distinguish these wavelike and other submesoscale phenomena (Callies and Ferrari, 2013), which is a major challenge for present and planned observations and data assimilating submesoscale models.

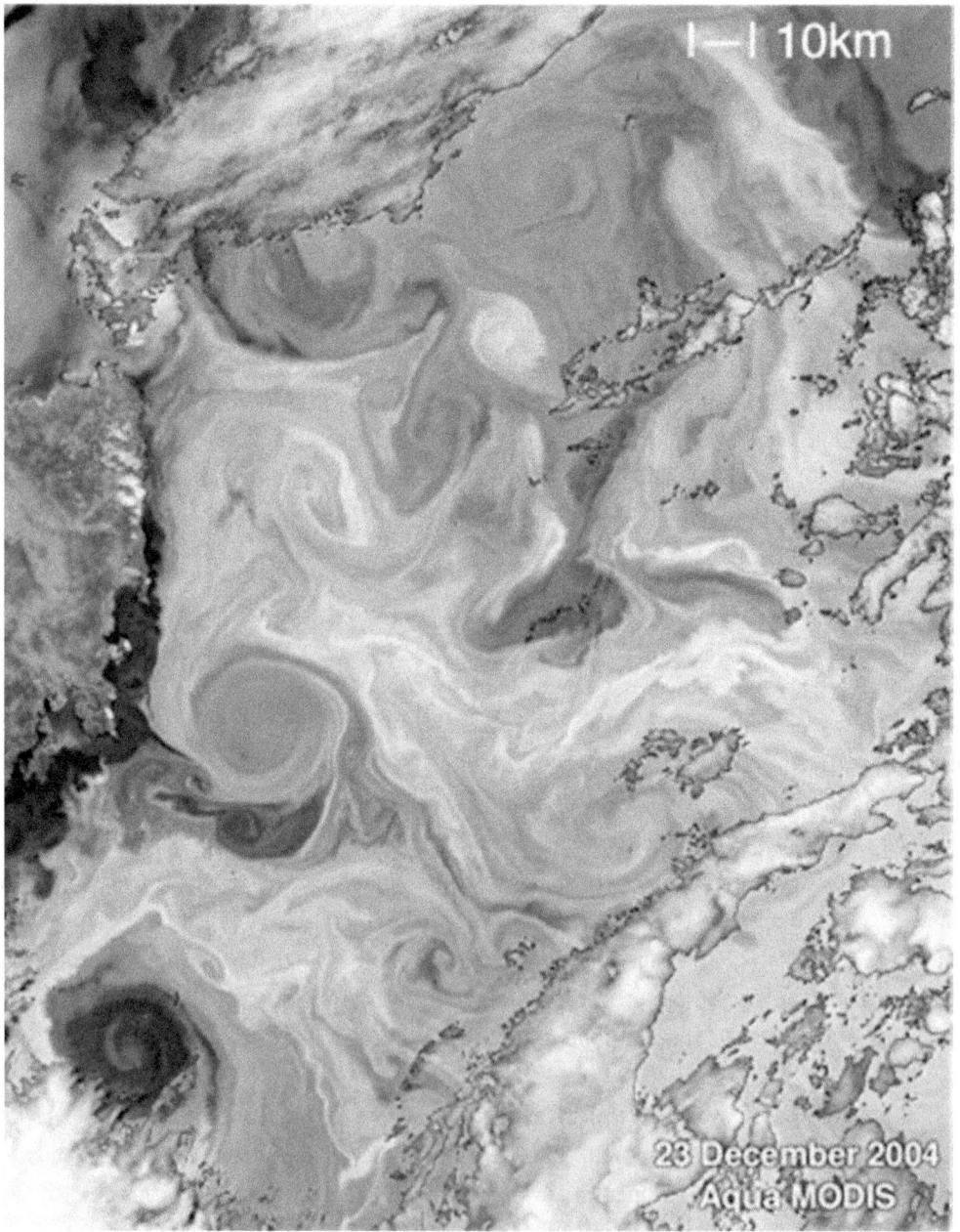

Figure 2.3. Taken from the NASA Goddard Space Flight Center Gallery: Mesoscale and submesoscale features in the Tasman Sea appear in this false color image of ocean color. The true colors from blue to green are expanded here, revealing the large mesoscale eddies (> 10 km diameter) surrounded by submesoscale fronts, filaments, and smaller (< 10 km diameter) submesoscale meanders.

A few parameterizations of submesoscale phenomena have been developed. A commonly used one seeks to capture the restratification effect of submesoscales (Fox-Kemper et al., 2008; Fox-Kemper et al., 2011) formulated as an advection similar to Gent and McWilliams (1990) with a preference for along-isopycnal transport. However, unlike the mesoscale parameterizations, which emphasized the horizontal processes, this parameterization emphasizes the vertical restratification aspect of the motions. Extensions of this parameterization and variants on these ideas remain to be implemented and tested in operational systems (Bachman and Fox-Kemper, 2013; Bachman et al., 2017c). Other submesoscale processes have been studied for parameterization purposes, such as the interaction between Ekman flow and fronts (Thomas and Lee, 2005), frontogenesis and turbulent thermal wind scalings (McWilliams, 2016), and symmetric instabilities (Bachman et al., 2017b).

Boundary layer and finescale turbulence

Long before the interest in submesoscale-convection/mixing interactions, the turbulence that mixes the upper and lower boundary layers of the ocean was recognized as important. Similarly, turbulent mixing in the ocean interior driven by breaking internal waves and tides over rough topography plays an important role in setting the stratification and overturning circulation of the abyss (Munk and Wunsch, 1998). This turbulence is similar, on its smaller scales, to isotropic three-dimensional turbulence. The aspect of this turbulence that affects larger scales and climate is its ability to overturn and mix stratification and momentum, and this overturning is naturally connected to nonhydrostatic accelerations. Thus, when this scale is reached, the final approximation to the Boussinesq equations in (35)-(40) must be abandoned (although on these small scales, sometimes planetary and Coriolis effects are neglected).

In the ocean interior away from boundary layers, mixing against stratification is a theme of the categorization of turbulence. It is the Thorpe scale–the vertical scale of the tallest well-mixed regions–and the Ozmidov scale–the largest scale at which turbulence in a stratified fluid can persist before giving way to larger-scale internal waves–that categorize turbulence.

In the boundary layers, turbulence tends to be quantified using forcing mechanisms. Wind-driven and bottom stress-driven mixing can be quantified with the friction velocity $u^* = (\tau/\rho)^{1/2}$. Langmuir turbulence, which also receives energy from surface gravity waves as well as wind, is quantified with the friction velocity and the turbulent Langmuir number which is the square root of the ratio of u^* to the Stokes drift (McWilliams et al., 1997). A third important scale is the buoyancy forcing velocity $w^* = (Bsh)1/3$, which measures the potential of cooling or heating at the surface to produce or suppress mixing. There are many examples of parameterizations of boundary layer mixing that take these forcings and convert them into predictions of bulk mixed layer depth (Kraus and Turner, 1967; Price et al., 1986), or predictions of vertical mixing rates (Pacanowski and Philander, 1981; Large et al., 1994; Van Roekel et al., 2012), or predictions of budgets for second moments such as variances and covariances (Mellor and Yamada, 1982; Kantha and Clayson, 2004; Harcourt, 2013), or bottom boundary layer mixing (Jayne and St Laurent, 2001; Simmons et al., 2004).

Boundary layer mixing is critically important for climate and weather modeling, as it is through these layers that the air-sea exchanges of energy, momentum, gasses (including greenhouse gasses) and moisture are exchanged. Some of the key questions that remain in the boundary layer problem are the role of waves in driving turbulence (both through breaking waves and Langmuir mixing), whether there are equation sets simpler than resolving the free surface including spray, bubbles, etc., and the details of air-sea coupling at high wind speeds where it is difficult to observe, perform experiments, and model.

Surface gravity waves

Even though surface gravity waves are (inviscid, irrotational) solutions to the Boussinesq equations with surface boundary conditions (7)-(10), for some decades wave modeling has been a separate activity from ocean interior modeling, resulting in typically separate codes, numerics, and scientific communities. However, recently these dynamics have begun to be incorporated into weather (Janssen et al. 2002) and climate (Fan and Griffies, 2014; Li et al., 2016) models, which is a step toward including their effects into a holistic portrayal of the earth system (Cavaleri et al., 2012).

Surface waves that are triggered by seismic activity or seafloor shifts are among the fastest propagating signals in the oceans. These tsunami travel at a speed of $\sqrt{gH} \approx 200$ m s^{-1}, and cross entire ocean basins in less than a day. The more common wind waves are slower and tend to adhere to nearly universal power spectra shapes governed by a combination of source functions and wave-wave interactions. Over time, wind wave energy is transferred into lower frequency swell waves, with typical wavelengths of 30 m and periods of 6 s. This assortment of waves makes for difficult modeling, but it can be combined into a prediction of the Stokes drift which is the key link between the waves, turbulence, and currents as will be described below (Webb and Fox-Kemper, 2011; 2015). Many operational oceanography systems contain a wave prediction component, but it is less common to utilize these wave predictions to force aspects of the currents and turbulence.

Examples of Reynolds-Averaged and Large Eddy Simulation Closures

The preceding sections illustrated how the governing equations can be made simpler and the phenomena one might choose to model with a tera-grid. The Boussinesq approximation eliminated sound waves, and the hydrostatic, Boussinesq equations are easier to code and more efficient than the nonhydrostatic Boussi- nesq equations. Going still further, the quasigeostrophic equations were a powerful tool in discovering the mesoscale dynamics. Just as the Boussinesq equations increase efficiency over the compressible equations by removing sound waves, the quasigeostrophic equations are vastly more efficient than the hydrostatic, Boussinesq equations. Some of the earliest proto-operational oceanography was done with the quasigeostrophic equations (e.g., Capotondi et al., 1995).

Yet, oceanographers are moving away from using these equations outside of theoretical studies. There are many reasons for this, a primary one being the availability of computing power capable of handling more complex equation sets. However, a bigger reason is that it is not only the internal waves that are missing from the quasigeostrophic equations. Quasigeostrophy limits the vertical stratification to dominate the horizontal, and the vertical stratification is not usually allowed to change within the model. This means that eddy-driven restratification, outcropping fronts, deep convection, etc., are untenable with the quasigeostrophic equations. Furthermore, converting surface forcing and topography into a meaningful form for quasigeostrophic requires a sophisticated appreciation of the asymptotic behavior of the model. As these deficits became more clear, hydrostatic Boussinesq models have become more popular, even when the dynamics at the gridscale is asymptotically consistent with quasigeostrophy (e.g., Pearson et al.,2017).

Numerical models over a limited range of scales must represent the averaged effects of smaller-scale turbulence on the resolved scales. This process can involve either Reynolds-averaging or Large Eddy Simulation style averaging (Fox-Kemper and Menemenlis, 2008). In either approach, parameterizations can be made scale-aware–that is, parameterizations adjust to the gridscale and model conditions to respond. For example, Hallberg (2013) suggests that models can evaluate the deformation radius and compare it to the grid scale. If the deformation radius and thus expected baroclinic instabilities will be resolved locally, then the Reynolds-averaged parameterization of these eddies can be turned off. If they aren't resolved locally, then leave the parameterization on. As grid scale often changes with latitude and Coriolis parameter and stratification vary, making the model aware of the dynamical scales to expect is a powerful idea. A closely related concept is flow-awareness. In this approach, the resolved flow in the model is used to inform the parameterization. Both approaches are common in Large Eddy Simulation modeling, but are not uniformly applied over all oceanographic applications, largely because parameterization systems are not well-understood for many of the tera-grids that might be chosen from Fig. 2.2.

To illustrate some of these ideas, a few example systems will be presented. The Smagorinsky (1963) classic parameterization started the idea of flow-aware, scale-aware modeling. A recent application and extension of the Leith (1996) scheme to quasigeostrophic dynamics from Bachman et al. (2017a); Pearson et al. (2017) will be used to show how important subgrid schemes can be *even at high resolution*. Finally, a discussion of the wave-averaged equations, a good example of multi-scale dynamics used to study Langmuir turbulence, will be used to motivate the inclusion of wave-current coupling in a broader range of models.

Scale-aware parameterizations

Kolmorogov & Smagorinsky 3D turbulence

Three-dimensional, nonrotating, unstratified turbulence occurs on small, fast scales in the limit of $Re_* \gg 1$, $Fr_* \gg 1$, $\epsilon \gg 1$, $\Delta z \sim \Delta s$, in which case we have equations (35)-(40). If we were to make a nonhydrostatic, Boussinesq model, a scale-aware LES parameterization (Kolmogorov, 1941; Smagorinsky, 1963, 1993) would be useful. The theory stems from the budget of total kinetic energy, which is the domain-average of the dot product of the momentum equations with the velocities and equal to the integral of the power spectrum $E(k)$.

$$\bar{E} = \frac{1}{2}\langle \mathbf{v} \cdot \mathbf{v}\rangle = \int_0^\infty E(k) \mathrm{d}k. \qquad (52)$$

If a steady-state situation where energy is produced or otherwise injected at a rate ε on a scale far larger than the scale at which it is dissipated, then between these scales Kolmogorov (1941) argues that the energy spectrum $E(k)$ can only depend on the local scale k and ε. Dimensional analysis then demands $E(k) \propto \varepsilon^{2/3} k^{-5/3}$ and the dissipation scale is $k_d = \varepsilon^{1/4} \nu^{-3/4}$. Smagorinsky (1963, 1993) argues that in a limited resolution model, i.e., not DNS, resolved dissipation should balance the energy production, because the energy transfer through every scale in the cascade from large to small should be equal. An isotropic viscosity ν_* that can be applied to the resolved flow that has this property is

$$\nu_* = \left(\frac{\Upsilon}{k_*}\right)^2 |D_*| = \left(\frac{\Upsilon \Delta s}{\pi}\right)^2 |D_*|, \qquad (53)$$

$$|D_*| = \sqrt{\frac{1}{4}\left(\frac{\partial u_{*i}}{\partial x_j} + \frac{\partial u_{*j}}{\partial x_i}\right)\left(\frac{\partial u_{*i}}{\partial x_j} + \frac{\partial u_{*j}}{\partial x_i}\right)}. \qquad (54)$$

Here Υ is a dimensionless constant with a value near 1, and note that u_* is the resolved model velocity not the friction velocity u^*. Examining these equations, it is clear that even though the goal was to represent unresolved processes, there all aspects of this viscosity are known from only resolved fields–it is flow-aware. There are, of course, oversimplifications made in arriving at this simple form, but it is a robust approximation based on known behavior. Also note that the grid scale Δs appears explicitly in the formula, thus as the grid scale changes the viscosity changes automatically – it is scale-aware.

Kraichnan and Leith 2D turbulence

The eddies of the mesoscale and submesoscale are not 3D isotropic, in fact they are closer to 2D turbulence. They are much wider than they are tall, and they tend to prefer to exchange properties along isopycnals. They also conserve energy, and like 2D turbulence they conserve a form of enstrophy (vorticity squared, with power spectrum $Z(k)$). Proceeding along the same lines as Kolmogorov (1941), Kraichnan (1967) finds two different inertial ranges are consistent with these behaviors: a forward enstrophy cascade with rate η and an inverse energy cascade. If the scale of mesoscale eddy formation is resolved, then the grid scale is expected to be smaller and in the forward cascade of enstrophy, where $E(k) \propto \eta^{2/3} k^{-3}$, $Z(k) \propto \eta^{2/3} k^{-1}$, $k_d = \eta^{1/6} \nu^{-1/2}$, and energy does not cascade.

Leith (1996) follows Smagorinsky (1963, 1993) to find a viscosity to halt the enstrophy cascade:

$$\nu_* = \left(\frac{\Lambda}{k_*}\right)^3 |\nabla_h q_{z*}| = \left(\frac{\Lambda \Delta s}{\pi}\right)^3 |\nabla_h (f\mathbf{z} + \nabla \times \mathbf{v}_{h*})|. \qquad (55)$$

Here Λ is a dimensionless constant with a value near 1. Fox-Kemper and Menemenlis (2008) propose a modification to (55) to remove noisy horizontally-divergent motions.

Charney Quasigeostrophic turbulence

Quasigeostrophic theory also has an inverse energy cascade and a forward (potential) enstrophy cascade (Charney, 1971). Fig. 2.4 schematizes and illustrates the dual cascades of mesoscale turbulence in the quasigeostrophic regime.

Fox-Kemper and Menemenlis (2008) and Bachman et al. (2017a) extend the approach of Leith (1996) to apply to the potential enstrophy cascade and find,

$$\nu_* = \left(\frac{\Lambda}{k_*}\right)^3 |\nabla_h q_{q_*}| = \left(\frac{\Lambda \Delta s}{\pi}\right)^3 \left|\nabla_h(f\mathbf{z}) + \nabla_h\left(\nabla \times \mathbf{v}_{h*}\right) + \partial_z \frac{f}{N^2}\nabla_h b\right|. \quad (56)$$

As in 2D, (56) requires the extra damping of spurious divergent modes (Fox-Kemper and Menemenlis, 2008).

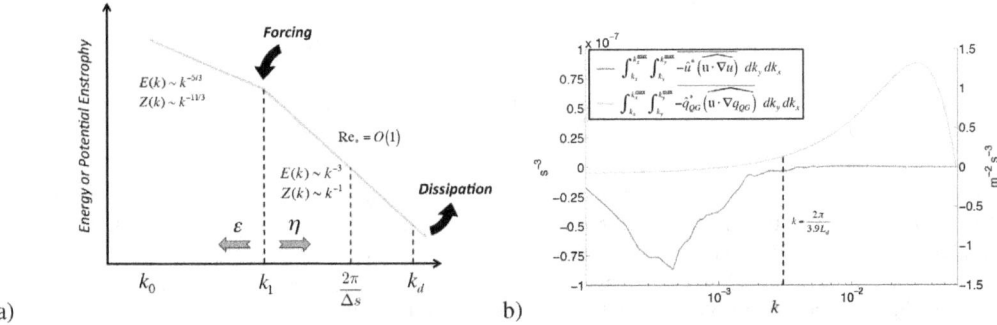

Figure 2.4. Taken from Bachman et al. (2017a): a) Schematic of the cascades of energy and potential enstrophy in quasigeostrophic flow, with spectral slopes from Smith et al. (2002). Energy and enstrophy are produced by instabilities near wavenumber k_1, leading to an inverse cascade of energy toward at the domain scale k_0 and a forward cascade of potential enstrophy toward the grid wavenumber $2\pi/\Delta s$. For a statistically steady state, a model must dissipate this energy and enstrophy appropriately. b) Spectral fluxes of (purple) energy and (blue) potential enstrophy as a function of wavelength λ, taken from a simulation with $\Delta s = L_d/10.0$, $Bu_* = 100$. Negative values indicate a flux toward larger scales; positive values indicate fluxes toward smaller scales. The two cascades originate near the production scale, estimated here using the Eady (1949) fastest growing baroclinic instability wavenumber (black dashed vertical line).

Comparing the schemes

Figure 2.5 compares the average viscosity that results in the different simulations with different resolutions and subgrid models. 2D Leith tends to have the lowest average viscosity, while Smagorinsky tends to have the largest. QG Leith is systematically larger than 2D Leith, revealing that the last "stretching" term in (56) has an important effect.

Not only do the viscosities vary, the solutions vary. Fig. 2.6 shows that not only does the horizontal dissipation decrease when different subgrid schemes are used, but bottom drag increases– even though the bottom drag coefficient was unchanged. Thus, even high-resolution solutions are sensitive to changes in their parameterization of small scales.

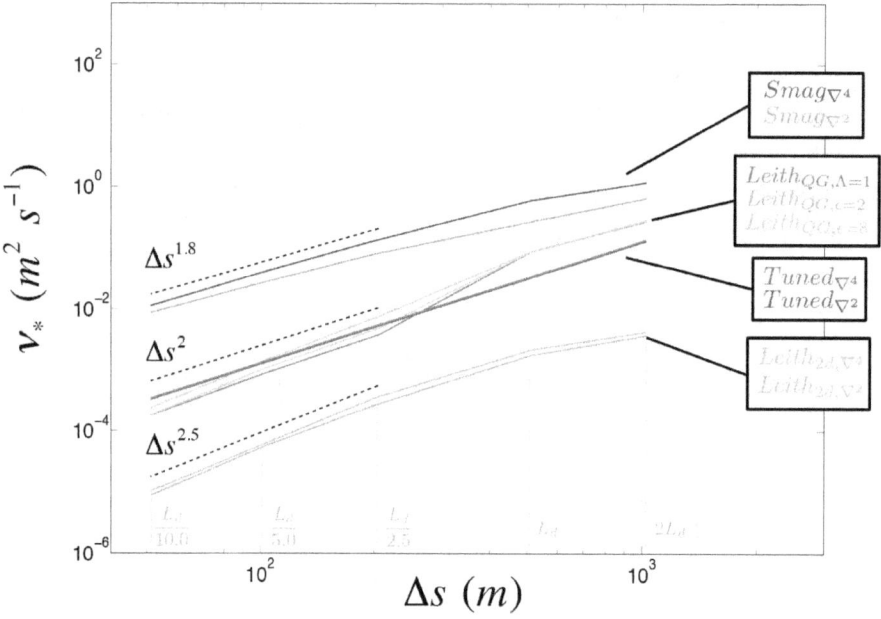

Figure 2.5. From Bachman et al. (2017a): A wide range for average viscosity (ν_*) results from using the different closures as a function of horizontal resolution (Δs). Biharmonic viscosities are shown here using the conversion factor $\nu_{2*} \approx \nu_{4*}8/\Delta s^2$ (Griffies and Hallberg, 2000; Fox-Kemper and Menemenlis, 2008). The label above each dashed black line is its slope, which shows how the viscosity scales with resolution. The dashed grey lines indicate the resolution at which each simulation was run.

Wave-averaged equations

Not only small-scale turbulence affects the larger scales–waves do as well. Just as scales that are averaged over can contribute to larger scales through mixing, waves that are averaged over (by a period or a wavelength) can contribute to larger scales. As the discussion above indicated, one key way that spatially-dependent waves affect averaged quantities is through the Stokes drift: the difference between the Lagrangian and Eulerian velocities. Because models usually solve for the Eulerian velocities, yet conserved tracers are carried around by the Lagrangian velocity, it is important to include the Stokes drift in tracer advection. However, the Lagrangian velocity also affects the momentum equation, as it experiences a Coriolis force and a net transfer of momentum and energy from waves to currents when they are both present.

Since Craik and Leibovich (1976) showed that the Lagrangian-averaged effects of tilting vorticity by the Stokes drift shear could produce Langmuir cells, this system has been used for analysis and modeling. Notably, Holm (1996) and Gjaja and Holm (1996) improved the theoretical underpinnings by being more explicit about Lagrangian averaging, while Skyllingstad and Denbo (1995) and McWilliams et al. (1997) show that Large Eddy Simulations using the Stokes forces result in a Langmuir turbulence that qualitatively resembles observations. Even though additional work is required to add in the effects of breaking waves (Kukulka et al., 2007; Sullivan et al., 2007), D'Asaro et al. (2014) were able to verify that an LES-based scaling from Harcourt and D'Asaro (2008) does indeed seem to be a good fit to observations of vertical velocity over varying wave strengths. Other approaches to wave-driven mixing (Huang and Qiao, 2010; Babanin, 2006) based on using the wave orbital velocities to infer a Reynolds number have mixed support (Babanin and Haus, 2009; Kantha et al., 2014; Fan and Griffies, 2014). Other recent Langmuir work has broadened the scope of forcing scenarios and analysis and demonstrated that Langmuir turbulence plays an important role in the global climate system through deepening the upper ocean boundary layer (Kantha and Clayson, 2004; Van Roekel et al., 2012; Belcher et al., 2012; Fan and Griffies, 2014; Harcourt, 2013; Li et al., 2016; Li et al., 2017; Li and Fox-Kemper, 2017). These scalings for Langmuir mixing are available for operational oceanographic or climate modeling use.

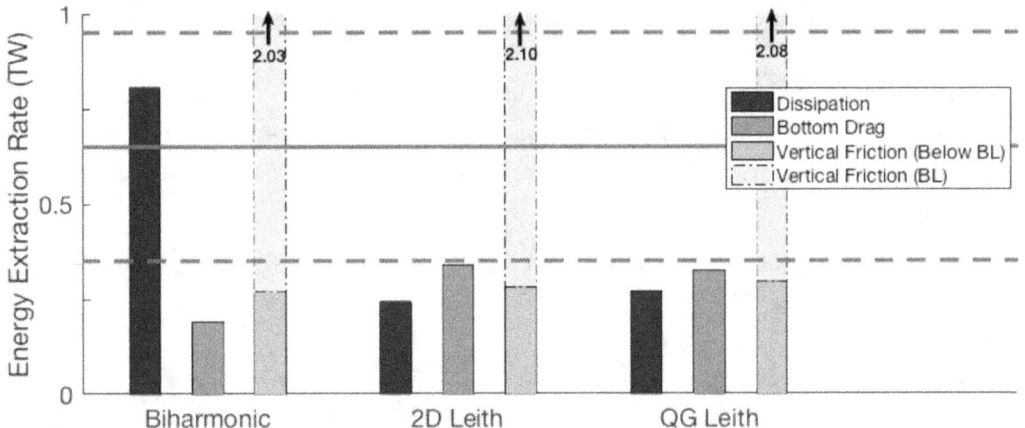

Figure 2.6. From Pearson et al. (2017): A comparison of the global integrated kinetic energy extraction by horizontal dissipation, bottom drag, vertical friction below the upper boundary layer, and vertical friction within the boundary layer in three realistically-forced, global mesoscale-resolving models.

However, Langmuir effects are not the only effect of Stokes forces on currents and turbulence. Waves have been known to affect nearshore currents and rips (e.g., Longuet-Higgins and Stewart, 1964; Uchiyama et al., 2009), to play a role in the Ekman layer (McWilliams et al., 2012), fronts (McWilliams and Fox-Kemper, 2013), submesoscale instabilities (Haney et al., 2015), and frontogenesis (Suzuki et al., 2016). The Stokes forces don't affect just Langmuir turbulence!

There are at least four schools of thought as to how to incorporate these effects. Radiation stresses are the traditional manner (Longuet-Higgins and Stewart, 1964), and the approach by Mellor (2003) is intended to simplify this approach. Fan et al. (2010) and Janssen (2004) consider the wave layer as a boundary condition capable of absorbing and releasing momentum, essentially by altering the boundary condition for wave storage. Ardhuin et al. (2008) describe the wave effect using the generalized Lagrangian mean (Andrews and McIntyre, 1978; Bühler, 2014). Another chapter in this volume addresses some of these points (Ardhuin and Orfila, 2018). Finally, the "Stokes vortex" implementation (McWilliams et al., 1997; McWilliams et al., 2004; Lane et al., 2007) has recently been shown to be inconsistent with the Mellor (2003) in theory (Ardhuin et al., 2017) and model simulations (Bennis et al., 2011; Wang et al., 2017). These discussions imply that including these forces will be an important effect, yet Breivik et al. (2015) found minimal effects when using a coarse resolution version of the NEMO ocean model. For this reason it is useful to consider the dimensionless form of the equations to see when the effects are expected to be large.

Here I will present the formulation of the Stokes forces from Suzuki and Fox-Kemper (2016), because I think they are the easiest to understand. These equations are analytically identical to the Stokes vortex formalism (McWilliams et al., 2004). Within the framework of the dimensionless equations here, it is easy to see how these forcing terms can be included into a nonhydrostatic or hydrostatic model. The wave-averaged Boussinesq momentum equation in the Suzuki and Fox-Kemper (2016) form, which solves for the Eulerian velocity **u** *averaged over wave phase in time*, are

$$\frac{\partial \mathbf{u}}{\partial t} + \underbrace{(\mathbf{u}^L \cdot \nabla)\mathbf{u}}_{\text{Lagrangian advection}} = \frac{1}{\rho_0}\nabla \cdot (-\pi \mathbf{I} + \sigma^{\text{mol}}) - \underbrace{2\mathbf{\Omega} \times \mathbf{u}^L}_{\text{Lagrangian Coriolis}} + b\mathbf{k} - \underbrace{u_j^L \nabla u_j^S}_{\text{Stokes shear force}}, \quad (57)$$

This form of the wave-averaged Boussinesq equations frames the wave effect in terms of three forces which have distinct roles in the dynamics and energetics of wave-influenced flows; namely, the Lagrangian advection $((\mathbf{u}^L \cdot \nabla)\mathbf{u})$, the Lagrangian Coriolis force $(\mathbf{f} \times \mathbf{u}^L)$, and the Stokes shear force $(-u_j^L \nabla u_j^S)$.

The dimensionless form of the Boussinesq equations, now taken to represent the wave-averaged Boussinesq equations, are

$$\text{Ro}_* \left[\partial_t \mathbf{v}_h + \mathbf{v}_h^L \cdot \nabla \mathbf{v}_h + \epsilon w^L \partial_z \mathbf{v}_h + \omega_h u_j^L \partial_z u_j^S\right] = \underbrace{-\left(1 + \frac{y\text{Pl}_*}{\Delta y}\right)\mathbf{z} \times \mathbf{v}_h^L - M_{R_*}\nabla_h \pi}_{\text{Lagrangian geostrophic}} + \frac{\text{Ro}_*}{\text{Re}_*}\nabla_i \sigma_{ih}, \quad (58)$$

$$\text{Fr}_*^2 \frac{\Delta z^2}{\Delta s^2}\left[\partial_t w + \mathbf{v}_h^L \cdot \nabla w + \epsilon w^L \partial_z \mathbf{v}_h\right] = \underbrace{-\partial_z \pi + b - \omega_z \mathbf{u}_j^L \partial_z u_j^S}_{\text{wavy hydrostatic}} + \frac{\text{Fr}_*^2 \Delta z^2}{\text{Re}_* \Delta s^2}\nabla_i \sigma_{iz}, \quad (59)$$

$$\partial_t S + \mathbf{v}_h^L \cdot \nabla S + \epsilon w^L \partial_z S + w^L \partial_z \bar{S} = \frac{1}{\text{Pe}_*}\nabla \cdot \mathbf{I}_S^{\text{all}}, \quad (60)$$

$$\partial_t \Theta + \mathbf{v}_h^L \cdot \nabla \Theta + \epsilon w^L \partial_z \Theta + w^L \partial_z \bar{\Theta} = \frac{1}{\text{Pe}_*}\nabla \cdot \mathbf{I}_\theta^{\text{all}}, \quad (61)$$

$$\partial_t b + \mathbf{v}_h^L \cdot \nabla b + \epsilon w^L \partial_z b + w^L \partial_z \bar{b} = \frac{1}{\text{Pe}_*}\nabla \cdot \left(\alpha \mathbf{I}_\theta^{\text{all}} - \beta \mathbf{I}_S^{\text{all}}\right), \quad (62)$$

$$\nabla \cdot \mathbf{v}_h + \epsilon \partial_z w = 0, \quad (63)$$

$$M_{R_*} \equiv \max(1, \text{Ro}_*), \quad \epsilon \equiv \frac{\text{Fr}_*^2}{\text{Ro}_*}M_{R_*} = \begin{cases} \text{Fr}_*^2 & \text{Ro}_* \geq 1, \\ \text{Ro}_*\text{Bu}_*^{-1} & \text{Ro}_* < 1 \end{cases} \quad (64)$$

The wave parameters $\omega_h = \frac{v^S}{v^L}\frac{\Delta s}{L^S}$ and $\omega_z = \frac{v^S}{v^L}\frac{\Delta z}{H^S}\min(1, \text{Ro}_*)$ categorize the strength of the Stokes shear force in the horizontal and vertical momentum equations, respectively. Generally, ω_h is small, as Stokes drift is slightly weaker than the Lagrangian velocity and varies slowly in the horizontal.

However, ω_z is $O(1)$ for Langmuir, submesoscale, and strong mesoscale flows, which means that not only does the Stokes force drive Langmuir circulations, but *the Stokes shear force contributes as much as the buoyancy in forcing the strengthening of fronts by enhancing the downward part of their frontogenetic secondary circulation* (Suzuki and Fox-Kemper, 2016). The essential example of this force is shown in Fig. 2.7, where a surface jet (in the Langmuir case) or front (in the submesoscale case) is aligned with the direction of the surface Stokes drift. As the

Stokes drift decays with depth, the Stokes shear force is downward in the center of the jet or front. A downward force at this point tends to enhance the strength of the jet or front, which in turn leads to more Stokes shear force, etc. This mechanism is the same mathematically as the primary mechanism explained by Craik and Leibovich (1976) for Langmuir circulations, but it is simpler and also applies equally well to the frontal case. Furthermore, as the Stokes shear force depends on the match in direction between fronts and the Stokes drift, it predicts that Langmuir cells will orient downwind and that up-Stokes fronts will be weakened while down-Stokes fronts are enhanced. Suzuki et al. (2016) show that in a submesoscale-resolving simulation including Stokes drift, this effect is strong, accounting for the equivalent of 40% of the buoyancy effects in forming a front. Even in hydrostatic models, the Stokes shear force can be included, and it will affect the pressure field of currents that are aligned with the wave direction. It is easy to include, by including it wherever the buoyancy appears. Fig. 2.7 schematizes the effect of this term.

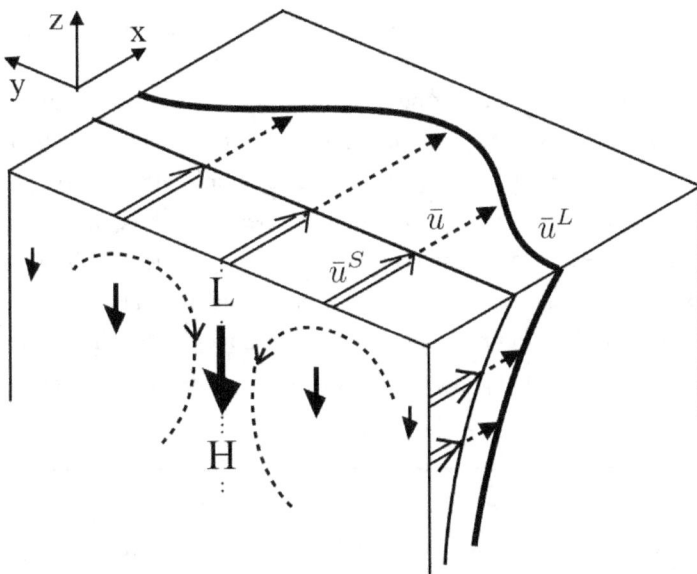

Figure 2.7. From Suzuki and Fox-Kemper (2016): This schematic illustrates how the Stokes shear force can drive, defeat, or enhance a secondary circulation related to a surface jet (i.e., a Langmuir cell) or front (and its associated secondary circulation). In this example, the jet or front is oriented along the direction of the Stokes shear, causing a strong downward acceleration in the middle of the feature. To either side, the velocity is weaker and thus the downward Stokes shear force is weaker, even though the Stokes drift is horizontally uniform.

One final comment on the wave averaged equations. Sometimes the effects of Stokes drift are estimated by only including Stokes advection without adding Stokes Coriolis or Stokes shear force to the momentum equation (e.g., McWilliams and Restrepo, 1999; Breivik et al., 2015; Curcic et al., 2016). This is almost certainly an overestimate of the Stokes effects, because the Stokes forces frequently act to make the Lagrangian advection similar to the Eulerian advection when no Stokes forces are present (Monismith et al., 2007; Lentz and Fewings, 2012; McWilliams and Fox-Kemper, 2013).

Conclusions

Although the oceans are vast and diverse, like a giant aquarium filled with interesting dynamical phenomena, much progress has been made in modeling the variability. Although computation is limited to roughly a few tera-grids and will expand only modestly during our lifetimes, there is much that can be done with this amount of computation. Rather than solving the fundamental equations of compressible fluid dynamics with molecular effects, approximations to these equations can be made.

The Boussinesq approximation filters sound waves, allowing longer time steps. The quasigeostrophic equations filter gravity waves for longer time steps still, although these equations are limited to use on the extratropical mesoscales only. The wave-averaged equations are less expensive than resolving waves, but require Stokes forces be introduced as a parameterization of leading wave effects. These wave effects are at the root of Langmuir turbulence, but also have profound effects in coastal regions and on submesoscale dynamics.

Other parameterizations can be formulated–for mesoscale eddies, submesoscale eddies, boundary layer mixing, etc. It is particularly helpful when these parameterizations are scale-aware and flow-aware, as this means that they do not require tuning and can respond differently and appropriately in different regimes across the globe. Dimensionless equations and dimensionless parameters can help decide how and where to make these adjustments. The Smagorinsky, Leith, and QG Leith schemes demonstrate how different dynamics at the gridscale affect the parameterization choice that is optimal, and comparing matching simulations varying only the choice of parameterizations demonstrates the influence of this choice even in high-resolution simulations.

References

Andrews, D. G. and M. E. McIntyre: 1978, An exact theory of nonlinear waves on a Lagrangian-mean flow. *Journal of Fluid Mechanics*, **89**, 609–646.

Ardhuin, F. and A. Orfila: 2018, Wind waves. *GODAE Book*.

Ardhuin, F., N. Rascle, and K. A. Belibassakis: 2008, Explicit wave-averaged primitive equations using a generalized Lagrangian mean. *Ocean Modelling*, **20**, 35–60.

Ardhuin, F., N. Suzuki, J. C. McWilliams, and H. Aiki: 2017, Comments on "a combined derivation of the integrated and vertically resolved, coupled wave-current equations". *Journal of Physical Oceanography*, **47**(9):2377–2385, URL http://dx.doi.org/10.1175/JPO-D-17-0065.1.

Babanin, A.: 2006, On a wave-induced turbulence and a wave-mixed upper ocean layer. *Geophysical Research Letters*, **33**, L20605, URL http://dx.doi.org/10.1029/2006GL027308.

Babanin, A. V. and B. K. Haus: 2009, On the existence of water turbulence induced by nonbreaking surface waves. *Journal of Physical Oceanography*, **39**.

Bachman, S. and B. Fox-Kemper: 2013, Eddy parameterization challenge suite. I: Eady spindown. *Ocean Modelling*, **64**, 12–28. URL http://dx.doi.org/10.1016/j.ocemod.2012.12.003

Bachman, S. D., B. Fox-Kemper, and B. Pearson: 2017a, A scale-aware subgrid model for quasigeostrophic turbulence. *Journal of Geophysical Research–Oceans*, **122**, 1529–1554. URL http://dx.doi.org/10.1002/2016JC012265

Bachman, S. D., B. Fox-Kemper, J. R. Taylor, and L. N. Thomas: 2017b, Parameterization of frontal symmetric instabilities. I: Theory for resolved fronts. *Ocean Modelling*, **109**, 72–95. URL http://dx.doi.org/10.1016/j.ocemod.2016.12.003

Bachman, S. D., D. P. Marshall, J. R. Maddison, and J. Mak: 2017c, Evaluation of a scalar eddy transport coefficient based on geometric constraints. *Ocean Modelling*, **109**, 44–54. URL http://dx.doi.org/10.1016/j.ocemod.2016.12.004

Barkan, R., J. C. McWilliams, M. J. Molemaker, J. Choi, K. Srinivasan, A. F. Shchepetkin, and A. Bracco: 2017, Submesoscale dynamics in the northern gulf of mexico. Part II: Temperature-salinity relations and cross shelf transport processes. *Journal of Physical Oceanography*.

Bates, S. C., B. Fox-Kemper, S. R. Jayne, W. G. Large, S. Stevenson, and S. G. Yeager: 2012, Mean biases, variability, and trends in air-sea fluxes and SST in the CCSM4. *Journal of Climate*, **25**, 7781–7801. URL http://dx.doi.org/10.1175/JCLI-D-11-00442.1

Belcher, S. E., A. A. L. M. Grant, K. E. Hanley, B. Fox-Kemper, L. Van Roekel, P. P. Sullivan, W. G. Large, A. Brown, A. Hines, D. Calvert, A. Rutgersson, H. Petterson, J. Bidlot, P. A. E. M. Janssen, and J. A. Polton: 2012, A global perspective on Langmuir turbulence in the ocean surface boundary layer. *Geophysical Research Letters*, **39**, L18605, 9pp. URL http://dx.doi.org/10.1029/2012GL052932

Bennis, A.-C., F. Ardhuin, and F. Dumas: 2011, On the coupling of wave and three-dimensional circulation models: Choice of theoretical framework, practical implementation and adiabatic tests. *Ocean Modelling*, **40**, 260–272.

Boccaletti, G., R. Ferrari, and B. Fox-Kemper: 2007, Mixed layer instabilities and restratification. *Journal of Physical Oceanography*, **37**, 2228–2250. URL http://dx.doi.org/10.1175/JPO3101.1

Boussinesq, J.: 1897, *Théorie de l'e´coulmnent tourbillonnant et tumultuex des liquides dans les lits rectilignes à grande section*, volume 1. Gauthier-Villars.

Brannigan, L., D. P. Marshall, A. Naveira-Garabato, and A. G. Nurser: 2015, The seasonal cycle of submesoscale flows. *Ocean Modelling*, **92**, 69–84.

Breivik, O., K. Mogensen, J.-R. Bidlot, M. A. Balmaseda, and P. A. Janssen: 2015, Surface wave effects in the NEMO ocean model: Forced and coupled experiments. *Journal of Geophysical Research: Oceans*.

Bühler, O.: 2014, *Waves and mean flows*. Cambridge monographs on mechanics, Cambridge University Press, Cambridge, United Kingdom, second edition.

Caldwell, D. and S. Eide: 1981, Soret coefficient and isothermal diffusivity of aqueous solutions of five principal salt constituents of seawater. *Deep Sea Research Part A. Oceanographic Research Papers*, **28**, 605–1618.

Callies, J. and R. Ferrari: 2013, Interpreting energy and tracer spectra of upper-ocean turbulence in the submesoscale range (1–200 km). *Journal of Physical Oceanography*, **43**, 2456–2474.

Callies, J.: 2018, Baroclinic instability in the presence of convection. *Journal of Physical Oceanography*, **48**, 45–60.

Callies, J., R. Ferrari, J. M. Klymak, and J. Gula: 2015, Seasonality in submesoscale turbulence. *Nature communications*, **6**, 6862.

Callies, J., G. Flierl, R. Ferrari, and B. Fox-Kemper: 2016, The role of mixed layer instabilities in submesoscale turbulence. *Journal of Fluid Mechanics*, **788**, 5–41. URL http://dx.doi.org/10.1017/jfm.2015.700

Capet, X., J. C. Mcwilliams, M. J. Mokemaker, and A. F. Shchepetkin: 2008a, Mesoscale to submesoscale transition in the California current system. Part I: Flow structure, eddy flux, and observational tests. *Journal of Physical Oceanography*, **38**, 29–43.

Capet, X., J. C. Mcwilliams, M. J. Molemaker, and A. F. Shchepetkin: 2008b, Mesoscale to submesoscale transition in the California current system. Part II: Frontal processes. *Journal of Physical Oceanography*, **38**, 44–64.

Capet, X., J. C. McWilliams, M. J. Molemaker, and A. F. Shchepetkin: 2008c, Mesoscale to submesoscale transition in the California current system. part III Energy balance and flux. *Journal of Physical Oceanography*, **38**, 2256–2269.

Capotondi, A., P. Malanotte-Rizzoli, and W. R. Holland: 1995, Assimilation of altimeter data into a quasi-geostrophic model of the gulf-stream system .1. dynamical considerations.

Cavaleri, L., B. Fox-Kemper, and M. Hemer: 2012, Wind waves in the coupled climate system. *Bulletin of the American Meteorological Society*, **93**, 1651–1661. URL http://dx.doi.org/10.1175/BAMS-D-11-00170.1

Charney, J. G.: 1955, The Gulf Stream as an inertial boundary layer. *Proceedings of the National Academy of Sciences*, **41**, 731–740.

— 1971, Geostrophic turbulence. *Journal of the Atmospheric Sciences*, **28**, 1087–1095.

Charney, J. G., R. Fjörtoft, and J. V. Neumann: 1950, Numerical integration of the barotropic vorticity equation. *Tellus*, **2**, 237–254.

Chelton, D., M. Schlax, R. Samelson, and R. de Szoeke: 2007, Global observations of large oceanic eddies. *Geophysical Research Letters*, **34**, L15606.

Chelton, D. B., R. A. Deszoeke, and M. G. Schlax: 1998, Geographical variability of the first baroclinic Rossby radius of deformation. *Journal of Physical Oceanography*, **28**, 433–460.

Chelton, D. B. and M. G. Schlax: 1996, Global observations of oceanic Rossby waves. *Science*, **272**, 234–238.

Chen, C., I. Kamenkovich, and P. Berloff: 2016, Eddy trains and striations in quasigeostrophic simulations and the ocean. *Journal of Physical Oceanography*, **46**, 2807–2825.

Corrsin, S.: 1951, On the spectrum of isotropic temperature fluctuations in an isotropic turbulence. *Journal of Applied Physics*, **22**, 469–473.

Craik, A. D. D. and S. Leibovich: 1976, Rational model for Langmuir circulations. *Journal of Fluid Mechanics*, **73**, 401–426.

Curcic, M., S. S. Chen, and T. M. Ö̈zgökmen: 2016, Hurricane-induced ocean waves and stokes drift and their impacts on surface transport and dispersion in the Gulf of Mexico. *Geophysical Research Letters*, **43**, 2773–2781.

D'Asaro, E. A., A. Y. Shcherbina, J. M. Klymak, J. Molemaker, G. Novelli, C. M. Guigand, A. C. Haza, Haus, B. K., E. H. Ryan, G. A. Jacobs, et al.: 2018, Ocean convergence and the dispersion of flotsam. *Proceedings of the National Academy of Sciences*, 201718453.

D'Asaro, E. A., J. Thomson, A. Y. Shcherbina, R. R. Harcourt, M. F. Cronin, M. A. Hemer, and B. Fox-Kemper: 2014, Quantifying upper ocean turbulence driven by surface waves. *Geophysical Research Letters*, **41**, 102–107. URL http://dx.doi.org/10.1002/2013GL058193

Deardorff, J.: 1970, A numerical study of three-dimensional turbulent channel flow at large Reynolds numbers. *Journal of Fluid Mechanics*, **41**, 453–480.

Deser, C., A. Capotondi, R. Saravanan, and A. Phillips: 2006, Tropical Pacific and Atlantic climate variability in CCSM3. *J. Clim.*, **19**, 2451–2481.

Eady, E. T.: 1949, Long waves and cyclone waves. *Tellus*, **1**, 33–52.

Ekman, V. W.: 1905, On the influence of the Earth's rotation on ocean currents. *Arkiv. Mat. Astron. Fysik.*, **2**, 1–53.

Euler, L.: 1757, Principes generaux du mouvement des fluides. *Mémoires de l'Academie des Sciences de Berlin*, **11**, 274–315.

Fan, Y., I. Ginis, and T. Hara: 2010, Momentum flux budget across the air–sea interface under uniform and tropical cyclone winds. *Journal of Physical Oceanography*, **40**, 2221–2242.

Fan, Y. and S. M. Griffies: 2014, Impacts of parameterized langmuir turbulence and non-breaking wave mixing in global climate simulations. *Journal of Climate*, in press.

Fermi, E.: 1956, *Thermodynamics*. Dover, new edition, 176 pp. Fick, A.: 1855, Ueber diffusion. *Annalen der Physik*, **170**, 59–86.

Fofonoff, N. P.: 1954, Steady flow in a frictionless homogenous ocean. *Journal of Marine Research*, **13**, 254–262.

Fourier, J.: 1822, *Theorie analytique de la chaleur, par M. Fourier*. Chez Firmin Didot, père et fils.

Fox-Kemper, B., S. Bachman, B. Pearson, and S. Reckinger: 2014, Principles and advances in subgrid modeling for eddy-rich simulations. *CLIVAR Exchanges*, **19**, 42–46. URL http://bit.ly/1qSMTzA

Fox-Kemper, B., G. Danabasoglu, R. Ferrari, S. M. Griffies, R. W. Hallberg, M. M. Holland, M. E. Maltrud, S. Peacock, and B. L. Samuels: 2011, Parameterization of mixed layer eddies. III: Implementation and impact in global ocean climate simulations. *Ocean Modelling*, **39**, 61–78. URL http://dx.doi.org/10.1016/j.ocemod.2010.09.002

Fox-Kemper, B. and R. Ferrari: 2009, An eddifying Parsons model. *Journal of Physical Oceanography*, **39**, 3216–3227. URL http://ams.allenpress.com/perlserv/?request=get-abstract&doi=10.1175%2F2009JPO4104.1

Fox-Kemper, B., R. Ferrari, and R. Hallberg: 2008, Parameterization of mixed layer eddies. Part I: Theory and diagnosis. *Journal of Physical Oceanography*, **38**, 1145–1165. URL http://dx.doi.org/10.1175/2007JPO3792.1

Fox-Kemper, B. and D. Menemenlis: 2008, Can large eddy simulation techniques improve mesoscale-rich ocean models? *Ocean Modeling in an Eddying Regime*, M. Hecht and H. Hasumi, eds., AGU Geophysical Monograph Series, volume 177, 319–338.

Fox-Kemper, B. and J. Pedlosky: 2004, Wind-driven barotropic gyre I: Circulation control by eddy vorticity fluxes to an enhanced removal region. *Journal of Marine Research*, **62**, 169–193. URL http://www.ingentaselect.com/rpsv/cgi-bin/cgi?body=linker&reqidx=0022-2402(20040301)62:2L.169;1-

Frankignoul, C. and K. Hasselmann: 1977, Stochastic climate models. II: Application to sea surface temperature variability and thermocline variability. *Tellus*, **29**, 284–305.

Frenger, I., N. Gruber, R. Knutti, and M. Münnich: 2013, Imprint of southern ocean eddies on winds, clouds and rainfall. *Nature Geoscience*, **6**, 608–612.

Gebbie, G. and P. Huybers: 2012, The mean age of ocean waters inferred from radiocarbon observations: sensitivity to surface sources and accounting for mixing histories. *Journal of Physical Oceanography*, **42**, 291–305.

Gent, P. R. and J. C. McWilliams: 1990, Isopycnal mixing in ocean circulation models. *Journal of Physical Oceanography*, **20**, 150–155.

Gjaja, I. and D. Holm: 1996, Self-consistent hamiltonian dynamics of wave mean-flow interaction for a rotating stratified incompressible fluid. *Physica D*, **98**, 343–378.

Gnanadesikan, A.: 1999, A simple predictive model for the structure of the oceanic pycnocline. *Science*, **283**, 2077–2079.

Griffies, S. and R. Hallberg: 2000, Biharmonic friction with a Smagorinsky-like viscosity for use in largescale eddy-permitting ocean models. *Monthly Weather Review*, **128**, 2935–2946.

Griffies, S. M. and R. J. Greatbatch: 2012, Physical processes that impact the evolution of global mean sea level in ocean climate models. *Ocean Modelling*, **51**, 37–72.

Grist, J. P. and S. A. Josey: 2003, Inverse analysis adjustment of the SOC air–sea flux climatology using ocean heat transport constraints. *Journal of Climate*, **16**, 3274–3295.

Grooms, I., K. Julien, and B. Fox-Kemper: 2011, On the interactions between planetary geostrophy and mesoscale eddies. *Dynamics of Atmospheres and Oceans*, **51**, 109–136. URL http://dx.doi.org/10.1016/j.dynatmoce.2011.02.002

Haidvogel, D., E. Churchitser, S. Danilov, and B. Fox-Kemper: 2017, Multiscale multi-physics ocean modeling: Numerics (invited). *The Sea*, Journal of Marine Research, in press. URL http://bit.ly/2BnSdaQ

Hallberg, R.: 2013, Using a resolution function to regulate parameterizations of oceanic mesoscale eddy effects. *Ocean Modelling*, **72**, 92–103.

Hamlington, P. E., L. P. Van Roekel, B. Fox-Kemper, K. Julien, and G. P. Chini: 2014, Langmuir-submesoscale interactions: Descriptive analysis of multiscale frontal spin-down simulations. *Journal of Physical Oceanography*, **44**, 2249–2272. URL http://dx.doi.org/10.1175/JPO-D-13-0139.1

Haney, S., B. Fox-Kemper, K. Julien, and A. Webb: 2015, Symmetric and geostrophic instabilities in the wave-forced ocean mixed layer. *Journal of Physical Oceanography*, **45**, 3033–3056. URL http://dx.doi.org/10.1175/JPO-D-15-0044.1

Hansen, J., M. Sato, P. Kharecha, and K. v. Schuckmann: 2011, Earth's energy imbalance and implications. *Atmospheric Chemistry and Physics*, **11**, 13421–13449.

Harcourt, R. R.: 2013, A second-moment closure model of langmuir turbulence. *Journal of Physical Oceanography*, **43**.

Harcourt, R. R. and E. A. D'Asaro: 2008, Large-eddy simulation of Langmuir turbulence in pure wind seas. *Journal of Physical Oceanography*, **38**, 1542–1562.

Hasselmann, K.: 1976, Stochastic climate models. Part I: Theory. *Tellus*, **28**, 473–485.

Held, I. M. and T. Schneider: 1999, The surface branch of the zonally averaged mass transport circulation in the troposphere. *Journal of the Atmospheric Sciences*, **56**, 1688–1697.

Holm, D.: 1996, The ideal Craik-Leibovich equations. *Physica D*, **98**, 415–441.

Hoskins, B. J.: 1975, The geostrophic momentum approximation and the semi-geostrophic equations. *Journal of the Atmospheric Sciences*, **32**, 233–242.

Huang, C. J. and F. Qiao: 2010, Wave-turbulence interaction and its induced mixing in the upper ocean. *Journal of Geophysical Research: Oceans*, **115**.

Huang, R. X. and G. R. Flierl: 1987, Two-layer models for the thermocline and current structure in subtropical/subpolar gyres.

Jansen, M. F. and I. M. Held: 2014, Parameterizing subgrid-scale eddy effects using energetically consistent backscatter. *Ocean Modelling*, **80**, 36–48.

Janssen, P.A.E.M., J. D. Doyle, J. Bidlot, B. Hansen, L. Isaksen, and P. Viterbo. Impact and feedback of ocean waves on the atmosphere. In W. Perrie, editor, *Adv. Fluid. Mech.*, volume I of *Atmosphere–Ocean Interactions*. 2002.

Janssen, P.: 2004, *The interaction of ocean waves and wind*. Cambridge University Press.

Jayne, S. and L. St Laurent: 2001, Parameterizing tidal dissipation over rough topography. *Geophysical Research Letters*, **28**, 811–814.

Johnson, G. C. and H. L. Bryden: 1989, On the size of the Antarctic Circumpolar Current. *Deep-Sea Research Part A-Oceanographic Research Papers*, **36**, 39–53.

Julien, K., E. Knobloch, R. Milliff, and J. Werne: 2006, Generalized quasi-geostrophy for spatially anisotropic rotationally constrained flows. *Journal of Fluid Mechanics*, **555**, 233–274.

Kantha, L. and C. Clayson: 2004, On the effect of surface gravity waves on mixing in the oceanic mixed layer. *Ocean Modelling*, **6**, 101–124.

Kantha, L., H. Tamura, and Y. Miyazawa: 2014, Comment on "wave-turbulence interaction and its induced mixing in the upper ocean" by Huang and Qiao. *Journal of Geophysical Research: Oceans*, **119**, 1510–1515.

Karsten, R. H. and J. Marshall: 2002, Constructing the residual circulation of the acc from observations. *Journal of physical oceanography*, **32**, 3315–3327.

Kolmogorov, A. N.: 1941, The local structure of turbulence in incrompressible viscous fluid for very large reynolds number. *Dokl. Akad. Nauk. SSSR*, **30**, 9–13.

Kourafalou, V., P. De Mey, M. Le He′naff, G. Charria, C. Edwards, R. He, M. Herzfeld, A. Pascual, E. Stanev, J. Tintore′, et al.: 2015, Coastal ocean forecasting: system integration and evaluation. *Journal of Operational Oceanography*, **8**, s127–s146.

Kraichnan, R. H.: 1967, Inertial ranges in two-dimensional turbulence. *Physics of Fluids*, **16**, 1417–1423.

Kraus, E. and J. Turner: 1967, A one-dimensional model of the seasonal thermocline. II: The general theory and its consequences. *Tellus*, **19**, 98–106.

Kukulka, T., T. Hara, and S. E. Belcher: 2007, A model of the air-sea momentum flux and breaking-wave distribution for strongly forced wind waves. *Journal of Physical Oceanography*, **37**, 1811–1828.

Lane, E. M., J. M. Restrepo, and J. C. McWilliams: 2007, Wave-current interaction: A comparison of radiation-stress and vortex-force representations. *Journal of Physical Oceanography*, **37**, 1122–1141.

Laplace, P. S., N. Bowditch, and N. I. Bowditch: 1829, *Mécanique céleste*. Hillard, Gray, Little, and Wilkins, Boston.

Large, W. G., J. C. McWilliams, and S. C. Doney: 1994, Oceanic vertical mixing a review and a model with a nonlocal boundary-layer parameterization. *Reviews of Geophysics*, **32**, 363–403.

LeBlond, P. H. and L. A. Mysak: 1978, *Waves in the Ocean*. Number 20 in Elsevier Oceanography, Elsevier Scientific Publishing Company, New York.

Leith, C. E.: 1996, Stochastic models of chaotic systems. *Physica D*, **98**, 481–491.

Lentz, S. J. and M. R. Fewings: 2012, The wind-and wave-driven inner-shelf circulation. *Annual review of marine science*, **4**, 317–343.

Li, Q., B. Fox-Kemper, O. Breivik, and A. Webb: 2017, Statistical modeling of global Langmuir mixing. *Ocean Modelling*, **113**, 95–114. URL http://dx.doi.org/10.1016/j.ocemod.2017.03.016

Li, Q. and B. Fox-Kemper: 2017, Assessing the effects of Langmuir turbulence on the entrainment buoyancy flux in the ocean surface boundary layer. *Journal of Physical Oceanography*, in press. URL http://bit.ly/2otyrUT

Li, Q., A. Webb, B. Fox-Kemper, A. Craig, G. Danabasoglu, W. G. Large, and M. Vertenstein: 2016, Langmuir mixing effects on global climate: WAVEWATCH III in CESM. *Ocean Modelling*, **103**, 145– 160. URL http://dx.doi.org/10.1016/j.ocemod.2015.07.020

Lilly, D. K.: 1983, Stratified turbulence and the mesoscale variability of the atmosphere. *Journal of the Atmospheric Sciences*, **40**, 749–761.

Longuet-Higgins, M. S. and R. W. Stewart: 1964, Radiation stresses in water waves: A physical discussion, with applications. *Deep Sea Res.*, **11**, 529–562.

Lozier, M. S.: 2010, Deconstructing the conveyor belt. *Science*, **328**, 1507–1511.

Luyten, J. R., J. Pedlosky, and H. Stommel: 1983, The ventilated thermocline. *Journal of Physical Oceanography*, **13**, 292–309.

Maltrud, M. E. and J. L. McClean: 2005, An eddy resolving global 1/10° ocean simulation. *Ocean Modelling*, **8**, 31–54.

Marangoni, C.: 1865, *Sull'espansione delle goccie d'un liquido galleggianti sulla superfice di altro liquido*. Fratelli Fusi, Pavia, Italy.

McDougall, T.: 2003, Potential enthalpy: A conservative oceanic variable for evaluating heat content and heat fluxes. *Journal of Physical Oceanography*, **33**, 945–963.

McDougall, T. J. and P. M. Barker: 2011, Getting started with teos-10 and the gibbs seawater (gsw) oceanographic toolbox. *SCOR/IAPSO WG*, **127**, 1–28.

McWilliams, J. and J. Restrepo: 1999, The wave-driven ocean circulation. *Journal of Physical Oceanography*, **29**, 2523–2540.

McWilliams, J., J. Restrepo, and E. Lane: 2004, An asymptotic theory for the interaction of waves and currents in coastal waters. *Journal of Fluid Mechanics*, **511**, 135–178.

McWilliams, J. C.: 1985, A uniformly valid model spanning the regimes of geostrophic and isotropic, stratified turbulence: Balanced turbulence. *Journal of the Atmospheric Sciences*, **42**, 1773–1774.

— 2016, Submesoscale currents in the ocean. *Proc. R. Soc. A*, The Royal Society, volume 472, 20160117.

McWilliams, J. C. and B. Fox-Kemper: 2013, Oceanic wave-balanced surface fronts and filaments. *Journal of Fluid Mechanics*, **730**, 464–490. URL http://dx.doi.org/10.1017/jfm.2013.348

McWilliams, J. C., J. Gula, M. J. Molemaker, L. Renault, and A. F. Shchepetkin: 2015, Filament frontogenesis by boundary layer turbulence. *Journal of Physical Oceanography*, **45**, 1988–2005.

McWilliams, J. C., E. Huckle, J.-H. Liang, and P. P. Sullivan: 2012, The wavy Ekman layer: Langmuir circulations, breaking waves, and reynolds stress. *Journal of Physical Oceanography*, **42**, 1793–1816.

McWilliams, J. C., P. P. Sullivan, and C.-H. Moeng: 1997, Langmuir turbulence in the ocean. *Journal of Fluid Mechanics*, **334**, 1–30.

Meehl, G. A., L. Goddard, G. Boer, R. Burgman, G. Branstator, C. Cassou, S. Corti, G. Danabasoglu, F. Doblas-Reyes, E. Hawkins, et al.: 2014, Decadal climate prediction: an update from the trenches. *Bulletin of the American Meteorological Society*, **95**, 243–267.

Mellor, G.: 2003, The three-dimensional current and surface wave equations. *Journal of Physical Oceanography*, **33**, 1978–1989.

Mellor, G. L. and T. Yamada: 1982, Development of a turbulent closure model for geophysical fluid problems. **20**, 851–857.

Mensa, J. A., Z. Garraffo, A. Griffa, T. M. O¨ zgo¨kmen, A. Haza, and M. Veneziani: 2013, Seasonality of the submesoscale dynamics in the gulf stream region. *Ocean Dynamics*, **63**, 923–941.

Milankovitch, M.: 1930, Mathematische klimalehre und astronomische theorie der klimashwankungengebruder borntraeger.

Monismith, S. G., E. A. Cowen, H. M. Nepf, J. Magnadaudet, and L. Thais: 2007, Laboratory observations of mean flows under surface gravity waves. *Journal of Fluid Mechanics*, **573**, 131–147.

Moore, G. E.: 1965, Cramming more components onto integrated circuits. *Electronics*, **38**, 114–117. Mu¨ller, P.: 2006, *The equations of oceanic motions*. Cambridge University Press.

Munk, W. and C. Wunsch: 1998, Abyssal recipes ii: energetics of tidal and wind mixing. *Deep-Sea Research Part I-Oceanographic Research Papers*, **45**, 1977–2010.

Munk, W. H.: 1950, On the wind-driven ocean circulation. *Journal of Meteorology*, **7**, 79–93.

Navier, C.-L.: 1822, Me´moire sur les lois du mouvement des fluides. *Mem. de l'Acad. des Sciences*, **389**.

Nazarenko, S.: 2011, *Wave turbulence*, volume 825. Springer Science & Business Media.

NOAA, N.: 2018, State of the climate: Global climate report for annual 2017. URL https://www.ncdc.noaa.gov/sotc/global/201713

Obukhov, A.: 1949, Structure of the temperature field in turbulent flow. *Izvestiia Akademii M!auk S.S.S.R.*, **13**, 58–69, translation (from Russian) No. 334 by Army Biological Labs, 1968.

Olbers, D. and C. Eden: 2003, A simplified general circulation model for a baroclinic ocean with topography. part I: Theory, waves, and wind-driven circulations. *Journal of Physical Oceanography*, **33**, 2719–2737.

Onsager, L.: 1931a, Reciprocal relations in irreversible processes. i. *Physical review*, **37**, 405.

— 1931b, Reciprocal relations in irreversible processes. ii. *Physical review*, **38**, 2265.

Pacanowski, R. and S. Philander: 1981, Parameterization of vertical mixing in numerical models of tropical oceans. *J. Phys. Ocean.*, **11**, 1443–1451.

Pauluis, O.: 2011, Water vapor and mechanical work: A comparison of Carnot and steam cycles. *Journal of the Atmospheric Sciences*, **68**, 91–102.

Pearson, B. and B. Fox-Kemper: 2018, Lognormal turbulence dissipation in global ocean models. *Physical Review Letters*, in press. URL http://bit.ly/2A11PnD

Pearson, B., B. Fox-Kemper, S. D. Bachman, and F. O. Bryan: 2017, Evaluation of scale-aware subgrid mesoscale eddy models in a global eddy-rich model. *Ocean Modelling*, **115**, 42–58. URL http://dx.doi.org/10.1016/j.ocemod.2017.05.007

Peixoto, J. P. and A. H. Oort: 1992, *Physics of Climate.*. American Institute of Physics, New York, 520pp, 520 pp.

Penney, J. and M. Stastna: 2016, Direct numerical simulation of double-diffusive gravity currents. *Physics of Fluids*, **28**, 086602.

Price, J. F., R. A. Weller, and R. Pinkel: 1986, Diurnal cycling: Observations and models of the upper ocean response to diurnal heating, cooling, and wind mixing. *Journal of Geophysical Research-Oceans*, **91**, 8411–8427.

Radko, T. and J. Marshall: 2006, The Antarctic Circumpolar Current in three dimensions. *Journal of Physical Oceanography*, **36**, 651–669.

Redi, M. H.: 1982, Oceanic isopycnal mixing by coordinate rotation. *Journal of Physical Oceanography*, **12**, 1154–1158.

Reynolds, O.: 1895, On the dynamical theory of incompressible viscous fluids and the determination of the criterion. *Philosophical Transactions of the Royal Society of London. A*, **186**, 123–164.

Rhines, P. B.: 1979, Geostrophic turbulence. Annual Review of Fluid Mechanics.

Richter, F. M.: 1986, Kelvin and the age of the Earth. *The Journal of Geology*, **94**, 395–401.

Rocha, C. B., T. K. Chereskin, S. T. Gille, and D. Menemenlis: 2016a, Mesoscale to submesoscale wavenumber spectra in Drake Passage. *Journal of Physical Oceanography*, **46**, 601–620.

Rocha, C. B., S. T. Gille, T. K. Chereskin, and D. Menemenlis: 2016b, Seasonality of submesoscale dynamics in the kuroshio extension. *Geophysical Research Letters*, **43**.

Samelson, R. M. and G. K. Vallis: 1997, A simple friction and diffusion scheme for planetary geostrophic basin models. *Journal of Physical Oceanography*, **27**, 186–194.

Schlottke, J. and B. Weigand: 2008, Direct numerical simulation of evaporating droplets. *Journal of Computational Physics*, **227**, 5215–5237.

Shevchenko, I. and P. Berloff: 2015, Multi-layer quasi-geostrophic ocean dynamics in eddy-resolving regimes. *Ocean Modelling*, **94**, 1–14.

Simmons, H., S. Jayne, L. St Laurent, and A. Weaver: 2004, Tidally driven mixing in a numerical model of the ocean general circulation. *Ocean Modelling*, **6**, 245–263.

Skyllingstad, E. D. and D. W. Denbo: 1995, An ocean large-eddy simulation of Langmuir circulations and convection in the surface mixed-layer. *Journal of Geophysical Research-Oceans*, **100**, 8501–8522.

Smagorinsky, J.: 1963, General circulation experiments with the primitive equations I: The basic experiment. *Monthly Weather Review*, **91**, 99–164.

— 1993, Some historical remarks on the use of nonlinear viscosities. *Large Eddy Simulation of Complex Engineering and Geophysical Flows*, B. Galperin and S. A. Orszag, eds., Cambridge University Press, 3–36.

Smith, K. S., G. Boccaletti, C. C. Henning, I. Marinov, C. Y. Tam, I. M. Held, and G. K. Vallis: 2002, Turbulent diffusion in the geostrophic inverse cascade.

Stacey, F. D.: 2000, Kelvin's age of the Earth paradox revisited. *Journal of Geophysical Research: Solid Earth*, **105**, 13155–13158.

Stamper, M. A. and J. R. Taylor: 2017, The transition from symmetric to baroclinic instability in the eady model. *Ocean Dynamics*, **67**, 65–80.

Stokes, G. G.: 1845, On the theories of the internal friction of fluids in motion, etc. *Trans. Camb. Philos. Soc.*, **8**, 287–319.

Stommel, H. and A. B. Arons: 1960, On the abyssal circulation of the world ocean .2. An idealized model of the circulation pattern and amplitude in oceanic basins. *Deep-Sea Research*, **6**, 217–233.

Stommel, H. M.: 1948, The westward intensification of wind-driven ocean currents. *Transactions, American Geophysical Union*, **29**, 202–206.

Straub, D. N. and B. T. Nadiga: 2014, Energy fluxes in the quasigeostrophic double gyre problem. *Journal of Physical Oceanography*, **44**, 1505–1522.

Sullivan, P. P. and J. C. McWilliams: 2018, Frontogenesis and frontal arrest of a dense filament in the oceanic surface boundary layer. *Journal of Fluid Mechanics*, **837**, 341–380.

Sullivan, P. P., J. C. McWilliams, and W. K. Melville: 2007, Surface gravity wave effects in the oceanic boundary layer: large-eddy simulation with vortex force and stochastic breakers. *Journal of Fluid Mechanics*, **593**, 405–452.

Sun, D.-Z. and Z. Liu: 1996, Dynamic ocean-atmosphere coupling: a thermostat for the tropics. *Science*, **272**, 1148–1150.

Suzuki, N. and B. Fox-Kemper: 2016, Understanding Stokes forces in the wave-averaged equations. *Journal of Geophysical Research–Oceans*, **121**, 1–18. URL http://dx.doi.org/10.1002/2015JC011566

Suzuki, N., B. Fox-Kemper, P. E. Hamlington, and L. P. Van Roekel: 2016, Surface waves affect frontogenesis. *Journal of Geophysical Research–Oceans*, **121**, 1–28. URL http://dx.doi.org/10.1002/2015JC011563

Sverdrup, H. U.: 1947, Wind-driven currents in a baroclinic ocean; with appplication to the equatorial currents of the eastern Pacific. *Proc. National Acad. Sci.*, **33**, 318–326.

Talley, L. D.: 2008, Freshwater transport estimates and the global overturning circulation: Shallow, deep and throughflow components. *Progress In Oceanography*, **78**, 257–303.

Tandon, A. and C. Garrett: 1995, Geostrophic adjustment and restratification of a mixed layer with horizontal gradients above a stratified layer. *Journal of Physical Oceanography*, **25**, 2229–2241.

Taylor, J. R. and R. Ferrari: 2010, Buoyancy and wind-driven convection at mixed layer density fronts. *Journal of Physical Oceanography*, **40**, 1222–1242.

Thomas, L. N. and C. M. Lee: 2005, Intensification of ocean fronts by down-front winds. *Journal of Physical Oceanography*, **35**, 1086–1102.

Thomas, L. N., A. Tandon, and A. Mahadevan: 2008, Submesoscale processes and dynamics. *Ocean Modelling in an Eddying Regime*, M. Hecht and H. Hasumi, eds., AGU Geophysical Monograph Series, volume 177, 17–38.

Thomas, L. N., J. R. Taylor, R. Ferrari, and T. M. Joyce: 2013, Symmetric instability in the Gulf Stream. *Deep Sea Research Part II: Topical Studies in Oceanography*, **91**, 96–110.

Trenberth, K. and J. Caron: 2001, Estimates of meridional atmosphere and ocean heat transports. *Journal of Climate*, **14**, 3433–3443.

Trenberth, K. E. and J. T. Fasullo: 2009, Changes in the flow of energy through the Earth's climate system. *Meteorologische Zeitschrift*, **18**, 369–377.

— 2010, Climate change: Tracking Earth's energy. *Science*, **328**, 316–317.

Uchiyama, Y., J. C. McWilliams, and J. M. Restrepo: 2009, Wave-current interaction in nearshore shear instability analyzed with a vortex force formalism. *Journal of Geophysical Research-Oceans*, **114**.

Vallis, G. K.: 2006, *Atmospheric and Oceanic Fluid Dynamics: Fundamentals and Large-Scale Circulation*. Cambridge University Press, Cambridge. URL http://bit.ly/SDSMSK

Van Roekel, L. P., B. Fox-Kemper, P. P. Sullivan, P. E. Hamlington, and S. R. Haney: 2012, The form and orientation of Langmuir cells for misaligned winds and waves. *Journal of Geophysical Research–Oceans*, **117**, C05001, 22pp. URL http://dx.doi.org/10.1029/2011JC007516

Veronis, G.: 1966, Wind-driven ocean circulation–Part I. Linear theory and perturbation analysis. *Deep-Sea Research*, **13**, 17–29.

Von Schuckmann, K., M. Palmer, K. Trenberth, A. Cazenave, D. Chambers, N. Champollion, J. Hansen, S. Josey, N. Loeb, P.-P. Mathieu, et al.: 2016, An imperative to monitor Earth's energy imbalance. *Nature Climate Change*, **6**, 138–144.

Wang, P., J. Sheng, and C. Hannah: 2017, Assessing the performance of formulations for nonlinear feedback of surface gravity waves on ocean currents over coastal waters. *Continental Shelf Research*.

Waterhouse, A. F., J. A. MacKinnon, J. D. Nash, M. H. Alford, E. Kunze, H. L. Simmons, K. L. Polzin, L. C. St. Laurent, O. M. Sun, R. Pinkel, et al.: 2014, Global patterns of diapycnal mixing from measurements of the turbulent dissipation rate. *Journal of Physical Oceanography*, **44**, 1854–1872.

Webb, A. and B. Fox-Kemper: 2011, Wave spectral moments and Stokes drift estimation. *Ocean Modelling*, **40**, 273–288.

Webb, A. and B. Fox-Kemper: 2015, Impacts of wave spreading and multidirectional waves on estimating Stokes drift. *Ocean Modelling*, **96**, 49–64. URL http://dx.doi.org/10.1016/j.ocemod.2014.12.007

Whitt, D. B. and J. R. Taylor: 2017, Energetic submesoscales maintain strong mixed layer stratification during an autumn storm. *Journal of Physical Oceanography*, **47**, 2419–2427.

Wunsch, C. and R. Ferrari: 2004, Vertical mixing, energy and the general circulation of the oceans. *Annual Review of Fluid Mechanics*, **36**, 281–314.

Wunsch, C. and D. Roemmich: 1985, Is the North Atlantic in Sverdrup balance? *Journal of Physical Oceanography*, **15**, 1876–1880.

Yeager, S.: 2015, Topographic coupling of the Atlantic overturning and gyre circulations. *Journal of Physical Oceanography*, **45**, 1258–1284.

Young, W. R.: 2010, Dynamic enthalpy, conservative temperature, and the seawater Boussinesq approximation. *Journal of Physical Oceanography*, **40**, 394–400.

CHAPTER 3

The Global Ocean Observing System

Bernadette M. Sloyan[1], Moninya Roughan[2,3], and Katherine Hill[4]

[1]CSIRO Oceans and Atmosphere, Hobart Tasmania 7000, Australia; [2]MetOcean Solutions, Raglan, 3225, New Zealand; [3]School of Mathematics and Statistics and Biological and Earth Sciences, University of New South Wales, Sydney NSW, 2052, Australia; [4]World Meteorological Organization, Geneva, CH-1211, Switzerland

In the last 20-30 years, much progress has been made in the deployment of sustained, nationally and internationally coordinated ocean observing programs. These include the Argo array of profiling floats, ocean glider missions, the global drifter array, Ships of Opportunity (SOOP) eXpendable Bathythermograph (XBT) lines, deep water moorings, Global Ocean Ship-based Hydrographic Investigation Program (GO-SHIP), and new pilot studies extending to boundary currents, the deep ocean, and the marginal ice zones. In general, many of the observing systems were originally designed to resolve ocean variability at timescales from sub-seasonal to longer; however, the data are now also essential for ocean forecasting and prediction projects. Improved satellite technologies and telecommunications has enabled much of the ocean observations data to be recovered in real-time. The advent of rapid and timely access to data has led to the increasing use of ocean data in operational oceanography systems for the purpose of providing increasing accurate and reliable global and regional ocean (eddy resolving) forecasts. In this chapter, we provide an overview of the global in situ ocean observing systems for measuring physical Essential Ocean Variables (EOVs), including all the major platforms, and the efforts of the international coordinating programs.

Physical Ocean Observations

Introduction

In situ observations of the ocean remained relatively sparse and regionally-focused until the 1990s (Gould et al., 2013). The success of large multinational projects — Tropical Ocean Global Atmosphere (TOGA), TOGA-Coupled Ocean Atmosphere Response Experiment (TOGA-COARE), World Ocean Circulation Experiment (WOCE), and Joint Global Ocean Flux Study (JGOFS) — provided the impetus for coordinated global-scale in situ observations. This, in part, led to the Intergovernmental Oceanographic Commission (IOC) creating the Global Ocean Observing System (GOOS, www.goosocean.org) in 1991.

The GOOS is a highly collaborative, multidisciplinary system of observation networks built around nationally-managed and nationally-funded observing elements (satellites, buoys, scientists, etc.). GOOS provides a coordination mechanism for national contributions to come together to deliver sustained observations of the global ocean and relies on its expert stakeholders to represent concerns of their home nations. GOOS utilizes the Framework for Ocean Observing (http://www.oceanobs09.net/foo/) to guide its implementation of an integrated and sustained ocean

---

Sloyan, B.M., M. Roughan, and K. Hill, 2018: The global ocean observing system. In "*New Frontiers in Operational Oceanography*", E. Chassignet, A. Pascual, J. Tintoré, and J. Verron, Eds., GODAE OceanView, 75-90, doi:10.17125/gov2018.ch03.

observing system. This approach is designed to enable the ocean observing system to build flexible networks that can adapt to evolving technological developments as well as scientific and user requirements.

Essential Ocean Variables

Following the Global Climate Observing System's (GCOS) successful development and advocation for measuring Essential Climate Variables (Bojinski et al., 2014), GOOS has defined the Essential Ocean Variables (EOVs) required to meet the needs of the diverse user community. This user community includes climate and climate variability researchers, operational services and ocean health application areas. The adoption of EOVs is not intended to replace existing ocean observing networks and international coordination groups, but to provide a way to bring them together to develop a holistic ocean observing system and strengthen the ocean observing system's ability to meet all current and future user requirements.

EOVs are identified by the GOOS Physics and Climate, Biogeochemical, and Biology Panels (the GOOS Expert Panels), based on the following criteria:

Relevance: The variable is effective in addressing the overall GOOS Themes – Climate, Operational Ocean Services, and Ocean Health.

Feasibility: Observing or deriving the variable on a global scale is technically feasible using proven, scientifically understood methods.

Cost effectiveness: Generating and archiving data on the variable is affordable, mainly relying on coordinated observing systems using proven technology and taking advantage, whenever possible, of historical datasets.

The EOVs were determined using these criteria. However, it was recognized that not all EOVs are at the same level of implementation in the global observing system. A readiness level has been developed to offer an indication of our ability to provide a variable to meet the users' requirements. The readiness level considers: (1) **requirement processes**: technological maturity, adequate sampling frequency, measurement precision and quality control, satisfaction of multiple user needs, and ongoing international community support; (2) **coordination of observations elements**: Global and sustained observations, periodic review process, availability of specifications and documentation, and (3) **data management and information products**: Standardized and interoperable data outputs, global availability of useful data, data management and distribution policies. Each EOV is assessed against the three readiness elements and categorized as either concept, pilot, or mature.

The list of EOVs and their readiness level are shown in Figure 3.1. All of the physical and biogeochemical ocean variables are classed as mature, except for ocean surface heat flux, while many of the biological/marine ecosystem EOVs are either pilot or concept.

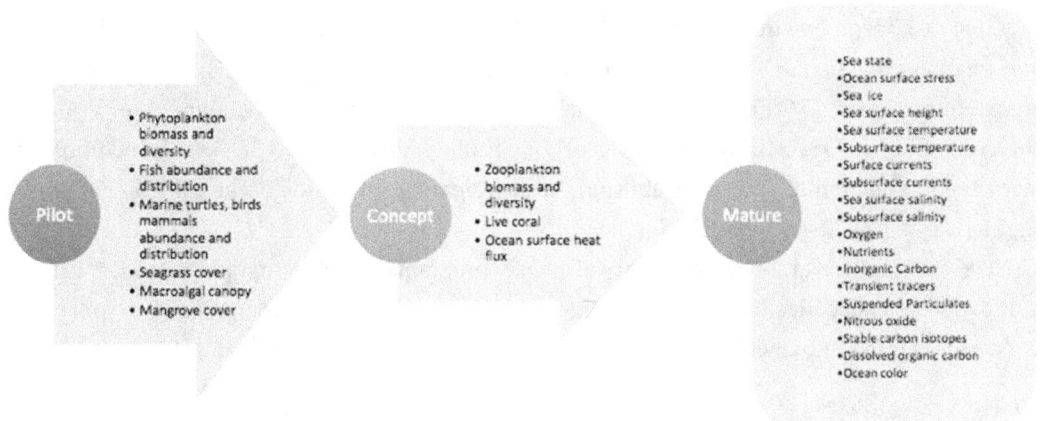

Figure 3.1. List of GOOS Essential Ocean Variables (EOVs) grouped as Pilot, Concept, or Mature.

International Ocean Observing Networks.

The EOVs are collected using a wide array of platforms and, in some cases, instruments and technologies. The technologies comprising the physical ocean in situ observing system include moored surface buoys, surface drifters, sub-surface moorings, satellites, floats, gliders, research and vessels of opportunity, and tidal stations (Figure 3.2).

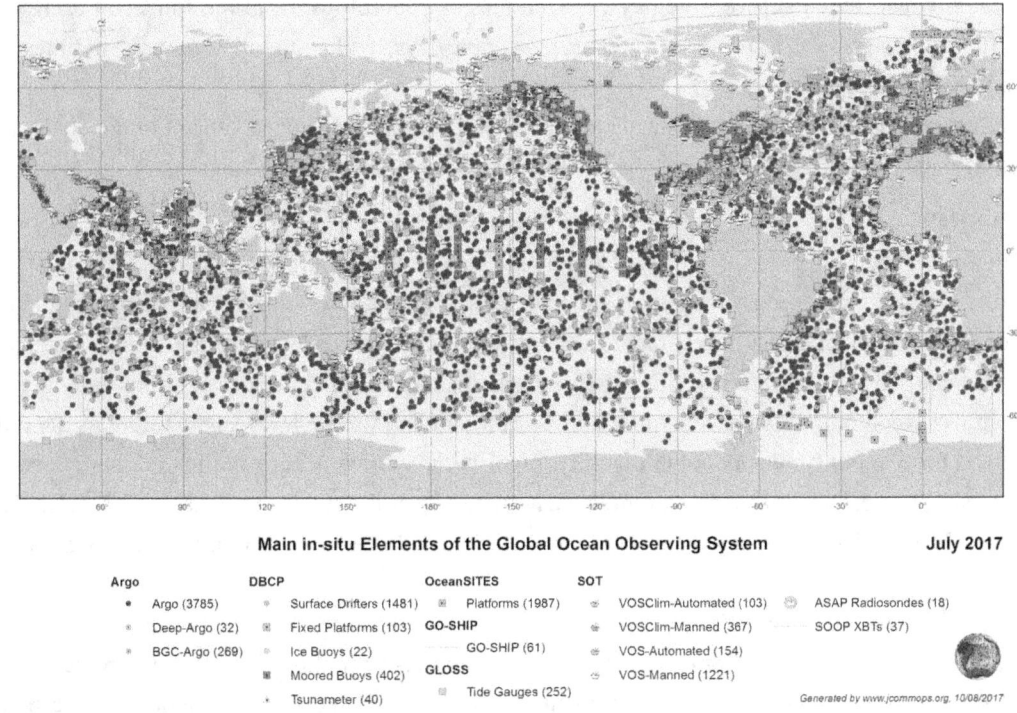

Figure 3.2. Components of the global ocean observing system.

The six major programs that currently comprise the global ocean observing system are Argo, Data Buoy Cooperation Group (DBCP), OceanSITES, the Global Ocean Ship-based Hydrographic Investigation Program (GO-SHIP), the Global Sea Level Observing System (GLOSS), and the Ship Observations Team (SOT), which consists of several programmes including the Volunteer Observing Ship (VOS) scheme and Ship of Opportunity Programme (SOOP). In addition, the international community is now establishing a coordinating group for ocean gliders and animal tagging.

Here we provide a brief overview of these programs. Each program provides detailed information on data collection, data processing, and data availability via comprehensive web pages. The reader is referred to these for detailed information and recent updates.

Argo

The broad-scale global array of temperature/salinity (T/S) profiling floats, known as Argo (http://www.argo.ucsd.edu/), is a major component of the ocean observing system, complementing satellite observations of sea surface height. Argo exemplifies international collaboration and data management, and offers a new paradigm for data collection. Float deployments began in 2000 and continues at the rate of about 800 floats per year. The design of the Argo network is based on experience from the present observing system, knowledge of ocean variability observed by satellite altimeter, and the requirements for climate and high-resolution ocean models. The array of almost 4000 floats provides 140,000 T/S profiles and velocity measurements each year distributed over the global ocean at an average 3-degree spacing, including the seasonal ice zone. Argo park depth is 1000 db where they drift for nine days, descending to 2000 db to begin the full 2000 db ascent profile. Floats will cycle to 2000 m depth every ten days, with four- to five-year lifetimes for individual instruments.

Pilot projects or design experiments are underway for enhanced observations in the equatorial and boundary current regions, as well as a Deep Argo array, and a Biogeochemical Argo array.

GO-SHIP

Building on previous global-scale hydrography efforts (WOCE, JGOFS, CLIVAR), GO-SHIP (the Global Ocean Ship-based Hydrographic Investigation Program; www.go-ship.org) is the systematic and global survey of select hydrographic sections being carried out by an international consortium of 16 countries and laboratories. GO-SHIP sections span all of the major ocean basins and the full-depth water column. Ship-based hydrography, at present and for the foreseeable future, is the only method for obtaining coincident high-quality, high spatial and vertical resolution measurements of a suite of physical, chemical, and biological parameters over the full water column. GO-SHIP data are freely available in a timely manner to the scientific and general community from a number of data servers.

GO-SHIP's unique contributions to the observing system are: coast-to-coast, top-to-bottom, near-synoptic section observation resolving boundary currents; highly accurate measurements of a full suite of water properties (salinity, oxygen, nutrients, carbon parameters, chlorofluorocarbons, isotopes, turbulence, and more) to detect subtle changes in the full ocean depth; provision of high-quality reference observations to other components of global ocean observing system that use autonomous observing platforms (e.g., Argo and SOOP), and supports validation and development of regional and global climate models, and a platform for testing new ocean observing technologies.

GO-SHIP provides changes in inventories of heat, freshwater, carbon, oxygen, nutrients, transient tracers, and other ocean properties at approximately decadal resolution. The GO-SHIP data are used for major assessments of the role of the ocean in mitigating climate change, research publications, atlases and other climate products, and outreach materials.

Moored/ship-based time series (OceanSITES)

OceanSITES (http://www.oceansites.org/) oversees a worldwide system of long-term, open-ocean time series stations measuring dozens of variables and monitoring the full depth of the ocean from air-sea interactions down to the seafloor. It is a network of stations or observatories measuring many aspects of the ocean's surface and water column using, where possible, automated systems with advanced sensors and telecommunications systems and yielding high time resolution, often in real-time, while building a long record. Observations cover meteorology, physical oceanography, transport of water, biogeochemistry, and parameters relevant to the carbon cycle, ocean acidification, the ecosystem, and geophysics.

OceanSITES comprises air-sea flux moorings, transport arrays, the Tropical Moored Buoy[2] array, and multidisciplinary time series sites. In some cases, these are individual moorings in a region of high interest. In other cases, multiple moorings are used in an array to measure transport, for example, to observe boundary current transport or to observe basin-scale meridional transports.

While most moorings carry instrumentation that record data internally, technical advances are increasing the real-time availability of OceanSITES data. Surface buoys allow satellite data telemetry, and subsurface data are brought to the surface by inductive, acoustic, or hardwire links. At some sites, ocean gliders are now used to acquire the data from subsurface moored instrumentation via acoustic modems and then transmit the data via iridium when they surface. Other sites use data capsules that are periodically released from the mooring to float to the surface and pass on subsurface data.

DBCP – Meteorological moorings

Marine meteorological moored buoys (http://www.jcommops.org/dbcp/) are deployed, operated, and maintained by various National Meteorological and Hydrological Services (NMHSs) under the World Meteorological Organization (WMO) framework and complement other sources of synoptic surface marine meteorological observations in coastal areas and the high seas. They provide data in support of marine services such as marine weather (and wave) forecasts, provision of maritime safety information to end users, and are assimilated into high-resolution and global numerical weather prediction models. Capabilities vary from country to country, with most (if not all) buoys measuring meteorological variables and some networks also measuring oceanographic variables. Many of these networks have been in place for 20 years or so and deliver data for weather and ocean state prediction.

---

[2] TAO/TRITON (in the Pacific), RAMA (in the Indian Ocean), PIRATA (in the Atlantic).

DBCP – Global Drifter Program

The objectives of the Global Drifter Program (http://www.aoml.noaa.gov/phod/dac/index.php), formerly known as the Surface Velocity Program (SVP), are to maintain a global 5° x 5° array of satellite-tracked surface drifting buoys (excluding marginal seas, latitudes > 60°N/S and those areas with high drifter 'death' rates) to meet the need for an accurate and globally dense set of in situ observations of mixed layer currents, sea surface temperature (SST), atmospheric pressure, winds and salinity, and provide a data processing system to deliver the data to operational and research users. Data from the Global Drifter Array make a valuable contribution to short-term numerical weather prediction (NWP), longer-term (seasonal to inter-annual) climate predictions, as well as climate research and monitoring. They are also used to validate satellite-derived SST products.

Ocean surface drifters were standardized in 1991, with drogues centered at 15 m below the surface. In 1993, drifters with barometer ports (called SVPB drifters) were tested in the high seas and proven reliable. Recent analysis has refined the array target to improve global coverage, including marginal seas. More accurate thermistors for high-resolution SST have been deployed and are being evaluated for impact on satellite SST calibration and validation. The archive of quality-controlled drifter data is updated quarterly.

SOT- Underway Observations from Ships (Ship Observations Team – VOS and SOOP)

Observations (http://www.aoml.noaa.gov/phod/soop/index.php) are being taken aboard underway vessels for a variety of observation programmes, which requires different levels of engineering and human intervention on the ship. Vessels used include commercial ships, ferries, as well as research and supply vessels. Some of the programmes require repeat transect observations, while others are focused on broader-scale observations. Research vessels and those servicing moorings (DBCP and OceanSITES moorings) provide the added benefit of delivering comprehensive high quality underway observations.

SOT - Voluntary Observing Ships (VOS)

Voluntary Observing Ships (VOS, http://sot.jcommops.org/vos/) are recruited and operated by NMHSs under the framework of the Joint WMO-IOC Technical Commission for Oceanography and Marine Meteorology (JCOMM) Ship Observations Team (SOT) to complement other sources of synoptic surface marine meteorological observations in coastal areas and the high seas. They provide essential support for global numerical weather prediction, climate applications, and marine services activities such as marine forecasting and the provision of maritime safety information to the maritime industry and port authorities. VOS data are also used in climate research and reanalysis. VOS provide most of the air temperature and humidity observations over the ocean.

While observations have increased close to the coast, there has been a decline in the number of observations in the open ocean due to changes in ship operations. VOS objectives are to sustain a network of vessels that provides weather and ocean observations via automated systems and human (manual) observations. There are currently over 3,000 active VOS ships that submit nearly two million observations each year.

SOT - VOS underway thermosalinograph observations

Thermosalinographs collect underway temperature and salinity data from the engine intake on vessels, usually complementary to other data streams such as underway CO_2 observations. These observations are collated and quality-controlled as part of the Global Ocean Surface Underway Data (GOSUD) Project.

SOT - SOOP XBT

An eXpendable BathyThermograph (XBT) is a probe that is dropped from a ship and measures the temperature as it falls through the water to a depth of approximately 800 m. The core XBT mission is to obtain multi-decadal upper ocean temperature profile data along specific transects that typically span ocean basins. The XBT observations constitute a large fraction of the archived ocean thermal data between 1970-1992. Until the full implementation of the Argo array, XBTs constituted 50% of the global ocean thermal observations, providing sampling initially during regional research cruises and later along major shipping lines but with a broad-scale spatial sampling strategy. Currently, XBT observations represent approximately 15% of temperature profile observations and they are the main practical system used for monitoring transports in boundary currents, eddies, and fronts by repeat sampling across fixed transects, some of which now have 30-year time series.

XBT observations are complementary to other ocean observation systems, and transects are maintained in locations that maximize the scientific value of the observations. Fixed transects (30-35) are maintained by the scientific community in either high density or frequently repeated modes. High density transects (occupied at least four times per year, with profiles at approximately 25-50 km intervals along the ship track and finer resolution of approximately 10 km across the equator and in-boundary currents), enable the calculation of heat and mass fluxes of boundary currents and the closing of gyre-scale heat and mass budgets of ocean basins. Frequently repeated transects (12-18 times per year, 100-150 km intervals) are positioned in areas of high temporal variability and enable studies of long-term means, seasonal cycles, and large-scale ocean circulation.

Tide gauge network (GLOSS)

The Global Sea Level Observing System (GLOSS; http://www.psmsl.org/gloss/) maintains high quality global and regional sea level observations. The network is comprised of approximately 300 sea level/tide gauge stations around the world for long-term climate change and oceanographic sea level monitoring that conform to requirements for representativeness of regional conditions, a core set of observations, and data delivery/availability. The core network is designed to provide an approximately evenly-distributed sampling of global coastal sea level variations. The final repository for GLOSS data is delivered to the Permanent Service for Mean Sea Level (PSMSL) repository, which is the preeminent global data bank for long-term sea level change information from tide gauges.

In addition to these well-established programs, GOOS is helping to establish coordinating programs for new technologies, which are increasingly becoming a part of the global ocean observing system (e.g., animal tagging).

EOV	Profiling Floats (Argo)	Repeat Hydrography (GO-SHIP)	Time series Sites - Moored/ Ship	Metocean moorings (DBCP)	Drifters – including buoys on ice (DBCP)	Voluntary Observing Ships (VOS)	Ships of Opportunity (SOOP)	Tide Gauges (GLOSS)	Ocean Gliders	Tagged Animals
Temp. (surface)	X	X	X	X	X	X	X		X	X
Temp. (subsurface)	X	X	X				X		X	X
Salinity (surface)	X	X	X	X	X	X	X		X	X
Salinity (subsurface)	X	X	X						X	X
Currents (surface)		X	X		X					
Currents (subsurface)	X	X	X						X	
Sea level	X		X					X		
Sea state			X	X	X	X				
Sea ice				X	X	X				
Ocean surface stress (OSS)		X	X	X		X				
Ocean surface heat flux (OSHF)		X	X			X				
Atmospheric Surface Variables										
Air temperature			X	X		X				
Wind speed and direction			X	X	X	X				
Water vapour			X	X		X				
Pressure			X	X	X	X				
Precipitation			X	X		X				
Surface radiation budget			X	X						

Table 3.1. Relationship between physical EOVs and Observing Platforms/Networks. Also included are atmospheric observation collected by the observing platforms/networks (GCOS, 2016).

Ocean gliders

Autonomous underwater glider technology has developed significantly over the last decade and gliders are now operated routinely, providing sustained fine-resolution observations of the coastal ocean, from the shelf to the open ocean. Long-term repeat sections can be carried out with gliders, considered steerable profiling floats to maintain oceanic measurements over the water column in regions of interest. A global glider program (http://www.ego-network.org/dokuwiki/doku.php) is being established as part of GOOS that will provide international coordination and scientific oversight to consider the role of gliders in the sustained observing system; the focus will likely be on the ocean boundary circulation area that links the coastal ocean and the open sea.

Animal tagging

Tagged animals (particularly conductivity, temperature, and depth [CTD]-tagged pinnipeds such as seals and sea lions) fill a critical gap in the observing system by providing profile data in the high-latitude ocean, including under the ice. Activity peaked during the International Polar Year (2007–2009). The primary motivation for tagging pinnipeds is for ecosystem monitoring, so coordination is needed to ensure that deployments provide information for biological and physical applications. Coordination is generally regional-/project-based but it would be beneficial to move towards global coordination of observations (including tagging locations, species, and their ranges), and particularly in the coordination of data assurance and quality procedures. Such global coordination would also facilitate systematic expansion and integration of T/S profile collection from other species. As a start toward global coordination, the Marine Mammals Exploring the Oceans Pole to Pole (MEOP) consortium (www.meop.net) brings together several national programmes to produce a comprehensive, quality-controlled database of oceanographic data obtained in polar regions from instrumented marine mammals.

International Coordination and Monitoring of the Observing Network

The complex relationship amongst EOVs and the major ocean observing programs is summarized in Table 3.1. This table clearly shows that the ocean physical EOVs requirements are satisfied in a number of ways using a variety of sensors and platforms. To provide resilience to the observing system, continued coordination through global networks organized around a particular platform or observing approach (e.g., satellites, surface drifters, profiling floats, mooring, time series sites, research vessels, and volunteer ships) is warranted and will benefit from evaluation studies considering combined multi-platform systems. It is the role of the GCOS-GOOS-WCRP Physics and Climate Panel (the Ocean Observations Panel for Climate, OOPC) to evaluate and set requirements, as well as to foster agreement on multiplatform design of the observing system and to evaluate observing system performance against these requirements. The coordination and monitoring of the ocean system is then undertaken by the JCOMM via two connected programs the Ocean Coordination Group (OCG) and JCOMM Operations (JCOMMOPS). The OCG and JCOMMOPS work with the individual observing networks to track the implementation performance of the ocean observing system against targets (e.g., deployments, coverage, data delivery).

The OCG's main task is to work across the observing networks to improve the performance of the observing system while also building on strengths and synergies. The OCG workplan focuses on responding to requirements (in consultation with OOPC and the WMO Integrated Global Observing System), observing system implementation (including roll-out of new technology and performance metrics), standards and best practices, and data and integration. JCOMM OCG is also the interface with data management coordination activities under JCOMM, along with the WMO Information System (WIS) and the Intergovernmental Oceanographic Commission International Ocean Data and Information Exchange (IODE).

The JCOMMOPS (www.jcommops.org) provides the ability to track implementation of the observing networks, systems monitoring, and data flow. From this metadata, we can monitor the key performance indicators of each of the observing networks (Figure 3.3)

	Argo			DBCP				SOT		GLOSS
	Argo Core	Argo Global	Argo BioGeoChemical	Global Drifter Program	Tropical Moored Buoy	Coastal/National MB	Tsunami Buoy	VOS	SOOP XBT	
Implementation										
Activity Global Ocean	85.16% 8/2017 ↗	97.42% 8/2017 ↗	28.79% 8/2017 ↘	115.06% 7/2017 ↗	65.55% 7/2017 ↘	108% 7/2017 ↘	52.5% 7/2017 ↘	92.25% 7/2017 ↗		84% 3/2017
Coverage (Monthly) Global Ocean		50.92% 8/2017 ↗		42.21% 7/2017 ↘				1.24% 7/2017 ↘		
Coverage (Yearly) Global Ocean	58.61% 2016 ↗	66.04% 2016 ↗								
Density (Monthly) Global Ocean	83.5% 8/2017 ↘	86% 8/2017 ↘	39.04% 8/2017 ↗	84% 9/2017 ↘						
Intensity Global Ocean	76.45% 8/2017 ↗	93.87% 8/2017 ↗	44.59% 8/2017 ↗	96.69% 7/2017						
Data Flow										
Delivery Global Ocean	96.56% 8/2017	95.89% 8/2017 ↗	89.8% 8/2017 ↗					61.74% 7/2017 ↘		
Timeliness (GTS FR) Global Ocean	95.57% 8/2017 ↗	95.62% 8/2017 ↗	93.62% 8/2017 ↗	78.98% 7/2017 ↗						
International										
Diversity (National) Global Ocean	19 2016 ↘	20 2016 →	11 2016 ↗	10 7/2017	13 7/2017 →			27 2016 ↗	9 2016 ↗	

Figure 3.3. Example of assessment of key performance indicators of components of the ocean observing network.

Data Management, Distribution and Integration

Each observing platform has its own data management system that, in association with OCG and JCOMMOPS, is used to closely monitor group collaboration across activities (e.g., observation collection, metadata and data assembly using community accepted standards, and quality assurance and control). Observing networks may provide data as real-time and delayed mode, or only as delayed mode. Real-time data reach operational ocean and climate forecast/analysis centers via the Global Telecommunications System (GTS). The target is for these real-time data to be available within approximately 24 hours of their transmission from the observing platform. Delay-mode data are the climate-standard, quality-controlled data. The production of delayed mode data is, for most networks, the responsibility of individual research in each country following the program/network standards and procedures practices. In general, delayed mode data is available from network data centers within six months of collections of data.

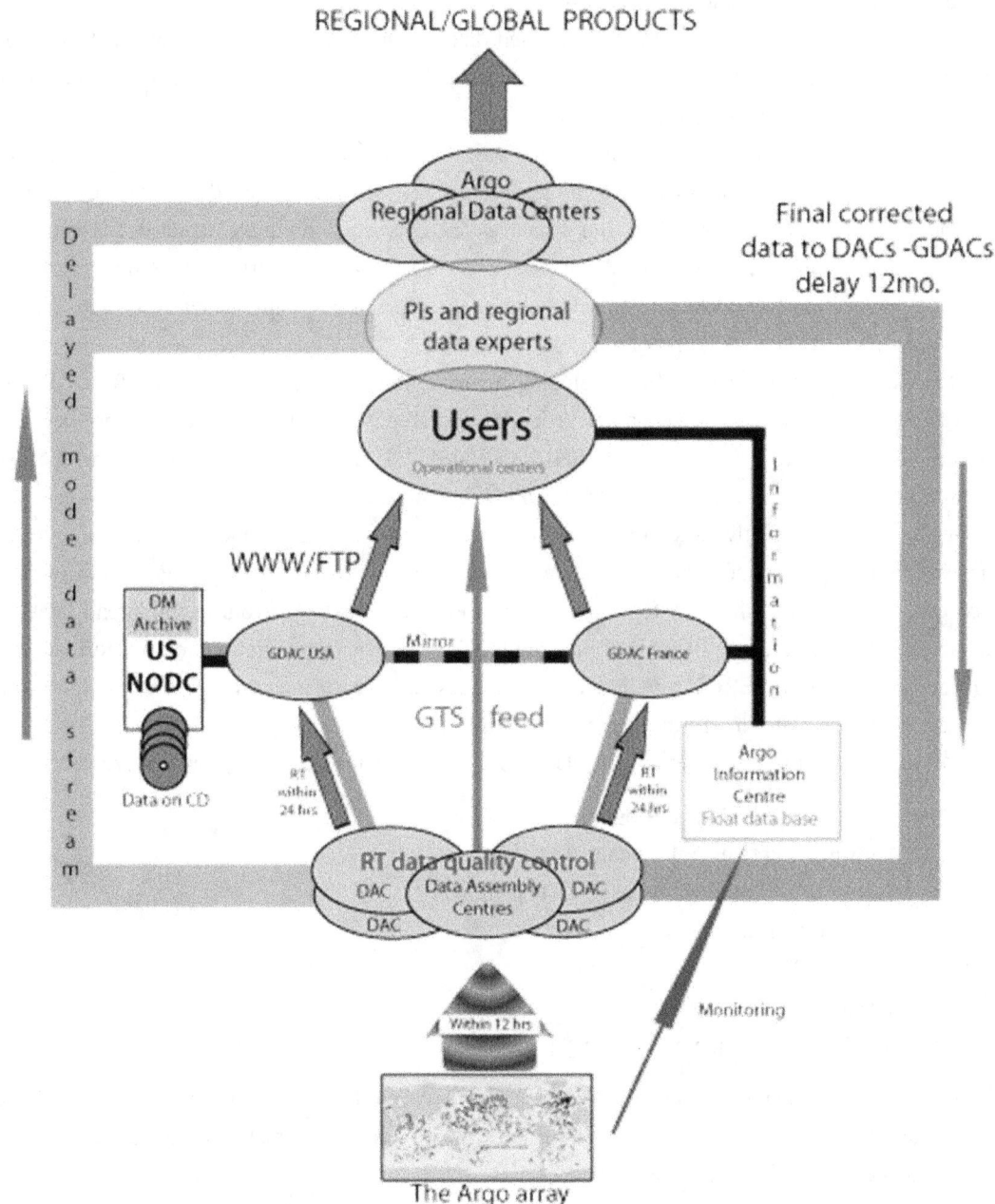

Figure 3.4. Schematic of the real-time and delayed mode data flow from the Argo array (from www.argo.ucsd.edu).

Some ocean observing networks are well developed and are largely successful in all these data management functions, while others are challenged to operate consistently due to varying data policies and submission requirements and sometimes due to a lack of sufficient resources for all the needed experienced staff and cyber-infrastructure for data services and preservation.

Argo provides an example of a well-defined data management system for both real-time and delayed mode data (Figure 3.4). Many observing networks that provide both real-time and delayed mode data follow similar data flows patterns.

The JCOMM OCG observing programs are working to build data systems that enable interoperability between the platforms. Interoperability serves the routine data exchanges within and amongst the networks, as well as user discovery and access. Community standards for metadata, data formats, communication protocols, and data server software infrastructure are the foundation for interoperability. The technical aspects have been demonstrated and successfully deployed in limited regions and specific parts of the global networks.

For example, a software system developing interoperability of the GOOS is the ERDDAP data server (http://coastwatch.pfeg.noaa.gov/erddap). ERDDAP functions as a broker between observational platforms (e.g, Argo, GO-SHIP, OceanSITES, Global Drifter Network) and users of data (Figure 3.5). Because it implements the OPeNDAP protocol and supports modern web services architecture, it is perfectly suited to pair with the various platform networks to provide integrated access to the real-time and delayed mode data streams. From the ingest side, it allows scientists to continue to work with the data formats they are most familiar with, from excel spreadsheets to database tables to netCDF files. From the data user's perspective, the various output formats and protocols supported by ERDDAP allow them to access and use the data through the clients of their choice, without having to reformat desired data.

An example of the application of ERDDAP is shown in Figure 3.6, using temperature data from:

- the Global Data Drifter Program,
- Argo-gridded temperature data,
- GO-SHIP and other ship-based CTD profiles,
- Pacific glider data,
- OceanSITES,
- SOT, and
- DBCP – Meteorological moorings.

The 2015 zonal mean temperature section was constructed for the Pacific Ocean between 30°N and 30°S. Here ERDDAP was used as the data framework to unite those platforms as part of an OGC pilot project to show improved integration of data.

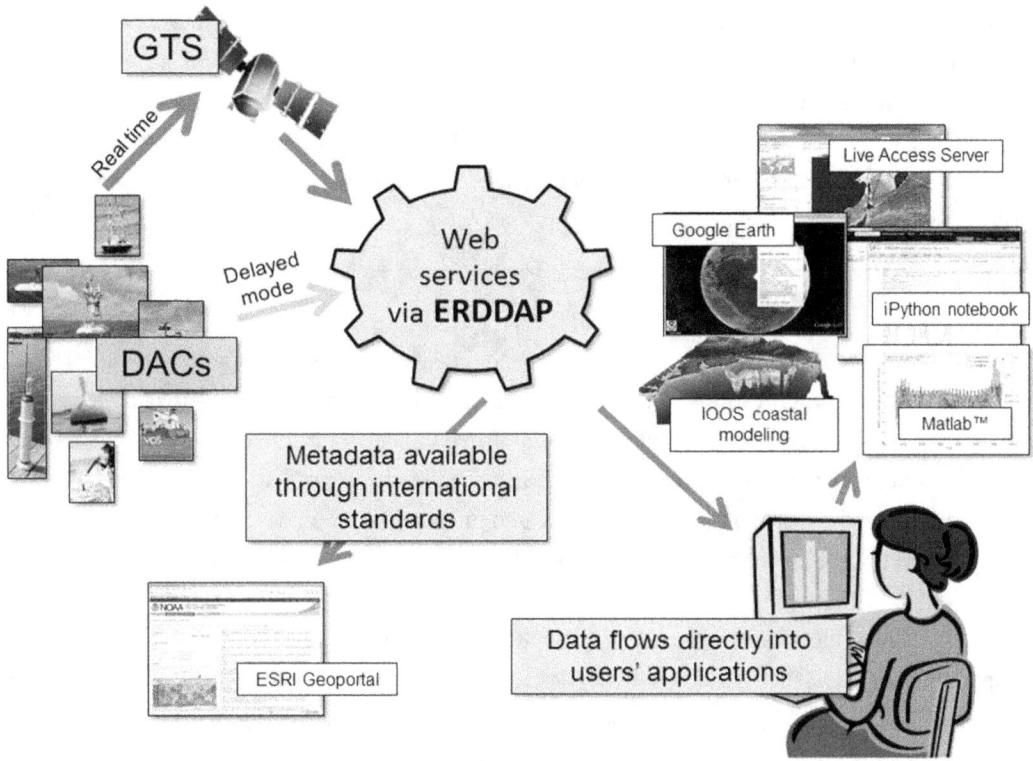

Figure 3.5. A schematic of the functionality of ERDDAP to combine data from various observations platforms to a user based on their requirements and applications.

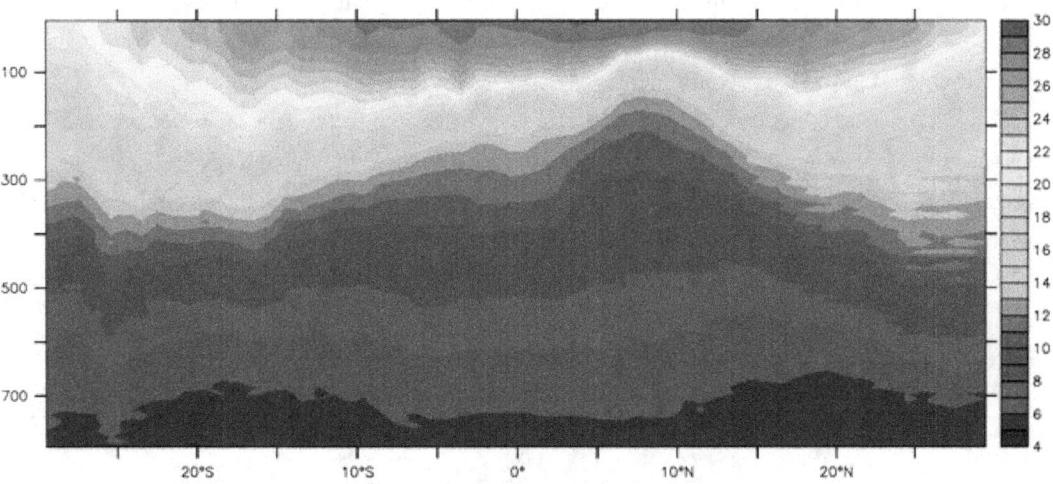

Figure 3.6. ERDDAP compiled 2015 Pacific Ocean zonal mean temperature section between 30ºS and 30ºN from numerous observations systems that collect temperature measurements

Expanding on these successes is important and is being guided by various ocean observing programs (both national and international) as well as by coordinating organizations – JCOMM OCG, JCOMMOPS, and GOOS. Improved interoperability across the observing system networks will enable us to create integrated datasets for the EOVs that include data from all observing networks. Integrated datasets have the potential to change the way operational oceanography utilizes the observational systems.

Future Developments

New or improved ocean observing satellites and in situ sensors and platforms, coupled with advances in telecommunications, are enabling us to close gaps and expand into new areas, measure new variables, lower cost per observation, and improve the impact of the observing system relative to investment (Sloyan et. al., 2017). During the last decade, the use of autonomous in situ platforms has revolutionized the ocean observing system, and the fast technological advance on platforms and sensors (including biogeochemical sensors) will continue to improve our ability to observe the ocean.

These sensor developments and improved telecommunications will make the concurrent observations of ocean physical, biogeochemistry, and biological/ecosystem observations realizable. This will lead to exciting developments towards multidisciplinary observing systems that meet requirements for climate, real-time services and ocean health. Indeed, the ocean community is strongly heading in this direction with the development of the Tropical Pacific Observing System 2020 Project (TPOS, www.tpos2020.org), the Deep Ocean Observing Strategy (www.deepoceanobserving.org), AtlantOS (www.atlantos-h2020.eu), and the related Atlantic Ocean Observing Blueprint (atlanticblueprint.net/). While these projects have a particular science/regional focus, the common thread is building an integrated, multi-disciplinary observing system that will have the ability to provide an efficient and comprehensive set of observations to meet current and future user requirements.

In addition, it is imperative to extend the focus of EOVs into the coastal zones and marginal seas, where societal impacts of the oceans are mostly keenly evident through sea level rise, extreme events, loss of ecosystem services, and impacts on coastal infrastructure. This also requires special attention to the integration of the ocean observing system – physical, biogeochemical, and biological – due to the variability of this region and immediate societal impacts. Coastal/open-ocean exchange processes are also key controllers of coastal ocean water properties, thus strong integration of the coastal and open-ocean observing systems is required.

Conclusion

The ocean observing system is used by an increasingly diverse user group for fundamental underpinning ocean research, as well as for real-time numerical weather forecast and near-term climate prediction services. The sampling strategy of the ocean observing system will evolve as we improve our understanding of the spatial and temporal scales that need to be resolved, technology advances, and experience expands from the user community. An ongoing challenge for the stakeholders of the ocean observing system is designing, implementing, and sustaining the critical multidisciplinary ocean observations required by the user groups.

Improved models are essential for guiding national and international policies that relate to resources (such as fisheries, agriculture, and water supply) that are impacted by climate variability and change, and support efforts aimed at mitigating long-term climate change. As the forecast range increases, processes in the ocean become increasingly important. It is anticipated that improved model skill will result from the addition of sub-surface ocean information, particularly at the seasonal to multi-year timescale (Kirtman et al., 2013). Therefore, the reliability and timeliness of weather forecasts and climate predictions provided to the public, industry, government (including emergency services), and policy makers relies on a comprehensive and timely set of ocean observations.

Acknowledgments

We thank Kevin O'Brien for providing ERDDAP information and JCOMMOPS that enable the tracking of ocean network performance. We acknowledge the vast number of people who contribute to the collection of in situ ocean observations, quality control of data, submission, and serving of data at the various national and international data portals, as well as the coordination and oversight committees. The global ocean observing system would not be possible without their commitment and dedication.

References

Bojinski, S., Verstraete, M., Peterson, T.C., Richter, C., Simmons, A. and Zemp, M.A., 2014. The concept of essential climate variables in support of climate research, applications, and policy. Bulletin of the American Meteological Society, **95**, 1431-1441.

Gould, J., Sloyan, B. M., and Visbeck, M., 2013. In-situ Ocean Observations: A brief history present status and future directions. In Gerold Siedler, Stephen Griffies, John Gould and John Church (eds) Ocean Circulation and Climate: A 21st Century Perspective, 2nd Ed. Sydney, Academic Press, 59-79 (International Geophysics, 103).

Kirtman, B. and Stockdale, T. and Burgman, R., 2013. The Ocean's role in modeling and predicting seasonal-to-interannual climate variations. In Gerold Siedler, Stephen Griffies, John Gould and John Church (eds) Ocean Circulation and Climate: A 21st Century Perspective, 2nd Ed. Sydney, Academic Press, 625-643.

Sloyan, B., Craw, P., King, E., Neill, C., Kloser, R., and Bodrossy, L., 2017. Future technologies. In B. Mapstone (Ed) Oceans: science and solutions for Australia, Clayton South, Australia, CSIRO Publishing, 179-186.

CHAPTER 4

Shelf and Coastal Ocean Observing and Modeling Systems: A New Frontier in Operational Oceanography

Moninya Roughan[1,2,3], Colette Kerry[2], and Peter McComb[1]

[1]*MetOcean Solutions and New Zealand MetService New Zealand;* [2]*School of Mathematics and Statistics, University of New South Wales, Sydney NSW, 2052, Australia;* [3]*School of Biological, Earth and Environmental Sciences, University of New South Wales, Sydney NSW, 2052, Australia*

One of the new frontiers in operational oceanography includes progress in observing and modeling the coastal ocean. In addition to this, it is becoming increasingly important to bring operational oceanographic data to industry in a usable format. We use the case study of Australia's Integrated Marine Observing System (IMOS) and its application to the East Australian Current to introduce some of the latest ideas about shelf and coastal ocean observing and modeling, and its applications to operational oceanography.

Introduction and Motivation

Coastal oceans are among the most productive ecosystems on the planet, providing an array of services that directly and indirectly support economic activity and growth. In addition to the plethora of ecosystem services, social and environmental benefits, our coastal oceans provide services of great economic value including: protection from natural hazards; weather regulation; shoreline stabilization; carbon sequestration; wild-catch fisheries; energy from wind, waves and offshore oil; sea bound trade; and tourism. These services, along with many others, provide the foundation for an estimated $3–$5 trillion dollars in annual global ocean economic activity (http://www.pemsea.org/our-work/blue-economy).

Indeed, in some Asian countries it is estimated that up to 20-30% of their Gross Domestic Product (GDP) comes from the ocean (http://www.pemsea.org/our-work/blue-economy), while Australia has one of the world's largest Exclusive Economic Zones (EEZs) providing a total value of over $70 billion and more than 400,000 jobs (2013-14, http://www.aims.gov.au/aims-index-of-marine-industry). Continental shelf regions are even more important economically. Even though only a small percentage of the surface area of the world oceans is comprised of these regions, they provide between 15 and 30 % of the entire oceanic primary production (Yool and Fashman, 2001).

Presently about 40% of the world's population lives within 100 km of the coast, and in some countries, such as Australia, this figure can be as high as 80%

Roughan, M., C. Kerry, and P. McComb, 2018: Shelf and coastal ocean observing and modeling systems: A new frontier in operational oceanography. In *"New Frontiers in Operational Oceanography"*, E. Chassignet, A. Pascual, J. Tintoré, and J. Verron, Eds., GODAE OceanView, 91-116, doi:10.17125/gov2018.ch04.

(http://www.un.org/esa/sustdev/natlinfo/indicators/methodology_sheets/oceans_seas_coasts/pop_coastal_areas.pdf). As population density and economic activity in coastal zones increases, pressures on coastal ecosystems increase.

Moreover, a country's EEZ extends up to 200 nautical miles from its coastline, thus individual nations and broader groups such as the European Union (EU) have an impetus to understand their coastal ocean domains. An adequate understanding of the coastal marine environment is fundamental to manage it sustainability. However, we are presently limited in our ability to measure and model our marine environments, particularly at scales sufficient for resolving the dynamics of the coastal ocean.

In recognition of the significance of our coastal and shelf regions, a range of long-term observing initiatives have been developed over the past 1-2 decades. In the U.S., coastal ocean observing programs emerged through a bottom-up approach, where local counties, universities, and organizations began sustained observing. In the absence of regional coordination, these groups grew into a network of regional alliances (e.g., see Chapter 21 by Wilkin et al.) brought together into a nationally coordinated way under the Integrated Ocean Observing System (IOOS) framework. In Australia, sustained ocean observing was instigated through a series of tranches of federal funding to support the Integrated Marine Observing System (IMOS, www.imos.org). This resulted in a nationally coordinated network of oceanographic observations for the coastal and deep ocean.

In the ocean modeling space, the experiences of the U.S. and Australian regional alliances also differ substantially. The U.S. has led the way in ocean modeling, both in development of open source ocean models (such as POM, ROMS, and FVCOM) as well as in implementation in operational oceanography and the application to regional alliances. In contrast, Australian modeling initiatives were not funded through the "ocean observing" program, which has meant that the application of coastal observations in an operational ocean hindcasting or forecasting framework has been more ad hoc.

To date, Australia's ocean modeling needs have been serviced by the BlueLink project, which provides circulation at 10 km resolution (Oke et al., 2008). However, near-global models, such as these with 7–10 km resolution, are insufficient for resolving flow in the coastal ocean (e.g., along southeastern Australia where the shelf is <30 km wide in many places). Notably, in addition to insufficient resolution, near-global models, including BlueLink, exclude important coastal processes, such as tidal forcing, local winds, and river input.

There are very few countries globally that have high resolution operational ocean forecasting capability. Presently, the Australian Bureau of Meteorology and the Commonwealth Scientific and Industrial Research Organisation (CSIRO) have coastal capability in only two regions (the Great Barrier Reef and the Great Australian Bight) driven by end user and stakeholder needs. Despite this, significant efforts are being undertaken in the ocean observing space in the coastal ocean, and it is the integration of these modeling and observing efforts that are the leading edge in operational oceanography.

Here we present the example of the New South Wales node of the Australian Integrated Marine Observing System (NSW-IMOS) and showcase the benefit of sustained ocean observations in the coastal ocean as a pathway for operational oceanography in our shelf sea. In the context of operational oceanography, there is unlimited potential to develop coastal operational hindcasting and prediction systems, which is clearly one of the new frontiers in operational oceanography and coastal modeling.

Observing the Coastal Ocean: Australia's Integrated Marine Observing System (IMOS)

Boundary currents (BCs) are highly energetic regions of our ocean basins that redistribute water, heat, and salt around the globe. As such, BCs play a major role in regulating the global climate system and they have a profound influence on local ocean and weather processes. Yet monitoring the multi-space and timescales of the energetic dynamic flows of boundary currents can be complicated.

One of the next frontiers in operational oceanography is development and implementation of a global network designed for sustained monitoring of our BCs in order to understand their full multi-scale variability from time and space scales that span from sub-seasonal to multi-decadal and from turbulent to basin scales; moreover we need to understand the interaction of BCs with marginal seas, frontal processes that aid mixing and cross front flows, and air-sea interaction, as well as their impacts on marine ecosystems.

Along the east coast of Australia, significant effort has been put into sustained monitoring of the East Australian Current (EAC, the Western boundary current of the South Pacific Ocean) and its interaction with shelf waters (Fig. 4.1). This has been facilitated since 2006 through IMOS (www.imos.org.au). In the IMOS framework, funding flows through 'centralized facilities' that are responsible for the deployment of the ocean observing "kit" such as high frequency (HF) radar, autonomous gliders, and moorings (deep and shallow). Whereas the science is driven by regional "nodes" that are a loose grouping of scientists united under a coordinated science and implementation plan (http://imos.org.au/plans.html). There are three IMOS nodes that contribute to our understanding of the EAC: two coastal nodes that together cover the entire east coast of Australia (Q-IMOS and NSW-IMOS) and a blue water and climate node which encompasses the offshore regions around Australia. Over the past ten years, a comprehensive ocean observing network has been deployed and refined with the purpose of understanding the dynamics of the WBC, its impact on shelf circulation, and biological productivity.

Since 2006, the Australian federal government has invested more than $190 million in IMOS, with an additional co-investment of 140% from industry, universities, stakeholders, state and federal agencies. The main goal is to provide a multidisciplinary, multi-institutional approach to enhance the observation and understanding of the oceans around Australia. The funding flows through ten centrally coordinated "infrastructure" facilities that are responsible for deploying ocean observing equipment in both the coastal and deep ocean, to measure a number of physical, chemical, and biological variables. These facilities are supported by the national IMOS office and an eleventh data facility responsible for management and free distribution of the data, the Australian Ocean Data Network (AODN, www.aodn.org.au).

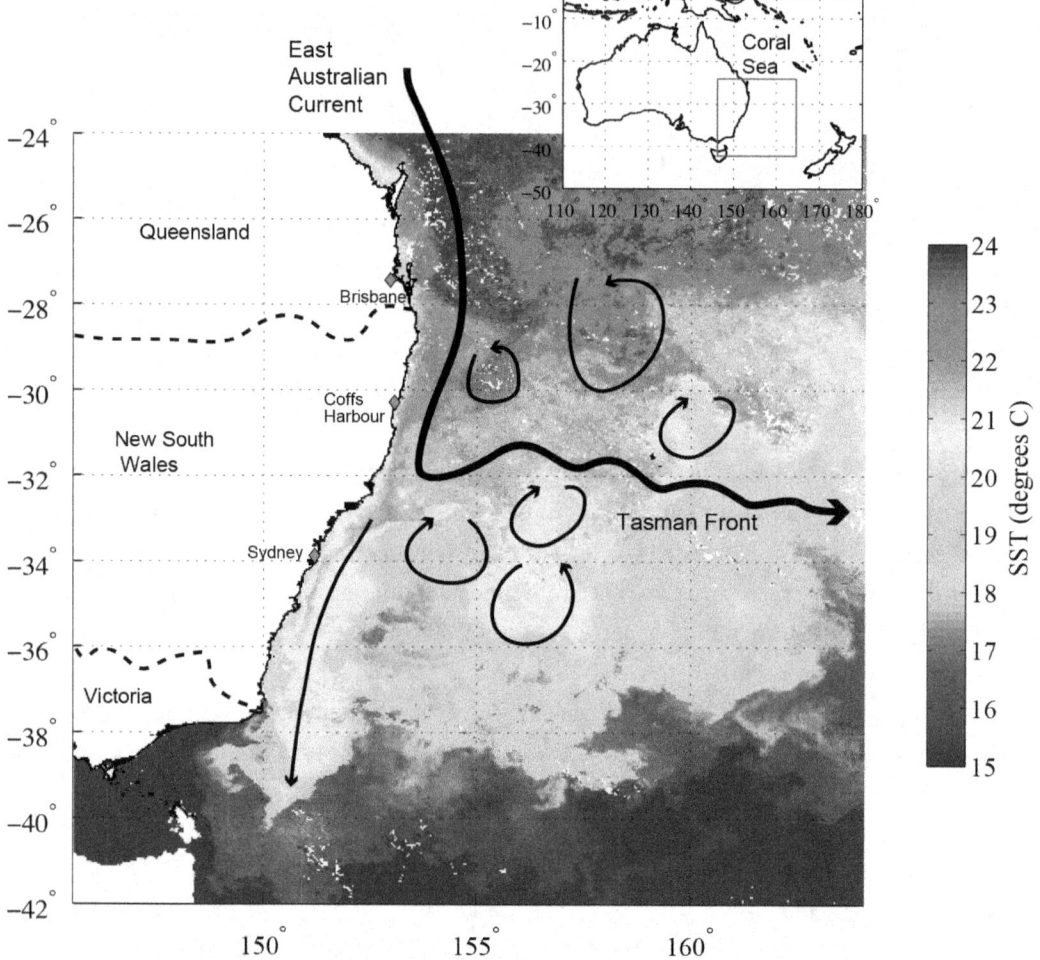

Figure 4.1. Sea surface temperature (SST) image (AVHRR SST L3 for 7 June 2015) showing the East Australian Current flowing poleward along the east coast of Australia. Color indicates SST and identifies the WBC jet, the Tasman front and the EAC eddy field.

Science objectives

The infrastructure deployment around Australia has been driven by a series of science questions that are documented in the IMOS Node Science and Implementation Plans (http://imos.org.au/about/about-imos/plansreports/). Through this process, a number of unifying national overarching goals were identified: 1. multi-decadal ocean change, 2. climate variability and weather extremes, 3. major boundary currents and inter-basin flows, 4. continental shelf and coastal processes, and 5. the ecosystem response. In the context of coastal oceanography, there are two main objectives:

1. To investigate the East Australian Current, its separation from the coast and the resultant eddy field in order to:
   - determine the statistical, dynamic, and kinematic properties of EAC eddies, and the frequency and dynamical drivers of eddy shedding;
   - quantify the impact of key physical processes driven by the EAC such as onshore encroachment, shelf circulation driven by the EAC, cross shelf processes, and internal waves; and
   - understand air-sea interactions, particularly to determine the development of east coast lows and severe winter storms in relation to warm core eddies and oceanic conditions.
2. To quantify oceanographic processes on the continental shelf and slope of southeastern Australia in order to:
   - examine the coastal wind and wave climate in driving nearshore currents and the northward sediment transport;
   - quantify the biogeochemical cycling of carbon (nutrients and phytoplankton composition); and
   - determine the transport and dispersal of passive particles (e.g., larvae, eggs, spores) and the degree of along-coast connectivity and trophic linkages.

Ocean observing infrastructure

To provide insight into these science objectives in the context of coastal Australia, three coastal focal regions were chosen: Coffs Harbour (30°S), Sydney (34°S), and Narooma (36°S). Coffs Harbour lies approximately 550 km to the north of Sydney, while Narooma lies approximately 350km to the south. In addition, a full-depth transport resolving array was deployed from the continental shelf off Brisbane to 150 km offshore at about 27.5°S. These are the only permanent mooring along an approximately 2000 km stretch of coastline, and not a single one has a surface expression for real-time or air-sea fluxes and over-ocean wind measurements. These arrays are described below and outlined in Fig. 4.2.

Deep water moorings

In an effort to resolve the full-depth transport of the EAC, a mooring array was deployed across 27.5°S where the EAC has been shown to be most coherent. Seven moorings were deployed in water depths of 200 to ~4800 m over a distance of ~153 km from the continental shelf to the abyssal plain (See Fig. 4.2 and Sloyan et al., 2016, Table 1). Q-IMOS, the coastal node of IMOS responsible for observing northern Australia, deployed the shelf moorings in 200 and 400 m of water. The deep-water moorings were deployed shore normal to this array across the continental slope and abyssal moorings by the IMOS deep water mooring program as part of the Bluewater and Climate Observing node (BWC). The moorings were instrumented with Teledyne RD Instrument (TRDI) acoustic Doppler current profilers (ADCPs) of various frequencies (e.g., 75, 150, and 300 kHz), which provided velocity profiles at varying vertical resolution from 4 to 16 m, in the upper 1000 m in a range of configurations. In addition, point-source velocity data were obtained from Nortek

Aquadopp instruments at varying vertical resolution of 500 to 1000 m. In addition, 55 point-source temperature sensors and 25 point-source salinity sensors (Sea-Bird Electronics) were distributed across the seven moorings. In total, 143 instruments were deployed across the array. See Sloyan et al. (2016) for further details of the mooring program and instrumentation.

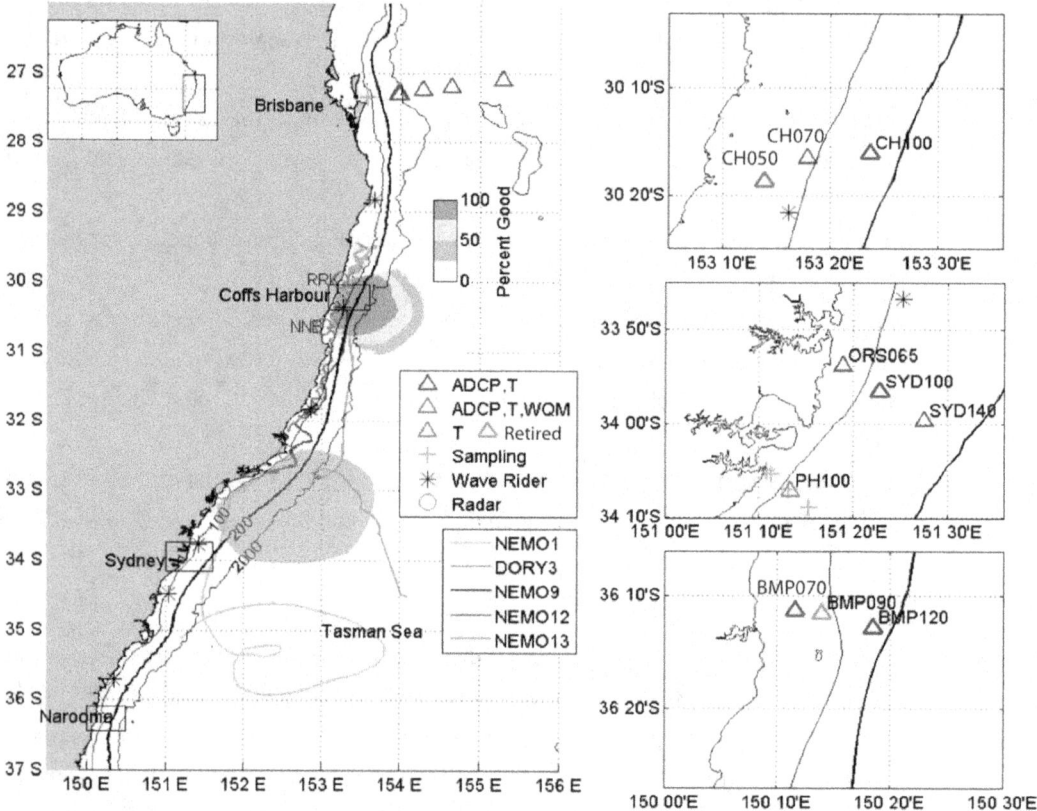

Figure 4.2. Location of the observing infrastructure along the coast of southeastern Australia. This includes deep water moorings (off Brisbane at ~27.5°S) and shelf moorings, HF coastal radar, biogeochemical sampling, wave rider buoys and example tracks from autonomous ocean gliders (where Nemo are shelf missions < 200m, and Dory are deep missions < 1000m). Insets from top to bottom (right) show the moorings at Coffs Harbour, Sydney and Narooma. Not shown are satellite remote sensed data, Argo floats and repeat XBT lines (shown in Fig. 4.4). 100 m, 200 m (bold) and 2000 m contours are shown. Note the HF radar array at 33°S shows representative coverage only.

The mooring array was deployed for an 18-month period from April 2012 to August 2013 (Sloyan et al., 2016). Then, due to funding constraints, was removed for a two-year period. The array was consolidated into six moorings and redeployed in May 2015.

These data were used to understand the mean velocities, coherence of the EAC, and the transport of heat and mass. Mean velocities calculated from the 18-month time series show poleward velocity from the 0 to 1500 m adjacent to the continental shelf, with a recirculation feature offshore. While the EAC has been shown to exhibit variability on a range of timescales and a degree of incoherence, this dataset showed that the EAC was coherent during this time at this latitude, with an eddy kinetic to mean kinetic energy ratio of less than 1.

Moreover, they calculated the 18-month mean, poleward-only mass transport above 2000 m to be 22.1 ± 7.5 Sv (where 1 Sv is 10^6 m^3 s^{-1}). The mean, poleward-only heat transport and flow-weighted temperature above 2000 m were −1.35 ± 0.42 PW and 15.33°C, respectively.

Using the results of a complex empirical orthogonal function (EOF) analysis of the along-slope velocity anomalies, Sloyan et al. (2016) further showed that velocity variance is dominated by two modes of variability at periods of approximately 60 days (Mode 1) and 120 days (Mode 2), and together they explain more than 70% of the velocity variance.

Shelf moorings

A network of up to ten coastal moorings has been deployed in shelf waters from 65-140 m deep along the coast of southeastern Australia since 2008 (Fig. 4.2). The moorings were deployed in shore-normal arrays in each of the three coastal focal regions: Coffs Harbour (30°S), Sydney (34°S), and Narooma (36°S). The moored array was designed to build upon existing operational observations e.g., a hydrographic mooring supported by Sydney Water Corporation off Sydney (ORS065) since 1991, and a network of wave rider buoys measuring wave height and direction funded by NSW state government (Fig. 4.2). The ORS065 dataset has been collected operationally on behalf of Sydney Water Corporation as part of their license to operate an offshore sewage ocean outfall since 1991. This is one of the longest moored full water column temperature records available around Australia, and it exists entirely due to operational necessity.

The moorings have existed in a number of different configurations over the years, however, presently the moorings consist of a bottom-mounted TRDI 300kHz ADCP and a string of Aquatech 520 temperature and temperature/pressure loggers at 8 m intervals through the water column (Table 1, Roughan et al., 2013, 2015). The line of thermistors is supported by a float, approximately 20 m below the surface. Below the sub-surface float at PH100 is a Wetlabs water quality meter (WQM) that consists of a SeaBird CTD, as well as measurements of dissolved oxygen, fluorescence, and turbidity (Wetlabs FLNTU). The WQM is being phased out and replaced with a SeaBird SBE37 CTD at the bottom and below the surface (See Roughan et al., 2015, 2013, 2011 for further details).

Needless to say, over-ocean winds are an important component of any observing system as they provide validation for atmospheric forcing products. In this region, very limited over-ocean wind observations have been available historically, however, modeled products perform reasonably well (Wood et al., 2012). Another alternative for broad-scale wind patterns are satellite-derived observations and products, which have been shown to perform well in this region (Rossi et al., 2014).

In regions of strong currents (such as the EAC), it is challenging to deploy moorings with a surface expression that would hold atmospheric sensors; however, this is becoming more feasible as technology improves. Associated with a mooring that has a surface expression is the ability to telemeter data in real time. Thus, development of a real-time data delivery system is the next obvious improvement to the array, which will support operational purposes as well as real-time validation of model output.

Autonomous ocean gliders

Between 2008 and 2017, 32 autonomous ocean glider missions were undertaken along the coast of southeastern Australia. Typically, these missions spanned the continental shelf on the inshore edge of the EAC from 29.5–33.5°S. The region is both physically and biologically significant, and it is in a hotspot of ocean warming. This comprehensive dataset of over 40,000 new CTD profiles from the surface to within 10 m of the bottom in water depths ranging 25–200 m provides unprecedented high resolution observations of the properties of the continental shelf waters adjacent to a western boundary current, straddling the region where it separates from the coast. The dataset consists of temperature, salinity, and density, as well as dissolved oxygen and chlorophyll-a fluorescence (indicative of phytoplankton biomass).

The data have been released publically in a gridded dataset (Schaeffer et al., 2016a) and are an invaluable resource with which to understand shelf circulation dynamics (Schaeffer and Roughan 2015), stratification (Schaeffer et al., 2016b, Schaeffer and Roughan 2015), biophysical and biogeochemical interactions (Baird et al., 2011; Everett et al., 2015; Schaeffer et al., 2016b). They are also useful for data assimilation into and validation of high-resolution ocean models (Kerry et al., 2016). Moreover, the data are being used as teaching material for ocean observing classes introducing students to the newest methodology and procedures in ocean observing.

Using the combined gridded dataset, Schaeffer et al. (2016b) calculated the spatial scales of variability on the continental shelf—showing the anisotropy in the EAC in both the physical and biogeochemical datasets. These data are useful in the context of ocean data assimilation, where errors need to be prescribed to the data prior to assimilation (see following section and Kerry et al., 2016).

High frequency coastal radar

A pair of high frequency (HF) coastal radars (Wellan Radar) were installed at 30°S off southeastern Australia in February 2012. The phased array system remotely measures surface currents in the top 0.9 m off Coffs Harbour (Eastern Australia, 30°S–31°S; Fig. 4.2). This HF system operates at 13.92MHz frequency, with radial and azimuthal resolutions of 1.5 km and 10.4°, respectively (Wyatt et al., 2017). It acquires data every 10 minutes with an offshore range of up to 150 km (see Wyatt et al., 2017; Mantovanelli et al., 2017; Archer et al., 2017, and Schaeffer et al., 2017 for more details of the system deployment and data validation).

Cyclonic eddies that form on the inside edge of the WBC have been shown to entrain coastal waters (Everett et al., 2015; Macdonald et al., 2016; Roughan et al., 2017). Schaeffer et al. (2017) presented results from an automated eddy detection algorithm applied to the radar surface currents over a one-year period and showed that cyclonic eddies are generated on average every seven days on the inside edge of the EAC.

High frequency measurements of over-ocean wind direction and surface currents were used by Mantovanelli et al. (2017) to reveal the influence of the short-term, small-scale wind forcing on the surface circulation, enhancement of horizontal shear, frontal jet destabilization, and the generation and decay of a cyclonic eddy.

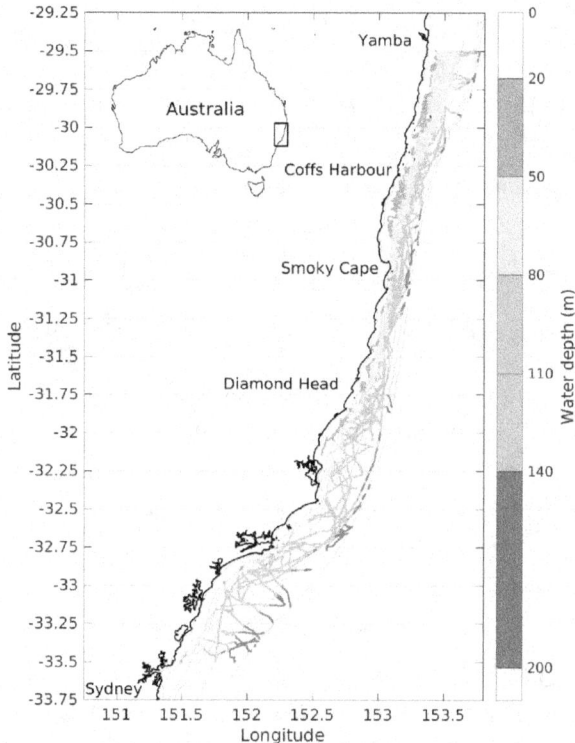

Figure 4.3. Glider paths from the 26 shelf glider deployments along the coast of southeastern Australia between 2008 and 2015, adapted from Figure 1 of Schaeffer et al. (2016) and Schaeffer and Roughan (2015). The gliders were generally deployed in Yamba (29.5°S) and retrieved as far south as possible. Color shading indicates the depth of the water that the gliders traversed, and typically represents the dive depth.

Using four years of data Archer et al. (2017) were able to identify periodicity in the EAC meandering, core velocity and width, and eddy kinetic energy. Their results showed periodicities at 65-100 days, similar to the results of Schaeffer et al. (2013, 2014a,b) who identified periodicity in cross-shore transport and eddy encroachment and variability. This periodicity differs somewhat from that of Sloyan et al. (2016) who found dominant modes of 60 and 120 days located approximately 200 km to the north of this site (using a shorter dataset).

Interestingly, using a flow following technique, Archer et al. (2017) were able to show that the EAC had a distinct annual cycle in core velocity, a result that had remained elusive using data from fixed moorings because the results were aliased as the core of the EAC meandered onshore and offshore across the mooring array. These results highlight the benefit of HF radar for spatial mapping of surface flows.

Boat-based sampling

In addition to the moored data, boat-based biogeochemical sampling has been undertaken off Sydney (34°S, PH100 in Fig. 4.2) since the 1940s for a range of variables (see Roughan et al., 2011, 2013 and Lynch et al., 2014 for full details). These data provide an invaluable resource of more than 70 years of weekly to monthly temperature data. Among other studies, the data have been used to show that sea surface temperatures (SSTs) have warmed at a rate of 0.746°C per century

(Thompson et al., 2009) off Port Hacking, Sydney, Australia. More recently, Schaeffer and Roughan (2017) used the data to create a daily temperature climatology for the region, which they used in conjunction with the high resolution mooring data to investigate the sub-surface structure and dynamics of marine heatwaves (MHWs) over the past 20 years. Their results showed that MHWs are subsurface intensified and are longer below the surface. These data are becoming increasingly essential and make a valuable contribution to operational oceanography, model verification and validation (discussed further below).

Model-Data Integration to Improve Operational Outcomes

Without a doubt, the next ten years will see a range of advances in ocean modeling and the integration of observational data at increasing resolution, particularly in coastal areas. As the coastal zone is where societal necessities are greatest, there is a pressing need for accurate, detailed, and timely information on the marine environment at increasing resolutions. In Australia, the Australian Bureau of Meteorology/CSIRO presently provides coastal forecasts for the Great Barrier Reef at 4 km resolution (www.ereefs.org.au). However, like most countries, Australia currently depends on a global operational ocean forecasting system of ~10 km (1/12°) resolution (BlueLink, Oke et al., 2008) for forecasts along the majority of its coastline. Many developed countries, such as New Zealand, do not have their own ocean forecasting system and thus they rely on products developed by other nations, also at about 10 km resolution (e.g., the U.S. Navy implementation of Global HYCOM, Chassignet et al. (2009), and Mercator Ocean of France, www.mercator-ocean.fr). The relatively coarse resolution of global models, required at present due to computational limitations, makes them inappropriate for coastal-scale prediction where they are unable to resolve the complex bathymetry and submesoscale processes. Along the coast of southeastern Australia, for example, the continental shelf is narrow (~15-30 km wide) and is not well-resolved at 10 km resolution.

While the global operational ocean forecasts presently available resolve the typically slowly evolving, synoptic, mesoscale flows, regional and coastal models resolve smaller temporal and spatial scales that are more rapidly evolving and have shorter decorrelation timescales. For slowly evolving flows, data assimilation schemes that assume all data to be at the analysis time can be effectual; the global BlueLink reanalysis and forecast system uses ensemble optimal interpolation (Oke et al., 2008; Oke et al., 2013), while HYCOM uses 3D-Var (Cummings, 2005). Mercator Ocean of France uses a data assimilation system based on the singular evolutive extended Kalman (SEEK) filter (Brasseur and Verron, 2006). These data assimilation techniques proceed only sequentially and do not optimize the fit between observations and model over the full extent of the assimilation window. This makes them ill-suited to highly intermittent flows, typical of non-synoptic scales in the coastal ocean, with irregularly sampled observations. Indeed, these techniques are no longer used in numerical weather prediction (e.g., Lorenc and Rawlins, 2005). The two techniques that are now the most promising in numerical weather prediction are 4D-Var and the ensemble Kalman filter (EnKF), which consider the time-evolving model state in the data assimilation methodology. 4D-Var and EnKF are considerably more computationally expensive but are becoming viable for regional ocean forecasting. In the ocean, 4D-Var is used by select groups for regional forecasts (e.g., the Pacific Islands Ocean Observing System, the Mid-Atlantic Coastal Ocean Observing System, and the Central and Northern California Ocean Observing System, all in the U.S.), while the EnKF is used operationally in the Great Barrier Reef, Australia (www.ereefs.org.au).

With the step change in observations available in the coastal ocean, it behooves us to extend our operational models into the coastal ocean. The coastal and continental shelf region of southeastern Australia, the country's most densely populated region, has been well-observed over the past ten years. While these observations have been very useful in understanding coastal and continental shelf processes, such as cross-shelf dynamics (Schaeffer et al., 2013), upwelling (Schaeffer et al., 2014; Schaeffer and Roughan, 2015) and frontal eddies (Schaeffer et al., 2017). They have largely not been exploited for operational forecasting purposes. To make use of southeastern Australia's coastal observing system and provide coastal forecasts we need to develop a high resolution data assimilating model of the coastal ocean.

A high-resolution reanalysis of the East Australian Current

As a first step towards integrating the array of observations described above into an operational model, we perform a two-year case study assimilating all of the available observations into a shelf-resolving model. For the region dominated by the EAC, we combine a high-resolution state-of-the-art numerical ocean model with a variety of traditional and newly available observations using an advanced variational data assimilation scheme. The southeast Australian coastal ocean provides an ideal study site for this research as it exhibits complex shelf dynamics, is impacted by an intense WBC, and there are a large number of observations on the shelf.

We begin by configuring a numerical model of the oceanic region off southeastern Australia (domain shown in Figs. 4.4a and 4.4b) that is capable of representing the mean ocean circulation and its eddy variability. We configure the model using the Regional Ocean Modeling System (ROMS 3.4) and provide boundary forcing from the BlueLink ReANalysis (BRAN3, Oke et al., 2008, 2013), the Australian operational hindcast. The model has a variable horizontal resolution in the cross-shore direction, with 2.5 km over the continental shelf and slope that gradually increases to 6 km in the open ocean and a resolution of 5 km in the along-shore direction. The variable resolution allows us to resolve the continental shelf, which is 15 km wide at its narrowest point, while avoiding excessive computational expense. This work is described in detail in Kerry et al. (2016).

In order to correctly represent the spatial and temporal evolution of the eddy field, we need to constrain the model with observations. For the data assimilation, we use an Incremental Strong-Constraint 4-Dimensional Variational (IS4D-Var) scheme (Moore et al., 2004, 2011a). This technique uses the (linearized) model dynamics to compute increments in the initial conditions, atmospheric forcing, and boundary conditions, such that the difference between the new model solution and the observations is minimized (in a least squares sense) over an assimilation window, given prior assumptions of the uncertainties in the observations and the model background state.

The new modeled ocean state (the reanalysis) better fits and is in balance with the observations. The assimilation is performed over five-day windows as, for this model configuration, the linear assumption remains acceptable for typical perturbations over five days.

The reanalysis is configured for the two-year period of 2012-2013 because of the availability of significant observational resources during this time; in particular, a deep water mooring array deployed to capture the transport of the EAC (described above and in Sloyan et al., 2016). In addition to the traditional data streams (satellite derived sea surface height (SSH) and SST, Argo profiling floats, and XBT lines), we exploit many of the newly available observations that were collected as part of IMOS. These include velocity and hydrographic observations from the above-mentioned deep water mooring array and several moorings on the continental shelf, surface radial current observations from a HF radar array, and hydrographic observations from ocean gliders (Fig. 4.4). The SSH observations are from the Archiving, Validation and Interpretation of Satellite Oceanographic Data (AVISO) daily gridded Sea Level Anomaly product and the SST data is from the U.S. Naval Oceanographic Office Global Area Coverage Advanced Very High Resolution Radiometer level-2 product. For details of the observations and their processing before integration into the assimilation system, refer to Kerry et al. (2016).

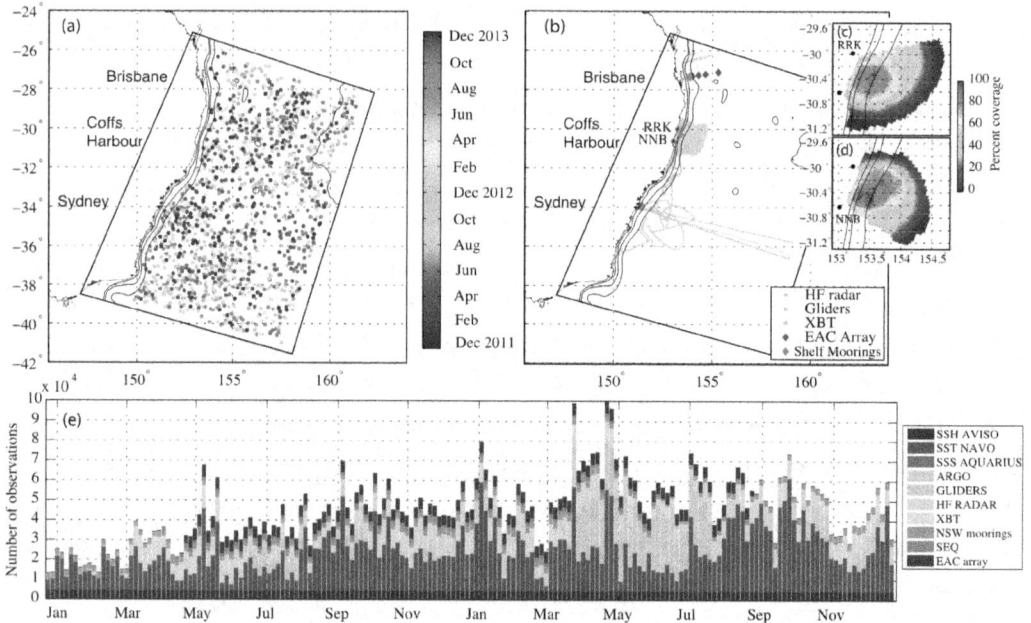

Figure 4.4. Observations used in the two-year reanalysis from 2012-2013. Argo observations colored by time of occurrence, (a), and all other observations, with the exception of satellite-derived SSH and SST, (b). 100m, 200m and 2000m contours are shown. HF radar sites Red Rock (RRK) and North Nambucca (NNB) are shown with black asterisks in (b) and zooms showing the percent coverage of radial data for the two stations are shown in (c) and (d). Number of observations (after processing) from each observation platform used in each five-day assimilation window are shown in (e). Adapted from Kerry et al. (2016).

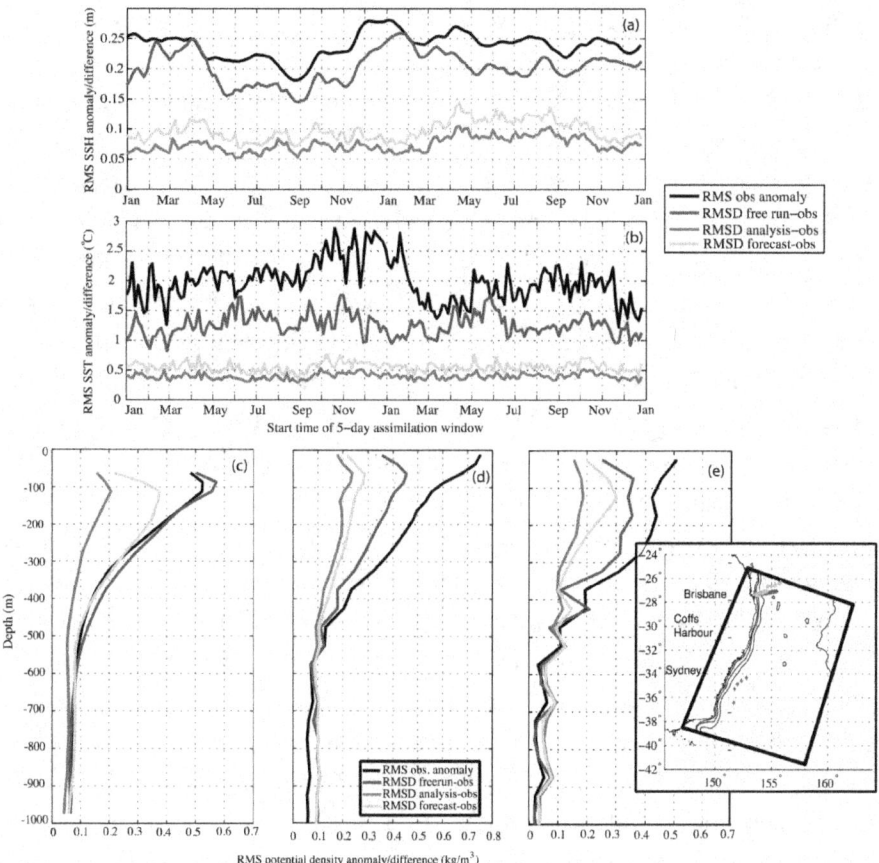

Figure 4.5. Time series of spatially-averaged RMS SSH observation anomaly, RMS SSH difference between the free run and observations, RMS SSH difference between the analysis and observations, and RMS SSH difference between the forecast and observations, for each assimilation window (a). As above but for SST (b). RMS potential density observation anomaly and RMS difference between the free run and observations, the analysis and observations, and the forecast and the observations for glider observations (c). As above but for Argo observations (d) and for independent CTD cast observations (e). Locations of the CTD casts are shown in the inset in (e). The blue, green and magenta dots represent CTD casts taken on different cruises over the two-year period. Adapted from Kerry et al. (2016).

The reanalysis is shown to represent both assimilated and non-assimilated observations well. The system achieves mean spatially-averaged Root Mean Squared (RMS) residuals between the analysis and the observations of 7.6 cm for SSH and 0.39°C for SST over the assimilation period. The mean RMS residual between the forecast and the observations are 9.8 cm for SSH and 0.55°C for SST. Figs. 4.5a and 4.5b show the time series of the spatially-averaged RMS observation anomalies and RMS residuals between the free-running model and the observations, the analysis and the observations and the forecasts and the observations for each 5-day window over the 2-year reanalysis for SSH and SST, respectively. For SSH, the free-running model's residuals with the observations are of similar order to the observation anomalies (which represent the SSH variability), indicating that the model has very little skill in predicting SSH without data assimilation. For SST, the free-running model has some skill; the root-mean-square deviations (RMSDs) between the free run and the observations are lower than the SST observation anomalies. For both SSH and SST, the analyses and forecasts show significantly improved representation of the observations, compared to the free-running model.

Both Argo floats and ocean gliders measure subsurface temperature and salinity, allowing potential density to be computed. Figs. 4.5c and 4.5d show profiles of the RMS potential density observation anomaly and RMS difference between the free run and observations, the analysis and observations, and the forecast and the observations for glider observations and Argo float observations, respectively. For Argo, the peak in the RMS residual between the analysis and the observations occurs at about 60-100 m and is 0.23 kgm^{-3}, with an increase to 0.28 kgm^{-3} for the residual between the forecast and the observations. The majority of the glider observations are taken on the continental shelf and shelf slope, and the forecasts degrade more rapidly from the analyses (compared to the Argo observations, which mostly sample the offshore deep water region). At their peak (100 m depth), the RMS residual between the analysis and the glider observations is 0.21 kgm^{-3}, and the residual with the forecasts is 0.37 kgm^{-3}.

Comparison with independent (non-assimilated) shipboard CTD cast observations shows a marked improvement in the representation of the subsurface ocean in the reanalysis, with the RMS residual in potential density reduced to about half of the residual with the free-running model in the upper eddy-influenced part of the water column (Fig. 4.5e). This shows that information is successfully propagated from observed variables to unobserved regions as the assimilation system uses the model dynamics to adjust the model state estimate.

Velocities at several offshore and continental shelf moorings are well-represented in the reanalysis, with complex correlations between 0.8-1 for all observations in the upper 500 m (not shown). Surface radial velocities from the HF radar array are assimilated and the reanalysis provides surface velocity estimates with complex correlations with observed velocities of 0.8-1 across the radar footprint (not shown). An example of the improvement in surface current representation achieved by assimilation of the HF radar data is shown in Fig. 4.6 for the assimilation window beginning on March 14, 2012. The surface velocities computed from the assimilated HF radar observations of the radial components of the surface currents (Fig. 4.6a) show a recirculation feature with northward flow inshore of the dominant southerly flow. This is not represented in the forecast (Fig. 4.6b), but is present upon assimilation (Fig. 4.6c). The corresponding SSH observations, model forecast, and analysis are shown in Figs. 4.6d-f. The anticyclonic eddy that is in balance with the observed currents is not present in the assimilated SSH observations (which are only used further than 100 km from the coast), but becomes present in the analysis SSH field.

This is the first study to generate a reanalysis of the region at such a high resolution, making use of an unprecedented observational dataset and using an assimilation method that uses the time-evolving model physics to adjust the model in a dynamically consistent way. The reanalysis provides a good representation of the ocean state, as measured by both the assimilated and non-assimilated observations. The RMS residuals with the observations remain low in the subsequent five-day forecasts for the SSH, SST, and Argo float observations. However, the forecasts have considerably greater residuals for the glider observations, most of which are taken close to or over the continental shelf. Likewise, as seen in the example in Fig. 4.6, small-scale circulation features on the shelf are often not captured in the forecasts.

From an operational forecasting perspective, this system is likely to be effective for mesoscale type forecasts, but would require further tuning and/or downscaling to provide accurate predictions close to the coast or on the shelf and shelf slope. Some of the considerations when downscaling to increasingly higher resolutions are discussed in the next section. Also, many of the observations used are not available in real time. Understanding which observations are important in informing the model, and therefore are worth making available operationally, is a key step towards developing an effective forecasting system for this region.

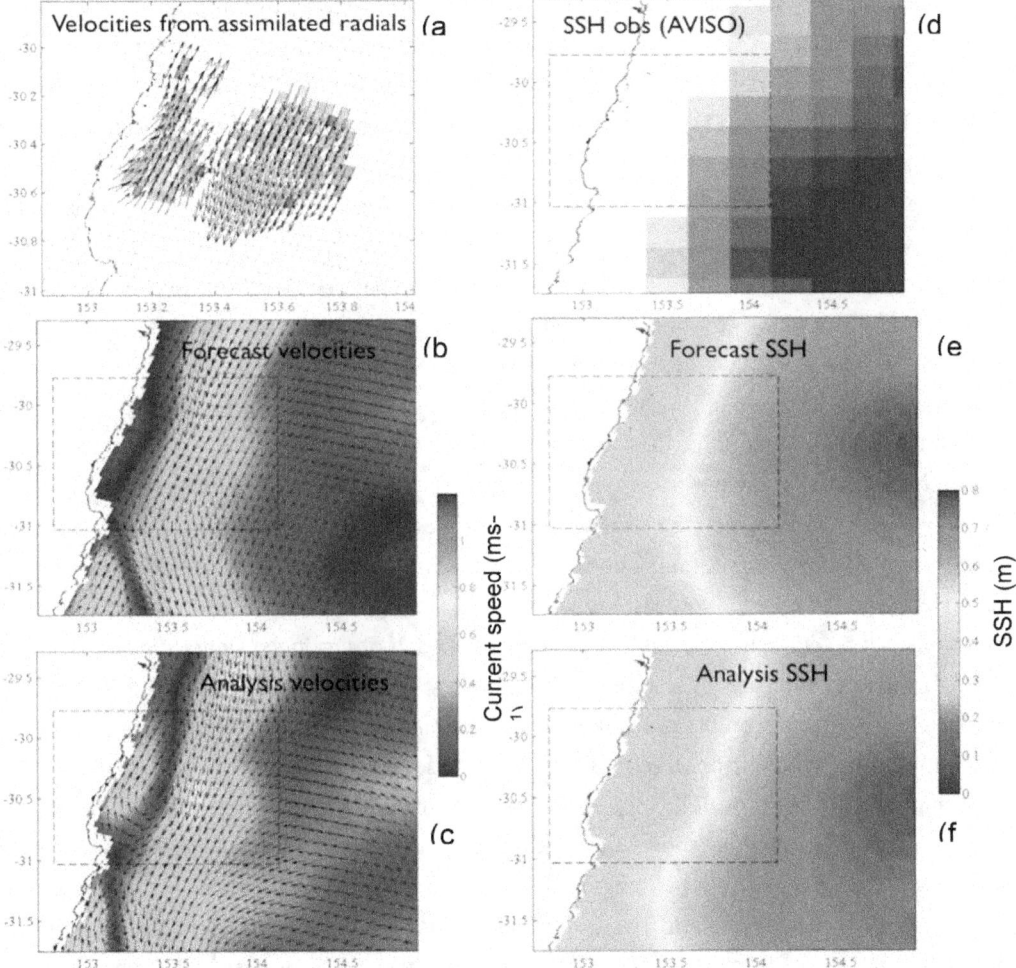

Figure 4.6. Example of one particular assimilation window (March 14, 2012) showing the impact of the assimilation of the HF radial component of surface velocity. Surface velocity vectors computed from assimilated surface radial currents, with current speed shown in color bar (a), model forecast velocities (b) and analysis velocities (c). SSH-assimilated observations (d), model forecast SSH (e) and analysis SSH (f). The box represented by the dashed line in (b)-(f) shows the area of (a), which has different axis limits to show the surface velocities computed from the observed radial velocities more clearly.

Understanding observation impact

Given the significant investment required to install and sustain ocean observations, and to make the data available in real time, it is important to understand which observations (locations and variables) are most useful in informing our models. This allows one to assess the cost–benefit of each observation stream and design an observing system that can be applied for operational purposes to provide effective forecasts.

Various methods have been used to assess the impact of observations on model predictions. Oke and Sakov (2012) used satellite-derived SSH and SST and model output to assess the footprint of mooring observations around Australia based on zero time-lag correlations. This general approach is not dependent on a specific model or data assimilation configuration and provides a useful assessment of the moorings' effectiveness in monitoring the circulation, but is limited in that it does not consider multivariate correlations or time-lags. Observing System Experiments (e.g. Oke and Schiller, 2007; Oke et al., 2015), in which observations are systematically withheld, are useful to assess the relative importance of various observations; however, these experiments are unable to quantify the value of the observations given the complete assimilation system, as when some observations are withheld the value of the remaining observations changes.

A particularly useful method of assessing observation impact is Observing System Simulation Experiments (OSSEs, e.g., Schiller et al., 2004; Tranchant et al., 2008). By defining a given model solution as the *true state*, these experiments allow us to assimilate synthetic observations (i.e., simulated observation types and locations that are constructed from model output to assess their utility) and investigate predictive skill based on a known ocean state. The goal is to assimilate observations extracted from the *true state*, with realistic errors applied, into a predictive model allowing one to assess how well the assimilation improves the model estimates. The *true state* can be chosen as a different model realization of the same time period, a coarser model, or even the same model, in which case the forecast model is initialized from a perturbed initial state. Synthetic observations can be withheld to establish the impact of each observing system. Such experiments are useful as we can compare our estimates to a known state, as compared to prediction studies of the real ocean using actual observations in which the ocean state is largely unknown.

The influence of synthetic observations can also be assessed using a representer-based approach (e.g., Zhang et al., 2010; Powell, 2017), which requires the adjoint and tangent-linear versions of the ocean model. This method shows how individual observations project onto the ocean state, which reveals the area and relative magnitude of the observations' influence. For example, this method was used by Zhang et al. (2010) to study the influence of observations on circulation in the New York Bight and show that regular glider transects influence a greater area than a profiling mooring, but the mooring has a stronger influence at its particular location.

A unique advantage of variational data assimilation methods is that they can reveal the dynamical connections between model fields through the use of the tangent-linear and adjoint models (Moore et al., 2004). In solving the state estimation problem with 4D-Var, one computes the dynamical covariance between the observations and the model that allows direct computation of the impact of each observation on the change in circulation estimate between the forecast and the analysis (as described in Moore et al., 2011b). This method was used by Powell (2017) to show that ocean glider observations are particularly impactful in constraining transport in the Hawai'i Lee Countercurrent. Moore et al. (2011b) used 4D-Var to assess observation impact in the California Current and described a method to determine the sensitivity of circulation estimates to changes in the observations, providing a way of estimating the change in the analysis due to degradation or failure of an observation platform.

In order to make specific decisions on observing system design, we need to undertake targeted studies specifically designed to determine optimum observations given specific prediction goals. Such experiments should quantify the potential benefits of making different observations and make recommendations for optimizing the deployment of expensive assets in our present observing system.

Downscaling and data assimilation at increasingly higher resolution

Fine-scale predictability of the coastal ocean is a new frontier in operational oceanography. With increased computational resources and model efficiency, there is an increasing push towards providing forecasts that resolve even finer scale features in coastal regions, such as submesoscale dynamics, coastal upwelling, shelf fronts, and filaments. However, the predictability of the coastal ocean is not yet well-understood.

Errors in ocean modeling can arise from uncertainties in initial conditions and model error, which is composed of a number of factors including grid resolution, numerics, atmospheric forcing, boundary conditions, and parameterization of physical processes. While errors in initial conditions dominate at the beginning of the forecast window, as the forecast length grows model error dominates the forecast skill. Operational ocean forecasts available at present typically resolve the slowly evolving mesoscale circulation (with model resolutions of about 10 km), which is highly sensitive to the initial state, and data assimilation methods typically focus on minimizing the error in the initial state at the beginning of each analysis window. However, with improved model resolution and physics, model error increases more rapidly. This so-called "curse of resolution" stems from the fact that finer grid resolution and numerics increase the model error by resolving finer scales of short period turbulence that have faster error growth. As such, prediction of finer-scale oceanic processes presents additional challenges as the modelled circulation is likely to be more rapidly decoupled from the initial state and depend strongly on surface and boundary forcing and model parameters.

In order to develop accurate forecasting systems, we need to first understand the sensitivity of the forecast skill to uncertainty in the initial conditions, model forcing, and model parameters. Indeed, for a downscaled coastal model where processes strongly depend on boundary conditions and surface forcing, using data assimilation to update the initial conditions may provide little value. Configuring a well-tuned, free-running model and using data assimilation for parameter estimate may be more effective.

Framework for model inter-comparison

One of the major limitations in comparing model performance internationally is that there is no single set of metrics nor is there a consistent observational framework with which to assess models. This is further complicated by different data formats, e.g., mean dynamic topographies, time averaging (daily means etc.), and no consistent file naming convention. This has been recognized as a limitation in the GODAE global ocean data assimilation experiment task team who have proposed an inter-comparison framework as documented in Martin et al. (2013). To encourage global best practices, we suggest adopting the GODAE guidelines to ensure international operability from the outset. Furthermore, this will ensure future modeling efforts will feed back into global assessment programs. Importantly, coastal observations are needed for model assessment and validation. For example, assessment can be against observations that have been withheld (such as shipboard CTD casts, glider surveys, HF radar velocities, etc.) depending on the experiment. Thus, good dialogue is required between the observing and modeling communities.

The Need for Operational Oceanography in the Coastal Ocean: Case Study of Marine Heatwaves and Aquaculture.

In this section, we introduce new research into marine heatwaves (MHWs), a very recent area of research in coastal oceanography. This research has been facilitated by access to long-term datasets of ocean temperature, which in this case were collected originally for operational purposes at the Ocean Reference Station (see Fig. 4.2, Sydney insert). We then introduce an aquaculture example and explore operational considerations for fish farming in the coastal ocean, in the context of ocean warming.

Marine heatwaves

Analogous to atmospheric heatwaves, MHWs are defined as discrete, prolonged anomalously warm water events, lasting more than five days (Hobday et al., 2016). Extremely hot SSTs have become more common in one-third of the world's coastal areas over the past 30 years (Lima and Wethey, 2012). In recent years, MHWs have become more common in the coastal ocean and they are having unprecedented biological impacts including mass mortality and habitat shifts.

In response to severe thermal stress, marine communities have to either acclimatize or move to more suitable (cooler) habitat either further poleward or deeper. This includes not only shallow communities but also benthic species, some coral communities, and seaweed below the surface mixed layer (Schaeffer and Roughan, 2017). Increasingly, habitat-forming species are dying during MHWs and they are not able to recover after an MHW has subsided (Wernberg et al., 2016). In order to sustainably manage our coastal oceans into the future, we need to understand both long-term trends in ocean warming, as well as the characteristics of events such as MHWs.

Off western Australia, a severe MHW event occurred during the 2010-2011 austral summer, lasting over a month; with anomalies up to 5°C above the climatological mean (Pearce and Feng,

2013). Wernberg et al. (2016) showed that two years after the event the cool water kelp forests had not recovered, and there was a community-wide shift towards warm water seaweeds and fish. However, little is known about the characteristics of MHWs due to the lack of long-term in situ observations.

While satellite derived data such as SST can be used to identify broad spatial patterns in MHWs, the time series is only 30 years and the data are only representative of the sea surface. Because Schaeffer and Roughan (2017) had access to long-term high resolution moored temperature information (5-minute data at 4-8 m resolution through the water column, since 1991 off Sydney at the ORS065 site as described above, see also Fig. 4.2), they were able to characterize the MHWs that have occurred in the region over the past 20 years.

Using this valuable dataset in conjunction with temperature climatology for the region, Schaeffer and Roughan (2017) were able to show that MHWs are sub-surface intensified, and appear to be driven by local wind forcing. Downwelling winds force the warm water downward, resulting in a maximum peak right below the thermocline. In contrast, upwelling favorable winds are able to arrest the temporal evolution of the MHW. The data also showed that SST underestimates the peak intensity and duration of MHWs off the coast of southeastern Australia because of their sub-surface intensification. Little is known about temperature extremes at depth due to the lack of long-term in situ measurements so this is an area that needs more observational effort.

Salmon aquaculture

From an operational perspective, one of the industries that is presently being impacted by ocean warming and MHWs is the salmon aquaculture industry. Salmon farming is a highly lucrative industry, with salmon products in increasingly high demand. In Australia, the Tasmanian salmon farming industry is worth ~$550 million AUS per year, and is projected to be a billion-dollar industry within 20 years as it is rapidly surpassing other fisheries. Chile is the second largest exporter of salmon and their industry is worth over $4.5 billion (USD) p.a., and in New Zealand, salmon aquaculture is worth about $50-$70 million pa.

Salmon cannot regulate their body temperature and thrive when water is between 12-17°C. When water temperatures rise, salmon cardiovascular capacity is lowered, fish become more vulnerable to disease and predators, and if temperatures get too high it can be fatal. Less catastrophic, but still an operational consideration, is the impact of water temperature on consumption. Salmon slow down their eating when temperatures rise, and increase when water is cooler. Thus, water temperature has an impact on how much feed a company requires.

Globally rising temperatures and MHWs are having severe consequences on the salmon aquaculture industry. During early 2016, the salmon industry in Chile experienced $800 million (USD) in losses (100,000 tons lost) with sea temperatures 2-4°C above average for a prolonged period of time and recurrent harmful algal blooms.

In the austral summer of 2015, water temperatures stayed above 18°C for three months in the Marlborough Sounds of central New Zealand. Large numbers of salmon died in the Marlborough Sounds creating a "multimillion-dollar problem" for the New Zealand salmon industry.

Operational considerations of MHWs and the salmon aquaculture industry

In recent years, ocean temperatures off Tasmania (southern Australia) and New Zealand have been regularly exceeding 18°C in summer. This raises a number of questions with operational implications, such as: What are the long-term temperature trends? What is driving the increase in temperature? Is it in situ heating, anomalous air sea fluxes, advection, reduced mixing or a combination of mechanisms?

Once the in situ temperature datasets are long enough, we can begin to assess trends in MHWs and correlate climatology with salmon losses. There is a pressing need to understand if MHWs are increasing in frequency, duration and or intensity. The results of Schaeffer and Roughan (2017) showed that MHWs were sub-surface intensified off the east coast of Australia. However, we don't know the implication of sub-surface intensification on the salmon aquaculture industry. What we do know is that the salmon pens in Marlborough Sound New Zealand extend ~24 m below the surface, which may impact salmon health if MHWs are sub-surface intensified in this region.

Finally, off southeastern Australia, Schaeffer and Roughan (2017) showed that MHWs had an average duration of 8-12 days but could last for over month. This is sufficient time to have an impact on marine life, particularly on organisms that live below the surface. However, the question remains: how long is too long and how do these timeframes differ from organisms such as microbes and plankton to larger more developed organisms such as fish?

Operational decision timescales

For operational purposes, there are three main timescales on which decisions need to be made; therefore, pertinent information is required within these three timeframes in mind to facilitate the decision-making process. The timeframes are:

1. Weather timescales, typically 1–7 days. With minimal warning, typically the only management option is reactive management.
2. Seasonal timescales, 2 weeks–9 months. This timeframe gives an early window for implementation of strategies with which to minimize impacts and maximize opportunities.
3. Climate forecasting, 10–100s of years. This timeframe serves for long-term planning purposes and future proofing of the industry.

The seasonal timescale is the most useful for proactive management. Business performance and industry resilience could be improved with prediction about future conditions.

Benefits of public–private partnerships

There is no doubt that there have been many benefits from the public–private partnerships in sustained ocean observing. One example along the coast of southeastern Australia is the long ocean temperature time series at the Ocean Reference Station (ORS) provided by the Sydney Water Corporation (SWC, see Fig. 4.2). This data is collected on behalf of SWC as part of their license to operate a deep water ocean sewage outfall. SWC requires information on the ocean stratification and velocities to drive a plume dispersion model. This data provider has strict key performance indicators against it, as such the data quality are extremely high including, on average, a more than 90% data return over the past ten years. SWC has generously made this data available to the scientific community through the ocean data portal, which has facilitated a whole range of research opportunities. In addition, by opening up the dataset the research community has been able to build on their efforts by augmenting the data collection through two additional moorings forming a cross shelf array, anchored in the near shore by the ORS. In other instances, where data has not been released publically, data collection efforts are often duplicated, sometimes knowingly.

Operational Data Delivery

Increasingly, industry end users are requiring sophisticated tools to access display and analyze meteorological and oceanographic data. Moreover, many industry end users of oceanographic data prefer a one stop shop data and model display platform to streamline their operations. As coastal hindcast and forecast modeling becomes more sophisticated with assimilation at high resolution, and more real-time data streams become available, these platforms will be necessary to deliver real-time operational data streams to users.

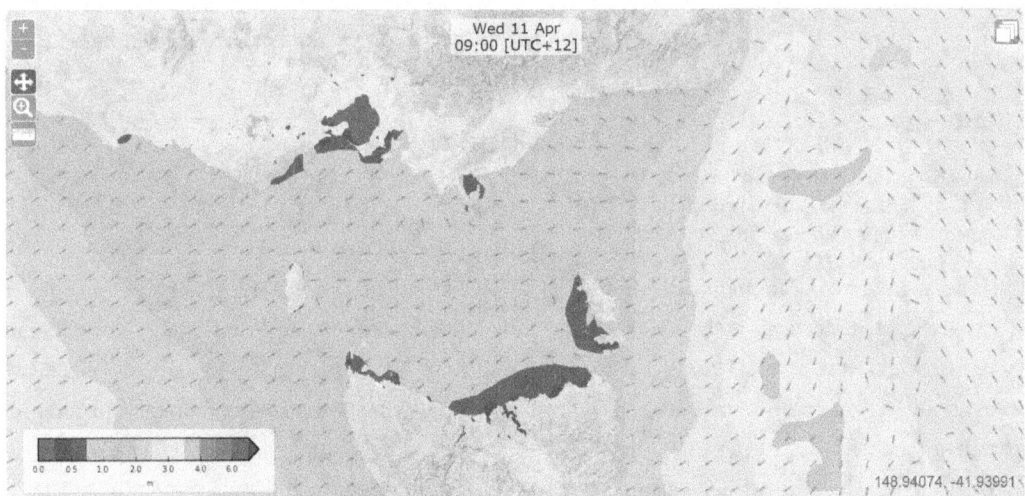

Figure 4.7. Example screen shot from the MetOceanView (MOV) platform showing wave heights (m) through Bass Straight (southern Australia) for operational purposes. Note the complex scenario through the straight where waves are coming from both directions and are interacting with the topography.

Traditionally, the barriers to this path have been timely access to data and the experience of the people involved. When an expert (e.g., a weather forecaster) is inserted into the decision-making process, the operational context is often lost and numerous simplifications need to be made. The net result can be conservative and insufficiently quantitative for modern logistics, with the justification being that people are busy and need only the distilled result without all the detail. The contrary view is that people are intelligent and can make wise decisions if provided the correct information and empowered with tools to create project-specific knowledge.

Recognition for improved operational decision-making is at the core of the global move toward providing the common operational picture at an organizational level. Typically delivered within GIS architecture, these systems display assets alongside the spatial gradients in the important environmental variables. Time can be applied as a fifth dimension, allowing forecast or hindcast data to be accessed seamlessly. Further, with or without a common operational picture framework, many industry end users still require a web-delivered service to access, display, and analyze meteorological and oceanographic data. The goal is a one-stop platform to streamline their operations.

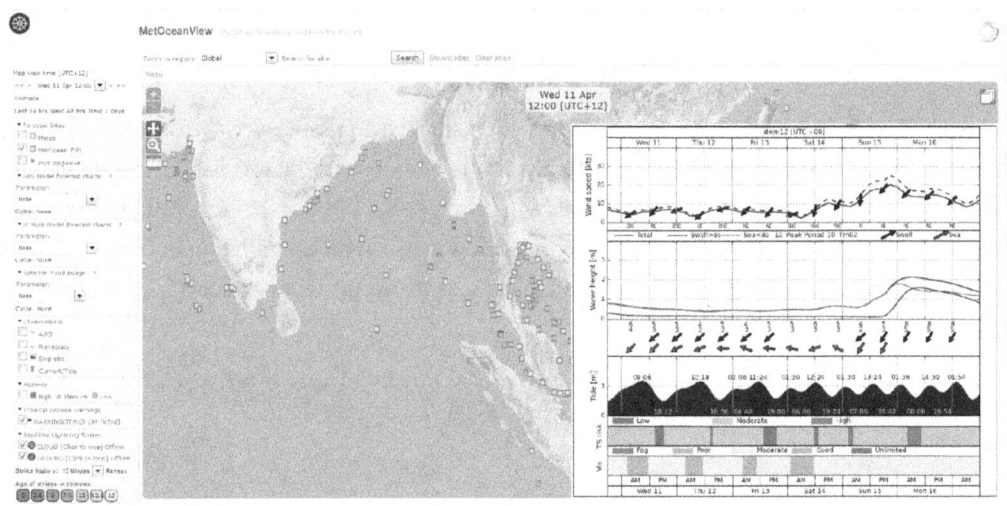

Figure 4.8. Screen shot from the MOV data display platform showing real-time weather observations displayed alongside the forecast predictions (left), and presentation of the full population of forecast ensembles with customized probabilistic measures (right).

There are a growing number of examples of services that cater to this need, and one of the more mature is the MetOceanView (MOV) platform developed by MetOcean Solutions in New Zealand (www.metoceanview.com). This is a web-delivered, map-based tool that allows customized user access to a wide range of marine weather variables along with some tools to display and integrate those data (see Fig. 4.7 for an example). The underlying architecture of MOV is a series of API micro-services, which allow a spectrum of service delivery levels – either via the dedicated MOV web service or integration into an existing dynamic GIS platform or static monitoring display system.

The original concept for MOV was first developed in 2006, with the aim of creating an offshore construction project resource where users would have unlimited and instant access to all the relevant marine observations, forecasts, and hindcasts in one place. By opening up all the datasets and displaying them to everyone on the project, the project team members became highly skilled users over time. Their feedback led to the development of a suite of tools and ways of displaying data for rapid inclusion in the decision-making process. Two good examples are displaying real-time weather observations alongside the forecast predictions (Fig. 4.8, left) and presentation of the full population of forecast ensembles with customized probabilistic measures (Fig. 4.8, right).

The coastal zone is the area of greatest interest to industry users. Here too, the spatial gradients in oceanographic variables are typically the highest and their temporal changes of larger consequence. Coastal modeling is meeting those needs with more sophisticated solutions; unstructured model domains, and real-time data assimilation at high resolution are becoming operational norms. The challenge is how to let users engage meaningfully with these new and important data sources. Without question, it requires dialogue and deep involvement between the scientists, developers, and the end users. Finally, a key consideration is to ensure delivery platforms are not built as monolithic structures unable to cope with new observational datasets and evolving model outputs. Keeping systems flexible will allow for future integration and ingestion of new and increasingly larger datasets. In this way, we ensure that end users have access to the data needed to ensure safe and efficient operations.

Acknowledgments

MOV figures provided by Simon Weppe. The operational decision timescales were adapted from C. Spillman personal communication. IMOS is supported by the Australian Government through the National Collaborative Research Infrastructure Strategy and the Super Science Initiative and the Education Infrastructure Fund. CK is partially funded through and Australian Research Council Discovery Project to MR (ARC DP140102337). We acknowledge the vast contribution made by the IMOS technical staff and fieldwork teams including the NSW-IMOS moorings team, the ANMN, ANFOG, ACORN, and ABOS facilities.

References

Archer, M. R., Roughan, M., Keating, S. R., Schaeffer, A., 2017: On the variability of the East Australian Current: Jet structure, meandering, and influence on shelf circulation. J. Geophys. Res. Oceans, **122**, 8464–8481. https://doi.org/10.1002/2017JC013097.

Baird, M.E., I. M. Suthers, D. A. Griffin, B. Hollings, C. Pattiaratchi, J. D. Everett, M. Roughan, K. Oubelkheir, M. Doblin. 2011: The effect of surface flooding on the physical-biogeochemical dynamics of a warm-core eddy off southeast Australia. Deep Sea Research II, **58**, 592–605.

Brasseur, P. and Verron, J., 2006: The SEEK filter method for data assimilation in oceanography: a synthesis. Ocean Dynamics, **56**, 650-661.

Chassignet, E.P., H.E. Hurlburt, E.J. Metzger, O.M. Smedstad, J., G.R. Halliwell, R. Bleck, R. Baraille, A.J. Wallcraft, C. Lozano, and others. 2009: US GODAE: Global Ocean Prediction with the HYbrid Coordinate Ocean Model (HYCOM). Oceanography, **22**(2), 64–75.

Cummings J A., 2005: Operational Multivariate Ocean Data Assimilation. Q. J. R. Meteorol. Soc., **131**, 3583–3604

Everett, J. D., H. S. Macdonald, M.E. Baird, J. Humphries, M. Roughan, I. M. Suthers, 2015: Cyclonic entrainment of pre-conditioned shelf waters into a Frontal Eddy. J. Geophys. Res. Oceans, **120**, doi:10.1002/2014JC010301.

Hobday, A. J., et al. 2016: A hierarchical approach to defining marine heatwaves. Prog. Oceanogr., **141**, 227–238, doi:10.1016/j.pocean.2015.12.014.

Kerry, C. G., B. Powell, M. Roughan and P. Oke 2016: Development and evaluation of a high-resolution reanalysis of the East Australian Current region using the Regional Ocean Modeling System (ROMS 3.4) and Incremental Strong-Constraint 4-Dimensional Variational data assimilation (IS4D-Var). Geosci. Model Dev., **9**, 3779-3801, doi:10.5194/gmd-2016-44.

Lima, F., and D. Wethey 2012: Three decades of high-resolution coastal sea surface temperatures reveal more than warming, Nat. Commun., **3**, 1–13, doi:10.1038/ncomms1713.

Lynch, T.P., Morello, E.B., Evans, K., Richardson, A., Rochester, W., Steinberg, C. R., Roughan, M., Thompson, P., Middleton J. F., Feng, M., Sherrington, R., Brando, V., Tilbrook B., Ridgway, K., Allen, S., Doherty, P., Hill, K., Moltmann, T.C.Z. 2014: IMOS National Reference Stations: A continental scale physical, chemical, biological coastal observing system. PLoS ONE, **9**(12): e113652, doi:10.1371/journal.pone.0113652.

Lorenc A. C. & Rawlins F. 2005: Why does 4D-Var beat 3D-Var? Quarterly Journal Royal Meteorological Society, **131**, 3247-3257, doi: 10.1256/qj.05.85.

Macdonald, H. S., Roughan, M., Baird, M.E. and Wilkin, J. 2016: The formation of a cold-core eddy in the East Australian Current. Continental Shelf Res., doi:10.1016/j.csr.2016.01.002.

Mantovanelli, A., Keating, S., Wyatt, L., Roughan, M. and Schaeffer, A. 2017: Lagrangian and Eulerian characterization of two counterrotating submesoscale eddies in a western boundary current, J. Geophys. Res. Oceans, **122**(6), doi: 10.1002/2016JC011968.

Martin M., F. Hernandez and A. Sellar 2013: Inter-comparison of forecast metrics. https://www.godae-oceanview.org/files/download.php?m=documents&f=130312134130-intercomparisonproposal.doc

Moore, A. M., Arango, H. G., Broquet, G., Powell, B. S., Zavala-Garay, J., and Weaver, A. T. 2011a: The Regional Ocean Modeling System (ROMS) 4-dimensional variational data assimilation systems: Part I – System overview and formulation, Prog. Oceanogr., **91**, 34–49, doi:10.1016/j.pocean.2011.05.004

Moore, A. M., H. G. Arango, G. Broquet, C. Edwards, M. Veneziani, B. S. Powell, D. Foley, J. Doyle, D. Costa, and P. Robinson 2011b: The Regional Ocean Modeling System (ROMS) 4-dimensional variational data assimilation systems: Part III Observation impact and observation sensitivity in the California Current System, Prog. Oceanog., **91**, 74-94, doi:10.1016/j.pocean.2011.05.005.

Moore, A. M., H. G. Arango, E. Di Lorenzo, B. D. Cornuelle, A. J. Miller, and D. J. Neilson 2004: A comprehensive ocean prediction and analysis system based on the tangent linear and adjoint of a regional ocean model, Ocean Modelling, **7**, 227-258.

Mourre, B., and A. Alvarez 2012: Benefit assessment of glider adaptive sampling in the Ligurian Sea, Deep-Sea Res. I, **68**, 68-78.

Oke, P. R., Brassington, G. B., Griffin, D. A., and Schiller, A. 2008: The BlueLink ocean data assimilation system (BODAS), Ocean Modelling, **21**, 46–70.

Oke, P. R., G. Larnicol, E. Jones, V. Kourafalou, A. Sperrevik, F. Carse, C. Tanajura, B. Mourre, M. Tonani, G. Brassington, M. L. Hena_, G. H. Jr., R. Atlas, A. Moore, C. Edwards, M. Martine, A. Sellare, A. Alvarez, P. DeMey, and M. Iskandaranic 2015: Assessing the impact of observations on ocean forecasts and reanalyses: Part 2, Regional applications, J. Operational Oceanogr., **8**(S1), s63-s79.

Oke, P. R., and P. Sakov 2012: Assessing the footprint of a regional ocean observing system, J. Mar. Res., **105-108**, 30-51.

Oke, P., Sakov, P., Cahill, M. L., Dunn, J. R., Fiedler, R., Griffin, D. A., Mansbridge, J. V., Ridgway, K. R., and Schiller, A. 2013: Towards a dynamically balanced eddy-resolving ocean reanalysis: BRAN3, Ocean Modell., **67**, 52–70.

Oke, P. R., and A. Schiller 2007: Impact of Argo, SST, and altimeter data on an eddy-resolving ocean reanalysis, Geophys. Res. Lett., **34**(L19601), 1-7.

Pearce, A. F., and M. Feng 2013: The rise and fall of the "marine heatwave" off Western Australia during the summer of 2010/2011, J. Mar. Syst., **111-112**, 139–156, doi:10.1016/j.jmarsys.2012.10.009.

Powell, B. S. 2017: Quantifying how observations inform a numerical reanalysis of Hawaii, J. Geophys. Res. Oceans, **122**, 1-18, doi: 10.1002/2017JC012854.

Rossi, V., Schaeffer, A., Wood, J., Galibert, G., Morris, B., Sudre, J., Roughan, M. & Waite, A. 2014 Seasonality of sporadic physical processes driving temperature and nutrient high-frequency variability in the coastal ocean off southeast Australia. J. Geophys. Res. Oceans. **119**, 1-19, doi:10.1002/2013JC009284.

Roughan, M., Keating, S. R., Schaeffer, A., Cetina Heredia, P., Rocha, C., Griffin, D., Robertson, R. and Suthers, I.M. 2017: A tale of two eddies: The biophysical characteristics of two contrasting cyclonic eddies in the East Australian Current System. J. Geophys. Res. Oceans. **122**, doi:10.1002/2016JC012241.

Roughan, M, Schaeffer, A. Suthers, I.M, 2015: *Sustained ocean observing along the coast of southeastern Australia:* NSW-IMOS 2007 -2014 (Eds, Y. Lui, H, Kerling, Weisberg). Coastal Ocean Observing Systems, Elsevier ISBN: 9780128020227

Roughan, M., Schaeffer, A., Kioroglou, S. 2013: Assessing the design of the NSW-IMOS Moored Observation Array from 2008-2013, Recommendations for the future. In Proceedings of MTS/IEEE Oceans 2013, San Diego USA, Sept 2013

Roughan, M. and Morris, B.D., 2011: Using high-resolution ocean time series data to give context to long term hydrographic sampling off Port Hacking, NSW, Australia, In Proceedings of MTS/IEEE Oceans 2011 Kona USA.

Roughan, M., Morris, B.D. and Suthers, I.M. 2010: NSW-IMOS, An integrated marine observing system for Southeastern Australia. IOP Conf. Ser.: Earth Environ. Sci. 11012030, doi:10.1088/1755-1315/11/1/012030.

Schaeffer, A., and M. Roughan 2017: Subsurface intensification of marine heatwaves off southeastern Australia: The role of stratification and local winds. Geophys. Res. Lett., **44**, 5025–5033, doi:10.1002/2017GL073714.

Schaeffer, A. and M. Roughan 2015: Influence of a Western Boundary Current on shelf dynamics and upwelling from repeat glider deployments. Geophys. Res. Lett. **42**, 121-128, doi:10.1002/2014GL062260.

Schaeffer, A. S., M. Roughan, T. Austin, J.D. Everett, D. Griffin, B. Hollings, E. King, A. Mantovanelli, S. Milburn, B. Pasquer, C. Pattiaratchi, R. Robertson, D. Stanley, I. Suthers, D. White, 2016a: Mean hydrography on the continental shelf from 26 repeat glider deployments along Southeastern Australia. Sci. Data, **3**, 160070 doi: 10.1038/sdata.2016.70.

Schaeffer, A. S, M. Roughan, M. E. Jones, D. White. 2016b: Physical and biogeochemical spatial scales of variability in the East Australian Current separation from shelf glider measurements. Biogeosciences, **13**, 1967-1975, doi:10.5194/bgd-12-1-2015.

Schaeffer, A., M.Roughan,and J. E.Wood 2014: Observed bottom boundary layer transport and uplift on the continental shelf adjacent to a western boundary current, J. Geophys. Res. Oceans, **119**, doi:10.1002/2013JC009735.

Schaeffer, A., Gramoulle, A., Roughan, M. and Mantovanelli, A. 2017: Characterizing frontal eddies along the East Australian Current from HF radar observations. J. Geophys. Res. Oceans, **122**, doi: 10.1002/2016JC012171.

Schaeffer A.S and M. Roughan 2017: Sub-surface intensification of Marine Heatwaves off southeastern Australia: the role of stratification and local winds. Geophys. Res. Lett., **44**, 1–9, doi:10.1002/2017GL073714.

Schaeffer, A., Roughan, M., Morris, B.D. 2013: Cross-shelf dynamics in a Western Boundary Current regime: Implications for Upwelling. J. Phys. Oceanogr., **43**, 1042–1059, doi:10.1175/JPO-D-12-0177.1

Schiller, A., S. E. Wijffels, and G. A. Meyers 2004: Design Requirements for an Argo float array in the Indian Ocean inferred from observing system simulation experiments, J. Atmos. Ocean. Technol., **21**, 1598-1620.

Sloyan, B. M., K. R. Ridgway, and R. Cowley 2016: The East Australian Current and Property Transport at 27°S from 2012-2013., J. Phys. Oceanogr., **46**, 3.

Smith, R. N., Y. Chao, P. P. Li, D. A. Caron, B. H. Jones, and G. S. Sukhatme 2010: Planning and Implementing Trajectories for Autonomous Underwater Vehicles to Track Evolving Ocean Processes Based on Predictions from a Regional Ocean Model. The International Journal of Robotics Research, **29**, 12, 1475-1497.

Thompson P. A, Baird M. E, Ingleton T, Doblin M. A. 2009: Long-term changes in temperate Australian coastal waters: implications for phytoplankton. Mar. Ecol. Prog. Ser., **394**, 1-19, doi:10.3354/meps08297.

Tranchant, B., C.-E. Testut, L. Renault, N. Ferry, F. Birol, and P. Brasseur 2008: Expected impact of the future SMOS and Aquarius Ocean surface salinity missions in the Mercator Ocean operational systems: New perspectives to monitor ocean circulation. Rem. Sensing Environ., **112 (4)**, 1476-1487.

Wernberg, T., et al., 2016: Climate-driven regime shift of a temperate marine ecosystem, Science, **353**(6295), 169–172, doi:10.1126/science.aad8745.

Wood, JE, Roughan, M & Tate, PM, 2012: Finding a proxy for wind stress over the coastal ocean. Marine and Freshwater Research, **63 (6)** ,528 – 544, doi:10.1071/MF11250.

Wyatt, L.R, Mantovanelli, A., Heron, M.L., Roughan, M. and Steinberg, C.R. 2017 Assessment of surface currents measured with high-frequency phased-array radars in two regions of complex circulation. IEEE, J. Ocean Engineering, **99**, 1-22, doi:10.1109/JOE.2017.2704165.

Yool A. Fashman M.J.R. 2001: An examination of the 'continental shelf pump' in an open ocean general circulation model. Global Biogeochemical Cycles, **15**(4), 831-844.

Zhang, W. G., J. L. Wilkin, and J. C. Levin 2010: Towards an integrated observation and modeling system in the New York Bight using variational methods. Part II: Representer-based observing strategy evaluation, Ocean Modell., **35**, 134-145.

CHAPTER 5

Multi-Platform Observations and Numerical Simulations to Understand Meso and Submesoscale Processes: A Case Study of Vertical Velocities in the Western Mediterranean

Simón Ruiz[1], Amala Mahadevan[2], Ananda Pascual[1], Mariona Claret[3], Joaquín Tintoré[1,4], and Evan Mason[1]

[1]*Instituto Mediterráneo de Estudios Avanzados (CSIC-UIB), Esporles, Spain;* [2]*Wood Hole Oceanographic Institution, Woods Hole, MA, USA;* [3]*Joint Institute for the Study of the Atmosphere and Ocean, University of Washington, Seattle, WA, USA;* [4]*Balearic Islands Observing and Forecasting System (SOCIB), Palma de Mallorca, Spain*

In this chapter we provide a description of an intense ocean front and an anomalous anticyclonic eddy in the Western Mediterranean. We use observations from two multi-platform experiments carried out in the eastern Alboran Sea and in the northern Balearic Islands. We diagnose mesoscale vertical motion (± 1-10 m/day) associated with these ocean structures using quasi-geostrophic dynamics. A unique characteristic of both field experiments is the combination of conventional in situ measurements from ships with high-resolution observations using autonomous underwater vehicles (gliders). For the eastern Alboran Sea, we also use a high-resolution numerical model that is initialized with hydrographic data (0.5-1 km resolution) from gliders. Numerical simulations show that lateral buoyancy gradients are large enough to trigger submesoscale mixed layer instabilities. Results from the model illustrate that a mixed layer tracer subducts to form vertical intrusions extending to depths of 80-90 m, which is in agreement with remarkable subduction events of chlorophyll and oxygen captured by ocean gliders.

Introduction

Mesoscale (10–100 km) and submesoscale (1–10 km) ocean structures play a major role in the redistribution of properties such as heat, salt, and biochemical tracers, with a significant impact on the ocean's primary productivity (Levy et al., 2001; Ramachandran et al., 2004; Omand et al., 2015). The dynamics associated with these features result in enhanced vertical velocities and mixing, as well as stratification, on time scales that range from a few days to several months and from a few kilometers to 100 km (Klein and Lapeyre, 2009; Ruiz et al., 2009a; Pascual et al., 2015; McWilliams, 2016). In the last 20 years, satellite altimetry has

---

Ruiz, S., et al., 2018: Multi-platform observations and numerical simulations to understand meso and submesoscale processes: A case study of vertical velocities in the Western Mediterranean. In "*New Frontiers in Operational Oceanography*", E. Chassignet, A. Pascual, J. Tintoré, and J. Verron, Eds., GODAE OceanView, 117-130, doi:10.17125/gov2018.ch05.

helped improve our understanding of surface ocean circulation (e.g., Le Traon, 2013), leading to major breakthroughs such as the quantification of eddy kinetic energy (Pascual et al., 2006) and eddy identification and tracking (Chelton et al., 2011; Mason et al., 2014) at scales >100 km. However, the present constellation of altimeters still lacks enough resolution to cover scales shorter than 100 km, typical of ubiquitous mesoscale and submesoscale features such as fronts, meanders, eddies, and filaments. Increasing our knowledge about the relationship between the physical, chemical, and biological processes in the upper ocean is essential for understanding and predicting the ocean and the functioning of marine ecosystems. High-resolution observations (both in situ and satellite) and multi-sensor approaches (e.g., Shcherbina et al., 2015) are necessary to advance our knowledge and prediction of the ocean.

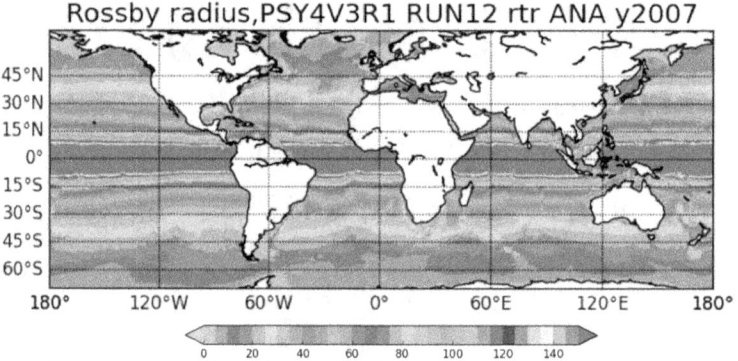

Figure 5.1. Rossby radius of deformation (km) from global operational model (Courtesy of Angélique Melet, Mercator-Ocean).

The Mediterranean Sea is considered an accessible and reduced-scale ocean laboratory where most of the processes that are present in different regions of the global ocean can be investigated at a smaller scale, including deep convection (Herrmann et al., 2009), shelf-slope exchanges (Bethoux and Gentili 1999), thermohaline circulation and water mass interaction (Schröder et al., 2006), and mesoscale and sub-mesoscale dynamics (Bosse et al., 2017). In the context of ocean variability in the Mediterranean Sea, we see that the Rossby radius of deformation is smaller in this semi-closed sea than in other areas of the global ocean located at a similar latitude (Fig. 5.1).

Mesoscale and submesoscale features such as eddies and filaments (Fig. 5.2) interact with each other on the sub-basin and basin scales (Allen et al., 2001; Ruiz et al., 2009a). This amalgam of intricate processes requires high-resolution and comprehensive observations to be fully understood (Pascual et al., 2017).

Meso and Submesoscale Processes: Multi-platform Observations and Numerical Simulations

Methods and data

The vertical velocity (w) in the ocean is generally four orders of magnitude smaller than horizontal velocity and, therefore, it is difficult to obtain direct measurements of vertical velocity (Klein and

Lapeyre, 2009). Vertical velocity can be diagnosed resolving the continuity equation in primitive equation numerical models; however, this approach cannot be followed using observations since instrumental error is as large as the magnitude of the variable (w) to estimate. In the 1990s and the beginning of the 21st century, a common approach to diagnose the vertical motion associated with mesoscale eddies was the use of simplified diagnostic models from two-dimensional (2D) and/or three-dimensional (3D) gridded density and geostrophic fields (Tintoré et al., 1991; Pollard and Regier, 1992; Pinot et al., 1996). The use of the quasi-geostrophic (QG) approximation has led to a better understanding dynamics of physical mesoscale processes in the upper ocean (Rudnick, 1996; Buongiorno-Nardelli et al., 2012; Ruiz et al., 2014) and the coupling between physical and biogeochemical processes (Rodríguez et al., 2001; Allen et al., 2005).

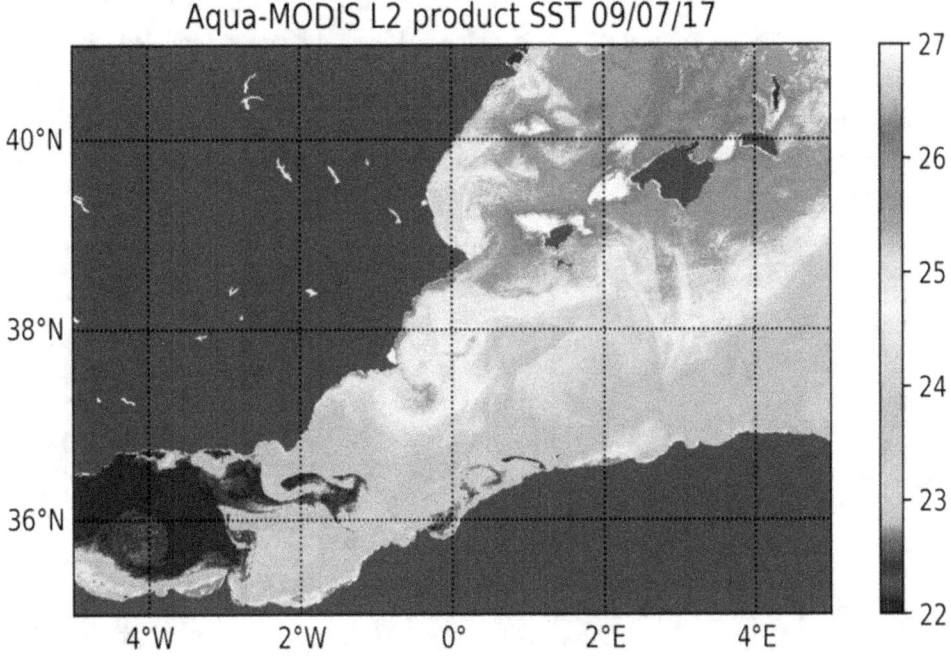

Figure 5.2. Sea surface temperature (SST) image (°C) corresponding to 09 July 2017 covering the western Mediterranean Sea. Fine-scale features are observed due to the interactions of the entrance of fresh and cold Atlantic water (blue) and the more saline and warmer resident Mediterranean water (green).

The diagnostic QG omega equation is directly obtained from the QG vorticity and thermodynamic equation (Hoskins et al., 1978):

$$f^2 \frac{\partial^2 w}{\partial z^2} + \left(\frac{\partial^2}{\partial x^2} + \frac{\partial^2}{\partial y^2}\right)(N^2 w) = \nabla_h \cdot \mathbf{Q} \tag{1}$$

where

$$\mathbf{Q} = \left[2f\left(\frac{\partial V}{\partial x}\frac{\partial U}{\partial z} + \frac{\partial V}{\partial y}\frac{\partial V}{\partial z}\right), -2f\left(\frac{\partial U}{\partial x}\frac{\partial U}{\partial z} + \frac{\partial U}{\partial y}\frac{\partial V}{\partial z}\right)\right]$$

and (U,V) are the geostrophic velocity components, f is the Coriolis parameter, and N is the buoyancy frequency. The vertical velocity (w) can be estimated considering a set of boundary conditions for w and using a 3D snapshot of the density field. In the examples shown in this chapter,

$w=0$ at the upper and lower boundaries and Neumann conditions at the lateral boundaries. The 3D density field (2 x 2 km horizontal resolution and every 5 m in the vertical) is obtained applying an optimal statistical interpolation (OSI) to hydrographic data.

The AlborEx experiment: An intense front

A major intensive, multi-platform and multidisciplinary experiment was completed in May 2014 as a part of PERSEUS (Policy-oriented marine Environmental Research for the Southern European Seas) EU-funded project, led by the Consejo Superior de Investigaciones Científicas (CSIC) and with strong involvement of the Balearic Islands Coastal Observing and Forecasting System (SOCIB), the Instituo Nazionale di Oceanografia e di Geofisica Spermentale (OGS), the Consiglio Nazionale delle Ricerche (CNR), Woods Hole Oceanographic Institution (WHOI), and McGill University. The multi-platform AlborEx experiment (Fig. 5.3) conducted over eight days, included 25 drifters, two gliders, three Argo floats, one ship, and 50 scientists. The week-long experiment was designed to capture the intense but transient vertical motion associated with mesoscale and submesoscale features in order to fill gaps in our knowledge connecting physical processes to ecosystem response (Ruiz et al., 2015).

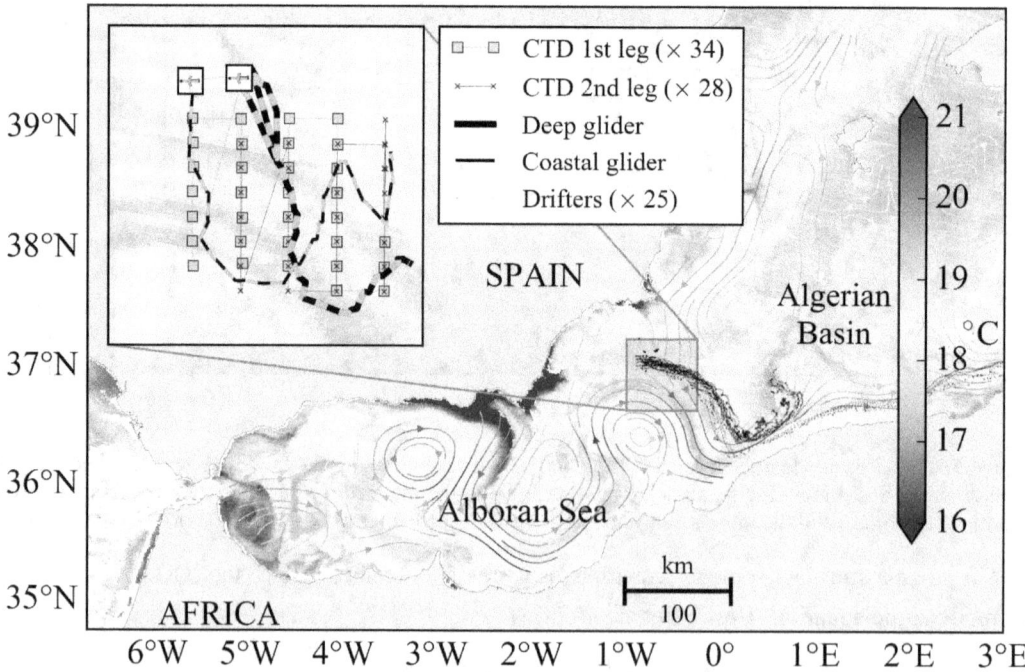

Figure 5.3. SST (MODIS-Aqua) for 29 May 2014 with conductivity, temperature, and depth (CTD) casts, glider and drifter paths during AlborEx experiment. The gray lines correspond to streamlines computed from absolute dynamic topography from gridded altimeter fields (Copernicus Marine Environment Monitoring Service or CMEMS-SLTAC). (From Pascual et al., 2017).

In the Western Mediterranean, the transition region between the Alboran Sea and the Algerian sub-basin to the east is characterized by strong fronts and mesoscale anticyclonic eddies. Transient fronts, such as the Almería-Orán front, separate Atlantic water flowing into the Mediterranean Sea and resident Mediterranean Water that intrudes southwestward along the Spanish coast. Ruiz et al. (2009a) reported mesoscale (~100 km) quasi-geostrophic vertical velocities of the order ±1 m/day estimated from a combination of altimetry and glider observations south of Cartagena, although higher velocities (up to ±20-25 m/day) can be assumed for smaller meso- and submesoscale structures embedded within the front (Allen et al., 2001).

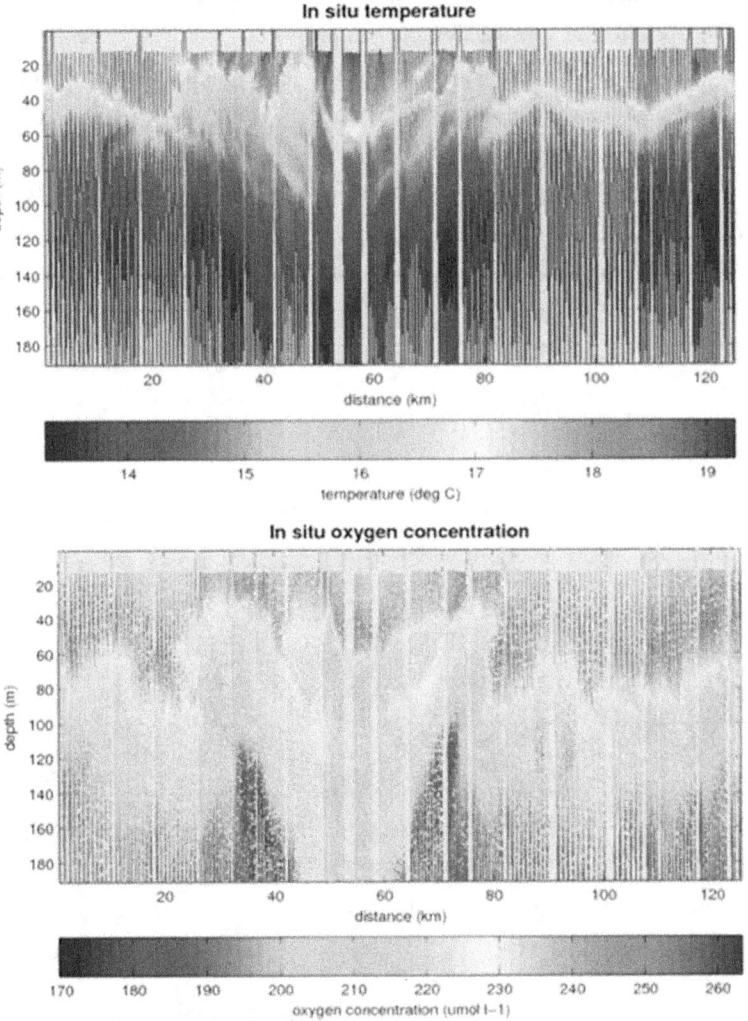

Figure 5.4. Vertical section of temperature and oxygen from the coastal glider.

Autonomous underwater vehicles

Two ocean gliders were able to sample at high-resolution the frontal zone. The coastal glider reached a maximum depth of 200 m and was configured to collect hydrographic and biochemical data at about 0.5 km, while resolution of data from the deep glider was at a depth of ~1 km along-track (data not shown) and it collected data from the surface to 500 m. Fig. 5.4 shows the temperature and oxygen data from the coastal glider. Subduction of small filaments (less than 10 km wide) was observed in different parts of the sampled area suggesting exchanges of properties from the upper layer to the ocean interior

Ship CTDs and a Vessel-Mounted Acoustic Doppler Current Profiler (VM-ADCP)
Conductivity, temperature, and depth (CTD) data along with data from an Acoustic Doppler Current Profiler (ADCP) were gathered from the R/V SOCIB. All physical in situ data were quality-controlled (see Ruiz et al., 2015, for further details). The exact location of the CTD stations was fixed based on the presence of mesoscale and submesoscale features in SST data from remote sensing. The maximum depth reached during all of the CTD casts was 600 m and water samples were collected at each station at the following depths: 5, 20, 40, 60, 90, 100, 120, 150 m. Salinity samples were collected at different depths in one of two stations. An additional sample at 350 m was collected at certain stations for salinity calibration. The first CTD survey consisted of 34 CTD casts distributed over five north/south legs, performed 26-27 May 2014 (Fig. 5.3). During the second survey, 28 CTD casts were conducted on 29-30 May 2014 in almost the same positions as those performed during the first survey (Fig. 5.3).

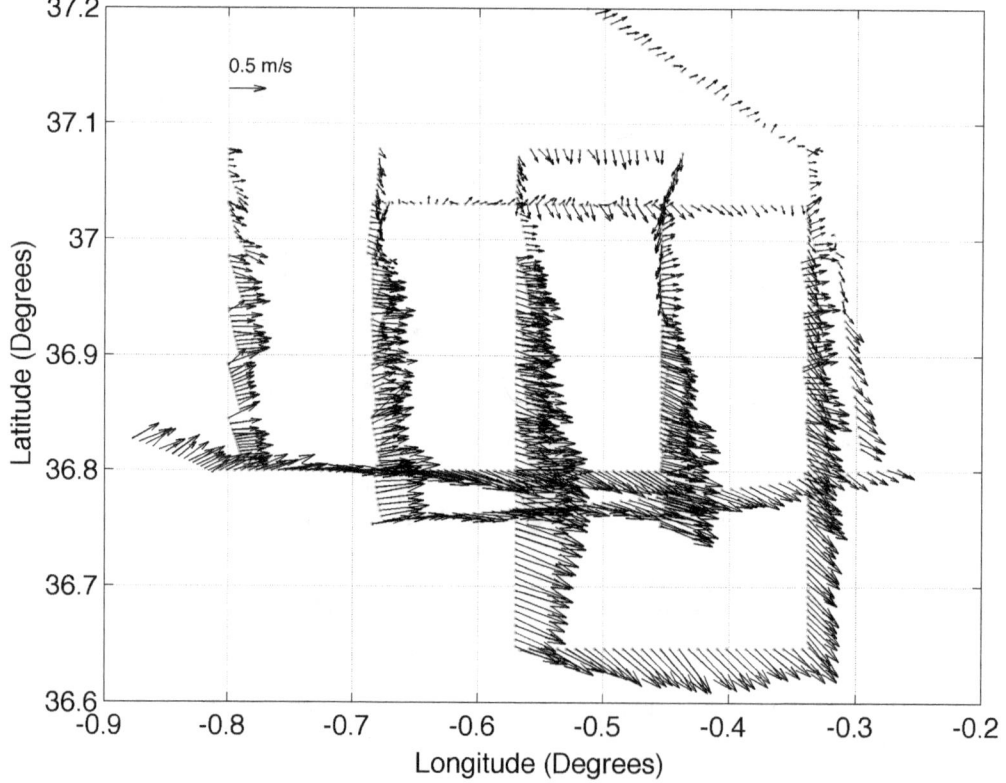

Figure 5.5. Velocity field (cm/s) from VM-ADCP at 48 m depth.

Direct current measurements were obtained from a VM-ADCP 153kHz, with an accuracy of about 1 cm/s. Original profiles were collected every two minutes resulting in approximately one profile every 0.5 km. In the vertical, data were also averaged over 8 m depth bins. Velocity from the first bin (16 m depth) reached values near 1 m/s (Fig. 5.5). It is worth noting that mean geostrophic flow at 50 m depth (Fig. 5.6) was smaller (~40 cm/s) than the velocity from the VM-ADCP, which is about 50 cm/s (not shown). That is to be expected since the VM-ADCP measured actual velocity, including all velocity components.

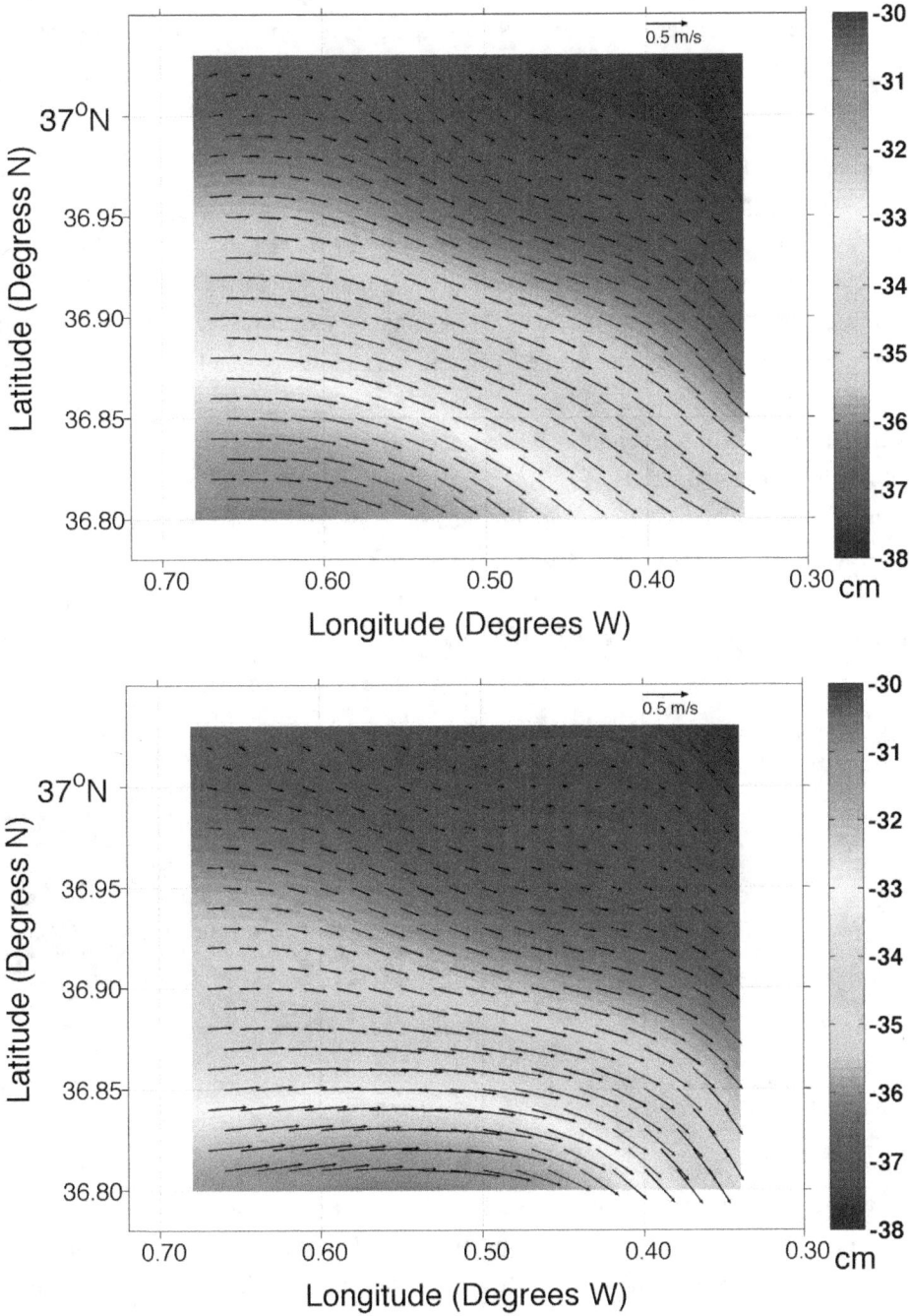

Figure 5.6. Dynamic height (cm, background colour) and geostrophic velocity (cm/s, vectors) at 50 m depth from ship-CTD survey 1 (left) and ship-CTD survey 2 (right). A reference level of no motion is assumed at 600 m depth (From Pascual et al., 2017).

Using the ship CTD data, a 3D density field was diagnosed and used to estimate the vertical velocity following the quasi-geostrophic approach described above (Eq. 1). On average, the magnitude of qg-ω from observations is about ±10 m/day. The pattern is coherent having upwelling/downwelling upstream/downstream of the flow, however these vertical velocities are associated to structures larger than 20 km (Ruiz et al., pers. comm.). Potential contributions from smaller structures that were revealed by gliders (Fig.5.4) have been filtered out in the QG analysis using a low-pass filter (Pedder, 1993).

Numerical modelling

To explore the role of fine-scale features (< 20 km) in enhancing vertical transport at the front, a Process Ocean Study Model (PSOM, https://github.com/PSOM, Mahadevan et al., 1996) was used. Atmospheric forcing is not included in the model configuration in order to isolate and evaluate the particular role of the lateral density gradients. Flow remains statically, inertially, and symmetrically stable during the time period of the analyses. Domain is a periodic channel along the zonal direction (120 km) meridional dimension (200 km), lateral resolution (500 m). The vertical extent is 550 m (~3 m vertical resolution). A front defined by an encounter with Mediterranean and Atlantic waters is initialized in thermal-wind balance using a hydrographic glider section extended with 2 km-resolution output from the Western Mediterranean Operational forecasting system (Juza et al., 2016). A passive tracer was implemented homogeneous within the mixed layer, the depth of which is defined by a density criterion (a difference of 0.01 kg m^{-3} from the surface) in order to investigate vertical intrusions and subduction events below it.

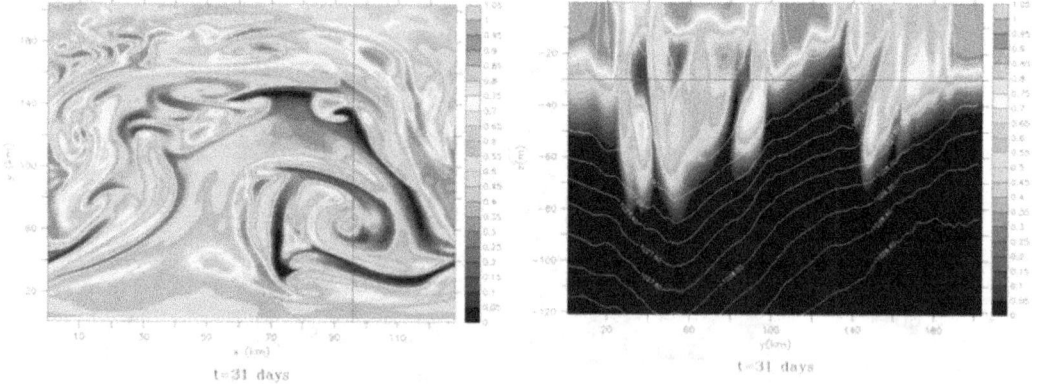

Figure 5.7. Horizontal section at 30m depth (top) and vertical section (bottom) of tracer after 31 days of simulation from PSOM model (resolution 500 m). Contours correspond to isopycnals.

Numerical simulations show the presence of mesoscale feature and fine-scale filaments. Mixed layer tracer subducts form vertical intrusions extending to depths of 80-90 m that become narrow filaments in the horizontal flow (Fig. 5.7). The intrusion of surface water occurs predominantly in regions of weak stratification (Ruiz et al., 2016, pers. comm.). Upwelling of tracer-free water occurs alongside downwelling and is less intense but highlights the subducted tracer.

SINOCOP Experiment: A moderate anticyclonic eddy

The SINOCOCP (Towards an integrated System of Coastal Operational Oceanography) experiment took place 12–18 May, 2009 covering a study area of 50 x 40 km^2. The multi-platform experiment consisted of two Slocum gliders operated simultaneously and in combination with drifters, CTDs from ships, and remote sensing north of Mallorca Island (Fig. 5.8).

The scientific motivation of this experiment was to develop new methodologies to diagnose the 3D dynamics of the ocean using a multi-sensor approach in combination with numerical models. Particular attention was paid to the formation, evolution, and decay of mesoscale variability (and fine-scale features) associated with instabilities of the Balearic current, the main oceanographic feature that is found along the northwest coast of Mallorca Island (Western Mediterranean).

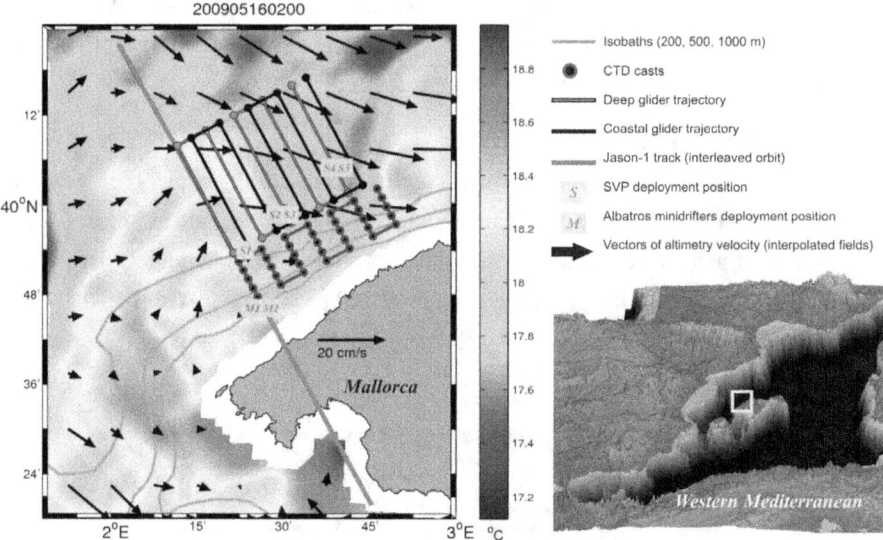

Figure 5.8. Study area during the multi-platform SINOCOP experiment.

Gliders and ship CTDs

Two Slocum gliders, one shallow and one deep, were used. The shallow glider was configured to operate between the surface and 200 m, while the deep glider was set up to reach 600 m in order to meet the scientific objective of the experiment. The net horizontal speed of these vehicles is about 1 km/h, taking into account data transmission when they are at surface. During one week, 811 hydrographic (temperature and salinity) and biogeochemical measurements (turbidity, oxygen and chlorophyll) were registered. Gliders were set up to gather data when descending and ascending, resulting in a final horizontal resolution along-track of 0.5 km for the coastal glider, around 1.1 km for the deep glider, and at about 4 km between adjacent tracks. Final profiles were averaged in the vertical every 1 m. Regarding the glider data processing, special attention was paid to the thermal lag correction, which was applied following a specific procedure developed for gliders (Garau et al., 2010).

Twenty-four CTD casts (SeaBird-19 probe) were performed from IMEDEA coastal ship. The sampling covered the coastal zone between the isobaths of 200 m and 1000 m. The distance between stations was 2.5 km along the same transect and 6.5 km between consecutive transects. Standard CTD processing was applied and final profiles were 1 m averaged in the vertical (see Pascual et al., 2009 for further details).

Drifters

In order to measure the surface currents, five Surface Velocity Program (SVP) ClearWater drifters were deployed in the study area. Each drifter was composed of a surface buoy with a subsurface drogue attached and centered at 15 m depth (Lumpkin and Garzoli, 2005), which guaranteed the flow of the drifter with the ocean currents (minimizing the wind effect). Positioning time series from drifters were linearly interpolated every six hours and a low-pass filter with a 36-hour cut-off frequency was applied in order to remove inertial oscillations. A velocity following the path of the drifters was computed by time differencing the processed 6-hourly positions. Finally, the time series were sub-sampled every 24 hours to obtain a daily product.

Remote sensing: altimetry and SST

Two-dimensional interpolated gridded fields, currently delivered by the AVISO web server, were used in this experiment. The gridded fields are specific to the Mediterranean Sea and are computed on a regular 1/8° grid using a suboptimal space/time optimal interpolated analysis. Regarding SST satellite data, raw images at 1.2 km resolution acquired and processed by GOS-ISAC(CNR) as well as 6-hour averaged images at 2 km spatial resolution from EUMETSAT web server were available during the experiment.

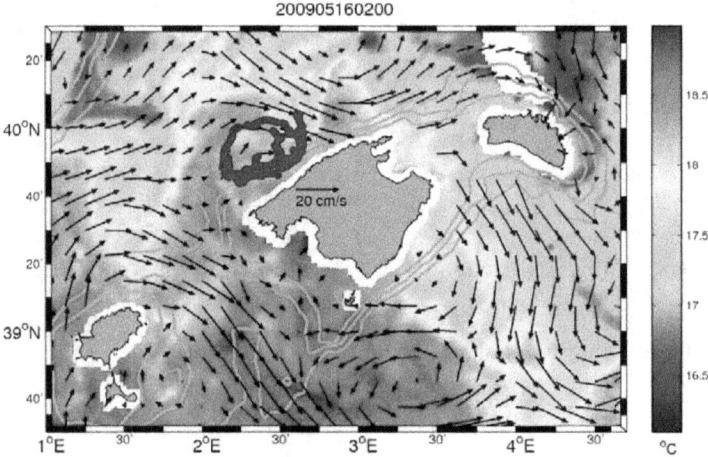

Figure 5.9. Drifter trajectories (blue). Arrows correspond to absolute geostrophic currents from altimetry and the color bar is SST. Isobaths (gray lines) are 200 m, 500 m, and 1000 m.

Relative fresh water from Atlantic origin (37.4) was not detected near the coast, as was found in the previous studies (Ruiz et al., 2009b). Instead, Mediterranean water (with salinity of about 38.4) was dominant in the study area. Analysis of in situ and remotely sensed observations revealed the presence of an anomalous anticyclonic eddy ~ 40 km in diameter near the northwest coast of Mallorca Island (Fig. 5.9). This structure blocked the usual path of the Balearic Current along the coast, deflecting the main northeastward flow to the north. From drifter data analysis, horizontal velocities associated with the eddy were estimated to be about 20 cm/s. Comparisons of drifter and altimetry data revealed that altimeter geostrophic currents derived from gridded products do not have sufficient resolution for the detection of these kinds of small mesoscale and submesoscale features (Fig. 5.9).

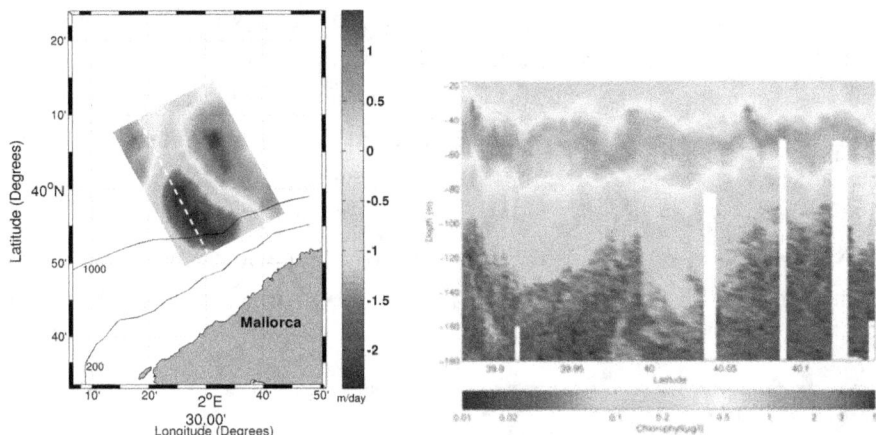

Figure 5.10. Left: QG vertical velocity (m/day) at 100 m depth estimated from hydrographic data (ship-CTD and glider-CTD). The white dashed line indicates the location of vertical chlorophyll (ug/l) section (right) from the first transect of coastal glider.

Hydrographic data from the ship's CTD and derived geostrophic velocities were interpolated in a regular grid at 2 km resolution using an optimal interpolation scheme. Then, the Omega equation (Eq. 1) was solved setting $w = 0$ at the upper and lower boundaries and Neumann conditions at the lateral boundaries. The anticyclonic eddy sampled had associated a vertical motion of about ±2 m/day (Fig. 5.10). Preliminary analysis showed a sinking motion in the center of the eddy that may indicate an early stage of formation of the anticyclonic eddy. A more complete analysis of the whole dataset is in progress, including the modelling of this kind of mesoscale structure.

Summary and Perspectives

The AlborEx and SINOCOP multi-platform and multidisciplinary experiments are examples of integrated multi-platform approaches to investigate the three-dimensional state of the ocean. Although there are still some limitations in terms of spatial coverage (e.g., ocean gliders)—the lack of repeated high-resolution in situ measurements or low resolution of altimetry—the examples provided in this chapter represent a step forward in the direction of combining multiple sensors to further improve our understanding of ocean dynamics (D'Asaro et al., 2011). We have shown that for diagnosis of the vertical motion, which is a key variable for the ecosystem functioning, both field and numerical studies are necessary. Vertical velocities associated with mesoscale eddies derived from observations are in the range of 1–10 m/day and can be responsible for the spatial distribution of biogeochemical parameters such as chlorophyll (Pascual et al., 2015; Olita et al.,

2017; Mason et al., 2017). Fine-scale features can enhance the vertical exchange up to 20–30 m/day as was demonstrated in Mahadevan et al. (2006). Finally, and no less important, the AlborEx and SINOCOP experiments are examples of collaborative joint efforts between scientists, students, and technicians that encourage knowledge exchanges at national and international level.

To better evaluate the vertical and horizontal exchanges and interactions at meso- and submesoscale, we need to improve the coverage and resolution of in situ and remote sensing observational data (e.g., the Surface Water and Ocean Topography or SWOT altimeter mission) while increasing the resolution of numerical models capable of resolving fine-scale processes. The SWOT mission is scheduled to be launched in 2020 and the objectives of the altimeter mission are to characterize the ocean mesoscale and submesoscale circulation determined from ocean surface topography at spatial resolutions of 15 km. This represents a resolution one order of magnitude higher than can be resolved with present-day altimeters (Fu and Ferrari, 2008; Gómez-Navarro et al., 2018). New in situ autonomous platforms are also contributing to the measurement of fine-scale velocities in the ocean (Shcherbina et al., 2018; Jaffe et al. 2016) and exploration of submesocale dynamics.

Acknolwedgements

The AlborEx experiment was conducted in the framework of PERSEUS EU-funded project (Grant agreement no: 287600) with substantial support from SOCIB. Glider operations were partially funded by JERICO FP7 projects. We would like to thank all the crew on board R/V SOCIB for their efficient collaboration during the Alborex experiment and all scientist that have contributed to the discussion of results (Charles Troupin, Arthur Capet, Antonio Olita, Antonio Tóvar-Sánchez, Pierre-Marie Poulain, John Allen). SINOCOP project is an Intramural project funded by Spanish National Research Council. We would like to extend especial thanks to Bruno Buongiorno Nardelli, Miguel Martínez-Ledesma, Bartolomé Garau, Kristian Sebastian, Enrique Vidal, Jérôme Bouffard and Romain Escudier for their efficient work during the SINOCOP experiment. E. Mason contract was partially funded by the Copernicus Marine Environment Monitoring Service (CMEMS) MedSUB project. This study is also a contribution to the PRE-SWOT project (CTM2016-78607-P) funded by the Spanish Research Agency and the European Regional Development Fund (AEI/FEDER, UE).

References

Allen, J. T. et al. (2005), Diatom carbon export enhanced by silicate upwelling in the northeast atlantic. Nat. 728–732. doi:10.1038/nature03948.

Allen, J. T., D. A. Smeed, J. Tintoré, and S. Ruiz (2001), Mesoscale subduction at the Almeria-Oran front. Part 1: Agesotrophic flow, J. Mar. Syst., 30, 263– 285, doi:10.1016/S0924-7963(01)00062-8.

Bethoux, J.P., Gentili, B. (1999), Functioning of the Mediterranean Sea: past and present changes related to freshwater input and climate changes. J. Mar. Syst. 20, 33–47.

Bosse, A., Testor, P., Mayot, N., Prieur, L., D'Ortenzio, F., Mortier, L., Le Goff, H., Gourcuff, C., Coppola, L., Lavigne, H., Raimbault, P. (2017), A submesoscale coherent vortex in the Ligurian Sea: From dynamical barriers to biological implications, J. Geophys. Res. Oceans, 122, 6196–6217, doi:10.1002/2016JC012634.

Buongiorno Nardelli, B., Guinehut, S., Pascual, A., Drillet, Y., Ruiz, S., & Mulet, S. (2012), Towards high resolution mapping of 3-D mesoscale dynamics from observations. Ocean Science, 8(5), 885–901, doi:10.5194/os-8-885-2012.

Chelton, D. B., P. Gaube, M. G. Schlax, J. A. Early, and R. M. Samelson (2011), The influence of nonlinear mesoscale eddies on near-surface oceanic chlorophyll, Science, 334(6054), 328–332, doi:10.1126/science.1208897.

D'Asaro, E., Lee, C., Rainville, L., Harcourt, R., Thomas, L. (2001), Enhanced turbulence and energy dissipation at ocean fronts, Science 332, 318 (2011), doi: 10.1126/science.1201515.

Fu, L.-L. and R. Ferrari, (2008), Observing Oceanic Submesoscale Processes from Space. Eos Trans. AGU, 89(48):488-488.

Garau, B., S. Ruiz, W. G. Zhang, A. Pascual, E. E. Heslop, J. Kerfoot, and J. Tintoré (2011), Thermal lag correction on Slocum CTD glider data, J. Atmos. Oceanic Technol., 28, 1065–1071.

Gómez-Navarro, L., Fablet, R., Mason, E., Pascual, A., Mourre, B., Cosme, E., Le Sommer, J. (2018), SWOT Spatial Scales in the Western Mediterranean Sea derived from pseudo-observations and and ad hoc filtering. Remote Sens., 10, 599, doi:10.3390/rs10040599.

Herrmann, M., J. Bouffard, and K. Béranger (2009), Monitoring open-ocean deep convection from space, Geophys. Res. Lett., 36, L03606, doi:10.1029/2008GL036422.

Hoskins, B. J., Draghici, I. & Davies, H. C. (1978), A new look at the w -equation. Q. J. Royal Meteorol. Soc. 104, 31–38, doi:10.1002/qj.49710443903.

Jaffe, J., Franks, J.S., Roberts, P., Mirza, D., Schurgers, C., Kastsner, R., Boch, A. (2016), A swarm of autonomous miniature underwater robot drifters for exploring submesoscale ocean dynamics, Nat. comm, 8, 14189, doi:10.1038/ncomms14189.

Juza, M., B. Mourre, L. Renault, S. Gómara, K. Sebastián, S. Lora, J. P. Beltran, B. Frontera, B. Garau, C. Troupin, M. Torner, E. Heslop, B. Casas, R. Escudier, G. Vizoso, and J. Tintoré (2016), Socib operational ocean forecasting system and multi-platform validation in the western mediterranean sea, Journal of Operational Oceanography, 9 (sup1), s155–s166, doi:10.1080/1755876X.2015.1117764.

Klein, P. & Lapeyre, G (2009), The oceanic vertical pump induced by mesoscale and submesoscale turbulence. Annu. Rev. Mar. Sci. 1,351–375, doi:10.1146/annurev.marine.010908.163704.

Le Traon, P.Y. (2013). From satellite altimetry to Argo and operational oceanography: three revolutions in oceanography. Ocean Sci. 9, 901–915, doi:10.5194/os-9-901-2013.

Lévy, M., P. Klein, and A.-M. Treguier (2001), Impact of sub-mesoscale physics on production and subduction of phytoplankton in an oligotrophic regime., J. Mar. Res., 59,535–565.

Lumpkin, R., and S. L. Garzoli, (2005), Near-surface circulation in the tropical Atlantic Ocean, In Deep Sea Research Part I: Oceanographic Research Papers, 52, 3, 495-518, doi:10.1016/j.dsr.2004.09.001.

Mahadevan, A., Oliger, J. & Street, R. A (1996), Nonhydrostatic mesoscale ocean model. Part II: Numerical implementation. J. Phys. Oceanogr. 26, 1881–1900.

Mahadevan, A., A. Tandon, (2006), An analysis of mechanisms for submesoscale vertical motion at ocean fronts. Ocean Model. 14, 241.

Mason, E., A. Pascual, and J.C. McWilliams (2014), A new sea surface height–based code for oceanic mesoscale eddy tracking. J. Atmos. Oceanic Technol., 31, 1181–1188, doi:10.1175/JTECH-D-14-00019.1

Mason, E., A. Pascual, P. Gaube, S. Ruiz, J. L. Pelegrí, and A. Delepoulle (2017), Subregional characterization of mesoscale eddies across the Brazil-Malvinas Confluence, J. Geophys. Res. Oceans, 122, 3329–3357, doi:10.1002/2016JC012611.

McWilliams, J. C. (2016), Submesoscale currents in the ocean. Proceedings of the Royal Society of London A: Mathematical, Physical and Engineering Sciences, 472 (2189), doi:10.1098/rspa.2016.0117.

Olita, A., Capet, A., Claret, M. et al. Ocean Dynamics (2017), Frontal dynamics boost primary production in the summer stratified Mediterranean Sea. Oce. Dyn., 67, 767–782. DOI 10.1007/s10236-017-1058-z.

Omand, M. M., E. A. D'Asaro, C. M. Lee, M. J. Perry, N. Briggs, I. Cetinic, and A. Mahadevan (2015), Eddy-driven subduction exports particulate organic carbon from the spring bloom, Science, doi:10.1126/science.1260062.

Pascual, A., Faugere, Y., Larnicol, G., and Le Traon, P. Y. (2006), Improved description of the ocean mesoscale variability by combining four satellite altimeters, Geophys. Res. Lett., 33, L02611, doi:10.1029/2005GL024633.

Pascual, A., Ruiz, S., Martínez-Ledesma, M., Garau, T., Casas, B. Bonet, M., Vidal-Vijande, E., Bouffard, J., Sole, J., Vizoso, G., Orfila, A., Sebastian, K., Buongiorno Nardelli, B., Escudier, R., Schaff, E., Zavala, J., Tintoré, J., (2009), SINOCOP EXPERIMENT: BALEARIC SEA 2009, "A multi-sensor approach"., Technical Report, IMEDEA (CSIC-UIB), 48pp.

Pascual, A., S. Ruiz, B. Buongiorno Nardelli, S. Guinehut, D. Iudicone, and J. Tintoré (2015), Net primary production in the Gulf Stream sustained by quasi geostrophic vertical exchanges, Geophys. Res. Lett., 42, doi:10.1002/2014GL062569.

Pascual, A., S. Ruiz, A. Olita, C. Troupin, M. Claret, B. Casas, B. Mourre, P.-M. Poulain, A. Tovar-Sanchez, A. Capet, E. Mason, J. T. Allen, A. Mahadevan, and J. Tintoré (2017), A multiplatform experiment to unravel meso- and submesoscale processes in an intense front (AlborEx). Front. Mar. Sci. 4, 39, doi:10.3389/fmars.2017.00039.

Pedder, M. A. (1993), Interpolation and filtering of spatial observations using successive corrections and gaussian filters. Mon. Weather. Rev. 121, 2889–2902.

Pinot, J.-M., Tintoré, J. & Wang, D.-P. (1996), A study of the omega equation for diagnosing vertical motions at ocean fronts. J. Mar. Res. 54, 239–259.

Pollard, R. T., and L. A. Regier (1992), Vorticity and vertical circulation at an ocean front, J. Phys. Oceanogr., 22, 609–625, doi:10.1175/1520- 0485(1992)022<0609:VAVCAA>2.0.CO;2.

Ramachandran, S., A. Tandon, and A. Mahadevan (2014), Enhancement in vertical fluxes at a front by mesoscale-submesoscale coupling, Journal of Geophysical Research: Oceans, 119(12), 8495-8511, doi:10.1002/2014JC010211.

Rodriguez, J., J. Tintoré, J. Allen, J. Blanco, D. Gomis, A. Reul, J. Ruiz, V. Rodríguez, F. Echevarría, and F. Jiménez-Gómez (2001), Mesoscale vertical motion and the size structure phytoplankton in the ocean. Nat. 410, 360–363, doi:10.1029/35066560.

Rudnick DL (1996) Intensive surveys of the Azores front: 2. Inferring the geostrophic and vertical velocity fields. J Geophys Res Oceans 101:16291–16303.

Ruiz, S., Pascual, A., Garau, B., Pujol, I. & Tintoré, J. (2009a), Vertical motion in the upper ocean from glider and altimetry data. Geophys. Res. Lett., 36., doi:10.1029/2009GL038569. L14607.

Ruiz S., Pascual A., Garau B., Faugere Y., Alvarez A., Tintoré J. (2009b), Mesoscale dynamics of the Balearic front integrating glider, ship and satellite data. J. Mar. Syst. 78: S3-S16.

Ruiz, S., J. L. Pelegrí, M. Emelianov, A. Pascual, and E. Mason (2014), Geostrophic and ageostrophic circulation of a shallow anticyclonic eddy off Cape Bojador, J. Geophys. Res. Oceans, 119, 1257–1270, doi:10.1002/2013JC009169.

Ruiz, S., Pascual A., Casas B., Poulain P., Olita A., Troupin C., Torner M., Allen J.T., Tovar-Sánchez A., Mourre B., Massanet A., Palmer M., Margirier F., Balaguer P., Castilla C., Claret M., Mahadevan A., and Tintoré J. (2015), On operation and data analysis from multi-platform synoptic intensive experiment (alborex). Tech. Rep. 120pp, doi:10.13140/RG.2.1.3730.4721.

Ruiz, S., Pascual, A., Mahadevan, A., Claret, M., Olita, A., Troupin, C., Tintoré, J., Poulain,, P., Tovar-Sánchez, A., Mourre, B., Capet, A. (2016), Intense ocean frontogenesis inducing submesoscale processes and impacting biochemistry, Pers. Comm., International Liège Colloquium on Submesoscale Processes: Mechanisms, Implications And New Frontiers, 23-27 May 2016, Liège, Belgium.

Schroeder, K., Gasparini, G. P., Tanghelini, M., and Astraldi, M. (2006), Deep and intermediate water in the western Mediterranean under the influence of the Eastern Mediterranean Transient, Geophys. Res. Lett., 33, L21607, doi:10.1029/2006GL027121.

Shcherbina, A.Y., M.A. Sundermeyer, E. Kunze, E. D'Asaro, G. Badin, D. Birch, A.E. Brunner-Suzuki, J. Callies, B.T. Kuebel Cervantes, M. Claret, B. Concannon, J. Early, R. Ferrari, L. Goodman, R.R. Harcourt, J.M. Klymak, C.M. Lee, M. Lelong, M.D. Levine, R. Lien, A. Mahadevan, J.C. McWilliams, M.J. Molemaker, S. Mukherjee, J.D. Nash, T. Özgökmen, S.D. Pierce, S. Ramachandran, R.M. Samelson, T.B. Sanford, R.K. Shearman, E.D. Skyllingstad, K.S. Smith, A. Tandon, J.R. Taylor, E.A. Terray, L.N. Thomas, and J.R. Ledwell (2015), The LatMix Summer Campaign: Submesoscale Stirring in the Upper Ocean. Bull. Amer. Meteor. Soc., 96, 1257–1279, doi :10.1175/BAMS-D-14-00015.1.

Shcherbina, A.Y., E.A. D'Asaro, and S. Nylund (2018), Observing finescale oceanic velocity structure with an autonomous Nortek Acoustic Doppler Current Profiler. J. Atmos. Oceanic Technol., 35, 411–427, doi:10.1175/JTECH-D-17-0108.1.

Tintoré, J., D. Gomis, S. Alonso, and G. Parrilla (1991), Mesoscale dynamics and vertical motion in the Alboran Sea, J. Phys. Oceanogr., 21, 811–823, doi:10.1175/1520-0485(1991)021<0811:MDAVMI>2.0.CO.

CHAPTER 6

Biogeochemical In Situ Observations – Motivation, Status, and New Frontiers

Maciej Telszewski[1], Artur Palacz[1], and Albert Fischer[2]

[1]*International Ocean Carbon Coordination Project, Institute of Oceanology of Polish Academy of Sciences, Sopot, Poland;* [2]*Intergovernmental Oceanographic Commission of UNESCO, Paris, France*

We begin this chapter on in situ biogeochemical observations by presenting the three major areas of societal benefit related to ocean observations: climate, operational ocean services, and ocean health. Biogeochemistry constitutes a varying proportion of each of these areas, while climate and ocean health benefit more from sustained flow of accurate information than operational ocean services. Once the societal drivers are presented, we focus on identifying the relevant phenomena that need quantifying. These phenomena are closely related to the scientific dimension, which helps to establish specific observing targets and observing system design. Scales, seasonality, and geographic limitations are briefly discussed. Consideration is also given to the fact that often a given biogeochemical phenomenon is primarily driven by physical processes (e.g., ventilation, air-sea fluxes) or biological and ecosystem mechanisms (e.g., organic matter cycling, eutrophication) and, therefore, parameters across all three disciplines ought to be measured. Next, we provide an overview of the current capabilities of the global ocean observing system (GOOS) for biogeochemistry. The capacity is considered as an ability (or lack thereof – a gap in capacity) to address the requirements stated in the earlier part of the chapter. A holistic approach to thinking about platforms and sensors is presented. In the following section, the data quality requirements and efforts, as well as data management practices are briefly explained. There has been a strong, long-standing effort among the carbon and biogeochemical observationalists to make biogeochemistry data not only freely available, but also quality-controlled and inter-comparable. These grassroots efforts eventually led to the successful creation of two information products: SOCAT and GLODAP, which are predominantly carbon-focused and represent almost exclusively ship-based, benchtop instrument-based observations. We also discuss an urgent need to expand biogeochemical data availability, quality control, and inter-comparability beyond carbon parameters and onto a wider suite of available platforms and observing techniques (sensors). Finally, to the extent possible, a perspective on existing and planned prototype technology is provided.

Global Ocean Observing System and the Framework for Ocean Observing

The ocean covers 70% of the Earth's surface and is the natural system that ultimately provides most of the air we breathe and the fresh water we drink. The ocean is the primary controller of the global climate that makes this planet habitable for humankind. The ocean is also the pathway for 90% of global trade, it provides 17% of the animal protein consumed by the world's human population, it is a huge draw globally for tourism, and it hosts 99% of the habitable space for our planet's animal and plant life, much of which we believe has not yet even been discovered. Sustainable use and development of the ocean, or the "blue economy", has tremendous

---

Telszewski, M, A. Palacz, and A. Fischer, 2018: Biogeochemical in situ observations – Motivation, status, and new frontiers. In "*New Frontiers in Operational Oceanography*", E. Chassignet, A. Pascual, J. Tintoré, and J. Verron, Eds., GODAE OceanView, 131-160, doi:10.17125/gov2018.ch06.

potential to create jobs and economic value. The ocean is also changing. Shrinking ice caps, sea level rise, ocean acidification, degradation of coastal and open-ocean habitats, plastics and other pollutants, over-exploitation of fish populations, the death of coral reefs and other declines in biodiversity, extreme storms, and coastal flooding — these all pose increasing risks. Understanding, forecasting, and adapting to these growing risks urgently requires that more ocean information be collected, processed, and made available in better ways to support multiple users — governments for policy-making, businesses for safe and efficient operations on the seas and coastlines, scientists for greater understanding of ocean processes, and coastal citizens who are increasingly dependent on forecasts and warnings to protect them from local disasters (GOOS, 2018).

The Global Ocean Observing System (GOOS) and the Global Climate Observing System (GCOS) coordinate sustained observations around the global ocean for three critical themes: climate, ocean health, and real-time services. These themes correspond to the GOOS/GCOS mandate to contribute to the:

- United Nations (UN) Framework Convention on Climate Change (UNFCCC),
- UN Convention on Biological Diversity (CBD),
- Intergovernmental Oceanographic Commission of UNESCO (IOC-UNESCO), and
- World Meteorological Organization (WMO).

This mandate includes observations used for operational ocean services. Historically, these services were based mainly on information related to physical oceanography; however, over the past decade or so our needs for near real-time operational knowledge related to marine biogeochemistry has increased significantly, as did our overall capacity to deliver the required information. The ocean observing community has realized that many observational disciplines are needed to quantify the simultaneous impacts of multiple stressors on ocean ecosystems. This required a re-thinking of many observing strategies, and some compromises (within and across disciplines) have to be made in order to achieve a fit-for-purpose global ocean observing system

There are three broad areas where multidisciplinary sustained ocean observations can bring societal benefit:

- Climate. The ocean is a key component of the climate system and influences its evolution and change through the energy, water, and element cycles. Better monitoring and knowledge will inform both mitigation and adaptation to climate change as well as improved climate services.
- Operational ocean services. Coastal populations and infrastructure are growing and are increasingly exposed to ocean-related hazards. Also, the major marine industries and other ocean users continue to grow. Ocean forecasts and early warning systems can help manage risk and improve business efficiency.
- Ocean health. Ocean ecosystems are coming under increasing pressure from anthropogenic influences, through climate change (warming), acidifying and changing oxygen distributions, and direct human impact. Better monitoring and knowledge will help in sustaining livelihoods and ecosystem services from the ocean.

The impact of such multidisciplinary, sustained ocean observing system derives from a value chain (Figure 6.1) that links research and technology innovation; sustained observing systems; data management systems, analyses, syntheses, and information products; ocean forecast systems and scientific analysis; and operational services and scientific assessments to societal benefit.

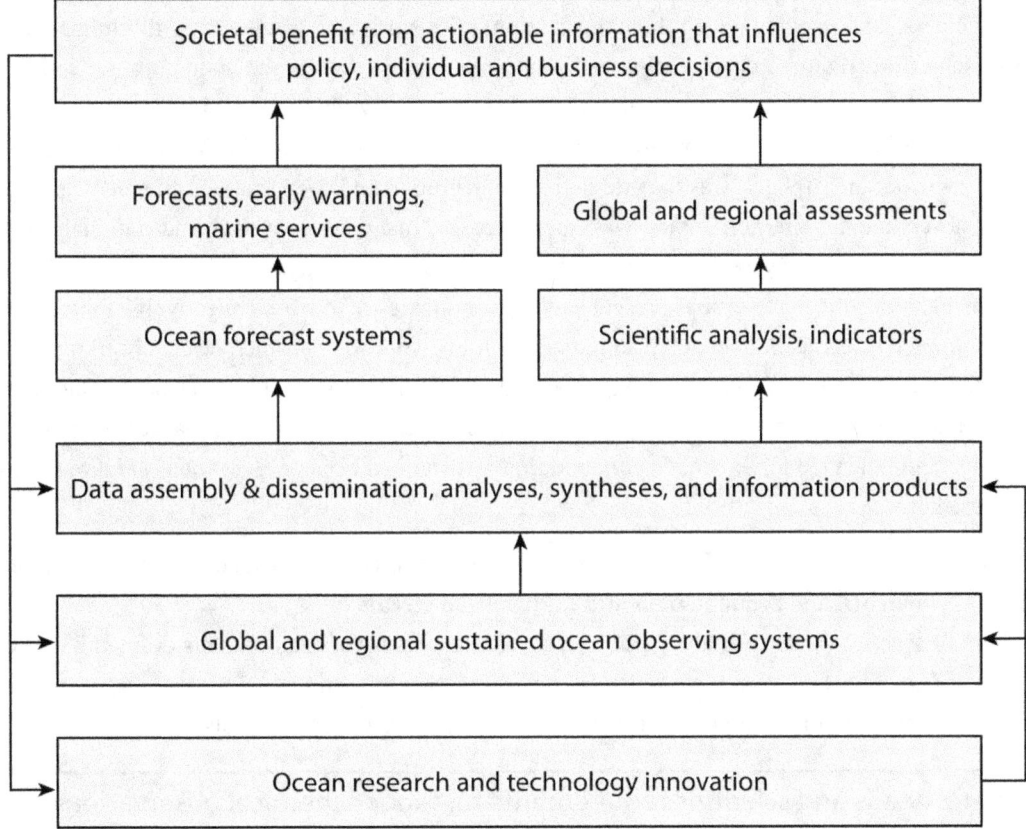

Figure 6.1. An illustration of the value chain linking sustained ocean observations with societal benefit. *Adapted from the G7 Ocean Expert Group think piece, May 2016.*

Improving the feasibility of sustained ocean observations benefits from a close alliance between ocean research and technology innovation, which has always been the source of the observing platforms and sensors that comprise sustained observing systems. Data from the observing system must then flow into systematic data assembly and dissemination centers where they can also be transformed into analyses, syntheses, and information products for use in scientific research or for direct input into indicator frameworks such as 2030 Agenda for Sustainable Development or ocean forecast systems. Here the value chain diverges somewhat into two paths based on our readiness to accurately model the phenomena in question: one through ocean and climate forecasting systems into forecasts, early warnings, and marine services that allow individuals and businesses to make decisions; and the second through scientific analysis or indicator frameworks to global and regional assessments and policy briefs that can inform government or business decisions and policy. This reinforces the dual purpose of a sustained infrastructure for operational benefits as well as for scientific research. The primary concern for operational services will be to estimate the state of the ocean and for scientific analysis and assessments to provide sustained observational infrastructure to understand phenomena and build knowledge.

In 2009, the global observing community agreed to develop and implement the global observing system that could deliver to the above value chain, incorporating information gathered across disciplinary-focused communities: physical oceanographers, marine biogeochemists, and biological oceanographers. Such homogenization of historically fragmented, discipline-based observing efforts required developing a fit-for-purpose strategy, where the main purpose would be addressing short- to long-term societal needs while preserving the well-being of the ocean ecosystem. A key recommendation from the OceanObs'09 Conference held in Venice in September 2009 (www.oceanobs09.net) was the international integration and coordination of interdisciplinary ocean observations. The conference was sponsored by many international and national ocean agencies and attended by representatives of ocean observation programs worldwide. Based on agreement among the many groups present and their strong desire to work collectively, the sponsors commissioned a task team to develop an integrated framework for sustained ocean observing.

The *Framework for Ocean Observing* (FOO, 2012) identifies lessons learned from the successes of previously existing ocean observing efforts and provides an internationally-accepted common language and guidance for expanded collaboration in sustained ocean observations (Figure 6.2). It is focused on:

- delivering a system based on common **requirements**, coordinated ocean **observing elements**, and common **data and information streams**;
- Essential Ocean Variables (EOVs), a common focus for requirements defined based on *feasibility* and *impact* on societal and scientific drivers; and
- evaluation of "readiness levels" for each of these system components.

Societal needs and scientific requirements for biogeochemical observations

Surface fluxes are commonly thought of as the transfer of something (e.g., momentum, heat, moisture, a gas, or particulates) from the ocean to the atmosphere or vice versa. For such a transfer to be maintained, the fluxes must also apply on both sides of the air-sea interface, otherwise there will be a buildup or a loss at the ocean surface. For example, the latent heat flux is related to evaporation at the ocean surface, but is also the vertical transport of this energy in the atmospheric boundary layer. Stress is the vertical transport of momentum at the air-sea interface, but there must also be an identical transport of momentum in the atmospheric and oceanic boundary layers unless that momentum is released or transported away by waves in the process of crossing the air-sea interface. Consequently, we realize that air-sea fluxes modify the air-sea interface and could plausibly be measured from satellite. Again, understanding the geophysical processes that modify the surface and hence modify the electromagnetic characteristics of the surface can be useful in developing more accurate retrievals and retrievals of new variables.

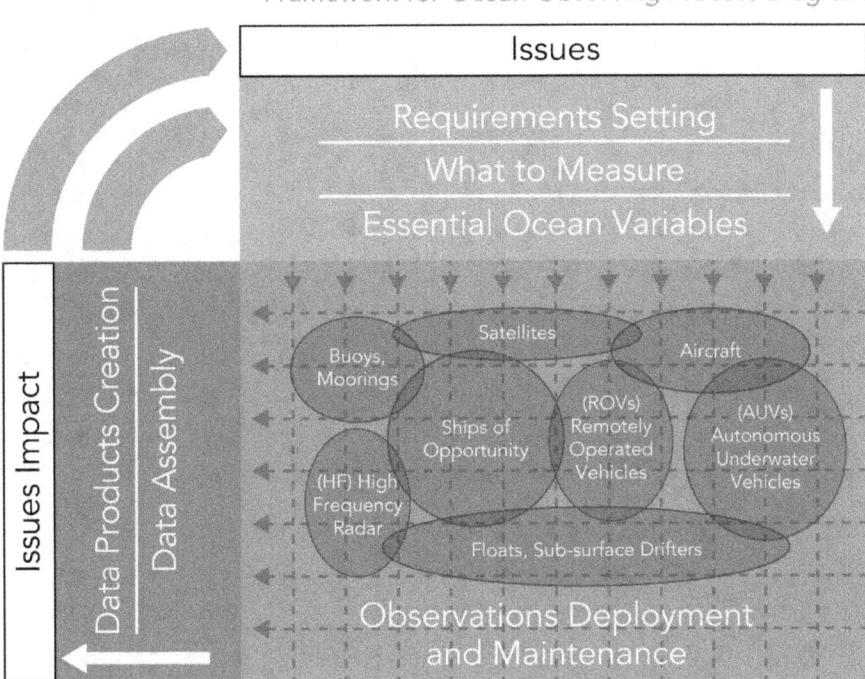

Figure 6.2. Processes in the *Framework for Ocean Observing,* with feedback loops in the definition of requirements and the outputs of the observing system, and a check for fitness-of-purpose of these outputs against societal drivers.

Experts representing a wide array of programs, institutions, national agencies, and intergovernmental and non-governmental organizations were consulted in order to develop a set of presently most pressing requirements related to marine biogeochemistry. Identifying a common approach to requirements across the stakeholders of multidisciplinary global sustained ocean observing systems facilitated developing a common understanding of the societal and scientific needs that facilitates the joint investment needed to build and maintain such an integrated multidisciplinary system. It also encourages integration of the existing elements of the observing system, ensuring that we can develop an infrastructure for both operational services and research. Several workshops, individual consultations with experts representing specific and narrow scientific fields, as well as open-access consultations during large-scale conferences led to agreement upon the following three overarching requirements, each of which is divided into two main questions:

The role of ocean biogeochemistry in climate

The oceans play a critical role in the cycling of many greenhouse gases. It is responsible for taking up and storing about 50% of the anthropogenic emissions of carbon dioxide since the pre-industrial era, thereby buffering (or mitigating) the rate of climate change. Other biologically active elements, such as nitrogen and oxygen, also play an important role in regulating the climate and its effects on how habitable our planet is, most notably through ocean ventilation and the so-called biological pump — the process of uptake of carbon by phytoplankton and its export to the ocean interior and sediments. Marine biogeochemical processes contribute to the complexity of ocean-atmosphere interactions and feedback mechanisms. Constraining the seasonal, regional, and global patterns of fluxes of carbon and other biologically active elements in and out of the ocean is critical to understanding the natural cycle of carbon in the ocean, and thus enhancing the capacity to predict how it might change in the future and what its effects on climate will be.

Key Questions:

How is the ocean carbon content changing?
As the biggest mobile reservoir of carbon in the earth system, any change in the ocean's ability to uptake and store anthropogenic carbon will have a direct impact on rates of atmospheric carbon dioxide (CO_2) concentrations, and hence on climate. Therefore, an observing framework that allows quantification and detection of change of both anthropogenic and total ocean carbon storage and uptake is critical (e.g., for setting emission targets, carbon accounting, model predictions, etc.). Additionally, understanding ocean oxygen fluxes and inventories are important indicators of ocean ventilation and respiration that are needed for more accurate carbon budgets.

How does the ocean influence cycles of non-CO_2 greenhouse gases?
The ocean is a key unknown in the cycling of many other non-CO_2 greenhouse gases such as ozone, depleting, e.g., halocarbons (methyl bromide, bromoform), methane, nitrous oxide (N_2O), and dimethyl sulfate. Ocean measurements are essential for closing the budgets of these gases, which are potentially strong amplifiers of climate change. Furthermore, an ocean observing system that allows for early detection would serve as a warning system alerting us to the risk of passing key tipping points in the climate system.

Human impacts on ocean biogeochemistry
Human activities, such as fossil fuel burning and industrial fertilizer production, have perturbed the global elemental cycles of carbon and nitrogen and significantly impact ocean chemistry. For example, shifts in the carbon chemistry of seawater, as well as changes in nitrogen and oxygen in both coastal and open ocean waters, have been widely recorded. These induce a variety of shifts in marine resources, the full impact of which we still don't understand. The rates at which these changes occur often exceed the recent geological record and highlight the need for a more comprehensive, multivariable approach to ocean biogeochemical analyses in order to better track and predict changes and impacts on marine ecosystems.

Key questions:

How large are the ocean's "dead zones" and how fast are they changing?
The oxygen content of the ocean is decreasing in many areas and, in particular, oxygen minimum zones (OMZ) are growing, likely due to combined effects of changes in circulation and rates of biological oxygen consumption. Oxygen is a strong habitat constraint for most marine animals, and OMZs are areas of highly reduced animal diversity. Low oxygen concentration leads to significant changes in biogeochemistry such as reduction of available nitrate, which can impact ocean productivity.

What are rates and impacts of ocean acidification?

Ocean acidity (i.e., the activity of hydrogen protons, H^+) has increased by 30% since the pre-industrial period. This acidification will likely have significant effects on all levels of the trophic chain (e.g., reproduction, ecosystem structure, physiology), directly impacting future food security. Changes and impacts are expected to be heterogeneous and more severe in the coastal ocean.

Ocean ecosystem health

Changes in ocean chemistry will directly impact the health of marine ecosystems and, as a result, affect humans that rely on marine resources for ecosystem services (e.g., food security, aquaculture).

Key Questions:

Is the biomass of the ocean changing?

Quantifying the magnitude of changes in ocean biomass and productivity, and separating natural variability and secular trends is crucial for understanding and mitigating future impacts on fisheries. Changes in nutrient supply and distribution of macro- and micronutrients are key drivers of primary productivity, which will be impacted by changes in the nitrogen cycle (e.g., nitrogen [N_2] fixation, denitrification). Understanding biogeochemistry changes is key to predicting potential impacts on food webs.

How does eutrophication and pollution impact ocean productivity and water quality?

Land-based sources of nutrients (macro and micro) and carbon (organic and inorganic) into the coastal ocean increasingly lead to eutrophication and hypoxia directly impacting productivity and leading to deleterious effects such as harmful algal blooms. Furthermore, human pollution caused by the use of persistent organic pollutants (POPs), plastics, and dioxins can adversely impact ecosystem health.

Phenomena and essential ocean variables

Minimum set of biogeochemical phenomena to quantify

In order to provide relevant quantifiable information that will allow us to address the above questions, we must develop or adapt an observing system so it is capable of monitoring a comprehensive list of relevant phenomena that control the ocean variability specifically in the context of these drivers. Defining biogeochemical phenomena is challenging, mainly due to the fact that many are either primarily driven by physical (e.g., ventilation, air-sea fluxes) or biological and ecosystem mechanisms (organic matter cycling, eutrophication). In addition to identifying the key phenomena of interest, which should actively drive the design of the global observing system, the spatial and temporal scales on which these phenomena operate need to be well understood. Each phenomenon operates on a variety of spatial and temporal scales that must be considered when choosing the most efficient measuring platform and sensor/instrument. Although the current observing system setup is not centered around phenomena, but rather around observing approaches and/or individual EOVs, such a paradigm shift is much needed to move from a fragmented towards an integrated multiplatform and multidisciplinary observing system.

Here, we will define a phenomenon as follows (adopted from GOOS): *A **phenomenon** is an observed process, event, or property, with characteristic spatial and timescale(s), measured or derived from one or a combination of EOVs, and needed to answer at least one of the scientific questions asked in order to address relevant societal need.*

The following list of phenomena, with short descriptions given in terms of their biogeochemical relevance, has been adopted by GOOS. Whether in the context of operational oceanography, meaning short-term monitoring, or climate variability understanding, which requires decades of observations, these phenomena have to be quantified in parallel with "auxiliary" physical and biological measurements to enable the delivery of information required to address today's societal needs and scientific questions:

- Ventilation (water mass age)
- Air-sea fluxes
- Cross-shelf interactions
- Anthropogenic carbon sequestration
- Ocean acidity
- Inorganic nutrient cycling
- Organic matter cycling
- Hypoxia
- Eutrophication
- Contamination/pollution

Ventilation (water mass age)

Ocean ventilation describes the rate and pathways by which surface waters are carried into the interior of the ocean (e.g., Church et al., 1991), and it is the key physical process determining water mass age. Documenting long-term changes in water mass age and ventilation rates is a key requirement for understanding the role of biogeochemical cycling in climate. Ventilation occurs via a number of downward physical transport mechanisms, such as deep water formation, subduction processes (mode water formation), seasonal mixed layer dynamics, and diffusive fluxes.

Ventilation rates strongly affect biogeochemical cycling of elements, especially carbon, oxygen, and nutrients in the ocean (Sarmiento et al., 2004; Le Quere et al., 2007; Gille, 2008). Understanding the response of ventilation processes is critical now that we have begun to understand feedbacks between climate change and the rate of uptake of anthropogenic CO_2 (C_{ant}) by the ocean (Fung et al., 2005). Monitoring changes in ventilation strength through biogeochemical measurements helps answer the question of how is the ocean carbon content changing. Observing variability in ventilation strength on adequate spatial and temporal scales will greatly improve the constraints on modelling the role of ocean biogeochemistry in climate.

Spatial scales over which ventilation occurs range from local- to basin-scale, i.e., from 100 to 10,000 km. Timescales of ocean ventilation span from sub-annual to millennial, corresponding to water mass ages ranging from zero to 1,000 years. Detecting short-term changes in water mass age and associated changes in biogeochemical properties is important in the context of seasonal to interannual variability in biological production and carbon content in the ocean. However, the observing system should remain focused on documenting changes in ocean ventilation and resultant water mass ages on decadal and longer timescales. This variability is necessary to quantify long-term trends in C_{ant} ocean uptake or oxygen and nutrient storage in the ocean interior.

Air-sea fluxes

The surface turbulent fluxes of momentum, heat, and moisture at the interface between the atmosphere and the oceans represent an exchange of energy between the two. The air-sea fluxes are also an important phenomenon responsible for cycling numerous biogeochemical elements. Biogeochemical observations in the Atlantic focus on several types of air-sea fluxes: (i) air-sea fluxes of CO_2, (ii) air-sea fluxes of O_2, (iii) N_2O flux to the atmosphere, and (iv) dust deposition; observations of which help answer one or more of the GOOS scientific questions.

Air-sea fluxes of CO_2: Although currently not as important as ocean circulation and mixing, air-sea CO_2 gas exchange is one of the mechanisms controlling C_{ant} uptake by the ocean (Talley et al., 2016). For any particular location, the flux of CO_2 between the air and the sea is the product of two principal factors: 1) the difference in partial pressure of CO_2 between the air and the bulk water (ΔpCO_2), which can be considered as the thermodynamic driving force, and 2) the gas exchange rate or "transfer velocity" (k_w), which is the kinetic parameter. The transfer velocity incorporates both the diffusivity of the gas in water (which varies with temperature and between different gases), as well as the effect of physical processes within the water boundary layer.

The rate of CO_2 exchange is determined by the transfer across the water boundary layer; thus, the flux is obtained by multiplying the difference between the air and water pCO_2 (partial pressure of CO_2 in air, which is in equilibrium with the water) by the solubility "K_0" in mol l^{-1} atm.

Air-sea fluxes of oxygen (O_2): Two separate mechanisms contribute to trends in dissolved oxygen storage in the ocean. First, the air-sea flux component corresponds directly to the amount of oxygen (O_2) that the ocean is losing or gaining from the atmosphere, i.e., it is that part of the marine O_2 that leaves an imprint on atmospheric oxygen (Gruber et al., 2001; Keeling and Garcia, 2002). It is important to separate this component from the second mechanism that includes all processes at the surface and interior not associated with the exchange of O_2 across the air-sea interface, i.e. as a result of biological consumption and production of O_2 in the surface ocean and through physical transport and mixing (Stendardo and Gruber, 2012, and references therein). To quantify the contributions of these mechanisms driving the changes in oxygen, one needs to calculate trends in the saturation concentration of O_2, trends in the apparent oxygen utilization and trends in the quasi-conservative tracer O^*_2 derived from dissolved oxygen and phosphorus concentrations (Keeling and Garcia, 2002). Dissolved oxygen tends to respond very sensitively to climate variability and change because any perturbation in sea surface temperature not only changes the solubility of dissolved oxygen, but also alters upper ocean stratification in a way that tends to amplify the solubility effect (e.g., Najjar and Keeling, 2000; Keeling et al., 2010). This high sensitivity to climate forcing makes oxygen one of the best candidates for detecting, and thus better understanding the link between global warming and the resulting biogeochemical and physical changes in the ocean (e.g., Joos et al., 2003; Keeling et al., 2010).

N_2O flux to the atmosphere: Because of the ongoing decline of chlorofluorocarbons and the continuous increase of N_2O in the atmosphere (e.g., Machida et al., 1995; IPCC AR4; IPCC AR5), the contributions of N_2O to both the greenhouse effect and ozone depletion will be even more pronounced in the 21st century. The oceans — including coastal areas such as continental shelves, estuaries, and upwelling areas — are a major source of N_2O and contribute about 30% to the atmospheric N_2O budget. Oceanic N_2O is mainly produced as a by-product during archaeal nitrification (i.e., ammonium oxidation to nitrate), whereas bacterial nitrification seems to be of minor importance as a source of oceanic N_2O. N_2O also occurs as an intermediate during microbial denitrification (nitrate reduction via N_2O to dinitrogen, N_2). Nitrification is the dominating N_2O production process, whereas denitrification contributes only 7-35% to the overall N_2O water column budget in the ocean. The amount of N_2O produced during both nitrification and denitrification strongly depends on the prevailing dissolved O_2 concentrations and is significantly enhanced under low (i.e., suboxic) O_2 conditions.

Atmospheric dust deposition: Dust is produced primarily in desert regions and transported long distances through the atmosphere to the oceans. Once deposited, dust dissolution can be an important source of a range of nutrients, particularly N_2 and iron, to microbes living in open ocean surface waters (Jickels and Moore, 2015). Direct measurements of N_2 from dust deposition are difficult. The majority of flux estimates used come from particle tracking models, with assumptions about dust deposition solubility and bioavailability. Ship-based oceanic measurements of dissolved nutrients are needed in parallel with atmospheric measurements of dust deposition rates to further validate these models.

Cross-shelf interactions

This phenomenon describes the biogeochemical exchanges with shelf and marginal seas. In many coastal circulation regimes, the proximity of energetic boundary currents in deep water at the shelf edge is a key dynamic in mediating shelf/open ocean exchange. On coasts for which estimates exist, fluxes of nutrients and carbon across this boundary are leading order terms in the N_2 and carbon budgets of shelf ecosystems. The exchange at the ocean boundary and shelf edge dynamics immediately impact ecosystem function and productivity on weekly-to-seasonal timescales; they can also drive multi-decadal changes in ecosystem structure via effects on habitat ranges and biodiversity.

Direct observations of biogeochemical and physical exchanges across the shelf-open ocean boundary have not been sustained to the extent required to fully complement observations within the ocean interior. In large part, this is due to the particular challenges of maintaining observing networks within energetic regimes, and capturing the significantly shorter time and space scales of variability there.

Quantifying nutrient fluxes across the shelf-open ocean boundary, often occurring in pulse form in response to passing fronts and eddies, is essential to inform, calibrate, and validate biogeochemical models, which can enhance the capacity for harmful algal bloom forecasting. Measurements of carbon and oxygen fluxes are another important application.

Anthropogenic carbon sequestration

The term "carbon sequestration" here describes only the natural oceanic processes by which CO_2 is removed from the atmosphere and stored in the ocean interior or buried in marine sediments. Given the accelerating pace of emissions related to human activities, it is important to explicitly detect changes in the uptake and storage of the anthropogenic component of CO_2 in the atmosphere, in addition to quantifying the ocean's role as a sink in the global carbon budget. By continuing to take up a substantial fraction of the C_{ant} emissions from fossil fuel combustion and net land-use change, the ocean is a major mediator of global climate change. Therefore, observing the phenomenon of anthropogenic carbon sequestration helps answer the question 'How is the ocean carbon content changing?'

Mechanisms of carbon sequestration are either physicochemical (through the solubility pump) or biological (through the biological pump), and are no different regardless of whether the sources of carbon are natural or anthropogenic (Raven and Falkowski, 1999). Several methods have been developed and tested to accomplish the separation between changes in dissolved inorganic carbon due to C_{ant} uptake and those due to natural variations in circulation and organic matter remineralization (Friis et al., 2005; Locarnini et al. 2013; Zweng et al. 2013). The physicochemical mechanism is closely related to two other phenomena discussed in this chapter, i.e., ventilation and air-sea fluxes. On the other hand, the biological mechanism relates to organic matter cycling. As a result, setting biogeochemical observing targets with respect to C_{ant} sequestration phenomenon should not be done in isolation from considering requirements for observations of these other phenomena.

Ocean acidity

Acidity is hydrogen ion (H+) concentration in a liquid, and pH is the logarithmic scale on which this concentration is measured. Ocean acidification is a progressive increase in the acidity (decrease in pH) of the ocean over an extended period, typically decades or longer, which is caused primarily by uptake of CO_2 from the atmosphere. It can also be caused or enhanced by other chemical additions or subtractions from the ocean. Acidification can be more severe in areas where human activities and impacts, such as acid rain and nutrient runoff, further increase acidity.

The pH of the open ocean surface layer is unlikely to ever become acidic (i.e., drop below pH 7.0) because seawater is buffered by dissolved salts. However, ocean acidification is changing seawater carbonate chemistry. The concentrations of dissolved CO_2, hydrogen ions, and bicarbonate ions are increasing, and the concentration of carbonate ions is decreasing. Changes in pH and carbonate chemistry force marine organisms to spend more energy regulating chemistry in their cells. For some organisms, this may leave less energy for other biological processes such as growing, reproducing, or responding to other stresses.

Many shell-forming marine organisms are very sensitive to changes in pH and carbonate chemistry. Corals, bivalves (such as oysters, clams, and mussels), pteropods (free-swimming snails), and certain phytoplankton species fall into this group. But other marine organisms are also stressed by the higher CO_2 and lower pH and carbonate ion levels associated with ocean acidification. The biological impacts of ocean acidification will vary because different groups of marine organisms have a wide range of sensitivities to changing seawater chemistry.

Aragonite saturation state is another indicator of change in ocean acidity. Aragonite is a mineral form of calcium carbonate, a basic building block of corals and many forms of zooplankton. The aragonite saturation state decreases with increasing acidity of ocean water. Below a certain threshold, calcifying organisms using aragonite cannot produce shells or skeletons effectively. Thus, changes in aragonite saturation state will become a broad-scale ocean ecosystem stressor that will affect a large set of organisms.

Aragonite saturation state in surface and subsurface waters is calculated from dissolved inorganic carbon and total alkalinity data. According to Jiang et al. (2015), through the year 2012 surface aragonite saturation state in the open ocean was always supersaturated ($\Omega > 1$), ranging between 1.1 and 4.2. It was above 2.0 (2.0–4.2) between 40°N and 40°S, but decreased toward higher latitude to below 1.5 in polar areas.

Inorganic nutrient cycling

Nitrogen in various forms occurs naturally in the environment, including inorganic species such as ammonium (NH_4^+), nitrate (NO_3^-), nitrite (NO_2^-), and nitrogen gas (N_2), and organic forms such as amino acids, proteins, DNA, and RNA; the latter forms occur both as particulate and dissolved fractions. The phosphorus and silicon cycles are less complex but are equally important in controlling ocean biomass and carbon content changes.

The following processes reflect either sources or sinks of inorganic nitrogen in the ocean, and they require support from biogeochemical observations and modelling: (i) denitrification, nitrification and anaerobic ammonium oxidation (anammox), (ii) N_2 fixation, and (iii) non-point and point source nutrient fluxes.

Denitrification, nitrification, and anammox processes determine the availability of macronutrients for phytoplankton uptake. Denitrification reduces nitrate, first to nitrite and then to ammonia, burning oxygen. we often observe predominance of denitrification in OMZs and in coastal eutrophicated waters, thus further augmenting the loss of oxygen and leading to temporary or persistent hypoxia (i.e., oxygen reduced in concentration to a point where it becomes detrimental to aquatic organisms). Anammox is a significant component of the biogeochemical nitrogen cycle, whereby ammonia and nitrite are converted directly into nitrogen gas. Globally, this process may be responsible for 30-50% of the N_2 gas produced in the oceans (Devol et al., 2003). Therefore, it is a major sink for fixed nitrogen and it limits oceanic primary productivity. In the Atlantic Ocean, for example, anammox appears to be closely coupled to denitrification, with regional differences observed between the shelf and continental slope areas (e.g., Trimmer and Nichols, 2009).

Nutrient cycling on scales from hours (ammonia production) to seasons (nitrate replenishment in surface waters) affects patterns of phytoplankton community composition and primary productivity on scales from days to months. In combination with changes in the physical transport of new and regenerated nutrients, interannual or decadal changes in nutrient availability also affect the long-term variability in organic matter cycling.

N_2 fixation is a biologically mediate process in which atmospheric N_2 gas is converted into ammonia by prokaryotes called diazotrophs. N_2 fixation provides a means of fixing organic matter under reduced nitrogen availability conditions. Nevertheless, diazotrophs have strong requirements for phosphorus and iron. N_2 fixation is not directly measured in the ocean, but can be derived from measurements of available phosphate. The importance of this process increases as we move towards the equator and decreases towards the higher latitudes in the Atlantic, with some regional enhancements associated with boundary current regions.

Organic matter cycling

Organic matter cycling refers to a group of processes that either biologically transform or physically transport organic matter between the surface and interior ocean or across the water-sediment interface. Biological transformations of organic matter include gains due to fixation of atmospheric CO_2 and inorganic nutrients into particulate organic matter, as well as losses due to grazing and respiration that transform particulate into dissolved organic matter and organic carbon and nutrients back into their inorganic forms.

Organic matter fixation is particularly important with respect to the biological component of anthropogenic CO_2 uptake that results from net community production, defined as the gross primary production by autotrophs minus the total respiration by phytoplankton, zooplankton, and the resident microbial community. Globally, the magnitude of the net community production signal is estimated to be in the range of 5-15 Pg-C per year. In the Atlantic, the North Atlantic spring bloom has gained a lot of attention due to the uptake and potential sequestration of CO_2 via rapid growth of large phytoplankton cells (i.e., diatoms) and subsequent vertical export of assimilated carbon to depth following nutrient limitation. The highly seasonal nature of the diatom bloom, specifically the rapid growth and equally rapid export of cells during bloom termination, as well as the patchy nature of the bloom makes this phenomenon particularly difficult to study. Higher-resolution sampling in time and space is required to capture the smaller, episodic events describing the coupled physical and biogeochemical dynamics of the spring bloom.

The oligotrophic waters of subtropical ocean gyres occupy >40% of the Earth's surface; thus, even relatively low carbon exports in these immense ocean provinces may significantly contribute to the global carbon budget. While surface ocean biology is very similar between two Atlantic central gyres, there is an established spatio-temporal variability in their upper ocean biogeochemistry. Presence of such a signature in the deep ocean has long been unknown due to a lack of direct measurements.

Changes in organic matter fixation and remineralization, and resultant amount of particulate carbon export into the ocean interior, occur over a broad range of timescales, from weekly to interannual and longer. However, as shown by Henson et al. (2016), detecting long-term climate-driven trend in changing biological production and export might require temporal records substantially longer than what is available beyond a few existing time series stations. Except for a very few regions where natural variability signal is relatively weak, in most areas of the global ocean, records between 25 and 50 years might be necessary to detect climate-driven trends in these processes. This is also significantly longer than chlorophyll a (Chl-a) observations available from remote sensing.

Considering a limited number of open ocean fixed-point observatories and their limited spatial footprint of 42-43 10^6 km^2 (defined as the area over which a station is representative of a broader region; Henson et al., 2016), this represents a significant gap in the observing system capacity to answer the question of how does the ocean carbon content change, at least on long-term scales.

Dissolved organic carbon (DOC) is one of the largest bioreactive pools of carbon in the ocean (Hansell et al., 2009; 2012). The inventory of oceanic DOC is estimated to be ~662 ± 32 Pg C, 200 times the mass of the organic carbon in suspended particles but approximately 1/50th of the total dissolved inorganic carbon inventory (Hansell et al., 2009). The majority of the newly produced DOC is rapidly remineralized by heterotrophic bacterioplankton within the ocean's surface layer (Carlson and Hansell, 2015). However, ~20% of global net community production (~1.9 Pg C year−1) escapes rapid microbial degradation for periods long enough to be exported from the euphotic zone via convective mixing or isopycnal exchange into the ocean's interior (e.g., Hansell et al., 2009). DOC export occurs with deep- and mode-water formation as mid-latitude, warm, DOC-enriched surface waters are transported with surface currents to subpolar and high latitudes. Here, convective overturn transports the DOC deep into the interior, where it is slowly removed through southward flow.

Hypoxia

Hypoxia is the result of a process of decomposing organic matter (mostly phytoplankton) biomass that uses up dissolved oxygen in the water in equilibrium with a certain rate of oxygen flux by ocean transport processes (phenomena: circulation, mesoscale transport). It can occur naturally, or be stimulated by anthropogenic nutrient pollution through non-point and point sources (see also eutrophication below).

Changes in oxygen content in the ocean have a direct impact on both the climate and ecosystem health. The concentration of dissolved oxygen is a major determinant of the distribution and abundance of marine species globally. Open ocean deoxygenation has already been recorded in nearly all ocean basins during the second half of the 20th century. Increased temperatures are responsible for approximately 15% of the observed change, and the remaining 85% is due to reduced O_2 supplies from increased ocean stratification and deep-sea microbial respiration (IOC-UNESCO and UNEP, 2016).

Although deoxygenation is the predominant direction of change in the ocean, this trend is not uniform across ocean basins. In the North Atlantic, for example, although the upper, mode, and intermediate waters are indeed losing oxygen because of changes in solubility, the deeper waters actually gained oxygen over the last 50-year period owing to changes in circulation and ventilation (Stendardo and Gruber, 2012). Oxygen decline is found most consistently in the OMZs

Although seasonal or persistent OMZs occur on local (up to 100 km) scales frequently in the coastal ocean, in the open ocean Atlantic the main areas of interest with respect to the observing system are upwelling regions. These regions present a very complex case where several phenomena are inter-linked with each other, through a myriad of physical, biogeochemical, and biological processes. The schematic on Figure 6.3 illustrates the complex interaction between coastal hypoxia and eutrophication and open ocean oxygen conditions only found in some upwelling regions (Doney, 2010).

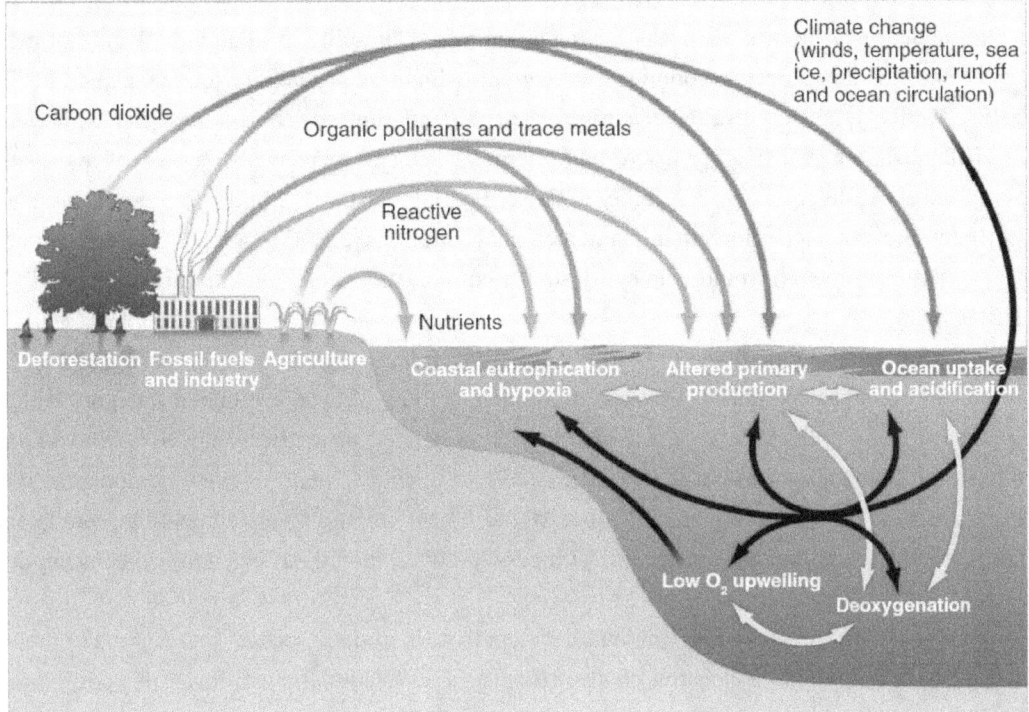

Figure 6.3. Schematic of human impacts on ocean biogeochemistry either directly via fluxes of material into the ocean (colored arrows) or indirectly via climate change and altered ocean circulation (black arrows). Source: Doney, 2010.

Eutrophication

Monitoring and predicting the eutrophication status of waters is an important part of determining whether an ecosystem is in a healthy state or not. The eutrophication is driven by a surplus of the nutrients nitrogen and phosphorus in the sea, caused mainly by agriculture activities (fertilizer use and wastes from livestock) and urban wastewater. Nutrient over-enrichment, also referred to as nutrient pollution, causes elevated levels of algal and some plant growth, increased turbidity, oxygen depletion, changes in species composition, and increase in toxins (e.g. through incidence of harmful algal blooms). As these properties and events are directly measurable or are easily derived from such measurements, eutrophication is observed based on its impacts on biogeochemistry and biology, rather than being observed based on the state of its pressures and drivers such as excess concentration of nutrients in river run-off or atmospheric deposition. This has implications for selecting core indicators of eutrophication and designing an observing system adequate for monitoring changes in eutrophication status. Observing changes in eutrophication is tightly coupled with three other phenomena: inorganic nutrient cycling, hypoxia, and organic matter cycling.

Contamination/pollution

Marine water pollution is a significant concern for ocean ecosystem health. The following contaminants are generally identified: plastics, organic contaminants, heavy metals, hydrocarbons, and underwater noise.

Plastic debris or litter in the ocean is now ubiquitous. Durability is a common feature of most plastics and it is this property, combined with an unwillingness or inability to manage end-of-life plastic effectively, that has resulted in marine plastics and microplastics becoming a global problem. We distinguish between floating macro and microplastics, and there is also a pool of mid-water microplastics. At the moment, our ability to detect floating plastics is limited to presence/absence data, but future sustained efforts to measure their concentrations (e.g., through underway automated data capture instruments) would help constrain the current, very large level of uncertainty on their distribution.

Persistent, bioaccumulating, and toxic organic compounds (PBTs) are also ubiquitous in the marine environment, primarily because of human activity. These include a category called persistent organic pollutants (POPs), which are either banned or restricted under the Stockholm Convention. Some are hydrophilic and others are hydrophobic. Many of these compounds have chronic impacts on marine organisms, especially at higher trophic levels amongst top predators. There are human populations, particularly at higher latitudes, directly affected due to consumption of traditional foodstuffs.

There has been extensive research and development to produce sensors that can make in situ sampling by employing a concentration stage, using passive samplers with lower detection limits based on a variety of gel and films (e.g., Burgess et al., 2015).

Floating plastics, PBTs, and POPs remain in the natural system on similar temporal scales, from weeks to decades. Due to their wide-spread transport in the ocean, they affect the marine system in spatial scales from meters to thousands of kilometres.

Atmospheric deposition is the main source of dissolved heavy metals in the ocean (cadmium, copper, lead, zinc, cobalt, zinc, mercury, methyl mercury). The partitioning of atmospheric inputs between dissolved and particulate phases within the surface layer strongly determines the behaviour of trace metals and their involvement in biogeochemical cycles (Cossa et al., 2009). Basically, the assimilation of metals by biota may be constrained by their solubilisation, resulting from physicochemical and biological processes (dissolution through zooplankton intestines, for example). Riverine fluxes are the main source of heavy metal particulates. Distribution of heavy metals changes in annual scales, and spatially, from 100-3000 km when affected by atmospheric deposition and from 1-1000 km when driven by river inputs.

Minimum set of biogeochemical variables to measure

EOVs are the *Framework for Ocean Observing* concept of the fundamental physical, biogeochemical, and biological measurements needed for the scientific understanding of ocean phenomena and the provision of applications in support of societal needs. What makes these observables essential is that they are the minimum subset required; they are not replaceable by other variables. Their essential nature is defined by both a:

- high *feasibility* of sustained observation, based on the platforms and sensors that can observe this variable at the space and timescales and accuracy needed to capture the required phenomena, and
- high *impact* of the observation in creating the application or contributing to the needed scientific knowledge, and therefore providing societal benefit.

EOVs are basic variables observable by one or more practical instrumentation systems. Measuring the whole suite of defined EOVs is necessary to provide the data to quantify all the phenomena in support of scientific research and address societal issues related to the ocean and climate. EOV specification sheets linking EOVs to societal benefits describe: their importance in scientific phenomena, present observation strategies (e.g., spatial and temporal resolution, accuracy, technological readiness level), required complementary variables, derived variables, and the observation programmes and networks measuring the data. Specification sheets for marine biogeochemistry EOVs briefly described below can be found at http://www.ioccp.org/index.php/foo. There are nine biogeochemical EOVs (listed below).

Dissolved Oxygen

Measuring and understanding the large (mostly) decreasing trends in the concentrations of dissolved oxygen in the ocean over the last few decades has important implications for our understanding of anthropogenic climate change. Sub-surface oxygen concentrations in the ocean reflect a balance between supply through circulation and ventilation and consumption by respiratory processes. An observing network of dissolved oxygen, among other things, results in the following products: (i) improved constraint on the ocean-land-partitioning of anthropogenic CO_2, (ii) determination of the seasonal to interannual net remineralization rates as a proxy for the amount of organic matter exported from the surface ocean, (iii) better interpretation of variations in water mass ventilation strength, (iv) and increased availability of crucial data (initial conditions, evaluation) for ocean biogeochemistry models.

Nutrients

The availability of inorganic macronutrients (NO_3, phosphate [PO_4], silicon [Si], ammonium (NH_4), nitrogen dioxide (NO_2)) in the upper ocean frequently limits and regulates the amount of organic carbon fixed by phytoplankton, thereby constituting a key control mechanism of carbon and biogeochemical cycling. There is a number of biogeographic regions in the open ocean characterized by different macronutrient regimes, either permanently or seasonally limiting the growth of phytoplankton. Measuring changes in macronutrient concentrations is essential to constraining net biological production and export fluxes, detecting shifts in biogeographic regimes, but also monitoring eutrophication and pollution phenomena.

Inorganic Carbon

The observations required to constrain the carbon system at a point in space and time are any two of dissolved inorganic carbon, total alkalinity (T_{ALK}), partial pressure of carbon dioxide (pCO_2) and pH, and associated physical variables (temperature and salinity). High resolution and long-term observations of the carbonate system are essential for distinguishing the climate change-driven trends from the strong seasonal to decadal variability signal in net biological production and export flux, in particular in the high-latitude spring bloom systems. The carbon system is in a delicate balance such that high quality observations and predictions of the carbonate system will continue to be required to have a mechanistic understanding and ability to predict the changes in the anthropogenic carbon flux and storage in the interior ocean, and ocean acidification rates.

Transient Tracers

Transient tracers are a group of (chemical) compounds that can be used in the ocean to quantify ventilation strength, transit time distribution and transport time-scales. These compounds are all conservative in sea-water, or have well-defined decay-functions, and a well-established source function over time at the ocean surface. Measurement of transient tracers in the interior ocean thus provides information on the time-scales since the ocean was ventilated, i.e. in contact with the atmosphere. Knowledge of the transit time distribution of a water-mass allows for inference of the concentrations or fates of other transient compounds, such as anthropogenic carbon or N_2O. Commonly measured transient tracers are the chlorofluorocarbons (CFCs) 11 and 12, although in the past also CFC-113 and CCl_4 have been measured. More recently, measurement of transient tracers includes Sulphur hexafluoride (SF_6), radioactive isotopes 14C, tritium (decaying to stable 3He), and argon isotope 39Ar.

Particulate Matter

Particulate matter include the variables referred to as particulate organic matter (i.e. particulate organic carbon and particulate organic nitrogen; but also particulate inorganic carbon and biogenic silica (BSi); as well as the vertical transport (export) flux of all particulates. Observation of particulate organic matter within a global observing system would directly address the question of whether the ocean's biomass and productivity are changing. Changes in particulate organic matter could be important indicators of deteriorating water quality due to eutrophication in coastal regions, and of declines in primary production that could potentially translate up the food chain negatively impacting fisheries. Observation of particulate inorganic carbon would directly address the question of what impacts does ocean acidification have on calcareous organisms and, thus, community structure. Export production gradients occur over a multitude of spatial and temporal scales, therefore high spatial resolution measurements are needed (for example, in upwelling areas in the eastern boundary currents), while high temporal resolution measurements are needed, particularly in polar regions where spring blooms can be highly pulsed and the bulk of annual export rates occur often over only a few weeks' time.

Nitrous Oxide (N_2O)

The oceans, including its coastal areas such as continental shelves, estuaries, and upwelling areas, are a major source of N_2O and contribute about 30% of this important climate-relevant trace gas to the atmospheric budget. Because of the ongoing decline of chlorofluorocarbons and the continuous increase of N_2O in the atmosphere, the contributions of N_2O to both the greenhouse effect and ozone depletion will be even more pronounced in the 21st century. A ship-based observing network in the

Atlantic Ocean not only helps estimate global N_2O emissions but also helps capture information about ocean phenomena such as deoxygenation, eutrophication, and upwelling.

Stable Carbon Isotopes

Recent improvements in measuring the carbon-13 to carbon-12 isotope ratio (13C/12C) and concentration of CO_2 gas dissolved in seawater using field portable spectrometers open up the possibility of underway 13C/12C observations across large portions of the surface ocean. Such datasets would substantially improve δ13C-based estimates of organic matter export rate and of the air-sea $13CO_2$ flux. The latter term can be compared to depth-integrated $13CO_2$ inventory changes in the water column to provide a separation of anthropogenic CO_2 change due to air-sea CO_2 flux versus change due physical transport by ocean circulation. One example of a recent application of this approach in the North Atlantic indicates that 50% of the anthropogenic CO_2 increase in this ocean basin is a result of transport from the South Atlantic as part of the meridional overturning circulation.

Dissolved Organic Carbon (DOC)

DOC exceeds the inventory of organic particles in the oceans by 200-fold, making it one of the largest of the bioreactive pools of carbon in the ocean, second only to dissolved inorganic carbon. The size of the reservoir (comparable to that of atmospheric CO_2), as well as its role as a sink for autotrophically-fixed carbon, a substrate to heterotrophic microbes, and a sink/source of carbon involved in climate variations over long timescales, highlights its importance in the ocean carbon and nitrogen cycles. DOC exported from the epipelagic zone contributes around 20% to the biological pump via meridional overturning circulation.

Ocean Colour

Ocean colour radiance is the wavelength-dependent solar energy captured by an optical sensor looking at the sea surface. The spectral distribution of the water-leaving radiance contains information on the ocean albedo and optical constituents of the seawater, in particular the concentration of the phytoplankton pigment chlorophyll-a (a proxy for phytoplankton biomass). Deriving ocean colour products is not easy because the water-leaving radiance signal is relatively weak at the altitude of a satellite sensor (only 5-15% of incident solar radiation, with the remaining light having an atmospheric origin).

Ocean colour radiometry observations from space have revealed decadal-scale changes in the ocean biosphere. Passive ocean colour sensors observe only the first (top) optical depth of the ocean (40-60 m in the open ocean to less than 1 m in turbid coastal waters). However, when coupled with in situ observations and numerical models, these space-based observations provide a three-dimensional understanding of ocean processes, their complexity, and their interactions with other parts of the Earth system. Therein, enhanced in situ sampling of ocean colour and ecosystem variables is a requirement and a complement to satellite-based data.

Societal Drivers	Scientific Questions	Biogeochemical Phenomena to Capture	EOVs
The role of ocean biogeochemistry in climate	How is the ocean carbon content changing?	Ventilation (water mass age)	Transient tracers, oxygen, stable carbon isotopes
		Air-sea fluxes	Oxygen, inorganic carbon, N_2O, nutrients
		Anthropogenic carbon sequestration	Inorganic carbon, transient tracers
		Organic matter cycling	Oxygen, inorganic carbon, nutrients, suspended particulates, DOC, transient tracers
		Cross-shelf interactions	Oxygen, nutrients, inorganic carbon, suspended particulates, DOC
	How does the ocean influence cycles of non-CO_2 greenhouse gases?	Air-sea fluxes	N_2O, oxygen
Human impacts on ocean biogeochemistry	How large are the ocean's "dead zones" and how fast are they changing?	Hypoxia	Oxygen
	What are the rates and impacts of ocean acidification?	Ocean acidity	Inorganic carbon
Ocean ecosystem health	Is the biomass of the ocean changing?	Organic matter cycling	Oxygen, nutrients, inorganic carbon, Suspended particulates, DOC
		Inorganic nutrient cycling	Nutrients
	How does eutrophication and pollution impact ocean productivity and water quality?	Eutrophication	Oxygen, nutrients, suspended particulates, DOC
		Hypoxia	Oxygen
		Contamination/pollution	NA

Table 6.1. Links between societal drivers, scientific questions, and corresponding identified biogeochemical EOVs.

Table 6.1 summarizes the links between societal needs (drivers), scientific questions, and corresponding identified biogeochemical EOVs. This table indicates the central role of phenomena in the process of implementing the GOOS and provides a synthetic view of why and what needs to be observed in terms of marine biogeochemistry. The following two sections focus on the question: How?

Era of Biogeochemical Observations – New Frontier for Operational Oceanography

Current and expanding capacity for biogeochemical observations

All observing elements (platforms) organized into observing networks are characterized by time and space sampling, which are ultimately set by the technology. By combining the observing elements in the integrated GOOS, the sampling limitations of the individual observing elements can be minimized and a more comprehensive time and space sampling can be achieved; thus, addressing the observing objectives in a more complete way. The combination of the observing elements in a system is closely related to the capacities of an integrated GOOS. Despite the integration, unresolved time, space, and parameters remain – a fact that is closely related to the gaps of the system and which might be overcome by technological advancement and engineering. The maturity of the observing elements, particularly that of sensors, can be categorized in different levels of technical readiness. The Framework of Ocean Observations categorizes nine levels in three categories: requirement setting, observing capacity and data and information products delivery (Figure 6.4 below).

Biogeochemical EOV specification sheets indicate technical readiness levels for each category specifically for each observing element. Overall, a global ocean observing system for biogeochemistry based on the major networks described below has reached a satisfactory technical readiness level to address the current scientific questions globally. That might not be the case regionally and several processes remain unresolved due to technological shortcomings.

<u>Ship-based repeat hydrography</u>
Despite numerous technological advances over the last several decades, ship-based hydrography remains the only method for obtaining high-quality, high spatial and vertical resolution measurements of a suite of physical, chemical, and biological parameters over the full water column. Ship-based hydrography is essential for documenting ocean changes throughout the water column, especially for the deep ocean below 2 km.

The Global Ocean Ship-Based Hydrographic Investigations Program (GO-SHIP; http://www.go-ship.org/) helps develop a globally-coordinated network of sustained hydrographic sections (i.e., repeat hydrography) as part of the global ocean/climate observing system, providing information on physical oceanography, the carbon cycle, marine biogeochemistry, and ecosystems. GO-SHIP provides approximately decadal resolution of the changes in inventories of biogeochemical variables (such as carbon, oxygen, nutrients, and transient tracers), covering the Atlantic and other ocean basins from coast to coast and full depth (top to bottom), with measurements of the highest required accuracy to detect these changes.

The ship-based hydrography observing network is critical to addressing questions of how the ocean will respond to an increase in dissolved inorganic carbon, a decrease in pH, and changes in ventilation strength processes. Repeated decadal since the 1970s, these observations are a crucial resource for documenting baselines and patterns of long-term variability in many of the Atlantic

Ocean biogeochemical phenomena considered: ocean ventilation, anthropogenic carbon sequestration, organic matter cycling, hypoxia, and ocean acidity, both regionally and on basin-scale. GO-SHIP data also provide reference data to calibrate autonomous platform sensors that cannot be recovered, and cruises provide a stage for the deployment of many autonomous platforms as well.

Considering only the status of decadal full GO-SHIP lines, this observing network has lines distributed every approximately 20°, providing along-section sampling every 30 nautical miles. Such a sampling design is adequate for studying interannual to decadal variability in basin-scale signals for many of the key biogeochemical phenomena discussed above. Considering the availability of GO-SHIP data since the 1970s, these records are also long enough to detect climate-driven trends in key biogeochemical EOVs.

FRAMEWORK PROCESSES BY READINESS LEVELS

Readiness Levels	Requirements Processes	Coordination of Observational Elements	Data Management & Information Products
Mature			
Level 9 "Sustained"	Essential Ocean Variable: • Adequate sampling specifications • Quality specifications	System in Place: • Globally • Sustained indefinitely • Periodic review	Information Products Routinely Available: • Product generation standardized • User groups routinely consulted
Level 8 "Mission qualified"	Requirements "Mission Qualified:" • Longevity/stability • Fully scalable	System "Mission Qualified:" • Regional implementation • Fully scalable • Available specifications and documentation	Data Availability: • Globally available • Evaluation of utility
Level 7 "Fitness for purpose"	Validation of Requirements: • Consensus on observation impact • Satisfaction of multiple user needs • Ongoing international community support	Fitness-for-Purpose of Observation: • Full-range of operational environments • Meet quality specifications • Peer review certified	Validation of Data Policy • Management • Distribution
Pilot			
Level 6 "Operational"	Requirement Refined: • Operational environment • Platform and sensor constraints	Implementation Plans Developed: • Maintenance schedule • Servicing logistics	Demonstrate: • System-wide availability • System-wide use • Interoperability
Level 5 "Verification"	Sampling Strategy Verified: • Spatial • Temporal	Establish: • International commitments and governance • Define standardized components	Verify and Validate Management Practices: • Draft data policy • Archival plan
Level 4 "Trial"	Measurement Strategy Verified at Sea	Pilot project in an operational environment	Agree to Management Practices: • Quality control • Quality assurance • Calibration • Provenance
Concept			
Level 3 "Proof of concept"	Proof of Concept via Feasibility Study: • Measurement strategy • Technology	Proof of Concept Validated: • Technical review • Concept of operations • Scalability (ocean basin)	Verification of Data Model with Actual Observational Unit
Level 2 "Documentation"	Measurement Strategy Described • Sensors • Sensitivity • Dependencies	Proof of Concept: • Technical capability • Feasibility testing • Documentation • Preliminary design	Socialization of Data Model • Interoperability strategy • Expert review
Level 1 "Idea"	Environment Information Need and Characteristics Identified: • Physical • Chemical • Biological	System Formulation: • Sensors • Platforms • Candidate technologies • Innovative approaches	Specify Data Model • Entities, Standards • Delivery latency • Processing flow

Figure 6.4. A detailed description of varying readiness levels defined by the Framework for Ocean Observing. Source: FOO, 2012.

Ship-based underway observations

The Joint Technical Commission for Oceanography and Marine Meteorology Ship Of Opportunity Programme (SOOP) makes use of volunteer merchant ships that routinely transit strategic shipping routes. A number of biogeochemical EOV measurements depends on the SOOP network coverage. So-called 'underway' measurements of pCO_2 in surface sea water and in the air are made routinely by the SOOP network with high accuracies achieved. The SOOP pCO_2 data, potentially supplemented in the near future with underway measurements of pH, dissolved inorganic carbon, or total alkalinity, will be vital in describing basin-wide changes in the carbonate system, thereby improving seasonal and inter-annual climate predictions, and better constraining annually updated calculations of the global carbon budget.

Underway surface pCO_2 observations provide the capacity to constrain the air-sea CO_2 fluxes in key regions of the global ocean, on timescales from monthly to decadal. Combined with ocean interior observations collected on GO-SHIP lines, these measurements enable well-constrained estimates of carbon storage in the ocean.

The biggest limitation of using the underway ship-based observations from volunteer/commercial ships is the fact that we cannot directly alter the sampling scheme. Therefore, any spatial gaps (e.g., the South Pacific) or seasonal biases (e.g., boreal winter measurements in deep water formation regions in the North Atlantic) cannot be alleviated from the level of observing system design, but instead require installing a new pCO_2 line on an existing commercial vessel.

Fixed-point observatories

Single-point time series stations have increased understanding of the patterns of temporal variability but, by their nature, remain limited in the spatial domain. Henson et al. (2016) attempted to analyse the spatial footprint of moored fixed-point observatories with biogeochemical EOVs measured on them. They concluded that depending on the variable of interest, these footprints account for only 10-15% of the global ocean.

The fact that existing time series stations are representative of relatively large surface areas of the ocean confirms their role in estimating sub-basin scale patterns of variability. However, it is clear that much of the Atlantic, as well as the global ocean, remains under-sampled. There is a need to build and maintain a basin-wide and global network of multi-disciplinary, fixed-point surface and subsurface time series using mooring, ship and other fixed instruments, and to establish a coordinated network of ship-based multidisciplinary time series that is geographically representative.

Implementing this target will depend on the development of units attempting to coordinate fixed-point observatories. The Fixed-point Open Ocean Observatory (FixO3) network seeks to integrate European open-ocean, fixed-point observatories and to improve access to these key installations for the broader community. A similar mission, but on the global scale is adopted by OceanSITES, the goal of which is to collect, deliver, and promote the use of high-quality data from long-term, high-frequency observations at fixed locations in the open ocean. OceanSITES typically aims to collect multidisciplinary data worldwide from the full-depth water column as well as the overlying atmosphere.

Another key aspect of enhancing the fixed-point observatory capacity is the need to provide accurate information with which moorings can actually measure any of the biogeochemical EOV and, in the long-run, expand on the number of biogeochemical measurements performed routinely at time series stations.

In terms of the key geographic regions samples by fixed-point observatories, there is a gap in such measurements in places of deep water formation. Although resolving sub-decadal variability in basin-scale ventilation is not a primary goal for the observing system, weekly-to-seasonal variability in air-sea fluxes, inorganic nutrient cycling, and organic matter cycling are all crucial to understanding how the ocean carbon content and biomass are changing. While there transport mooring arrays exist in these key locations, there is a lack of coincident biogeochemical observations.

Profiling floats

Although Argo profiling floats can be considered a mature observing approach as far as technological readiness level is concerned, the newly-formed global biogeochemical Argo array as a coordinated observing network remains in pilot stage. There is currently a limited number of floats with biogeochemical sensors deployed.

Biogeochemical-Argo is set to enable direct observation of the seasonal- to decadal-scale variability in net community production, the supply of essential inorganic nutrients transported from deep waters to the sunlit surface layer, ocean acidification, hypoxia, and ocean uptake of CO_2. Bio-optical sensors would supplement ocean colour satellite observations by providing measurements of chlorophyll, light, and light scattering deep into the ocean interior throughout the year, in cloud- and ice-covered areas, or during the dark of polar winter.

The regional profiling float arrays equipped with biogeochemical sensors provide a sampling of ocean conditions around the world designed to produce an integrated dataset that can be used to address questions related to physical-biogeochemical coupling in eddies, phytoplankton phenology (cyclic and seasonal phenomena), nutrient supply, and climate effects on ocean carbon cycling in selected regions. Some of these arrays include:

- the Southern Ocean Carbon and Climate Observations and Modelling (SOCCOM) project,
- the Remotely Sensed Biogeochemical Cycles in the Ocean (remOcean) project in the North Atlantic subpolar gyre,
- the Novel Argo Ocean Observing System in the Mediterranean Sea (NAOS),
- the Integrated Physical-Biogeochemical Ocean Observation Experiment (INBOX) in the Kuroshio region of the North Pacific, and
- the Australia-India Joint Indian Ocean Bio-Argo Project (IO Bio-ARGO).

When setting targets for and optimizing the Biogeochemical-Argo array of floats, there is an inherent trade-off between meeting the requirements for observing relevant phenomena on adequate spatial and temporal scales, and the cost of maintaining a sustained observing network. Currently, the set target of deploying 1,000 biogeochemical floats is based on the results of a series of observing system simulation experiments performed by the SOCCOM team. That experiment revealed that the biggest gain, in terms of reducing the air-sea CO_2 flux reconstruction error, is between 500 and 1,000 randomly distributed floats. Hence, the target is currently 1000 floats (Johnson and Claustre, 2016)

The regional profiling float observation programs are also building the expertise needed to operate a global network that interacts with other components of the GOOS, including satellites (IOCCG, 2011), shipboard programs such as GO-SHIP, and various time series stations.

Currently, the size of the Biogeochemical-Argo array is insufficient to resolve many of the phenomena on basin-scale. Until a denser network is developed, they should be viewed as providing high spatial and temporal data on local to regional scales (1–1000 km), which are complementary to the basin-scale, decadal-scale ship-based repeat hydrography observations.

Gliders

Underwater gliders have enhanced capabilities, when compared with profiling floats. They provide some level of manoeuvrability and, hence, position control. The gliders perform sawtooth trajectories from the surface to depths of 1000-1500 m, along reprogrammable routes (using two-way communication via satellite), and they can be operational for several months. Their role in the integrated observing system is to fill the gaps left by other observing platforms. The mission of the EGO (Everyone's Gliding Observatories; http://www.egonetwork.org/) underwater glider network is to develop a new observational capacity for process studies and operational monitoring of the ocean physics and biogeochemistry with gliders, and thereby go beyond the marine sciences frontiers. In particular, gliders could be deployed to sample most of the western and eastern boundary circulations and the regional seas, which are not well-covered by the present ocean observing system, as well as in the vicinity of fixed-point time series stations. Gliders can operate at higher resolution than the ca. 300 km/10-day float in the Argo profiling float network, and the even sparser ship-based observations. Therefore, glider-based observations have great potential to address regional and coastal issues, which are so important for societal applications.

Remote sensing observations

The space-based observing system is an important component of the GOOS. An array of geostationary and polar-orbiting satellites operated by the National Aeronautics and Space Administration (NASA) and the European Space Agency (ESA) sample the Atlantic and global surface ocean on unprecedented spatial and temporal scales, weaving together the requirements for observing surface signatures of key biogeochemical phenomena on short- and long timescales, providing basin-wide coverage with a simultaneous high spatial resolution on the order of kilometers.

Many products are derived from remote sensing observations that provide often unique information on a number of sub-variables listed under the particulate matter and dissolved organic carbon biogeochemical EOVs. Observations provided by the Ocean Colour Radiometry Virtual Constellation, and recently also by the LIDAR (light detection and ranging) method, enable near real-time monitoring of phenomena such as organic matter cycling and eutrophication. On the other hand, the focus of remote sensing observations is often on the physical and biological phenomena.

There are often very few algorithms available to estimate biogeochemical properties of interest, e.g., DOC or particulate inorganic carbon. Promoting new and alternative algorithms should be a goal that will lead to decreasing the high uncertainty of satellite-based measurements of biogeochemical variables.

Remote sensing observations are essential to studying surface processes related to organic matter cycling in key regions of the global ocean such as boundary currents and upwelling regions. Although much more challenging and associated with very large uncertainties, coastal satellite observations also provide information about changing carbon content on the continental shelf and in marginal seas regions.

The table below (Table 6.2) summarizes the links between the major observing elements and the biogeochemical EOVs that they measure (green indicates autonomous, blue indicates ship-based, and orange indicates remote sensing). Spatial and temporal scales captured by the observing elements measuring individual biogeochemical EOVs are described in the respective biogeochemical EOV specification sheets

Observing element	EOVs (sub-variables measured)
Profiling floats	Oxygen, nutrients (NO_3), inorganic carbon (pH), particulate matter
Gliders	Oxygen, nutrients (NO_3), inorganic carbon (pCO_2, pH), particulate matter
Moorings	Oxygen, nutrients (NO_3), inorganic carbon (pCO_2, pH), particulate matter
Drifting buoys	Nutrients (NO_3), inorganic carbon (pCO_2, pH)
Sediment traps	Particulate matter
Ship-based hydrography (including repeat hydrography)	Oxygen, nutrients (NO_3, PO_4, Si), inorganic carbon, transient tracers, particulate matter, N_2O, stable carbon isotopes, dissolved inorganic carbon
Ship-based time series^	Oxygen, nutrients (NO_3, PO_4, Si), inorganic carbon, N_2O, stable carbon isotopes, dissolved inorganic carbon
Ship-of-Opportunity	Oxygen, nutrients (NO_3, PO_4, Si), inorganic carbon (pCO_2, dissolved inorganic carbon*, pH*), particulate matter, N_2O*, stable carbon isotopes
Satellites	Particulate matter

Table 6.2. Links between major observing elements and biogeochemical EOVs measured by these elements.

Data quality and availability

There has been a strong, long-standing effort among the marine biogeochemistry observing and modelling community to make biogeochemistry EOV data not only freely available, but also quality-controlled and inter-comparable. These grassroots efforts eventually led to the successful creation of two information products: Surface Ocean CO_2 Atlas (SOCAT; Bakker et al., 2016) and GLODAP (Lauvset et al., 2016; Olsen et al., 2016). However, the data and synthesis products handled by SOCAT and GLODAP are predominantly carbon-focused and represent almost exclusively ship-based observations. There is an urgent need to expand biogeochemical data availability, quality control and inter-comparability beyond carbon parameters and onto a wider suite of available observing elements.

Vast number of metadata queried in, for example, European Union databases (e.g., SeaDataNet) is in fact under restricted access. Any access restriction, even as minimal as registration on the website through which data is to be acquired, prevents such data from being considered open-access and inevitably hinders data sharing. One example is the availability of in situ chlorophyll-*a* (Chl-*a*) and other pigment data from the British Atlantic Meridional Transect (AMT) cruises. These data, being of fundamental importance to calibration and validation of remote sensing-derived ocean product algorithms (for Chl-*a*, plankton size classes, plankton functional types, etc.), appear in SeaDataNet under restricted access. However, the same data is freely available through the US-based source, NOMAD: the NASA bio-Optical Marine Algorithm Dataset, available for download from: https://seabass.gsfc.nasa.gov/wiki/NOMAD.

N_2O is an example of an EOV observation for which a comprehensive, global database, MEMENTO (MarinE MethanE and NiTrous Oxide; https://memento.geomar.de/home) exists. While MEMENTO data is freely usable, access to the database is restricted (i.e., granted upon request via email). Although this could be considered a 'light' restriction, it is responsible for a gap in the observing system from the perspective of open access to information products that directly answer one of the key biogeochemical scientific questions: 'How does the ocean influence cycles of non-CO_2 greenhouse gases?'.

Future Prospects

The main focus for enhancement of the current biogeochemistry observing system is on sensor technology development. The sensor suite that is now available and tested throughout the array of autonomous and moored platforms (oxygen, pH, nitrate, chlorophyll fluorescence, and backscattering and downwelling irradiance sensors) is hardly sufficient to address the needs and questions driving the requirements for observations described above. These biogeochemical sensors are a relatively recent development and reflect the rapid expansion of technological capabilities that has resulted from the rapid development of electronics and optics over the past decade. It is likely that yet-to-be-developed sensors will enable significant extensions to the current capabilities. It is also possible that the existing sensors may be improved significantly. This has already happened for oxygen, for example. Early biogeochemical floats and gliders used oxygen sensors based on Clark-type oxygen electrodes (Edwards et al., 2010; Riser and Johnson, 2008). These have been replaced with optical sensors based on fluorescence lifetimes, which have improved stability and a capability for calibration in air (Körtzinger et al., 2005). New sensors that might be available in the future include pCO_2 sensors. The current state of optode-based pCO_2 sensors (Atamanchuk et al.,

2015) has not yet reached deployment-readiness. However, the pCO$_2$ sensor is expected to enable measurements of a prospective second carbon system parameter that may mature quickly and become an alternative to the pH sensor, allowing for direct observation of the CO$_2$ saturation state at the sea surface. This sensor, when installed on autonomous platforms, would also link directly with the pCO$_2$ measurements provided by the global ship-based networks, adding the much-needed vertical dimension. Other examples might include the development of particulate inorganic carbon sensors (Guay and Bishop, 2002) or fast repetition rate fluorometers. The highest accuracy pH measurements are generally made by spectrophotometry using well-characterized indicator dyes (Clayton and Byrne, 1993). Spectrophotometric pH profiles have been measured in situ (Liu et al., 2006) and such systems may become alternate approaches for pH determination if they are proven to have the appropriate performance needed for long-term deployments. Also, electrochemical sensors for N$_2$O, phosphate, oxygen, and silicate, as well as new optodes for pH, CO$_2$, O$_2$ and ammonia have been developed and are being tested. As new sensors are proven robust and effective, they may be considered for addition to the system based on performance, cost, and scientific merit.

In addition to new sensors, the performance of existing sensors may also be extended. One example might be application of the ultraviolet nitrate sensor to also observe dissolved nitrite in OMZ regions. The nitrite ion has a ultraviolet spectrum that is moderately distinct from that of nitrate (Johnson and Coletti, 2002) and quantification of the higher nitrite concentrations that result from denitrification in OMZ regions may be feasible. Such a capability would greatly add to interpretation of nitrate loss processes in OMZ regions.

Teams across the globe continue to push the sensors along the technology readiness levels, seeking to make their sensors commercially available as soon as possible. Issues such as power and resource management, internal and external communication, and expanding from single-parameter sensors to multi-parameter sensors are just a few of the myriad of challenges that need to be addressed.

Conclusion

Marine biogeochemistry observations are well-justified, well-designed and well-implemented, more so than ever before. Our understanding of the inter-connectedness of ocean processes requires a multidisciplinary and multi-parameter approach, whether we focus on operational oceanography or climate variability. There is still a large space for improvements in delivering integrated information products across disciplines, platforms, and parameters for the benefit of the informed ocean management. Potential benefits cannot be overstated and the efforts to operationalize all ocean observations, no matter how demanding they might initially seem, should not be neglected.

Acknowledgments

MT was funded via a US National Science Foundation grant (OCE-124 3377) to the Scientific Committee on Oceanic Research and AP was funded by the AtlantOS Project, the European Union's Horizon 2020 research and innovation programme under grant agreement n° 633211.

References

Biogeochemical-Argo Planning Group. 2016. "The scientific rationale, design and Implementation Plan for a Biogeochemical-Argo float array." doi:10.13155/46601.

Church, John A., J. Stuart Godfrey, David R. Jackett, and Trevor J. McDougall. 1991. "A Model of Sea Level Rise Caused by Ocean Thermal Expansion." *Journal of Climate* 4: 438-456, doi:10.1175/1520-0442(1991)004<0438:AMOSLR>2.0.CO;2.

Clayton, Tonya D., and Robert H. Byrne. 1993. "Spectrophotometric seawater pH measurements: total hydrogen ion concentration scale calibration of m-cresol purple and at-sea results." *Deep Sea Research Part I:* Oceanographic Research Papers 40: 2115-2129, doi:10.1016/0967-0637(93)90048-8.

Cossa, Daniel, Bernard Averty, and Nicola Pirrone. 2009. "The origin of methylmercury in open Mediterranean waters." *Limnology and Oceanography* 54: 837-844.

Devol, Allan H. 2003. "Nitrogen cycle: solution to a marine mystery." *Nature* 422: 575-576.

Edwards, D. Murphy, C. Janzen, and N. Larson. 2010. "Calibration, Response, and Hysteresis in Deep-Sea Dissolved Oxygen Measurements." *Journal of Atmospheric and Oceanic Technology* 27: 920-931, doi:10.1175/2009jtecho693.1.

FOO. 2012. "A Framework for Ocean Observing. By the Task Team for an Integrated Framework for Sustained Ocean Observing" (Eric Lindstrom, John Gunn, Albert Fischer, Andrea McCurdy, L.K. Glover), doi:10.5270/OceanObs09-FOO.

Friis, K., Arne Körtzinger, J. Pätsch, and Douglas W. R. Wallace. 2005. "On the temporal increase of anthropogenic CO_2 in the subpolar North Atlantic." *Deep Sea Research Part I*: Oceanographic Research Papers 52: 681-698.

Fung, Inez Y., Scott C. Doney, Keith Lindsay, and Jasmin John. 2005. "Evolution of carbon sinks in a changing climate." *Proceedings of the National Academy of Sciences of the United States of America* 102: 11201-11206.

GOOS. 2018. "Global Ocean Observing System (GOOS) website."

Gruber, Nicolas, Manuel Gloor, Song-Miao Fan, and Jorge L. Sarmiento. 2001. "Air-sea flux of oxygen estimated from bulk data: Implications for the marine and atmospheric oxygen cycles." *Global Biogeochemical Cycles* 15: 783-803.

Guay, Christopher K. H., and James K. B. Bishop. 2002. "A rapid birefringence method for measuring suspended $CaCO_3$ concentrations in seawater." *Deep Sea Research Part I:* Oceanographic Research Papers 49: 197-210.

Hansell, Dennis A., Craig A. Carlson, Daniel J. Repeta, and Reiner Schlitzer. 2009. "Dissolved organic matter in the ocean: A controversy stimulates new insights." *Oceanography* 22.

IOC-UNESCO, and UNEP. 2016. "Open Ocean: Status and Trends, Summary for Policy Makers."

IPCC. 2007. Climate Change 2007: The Physical Science Basis. Contribution of Working Group I to the Fifth Assessment Report of the Intergovernmental Panel on Climate Change. Edited by D. Qin M. Manning Z. Chen M. Marquis K. B. Averyt M. Tignor Solomon and H. L. Miller. Cambridge University Press, Cambridge, United Kingdom and New York, NY, USA, 996 pp, doi:10.1017/CBO9781107415324.

—. 2013. Climate Change 2013: The Physical Science Basis. Contribution of Working Group I to the Fifth Assessment Report of the Intergovernmental Panel on Climate Change. Edited by D. Qin G.-K. Plattner M. Tignor S. K. Allen J. Boschung A. Nauels Y. Xia V. Bex Stocker and P. M. Midgley. Cambridge University Press, Cambridge, United Kingdom and New York, NY, USA, 1535 pp.

Jiang, Li-Qing, Richard A. Feely, Brendan R. Carter, Dana J. Greeley, Dwight K. Gledhill, and Krisa M. Arzayus. 2015. "Climatological distribution of aragonite saturation state in the global oceans." *Global Biogeochemical Cycles* 29: 1656-1673.

Jickells, Tim, and C. Mark Moore. 2015. "The importance of atmospheric deposition for ocean productivity." *Annual Review of Ecology, Evolution, and Systematics* 46: 481-501.

Johnson, Kenneth S., and Luke J. Coletti. 2002. "In situ ultraviolet spectrophotometry for high resolution and long-term monitoring of nitrate, bromide and bisulfide in the ocean." *Deep Sea Research Part I*: Oceanographic Research Papers 49: 1291-1305, doi:10.1016/S0967-0637(02)00020-1.

Joos, Fortunat, Gian-Kasper Plattner, Thomas F. Stocker, Arne Körtzinger, and Douglas W. R. Wallace. 2003. "Trends in marine dissolved oxygen: Implications for ocean circulation changes and the carbon budget." *Eos, Transactions American Geophysical Union* 84: 197-201.

Körtzinger, Arne, Jens Schimanski, and Uwe Send. 2005. "High quality oxygen measurements from profiling floats: A promising new technique." *Journal of Atmospheric and Oceanic Technology* 22: 302-308.

Keeling, Ralph F., Arne Körtzinger, and Nicolas Gruber. 2010. "Ocean deoxygenation in a warming world." *Annual Review of Marine Science* 2: 199-229.

Lauvset, Siv K., Robert M. Key, and Fiz F. Perez. 2016. "A new global interior ocean mapped climatology: the 1°× 1° GLODAP version 2." *Earth System Science Data* 8: 325.

Le Quéré, Corinne, Christian Rödenbeck, Erik T. Buitenhuis, Thomas J. Conway, Ray Langenfelds, Antony Gomez, Casper Labuschagne, et al. 2007. "Saturation of the Southern Ocean CO_2 sink due to recent climate change." *Science* 316: 1735-1738.

Locarnini, R. A., A. V. Mishonov, J. I. Antonov, T. P. Boyer, H. E. Garcia, O. K. Baranova, M. M. Zweng, and D. R. Johnson. 2006. "World Ocean Atlas 2005, Vol. 1: Temperature."

Olsen, A., R. M. Key, S. Heuven, S. K. Lauvset, A. Velo, X. Lin, C. Schirnick, et al. 2016. "The Global Ocean Data Analysis Project version 2 (GLODAPv2) -- an internally consistent data product for the world ocean." *Earth System Science Data* 8: 297-323, doi:10.5194/essd-8-297-2016.

Raven, J. A., and P. G. Falkowski. 1999. "Oceanic sinks for atmospheric CO_2." *Plant, Cell & Environment* 22: 741-755.

Stendardo, I., and N. Gruber. 2012. "Oxygen trends over five decades in the North Atlantic." *Journal of Geophysical Research: Oceans* 117: 2156-2202, doi:10.1029/2012JC007909.

Talley, L. D., R. A. Feely, B. M. Sloyan, R. Wanninkhof, M. O. Baringer, J. L. Bullister, C. A. Carlson, et al. 2016. "Changes in Ocean Heat, Carbon Content, and Ventilation: A Review of the First Decade of GO-SHIP Global Repeat Hydrography." *Annual Review of Marine Science* 8: 185-215, doi:10.1146/annurev-marine-052915-100829.

Trimmer, Mark, and Joanna Claire Nicholls. 2009. "Production of nitrogen gas via anammox and denitrification in intact sediment cores along a continental shelf to slope transect in the North Atlantic." *Limnology and Oceanography* 54: 577-589.

Zweng, M. M., J. I. Antonov, J.R. Reagan, A. V. Mishonov, R.A. Locarnini, H. E. Garcia, T.P. Boyer, D. R. Johnson, O.K. Baranova, and M. M. Biddle. 2006. "World Ocean Atlas 2013, Volume 2: Salinity."

CHAPTER 7

Satellites and Operational Oceanography

Pierre-Yves Le Traon[1,2]

[1]*Mercator Océan, Parc Technologique du Canal, Ramonville-Saint-Agne, France;* [2]*Ifremer, Plouzané, France*

The chapter starts with an overview of satellite oceanography, its role and use for operational oceanography. Main principles of satellite oceanography techniques are summarized. We then describe key techniques of radar altimetry, sea surface temperature, and ocean colour remote sensing. This includes measurement principles, data processing issues, and the use of data for operational oceanography. Synthetic aperture radar, scatterometry, sea ice and sea surface salinity measurements are also briefly described. Techniques used to assess the impact of present and future satellite observations for ocean analysis and forecasting are reviewed. We also discuss future requirements for satellite observations. Main prospects are given in the conclusion.

Introduction

There are very strong links between satellite oceanography and operational oceanography. The development of operational oceanography has been mainly driven by the development of satellite oceanography capabilities. The ability to observe the global ocean in near real-time at high space and time resolution is indeed a prerequisite for the development of global operational oceanography and its applications. The first ocean parameter to be globally monitored from space was sea surface temperature by meteorological satellites in the late 1970s. It was, however, the advent of satellite altimetry in the late 1980s that led to the development of ocean data assimilation and global operational oceanography. In addition to providing all kinds of weather observations, sea level from satellite altimetry is an integral of the ocean interior and provides a strong constraint on the 4D ocean state estimation. The satellite altimetry community was also keen to develop further the use of altimetry, and this required an integrated approach merging satellite and in situ observations with models. Thus, the GODAE demonstration was phased to coincide with the Jason-1 and ENVISAT altimeter missions (Smith and Lefebvre, 1997). Satellite oceanography is now a major component of operational oceanography. Data are usually assimilated in ocean models but they can also be used directly for applications.

An overview of satellite oceanography will be provided here, focusing on the most relevant issues for operational oceanography. The chapter is organized as follows. First is an overview of satellite oceanography, its role and use for operational oceanography. Main operational oceanography requirements are summarized. The complementary role of in situ observations is also emphasized. Next, main principles of satellite oceanography and general data processing issues are described. We then detail key techniques of radar altimetry, sea surface temperature, and ocean

---

Le Traon, P.-Y., 2018: Satellites and operational oceanography. In "*New Frontiers in Operational Oceanography*", E. Chassignet, A. Pascual, J. Tintoré, and J. Verron, Eds., GODAE OceanView, 161-190, doi:10.17125/gov2018.ch07.

colour remote sensing. This includes measurement principles, data processing issues, and the use of these data for operational oceanography. Synthetic aperture radar (SAR), scatterometry, sea ice and sea surface salinity measurements are briefly described next followed by a review of tools used to quantify the impact of present and future satellite observations for ocean analysis and forecasting. Finally, future challenges and requirements for satellite observations are discussed. Main prospects are given in the chapter conclusion.

Role of Satellites in Operational Oceanography

The global ocean observing system and operational oceanography

Operational oceanography critically depends on the near real-time availability of high quality in situ and remote sensing data with sufficiently dense space and time sampling. The quantity, quality, and availability of data sets directly impact the quality of ocean analyses and forecasts and associated services. Observations are required to constrain ocean models through data assimilation and also to validate them. Products derived from the data themselves can also be directly used for applications.

This requires an adequate and sustained global ocean observing system. Climate and operational oceanography applications share the same backbone system (i.e., GOOS, GCOS, and JCOMM). Operational oceanography has, however, specific requirements for availability of high space and time resolution measurements and for near real-time measurements.

The unique contribution of satellite observation

Satellites provide long-term, continuous, global, high space and time resolution data for key ocean parameters: sea level and ocean circulation, sea surface temperature, ocean colour, sea ice, waves, and winds. These are the observational core variables required to constrain global, regional and coastal ocean monitoring and forecasting systems. They are also needed to validate them. Only satellite measurements can, in particular, provide observations at high space and time resolution to partly resolve the mesoscale and coastal variability. Satellite data can also be directly used for applications (e.g., SAR for sea ice and oil pollution monitoring, ocean colour for water quality monitoring). Sea surface salinity is a new and important parameter that could be operationally monitored from space; the feasibility has been demonstrated with the European Space Agency's Soil Moisture and Ocean Salinity (SMOS) mission and the NASA/Comisión Nacional de Actividades Espaciales (CONAE) Aquarius mission.

Main requirements

Operational oceanography requirements have been presented in the GODAE strategic plan and by Le Traon et al. (2001). They have been further detailed in Clark and Wilson (2009), Drinkwater et al. (2010) and Le Traon (2011). Sea level, sea surface temperature (SST), surface geostrophic currents, ocean colour, sea surface salinity (SSS), waves, sea ice, and winds form the core operational satellite observations required for global, regional, and coastal ocean monitoring and forecasting systems. To deliver sustained, high resolution observations while meeting operational constraints such as near-real-time data distribution and redundancy in the event of satellite or instrument failure requires international cooperation and the development of virtual constellations as promoted by the Committee on Earth Observation Satellites (CEOS; e.g. Bonekamp et al., 2010).

Specific requirements for operational oceanography are as follows:

- In addition to meteorological satellites, a high precision (Advanced Along-Track Scanning Radiometer - AATSR-class) SST satellite mission, is needed to give the highest absolute SST accuracy. A microwave mission is also needed to provide an all-weather global coverage.
- At least three or four altimeters are required to observe the mesoscale circulation. This would also useful for significant wave height measurements. A long-term time series of a high accuracy altimeter system (Jason satellites) is needed to serve as a reference for the other missions and for the monitoring of climate signals.
- Ocean colour is increasingly more important, particularly in coastal areas. At least two satellites are required.
- Two scatterometers are required to globally monitor the wind field and sea ice at high spatial resolution.
- Two SAR satellites are required for waves, sea ice characteristics, and oil slick monitoring.

These minimum requirements have been only partly met over the past ten years (see Le Traon et al., 2015, for a recent review). Long-term continuity and transition from research to operational mode remains a major challenge.

Role of in situ data

Satellite observations need to be complemented by in situ observations. First, in situ data are needed to calibrate satellite observations. Most algorithms used to transform satellite observations (e.g., brightness temperatures) into geophysical quantities are partly based on in situ/satellite match up data bases. The in situ data are then used to validate satellite observations and to monitor the long-term stability of satellite observations. The stability of the different altimeter missions is, for example, commonly assessed by comparing the altimeter sea surface height measurements with those from tide gauges (Mitchum, 2000). Other examples include the validation of altimeter velocity products with drifter data (e.g., Bonjean and Lagerloef, 2002; Pascual et al., 2009), the systematic validation of satellite SST with in-situ SST from drifting buoys and the use of dedicated ship mounted radiometers to quantify the accuracy of satellite SST (Donlon et al, 2008). Comparison of in situ and satellite data can also provide indications of the quality of in situ data (e.g., Guinehut et al., 2009). The comparison of in situ and satellite data is also useful to check the consistency and information content between the different data sets (e.g., satellite sea level versus in situ dynamic height measurements) before they are assimilated in an ocean model (e.g., Guinehut et al., 2006).

Most importantly, in situ data are mandatory (and this is their main role) to complement satellite observations and to provide measurements of the ocean interior. Only the joint use of high resolution satellite data with precise (but sparsely available) in situ observations of the ocean interior has the potential to provide a high resolution description and forecast of the ocean state. The development of the Argo array of profiling floats and their integration with satellite altimetry and operational oceanography is an outstanding example of the value of an integrated ocean observing system (see discussion in Le Traon, 2013).

Data processing issues

Satellite data processing takes place in steps: level 0 and level 1 (from telemetry to calibrated sensor measurements), level 2 (from sensor measurements to geophysical variables), level 3 (space/time composites of level 2 data), and level 4 (merging of different sensors, data assimilation). Processing from level 0 to level 2 is generally carried out as part of the satellite ground segments.

Assembly of level 2 data from different sensors, intercalibration of level 2 products, and higher level data processing is usually done by specific data processing centers or thematic assembly centers. The role of data processing centers is to provide modelling and data assimilation centers with the real-time and delayed mode data sets required for validation and data assimilation. This includes uncertainty estimates, which are critical to an effective use of data in modelling and data assimilation systems. Links with ocean forecasting centers are needed, in particular, to organize feedback on: the quality control performed at the level of ocean forecasting centers (e.g., comparing an observation with a model forecast); the impact of data sets and data products in the assimilation systems; and new or future requirements.

High level data products (level 3 and 4) are also needed for applications (e.g., a merged altimeter surface current product for marine safety or offshore applications) and can be used to validate data assimilation systems (e.g., statistical versus dynamical interpolation) and complement products derived through modelling and data assimilation systems. It is important, however, to be fully aware of the limitations (e.g., mapping errors, limited effective space/time resolution) of high level satellite products (e.g., gridded sea surface temperature or sea level data sets) when using them.

Use of satellite data for assimilation into ocean models

The use of satellite data for assimilation into ocean models is discussed at length in other chapters of this book. Three important issues are emphasized in this chapter:

1. There can be large differences in data quality between real-time and delayed mode (reprocessed) data sets. Depending on applications, trade-offs between time delay and accuracy often need to be considered.
2. Error characterisation is mandatory for data assimilation, and a proper characterisation of error covariance can be quite complex for satellite observations. Data error variance (and covariance, but this is much more challenging) should always be tested and checked as part of the data assimilation systems.

In theory and for advanced assimilation schemes, it is much better to use raw data (level 2 or in some cases level 1 when the model can provide data needed for level 1 processing). The data error structure is generally more easily defined. The model and the assimilation scheme should also deliver a better high level processing (e.g., a model forecast should provide a better background than climatology or persistence). However, in practice this is not always true. Some high level data processing is often needed (e.g., correcting biases or large-scale errors, intercalibration of satellite missions) as it cannot be easily accomplished within complex data assimilation systems.

Overview of Satellite Oceanography Techniques

Passive/active techniques and choice of frequencies

There are two main types of satellite techniques to observe the ocean[1]. Passive techniques measure the natural radiation emitted from the sea or from reflected solar radiation. Active or radar techniques send a signal and measure the signal received from its reflection at the sea surface. In both cases, the propagation of the signal through the atmosphere and the emission from the atmosphere itself must be taken into account in order to extract the sea surface signal. The intensity and frequency distribution of the radiation that is emitted or reflected from the ocean surface allows the inference of its properties. The polarization of the radiation is also often used in microwave remote sensing.

Satellite systems operate at different frequencies depending on the signal to be derived. Visible (400 – 700 nm) and infrared (0.7 – 20 μm) frequencies are used for ocean colour and sea surface temperature measurements. Passive (radiometry) microwave systems (1 cm-30 cm) are used for sea surface temperature measurements in cloud situations, wind, sea ice and sea surface salinity retrievals. Radars operate in the microwave bands and provide measurements of sea surface height, wind speed and direction, wave spectra, sea ice cover, and types and surface roughness. Radar pulses are emitted obliquely (15 to 60°) (SAR, scatterometer) or vertically (altimetry).

The choice of frequencies is limited by other usages (e.g., radio, cellular phones, military and civilian radars, satellite communications). This is particularly important at microwave frequencies

---

[1] Gravimetry satellites (e.g. GRACE, GOCE), which measure the earth gravity field and its variations, are not included in these two categories.

in the range 1-10 GHz, which limits the frequencies used for earth remote sensing. The atmosphere also greatly affects the transmission of radiation between the ocean surface and the satellite sensors. The presence of fixed concentrations of atmospheric gases (e.g., O_2, CO_2, and O_3) and of water vapour means that only a limited number of windows exist in the visible, infrared and microwave for ocean remote sensing. Even at these frequencies, the propagation effects through the atmosphere (from the troposphere to the ionosphere) must be taken into account and corrected for. Clouds are a strong limitation of visible and infrared measurements.

There are also technological constraints for the choice of frequencies. The resolution of a given sensor is generally related to the ratio between the observed wavelength (λ) and the antenna diameter (D). For antenna diameters of a few meters, typical resolution at around 1 GHz (wavelength of 30 cm) is about 100 km, while at 30 GHz (wavelength of 1 cm) resolution is about 10 km. Radar altimeters use pulse limited techniques (which are much less sensitive to mispointing errors). Their footprint size is related to the pulse duration and is much smaller than that of a beam limited sensor. SAR uses the motion of the satellite to generate a very long antenna (e.g., 20 km for the ENVISAT Advanced Synthetic Aperture Radar (ASAR)) and thus to provide very high resolution measurements (up to a few meters).

Satellite orbits and measurement characteristics

Orbits for ocean satellites are geostationary, polar, or inclined. A geostationary orbit is one in which the satellite is always in the same position with respect to the rotating Earth. The satellite orbits at an elevation of approximately 36,000 km because that produces an orbital period equal to the period of rotation of the Earth. By orbiting at the same rate, in the same direction as Earth, the satellite appears stationary. Geostationary satellites provide a large field of view (up to 120°) at very high frequency, enabling coverage of weather events. Because of the high altitude, spatial resolution is on the order of a few kilometers, while it is 1 km or less for polar orbiting satellites. Because a geostationary orbit must be in the same plane as the Earth's rotation (i.e., the equatorial plane), it provides distorted images of the polar regions. Five or six geostationary meteorological satellites can provide a global coverage of the earth (for latitudes below 60°) (e.g., Martin, 2004).

Polar-orbiting satellites provide a more global view of Earth by passing from pole to pole, observing a different portion of the Earth with each orbit due to the Earth's own rotation. Orbiting at an altitude of 700 to 800 km, these satellites have an orbital period of approximately 90 minutes. They usually operate in a sun-synchronous orbit. At the same local solar time each day, the satellite passes the equator and any given latitude. Inclined orbits have an inclination between 0 degrees (equatorial orbit) and 90 degrees (polar orbit) and are used to observe tropical regions (e.g., Tropical Rainfall Measuring Mission (TRMM) Microwave Imager). High accuracy altimeter satellites such as TOPEX/Poseidon and Jason use higher altitude and non-synchronous orbits to reduce atmospheric drag and (mainly) to avoid aliasing of the main tidal signals.

The sampling pattern of a given satellite will be different depending on instrument types (along-track, imaging, or swath), frequencies, and antennas (see above). In addition, in the visible and infrared frequencies, cloud cover can strongly reduce the effective sampling.

Radiation laws and emissivity

Radiation from a blackbody

Planck's law describes the rate of energy emitted by a blackbody as a function of frequency or wavelength. A blackbody absorbs all the radiation it receives and emits radiation at a maximum rate for its given temperature. Planck's law gives the intensity of radiation L_λ emitted by unit surface area into a fixed direction (solid angle) from the blackbody as a function of wavelength (or frequency). The law can be expressed through the following equation:

$$L_\lambda = 2hc^2 / \lambda^5 \, [exp\,(hc/\lambda kT)-1]$$

where T is the temperature, c the speed of light ($2.99\ 10^{-8}$ m s^{-1}), h the Planck's constant (6.63×10^{-34} J s), k the Boltzmann's constant ($1.38\ 10^{-23}$ J °K^{-1}), and L_λ the spectral radiance per unit of wavelength and solid angle in W m^{-3} sr^{-1}.

Planck's law gives a distribution that peaks at a certain wavelength; the peak shifts to shorter wavelengths for higher temperatures. Wien's law gives the wavelength of the peak of the radiation distribution ($\lambda_{max} = 3\ 10^7/T$), while the Stefan-Boltzmann law gives the total energy E being emitted at all wavelengths by the blackbody ($E = \sigma T^4$). Thus, Wien's law explains the shift of the peak to shorter wavelengths as the temperature increases, while the Stefan-Boltzmann law explains the growth in the height of the curve as the temperature increases. Notice that this growth is very abrupt, since it varies as the fourth power of the temperature.

The Rayleigh-Jeans approximation ($L\lambda = 2kcT/\lambda^4$) holds for wavelengths much greater than the wavelength of the peak in the black body radiation form. This approximation is valid over the microwave band.

Graybodies and emissivity

Most bodies radiate less efficiently than a blackbody. The emissivity e is defined as the ratio of graybody radiance to the blackbody. It has a non-dimensional unit and its value is comprised between 0 and 1. The emissivity (e) generally depends on wavelength (λ) and polarization and has a directional dependence; it can be considered as a physical surface property and is a key quantity for ocean remote sensing. A graybody absorbs only part of the energy it receives and the remaining part is reflected and/or transmitted. The absorptivity is equal to the emissivity, as a surface in equilibrium must absorb and emit energy at the same rate (Kirchoff's law). Similarly, the reflectivity is equal to $1 - e$.

Retrieval of geophysical parameters for microwave radiometers

The brightness temperature (BT) is defined as BT = eT where T is the (physical) temperature. In the microwave band, it is proportional to the radiation L_λ. Brightness temperature is a measure of the intensity of emitted radiation. It is the physical temperature a blackbody would have to yield the same observed intensity of radiation emitted by a graybody.

The brightness temperature is an integrated measurement that includes all surface and atmosphere emitted power. Depending on frequency, it is more sensitive to a given parameter. Physical retrieval algorithms for geophysical parameters, such as the sea surface temperature, sea surface wind speed, sea ice or sea surface salinity are derived from a radiative transfer model, which computes the brightness temperatures that are measured by the satellite as a function of these variables. The radiative transfer model is based on a model for the sea surface emissivity and a model of microwave absorption in the Earth's atmosphere. The ocean sea surface emissivity (or reflectivity, see above) depends on the dielectric constant ε (which is a function of frequency, water temperature, and salinity), small-scale sea surface roughness, foam, as well as viewing geometry and polarization. The retrieval of a given parameter is possible through the inversion of a set of brightness temperatures measured at different frequencies and/or at different incidence angles. Inversion methods minimize the difference between measured and simulated (through a radiative transfer model) brightness temperatures. And given uncertainties in radiative transfer models, statistical or empirical inversions are also often used. These use a regression formalism (e.g., parametric, neural network) to find the best relation between brightness temperatures and the geophysical parameter to be retrieved.

Altimetry

Overview

Satellite altimetry is the most essential observing systems required for global operational oceanography (see Le Traon et al., 2017a for a recent review). It provides global, real-time, all-weather sea level measurements (sea surface height or SSH) with high space and time resolution. Sea level is directly related to ocean circulation through the geostrophic approximation. Sea level is also an integral of the ocean interior (density) and a strong constraint for inferring the 4D ocean circulation through data assimilation. Altimeters also measure significant wave height, which is essential for operational wave forecasting. High resolution from multiple altimeters is required to adequately represent ocean eddies and associated currents (aka the "ocean weather") in models. Only altimetry can constrain (through data assimilation) the 4D mesoscale circulation in ocean models that is required for most operational oceanography applications.

Measurements principles

An altimeter is active radar that sends a microwave pulse towards the ocean surface. A precise clock on board measures the return time of the pulse from which the distance or range (d) between the satellite and the sea surface is derived ($d=t/2c$). The range precision is within a few centimeters for a distance of 800 to 1300 km. The altimeter also measures the backscatter power related to surface roughness and wind and significant wave height.

An altimeter mission generally includes a bifrequency altimeter radar (usually in Ku and C or S Band) for ionospheric corrections, a microwave radiometer for water vapour correction, and a tracking system for precise orbit determination (Laser, GPS, Doris) that provides the orbit altitude relative to a given earth ellipsoid.

The main measurement for an altimeter radar is the SSH relative to a given earth ellipsoid. The SSH is derived as the difference between the orbit altitude and the range measurement. SSH precision depends on orbit and range errors. Altimeter range measurements are affected by a large number of errors (propagation effects in the troposphere and ionosphere, electromagnetic bias, errors due to inaccurate ocean and terrestrial tide models, inverse barometer effect, residual geoid errors). Some of these errors can be corrected with dedicated instrumentation (e.g. dual frequency altimeter, radiometer).

For a comprehensive description of altimeter measurement principles and measurement errors, the reader is referred to Chelton et al. (2001) and Escudier et al. (2017).

Geoid and repeat-track analysis

The altimeter missions provide along-track measurements every 7 km along repetitive tracks (e.g., every 10 days for the TOPEX/Poseidon and Jason series and 35 days for ERS, ENVISAT, and SARAL/Alti-Ka in its repeat-track phase). The distance between tracks is inversely proportional to the repeat time period (e.g., about 315 km at the equator for Jason and 90 km for ERS/ENVISAT/SARAL).

The sea surface height $SSH(x,t)$ measured by altimetry can be described by:

$$SSH(x,t) = N(x) + \eta(x,t) + \varepsilon(x,t)$$

where N is the geoid, η the dynamic topography and ε are measurement errors. The quantity of interest for the oceanographer is the dynamic topography (see next subsection). Geoids are not accurate enough to estimate globally the absolute dynamic topography η at all wavelengths.

The variable part of the dynamic topography η' ($\eta - <\eta>$) (or sea level anomaly, SLA) is, however, easily extracted using the so-called repeat track method. For a given track, η' is obtained by removing the mean profile over several cycles, which contains the geoid N and the mean dynamic topography $<\eta>$:

$$SLA(x,t) = SSH(x,t) - <SSH(x)>_t = \eta(x,t) - <\eta(x)>_t + \varepsilon'(x,t)$$

To get the absolute signal, a climatology or existing geoids must be used together with altimeter Mean Sea Surface (MSS), or both. A model mean can also be relied upon. Gravimetric missions (GRACE, GOCE) are now providing much more accurate geoids. Even with GOCE, however, repeat-track analysis is still needed because the small scales of geoid (below 50 to 100 km) will not be precisely known. GOCE is used with an altimetric MSS to derive $<\eta>_t$ that can then be added to η' (see next subsection).

High-level data processing issues and products

The SSALTO/DUACS system is the main multi-mission altimeter data center used today for operational oceanography. It aims to provide directly usable, high quality near real-time and delayed mode (for reanalyses and research users) altimeter products to the leading operational oceanography and climate centers in Europe and worldwide. The main processing steps are product homogenization, data editing, orbit error correction, reduction of long wavelength errors, production of along-track data, and maps of sea level anomalies. Major progress has been made with higher level processing issues such as orbit error reduction (e.g., Le Traon and Ogor, 1998), intercalibration, and merging of altimeter missions (e.g., Le Traon et al., 1998; Ducet et al., 2000; Pascual et al., 2006). The SSALTO/DUACS weekly production moved to daily production in 2007 to improve timeliness of data sets and products. A new real-time product was also developed for specific real-time mesoscale applications. A review of the SSALTO/DUACS processing is given in Dibarboure et al. (2011). Recent evolutions of the system are detailed on the DUACS website (https://duacs.cls.fr/)

Accurate knowledge of the marine geoid is a fundamental element for the full exploitation of altimetry for oceanographic applications and, in particular, for assimilation into operational ocean forecasting systems. SSH measured by an altimeter is the sea level above the ellipsoid, which is the sum of the Absolute Dynamic Topography (ADT or η) and the geoid height (N). The ADT is usually obtained by estimating a Mean Dynamic Topography (MDT or $<\eta>$) and adding it to the altimetric SLAs (η'). The MDT is obtained, at spatial scales where the geoid is known with sufficient accuracy, as the difference between an altimeter Mean Sea Surface Height (MSSH=$<SSH>$) and a geoid model.

Thanks to the recent dedicated space gravity missions of GRACE and GOCE, the knowledge of the geoid at scales of around 100-150 km has greatly improved in the past years, so that the ocean MDT is now resolved at those scales with centimetre accuracy. However, the true ocean MDT over a given period (e.g., 10 – 20 years) contains scales shorter than 100-150 km, which are not resolved in geoid models based on remote sensing. To compute higher resolution MDT, space gravity data can be combined with altimetry and oceanographic in situ measurements such as hydrological profiles from the Argo array and velocity measurements from drifting buoys. This approach was used by Rio et al. (2014) to compute the CNES-CLS13 MDT, which is used in several global data assimilation systems.

Operational oceanography requirements

Le Traon et al. (2006) have defined the main priorities for altimeter missions in the context of the European Copernicus Marine Service. Tables 7.1 and 7.2 from this paper give the requirements for different applications of altimetry and characteristics of altimeter missions.

	Application area	Accuracy*	Spatial resolution	Revisit Time	Priority
1	Climate applications and reference mission	1 cm*	300-500 km	10-20 days	High
2	Ocean nowcasting/ forecasting for mesoscale applications	3 cm*	50-100 km	7-15 days	High
3	Coastal/local	3 cm*	10 km	1 day	Low**

Table 7.1: User requirements for different applications of altimetry (*for the given resolution; **limited by feasibility).

Class	Orbit	Mission characteristics	Revisit interval	Track separation at the Equator
A	Non-sun synchronous	High accuracy for climate applications and to reference other missions	10-20 days	150-300 km
B	Polar	Medium-class accuracy	20 - 35 days	80 - 150 km

Table 7.2: Altimeter mission characteristics.

The main operational oceanography requirements for satellite altimetry can be summarized as follows:

1. There is a strong need to maintain a long time series of a high accuracy altimeter system (e.g., the Jason series) to serve as a reference mission and for climate applications. It requires one class A altimeter with an overlap between successive missions of at least six months.

2. The main requirement for medium to high resolution altimetry would be to fly three class B altimeters in addition to the Jason series (class A). Most operational oceanography applications (e.g., marine security, pollution monitoring) require high resolution surface currents that cannot be adequately reproduced without a high resolution altimeter system. Studies (e.g., Pascual et al., 2006) have shown that at least three, but preferably four, altimeter missions are needed for monitoring the mesoscale circulation. This is particularly desirable for real-time nowcasting and forecasting. Pascual et al. (2009) showed that four altimeters in real time provide similar results as two altimeters in delayed mode. Such a scenario would also provide improved operational reliability. Moreover, it would enhance the spatial and temporal sampling for monitoring and forecasting significant wave height.

In parallel, there is a need to develop and test innovative instrumentation (e.g., wide swath altimetry with the NASA/CNES SWOT mission) to better answer existing and future operational oceanography requirements for high to very high spatio/temporal resolution (e.g., mesoscale/submesoscale and coastal dynamics). There is also a need to improve nadir altimetry technology (e.g., increase resolution, reduce noise) and to develop smaller and cheaper instruments that could be embarked on a constellation of small satellites. For instance, the use of the Ka band (35 Ghz) allows for a major reduction in the size and weight of the altimeter and improved performances (Verron et al., 2018). The new generation of nadir altimeters provide enhanced capability thanks to a SAR mode (also known as Delay-Doppler processing mode) that allows reducing significantly the measurement noise level compared to conventional pulse-limited altimeters (Dibarboure et al., 2014; Boy et al., 2017). Noise level for 1 Hz sampling (about 7 km) is thus about 1 cm RMS for SAR altimeters compared to about 3 cm RMS for conventional pulse limited altimeters. More information is given in Chapter 8 (Morrow et al.).

Sea Surface Temperature

Sea surface temperature measurements and operational oceanography

Sea surface temperature (SST) is a key variable in operational oceanography and for assimilation into ocean dynamical models. SST is strongly related to air-sea interaction processes and provides a means to correct for errors in forcing fields such as heat fluxes and wind. It also characterizes the mesoscale variability of the upper ocean, resolving eddies and frontal structures, at very high resolution (a few km). SST data are often directly used for operational oceanography applications. They provide useful indices (e.g., climate changes, upwelling, thresholds). SST data can also be used to derive high resolution velocity fields (e.g., Bowen et al., 2002). Accurate, stable, well-resolved maps of SST are essential for climate monitoring and climate change detection. They are also central for numerical weather prediction, for which the role of high resolution SST measurements has been clearly evidenced (e.g., Chelton, 2005).

Measurements principles

Infrared radiometers (IR) operate at wavebands around 3.7, 10.5 and 11.5 μm where the atmosphere is almost transparent. The brightness temperature measured from infrared radiometers differs from the actual temperature of the observed surface because of non-unit emissivity and the effect of the atmosphere. Emissivity at IR frequencies is between 0.98 and 0.99 (close to a blackbody). Atmospheric correction is based on a multispectral approach, when the differences between brightness temperatures measured at different wavelengths are used to estimate the contribution of the atmosphere to the signal. At 10 μm, the solar irradiance reaching the top of the atmosphere is about 1/300 of the sea surface emittance. At 3.7 μm, the incoming solar irradiance is on the same order as the surface emittance. As a result, this wavelength can be used during nighttime only. Different algorithms are thus used for nighttime and daytime.

There is no IR way of measuring SST below clouds. Thus, the first priority is to detect cloud through a variety of methods. For cloud detection, the thermal and near-infrared waveband thresholds are used, as well as different spatial coherency tests. Poor cloud detection biases the SST low in climatic averages, and "false hits" of cloud can hide frontal and other dynamical structures. Geostationary infrared sensors can see whenever the cloud breaks.

Microwave sensors operate at several frequencies. Retrieval of SST is done at 7 and/or 11 GHz. Higher frequency channels (19 to 37 GHz) are used to precisely estimate the attenuation due to oxygen, water vapour, and clouds. The polarization ratio (horizontal versus vertical) of the measurements is used to correct for sea surface roughness effects. The great advantage of microwave measurements compared to infrared measurements is that SST can be retrieved even through non-precipitating clouds, which is very beneficial in terms of geographical coverage.

SST infrared and microwave sensors

Infrared radiometers, such as the Advanced Very High Resolution Radiometer (AVHRR), on board operational meteorological polar orbiting satellites offer a good horizontal resolution (1 km) and potentially global coverage, with the important exception of cloudy areas. However, their accuracy (0.4 to 0.5°K derived from the difference between collocated satellite and buoy measurements) is limited by the radiometric quality of the AVHRR instrument and the correction of atmospheric effects. Geostationary satellites (e.g., the GOES and MSG series) are carrying radiometers with infrared window channels similar to the ones on the AVHRR instrument. Their horizontal resolution is coarser (3-5 km), but their high temporal resolution sampling provides an advantage. Advanced measurements of SST suitable for climate studies include the Along Track Scanning Radiometer (ATSR) series of instruments, which have improved on-board calibration and make use of dual views at nadir and 55° incidence angle. The along track scanning measurement provides an improved atmospheric correction leading to an accuracy better than 0.2°K (O'Carroll et al, 2008). The main drawback of these instruments is their limited coverage due to a much narrower swath than the AVHRR instruments. Several microwave radiometers have also been developed and flown over the last 10 years (e.g., AMSR, TMI). The horizontal resolution of these products is around 25 km and their accuracy around $0.6 - 0.7$°K.

Key developments in SST data processing

During the past ten years, there has been a concerted effort to understand satellite and in situ SST observations, revolutionizing the way we approach the provision of SST data to the user community. GODAE, recognizing the importance of high resolution SST data sets for ocean forecasting, initiated the GODAE High Resolution SST Pilot Project (GHRSST-PP) to capitalize on these developments and create a set of dedicated products and services. There have also been key advances in the data processing of SST data sets over the last 10 years. As a result, new or improved products are now available. A full description of the GHRSST-PP is provided by Donlon et al. (2009). Data processing issues are summarized by Le Traon et al. (2009).

A satellite measures the so-called skin temperature, i.e., from a few tens of microns (infra-red) up to only a few mm (microwave). Diurnal warming changes the SST over a layer of 1 to 10 meters. The effect can be particularly large in regions of low wind speed and high solar radiation. GHRSST has defined the foundation SST as the temperature of the water column free of diurnal temperature variability. A key issue in SST data processing is to correct satellite SST measurements for skin and diurnal warming effects to provide precise estimations of the foundation SST. Night and day SST data from different satellites can then be merged through an optimal interpolation or a data assimilation system.

Several new high resolution SST products have been produced specifically in the framework of GHRSST-PP. These high resolution data sets are estimated by optimal interpolation methods merging SST satellite measurements from both infrared and microwave sensors. The pre-processing consists mainly of screening and quality control of the retrieved observations from each single data set and then constructing a coherent merged multi-sensor set of the most relevant and accurate observations (level 3). The merging of these observations requires a method for bias estimation and correction (relative to a chosen reference, currently AATSR). Finally, the gap-free SST foundation field is computed from the merged set of selected observations using an objective analysis method. The guess is either climatology or a previous map.

Operational oceanography requirements

Table 7.3 from Le Traon et al. (2006) summarises weather, climate, and operational oceanography requirements for SST. No single sensor is adequate meets the key requirements for SST. To remedy this, GHRSST-PP has established an internationally accepted approach to blending SST data from different sources that complement each other (refer to previous subsection). For this to work effectively, there must be an assemblage of four distinct types of satellite SST missions in place at any time, as defined in Table 7.4 (from Le Traon et al., 2006).

	Application area	Temperature accuracy [K]	Spatial resolution [km]	Revisit Time	Priority
1	Weather prediction	0.2 – 0.5	10 – 50	6 – 12 hrs	High
2	Climate monitoring	0.1	20 – 50	8 d	High
3	Ocean forecasting	0.2	1 – 10	6 – 12 hrs	High

Table 7.3: User requirements for SST provision.

The priority expressed by the international SST community, through GHRSST, is to continue to provide a type B (ATSR class) sensor. Its on-board calibration system, especially its dual-view methodology, allow AATSR to deliver the highest achievable absolute accuracy of SST, robustly independent of factors that cause significant biases in other infrared sensors such as stratospheric aerosols from major volcanic eruptions or tropospheric dust. Because its absolute calibration (for dual view) is better than 0.2 K it is used for bias correction of the other data sources before assimilation into models or analyses. A type C sensor (microwave) is also required beyond AMSR-E on Aqua.

Ocean Colour

Ocean colour measurements and operational oceanography

Over the last decade, the applications of satellite-derived ocean colour data have made important contributions to biogeochemistry, physical oceanography, ecosystem assessment, fisheries oceanography, and coastal management (IOCCG, 2008). Ocean colour measurements provide a global monitoring of chlorophyll (phytoplankton biomass) and associated primary production. They can be used to calibrate and validate biogeochemical, carbon, and ecosystem models. Progress towards assimilation of ocean colour data is less mature than for SST or SSH, but there are already convincing examples of assimilation of chlorophyll-a (Chla) in ocean models.

Data products needed to support ocean analysis and forecasting models of open ocean biogeochemical processes include the concentration of chlorophyll-a (Chla), total suspended material (TSM), the optical diffuse attenuation coefficient (K), and the photosynthetically available radiation (PAR). Use of K and PAR is needed to define the in-water light field that drives photosynthesis in ocean ecosystem models and that is required to model and forecast the ocean surface temperature. Ocean colour is a tracer of dynamical processes (mesoscale and submesoscale) and this is of great value for model validation. It also plays a role in air-sea CO_2 exchange monitoring.

	SST mission type	Radiometer wavebands	Nadir resolution	Swath width	Coverage / revisit
A	Two polar orbiting meteorological satellites with infrared radiometers. Generates the basic global coverage	3 thermal IR (3.7, 11, 12 µm) 1 near-IR, 1 Vis	~1 km	~2500 km	Day and night global coverage by each satellite
B	Polar orbiting dual-view radiometer. SST accuracy approaching 0.1K, used as reference standard for other types.	3 thermal IR (3.7, 11, 12 µm) 1 near-IR, 1 vis, each with dual view	~1 km	~500 km	Earth coverage in ~4 days
C	Polar orbiting microwave radiometer optimised for SST retrieval. Coarse resolution coverage of cloudy regions	Requires channels at ~7 and ~11 GHz	~50 km (25 km pixels)	~1500 km	Earth coverage in 2 days
D	Infrared radiometers on geostationary platforms. Spaced around the Earth	3 thermal IR (3.7, 11, 12 µm) 1 near-IR, 1 Vis	2 - 4 km	Earth disk from 36000 km altitude	Sample interval < 30 min

Table 7.4: Minimum assemblage of missions required to meet the need for operational SST.

At regional and coastal scales, there are many applications that require ocean colour measurements: monitoring of water quality, measurement of suspended sediment, sediment transport models, measurement of dissolved organic material, validation of regional/coastal ecosystem models (and assimilation), detection of plankton and harmful algal blooms, and monitoring of eutrophication. Use of ocean colour data in coastal seas is, however, more challenging as explained below.

Measurements principles

The sunlight is not merely reflected from the sea surface. The colour of water surface results from sunlight that has entered the ocean, been selectively absorbed, scattered and reflected by phytoplankton and other suspended material in the upper layers, and then backscattered through the surface. The subsurface reflectance $R(\lambda)$ (ratio of subsurface upwelled or water-leaving radiance on incident irradiance) that is the ocean signal measured by a satellite is proportional to $b(\lambda)/[a(\lambda)+b(\lambda)]$ or $b(\lambda)/a(\lambda)$ where $b(\lambda)$ is the backscattering and $a(\lambda)$ the absorption of the different water constituents.

Sunlight backscattered by the atmosphere (aerosols and molecular/Rayleigh scattering) contributes to more than 80% of the radiance measured by a satellite sensor at visible wavelengths. Atmospheric correction is calculated from additional measurements in the red and near-infrared spectral bands. Ocean water reflects very little radiation at these longer wavelengths (the ocean is close to a blackbody in the infrared) and the radiance measured is thus due almost entirely to scattering by the atmosphere.

Unlike observations in the infrared or microwave frequencies for which emission is from the sea surface only, ocean colour signals in the blue-green can come from depths as great as 50 m.

Sources of ocean colour variations include:

- Phytoplankton and its pigments
- Dissolved organic material
    - Coloured Dissolved Organic Material (CDOM or yellow matter) is derived from decaying vegetable matter (land) and phytoplankton degraded by grazing or photolysis.
- Suspended particulate matter (SPM)
    - The organic particulates (detritus) consist of phytoplankton and zooplankton cell fragments and zooplankton fecal pellets.
    - The inorganic particulates consist of sand and dust created by erosion of land-based rocks and soils (from river runoff, deposition of wind-blown dust, wave or current suspension of bottom sediments).

Colour can tell us about relative and absolute concentrations of those water constituents that interact with light. Hence we measure chlorophyll, yellow substance, and sediment load. It is difficult to distinguish independently varying water constituents:

- Case 1 waters are where the phytoplankton population dominates the optical properties (typically open sea). Only one component modulates the radiance spectrum backscattered from the water (phytoplankton pigment). Concentration range is 0.03 – 30 mg m^{-3}. Water in the near IR is nearly black for blue water. Atmospheric correction that is based on IR frequency measurements is thus relatively simple. Using green/blue ratio algorithms for chlorophyll, of the form Chla = A(R550/R490), provides an accuracy for Chla of ~ ±30% in open ocean.
- Case 2 waters are where other factors (CDOM, SPM) are also present. There are multiple independent components in water, which have an influence on the backscattered radiance spectrum. The retrieval procedure has to deal with these multiple components, even if only one should be determined. At high total suspended matter concentrations, problems also occur with atmospheric correction. Therefore, more complex algorithms (e.g., neural network) and more frequencies are required. Although this remains a challenging task, much progress has been made over the past five years. Useful estimations of Chla and SPM can thus be obtained in the coastal zone (e.g., Gohin et al., 2005).

Ocean colour can also provide information on phytoplankton functional types as changes in phytoplankton composition can lead to changes in absorption and backscattering coefficients (IOCCG, 2014). This is an area of active research with important implications for the assimilation of ocean colour data in ocean models.

An ocean colour satellite should have a minimum number of bands from 400-900nm. The role of the various bands is:

- 413 nm - Discrimination of CDOM in open sea blue water.
- 443, 490, 510, 560 nm - Chlorophyll retrieval from blue-green ratio algorithms.
- 560, 620, 665 nm and others - Potential to retrieve water content in turbid Case 2 waters using new red-green algorithms.
- 665, 681, 709 nm and others - Use of fluorescence peak for chlorophyll retrieval.
- 779, 870 nm for atmospheric correction plus another above 1000 nm to improve correction over turbid water.

Processing issues

The processing transforms the level 1 data, normalized radiances observed by the ocean colour radiometer, into geophysical properties corrected from atmospheric effects. Level 2 products include water leaving radiances at different wavelengths, chlorophyll-a concentration of the surface water (usually with case 1 and case 2 algorithms), total suspended matter (TSM), coloured dissolved and detrital organic materials (CDOM), diffuse attenuation coefficient (K) and photosynthetically available radiation (PAR).

Merging of several ocean colour satellites is needed to improve the daily ocean coverage. This requires combining data from individual sensors with different viewing geometries, resolution, and radiometric characteristics (e.g. IOCCG 2007; Le Traon et al., 2015). The availability of merged datasets allows the users to exploit a unique, quality-consistent, time series of ocean colour observations, without being concerned with the performance of individual instruments.

Category	Category of use	Optical class of water	Minimum set of satellite-derived variables needed	Accuracy [%]	Spatial resolution [km]	Revisit Time
1	Assimilation into operational open ocean models	Case 1	Chlor K PAR $L_w(\lambda)$	30% 5% 5% 5%	2 - 4	1 – 3 days
2	Ingestion in operational shelf sea & local models	Case 2	K - PAR $L_w(\lambda)$ Chlor TSM CDOM	5% 5% 5% 30% 30% 30%	0.5 - 2	1 day desired, but 3-5 days useful
3	Data products used directly by marine managers in shelf seas	Case 2	K PAR $L_w(\lambda)$ Chlor TSM CDOM	5% 5% 5% 30% 30% 30%	0.25 - 1	1 day desired, but 3-5 days useful
4	Global ocean climate monitoring	Case 1	Chlor K PAR	10 – 30% 5% 5%	5 - 10	8 d average
5	Coastal ocean climate monitoring	Case 2	Chlor TSM CDOM K PAR K	10 – 30% 10 – 30% 10 – 30% 5% 5%	5	8 day average
6	Coastal and estuarine water quality monitoring	Case 2	$L_w(\lambda)$	5%	0.1 - 0.5	0.5 – 2 hrs

Table 7.5: User requirements for ocean colour data products

Operational oceanography requirements

The needs and the broad classes of colour sensor are summarised in Tables 7.5 and 7.6 from Le Traon et al. (2006). They distinguish categories of use between the needs of the open ocean forecasting models, the finer scale shelf sea and local models, and those operational end users who analyse the data directly rather than through assimilation into a model system. There is a variety of additional products desired in coastal waters depending on the local water character. These include the CDOM and the discrimination of different functional groups of phytoplankton.

Some operational users prefer to use directly the atmospherically corrected water leaving radiance, $L_w(\lambda)$ (defined over the spectrum of given wavebands), applying their own approach for deriving water quality information or for confronting a model. Climate applications (categories 4 and 5) are envisaged to be derived from the operational categories 1 and 2, respectively, trading spatial and temporal resolution for improved accuracy. Category 6 is included in Table 7.5 to represent those users needing to monitor estuarine processes in fine spatial detail and to resolve the variations within the tidal cycle. This is a much more demanding category than the others.

A Class A simple SeaWiFS-like instrument with a resolution of 1 km and a set of 5 or 6 wavebands would be adequate for user Categories 1 and 4, to monitor global chlorophyll for assimilation into open ocean ecosystem models and for monitoring global primary production. It would fail to meet the main requirement to monitor water quality in coastal and shelf seas represented by user categories 2 and 3. These require a Class B imaging spectrometer sensor.

In order to satisfy the ocean colour measurement requirements for operational oceanography, the minimum requirement is for one Class B sensor and at least one other sensor (Class A, B or C). The Class C sensor corresponds to an imaging spectrometer on a geostationary platform. As well as uniquely serving the user Category 6 by resolving variability within the tidal cycle, it also serves other user categories in cloudy conditions by exploiting any available cloud windows that occur during the day.

Class	Orbit	Sensor type	Revisit Time	Spatial resolution	Priority
A	Polar	SeaWiFS type multispectral scanner, 5-8 Vis-NIR wavebands	3 days	1 km	High
B	Polar	Imaging spectrometer (MERIS/MODIS type)	3 days	0.25 – 1 km	High
C	Geostationary	Radiometer or spectrometer - feasibility to be determined	30 min	100 m – 2 km	Medium

Table 7.6: Classes of ocean colour sensor

Other Techniques

Synthetic aperture radar

SAR is an active instrument that transmits and receives electromagnetic radiation. It operates at microwave (or radar) frequencies. Wavelengths are in the range of 2 cm to 30 cm corresponding to frequencies in the range of 15 GHz – 1 GHz. It works in the presence of clouds, day and night. Synthetic aperture principle is to generate a very long antenna through the motion of the platform. For ASAR, the length of the synthetic antenna is approximately 20 km. This leads to very high resolution.

The surface roughness is the source for the backscatter of the SAR signal. The signal that arrives at the antenna is registered both in amplitude and phase. Although the SAR sees only the Bragg waves ($\lambda_B = \lambda/2 \sin \theta$, where θ is the incidence angle, λ the radar wavelength and λ_B the resonant Bragg wavelength), these waves are modulated by a large number of upper ocean and atmospheric boundary layer phenomena. This is the reason why SAR images express wave field, wind field, currents, fronts, internal waves, and spilled oil. They also provide high resolution images of sea ice (see next sub section).

Sea ice

Low resolution passive microwave sensors (Special Sensor Microwave/Imager (SSM/I) series of the US Defense Meteorological Satellite Program) have provided essential sea ice extent and concentration data from 1979 to present. These data can be extended and combined with scatterometer data. Moreover, sea ice drift estimates from scatterometers and radiometers are widely used for model validations and contribute importantly to the long-term sea ice monitoring. Sea ice thickness observations are needed to enable accurate estimations of the total sea ice volume. Great expectations were given to the CryoSat-2 launched in April 2010 with an altimeter designed for sea ice freeboard measurements. Thanks to the careful inter-comparison to the ICESAT laser altimeter mission data the retrieval accuracy has been reliably assessed (e.g. Laxon et al., 2013). Thickness of thin sea ice (<30 cm) for the Arctic freeze-up period can also be derived from the SMOS mission, which make these data highly complementary to the Cryosat-2 estimates of the thicker sea ice. From 2016, the Sentinel-1 A/B satellites have delivered high resolution (better than 30 m) SAR data in wide swath mode and simultaneous co- and cross-polarisation. This allows estimations of sea ice drift with a daily temporal resolution. Studies are now focusing on sea ice thickness, drift, and deformation analysis combining satellite data (like SMOS, CryoSat-2) with new data from the Sentinel-1 SAR missions.

Winds and waves

Scatterometers (e.g., Seawinds/QuikSCAT, ASCAT/MetOp) are radars operating at C or Ku bands. The main ocean parameters measured are the wind speed and direction. They also provide useful information on sea ice roughness. The principle is based on the resonant Bragg scattering. For a smooth surface, oblique viewing of the surface with active radar yields virtually no return. When wind increases, so does surface roughness and the reflected signals towards the satellite sensor. The wind direction can be derived because of the azimuthal dependence of the reflected signal with respect to the wind direction.

Over the past five years, there have been significant advances to develop and make accessible a harmonized set of altimeter and SAR wave products. These have been invaluable for numerical wave modelling and for applications. In particular, multiple altimeter missions have continued to provide precise significant wave height observations with a global coverage, which are essential to calibrate and validate numerical wave models and improve their forecasting skills through data assimilation. The sequence of ESA SAR C- and X-band instruments continuously operated on the ERS-1, ERS-2, ENVISAT RADARSAT, and TerraSAR-X satellites from 1991–2015 has also had a valuable impact on ocean wave observation and modelling, especially with regards to adequate determination of the swell attenuation over large distances. The future CFOSAT mission (to be launched in 2018) is expected to provide significant advances for winds and waves monitoring from space.

A new challenge: estimate sea surface salinity from space

At L-band (1.4 GHz), brightness temperature (BT) is mainly affected by ocean surface emission (atmosphere is almost transparent):

$$BT = e \, SST = (1-R) \, SST$$

where BT is brightness temperature and e is the sea surface emissivity. R (θ, SSS, SST, U...) is the reflection coefficient (see section 3.3). R depends on sea water permittivity and thus on sea surface salinity. Sensitivity is maximum at L-band, however it is very low (0.2 - 0.8 °K/psu) and increases with sea surface temperature.

The SMOS satellite was launched in November 2009. It is an L-band radiometer that measures BT at different incidence angles (0-60°). SMOS is a synthetic aperture radiometer which provides a high spatial resolution (~40 km, precision 1 PSU). SSS accuracy of 0.1-0.2 PSU over 10-day 200 km x 200 km areas is achieved through averaging of individual measurements. The Aquarius satellite was launched in 2010. It is a conventional L-band radiometer operating at three incident angles. Aquarius includes an L-band scatterometer to correct for sea surface roughness effects. These missions provide global SSS at spatial resolution varying from 50 km (SMOS) to 100 km (Aquarius) on weekly to monthly time scales. Many scientific studies have revealed the high potential of these new data sets (see Lagerloef et al., 2014 and Reul et al., 2014 for a review). Efforts to demonstrate the impact of satellite SSS data assimilation for ocean analysis and forecasting is an ongoing activity (see chapter by T. Lee). Measurement errors remain an issue, but more work should be carried out in the coming years thanks to improved data sets and products.

Assessing the Impact of Satellite Observations in GOV models

GOV (GODAE OceanView) systems have been used to investigate the impact of satellite data within their data assimilation and forecasting framework in a number of global and regional studies summarized in Oke et al. (2015a) and Oke et al. (2015b). This is also the focus of the GOV OSEval (Observing System Evaluation) Task Team. Most of these studies demonstrate the impact of the satellite data in the context of the other observing systems using Observing System Experiments (OSEs) whereby an experiment is run assimilating all available data, and a parallel run of the system is carried out assimilating all the data except for the data type to be investigated. The difference between the two runs shows the impact of the withheld data in the context of all the other data, and the two runs can be assessed by comparing the outputs with assimilated and independent data. Le Traon et al. (2017) provided a review on the use of OSEs to assess the impact of the altimeter constellation on GOV systems. A complementary approach for estimating the influence of the observations on the analysis is the computation of the so-called "degrees of freedom for signal" (DFS), which represents the equivalent number of independent observations that constrain the model analysis at the observation point (Cardinali et al., 2004). DFS quantifies the influence of observations on analyses without having to run dedicated experiments withholding some data.

Impact of future observations can be assessed using OSSEs (Observing System Simulation Experiments). OSSEs typically use two different model runs (from two different models or from the same model with different resolution, parametrization, or forcing). One model is used to perform a "truth" run and it is treated as if it is the real ocean. The truth run is sampled in a manner that mimics either an existing or future observing system, yielding synthetic observations. A measurement error (and often a representativity error) is then added to the synthetic observations that are then assimilated into the second model. The model performance is evaluated by comparing it against the truth run. OSSEs need a careful design so that results are representative of the actual ocean. Calibration of OSSEs with OSEs (i.e., verifying that, for an existing observing system, an OSSE yields the same result as an OSE) should be systematically applied. For example, Mercator Ocean has performed first OSSEs of SWOT observations using a 1/12° regional model of the Iberian Biscay region that includes tidal forcing (Benkiran et al., 2017). The truth run was derived from a 1/36° model run over the same region. SWOT errors were derived using the JPL SWOT Simulator. This first study quantified the highly significant improvement of SWOT data with respect to existing conventional nadir altimeters to constrain ocean models.

Future Satellite Requirements

Challenges for the next decade

A complete overview of the evolution (past, present, future) of the satellite ocean mission details is maintained within the CEOS Earth Observation handbook (http://www.eohandbook.com/) and/or the World Meteorological Organization's Observing Systems Capability Analysis and Review (OSCAR) tool (http://www. wmo-sat.info/oscar/).

The satellite-based data record lengths now exceed 25 years (altimetry, SAR, scatterometry) and 35 years (radiometry). In the 2020-2030 timeframe, sustained satellite observations will be common for almost all the variables addressed in the previous section, except for the sea surface salinity. Altogether, this ensures that operational oceanography will be supplied with a rich amount of highly important satellite observations. In addition to securing continuity in the observations, the retrieval accuracies have also gradually improved thanks to advancing sensor technology and retrieval algorithm performances.

Still, many key ocean phenomena are undersampled and require much higher space and time resolution. Model resolutions are regularly increasing, but our observation capabilities are not. Coastal regions, which are of paramount importance for operational oceanography, are characterized by small spatial and temporal scales. There is also growing evidence that we need to better observe the submesoscale ocean dynamics. Hence, a major challenge for satellite oceanography is to address the ongoing need for improved resolution. This challenge is partly addressed through the development of virtual constellations, but it also requires developing new observing capabilities (e.g., swath altimetry with SWOT, geostationary ocean colour missions). New satellite observing capabilities (e.g., for surface currents) also need to be developed (e.g., SKIM, see Ardhuin et al.,

2017). Fusing of different types of observations to extract better information is another complementary approach.

Future requirements: The Copernicus Marin Service perspective

The Copernicus Marine Environment Monitoring Service (CMEMS) is one of the six pillar services of the Copernicus program (see Le Traon et al., 2017 and chapter by Drevillon et al.). CMEMS includes most of the European contributions to GOV. It provides regular and systematic reference information on the physical state, variability, and dynamics of the ocean, ice, and marine ecosystems for the global ocean and the European regional seas. Copernicus Marine Service perspectives with respect to the long-term evolution of satellite observing systems are given hereafter.

First, continuity of the present Copernicus satellite observing system should be guaranteed as this is mandatory for maintaining the CMEMS service. This is particularly relevant to the Sentinel 6 altimeter reference mission (follow-on of Jason-3) and the two satellite constellation of Sentinel 3 (altimetry, sea surface temperature, and ocean colour) and Sentinel 1 (SAR).

In the post 2025 time period, Copernicus Marine Service model resolutions will be increased by a factor of at least 3 (e.g. global 1/36°, regional 1/108°) compared to the present and more advanced data assimilation methods will be available. The objective will be to describe at fine scale the upper ocean dynamics to improve our capabilities to describe and forecast the ocean currents and provide better boundary conditions for very high resolution coastal models (up to a few hundred meters). This is essential for key applications such as maritime safety, maritime transport, search and rescue, fish egg and larvae drift modelling, riverine influence in the coastal environments, pollution monitoring, and offshore operations. When moving to higher resolution, it will be necessary to constrain CMEMS models with new observations. The most important satellite-based observation is SSH from altimetry. As explained in previous sections, the SSH is an integral of the ocean interior properties and is a strong constraint for inferring the 4D ocean circulation through data assimilation. Multiple nadir altimeters (at least 4 altimeters) are required to adequately represent ocean eddies and associated currents in models. Much higher space/time resolution (e.g., 50 km / 5 days) will be needed in the post-2025 time period. This can be achieved through a combination of swath altimetry (to be demonstrated during the SWOT mission to be launched in 2021) with nadir SAR altimetry.

Monitoring the ecosystems in European seas is also a fundamental need, both for European policies (Marine Strategy) to monitor the health of the seas and coastal waters and to support sustainable fishery and aquaculture industries. Specifically, much more frequent observations of the highly variable biological parameters of the European regional seas are urgently needed. This is required to monitor the ocean ecosystem functioning at the diurnal scale and to monitor rapidly evolving phenomena (e.g. river outflows, phytoplankton and harmful algae blooms, sub-mesoscale features) and to constrain coupled biological-physical 3D models in regional seas and coastal zones. An ocean colour geostationary satellite would provide unique capabilities to provide such a monitoring. It should be complemented with new in situ biogeochemical observations (e.g., BGC Argo, Gliders, FerryBoxes). The development of hyperspectral sensor capabilities would also be relevant to

CMEMS (e.g., to improve the quality of Ocean Colour products in coastal zones and to differentiate between the types of phytoplankton in the ocean) (e.g., the NASA PACE and ASI PRISMA missions).

Sustainable passive microwave SST and sea ice observations are also very important in the global ocean and in polar regions. Such observations are available in all weather conditions, while infrared SST observations are available in cloud-free conditions only. Passive microwave SST and sea ice are a crucial contribution to weather forecasting and CMEMS ocean and analysis and forecasting models. CMEMS also requires specific observations for polar regions. In addition to passive microwave observations, one of the most important short term priorities is the continuation (including a few enhancements such as additional use of Ka band, more optimized orbit configuration, real-time capabilities) of the Cryosat-2 mission to monitor sea ice thickness, continental ice shelves elevation changes and contribute to the observation of the ocean surface topography in ice free regions.

Other requirements (e.g., surface current, SSS, wave, improved geoid, wind) should also be considered. Today, they do not have today the same (due to maturity of satellite technology or foreseen impact on the service). Surface current and SSS are two very important variables required for CMEMS. The potential impact is high but this requires developments or improvements in satellite technology. Thus, research and development should be done to further advance our capabilities to observe SSS from space building on SMOS achievements. There is also a strong need to test new mission concepts allowing the direct measure of surface currents at high resolution and to develop capabilities to fully use these observations to constrain ocean analysis and forecasting systems. Meanwhile, use of imaging SAR-based range Doppler retrievals (S1) should be reinforced.

Waves are observed today through altimeter (S3, S6) and SAR (S1) missions. New satellite concepts (CFOSAT) will soon allow for a better retrieval of directional wave spectra. This could lead to an improved design of future Sentinel missions. Finally, new gravity missions could be required to improve the geoid at small scales (and derived mean dynamic topography) and to monitor large-scale mass changes in the ocean. However, the main priority at present is to further develop the exploitation of GOCE data and to derive new mean dynamic topographies from the merging of GOCE, altimetry, and in situ observations. The Ocean Surface Vector Wind (OSVW) measurements through scatterometers are also important to improve NWP forcing fields. Europe, through the Eumetsat MetOP series, provides a unique contribution to the international CEOS OSVW virtual constellation. This should be pursued and coordination of the CEOS OSVW should be reinforced to optimize the existing and future scatterometer constellation.

Concluding Remarks

The chapter provides a brief introduction to ocean remote sensing measurement principles and the use of satellite observations for operational oceanography. The different techniques will be detailed further in the other chapters. More information can also be found in Fu and Cazenave (2001), Robinson (2004), Martin (2004), Emery and Camps (2017), and Stammer and Cazenave (2017).

Satellite data plays a fundamental role for operational oceanography. There have been important achievements over the past 10 years to ensure real-time availability of high quality satellite data and to develop the use of satellite observations for operational oceanography. Multiple mission high resolution altimeter products are now readily available. There have been many improvements (timeliness, new products) and major efforts were undertaken to include new missions in the operational data stream in a very limited time. New MDTs from GRACE and GOCE have a major impact on data assimilation systems and further improvements are expected. Thanks to GHRSST, major improvements in SST data processing techniques and use of different types of sensors have occurred. The use of ocean colour data in operational oceanography has become a reality and there has been continuous progress in data access, data processing, and data assembly systems. SMOS and Aquarius have demonstrated the feasibility and utility of measuring SSS from space. The ocean community and GOV now need to fully invest in the critical assessment and application of the data.

In situ data are mandatory to calibrate and validate and to complement satellite observations. Although the consolidation of the Argo in situ observing system and its integration with satellite altimetry and operational oceanography (e.g., Le Traon, 2013) was an outstanding advancement, the evolution of the in situ observing system remains a strong concern. The potential of satellite observations is not and will not be fully realized without a sustained in situ observing system.

Improvements in models and data assimilation techniques have resulted in a better use of satellite observing capabilities. However, the information content of satellite observations is not being fully exploited. Use of new theoretical frameworks to better exploit high resolution information from satellite data is required and further improvements in data assimilation schemes are needed to better take into account observations (e.g., towards L1B and L2 assimilation, correlated errors, biases, representativity errors). The potential of ocean colour data to calibrate or improve biogeochemical models is considerable. But this is a complex issue and, as such, development lags behind other remote sensing techniques. This is a challenging and high priority research topic for operational oceanography.

There is a need to develop further OSE/OSSE activities (GOV Task Team). This is essential to define needs, quantify impacts, and to improve data assimilation systems.

Finally, new satellite missions such as high resolution altimetry (SWOT) missions will likely have a major impact on operational oceanography. There is also great potential for satellite missions to improve the monitoring of waves and winds (e.g., CFOSAT, see Hauser et al., 2016) or for directly measuring surface currents from space (e.g., SKIM, see Ardhuin et al., 2017).

References

Ardhuin, F., Aksenov, Y., Benetazzo, A., Bertino, L., Brandt, P., Caubet, E., Chapron, B., Collard, F., Cravatte, S., Dias, F., Dibarboure, G., Gaultier, L., Johannessen, J., Korosov, A., Manucharyan, G., Menemenlis, D., Menendez, M., Monnier, G., Mouche, A., Nouguier, F., Nurser, G., Rampal, P., Reniers, A., Rodriguez, E., Stopa, J., Tison, C., Tissier, M., Ubelmann, C., van Sebille, E., Vialard, J., and Xie, J. (2017). Measuring currents, ice drift, and waves from space: the Sea Surface KInematics Multiscale monitoring (SKIM) concept. Ocean Sci. Discuss., doi:10.5194/os-2017-65.

Benkiran M., E. Remy, E. Greiner, Y. Drillet and P.Y. Le Traon (2017). An Observing System Simulation Experiment to evaluate the impact of SWOT in a regional data assimilation system. Remote Sensing Environment (in press).

Bonekamp, H. & Co-Authors (2010). Transitions towards operational space based ocean observations: from single research missions into series and constellations in Proceedings of OceanObs'09: Sustained Ocean Observations and Information for Society (Vol. 1), Venice, Italy, 21-25 September 2009, Hall, J., Harrison, D.E. & Stammer, D., Eds., ESA Publication WPP-306, doi:10.5270/OceanObs09.pp.06.

Bowen, M., W. J. Emery, J. Wilkin, P. Tildesley, I. Barton and R. Knewtson (2002). Extracting multi-year surface currents from sequential thermal imagery using the Maximum Cross Correlation technique, Journal of Atmospheric and Oceanic Technology, 19, 1665-1676.

Boy, F., J.D. Desjonquères, N. Picot, T. Moreau, M. Raynal (2017). CryoSat-2 SAR-Mode Over Oceans: Processing Methods, Global Assessment, and Benefits. IEEE Transactions on Geoscience and Remote Sensing, 55, 1, 148 - 158, doi:10.1109/TGRS.2016.2601958

Cardinali C., S. Pezzulli and E. Andersson (2004). Influence-matrix diagnostic of a data assimilation system. Quarterly J R Meteorol Soc. 130:2767–2786. doi: 10.1256/qj.03.205.

Chapron, B., F. Collard, and F. Ardhuin (2005). Direct measurements of ocean surface velocity from space: Interpretation and validation, J. Geophys. Res., 110, C07008, doi:10.1029/2004JC002809.

Chelton, D. B. (2005). The impact of SST specification on ECMWF surface wind stress fields in the eastern tropical Pacific. J. Climate, 18, 530–550.

Chelton, D.B., J.C. Ries, B.J. Haines, L.L. Fu, P. Callahan (2001). Satellite Altimetry, Satellite altimetry and Earth sciences, L.L. Fu and A. Cazenave Ed., Academic Press.

Choi, J. K., Park, Y. J., Ahn, J. H., Lim, H. S., Eom, J., & Ryu, J. H. (2012). GOCI, the world's first geostationary ocean color observation satellite, for the monitoring of temporal variability in coastal water turbidity. Journal of Geophysical Research: Oceans (1978-2012), 117(C9).

Clark C. and W. Wilson (2009). An overview of global observing systems relevant to GODAE. Oceanography Magazine, Vol. 22, No. 3, Special issue on the revolution of global ocean forecasting—GODAE: ten years of achievements.

Dibarboure, G., M.I. Pujol, F. Briol, P.-Y. Le Traon, G. Larnicol, N. Picot, F. Mertz and M. Ablain (2011). Jason-2 in DUACS: Updated system description, first tandem results and impact on processing and products. Marine Geodesy, 34(3-4), 214-241.

Donlon C., N. Rayner, I. Robinson, D. J. S. Poulter, K. S. Casey, J. Vazquez-Cuervo, E. Armstrong, A. Bingham, O. Arino, C. Gentemann, D. May, P. LeBorgne, J. Piollé, I. Barton, H. Beggs, C. J. Merchant, S. Heinz, A. Harris, G. Wick,B. Emery, P. Minnett, R. Evans, D. Llewellyn-Jones, C. Mutlow, R. W. Reynolds, H. Kawamura (2007). The Global Ocean Data Assimilation Experiment High-resolution Sea Surface Temperature Pilot Project, Bulletin of the American Meteorological Society, Volume 88, Issue 8 (August 2007) pp. 1197-1213 doi: http://dx.doi.org/10.1175/BAMS-88-8-1197

Donlon, C., I.S. Robinson, M. Reynolds, W. Wimmer, G. Fisher, R. Edwards, and T.J. Nightingale (2008). An Infrared Sea Surface Temperature Autonomous Radiometer (ISAR) for Deployment aboard Volunteer Observing Ships (VOS), J. Atmos. Oceanic Technol., 25, 93–113.

Drinkwater, M. & Co-Authors (2010). Status and Outlook for the Space Component of an Integrated Ocean Observing System in Proceedings of OceanObs'09: Sustained Ocean Observations and Information for Society (Vol. 1), Venice, Italy, 21-25 September 2009, Hall, J., Harrison, D.E. & Stammer, D., Eds., ESA Publication WPP-306, doi:10.5270/OceanObs09.pp.17.

Ducet, N., P.Y. Le Traon and G. Reverdin (2000). Global high resolution mapping of ocean circulation from the combination of TOPEX/POSEIDON and ERS-1/2. Journal of Geophysical Research, 105, C8, 19,477-19,498.

Emery, B. and A. Camps (2017). Introduction to satellite remote sensing. Atmsophere, Ocean, Land, Cryosphere applications. Elsevier, pp 843.

Escudier, P., A. Couhert, F. Mercier, A. Mallet, P. Thibaut, N. Tran, L. Amarouche, B. Picard, L. Carrère, G. Dibarboure, M. Ablain, J. Richard, N. Steunou, P. Dubois, M. H. Rio, and J. Dorandeu. Satellite radar altimetry: principle, accuracy, and precision in Satellite Altimetry Over Oceans and Land Surfaces, CRC Press, Editors: Stammer and Cazenave.

Gohin F., Loyer S., Lunven M., Labry C., Froidefond J.M., Delmas D., Huret M., and Herbland A. (2005). Satellite-derived parameters for biological modelling in coastal waters: Illustration over the eastern continental shelf of the Bay of Biscay. Remote Sensing of Environment, 95 (1): 29-46.

Guinehut, S., C. Coatanoan, A.-L Dhomps, P.-Y. Le Traon and G. Larnicol (2008). On the use of satellite altimeter data in Argo quality control. Journal of Atmospheric and Oceanic Technology, 26(2), 395-402.

Guinehut, S., P.-Y. Le Traon and G. Larnicol (2006). What can we learn from global altimetry/hydrography comparisons? Geophysical Research Letters, 33, L10604, doi:10.1029/2005GL025551.

Hauser, D., C. Tison, T. Amiot, L. Delaye, A. Mouche, G. Guitton, L. Aouf and P. Castillan (2016). CFOSAT: a new Chinese-French satellite for joint observations of ocean wind vector and directional spectra of ocean waves. Proc. SPIE 9878, Remote Sensing of the Oceans and Inland Waters: Techniques, Applications, and Challenges, 98780T, doi: 10.1117/12.2225619.

IOCCG (2007). Ocean Colour Data Merging. Gregg W.W. (Ed.), with contribution by W. Gregg, J. Aiken, E. Kwiatkowska, S. Maritorena, F. Mélin, H. Murakami, S. Pinnock, and C. Pottier. IOCCG Monograph Series, Report #6, 68pp.

IOCCG (2008). Why Ocean Colour? The societal benefits of Ocean-Colour Technology, Platt T., N. Hoepffner, V. Stuart, and C. Brown (Eds.), Reports of the International Ocean-Colour Coordinating Group, No. 7, IOCCG, Dartmouth, Canada, 141pp.

IOCCG (2012). Ocean-Colour Observations from a Geostationary Orbit. Antoine, D. (ed.), Reports of the International Ocean-Colour Coordinating Group, No. 12, IOCCG, Dartmouth, Canada.

IOCCG (2014). Phytoplankton Functional Types from Space. Sathyendranath, S.(ed.), Reports of the International Ocean-Colour Coordinating Group, No. 15, IOCCG, Dartmouth, Canada.

Lagerloef, G., F. Wentz, S. Yueh, H.-Y. Kao, G. C. Johnson, and J. M. Lyman (2012). Aquarius satellite mission provides new, detailed view of sea surface salinity, State of the Climate in 2011. Bull. Am. Meteorol. Soc., 93(7), S70-71.

Laxon S. W., K. A. Giles, A. L. Ridout, D. J. Wingham, R. Willatt, R. Cullen, R. Kwok, A. Schweiger, J. Zhang, C. Haas, S. Hendricks, R. Krishfield, N. Kurtz, S. Farrell and M. Davidson (2013). CryoSat-2 estimates of Arctic sea ice thickness and volume. Geophys. Res. Lett., 40, 732-737.

Le Traon P.Y. (2011). Satellites and Operational Oceanography. In Operational Oceanography in the 21st Century (Springer-verlag Berlin). http://archimer.ifremer.fr/doc/00073/18383/

Le Traon P.Y. (2013). From satellite altimetry to Argo and operational oceanography: three revolutions in oceanography. Ocean Science, 9(5), 901-915, doi:10.5194/os-9-901-2013.

Le Traon, P.Y. G. Dibarboure, G. Jacobs, M. Martin, E. Remy and A. Schiller (2017a). Use of satellite altimetry for operational oceanography in Satellite Altimetry Over Oceans and Land Surfaces, CRC Press, Editors: Stammer and Cazenave.

Le Traon, P.Y. (2013). From satellite altimetry to Argo and operational oceanography: three revolutions in oceanography. Ocean Science 9(5): 901-915, doi:10.5194/os-9-901-2013.

Le Traon, P.Y. J. Johannessen, I. Robinson, O. Trieschmann (2006). Report from the Working Group on Space Infrastructure for the GMES Marine Core Service. GMES Fast Track Marine Core Service Strategic Implementation Plan. Final Version, 24/04/2007.

Le Traon, P.Y., Antoine D., Bentamy A., Bonekamp H., Breivik L.A., Chapron B., Corlett G., Dibarboure G., Digiacomo P., Donlon C., Faugere Y., Font J., Girard-Ardhuin F., Gohin F., Johannessen J., Kamachi M., Lagerloef G., Lambin J., Larnicol G., Le Borgne P., Leuliette E., Lindstrom E., Martin M.J., Maturi E., Miller L., Mingsen L., Morrow R., Reul N., Rio M., Roquet H., Santoleri R., and J. Wilkin (2015). Use of satellite observations for operational oceanography: recent achievements and future prospects. Journal of Operational Oceanography, 8(supp.1), s12-s27, doi:10.1080/1755876X.2015.1022050.

Le Traon, P.Y., M. Rienecker, N. Smith, P. Bahurel, M. Bell, H. Hurlburt, and P. Dandin, (2001). Operational Oceanography and Prediction – a GODAE Perspective, in Observing the Oceans in the 21st Century, edited by C.J. Koblinsky and N.R. Smith. GODAE project office, Bureau of Meteorology, 529-545.

Le Traon, P.Y., Nadal F. and N. Ducet (1998). An improved Mapping Method of Multisatellite Altimeter Data. Journal of Atmospheric and Oceanic Technology, 15, 522-533.

Le Traon, P.Y., G. Larnicol, S. Guinehut, S. Pouliquen, A. Bentamy, D. Roemmich, C. Donlon, H. Roquet, G. Jacobs, D. Griffin, F. Bonjean, N. Hoepffner, and L.A. Breivik (2009). Data assembly and processing for operational oceanography:10 years of achievements, Oceanography Magazine, Vol. 22, No. 3, Special issue on the revolution of global ocean forecasting—GODAE: ten years of achievement.

Le Traon P.Y., A. Ali, E. Alvarez Fanjul, L. Aouf, L. Axell, R. Aznar, M. Ballarotta, A. Behrens, M. Benkiran, A. Bentamy, L. Bertino, P. Bowyer, V. Brando, L. A. Breivik, B. Buongiorno Nardelli, S. Cailleau, S. A. Ciliberti, E. Clementi, S. Colella, N. Mc Connell, G. Coppini, G. Cossarini, T. Dabrowski, M. de Alfonso Alonso-Muñoyerro, E. O'Dea, C. Desportes, F. Dinessen, M. Drevillon, Y. Drillet, M. Drudi, R. Dussurget, Y. Faugère, V. Forneris, C. Fratianni, O. Le Galloudec, M. I. García-Hermosa, M. García Sotillo, P. Garnesson, G. Garric, I. Golbeck, J. Gourrion, M. L. Grégoire, S. Guinehut, E. Gutknecht, C. Harris, F. Hernandez, V. Huess, J. A. Johannessen, S. Kay, R. Killick, R. King, J. de Kloe, G. Korres, P. Lagemaa, R. Lecci, J.F. Legeais, J. M. Lellouche, B. Levier, P. Lorente, A. Mangin, M. Martin, A. Melet, J. Murawski, E. Özsoy, A. Palazov, S. Pardo, L. Parent, A. Pascual, A. Pascual, J. Paul, E. Peneva, C. Perruche, D. Peterson, L. Petit de la Villeon, N. Pinardi, S. Pouliquen, M. I. Pujol, R. Rainaud, P. Rampal, G. Reffray, C. Regnier, A. Reppucci, A. Ryan, S. Salon, A. Samuelsen, R. Santoleri, A. Saulter, J. She, C. Solidoro, E. Stanev, J. Staneva, A. Stoffelen, A. Storto, P. Sykes, T. Szekely, G. Taburet, B. Taylor, J. Tintore, C. Toledano, M. Tonani, L. Tuomi, G. Volpe, H. Wedhe, T. Williams, L. Vandendbulcke, D. van Zanten, K. von Schuckmann, J. Xie, A. Zacharioudaki, and H. Zuo (2017b). The Copernicus Marine Environmental Monitoring Service: Main Scientific Achievements and Future Prospects. Special Issue Mercator Océan Journal #56, doi:10.25575/56

Lea, D. J., Martin, M. J. and P. R. Oke (2014). Demonstrating the complementarity of observations in an operational ocean forecasting system. Q.J.R. Meteorol. Soc. 140: 2037-2049. doi: 10.1002/qj.2281.

Martin S. (2004). An introduction to ocean remote sensing, Cambridge University Press. ISBN-13: 9780521802802 | ISBN-10: 0521802806.

Mélin, F., and G. Zibordi (2007). An optically-based technique for producing merged spectra of water leaving radiances from ocean colour remote sensing. Applied Optics, 46, 3856-3869.

Mitchum, G. T. (2000). An improved calibration of satellite altimetric heights using tide gauge sea levels with adjustment for land motion. Marine Geodesy, 23, 145-166.

O'Carroll, A.G., J.R. Eyre, and R.W. Saunders, (2008). Three-Way Error Analysis between AATSR, AMSR-E, and In Situ Sea Surface Temperature Observations. Journal of Atmospheric and Oceanic Technology, 25, 1197–1207.

Oke, P.R., G. Larnicol, E.M. Jones, V. Kourafalou, A.K. Sperrevik, F. Carse, C.A.S. Tanajura, B. Mourre, M. Tonani, G.B. Brassington, M. Le Henaff, G.R. Halliwell Jr., R. Atlas, A.M. Moore, C.A. Edwards, M.J. Martin, A.A. Sellar, A. Alvarez, P. De Mey, and M. Iskandarani (2015b). Assessing the impact of observations on ocean forecasts and reanalyses: Part 2, Regional applications, Journal of Operational Oceanography 8:sup1, s63-s79, doi:10.1080/1755876X.2015.1022080

Oke, P.R., G. Larnicol, Y. Fujii, G.C. Smith, D.J. Lea, S. Guinehut, E. Remy, M. Alonso Balmaseda, T. Rykova, D. Surcel-Colan, M.J. Martin, A.A. Sellar, S. Mulet, and V. Turpin (2015a). Assessing the impact of observations on ocean forecasts and reanalyses: Part 1, Global studies, Journal of Operational Oceanography, 8:sup1, s49-s62, doi:10.1080/1755876X.2015.1022067.

Pascual, A., C. Boone, G. Larnicol, P.Y. Le Traon (2009). On the quality of real time altimeter gridded fields: comparison with in situ data. Journal of Atmospheric and Oceanic Technology 26, 556–569.

Pascual, A., Faugere, Y., G. Larnicol, P.Y. Le Traon (2006). Improved description of the ocean mesoscale variability by combining four satellite altimeters. Geophysical Research Letters, 33 (2): Art. No. L02611.

Reul N., Fournier S., Boutin J., Hernandez O., Maes C., Chapron B., Alory G., Quilfen Y., Tenerelli J., Morisset S., Kerr Y., Mecklenburg S. and S., Delwart S. (2014). Sea Surface Salinity Observations from Space with the SMOS Satellite: A New Means to Monitor the Marine Branch of the Water Cycle. Surveys in Geophysics, 35(3), 681-722, doi:10.1007/s10712-013-9244-0

Rio, M. H., S. Mulet and N. Picot (2014). Beyond GOCE for the ocean circulation estimate: Synergetic use of altimetry, gravimetry, and in situ data provides new insight into geostrophic and Ekman currents. Geophysical Research Letters, 41(24), 8918-8925. Robinson, I. (2004). Measuring the Oceans from Space: The principles and methods of satellite oceanography, Springer, 669 pp.

Robinson, I. (2004). Measuring the Oceans from Space: The principles and methods of satellite oceanography, Springer, 669 pp.

Smith, N., and M. Lefebvre (1997). The Global Ocean Data Assimilation Experiment (GODAE). Paper presented at Monitoring the Oceans in the 2000s: An Integrated Approach, Biarritz, France, October 15–17.

Verron, J., Bonnefond, P., Aouf, L., Birol, F. Bhowmick, S.A., Calmant, S., Conchy, T., Crétaux, J.-F., Dibarboure, G., Dubey, A.K., Faugère, Y., Guerreiro, K., Gupta, P.K., Hamon, M., Jebri, F., Kumar, R., Morrow, R., Pascual, A., Pujol, M.I., Rémy, E., Rémy, F., Smith, W.H.F., Tournadre, J., and Vergara, O. (2018). The Benefits of the Ka-Band as Evidenced from the SARAL/AltiKa Altimetric Mission: Scientific Applications. Remote Sens. 2018, 10, 163.

CHAPTER 8

Fine-scale Altimetry and the Future SWOT Mission

Rosemary Morrow[1], Denis Blumstein[1,2], and Gerald Dibarboure[2]

[1]LEGOS/CNRS/IRD/CNES/University Toulouse III, Toulouse, France; [2]CNES, Toulouse, France

This chapter describes recent advances in improving altimetry observations over the ocean for the detection of fine-scale ocean dynamics. The first section gives an overview of the different satellite radar altimetry techniques being used today at high-resolution over the open and coastal oceans: from conventional along-track nadir altimetry to along-track Synthetic Aperture Radar (SAR) at nadir. We present the advantages of the measurement techniques in conventional Ku-band (Jason) and Ka-band (Saral), and in global SAR mode (Sentinel-3). We show how the along-track errors are estimated, how they vary geographically and seasonally, and how they limit the sea surface height (SSH) scales resolved. We also address various mapping techniques being used to derive gridded SSH data and the issues for observing fine-scale ocean dynamics from altimeter data in the coastal zone. The second section addresses the future global SAR-interferometry mission, Surface Water Ocean Topography (SWOT), which aims to measure terrestrial surface waters and ocean SSH over a wide swath. We concentrate on the ocean component of this mission, which will provide the first two-dimensional (2D) observations of SSH on a 1-2 km grid. The low noise level of the SWOT observations should allow us to observe physical processes in the open and coastal oceans with wavelength scales down to 15-20 km. We present the SWOT SAR-interferometry technique, as well as the mission's sampling characteristics and error budget. Of particular interest is the range of ocean dynamics that have a SSH signature in the wavelength scale of 15-200 km, including small mesoscale structures, larger submesoscale fronts and filaments, internal tides, and internal gravity waves. These are difficult to observe with the present altimeter constellation due to the along-track altimetric sampling and higher noise levels. The chapter addresses how these fine-scale dynamics will be observed with the future SWOT SAR-interferometric altimetry technology, the challenges in mapping the SWOT swath SSH observations, and the preparation to assimilate SWOT 2D SSH images into operational ocean models.

Fine-scale Ocean Dynamics from Conventional and SAR Altimetry

Introduction

Spaceborne altimetry measurements were first demonstrated in the 1970s (Skylab, GEOS-3, Seasat) with a sea surface height (SSH) accuracy of around 1 m, mainly due to orbit determination errors. In the late 1980s, the Geosat mission (US Navy, 1985-1990) demonstrated the ability to monitor the ocean from space, with an 18-month geodetic phase to improve the Earth's gravity and marine geoid estimates, followed by a three-year ocean circulation phase. The era of high-precision satellite altimetry started with the launch of Topex/Poseidon in 1992, a joint NASA–CNES mission that provided the first estimates of SSH with 2–3 cm accuracy

---

Morrow, R., D. Blumstein, and G. Dibarboure 2018: Fine-scale altimetry and the future SWOT mission. In "*New Frontiers in Operational Oceanography*", E. Chassignet, A. Pascual, J. Tintoré, and J. Verron, Eds., GODAE OceanView, 191-226, doi:10.17125/gov2018.ch08.

on a 10-day repeat cycle. Since then, several high-precision altimetry missions have been launched: Jason-1, Jason-2, Jason-3, as successors to Topex/Poseidon. Complementary space–time coverage over all surfaces was available from ERS-1&2 and Envisat (missions from ESA), SARAL/AltiKa (joint mission from CNES and ISRO), and HY-2 (CNES and CSA-China). All of these missions are in conventional "low-resolution" mode (LRM).

A new generation of altimeters is starting to use nadir Synthetic Aperture Radar (SAR) and SAR interferometry modes, including CryoSat-2 (ESA) and the first global SAR altimetry mission from Sentinel-3A (ESA). In the future, global SAR nadir missions will continue with Sentinel-3B (ESA) and the reference climate mission, Jason-CS/Sentinel 6A (ESA, EUMETSAT, NOAA, CNES, and NASA), part of the operational Copernicus program of the European Union.

In this first section, we detail some of the characteristics and advances of the most recent LRM missions in Ku- and Ka-band, as well as their differences from the SAR along-track missions. In the second section, we present the wide-swath SAR interferometric altimetry mission (Surface Water Ocean Topography or SWOT) able to study the two-dimensional (2D) fine-scale ocean circulation and terrestrial surface waters, jointly developed by NASA and CNES. The different radar altimeter missions providing data from 1992 onwards are shown in Table 8.1, after Frappart et al. (2017).

Mission	Topex/Poseidon Jason-1/2/3	ERS-2 ENVISAT	CryoSat-2	SARAL	Sentinel-3 A, B	Jason-CS/ Sentinel-6	SWOT
Instrument	NRA/SSALT Poseidon-2 Poseidon-3 Poseidon-3B	RA RA-2	SIRAL	AltiKa	SRAL	Poseidon-4	KaRIN Poseidon-3 like
Space agency	CNES, NASA	ESA	ESA	CNES, ISRO	ESA	ESA, EUMETSAT, NOAA, CNES	CNES, NASA
Operation	1992-2005 2001-2013 Since 2008 Since 2016	1995-2003 2002-2012	Since 2010	Since 2013	Since 2016 Expected in 2017	Expected in 2020	Expected in 2021
Acquisition mode	LRM	LRM	LRM, SAR, InSAR	LRM	LRM, SAR	LRM, SAR	InSAR LRM
Acquisition	Along-track	Along-track	Along-track	Along-track	Along-track	Along-track	120 km swath Along-track
Frequency (GHz)	13.575 (Ku) 5.3 (C)	13.8 (Ku)[b] 13.575 (Ku)[c] 3.2 (S)[c]	13.575 (Ku)	35.75 (Ka)	13.575 (Ku) 5.41 (C)	13.575 (Ku) 5.3 (C)	35.6 (Ka) 13.575 (Ku) 5.3 (C)
Altitude (km)	1315	800	717	800	814.5	1315	890
Orbit inclination (°)	66	98.55	92	98.55	98.65	66	77.6
Repetitivity[a] (days)	9.9156	35	369	35	27	9.9156	21
Equatorial cross-track separation* (km)	315	75	7.5	75	104	315	133.1

[a] On nominal orbit
[b] RA onboard ERS-2
[c] RA-2 onboard ENVISAT

Table 8.1. Major characteristics of the high-precision radar altimetry missions since 1992

Conventional low-resolution mode (LRM) altimetry; examples from the Jason series and Saral

The satellite radar altimetry measurement technique is based on an active radar emission at microwave frequencies from the satellite platform. A detailed description can be found in Chelton et al. (2001) and others (e.g., Robinson, 2004; Frappart et al., 2017; Escudier et al., 2017). We provide here a simplified view that allows a user to understand how the measurement is formed.

The radar sensor emits an electromagnetic pulse in the nadir direction (vertically downwards) and precisely measures the two-way travel time of the signal (Δt) between the satellite and the reflecting surface. Given the known speed of light in a vacuum, c, the distance or range, R, between the sensor the reflecting surface, such as the ocean, can be given by:

$$R = \frac{c \Delta t}{2} - \sum_i \Delta R_i \quad (1)$$

where ΔR_i represent the sum of the corrections in the delay of the round trip travel time, due to atmospheric, instrument, or surface effects.

In reality, to obtain a height resolution of around 1 cm, we would need a pulse of a fraction of a nanosecond (with a pulse bandwidth of > 3 GHz, which is not possible). So altimetry uses a technique called pulse compression. The emitted signal has a frequency that is linearly modulated in time (a chirp). When the return signal (echo) is received onboard, the radar passes the signal through an inverse filter to compress the chirp back into a short pulse and recover the required range resolution (for details see Chelton et al., 2001).

Over an ocean surface, the altimeter sends out the radar energy, which is reflected back from the sea surface. Even if the emitted signal has an infinitely short duration, the return echo has a longer duration because it is scattered back by many reflectors on the surface that are at different distances from the radar (see Figure 8.1). The power of the return signal has a resolution of many tens of centimeters (about 47 cm for altimeters of the Jason series), which is not sufficient to study cm-level ocean SSH signals. However, if we apply a theoretical model, we can analyze how the power is returned from a series of gates/bins, each at a slightly different time (see Figure 8.1). This return power over time is called a "waveform" and it has a characteristic shape that can be described analytically (for example, over the open ocean by the Brown [1977] model) or numerically. The waveform shape and amplitude provide information on the nature of the surface and the geophysical parameters we wish to derive (SSH, significant wave height, wind speed, antenna mispointing). In the Brown model, the altimeter range (R) corresponds to the time (τ or epoch at mid-height) when the received power reaches the middle of the leading edge (see Figure 8.1). Using the epoch given by the Brown model allows us to obtain the centimeter precision from the original signal with a sampling step of 47 cm (2%). In addition, the power of the return signal gives information on the backscattering coefficient, from which the wind speed (amplitude, not direction) can be derived using empirical algorithms. The significant wave height is derived from the slope of the waveform's leading edge, and the antenna mispointing from the trailing edge slope.

The Brown model's validity is limited to conditions where the characteristics of the water surface and the transmission through the atmosphere are very uniform. In the open ocean, this is generally true but, in some cases, the validity conditions are not met and the accuracy of the Brown model's range retrievals deteriorates. This occurs, for example, when the ocean backscatter or surface roughness varies rapidly (e.g., across low wind patches or near the coast when the ocean signal is polluted by the nearby land) or in cases of heavy rain cells. The occurrence of such events can be monitored and edited using the backscatter coefficient (e.g., Quartly et al., 2001; Dibarboure et al., 2014). Conventional radar altimeters provide measurements at a frequency around 10 Hz (T/P, ERS), 20 Hz (Jason, Envisat, CryoSat2), and 40 Hz (AltiKa). Each of these measurements is the (incoherent) sum of about 100 individual echos, and the summing helps reduce the measurement noise.

Figure 8.1 also shows a schematic of the pulses' "round" footprint on the ocean surface. For the Jason class altimeters, flying at 1336 km altitude, and for the pulses averaged over 1 second, this footprint is more oval in shape, being around 10 km along-track and 5 km cross-track (Chelton et al., 2001). The SSH signal and any noise effects are integrated over the original footprint domain (50-70 km^2), and then averaged over the 1-second time. The footprint size is determined by the pulse duration (3 ns <-> 2 km for calm seas) but also depends on the sea state conditions, the altimeter frequency band, the antenna pattern and the satellite altitude. As an example, for a 3 m high significant wave height sea, the footprint cross-track radius is around:

- 3 km for Ka-band Saral at ~800 km altitude;
- 4.4 km for Ku-band Envisat at ~800 km altitude
- 5.5 km for Ku-band Jason at 1336 km altitude.

The footprint size also increases in high seas as the sea-state increases. For example, 10 m high significant wave height increases the footprint for Envisat to 7.7 km, and to 9.6 km for Jason (Chelton et al., 1989).

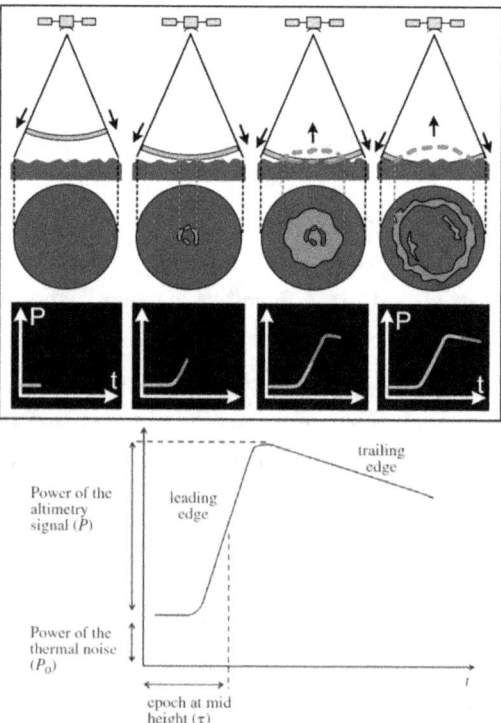

Figure 8.1. Schematic of a time series of a radar pulse. In green, the outgoing radar pulse before it reaches the surface, in red, the pulse reflected from an ocean surface with waves received back at the sensor. The bottom panels show the radar power received at the sensor over time. The main waveform parameters marked are derived from the Brown model, used to determine the different altimetric geophysical parameters. (Credits: CNES)

To obtain a precise estimate of the sea surface topography, the range has to be adjusted for instrument corrections, atmospheric propagation delays (as the radar pulse passes through the wet and dry atmosphere, or due to interactions with the electron content in the ionosphere), and surface geophysical corrections. The details of this processing can be found in Chelton et al. (2001), Pujol et al. (2016), and Escudier et al. (2017). Surface geophysical corrections need to be applied to remove the effects of tides (ocean tides, solid earth tides, loading tides, and the pole tide), rapid atmospheric dynamical effects from wind and pressure forcing that are not resolved by the altimeter temporal sampling, and corrections for the geoid or mean sea surfaces.

The Saral/AltiKa altimeter is the first altimeter to use a high single frequency in Ka-band (35.75 GHz). Ka-band is much less attenuated by the ionosphere than the Ku-band, and is designed to have greater performance in terms of vertical resolution, time decorrelation of echoes, spatial resolution and range noise (Verron et al., 2015). Theoretically, Ka-band pulses are sensitive to rain; however, Saral does not exhibit large missing data during rain events, except for strong rain rates. The Ka-band wavelength is better suited to describing the slopes of small facets on the sea surface (capillary waves, etc.) and gives a more accurate measurement of the backscatter coefficient over calm or moderate seas. Saral also has a smaller footprint and higher resolutions (40 Hz as opposed to 20 Hz for other altimeters), leading to a noise reduction of a factor of two compared to Jason-class altimeters for wave heights greater than 1 m. The increased spatial resolution and lower noise are associated with a more focused cross-track footprint for Saral, being nearly half the size of the Jason-class altimeters.

Along-track SAR altimetry, Cryosat-2, Sentinel-3

We have seen that the LRM altimetry is based on the pulse compression of a series of individual radar pulses and, when averaged over 1000s of pulses (at 1Hz), the footprint size is expanded beyond the normal pulse-limited diameter. The main interest of the SAR processing is to provide a greatly reduced footprint (less than 5 km^2 for SARM) compared to conventional LRM measurements, which have a footprint size of more than 200 km^2. This gain in spatial resolution is in the direction of displacement of the satellite (along-track). Delayed Doppler/SAR altimetry exploits a coherent processing of groups of transmitted pulses in order to zoom in on a smaller along-track bin (Raney, 1998; Boy et al., 2017; Frappart et al., 2017; Escudier et al., 2017). Although the SAR processing is complex, the basic principle is based on the Doppler effect due to the moving satellite.

Consider the frequency shift caused by the Doppler effect generated by the moving satellite, which affects the signal coming from a given scattering point within the footprint on the ground. If the satellite flies horizontally, the Doppler shift is zero for the scattering point at nadir, but it increases almost linearly with the along-track distance between the scatterer and the nadir point. Figure 8.2d shows a schematic of this effect. The straight lines correspond to contours of equal linear Doppler shift and the circles are a selection of contours of equal range seen from the radar. The satellite is moving to the right along the track in this schematic. In essence, the SAR processing identifies the exact position of the return power and removes the power coming from all the vertical bands shown on Figure 8.2 except from the central band (called the zero Doppler band), whose width is typically 340 m. This is then repeated for the adjacent Doppler bands to form a stack of waveforms. This is analogous to the pulse compression technique used for range compression. For the pulse compression in range (cross-track or perpendicular to the flight direction), the linear variation of frequency in time is imposed by the instrument, whereas it is imposed by the Doppler frequency shift for the pulse compression in azimuth (along-track).

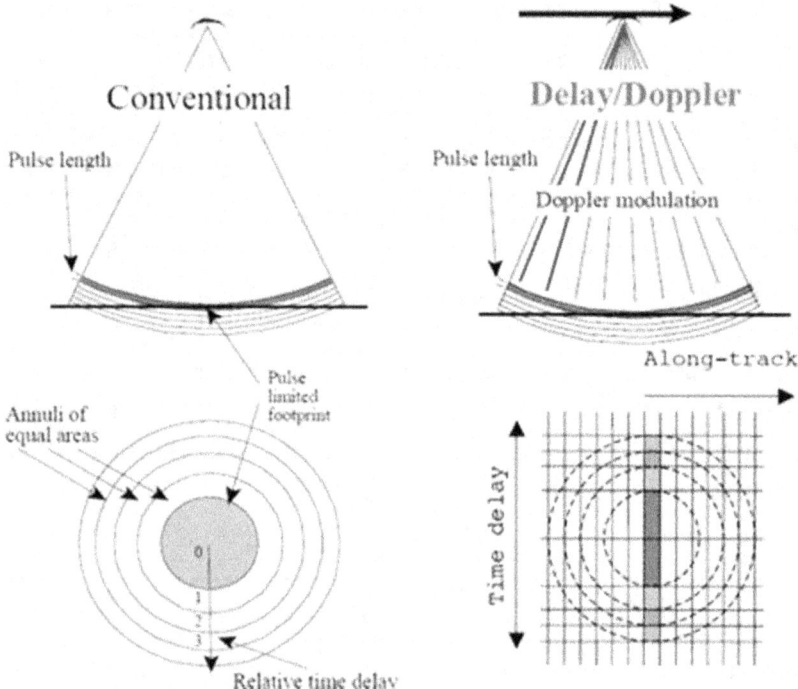

Figure 8.2. (left panels): Schematic of conventional LRM pulse-limited altimetry showing the concentric annuli of pulses being reflected back to the satellite, and contributing to the steady increase in power over the Brown waveform. (right panels): SAR uses a combination of the range information and the time delay. Here the circles represent the iso-range contours and vertical lines the isodoppler contours). The horizontal line is the trajectory of the satellite. (Credits: R.K. Raney, Johns Hopkins University Applied Physics Laboratory)

As the satellite moves along-track, the same vertical band in Figure 8.2d can be observed many times from the front of the round footprint, then the centre, then the back. A multi-looking technique is used in the SAR processing, taking the sum of the contribution of many co-located vertical bands during a 2 sec period, instead of just analyzing the millisecond observation from individual pulses. The advantage of this "stacking" within the SAR processing is that the averaged return signal (and integrated noise) are concentrated in the smaller central zero Doppler band, being 340 m along-track and ~5 km across track. In comparison, the convention LRM mode has the signal and noise integrated over the spherical diameter surface (Figure 8.2), whose footprint size varies depending on frequency, altitude, and surface roughness conditions. The SAR processing gives a more focused signal, and a much smaller noise level, since the noise is integrated over a smaller surface area with less sensitivity to sea-state. It can also approach closer to the coast or islands or sea-ice, with fewer perturbations to the radar signal when the ground track is perpendicular to the coast/sea-ice boundary.

Improvements in along-track data and mapping capabilities

Over the last decade, there have been important advances in the processing and reprocessing of the along-track altimeter data from all missions, as well as in the quality of the multi-mission mapping by DUACS/AVISO, which is used in most oceanographic studies. Although the mesoscale band is the most energetic component of sea level, small-scale errors and noise are two of the factors limiting our observation of the smaller mesoscales. The choice of editing, filtering, or mapping can also impact on the ocean signals we can observe. Here, we give a brief overview of the recent improvements in along-track data processing and multi-mission mapping.

Reprocessing of along-track data

Many studies use the along-track sea level anomaly (SLA) data directly (for example, for calculating along-track wavenumber spectra, for calculating fine-scale currents in a local area, or for assimilation in models). There have been many improvements in the quality of these along-track data, described in detail by Dibarboure et al. (2011) and aided by the ESA Climate Change Initiative effort (Ablain et al., 2015). The most recent altimeter datasets have been reprocessed using the latest standards for climate studies and applied homogeneously to all missions. The radar waveforms have been reprocessed, rigorous editing and selection processes have been applied, especially important in coastal and high-latitude regions, and the most up-to-date standards applied for orbits and atmospheric forcing (Ablain et al., 2015). This has improved the signal-to-noise for each mission.

The next step is to cross-calibrate the different altimeter missions within the constellation. This is done first between the reference missions (T/P, J1, J2, J3) to reduce the long wavelength, geographically correlated errors. A second cross-calibration is then performed to reduce the long wavelength errors between the reference missions and the other altimeter missions. The details are described in Dibarboure et al. (2011) and Pujol et al. (2016). The long wavelength reduction removes orbit errors, as well as errors in the tidal corrections, atmospheric forcing corrections, or in the mean sea surface for non-repeat altimeter missions. This can potentially improve the quality of the mapped mesoscale signals.

Once the missions are cross-calibrated, SLAs are calculated by removing a reference surface: either a precise time-mean along-track profile (for the missions on a long-term repeating track) or a gridded 2D mean sea surface product (for the new missions or non-repeating missions). There has been recent progress in improving the time-mean along-track profile calculation over a 20-year period, especially in coastal and high-latitude regions (Pujol et al., 2016), and a new generation of 2D mean sea surface products are being developed using more geodetic altimeter data from Cryosat-2 and Jason-1 (DTU2015; O. Anderson; CNES-CLS-2015; P. Schaeffer; pers. Communication). These improvements in the mean sea surface have a direct impact on the mesoscale SLAs, as shown by Pujol et al. (2016) and Dufau et al. (2016).

Finally, due to the limited temporal sampling of the altimeters, rapid barotropic signals are not well-resolved. In the mapped and along-track SLA data, their high-frequency component < 20 days is removed using a dynamical atmospheric correction (DAC) based on a barotropic model forced by wind and atmospheric pressure (Carrère and Lyard, 2003). This is a relatively large-scale correction and should not impact the barotropic component of the mesoscale dynamics, except in the coastal zone where the DAC correction has higher resolution and resolves smaller scales. In general, the corrected altimetric data represents both barotropic and baroclinic motions at mesoscales.

Multi-mission mapping

Since 1992, the altimeter constellation has varied over time, having two to four altimeter missions available that can be combined for better spatial-temporal resolution of the mesoscale field (Table 1). Jason-class altimeters follow a repeat cycle of ten days designed to monitor ocean variations so they pass over the same points fairly frequently, but their ground tracks are some 315 km apart at the equator – more than the average span of an ocean eddy. On the other hand, ERS/Envisat and Saral only revisit the same point on the globe every 35 days, but the maximum distance between two tracks at the equator is just 80 kilometers. Figure 3.2.3 shows an example of the spatial coverage of the altimeter constellation in June 2015 in the Mediterranean Sea, with all tracks cumulated over one day and ten days. Combining these different sampling missions gives a much better spatial-temporal evolution of the ocean circulation. Even after ten days, some regions are well-sampled, but others still have gaps of ~150 km.

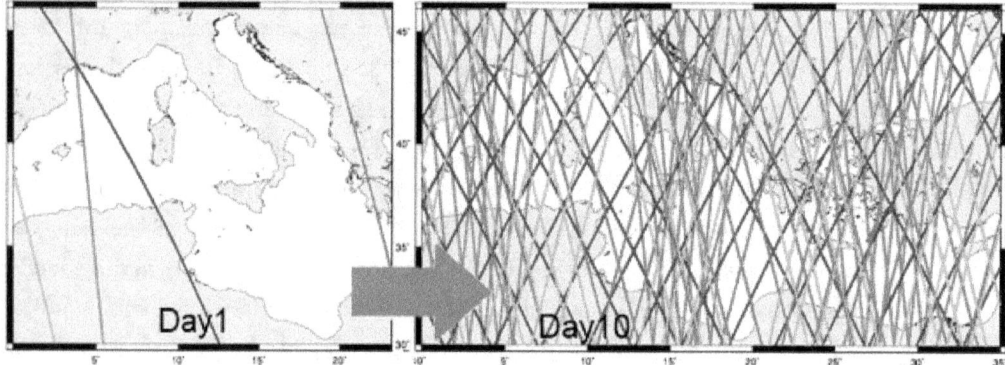

Figure 8.3. Grounds tracks for the four operational satellites on June 2015, over the Mediterranean Sea: Jason-2 ground tracks are blue, Saral/AltiKa are green, Cryosat-2 are red, and HY-2A are pink. (Credits: AVISO)

Multi-mission merged SLA maps are used for ocean studies, where the along-track data are sub-optimally interpolated onto a fixed grid using Gaussian spatial and temporal decorrelation functions (Dibarboure et al., 2011). Two series of maps exist. The first is a two-satellite or "reference" series of maps that has consistent sampling over the entire period from 1992 to 2015 using data from T/P-J1-J2 and ERS-Envisat-Saral (Cryosat-2 data is used in the small gap between Envisat and Saral). This allows consistent sampling for long-term studies of interannual variability. A second "all-satellite" series of maps includes data from all available missions, with increased spatio-temporal sampling. This is available from 2000 onwards, and provides better anisotropic structure for mesoscale studies (Pascual et al., 2006) and improved sea level and velocity variances, which are used to evaluate the realism of ocean circulation models (Le Traon and Dibarboure, 2002). An example of the mapping for the two series is given for the Mediterranean Sea for one date, in Figure 8.4.

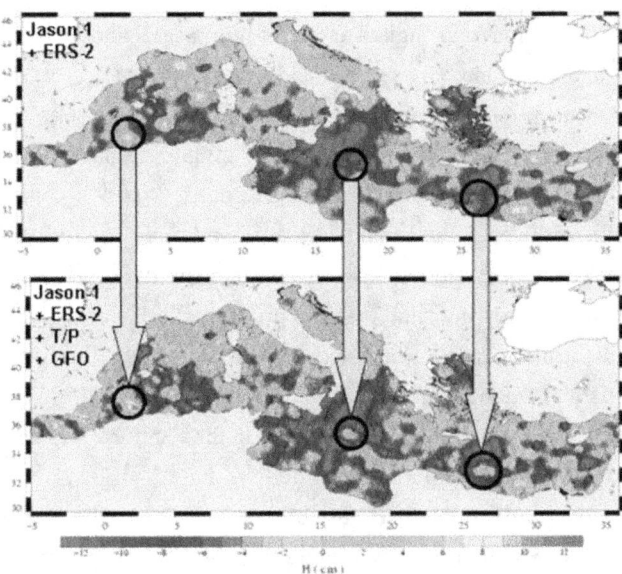

Figure 8.4. SLA maps over the Mediterranean on June 11, 2003, made from Jason-1 + ERS-2 (top) and Jason-1+ERS-2+T/P+GFO (bottom). Merging of the four satellite data shows eddies (circles) that are invisible or just barely visible with two satellites and much better sorted out with four of them. (Credits: A. Pascual, IMEDEA).

The latest version of the DUACS DT2014 gridded products are described in (https://www.aviso.altimetry.fr/en/data/products/sea-surface-height-products/global.html) and have a number of advantages for mesoscale applications (Pujol et al., 2016). The data are referenced to a 20-year period, which provides better interannual variability; the editing has been improved and homogeneous standards applied to all missions; the along-track data input in the system have less filtering, maintaining scales down to 65 km wavelength (older versions had along-track filtering to 250 km in the tropics, decreasing to higher latitudes); the gridded data are available daily on a ¼° Cartesian grid, reducing interpolation errors for users converting from the original 1/3° Mercator grid; correlation scales vary regionally (correlation scale maps are given in Dibarboure et al., 2011); and the error budget varies regionally and between missions depending on data quality. These processing changes allow additional mesoscale signals at wavelengths less than 250 km, and the global SLA variance has increased by 5%; the EKE variance by 15%. This has particular benefits for mesoscale studies in the boundary region, as noted by Capet et al. (2014) for the eastern boundary upwelling systems.

The mean wavelength resolution of these gridded products remains 2°, i.e., 200 km at mid latitudes (Chelton et al., 2011). A comparison with the along-track data shows that in the 65-300 km mesoscale band, around 40% of the mesoscale variability is missing in these gridded products, mainly associated with the smaller mesoscale signals (this depends very much on latitude) (Pujol et al., 2016). A second consequence of the reduced along-track filtering is that additional signals are now observed, associated with residual internal tide signals in both the along-track and mapped products (Ray and Zaron, 2015; Dufau et al., 2016). As we resolve smaller-scale structures, the separation and identification of internal tides and internal waves becomes more critical. Despite the progress in improving the along-track signal-to-noise, mesoscale studies today are still limited by this along-track noise and the altimeter ground track sampling.

The improved coverage in recent years has allowed a number of regional gridded products to be developed in the Mediterranean Sea (Pujol and Larnicol, 2005; Pascual et al., 2005), the Black Sea, the Mozambique region, the Arctic Ocean, and other regions (see https://www.aviso.altimetry.fr/en/data/products/sea-surface-height-products/regional.html).

Regional studies have created gridded altimeter products with correlation scales that are better tuned to the local dynamics, allowing smaller-scale processes in the coastal zone (Dussurget et al., 2011) or including anisotropic scales associated with bathymetry (Escudier et al., 2012). As an alternative to the statistical optimal interpolation, Ubelmann et al. (2015) recently developed a simple dynamical interpolation method to retrieve SSH in the temporal gap between two SSH fields separated by up to 20 days. This experimental method, based on the conservation of potential vorticity and assuming that most of the SSH variability is carried by the first baroclinic mode, was able to reconstruct gridded SSH maps with wavelengths smaller than 150 km. A more recent version uses the simple dynamical model to introduce dynamical time-evolving structure into covariance functions, which are then applied in an optimal interpolation framework using along-track altimeter data (Ubelmann et al., 2016). Simulations in the Gulf Stream region show reduced mapping errors and a better representation of the smaller-scale eddies.

Global along-track wavenumber spectra of altimetry data: quantifying errors and observable scales

Altimeters fly at 7 km/s and cross ocean basins in a few minutes, providing a "snapshot" of the interaction of different spatial scales of ocean variability. Many altimeter missions on repeat tracks have these snapshots repeated every ten to 35 days, over decadal time periods. This global and decadal coverage from the series of repeating satellite altimetry missions with high along-track resolution provides a unique framework for estimating frequency-wavenumber spectra of ocean SSH variability. Figure 8.5 shows the globally-averaged wavenumber spectrum of SLA from different altimeter missions with different technology at fine resolution: Jason Ku-band LRM (20 Hz), Saral Ka-band LRM (40 Hz), and Sentinel-3 SAR (20 Hz). The wavenumber spectrum of SSH or SLA is generally 'red,' with the largest SSH fluctuations on the largest spatial scales and increasingly smaller fluctuations at smaller scales. This decrease of SSH signal with decreasing spatial scale means that the smallest spatial scales that can be resolved are limited by the noise floor of the altimeters.

Figure 8.5a shows that these three missions with different ground tracks are observing similar ocean SLA power spectral density at large scales and the same cascade of spectral energy down to scales of 100 km (wavenumber of 10^{-2} cycles per km - cpkm). At smaller scales, from 100-10 km (wavenumber of 10^{-2} to 10^{-1} cpkm), the three missions observe very different SLA energy. Jason-2 (in green) has a large spectral "bump" from 10-30 km wavelength due to inhomogeneous sea-state perturbations within the altimeter footprint (wind, waves, rain cells, ...; Dibarboure et al., 2014). Since the Jason-2 altimeter footprint is relatively large, these conditions add more noise, masking the ocean signal out to 70-100 km wavelength. Saral and Sentinel-3 SAR mode have smaller footprints and are designed to have lower noise. Saral still shows a spectral "bump" at smaller scales, mainly due to sea-state effects that mask the ocean signal out to scales of 30-50 km. Sentinel-3 SAR technology shows a different red noise slope at small wavelengths and a greatly reduced noise level compared to the Jason-class of altimeters. Work is ongoing to understand the spectral shape of SAR instrument noise in varying wind-wave-swell conditions and the impact of small-scale ocean signals such as internal waves on this red spectral slope.

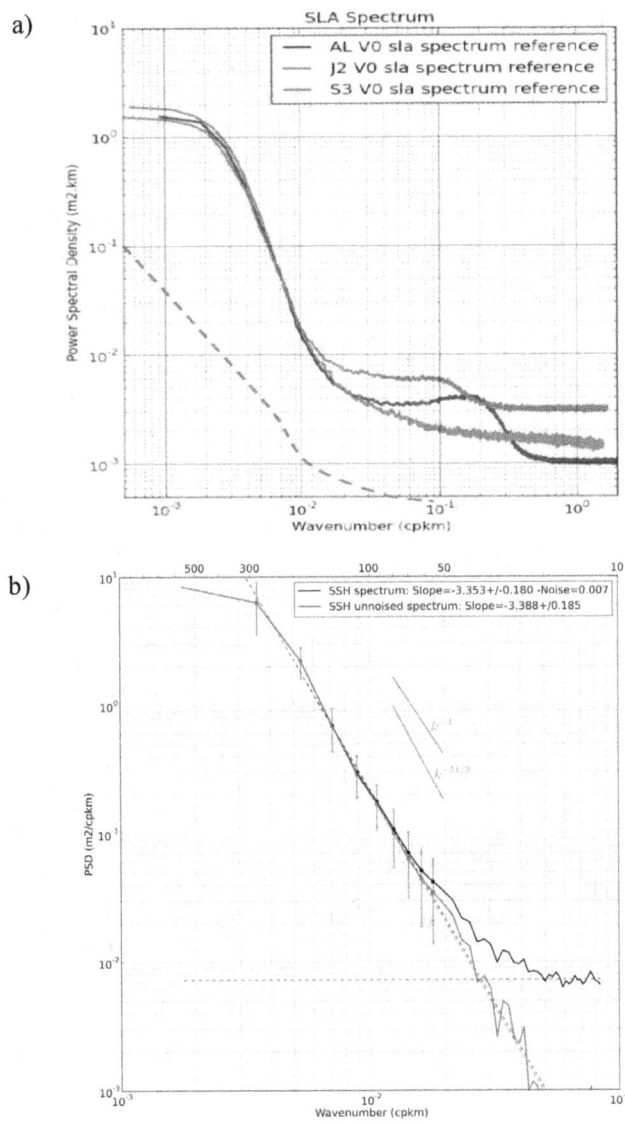

Figure 8.5. a) Global mean SLA wavenumber spectra from high-resolution along-track altimetry data from Jason-2-LRM (in green), Saral-LRM (in blue) and Sentinel-3 SAR (in red). 20 Hz data are used for Jason and Sentinel-3, Saral is at 40 Hz. The dashed line shows the projected SWOT error levels. b) An example of mean SSH wavenumber spectra from 1 Hz along-track Jason-2-LRM in a 10x10° box in the Kuroshio Current centered on 294°E, 39°N. The red curve is the unbiased spectra with the constant noise level removed (horizontal dashed line) from the original spectrum (black curve). (Credits: CLS/CNES and Dufau et al., 2016)

If we look at the averaged 1-Hz, 1-sec data, the spectral "bump" is absorbed into a white noise floor, as shown in the example for Jason in Figure 8.5b (black line) for the energetic Kuroshio Current region. If this noise floor is stable, it can be subtracted as a constant value from the global spectra to reveal the unbiased spectra, showing an ocean signal that decreases more smoothly to smaller scales (red line). Geostrophic turbulence theory predicts a particular slope of the power-law dependence of the spectrum on wavenumber, k, in a log-log plot, being k^{-5} in SSH or SLA, and k^{-3} in surface velocity. Figure 8.5b shows that over the Kuroshio, the SSH spectra are flatter than predicted by geostrophic turbulence.

Different studies have used this technique to estimate how the unbiased spectral slopes vary from one region to another. Figure 8.6 shows the global distribution of spectral slopes of unbiased SSH wavenumber spectra from Xu and Fu (2012) that have a large regional variation. Spectral slopes in the more energetic western boundary regions and the Antarctic Circumpolar current can reach k^{-4} or $k^{-4.5}$, but never the k^{-5} of geostrophic turbulence theory. Spectral SSH wavenumber slopes are much flatter in the eastern basins and the tropics. The reasons for this discrepancy are being actively explored, but the presence of high-frequency internal tides and internal waves may increase the spectral SSH energy at short wavelengths and flatten the regional spectra (Richman et al., 2012; Rocha et al., 2016; Dufau et al., 2016).

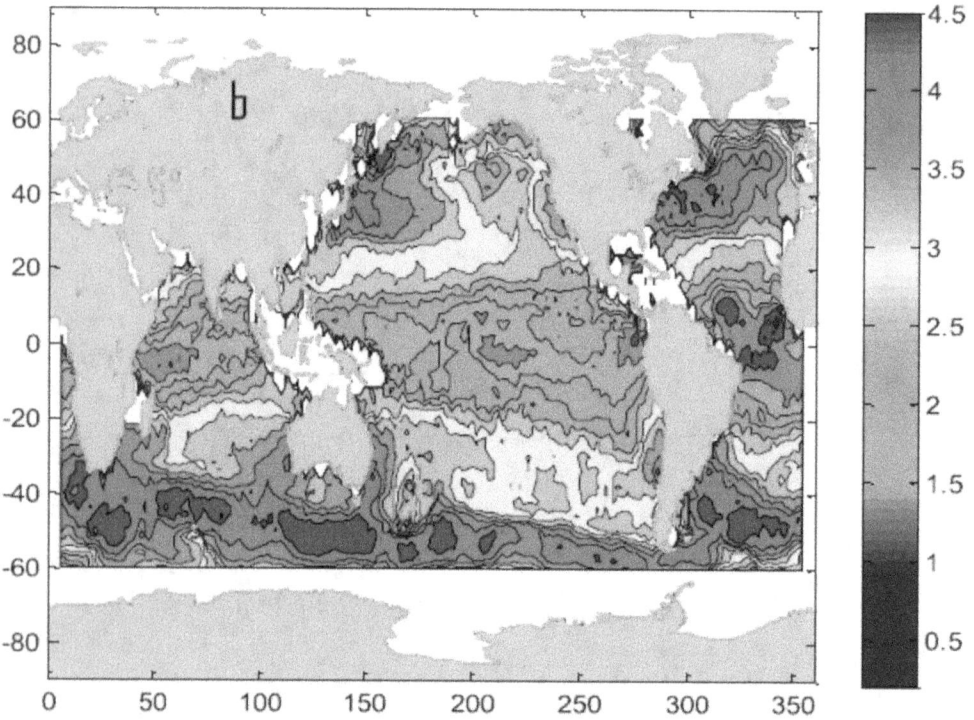

Figure 8.6. Global distribution of spectral slopes of unbiased SSH wavenumber spectra over the "mesoscale" band of 70-250 km estimated from Jason-1, after removing the noise. After Xu and Fu (2012).

Major measurement error contributions to SSH at different space-time scales

In the previous section, we have seen how the global altimetric spectral noise levels vary from one mission to another. The analysis of along-track SSH wavenumber spectra also allow us to investigate the seasonal and regional patterns of the 1 Hz noise level for each satellite.

Figure 8.7 shows an example of how the altimetric noise levels vary regionally (left) and seasonally (right), based on Jason-2, Saral and Sentinel-3 data (O. Vergara, personal communication). The 1 Hz altimetric noise is estimated from a horizontal straight line fit to the averaged wavenumber spectra from 10-25 km wavelength (see Figure 8.5b). The averaged wavenumber spectra are calculated from all of the edited along-track tracks occurring in 15°x15° boxes over the entire period (left) or each season (right). The decreasing instrument noise from Jason-2 to Saral to Sentinel-3 is clearly shown in the zonally averaged values (middle panel). The right panel in Figure 8.7 shows that the highest noise levels occur in winter in both hemispheres. Regions with high noise levels also have a high hump artifact (Dibarboure et al., 2014), i.e., where the altimeter waveforms include inhomogeneity within the footprint due to surface roughness changes, rain cells, etc., that increase the altimeter error level for wavelength between 10 and 25 km. Different techniques have been proposed to reduce this effect in the future using a dedicated data selection (Dibarboure et al., 2014) or by applying a two-step retracking algorithm on Cryosat-2 LRM data (Garcia et al., 2014).

There is a strong seasonal variation in the SSH error level since all factors contributing to the 1 Hz error (significant wave height, rain wells, wind streaks, and ocean slicks) vary seasonally and regionally. Figure 8.7 shows that the maximum SSH error level occurs during winter (December-January-February for the Northern hemisphere, June-July-August for the Southern hemisphere) with higher values in the south (3 cm rms compared to 2.5 cm rms for the north). The dominant contributor to the increased noise in winter is the higher sea-state. Around the Equator, a dipole in the SSH error level is present during winter and summer, with strongest values in the Inter Tropical Convergence Zone (ITCZ) and weak values in the trades wind regions on either side. These higher values are particularly due to rain cells and so-called sigma-0 bloom patches in the ITCZ (Dibarboure et al., 2014).

Another source of errors comes from the mean sea surface (MSS) used, particularly for non-repeat or new orbit missions where the SLAs are calculated with respect to gridded MSS products. Although these MSS products have greatly improved over the last years, including recent geodetic or drifting altimeter missions, MSS errors still impact the altimetric signal over wavelengths less than 100 km. Pujol et al. (2018) showed that MSS errors can explain 30% of the global SLA variance on new or drifting missions (e.g., Sentinel-3) and the error can be 2.5 times higher for unchartered tracks over rough bathymetry.

Many altimetry users apply an altimeter error for each mission that is constant in space and time. In the future, it could be beneficial for users to take into account how the SSH error level is modulated in space and time (e.g., for the assimilation of along-track altimetry in ocean models).

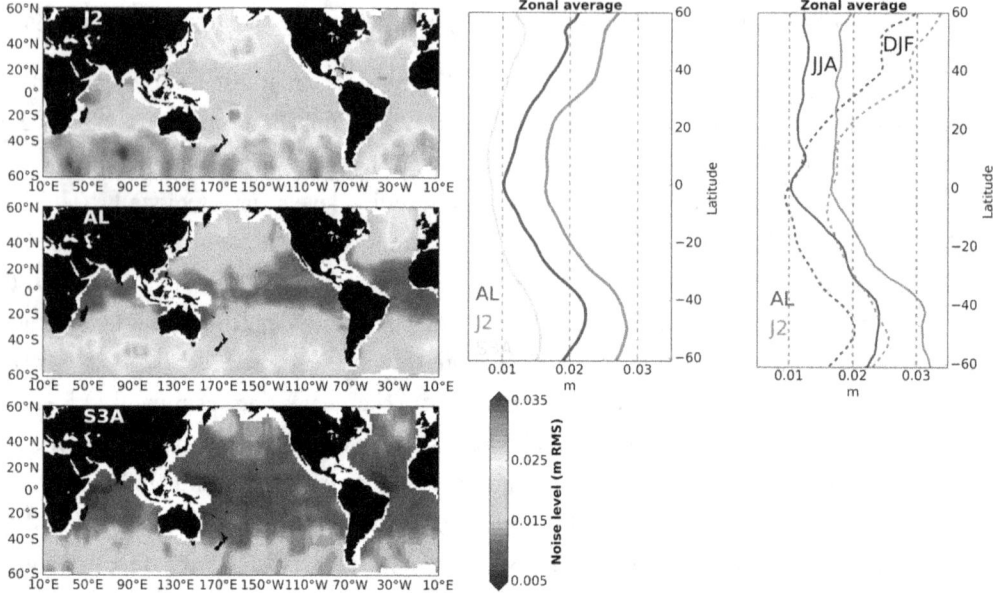

Figure 8.7. Left panel: 1 Hz noise levels (in m rms) from different altimeter missions based on mean SSH wavenumber spectra calculated in 15°×15° boxes from (top) Jason-2, (middle) Saral, and (bottom) Sentinel-3A. Middle panel: Zonally-averaged noise levels for each mission. Right panel: Seasonal variations in zonally-averaged noise levels for the Jason-2 (red) and Saral (blue) missions for Jun-Jul-Aug (JJA solid) and Dec-Jan-Feb –DJF, dashed). (Credits: O. Vergara, LEGOS)

Fine-scale altimetry near the coast

A key issue in satellite altimetry is the monitoring of coastal oceans (Vignudelli, et al., 2011). The coastal ocean dynamics have shorter space and time scales and larger SSH amplitudes due to their shallower bathymetry. Narrow coastal currents and tides need to be monitored due to their important socio-economic effects (pollution control, biomass advection, marine safety and shipping, …). Changing sea level at the coast impacts coastal morphology and erosion occurring over a large range of time scales, including extreme events via storm surges.

The excellent monitoring of the open-ocean dynamics from altimetry is mainly due to the fact that we can resolve the scales of the most energetic mesoscale features (150-200 km). The scales resolved by these gridded maps are not well-adapted to the smaller, faster coastal dynamics. For the along-track data, major problems exist – the footprint size of 200 km^2 for conventional altimetry is very large and averages over different dynamical regimes close to the coast (SARM with 5 km^2 footprint size is better adapted, especially when the ground tracks are perpendicular to the coast). The distance between altimetric ground tracks is also large compared to the smaller space and time scales of the coastal dynamics. In addition, as the along-track altimeter data approaches the coast, the data returned from the radar footprint becomes perturbed by the coast or islands/shallow banks/sea-ice (see Figure 8.8). The standard Brown waveform retracking fails and data is flagged around 10 km from the coast.

Figure 8.8. Schematic of the along-track radar waveforms being perturbed by the presence of land in the footprint. (Credits: ESA)

Additional problems affect the availability and quality of the coastal altimeter data. The radiometer measurements used to derive the wet-troposphere correction becomes perturbed even further from the coast due to their larger footprint at different frequencies, and standard data is lost up to 25-50 km from the coast. Other corrections such as the ionospheric correction or the ocean tides and dynamic atmospheric correction require specific processing or models for the coastal band.

Over the last decade, major advances have extended the capabilities of satellite altimetry for the observation of the coastal circulation and the complex interactions between the coastal and the open ocean circulation. Progress has been made in reducing the altimeter footprints and noise to approach closer to the coast (CryoSat-2- SAR, Saral/AltiKA, Sentinel-3-SAR), but also in processing algorithms, corrections, and products for coastal applications (Vignudelli et al., 2011; Obligis et al., 2011; Fernandes et al., 2014, Birol et al., 2017). The improved coastal altimeter data and their derived sea level and wind–wave data are being integrated into coastal observing systems. These data provide essential monitoring for both research and operational applications in these coastal regions where in situ measurements are sparse.

Future SWOT Mission – 2D Swath SSH Observations

The last 30 years of satellite altimetric studies have greatly advanced our understanding of the large-scale ocean circulation and its interaction with the larger mesoscale dynamics (Fu and Cazenave, 2001; Morrow et al., 2017). We are entering a new period where ocean models are evolving at high-resolution and global models, with and without tides, are now available (HYCOM at 1/25° - Chassignet and Xu, 2017; 1/48° MITgcm, 1/60° NEMO NATL, etc.). However, the dynamics of these models cannot be validated due to the lack of global observations at these finer scales. The future SWOT SAR-interferometry wide-swath altimeter mission is designed to provide global 2D SSH data resolving spatial scales down to 15-20 km, for a 2 m high significant wave height (Fu et al., 2012; Fu and Ubelmann, 2014); these values vary depending on the sea-state. These observations will fill the gap in our knowledge of the 15-200 km 2D SSH dynamics, important for the ocean horizontal circulation and kinetic energy budget, but also for driving energetic vertical velocities and tracers' transports (Levy et al., 2012).

As the name suggests, the Surface Water and Ocean Topography (SWOT) mission will bring together two scientific communities – oceanographers and hydrologists. For the hydrology community, SWOT SAR interferometric data will enable the observation of the surface elevation of lakes, rivers and flood plains, and will provide a global estimate of discharge for rivers > 100 m wide, and water storage for lakes > 250 m^2. SWOT will also provide unprecedented observations in the coastal and estuarine regions, of interest to both communities. The science objectives covering all disciplines are outlines in the SWOT Mission Science Document (Fu et al., 2012).

Here, we provide an overview of the SAR-interferometry measurement technique, the chosen orbits and space-time sampling, the ocean science objectives and applications, and the ocean error budget including the impact of waves. We present some of the research studies aiming to understand the new 2D SSH dynamics that will be observed at scales from 15-200 km. In particular, there will be a significant SSH contribution from high-frequency contributions such as tides, internal tides, and internal gravity waves, which are not in geostrophic balance. Observing their regional and seasonal signals and their interactions will be a great achievement for SWOT. A challenge is in separating the balanced motions from the high-frequency signals in order to calculate geostrophic velocity or vorticity. The spatial sampling is excellent, but weaker temporal sampling presents a challenge in mapping the high-resolution 2D "snapshots". We discuss the new 2D and 3D mapping techniques being developed for SWOT swath data. Finally, we present some early studies on assimilating SWOT simulated SSH data in operational ocean models.

Principle of SARin SWOT measurements

Interferometric Synthetic Aperture Radar (InSAR) has been used extensively during the past three decades for fine-resolution mapping and measuring changes of the surface of the Earth. The principle is to combine two SAR images of the same place on the ground taken from two distinct positions of the radar A_1 and A_2 separated by a distance B called the baseline (see Figure 8.9). Figure 8.9 is drawn in the plane orthogonal to the satellite speed vector, the line A_1, A_2 is then in the across-track direction. This explains the name cross-track interferometry.

In the past, the two images were standard SAR images taken at two different times for successive overflights of the same region by the same radar. The incidence angle (or look angle θ shown on Figure 8.9) was typically between 20 and 40 degrees. In the case of SWOT, the satellite carries two antennas and then the two images can be obtained simultaneously. One other important difference is that the look angle theta is much lower (between 0.6 and 4° for SWOT), so these measurements are quasi-nadir SAR interferometry. This has some noticeable effects that are discussed at the end of this section.

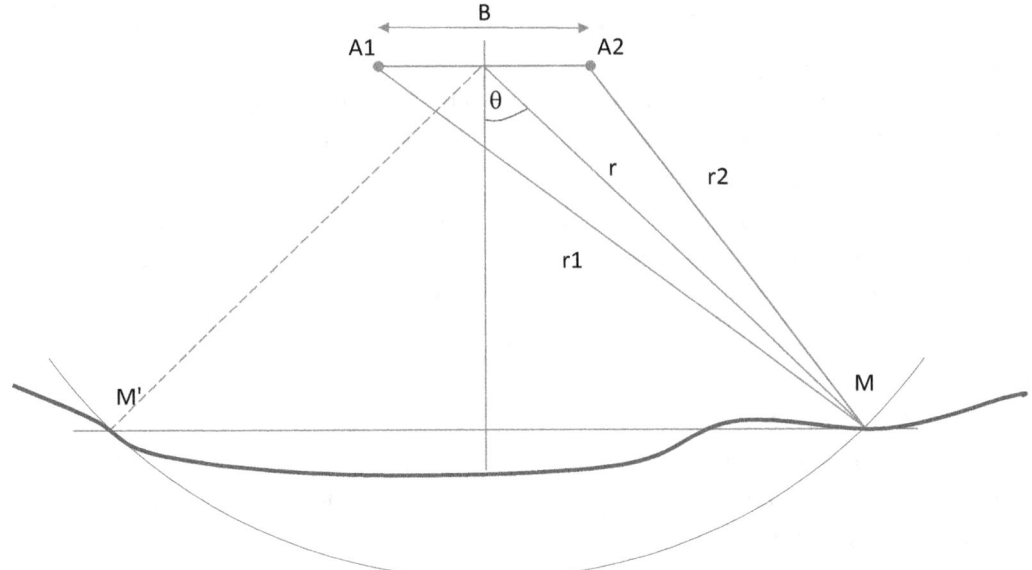

Figure 8.9. Principle of cross-track interferometry. After Frappart et al. (2017).

The two radar measurements can be used to determine the range r_1 and r_2 with a range resolution of around one meter (0.75 meter for SWOT). However, the instruments also measure the phase of the signal, and the interferometric phase Φ can be computed for each bin of the interferogram line (which is analogous to the waveform defined in nadir altimetry).

$$\Phi = \Phi_1 - \Phi_2 = 2\pi \frac{r_1 - r_2}{\lambda} \ (2\pi) \tag{2}$$

where λ is the wavelength of the electromagnetic wave emitted by the altimeter.

This measurement is made with great accuracy but the phase difference between the two return signals has an ambiguity of 2π, which is removed by a processing called phase unwrapping (Rodriguez et al., 2017). After phase unwrapping, the absolute phase Φ_a is known and the differential range r_1-r_2 can be deduced from it.

$$\Phi_a = 2\pi \frac{r_1 - r_2}{\lambda} \tag{3}$$

The incidence angle θ can also be computed from the knowledge of Φ_a.

$$\Phi_a = \frac{2\pi}{\lambda} B \sin(\theta) \tag{4}$$

Then, knowing the range r and the angle θ, the position in 3D space of the point M can be obtained. This process is called geolocation and height retrieval. This is a very simplified description; again, further details are given in Rodriguez et al. (2017).

When the surface is flat enough (such as over the ocean) and the backscattering is almost uniform, the SWOT measurements can be used to retrieve water heights from each line of the interferogram. This provides a height profile in the plane of Figure 8.9. The measurement is repeated with a high frequency, and assembling the successive profiles allows the processor to build an image of the surface height over the swath width of 50 km (see Figure 8.10).

The processing of the image formation is done independently for the left and the right swaths of SWOT. Note that if the point M' is symmetrical with M in respect to the vertical line passing by the middle of the two antennas, then M' is measured at the same range as the signal coming from M (Figure 8.9). This leads to an ambiguity in the detection of signals coming from the left and right swath that has to be removed at the instrument level. For SWOT, this is done using the right and left antenna patterns and using different polarization of the signals (one pulse is emitted in H-polarisation, the next in V-polarisation, etc.). For InSAR on CryoSat-2 used over the polar ice-caps, the goal was to provide measurements at nadir and there is no device to remove the left-right ambiguity. So, Cryosat-2 InSAR observations are not well-adapted for flat ocean observations. However, if the incidence angle on the ground is very different on the left and right sides (e.g., in the presence of steep terrain slopes in the cross-track direction on a polar ice-cap), then the radar-returned power on one side is close to zero. Consequently, the return signal is considered to come only from one side only, removing the left/right ambiguity. In these cases, the near-nadir CryoSat-2 SAR interferometer measurement can identify the direction of the main signal and then obtain both a height measurement and position.

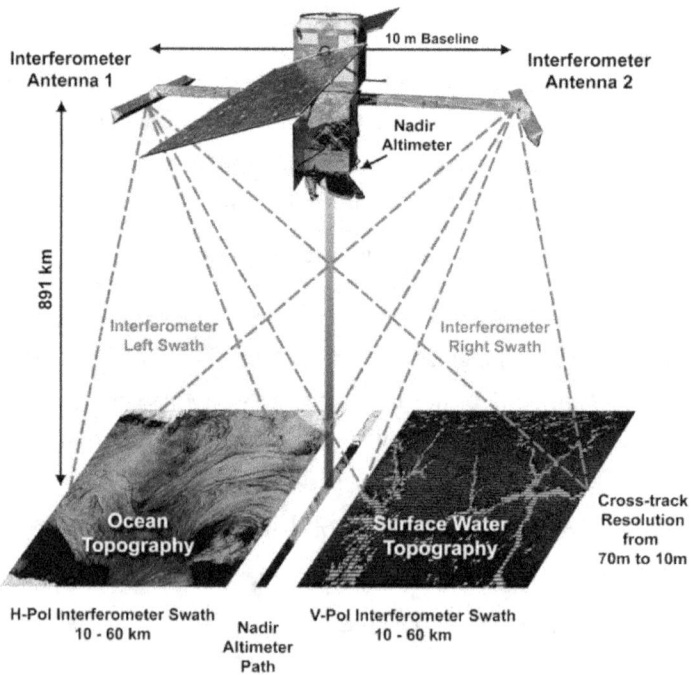

Figure 8.10. Schematic of the SWOT measurement technique using the KaRIn instrument for SAR-interferometry over the two swaths, and a Jason-class nadir altimeter in the gap. (Credits: NASA-JPL)

SWOT uses SAR processing to refine the along-track resolution of the return signal, as explained in the second section of this chapter. The interferometric processing refines the cross-track resolution. The SWOT SAR-KaRIn instrument provides a basic measurement resolution of

2.5 m along-track, and ranges from 70 m in the near-nadir swath to 10 m in the far swath. Over the terrestrial water surfaces, an onboard pre-summing is performed and data are downloaded at 5.5 m resolution along-track, with the maximum cross-track resolution. Full interferometric processing is then performed on the ground. Over the 70% of the Earth's surface covered by the oceans, the huge amount of data being produced cannot be downloaded from the satellite. Instead, so-called "low-resolution" data are pre-processed onboard and building blocks of nine interferograms at 250 m posting (and 500 m resolution) are downloaded from each antenna. Other parameters that are useful for the surface waves and front detection, such as the 250 m resolution backscatter images, are also downloaded. These data are combined through the geolocation and height calculation into a 250 m x 250 m expert SSH product with higher errors in swath co-ordinates, and a standard 2 km x 2 km product in along-swath and geographically fixed coordinates. The error estimate of these averaged data is 2.25 cm/km^2, more than an order of magnitude smaller than conventional Jason-class measurements.

SWOT Orbit coverage and space-time sampling

The SWOT Science Definition Team investigated the SWOT nominal orbit coverage in detail. Any orbit choice is a trade between good spatial coverage and temporal coverage, but not both. The two major communities using SWOT observations had different objectives – the hydrologists needed global coverage of the smallest lakes and rivers, including at the Equator, on a monthly time scale. The oceanographers needed to cover the small ocean scales evolving rapidly – yet full space and time coverage is not possible with one satellite. The final orbit was chosen to cover most of the terrestrial surface waters and oceans up to 78°N and S (inclination of 77.6°) at 890.6 km altitude on a 21-day repeat orbit, allowing near-global coverage after the full cycle (see Figure 8.11b). SWOT has a non-sun-synchronous orbit and its inclination and repeat sampling were specifically chosen to resolve the major tidal constituents, particularly important to resolve the coastal tides, the high-latitude tides, and to advance on the observation of internal tides.

The orbit has a one- and ten-day sub-cycle, with a westward coverage of tracks. Figure 8.11a shows how the tracks are laid down after three days. One day of coverage loosely covers the globe with big gaps between the swaths (one-day sub-cycle), then successive tracks are laid down to the west leaving a gap of approximately one swath at the equator, but helping to monitor the predominant westward propagation of mesoscale instabilities. After ten days, the whole globe is covered (the ten-day sub-cycle) and over the next ten days, the entire pattern is laid down again, but shifted westward to fill in the small holes between tracks. This satisfies the hydrologists with complete global coverage and gives the oceanographers good global coverage every ten days. Given the overlapping swaths, local coverage at certain latitudes is observed more frequently (two to four times at mid-latitude, up to seven times at high latitude).

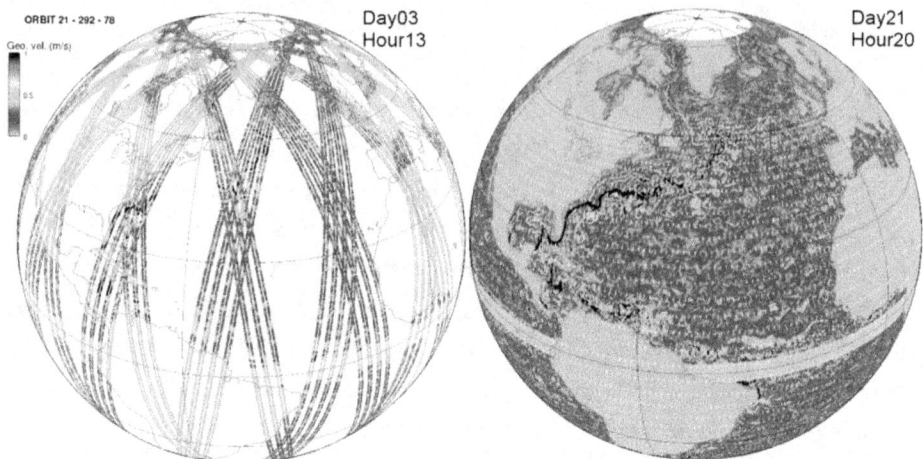

Figure 8.11. SWOT's nominal orbit coverage up to 78°N and S after a) 3-days and b) the full 21-days of a complete cycle. Color shows the evolving simulated along-track geostrophic currents for each track. (Credits: C. Ubelmann, CLS)

After the launch planned for 2021, SWOT will spend six months in the so-called "Calibration orbit," where the satellite passes over the same site every day to calibrate the satellite parameters. The first three months of this orbit will be to adjust and calibrate the instrument parameters, the second three months will be available for science studies, including studies of rapidly evolving small-scale ocean dynamics. SWOT will then continue in the nominal 21-day repeat orbit for three years, from 2021 to 2024. Details of the nominal and calibration orbits in different formats are given on the AVISO website: https://www.aviso.altimetry.fr/en/missions/future-missions/swot/orbit.html.

SWOT error budget

As discussed in the previous section, the noise level is important as it will limit the observation of small scale ocean dynamics. The SWOT SAR-interferometry measurement is designed to have small instrument noise to meet the stringent requirement of 2.25 cm^2/km^2 noise over the oceans at short wavelengths, for a 2 m SWH average sea-state. This is calculated over 1 km x 7.5 km averages (to resolve the 15 km Nyquist wavelength). For comparison, the noise level for the Jason series is around 100 cm^2/km^2 at 1 Hz, i.e., averaged over the 1 sec oval footprint or roughly 100 km^2. The SWOT instrument measurement will be very, very precise! For the first time for an altimetric mission, the error budget is also set in terms of wavenumber, so the instrument design must meet long wavelength and short wavelength goals. SWOT needs to account for standard altimetric SSH range errors, but in wavenumber space. SWOT also has specific errors associated with the interferometric calculation, which are detailed in the SWOT Error Budget and Performance document (Esteban Fernandez, 2017) and in Rodriguez et al. (2017). These include:

- Roll errors. The largest error source for swath altimeters, due to the platform roll, which is minimized by having a stiff mast. This error is linear in the cross-track direction and is estimated by crossover calibration.

- Phase error. This has the effect of moving the estimated height along an iso-range line, and is similar to the roll error but with less magnitude. Phase errors can be random, e.g., due to thermal noise, or systematic; systematic errors can be induced by mismatches in the path lengths of the two channels, temperature changes, or differences induced by the antenna patterns and baseline structures.
- Range errors. These errors also occur in along-track altimetry, including orbit errors; atmospheric propagation errors, including the wet and dry tropospheric errors, ionospheric errors; errors in the geophysical corrections: tides, dynamical atmospheric correction, sea state bias, etc.)
- Baseline errors. An error in the baseline (mast) length results in a height error. This is estimated to be very small.
- Radial velocity errors. To have good geolocation with SAR processing, one needs good knowledge of the radial component of the relative velocity between the satellite and the surface. Any mean error in this radial velocity will induce a horizontal displacement of the moving target and an error in the position (and geolocation).
- Wave effects. The combination of the near-nadir incidence of SWOT and moving surface waves leads to several effects, which are detailed in Peral et al (2015) and Rodriguez et al; (2017). These include a "surfboard" effect, where the waves are sampled multiple times leading to distortion of the waves' height and spatial distribution. A second "wave bunching" effect is due to the wave scatterers on a moving surface sending back different magnitudes and signs, leading to a non-linear distortion of the surface. Interferogram averaging can help suppress the wave-bunching effect from leaking into the sub-mesoscale range.
- EM bias. This bias is induced by modulations in the surface brightness that lead to a net lowering of the mean surface

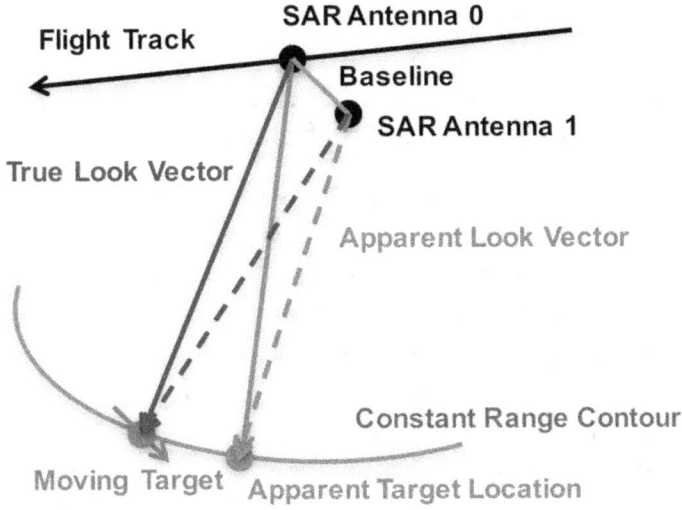

Figure 8.12. Schematic of the apparent shift of a moving target on the ocean along constant range lines, causing a small error in the interferometric height estimate. After Rodriguez et al. (2017).

An example of the impact of radial velocity effects from a moving target (such as surface waves) is shown in the schematic in Figure 8.12. Any moving targets on the ocean surface appear shifted in SAR images. Without knowledge of the target motion (and a correction for it), the difference between the interferometric phase between the target's true and apparent position becomes an interferometric height error. For SWOT over the oceans, the pointing control error is small and the velocity errors from surface waves are thus small, and are included in the error budget.

The breakdown of all of these errors in terms of their wavenumber spectral density is shown in Figure 8.13, from the SWOT Error Budget and Performance document by Esteban-Fernandez (2017). The largest error source is from the KaRIn instrument errors, the random errors dominating at scales < 100 km, gyro and systematic errors dominating the large scales.

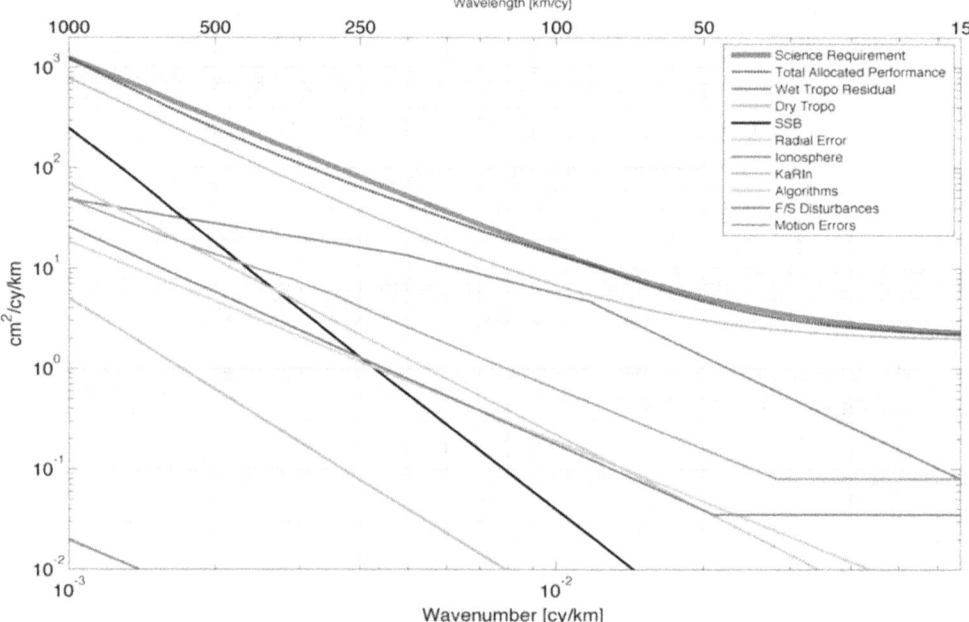

Figure 8.13. Wavenumber spectral estimates of the different components of the SWOT error budget for wavelengths < 1000 km. After Esteban-Fernandez (2017).

New 2D SSH dynamics observed at scales from 20-200 km

The primary oceanographic objective of the SWOT mission is to characterize the ocean mesoscale and submesoscale circulation determined from the ocean surface topography at spatial resolutions down to 15 km wavelength. Current altimeter constellations can only resolve the 2D ocean circulation at resolutions larger than 200 km.

Although the larger mesoscale eddies have been tracked and analyzed with standard altimetry maps (e.g., Chelton et al., 2011), these eddies "spontaneously" appear and disappear in mid-ocean and, at present, we cannot observe the smaller-scale processes that create the larger eddies, nor their cascade down to smaller dissipative scales. Recent high-resolution modelling has highlighted the importance of the smaller scales generated, for example, by energetic instabilities in the deep winter mixed layers. Figure 8.14 shows a comparison of winter and summer surface relative vorticity (the

Laplacian of SSH – upper panel) and upper ocean vertical velocities (lower panel), derived from the 1/48° Earth Simulator model for the North Pacific, with no tides (Sasaki et al., 2014). The late winter surface relative velocity shows a myriad of small-scale structures and filaments associated with strong vertical velocities at small-scales in the deep winter mixed layer, and some injection into the sub-surface layers (left panels). In late summer, when the mixed layer is rather shallow, the near-surface vertical velocities are quite weak, and the surface relative vorticity is less energetic and at larger scales (right panel). These processes are mainly in geostrophic balance and have a SSH signature that should be detected by SWOT.

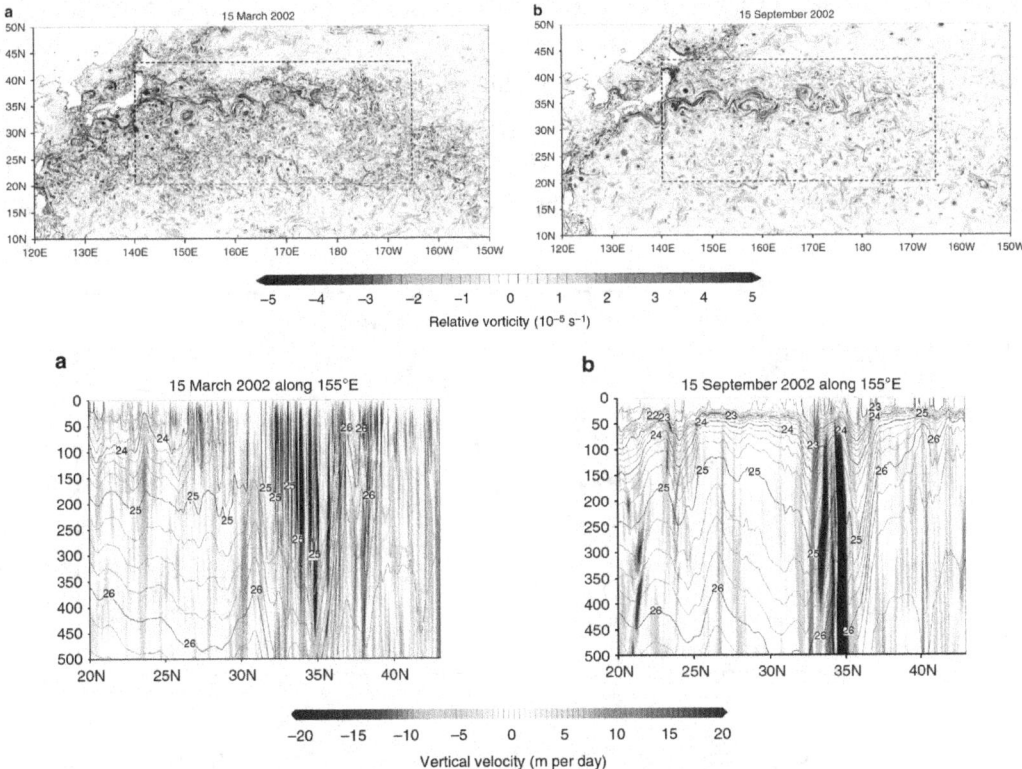

Figure 8.14. Upper panel: Surface relative vorticity in the Northwestern Pacific on (left) March 15, in winter and (right) September 15, 2002, in late summer. Lower panel: model vertical velocity along 155°E for both dates. The depth of the mixed layer is marked by the green contour. After Sasaki et al. (2014).

Kinetic energy and tracer transports

Global study of the circulation at scales of 15 -200 km is essential for quantifying the kinetic energy of ocean circulation and the ocean uptake of heat and carbon that are key factors in climate change. Exchange of heat and carbon between the ocean and the atmosphere is regulated by the large-scale mean circulation, as well as by the mesoscale and submesoscale eddies. Traditional altimeters combined with in situ data have revealed the fundamental role of mesoscale eddies in the horizontal transport of heat and carbon (Dong et al., 2014). The uptake of heat and carbon by the ocean is complete only after the vertical transport process from the surface turbulent boundary layer into the ocean interior is accomplished. The vertical transport is mostly accomplished by the submesoscale fronts with horizontal scales 15-50 km (e.g., Lévy et al., 2012). The SWOT mission will open a new window for studying the SSH signature of these processes.

The transfer of kinetic energy between the different dynamical scales is also of interest. Figure 8.15 also from Sasaki et al. (2014) shows the temporal evolution of the surface kinetic energy (KE) field, which is mainly in geostrophic balance at these scales. The KE field has been filtered to reveal the small-scale motions less than 200 km (in black) and the larger scales in red and blue, which are resolved today by gridded altimetry data. The peak in KE in winter is associated with the energetic small-scale mixed layer instabilities seen in Figure 8.14. Calculations on the transfer of energy reveal that the small-scale processes < 100 km wavelength peak in late winter Mar-Apr; these then feed energy up to the larger mesoscales that develop over time, with scales of 100-200 km peaking in Apr-May and 200-300 km peaking in early summer June (in red). With conventional altimetry maps, only the scales greater than 200 km are observed and the seasonal cycle is biased towards the summer peak, with no observations of the small-scale generating mechanisms in winter. These model results need to be validated by 2D observations, which will be available from the SWOT mission.

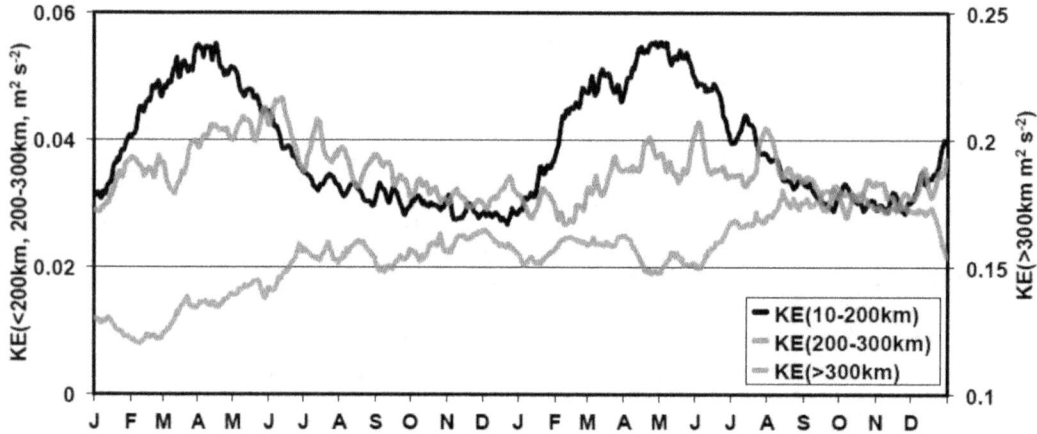

Figure 8.15. Kinetic energy in the 10-200 km band (black), in the 200-300 km band (red) and for scales > 300 km (blue) calculated over a two-year period in the box 20-43°N, 148°E-168°W from the Earth Simulator. After Sasaki et al. (2014).

Climate change and ocean circulation

The new knowledge of the KE of the ocean circulation and the vertical transport of carbon and heat is crucial for understanding the role of the ocean in regulating climate change through the interaction of the mesoscale and submesoscale variability with the large-scale circulation. Accurate knowledge of the large-scale circulation is thus also required to achieve these objectives, posing a requirement on measurement accuracy from the submesoscale to the global scale. For this, SWOT will carry a nadir altimeter positioned between the two swaths (see Figure 8.10), providing accurate SSH monitoring of the larger mesoscales and large scale circulation simultaneously with the 2D KaRIn observations of the smaller scales.

Coastal ocean dynamics

Coastal ocean dynamics are important for many societal applications. As mentioned earlier, they have smaller spatial and temporal scales than the dynamics of the open ocean and require finer-scale monitoring. SWOT will provide global, high-resolution observations of 250 m to 2 km resolution right up to the coasts, for observing coastal currents and storm surges. While SWOT is not designed to monitor the fast temporal changes of the coastal processes, the swath coverage will allow us to characterize the spatial structure of their dynamics when they occur within the swath.

Figure 8.16 shows the winter-time relative vorticity for the MARS model in the Bay of Biscay, with energetic fine-scale structures near the coasts and at the shelf edge, under the influence of density gradient instabilities from the freshwater river plumes and tidal mixing. The continuum from the regional seas to the shelf, to the nearshore region and into the estuaries involves complex coastal dynamics, and SWOT will help monitoring these exchanges over the 3.5 years of observations.

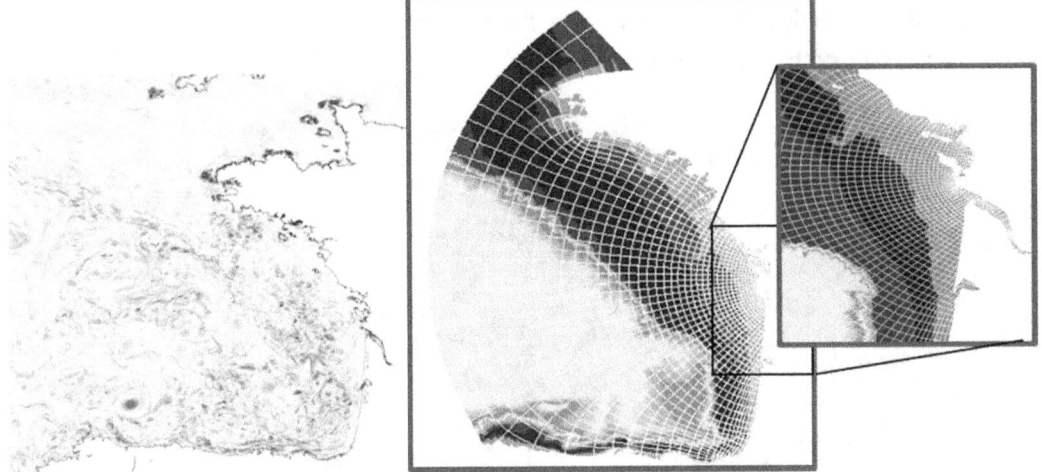

Figure 8.16. (left) winter-time relative vorticity in the Bay of Biscay from the MARS ocean model. (right) a hierarchy of embedded models using the Symphony code, with increased resolution towards the coast and into the estuaries. (Credits: N. Ayoub, LEGOS)

High-frequency contributions to SSH at 20-200 km: coastal tides, internal tides and internal gravity waves

The choice of orbit for SWOT is specifically designed to resolve the eight major tidal constituents. Although the Jason series of satellite altimeters has greatly improved our observations (and assimilation into tide models) of the open ocean barotropic tides (Stammer et al., 2014), the short spatial scales of the coastal tides and their non-linear constituents have been harder to monitor, since we miss information in between the Jason tracks. SWOT with its 250 m resolution 2D coverage can be used as a tide gauge at each grid point, with a time series over a three-year period. This should allow unprecedented coverage of the coastal tides and open-ocean barotropic tides, extended up to high latitudes of 78°N and S. Note that the precise historical Jason tidal observations only reach 66°N and S; and the sun-synchronous orbits of the higher-latitude altimeters (ERS, Envisat, Cryosat, Saral) cannot accurately resolve all of the tidal constituents, especially S2.

The new capability of mapping SSH down to 15 km scales will greatly improve our observation and understanding of internal tides that have a small signature in SSH. Internal tides result from the interaction of the barotropic tide with strongly sloping bathymetry in a stratified ocean. They create large amplitude internal waves at the thermocline (10-30 m), leading to small SSH signatures (1-3 cm). They can already be detected in one direction from along-track altimeter data (Ray and Zaron, 2016). SWOT with its lower noise and 2D coverage should provide the first global observations of these internal tides. Figure 8.17 shows the global internal tide M2 amplitude estimated from 20 years of along-track T/P-Jason data and the HYCOM model forced by the M2 tide. Although we have a good idea on how and where these internal tides are generated (Figure 8.17), where they are dissipated remains a crucial question. The interaction of the internal tide with the ocean circulation and currents has been shown to be complex, with ocean currents refracting and dissipating the tide (Ponte and Klein, 2015). The dissipation of the internal tide is estimated to have an important influence on the ocean's energy budget and the mixing of water masses (e.g., Munk, 1966), but we lack observations to validate this. Knowledge of the 2D energy fluxes in an ocean with internal tides is a key issue that may be addressed with SWOT.

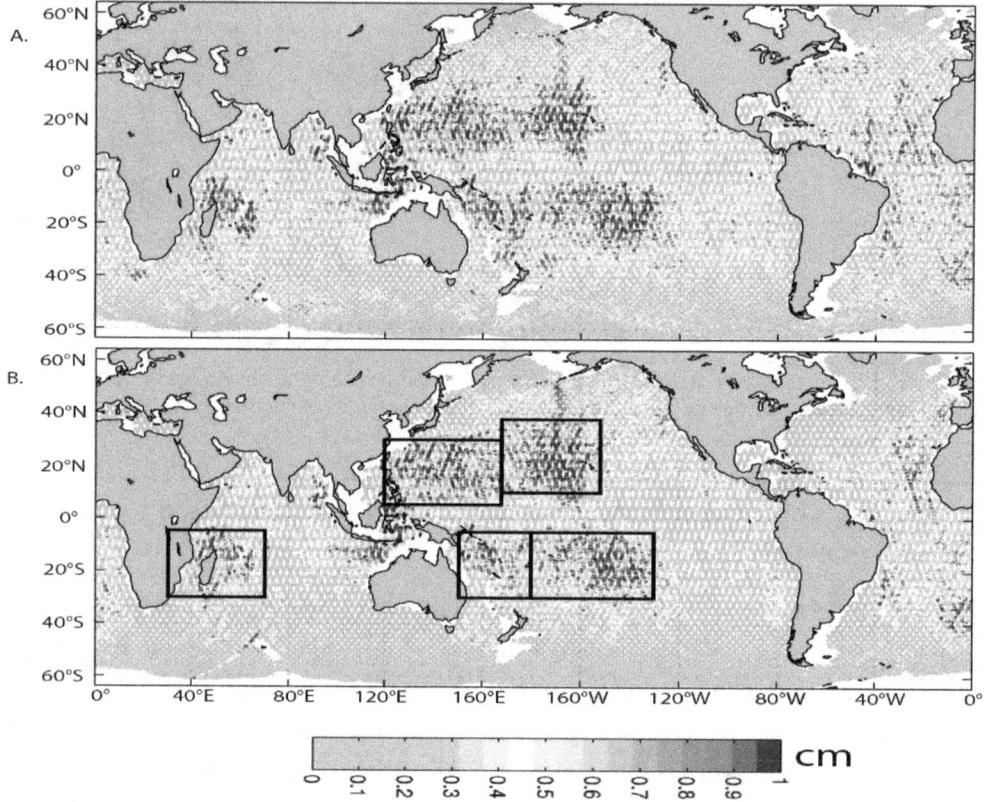

Figure 8.17. Amplitudes (top, altimetric and bottom, HYCOM model) of the M2 internal tide signature in sea surface elevation along altimeter tracks. HYCOM output is interpolated to altimeter tracks for comparison. HYCOM simulation is a 32-layer, wind-, buoyancy-, and M2- forced simulation. For both subplots, the internal (baroclinic) tide amplitudes are computed after removing the barotropic M2 signal. After Shriver et al. (2012).

So, these new SWOT observations will not only be crucial for achieving the ocean circulation objectives by separating barotropic and internal tidal signals from circulation signals, but will also be important for applications in both coastal and open oceans (such as navigation, surface drift, pollution control, etc.) and improved understanding of ocean mixing in the coastal and open ocean.

In addition to internal tides, internal gravity waves can be generated in a stratified ocean by the interaction of currents with bathymetry or by localized atmospheric forcing (Garrett and Munk, 1979). Conventional along-track SSH measurements are too noisy to allow direct estimation of the SSH wavenumber spectrum in the 10-80 km wavelength range where internal waves and submesoscale variability co-exist. Both should be apparent in the along-track SAR and 2D SWOT data, and disentangling the two is an important and interesting research challenge. Numerical simulations are starting to produce realistic internal-wave fields (e.g., Richman et al., 2012; Muller et al., 2015; Rocha et al., 2016) that can be used to analyze the relative contributions of internal waves and lower-frequency variability.

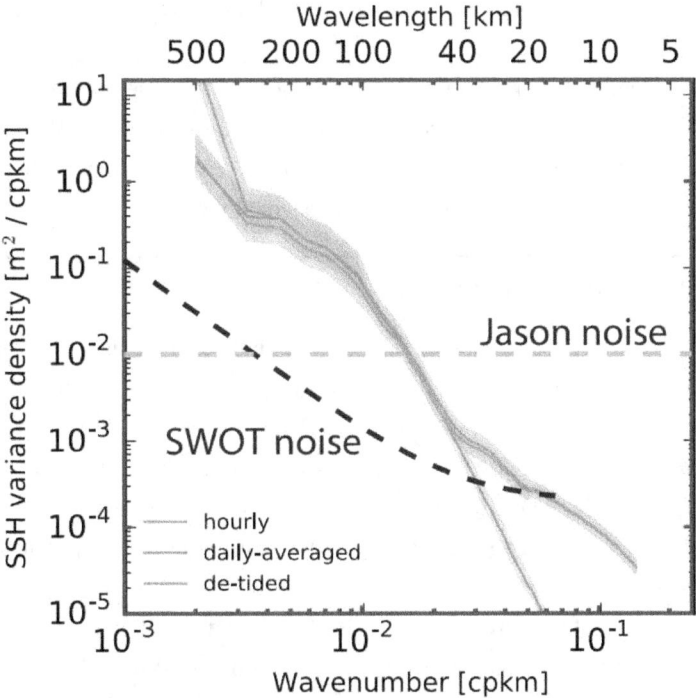

Figure 8.18. Wavenumber spectra of SSH in the Drake Passage region from a 1/48° ocean MITGCM model that represents internal waves and geostrophic variability down to scales of 5 km. The difference between SSH wavenumber spectra of hourly model output (red line) and daily-averaged output (blue line) gives an indication of the contribution of variability at periods of two hours to two days, which is expected to be largely due to internal waves. At wavelengths larger than 50 km in Drake Passage, the low-frequency variability dominates the SSH variability (and hence the blue curve and the red curve are almost the same); at scales smaller than 40 km, the high frequency variability from internal waves becomes larger than the low-frequency variability. After Morrow et al. (2017), modified after Rocha et al. (2016).

Figure 8.18 shows wavenumber spectra of SSH in the Drake Passage region from a 1/48° ocean general circulation model that represents internal waves and geostrophic variability down to scales as small as 5 km (figure modified from Rocha et al., 2016). The model output was from the MITgcm forced by realistic high-frequency wind and tidal forcing (see Rocha et al., 2016). By filtering the model output in time, one can crudely separate the variability at internal wave frequencies (from minutes up to the inertial period, which is about 14 hours in Drake Passage) from the lower frequency variability that is expected to be in geostrophic balance. Comparison of the SSH wavenumber spectra of hourly model output and daily-averaged output gives an indication of the contribution of variability at periods of two hours to two days, which is expected to be largely due to internal waves (Figure 8.18). In Drake Passage, on wavelengths larger than 50 km, the low-frequency variability dominates the SSH variability; at scales smaller than 40 km, the high frequency variability dominates. Figure 8.18 also shows the estimated noise levels for Jason (100 cm^2/cycle/km, e.g., Fu and Ubelmann, 2014) and SWOT (from the SWOT Science Requirements)—the dominance of internal waves begins to occur at the wavelengths that will be newly accessible with the much lower noise levels of SWOT. Recent studies by Qiu et al (2017) analyzing Acoustic Doppler Current Profiler data in the Pacific show that the scales where internal waves become important is geographically variable, and longer wavelengths are affected in the tropics and in low eddy energy regions.

SWOT has the potential of providing the first global SSH observations of the combined geostrophically-balanced flow and the internal wave field, and how they vary both geographically and seasonally and how they interact. Disentangling the two contributions will also be a major challenge, for the future SWOT data and also the fine-scale along-track data.

New 2D and 3D mapping techniques for SWOT swath data

The small-scale features we are starting to detect with the improved along-track altimetry data and with the future SWOT mission are also characterised by their rapid temporal evolution. This presents a challenge since the along-track SAR or SWOT altimetry are providing high spatial resolution but poor temporal resolution (ten-, 20-, 35-day repeats for along-track data; ten-day swath coverage for SWOT). The orbit characteristics of SWOT shown in Figure 8.11 highlight that small, rapid structures will be well sampled for a few days, then re-sampled ten days later, but we cannot observe their temporal evolution.

This dilemma can be illustrated from Figure 8.19. On the left panel, we see 21-days of SWOT swath coverage off the California coast when the simulated noisy SWOT data are just combined, with no interpolation or model advection. The right panels show a series of images based on optimal interpolation (OI) of a true modelled scene of SSH (upper panels) and velocity (lower panels), and a reconstructed scene at the same date from SWOT-like sampling and along-track nadir sampling (Gaultier et al., 2016). In both cases, the distance between the neighbouring tracks or swaths is the limiting factor if we want to ensure a smooth evolution of the flow field. When standard OI is applied after noise removal, we see that the SWOT sampling still retains more anisotropic structure in the SSH field, which gives more energetic currents than with OI applied to the nadir sampling. However, the detailed small-scale eddies and filaments have been smoothed away by the interpolation.

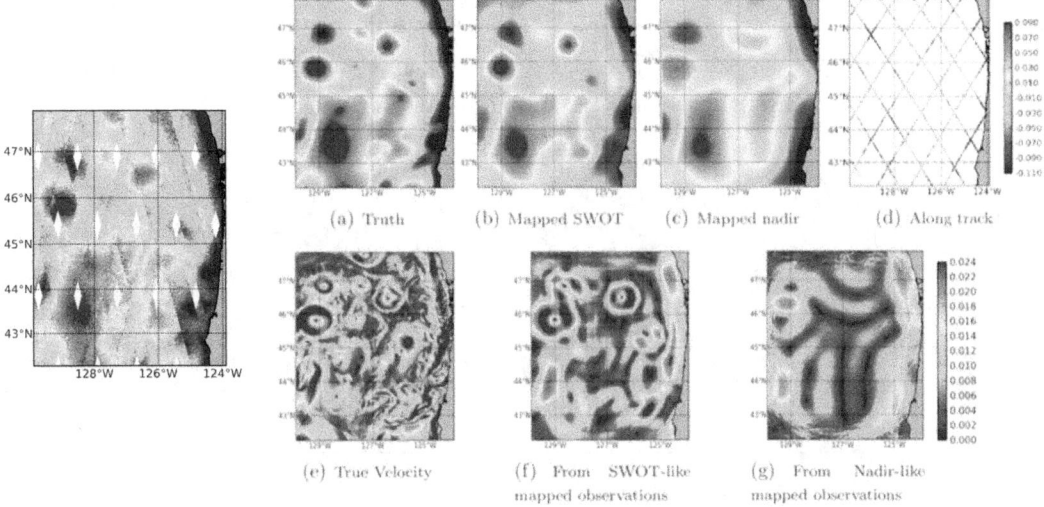

Figure 8.19. (Left): SWOT swath coverage off the Californian Coast with simulated SSH and noise over all tracks during a 21-day period. The SWOT modelled SSH is from the MITGCM. (Right): Upper panels: SSH on one date from MITGCM (Truth), interpolated using SWOT-like sampling, using nadir-like sampling, and along-track nadir SSH coverage. SSH in m. Lower panels: same fields in geostrophic velocity (in m/s). After Gaultier et al. (2016).

The first issue to address before mapping the swath data is the noise removal. Figure 8.19 (left) shows simulated SWOT data with the noise from the different error sources added. Simulated SWOT sampling and noise levels are available for ocean studies using a portable tool (not the full instrument simulator) in open source at: https://github.com/SWOTsimulator/swotsimulator.git. Since the total SWOT error has different causes and space-time structure, techniques are being developed to estimate the noise using cross-spectral methods (Ubelmann et al., JAOT, submitted) or advanced denoising techniques (Cosme et al., 2017, submitted). This is needed since any ocean studies requiring velocity or vorticity calculations will amplify the small-scale noise when taking the first or second derivatives of SSH.

The second issue is whether a linear statistical model (such as OI) is appropriate to describe the rapid evolution of the smaller eddies and fronts. An example of this problem is shown in Figure 8.20. The upper panels show the SSH evolution over a 300 km x 300 km box covering a large meander, but with several small eddies embedded that should be resolved with SWOT. These eddies move rapidly over this four-day period around the meander. If a linear interpolation is used between days zero and four (equivalent to OI over short time scales), the reconstructed field at day two (lower panels, left) places the large slow meander well, but the small eddy marked with a '+' has not been carried with the flow. Instead it is split into two weak eddies at the original and final positions with half the energy. Ubelmann et al. (2015) proposed instead to use a simple dynamical interpolation based on a nonlinear quasi-geostrophic model with one active layer and the conservation of potential vorticity. For most of the ocean, the first baroclinic mode explains most of the SSH evolution (Wunsch, 1997), so one active layer at fine-scale allows a good reconstruction over short time scales (< ten days). The bottom panels (right) show that the large meander and the small eddy have been correctly positioned with the dynamical interpolation, with greatly reduced errors. This simple dynamical interpolation works well when the lateral advection dominates other forcing terms, such as air-sea fluxes, and is able to retrieve gridded maps with wavelengths smaller than 100 km over a ten-day window in a mid-energy region (Ubelmann et al; 2015), and down to 60 km in the Mediterranean with weaker energy and smaller scales (Roge et al, 2017).

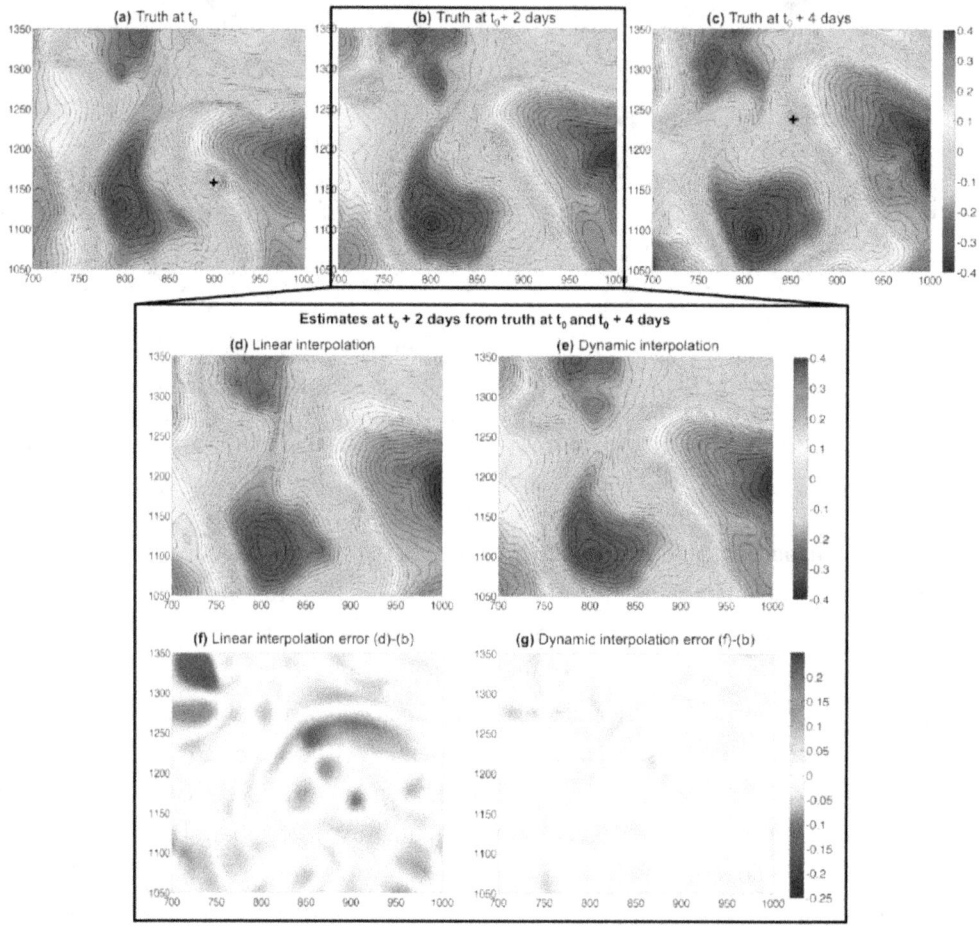

Figure 8.20. a,b,c: "True" SSH over a 300 km x 300 km domain at two-day intervals. d: linear SSH estimation of (b) from (a) and (c). e: dynamic SSH estimation of (b) from (a) and (c). Error fields are the difference f: between (d) and (b), and g: between (e) and (b). Units in meters. After Ubelmann et al. (2015).

In addition to the 2D reconstruction, techniques are being developed to try and diagnose the 3D dynamics in the ocean interior based on characterizing the potential vorticity (PV) field. This can allow an estimate of the 3D balanced ocean dynamics (including the vertical velocity field) using dynamical frameworks such as surface quasi-geostrophy (e.g., Lapeyre and Klein, 2006; LaCasce and Mahadevan, 2006; Wang et al., 2013; Ponte and Klein 2013; Berti and Lapeyre, 2014). The first step is to estimate surface PV at the highest possible resolution from satellite observations, either from high resolution SWOT SSH or SST as a proxy for surface density. The next step is to relate surface PV to interior PV. This can be done through a structure function that is depth-dependent (see Wang et al., 2013; Ponte and Klein, 2013) and can be estimated from climatology or the ARGO dataset if they resolve the critical layers at depth.

This surface-to-subsurface dynamical projection can be enhanced by other properties. PV is conserved along a Lagrangian trajectory and experiences a direct cascade. This means that Lagrangian techniques can allow us to recover smaller-scale structures at the surface and at depth. These techniques have proved to be very efficient. Lagrangian studies based on the temporal evolution of gridded altimetry maps have already allowed us to reconstruct ocean features with finer scales, including dynamical transport barriers aligned with the larger fronts (D'Ovidio et al., 2009, LeHahn et al., 2007) as well as finer-scale tracer fields (Despres et al., 2011; Dencausse et al., 2013). The combination of surface QG vertical projection techniques and Lagrangian advection is currently being explored (Berti and Lapeyre, 2014).

Challenges of assimilating SWOT-simulated SSH data in operational ocean models

More sophisticated assimilation schemes are being considered to ingest SWOT-simulated SSH data into operational ocean models. Due to the large volume of SWOT data (1-2 Go/day), different direct or Lagrangian approaches are being explored. A question is how the high-resolution observations available locally along the swath will constrain the model, and what the impact on the operational analysis and forecast fields will be. The noise reduction schemes developed for 2D SSH mapping are also of use in pre-processing the SWOT data before assimilation.

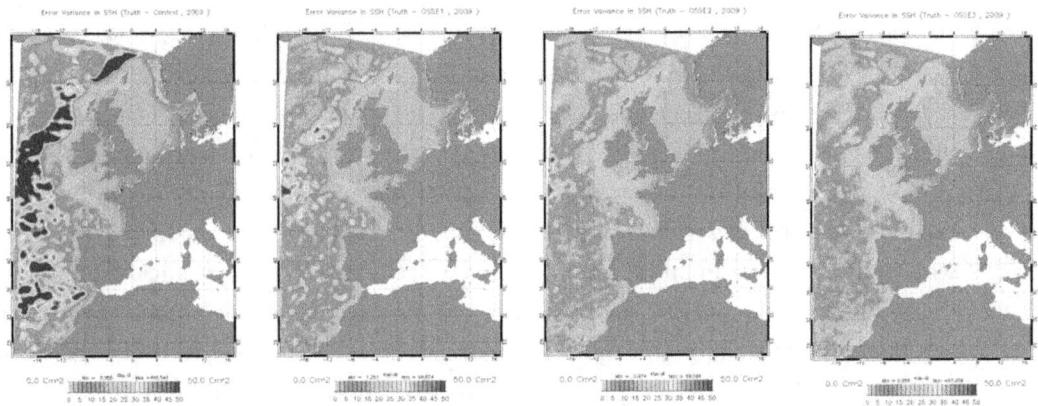

Figure 8.21. SSH Error variance (in cm^2) averaged over one year – 2009 for different assimilation experiences compared to the "truth" 1/36° model. Error variance for a) the Free model run (1/12°) with no assimilation, b) OSSE1 assimilation with along-track Jason-2, Jason-1 and Envisat sampling, c) OSSE2 with SWOT sampling, d) OSSE3 with SWOT and along-track J2-J1-Env sampling. (Credits: M. Benkiran, Mercator-Ocean)

At Mercator-Ocean, simulated SWOT SSH data with realistic position and errors have been generated from a 1/36° model with tides and assimilated into a regional 1/12° model with tides as a testbed. At mid latitudes, the 1/36° "truth" data resolve scales of around 15 km, similar to SWOT, the 1/12° model used as the control run resolves scales of around 50 km. In these early tests, the analyzed fields have the tide and other high-frequency signals removed using a 25-h mean, to concentrate on the internal dynamics. Three cases of Observing System Simulation Experiments (OSSEs) are considered: OSSE1 with a typical sampling of three along-track altimeters based on Jason-2, Jason-1 in its geodetic phase, and Envisat tracks; OSSE2 with the SWOT 21-day sampling;

and OSSE3 with SWOT and the three altimeter sampling. Figure 8.21 shows how the error variance in the open ocean is well constrained by the SWOT sampling. Interestingly, the along-track altimeter assimilation tends to add in coherent eddy signals that evolve slowly when there are no new data assimilated. In contrast, the strong anisotropy inherent in the SWOT SSH images add a non-linear observed flow field that evolves more rapidly and more realistically during the data gaps. Due to the improved anisotropy and small-scale gradients, the relative vorticity and the sub-surface horizontal and vertical velocities are better constrained with the SWOT observations. Work is ongoing to test the assimilation with higher resolution (1/60° "truth" assimilated into the 1/36° model) and to explore the representation and observability of the internal wave field.

Other techniques are being explored to use the complementary information from high-resolution surface tracer images (SST or ocean colour) and SWOT SSH maps to control the ocean circulation. Indeed, altimetric SSH has more energy at the larger mesoscales, whereas at smaller scales, tracer images can help control the more complex non-linear structures. Gaultier et al (2013) derived Lagrangian transport barriers from high-resolution SSH data and inverted tracer fields and assimilated these Lagrangian structures, rather than the original fields, to constrain the surface dynamics.

Conclusions

Even before the SWOT launch, the promise of observing a new 2D SSH field over scales from 15-200 km is opening new research domains. Exciting questions are being explored on the role of small mesoscale and sub-mesoscale dynamics in the ocean circulation, and their impact on the energy budget, on mixing and dissipation, on the generation of larger-scale dynamics, and on the vertical exchange between the surface and deeper layers. Similar questions exist for the role of internal tides and internal waves interacting with the "balanced" ocean circulation, and modifying the eddy energy, evolution and mixing. In this chapter, we have concentrated on the dynamical fields being resolved by altimetry. Yet improving the horizontal and vertical flow at small scales will have a huge impact on the exchange of heat, carbon, and nutrients across the oceans, as well as between the surface and deeper layers, with a big impact on biogeochemical cycles and biomass evolution. Improved small-scale circulation from altimetry will be analyzed in parallel with fine-scale tracer data (SST, ocean colour) and surface parameters (surface roughness, sun-glitter, etc.), which are strongly modified across fronts and filaments, to link the deeper dynamics with the surface fronts. Finally, if SWOT data will help us validate the next generation of global, high resolution ocean models, SWOT also needs to be validated against high spatial resolution in situ data. Exploring the overlapping dynamics from small-scale internal waves and internal dynamics from existing in situ and satellite data and models will occupy a lot of our energy over the coming years.

References

Ablain M., A. Cazenave, G. Larnicol, M. Balmaseda, P. Cipollini, Y. Faugère, M. J. Fernandes, O. Henry, J. A. Johannessen, P. Knudsen, O. Andersen, J. Legeais, B. Meyssignac, N. Picot, M. Roca, S. Rudenko, M. G. Scharffenberg, D. Stammer, G. Timms, J. Benveniste, (2015). Improved sea level record over the satellite altimetry era (1993–2010) from the Climate Change Initiative project. *Ocean Science*, 11, 67-82, doi:10.5194/os-11-67-2015.

Berti S and G Lapeyre (2014). Lagrangian reconstructions of temperature and velocities at sub-mesoscales. *Ocean Modelling*, 76, 59—71

Birol, F., N. Fuller, F. Lyard, M. Cancet, F. Niño, C. Delebecque, S. Fleury, F. Toublanc, A. Melet, and M. Saraceno. (2017). Coastal applications from nadir altimetry: Example of the X-TRACK regional products. *Adv. Space Res*, 59(4), 936–953.

Boy, F., J. D. Desjonquères, N. Picot, T. Moreau and M. Raynal, (2017). "CryoSat-2 SAR-Mode Over Oceans: Processing Methods, Global Assessment, and Benefits," in *IEEE Transactions on Geoscience and Remote Sensing*, 55(1), 148-158, doi: 10.1109/TGRS.2016.2601958

Brown, G. (1977). The average impulse response of a rough surface and its applications. *IEEE transactions on antennas and propagation*, 25(1), 67-74.

Capet A., E. Mason, V. Rossi, C. Troupin, Y. Faugere, M.-I. Pujol, A. Pascual, (2014). Implications of a Refined Description of Mesoscale Activity in the Eastern Boundary Upwelling Systems. *Geophys. Res. Lett.*, 41, doi:10.1002/2014GL061770.

Carrere, L. and F. Lyard, (2003). Modeling the barotropic response of the global ocean to atmospheric wind and pressure forcing - comparisons with observations. *Geophys. Res. Lett*. ISSN: 0094-8276, 30, n. 6, 1275-1278.

Chassignet, E.P. and X. Xu, (2017). Impact of horizontal resolution (1/12° to 1/50°) on Gulf Stream separation, penetration, and variability. *J. Phys. Oceanogr.*, 47, 1999-2021, doi:10.1175/JPO-D-17-0031.1.

Chelton,D.B, J. C. Ries, B. J. Haines, L.-L. Fu, and P. S. Callahan, (2001). In Satellite Altimetry and Earth Sciences: A Handbook of Techniques and Applications; Fu, L. L.; Cazenave, A., Eds.; Academic Press: San Diego, 1–131.

Chelton DB, Gaube P, Schlax MG, Early JJ, Samelson RM. (2011). The influence of nonlinear mesoscale eddies on near-surface oceanic chlorophyll. *Science*, 334(6054):328-32.

Dencausse G., R. Morrow, M. Rogé and S. Fleury (2014). Lateral stirring of large-scale tracer fields by altimetry. *Ocean Dynamics*, 64, 61-78.

Despres, A., G. Reverdin, and F. D'Ovidio, (2011). Mechanisms and spatial variability of mesoscale frontogenesis in the northwestern North Atlantic Subpolar gyre. *Ocean Modelling*, 39, 97-113, doi:10.1016/j.ocemod.2010.12.005

Dibarboure, G., M.-I. Pujol, F. Briol, P.-Y. Le Traon, G. Larnicol, N. Picot, F. Mertz and M. Ablain, (2011). Jason-2 in DUACS: first tandem results and impact on processing and products. *Marine Geodesy*, 34, 214-241. DOI, 10.1080/01490419.2011.

Dibarboure G, Boy F, Desjonqueres JD, Labroue S, Lasne Y, Picot N, Poisson JC, Thibaut P., (2014). Investigating short wave-length correlated errors on low-resolution mode altimetry. *J. Atmos. Ocean Technol.*, 31(6):1337–1362.

Dong, C., J. C. McWilliams, Y. Liu, and D. Chen, (2014). Global heat and salt transports by eddy movement, *Nat Commun*, 5, http://dx.doi.org/10.1038/ncomms4294

d'Ovidio, F., J. Isern-Fontanet, C. López, E. Hernández-García, and E. García-Ladona (2009), Comparison between Eulerian diagnostics and finite-size Lyapunov exponents computed from altimetry in the Algerian basin. *Deep Sea Res.*, Part I, 56(1), 15–31.

Dufau, C., M. Orsztynowicz, G. Dibarboure, R. Morrow, P.Y. Le Traon, (2016). Mesoscale Resolution Capability of altimetry: present and future. *J. Geophys. Res. Oceans*, 121, 4910–4927, doi:10.1002/2015JC010904.

Dussurget, R., F Birol, R Morrow and P De Mey, (2011): Fine Resolution Altimetry Data for a Regional Application in the Bay of Biscay. *Marine Geodesy*, 34:3-4, 447-476.

Escudier, P. , A. Couhert, F. Mercier, A. Mallet, P. Thibaut, N. Tran, L. Amarouche, B. Picard, L. Carrère, G. Dibarboure, M. Ablain, J. Richard, N. Steunou, P. Dubois, M. H. Rio, J. Dorandeu (2017). Satellite radar altimetry: principle, geophysical correction and orbit, accuracy and precision. In: "Satellite altimetry over oceans and land surfaces". Chapter 1, Eds D. Stammer and A. Cazenave. Taylor and Francis (in press)

Escudier, R., J. Bouffard, J., A. Pascual, P.-M. Poulain, M.-I. Pujol (2012): Improvement of coastal and mesoscale observation from space: Application to the Northwestern Mediterranean Sea. *Geophys. Res. Lett.*, 40 (10), 2148-2153

Esteban-Fernandez, D. (2017). SWOT mission performance and Error Budget. JPL Publication D-79084.

Fernandes, J., C. Lázaro, A. L. Nunes, and R. Scharroo (2014). Atmospheric Corrections for Altimetry Studies over Inland Water. *Remote Sens.*, 6(6), pp 4952–4997.

Frappart, F., D. Blumstein, A. Cazenave, G. Ramillien, F. Birol, R. Morrow, F. Remy (2017). Satellite Altimetry: Principles and Applications in Earth Sciences, J. Webster (ed.), Wiley Encyclopedia of Electrical and Electronics Engineering. Copyright 2017 John Wiley & Sons, Inc. DOI: 10.1002/047134608X.W1125.pub2

Fu, L.L and A. Cazenave. (2001). Satellite Altimetry and Earth Sciences: A Handbook of Techniques and Applications. Academic Press: San Diego.

Fu, LL., E Rodriguez, D Alsdorf, R. Morrow (eds). (2012). The SWOT Mission Science document. https://swot.jpl.nasa.gov/files/swot/SWOT_MSD_1202012.pdf

Fu L.L. and C. Ubelmann (2014). On the transition from profile altimeter to swath altimeter for observing global ocean surface topography. *J. Atmos. Oceanic. Tech.*, https://doi.org/10.1175/JTECH-D-13-00109.1

Garcia, E. S., D. T. Sandwell, and W. H. F. Smith. (2014). Retracking CryoSat-2, Envisat, and Jason-1 radar altimetry waveforms for improved gravity field recovery. *Geophysical Journal International*, 196(3):1402–1422.

Garrett, C. and W Munk (1979). Internal waves in the ocean. *Ann. Review Fluid Mech.*, 11:1, 339-369

Gaultier L., Verron J., Brankart J.-M., Titaud O. and Brasseur P., (2013). On the use of submesoscale tracer fields to estimate the surface ocean circulation. *Journal of Marine Systems*, 126, 33–42

Gaultier, L., C. Ubelmann, and L. Fu, (2016): The Challenge of Using Future SWOT Data for Oceanic Field Reconstruction. *J. Atmos. Oceanic Technol.*, 33, 119–126, https://doi.org/10.1175/JTECH-D-15-0160.1

LaCasce, J. and A. Mahadevan, (2006): Estimating subsurface horizontal and vertical velocities from sea surface temperature. *J. Marine Res.*, 64, 695–721.

Lapeyre G, Klein P (2006). Dynamics of the upper oceanic layers in terms of surface quasigeostrophy theory. *J. Phys. Oceanogr.*, 36(2), 165-176. http://dx.doi.org/10.1175/JPO2840.1

Lehahn Y., F. d'Ovidio, M. Lévy, Y. Amitai and E. Heifetz (2011). Long range transport of a quasi-isolated chlorophyll patch by an Agulhas ring. *Geophys. Res. Lett.*, 38, L16610, doi:10.1029/2011GL048588.

Le Traon, P.Y. and G. Dibarboure, (2002). Velocity mapping capabilities of present and future altimeter missions: the role of high frequency signals. *J. Atm. Ocean Tech.*, 19, 2077-2088.

Levy M., Iovino D., Resplandy L., Klein P., Madec G., Treguier A.-M., Masson S., Takahashi K. (2012). Large-scale impacts of sub-mesoscale dynamics on phytoplankton: Local and remote effects. *Ocean Modelling*, 43-44, 77-93. http://dx.doi.org/10.1016/j.ocemod.2011.12.003

Morrow, R., L-L Fu, T. Farrar, H. Seo, P-Y Le Traon (2017). Ocean eddies and mesoscale variability. In "Satellite altimetry and its use for earth observation", in 'Earth Observation of Global Changes' ed. CRC Press.

Müller, M., B. K. Arbic, J. G. Richman, J. F. Shriver, E. L. Kunze, R. B. Scott, A. J. Wallcraft, L. Zamudio (2015), Toward an internal gravity wave spectrum in global ocean models. *Geophys. Res. Lett.*, 42, 3474–3481. doi: 10.1002/2015GL063365.

Munk, W., (1966). Abyssal recipes. *Deep-Sea Res.*, 13 (1966), pp. 707-730

Obligis, E., C. Desportes, L. Eymard, M. J. Fernandes, C. Lázaro, and A. Nunes (2011). In Coastal Altimetry; Springer-Verlag Berlin Heidelberg, 578, doi:10.1007/978-3-642-12796-0

Pascual A., M.I. Pujol, G. Larnicol, P.Y. Le Traon and M.H. Rio (2005). Mesoscale Mapping Capabilities of Multisatellite Altimeter Missions: First Results with Real Data in the Mediterranean Sea. *J. Mar. Systems*, 65, 190–211, doi:10.1016/j.jmarsys.2004.12.004.

Pascual, A., Faugere, Y., G. Larnicol, P.Y. Le Traon, (2006). Improved description of the ocean mesoscale variability by combining four satellite altimeters. *Geophys. Res. Letters*, 33 (2): Art. No. L02611.

Peral, E. E. Rodriguez, D. Esteban-Fernandez (2015). Impact of surface waves of SWOT's projected ocean accuracy *Remote Sens.* 7(11), 14509-14529; doi:10.3390/rs71114509

Ponte A, Klein P (2013). Reconstruction of the upper ocean 3D dynamics from high-resolution sea surface height. *Ocean Dynamics*, 63(7), 777-791. http://dx.doi.org/10.1007/s10236-013-0611-7

Ponte A, Klein P (2015) Incoherent signature of internal tides on seal level in idealized numerical simulations. *Geophys. Res. Lett.*, 42: 1520–1526. doi: 10.1002/2014GL062583.

Pujol, M. I. Larnicol, G, (2005). Mediterranean Sea eddy kinetic energy variability from 11 years of altimetric data, *J. Mar. Systems*, 58, 121-142.

Pujol, M.-I., Y. Faugère, G. Taburet, S. Dupuy, C Pelloquin, M. Ablain, N. Picot (2016). DUACS DT2014: the new multi-mission altimeter dataset reprocessed over 20 years. *Ocean Sci. Discuss.*, doi:10.5194/os-2015-110.

Pujol, M.-I., P. Schaeffer, Y. Faugère, M. Raynal, G. Dibarboure, N. Picot (2018). Gauging the improvement of recent mean sea surface models: a new approach for identifying and quantifying their errors. *J. Geophys. Res.*, submitted.

Qiu B, T Nakano, S Chen, P Klein (2017). Submesoscale transition from geostrophic flows to internal waves in the northwestern Pacific upper ocean. *Nat Comm.*, 8:14055. doi:10.1038/ncomms14055

Quartly, G. D., M. A. Srokosz, and A. C. McMillan, (2001): Analyzing altimeter artifacts: Statistical properties of ocean waveforms. *J. Atmos. Oceanic Technol.*, 18, 2074–2091, doi:10.1175/1520-0426(2001)018<2074:AAASPO>2.0.CO;2

Raney, R.K. (1998) "The delay/Doppler radar altimeter" IEEE Trans. Geosci. Remote Sens vol. 36 no. 5 pp. 1578-1588.

Ray R.D. and Zaron E.D., (2016). M2 internal tides and their observed wavenumber spectra from satellite altimetry, *J. Phys. Oceanogr.*, 46:1, 3-22, doi:10.1175/JPO-D-15-0065.1.

Richman, J. G., B. K. Arbic, J. F. Shriver, E. J. Metzger, and A. J. Wallcraft, (2012): Inferring dynamics from the wavenumber spectra of an eddying global ocean model with embedded tides. *J. Geophys. Res.*, 117, C12012, doi:10.1029/2012JC008364.

Robinson, I.S., (2004). Measuring the oceans from space. Praxis-Springer-Verlag, 669pp.

Rocha, C.B., T. K. Chereskin, S. T. Gille, and D. Menemenlis (2016) Mesoscale to Submesoscale Wavenumber Spectra in Drake Passage. *J. Phys. Oceanogr.*, 46:2, 601-620. doi:10.1175/JPO-D-15-0087.1

Rodriguez, E, D Esteban Fernandez, E Peral, C W Chen, J-W De Bleser, B Williams (2017). Wide-swath altimetry: A review. In: "Satellite altimetry over oceans and land surfaces", Book Chapter 2. Eds D. Stammer & A. Cazenave, Taylor and Francis.

Rogé, M., Morrow, R., Ubelmann, C., and Dibarboure, G. (2017). Using a Dynamical Advection to Reconstruct a Part of the SSH Evolution in the Context of SWOT, Application to the Mediterranean Sea. *Ocean Dynamics*, 1-20, doi:10.1007/s10236-017-1073-0.

Sasaki H., Klein P., Qiu B., Sasai Y. (2014). Impact of oceanic-scale interactions on the seasonal modulation of ocean dynamics by the atmosphere. *Nature Comm*, 5, 1-8.

Shriver, J. F., B. K. Arbic, J. G. Richman, R. D. Ray, E. J. Metzger, A. J. Wallcraft, and P. G. Timko (2012), An evaluation of the barotropic and internal tides in a high-resolution global ocean circulation model. *J. Geophys. Res.*, 117, C10024, doi:10.1029/2012JC008170.

Stammer, D., R.D. Ray, O.B. Andersen, B.K. Arbic, W. Bosch, L. Carrere, Y. Cheng, D.S. Chinn, B.D. Dushaw, G.D. Egbert, S.Y. Erofeeva, H.S. Fok, J.A.M. Green, S. Griffiths, M.A. King, V. Lapin, F.G. Lemoine, S.B. Luthcke, F. Lyard, J. Morison, M. Müller, L. Padman, J.G. Richman, J.F. Shriver, C.K. Shum, E. Taguchi, and Y. Yi, (2014): Accuracy assessment of global barotropic ocean tide models. *Reviews of Geophysics*, 52, 243-282, doi:10.1002/2014RG000450.

Ubelmann, C., P. Klein, L.-L. Fu (2015) Dynamical interpolation of sea surface height and potential applications for future high-resolution altimetry mapping. *J. Atmos. Ocean. Tech.*, 32, 177-184, doi:10.1175/JTECH-D-14-00152.1

Ubelmann, C., B. Cornuelle, and L. Fu (2016) Dynamic Mapping of Along-Track Ocean Altimetry: Method and Performance from Observing System Simulation Experiments. *J. Atmos. and Oceanic Tech.*, 33:8, 1691-1699. doi:10.1175/JTECH-D-15-0163.1

Verron, J., Sengenes, P., Lambin, J., Noubel, J., Steunou, N., Guillot, A., Picot, N., Coutin-Faye, S., Sharma, R., Gairola, R. M., Raghava Murthy, D. V. A., Richman, J. G., Griffin, D., Pascual, A., Rémy, F., and Gupta, P.K., (2015): The SARAL/AltiKa altimetry satellite mission. *Mar. Geod.*, 38, 2–21, doi:10.1080/01490419.2014.1000471.

Vignudelli S, Kostianoy AG, Cipollini P, Benveniste J. (eds.) (2011). Coastal Altimetry, Springer-Verlag Berlin Heidelberg, 578, doi:10.1007/978-3-642-12796-0

Wang, J., G.R. Flierl, J.H. LaCasce, J.L. McClean and A. Mahadevan, (2013). Reconstructing the ocean's interior from surface data. *J. Phys Oceanogr.*, 43, 1611-26, doi:10.1175/JPO-D-12-0204.1

Wunsch C (1997). The vertical partition of oceanic horizontal kinetic energy. *J. Phys. Oceanogr.*, 27(8):1770–1794. doi:10.1175/1520-0485(1997)027<1770:TVPOOH>2.0.CO;2

Xu, Y., and L-L. Fu, (2012): The effects of altimeter instrument noise on the estimation of the wavenumber spectrum of sea surface height, *J. Phys. Oceanogr.*, 42, 2229-2233. doi:10.1175/JPO-D-12-0106.1

CHAPTER 9

An Operational Interpolated Ocean Colour Product in the Mediterranean Sea

Gianluca Volpe[1], Bruno Buongiorno Nardelli[1,2], Simone Colella[1], Andrea Pisano[1], and Rosalia Santoleri[1]

[1]*Istituto di Scienze dell'Atmosfera e del Clima, CNR, Roma, Italy,* [2]*Istituto per l'Ambiente Marino Costiero, CNR, Napoli, Italy*

A novel technique to interpolate satellite ocean colour data has been developed and calibrated in the framework of the European MyOcean2 project and successively implemented within the Copernicus Marine Environment Monitoring Service (CMEMS) specifically for the Mediterranean Sea products. The methodology is based on the Data Interpolating Empirical Orthogonal Functions technique, which interpolates data voids from Empirical Orthogonal Function (EOF) modes iteratively estimated as characteristic spatial patterns. Here, this method is extended to take into account the temporal correlation between the observations. A higher-dimensional approach is followed by using a temporal sequence of daily images to build the state vector and thus the observation matrix used to compute the EOFs. An ad-hoc smoothing procedure is also applied to resulting 2-dimensional fields to filter out spurious signals and provide consistent spatial reconstructions. Several tests are performed on a dataset at 4 km resolution to calibrate the technique and to assess, among other issues, the most convenient number of images to be included in the state vector. The final CMEMS product at 1 km resolution is then validated with the independent chlorophyll data collected during dedicated oceanographic surveys between 1997 and 2015 across the entire Mediterranean basin.

Introduction

Numerous marine environmental and scientific applications require complete data time series of bio-geophysical parameters measured at fixed locations with a high spatial coverage, such as those provided by satellite optical remote sensing (e.g., Ferreira et al., 2011). They include, for example, the assimilation of observations in numerical biogeochemical models, the operational detection of harmful algae blooms, as well as the monitoring and assessment of marine ecosystem status by national and international institutions in order to comply with international rules/legal acts (e.g., for implementation of the European Marine Strategy Framework Directive) and/or to investigate the ecosystem functioning and its response to human and natural pressures (e.g., Volpe et al., 2012). However, cloud cover, low repetitiveness of satellite passes, and the relative narrow satellite swaths all prevent a full exploitation of optical satellite measurements.

Volpe, G., et al., 2018: An Operational interpolated ocean colour product in the Mediterranean Sea. In *"New Frontiers in Operational Oceanography"*, E. Chassignet, A. Pascual, J. Tintoré, and J. Verron, Eds., GODAE OceanView, 227-244, doi:10.17125/gov2018.ch09.

As a consequence, optical sensors can only provide a partial view of sea surface, and interpolation techniques are needed to overcome the sparseness and uneven temporal coverage of satellite data.

The most common approach to operationally interpolate sea surface temperature (SST) or sea surface height (SSH) data is the optimal interpolation (OI) or analogous approaches (e.g., Martin et al., 2012). These are implemented in several processing centres for data production at both global and regional scales and within real-time operational chains and for offline processing (Le Traon et al., 1998; Reynolds et al., 2007; Buongiorno Nardelli et al., 2010, 2013; Donlon et al., 2012; Roberts-Jones et al., 2012). However, different approaches have been proposed to fill in the gaps present in ocean colour imagery (e.g., Chlorophyll-a, Total Suspended Matter, etc.), either based on kriging/optimal interpolation (e.g., Saulquin et al., 2011) or iterative methods (Alvera-Azcárate et al., 2009, 2015; Sirjacobs et al., 2011). In particular, the Data Interpolating Empirical Orthogonal Functions (DINEOF) method has been proposed and tested for the interpolation of optical satellite data on regular high-resolution spatial grids, even at very high temporal sampling (Beckers and Rixen, 2003; Miles and He, 2010; Sirjacobs et al., 2011; Alvera-Azcárate et al., 2015; Liu and Wang, 2016). This method allows the extraction of the dominant spatial patterns observed in a data time series through an iterative approach, while simultaneously filling in the missing data. DINEOF presents some interesting advantages as compared to more classical approaches (such as optimal interpolation), especially when working with ocean colour data. In fact, different scales of variability and background concentrations are generally associated with the coastal and open sea domains, along with their respective seasonal cycles. The underlying hypothesis of the standard OI techniques for the estimation of the field covariance is that the field is stationary and generally characterized by isotropy (Bretherton et al., 1976). This assumption may potentially lead to artefact propagations of coastal signals offshore, in the presence of extended cloud cover or when analysing different seasons. Adapting the OI covariance parameter estimation to bypass these assumptions would then require unpractical computational steps. These hypotheses are automatically relaxed when using DINEOF, as this technique directly identifies the dominant patterns and the main sources of variability through the calculation of the field principal components. In its original formulation, DINEOF is based on a purely spatial state vector (Beckers and Rixen, 2003). Consequently, the reconstruction of long time series by projecting sparse observations on dominant spatial modes may eventually lead to temporal discontinuities when large fractions of the original images are empty. To address this issue, a filtering of the covariance matrix has been suggested allowing for the reduction of spurious signals and to obtain more realistic reconstructions (Alvera-Azcárate et al., 2009). Here we propose a different approach, namely to augment the DINEOF technique by using a time-lagged extended Empirical Orthogonal Function (EOF) analysis (e.g., Weare and Nasstrom, 1982) instead of standard EOF in the iterative processing. This technique has been developed in the framework of the European project MyOcean2 and successively implemented within the Copernicus Marine Environment Monitoring Service (CMEMS) to provide both Near Real-Time/Delayed Time (NRT/DT) and reprocessed datasets (REP) over the Mediterranean Sea.

Operational applications often require NRT data, but the accuracy of NRT ocean colour data suffers the lack of up-to-date ancillary information, such as meteorological and ozone data, which are crucial for the atmospheric correction and are generally available only after a few days from the satellite overpass. Despite the lower quality of NRT data, they still represent the only available observations of the sea state and, therefore, they are routinely produced. Once the ancillary data are available, NRT data are reprocessed and the DT data are produced with a more reliable scientific quality. Several experiments have thus been carried out to define the optimal configuration for both products. These have been performed on a test dataset, varying the length of the data time series used to build the state vector, the lag considered to interpolate the single daily image and the number of EOF modes used to reconstruct the field.

A complete description of the DINEOF algorithm calibration, as well as of the validation of the final product with independent measurements, is provided in the next section (Data and Methods). As a matter of comparison, the next section includes an overview of the OI procedure currently used to produce daily gap-free (Level-4) SST data over the Mediterranean Sea and available through CMEMS. In the second subsection, the method operationally used to interpolate ocean colour data in the context of CMEMS is presented, while the different tests are shown and discussed in the final subsection.

Data and Methods

This section provides the details of the methodology for field interpolation using both the OI and the DINEOF approaches, with a special focus on the procedures followed to obtain both DINEOF input and output. The pre-processing of the Level-3 original data is a crucial part of the method and is essential to building the data matrix input to the interpolation procedure. Figure 9.1 summarizes the entire interpolation scheme.

Optimal Interpolation

The OI procedure is a mature and consolidated technique to reconstruct gap-free, 2-dimensional fields for oceanographic variables, such as SST, SSH, SSS (sea surface salinity), and ocean colour, as demonstrated by the growing body of literature on the topics. This section deals with the description of the regional and fine-tuned OI scheme, which is operationally used in the context of CMEMS to produce daily SST data at 1 km resolution over the Mediterranean Sea. The same OI parameterization and algorithm used for SST data interpolation are used here for the ocean colour chlorophyll images under the assumption that they (ocean colour and SST) show the same scales of spatial and temporal variability (Volpe et al., 2012). In particular, we use a space-time OI procedure to fill in data voids.

Within OI, the single interpolated pixel is obtained as a linear combination of the observations (e.g., chlorophyll anomalies with respect to the first guess) directly weighted with their correlation (or equivalently their covariance) to the interpolation pixel and inversely with their cross-correlation and error. The estimation of the covariance of all observations is computationally very demanding and operationally unfeasible. To overcome these limitations, the covariance is generally assumed to be well represented by a characteristic functional form. Furthermore, only a subset of the available observations concurs to the estimation of the interpolated value, depending on the spatial and temporal *influential radius* to the interpolation point, i.e., the maximum distance (in both space and time) over which the observations are deemed useful to the interpolation. To this aim, we use an *influential spatial radius* of 20 km and a *temporal window* of 10 days centered on the day that has to be interpolated (as defined in Buongiorno Nardelli et al., 2013).

Following Marullo et al. (2007) and Buongiorno Nardelli et al. (2013), the covariance, C, is directly estimated through the functional form:

$$C(\Delta r, \Delta t) = e^{-\frac{\Delta t}{\tau}} e^{-\frac{\Delta r}{\lambda}},$$

with Δr being the relative spatial distance between observation and interpolation pixels and Δt their time difference; τ and λ are the temporal and spatial de-correlation lengths and are set as three days and 5 km, respectively. These space (5 km) and time (three days) decorrelation lengths are, in turn, defined by the field specific temporal and spatial scales of variability (see Figure 9.4a for the temporal scale and Buongiorno Nardelli et al. (2013) for the spatial scale). When no direct observations are available to the interpolation, for example due to prolonged cloud cover, the first guess is used as interpolation value to fill in the data gaps. Here, as first guess we use the daily climatology at the same spatial resolution as the ocean colour data, as derived from the Sea-Viewing Wide Field-of-View Sensor (SeaWiFS, McClain et al., 2004).

DINEOF Interpolation

Processing modes

Here, the two configurations (NRT and DT) for interpolating the daily chlorophyll data are described. For filling in missing data, both schemes require a data time series to get geophysical information from. Intuitively and from the biogeochemical point of view, the optimum would be to center the data to be interpolated within the time series, e.g., half of the data time series in the future and half in the past with respect to the day that has to be interpolated. This approach should ensure all the data time series to be relevant for the reconstruction of the field; this configuration is hereafter referred to as OPT and is used within CMEMS for the REP processing mode. In the context of operational oceanography, the main goal is to minimize the time required to produce the gap-free daily field while keeping its scientific robustness and reliability. Time minimization means shifting the data that has to be interpolated towards the future end of the time series, in order to provide users (e.g., data assimilation into ecosystem models) with a reasonably recent field. The main difference between NRT and DT is that NRT requires a data time series, with the most recent day being the current day that needs to be interpolated. While in the DT configuration, the same input data is reprocessed after a defined time lag (TL) so that it not only benefits from the information contained in these last days but also from the better quality of the data processed with updated ancillary information. Another important difference between the two approaches is the way the input data matrix is built and given to the interpolation procedure. Moreover, the structure of the input data matrix becomes crucial to accounting for both the spatial and the temporal variability of the data time series and to include them into the interpolated field. In the input data matrix, the rows are used to identify the spatial variability, and they are used by the interpolation procedure to build the EOF spatial patterns.

DINEOF input data

Two satellite datasets are involved in this work, the testing and operational datasets at 4 km and 1 km spatial resolution, respectively. Since computational time depends on the dimension of the data, the 4 km-resolution data are used for testing purposes. The full resolution dataset is then used to implement the final configuration in order to provide users with appropriate statistics associated with the operational product. The 4 km dataset is made from the SeaWiFS, while the 1 km-resolution data are derived from the CMEMS operational processing chain. Each of the two datasets (testing and operational) has its own reference climatology at 4 and 1 km spatial resolution, respectively.

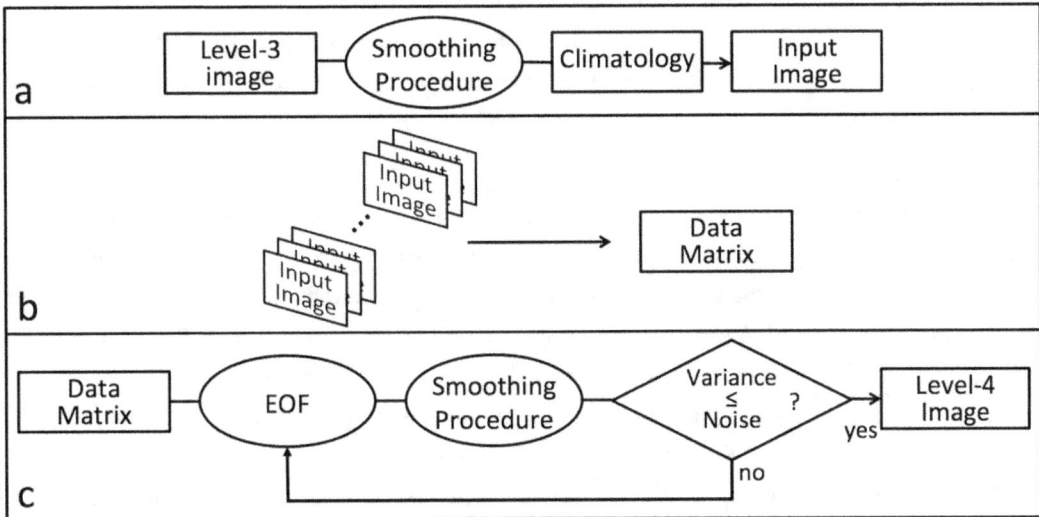

Figure 9.1. Interpolation scheme. Panel a shows that the original Level-3 image has to be merged with the climatology through a smoothing procedure to obtain the input image, which is then used to build the data matrix (panel b), input to the interpolation procedure. The data matrix enters the EOF calculations, and spatial modes are merged with the original data matrix via the smoothing procedure until the explained variance does not exceed that of the noise.

The testing dataset derives from the SeaWiFS Level-0 data that have been processed up to Level-3 with the SeaWiFS Data Analysis System (SeaDAS) software package version 7.0 available from NASA website (http://seadas.gsfc.nasa.gov). The Mediterranean- and sensor-specific Ocean Colour 4-band algorithm (MedOC4, Volpe et al., 2007) for chlorophyll retrieval in Case-1 waters is then applied to the resulting remote sensing reflectance. To ensure the most possible reliability to data, apart from the shallow water and turbid water flags, all masking criteria provided by the SeaDAS software package are applied (McClain et al., 1995). Single chlorophyll swath maps are remapped at a nominal spatial resolution of 4 km on an equirectangular grid covering the Mediterranean domain (30°N–46°N; 6°W–36.5°E).

The operational dataset is the multi-sensor data derived from CMEMS over the Mediterranean Sea that merges all available sensors at any given time (Volpe et al., 2017;

http://marine.copernicus.eu/documents/QUID/CMEMS-OC-QUID-009-038to045-071-073-78-079-095-096.pdf). With the merged product, the number of daily valid observations increases by roughly 20% with respect to their single-sensor contributors, without the introduction of any significant source of uncertainty (Volpe et al., 2017).

Figure 9.1a shows the schematics of the pre-processing that any satellite images undergo before being ingested into the interpolation processor. Daily images are transformed into their base-10 logarithm to account for chlorophyll lognormal distribution. To avoid spurious and noisy signals in the DINEOF output due to the low quality of the input images, a filtering procedure is routinely applied to daily input images. The filtering procedure is made of two parts. It first checks for the existence of isolated pixels and, if they are missing, set them to a predefined missing value. Afterwards and with the assumption that ocean colour data do not vary much on the pixel distance level, the procedure checks for the existence of isolated missing pixels to fill them using the median value of its surrounding good pixels. As EOF requires complete input data time series, missing values within this filtered daily image are set to respective climatological values via a procedure that smoothens out spurious spatial gradients. These images constitute the single building blocks of the input data matrix.

The input data matrix

As mentioned, regardless of the processing mode and hence of the input data structure, the interpolation requires a data time series to get geophysical information from. This paragraph explains the differences in the structure of the input data matrix among the three processing modes that are operationally used in the CMEMS processing chain: NRT, DT, and OPT. Under the NRT configuration, the input data matrix is built such that each row, the state vector, corresponds to the single daily sea domain. In this way, the columns represent the single sea pixel time series; the NRT represents the standard DINEOF configuration, which only accounts separately for space and time covariance (meaning that dominant modes are identified exclusively either as spatial patterns or as temporal patterns). Conversely, extended EOFs used in DT and OPT configurations consist of a temporal sequence of spatial patterns and therefore effectively account for the full space-time covariance. Thus, the DT configuration differs from the NRT in such a way that each row, rather than being one single daily sea domain, is made up of multiple subsequent daily sea domains (Figure 9.2). This configuration enables the temporal variability to be taken into account when computing the EOF modes. Starting from the top row, the first image on the right is the image that has to be interpolated and refers to time T0; all other images in this row are more recent than that, referring to times T+1, T+2, up to T+TL. When seeking data, the operational procedure first looks for the correct L3 product. If, for any reason, this product is missing the climatology is taken.

a) NRT		b) DT					c) OPT										
T0		T+4	T+3	T+2	T+1	T0	T+10	T+9	T+8	T+7	T+6	T+5	T+4	T+3	T+2	T+1	T0
T-1		T+3	T+2	T+1	T0	T-1	T+9	T+8	T+7	T+6	T+5	T+4	T+3	T+2	T+1	T0	T-1
T-2		T+2	T+1	T0	T-1	T-2	T+8	T+7	T+6	T+5	T+4	T+3	T+2	T+1	T0	T-1	T-2
T-3		T+1	T0	T-1	T-2	T-3	T+7	T+6	T+5	T+4	T+3	T+2	T+1	T0	T-1	T-2	T-3
T-4		T0	T-1	T-2	T-3	T-4	T+6	T+5	T+4	T+3	T+2	T+1	T0	T-1	T-2	T-3	T-4
T-5		T-1	T-2	T-3	T-4	T-5	T+5	T+4	T+3	T+2	T+1	T0	T-1	T-2	T-3	T-4	T-5
T-6		T-2	T-3	T-4	T-5	T-6	T+4	T+3	T+2	T+1	T0	T-1	T-2	T-3	T-4	T-5	T-6
T-7		T-3	T-4	T-5	T-6	T-7	T+3	T+2	T+1	T0	T-1	T-2	T-3	T-4	T-5	T-6	T-7
T-8		T-4	T-5	T-6	T-7	T-8	T+2	T+1	T0	T-1	T-2	T-3	T-4	T-5	T-6	T-7	T-8
T-9		T-5	T-6	T-7	T-8	T-9	T+1	T0	T-1	T-2	T-3	T-4	T-5	T-6	T-7	T-8	T-9
T-10		T-6	T-7	T-8	T-9	T-10	T0	T-1	T-2	T-3	T-4	T-5	T-6	T-7	T-8	T-9	T-10

Figure 9.2. Graphical representation of the input data matrix for a) NRT, b) DT, and c) OPT processing modes. Single squares refer to the single daily sea pixel domain. The image that has to be interpolated is referred to as T0 and is highlighted in grey for each configuration. All squares marked with T-1 refer to data images of one day earlier than the one that has to be interpolated. It is possible to see that a single data file is used more than once with the purpose of accounting for the data temporal variability. In all three processing modes, each row represents the state vector of the input data matrix.

One aspect that must be taken under consideration is the position of the image that has to be interpolated within the input data matrix. Here, two possibilities are examined: one in which the image that has to be interpolated is centered in the state vector (red square in Figure 9.2b, and referred to as DT_C), and the other as shown by the grey square of Figure 9.2b. The latter constitutes the operational way of interpolating the Level-3 fields in DT mode and is referred to as DT_L. However, it must be noted that since the method is based on a statistical approach, the most appropriate configuration is linked to the availability of observations within the input data matrix, and, therefore, cannot be determined a priori. The operational configuration assumes the data voids to be more or less equally distributed within the input data matrix. A future version of this approach should dynamically individuate the best configuration aimed at minimizing the distance between T0 and the most clear sky days to allow the contribution of the observations to be more relevant than that of the climatology.

The procedure applied to build the input data matrix is exactly the same, whether the input data are from SeaWiFS (4 km) or from CMEMS (1 km). This means that all results obtained from the testing phase, and hence about the fine-tuning of the methodology, can be realistically applied to the operational context.

Climatology

The general scope of earth observation is to know the value of a parameter (e.g., the chlorophyll concentration, the diffuse attenuation coefficient of light) at any time and location. However, when observing ocean colour, clouds may prevent the single pixel to be seen for long time periods, especially during winter when the interpolation becomes even more important. In this context, the only source of information relevant for the field interpolation is the climatology, which turns out to be more reliable than the data referring to periods too far away in time from the one that has to be interpolated. It is known that the decorrelation timescale in ocean colour data in the Mediterranean Sea is around a few days. Thus, it is generally safe to assume that the most plausible expectation of such a value will be the average (or the median value) of that parameter from past observations, i.e., the climatology field, with a certain degree of variability (e.g., within a few standard deviations from the average). The climatology is obtained from the 13 years of SeaWiFS data using the MedOC4 regional algorithm for chlorophyll (Volpe et al., 2007). As mentioned, this daily field has the same cylindrical projection and spatial resolution, 1 km and 4 km, as the testing and the operational fields, respectively. To reduce the impact of the short-scale variability, these climatology maps are created using all data falling into a moving temporal window of ± 5 days. One of the main purposes of a climatology field is to serve as a reference, and therefore it is expected to be as reliable as possible, thus avoiding biases caused by single incorrect pixel values. To overcome these possible biases, a filtering procedure is applied to the entire SeaWiFS time series by removing all isolated pixels and by filling in all isolated missing pixels using the near-neighbourhood approach. The resulting daily climatology time series includes the pixel-scale standard deviation, the average, the median, the modal, the minimum, and the maximum values.

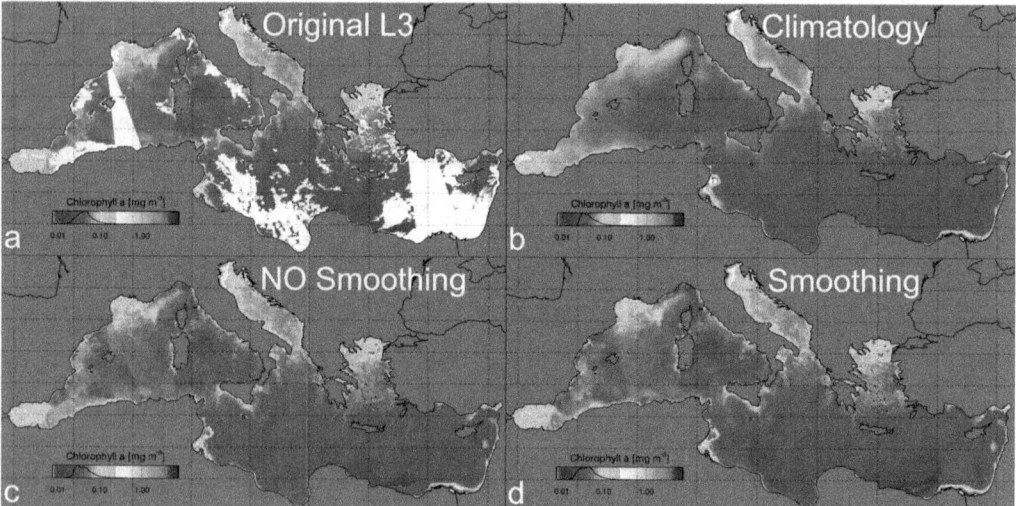

Figure 9.3. Example of the impact of the smoothing procedure used to merge the original Level-3 data (panel a) downloaded from the CMEMS website and referring to the merged product between MODIS-Aqua and NPP-VIIRS collected on the 28 July, 2017, and relevant daily climatology (panel b). The output image when the two fields are merged without (panel c) any smoothing procedure and by applying SP (panel d). The colour palettes contain the distribution histogram of each image. In panel a, white areas refer to missing data either because of the clouds or because out of the orbits of the satellites.

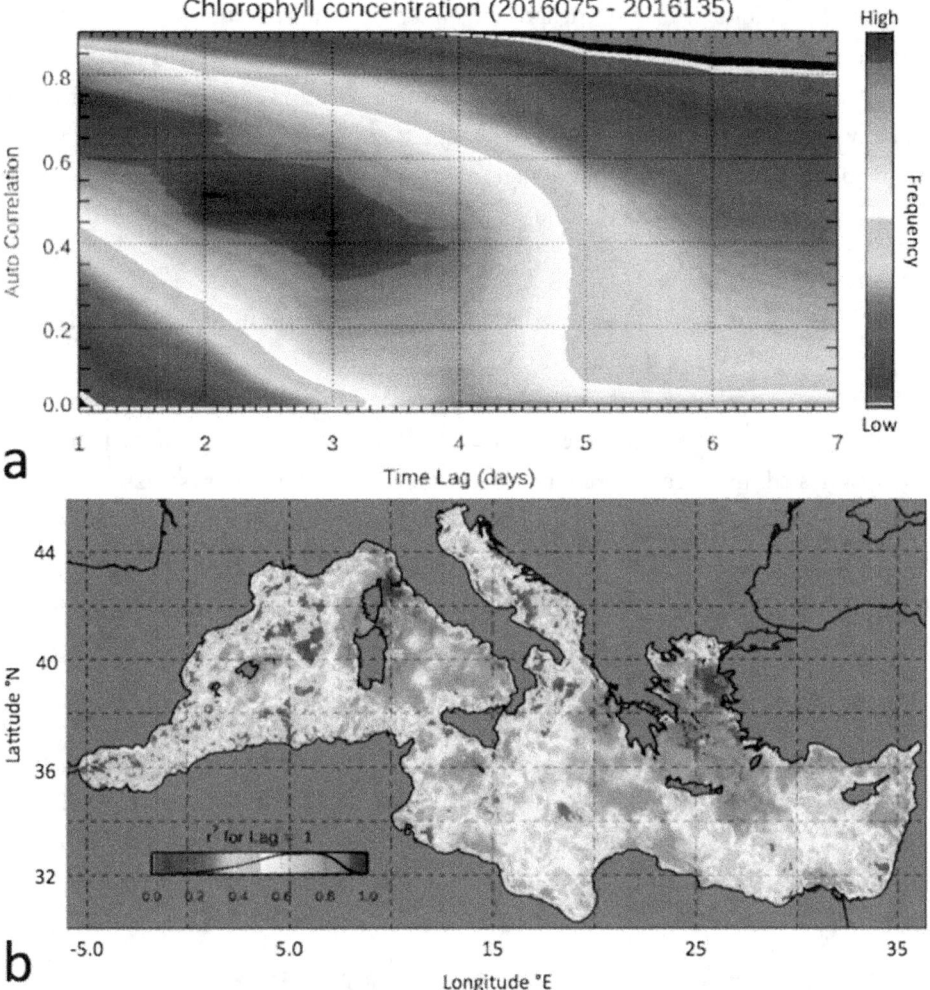

Figure 9.4. Panel a: Frequency distribution of the chlorophyll autocorrelation (r2) as a function of time lag, computed with two months of L4 data. Panel b: autocorrelation map computed with the same two-month data with a time lag of one day.

Smoothing procedure

Smoothing procedure (SP) allows for minimizing the occurrence of spurious gradients due to the merging of two fields of different origin (e.g., observations and climatology) that might contain information at different temporal and spatial scales. SP is applied anytime two fields need to be merged together: it is applied when L3 and climatology are merged together, in correspondence with the first EOF run or for the subsequent runs when the holes in the original L3 daily maps are filled in with the EOF modes. The procedure smooths out the differences computed in correspondence with common observations, that is, where both observations and climatology (or the reconstructed field) co-occur. This procedure allows for filling gaps in while keeping the original observations and without introducing any spurious gradients. Figure 9.3 shows an example of a typical two-dimensional field with and without the application of the smoothing procedure. The impact of using SP is particularly evident south of the Balearic Islands, where the original L3 and the reference climatology differ significantly giving origin to unrealistic spatial gradients that are efficiently erased by the application of the smoothing function. Moreover, the distribution histogram of the SP output (Figure 9.3d, inside the colour palette) is much closer in shape than the original L3, whereas the one in Figure 9.3c (without the application of SP) is very similar to the one of the climatology (Figure 9.3b). This highlights the importance of using a procedure that does not significantly alter the original data distribution.

DINEOF interpolation

The reconstruction of missing data is performed using the DINEOF method first developed by Beckers and Rixen (2003) and later used by Volpe et al. (2012). Volpe et al. (2012) provided some insight, in the ocean colour context, into the advantages of using the DINEOF approach to fill in missing data with respect to optimal interpolation. All three versions of this product (NRT, DT, and OPT) are here tested. In all configurations, the technique works as follows: starting from the day that has to be interpolated (grey square in Figure 9.2), the time series of the previous 10-day data is used to build a data matrix (Figure 9.2). Ten days are deemed to be a good compromise between the need for observations that tend to include as many images as possible and the chlorophyll temporal decorrelation scale in the Mediterranean Sea, to reduce the contribution of uncorrelated observations to the reconstruction of the missing data.

In this respect, Figure 9.4a shows an example of the autocorrelation temporal scales of variability at basin-scale computed with a two-month data time series of interpolated fields: correlation sharply decreases after a few days, reaching the null value in five days. Similarly, Figure 9.4b shows that there is a considerable spatial patchiness due to the short temporal data time series used to compute the statistics, thus it is not entirely representative of all of the scales that can be encountered in the basin.

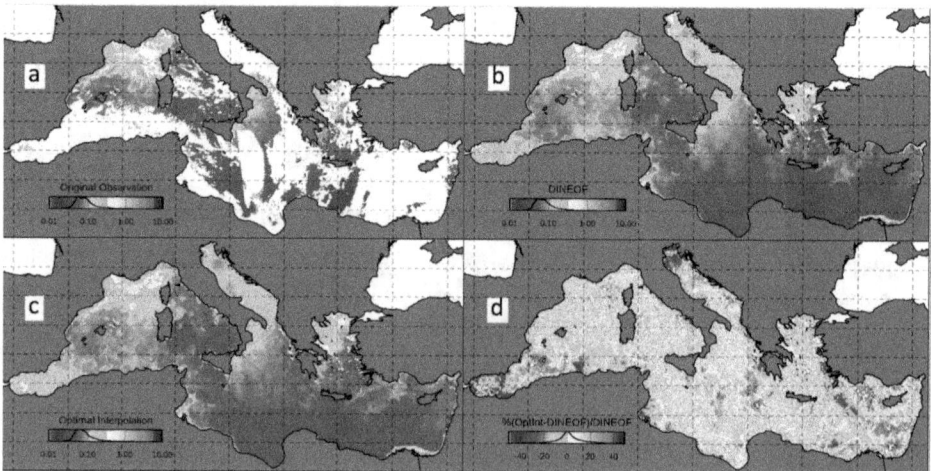

Figure 9.5. Panel a shows the daily merged product from MODIS-AQUA and NPP-VIIRS on 1 April, 2012. Panel b shows the Level-4, in DT mode. The entire data time series of both Level-3 and Level-4 is available on the CMEMS website (http://www.marine.copernicus.eu) and at http://gos.artov.isa.cnr.it. Panel c shows the same interpolated chlorophyll image using the OI procedure, with the settings shown in the second section. Panel d shows the difference (%) between the two interpolated fields.

All missing data are replaced with respective daily climatological sea pixels, and a mask matrix, namely "holes", is built with zeros in correspondence with effective observations and ones where the climatology is used. This data matrix constitutes the input to the iterative EOF procedure, which uses the singular-value decomposition approach (SVD from the LAPACK library under the IDL environment and is equivalent to the EOF). After each iteration, the input data matrix to the next SVD iteration is built with original observations for holes=0, and with the field reconstructed from the SVD output of the previous iteration for holes=1. The reconstruction is carried out using the number of modes corresponding to the iteration number. Thus, after the first iteration the climatology is replaced with the first EOF mode, which is then replaced with the field reconstructed with the first two modes after the second iteration, and so on. To determine the number of modes that can be used effectively to reconstruct the final field, the cumulative variance is computed at each iteration and compared with the one determined by an independent EOF run performed using a data matrix of the same dimensions as the input data matrix and filled with random numbers. The variance explained by the first mode of this EOF identifies the noise. The interpolation procedure stops when the variance explained by the current iterative mode equals that of noise. It is important to stress that the number of modes used to reconstruct the final field can vary from one day to another or depending on the configuration (NRT, DT, OPT), because of the intrinsic variability of the observations used to build the input data matrix.

Moreover, as it will detailed further in a section below, the procedure smooths out differences between original observations and a reconstructed field the same way, as shown in Figure 9.3. This scheme prevents artificial gradients from being created and is particularly effective during periods of scant data availability, (e.g., winter, when the cloudy pixels are much more numerous than the clear sky pixels). Figure 9.5 shows an example of the CMEMS daily merged chlorophyll product and of its interpolation in DT mode. As a means of comparison and to show how robust the interpolation procedures used here are, the same image is provided applying the OI technique. Figure 9.5d shows that less than 15% of the pixels exceed 25% difference.

Results and Discussion

This section discusses how the final configuration described above is selected from among the many available options. As mentioned, the various tests performed are all aimed at addressing a series of different issues concerning both the way the input data matrix is built and how it is processed to achieve the Level-4. All tests are repeated any time a new issue is identified. In general, all tests are meant to improve the quality of the operational product so that when one specific test clearly tackles the issue that it was meant to address, the operational configuration is changed accordingly. The issues that need to be sorted out before the operational product can be regularly delivered to users are: 1) the number of EOF modes used to reconstruct the final field, 2) the number of days used to build the data matrix, 3) the structure of the data within the input matrix to account for the temporal variability, and 4) the position of the image that has to be interpolated within the input matrix.

The first of these issues has to do with determining the most appropriate number of EOF modes to be used in a field reconstruction. In theory, one would correctly expect more EOF modes to correspond to more information in the final reconstructed field. However, the use of higher EOF modes explaining lower variance could correspond to lower signal-to-noise information, which translates into a noisier final output (i.e., salt-and-pepper noise). With this in mind, we test the reconstruction of the final field with a number of EOF modes such that the variance can be explained by the higher mode to be lower than the one explained by the noise, or with a number of modes such that their cumulative variance can be explained to be at least 90%, 95%, or 99% of total variance, and lastly using all available EOF modes. The result (not shown) is that the root mean square error and bias between the original fields and the interpolated fields increases with the number of modes used to reconstruct the final field. Stopping the EOF iterations in correspondence with the variance explained by the noise becomes an operational requisite and it is thus implemented in the CMEMS operational processing chain.

The standard DINEOF approach only considers the spatial variability, and one of the novelties of this work is the fact that it aims at including the temporal variability in the interpolation scheme. As already discussed in a previous section, this is achieved by building the state vector as a sequence of daily images. The issue then becomes the number of daily data to be used for building the data matrix, input to the interpolation procedure. The rationale for such a question is intuitive and has to do with the fact that the EOF procedure needs the observed fields to get information from. The question becomes particularly relevant under persistent cloud cover that prevents the retrieval of the geophysical information. One would increase the length of the time series until enough information is available for the interpolation procedure. However, one constraint is that the temporal de-correlation scale of the field has to be interpolated. In the Mediterranean Sea, chlorophyll autocorrelation sharply decreases within a few days, so that the input data matrix should not contain data newer or older than one or two weeks of the data that has to be interpolated. In winter, when the number of available pixels dramatically decreases due to persistent cloud cover, the two-week limit may result in no available local data. To address this issue, daily observations and climatology are operationally blended together via the smoothing procedure (see next section and Figure 9.3). The only drawback to this approach is the different space and time scales of variability of the two kinds of data that, in correspondence with persistent and spatially-diffused cloud cover, results in images with lower patterns of variability.

Another issue is represented by the position of the data that has to be interpolated within the input data matrix. Figure 9.2 shows the input data matrices for the three operational processing modes; these are the result of the various tests in which the position of data to be interpolated is allowed to change within the state vector, from the most recent (NRT, the upper left square in any of the configurations shown in Figure 9.2) to the one centred in the state vector (DT$_C$, centred in the upper row in any of the configurations of Figure 9.2) or the one in the last position of the state vector (DT$_L$, the upper right square in any of the configurations of Figure 9.). These issues and their impact on the quality of the interpolated field are investigated through a matchup analysis against in situ observations (Figure 9.5).

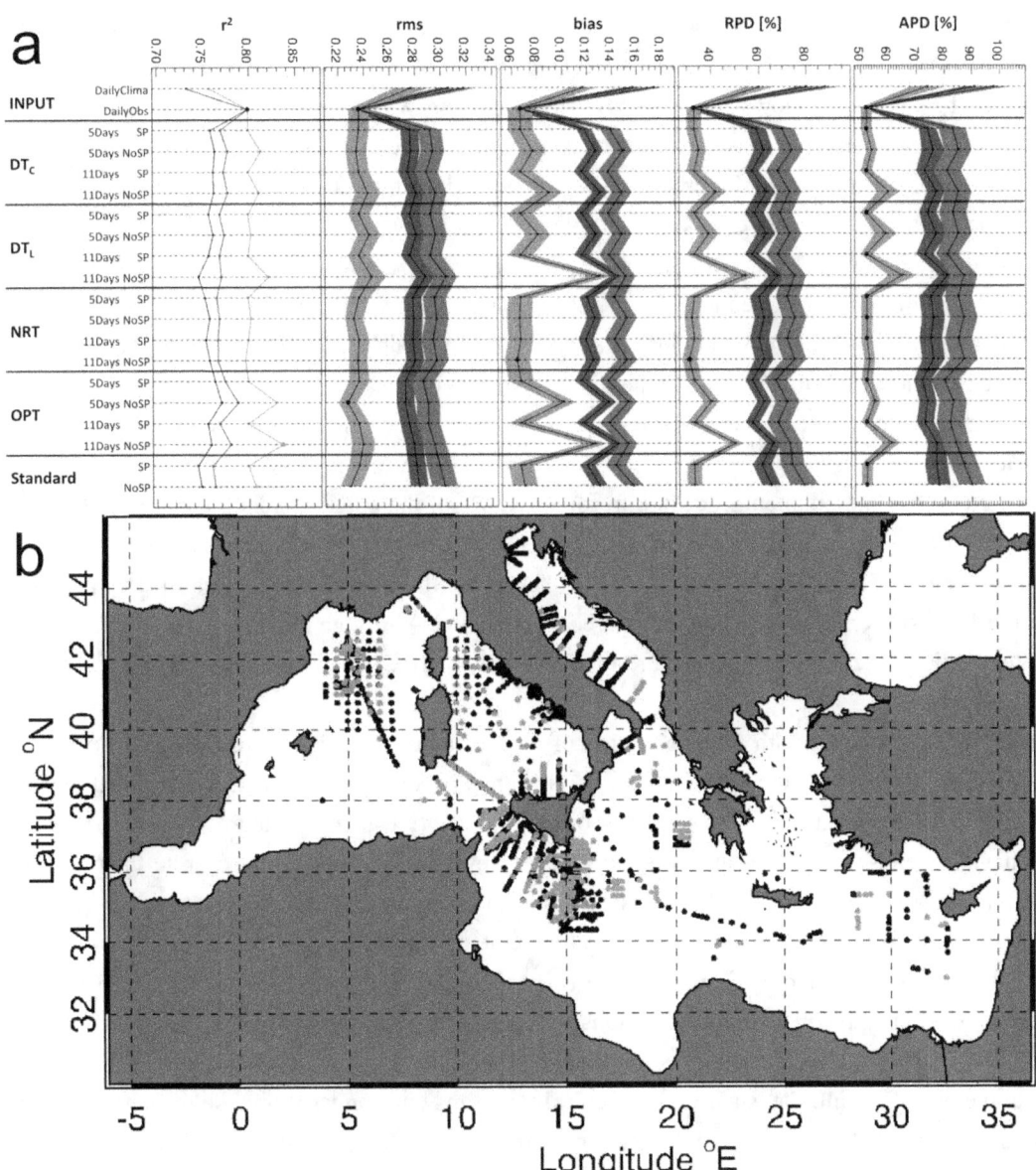

Figure 9.6. Panel a: Shaded areas correspond to ±1STD, computed via bootstrap method, provide the confidence level of the associated statistics. The three coloured shaded areas refer to all matchups (blue), matchups in correspondence with missing original pixels only (grey), and in correspondence with valid original pixels only (red). The five plots are the correlation coefficient, r2, the root mean square error, rms, the bias, and the relative and absolute differences (%) between satellite-derived and in situ chlorophyll concentration. For details on the various tests shown on the y-axis see the main text. The labels 5Days and 11Days refer to the length of the state vector, whereas SP and NoSP refer to the use or non-use of the smoothing procedure. The number of the three matchup datasets is 1139, 504 and 1643, respectively, and their spatial distribution is shown in panel b. Red points in panel b refer to matchups in correspondence with valid original pixels only and black points indicate those with interpolated pixels only.

Figure 9.5b shows the spatial distribution of the matchup points, which consists of two sub-datasets with one (1139 stations) made with the in situ data in correspondence with missing initial observations, and the other (504 stations) with in situ data in correspondence with valid initial observations. This separation allows discerning, above all, the added value of the interpolation procedure. Figure 9.5 shows the results of the matchup analysis in terms of correlation coefficient, root mean square error, bias, and the relative and absolute differences (%) between in situ observations and the interpolation outputs from the various tests. Both the input and output of the interpolation procedure are compared with the in situ observations, and this provides a means for precisely quantifying the impact of the method. Moreover, to address the significance of the results, all calculations are performed using the bootstrapping method in which the statistics are computed 1000 times over half of the validation dataset only, at any time. This method allows the estimation of the statistics variability providing their confidence level.

The first important and non-trivial result is the fact that in all configurations the interpolated fields behave better than the climatology. This is not straightforward, as the climatology enters as first guess the interpolation procedure in roughly 70% of the entire matchup dataset (1139 initial missing pixels against the total number of matchups of 1643). Apart from a few tests, e.g., DT_L 11Days NoSP, OPT 5Days NoSP and OPT 11Days NoSP, all other tests exhibit pretty much the same level of performance when compared with their in-situ counterparts. This means that the interpolation scheme is able to capture the input data matrix variability and to project it into the final output.

Results associated with the application of the smoothing procedure are generally better than when the procedure is not applied (all the NoSP runs). Moreover, the statistics associated with the matchup dataset built only with the valid original pixels (red lines in Figure 9.5) show a similar behavior to those of the missing original pixels (grey lines in Figure 9.5), despite their absolute difference. This highlights the importance of the smoothing procedure, which became a requisite of the operational processing and is currently implemented in the CMEMS chain. The uncertainty contribution of the interpolation procedure is on average less than 30% (Figure 9.5). Despite the above-mentioned theoretical considerations, Fig. 9.6 does not provide definitive evidence that the DT and OPT configurations are unequivocally better than the NRT or standard DINEOF, showing that the use of a multi-day state vector (basically all runs but standard) only slightly improves the statistics. Thus, what surely drives the quality of the output is the amount and quality of the original observations in the input data matrix.

The use of the multi-image state vector, along with the position of the image that has to be interpolated within the input data matrix, are the two elements that mostly contribute to the variability among the three configurations (NRT, DT, OPT). In fact, Figure 9.6 shows, in correspondence with the areas where the original Level-3 presents data voids (panel a), that the impact of the climatology (panel b) on the interpolation outputs decreases from the NRT (panel d) to the DT (panel e) and to the OPT (panel f) configurations, even if its features are clearly visible in the input image (panel c). The three configuration outputs mainly differ in their correspondence with the Gulf of Lions, where the area of the bloom is represented by two distinct patches in the

NRT and DT and by a more uniform and coherent patch in the OPT run. This result is due to the fact that, in the OPT configuration, the space-time variability benefits of as many days in the future as they are in the past. There are several little features where the three outputs differ, such as the one in correspondence with the Rhodes Gyre, east of the Island of Crete, which varies in shape and to a lesser extent in its background intensity. Analogously, the Bonifacio Gyre visible in the input image (Figure 9.6c) and clearly derived from the climatology field is visible in the NRT and DT runs, but not in the OPT run (Figure 9.6f).

The operational product

The previous section showed that the best performing configurations for the three processing modes are those schematically drawn in Figure 9.2, by smoothing differences anytime two fields of different origin need to be merged together, and by stopping the iterations before the variance explained by the current mode reaches the level of the noise. As mentioned, this configuration is operationally implemented into the CMEMS processing chain through which the entire data archive has recently been reprocessed.

Processing Level	r^2	RMS	Bias	RPD	APD	N
Level-3	0.720	0.272±0.009	-0.029±0.009	17±6	56±5	784
Level-4	0.720	0.288±0.006	-0.062±0.006	8±2	53±2	2005
Level-4*	0.718	0.286±0.005	-0.061±0.006	8±2	53±2	1967
Optimal Interpolation	0.724	0.281±0.006	-0.040±0.006	16±4	58±3	1967

Table 9.1. Statistics associated with the matchup exercise performed over the entire Mediterranean Sea in situ dataset collected in several cruises carried out by CNR from 1997 to 2015. First and second rows refer to the entire operational Level-3 and Level-4 datasets. Statistics associated with matchups in correspondence with valid DINEOF and OI data are shown for comparison in the third and fourth rows.

Figure 9.7 and Table 9.1 show that the comparison of both Level-3 and Level-4 data with in situ observations does yield similar agreement. They also confirm the importance of the data interpolation in terms of field coverage (the number of matchups for the Level-4 is nearly triple as compared with the Level-3), without introducing significant sources of uncertainty. This result is quite different than the one achieved within the testing analysis, in which there is a net increase of about 30% uncertainty from Level-3 (red statistics in Figure 9.5a) and Level-4 (blue statistics in Figure 9.5a), while in the operational context the uncertainty difference between the two is of only a few percentages. As a means of comparison, days having corresponding in situ measurements are also processed with the OI method; the result of this matchup exercise, shown in Table 9.1, highlights that the two methods are absolutely equivalent at the scale of the single pixel statistics. In cases of prolonged cloud cover, however, the OI method is more susceptible to propagating features offshore giving rise to unphysical oceanographic structures (results not shown here). On the other hand, we observe that the DINEOF method, at least the way it is implemented here, is largely dependent on the climatology. This results in the two-dimensional output sometimes being unable to reproduce the small-scale processes, which can be better inferred from the OI technique if observations falling within the spatial and temporal influential radius are available.

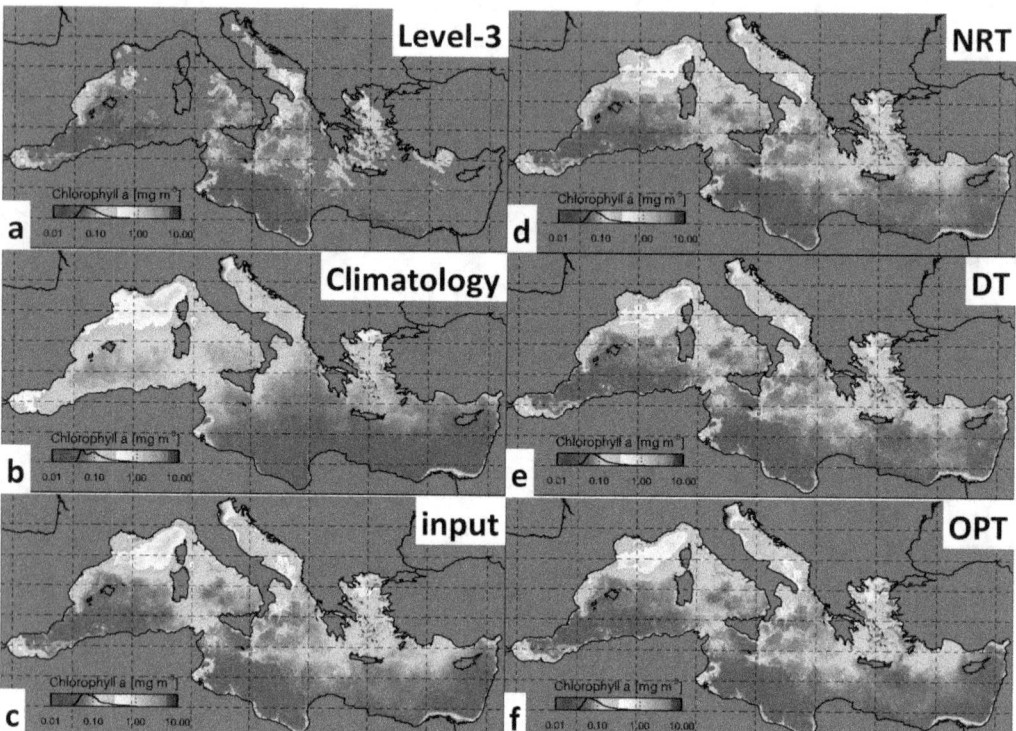

Figure 9.7. Example of the full chain that transforms the original observations (panel a), using the daily climatology (panel b), into the data that constitutes the single image (panel c) for the input data matrix. The outputs were determined using the input data matrix schemes of Figure 9.2a for NRT (panel d), Figure 9.2b for DT (panel e), and Figure 9.2c for OPT (panel e). These images refer to SeaWiFS daily products collected on April 3, 1998.

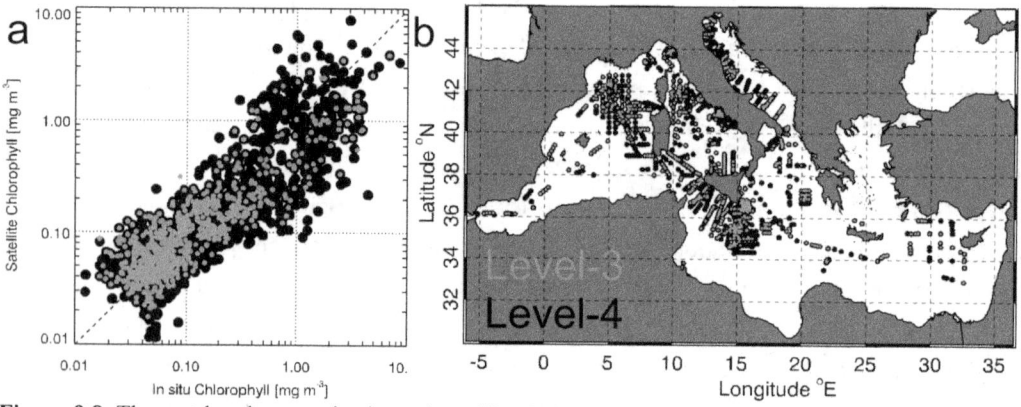

Figure 9.8. The matchup between in situ and satellite derived chlorophyll spans from 1997 to 2015, for both Level-3 (red) and Level-4 (black).

Conclusions

The aim of this work is to describe and fully characterize the steps needed to achieve an operational ocean colour interpolated product over the Mediterranean Sea. This product is operationally provided to the user community through the CMEMS web portal. An important aspect that needs to be pointed out is the fact that the method described here is totally independent from the region of interest and from the product itself; it is general and it could reasonably be applied for the interpolation of other fields over other regions including the global ocean. An example is the L4 chlorophyll field over the Black Sea[2], which is operationally produced with this approach, in the context of CMEMS. One aspect that surely needs to be considered is the de-correlation temporal and spatial scales that should drive the length of the data time series used as input to the interpolation procedure.

The DINEOF method has been demonstrated to perform equally well as, and under given circumstances even better, than the *standard* OI procedure, which is widely used to fill in different oceanographic variables such as SST, SSS, and SSH.

To avoid the use of data time series that are too long, because they might be poorly correlated with the data that has to be interpolated, the input data matrix is built starting from relatively recent original observations with holes filled in with daily climatology, as first guess. This allows for reducing the length of the data time series, thus significantly saving computational resources without any appreciable drawbacks. The impact of the climatology fields is demonstrated to be negligible, as all configurations return statistics better than those obtained with climatology.

The use of climatology is made possible by the *ad hoc* smoothing procedure, which efficiently enables the merging of fields with different temporal and spatial scales without introducing unrealistic features. The overall impact of the smoothing procedure is that it allows the original observations to be kept in the final interpolated field. As a result, the interpolated field behaves exactly the same way as the original in correspondence with valid initial observations. An important point is that, since the original observations are kept as they are, their quality needs to be very carefully checked before entering the interpolation procedure.

The multi-image state vector is efficient when including the data space-time variability in the final interpolated fields and decreasing the weight that climatology has in determining the space-time distribution of the interpolated fields.

Acknowledgements

This work was funded by the European Integrated project MyOcean2 and by the Italian Flagship project, RITMARE. The results of this work are operationally implemented in the CMEMS processing chain. A special thanks to Clifford Watkins, Virginie van Dongen, Dakui Wang, and Natalia Tilinina for their early review of this work. An anonymous reviewer is also thanked for the useful comments that helped improve the work.

---

[2] http://marine.copernicus.eu/services-portfolio/access-to-products/?option=com_csw&view=details&product_id=OCEANCOLOUR_BS_CHL_L4_NRT_OBSERVATIONS_009_045

References

Alvera-Azcárate, A., A. Barth, D. Sirjacobs, and J.-M. Beckers (2009), Enhancing temporal correlations in EOF expansions for the reconstruction of missing data using DINEOF, Ocean Sci. Discuss., 6(2), 1547–1568, doi:10.5194/osd-6-1547-2009.

Alvera-Azcárate, A., Q. Vanhellemont, K. Ruddick, A. Barth, and J. M. Beckers (2015), Analysis of high frequency geostationary ocean colour data using DINEOF, Estuar. Coast. Shelf Sci., 159, 28–36, doi:10.1016/j.ecss.2015.03.026.

Beckers, J., and M. Rixen (2003), EOF Calculations and Data Filling from Incomplete Oceanographic Datasets*, J. Atmos. Ocean., 1–32.

Bretherton, F., R. Davis, and C. Fandry (1976), A technique for objective analysis and design of oceanographic experiments applied to MODE-73, Sea Res. Oceanogr., 23.

Buongiorno Nardelli, B., S. Colella, R. Santoleri, M. Guarracino, and A. Kholod (2010), A re-analysis of Black Sea surface temperature, J. Mar. Syst., 79(1–2), 50–64, doi:10.1016/j.jmarsys.2009.07.001.

Buongiorno Nardelli, B., C. Tronconi, a. Pisano, and R. Santoleri (2013), High and Ultra-High resolution processing of satellite Sea Surface Temperature data over Southern European Seas in the framework of MyOcean project, Remote Sens. Environ., 129, 1–16, doi:10.1016/j.rse.2012.10.012.

Donlon, C. J., M. Martin, J. Stark, J. Roberts-Jones, E. Fiedler, and W. Wimmer (2012), The Operational Sea Surface Temperature and Sea Ice Analysis (OSTIA) system, Remote Sens. Environ., 116, 140–158, doi:10.1016/j.rse.2010.10.017.

Ferreira, J. G. et al. (2011), Overview of eutrophication indicators to assess environmental status within the European Marine Strategy Framework Directive, Estuar. Coast. Shelf Sci., 93(2), 117–131, doi:10.1016/j.ecss.2011.03.014.

Liu, X. and M. Wang, (2016), Analysis of ocean diurnal variations from the Korean Geostationary Ocean Color Imager measurements using the DINEOF method, Estuarine Coastal Shelf Sci., 180, 230-241. doi:10.1016/j.ecss.2016.07.006

Martin, M. et al. (2012), Group for High Resolution Sea Surface temperature (GHRSST) analysis fields inter-comparisons. Part 1: A GHRSST multi-product ensemble (GMPE), Deep Sea Res. Part II Top. Stud. Oceanogr., 77–80, 21–30, doi:10.1016/j.dsr2.2012.04.013.

McClain, C. R., Feldman, G. C., & Hooker, S. B. (2004). An overview of the SeaWiFS project and strategies for producing a climate research quality global ocean bio-optical time series. Deep-Sea Research Part II: Topical Studies in Oceanography, 51(1–3), 5–42. http://doi.org/10.1016/j.dsr2.2003.11.001

Miles T.N., He R., (2010), Temporal and spatial variability of Chl-a and SST on the South Atlantic Bight: Revisiting with cloud-free reconstructions of MODIS satellite imagery, Continental Shelf Research, 30 (18), pp. 1951-1962.

Reynolds, R. W., T. M. Smith, C. Liu, D. B. Chelton, K. S. Casey, and M. G. Schlax (2007), Daily High-Resolution-Blended Analyses for Sea Surface Temperature, J. Clim., 20(22), 5473–5496, doi:10.1175/2007JCLI1824.1.

Roberts-Jones, J., E. K. Fiedler, and M. J. Martin (2012), Daily, Global, High-Resolution SST and Sea Ice Reanalysis for 1985–2007 Using the OSTIA System, J. Clim., 25(18), 6215–6232.

Saulquin, B., F. Gohin, and R. Garrello (2011), Regional objective analysis for merging high-resolution MERIS, MODIS/Aqua, and SeaWiFS chlorophyll-a data from 1998 to 2008 on the european atlantic shelf, IEEE Trans. Geosci. Remote Sens., 49(1 PART 1), 143–154, doi:10.1109/TGRS.2010.2052813.

Sirjacobs, D., A. Alvera-Azcárate, A. Barth, G. Lacroix, Y. Park, B. Nechad, K. Ruddick, and J.-M. Beckers (2011), Cloud filling of ocean colour and sea surface temperature remote sensing products over the Southern North Sea by the Data Interpolating Empirical Orthogonal Functions methodology, J. Sea Res., 65(1), 114–130, doi:10.1016/j.seares.2010.08.002.

Le Traon, P. Y., F. Nadal, and N. Ducet (1998), An Improved Mapping Method of Multisatellite Altimeter Data, J. Atmos. Ocean. Technol., 15(2), 522–534.

Volpe, G., B. Buongiorno Nardelli, P. Cipollini, R. Santoleri, and I. S. Robinson (2012), Seasonal to interannual phytoplankton response to physical processes in the Mediterranean Sea from satellite observations, Remote Sens. Environ., 117, 223–235, doi:10.1016/j.rse.2011.09.020.

Volpe, G., S. Colella, V. Brando, V. Forneris, F. La Padula, J. Pitarch Portero, A. Di Cicco, M. Bragaglia, F. Artuso, R. Santoleri (2017), The Mediterranean component of the Copernicus Ocean Colour Thematic Assembly Centre: Level 2 to Level 3 processing, Remote Sens. Environ., submitted.

Weare, B. C., and J. S. Nasstrom (1982), Examples of Extended Empirical Orthogonal Function Analyses, Mon. Weather Rev., 110, 481–485, doi:10.1175/1520-0493(1982)110<0481:EOEEOF>2.0.CO;2.

CHAPTER 10

Surface Heat Fluxes and Wind Remote Sensing

Mark A. Bourassa and Paul J. Hughes

Dept. of Earth, Ocean and Atmospheric Science and Center for Ocean-Atmosphere Prediction Studies, Florida State University, Tallahassee, FL, USA

The exchange of heat and momentum through the air-sea surface are critical aspects of ocean forcing and ocean modeling. Over most of the global oceans, there are few in situ observations that can be used to estimate these fluxes. This chapter provides background on the calculation and application of air-sea fluxes, as well as the use of remote sensing to calculate these fluxes. Wind variability makes a large contribution to variability in surface fluxes, and the remote sensing of winds is relatively mature compared to the air sea differences in temperature and humidity, which are the other key variables. Therefore, the remote sensing of wind is presented in greater detail. These details enable the reader to understand how the improper use of satellite winds can result in regional and seasonal biases in fluxes, and how to calculate fluxes in a manner that removes these biases. Examples are given of high-resolution applications of fluxes, which are used to indicate the strengths and weakness of satellite-based calculations of ocean surface fluxes.

Introduction

Satellite remote sensing works by receiving an electromagnetic signal that is dependent on the geophysical variable that is being measured. For example, ocean color is determined through measuring wavelengths of reflected light in the visible band; temperature is determined from measurements of emitted radiation, and wind speed can be determined through a wide variety of approaches including emissions at multiple bands or the fraction of radar energy that interacts with the surface and returns to the satellite. The geophysical variable determined in this manner is called a retrieval (e.g., wind speed retrievals). Passive remote sensing usually measures how much energy (over a band of wavelengths) is emitted, reflected, or scattered from an object (e.g., the ocean surface). Active satellites emit a signal and measure how much of this emitted signal is reflected or scattered from an object and returned to the satellite or a different satellite. For remote sensing to be effective, the geophysical variable that is retrieved must modify the measured electromagnetic signal. It is easy to imagine how wind modifies the ocean surface (e.g., water waves), making the remote sensing of wind an obvious candidate for observation from space. A better understanding of the geophysical processes that modify the surface can be used to develop better retrievals and to more carefully define the variable being measured.

Bourassa, M.A., and P.J. Hughes, 2018: Surface heat fluxes and wind remote sensing. In "*New Frontiers in Operational Oceanography*", E. Chassignet, A. Pascual, J. Tintoré, and J. Verron, Eds., GODAE OceanView, 245-270, doi:10.17125/gov2018.ch10.

There is a long history of efforts to use satellite observations to determine surface turbulent and radiative fluxes. Surface fluxes are commonly thought of as the transfer of something (e.g., momentum, heat, moisture, a gas, or particulates) from the ocean to the atmosphere or vice versa. For example, the latent heat flux is related to evaporation at the ocean surface, but it is also the vertical transport of this energy (stored in water vapor) in the atmospheric boundary-layer. Stress is the vertical transport of momentum at the air/sea interface, but there must also be an identical transport of momentum in the atmospheric and oceanic boundary-layers, unless that momentum is released or transported away by waves in the process of crossing the air/sea interface. Air-sea fluxes transfer momentum, heat, or material through the ocean surface and the layers near the surface, in the process modifying the surface and materials near the surface (e.g., modifying the motions, temperatures near the surface). A variety of approaches can be used to measure information at slightly different depths, and hence the changes caused by fluxes can be used to measure fluxes. Consequently, we realize that air/sea fluxes modify the air/sea interface and could plausibly be measured from satellite. Understanding the geophysical processes that modify the surface and thus modify the electromagnetic characteristics of the surface can be useful in developing more accurate retrievals and retrievals of new variables.

The above considerations are extremely useful for the developers of remote sensing retrievals, but are not relevant to most users of the retrieved geophysical data. That said, because the above considerations have been used to improve retrievals and quantify error characteristics, the descriptions of what is retrieved is often very carefully defined and often is not quite consistent with the assumptions of a casual user of the data. In some cases, the consequences of the errors associated with ignoring these definitions are small compared to other assumptions made in the application; however, in many cases they are not small and can result in a very misleading result. For example, satellite winds are 'equivalent neutral winds' (defined below, which differ slightly from traditional winds. If this slight difference is ignored, the average impact on mid-latitude fluxes is approximately a 10 Wm^{-2} bias, in addition to further regional and seasonal biases. It is very useful to know enough about the physical interpretation of retrievals to be able to assess if the assumptions made in using them will seriously alter the interpretation resulting from using the satellite data.

An introductory discussion of how air-sea coupling processes cause responses in the ocean and atmosphere provides relatively simple examples that show the complexity of coupling in the real world. These example illustrate problems that can be examined with uncoupled and coupled models. Satellite data can be used to assess the realism of the modeled near surface conditions, but a new generation of much finer resolution observations would be much more useful in determining the accuracy of models.

This chapter is designed to provide novices to remote sensing a very general idea of how remote sensing works, with a focus on surface wind and stress. It includes the basic physics and parameterizations of air/sea fluxes and how these fluxes modify the ocean surface in ways that can be used to measure air/sea fluxes. The goals are that readers will be able to better understand satellite remote sensing, gain an appreciation of the physics that impacts air/sea fluxes and satellite retrievals, and learn the types of questions that should be asked when applying any satellite retrieval.

Basics of Remote Sensing of Winds

Remote sensing relies on measuring electromagnetic radiation that is either emitted by the surface or emitted off the surface and interacts with the surface. If the emission or interaction of the electromagnetic radiation changes enough as a function of wind speed, then the observations can be used to determine wind speed. In addition, if the emissions or interactions are non-isotropic (directionally-dependent), then the observations can be used to determine the wind direction. The measurement of wind direction requires observations of the electromagnetic signal from several different directions. One further point is that if the signal being measured is non-isotropic, then the measurement of speed will be more accurate if the direction is determined. In other words, not knowing or measuring the direction can result in a greater uncertainty in the measurement of speed. These observations can be made more accurate by better accounting for how the electromagnetic radiation interacts with the atmosphere or by using wavelengths that are not heavily altered or modified by the atmosphere. Measurements of several wavelengths or polarizations can be used to learn a lot more about the atmosphere and have been shown to be effective in retrieving the near surface air temperature and humidity (Jackson et al., 2006, 2009; Jackson and Wick, 2010; Roberts et al., 2010; Bourassa et al., 2010; Smith et al., 2012), which are very important for determining surface fluxes of heat and moisture. The complexities of interaction with the atmosphere will not be discussed in this chapter, other than as warnings about when the atmospheric conditions cause trouble with the accuracy of remotely sensed variables.

Active remote sensing of surface winds

Radar is an example of active remote sensing. An electromagnetic signal, often microwaves, is generated and aimed at the surface. The signal that is reflected or scattered from the surface is measured and the ratio of the input to the output signal provides information about the surface. Radars use wavelengths that respond to the shape or roughness of the ocean surface. The term roughness is used for electromagnetic interactions and aerodynamic interactions; these do not mean the same thing. Electromagnetic waves interact with water waves of similar wavelengths to the electromagnetic wavelengths. For example, a radar wavelength of 6 cm moving roughly parallel to the surface would interact with surface waves of around 6 cm in wavelength. If the electromagnetic waves approach the surface at a greater angle, then they interact with shorter wavelengths. For example, scatterometers (described below) tend to operate at Ku-band interacting with longer ultragravity waves or at C-band interacting with very short gravity waves. These wavelengths are desired because these waves respond very quickly to changes in the surface winds. However, shorter radar wavelengths are more adversely impacted by the atmosphere and rain, making longer wavelengths, such as those used to measure ocean salinity, preferred for observations in rainy conditions.

Scatterometers and synthetic aperture radars

Scatterometers and synthetic aperture radars (SARs) are active sensors that use short microwaves (specifically Ka- to C-band) to measure the surface roughness. This roughness is a function of wind stress, which is closely related to wind speed (see the section on air-sea fluxes for details). The backscatter at these wavelengths has a directional dependence that allows direction to be retrieved if there are observations from multiple directions or the direction can be assumed from features in the data. Scatterometers retrieve direction by observing the surface from multiple directions as the satellite moves over the surface. This approach requires that the satellite move relative to the ocean surface. Therefore, we will never see a scatterometer in geosynchronous orbit. Similarly, SARs move relative to the surface and take observations from slightly different locations to effectively increase the size of the antenna and allow for much finer resolution observations. This resolution is used to identify bands of higher winds from rolls or other features to estimate the wind direction. SAR directions are less accurate than scatterometer directions and can, in both cases, have substantial biases in areas because of incorrect assumptions.

It is clear that scatterometer and SAR observations respond to stress rather than wind (Bourassa et al., 2010); however, surface wind datasets are produced. This is in part because there are far more wind observations than stress observations with which to calibrate the satellites, and in part because it is believed that operational weather centers will be able to more easily assimilate wind than surface stress.

The strengths of scatterometer and SAR observations are that they are vector observations (i.e., speed and direction), which is much more useful than scalar winds (i.e., speed only) for many ocean and atmospheric applications. Vector observations allow for the calculation of curl and divergence, and have much more dynamical impact during data assimilation in weather models. The vector observations also usually come from active systems, which can penetrate cloud cover and rain far better than passive observations. However, in both cases rain can cause serious complications in retrieving accurate surface winds. In the case of scatterometers, those that use Ku-band (e.g., QuikSCAT) are more sensitive to rain than C-band instruments (e.g., ASCAT), which have longer wavelengths relative to rain drop sizes. The problems are greatest when there are light winds and heavy rain, and negligible for most applications when there are high winds and light rain. Perhaps the greatest weakness of any single wind sensing satellite is the temporal sampling. There is a global average of two overpasses per day from a QuikSCAT-like instrument (1800 km swath width), with about 1.5 observations per day in the equator and four or more (in two clusters) near the poles where adjacent swaths overlap. An example of the coverage for one day is shown below. This limited coverage means that it is highly desirable to use data from a constellation of intercalibrated instruments. While intercalibration is very good for commonly occurring wind speeds (4–17 ms^{-1}), the two popular calibrations (from KNMI and Remote Sensing Systems) diverge outside this range (Verspeek et al., 2012; Chakraborty et al., 2013; Wentz et al., 2017; Holbach and Bourassa, 2017).

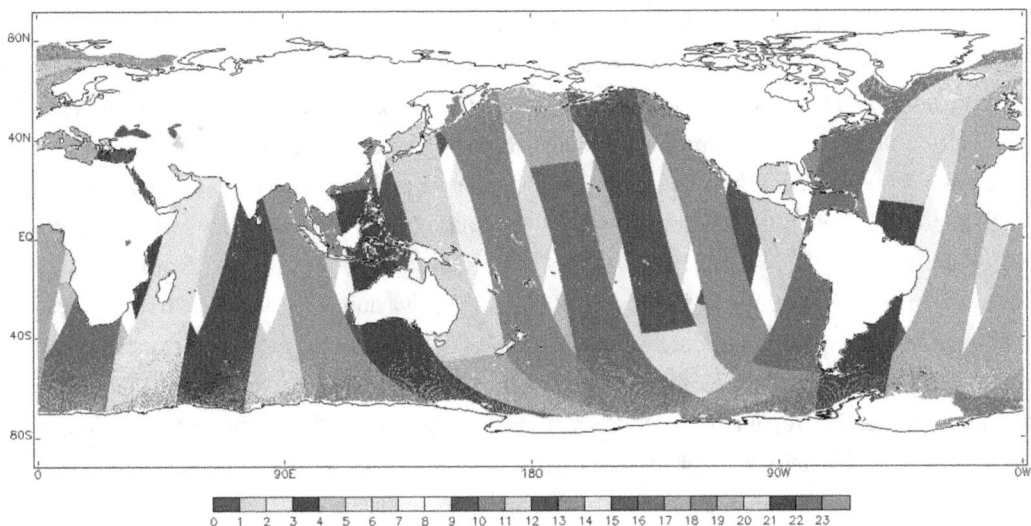

Figure 10.1. Example of daily coverage from QuikSCAT, where the color indicates the hour of the overpass. Any satellite in low earth orbit takes roughly 100 minutes to circle the Earth. The swath width and orbit determine the temporal sampling.

GPS reflectometry

Surface winds can also be estimated from the reflection of signals from the Global Positioning System (GPS). This approach is called GPS reflectometry (Zavorotny and Voronovich, 2000) and it requires a separate satellite to measure the signal emitted from the GPS satellite. The footprint from this technique is currently rather large, making it difficult to observe extreme features, which usually are much smaller than the footprint (e.g., fonts, Tropical Cyclone eyewalls, and near-shore features). NASA has recently developed the CYGNSS observing system (Ruf et al., 2013), which is a set of eight satellites flown in formation to improve the coverage. This observing system is focused on the tropics, providing improved sampling in time. However, the pattern in space is not well suited for spatial derivatives (e.g., calculation of curl and divergence, in addition to gradients of speed). The calibration of the CYGNSS system is an ongoing effort. It is expected to provide scalar wind speeds only, with much more sampling needed to provide vector observations, which require looks at the same surface location from difference azimuthal angles (angles relative to true north). One advantage of this system is that the GPS wavelengths are very long and interact very weakly with rain. Therefore, this system is expected to be useful for activities (operational and research) involving tropical cyclones.

Passive remote sensing (radiometry) of surface winds

Passive systems for observing surface winds measure the brightness temperature of the ocean surface. This is the temperature that the ocean would have if the observed radiance came from a black body. The ocean is not a black body, therefore this temperature is less than the ocean's surface temperature (Petty, 2006). This radiance is a function of temperature and surface wind speed (technically stress – see the section on air-sea fluxes) to understand this distinction) with greater radiance coming from whitecaps (Uhlhorn and Black, 2003; Petty, 2006; Paget et al., 2015). As wind stress increases there is greater breaking and a larger fraction of the ocean surface is covered by whitecaps. Combined with an estimate of the ocean temperature, the excess emissivity (above that expected from the temperature and a flat water surface) can be determined (Uhlhorn and Black, 2003; Petty, 2006). This temperature estimate is usually made from the same satellite used to estimate wind speed. There is a wind direction dependence in these observations, which is usually not considered. However, the WindSat mission uses polaromitry (Wentz, 1983; Gordon and Wang, 1994; Yueh et al., 1995; Wentz, 1997; Smith, 1998; Gordon and Voss, 1999; Rose et al., 2002; Gaiser et al., 2004; Anguelova and Webster, 2006; Anguelova and Gaiser, 2012, 2013) to measure this directional dependence and determine wind direction. In general, radiometers can accurately measure surface wind speeds >3 ms^{-1}; however, those designed to measure hurricane winds are typically not effective for wind speeds <8 ms^{-1}. WindSat can usefully determine directions for speeds >8 ms^{-1}. Radiometry is much more sensitive to rain than scatterometry and SAR, and usually does not provide wind data coincident with rain, whereas scatterometry (particularly C-band) and SAR can work in rain rates up to a wind speed dependent threshold (Draper and Long, 2004).

Air/Sea Fluxes

Air/sea fluxes refer to the rate of transfer, per unit area, through the air/sea interface and the ocean and atmosphere near this interface. Physicists refer to a transfer that includes the units of 'per unit area' as flux densities, but meteorologists and oceanographers drop the word 'density' and simply say 'fluxes.' Similar vertical rates of turbulent transfer exist in the atmospheric boundary-layer. Stress in the near-surface ocean is slightly reduced because some stress is lost to surface water waves. Heat fluxes are radiative fluxes and turbulent heat fluxes (sensible and latent heat). Radiative fluxes are similar in the lower atmosphere, but change rapidly with depth in the ocean. Radiative fluxes refer to the rate of transfer of energy per unit area in the form of electromagnet radiation, rather than the much more scary things that people normally associate with the word radiation. Fluxes are associated with motion and the net transport of something like heat and momentum. Most people think of conduction (heat transferred through the motion of atoms and molecules) as a good method for transporting energy, but that is an extremely poor assumption for the atmosphere. For example, Styrofoam coolers work very well not because of the foam, but because air is trapped in the Styrofoam and air is a terrible conductor. Similarly, Brownian motion is a very slow way of transporting energy compared to turbulent transport and radiative transport. Turbulence, where it exists, is much more efficient than conduction and Brownian motion.

In simple and somewhat recursive terms, turbulence transport is a transport due to turbulent motion. This motion comes about from vertical shear in horizontal motion and from unstable stratification. In contrast, stable stratification inhibits turbulent motion. Turbulent motion can be thought of as due to local changes of the speed of air or water associated with friction on a boundary and buoyancy. For example, any broad surface that has friction will have slower wind speeds near this surface and larger wind speeds away from the surface. These wind speeds increase logarithmically with height in the atmospheric boundary-layer. Increases in mechanical mixing tend to reduce the impacts of buoyancy, and increased buoyancy (either upward or downward) tends to reduce vertical shear except very near a boundary (e.g., the ocean surface). Turbulent fluxes (described in detail below) include stress (the flux of momentum), sensible heat, latent heat, and mass fluxes such as moisture or gas fluxes.

Turbulent fluxes are proportional to the covariance of vertical perturbations in vertical winds and perturbations in the quantity being transported. The example in Fig. 10.2 shows how a vertical perturbation of an air parcel is associated with a vertical motion (w'), and that when that air parcel moves up (positive w') it is slower than the surrounding air (a negative perturbation in momentum, u'). If the air parcel moves down it has negative w' and a positive u'. Except for the very rare cases when the current is moving faster than the air above, a wind profile will always have this shape. Consequently, the covariance of u' and w' is always negative. Hence the stress is proportional to $-\overline{u'w'}$.

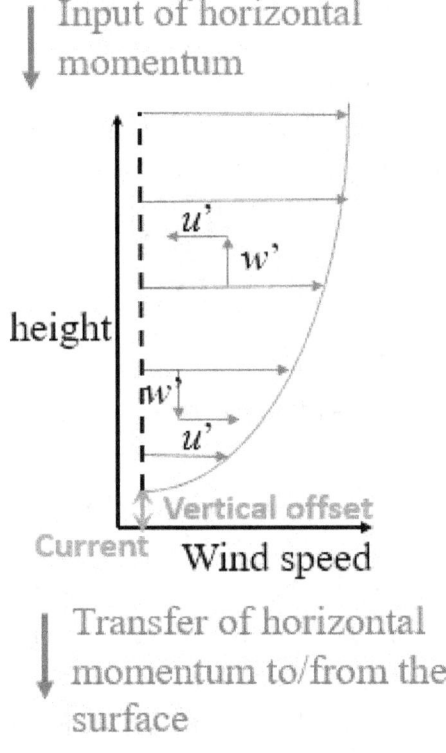

Figure 10.2. Log-wind profile with vertical perturbations in the wind (w') and the corresponding perturbation in horizontal velocity, which is proportional to momentum.

Stress

If the boundary layer has neutral stratification, meaning that force related to buoyancy is not contributing to turbulence, then stress is a shear stress, which is due to a change in velocity with height (Fig. 10.2). Near the surface, in the part of the atmospheric boundary layer known as the log layer (because the variables have a logarithmic function of height above the surface), all fluxes are independent of height within that layer. Very near the surface, part of the momentum flux can change to a horizontal pressure on the waves but the net momentum flux remains constant.

The top of the ocean's mixed layer feels this same stress as a 'surface stress,' reduced by any net rate of loss of momentum into surface waves. If the waves are in balance with the wind (i.e., the time and space averaged distributions of wave characteristics are not changing thus waves are not growing higher to reach equilibrium with wind that is stronger than needed for equilibrium, and waves are becoming shorter and longer because the wind has reduced in speed), the stress experienced by the ocean is a few percent less than the stress in the atmosphere because some of the momentum from the waves maintains the current. Stress is particularly important for oceanography because it can be closely related to wind-driven currents and vertical motions in the upper ocean (Ekman transport; Knauss, 2005) as well as horizontal motions in the deep ocean (Sverdrup transport; Sverdrup, 1947). Stress is also related to the generation of eddy kinetic energy, which is the dot product of surface stress and surface current.

Stress (τ) can be defined in terms of perturbations as $-\rho \overline{u'w'}$, where ρ is the density of the fluid. This is a great definition when measuring fluxes, but is not very useful for modeling stress in atmospheric and ocean models because tenth of a second changes in winds must be resolved to determine stress this way. In such models it is more useful to define stress in terms of friction velocity ($\mathbf{u}_*$, Eq. 1; where bold text indicates a vector, and non-bold indicates a scalar), which is closely related to stress (Eq. 1) and a scaling parameter in the log-profile for wind speed (Eq. 2).

$$\tau = \rho |\mathbf{u}_*|\mathbf{u}_* \qquad (1)$$

$$\overline{\mathbf{u}}(z) - \overline{\mathbf{u}}_{\text{sfc}} = \frac{\mathbf{u}_*}{k_v}\left[\log\left(\frac{z-d}{z_o}+1\right) - \varphi_M(z, z_o, L)\right] \qquad (2)$$

In Eq. 2, $\overline{\mathbf{u}}$ is the mean horizontal wind speed at a height z above the displacement height d. $\overline{\mathbf{u}}_{\text{sfc}}$ is the mean ocean surface current, κ is von Kármán's constant, z_o is the momentum roughness length, φ_M is an atmospheric stability term (Stull 1988) for momentum, L is the Moniv-Obukhov scale length (Monin and Obukhov, 1954; Stull, 1988), and $\overline{\mathbf{u}}$ - $\overline{\mathbf{u}}_{\text{sfc}}$ is parallel u_* and stress. The displacement height (d) is a vertical offset in the log-profile, and it is the height at which the left side of the equation extrapolates to zero. As implied by the word 'extrapolate,' the log profile does not apply near the displacement height. For example, over crops the displacement height is roughly 79% the height of crops. Waves have a height-to-spacing ratio that is much less than crops, suggesting that d is much less than 70% of wave height. In fact, d is almost always assumed to be zero over water. The impact of considering displacement height has been tested in a wind/wave tank (Bourassa et al., 1999), where the stress estimated using a log profile technique was compared to a stress determined from eddy covariance. The impact of displacement height on open ocean fluxes is small when a displacement height of 80% of the wind driven significant wave height was found to optimize the fit to observed fluxes (Bourassa, 2006) for wind waves. A student project later revealed that a displacement height of zero works well for swell. The Monin-Obukhov scale length is a related to atmospheric stratification, and is also dependent on the shear stress. The friction velocity can be obtained from Eq. **Error! Bookmark not defined.**, given L and a relation between roughness length and friction velocity, such as Charnock's relation (Stull, 1988). Typically, friction velocity is calculated in the atmosphere, so the density is that of air rather than water. Using the density of water rather than that of air results in roughly a factor of 800 in stress.

Solving for friction velocity and stress as a function of wind speed is a slow iterative process. Therefore, many ocean and atmosphere models define stress in terms of a drag coefficient (C_D).

$$\tau = \rho \, C_D \, |\bar{\mathbf{u}} - \bar{\mathbf{u}}_{sfc}|(\bar{\mathbf{u}} - \bar{\mathbf{u}}_{sfc}) \qquad (3)$$

The drag coefficient can be parameterized in terms of z_o and φ, assuming the displacement height is negligible, or the impact of displacement height can be included empirically while fitting the drag coefficient to observations, analogous to what is done with the COARE flux model (Fairall et al., 2003).

$$C_D(z) = \left[\frac{1}{k_v} \left[\ln\left(\frac{z-d}{z_o}+1\right) - \varphi_M(z, z_o, L) \right] \right]^{-2} \qquad (4)$$

This combination of Eqs. (2) and (3) clearly shows that specifying a drag coefficient is analogous to specifying a roughness length. The roughness length is a combination of roughness lengths for multiple types of surfaces (Smith, 1988; Bourassa et al., 1999; Zheng et al., 2013): a smooth surface (Nikuradse, 1933; Kondo, 1975), capillary waves (Wu, 1968, 1994; Bourassa et al., 1999; Zheng et al., 2013) and gravity waves (Wu, 1980; Smith, 1988; Dobson et al., 1994; Taylor and Yellend, 1999; Drennan et al., 2005; Bourassa, 2006). The fitting parameters in the gravity wave roughness length parameterizations are highly sensitive to the surface current and displacement height (Bourassa, 2006). In all cases, the parameterizations for stress are tuned to observations. If stability is considered (and it should be), then the parameterizations become complicated. If sea state is known, it is almost always a good idea to use a flux parameterization that considers sea state. Some models greatly speed up the time it takes to calculate fluxes by using polynomial curve fits to stability dependent parameterizations (Kara, 2000).

Current topics in stress parameterization include the dependency on sea state and the atmospheric response to winds flowing over sea surface temperature (SST) gradients. Another open question is the importance of the atmospheric response to changes in the ocean. There is no question that the atmosphere responds to changes in the ocean, but the outstanding questions are *how much does it respond?* and *what are the key physical processes?* Similar questions are asked about the response of the ocean to changes in the atmosphere. These are important questions because they could have large impacts on the importance of two-way coupling between the ocean and atmosphere, which could add a great deal of computer processing requirements relative to the common practice of assuming that these feedbacks are small.

The ocean's mixed layer has a very thin log layer compared to the atmosphere. Ekman transport cause ocean responses to changes in the surface stress. Changes in the thermal forcing (heat budget) also cause an ocean response. If the winds are light and there is no wave breaking, the current profile is explained in terms of Stokes drift. However, if breaking occurs there should be more mixing at

Sensible heat

The sensible heat flux (H) is the rate at which thermal energy (associated with heating, but without a phase change) is transferred from the ocean to the atmosphere. In the tropics, the latent heat flux is typically an order of magnitude greater than the sensible heat flux; however, in the polar regions the H can dominate. The quantity used to describe the storage of thermal and potential energy is the potential temperature. The sensible heat flux can be measured as proportional to the covariance of vertical velocity and potential temperature (Θ). Hence the sensible heat flux is proportional to $\overline{\theta'w'}$.

Sensible heat flux (H) can be defined in terms of perturbations as $-\rho C_p \overline{\theta'w'}$, where C_p is the heat capacity of air. In such models it is more useful to define sensible heat in terms of the magnitude of friction velocity (u_*, Eq. 1), an analogous term in the log-temperature profile (θ_*, Eq. 6).

$$H = \rho C_p |\mathbf{u}_*|\theta_* \quad (5)$$

$$\overline{\theta}_{sfc} - \overline{\theta}(z) = \frac{\theta_*}{k_v}\left[\ln\left(\frac{z}{z_{o\theta}}+1\right) - \varphi_\theta(z, z_{o\theta}, L)\right] \quad (6)$$

In Eq. 6, the displacement height for potential temperature is believed to be zero, however, it could easily be argued that evaporation from sea spray would change the temperature profile and make this term non-zero. A non-zero value for the displacement height for momentum (Eq. 2) suggests that a non-zero displacement height for potential temperature is quite plausible. Such considerations would be difficult to separate from adjustments due to stability. The stability term for temperature and moisture (φ_θ) is different than the stability term for momentum. Since H is an atmospheric flux, stability is calculated based on the atmospheric profile, rather than the oceanographic profile. The sensible heat flux should be non-zero in the near surface ocean, provided the net surface heat flux is non-zero (which implies that there is a temperature difference between the ocean and atmosphere) and wave breaking is causing turbulent motion. Turbulence will be damped out rapidly with increasing depth, meaning that other forms of energy transport will be more important away from the surface. Consequently, sensible heat flux almost always refers to the atmospheric side of the interface, with the understanding that the net heat flux is maintained on the ocean side of the interface.

Many models define H in terms of a transfer coefficient (C_H).

$$H = \rho C_p C_H |\overline{\mathbf{u}} - \overline{\mathbf{u}}_{sfc}|(\overline{\theta}_{sfc} - \overline{\theta}) \quad (7)$$

The merging of Eqs. (5) and (7) shows that C_H is equal to the square root of the drag coefficient times and similarly structured term with velocities replaced by potential temperature. The sensible heat flux should be sensitive to changes in drag, but it is rarely parameterized as such. Another issue with both sensible and latent heat fluxes is that organized eddies, such as atmospheric rolls, arguably act to increase fluxes beyond the expectations of Eq. (7). Similarly, at low enough wind speeds (how low is debatable) these parameterizations (Eqs., 1, 3, 5, and 7) underestimate fluxes because they assume that the vast majority of horizontal transport is due to the mean flow rather than eddies. Both of these issues have been attempted to be accounted for by slightly increasing the wind shear (Fairall et al., 1996). However, if this approach is used, the wind, temperature, and humidity cannot be correctly adjusted to different heights. For example, a temperature measured at a height of 4 m cannot be correctly adjusted to 10 m. If this type of fix is not used, then something else must be done to increase fluxes at low wind speeds. It has also been argued that roughness and turbulence from capillary waves partially account for this underestimation at low wind speeds (Bourassa et al., 1999; Zheng et al., 2013). Both approaches have a similar impact on the fluxes, but the inclusion of roughness from capillary waves does not cause problems with height adjustments. Such parameterizations work very well with the temperature and moisture roughness length parameterizations developed by Clayson et al. (1996).

Latent heat and evaporation

The evaporative moisture flux is the rate, per unit area, at which moisture is transferred from the ocean to the air. The latent heat flux (E) is related to the moisture flux (L); it is the rate (per unit area) at which energy associated with the phase change of water is transferred from the ocean to the atmosphere or the rate at which this energy is vertically transported within the atmosphere. The latent heat flux can be measured as proportional to the covariance of vertical velocity and specific humidity (q): $\overline{q'w'}$.

The latent heat flux can be defined in terms of perturbations as $-\rho L_v \overline{q'w'}$, where L_v is the latent heat of vaporization for water, and C_E is the transfer coefficient analogous to the drag coefficient. In such models it is more useful to define latent heat flux (Eq. 8) in terms of friction velocity (u_*, Eq. 1), and an analogous term in the log-humidity profile (Eq. 9):

$$Q = \rho L_v |\mathbf{u}_*| q_* \tag{8}$$

$$\overline{q}_{\text{sfc}} - \overline{q}(z) = \frac{q_*}{k_v} \left[\ln\left(\frac{z}{z_{oq}} + 1\right) - \varphi_\theta\left(z, z_{oq}, L\right) \right] \tag{9}$$

The stability term for temperature and moisture (φ_θ) is identical to the stability term for temperature. There is no latent heat flux in the ocean, so the density is that of the air. Many models define E in terms of a transfer coefficient (C_E).

$$E = \rho L_v C_E |\mathbf{\overline{u}} - \mathbf{\overline{u}}_{sfc}|(\overline{q}_{sfc} - \overline{q}) \tag{10}$$

The merging of Eqs. (8) and (10) shows that C_E is equal to the square root of the drag coefficient times and similarly structured term with velocities replaced by potential temperature. Evaporation (E) is the flux of moisture (not including precipitation).

It is almost identical to Q except that the water vapor is not multiplied by L_v.

$$L = E/L_v = \rho\, C_E\, |\bar{\mathbf{u}} - \bar{\mathbf{u}}_{sfc}|(\bar{q}_{sfc} - \bar{q}) \tag{11}$$

Radiative fluxes

Radiative fluxes are normally subdivided as either solar or terrestrial, and further subdivided as upward or downward. Upward and downward simply refers to the direction of propagation. Solar usually refers to near infrared and more energetic frequencies (i.e., larger frequencies and shorter wavelengths). These are energies that are rarely generated in a natural terrestrial environment. Therefore, upward terrestrial radiation is from reflected (or scattered) downward solar radiation rather than emitted from something in the terrestrial environment. For example, a red shirt looks red because it is relatively efficient in reflecting red light. The shirt does not have the temperature of a hot burner required to emit red light! Terrestrial radiation is generated by any body at terrestrial temperatures, and the rate of output of terrestrial radiation is closely tied to the temperature of the emitter.

Fig. 10.3 shows one of the two key concepts in both the diel (daily) cycle of light and seasonal changes in light. The other key concept is that the length of daytime changes as a function of season and latitude. For the diel cycle, the rotation of the earth causes the shift in the angle θ. For the seasonal cycle, the change in tilt is caused by the tilt of the Earth relative to the sun and the location of the Earth in its orbit about the sun. High latitude summers can have a daily input of solar radiation that is similar to tropical input because of the long length of a summer day.

Fig. 10.3 illustrates the importance of the angle of the surface relative to the incoming sunlight. The amount of incoming solar radiation depends on the distance from the Sun (Earth's orbit is not very elliptical, so this doesn't change much), the angle of the surface relative to the sun (which is a function of latitude, season and the time of day), and the amount of absorbing, reflecting and scattering materials in the atmosphere. The solar radiation that makes it through the atmosphere without interaction is call direct solar radiation, and reaches the surface as indicated in Fig. 10.3. Indirect light was either scattered (e.g., the blue light from the sky) or reflected from a cloud. The amount of solar radiation directly reaching the ground is easy to calculate if the mass and type of atmospheric absorbers are known, and if the number and type of scatters are known. The indirect solar radiation is much harder to model.

The absorption and emission of terrestrial radiation can be an effective method of transferring heat. In forest fires, enough heat can be transferred through radiation to start fires in neighboring trees untouched by flame. Temperatures in the lower atmosphere and ocean are much less extreme (excepting volcanoes), nevertheless radiative transfer can be an effective mechanism for transferring heat. This mechanism should be much more effective than diffusion in the ocean, and is perhaps more important than conduction in areas with weak thermal gradients.

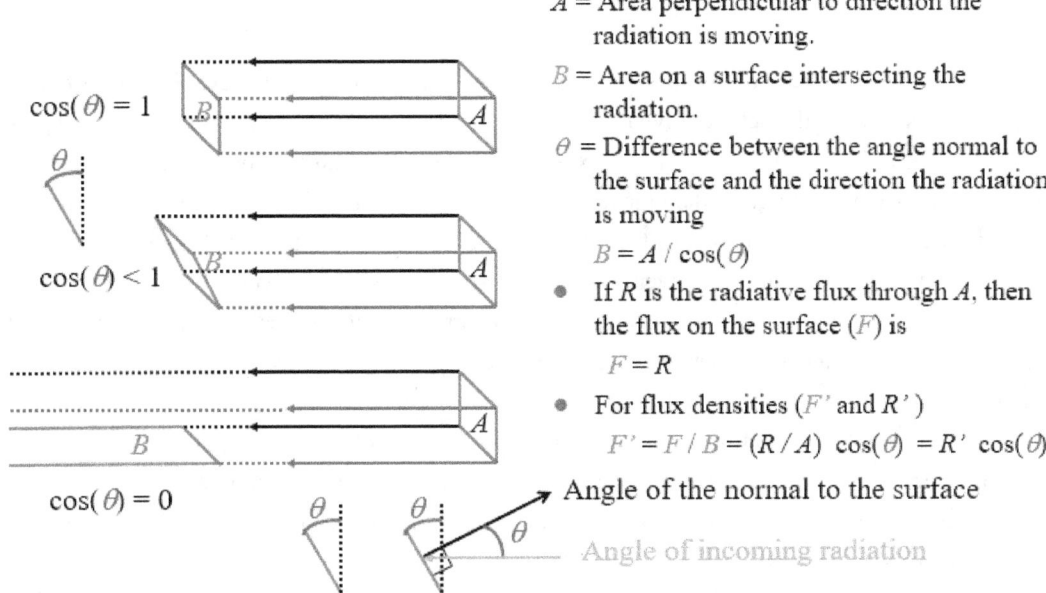

Figure 10.3. The flux of solar energy passing through the area A, which is perpendicular to light emitted from the sun, is constant in all three examples. This light strikes an ocean surface B that is tilted relative to A. In the top example, the tilt of A is zero relative to B, and the flux through A is equal to the flux through B. For the middle and bottom cases, the tilt is not zero, causing the energy passing through B to be spread over more area. The flux density through B is less than the flux density through A, even though the same amount of energy is passing through both areas. In the bottom case, the area of B is infinite, making the flux density through B equal to zero.

Terrestrial radiation is a function of the temperature and the emissivity (the fraction of energy emitted relative to a perfect emitter). The amount emitted is related to the surface area or mass of emitters. The atmosphere is a good absorber of most emitted terrestrial radiation, and the ocean is an excellent absorber. Emitted terrestrial radiation is usually absorbed and reemitted in all directions, contributing to the redistribution of heat in the earth system. The atmosphere also emits terrestrial radiation, and this contributes to heating the ocean's surface. Since the atmosphere is usually colder than the ocean, the terrestrial radiation emitted from the ocean tends to be larger than the input radiation. Cloud cover will increase the downward terrestrial radiation and block some of the solar radiation. In polar winter environments, most of the input radiation is terrestrial radiation emitted from clouds in the lower 200 m (Bourassa et al., 2013).

In the atmosphere, emissions at some frequencies can travel through the cloud-free atmosphere, and are very useful for remote sensing of the surface. Microwaves can travel through the atmosphere including clouds making them very effective for satellite observations of the Earth's surface. Longer wavelength microwaves can see through rain, whereas the shorter end of the microwave spectrum is largely blocked by rain. Therefore, microwaves are very effective for satellite observations of surface processes in the sense that they are rarely blocked by the atmosphere, with longer wavelengths being more effective. Unfortunately, the resolution of satellite observations is a function of wavelength, with finer resolution being much easier (i.e., less costly) to achieve with shorter wavelengths. This tradeoff between being able to see through rain and resolve smaller ocean features is a concern in the development of satellite missions.

Net fluxes

The net heat flux is critical to long-term energy budgets and to the diurnal cycle. The atmospheric net heat flux is equal to the net radiative fluxes (solar and terrestrial, downward minus upward) plus the sensible heat flux and the latent heat flux. The dominant mechanisms for transporting thermal energy in the ocean are different than those in the atmosphere, but in conditions of equilibrium the fluxes in the atmosphere must equal the fluxes in the ocean. While the whole integrated Earth system is in near-equilibrium when averaged over decades, equilibrium is very rare on regional and local scales. For example, more solar radiation is absorbed in the tropical oceans than at higher latitudes, resulting in a net transfer of energy from the tropics to the poles. The change in radiative forcing due to the seasonal cycle, and changes due to the diurnal cycle also change radiative forcing and cause non-equilibrium on regional, seasonal and sub-daily scales. In fact, modeling the Earth System assuming radiative equilibrium results in unrealistically extreme summers. Weather plays a large role in reducing the equator to pole imbalance of energy, but changes in weather cause large local imbalances of the energy budget.

Consider the diurnal cycle in ocean surface temperature, because this temperature is important to terrestrial radiative fluxes and turbulent heat fluxes. The diurnal cycle of solar radiative energy causes a large change in the solar radiative flux, most of which is absorbed in the top meter of the ocean. At night, the surface water temperature deceases, increasing the density of the water, and causing warmer sub-surface water to replace the surface water. This process greatly reduces the nighttime change in surface temperature. The daytime warming is more complicated. The stronger solar radiation heats the surface water, making it less dense, and inhibits downward mixing. At wind speeds above three or four m s^{-1}, the cooling due to the latent heat flux is sufficient to prevent the build-up of enough energy near the ocean surface, hence prevent increases in temperature of more than a few tenths of a degree Celcius (C°). At wind speeds <3 ms^{-1}, the latent heat flux and ocean mixing are too weak to prevent much larger increases in temperature (Weihs and Bourassa, 2016), which can easily exceed 0.6 C°. This heating is eventually disrupted by increased convection and by the duration of strong solar heating. This diurnal heating process is estimated to increase the neat heat flux by 10Wm-2 in the tropics and sub-tropics (Weihs and Bourassa, 2016). Modeling and explaining this change in temperature is a subject of ongoing research, and is clearly an important consideration in climate modeling but is a much less important consideration for short-term ocean and atmosphere forecasts.

Measurements of Stress and Example Practical Applications

Stress can be measured by buoys and research vessels simply by measuring horizontal and vertical winds frequently enough (roughly 10 Hz) to accurately determine the stress from the covariance of vertical and horizontal winds. However, because the spatial sampling is very poor for fluxes determined this way, these observations are not used to force ocean models. These observations are well suited for determining transfer coefficients (C_D, C_H and C_E) and roughness length parameterizations. The resulting flux models can then be used to determine fluxes from more typical ocean observations of wind speed, temperature, humidity and pressure which are much more commonly measured. Pressure is used to determine the air density and some humidity characteristics. The importance of wind speed, surface current, temperature and humidity are clear from Eqs. 3, 7 and 10. These equations are also dependent on the density of air, which is a function of pressure, temperature, and humidity. These transfer coefficients are also used in atmospheric models, which lack the 10 Hz temporal resolution needed to calculate fluxes through covariances, but do model the above list of flux-related variables. However, in situ based flux-related observations have insufficient space–time sampling to resolve fluxes needed to determine mesoscale fluxes without biases, and hence to determine basin-scale and global fluxes without biases without biases. For short-term weather forecasts, these biases are not a serious problem. For forecasts on monthly scales and longer, these biases contribute to model errors that must be someone compensated. Satellite observations and data assimilative numerical weather predictions can be used to reduce these biases, but will not be sufficient solutions unless the sufficiently small scales can be resolved. 'How small?' is a key question that is yet to be resolved, however, it appears that from the perspective of surface fluxes and coupled air-sea processes that a 2 or 3 km-scale must be resolved (Shi, 2017). A constellation of satellites in addition to in situ observations would be required to meet temporal sampling requirements (at least hourly sampling) over the global oceans. The daily coverage by a satellite (e.g., Fig. 1) is vast, particularly when it is realized that winds are separated by 12.5 km within each 1800 km-wide swath. However, a single satellite in low Earth orbit takes roughly 100 minutes to orbit the globe and does not come close to providing global coverage in that orbit. This explains the need for a constellation of satellites. The in situ observations are needed to test satellite calibration and to examine processes (including the subsurface) that are not well observed from satellite. In situ platforms are capable of making many more types of observations than would be available from satellites focused on surface fluxes and related variables.

Determining stress from satellite observations

Surface stress is particularly easy to determine from satellites because satellite winds are not quite winds but are equivalent neutral winds (Eq. 12; Kara et al. 2008). Such winds, when squared and multiplied by a neutral drag coefficient and an air density, are transformed into a stability-dependent stress (Eq. 13; Bourassa et al., 2010). This roughness is a function of the stress in a manner that is consistent with boundary layer stratification (Kara et al., 2008). Equivalent neutral winds were chosen as the calibration standard for satellite winds because (1) the satellites respond to ocean surface roughness rather than the wind at a 10-m height, (2) there are much more wind observations that stress observations to use for calibration, and (3) atmospheric models were already capable of assimilating surface winds.

$$\overline{\mathbf{u}}_{10EN} = \overline{\mathbf{u}}_{sfc} + \frac{\mathbf{u}_*}{k_v} \log\left(\frac{10}{z_o} + 1\right) \tag{12}$$

$$\tau = \rho\, C_{D10N}\, |\overline{\mathbf{u}}_{10EN}|\, \overline{\mathbf{u}}_{10EN} \tag{13}$$

Eq. 12 uses a friction velocity that is consistent with a stability-dependent stress. The great advantage of this approach is that it uses a neutral drag coefficient (one that is determined for neutral buoyancy, which means the boundary layer stability term in Eq. 4 is set to zero). This approach removes the need to know the temperature and humidity, which are often not available from observations and are relatively poorly modeled. The above neutral drag coefficient (C_{D10N}) calibrated to a height of 10 m.

While the flux-related variables listed above must be known to calibrate the satellite, it has long been argued that they do not need to be known to use the calibrated satellite products. However, satellite retrievals of stress have been shown to depend on air density (Bourassa et al., 2010), which was considered to be a minor error when the calibration approach was developed. Density-related errors are smaller than the accuracy requirements in the original mission planning. A stress-equivalent wind product, that accounts for this density dependence, is now produced by KNMI (De Kloe et al., 2017). It is clear that satellite winds are relative to the moving surface (Kelly et al., 2001; Cornillion and Park, 2001; Plagge et al., 2012), dependent on density (Bourassa et al., 2010), and dependent on boundary layer stability (personal communication, Jim Edson), all in the manner we expect for stress. Therefore, it is apparent that remotely-sensed winds are more directly sensitive to wind stress than wind at an arbitrary height. While this finding creates some difficulties for determining surface winds, it is excellent for determining surface stress from satellites.

The concept of a neutral equivalent wind works well because surface characteristics are relatively sensitive to wind speed, making an equivalent neutral wind similar to a wind speed, with most differences <0.5 ms^{-1}. It has proven much more challenging to provide similarly good calibrations for heat fluxes, with several groups publishing remarkable improvements in the last decade. Very recently calibrations for air/sea differences in temperature and humidity have become quite good, with random errors about 1.5 times as large as buoys, and about 25 km sampling from satellites (Jackson and Wick, 2010; Roberts et al., 2010; Bourassa et al., 2010; Smith et al., 2012). These techniques use satellite observations of air-sea temperature difference and air-sea humidity differences, applied in Eqs. 7 and 10.

Heat fluxes determined from equivalent neutral winds

Since satellite winds are referred to as 'winds' rather than equivalent neutral winds, there is an unsurprising tendency for satellite winds to be used in bulk formulas (Eqs. 7 and 10) to calculate sensible and latent heat fluxes. There are stability-dependent biases between equivalent neutral winds and winds, resulting in biased fluxes of sensible and latent heat. For mid-latitudes, this bias in heat flux is roughly 10 Wm^{-2} in mid-latitude fluxes, which can very safely be ignored for weather forecasts of several days or less. Such biases become important on scales of 10 to 14 days On seasonal scales, regional errors of this magnitude could alter the layering of ocean circulation. Bulk formulas can be revised to account for this error in interpretation, but that is rarely done. Fluxes using satellite winds should be calculated using Eqs. 5, 6, 8, 9, and 12. However, the resulting equations are computationally intensive, which makes them awkward to use for ocean and atmospheric modeling. Ideally, equations similar to 7 and 10 would be developed, using forms of C_H and C_E that account for equivalent neutral winds like Eq. 13.

Datasets of satellite-derived near surface potential temperature and humidity

At this time, there are several relatively high-quality datasets of satellite-derived potential temperature and humidity, which can be used to determine turbulent heat fluxes as described in above These are the Seaflux dataset (Roberts et al., 2010), which is available from Woods Hole Oceanographic Institution, the NOAA Earth System Research Laboratory (ESRL) products (Jackson et al., 2010), and the U.S. Naval Research Laboratory NFLUX products, which are not yet available to the public. The Ocean Heat Flux product has been produced by Abderrahim Bentamy at IFREMER. More information about these products is given in Table 10.1. These turbulent fluxes can be combined with radiative fluxes from Clouds and the Earth's Radiant Energy System (CERES) or Surface Radiation Budget (SRB) to determine net heat fluxes.

Product	Reference	Website
Seaflux	Roberts et al. (2010)	http://seaflux.org/
NOAA/ESRL	Jackson et al. (2009)	https://mdc.coaps.fsu.edu/data/noaa-cires-multi-satellite-air-temperature-and-humidity
NFLUX	May et al. (2017)	Not yet publicly available
IFREMER	None at this time	ftp://o1ef56:DeJd6uNv@eftp.ifremer.fr/oceanheatflux/data/third-party/fluxes/ifremerflux_v4
SRB		https://gewex-srb.larc.nasa.gov
CERES		https://ceres.larc.nasa.gov

Table 10.1. Modern satellite-derived near-surface temperature and humidity datasets.

Atmospheric response to ocean variability

It is well known that stress depends on sea state and that air/sea temperature differences are important for determining fluxes. Sea state and surface winds also respond to ocean surface currents and gradients of atmospheric stratification and SST. These concepts become more interesting and presumably important when examining air/sea interaction around strong ocean currents and eddies. For example, ocean surface currents (e.g., tides) moving in the same direction as the wind reduce wind shear, which reduces stress, which increases atmospheric stability departures from neutral conditions. Most ocean conditions have unstable stratification, thus currents moving in the wind direction tend to make stratification more unstable, increasing stress and more effectively stronger winds down from the top of the boundary layer. This series of adjustments represents a negative feedback. Stable stratification would lead to positive feedback in a manner that reduces fluxes and might eventually trigger a negative feedback through reduced wind speeds as described earlier in relation to diurnal changes in SST.

Coupling coefficients between stress and SST gradients

Winds and surface stress has been shown (detailed below) to be modified by gradients in SST. These changes imply changes in surface turbulent fluxes, as well. These changes in winds and stress have been identified in satellite data on scales <1000 km. Coupling coefficients are the regression coefficient between an SST gradient and a characteristic of wind or stress such as speed, a vector component, divergence and curl. If the focus is on surface fluxes, then links to wind and stress are clearly important. Changes in the divergence influence vertical motion in the atmosphere, and changes in curl influence vertical motion in the ocean.

There are several hypotheses concerning the mechanisms that fundamentally drive the surface wind and wind stress response to strong SST fronts: 1) stability-dependent adjustment of turbulent mixing of momentum from the upper atmosphere to the surface (e.g., Sweet et al., 1981; Hayes et al., 1989; Wallace et al., 1989; Wai and Stage, 1989; Liu et al., 2000; Hashizume et al., 2001, 2002; Tokinaga et al., 2005; Spall, 2007); 2) generation of hydrostatic sea level pressure gradients through changes in atmospheric baroclinicity across the SST front (e.g., Lindzen and Nigam, 1987; Wai and Stage, 1989; Hashizume et al., 2001; Small et al., 2003, 2005; Song et al., 2006; Spall et al., 2007; O'Neill et al., 2010b), and 3) a rapid change in turbulent mixing that results in an unbalanced Coriolis force in the vicinity of the SST front (Spall et al., 2007). A great deal of controversy surrounds these hypotheses. Most of these suggested mechanisms are related to surface fluxes of heat, and the winds and SSTs used in studying these relationships are often measured from satellites.

For the momentum-mixing mechanism, the modification of the surface wind and wind stress is attributed (see references above) to SST-induced changes in the stratification of the marine atmospheric boundary-layer. Over warm water, the stability of the marine atmospheric boundary-layer is reduced and the buoyancy-driven turbulent mixing is increased, which enhances the downward mixing of momentum from aloft to the surface. This decreases the vertical wind shear in the boundary layer and increases the surface winds. The converse is true over cool water, where the vertical turbulent mixing is suppressed by the stronger static stability, resulting in greater vertical wind shear and lighter surface winds.

The sea level pressure gradient mechanism attributes the modification of the surface wind field to the SST-induced changes in sea level pressure that develop across SST fronts (see references above). The resulting perturbation pressure gradient force tends to accelerate the surface wind toward locally warmer water. Therefore, the surface wind can be enhanced or reduced depending on its directional alignment with the SST-induced sea level pressure gradient vector. For example, Song et al. (2006) showed that the perturbation pressure gradient force enhanced the surface wind speed in conditions where the mean surface flow traversed the Gulf Stream from the cool to the warm side, and reduced the surface wind speed in the converse. Small et al. (2003) found similar results over tropical instability waves in the eastern equatorial Pacific, where the perturbation pressure gradient force was responsible for strengthening the trade winds over warm SST anomalies and weakening the trade winds over colder SST anomalies.

Satellite-based observations show that the wind stress divergence and curl fields are linearly related to the downwind and crosswind component of the SST gradient, respectively (see the review by O'Neill). The perturbation in wind stress divergence or convergence is locally maximized where the wind stress vector is oriented parallel to the SST gradient vector (across isotherms). Conversely, the perturbation in wind stress curl is locally largest where the wind stress vector is aligned perpendicular to the SST gradient vector (along isotherms). This linear relationship between surface wind stress and SST perturbations is observed remotely (to varying degrees) over the Gulf Stream (e.g., O'Neill, 2012), the Kuroshio Extension (e.g., O'Neill et al., 2010a, 2012b; Maloney and Chelton, 2006), the Agulhas Return Current (e.g., O'Neill et al., 2005, 2010a, 2010b, 2012; Maloney and Chelton, 2006), the Brazil-Malvinas Confluence in the South Atlantic (e.g., O'Neill et al., 2010a, 2012), the California Current System (e.g., Chelton et al., 2007), and the eastern equatorial Pacific (e.g., Xie et al., 1998; Liu et al., 2000; Chelton et al., 2007). O'Neill (2012) confirmed the linear dependence using moored buoys in the Gulf Stream and eastern equatorial Pacific, where the wind speed, 10 m equivalent neutral wind speed, and surface wind stress were found to respond linearly to SST differences.

The strength of the linear relationship or coupling between the perturbations in surface wind and wind stress, and SST is shown to exhibit large geographical variability (see aforementioned references). The prominence of each proposed mechanism or force balance is, in part, dependent on the local oceanic and atmospheric conditions. Some of the factors that may contribute to the spatial variability are: the strength and spatial extent of the SST front; the strength and directional steadiness of the prevailing wind; the marine atmospheric boundary-layer stratification; the state of equilibrium between marine atmospheric boundary-layer and underlying SST, capping inversions, latitude, and proximity to land (Hashizume et al., 2001; Chelton et al., 2007; Spall et al., 2007; O'Neill et al., 2010b). Therefore, knowledge of the relative dominance of each mechanism is important and should be investigated over an extensive range of oceanic and atmospheric conditions.

Responses to fluxes

This section provides a subset of near-surface responses to surface fluxes on timescales that are typically of less than one day, and describes how changes to the near surface ocean and atmosphere in turn cause changes in the fluxes. Take away points are that the ocean and atmosphere both respond to surface forcing on short timescales (as well as on timescales greater than one day, which are well described elsewhere). These changes alter surface fluxes, causing further changes, or countering changes to the near-surface ocean and atmosphere. This coupling is stronger near surface features with strong gradients, such as some eddies, ocean fronts, and strong currents, which tend to occur on scales <50 km. These are time- and space scales that are difficult to observe and model in a coupled system: observing, modeling, and understanding coupled processes on these scales will be a hot topic for the foreseeable future, and both air-sea fluxes and remote sensing will be important in understanding these processes.

Subsections address responses at the surface, in the ocean, and in the atmosphere. This topic is of interest for many applications, such as diurnal variability, satellite sampling of diurnal cycles, waves and currents changes with changing winds, thermodynamical coupling of the ocean and atmosphere, and air-sea coupling around ocean fronts. For example, if an ocean model is forced with atmospheric (e.g., heat, momentum, or CO_2) fluxes, it accumulates the influence of biases, often resulting in very unrealistic ocean characteristics such as surface temperature, salinity, current speed, and acidity. Thus, errors on sub-daily timescales are temporally integrated to cause longer term errors, which can lead to very unrealistic ocean and atmospheric conditions and weather in models. Of course, when errors get too large additional processes come into play to reduce biases. For example, excessive ocean surface temperatures are reduced to fluxes of terrestrial radiation, latent heat and sensible heat. Similarly, currents that are too strong are reduced by friction with surrounding slower moving water.

Models that are one-way forced (where an atmosphere is modeled with a specified SST, or where an ocean is forced with surface fluxes from an atmospheric model) cannot modify the surface forcing to reduce the impacts of biases. The impacts of these biases can be greatly mitigated by calculating surface fluxes based on atmospheric variables (wind, air temperature, humidity, and surface pressure) coupled with ocean variables (SST, currents, and sea state) based on Eqs. 4, 7, and 10. As biases in these flux-related variables occur, there are associated changes in surface fluxes (momentum, heat, and others) that reduce the biased fluxes that are causing the problems. This solution to reduce biases in fluxes might seem odd because the atmospheric data used to force an ocean model does not change in response to the ocean model. However, the changes in surface fluxes imply changes in the near surface atmosphere consistent with Eqs. 2, 6, and 9. This approach to reducing biases does not remove biases in the atmospheric variables (e.g., excessive humidity or winds) and it does result in errors in the ocean model (e.g., excess temperatures and currents), but it does reduce biases in fluxes and thereby reduced biases in the ocean model.

Full two-way coupling could be thought of as a further extension of calculating fluxes to reduce biases in forcing. Two-way coupling is more complicated because it allows the atmosphere to respond to the ocean including atmosphere-related changes to the ocean, and the ocean to respond to the atmosphere including ocean-related changes to the atmosphere. Currently there is considerable interest in evaluating the need for two-way coupling. If there is a substantial small-scale and short-term atmospheric response to ocean coupling, then ocean forcing will benefit from two way coupling. However, two-way coupling and forcing from atmospheric variables is sensitive to the physics included in the model. For example, including surface currents (without considering sea state and atmospheric response) results in a roughly 25% drop in the production of ocean eddy kinetic energy (Renault et al., 2016a, 2016b), which is clearly unrealistic. Preliminary results (Shi, 2017) suggest that there is a quite substantial atmospheric response over western boundary currents, and that this response compensates to cause a net impact of increased stresses. Considerably more work needs to be done to investigate what physics and resolutions are required to provide sufficiently accurate coupling.

Surface responses

One of the most well-known responses to fluxes is the changes in sea surface temperature and wind vectors associate with the diurnal cycle. The changes in wind vectors are best known in land-sea breezes. Currently, these are difficult to observe from satellite because most satellites cannot measure closer than roughly the width of a footprint from the coast. However, weather radars show the complicated interaction between surface temperature gradients, the horizontal pressure gradients and winds they induce, and atmospheric convection and outflow. In this example it is quite clear that the atmospheric response is substantial.

Large diurnal changes in SST are associated with low wind speeds, little mixing of the near-surface ocean, and daytime heating. There can be very large SST changes associated with strong advection of cold air, but these are not daily repeating changes. The low wind speeds are essential because they reduce the vertical transport of heat downward in the ocean and upward through latent heat. Consequently, the absorbed solar energy raises the water temperature, with greater absorption near the surface and hence greater heating near the surface. The resulting vertical gradient in density is stably stratified, which further reduces vertical transport in the ocean. Thus far, the process has a positive feedback, with strengthens the diurnal cycle (Weihs and Bourassa, 2014). The negative feedback is presumably due to increased surface winds associated with atmospheric convection and possibly associated with temperature gradients (discussed earlier). If there was no atmospheric feedback, the magnitude of the diurnal heating would be constrained only by the mean heat flux over a diurnal cycle.

Ocean responses

The case of diurnal heating is a simplified example of a thermodynamical ocean response to air sea coupling. In most cases or on large scales, the energy transfer in the ocean is dominated by vertical heat fluxes. Strong warm and cold currents are obvious exceptions, but the currents are not strong over much of the ocean, and when averaged over a large enough scale (several hundred kilometers), the impact currents are averaged down. However, around strong currents and eddies the horizontal transport of energy can be important. On small scales (mesoscale) there can be large gradients in surface stress, causing relatively strong vertical motions and vertical transport of energy. Surface-layer vertical motion is usually due to Ekman pumping, which is proportional to the curl of the surface stress. This makes resolution of the data and the complexity of the surface stress model more important. Surface stress parameterizations that are commonly applied consider the important of wind speed and (in more modern parameterizations) air-sea temperature gradients. Considering currents and sea state influences causes greater gradients and curls of stress, which makes vertical motion locally important. The importance of spatial scale and complexity of the flux parameterizations are topics of ongoing research. Satellite data are much easier to utilize in the sense that they account for additional physics; however, the sampling from satellite is sufficiently sparse to present additional challenges, and greater resolution of vector stresses is highly desired.

Atmospheric responses

As stated above, it is often assumed that there is no atmospheric response to the ocean; however, it is also pointed out that small-scale ocean variability imparts considerable forcing on the atmosphere. Since the atmosphere has relatively little inertia and thermal inertial compared to the ocean, it seems likely that atmospheric responses should occur. Such reasoning is strongly supported by the complex interaction of atmospheric convection and outflow associated with land-sea breezes. Clearly changes in surface stress associated with currents, waves, wave-current interaction, and horizontal temperature gradients (in the ocean and the atmosphere) should impact the lower atmospheric boundary layer. It has been argued that the Gulf Stream impacts the strength of cyclogeneses. This concept is strongly supported by models of atmospheric cyclogenesis; however, the limited time spent over the Gulf Stream greatly complicates analyses of cyclone evolution. Preliminary results indicated that observed coupling between SST gradients and stress can be used to test the sensitivity of coupled models to stress parameterizations (Shi, 2017). On smaller scales, it is much clearer how changes in atmospheric circulation modify air-sea fluxes and the ocean response. Such questions seem to be important for weather forecasts around western boundary currents (and downwind of them), to address biases in seasonal forecasting, to reduce biases in climate models, and, of more immediate interest, to improve operational ocean models.

Acknowledgements

This work was funded, in part, by the Jet Propulsion Laboratory, National Aeronautics and Space Administration under contract #1419699 and NNX15AD45G S02. This work was also funded, in part, by the Northern Gulf Institute (grant number 18-NGI3-42), National Oceanic and Atmospheric Administration, U.S. Department of Commerce under award number NA11OAR432019 and NA16OAR4320199.

References

Anguelova, M.D., and P.W. Gaiser, 2013: Microwave emissivity of sea foam layers with vertically inhomogeneous dielectric properties, *Remote Sensing of Environment*, **139**, 81-96, doi:10.1016/j.rse.2013.07.017.

Anguelova, M.D., and P.W. Gaiser, 2012: Dielectric and radiative properties of sea foam at microwave frequencies: Conceptual understanding of foam emissivity, *Remote Sensing*, **4**, 1162-1189, doi:10.3390/rs4051162.

Anguelova, M.D. and F. Webster, 2006: Whitecap coverage from satellite measurements: A first step toward modeling the variability of oceanic whitecaps, *J. Geophys. Res.*, **111**, C03017, doi:10.1029/2005JC003158.

Bourassa, M.A., E. Rodriguez, and R. Gaston, 2010: NASA's Ocean Vector Winds Science Team workshops. *Bull. Amer. Meteor. Soc.*, **91**, doi:10.1175/2010BAMS2880.1

Bourassa, M.A., 2006: Satellite-based observations of surface turbulent stress during severe weather, Atmosphere - Ocean Interactions, Vol. 2., ed., W. Perrie, Southampton, UK, Wessex Institute of Technology Press, 35 – 52 pp.

Bourassa, M.A., E. Rodriguez, and R. Gaston, 2010: NASA's Ocean Vector Winds Science Team workshops. *Bull. Amer. Meteor. Soc.*, **91**, doi:10.1175/2010BAMS2880.1

Bourassa, M.A., D.G. Vincent, W.L. Wood, 1999: A flux parameterization including the effects of capillary waves and sea state. *J. Atmos. Sci.*, **56**, 1123-1139.

Bourassa, M.A., S.T. Gille, D.L. Jackson, B.J. Roberts, and G.A. Wick, 2010: Ocean winds and turbulent air-sea fluxes inferred from remote sensing. *Oceanography*, **23**, 36-51.

Chakraborty, A., R. Kumar and A. Stoffelen, 2013: Validation of ocean surface winds from the OCEANSAT-2 scatterometer using triple collocation. *Remote Sensing Letters*, **4**(1), 84–93, doi:10.1080/2150704X.2012.693967.

Chelton, D.B., M. G. Schlax, and R.M. Samelson, 2007: Summertime coupling between sea surface temperature and wind stress in the California current system. *J. Phys. Oceanogr.*, **37**, 495–517.

Clayson, C.A., C. W. Fairall, and J.A. Curry, 1996: Evaluation of turbulent fluxes at the ocean surface using surface renewal theory. *J. Geophys. Res.*, **101**, 28503-28513.

Cornillon, P., Park, K.A., 2001: Warm core ring velocities inferred from NSCAT. *Geophys. Res. Letters*, **28**, 575–578, doi:10.1029/2000GL011487.

De Kloe, A. Stoffelen and A. Verhoef, 2017: Improved Use of Scatterometer Measurements by Using Stress-Equivalent Reference Winds. JSTARS, 10, doi:10.1109/JSTARS.2017.2685242

Dobson, F.W., S.D. Smith, and R.J. Anderson, 1994: Measuring the relationship between wind stress and sea state in the open ocean in the presence of swell. *Atmos.-Ocean*, **32**, 327-256.

Draper, D.W., Long, DG (2004), Evaluating the effect of rain on SeaWinds scatterometer measurements. *J. Geophys. Res.-Oceans*, **109**(C02005), doi:10.1029/2002JC001741.

Drennan, W.M., P.K. Taylor, and M.J. Yelland, 2005: Parameterizing the sea surface roughness. *J. Phys. Oceanogr.*, **35**, 835-848.

Fairall, C.W., E.F. Bradley, D.P. Rogers, J.B. Edson, and G.S. Young, 1996: Bulk parameterizations of air-sea fluxes for Topical Ocean-Global Atmosphere Coupled-Ocean Atmosphere Response Experiment. *J. Geophys. Res.*, **101**, 3747-3764.

Fairall, C., E.Bradley, J. Hare, A. Grachev, and J. Edson, 2003: Bulk parameterization of air–sea fluxes: updates and verification for the CORE algorithm. *J. Clim*, **16**, 571–591.

Gasier, P.W., and Coauthors, 2004: The WindSat spaceborne polarimetric microwave radiometer: Sensor description and early orbit performance. *IEEE Trans. Geosci. Remote Sens.*, **42**, 2347-2361.

Gordon, H.R., and M. Wang, 1994: Retrieval of water-leaving radiance and aerosol optical thickness over the oceans with SeaWiFS: A preliminary algorithm, *Appl. Opt.*, **33**, 443–452.

Gulev, K.S., S.A. Josey, M. Bourassa, L.-A. Breivik, M.F. Cronin, C. Fairall, S. Gille, E.C. Kent, C.M. Lee, M.J. McPhaden, P.M.S. Monterio, U. Schuster, S.R. Smith, K.E. Trenberth, D. Wallace, and S.D. Woodruff, 2010: Surface energy, CO_2 fluxes and sea ice. *Proceedings of the OceanObs'09: Sustained Ocean Observations and Information for Society Conference (Vol. 1)*, Venice, Italy, eds. J. Hall, D.E. Harrison and D. Stammer, ESA Publication WPP-306. doi:10.5270/OceanObs09.pp.19

Hansen, J., and Coauthors, 2005: Earth's energy imbalance: Confirmation and implications. *Science*, **308**, doi:10.1126/science.1110252.

Hashizume, H., S.-P. Xie, W. T. Liu, and K. Takeuchi, 2001: Local and remote atmospheric response to tropical instability waves: a global view from space. *J. Geophys. Res.*, **106**, 10173–10185.

Hashizume, H., S.-P. Xie, M. Fujiwara, M. Shiotani, T. Watanabe, Y. Tanimoto, W. T. Liu, and K. Takeuchi, 2002: Direct Observations of atmospheric boundary layer response to SST variations associated with tropical instability waves over the eastern Equatorial Pacific. *J. Climate*, **15**, 3379–3393.

Hayes, S. P., M. J. McPhaden, and J. M. Wallace, 1989: The influence of sea-surface temperature on surface wind in the Eastern Equatorial Pacific: weekly to monthly variability. *J. Climate*, **2**, 1500–1506.

Holbach, H.M. and M.A. Bourassa, 2017: Platform and across-swath comparison of vorticity spectra from QuikSCAT, ASCAT-A, OSCAT, and ASCAT-B Scatterometers. *IEEE Journal of Selected Topics in Applied Earth Observations and Remote Sensing*, **10**(5), 2205–2213, doi:10.1109/JSTARS.2016.2642583.

Jackson, D.L., and G.A. Wick, 2010: Near-surface air temperature retrieval derived from AMSU-A and sea surface temperature observations. *J. Atmos. Oceanic Technol.*, **27**, 1769-1776.

Jackson, D.L., G.A. Wick, and J.J. Bates, 2006: Near-surface retrieval of air temperature and specific humidity using multisensor microwave satellite observations, *J. Geophys. Res.*, **111**, D10306, doi:10.1029/2005JD006431.

Jackson, D.L., G.A. Wick, and F.R. Robertson, 2009: Improved multi-sensor approach to satellite-retrieved near-surface specific humidity observations. *J. Geophys. Res.*, 114, D16303, doi:10.1029/2008JD011341.

Kara, A.B., A.J. Wallcraft, and M.A. Bourassa, 2008: Air-Sea Stability Effects on the 10m Winds Over the Global Ocean: Evaluations of Air-Sea Flux Algorithms. *J. Geophys. Res.*, **113**, C04009, doi:10.1029/2007JC004324.

Kara, B., Kara A.B., Hurlburt H.E., Rochford P.A., 2000: Efficient and accurate bulk parameterizations of air–sea fluxes for use in general circulation models. *J. Atmos. Ocean. Technol.*, **17,** 1421-1438.

Kelly, K.A., S. Dickinson, M.J. McPhaden, and G.C. Johnson, 2001: Ocean currents evident in satellite wind data. *Geophys. Res. Letters*, **28**, 2469–2472, doi:10.1029/2000GL012610.

Knauss, J.A., 2005: *Introduction to Physical Oceanography*, Waveland Press. Second Edition. ISBN 978-1-57766-429-1

Kondo, J., 1975: Air-sea bulk transfer coefficients in diabatic conditions. *Bound.-Layer Meteor.*, **9**, 91-112.

Lindzen, R.S., and S. Nigam, 1987: On the role of sea surface temperature gradients in forcing low-level winds and convergence in the tropics. *J. Atmos. Sci.*, **44**, 2418–2436.

Liu, W.T., and W. Tang, 1996: Equivalent neutral wind. JPL Publ., 96–17, 8 pp.

Liu, W.T., X. Xie, P.S. Polito, S.-P. Xie, and H. Hashizume, 2000: Atmospheric manifestation of tropical instability wave observed by QuikSCAT and Tropical Rain Measuring Mission. *Geophys. Res. Lett.*, **27**, 2545–2548.

Maloney, E.D., and D.B. Chelton, 2006: An assessment of the sea surface temperature influence on surface wind stress in numerical weather prediction and climate models. *J. Climate*, **19**, 2743–2762.

May, J.C., C. Rowley, and C.N. Barron, 2017: NFLUX Satellite-Based Surface Radiative Heat Fluxes. Part I: Swath-Level Products. *J. Applied Met. Clim.*, **56**, 1025 – 1041, doi:10.1175/JAMC-D-16-0282.1

Monin, A.S. and A.M. Obukhov, 1954: Basic laws of turbulent mixing in the surface layer of the atmosphere. Tr. Akad. Nauk. SSSR Geophiz. Inst., **24** (151): 163–187.

Nikuradse, J., 1933: *Stromungsgesetze in rauben Rohren*. V. D. I. Forschungsheft 361, 22 pp.

O'Neill, L.W., 2012: Wind speed and stability effects on coupling between surface wind stress and SST observed from buoys and satellite. *J. Climate*, **25**, 1544–1569.

O'Neill, L.W., D.B. Chelton, and S.K. Esbensen, 2005: High-resolution satellite measurements of the atmospheric boundary layer response of SST variations along the Agulhas return current. *J. Climate*, **18**, 2706–2723.

O'Neill, L. W., D. B. Chelton, and S. K. Esbensen, 2010a: The effects of SST-indeced surface wind speed and direction gradients on midlatitude surface vorticity and divergence. *J. Climate*, **23**, 225–281.

O'Neill, L.W., D. B. Chelton, and S. K. Esbensen, 2010a: The effects of SST-indeced surface wind speed and direction gradients on midlatitude surface vorticity and divergence. *J. Climate*, **23**, 225–281.

O'Neill, L.W., S.K. Esbensen, N. Thum, R.M. Samelson, and D.B. Chelton, 2010b: Dynamical analysis of the boundary layer and surface wind responses to mesoscale SST perturbations. *J. Climate*, **23**, 559–581, doi:10.1175/2009JCLI2662.1.

Paget, A., M.A. Bourassa, and M.D. Anguelova, 2015: Comparing in situ and satellite-based observations of oceanic whitecaps. *J. Geophys. Res.*, **120**, 2826-2843, doi:10.1002/2014JC010328.

Petty, G. W., 2006: Thermal emission. A first course in atmospheric radiation. Second Edition, Sundog Publishing, 113–154.

Plagge, A.M., D. Vandemark and B. Chapron, 2012: Examining the impact of surface currents on satellite scatterometer and altimeter ocean winds. *J. Atmos. Oceanic Technol.*, **29**, 1776–1793, doi:10.1175/JTECH-D-12-00017.1.

Renault, L., M. J. Molemaker, J. Gula, S. Masson, and J. C. McWilliams, 2016a: Control and Stabilization of the Gulf Stream by Oceanic Current Interaction with the Atmosphere, *J. Phys. Oceanogr.*, **46**(11), 3439–3453, doi:10.1175/JPO-D-16-0115.1.

Renault, L., M.J. Molemaker, J.C. McWilliams, A.F. Shchepetkin, F. Lemarié, D. Chelton, S. Illig, and A. Hall, 2016b: Modulation of wind-work by oceanic current interaction with the atmosphere, *J. Clim.*, 0–52, doi:10.1175/JPO-D-15-0232.1.

Roberts, B., C.A. Clayson, F.R. Robertson, and D.L. Jackson, 2010: Predicting near-surface characteristics from SSM/I using neural networks with a first guess approach. *J. Geophys. Res.*, **115**, D19113, doi:10.1029/2009JD013099.

Rose, L., W. Asher, S. Reising, P. Gaiser, K. St Germain, D. Dowgiallo, K. Horgan, G. Farquharson, and E. Knapp, 2002: Radiometric measurements of the microwave emissivity of foam, *IEEE Trans. Geosci. Remote Sens.*, **40**(12), 2619–2625, doi:10.1109/TGRS.2002.807006.

Ruf, C., M. Unwin, J. Dickinson, R. Rose, D. Rose, M. Vincent, A. Lyons, "CYGNSS: Enabling the Future of Hurricane Prediction", *Geoscience and Remote Sensing Magazine IEEE*, vol. 1, pp. 52-67, 2013.

Small, R.J., S.-P. Xie, Y. Wang, S.K. Esbensen, and D. Vickers, 2005: Numerical simulation of boundary layer structure and cross-equatorial flow in the Eastern Pacific. *J. Atmos. Sci.*, **62**, 1812–1830.

Small, R.J., S.P. deSzoeke, S.P. Xie, L. O'Neill, H. Seo, Q. Song, P. Cornillon, M. Spall, and S. Minobe, 2008: Air–sea interaction over ocean fronts and eddies. *Dyn. Atmos. Oceans*, **45**, 274–319, doi:10.1016/j.dynatmoce.2008.01.001.

Shi, Q., 2017: Coupling ocean currents and waves with wind stress over the Gulf Stream. Ph.D Dissertation, Florida State University

Smith, P., 1998: The emissivity of sea foam at 19 and 37 GHz. *IEEE Trans. Geosci. Remote Sens.*, **26**, 541-547.

Smith, S. D., 1988: Coefficients for sea surface wind stress, heat flux, and wind profiles as a function of wind speed and temperature. *J. Geophys. Res.*, **93**, 15467-15472.

Smith, S.R., M.A. Bourassa, and D.L. Jackson, 2012: Supporting Satellite Research with Data Collected by Vessels. *Sea Tech.*, June 2012, 21 – 24.

Stull, R. 1988: An introduction to boundary layer meteorology. Netherlands: Springer. ISBN 978-94-009-3027-8.

Song, Q., P. Cornillon, and T. Hara, 2006: Surface wind response to oceanic fronts. *J. Geophys. Res.*, **111**, C12006, doi:10.1029/2006JC003680.

Spall, M.A., 2007: Midlatitude wind stress–sea surface temperature coupling in the vicinity of oceanic fronts. *J. Climate*, **20**, 3785–3801, doi:10.1175/JCLI4234.1.

Sweet, W., R. Fett, J. Kerling, and P. La Violette, 1981: Air–sea interaction effects in the lower troposphere across the North Wall of the Gulf Stream. *Mon. Wea. Rev.*, **109**, 1042–1052.

Sverdrup, H.U., 1947: Wind-driven currents in a baroclinic ocean; with application to the equatorial currents of the Eastern Pacific. *Proc. Natl. Acad. Sci. U.S.A.*, **33**(11): 318–26.

Taylor, P.K., and M.J. Yelland, 2001: The dependence of sea surface roughness onthe height and steepness of the waves. *J. Phys. Oceanogr.,* **31,** 572–590.

Tokinaga, H., Y. Tanimoto, and S.-P. Xie, 2005: SST-induced surface wind variations over the Brazil-Malvinas confluence: satellite and in situ observations. *J. Climate,* **18,** 3470–3482.

Ulaby, F.T. and D.G. Long, 2014: Microwave Radar and Radiometric Remote Sensing, *University of Michigan Press,* Ann Arbor, 984 pp., ISBN: 978-0-472-11935-6

Uhlhorn, E.W., and P.G. Black, 2003: Verification of remotely sensed sea surface winds in hurricanes. *J. Atmos. Oceanic Technol.,* **20,** 99–116, doi:10.1175/1520-0426(2003)020<0099:VORSSS>2.0.CO;2.

Wai, M. M.-K., and S.A. Stage, 1989: Dynamical analyses of marine atmospheric boundary layer structure near the Gulf Stream oceanic front. *Q. J. R. Meteorol. Soc.,* **115,** 29–44.

Wallace, J.M., T.P. Mitchell, and C. Deser, 1989: The influence of sea-surface temperature on surface wind in the Eastern Equatorial Pacific: seasonal and interannual variability. *J. Climate,* **2,** 1492–1499.

Weihs, R., and M.A. Bourassa, 2014: Modeled diurnally varying sea surface temperatures and their influence on surface heat fluxes. *J. Geophys. Res.,* **119**, 4101–4123, doi:10.1002/2013JC009489.

Wentz, F. (1983), A model function for ocean microwave brightness temperature, *J. Geophys Res.,* **88**, 1892-1908.

Wentz, F. (1997), A well-calibrated ocean algorithm for special sensor microwave.imager, *J. Geophys. Res.,* **102**, 8703-8718.

Wu, J., 1968: Laboratory studies of wind-wave interactions. J. Fluid Mech., 34, 91-111.

Wu, J., 1980: Wind-stress coefficients over sea surface near neutral conditions--A revisit. *J. Phys. Oceanogr.,* 10, 727-740.

Wu, J., 1994: The sea surface is aerodynamically rough even under light winds, *Bound..-Layer Meteor.,* **69,** 149-158.

Xie, S.P., Ishiwatari, M., Hashizume, H. and Takeuchi, K., 1998: Coupled ocean-atmospheric waves on the equatorial front. *Geophys. Res. Letters,* **25**(20), 3863-3866.

Yueh, S.H., W.J. Wilson, F.K. Li, S.V. Nghiem, W.B. Ricketts, 1995: Polarimetric measurements of sea surface brightness temperatures using an aircraft K-band radiometer. *IEEE Trans. Geosci. Remote Sens.,* **33,** 1, 85–92.

Zavorotny, V.U. and A.G. Voronovich, "Scattering of GPS signals from the ocean with wind remote sensing application", *IEEE Trans. Geosci. Remote Sensing,* vol. 38, pp. 951-964, 2000.

Zheng, Y., M.A. Bourassa, and P.J. Hughes, 2013: Influences of sea surface temperature gradients and surface roughness changes on the motion of surface oil: a simple idealized study. *J. Appl. Meteor. Clim.,* **52,** 1561–1575, doi:10.1175/JAMC-D-12-0211.1

CHAPTER 11

Satellite SST and SSS Observations and Their Roles to Constrain Ocean Models

Tong Lee[1] and Chelle Gentemann[2]

[1]Jet Propulsion Laboratory, California Institute of Technology, Pasadena, California, USA; [2]Earth and Space Research, Seattle, Washington, USA

Sea surface temperature (SST) and sea surface salinity (SSS) are important parameters of the ocean that influence ocean circulation, air-sea interactions, and biogeochemistry. In the past few decades, since SST measurements from space have become routine, they have been fundamental to ocean and climate research. In the past several years, satellite measurements of SSS have become available to strengthen research and applications for the oceans and the linkages with other elements of the Earth system. This chapter introduces the key principles and advantages of measuring SST and SSS from space, their complementarity use with other satellite and in situ observations, the past and current missions for these measurements, characteristics of uncertainties for the related data products, and the utility of these measurements in evaluating and constraining ocean model/assimilation systems and improving forecasts.

Importance of SST and SSS on Ocean and Air-sea Interaction Processes

Sea surface temperature (SST) and sea surface salinity (SSS) are important to ocean dynamics. These two parameters determine the surface density of the ocean. Surface density is essential to various ocean processes such as the formation of water masses at the ocean surface, ventilation, and mixed-layer dynamics. At high-latitude ocean where the vertical stratification is relatively weak, small variations in surface density due to SST and SSS changes can have significant impacts on vertical mixing and convection.

SST is a critical parameter for air-sea interactions. It affects the air-sea surface temperature difference, thereby contributing to local air-sea heat flux. Horizontal gradients of SST can also regulate ocean-atmosphere coupling, for example, through SST-wind feedback. A basin-scale example is the coupling of zonal SST gradients and zonal winds across the tropical Pacific Ocean associated with the El Niño-Southern Oscillation (e.g., Bjerknes 1969, Kumar and Hoerling, 1998). SST-wind coupling can also occur on smaller scales, such as those associated with tropical instability waves and ocean eddies (e.g., Chelton et al., 2001). SSS does not directly affect air-sea heat flux. However, it can influence SST and thus air-sea heat flux indirectly when there is SSS-related barrier layer (e.g., Lindstrom et al., 1987; Sprintall and Tomczak, 1992). The latter is a salinity-stratified layer within the isothermal layer. The barrier layer inhibits the vertical mixing

Lee, T., and C. Gentemann, 2018: Satellite SST and SSS observations and their roles to constrain ocean models. In "*New Frontiers in Operational Oceanography*", E. Chassignet, A. Pascual, J. Tintoré, and J. Verron, Eds., GODAE OceanView, 271-288, doi:10.17125/gov2018.ch11.

between the mixed layer above it and the thermocline beneath it, and thus indirectly affects SST and heat flux.

Given their importance in ocean and air-sea interaction processes, observations of SST and SSS and their use in ocean and coupled ocean-atmosphere model assimilation systems are central to the fidelity of the estimated ocean state as well as to ocean and coupled ocean-atmosphere forecasts across different timescales.

Advantages of Satellite SST and SSS and Complementarity with In situ Observations

Satellite and in situ measurements of SST and SSS are complementary to each other. Satellites provide more uniform spatiotemporal sampling than in situ systems to resolve oceanographic features. As such, they help monitor features with scales that are not resolved or adequately captured by in situ systems and they help decipher large-scale signals and smaller-scale features such as eddies and fronts. Additionally, satellites have synoptic coverage of coastal oceans, marginal seas, and high-latitude oceans, regions where in situ measurements are generally sparse. Satellite spatial sampling also enables the calculation of spatial derivative fields systematically, which is vital to studies of ocean dynamics and air-sea interaction. Moreover, synoptic coverage of global ocean SST and SSS by satellites facilitates the studies of large-scale teleconnections and impacts.

On the other hand, in situ SST and SSS measurements provide accurate ground truthing for calibration and validation of corresponding satellite measurements. In particular, the near-global Argo array and the tropical moored buoy array (TAO/TRITON, PIRATA, and RAMA) are important backbone datasets for evaluating and improving satellite retrievals of SST and SSS (Donlon et al., 2002; O'Caroll et al., 2008; Gentemann et al., 2004; Tang et al., 2014). High-frequency (e.g., hourly) measurements from moorings also help de-alias signals that may not be adequately sampled by satellites (e.g., diurnal signals). In situ measurements of temperature and salinity profiles are also important to the interpretation of satellite SST and SSS by providing measurements of the vertical structure. In addition, satellite and in situ data have been used synergistically to produce blended satellite/in situ products for SST (e.g., Reynolds et al., 2007) and SSS (e.g., Xie et al., 2014; Nardelli et al., 2016; Droghei et al., 2016). There have also been efforts that combined satellite SSS with the higher-resolution satellite SST to enhance the resolution of SSS products using multifractal fusion methods (e.g., Umbert et al., 2013; Olmeda et al., 2016).

Satellite Observations of SST

Satellite SST observing systems have decades of history. The first satellite program that measured SST was the TIROS weather satellite program in the late 1960s. The Advanced Very High Resolution Radiometers (AVHRR) onboard the NOAA series satellites from the early 1980s have brought the satellite SST observing system into a relatively mature stage for operational monitoring (Casey et al., 2010). There have been a series of technology advancements in satellite infrared (IR)

SST sensors, such as the European Space Agency's Along-Track Scanning Radiometer (ATSR) and Advanced Along-Track Scanning Radiometer (AATSR) (Llewellyn-Jones et al., 2001). Furthermore, a significant advancement in satellite SST observing systems has been the addition of passive microwave (PMW) sensors such as the Advanced Microwave Scanning Radiometer (AMSR) on JAXA's ADOES-II satellite (2002-2003), AMSR-E on the National Aeronautics and Space Administration's (NASA) EOS Aqua satellite (2002-2011) and AMSR-2 on JAXA's GCOM-W1 satellite (2012 onward) (Gentemann, 2014; Gentemann and Hilburn, 2015). International efforts have enabled a comprehensive constellation of SST-measuring satellites for past several years that will continue on into the next decade, both on polar and geostationary orbits (Figs. 11.1 and 11.2).

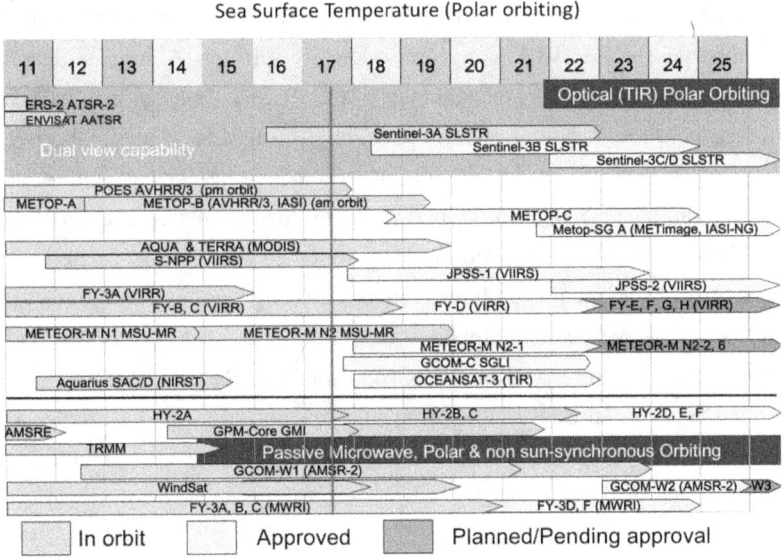

Figure 11.1. International constellation of polar-orbit SST satellites.

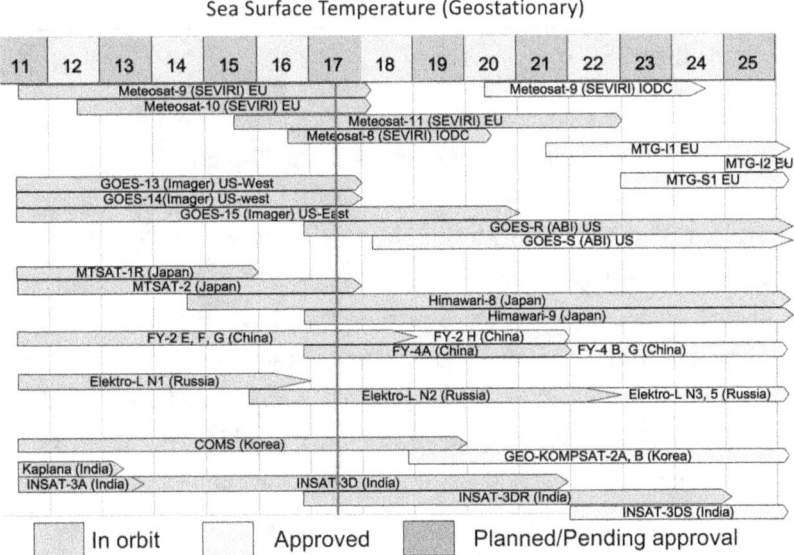

Figure 11.2. International constellation of geostationary SST satellites.

IR sensors measure SST in the top 10 microns of the ocean surface while PMW sensors measure SST in the top millimeters of the ocean, which can cause relative biases between IR and PMW SST especially due to diurnal heating and evaporative cooling, processes that are not well understood. This complicates the comparison and blending between IR and PMW SST. To address this, efforts have been made to convert the "skin SST" measured by IR and PMW sensors to "bulk SST" or "Foundation SST" (e.g., Reynolds et al., 2007; Donlon et al., 2007).

IR and PMW SST sensors have respective advantages and limitations. IR SST sensors have a long heritage and decades of service on record while PMW SST sensors have been in use for a much shorter time period (see above). IR SST sensors have much better spatial resolution (1-4 km) than PMW SST (~25 km). IR sensors are not able to measure SST through clouds and they are significantly affected by other atmospheric effects (e.g., water vapor and atmospheric aerosols). This is especially true for AVHRR-like single-view IR sensors. A large volcanic event could significantly impact the availability and accuracy of the IR SST (Reynolds, 1993). In comparison, PMW sensors can see through clouds and are relatively insensitive to atmospheric effects such as aerosols. However, PMW SST sensors are not able to retrieve SSTs near land or sea ice and they are influenced by rain, surface roughness, and radio frequency interference. The main source of radio frequency interference contamination is from reflected geostationary satellite transmissions, but satellite-to-satellite radio frequency interference is also a growing problem (Gentemann, 2014). A good graduate student-level resource that describes the differences between IR and PMW SST is http://www2.hawaii.edu/~jmaurer/sst, and the Group for High Resolution SST (GHRSST) Project (https://www.ghrsst.org) has produced a suite of high-resolution (< 10 km), level-4 SST products that synthesize SST observations from various satellites, some also including in situ SST measurements.

Salinity Remote Sensing: A New Frontier in Satellite Oceanography

Overview of satellite SSS missions

Three satellite missions have pioneered the measurements of SSS from space (Fig. 11.3). These are the Soil Moisture and Ocean Salinity (SMOS) Mission (2009-present) (Lagerloef and Font, 2010; Reul et al., 2012), the Aquarius/SAC-D Mission (June 2011-June 2015) (Lagerloef and Font, 2010; Lagerloef et al., 2013), and the Soil Moisture Active Passive (SMAP) Mission (January 2015-present) (Entekhabi et al., 2014; Tang et al., 2017). SMOS is a European Space Agency mission that measures both soil moisture and SSS. Aquarius/SAC-D is a joint mission between NASA and the Argentine National Space Activity Commission dedicated to SSS measurements. SMAP is a NASA mission with a main objective to measure soil moisture and the freeze/thaw state of the Earth. Even though SSS measurement is not one of the main objectives of the SMAP mission, the similarity in the designs of SMAP and Aquarius instruments allows SSS retrieval from SMAP. Likewise, soil moisture has also been retrieved from Aquarius even though the mission was dedicated to SSS measurements. In addition to soil moisture and SSS, thin sea ice thickness (up to approximately 50 cm) has also been retrieved from the measurements of these L-band these satellites.

All three missions operate at L-band radiometric frequencies (approximately 1.4 GHz), a frequency band where the surface brightness temperature has overall good sensitivity to changes in soil moisture and SSS. L-band is also a protected frequency band reserved for astronomy observations, thereby minimizing radio frequency interference by radar signals on the Earth. Aquarius and SMAP both have an active-passive design with a L-band radar scatterometer integrated with the L-band radiometer. SMOS' design is passive L-band radiometry. All three satellites are on sun-synchronous polar orbits with high inclinations, allowing for full coverage of polar oceans.

SMOS has a 33 km spatial resolution with a 10-day repeat and a three-day sub-cycle. Aquarius has a 100-150 km spatial resolution and a seven-day repeat. SMAP has a 40 km spatial resolution and an eight-day repeat. Therefore, all three missions provide synoptic measurements of SSS over the global ocean with spatial scales much finer than those afforded by the Argo array, and they resolve higher-frequency signals (e.g., tropical instability waves) that are difficult to capture using Argo floats.

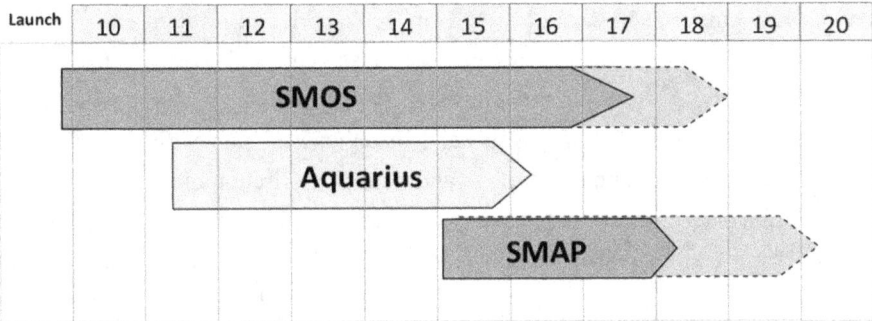

Figure 11.3. Satellite SSS missions.

Basic principles of measuring SSS from space

Measurements from L-band radiometers are used to retrieve SSS using a geophysical model function (Yueh et al., 2014). First, a set of procedures is used to estimate ocean surface brightness temperature (T_B) by removing atmospheric effects (e.g., water vapor, cloud, oxygen absorption) and galactic signals. Three major factors contribute to ocean surface T_B: SSS, SST, and surface roughness. SSS measurements are retrieved after correcting for the effects of SST and surface roughness. Because there is no onboard SST sensor on the L-band satellites, ancillary SST measurements from other satellites are used to remove the SST effect on T_B for all three missions. Surface roughness is generally a more important factor than SST, influencing T_B. Measurements from Aquarius' L-band radar have been very useful in removing the surface roughness effects. SMAP's radar failed after three months into the mission science operation. Therefore, ancillary wind products are used to estimate surface roughness and remove its effect from SMAP's T_B. SMOS does not have an onboard radar, so ancillary winds are also used to remove the effect of surface roughness on T_B. Splashing due to rain also roughens the ocean surface. Therefore, ancillary rain rate measurements are similarly helpful in removing the related surface roughness effect.

Calibration/validation and satellite SSS error characteristics

Evaluation of satellite SSS measurements are typically performed through comparisons with in situ near-surface salinity measurements obtained from Argo floats, tropical moorings, and ship-based CTD measurements (for level-2 satellite SSS), as well as with gridded maps based on these in situ salinity measurements (for level-3 gridded satellite SSS) (e.g., Tang et al., 2014; Tang et al., 2017). Various analyses of SSS from SMOS, Aquarius, and SMAP have demonstrated that satellite SSS measurements have better accuracy in the tropics and subtropics than at high latitudes. This is because L-band T_B is more sensitive to salinity in warmer waters than in waters colder than 5°C. Two important factors need to be taken into account when assessing the accuracy of satellite SSS using in situ measurements.

(1) Differences in spatiotemporal sampling between satellite and in-situ measurements.

Satellite SSS measurements represent the average SSS values within the satellite footprints (e.g., 33 km for SMOS, 40 km for SMAP, and 100-150 km for Aquarius). In situ measurements are point-wise observations. Evaluation of level-3 satellite SSS measurements is typically performed by "co-locating" in situ measurements within the satellite footprint and a certain time window (e.g., 7-10 days) to gather sufficient samples of in situ data. In regions of high variability (e.g., eddy-rich regions such as the western boundary currents and Antarctic Circumpolar Current, river plumes, tropical regions influenced by instability waves and transient and patchy rain), there could be actual differences between satellite SSS measurements within the footprint and point-wise in situ measurements because the sub-footprint variability is only partially sampled by the point-wise in situ measurements. Likewise, the presence of high-frequency variability can also cause differences between satellite SSS measurements averaged within the selected time window and snapshot in situ measurements. For evaluating level-3 satellite SSS measurements using gridded in situ salinity maps, the uncertainties of the in situ gridded maps due to potentially limited samplings and mapping errors also need to be considered. For example, Lee (2016) showed that in regions of high variability, such as those mentioned previously, the differences between satellite SSS and Argo products are also regions where significant differences exist between two Argo products, reflecting the effect of insufficient sampling by Argo to fully resolve the variability in those regions. Boutin et al. (2016) showed that these regions are in fact associated with substantial variability within satellite footprints as illustrated by ship-based, high-resolution themosalinograph data. Therefore, it is important to note that the differences between satellite SSS (level-2 or level-3) and in situ measurements (individual samples or gridded maps) not only reflect the uncertainties of the satellite SSS measurements, but also the actual differences due to the variances in sampling and mapping. In reality, this is a ubiquitous issue when evaluating satellite measurements using point-wise in situ measurements. It also has implications for the comparison of satellite or in situ measurements with model outputs of different spatial and temporal resolutions.

(2) Effect of near-surface salinity stratification

L-band satellites measure salinity in the top centimenters of the ocean. The shallowest measurement depths for in situ sensors are typically 5 m (for most Argo floats) and 1 m for tropical moorings. Recent measurements of near-surface salinity structures show that there are situations where actual salinity stratification exists above 1 m, especially in the tropics where the effect of transient rain is important (Drucker and Riser, 2014). The shallow, near-surface stratification depends on rain rate and wind speed (which affects vertical mixing) as well as advection by ocean currents. The physics for near-surface stratification is an area of active investigation (Boutin et al., 2016) that has important implications to near-surface vertical mixing and air-sea exchanges. NASA's field experiment Salinity Processes in the Upper Ocean Regional Study-2 (SPURS-2; https://ourocean3.jpl.nasa.gov/spurs2/index.php) is an example of recent efforts to understand near-surface salinity stratification and freshwater dispersal.

An international work group, the Satellite and in-situ Salinity Working Group (http://siss.locean-ipsl.upmc.fr/), brings the international salinity remote sensing community to further understand the two important issues discussed above. An ongoing project funded by the European Space Agency's SMOS Pilot Mission Exploitation Platform (SMOS Pi-MEP) is compiling various satellite SSS products along with in situ near-surface salinity measurements and ancillary data for validation, analysis, and applications of satellite SSS measurements (https://pimep-project.odl.bzh/home).

Scientific accomplishments of the L-band satellite SSS missions

Satellite SSS measurements from the SMOS, Aquarius, and SMAP missions have provided an unprecedented opportunity to map synoptic SSS in the global ocean. These measurements have brought new understanding of various ocean processes such as tropical instability waves (Lee et al., 2012, 2014; Yin et al., 2014), Rossby waves (Menezes et al., 2014), mesoscale eddies (Reul et al., 2014, Melnichenko et al., 2017), salinity fronts (Kao et al., 2014; Yu, 2015), hurricane haline wake (Grodsky et al., 2012), river plume variability (Gierach et al., 2013; Fournier et al., 2016a), and cross-shelf exchanges (Grodsky et al., 2017). They have also enhanced research about the relationship of SSS with climate variability, including research into the El Niño-Southern Oscillation, Indian Ocean Dipole, and Madden-Julian Oscillation (e.g., Grunseich et al., 2013; Guan et al., 2014; Qu and Yu, 2015; Du and Zhang, 2015), and the linkages of the ocean with different elements of the water cycle such as evaporation and precipitation and continental runoff (Yu, 2014; Fournier et al., 2016b).

Additionally, satellite SSS measurements are being used to constrain ocean state estimation and improve seasonal-to-interannual prediction (e.g., Hackert et al., 2014), which are discussed further in the next section. There are also emerging biogeochemical applications of satellite SSS measurements to study ocean's total alkalinity, ocean acidification, and air-sea CO_2 flux (e.g., Land et al., 2015; Fine et al., 2017). The L-band satellites provide complementary measurements to in situ platforms in that they provide global coverage, including the marginal seas and coastal oceans where in situ measurements are often sparse. Satellite SSS observations also capture scales that are not or inadequately resolved by in situ systems. Examples of these achievements are highlighted in this chapter.

Future challenges and ongoing technology development

The satellite salinity remote sensing community is actively pushing for the continuity and enhancement of satellite SSS observing systems. Enhancement includes two important aspects: (1) increasing spatial resolution to better resolve mesoscale features, which is particularly important to coastal ocean dynamics and biogeochemistry, and (2) improving the quality of satellite SSS observations, especially in high-latitude regions. The larger uncertainties of satellite SSS observation in high-latitude regions is to a large extent a result of the poor sensitivity of L-band T_B to salinity in cold (< 5°C) waters. In order to improve high-latitude satellite SSS measurement quality, ongoing technological developments are focused on combining L- and P-band radiometry. This is because P-band has 2.5-3 times better sensitivity to salinity than L-band for SST lower than 5°C (Fig. 11.4; Lee et al., 2016). The combined L- and P-band radiometry also help improve the uncertainties of seasonal sea ice thickness. Radar measurements of sea ice thickness have relatively large uncertainties for seasonal sea ice. Thus, the combined L-/P-band radiometry aims to fill a capability gap in sea ice thickness measurements. More accurate measurements of seasonal sea ice thickness also help improve the retrieval of T_B of the sea ice, which in turn reduce the errors of SSS measurements in marginal ice zones by removing the contamination of SSS signals by leakage of sea ice signal into the field of view of the radiometer over the ocean due to antenna side lopes.

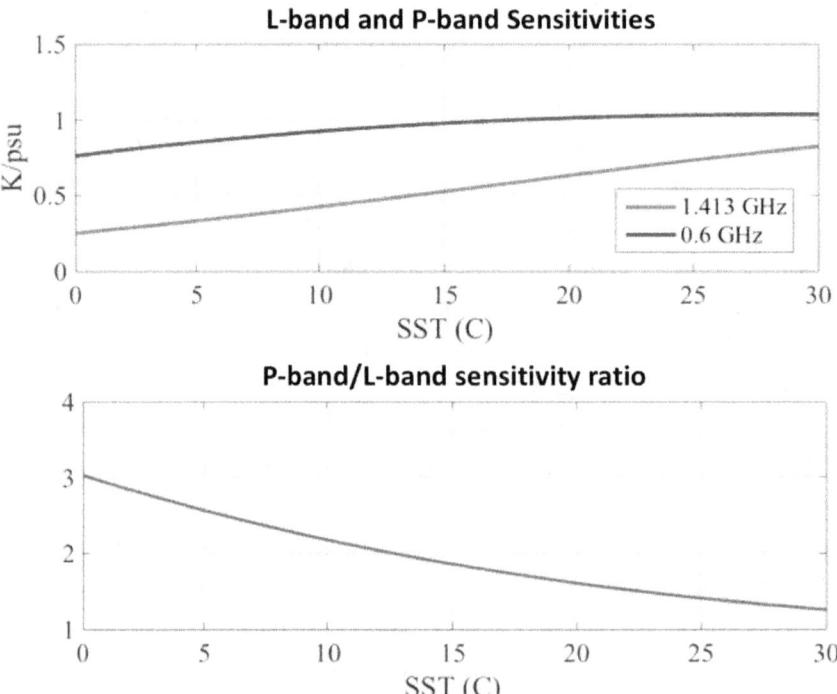

Figure 11.4. L-band and P-band brightness temperature (T_B) sensitivity to SST (upper) and their ratio (lower), showing 2.5-3 times better sensitivity for P-band than L-band for SST < 5°C. After Lee et al. (2016).

Use of Satellite SST and SSS for Constraining Ocean Models

SST data assimilation

Satellite SST products have long been a backbone dataset for most global and regional data assimilation systems due to the relative maturity and operational missions. All current global and regional assimilation systems integrate satellite SST (e.g., see the details under each system in the Ocean Synthesis/Reanalysis Directory http://icdc.cen.uni-hamburg.de/projekte/easy-init/easy-init-ocean.html or http://reanalyses.org/ocean/overview-current-reanalyses). These include many of the systems under the Global Ocean Data Assimilation (GODAE) OceanView Program.

Assimilation of SST aims to compensate for errors of surface heat flux forcing and to improve the representation of mixed-layer heat budget. For assimilation systems based on the adjoint (or four-dimensional variational) assimilation method such as that used in Estimating the Circulation and Climate of the Ocean (ECCO, http://www.ecco-group.org), the assimilation of SST also has the potential to improve surface heat flux estimation (inverse estimation) in combination with other ocean observations that are assimilated. Moreover, the improved estimates of the ocean state can be used to initialize short-term ocean forecasts and seasonal-to-interannual climate prediction. An example of the latter is the assimilation of SST data in JAMSTEC's SINTEX-F coupled ocean-atmosphere model, which demonstrated the positive impacts on the hindcasts of the El Niño-Southern Oscillation and Indian Ocean Dipole (Lou et al., 2005). An example of the encouraging hindcast skill is shown in Fig. 11.5 for Niño3.4 SST anomalies.

Note that SST observations are typically assimilated along with other measurements such as vertical profiles of temperature and salinity from Argo floats, ship-based CTDs, and moorings, as well as sea level anomalies from satellite altimetry to achieve a more comprehensive constraint on the model state. This can avoid misrepresentation of the mixed-layer heat budget through incorrect compensation of different term balance. Additional observational data assimilated typically improve the fidelity of the ocean state estimation (e.g., Fujii et al., 2015).

SSS data assimilation

Even though the satellite SSS data have relatively short records and the products continue to be improved, their values to improve ocean state estimation have been demonstrated. For example, Chakraborty et al. (2014) illustrated the improved representation of equatorial Pacific surface currents through the assimilation of Aquarius SSS data. Toyoda et al. (2015) showed that the assimilation of Aquarius SSS improved the representation of North Pacific mode water characteristics as well as salinity structure near the Maritime Continent, the Amazon plume, the eastern equatorial Pacific, and the Arctic Ocean. Lu et al. (2016) also documented improvements of surface and subsurface salinity fields through assimilation of SMOS SSS data. Köhl et al. (2014) used SMOS SSS to inversely constrain the estimate of the evaporation-precipitation (E-P) forcing field and found positive impacts of the assimilation of SMOS SSS in improving E-P estimates.

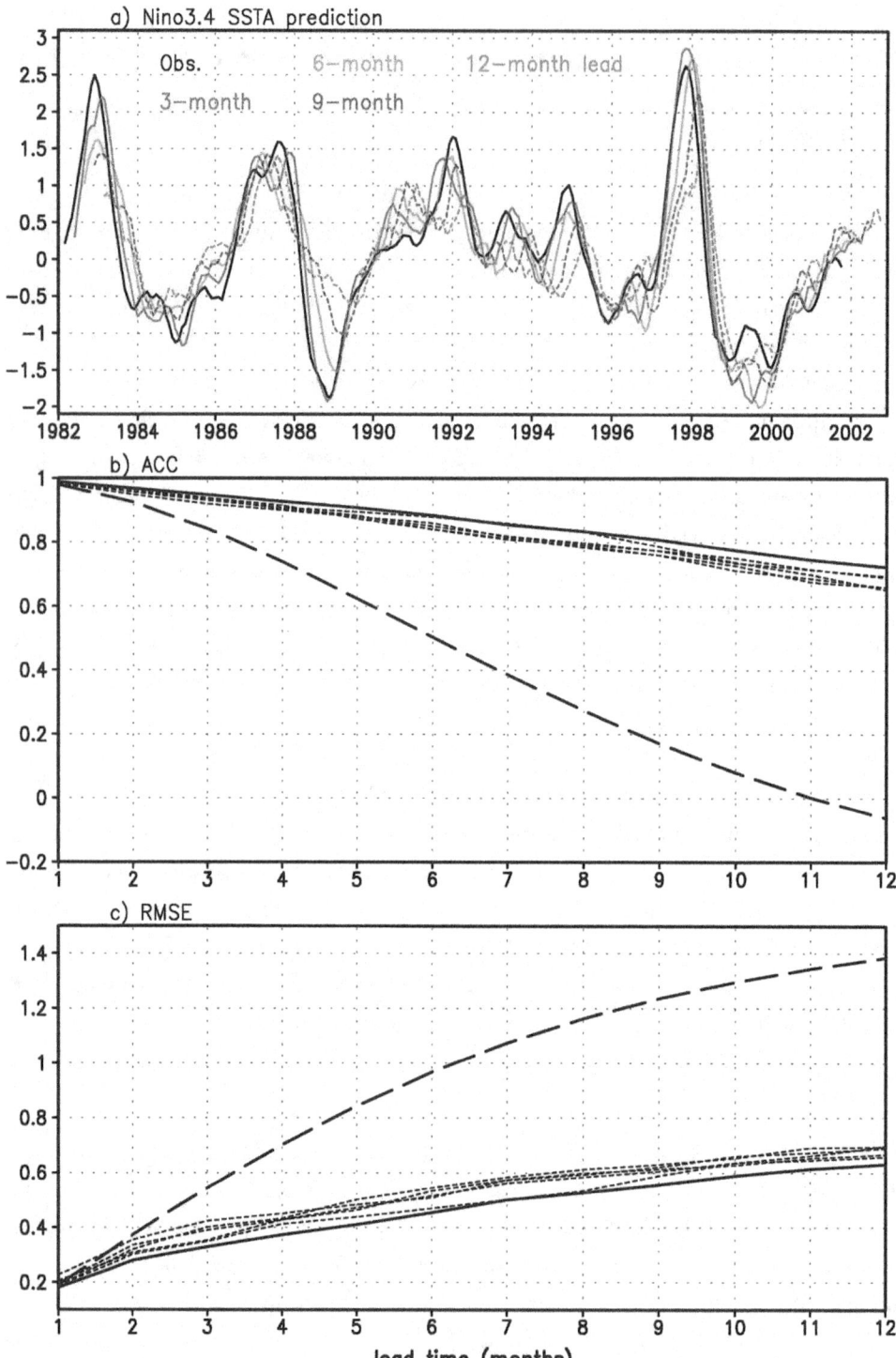

Figure 11.5. (a) Niño-3.4 SST anomalies (5°S–5°N, 170°–120°W) based on the NOAA/CDC observations (solid line) and model predictions at three- (red line), six- (green line), nine- (blue line), and 12-month (yellow line) lead times. Results have been smoothed with five-month running mean. (b), (c) ACC (Anomaly Correction Coefficient) scores and RMSEs (Root-Mean Square Errors) of the persistence (long dashed lines), ensemble mean (solid lines), and individual member forecasts (short dashed lines). After Lou et al. (2005).

Satellite SSS data have also been used to improve seasonal-to-interannual prediction. For example, Hacker et al. (2014) demonstrated the positive impacts of Aquarius SSS data assimilation to improve El Niño-Southern Oscillation hindcasts. In that effort, three hindcasts were performed for the period 2011-2014, initialized from three different initial states derived from ocean data assimilation. All three assimilation runs included the subsurface temperature profile data. In the first assimilation, no other data were assimilated. In the second assimilation, in situ salinity data were also assimilated. In the third assimilation, Aquarius SSS were assimilated instead of in situ salinity. The resultant hindcast skills initialized from these three assimilation runs show that the one with Aquarius SSS assimilated had the best hindcast skill for Niño3.4 SST anomalies, both in terms of correlation and root-mean-squared differences from observed Niño3.4 SST anomalies (Fig. 11.6). The improved hindcast skill was attributed to the better sampling of SSS measurements from Aquarius than the in situ measurements, which provides a better constraint on mixed-layer density variation on interannual timescales.

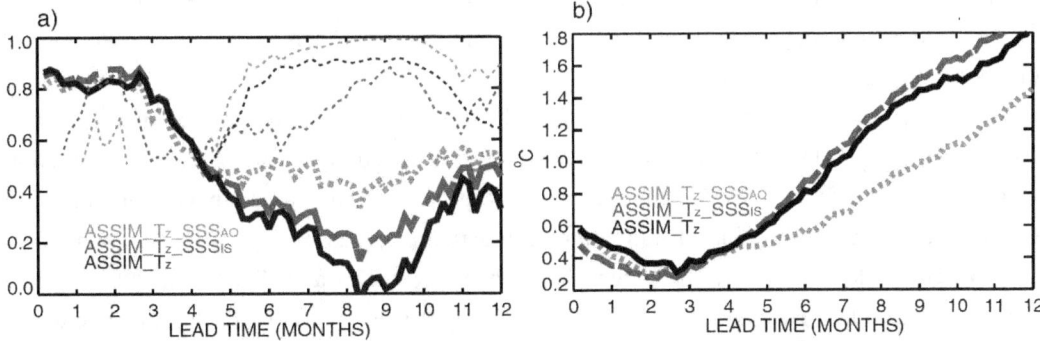

Figure 11.6. Validation of coupled model results for the Aquarius period, August 2011 to February 2014 using (a) correlation and (b) RMS versus observed NINO3 SST anomaly. The solid black curve is initialized using assimilation of subsurface temperature (ASSIM_Tz), the thick dotted red curve from Tz and Aquarius SSS (ASSIM_Tz_SSSAQ) and the dash blue curve from Tz and weekly OI of all available near-surface salinity (ASSIM_Tz_SSS$_{IS}$). The thin dotted lines show the significance of the differences assuming ASSIM_Tz_SSS$_{AQ}$ (red) and ASSIM_Tz_SSS$_{IS}$. After Hackert et al. (2014). (blue) are greater than ASSIM_Tz and ASSIM_Tz_SSSAQ is greater than ASSIM_Tz_SSSIS (black) using the Fisher Z test. Note that Fisher Z test is undefined (thus missing) when this condition fails.

Other ongoing efforts to employ satellite SSS assimilation include those by the Estimating the Circulation and Climate of the Ocean consortium (http://www.ecco-group.org), the NASA Goddard Space Flight Center, NOAA's Joint Center for Satellite Data Assimilation, the UK Meteorology Office, the French Mercator Ocean, and the Chinese National Marine Environmental Forecasting Center. In particular, the UK Meteorology Office and Mercator Ocean are conducting an observing system experiment to test the impacts of satellite SSS data assimilation on simulating the 2015-16 El Niño (https://www.godae-oceanview.org/projects/smos-nino15), an effort funded by the European Space Agency. The project is also part of the GODAE OceanView's Observing System Evaluation Task Team effort.

The fidelity of ocean models in presenting salinity has been not as good as for presenting temperature. There are three major contributing factors: (1) uncertainties of E-P forcing, (2) lack of discharge estimates for many rivers around the world, and (3) the lack of historical salinity observations (including SSS).

E-P is an important forcing for modelled SSS. In ocean models, E-P forcing is typically obtained from atmospheric models or analysis/reanalysis products that are subject to relatively large uncertainties, especially in terms of their global net balance. The errors in E-P forcing cause ocean models to deviate or drift away from reality. To alleviate this, a common practice in ocean modeling is to relax modelled SSS towards a seasonal climatology derived from historical in situ measurements (this is a crude way to compensate for the error of E-P forcing). In fact, a similar approach was used for SSTs (i.e., relaxing toward seasonal climatology) before the sustained development and improvement of the operational satellite SST observing system. Relaxation of modelled SSS to seasonal climatology tends to reduce the amplitudes of the non-seasonal SSS signals in ocean models. The development of the Argo system has provided broad-scale SSS measurements in the past decade. However, the Argo array is not global (it has limited to no coverage in many marginal seas, coastal oceans, and polar oceans). Satellite SSS observations thus provide a potentially important database for constraining ocean models to correct for errors in E-P and model representation of ocean dynamics.

In coastal oceans and marginal seas significantly affected by rivers, freshwater input from river discharge is an important forcing factor for SSS. However, due to the lack of accurate observations for the discharges of many rivers around the world (especially for interannually varying time series and near real-time data availability), a common practice in ocean models is to use estimates of seasonal climatology for river discharges (e.g., those from Dai and Trenberth, 2002 or other sources). SSS variations near river mouths in ocean models typically have little interannual variations even though satellite SSS data have revealed substantial interannual variations near some river mouths (e.g., Grodsky et al., 2014; Fournier et al., 2016a).

Therefore, the assimilation of satellite SSS data, especially in coastal oceans and marginal seas where in situ data tend to be sparse (or non-existent), can help improve the representation of salinity structure in ocean models. The improvement of salinity in these regions have significant implications to marine biogeochemistry.

Treatment of biases in observations and models in assimilation

An important issue in assimilating SST and SSS observations are potential biases in the data. Some assimilation schemes assume that the data are un-biased whereas in reality biases do exist in the data. For example, there are relative biases between SST datasets based on infrared and passive microwave sensors. Significant biases still exist in satellite SSS measurements. Biases can limit the effectiveness of the data in constraining the model representation of the temporal variability, especially when (and where) the magnitude of the biases are not small comparing to the magnitude of the temporal variability. Bias correction schemes have been employed in some assimilation systems to alleviate this issue. An example is the UK Meteorology Office's FOAM-NEMOVAR system (https://www.metoffice.gov.uk/binaries/content/assets/mohippo/pdf/t/e/frtr578.pdf) where bias correction was applied to the SST data by referencing the biased SST data to a reference set of observations that are known or assumed to be unbiased, similar to the procedure applied by Martin et al. (2007). Biases in observational data are not necessarily unique to SST and SSS and the related assimilation, but present in other observations as well.

Model biases also reduce the effectiveness of the assimilation to constrain the temporal variability in the model using temporal signals in the data. This is because the assimilation works hard to correct the model biases using the observations, thus limiting the influence of the temporal variability in the observations to constrain the temporal variability of the model. A widely-used approach to tackle this issue is to perform anomaly data assimilation, i.e., only use the temporal variability in the observations to constrain the counterpart in the model (e.g., Hackert et al., 2014). In the ECCO estimation system, the time mean state and temporal anomalies of the model are constrained by the corresponding values of the observations, respectively, by separating the constraint on the time mean and temporal anomalies separately in the cost function. Moreover, the control variables (e.g., surface forcing) are also separated into time mean and temporal anomalies, weighted by the respective a priori uncertainties.

Importance of understanding uncertainty characteristics of SST and SSS observations

It is also important to understand the uncertainty characteristics of the SST and SSS observations that assimilated into models beyond just the biases. There are three important sources of uncertainties for satellite SST observations: (1) effect of clouds on SST from infrared sensors, (2) ability to resolve diurnal variability, and (3) larger uncertainties in the Arctic Ocean. In particular, the GHRSST Project has identified the latter to be an important issue in terms of improving satellite SST data. The larger discrepancies among satellite SST products in the Arctic Ocean and differences from in situ SST data were reported by (Castro et al., 2016), who found large differences of various level-4, blended GHRSST products in the Arctic Ocean. The GHRSST Multi-Product Ensemble project produces median and standard deviation (Fig. 11.7) based on a dozen GHRSST products (https://www.ghrsst.org/latest-sst-map/) for the global ocean. The GHRSST Multi-Product Ensemble provides important resources to understand the consistency and uncertainties of the GHRSST products.

As discussed in earlier, L-band satellite SSS data generally have larger uncertainties in high-latitude oceans. This is in large part due to the poor sensitivity of L-band T_B to SSS in cold-water (<5 °C) environment. The larger uncertainty of SST (an ancillary data used in SSS retrieval) also contribute. As an example, the standard deviation values for various spatial scales between monthly, level-3 Aquarius Version-4 SSS and the 5 m salinity from an Argo-based monthly gridded product from the Scripps Institution of Oceanography is shown in Fig. 11.8 (Lee, 2016). The discrepancies between the Aquarius SSS and the Argo product in high-latitude oceans are larger than those at lower latitudes by a few times. Similarly, SMOS and SMAP SSS products have larger discrepancies with Argo products (not shown). This error structure is helpful to the assimilation of Aquarius SSS data.

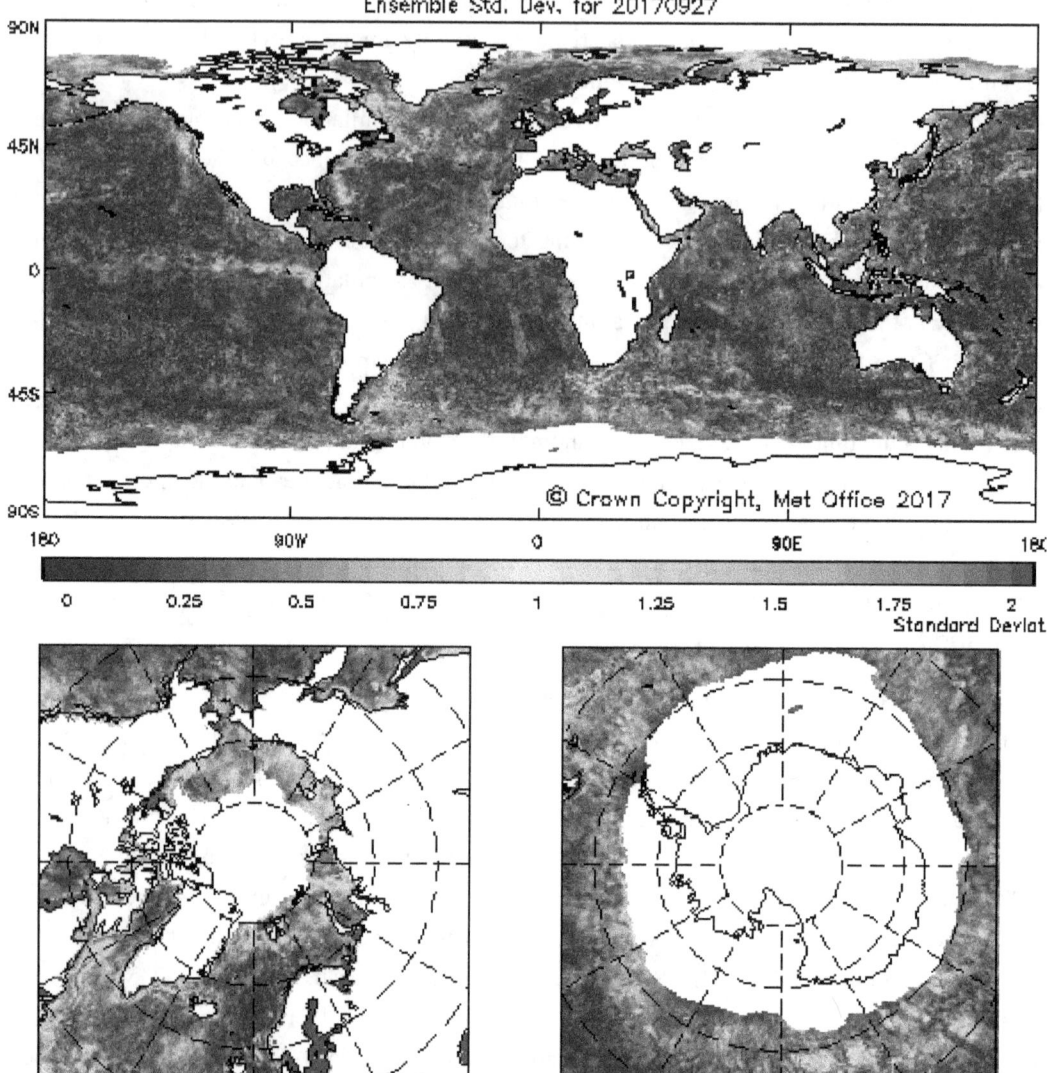

Figure 11.7. The spread among 11 SST products on 27 September 2017 (from https://www.ghrsst.org/latest-sst-map) as part of GHRSST's Multi-Product Ensemble comparison project, which is from the Copernicus Marine Environment Monitoring Service. © Crown copyright, Met Office.

It is important to note that the discrepancies between satellite SSS and the in situ-based gridded near-surface salinity products cannot be entirely contributed to the errors of the satellite SSS. The sampling differences between satellite SSS measurements (averages with the satellite footprint) and in situ (point-wise) measurements as well as the mapping errors also contribute. The nominal density of the Argo array is one profile 3°x3° at 10-day intervals. In areas with strong currents and divergence, the sampling of the Argo array may further reduce. As discussed earlier, in regions with high-frequency and small-scale variability, there will be larger sampling errors from Argo (e.g., insufficient sampling to accurately represent monthly averages on 1°x1° or 3°x3° scales. In fact, Figs. 11.8d and 11.8e illustrate relatively large standard deviation values between the Argo-Scripps product and that from University of Hawaii in regions of strong variability (e.g., tropical rain bands, river plumes, western boundary currents, Antarctic Circumpolar Current, etc.). The white-out areas in the third column of Fig. 11.8 show the locations where the differences between Argo-Scripps and Argo-University of Hawaii are larger than those between Aquarius and Argo-Scripps. If one excludes these areas, the areal average standard deviation between Aquarius and Argo is significantly smaller (the red values in the third column). The relative uncertainties of satellite and in situ SSS data due to the sampling differences need to be considered in assimilating salinity data in models with various resolutions. Even though the example above is for gridded satellite SSS and Argo products, the issue also applies for level-2 satellite SSS and individual in situ salinity measurements, as discussed earlier in this chapter.

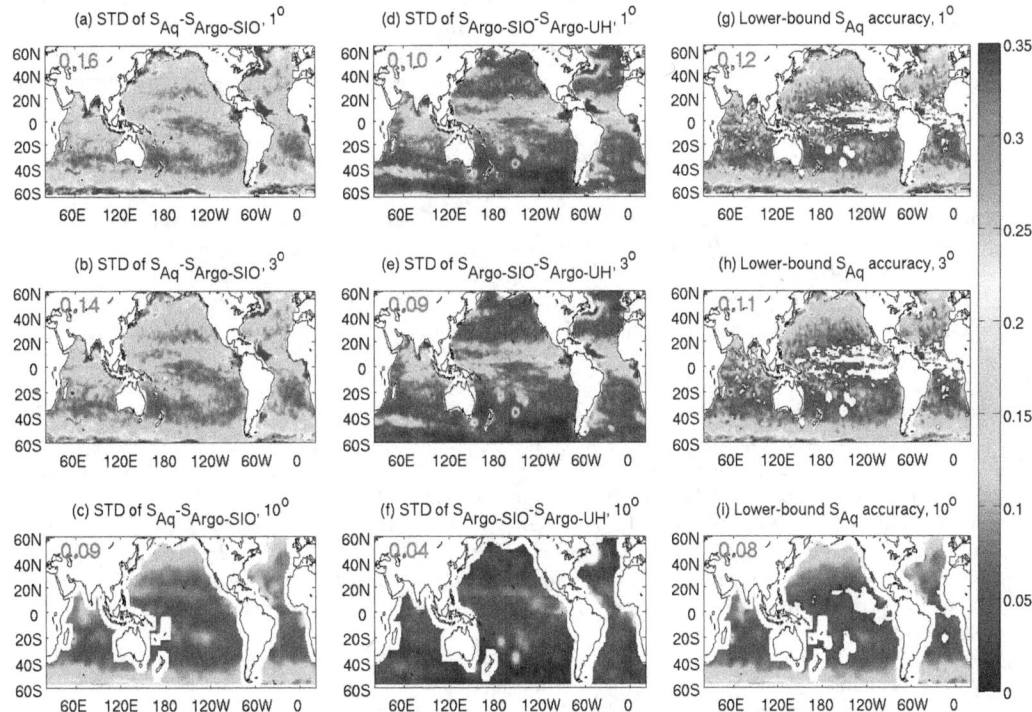

Figure 11.8. Standard deviation of SSS differences between Aquarius and Argo-Scripps on (a) 1°×1°, (b) 3°×3°, and (c) 10°×10° scales and between (d–f) Argo-Scripps and Argo-University of Hawaii on these scales. The values in Figs. 1a–1c represent upper-bound Aquarius SSS accuracies. (g–i) The estimated lower-bound Aquarius SSS accuracies obtained by removing the variance at each location in Figs. 1d–1f from that in Figs. 1a–1c then took the square root. The whited-out mid-ocean grid points in Figs. 1g–1i indicate where the discrepancies between the two Argo products are larger than those between Aquarius and Argo. The red numbers indicate the respective global averages of standard deviation. After Lee (2016).

Acknowledgments

This research was carried out in part at the Jet Propulsion Laboratory (JPL), California Institute of Technology, under a contract with NASA, supported by NASA Physical Oceanography Program.

References

Bjerknes, J, 1969: Atmospheric teleconnections from the equatorial Pacific. Mon. We. Rev., 97, 163-172.

Boutin, J., et al., 2016: Satellite and in situ salinity: understanding near surface stratification and sub-footprint variability. Bull. Amer. Meteorol. Soc., Vol.97, Issue 8, 1391-+, DOI: 10.1175/BAMS-D-15-00032.1.

Casey, K. S., Brandon, T. B., Cornillon, P., and Evans, R., 2010: The past, present, and future of the AVHRR Pathfinder SST program. In Oceanography from space (pp. 273-287). Springer Netherlands.

Castro, S.L., G.A. Wick, and M. Steele, 2016: Validation of satellite sea surface temperature analyses in the Beaufort Sea using UpTempO buoys. Remote Sensing of Environment, vol. 187, pp. 458-475.

Chakraborty, A., Kumar, R., Basu, S., and Sharma, R. (2015a). Improving Ocean State by Assimilating SARAL/AltiKa Derived Sea Level and Other Satellite-Derived Data in MITGCM, Mar. Geod., 38 (Sup 1), 328-338, doi: 10.1080/01490419.2014.1002142.

Chelton, D. B., S. K. Esbensen, M. G. Schlax, N. Thum, M. H. Freilich, F. J. Wentz, C. L. Gentemann, M. J. McPhaden, and P. S. Schopf, 2001: Observations of coupling between surface wind stress and sea surface temperature in the eastern tropical Pacific, Journal of Climate, 14(7), 1479-1498.

Donlon, C. J., Minnett, P. J., Gentemann, C., Nightingale, T. J., Barton, I. J., Ward, B., & Murray, M. J., 2002: Toward improved validation of satellite sea surface skin temperature measurements for climate research. Journal of Climate, 15(4), 353-369.

Donlon, C., Rayner, N., Robinson, I., Poulter, D. J. S., Casey, K. S., Vazquez-Cuervo, J., and May, D., 2007: The global ocean data assimilation experiment high-resolution sea surface temperature pilot project. Bulletin of the American Meteorological Society, 88(8), 1197-1213.

Droghei, R., B. Buongiorno Nardelli, and R. Santoleri, 2016: Combining in-situ and satellite observations to retrieve salinity and density at the ocean surface, J. Atmos. Ocean. Technol., 33, 1211–1223, doi:10.1175/JTECH-D-15-0194.1.

Drucker, R., & Riser, S. C., 2014: Validation of Aquarius sea surface salinity with Argo: Analysis of error due to depth of measurement and vertical salinity stratification. Journal of Geophysical Research: Oceans, 119(7), 4626-4637.

Du, Y. and Zhang, Y., 2015: Satellite and Argo Observed Surface Salinity Variations in the Tropical Indian Ocean and Their Association with the Indian Ocean Dipole Mode, J. Climate, 28 (2), 695-713, doi: 10.1175/JCLI-D-14-00435.1.

Entekhabi, D. et al., 2014: SMAP handbook soil moisture active passive: Mapping soil moisture and freeze/thaw from space, JPL CL14-2285, Jet Propulsion Laboratory, Pasadena, CA.

Fine, R.A., Willey, D.A., and Millero, F.J., 2017: Global Variability and Changes in Ocean Total Alkalinity from Aquarius Satellite Data, Geophys. Res. Lett., 44 (1), 261–267, doi: 10.1002/2016GL071712.

Fournier, S., T. Lee, and M. Gierach, 2016a: Seasonal and interannual variations of sea surface salinity associated with the Mississippi River plume observed by SMOS and Aquarius. Remote Sensing. Environ. 180, 431-439.

Fournier, S., J.T. Reager, and T. Lee et al., 2016b: SMAP observes flooding from land to sea: The Texas event of 2015. Geophys. Res. Lett., 43, 10.1002/2016GL068822.

Gentemann, C. L., Wentz, F. J., Mears, C. A., & Smith, D. K., 2004: In situ validation of Tropical Rainfall Measuring Mission microwave sea surface temperatures. Journal of Geophysical Research: Oceans, 109(C4).

Gentemann, C. L., 2014: Three-way validation of MODIS and AMSR-E sea surface temperatures. Journal of Geophysical Research: Oceans, 119(4), 2583-2598.

Gentemann, C. L., and Hilburn, K. A., 2015: In situ validation of sea surface temperatures from the GCOM-W1 AMSR2 RSS calibrated brightness temperatures. Journal of Geophysical Research: Oceans, 120(5), 3567-3585.

Gierach, M.M., J. Vazquez, T. Lee, and V. Tsontos, 2013: Aquarius and SMOS detect effects of an extreme Mississippi River flooding event in the Gulf of Mexico. Geophys. Res. Lett., 40, doi:10.1002/grl.50995.

Grodsky, S., N. Reul, G.S.E. Lagerloef, et al. 2012: Haline Hurricane Wake in the Amazon/Orinoco Plume: AQUARIUS/SAC-D and SMOS Observations, Geophys. Res. Lett., 39 (20), L20603, doi:10.1029/2012GL053335.

Grodsky, S.A., Reul, N., Chapron, B., Carton, J.A., and Bryan, F.O., 2017: Interannual Surface Salinity in Northwest Atlantic Shelf, J. Geophys. Res.-Oceans, 122 (5), 3638–3659, doi: 10.1002/2016JC012580.

Grodsky, S.A., Reverdin, G., Carton, J.A., and Coles, V.J., 2014b. Year-to-Year Salinity Changes in the Amazon Plume: Contrasting 2011 and 2012 Aquarius/SAC-D and SMOS Satellite Data, Remote Sens. Environ., 140, 14-22, doi: 10.1016/j.rse.2013.08.033.

Grunseich, G., Subrahmanyam, B., and Wang, B., 2013: The Madden-Julian Oscillation Detected in Aquarius Salinity Observations, Geophys. Res. Lett., 40 (20), 5461-5466, doi:10.1002/2013GL058173.

Guerrero, R.A., Piola, A.R., Fenco, H., Matano, R.P., Combes. V., Chao, Y., James, C., Palma, E.D., Saraceno, M., and Strub, P.T., 2014: The Salinity Signature of the Cross-Shelf Exchanges in the Southwestern Atlantic Ocean: Satellite Observations, J. Geophys. Res.-Oceans, 119 (11), 7794-7810, doi: 10.1002/2014JC010113.

Guan, B., T., Lee, T., D. Halkides, et al., 2014: Aquarius surface salinity and the Madden-Julian Oscillation: the role of salinity in surface layer density and potential energy, Geophys. Res. Lett., 41 (8), 2858-2869, doi:10.1002/2014GL059704.

Hackert, E., A. J. Busalacchi, and J. Ballabrera-Poy, 2014: Impact of Aquarius sea surface salinity observations on coupled forecasts for the tropical Indo-Pacific Ocean, J. Geophys. Res. Oceans, 119, 4045–4067, doi:10.1002/2013JC009697.

Kao, H.Y. and G.S.E. Lagerloef, G.S.E., 2015: Salinity Fronts in the Tropical Pacific Ocean, J. Geophys. Res.-Oceans, 120 (2), 1096-1106, doi:10.1002/2014JC010114.

Köhl, A., M. Sena Martins, and D. Stammer, 2014: Impact of assimilating surface salinity from SMOS on ocean circulation estimates, J. Geophys. Res. Oceans, 119, 5449–5464, doi:10.1002/2014JC010040.

Kumar A., and Hoerling M.P., 1998: Annual cycle of Pacific/North American seasonal predictability associated with different phases of ENSO. Journal of Climate 11: 3295–3308.

Lagerloef, G., and J. Font, 2010: SMOS and Aquarius/SAC-D Missions: The Era of Spaceborne Salinity Measurements is About to Begin, Oceanography from Space, 35-58, doi: 10.1007/978-90-481-8681-5_3.

Lagerloef, G., S.H. Yueh, and J. Piepmeier, 2013: Review & Forecast: NASA's Aquarius Mission Provides New Ocean View, Sea Technol., 54 (1), 26-29.

Land, P.E., Shutler, J.D., Findlay, H.S., Girard-Ardhuin, F., Sabia, R., Reul, N., Piolle, J.-F., Chapron, B., Quilfen, Y., Salisbury, J., Vandemark, D., Bellerby, R., and Bhadury, P., 2015: Salinity from Space Unlocks Satellite-Based Assessment of Ocean Acidification, Environ. Sci. Technol., 49 (4), 1987-1994, doi:10.1021/es504849s.

Lee, T., G. Lagerloef, M. M. Gierach, H.-Y. Kao, S. Yueh, and K. Dohan, 2012: Aquarius reveals salinity structure of tropical instability waves, Geophys. Res. Lett., 39, L12610, doi:10.1029/2012GL052232.

Lee, T., G. Lagerloef, H.-Y. Kao, M.J. McPhaden, J. Willis, and M.M. Gierach, 2014: The influence of salinity on tropical Atlantic instability waves. J. Geophys. Res., 10.1002/2014JC010100.

Lee, T., 2016: Consistency of Aquarius sea surface salinity with Argo products on various spatial and temporal scales. Geophys. Res. Lett., 43, 10.1002/2016GL068822.

Lee, T et al., 2016: Linkages of salinity with ocean circulation, water cycle, and climate variability. Community white paper in response to Request for Information #2 by the US National Research Council Decadal Survey for Earth Science and Applications from Space 2017-2027. http://surveygizmoresponseuploads.s3.amazonaws.com/fileuploads/15647/2604456/107-1abc9aa1a37ab7e77d91d86598954a50_LeeTong.pdf

Lindstrom, E., R. Lukas, R. Fine, et al. 1987: The western equatorial Pacific Ocean circulation study, Nature, 330, 533–538.

Llewellyn-Jones, D., Edwards, M. C., Mutlow, C. T., Birks, A. R., Barton, I. J., and Tait, H., 2001: AATSR: Global-change and surface-temperature measurements from Envisat. ESA bulletin, 105(10-21), 25.

Lu, Z., L. Cheng, J. Zhu, and R. Lin, 2016: The complementary role of SMOS sea surface salinity observations for estimating global ocean salinity state, J. Geophys. Res. Oceans, 121, 3672–3691, doi:10.1002/2015JC011480.

Luo, J.-J. et al., 2005: Seasonal Climate Predictability in a coupled OAGCM using different approach for Ensemble Forecasts. J. Climate, 18, 4474-4497. doi:10.1175/JCLI3526.1.

Martin, M., J.A. Hines, and M. J. Bell, 2007: Data assimilation in the foam operational short-range ocean forecasting system: a description of the scheme and its impact. Q. J. Roy. Meteorol. Soc., 133:981–995, 2007.

Melnichenko, O., A. Amores, N. Maximenko, et al., 2017: Signature of Mesoscale Eddies in Satellite Sea Surface Salinity Data, J. Geophys. Res.-Oceans, 122 (2), 1416–1424, doi:10.1002/2016JC012420.

Menezes, V.V., M.L. Vianna, and H.E. Phillips, 2014: Aquarius Sea Surface Salinity in the South Indian Ocean: Revealing Annual-Period Planetary Waves, J. Geophys. Res.-Oceans, 119 (6), 3883-3908, doi:10.1002/2014JC009935.

Nardelli, B., R. Droghei, and R. Santoleri, 2016: Multi-dimensional interpolation of SMOS sea surface salinity with surface temperature and in situ salinity data, Remote Sens. Environ., 180, 392–402, doi:10.1016/j.rse.2015.12.052.

O'Carroll, A. G., Eyre, J. R., & Saunders, R. W., 2008: Three-way error analysis between AATSR, AMSR-E, and in situ sea surface temperature observations. Journal of Atmospheric and Oceanic Technology, 25(7), 1197-1207.

Olmedo, E., Martínez, J., Umbert, M., Hoareau, N., Portabella, M., Ballabrera, J., Turiel, A., 2016: Improving time and space resolution of SMOS salinity maps using multifractal fusion. Remote Sensing of Environment 180, 246-263, doi:10.1016/j.rse.2016.02.038.

Qu, T.D. and J.-Y. Yu, 2014: ENSO Indices from Sea Surface Salinity Observed by Aquarius and Argo, J. Oceanogr., 70 (4), 367-375, doi:10.1007/s10872-014-0238-4.

Reul, N., J. Tenerelli, B. Chapron, D. Vandemark, Y. Quilfen, and Y. Kerr, 2012: SMOS satellite L-band radiometer: A new capability for ocean surface remote sensing in hurricanes, J. Geophys. Res., 117, C02006, doi:10.1029/2011JC007474.

Reul, N., B. Chapron, T. Lee, C. Donlon, J. Boutin, G. Alory, 2014: Sea surface salinity structure of the meandering Gulf Stream revealed by SMOS sensor. Geophys. Res. Lett., DOI: 10.1002/2014GL059215.

Reynolds, R. W., 1993: Impact of Mount Pinatubo aerosols on satellite-derived sea surface temperatures. Journal of climate, 6(4), 768-774.

Reynolds, R. W., T. M. Smith, C. Liu, et al., 2007: Daily high-resolution-blended analyses for sea surface temperature, J. Clim., 20, 5473–5496, doi:10.1175/2007JCLI1824.1.

Sprintall, J., and M. Tomczak (1992), Evidence of the barrier layer in the surface layer of the tropics, J. Geophys. Res., 97(C5), 7305-7316.

Tang, W., Yueh, S. H., Fore, A. G., and Hayashi, A., 2014: Validation of Aquarius sea surface salinity with in situ measurements from Argo floats and moored buoys. Journal of Geophysical Research: Oceans, 119(9), 6171-6189.

Tang, W., A. Fore, S. Yueh, T. Lee, A. Hayashi, A. Sanchez-Franks, B. King, D. Baranowski, and J. Martinez, 2017: Validating SMAP SSS with in-situ measurements. Remote Sensing Environ., doi:10.1016/j.rse.2017.08.021.

Toyoda, T., Fujii, Y., Kuragano, T., Matthews, J.P., Abe, H., Ebuchi, N., Usui, N., Ogawa, K., and Kamachi, M., 2015: Improvements to a Global Ocean Data Assimilation System Through the Incorporation of Aquarius Surface Salinity Data, Q. J. Roy. Meteor. Soc., 141 (692), 2750-2759, doi: 10.1002/qj.2561.

Umbert, M., N. Hoareau, A. Turiel, and J. Ballabrera-Poy, 2013: New blending algorithm to synergize ocean variables: The case of SMOS sea surface salinity maps, Remote Sens. Environ., doi:10.1016/j.rse.2013.09.018.

Xie, P., T. Boyer, E. Bayler, et al., 2014: An In Situ-satellite Blended Analysis of Global Sea Surface Salinity, J. Geophys. Res.-Oceans, 119 (9), 6140-6160, doi:10.1002/2014JC010046.

Yin, X., J. Boutin, and G. Reverdin, et al., 2014: SMOS sea surface salinity signals of tropical instability waves. J. Geophys. Res., 119, 7811–7826, doi:10.1002/2014JC009960.

Yu, L., 2014: Coherent Evidence from Aquarius and Argo for the Existence of a Shallow Low-salinity Convergence Zone Beneath the Pacific ITCZ, J. Geophys. Res.-Oceans, 119 (11), 7625-7644, doi: 10.1002/2014JC010030.

Yu, L., 2015: Sea-surface Salinity Fronts and Associated Salinity Minimum Zones in the Tropical Ocean, J. Geophys. Res.-Oceans, 120 (6), 4205-4225, doi:10.1002/2015JC010790.

Yueh, S., Tang, W., Fore, A., Hayashi, A., Song, Y. T., and Lagerloef, G., 2014: Aquarius geophysical model function and combined active passive algorithm for ocean surface salinity and wind retrieval. Journal of Geophysical Research: Oceans, 119(8), 5360-5379.

CHAPTER 12

Ocean Circulation Modeling for Operational Oceanography: Current Status and Future Challenges

Julien Le Sommer[1], Eric P. Chassignet[2], and Alan J. Wallcraft[2]

[1]*Univ. Grenoble Alpes, CNRS, IRD, IGE, Grenoble, France;* [2]*Center for Ocean-Atmospheric Prediction Studies (COAPS), Florida State University, Tallahassee, FL, USA*

This chapter focuses on ocean circulation models used in operational oceanography, physical oceanography and climate science. Ocean circulation models are a particular branch of ocean numerical modeling that focuses on the representation of ocean physical properties over spatial scales ranging from the global scale to less than a kilometer and time scales ranging from hours to decades. As such, they are an essential building block for operational oceanography systems and their design receives a lot of attention from operational and research centers.

Introduction

Ocean modeling is an important branch of operational oceanography. Ocean numerical models are an essential building block of global and regional operational oceanography systems. In this context, ocean circulation models are used in conjunction with data assimilation for extrapolating both in space and in time the available satellite and in situ oceanic observations in order to build a physically consistent estimate of the ocean state and its evolution. Ocean modeling is a relatively recent discipline in the field of oceanography. The underlying principles of the algorithmic formulation of ocean circulation models were first proposed in the 1960s by Bryan (1969; see also McWilliams [1996] for an historical review). Since then, the continuous increase in computing power has allowed ocean circulation models to provide ever more meaningful and comprehensive descriptions of the ocean circulation. Ocean modeling is now recognized as an essential supplement to more traditional scientific methodologies in oceanography. The variety of oceanic physical processes accounted for in ocean circulation models has also notably broadened over past decades. The original scope of ocean circulation models was to describe oceanic properties and physical processes at scales significantly larger than the mesoscale (horizontal scales on the order of 100 km and time scales on the order of three months). But the increase in computing power and the improved physical consistency of their formulation now allow ocean circulation models to resolve routinely oceanic flows down to the submesoscale (horizontal

---

Le Sommer, J., E.P. Chassignet, and A.J. Wallcraft, 2018: Ocean circulation modeling for operational oceanography: Current status and future challenges. In "*New Frontiers in Operational Oceanography*", E. Chassignet, A. Pascual, J. Tintoré, and J. Verron, Eds., GODAE OceanView, 289-306, doi:10.17125/gov2018.ch12.

scales on the order of 10 km; Chassignet and Xu [2017]) and to describe internal wave and internal tides at a global scale (Shriver et al., 2012).

The range of uses and applications of ocean circulation models has diversified over the past several decades. Ocean circulation models are now used standalone for simulating ocean circulation; when integrated in a data assimilation framework, they may be used for producing short-range ocean forecasts and for ocean reanalysis; coupled to atmospheric circulation models, they may be used for seasonal to decadal forecasts; and fully integrated into Earth System Models, they are crucial to climate modeling. Scientists also use ocean circulation models as the experimental tool of choice for improving our mechanistic understanding of the ocean.

The variety of oceanic physical processes accounted for in ocean circulation models has also broadened in recent decades. Ocean circulation models were initially used to describe oceanic properties and physical processes at scales significantly larger than the mesoscale (horizontal scales on the order of 100 km and time scales on the order of three months). But the increase in computing power and improved physical consistency of their formulation now allow ocean circulation models to resolve routinely oceanic flows down to the submesoscale (horizontal scales on the order of 10 km; Chassignet and Xu [2017]) and to describe internal wave and internal tides at the global scale (Shriver et al., 2012; Savage et al., 2017).

However, despite the maturity and broad range of applications of ocean circulation models achieved to date, ocean modelers still face notable challenges. Running ocean circulation models at increased model resolution does not solve the problem of subgrid scale closures (unresolved processes), and may even question some of the underlying assumptions and algorithms of current ocean circulation models (hydrostatic vs. non hydrostatic, for example). The modularity of modern geoscientific models requires the design of robust and physically rational approaches for coupling ocean circulation models with other model components. Most importantly, there is a growing concern about the necessity to describe explicitly how model uncertainty propagates in geoscientific modeling systems. This is why ocean circulation model design is still a very active field of research and will likely remain so well into the future.

In this chapter, we aim to: (i) briefly describe the principles that underpin the formulation of ocean circulation models, (ii) review current skills of ocean circulation models, and (iii) present what we believe are the new frontiers in ocean circulation model design. Obviously, given the complexity of the issues at stake and the amount of energy that is put into ocean model development, this chapter will only briefly touch upon the above objectives. Our goal is therefore not to cover these questions thoroughly, but rather to provide an entry point to the science of ocean models. The second section of this chapter focuses on the science and applications of ocean circulation models; the third section describes ongoing research avenues geared toward improving the representation of physical processes in ocean circulation models; and the fourth section describes the ongoing paradigm shift in ocean modeling towards a more probabilistic description of oceanic flows. More general considerations regarding the future of ocean modeling are presented in the conclusion.

Ocean Circulation Models: Scope, Usage, & Fundamentals

Ocean modeling is a branch of numerical modeling that focuses on representation of the physical mechanisms governing the evolution of ocean physical properties, namely T, S, u, v, and w, where T is temperature, S is salinity, and u, v, and w are the horizontal and vertical components of the velocity V. A good understanding of a range of physical processes is required in order to build a numerical model that is capable of faithfully representing the ocean circulation. Without this physical understanding, it is easy to derive erroneous conclusions since a numerical model is constructed using discretized equations of motion. Indeed, direct numerical simulations (DNS) of the ocean (i.e., numerical representation of the smallest turbulent scale, i.e. the Kolmogorov length scale, which is on the order of 1 cm (Smyth et al., 2001)), is not possible with present-day computers. The largest simulation achievable today is on scales of 10 m (Yeung et al., 2015).

At present, we cannot represent these small scales; we can only achieve a truncated representation of the ocean and this will remain the case for the foreseeable future. The spatial and temporal scales that one can currently represent strongly depends on the model configuration and application. High resolution operational oceanography requires accurate depiction of upper ocean structure and mesoscale features such as eddies and meandering fronts. Accurate sea level representation is crucial for coastal models (response to wind, tides, and surface pressure) and seasonal-to-interannual forecasts require a good representation of the upper ocean mass field and the coupling to an active atmosphere. On global and basin scales, high horizontal resolution ($1/10^o$ to $1/25^o$ and very rarely to $1/50°$; Chassignet and Xu [2017]) is mostly used for ocean "weather" and seasonal-to-decadal variability (Fig. 12.1). The emphasis is on short integrations (years to decades) and most models of this class are coupled to a sea-ice model but are stand-alone and use prescribed atmospheric fields. Coarser resolution ($1/4^o$ to 1°) is principally used for climate applications (Griffies et al., 2000). The emphasis for this class of models is on long integrations with fully coupled ocean-ice-atmosphere models.

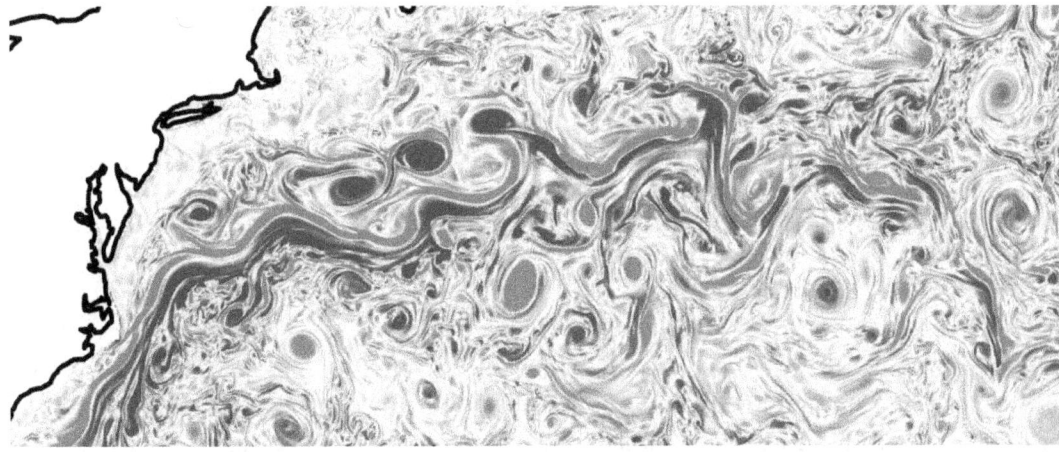

Figure 12.1. Example of submesoscale permitting modeling – March 1st surface vorticity over the Gulf Stream in a1/50° North Atlantic domain (~1.5 km horizontal grid spacing). Adapted from Chassignet and Xu (2017).

In practice, ocean models consist of a numerical solution to a set of partial differential equations (PDEs) describing the ocean dynamics. These PDEs are based on an approximated version of Navier-Stokes equations adapted to our regimes of interest (see the review papers by Griffies [2004] and Griffies and Treguier [2013] for details). An ocean numerical model typically consists of 20,000 to 200,000 lines of code, usually written in FORTRAN, and can easily take up to 10 years of community development to be fully functional.

The first ocean numerical model was put forward by Bryan (1969) following in the footsteps of numerical weather prediction. The first discretized equations for an atmospheric application were solved manually by Richardson (1922) and numerically by Charney et al. (1950). The latter consisted of a 15 by 18 grid ($\Delta x = 736$ km) and demonstrated that numerical weather prediction was possible. The latest configurations use horizontal resolution Δx on the order of 1 km (Chassignet and Xu, 2017). The sheer size of the ocean problem is such that it will always require the latest generation of supercomputers, and ocean/climate models have traditionally been one of the biggest users of computer resources. This means that close collaborations with computer scientists are essential to ensure that the numerical codes run efficiently and take full advantage of the computing architecture, as this changes constantly and ocean modelers need to be flexible and ready to adopt new approaches. The main limitation is not computational speed of the processors, but access to memory and latency in reading/writing on disk drives (I/O). This limitation is not likely to change in the near future since supercomputer development is closely linked to the performance of commodity chips, which are not well-adapted to ocean applications (i.e., GPUs) and do not facilitate memory access.

There is a wide range of applications for ocean models. Without data assimilation, they are mostly used to scientifically rationalize the observed ocean by testing mechanisms underlying observations via idealized or realistic configurations. With data assimilation, they are used to perform hindcasts in order to understand past evolution and to perform forecasts that can be used for societal applications (Chassignet and Verron, 2006; Dombrowsky et al., 2009; Schiller and Brassington, 2011; Bell et al., 2015). Observational data via data assimilation sets the stage for model state estimates and forecasts (Chassignet et al., 2009). The quality of the estimates and the forecast will depend on the ability of an ocean numerical model to faithfully represent the resolved dynamics of the ocean and the parameterized subgrid scale physics. It is therefore important to realize that even using an infinite amount of data to constrain the initial conditions of an ocean model will not necessarily improve the forecast when using a poorly performing ocean numerical model.

It is also important to realize that while numerical models are necessary to understand ocean dynamics, they cannot represent reality because of limitations in computational power, incomplete understanding of subgrid scale parameterizations, poorly known forcing fields, and poorly understood interactions with other components of the earth's system such as the atmosphere and sea ice. Observations are also not an accurate representation of reality because of the many space and time gaps in the observations that give us only limited information about the ocean's state, its variability, trends, and possible instabilities and regime shifts.

In summary, while numerical models allow for hypothesis testing and experimentation, one needs to understand their limitations in order to use them to their full potential. Good modelers are aware of the strengths AND weaknesses of their models, and the main difficulty is the quantification of the truncation errors introduced by the discretization of the Navier-Stokes equations.

Truncation errors arise from the discretization of PDEs derived from the Navier-Stokes equations, which usually assume that the ocean is incompressible and make the spherical approximation (assumes the earth is a sphere), the thin-shell approximation (allows to neglect variations of the local rotation rate with respect to depth, therefore simplifying the treatment of the Coriolis acceleration), the Boussinesq approximation (neglects variation of relative density in the horizontal momentum equations), and the hydrostatic approximation (allows to neglect the vertical acceleration).

The primitive Navier-Stokes equations are

$$\frac{\partial \mathbf{U}_h}{\partial t} = -[(\nabla \times \mathbf{U}) \times \mathbf{U} + \frac{1}{2}\nabla(\mathbf{U}^2)]_h - f\mathbf{k} \times \mathbf{U}_h - \frac{1}{\rho_0}\nabla_h p + \mathbf{D}^\mathbf{U} + \mathbf{F}^\mathbf{U}$$

$$\frac{\partial p}{\partial z} = -\rho g$$

$$\nabla \cdot \mathbf{U} = 0$$

$$\frac{\partial T}{\partial t} = -\nabla \cdot (T\mathbf{U}) + D^T + F^T$$

$$\frac{\partial S}{\partial t} = -\nabla \cdot (S\mathbf{U}) + D^S + F^S$$

with $\mathbf{U}_h$ is the horizontal velocity vector (u,v); $\mathbf{U}$, the three-dimensional velocity (u,v,w); f, the Coriolis parameter; p, the pressure; g, the gravity; ρ, the density; T, the temperature; S, the salinity; D, the dissipation; and F, the forcing. Temperature and salinity here implicitly refer to conservative temperature and absolute salinity as defined in the Thermodynamic Equation Of Seawater - 2010 (TEOS-10; McDougall and Barker, 2011). The equation of seawater relates density to temperature, salinity and pressure:

$$\rho = \rho(T, S, p)$$

Several other equations are required for setting boundary conditions at the ocean surface and at the seafloor. For instance, ocean circulation models enforce the following conditions at the ocean surface:

- Kinematic boundary condition

$$\frac{\partial \eta}{\partial t} + \mathbf{U}_h \cdot \nabla \eta = w_{|surf} + E - P$$

- Surface pressure condition

$$p_{|surf} = p_{atm}(x, y, t)$$

- Air-sea fluxes of freshwater and heat

$$K_v \frac{\partial T}{\partial z}\bigg|_{surf} = -\rho C_p Q \qquad K_v \frac{\partial S}{\partial z}\bigg|_{surf} = 0$$

- Air-sea fluxes of momentum

$$A_v \frac{\partial \mathbf{U}_h}{\partial z}\bigg|_{surf} = \frac{1}{\rho_0}(\tau_x, \tau_y)$$

where η is the sea surface height, E, the evaporation; P, the precipitation; τ, the wind stress; K, the diffusivity; A, the viscosity; and Q, the heat flux. Finally, one needs to use the turbulent closure hypothesis to close the system by assuming that the nonlinear terms involving correlations of variables at unresolved small scales can be fully prescribed from knowledge of the large scales.

Representation of Physical Processes in Ocean Circulation Models

In this section, we review some of the issues that arise as we strive to improve the representation of key physical processes in the ocean model component of operational systems. This includes broadening the spectrum of *explicitly represented processes* and improving the representation of *unresolved processes* in ocean circulation models.

Resolved versus unresolved physical processes in ocean circulation models.

As discussed in the previous section, ocean circulation models do not account *explicitly* for the entire range of scale interactions that control ocean circulation (Fig. 12.2). Because of the turbulent nature of oceanic flows, ocean circulation at a given scale is indeed fundamentally dependent on oceanic motions at scales ranging from global (of order 10,000 km) to dissipative (of order 1 cm; Fig. 12.3). But the finite grid resolution of a particular ocean model configuration constrains the spectrum of scales of motions that are explicitly represented in the model solution. Consider, for instance, ocean mesoscale eddies, which are known to play a fundamental role in shaping ocean circulation (McWilliams, 2008) and their representation in ocean models, which has motivated a large number of studies. Even the most high-end global ocean circulation models with grid resolutions down to just a few kilometers (Chassignet and Xu, 2017) are still not able to fully capture the dominant length scales of mesoscale variability at high latitudes (Hallberg, 2013). Scale interactions involving oceanic mesoscale eddies are therefore not explicitly represented at high latitudes in these models (Fig. 12.4).

OCEAN CIRCULATION MODELING FOR OPERATIONAL OCEANOGRAPHY: CURRENT STATUS AND FUTURE CHALLENGES

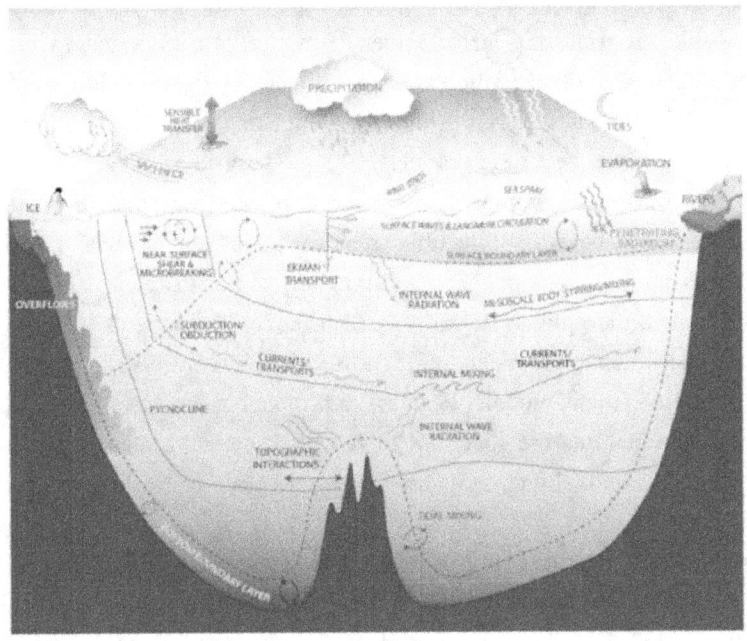

Figure 12.2. Range of ocean physical processes (adapted from Griffies and Treguier [2013]).

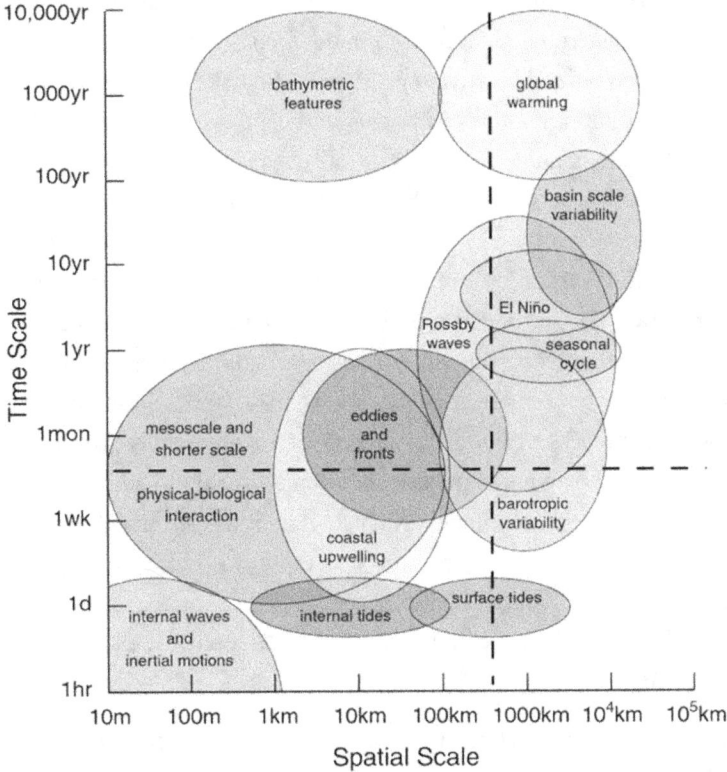

Figure 12.3. Space and time scale of ocean physical processes.

Therefore, a practical challenge for most applications of ocean circulation models is to describe the broadest possible spectrum of oceanic scales of motion at a given computational cost. The overall computational budget of an application is constrained by the available computing resources and by the expected time-to-solution, the latter constraint being particularly limiting in operational applications that deliver near real-time products. The model grid of ocean model components of operational systems is also usually constrained by other practical factors. The same model grid is often used for several years because of the effort required to tune the physical model configuration, calibrate the data assimilation components, and prepare the downstream data production chains. Given these constraints, the optimal design of ocean model components in operational systems requires a robust and rational understanding of a priori what controls the spectrum of resolved process in ocean circulation models besides the model grid resolution. Developing this understanding of what controls the spectrum of resolved scales in ocean model should rely on idealized process studies, model inter-comparison exercises and confrontation of model solutions with observations.

A commonly accepted practical criteria for deciding whether a flow feature is resolved in an ocean model is that there should be no less than 2π grid points spanning the feature (see, for instance, Griffies and Treguier [2013]). This is supported by the notion of effective resolution, defined by Skamarock (2004) as the smallest resolved scale that is not significantly affected by numerical dissipation due to discretization errors. The effective resolution of an ocean circulation model is about 6 to 10Δx, where Δx is the horizontal grid resolution, but the validity of the concept of effective resolution is generally restricted to model configurations with small enough grid resolutions and using high order discretization schemes (Soufflet et al., 2016).

Although effective resolution allows us to characterize the smallest scale of motion that could ideally be described without numerical error in an ocean model, it is generally impossible to distinguish unambiguously between resolved and unresolved *physical processes* in ocean circulation models. Indeed, only the physical processes with dynamics that do not depend on scales smaller than the effective resolution are unambiguously represented in a given ocean model configuration. In practice, a lot of physical processes of interest involve scales between the model grid scale and the model effective resolution. In such cases, although the model is a priori unable to fully account for the dynamics, the model solutions may still exhibit some properties of the misrepresented physical process. A good example of such a situation is discussed in a recent study by Uchida et al. (2017), who found signatures of submesoscale mixed layer instabilities in their 1/10° resolution model although their model grid resolution is a priori not able to adequately capture these dynamics.

How to broaden the spectrum of resolved processes in ocean circulation models?

Using computationally efficient numerical discretization schemes with good mimetic properties is a first approach for broadening the spectrum of resolved processes in ocean circulation models at a given computational cost. The notion of *computational efficiency* refers to the accuracy of model solution relative to the computational cost of numerical integration. The notion of *mimetic properties* refers to the ability of a discrete model to emulate the properties of the underlying continuous mathematical model, for instance as conservation laws and geometrical symmetries. Ocean circulation models with grid resolutions approaching 1 km can be sensitive to numerical discretization schemes (Ducousso et al., 2018). This is because ocean model solutions are generally more nonlinear and less controlled by diffusion in this range of resolution. The current trend in ocean circulation modeling is, therefore, to use high order discretization schemes in order to increase model effective resolution (Soufflet et al., 2016).

Multiscale modeling approaches coming from coastal ocean modeling also provide a framework for locally increasing the effective resolution according to a particular scientific or operational objective and at a given computational cost. Multiscale modeling methods, which allow us to locally refine the grid mesh, are now commonly used in large-scale ocean circulation modeling. Applications of multiscale approaches in ocean circulation modeling can be classified according to their grid mesh topology into *structured mesh* methods (for instance as orthogonal curvilinear grids used by most ocean circulation models), *unstructured mesh* methods (e.g., see Piggott et al., 2008; Danilov et al., 2017; Ringler et al., 2013) and *block structured mesh* methods (see Blayo and Debreu, 1999; Jablonowski et al., 2006). Although many research efforts have focused on unstructured mesh ocean circulation modeling over recent years, the design of robust discretization schemes for large-scale ocean models on triangular or hexagonal meshes remains challenging (Danilov, 2013). And it should also be noted that it is still unclear how to define objective criteria for adjusting meshes and, therefore, optimally use the potential for mesh refinement and mesh adaptability of multiscale modeling approaches in large-scale ocean circulation models.

Improving the performance of ocean circulation models on modern high performance computing (HPC) platforms is also essential for broadening the spectrum of resolved scales. The current trend in HPC is toward massively parallel machines with heterogeneous multicore architectures (Giles and Reguly, 2014). However, while modern HPC platforms can deliver a peak performance in the Petaflop/s range, existing ocean circulation models are unable to exploit this potential. The *computational intensity*, i.e., floating point operations per memory access, of ocean models is very low because they are dominated by stencil operations (typical of discretized PDEs). So, they typically run at ~5% of the system's peak speed. Also, ocean models have a very small vertical dimension, typically $O(10)$, so they scale more like 2-D domains than 3-D domains. In practice, $1/12°$ and $1/25°$ global ocean models might scale well to 8,000 and 30,000 cores, respectively. Scalability is eventually limited by the communication overhead, load imbalance, and latency inherent to spatial 2-D domain decomposition and by I/O overhead. Single processor performance depends on computational intensity, which can be improved by using higher order discretization schemes; but scalability can actually be reduced by better single processor performance because relatively more time is spent in communications.

In the future, overcoming single processor and scalability bottlenecks will require sustained collaborations between ocean modelers and computer scientists. Practical approaches will likely involve an increase in hybrid parallel programming in order to more efficiently exploit memory hierarchy and innovative algorithms for solving the set of PDEs that govern ocean dynamics, as for instance parallelization in time in addition to spatial domain decomposition (Schreiber et al., 2017).

Looking ahead, because of the multiscale nature of oceanic flows and the needs of end-users of operational oceanography, it is necessary for ocean circulation models to account explicitly for the broadest range of physical scales possible. But this broadening of the spectrum of resolved scales will not simply be a consequence of an increase in computing power. How efficiently ocean circulation models will use computing resources will depend on the efficiency of the algorithms used for translating ocean dynamical equations into practical computation. And foreseeing what approach will be most used in the future for broadening the spectrum of resolved scales is not straightforward, but it is arguable that HPC and algorithmic aspects will be crucial in future ocean model development. New advances will most likely rely on a high level of collaboration between ocean modelers, applied mathematicians, and computer scientists.

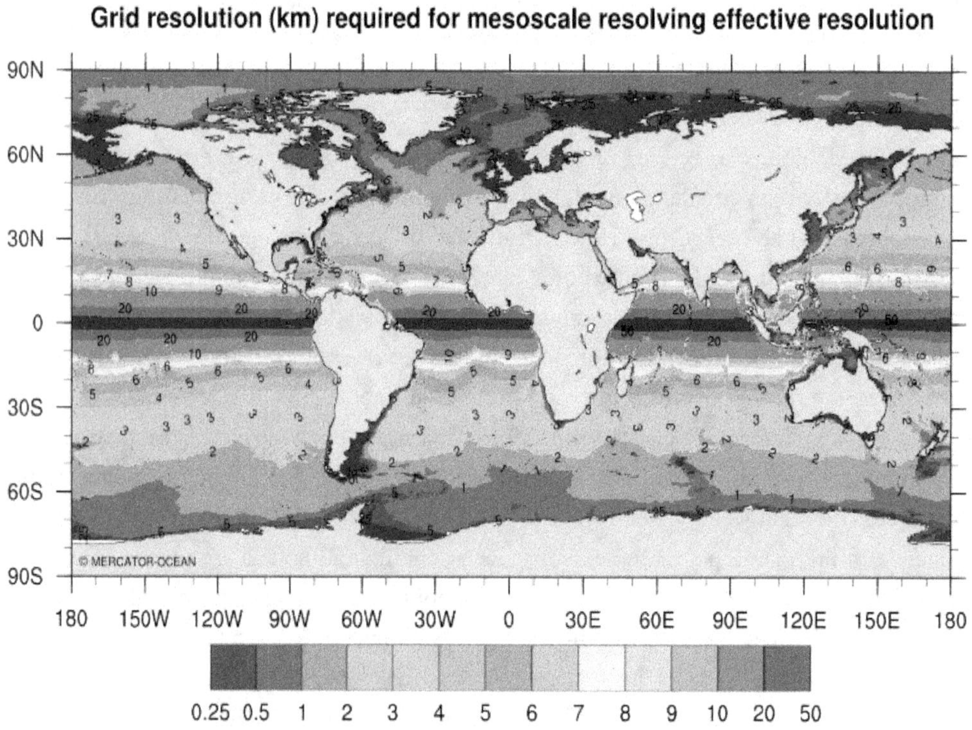

Figure 12.4. Grid resolution required to resolve the first baroclinic radius with 10 grid points. The figure shows the local grid resolution (in kilometers) divided by 10 as an estimate the grid resolution required for a model to effectively capture structures with wavelength close to the first baroclinic Rossby radius. The first baroclinic Rossby radius is estimated from a CMEMS 1/12° global reanalysis. The estimate of the number of points required for the effective resolution to match the first baroclinic Rossby radius is slightly more restrictive than that proposed by Hallberg (2013) following the work by Soufflet et al. (2016).

How to better account for unresolved processes in ocean circulation models

Ocean circulation models need to allow for the important physical processes that are not accounted for in their discrete approximation to the ocean dynamical equations. A common approach to the representation of unresolved processes in ocean circulation models is to use a subgrid scale (SGS) closure (aka physical parameterization). Typically, these do not focus on particular nonlinear terms in the primitive equations, but rather on particular physical processes (that may affect several terms). A rough classification of unresolved physical processes would include: (i) SGS closures for balanced turbulence (eddy closures), (ii) SGS closures for processes in the ocean surface boundary layer, (iii) SGS closures for interior mixing processes, and (iv) SGS closures for bottom boundary layer processes. The design of SGS closures associated with physical processes usually involves some combination of idealized process studies with models, realistic process studies with models, targeted field experiments at sea, and (more rarely) lab experiments. Importantly, a SGS closure should always target a particular ocean circulation model resolution because the information at the resolved scale depends on the model resolution.

Depending upon the available information from the scales assumed to be resolved in the target model, two categories of closures can be distinguished. The first is a RANS-type (cf Reynolds average) closure, implicitly based on some sort of ensemble average (in practice emulated through time average); for example, the Gent-McWilliams closure for mesoscale eddy transport (Gent, 2011). The other is a LES type closure (by analogy to Large Eddy Simulation) implicitly based on some spatial average; for example, Leith's (1968) momentum closure. LES-type subgrid closures can further be subdivided into functional and structural closures. Functional closures consider the bulk action of the subgrid terms on the resolved scales (e.g., Smagorinsly momentum closures [Griffies and Hallberg, 2000]). Structural closures aim at estimating the best local approximation of the unknown SGS term by constructing it from the known small-scale stuctures (e.g., closures defined by Taylor series expansion). A recent trend in the design of SGS closures in ocean circulation modeling deals with how to adapt them to the resolved physical scales of motions in the model solution (Bachman et al., 2017). There is also a growing concern in ocean models design regarding the complex interplay between numerical discretization errors and SGS closures, especially in the finest resolution applications (Campin et al., 2011; Lemarie et al., 2012). Another emerging approach to the representation of unresolved processes in ocean circulation models is to couple ocean circulation models with third-party code components, which are specifically designed to describe the evolution of a particular class of processes. Physical processes that are not explicitly accounted for in ocean models have been extensively studied (e.g., wave modelling) and estimates of the subgrid states associated with these unresolved processes must take into account non-local effects in space and time (e.g., propagation and dissipation of internal waves, cf. IDEMIX [Olbers and Eden, 2013]). It may also make sense to modularize, as much as possible, a complex closure so that it can be used in different ocean models and thus become an independent model library or component (e.g., https://github.com/CVMix). How to fully integrate multiple code components, each describing different aspect of the dynamics, in operational systems with data assimilation is likely to become a key research question in the future. A particularly important concern is associated with the representation of the fine-scale feedbacks between the atmospheric and oceanic boundary layers.

Towards Data-driven, Probabilistic Ocean Circulation Models

Stochastic parameterizations in ocean circulation models

Classical approaches in the design of SGS closures for ocean circulation models assumes that the bulk effect of SGS fluctuations can be deterministically predicted from the resolved scales of motions. A number of recent works have started to reconsider this assumption. These more recent studies allow the closure to be only weakly constrained by the resolved scales. Carefully designed *stochastic* parameterizations appear to be a promising approach to the representation of a large class of unresolved SGS processes in ocean circulation models (Berloff, 2005; Palmer and Williams, 2010; Williams, 2012; Grooms and Majda, 2013; Porta Mana and Zanna, 2014; Cooper and Zanna, 2015; Brankart et al., 2015; Andrejczuk et al., 2016). A certain amount of randomness is introduced with a stochastic procedure to account for the possible variability of fluxes at macroscales. Stochastic parameterizations seem particularly well-suited to the representation of the cross-scale exchanges of energy and momentum for SGS-balanced turbulence (in particular, energy backscatter). Notable improvement of Gulf stream dynamics has also been obtained through the stochastic representation of upscaling due to the nonlinear nature of the equation of the state of seawater (Brankart, 2013). More generally, the concept of introducing random perturbations in ocean models is also supported by another rationale coming from the theory of non-linear systems. The topology of the phase space of non-linear systems can indeed be rather complicated, showing multiple possible local energy minima. Adding perturbations could allow a system to escape from local potential wells and, therefore, explore more reliably the range of possible states.

The paradigm underpinning this idea of stochastic parameterization is actually not restricted to the design of SGS closures and could also allow us to account for more general sources of uncertainty in ocean circulation models. These include errors in their initial condition (particularly at depth), errors in the forcing function (uncertain atmospheric fields, uncertain parameter in bulk formula. etcetera), errors associated with subgrid closures (closures are imperfect models, with usually weakly constrained parameters), errors from the physical approximation use in for their continuous formulation (e.g., non-traditional Coriolis terms), and numerical discretization errors that tend to accumulate over time (even high order schemes have non-zero errors). Overall, uncertainty is a major property of ocean circulation models that physicists often tend to neglect, although some modeling frameworks are now tackling this problem more explicitly (e.g., see Brankart et al., 2015). At this point, it is arguable that further developing stochastic parameterizations for ocean circulation models will require a better understanding of their impact on the resolved scales of motion.

Toward probabilistic ocean circulation modeling through ensemble simulations

Ensemble modeling is a now common approach in operational forecast and climate modeling for accounting for the inherent uncertainty of geoscientific model solutions. Ensembles are, for instance, routinely used in operational systems to account for uncertainty in initial states, model formulation, or forcing. Because they sample the space of possible states given such uncertainty, ensemble approaches allow for a more objective comparison of model solutions with observations, which is key to operational systems. More recently, ensemble modeling has also become an experimental tool of choice for investigating the sources of oceanic variability in eddying regimes (Sérazin et al., 2017). It should be stressed that ensemble modelling is a reasonable strategy in terms of high performance computing, because in most cases ensemble runs are entirely independent of each other and therefore 100% scalable.

The combination of ensemble and stochastic methods allows us to deal more objectively with uncertainty in ocean models (Toth et al., 2003; Palmer, 2012; Brunton et al., 2016; Bessieres et al., 2017). The purpose of ongoing efforts is to attempt to explicitly sample the probability distribution of possible states given explicitly formulated uncertainties following a Bayesian approach. This shift towards probabilistic ocean circulation modeling is arguably a paradigm transition in our field. But it is still unclear how to deal with the daemon of dimensionality. Indeed, ensemble sizes are usually constrained to several tenth of members for practical reasons while the parameter space they are sampling is orders of magnitude larger. The emergence of probabilistic ocean circulation modeling could bring several benefits to ocean model design and usage in the future. But probabilistic ocean modeling also raises several technical and scientific challenges that ought to be addressed in the future.

Final Remarks

In this chapter, we have briefly described the fundamental principles that underpin the formulation of modern ocean circulation models, shown some of their recent achievements, and discussed what we believe are the future frontiers in ocean circulation model development. This chapter has, in particular, illustrated how a modeling strategy proposed in the 1960s for solving the primitive equations has yielded ocean circulation models that are now used for a wide range of applications and that form a building block of modern operational oceanography. We have also presented the view that future frontiers in ocean circulation modeling will depend upon (i) the computational performance of ocean models, (ii) their ability to represent scale interactions either explicitly or though parameterizations, and (iii) the representation of model uncertainty with stochastic and ensemble approaches.

This chapter illustrates how the field of ocean circulation model design has reached its maturity and now involves strong collaboration between different fields of expertise. Ocean circulation model design is a very active field of research with entire scientific teams dedicated to developing or improving ocean circulation models. The maturity of the field is arguably a consequence of the high level of collaboration and merging of efforts among different groups involved in different applications and aspects of ocean circulation models. A key driver for this collaborative approach to ocean model development is this notion of *seamless* geoscientific modeling, which suggests that the same numerical code can actually be used for a range of different applications covering a range of different resolutions and dynamical regimes (Hurrell et al., 2009). Overall, this approach has improved the robustness of ocean circulation models and the sustainability of the ocean model development process over recent decades.

Although not been discussed much in this chapter, we would like to stress the importance of ocean observing networks for improving ocean circulation models. Sustained ocean observations from satellites and in situ networks are critical for routinely assessing the skills and limitations of circulation ocean models over different timescales ranging from days to decades. There is also much to be learned from targeted field observations aimed at documenting specific oceanic processes in order to improve their representation in models. For instance, recent experiments documenting fine-scale ocean processes in the ocean surface boundary layer provided a wealth of information that can be used for improving surface processes in ocean circulation models (Shcherbina et al., 2015; Buckingham et al., 2016).

An aspect that we believe has also been critical in the continuous improvement of ocean circulation models over recent decades is the shared vision in the ocean modeling community that open source is the only sustainable approach to geoscientific model development. Ocean model developers have generally been early adopters of modern practices in software development (e.g., version control, unit testing, continuous integration). All the major ocean circulation codes are also distributed under open source licenses and therefore exposed to the scrutiny of other research groups. Arguably, workflows in ocean modeling could be improved and made more reproducible, but it is fair to recognize that this community has long been concerned with these issues. Building more open and transparent data processing chains for pre- and post-processing is probably an important next step to making ocean circulation modeling more robust and reproducible in the future (Stodden et al., 2016).

In conclusion, we would like to raise what is likely to become a key issue in the future development of ocean circulation models. With a broadening of scope and an increase in the number of users, ocean circulation models have also grown in complexity. But the actual size of the community of ocean model developers is still rather small. Furthermore, most of the research questions that have been raised in this chapter would require strong and sometimes new collaborations among ocean modelers, field oceanographers, process-oriented oceanographers, applied mathematicians, and computer scientists. Sustained and proactive initiatives from major funding agencies (one good example of such an initiative being the US CLIVAR Climate Process Team scheme) will certainly be vital to the success of these interdisciplinary collaborations and to bringing more early career scientists in to contribute to the science of ocean circulation models.

Acknowledgments

EPC and AJW are supported by the Office of Naval Research and the Naval Research Laboratory. The authors also thank R. Bourdallé-Badie for providing Fig. 12.4.

References

Andrejczuk, M., F.C. Cooper, S. Juricke, T.N. Palmer, A. Weisheimer, and L. Zanna, 2016: Oceanic stochastic parameterizations in a seasonal forecast system. *Monthly Weather Review*, **144 (5)**, 1867–1875.

Bachman, S.D., B. Fox-Kemper, and B. Pearson, 2017: A scale-aware subgrid model for quasi- geostrophic turbulence. *Journal of Geophysical Research: Oceans*, **122 (2)**, 1529–1554, doi: 10.1002/2016JC012265.

Bell M.J., A. Schiller, P.-Y. Le Traon, N.R. Smith, E. Dombrowsky, and K. Wilmer-Becker, 2015: An introduction to GODAE OceanView. *Journal of Operational Oceanography*, **8**:sup1, s2-s11, doi: 10.1080/1755876X.2015.1022041

Berloff, P. S., 2005: Random-forcing model of the mesoscale oceanic eddies. *Journal of Fluid Mechanics*, **529**, 71–95, doi:10.1017/S0022112005003393.

Bessieres, L., and coauthors, 2017: Development of a probabilistic ocean modelling system based on NEMO 3.5: Application at eddying resolution. *Geoscientific Model Development*, **10 (3)**, 1091–1106, doi:10.5194/gmd-10-1091-2017.

Blayo, E., and L. Debreu, 1999: Adaptive Mesh Refinement for Finite-Difference Ocean Models: First Experiments. *Journal of Physical Oceanography*, **29 (6)**, 1239–1250, doi: 10.1175/1520-0485(1999).

Brankart, J.-M., 2013: Impact of uncertainties in the horizontal density gradient upon low resolution global ocean modelling. *Ocean Modelling*, **66**, 64–76, doi:10.1016/j.ocemod.2013.02.004.

Brankart, J.-M., G. Candille, F. Garnier, C. Calone, A. Melet, P.-A. Bouttier, P. Brasseur, and J. Verron, 2015: A generic approach to explicit simulation of uncertainty in the NEMO ocean model. *Geoscientific Model Development*, **8 (5)**, 1285–1297, doi:10.5194/gmd-8-1285-2015.

Brunton, S.L., J.L. Proctor, and J.N. Kutz, 2016: Discovering governing equations from data by sparse identification of nonlinear dynamical systems. *Proceedings of the National Academy of Sciences*, **113 (15)**, 3932–3937, doi:10.1073/pnas.1517384113.

Bryan, K., 1969: A numerical method for the study of the circulation of the world ocean. *Journal of Computational Physics*, **4 (3)**, 347–376, doi:10.1016/0021-9991.

Buckingham, C.E., A. C. Naveira Garabato, A. F. Thompson, L. Brannigan, A. Lazar, D. P. Marshall, A. J. George Nurser, G. Damerell, K. J. Heywood, and S. E. Belcher, 2016: Seasonality of submesoscale flows in the ocean surface boundary layer. *Geophysical Research Letters*, **43**, 2118-2126, doi:10.1002/2016GL068009.

Campin, J.-M., C. Hill, H. Jones, and J. Marshall, 2011: Super-parameterization in ocean model- ing: Application to deep convection. *Ocean Modelling*, **36 (1-2)**, 90–101, doi:10.1016/j.ocemod. 2010.10.003.

Charney, J., A. Fjortoft, and J. von Neuman, 1950: Numerical integration of the barotropic vorticity equation. *Tellus*, **2**, 237-254, doi:10.1111/j.2153-3490.1950.tb00336.x.

Chassignet, E.P., and J. Verron (Eds.), 1998. *Ocean Modeling and Parameterization*. Kluwer Academic Publishers, 451 pp.

Chassignet, E.P., and J. Verron (Eds.), 2006: *Ocean Weather Forecasting: An Integrated View of Oceanography*. Springer, 577 pp.

Chassignet, E.P., H.E. Hurlburt, E.J. Metzger, O.M. Smedstad, J. Cummings, G.R. Halliwell, R. Bleck, R. Baraille, A.J. Wallcraft, C. Lozano, H.L. Tolman, A. Srinivasan, S. Hankin, P. Cornillon, R. Weisberg, A. Barth, R. He, F. Werner, and J. Wilkin, 2009: U.S. GODAE: Global Ocean Prediction with the HYbrid Coordinate Ocean Model (HYCOM). *Oceanography*, **22 (2)**, 64-75.

Chassignet, E.P. and X. Xu, 2017: Impact of horizontal resolution (1/12° to 1/50°) on Gulf Stream separation, penetration, and variability. *Journal of Physical Oceanography*, **47**, 1999-2021, doi:10.1175/JPO-D-17-0031.1.

Cooper, F. C., and L. Zanna, 2015: Optimisation of an idealised ocean model, stochastic parameterisation of sub-grid eddies. *Ocean Modelling*, **88**, 38–53, doi:10.1016/j.ocemod.2014.12.014.

Danilov, S., 2013: Ocean modeling on unstructured meshes. *Ocean Modelling*, **69**, 195– 210, doi:10.1016/j.ocemod.2013.05.005.

Danilov, S., D. Sidorenko, Q. Wang, and T. Jung, 2017: The Finite volumE Sea-ice Ocean Model (FESOM2). *Geoscientific Model Development*, **10 (2)**, 765–789, doi:10.5194/gmd-10-765-2017.

Dombrowsky, E., L. Bertino, G.B. Brassington, E.P. Chassignet, F. Davidson H.E. Hurlburt, M. Kamachi, T. Lee, M.J. Martin, S. Mei, and M. Tonani, 2009: GODAE systems in operation. *Oceanography*, **22**, 80–95.

Ducousso, N., J. Le Sommer, J.-M. Molines, and M. Bell, 2018: Impact of the Symmetric Instability of the Computational Kind on oceanic hindcasts at mesoscale and submesoscale permitting resolutions. *Ocean Modelling*, submitted.

Gent, P.R, 2011: The Gent–McWilliams parameterization: 20/20 hindsight. *Ocean Modelling*, **39**, 2-9. doi:10.1016/j.ocemod.2010.08.002.

Giles, M.B., and I. Reguly, 2014: Trends in high-performance computing for engineering calculations. *Philosophical Transactions of the Royal Society A: Mathematical, Physical and Engineering Sciences*, **372 (2022)**, 20130 319–20130 319, doi:10.1098/rsta.2013.0319.

Griffies, S.M., C. Boening, F.O. Bryan, E.P. Chassignet, R. Gerdes, H. Hasumi, A. Hirst, A.-M. Treguier, and D. Webb, 2000: Developments in Ocean Climate Models, *Ocean Modelling*, **2**, 123-192.

Griffies, S.M., and R.W. Halberg, 2000: Biharmonic friction with a smagorinsky-like viscosity for use in large-scale eddy-permitting ocean models. *Mon. Weather Rev.*, **128**, 2935–2946.

Griffies, S., 2004: *Fundamentals of Ocean Climate Models*. Princeton University Press.

Griffies, S.M., and A.M. Treguier, 2013: Ocean Circulation Models and Modeling. *International Geophysics*, Vol. 103, Elsevier, 521–551.

Grooms, I., and A.J. Majda, 2013: Efficient stochastic superparameterization for geophysical turbulence. *Proceedings of the National Academy of Sciences*, **110 (12)**, 4464–4469, doi: 10.1073/pnas.1302548110.

Hallberg, R., 2013: Using a resolution function to regulate parameterizations of oceanic mesoscale eddy effects. *Ocean Modelling*, **72**, 92–103, doi:10.1016/j.ocemod.2013.08.007.

Hurrell, J., G. A. Meehl, D. Bader, T. L. Delworth, B. Kirtman, and B. Wielicki, 2009: A Unified Modeling Approach to Climate System Prediction. *Bulletin of the American Meteorological Society*, **90 (12)**, 1819–1832, doi:10.1175/2009BAMS2752.1.

Jablonowski, C., M. Herzog, J. E. Penner, R. C. Oehmke, Q. F. Stout, B. Van Leer, and K. G. Powell, 2006: Block-structured adaptive grids on the sphere: Advection experiments. *Monthly weather review*, **134 (12)**, 3691–3713.

Leith, C.E., 1968: Diffusion approximation for two-dimensional turbulence. *Physics Fluids*, **10**, 1409–1416.

Lemarie, F., L. Debreu, A. Shchepetkin, and J. McWilliams, 2012: On the stability and accuracy of the harmonic and biharmonic isoneutral mixing operators in ocean models. *Ocean Modelling*, **52-53**, 9–35, doi:10.1016/j.ocemod.2012.04.007.

McDougall, T. J. and P. M. Barker, 2011: Getting started with TEOS-10 and the Gibbs Seawater (GSW) Oceanographic Toolbox, 28pp., SCOR/IAPSO WG127, ISBN 978-0-646-55621-5.

McWilliams, J.C., 1996: Modeling the oceanic general circulation. *Annual Review of Fluid Mechanics*, **28 (1)**, 215–248.

McWilliams, J.C., 2008: The nature and consequences of oceanic eddies. *Geophysical Monograph Series*, **177**, 5–15, doi:10.1029/177GM03.

Olbers, D. and C. Eden, 2013: A global model for the diapycnal diffusivity induced by internal gravity waves. *Journal of Physical Oceanography*, **43**, 1759–1779, doi:10.1175/JPO-D-12-0207.1.

Palmer, T., and P.D. Williams, Eds., 2010: *Stochastic physics and climate modelling*. Cambridge University Press, Cambridge, UK; New York.

Palmer, T. N., 2012: Towards the probabilistic Earth-system simulator: a vision for the future of cli- mate and weather prediction. *Quarterly Journal of the Royal Meteorological Society*, **138 (665)**, 841–861, doi:10.1002/qj.1923.

Piggott, M.D., G.J. Gorman, C.C. Pain, P.A. Allison, A.S. Candy, B.T. Martin, and M.R. Wells, 2008: A new computational framework for multi-scale ocean modelling based on adapting unstructured meshes. *International Journal for Numerical Methods in Fluids*, **56 (8)**, 1003–1015, doi:10.1002/fld.1663.

Porta Mana, P., and L. Zanna, 2014: Toward a stochastic parameterization of ocean mesoscale eddies. *Ocean Modelling* **79**, 1–20.

Richardson, L. F.,1922: *Weather Prediction by Numerical Process*. Cambridge University Press. (Reprinted by Dover Publications, New York, 1965, with a new introduction by S. Chapman, xvi+236; 2d Edn. by Cambridge University Press, 2007, with a new introduction by P. Lynch.).

Ringler, T.D., M. Petersen, R. Higdon, D. Jacobsen, P. Jones, and M. Maltrud, 2013: A multiresolution approach to global ocean modeling. *Ocean Modelling*, **69**, 211–232, doi:10.1016/j.ocemod.2013.04.010.

Schiller, A. and G.B. Brassington (Eds), 2011: *Operational oceanography in the 21st century*. Springer, ISBN 9789400703315.

Schreiber, M., S.P. Peixoto, T. Haut, and B. Wingate, 2017: Beyond spatial scalability limitations with a massively parallel method for linear oscillatory problems. *International Journal of High Performance Computing Applications*. doi:10.1177/1094342016687625.

Sérazin, G., A. Jaymond, S. Leroux, T. Penduff, L. Bessières, W. Llovel, B. Barnier, J.-M. Molines, and L. Terray: 2017. A global probabilistic study of the ocean heat content low-frequency variability: Atmospheric forcing versus oceanic chaos, *Geophysical Research Letters*, **44**, 5580–5589, doi:10.1002/2017GL073026.

Shcherbina, A.Y., M.A. Sundermeyer, E. Kunze, E.A. D'Asaro, G. Badin, D. Birch, A.-M. E.G. Brunner-Suzuki, J. Callies, B.T. Kuebel Cervantes, M. Claret, B. Concannon, J. Early, R. Ferrari, L. Goodman, R.R. Harcourt, J.M. Klymak, C.M. Lee, M.-P. Lelong, M.D. Levine, R.-C. Lien, A. Mahadevan, J.C.

McWilliams, M.J. Molemaker, S. Mukherjee, J.D. Nash, T. Özgökmen, S.D. Pierce, S. Ramachandran, R.M. Samelson, T.B. Sanford, R.K. Shearman, E.D. Skyllingstad, K. Schafer Smith, A. Tandon, J.R. Taylor, E.A. Terray, L.N. Thomas and J.R. Ledwell, 2015: The LatMix summer campaign: Submesoscale stirring in the upper ocean. *Bulletin American Meteorological Society*, 1257-1279.

Shriver, J.F., B.K. Arbic, J.G. Richman, R.D. Ray, E.J. Metzger, A.J. Wallcraft, and P. G. Timko, 2012: An evaluation of the barotropic and internal tides in a high-resolution global ocean circulation model. *Journal of Geophysical Research*, **117**, C10024, doi:10.1029/2012JC008170

Skamarock, W.C., 2004: Evaluating mesoscale NWP models using kinetic energy spectra. Monthly Weather Review, **132**, 3019–3032, doi:10.1175/MWR2830.1.

Smyth, W.D., J.N. Moum, and D.R. Caldwell, 2001: The efficiency of mixing in turbulent patches: Inferences from direct simulations and microstructure observations, *Journal of Physical Oceanography*, **31**, 1969–1992, doi:10.1175/1520-0485(2001)031<1969:TEOMIT>2.0.CO;2.

Soufflet, Y., P. Marchesiello, F. Lemarie, J. Jouanno, X. Capet, L. Debreu, and R. Benshila, 2016: On effective resolution in ocean models. *Ocean Modelling*, **98**, 36–50.

Stodden V, M. McNut, D.H. Bailey, et al, 2016: Enhancing reproducibility for computational methods. *Science*, 354:1240-1. doi:10.1126/science.aah6168.pmid:27940837

Toth, Z., O. Talagrand, G. Candille, and Y. Zhu, 2003: Probability and ensemble forecasts, in *Forecast Verification: a Practitioner's Guide in Atmospheric Science*, edited by: Jolliffe, I. and Stephenson, D. B., Wiley, UK, 137–163, 2003.

Uchida, T., R. Abernathey and K.S. Smith, 2017: Seasonality in ocean mesoscale turbulence in a high resolution global climate model. *Ocean Modelling*, **118**, 41-58.

Williams, P.D., 2012: Climatic impacts of stochastic fluctuations in air–sea fluxes. *Geophysical Research. Letters*, **39**, L10705, doi:10.1029/2012GL051813.

Yeung, P.K., Zhai, X.M., and Sreenivasan, K.R., 2015: Extreme events in computational turbulence. *Proceedings of the National Academy of Sciences of the United States of America*, **112**(41), 12633–12638, doi:10.1073/pnas.1517368112.

CHAPTER 13

A Primer on Global Internal Tide and Internal Gravity Wave Continuum Modeling in HYCOM and MITgcm

Brian K. Arbic[1,2], Matthew H. Alford[3], Joseph K. Ansong[1,4], Maarten C. Buijsman[5], Robert B. Ciotti[6], J. Thomas Farrar[7], Robert W. Hallberg[8], Christopher E. Henze[6], Christopher N. Hill[9], Conrad A. Luecke[1,3], Dimitris Menemenlis[10], E. Joseph Metzger[11], Malte Müller[12], Arin D. Nelson[1], Bron C. Nelson[6], Hans E. Ngodock[11], Rui M. Ponte[13], James G. Richman[14], Anna C. Savage[1,3], Robert B. Scott[15], Jay F. Shriver[11], Harper L. Simmons[16], Innocent Souopgui[5], Patrick G. Timko[1,+], Alan J. Wallcraft[14], Luis Zamudio[14], and Zhongxiang Zhao[17]

[1]*University of Michigan, Ann Arbor, Michigan, USA;* [2]*Currently on sabbatical at Institut des Géosciences de L'Environnement (IGE), Grenoble, France, and Laboratoire des Etudes en Géophysique et Océanographie Spatiale (LEGOS), Toulouse, France;* [3]*University of California San Diego, La Jolla, California, USA;* [4]*University of Ghana, Accra, Ghana;* [5]*University of Southern Mississippi, Stennis Space Center, Mississippi, USA;* [6]*NASA Ames Research Center, Mountain View, California, USA;* [7]*Woods Hole Oceanographic Institution, Woods Hole, Massachusetts, USA;* [8]*Geophysical Fluid Dynamics Laboratory/NOAA, Princeton, New Jersey, USA;* [9]*Massachusetts Institute of Technology, Cambridge, Massachusetts, USA;* [10]*Jet Propulsion Laboratory, California Institute of Technology, Pasadena, California, USA;* [11]*Naval Research Laboratory, Stennis Space Center, Mississippi, USA;* [12]*Norwegian Meteorological Institute, Oslo, Norway;* [13]*Atmospheric and Environmental Research, Lexington, Massachusetts, USA;* [14]*Florida State University, Tallahassee, Florida, USA;* [15]*Université de Bretagne Occidentale, Brest, France;* [16]*University of Alaska-Fairbanks, Fairbanks, Alaska, USA;* [17]*University of Washington, Seattle, Washington, USA; +Now at: Welsh Local Centre, Royal Meteorological Society, UK*

In recent years, high-resolution ("eddying") global three-dimensional ocean general circulation models have begun to include astronomical tidal forcing alongside atmospheric forcing. Such models can carry an internal tide field with a realistic amount of nonstationarity, and an internal gravity wave continuum spectrum that compares more closely with observations as model resolution increases. Global internal tide and gravity wave models are important for understanding the three-dimensional geography of ocean mixing, for operational oceanography, and for simulating and interpreting satellite altimeter observations. Here we describe the most important technical details behind such models, including atmospheric forcing, bathymetry, astronomical tidal forcing, self-attraction and loading, quadratic bottom boundary layer drag, parameterized topographic internal wave drag, shallow-water tidal equations, and a brief summary of the theory of linear internal gravity waves. We focus on simulations run with two models, the HYbrid Coordinate Ocean Model (HYCOM) and the Massachusetts Institute of Technology general circulation model (MITgcm). We compare the modeled internal tides and internal gravity wave continuum to satellite altimeter observations, moored observational records, and the predictions of the Garrett-Munk (1975) internal gravity wave continuum spectrum. We briefly examine specific topics of interest, such as tidal energetics, internal tide nonstationarity, and the role of nonlinearities in generating the modeled internal gravity wave continuum. We also describe our first attempts at using a Kalman filter to improve the accuracy of tides embedded within a general circulation model. We discuss the challenges and opportunities of modeling stationary internal tides, non-stationary internal tides, and the internal gravity wave continuum spectrum for satellite altimetry and other applications.

Arbic, B.K., et al., 2018: A primer on global internal tide and internal gravity wave continuum modeling in HYCOM and MITgcm. In "*New Frontiers in Operational Oceanography*", E. Chassignet, A. Pascual, J. Tintoré, and J. Verron, Eds., GODAE OceanView, 307-392, doi:10.17125/gov2018.ch13.

Introduction

This book chapter is about global modeling of oceanic internal tides and the oceanic internal gravity wave continuum. The chapter focuses on hydrodynamical modeling, rather than empirical modeling, of such motions. Due to the operational oceanography theme of the book in which this chapter resides, we focus on high-spatial-resolution numerical models run over relatively short time scales—i.e., simulations that could form the dynamical backbone of operational models—rather than on lower-resolution models run over decades or centuries for climate forecasting purposes. In this introductory section, after defining internal gravity waves and internal tides, we discuss the motivation for, requirements for, and history of global modeling of internal tides and the internal gravity wave continuum. A subsequent section focuses on the technical details underlying such models, such as atmospheric forcing, bathymetry, astronomical tidal forcing, self-attraction and loading, quadratic bottom boundary layer drag, parameterized topographic internal wave drag, shallow-water tidal equations, and a brief synopsis of internal wave theory. We then show results from the first solutions, obtained with the HYbrid Coordinate Ocean Model (HYCOM; Bleck, 2002; Chassignet et al., 2009), of a high-resolution three-dimensional global ocean model with simultaneous atmospheric and tidal forcing fields. Following that, we compare results from newer simulations of HYCOM, and similar simulations of the Massachusetts Institute of Technology general circulation model (MITgcm; Marshall et al., 1997), with satellite altimeter observations, observations from moored records, and linear wave theory. We then investigate some science questions, including the impact of wave drag on the energetics and accuracy of tidal models, internal tide nonstationarity, and the role of nonlinearities in the development of the modeled internal gravity wave continuum. We describe applications of the HYCOM and MITgcm simulations to swath satellite altimetry, and ongoing attempts to improve the accuracy of tides embedded in the HYCOM simulations. After summarizing our completed work, ongoing work, and challenges for the future, we provide a dedication for the first author's contributions to the work presented here. We then describe author contributions, acknowledge our financial support and intellectual debts, and provide an appendix on the linear internal gravity wave dispersion relation.

Definition of internal gravity waves

Internal gravity waves (often abbreviated hereafter as IGWs) are "internal" to the ocean. That is, their largest vertical perturbation signals, such as the perturbations of temperature and density isolines, occur well below the ocean surface. This behavior contrasts with, for instance, the behavior of surface wind waves, for which the largest vertical perturbation signals are at the surface of the ocean[1]. Let us idealize the ocean as having n layers, where the density ρ_n of each layer increases with depth (Fig. 13.1). In this layered system, IGWs are waves that exist on the interfaces between

---

[1] We shall see later that the perturbation surface elevation signals of internal tides and IGWs, though small, are not zero. This is important for satellite altimetry.

layers, and for which the restoring force is gravity[2]. Because IGWs cause undulations of layer interfaces, they bring deeper (usually colder) water upwards, and shallower (usually warmer) water downwards. Therefore, IGWs bring about high-frequency fluctuations in time series of temperature taken at a fixed depth.

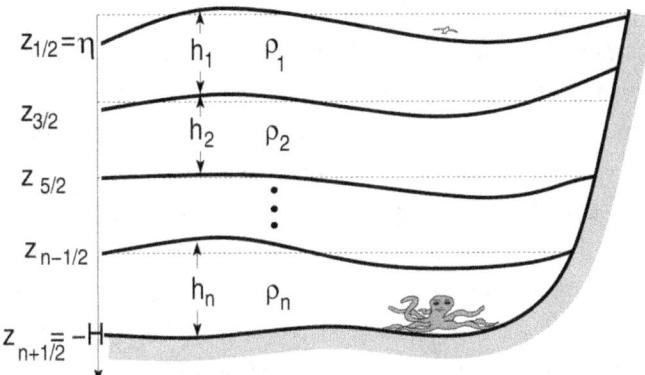

Figure 13.1. Schematic for a layered model of the ocean. There are n layers of density $\rho_1, \rho_2,...\rho_n$, with layer thicknesses $h_1, h_2,...h_n$. The sum of all layer thicknesses equals $H + \eta$, where H is the resting depth and η is the perturbation sea surface height. Internal gravity waves represent undulations of the layer interfaces from their resting positions, for which gravity is the restoring force. Reproduced from Figure 1 of Simmons et al. (2004b).

As demonstrated in the appendix, application of classical linear internal gravity wave theory shows that the frequency ω of a plane IGW obeys the inequality $|f| \leq \omega \leq N$, where f is the Coriolis frequency, also sometimes called the inertial frequency, and N is the buoyancy frequency[3]. Because IGWs are super-inertial (i.e., have frequencies exceeding the Coriolis frequency), they are not in geostrophic balance. IGWs with frequencies near the Coriolis frequency are called near-inertial waves. Near-inertial waves are primarily forced by fluctuations in surface winds (Pollard and Millard, 1970; D'Asaro, 1984; Silverthorne and Toole, 2009; Simmons and Alford, 2012). Because near-inertial waves are driven by fluctuating winds, they tend to be stronger in winter than in summer (D'Asaro, 1985). IGWs with tidal frequencies are known as internal tides, and they are generated by large-scale barotropic tidal flow over topographic features (Bell, 1975; Baines, 1982; among others). Topography will also be referred to as bathymetry throughout this chapter. The barotropic tidal flow over topographic features, including rough small-scale topography and topographic slopes along shelves, generates vertical motion, which, in a stratified fluid, implies that the interfaces between layers will oscillate vertically—i.e. that internal tides will be generated. Internal tides, also known as baroclinic tides, have smaller sea surface height signals and smaller horizontal scales than barotropic tides, large undulations at depth, and large baroclinic (vertically sheared) velocity fields. The vertical and horizontal structure of internal tides is intimately linked to the characteristics of the topographic features at which they are generated (St. Laurent and

---

[2] Note that there are other oceanic internal waves, for instance internal Rossby waves, for which the restoring force is not gravity.

[3] In the main text of this book chapter, we use ω to denote frequency, consistent with the notation used in our papers. In the appendix, we use σ for frequency, consistent with usage in the notes from Carl Wunsch that our appendix is based upon.

Garrett, 2002). High-vertical-mode internal tides, which have smaller horizontal scales, are generated by small-horizontal-scale topography characteristic of mid-ocean ridges, while low-vertical-mode internal tides, which have larger horizontal scales, are generated by larger horizontal-scale topographic features such as the Hawaiian Islands. High-mode IGWs are more likely to dissipate near their generation sites, while low-mode IGWs are more likely to propagate long distances, up to thousands of kilometers from their generation regions. The appendix outlines the Sturm-Liouville problem underlying the computation of vertical normal modes. At frequencies greater than tidal frequencies—hereafter often referred to as supertidal frequencies—there is a continuum of IGW energy, separate from the near-inertial and tidal peaks. This IGW continuum is described by the classical Garrett-Munk spectrum (Garrett and Munk, 1975). The Garrett-Munk spectrum is thought to arise from nonlinear interactions transferring energy out of the near-inertial and tidal frequencies and into the broadband continuum (e.g., Müller et al., 1986; Polzin, 2004).

Fig. 13.2 depicts internal wave generation mechanisms, including wind-generation of near-inertial flows, low-vertical-mode internal tide generation over larger-horizontal-scale topographic features, high-vertical-mode internal tide generation over smaller-scale rough topography, and generation of internal lee waves by low-frequency mesoscale eddy and current flows over topographic features. Fig. 13.2 also illustrates several important internal wave dissipation mechanisms, including scattering off of topographic features such as small-scale topography and continental shelf slopes, local dissipation of high-modes near their generation sites, and interactions amongst the waves themselves (wave-wave interactions) and between internal waves and mesoscale eddies and fronts. Understanding how internal tides and IGWs lose their energy is an important ongoing topic in physical oceanography. Global internal wave modeling both benefits from this conversation (in the form of improved parameterizations of internal wave damping) and contributes to it (by quantifying the long-range propagation of low-mode internal tides and hence providing some constraints on where they dissipate). As will be shown later, comparisons of global internal tide models with observations provide some clues about the locations and mechanisms of internal tide damping.

Several of the concepts described above manifest themselves in Fig. 13.3, which displays example frequency spectra[4] of temperature variance. Spectra from in-situ mooring data and three different global simulations of the MITgcm, with horizontal grid spacings of $1/12°$, $1/24°$, and $1/48°$, are given. Diurnal and semidiurnal tidal peaks near one and two cycles per day, respectively, can be seen in the mooring and model spectra. The internal gravity wave continuum (Garrett-Munk spectrum) seen at supertidal frequencies is better resolved as the model horizontal resolution increases, a theme we will return to later in this chapter. At frequencies below tidal frequencies (subtidal frequencies), the spectra also display a broadband continuum, with no visible peaks. The low-frequency broadband spectrum is dominated by mesoscale eddies and currents. Near-inertial peaks are not prominent in temperature variance spectra, but will be seen later in spectra of kinetic energy.

---

[4] Technically, what is shown in this figure, and in many other figures in this chapter, is spectral density. Like many authors, we will use the shorthand terms "spectra" or "spectrum" in place of "spectral density".

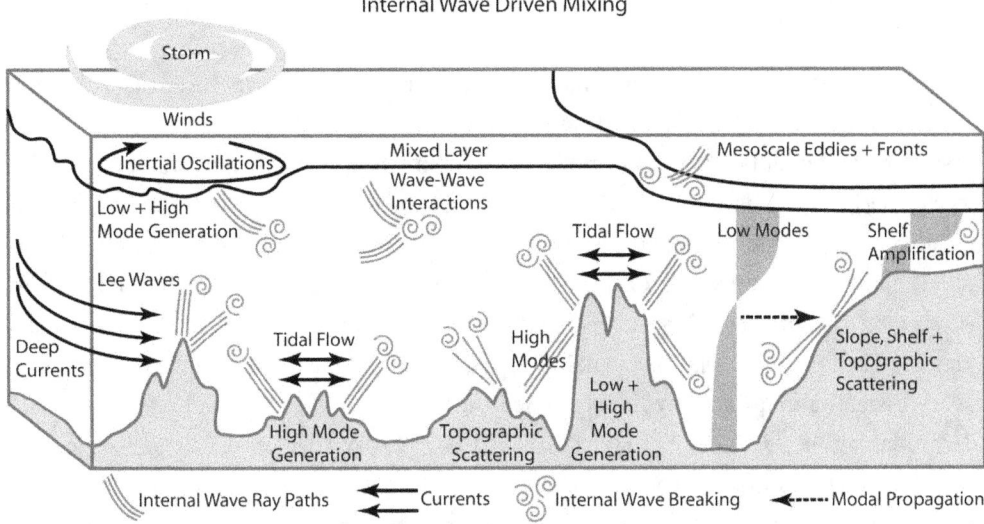

Figure 13.2. Schematic of open-ocean internal wave generation and dissipation processes that were considered as part of a Climate Process Team led by Jennifer MacKinnon of the University of California San Diego. Batropropic tidal flow interacts with topographic features to generate high-mode internal waves (e.g., at mid-ocean ridges) and low-mode internal waves (e.g., at tall steep ridges such as the Hawaiian Ridge). Deep low-frequency flows, including mesoscale eddies and currents, over topographic features can generate lee waves (e.g., in the Southern Ocean). Storms cause near-inertial oscillations in the mixed layer, which can generate both low- and high-mode internal waves (e.g., beneath storm tracks). In the open ocean these internal waves can scatter off of topographic features and potentially interact with mesoscale fronts and eddies, until they ultimately dissipate through wave-wave interactions that generate a nonlinear cascade to small-scale turbulence. Internal waves that reach the shelf and slope can scatter, dissipate via bottom boundary layer drag, or amplify as they propagate towards shallower water. Reproduced from Figure 1 of MacKinnon et al. (2017), ©American Meteorological Society, used with permission.

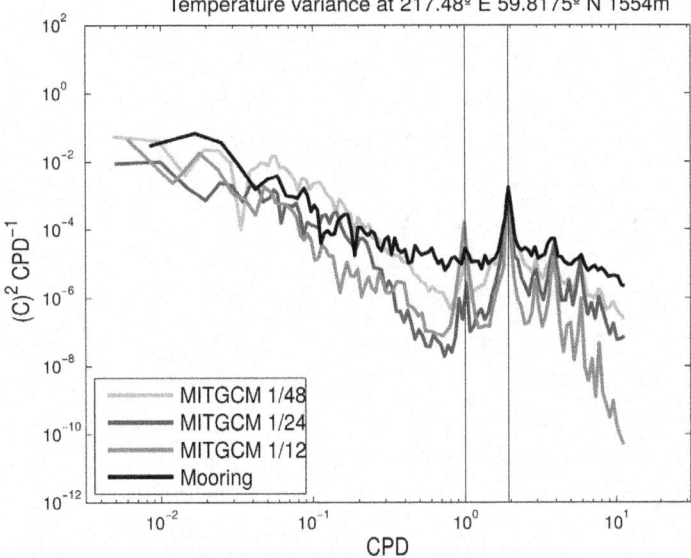

Figure 13.3. Frequency spectra of temperature variance at depth and geographical location given in plot title. Spectra are computed from observations in a historical archive (Scott et al. 2010), and from three different global simulations, with horizontal grid spacings of 1/12°, 1/24°, and 1/48°, of the MITgcm. "CPD" denotes "cycles per day." Extra vertical lines are drawn at the frequencies of M_2, the largest semidiurnal tidal line, and K_1, the largest diurnal tide.

Motivation

Why should the oceanographic community run global models of internal tides and the internal gravity wave continuum? The authors posit four important reasons.

First, global internal tide and IGW models provide important information on ocean mixing. In the ocean interior—e.g., in the open ocean, away from upper and lower boundaries—most of the diapycnal mixing[5] is due to breaking IGWs. A variety of theoretical and observational studies, including reports on observations of directly breaking IGWs (Moum et al., 2003, Klymak et al., 2008), provide support for this statement. However, the strongest evidence for the primacy of breaking IGWs in ocean interior diapycnal mixing comes from the remarkable agreement between observed dissipation rates and predictions based on IGW energy levels that assume a cascade to smaller (breaking) scales (Gregg, 1989; Polzin et al., 1995; Whalen et al., 2015; Kunze, 2017a). The breaking of IGWs is shown schematically in Fig. 13.2. It has been argued that mixing exerts an important control on the oceanic meridional overturning circulation (e.g., Munk and Wunsch, 1998; St. Laurent and Simmons, 2006; Kunze, 2017b). Indeed, the large-scale flow in ocean general circulation models is sensitive to parameterizations of tidal mixing (Simmons et al., 2004a). Global internal tide and IGW models cannot, of course, resolve IGW breaking. Nonetheless, by modeling the transport of energy by internal waves over long distances (Simmons et al., 2004b; Simmons and Alford, 2012), global internal tide and IGW models contribute importantly to the ongoing discussion of the space-time geography of internal wave breaking, a topic of great current interest in the oceanographic community, and one that is not fully settled; the fate of internal tide and IGW energy is not entirely known at present.

Second, internal tides and gravity waves are important components of any operational ocean model. IGW signals can be seen in oceanic measurements of temperature, density, and velocity. Because IGWs bring about vertical motions of water masses, they affect the speed of sound, which is a sensitive function of temperature and salinity. Modeling IGWs is therefore important for ocean acoustics. This is one reason for the enduring interest of the United States Navy in IGWs.

Third, internal tide and gravity wave modeling is of particular importance for the upcoming Surface Water Ocean Topography (SWOT) swath altimetry mission (Fu et al., 2012). Satellite altimetry has revolutionized operational oceanography, and other branches of physical oceanography, by providing accurate measurements of sea surface height (SSH) on a global scale, over decades, on repeated tracks (Fu and Cazenave, 2001). Current-generation nadir altimeters, such as the TOPEX/JASON series, provide SSH measurements along one-dimensional tracks. Examination of non-tidal phenomena—for instance El Niño, western boundary currents, and mesoscale eddies, among others—with altimeter data can only take place after the barotropic tides, which have large (of order 1000 km) horizontal scales in the open ocean, have been accurately removed. For this reason, it was recognized early on that highly accurate models of the large-scale barotropic tides were needed for successful altimeter missions.

---

[5] Diapycnal mixing is mixing across iospycnals, or density surfaces.

Barotropic tide models have improved greatly over the years. Cartwright and Ray (1990) proved that accurate barotropic tide models could be derived empirically from altimeter data—in their case, from GEOSAT data. The accuracy of the elevations due to the principal lunar semidiurnal tide M_2 in the Cartwright and Ray (1990) model was slightly better than the accuracy of M_2 elevations in the Schwiderski (1980) hydrodynamical one-layer (barotropic) shallow-water[6] tide model, which assimilated tide-gauge data and which served as the "go-to" tide model for more than a decade.[7] Shortly after the launch of TOPEX/POSEIDON, a number of accurate tide models were developed, including hydrodynamical barotropic shallow-water models that assimilated either altimeter or tide gauge data [e.g., the TPXO (Topex Poseidon Cross-Overs) model of Egbert et al. (1994) and Egbert and Erofeeva (2002), and the FES (Finite Element Solution) models of Le Provost et al. (1994) and Lyard et al. (2006)], and empirical models that estimated tides directly from altimeter data (e.g., Schrama and Ray, 1994; Ray, 1999). The accuracy of both empirical barotropic tide models based upon altimetry, and hydrodynamical barotropic tide models that assimilate altimeter data, was proven via comparison with independent tide-gauge data as well as other observations (Shum et al., 1997). Because many of these tide gauges were located at small islands, they were assumed to represent conditions in the open ocean, where altimetry-based tide models are more accurate. Barotropic tide models continue to improve. Stammer et al. (2014) provides a comprehensive recent overview of several state-of-the-art empirical and hydrodynamical barotropic tide models, including the new Taguchi et al. (2014) hydrodynamical model, and updated versions of the TPXO and FES models. Stammer et al. (2014) describes ongoing challenges for, and improvements in, barotropic tide models, including the need for more accurate tides in coastal and high-latitude regions, and the usage of new datasets, including tide data extracted from networks of bottom pressure sensors deployed after the 2004 Indian Ocean tsunami (Ray, 2013), to test tide models.

SWOT, which is due to launch in 2021, will measure SSH along high-resolution two-dimensional swaths, with pixel sizes of about 1 km^2. The instrument noise in SWOT is expected to be relatively low at smaller scales (wavelengths less than about 50 km). SWOT is therefore expected to allow examination of smaller-scale features than can be seen in present-generation nadir altimeter data. Because of SWOT's emphasis on smaller-scale features, internal tides and IGWs will be an important signal in SWOT, and will need to be accurately removed before smaller-scale features in low-frequency motions, such as mesoscale and submesoscale eddies, can be effectively examined. Stationary internal tides–that is, internal tides that can be described by amplitude and phase maps, as in equation (12)–have a high degree of predictability in theory, but in practice can be difficult to extract from altimeter data because of temporal sampling problems and "noise" from mesoscale eddy motions (e.g., Ray and Byrne, 2010; Shriver et al., 2012). Thus the stationary internal tides represent a significant challenge for SWOT. Some of the internal tide signal is non-stationary–that is, if one examines internal waves in the tidal band, a non-negligible fraction of the variance still remains after the stationary part has been predicted and removed. Non-stationary internal tides are

---

[6] The shallow-water equations will be introduced in a subsequent section.

[7] Parke and Hendershott (1980) also demonstrated that a hydrodynamical tide model could successfully assimilate tide-gauge data, but their model assimilated less data than Schwiderski's model did.

caused by internal tide interactions with the chaotic mesoscale eddy field, which has similar length and time scales as internal tides, and by temporal changes in stratification (e.g., Buijsman et al., 2017; among others). The non-stationary internal tides, and the IGW continuum, are inherently less predictable than stationary internal tides, and will therefore represent an even greater challenge for the SWOT mission. Global models of internal tides and IGWs are being used to predict the internal tide and IGW continuum SSH signals, and the impacts of internal tide and gravity wave motions on the SSH wavenumber spectrum, a critical characteristic of SWOT measurements. In some of our papers, and in other papers in the literature, the terms coherent/incoherent are used interchangeably with stationary/non-stationary. In this chapter, we will use the terms stationary/non-stationary, which many would argue are more precise, except when duplicating figures from past papers which used the coherent/incoherent terminology.

The fourth motivation for global internal tide and IGW modeling is that it is an interesting topic in and of itself. It is a new frontier in ocean modeling. Global IGW models have only existed since 2004, and evidence that global models can partially resolve the supertidal IGW continuum was only recently presented in 2015. Just as models with vigorous mesoscale eddy fields (e.g., Maltrud and McClean, 2005) have provided a new tool for understanding mesoscale eddies, global internal tide and IGW models provide a new tool for understanding higher-frequency motions. Global eddying models have developed to the point that they are used in operational forecast systems (e.g., Chassignet et al., 2009)[8]. Global internal tide and IGW models have not yet evolved to the point where they are used in forecasting systems, but that time is coming soon. It is worth noting that, just as the community has described models having resolutions of about 1/4° or better as "eddy-permitting" rather than "eddying" or "eddy-resolving", a similar distinction could be made for global internal tide and gravity wave modeling. Because the horizontal length scales of low-mode internal tides and mesoscale eddies are comparable, at least in midlatitudes, one might call, for instance, a 1/4° model with internal tides an "internal tide-permitting" model. It has been our experience that, as with mesoscale eddies, a horizontal resolution of at least 1/10° is needed for a fully vigorous low-mode internal tide field. A horizontal resolution of about 1/24° or finer is necessary for simulating a vigorous IGW continuum.

Requirements for global modeling of the internal gravity wave continuum

The classical paradigm for the generation of the IGW continuum spectrum is that winds produce near-inertial waves, barotropic tidal flow over topographic features creates internal tides, and nonlinear interactions fill out the IGW continuum. (The role of internal lee waves, generated by low-frequency flows over topography, in the IGW continuum is an interesting question). Based upon the classical paradigm, we might expect that the requirements for a resolved IGW spectrum in a model would be simultaneous atmospheric forcing and astronomical tidal forcing, together with horizontal and vertical resolution sufficiently high to allow nonlinear interactions to take place. The atmospheric forcing should contain both the low-frequency components that set up the background

---

[8] Following Hecht and Hasumi (2008), we will refer to models with horizontal resolutions of 1/10° or finer as "eddying," meaning that a vigorous mesoscale eddy field develops in such models.

large-scale circulation and stratification (Pedlosky, 1996), and the high-frequency components that drive near-inertial waves. Once a stratified flow is set up, the addition of astronomical tidal forcing will produce internal tides via barotropic tidal flow over topography. Because internal tide characteristics are linked to topographic scales, the resolution of small-scale bathymetric features in available global bathymetric datasets is an important topic, that we return to later.

History of global modeling of internal tides and the internal gravity wave continuum

This section is about the history of global internal tide and IGW continuum models. Some history of observations that have motivated such models is also presented. Prior to the work outlined in this chapter, global modeling of atmospherically-forced three-dimensional oceanic motions and global modeling of tides constituted separate fields of research. As computer power has increased, the frontiers of global operational modeling of the atmospherically-forced oceanic general circulation have moved toward ever higher horizontal and vertical resolution, meaning that the mesoscale eddy field has become better resolved in global and basin-scale models. Basin-scale models started to become eddying at the turn of the century (e.g., Paiva et al., 1999; Smith et al., 2000). Basin-scale models have now been run at resolutions of $1/50°$-$1/60°$ (e.g., Chassignet and Xu, 2017; Ducousso et al., 2017). Regional models are now run with grid spacings of order hundreds of meters, and have begun to resolve submesoscale eddies (e.g., Capet et al., 2008). The horizontal resolution of global tide models has also increased as a result of increasing computer power. One of the earliest global tide models, Hendershott (1972), employed a horizontal resolution of order $6°$ (Myrl Hendershott, personal communication, 2004). By the time of the TOPEX/POSEIDON launch, barotropic tide models were run on horizontal resolution grids of about $1°$ or less. As this book chapter was being written, global barotropic (one-layer) tide models are being run at horizontal resolutions of order $1/75°$ (Buijsman et al., paper in preparation); unstructured grids such as those used in finite-element models (Lyard et al., 2006) can go to much higher resolutions in selected areas, usually along shelves and coastlines. Along the way, larger computers were also used to run baroclinic tide models. The first baroclinic tide models run in realistic domains—e.g., not in idealized domains set up for process studies—were done on regional scales. Early regional internal tide models include Holloway (1996), who examined the Northwest Australian Shelf; Cummins and Oey (1997), who simulated the tides off of northern British Columbia; and several studies of the internal tides generated by the Hawaiian islands, for instance Kang et al. (2000) and Merrifield et al. (2001). See Carter et al. (2012) for a review of regional internal tide models, which continue to refine their resolutions with increasing computer power, and Nugroho et al. (2017), for an example of a recent regional internal tide modeling study. Nugroho et al. (2017) showed that inclusion of tides in a regional three-dimensional model of the Indonesian Seas yielded realistic levels of tidal dissipation, which in turn exerted important controls on water mass properties.

Both regional and global internal tide models have been motivated by evidence for stationary low-mode internal tides radiating over long distances in acoustic tomography data (Dushaw et al., 1995) and in altimeter data (Ray and Mitchum, 1996, 1997). Fig. 13.4, taken from Ray and Mitchum

(1996), displays M₂ tidal amplitudes estimated along a TOPEX/POSEIDON track (ragged curve), on top of M₂ amplitudes estimated from a barotropic tidal model along the same track (solid curve). As shown in the figure, the SSH signature of barotropic tides has much larger horizontal scales than the SSH signature of internal tides. Associated with the different length scales, the phase speeds also differ. The phase speed of barotropic tides in the open ocean is about 200 m s^{-1} while the open-ocean phase speed of vertical mode-1 internal tides is typically about 2 m s^{-1}. The higher-wavenumber "wiggles" in the ragged curve represent the M₂ internal tide signal. This altimeter evidence of stationary internal tides propagating over long distances, along with the tomography evidence, implies that tidal energy can be transported over long distances before dissipating, with profound implications for the three-dimensional geography of oceanic mixing.

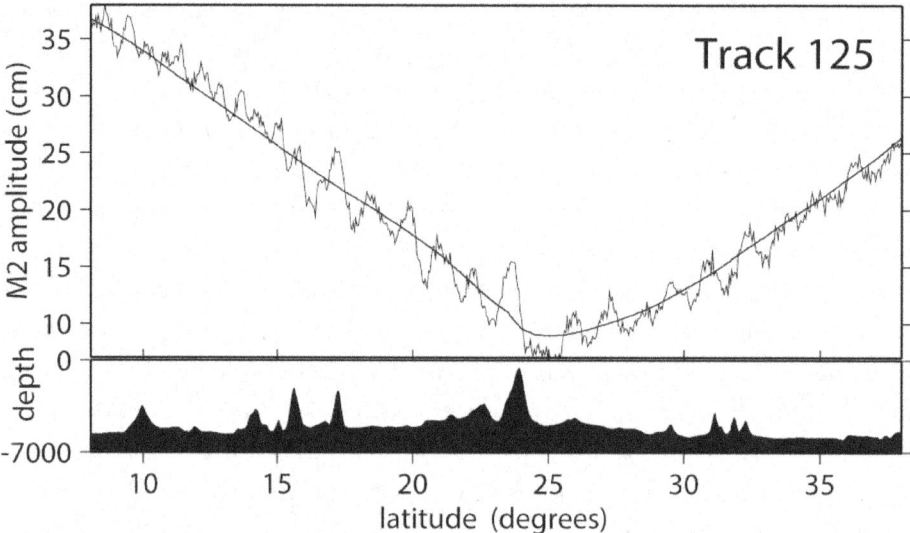

Figure 13.4. Ragged curve: M₂ tidal amplitudes along Topex/Poseidon track number 125. Smooth curve: M₂ tidal amplitudes estimated from a barotropic tide model. Bathymetry is shown in black. Reproduced from Figure 1 of Ray and Mitchum (1996), ©American Geophysical Union, Wiley Online Library, used with permission.

Further motivation for internal tide models arises from evidence that internal tide production over open-ocean topographic features accounts for about 25-30% of the energy lost in the global barotropic tidal energy budget. Egbert and Ray (2000, 2001, 2003) showed that about 1 TW of the total 3.5 TW barotropic tidal dissipation takes place in the open ocean, through generation of internal tides by barotropic flow over topography. Egbert and Ray's results implicate internal tide dissipation as an important contributor to the mixing that underpins the large-scale circulation (Munk and Wunsch, 1998; St. Laurent and Simmons, 2006; Kunze, 2017b). Fig. 13.5a displays energy lost from the barotropic tide, as inferred from an updated version of the TPXO model. The TPXO results display substantial energy dissipation over topographic features, such as the mid-Atlantic Ridge and the Hawaiian islands. The increased internal wave activity over open-ocean topographic features can also be seen in Fig. 13.5b, which displays the conversion of energy from barotropic tides to internal tides in simulations of the HYbrid Coordinate Ocean Model (HYCOM) computed by Buijsman et al. (2016). Further, enhanced dissipation of oceanic flows over topographic features has been seen in field campaigns, such as the Brazil Basin Tracer Release Experiment (BBTRE; e.g., Polzin et al., 1997; Fig. 13.6), the Diapycnal and Isopycnal Mixing Experiment in the Southern Ocean (DIMES; e.g., Gille et al., 2007; St. Laurent et al., 2012), and others, that include microstructure dissipation measurements. See MacKinnon et al. (2017), Waterhouse et al. (2014), and Whalen et al. (2012, 2015) for comprehensive discussions of observations indicating enhanced oceanic energy dissipation over rough topographic features. Along the shelves, considerable barotropic tidal dissipation takes place in regions with large coastal tides, such as the Hudson Strait, the Northwest European Shelf, the Bay of Fundy, and others. Collectively, the shelves dissipate about 70-75% of the tidal energy. For much of the 20th century, the shelves were thought to account for almost all of the tidal dissipation. The Egbert and Ray (2000, 2001, 2003) sequence of papers, and related work done by others, have brought about a paradigm shift in this important topic.

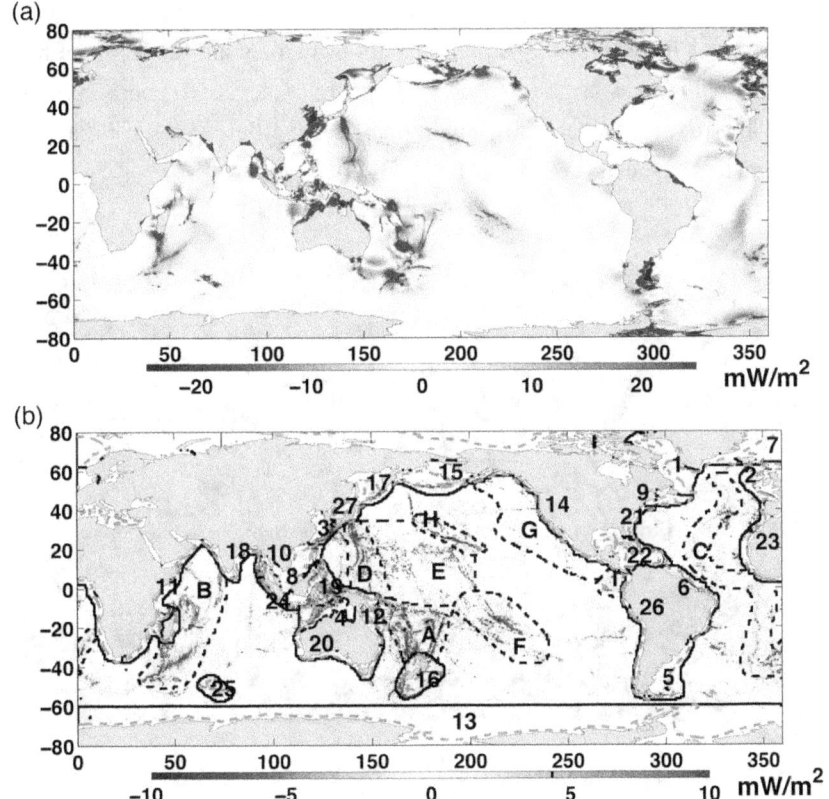

Figure 13.5. (a) Rate of energy lost from the M_2 barotropic tide (mW/m^2), estimated from TPXO8 global assimilation solution. (b) Baroclinic conversion rate (mW/m^2) for all semidiurnal constituents, computed from the HYCOM 18.5 solution (nominal resolution 1/12.5°; Buijsman et al., 2016). The conversion rate shown in (b) has been smoothed, and the plotting range has been reduced relative to that in panel [a], in order to enhance visibility of global scale patterns. Shallow and deep ocean areas used for localized dissipation and conversion calculations are defined by solid lines for shallow areas (numbered 1–27) and dashed lines for deep ocean areas (labeled A–I). The green dashed line is the more formally defined boundary between deep and shallow areas, as discussed in Ray and Egbert (2001). Reproduced from Figure 13.8 of Ray and Egbert (2017), which can be consulted for further details. Republished with permission of Taylor and Francis Group, conveyed through the Copyright Clearance Center.

The first basin-scale simulation of internal tides was performed by Niwa and Hibiya (2001). The first global internal tide simulations were performed in the companion papers Arbic et al. (2004) and Simmons et al. (2004b). These early global internal tide simulations were idealized in several respects. Atmospheric forcing was not included—the only forcing present was the astronomical tidal potential. Because the atmospheric forcing, which sets up the oceanic stratification, was lacking, the stratification profile was set to be horizontally uniform. The stratification profile was taken from observations in the subtropics, and was therefore not representative of equatorial or polar conditions. Figs. 13.7-13.9 display the interfacial height perturbations at three different times during the spin-up phase of the main two-layer simulation analyzed in Simmons et al. (2004b). The simplicity of this simulation allows for a nice illustration of the spin-up, but the exact values of the interfacial heights should not be taken too literally due to the unrealistic horizontally constant stratification. Internal wave generation at mid-ocean topographic features such as the Hawaiian and French Polynesian Islands is readily apparent. Tidal forcing is not required for global modeling of near-inertial waves, which are forced by rapid fluctuations in the wind fields. Examination of near-inertial waves in global models has been done by Furiuchi et al. (2008) and Simmons and Alford (2012). As with low-mode internal tides, low-mode near-inertial waves can carry energy over thousands of kilometers as they propagate (Alford, 2003).

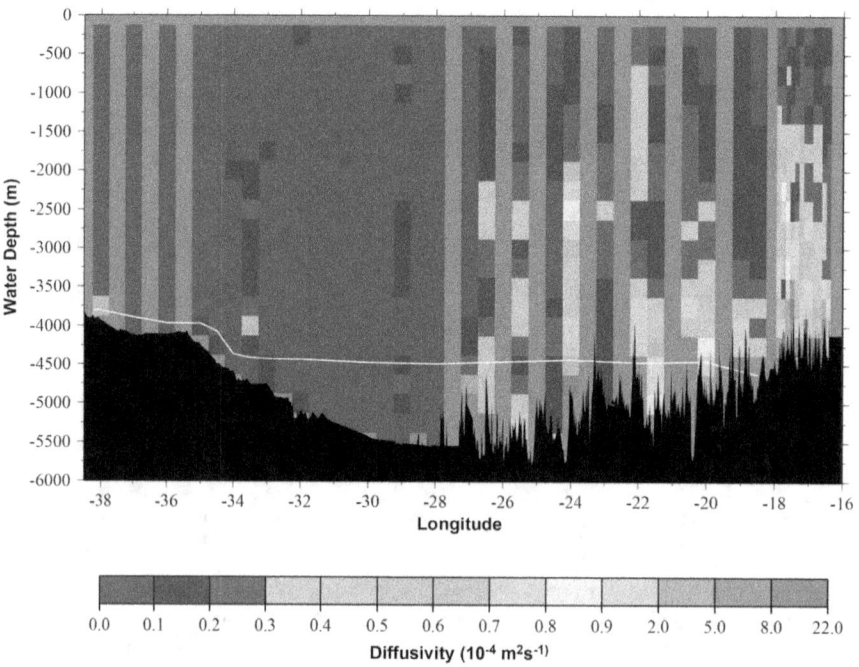

Figure 13.6. Depth-longitude section of cross-isopycnal (diapycnal) diffusivity in the Brazil Basin inferred from velocity micostructure observations. Note the nonuniform color scale. Microstructure data from two quasi-zonal transects have been combined without regard to latitude. Underway bathymetric data to 32°W is from an eastward track, and the balance comes from a westward track. The white line marks the depth of the 0.8°C interface. Reproduced from Figure 2 of Polzin et al. (1997). Reprinted with permission from AAAS. Except as provided by law, this material may not be further reproduced, distributed, transmitted, modified, adapted, performed, displayed, published, or sold in whole or in part, without prior written permission from the publisher.

In coastal modeling, and in global barotropic modeling, it has long been common to simultaneously resolve tides and atmospherically-driven motions. The earliest attempts to simultaneously include atmospheric and tidal forcing in three-dimensional global models (Thomas et al., 2001; Schiller and Fiedler, 2007; Müller et al., 2010) were done in simulations with horizontal resolutions of order 1°, in which neither mesoscale eddies nor internal tides are resolved; nonetheless, some interesting impacts of barotropic tides on the oceanic general circulation were found. The first global, three-dimensional, high-resolution simulation done with simultaneous atmospheric and tidal forcing was done using HYCOM and is described in Arbic et al. (2010). For a shorter and less technical overview of these HYCOM simulations, see Arbic et al. (2012a). Because of the high resolution, these HYCOM simulations resolved mesoscale eddies, western boundary currents, and internal tides, as well as the larger-scale barotropic tides. We will report on this HYCOM simulation as well as updated HYCOM simulations later. For now, we note that a small but growing number of "wind plus tides" simulations have been done in global three-dimensional high-resolution (1/10° or finer) models, for instance the German STORMTIDE model (e.g., Müller et al., 2012), the GFDL Generalized Ocean Layered Model (GOLD; e.g., Waterhouse et al., 2014), and the Massachusetts Institute of Technology general circulation model (MITgcm; e.g., Rocha et al., 2016a, 2016b). The lead author is collaborating with two modeling groups in France that are preparing to perform basin-scale and global high-resolution ("eddying") simulations with simultaneous atmospheric and tidal forcing. It is clear that concurrent atmospheric and tidal forcing will become increasingly common in global eddying ocean model runs of the future.

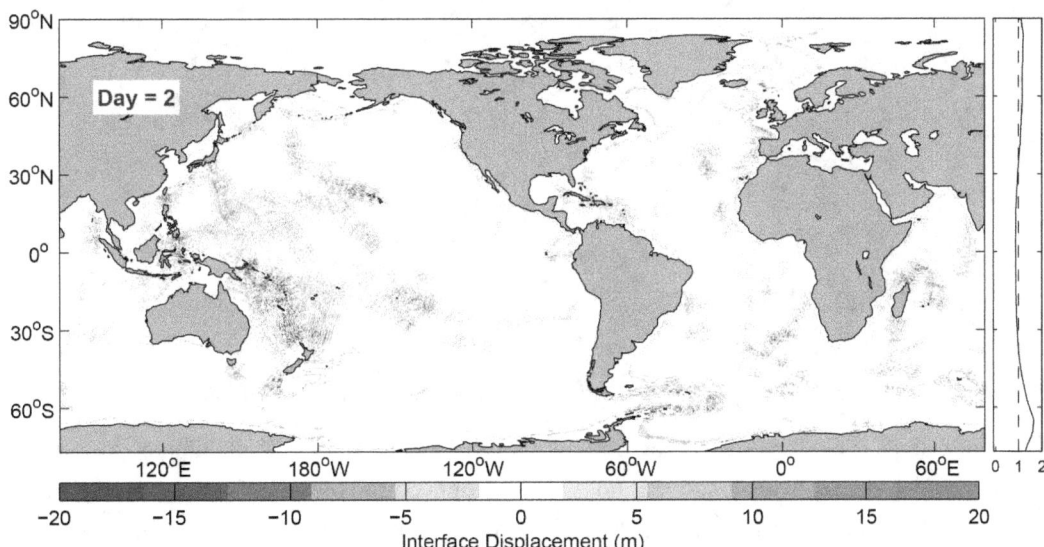

Figure 13.7. Interface displacement normalized according to Eq. (15) of Simmons et al. (2004b), on day 2 of spin-up of a two-layer M_2 simulation with simplified, horizontally uniform stratification. The resting depth of the interface is at 1100 m. The zonal mean of the normalization factor is shown on the right side of the plot. Reproduced from Figure 6 of Simmons et al. (2004b).

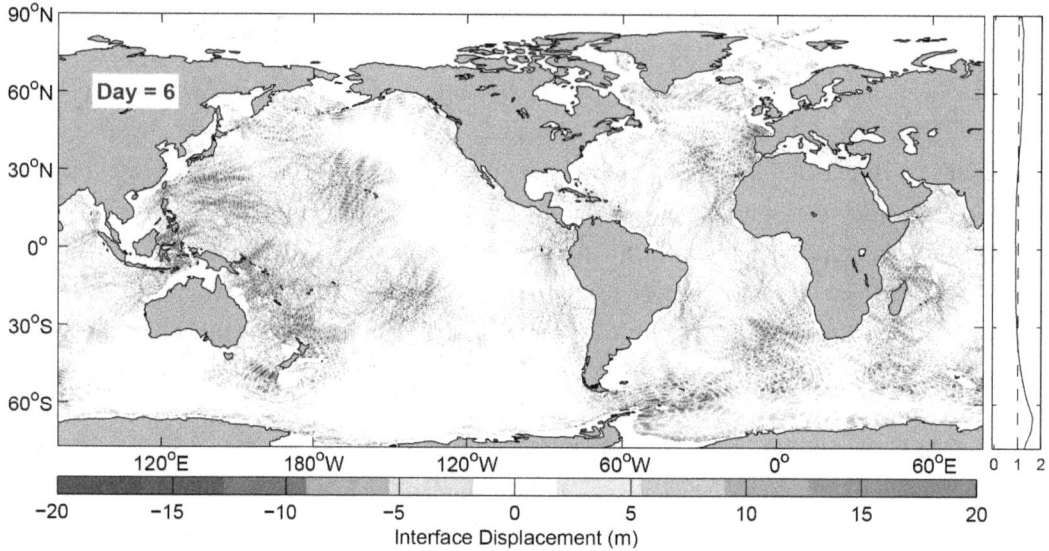

Figure 13.8. As in Fig. 13.7 but for day 6 of spin-up. Reproduced from Figure 7 of Simmons et al. (2004b).

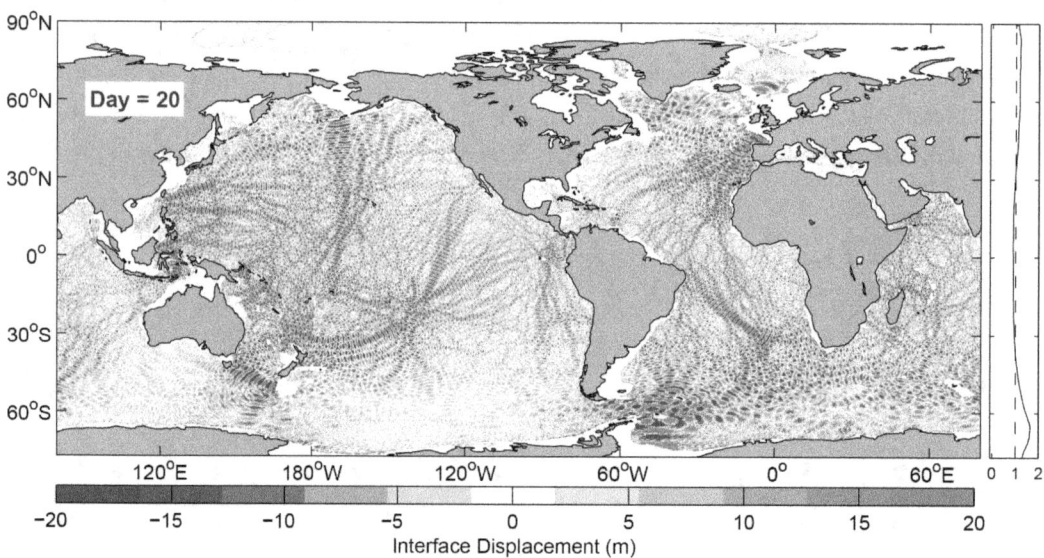

Figure 13.9. As in Fig. 13.7 but for day 20 of spin-up. Reproduced from Figure 8 of Simmons et al. (2004b).

It took some time to recognize the existence of a quasi-realistic IGW continuum spectrum in high-resolution models with concurrent tidal and atmospheric forcing. The first demonstration of a partial IGW continuum in global models, using HYCOM, was done in Müller et al. (2015). The realism of the HYCOM IGW kinetic energy spectra analyzed in Müller et al. (2015) increased as the horizontal grid spacing decreased from $1/12.5°$ to $1/25°$. Rocha et al. (2016a, 2016b) subsequently showed that MITgcm simulations also carry a partial IGW continuum. Savage et al. (2017a, 2017b) examined the IGW SSH fields and spectra in $1/12.5°$ and $1/25°$ simulations of HYCOM and in $1/12°$, $1/24°$, and $1/48°$ simulations of the MITgcm. The HYCOM and MITgcm simulations are being used extensively in preparation for the SWOT mission.

This book chapter focuses on global internal tide and wave modeling done with HYCOM and the MITgcm. HYCOM and MITgcm are full ocean general circulation models, with multiple layers, that include parameterizations for mixed-layer physics and other processes. They have been used as ocean general circulation models in many applications. Here we focus on high-resolution, three-dimensional simulations of these two models with simultaneous atmospheric and tidal forcing. The HYCOM "wind plus tides" simulations have existed longer than the MITgcm "wind plus tides" simulations, are associated with a longer publication list, have been compared with observations more frequently, and include a parameterized topographic internal wave drag, which is not included in the MITgcm simulations, or in the "wind plus tides" simulations performed thus far by other modeling groups. The parameterized wave drag (e.g., Garner, 2005) is meant to account for the breaking of internal waves throughout the water column, and for low-level turbulence, that are left unresolved in today's global models, even models run at the highest resolutions possible on current-generation supercomputers.

Because of the long history of the HYCOM simulations, they have gone through several iterations. The first HYCOM "wind plus tides" simulations used in Arbic et al. (2010) had some numerical problems (see the large spurious signals in bottom kinetic energy in the Gulf of Mexico, for example, in Figure 2b of that paper). These numerical problems were greatly reduced in the runs used in subsequent papers. Up until Ansong et al. (2015), the HYCOM simulations discussed in this chapter had 32 hybrid layers in the vertical direction; since Ansong et al. (2015) they have had 41. Ngodock et al. (2016) introduced a Kalman filter method to improve the accuracy of the modeled barotropic tides. All of the HYCOM simulations since then have employed the Kalman filter, which will be discussed in a subsequent section. The topographic drag scheme, and the scheme for self-attraction and loading, have also changed over time. Finally, we note that none of the HYCOM "wind plus tides" simulations used in our publications to date employ data assimilation acting on the mesoscale eddy field, although such runs are already being performed and will become more prevalent in the near future. The MITgcm "wind plus tides" simulations have not been around as long as the HYCOM "wind plus tides" simulations, and have not been vetted as extensively. One very significant advantage of the MITgcm simulations, however, is that they have been run with higher horizontal resolution, up to 1/48°, and with higher vertical resolution (90 z-levels, as opposed to 41 hybrid layers in HYCOM)[9]. The HYCOM simulations are run with horizontal grid spacings of 1/12.5° and 1/25° and are often referred to as HYCOM12 and HYCOM25 throughout this chapter. The MITgcm simulations are run with horizontal grid spacings of 1/12°, 1/24°, and 1/48° and are often referred to as MITgcm12, MITgcm24, and MITgcm48, respectively, throughout this chapter. The fact that a qualitatively similar IGW spectrum emerges in two different models, which

---

[9] Different modeling systems make use of different vertical coordinates. Some of the most common choices are z-level coordinates, which are employed in MITgcm, terrain-following coordinates, isopycnal (density-based) coordinates, and hybrid vertical coordinates. Hybrid coordinates, which can smoothly transition between the choices above, are used in HYCOM. Griffies et al. (2000) and Griffies (2005) can be consulted for detailed discussions of vertical coordinate choices in ocean models. Bleck (2002) describes the hybrid vertical coordinate approach used in HYCOM.

differ from each other in various respects including in their vertical coordinate systems, gives us more confidence in the results presented here.

Technical Details

This section describes some of the technical details involved in global internal gravity wave modeling, including atmospheric forcing, bathymetry, astronomical tidal forcing, self-attraction and loading, quadratic bottom boundary layer drag and parameterized topographic internal wave drag, and the shallow-water tidal equations. The section concludes with a brief synopsis of the Garrett and Munk (1975) description of the IGW continuum spectrum, and of linear dispersion relations for IGWs, both of which will be used to interpret our model results.

Atmospheric forcing

Atmospheric forcing of the ocean includes wind stress, evaporation minus precipitation (which impacts salinity at the ocean surface), and air-sea heat fluxes (which impact, and feed off of, sea surface temperature)—see Csanady (2001), Josey et al. (2013), and references therein for an overview. To obtain a realistically energetic near-inertial wave field, a model should be forced by winds that update frequently (e.g., about three hours or less). A component of atmospheric forcing that is often neglected in oceanic general circulation models is atmospheric pressure loading. There is a rich literature on oceanic motions forced by pressure loading (e.g., Ponte, 1994; Tierney et al., 2000a; Stammer et al., 2000; Carrère and Lyard, 2003). Because atmospheric pressure is broadband, it elicits a broadband response in the ocean. A special case, worth a brief mention here, is the atmospheric tide, which is thermally driven (Chapman and Lindzen, 1970). The predominant periods of the atmospheric thermal tide are 24 and 12 hours. The pressure loading of the 24-hour S_1 atmospheric tide is the predominant driver of the small S_1 tide in the ocean (Ray and Egbert, 2004). The pressure loading of the 12-hour S_2 atmospheric tide is an order 15% correction (e.g., Arbic, 2005) to the oceanic S_2 tide, which is primarily forced by the Sun's gravity and is the second largest tide in the ocean.

The atmospheric forcing for the HYCOM simulations presented here has evolved over time. The most recent HYCOM simulations include pressure loading as well as atmospheric buoyancy and wind forcing. The most recent HYCOM simulations are forced by the United States Navy atmospheric model, NAVGEM (Hogan et al., 2014), which recently replaced the NOGAPS model (Rosmond et al., 2002). The frequency of NAVGEM output has varied over time. Some of our recent HYCOM simulations have been forced by atmospheric fields updated every three hours, and some by atmospheric fields updated every hour. As noted, high-frequency atmospheric forcing is needed to represent the S_2 atmospheric tide (e.g., Ray and Ponte, 2003), to simulate the associated oceanic response (e.g., Arbic, 2005), and to simulate oceanic near-inertial waves. Our papers on the HYCOM tidal simulations, for instance Arbic et al. (2010), Shriver et al. (2012), and Savage et al. (2017b), can be consulted for more details on the atmospheric forcing fields, and their evolution over time.

The MITgcm simulations described here were forced by six-hourly atmospheric fields from the 0.14° European Center for Medium Range Weather Forecasts (ECMWF) operational atmospheric reanalysis. The ECMWF fields are converted to surface fluxes using the Large and Yeager (2004) versions of the bulk formulae. The MITgcm simulations are also forced by pressure loading in addition to atmospheric wind and buoyancy forcing.

Bathymetry

Bathymetry is a crucial consideration in global internal tide and IGW continuum models, because internal tides are generated by barotropic tidal flow over topographic features. Most global models use a version of the Smith and Sandwell (1997) bathymetry, which has been continually updated over time. The bathymetric database uses high-quality acoustic soundings where they are available. However, because of the military significance of acoustic sounding data in coastal waters, local governments often do not release them. In the open ocean, sounding data is rare; perhaps 10% of the ocean floor has been mapped acoustically (Wessel and Chandler, 2011), and most of the regions so mapped are in coastal shelf regions (Charette and Smith, 2010). Thus, in the open-ocean, Smith and Sandwell (1997) employ satellite altimetry data. The sea surface height signal is sensitive to local mass anomalies. The technique can resolve features down to about π times the ocean depth, in other words, about 10-20 km in the open-ocean. The altimeter inversion technique works poorly in shelf areas. In order to prevent numerical problems in ocean models, bathymetry must be smoothed to remove features with length scales less than the grid spacings employed in the models.

Bathymetry is, of course, a crucial control for other oceanic motions as well as for internal tides and IGWs. For example, deep-water masses often enter the ocean through bathymetric sills (Price and Baringer, 1994), and the paths of mesoscale eddies are steered by topography (LaCasce, 2000; Scott et al., 2008; Stewart et al., 2015). Bathymetry is also of primary importance for accurate barotropic tide modeling (Florent Lyard, personal communications over many years).

Astronomical tidal forcing

Tidal research has a long history, and it includes some of the greatest names in physics and geophysics, such as Isaac Newton, Pierre-Simon Laplace, George Biddell Airy, Lord Kelvin, and George Darwin. Cartwright (1999) and Pugh (1987) can be consulted for some of this fascinating history. Here we outline some of the fundamentals underlying astronomical tidal forcing, for those readers who might not have encountered them before.

The astronomical tidal potential is due to the differential gravitational forcing of a distant object across a body of finite size. Thus, for instance, the gravitational pull of the Moon is greater at points on the Earth facing the Moon than it is at the center of the Earth, and greater at the Earth's center than at points facing away from the Moon (Fig. 13.10). Due to the mutual gravitational attraction between the Earth and Moon, all points on Earth trace circles with a radius equal to the distance between the Earth's center and the center of mass of the Earth-Moon system. Therefore, all points on Earth experience a centripetal force equal to

$$F_{centripetal} = \frac{GM_{moon}}{r^2}, \quad (1)$$

where G is Newton's gravitational constant, M_{moon} is the mass of the Moon, and r is the distance between the center of the Moon and the center of the Earth. On the side of the Earth closest to the Moon, the gravitational pull of the Moon is given by

$$F_{gravitational} = \frac{GM_{moon}}{(r-a)^2}, \quad (2)$$

where a is the radius of the Earth. The tidal force is given by the difference of these forces,

$$F_{tidal} = F_{gravitational} - F_{centripetal} = GM_{moon}\left[\frac{1}{(r-a)^2} - \frac{1}{r^2}\right] \approx \frac{GM_{moon}}{r^2}\left[\frac{1}{1-\frac{2a}{r}} - 1\right] \approx \frac{2aGM_{moon}}{r^3}, \quad (3)$$

where we have used the fact that $a \ll r$. On the side of the Earth farthest from the Moon, we have

$$F_{tidal} = F_{gravitational} - F_{centripetal} = GM_{moon}\left[\frac{1}{(r+a)^2} - \frac{1}{r^2}\right] \approx \frac{GM_{moon}}{r^2}\left[\frac{1}{1+\frac{2a}{r}} - 1\right] \approx -\frac{2aGM_{moon}}{r^3}. \quad (4)$$

Thus to first order the tidal forces on the side of the Earth facing the Moon and the side farthest from the Moon are equal but opposite, yielding tidal bulges pointing outwards from the Earth in both cases. Note that the tidal force is proportional to Earth's radius a; hence, as anticipated, the finite size of Earth is a critical factor in tidal forcing. Because the gravitational force is proportional to $\frac{1}{r^2}$, the tidal force, being a difference and hence involving a derivative, is proportional to $\frac{1}{r^3}$.

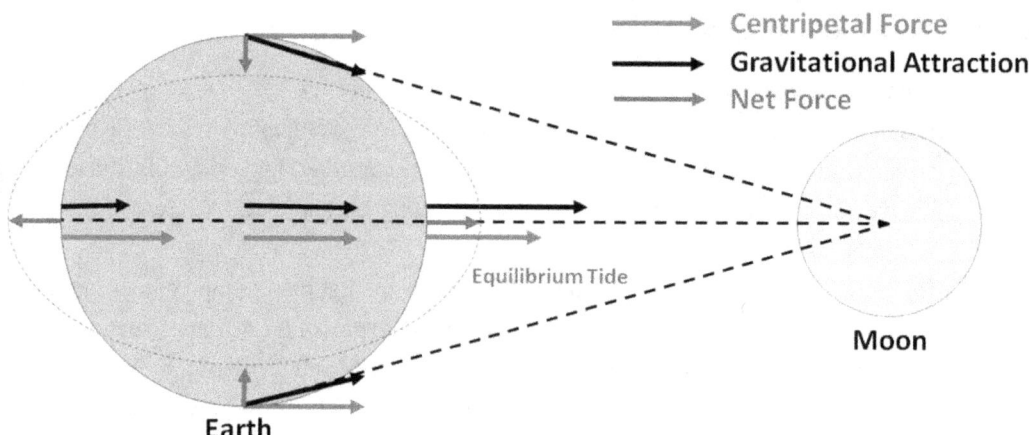

Figure 13.10. Tide-producing forces.

The same reasoning applies to the Sun's gravity. The Sun is 27,000,000 times more massive than the Moon, but it is 390 times farther away. This equates to the solar tidal force being 46% of the lunar tidal force; an important but smaller contribution.

Because the tidal forcing consists of two bulges, under which the Earth rotates, the period of the principal lunar semidiurnal tide is equal to half of a lunar day—the latter representing the time it takes for an observer on Earth to see the same point on the Moon two consecutive times as the Earth orbits. A lunar day is 24.84 hours, and the period of the principal lunar semidiurnal tide (M_2) is therefore 12.42 hours. Because a solar day—a day measured against the Sun—is 24 hours, the period of the principal solar semidiurnal tide (S_2) is 12 hours. The close but unequal frequencies of M_2 and S_2 yield a classic "beat" pattern in the tides. The tidal range, or difference between high and low tide, is especially large during spring tide, when the Earth, Moon, and Sun are aligned, and is especially small during neap tide, when the line between the Earth and Moon is at right angles to the line between the Earth and Sun. The tidal forcing is often referred to as the "equilibrium tide", a concept that is credited to Newton (Newton, 1687; Cartwright, 1999). Another commonly used term for tidal forcing is "astronomical tidal potential".

Tidal forcing is not just at semidiurnal frequencies. Because the Moon's orbit around the Earth is not in the Earth's equatorial plane, an observer on Earth sees two high tidal forcing peaks that are unequal during the course of a day (Fig. 13.11). As shown in the figure, a northern hemisphere observer sees a higher high tidal forcing at time B and a lower high tidal forcing at time A. At the same longitude, a southern hemisphere observer sees a lower high tide when the northern hemisphere observer sees a higher high tide and vice versa. This diurnal inequality yields an effective tidal forcing at periods close to once per day; in other words, diurnal tides. There are also diurnal tides that are caused by the Sun's gravity field.

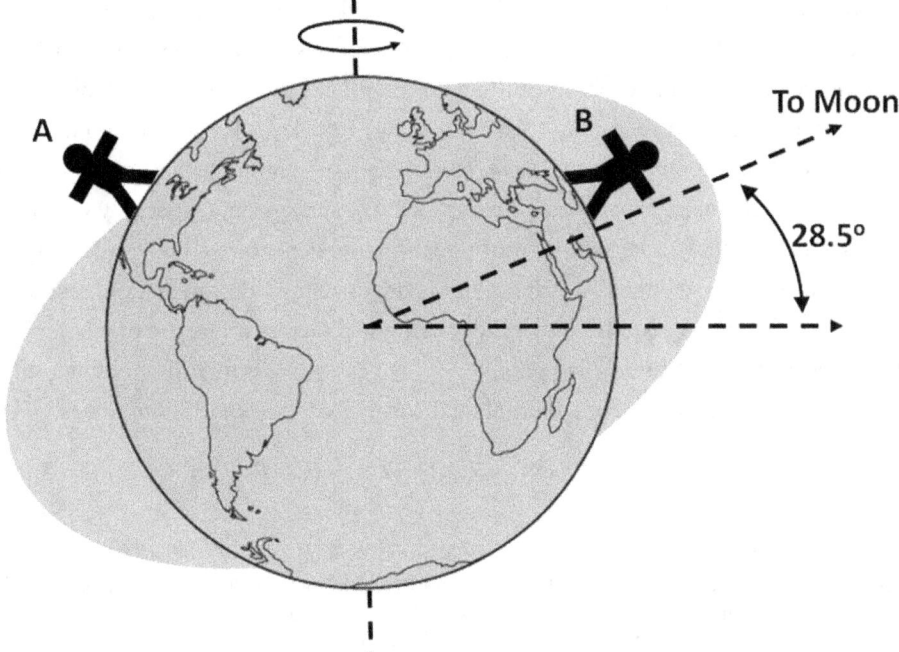

Figure 13.11. Declination and diurnal tides.

While the discussion of the astronomical tidal forcing given above is intuitive, a more complete treatment of the tidal forcing is done in terms of spherical harmonic expansions of the gravitational potential, where the potential is computed from a highly accurate trigonometric series expansion for the orbital positions of the Sun and Moon (Doodson, 1921; Cartwright and Tayler, 1971). This fuller treatment yields a multitude of terms having varying frequencies and amplitudes. The frequencies represent multiples, sums, and differences of the fundamental frequencies in the Earth-Moon-Sun system. The frequencies cluster around semidiurnal, diurnal, and longer periods, and the three types of tides correspond to the degree-two spherical harmonics.[10] These clusters are often referred to as "tidal species", a concept credited to Laplace (Cartwright, 1999). Cartwright and Tayler (1971)'s analysis netted approximately 400 terms. More recent treatments, e.g., Roosbeek (1996), have many thousands of terms. Most ocean tide studies include only some of these terms. See Williams and Boggs (2017) for a comprehensive treatment of the relationship between the dissipations of tidal constituents[11] and the secular changes in Earth's rotation, the semimajor axis of the Earth-Moon orbit, and other characteristics of the Earth-Moon system.

For a semidiurnal tidal constituent, the equilibrium tide, denoted by η_{EQ} below, is given by

$$\eta_{EQ} = Af(t_{ref})(1 + k_2 - h_2)\cos^2(\phi)\cos[\omega(t - t_{ref}) + \chi(t_{ref}) + \nu(t_{ref}) + 2\lambda], \quad (5)$$

where A and ω are constituent-dependent forcing amplitudes and frequencies, φ is latitude, λ is longitude, t is time, t_{ref} is a reference time (which could represent, for instance, the starting time of a particular model run), and $\chi(t_{ref})$ is the constituent-dependent astronomical argument referenced to t_{ref} (e.g., Schwiderski, 1980; Pugh, 1987). The Love numbers h_2 and k_2, which are frequency-dependent especially in the diurnal band (Wahr and Sasao, 1981), respectively account for the deformation of the solid earth resulting from the astronomical forcing, and the resulting alteration in the gravitational potential due to the redistributed mass within the solid Earth. The solid-earth deformations arising in response to the astronomical forcing are known as the "body tide". Solid-earth body-tide oscillations can reach several cm in amplitude, but they are not noticeable in common experience because of their long (planetary-scale) wavelengths. The nodal factors $f(t_{ref})$ and $\nu(t_{ref})$ respectively account for the slow modulation of amplitude and phase of a tidal constituent due to low-frequency changes in the Earth-Moon-Sun system (especially the 18.6 year nodal cycle—see Pugh, 1987). The inclusion of nodal factors depends on the application. Many tidal studies do not need to use them. On the other hand, accurate tidal prediction, and accurate analysis of ocean observations, requires either the use of nodal factors or the use of a large number of tidal lines over a record duration long enough ($\sim$ 19 years) to resolve the multitude of lines present in the actual tidal potential. For a diurnal tidal constituent, the equilibrium tide is given by

$$\eta_{EQ} = Af(t_{ref})(1 + k_2 - h_2)\sin(2\phi)\cos[\omega(t - t_{ref}) + \chi(t_{ref}) + \nu(t_{ref}) + \lambda], \quad (6)$$

---

[10] The astronomical tidal potential also includes degree-three and higher spherical harmonics. The higher-order harmonics are small and are neglected in many ocean tide models. See Cartwright (1975) for a demonstration that degree-three tides can be detected in observational records.

[11] As is explained in Egbert and Ray (2017), a "constituent" is a cluster of spectral lines, not a single isolated line. This point is often discussed loosely in the literature, including in some of our previous papers.

and for a long-period constituent, the equilibrium tide is given by

$$\eta_{EQ} = Af(t_{ref})(1 + k_2 - h_2)\left[\frac{1}{2} - \frac{3}{2}\sin^2(\phi)\right]\cos[\omega(t - t_{ref}) + \chi(t_{ref}) + \nu(t_{ref})]. \quad (7)$$

The diurnal tidal forcing is antisymmetric around the equator, consistent with the intuition developed from Fig. 13.11. The long-period constituents have relatively small amplitudes and will not be much discussed in this book chapter. To model many constituents, one must simply add the appropriate extra term to η_{EQ} each time a constituent is included.

Constituent	ω (10^{-4} s^{-1})	A (cm)	$1 + k_2 - h_2$	Period (solar days)
M_m	0.026392	2.2191	0.693	27.5546
M_f	0.053234	4.2041	0.693	13.6608
Q_1	0.6495854	1.9273	0.695	1.1195
O_1	0.6759774	10.0661	0.695	1.0758
P_1	0.7252295	4.6848	0.706	1.0027
K_1	0.7292117	14.1565	0.736	0.9973
N_2	1.378797	4.6397	0.693	0.5274
M_2	1.405189	24.2334	0.693	0.5175
S_2	1.454441	11.2743	0.693	0.5000
K_2	1.458423	3.0684	0.693	0.4986

Table 13.1. Constituent-dependent frequencies ω, astronomical forcing amplitudes A, and Love number combinations $1 + k_2 - h_2$ used to compute equilibrium tide η_{EQ}. The periods $2\pi/\omega$ are also given. Reproduced from Table 1 of Arbic et al. (2004).

Table 13.1 provides the frequencies, periods, amplitudes, and Love number combination $1 + k_2 - h_2$ for the four largest semidiurnal constituents, four largest diurnal constituents, and two largest long-period constituents. Note that the Love number combination is about 0.7, meaning that the body tide acts to reduce the amplitude of the equilibrium ocean tide forcing by 30%—a significant impact for an effect that sounds quite exotic when one first learns of it! Pugh (1987) and other sources can be consulted for computation of the astronomical arguments χ and nodal factors. As an alternative to modeling multiple tidal constituents using sums of the appropriate η_{EQ} terms, one could instead employ the full luni-solar tidal potential, as in Weis et al. (2008). In this approach, one would not include nodal factors, because one would instead be effectively modeling all tidal forcing frequencies. Employing the full luni-solar potential in its purest form would not allow for the frequency dependence of the body-tide Love numbers. To create such an accounting would require a harmonic analysis of the full potential, which would effectively mean writing the potential as a sum of many η_{EQ} terms, as in the more traditional approach outlined above.

Early HYCOM tides simulations (Arbic et al., 2010, 2012a; Shriver et al., 2012) included the four largest diurnal tidal constituents (K_1, O_1, P_1, and Q_1) and the four largest semidiurnal tidal constituents (M_2, S_2, N_2, and K_2). However, later simulations of HYCOM have included only the five largest tidal constituents— M_2, S_2, N_2, K_1, and O_1—due to the difficulties in separating the nearby $S_2 - K_2$ and $K_1 - P_1$ frequency pairs in relatively short model outputs. Savage et al. (2017b) can be consulted for a detailed description of the MITgcm simulations. Due to a misunderstanding amongst several co-authors, Savage et al. (2017b) incorrectly stated that the MITgcm tidal forcing included sixteen tidal constituents–eight long period tides, the four largest diurnal constituents, and the four largest semidiurnal constituents. In fact, the MITgcm simulations employed the full luni-solar tidal potential.

Self-attraction and loading

This section describes self-attraction and loading, an effect that those who are new to tidal research often find even more surprising than solid-earth body tides. In addition to its direct body-tide response to the astronomical tidal potential, the solid earth also compresses and expands due to the load of the ocean tide. Furthermore, the self-gravitation of both the ocean tide itself, and the load-deformed solid earth, alters the gravitational potential. Collectively, these effects are known as the self-attraction and loading (SAL) term (Hendershott, 1972; Ray, 1998). The η_{SAL} term is often computed in terms of a spherical harmonic expansion,

$$\eta_{SAL} = \sum_n \frac{3\rho_0}{\rho_{earth}(2n+1)}(1 + k'_n - h'_n)\eta_n, \qquad (8)$$

where $\rho_0 \approx 1035$ kg m^{-3} is the average density of seawater, $\rho_{earth} \approx 5518$ kg m^{-3} is the average density of the solid earth, n is an index of the spherical harmonics, and the η_n's are the nth spherical harmonics of the tidal elevation η. The load numbers h'_n and k'_n, introduced in Munk and MacDonald (1960), respectively account for solid-earth yielding and the resulting perturbation potential. We see from equation (8) that the SAL term η_{SAL} depends on the tidal elevations η themselves. The η_{SAL} term is more spatially complex than the η_{EQ} term because, as we will see shortly in Fig. 13.12, the ocean tide has a richer spatial structure than the degree-two spherical harmonic equilibrium tide. The ocean tide contains many spherical harmonic degrees, and the loading and self-gravitation responses are functions of these degrees via (8). The SAL term must therefore be evaluated either by decomposing the tides into spherical harmonics, or by using Green's functions, both of which are computationally expensive. Hendershott (1972) and Gordeev et al. (1977) demonstrated that the SAL effects are of first-order importance to tide modeling. It is not possible to model the global barotropic tides accurately without properly accounting for the SAL term.

Because spherical harmonics are computationally expensive, η_{SAL} is usually not computed inline in ocean models, with some notable exceptions such as Stepanov and Hughes (2004), Kuhlmann et al. (2011), and Vinogradova et al. (2015). To save computational expense, the SAL term is sometimes calculated with a "scalar approximation" (Accad and Pekeris, 1978; Ray, 1998) for which

$$\eta_{SAL} \approx \beta \eta, \qquad (9)$$

where β is a constant, usually taken to be about 0.09. As Ray (1998) and others have shown, the scalar approximation is not accurate enough for the most exacting tidal applications. A more accurate iterative method can be employed where resources allow. In the iterative method, one first estimates η_{SAL} with (9), runs the model out, and then uses the full spherical harmonic treatment given by (8) in a less expensive offline calculation. Finally, one takes advantage of the periodicity of tides to construct amplitude and phase maps, which are then used in the next iteration, and so on, until the results converge. Numerical devices are often employed to achieve convergence with a smaller number of iterations—see Arbic et al. (2004) and Egbert et al. (2004) for more discussion.

The early HYCOM "wind plus tides" simulations use a scalar approximation for the SAL term. More recent HYCOM simulations, beginning with the simulations described in Ngodock et al. (2016), have used the SAL maps from the Egbert et al. (1994) TPXO model.[12] The latter approach is generally the most accurate, but the iterative method is more self-consistent. The scalar approximation is the least accurate of the three approaches to SAL described above.

Quadratic bottom boundary layer drag and parameterized topographic internal wave drag

This section is about quadratic bottom boundary layer drag and parameterized topographic internal wave drag, the two main damping mechanisms that have been used thus far in global barotropic tide models and global internal tide and IGW continuum models. Taylor (1919), through examination of tidal dissipation in the Irish Sea, argued that bottom boundary layer drag should be modeled as quadratic in the velocity, with a drag coefficient of about 0.002. His formulation for quadratic bottom drag is commonly used today in ocean models, of both tidal and non-tidal motions. Because the energy budget equation for a model is derived through multiplication of the momentum equation by velocity, the quadratic bottom boundary layer drag yields a dissipation that is cubic in the velocity. Tidal velocities on shelves can be as large as 1 m s^{-1}, much larger than open-ocean tidal velocities, which are typically about 1-2 cm s^{-1}. Therefore, in global tidal models, dissipation by the quadratic bottom boundary layer drag term takes place primarily in coastal areas.

Motivated by Egbert and Ray's demonstration of barotropic tidal energy loss over topographic features, many recent barotropic tide models, beginning with the work of Jayne and St. Laurent

---

[12] Ngodock et al. (2016) displayed some prior solutions for which the SAL term was computed either iteratively, or using a scalar approximation. The ASEnKF solutions in Ngodock et al. (2016) employed SAL fields taken from the TPXO model, but the text did not make that clear. This omission led to Savage et al. (2017b) mistakenly stating that the HYCOM solutions they analyzed, taken from Ngodock's work, employed an iterative SAL.

(2001), have included a parameterized topographic internal wave drag term in the momentum equation. In the Jayne and St. Laurent (2001) formulation, the wave drag varies in space, as a function of the stratification at the bottom of the ocean, the RMS heights of small-scale topographic features, and the flow velocity. The horizontal length scales of topographic features also enter into the formulation, as a spatially constant tuning parameter. More complex formulations of wave drag, often based upon spatially varying tensors, are used in Arbic et al. (2004), Egbert et al. (2004), Nycander (2005), Lyard et al. (2006), and other studies. Arbic et al. (2004) employed a version of the atmospheric wave drag scheme of Garner (2005). See Nycander (2005) for development of a similar tensor formulation. These drag tensors depend on the local spectrum of topographic heights and horizontal length scales, and on stratification.

Employing wave drag, or other IGW damping parameterizations, in baroclinic tide models is more complex. First of all, the question arises as to whether one should parameterize IGW damping in a model that resolves some of the internal wave spectrum. Arbic et al. (2004), Arbic et al. (2010), Ansong et al. (2015), and Buijsman et al. (2016) have argued that wave drag (or, at least, some parameterized sink of internal wave momentum) is necessary in global baroclinic tide models, because such models do not resolve the actual breaking of internal tides and therefore require a parameterized momentum sink—which they take to be topographic wave drag—to make up for this deficiency. Some internal tide and wave modelers have not employed wave drag, either because they disagree with the point of view described above, or because they do not want to add another term into their models. If one does assume that a topographic wave drag or other IGW damping is needed, this introduces some numerical complexities. For instance, the question arises as to whether the wave drag should be applied to the bottom flow, barotropic flow, or some other flow. Parameterizations of upper-ocean wave-wave interactions, which have not been implemented in global internal wave models to date, would have to act on flows throughout the water column. In Arbic et al. (2004) and in the HYCOM simulations described here, a parameterized topographic wave drag is applied to the bottom flow, based upon the consideration that the bottom flow is the flow that actually interacts with topographic features. This follows the practice of atmospheric modelers, but raises another point of frequent discussion and argument.

Finally, there is the question of what to do about wave drag or other IGW damping in simulations, such as the ones described in this book chapter, that resolve both tidal and non-tidal motions. Non-tidal motions, such as low-frequency mesoscale eddies and currents, also likely generate an internal wave field that acts as an important momentum and energy sink (e.g., Nikurashin and Ferrari, 2011; Scott et al., 2011; Trossman et al., 2013, 2016). Wave drag acts on both tidal and non-tidal motions, but is quantitatively different in the two limits (Bell, 1975). The Trossman et al. (2013, 2016) papers employed the first inline insertion of wave drag into high-resolution models of the eddying general circulation. Trossman et al. (2013, 2016) found that the wave drag was a significant energy sink for the general circulation, and that the statistics of mesoscale eddies are significantly impacted by wave drag. The Trossman et al. (2013, 2016) simulations were done with HYCOM, but in order to avoid the problem discussed above—that wave drag acts differently on tidal and non-tidal flows—the Trossman et al. HYCOM simulations did not include tides. In the HYCOM simulations that include both tidal and non-tidal motions, such as the simulations that we focus on in this book chapter, the flow averaged over the bottom 500 meters is saved over a period of time of order 1-2 days (the exact number of hourly snapshots saved has been changed as the HYCOM tide simulations have evolved). Then a filter in time is employed to separate the tidal and non-tidal motions, and the wave drag is employed only on the tidal flow. The filter separates tidal and non-tidal motions imperfectly. See Arbic et al. (2010) for a description of the procedure, which produced some numerical artifacts in that first paper (see for instance the artificially large kinetic energy in the Gulf of Mexico and other regions in Figure 2b of that paper). In the subsequent HYCOM tides papers, beginning with Shriver et al. (2012), the artifacts have been greatly reduced. However, the question of how to emply wave drag in models containing both tidal and non-tidal motions is still a matter of active research.

As noted in Ansong et al. (2015) and MacKinnon et al. (2017), among others, in the actual ocean, internal tides likely lose their energy by a variety of mechanisms, including upper-ocean wave-wave interactions (e.g., McComas and Bretherton, 1977), interaction with mean flows and eddies (e.g., Dunphy and Lamb, 2014), and scattering into higher vertical modes and dissipation by bottom friction on continental shelves (e.g., Kelly et al., 2013). No parameterizations of these other internal tide damping processes have been incorporated into the HYCOM tides simulations, which for simplicity employ only a parameterized topographic wave drag meant to parameterize damping via breaking of high-mode motions. No parameterizations of wave drag or of any other internal wave damping mechanisms have been employed in the MITgcm tides simulations discussed here, or, to the best of our knowledge, in any of the other "wind plus tides" simulations performed thus far by modeling groups outside of the HYCOM modeling group. The wave drag scheme employed in earlier HYCOM papers was based on the Garner (2005) scheme. Buijsman et al. (2015) provides a detailed discussion of the performance of HYCOM barotropic tide simulations using various wave drag schemes. Buijsman et al. (2015) found that the simpler scheme of Jayne and St. Laurent (2001) produced errors with respect to TPXO that were comparable to the errors produced with more complex wave drag schemes. As a result, a decision was made to use the Jayne and St. Laurent (2001) scheme in our most recent HYCOM three-dimensional "wind plus tides" simulations.

Shallow-water tidal equations

We now describe the shallow-water tidal equations. Barotropic tides obey the shallow-water equations (Gill, 1982), which apply to motions having wavelengths that greatly exceed ocean depths. Shallow-water motions are hydrostatic, meaning that accelerations do not enter into the primary balance of forces in the vertical direction.

The barotropic tidal elevations do not equal the equilibrium tidal potential for a variety of reasons. First, for an ocean with an average depth of 4000 m, shallow-water waves travel about 200 m s^{-1}, which is not fast enough for the waves to stay under the Moon as it orbits the Earth.[13]

---

[13] Because waves in the solid earth travel rapidly, through both continental and oceanic crust, the solid-earth body tides do follow the equilibrium tidal forcing more closely. This is why the effects of solid-earth body tides represent a simple alteration of the astronomical tidal potential—multiplication by $1 + k_2 - h_2$.

Continents obstruct oceanic flows, another reason that the shallow-water waves cannot follow the equilibrium tidal forcing. Further complexities in the ocean tide response to equilibrium tidal forcing include the Coriolis force, frictional forces, and solid-earth body and load tides. The tides in the ocean represent a dynamical response, including all of the factors described above, to the equilibrium tidal forcing. Many authors have argued that the tidal response to astronomical forcing is in resonance (Wunsch, 1972; Garrett and Greenberg, 1977; Heath, 1981; Arbic et al., 2009), because the spatial structure and frequency of astronomical forcing is not dissimilar to the normal modes one obtains from the unforced shallow-water equations (Platzman et al., 1981; Platzman, 1991; Zahel and Müller, 2005; Müller, 2007). Numerous studies have argued that the large tides seen in some coastal regions around the globe (e.g., the Bay of Fundy, the Hudson Strait, the Northwest European Shelf, and other areas) are due to resonances in these regions (Garrett, 1972; Clarke, 1991; Arbic et al., 2007; Cummins et al., 2010; among others).

In the case of a one-layer (barotropic) shallow-water model, if we assume a tensor form of the wave drag, the governing momentum equation is

$$\frac{\partial \vec{u}}{\partial t} + \vec{u} \bullet \nabla \vec{u} + f\hat{k} \times \vec{u} = -g\nabla(\eta - \eta_{EQ} - \eta_{SAL})$$
$$+ \frac{\nabla \cdot [K_H(H+\eta)\nabla \vec{u}]}{H+\eta} - \frac{c_d|\vec{u}|\vec{u}}{H+\eta} + \frac{\overline{T}\vec{u}}{\rho_0(H+\eta)}, \tag{10}$$

and the governing mass conservation equation is

$$\frac{\partial \eta}{\partial t} + \nabla \bullet [(H+\eta)\vec{u}] = 0, \tag{11}$$

where $\vec{u}$ is the two-dimensional horizontal velocity vector, f is the Coriolis parameter, $\hat{k}$ is a unit vector in the vertical direction, g is gravitational acceleration, K_H is the horizontal eddy viscosity, H is the resting water depth, η is the perturbation tidal elevation, c_d is the quadratic drag coefficient (usually set to a value close to 0.0025), and $\overline{T}$ is the topographic internal wave drag tensor. The form of the one-layer shallow-water equations given in Arbic et al. (2004) is equivalent to the above, though written in a slightly different (flux-divergent) form. The form of the governing equations for two- and multi-layer shallow water models with only tidal forcing present are given in, for instance, Arbic et al. (2004) and Simmons et al. (2004b), respectively. A simplified form of the modern shallow-water equations lacking, for instance, the nonlinear advective terms and the damping terms, was written down by Laplace (Laplace, 1775, 1776). Laplace's equations are known as the "Laplace tidal equations." The study of tides has indeed contributed much to the development of ocean models.

The greater complexity of the ocean tide response, relative to the simple structure of the astronomical forcing, is illustrated in Fig. 13.12, which is a global map of the amplitudes and phases of the M_2 surface tidal elevations from Egbert et al. (1994). Tidal results are often displayed in amplitude and phase maps. A tidal harmonic analysis, which employs a least-squares fitting procedure, is generally used to extract tidal amplitudes and phases from observations or from model outputs (Foreman, 1977, 2004; Pawlowicz et al., 2002; Foreman et al., 2009). Amplitude and phase maps take advantage of the periodicity of the tides and assume that a scalar tidal variable V can be written as

$$V(\phi, \lambda) = Amplitude(\phi, \lambda) f(t_{ref}) cos[\omega(t - t_{ref}) + \chi(t_{ref}) + \nu(t_{ref}) - phase(\phi, \lambda)]. \quad (12)$$

The amplitudes and phases of tidal elevations can be used to compute the time-averaged discrepancy between two different estimates of the tides. The squared discrepancy D^2 between, for instance, modeled tidal elevations η_{MODEL} and observed elevations η_{OBS} at a point location is defined as

$$D^2 = <(\eta_{MODEL} - \eta_{OBS})^2> \quad (13)$$

where $<>$ denotes a time average, over an integer number of tidal periods. It is easily shown that D^2 can be written in terms of tidal amplitudes and phases, viz.

$$D^2 = \frac{1}{2}(A_{MODEL}^2 + A_{OBS}^2) - A_{MODEL}A_{OBS}cos(\phi_{MODEL} - \phi_{OBS}), \quad (14)$$

where A_{MODEL} and A_{OBS} are amplitudes of the model and observations, respectively, and ϕ_{MODEL} and ϕ_{OBS} are phases. Alternatively, one can write, as in Shriver et al. (2012),

$$D^2 = \frac{1}{2}(A_{MODEL} - A_{OBS})^2 + A_{MODEL}A_{OBS}[1 - cos(\phi_{MODEL} - \phi_{OBS})], \quad (15)$$

where the first term on the right-hand side is an amplitude error and the second is an amplitude-weighted phase error. Often an area-weighted D^2 is calculated in the tide literature. In such cases an area-weighted RMS error is taken as the square root of the area-weighted D^2, viz.

$$D = \sqrt{\frac{\int\int <(\eta_{MODEL} - \eta_{OBS})^2> dA}{\int\int dA}}. \quad (16)$$

Brief synopsis of internal wave theory

We briefly summarize two theoretical results that we will use later to interpret some of our modeled IGW results. First, as is shown in the appendix, the linear dispersion relation for plane IGWs can be written as

$$\omega^2 = f^2 + c_e^2 K^2, \quad (17)$$

where ω and K are the frequency and horizontal wavenumber, and c_e is the eigenspeed, of a particular vertical mode of interest.[14] Equation (17) implies that $|f| \leq \omega$. As can be seen from equation (59) in the appendix, the inequality $f \leq \omega \leq N$ holds in the special case of a constant N throughout the water column. Second, we note that the Garrett and Munk (1975) model for the IGW spectrum predicts that IGW spectra will fall off as ω^{-2} and as m^{-2}, where m is vertical wavenumber.

---

[14] Again, we remind the reader that frequency is denoted by σ in the appendix.

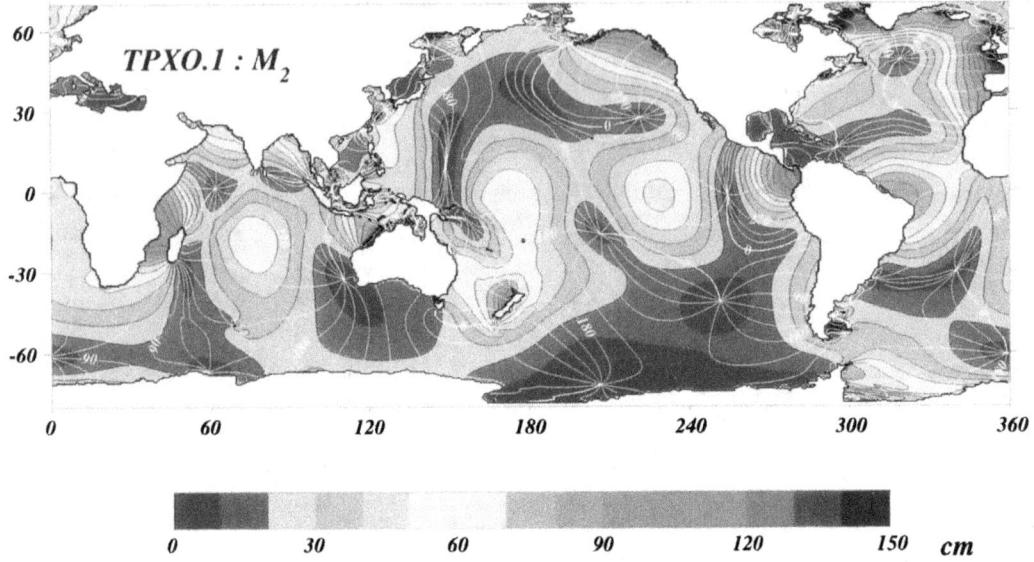

Figure 13.12. Global amplitude and phase maps of M_2 surface elevations, from TPXO model. Contour intervals are 10 cm for amplitude and 30° for Greenwich phase. Reproduced from Plate 3 of Egbert et al. (1994), ©American Geophysical Union, Wiley Online Library, used with permission.

Results from First HYCOM "Wind Plus Tides" Simulations

This section displays some results from the first HYCOM simulations forced concurrently by atmospheric fields and the astronomical tidal potential (Arbic et al., 2010); one result shown in this section is taken from the second round of HYCOM "wind plus tides" simulations (Arbic et al., 2012a; Shriver et al., 2012). Fig. 13.13a displays a global map of the amplitudes of the M_2 internal tide SSH signature (computed from steric SSH) in an experiment with two layers, a horizontally uniform stratification, and only M_2 tidal forcing present—in other words, in an experiment like those in Arbic et al. (2004) and Simmons et al. (2004b). Fig. 13.13b displays the M_2 internal tide SSH amplitude map in a simulation forced by both atmospheric fields and tides. There are some qualitative similarities between the two maps. For instance, internal tide generation regions such as Hawai'i, the French Polynesian Islands, and others, are clearly visible in both plots. There are quantitative differences throughout the globe, demonstrating (not surprisingly) that the internal tides are very different in a simulation that includes a realistic horizontally varying stratification. The differences are greatest in polar regions, where the oceanic stratification is weaker than the subtropical stratification employed globally in the simulation shown in Fig. 13.13a. Under the more realistic conditions in Fig. 13.13b, the internal tide signatures in polar regions are much weaker.

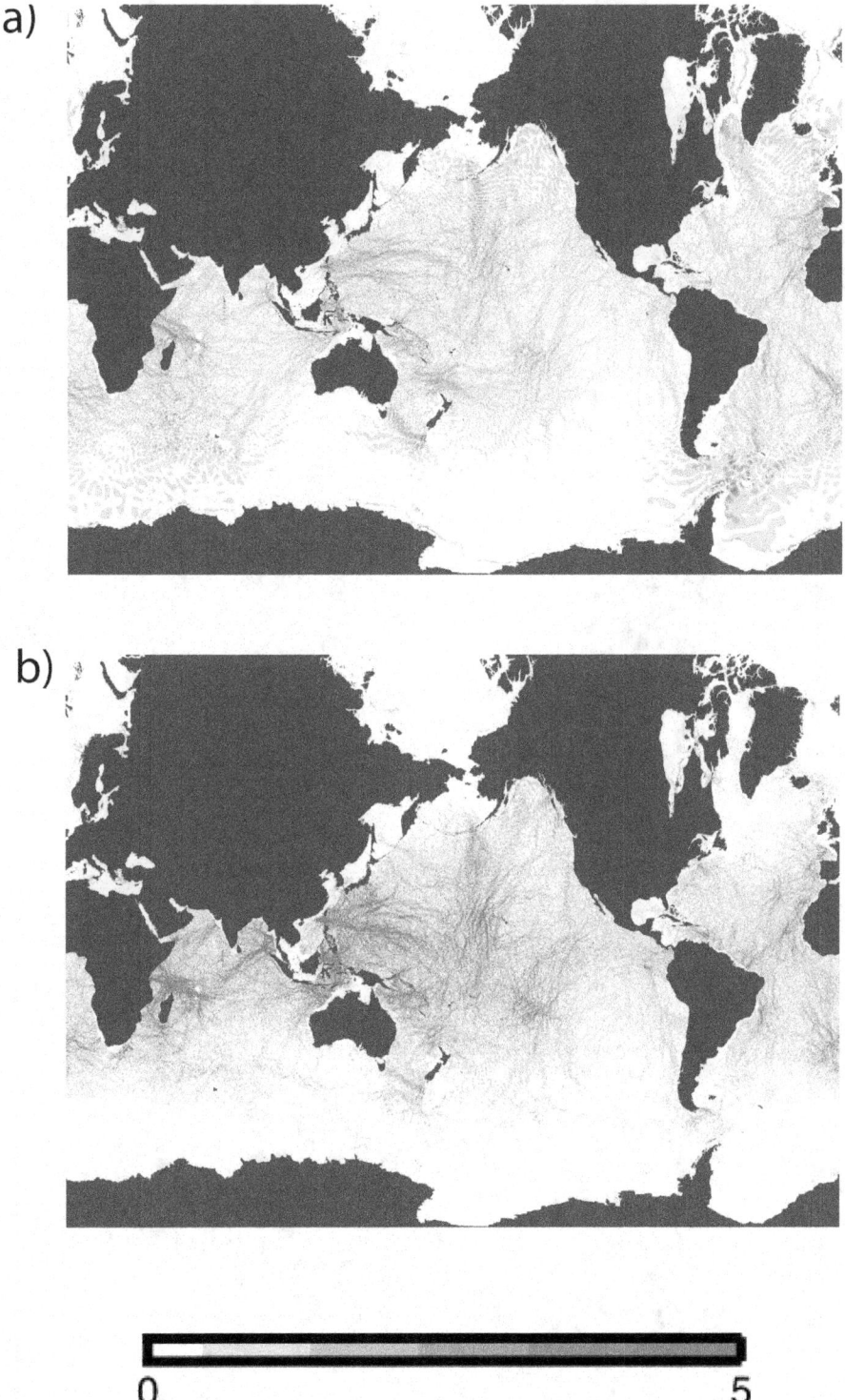

Figure 13.13. Amplitude (cm) of M_2 internal tide signature, computed from steric SSH, in (a) HYCOM simulation with two-layer, horizontally uniform stratification, and M_2 tidal forcing only, as in Arbic et al. (2004) and Simmons et al. (2004b), and (b) early HYCOM "wind plus tides" simulation, with a 32-layer, horizontally non-uniform stratification, forced by atmospheric fields as well as the M_2 astronomical tidal potential. Reproduced from Figure 4 of Arbic et al. (2010).

(a)

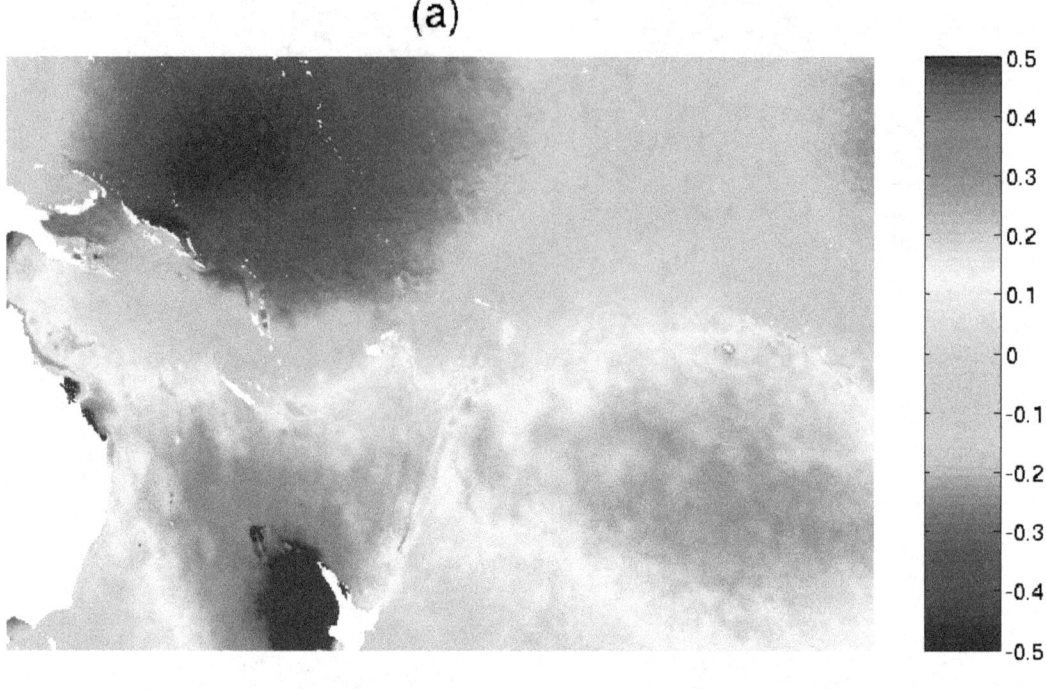

(b)

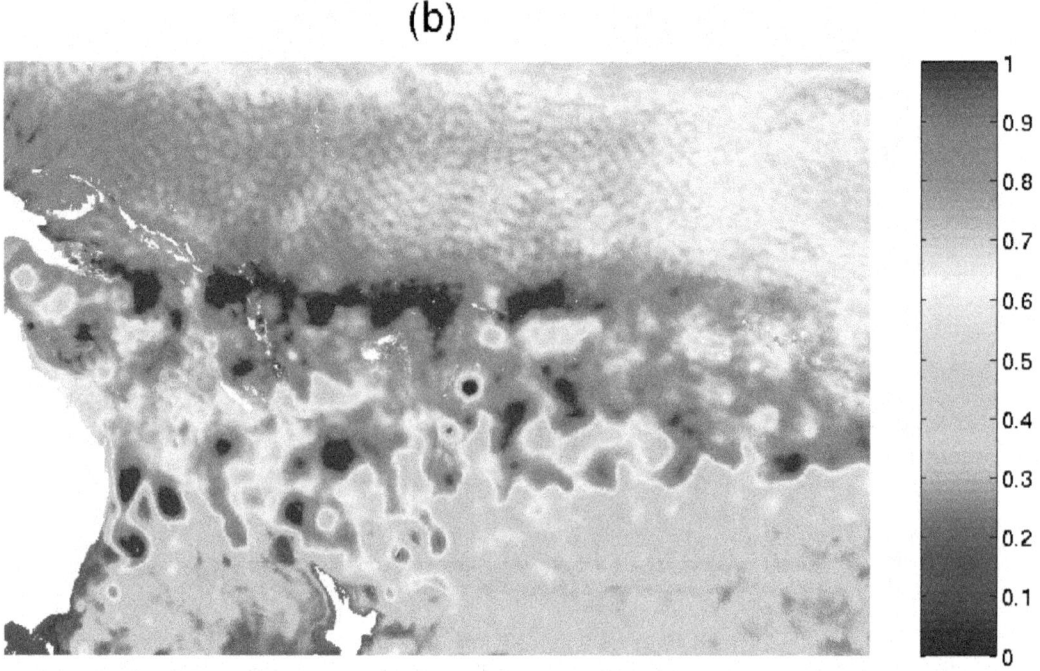

Figure 13.14. Snapshot of (a) non-steric and (b) steric sea surface heights (m) in the Southwest Pacific on June 30, 2006 at 00Z, from an early HYCOM simulation forced by both atmospheric fields and tides. Reproduced from Figure 8 of Arbic et al. (2010).

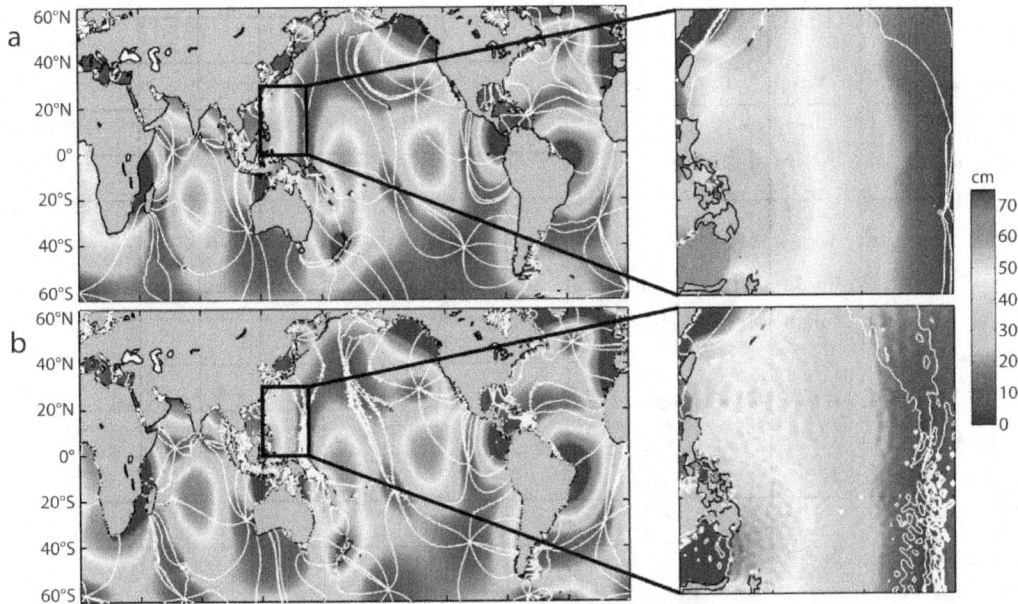

Figure 13.15. (a) Amplitude (cm) of M_2 surface tidal elevation in TPXO (Egbert et al., 1994), a highly accurate barotropic tide model constrained by satellite altimetry. (b) Amplitude (cm) of M_2 surface tidal elevation in early "wind plus tides" 32-layer HYCOM simulation with concurrent atmospheric and tidal forcing, which yields barotropic tides, internal tides, and mesoscale eddies. White curves in (a) and (b) are contours of constant phase. Insets to the right of (a) and (b) display amplitude and phase in the western Pacific region delineated by boxes in (a) and (b). Reproduced from Figure 1 of Arbic et al. (2012a).

The separation of HYCOM SSH fields into steric and non-steric components (see appendix of Savage et al., 2017a) has proven to be useful in visualizing different classes of motions simulated by the HYCOM "wind plus tides" simulations. Fig. 13.14 shows snapshots of non-steric and steric SSH in the South Pacific from the early simulations discussed in Arbic et al. (2010). The non-steric fields are dominated by the large-scale barotropic tides. The steric fields display both the internal tides—manifesting themselves in a small-scale "honeycomb" texture—and mesoscale eddies (which are more vortical). Animations of these fields reveal that the honeycomb patterns are high-frequency (because they are tidal) while the mesoscale eddies evolve more slowly.

Fig. 13.15 shows global maps of the amplitude and phase of M_2 surface elevations, in both HYCOM and in TPXO. The inset plots focus in on a region of the western Pacific where strong internal tides are present. The contrast between the inset of TPXO, which does not include internal tides, and the inset of HYCOM, illustrates the impact of internal tides upon the total (barotropic plus internal) surface tidal elevation. The internal tides, though of smaller amplitude, impart small-scale perturbations to both amplitude and phase contours, as in the altimeter data shown in Fig. 13.4. The amplitude perturbations have a "honeycomb" appearance, as in Fig. 13.14b.

Fig. 13.16 shows zonal velocities and isopycnal positions, in a section through Hawai'i, from the same HYCOM "wind plus tides" simulation. A snapshot is shown in (a), while a 25-hour mean is shown in (b). The snapshot reveals more structure in both the velocity and isopycnal position fields; this high-frequency structure is averaged out in the 25-hour mean.

Model Comparison with Observations and Theory

In this section, we compare the modeled internal tides and internal gravity wave continuum with observations. The observations used for comparison include satellite altimetry data, and observations from in-situ platforms such as moorings. For the purposes of this section, an altimeter-constrained model such as TPXO is considered to represent "observations". Because there have been many more model-observational comparisons with HYCOM than with MITgcm, we focus more on HYCOM than on MITgcm in this section. We compare the IGW spectra in both HYCOM and MITgcm with linear internal gravity wave theory.

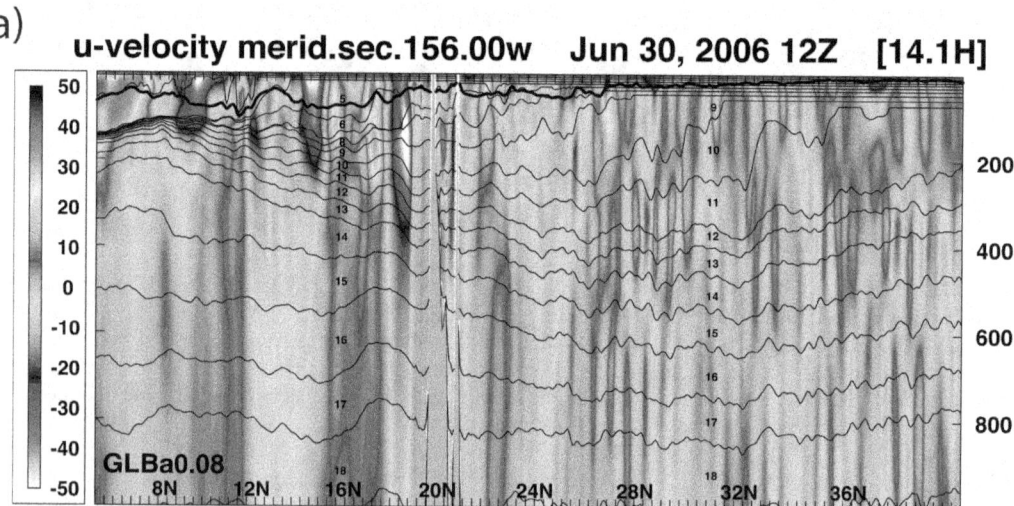

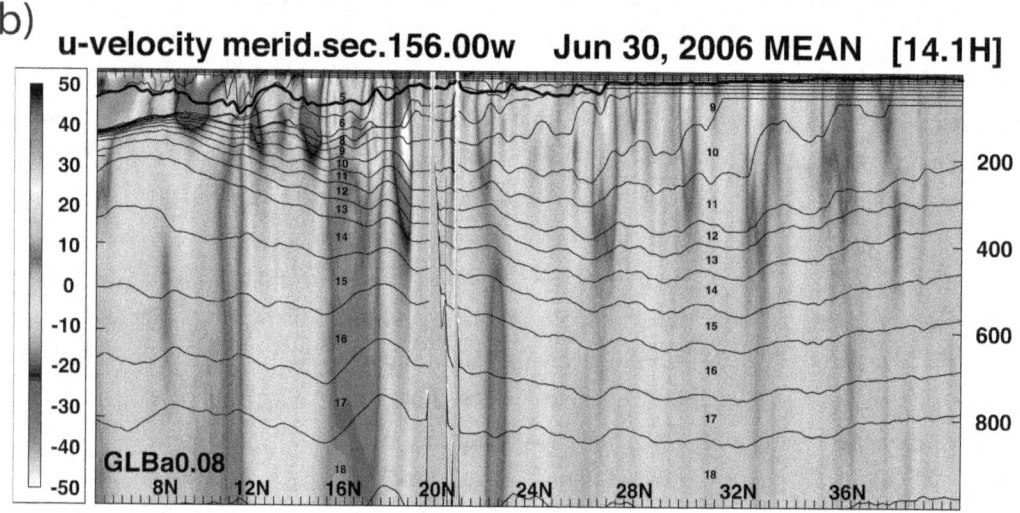

Figure 13.16. Zonal velocity (cm s^{-1}) in 156°W section through Hawai'i on June 30, 2006 from an early HYCOM simulation forced by both atmospheric fields and tides; (a) snapshot and (b) 25-hour mean. Isopycnal positions (black curves) shown versus depth in meters (right axes). Reproduced from Figure 10 of Arbic et al. (2010).

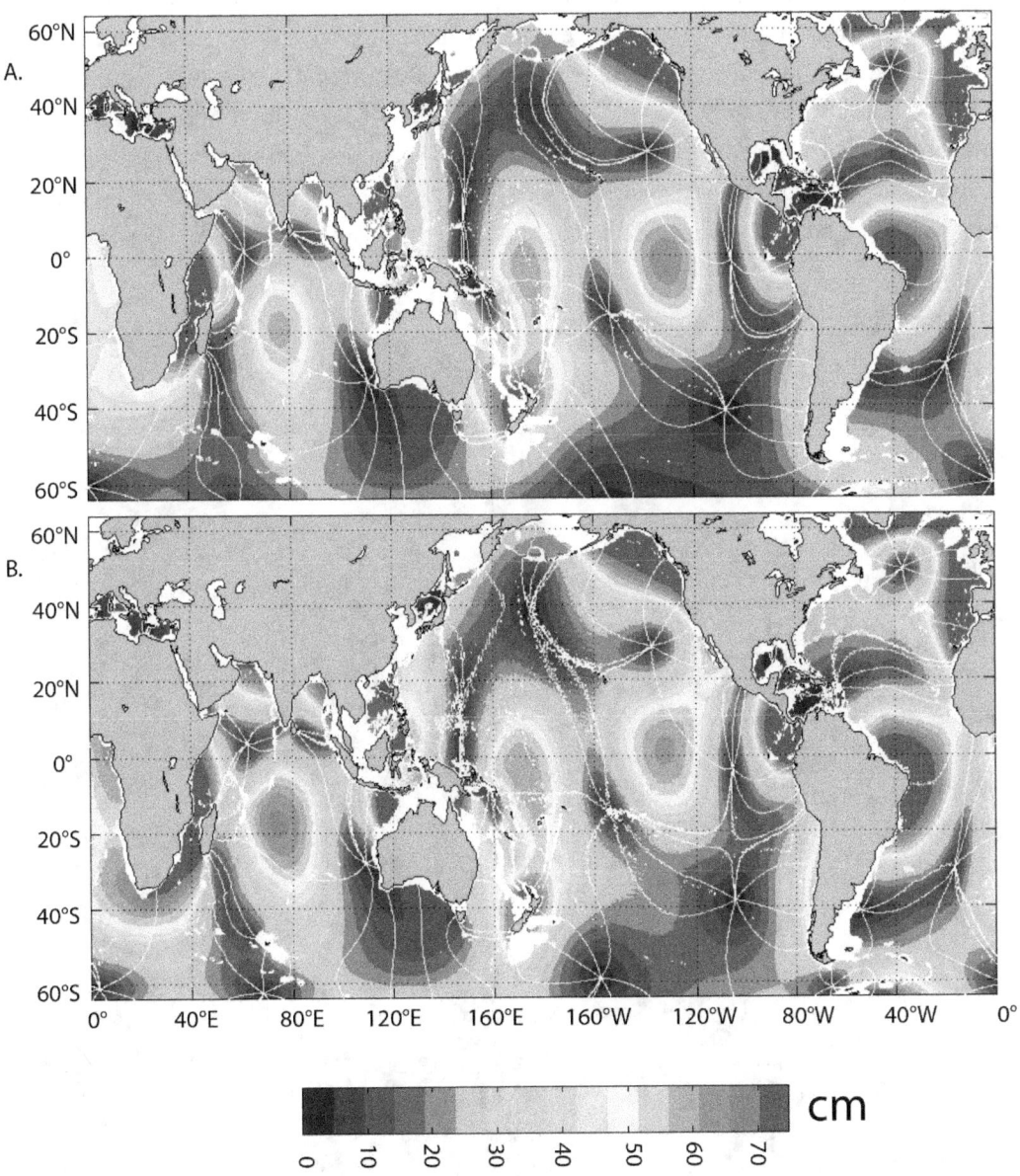

Figure 13.17. Amplitude (cm) of M_2 surface tidal elevation in (A) TPXO7.2 (an update to the model described by Egbert et al., 1994), a barotropic tide model constrained by satellite altimetry, and (B) 32-layer HYCOM simulations forced concurrently by atmospheric fields and the tidal gravitational potential. Lines of constant phase plotted every 45° are overlaid in white. Reproduced from Figure 2 of Shriver et al. (2012), ©American Geophysical Union, Wiley Online Library, used with permission.

Comparison with altimetry

Fig. 13.17 compares the M_2 surface tidal elevation amplitudes and phases in the altimeter-constrained model TPXO (A) and in HYCOM (B). Once again, small-scale structure, due to the presence of internal tides, is visible in HYCOM. The gross similarities between the TPXO and

HYCOM fields are clear, but there are also many differences in detail. For instance, the patterns in the eastern equatorial Pacific Ocean and equatorial Indian Ocean clearly differ. Later we will discuss methods that improve the accuracy of the HYCOM barotropic tides over what is seen in Fig. 13.17. Note that in Fig. 13.17 and in succeeding figures taken from Shriver et al. (2012), locations with water depths less than 1500 meters are whited out because the spatial high-pass filtering technique we used to separate barotropic and internal tides does not work well in shallow waters, where the horizontal length scales of barotropic tides are much smaller than in the open ocean.

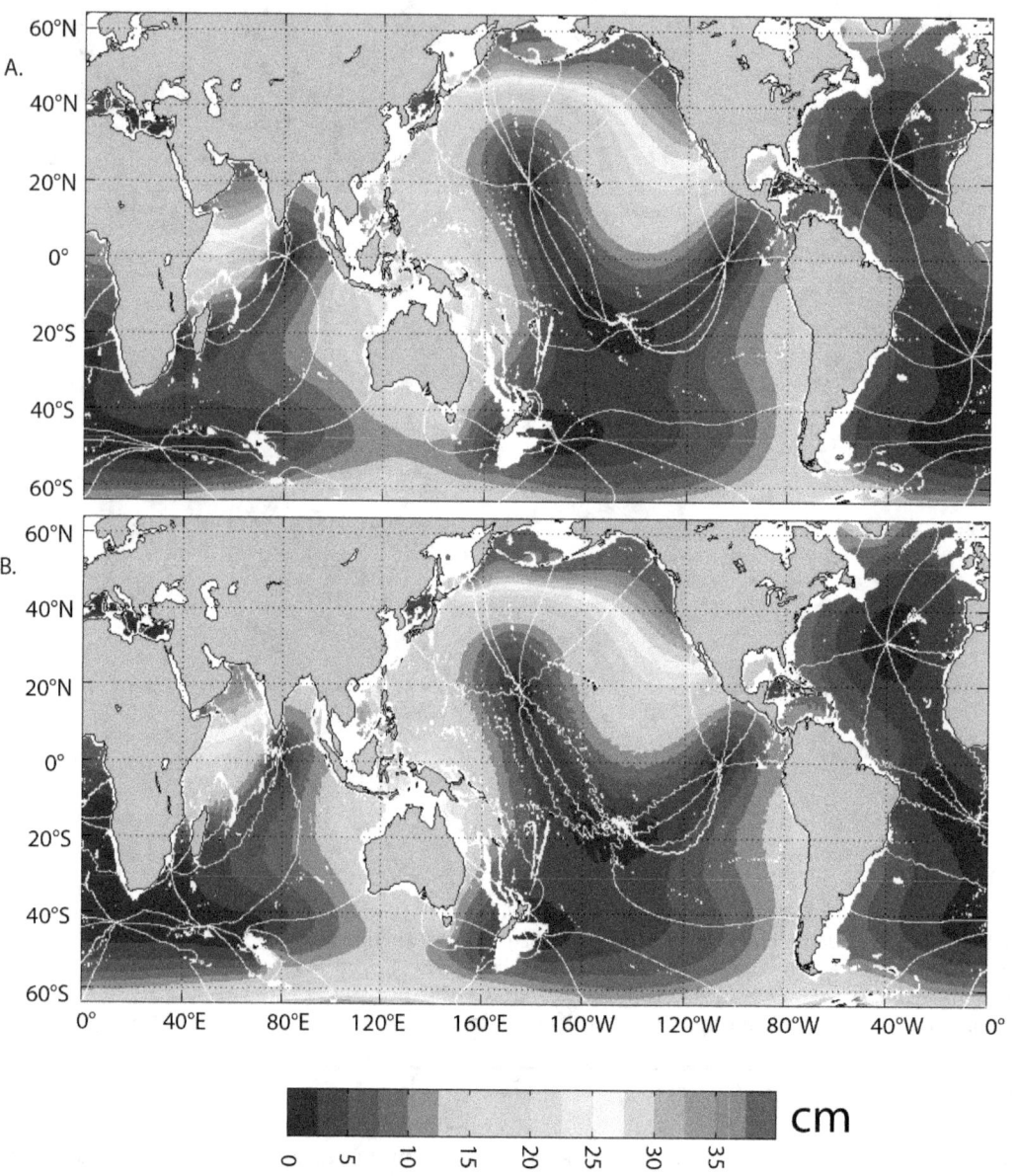

Figure 13.18. As in Fig. 13.17 but for the K_1 tidal constituent. Reproduced from Figure 3 of Shriver et al. (2012), ©American Geophysical Union, Wiley Online Library, used with permission.

Figure 13.18 compares the K_1 surface tidal elevation amplitudes and phases in TPXO (A) and HYCOM (B). We again note the small-scale structures in HYCOM, and the gross similarities and many quantitative differences between the HYCOM and TPXO fields.

Fig. 13.19 compares the stationary M_2 internal tide SSH elevation amplitudes in along-track altimeter data (A; from Ray and Byrne, 2010) and HYCOM (B). As noted in the discussion around Fig. 13.4, in the open ocean, the horizontal scales of internal tides are much less than the horizontal scales of barotropic tides. Therefore, we apply a spatial high-pass filter to the full sea surface height results shown in Fig. 13.17 to obtain the internal tides shown in Fig. 13.19. As with the barotropic tide comparisons above, the internal tide SSH comparison features gross similarities but also many differences in detail; differences are especially clear in the Atlantic Ocean. Both along-track altimetry and HYCOM reveal a small number of significant generation regions, or "hotspots", for instance, Hawai'i, the French Polynesian Islands, Northwest Australia, and others. Both subplots show a "dead zone" for stationary internal tides in the eastern equatorial Pacific. We will return to this "dead zone" in a subsequent section. Spatial averages of amplitudes computed from HYCOM over the hotspot regions outlined by black boxes in (B) agree with averages computed from along-track altimetry to within 15% for the four largest semidiurnal constituents (see Table 2 in Shriver et al., 2012). In later papers (e.g. Ansong et al., 2015), we realized that the duration of the model output impacts the average amplitude one attains, meaning that the Shriver et al. (2012) model values were slight overestimates.

Fig. 13.20 compares the stationary K_1 internal tide SSH elevation amplitudes in along-track altimeter data (A; from Ray and Byrne, 2010) and HYCOM (B). As with the M_2 internal tides, hotspot regions for the K_1 internal tides display gross agreement between HYCOM and along-track altimetry. Spatially-averaged amplitudes computed over the diurnal hotspot regions outlined by black boxes in (B) agree with averages computed from along-track altimetry to within 23% for the four largest diurnal constituents (see Table 3 in Shriver et al., 2012). The most obvious differences in the altimetric vs. HYCOM K_1 internal tides are seen in regions of strong mesoscale eddies and currents, such as the Gulf Stream, Kuroshio, Malvinas, Agulhas, and Antarctic Circumpolar Current regions (denoted by red ellipses in A). All of these regions are poleward of 30°, the latitude for which the K_1 frequency equals the Coriolis parameter f. By equation (17), we expect therefore that there should not be propagating diurnal IGWs poleward of 30°. The model results are consistent with this expectation, but the altimeter results are not. As discussed in Shriver et al. (2012) and in earlier references, the 10-day repeat times for the TOPEX/JASON series result in aliasing of tides into longer periods, that lie within the low-frequency broadband mesoscale eddy continuum. Thus, in regions of strong mesoscale eddies, it can be difficult to separate the eddies from internal tides, especially because the two types of motions have similar horizontal length scales. This leads to artificially high estimates of stationary internal tide amplitudes in altimeter data over regions of strong currents.

The global views displayed in the previous two figures do not allow for detailed comparison of individual features. Individual features can be seen in Fig. 13.21, which compares stationary M_2 internal tide amplitudes, in along-track altimetry vs. HYCOM, along ascending tracks in the North

Pacific. While the features in the model output and altimeter data are qualitatively similar, the individual peaks and troughs often do not line up closely. We are continuing to investigate the quality of internal tides in the HYCOM simulations, to see whether recent improvements to the HYCOM barotropic tides, to be discussed in a subsequent section, yield improvements in the model-observational comparison of internal tides.

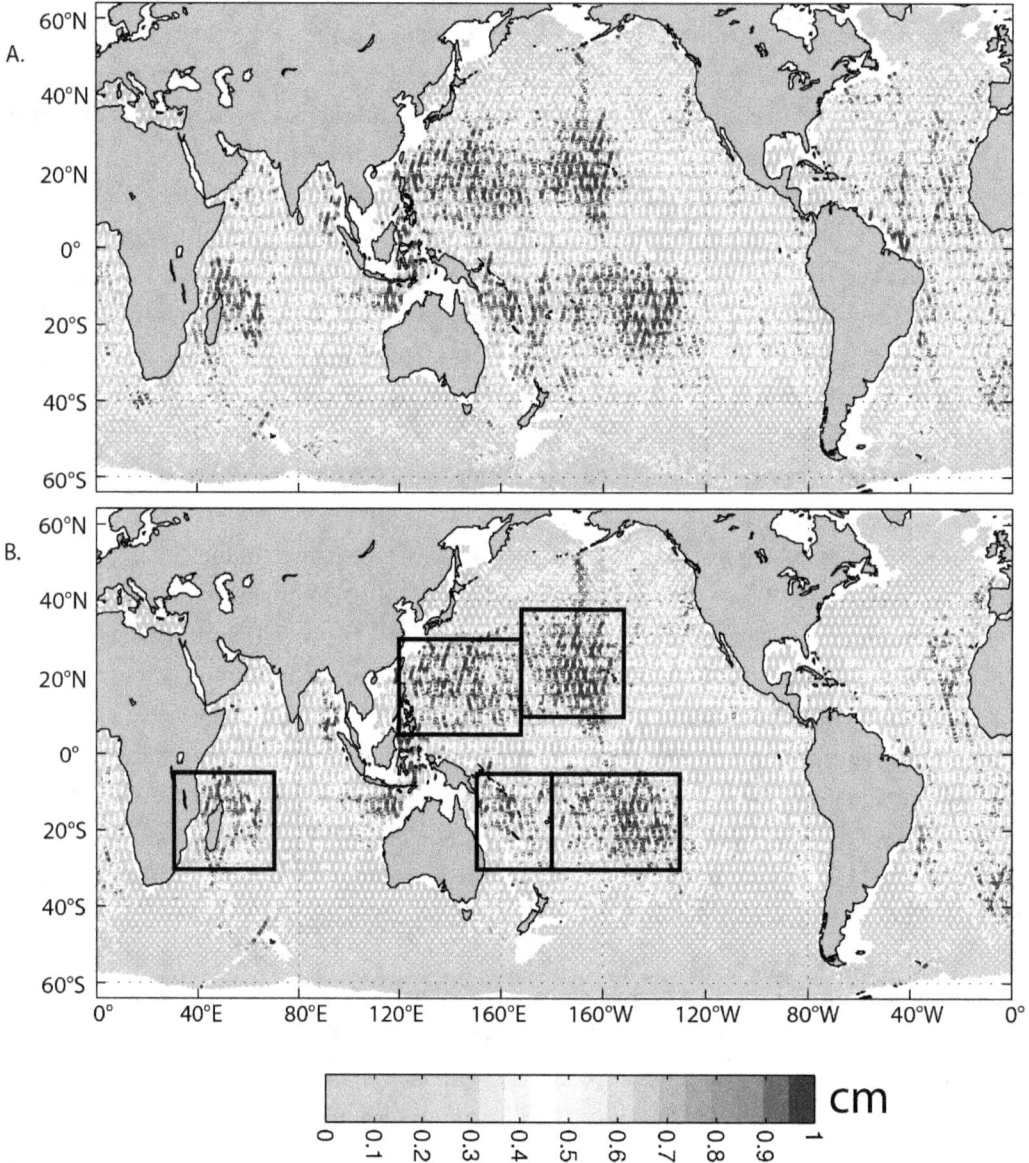

Figure 13.19. The stationary M_2 internal tide amplitude from the (A) along-track altimeter data as in Ray and Byrne (2010) and (B) analysis of HYCOM tidal model output. The five subregions denoted by black boxes in (B) are used to compute the area-averaged amplitudes in Table 2 of Shriver et al. (2012). Reproduced from Figure 7 of Shriver et al. (2012), ©American Geophysical Union, Wiley Online Library, used with permission.

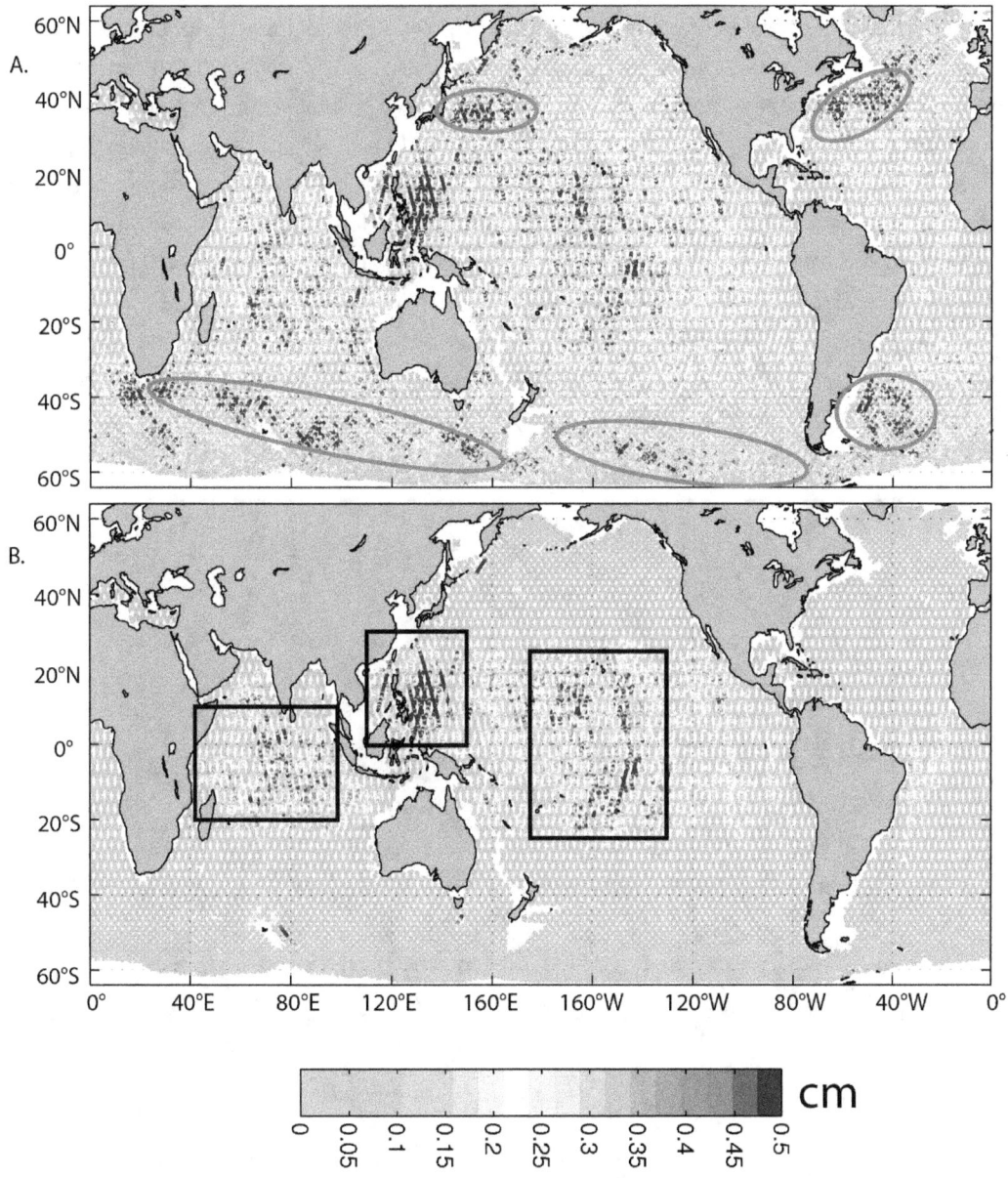

Figure 13.20. As in Fig. 13.19 but for the stationary K$_1$ internal tide. Reproduced from Figure 8 of Shriver et al. (2012), ©American Geophysical Union, Wiley Online Library, used with permission.

Comparison with moored observational records

We now display comparisons of results from global internal tide and IGW continuum models with results from an archive of historical moored instruments. Observations in the archive are in some ways the antithesis of altimeter records. Individual moored instruments in the archive provide high temporal resolution observations, often putting out measurement information at intervals of one hour or even less. This makes mooring data suitable for computing frequency spectra that encompass a range of oceanic motions, from low-frequency mesoscale eddies to high-frequency internal tides and the IGW continuum, as in Fig. 13.3. Variables commonly captured in mooring time series include velocity, temperature, and, in newer instruments for which the drift problems in older instruments have been reduced, salinity. We have used a database of archived temperature and velocity records (Scott et al., 2010; Wright et al., 2014), in several model-data comparison studies. Scott et al. (2010) compared the kinetic energy of low-frequency motions, such as mesoscale eddies and currents, in four different "eddying" models, none of which contained tides, vs. the archived current meter data. See Penduff et al. (2006), and Thoppil et al. (2011), for related efforts. Luecke et al. (2017) used the Scott et al. (2010) database to compare low-frequency temperature variance in non-tidal HYCOM simulations vs. moored observations, and Luecke et al. (paper in preparation; Fig. 13.3) compares both low- and high-frequency temperature variance and kinetic energy in HYCOM and MITgcm "wind plus tide" simulations with the archived historical mooring observations.

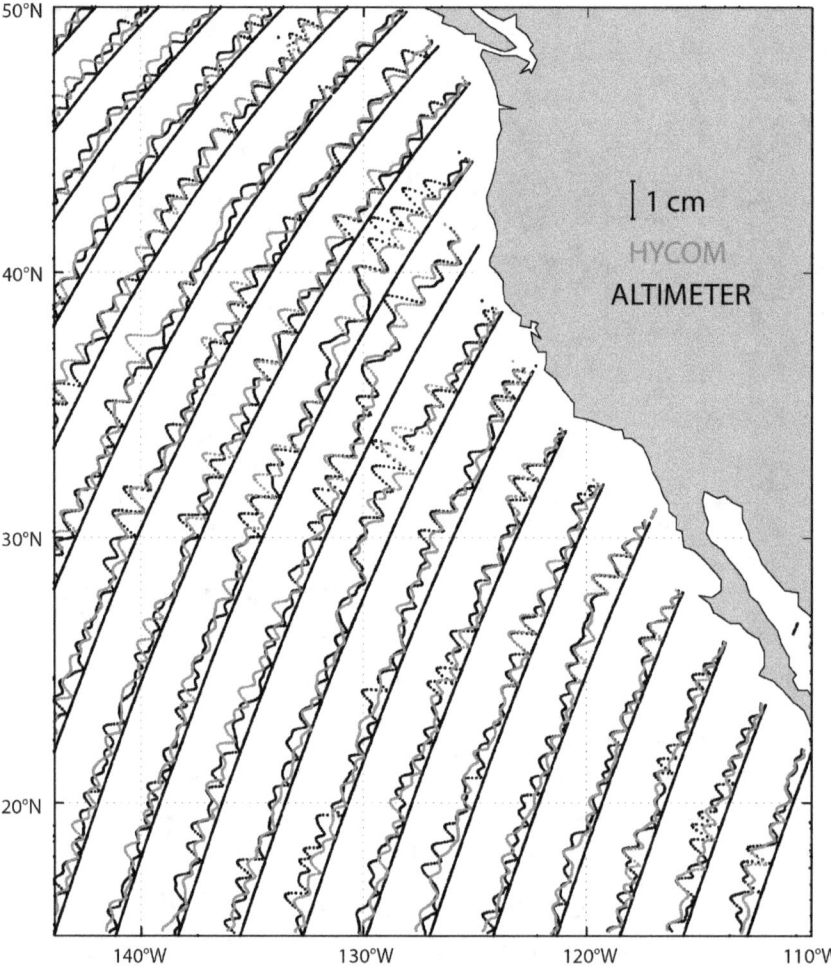

Figure 13.21. M_2 internal tide amplitude along ascending tracks from the HYCOM (red) and altimeter-based analysis (black). For each track, the line showing the coordinates of the track represents a zero amplitude for the tides on that track. The short-scale smoothness is due in part to the application of the band-pass filter and is not due to the response method used in the altimetric-based analysis. Reproduced from Figure 6 of Shriver et al. (2012), ©American Geophysical Union, Wiley Online Library, used with permission.

Timko et al. (2012, 2013) compared tidal currents in HYCOM with tidal currents inferred from the historical observations. Fig. 13.22 compares vertical profiles of the M_2 tidal current semi-major axis and phase at a particular location in the North Pacific, just south of the Aleutians. There is good general agreement throughout the water column, with the model tracking many of the changes seen with depth. However, not all locations show an agreement that is this close.

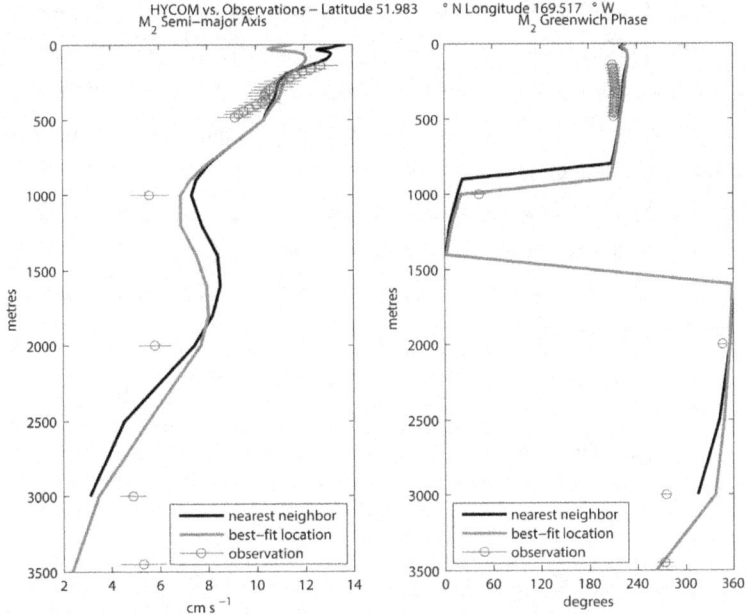

Figure 13.22. Comparison of M_2 semi-major axis and Greenwich phase in HYCOM and in a particular Acoustic Doppler Current Profiler (ADCP) record as a function of depth, at a location given in the plot title. The circles represent the value estimated from the observations and the line through the circle represents the 95% confidence intervals using harmonic tidal analysis. The black curve shows HYCOM values for the model grid point nearest the observation and the red curve shows the model values for the best-fit neighbor from a 9 point block of grid cells surrounding the observation. Reproduced from Figure 2 of Timko et al. (2013), ©American Geophysical Union, Wiley Online Library, used with permission.

Fig. 13.23 compares vertical profiles of M_2, S_2, K_1, and O_1 tidal kinetic energies, averaged over 5468 instrument records at 1618 geographical locations around the globe. As seen in the previous figure, Fig. 13.23 demonstrates that tidal currents vary significantly with depth, i.e. are quite baroclinic. Note that this vertical profile is not representative of a global average—rather it is the average profile over the particular geographical locations employed in the analysis. The model tracks the changes in kinetic energy with depth fairly well for M_2, but less well for S_2, K_1, and O_1. There are some plausible reasons for this. The simulation analyzed for Fig. 13.23 contained the four largest semidiurnal constituents and the four largest diurnal constituents. Storage limitations at the time this simulation was conducted prevented us from saving more than one month of three-dimensional model output. However, one month of output is not enough to separate S_2 from K_2, or to separate K_1 from P_1. In order to avoid this problem in more recent HYCOM simulations, we have employed two strategies. We save longer time series at locations where mooring data is available for comparison. We have also reduced the number of constituents employed, from 8 to 5. The five constituents employed— M_2, S_2, N_2, K_1, and O_1—can, according to the traditional Rayleigh criterion, be separated with just one month of model output.

Aside from the problem of separating nearby frequencies, another problem with the model output in Fig. 13.23 is that the topographic wave drag was tuned for M_2, not for the diurnal tides. However, theory (e.g., Bell, 1975) informs us that the drag on tidal flows will be frequency-dependent. Skiba et al. (2013) implemented a topographic wave drag tuned specifically for diurnal tides. In the Skiba et al. (2013) simulations, only the K_1 diurnal tide was present. In the HYCOM "wind plus tides" simulations, because both semidiurnal and diurnal tides are present, it is not a simple matter to tune a wave drag separately for semidiurnal and diurnal tides, and at present we continue to tune for the largest constituent M_2.

The variations throughout the water column seen in Figs. 13.22 and 13.23 make clear the limitations of predicting tidal currents with barotropic tide models, as is commonly done. On the other hand, multi-layer "wind plus tides" models will need more vetting before they are commonly used for predicting tidal currents.

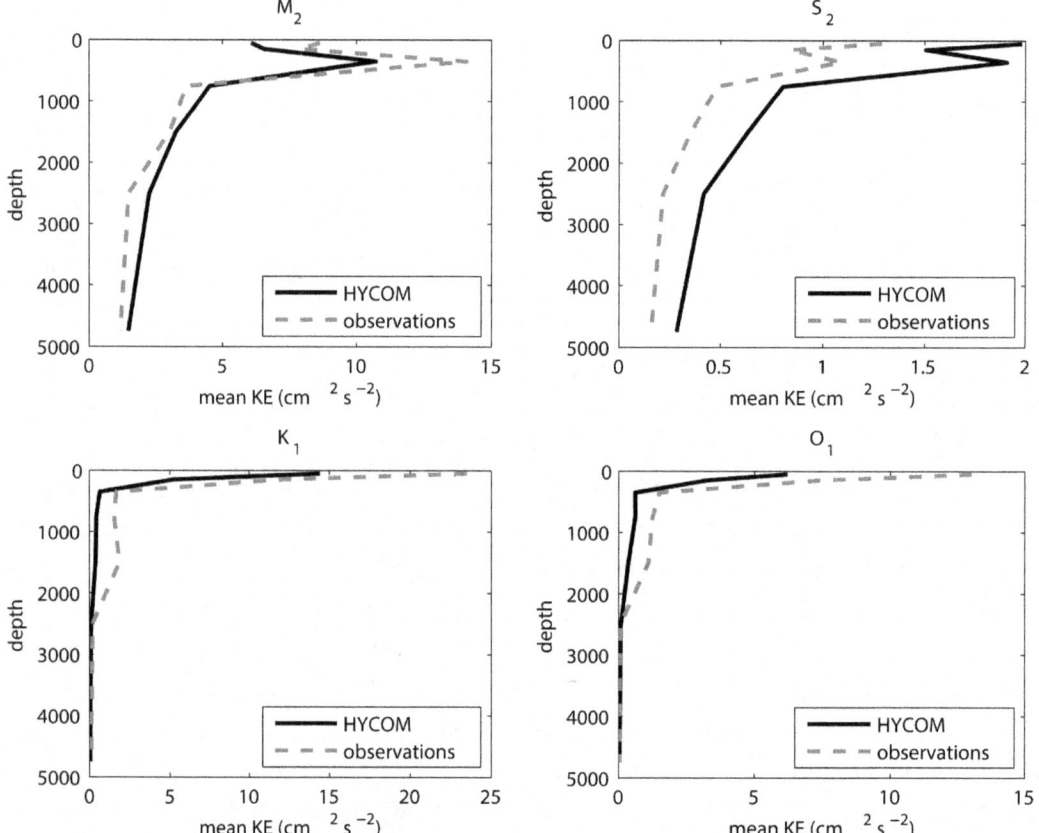

Figure 13.23. Vertical profiles of the HYCOM and observed tidal kinetic energy averaged over 5468 instrument records at 1618 geographic locations throughout the global ocean. The 5468 records, all in deep water (water column depth greater than 1000 m) are sorted into seven depth bins. Reproduced from Figure 5 of Timko et al. (2013), ©American Geophysical Union, Wiley Online Library, used with permission.

Internal tide energy fluxes can also be derived from moorings. Fig. 13.24 compares vertical mode-1 semidiurnal band energy fluxes in HYCOM vs. moorings from the Internal Waves Across the Pacific (IWAP) field campaign (Alford et al., 2007; Zhao et al., 2010; fluxes were taken from the latter paper). The agreement is visually close at 5 out of the 6 moorings. The 6 moorings shown in Fig. 13.24 are McLane profilers (Doherty et al., 1999), which provide very-high-resolution vertical coverage, in contrast to the sparse vertical coverage of traditional moorings. As is shown in Ansong et al. (2017), the model-observational agreement is not as close as in Fig. 13.24 when we use historical moorings, because sparse vertical coverage does not allow full vertical resolution of baroclinic signals.

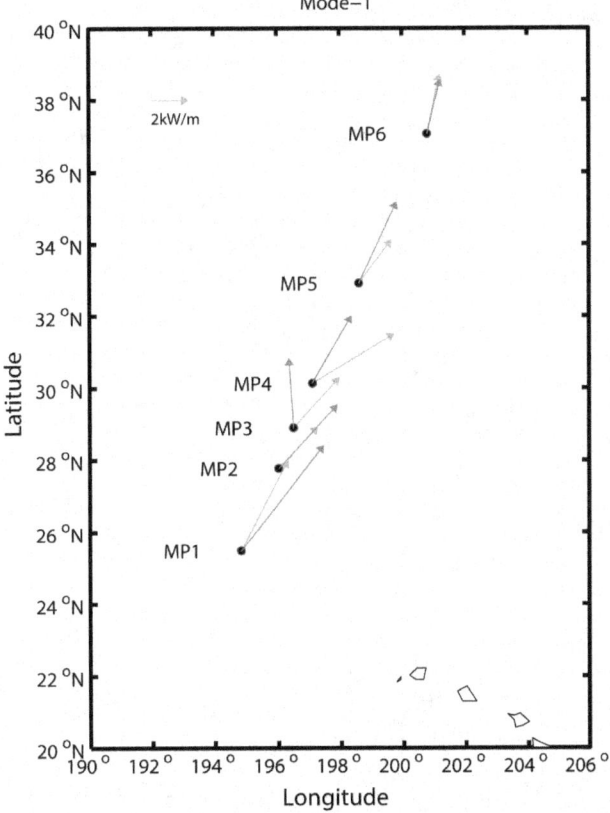

Figure 13.24. Map of depth-integrated vertical mode-1 semidiurnal energy fluxes computed for 1/25° HYCOM (red arrows) and observations (green arrows) from the IWAP experiment (Zhao et al., 2010). Arrow lengths are logarithmic and reference arrows are shown at the top left corner of the plot. Reproduced from Figure 11 of Ansong et al. (2017), ©American Geophysical Union, Wiley Online Library, used with permission.

Moorings provide information on the IGW continuum as well as on internal tides. Fig. 13.25 compares frequency spectra of surface kinetic energy in HYCOM vs. near-surface kinetic energy in historical North Pacific moorings (Schmitz, 1988). The low-frequency continuum, dominated by mesoscale flows, is seen alongside near-inertial peaks, semidiurnal tidal peaks, and the supertidal IGW continuum. The model IGW continuum approaches the observed continuum, and the Garrett and Munk (1975)/Cairns and Williams (1976) spectrum, more closely when horizontal resolution is increased from 1/12.5° to 1/25°.

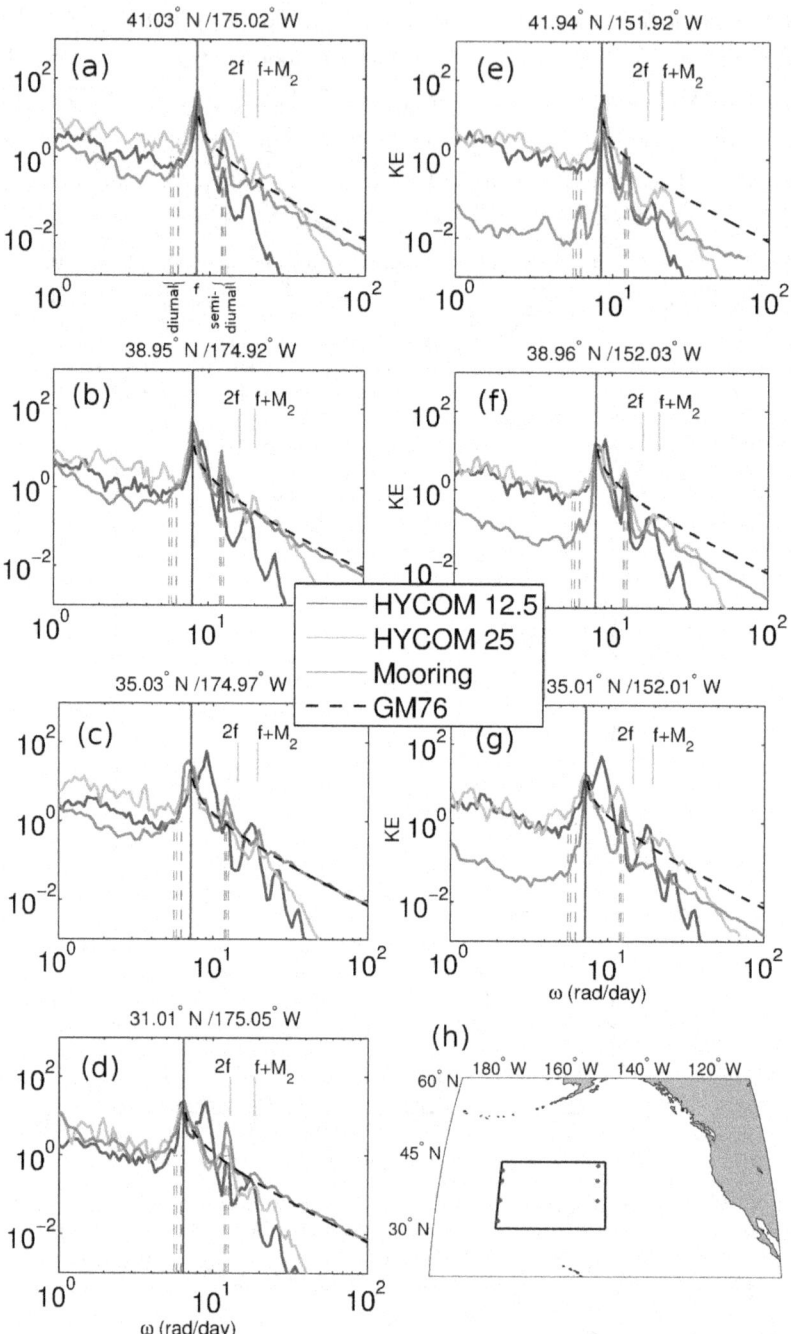

Figure 13.25. (a)-(g) Frequency spectra $E(\omega)$ of surface kinetic energy [(m/s)^2 / (rad/day)] from moored current meter observations (red), 1/12.5° HYCOM (HYCOM12; blue), and 1/25° HYCOM (HYCOM25; green) and the model spectrum of GM76 (dashed black; Cairns and Williams, 1976; built upon Garrett and Munk, 1975). Solid black and cyan vertical lines are drawn at the frequencies f, $2f$, and $f + M_2$, while dashed vertical lines are drawn at the frequencies of the four semidiurnal (M_2, S_2, N_2, and K_2) and four diurnal (K_1, O_1, P_1, and Q_1) tidal frequencies forced in the model. (h) The analysis region, showing the locations of the seven moorings in (a)-(g), in the northeast Pacific. The blue stars denote the locations of moored observations used in (a)-(g). Reproduced from Figure 1 of Müller et al. (2015),. ©American Geophysical Union, Wiley Online Library, used with permission.

Fig. 13.26 compares frequency spectra of dynamic height variance in global 1/25° HYCOM and 1/48° MITgcm vs. 9 McLane profilers. The broadband low-frequency continuum, semidiurnal peaks, and the IGW supertidal continuum are visible in all 9 subplots of Fig. 13.26. Diurnal peaks are visible in some of the subplots. The HYCOM and MITgcm spectra are similar in the subtidal and tidal bands. In the 0.2-1 cpd band, both HYCOM and MITgcm are generally deficient relative to observations, for reasons we do not understand. The supertidal IGW continuum is generally captured better by the MITgcm simulations, due to their higher horizontal and vertical resolution.

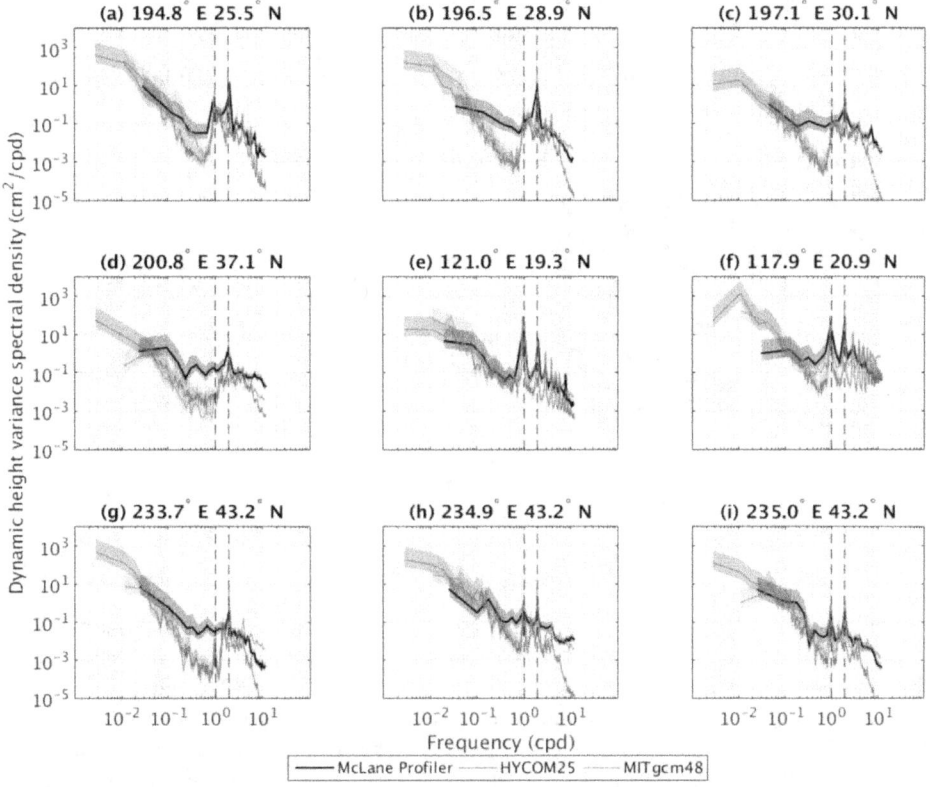

Figure 13.26. Dynamic height variance frequency spectra from McLane profilers and nearest neighbor gridpoints extracted from global simulations of 1/25° HYCOM (HYCOM25) and 1/48° MITgcm (MITgcm48). The dashed vertical lines denote K_1 diurnal and M_2 semidiurnal tidal frequencies. The shaded regions are the 95% confidence intervals that account only for random error in spectral density calculations. Reproduced from Figure 3 of Savage et al. (2017b), ©American Geophysical Union, Wiley Online Library, used with permission.

Comparison with linear internal gravity wave theory

In this section, we compute and display $K - \omega$ (horizontal wavenumber-frequency) spectra of surface kinetic energy and SSH variance, and examine whether the spectra follow the predicted linear IGW dispersion curves. Fig. 13.27 displays $K - \omega$ spectra of HYCOM surface kinetic energy computed over the North Pacific box shown in Fig. 13.25h. Dispersion relations for the first 3 vertical modes, computed from the Sturm-Liouville problem (see appendix) applied to model output, are drawn on the figure. Because both f and N change over the model gridpoints contained in the box, we computed the eigenspeeds at all model gridpoints along the southern and northern boundaries of the box, and we used the extreme values along each boundary to draw the dispersion curves—thus explaining why each of the 3 vertical modes is associated with two curves rather than just one. Near-inertial and semidiurnal peaks are clearly visible in spectra computed from both the 1/12.5° and 1/25° solutions. The kinetic energy increases with increasing resolution, especially along the vertical mode-1 and mode-2 dispersion curves. The bottom inset plots in Fig. 13.27 display the frequency spectra computed over all wavenumbers. The inset frequency spectra reveal sharp semidiurnal tidal peaks and a near-inertial peak that is broadened due to the range of latitudes represented in the box. The right-hand-side inset plots display the wavenumber spectra integrated over frequencies larger than the inertial frequency f_{29N} at 29°N. The wavenumber spectra show a peak associated with the first mode semidiurnal tide.

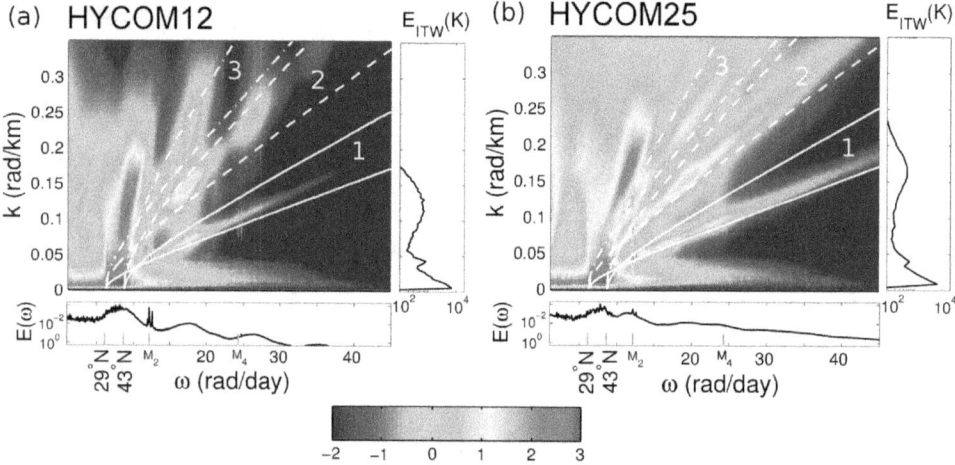

Figure 13.27. The $K - \omega$ spectra of kinetic energy $E(K, \omega)$, with units (m/s)² / [(rad/day) (rad/km)] on a log_{10} scale, for (a) HYCOM12 and (b) HYCOM25. The white curves represent bounding linear dispersion relations of the first three internal wave vertical modes (solid = mode 1; dashed = mode 2; and dash dotted = mode 3). The spectra $E(\omega)$ integrated over all horizontal wave numbers K, and $E_{ITW}(K)$ integrated over frequencies larger than the inertial frequency at 29°N ($\omega \geq f_{29N}$), are shown as bottom and right-hand-side insets, respectively. In the $E(\omega)$ spectra, cyan lines indicate the inertial frequencies at the bounding latitudes 29°N and 43°N, while M_2 and M_4 frequencies are also indicated. Reproduced from Figure 2 of Müller et al. (2015), ©American Geophysical Union, Wiley Online Library, used with permission.

The $K - \omega$ spectra of SSH variance in the North Pacific, in both HYCOM and MITgcm simulations, is displayed in Fig. 13.28. As with the IGW kinetic energy spectra, the SSH variance tends to lie along linear dispersion curves, and tends to increase with increasing horizontal resolution. Figure 5 in Savage et al. (2017b), and the supplementary figures in that paper, show that similar behaviors are seen over many different geographical regions; thus, the North Pacific region is not "special". In the higher-resolution simulations, the spectra fold back after reaching the Nyquist frequency of 12 cpd (associated with a 2 hour period). This folding in the spectra of the highest-resolution models is due to aliasing of energy in motions having sub-hourly periods. It is interesting that some global models now have so much high-frequency energy that aliasing problems arise with hourly snapshots.

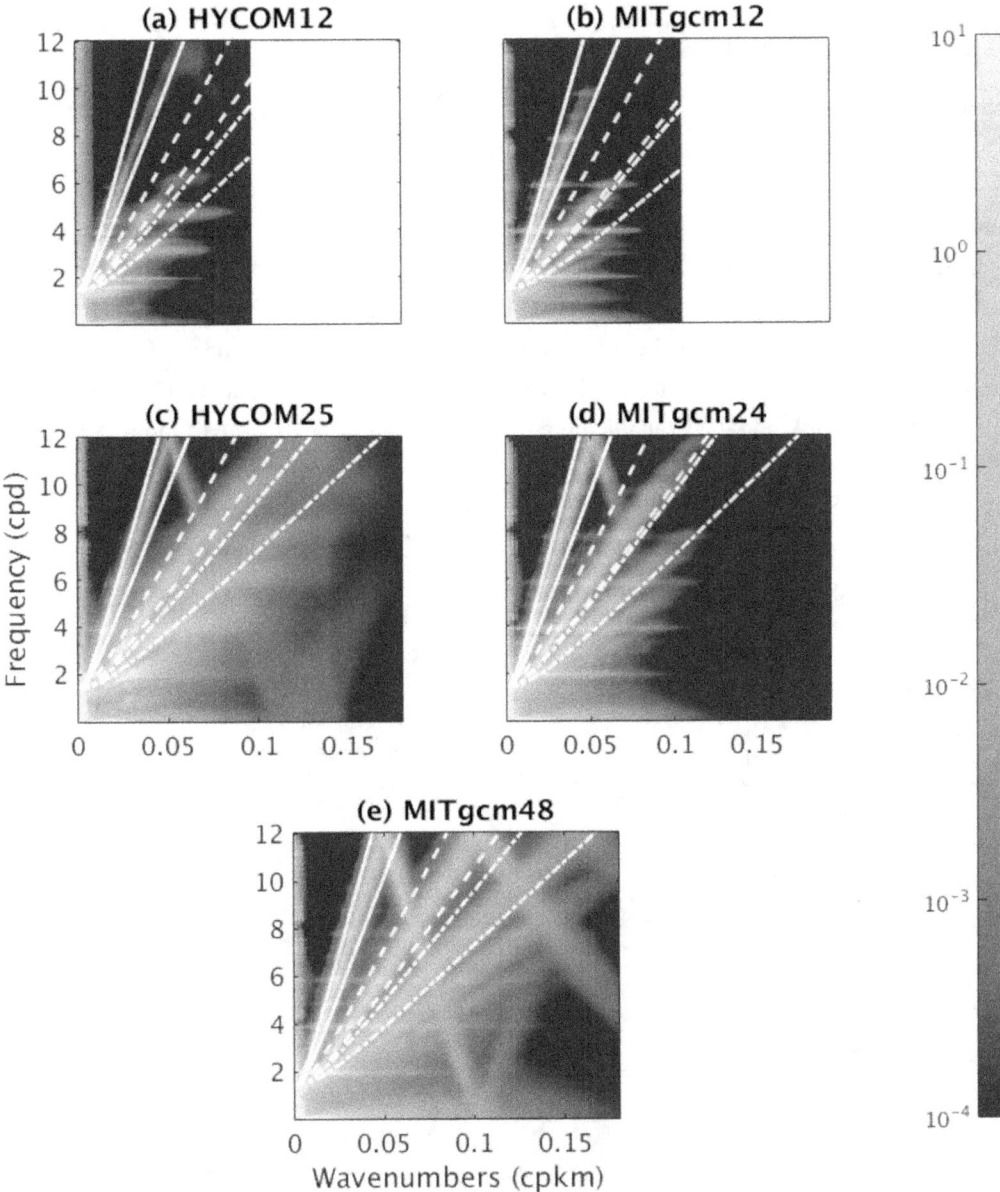

Figure 13.28. Sea surface height variance frequency-horizontal wave number spectra, in (cm^2/[(cpd cpkm)], computed in North Pacific region from HYCOM12, HYCOM25, MITgcm12, MITgcm24, and MITgcm48. Wave number axes are the same in each plot, and are set to the maximum wave number for the HYCOM25 calculation. White curves show theoretical IGW linear dispersion relations for first (solid), second (dashed), and third (dashed-dotted) vertical modes. Bounding curves for each vertical mode are computed from the maximal and minimal eigenspeeds along the northern and southern boundaries, as in Müller et al. (2015). Reproduced from Figure 6 of Savage et al. (2017b), ©American Geophysical Union, Wiley Online Library, used with permission.

Impact of Wave Drag on Tide Model Energetics and Accuracy

This section addresses the impact of parameterized topographic wave drag on tide model energetics and accuracy. Many studies have shown that inclusion of a parameterized topographic wave drag improves the accuracy of forward (unconstrained) barotropic tide models. Arbic et al. (2004) explored the reasons for this. Fig. 13.29A displays the globally integrated, temporally averaged M_2 barotropic kinetic and available potential energy, in a global barotropic tide model, as a function of wave drag strength, which is controlled with a multiplicative factor (see Arbic et al., 2004 for more explanation). Fig. 13.29B displays the globally averaged discrepancy D, defined in equation (13), of the model against the highly accurate altimeter-based barotropic tide model GOT99 (Schrama et al., 1994; Ray, 1999). When the available potential energy is either too large or too small relative to the value obtained from accurate altimeter-based values (extra horizontal line in Fig. 13.29a), the elevation error is relatively large. The minimum in the elevation error occurs when the modeled available potential energy is close to the observed value. Thus, because the wave drag strength controls tidal energy levels in models, it controls how close they lie to observations.

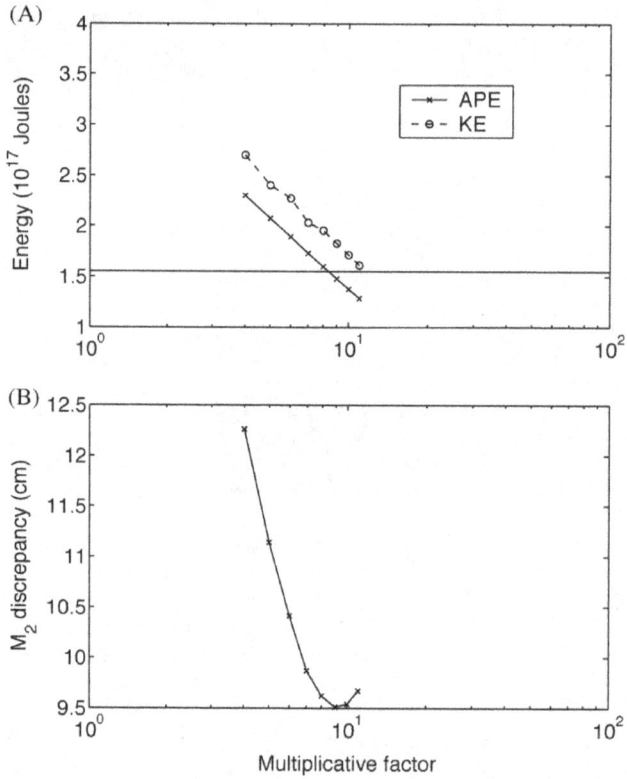

Figure 13.29. (A) Globally integrated, temporally averaged available potential energy (APE) and kinetic energy (KE) in 1/2° one-layer M_2-only simulations with a scalar approximation for the SAL term, run with a topographic wave drag scheme augmented by a variable multiplicative factor. The extra horizontal line represents the observed value of APE, taken from Tierney et al. (2000b). (B) Time- and area-averaged sea-level discrepancy D against GOT99.2, in waters deeper than 1000 m and equatorward of 66°. Reproduced from Figure 4 of Arbic et al. (2004).

The HYCOM tidal energy budget was thoroughly examined in Buijsman et al. (2016). Fig. 13.30 displays the energy dissipation of barotropic and baroclinic semidiurnal tidal flows by topographic wave drag and bottom drag. As expected, wave drag dissipates considerable energy over topographic features such as the mid-Atlantic ridge, while bottom drag dissipates considerable energy in coastal areas where tidal velocities are large. Buijsman et al. (2016) reported on a numerical instability in HYCOM, occurring in a region of the North Pacific. This instability is visible in high-frequency animations of HYCOM output, and can be seen in Fig. 13.30d (see anomalous patch just south of the Aleutians) as well as in other figures in Buijsman et al. (2016). Buijsman et al. (2016) noted that the instability is probably a thermobaric instability (Sun et al., 1999; Hallberg, 2005; Adcroft et al., 2008), a class of instability that manifests itself in weakly stratified regions of isopycnal models. Buijsman et al. (2016) found that the instability is confined to the North Pacific, and accounts for less than 2% of the globally integrated low-mode conversion. In the summary section we will briefly discuss ongoing efforts to eliminate this instability in HYCOM.

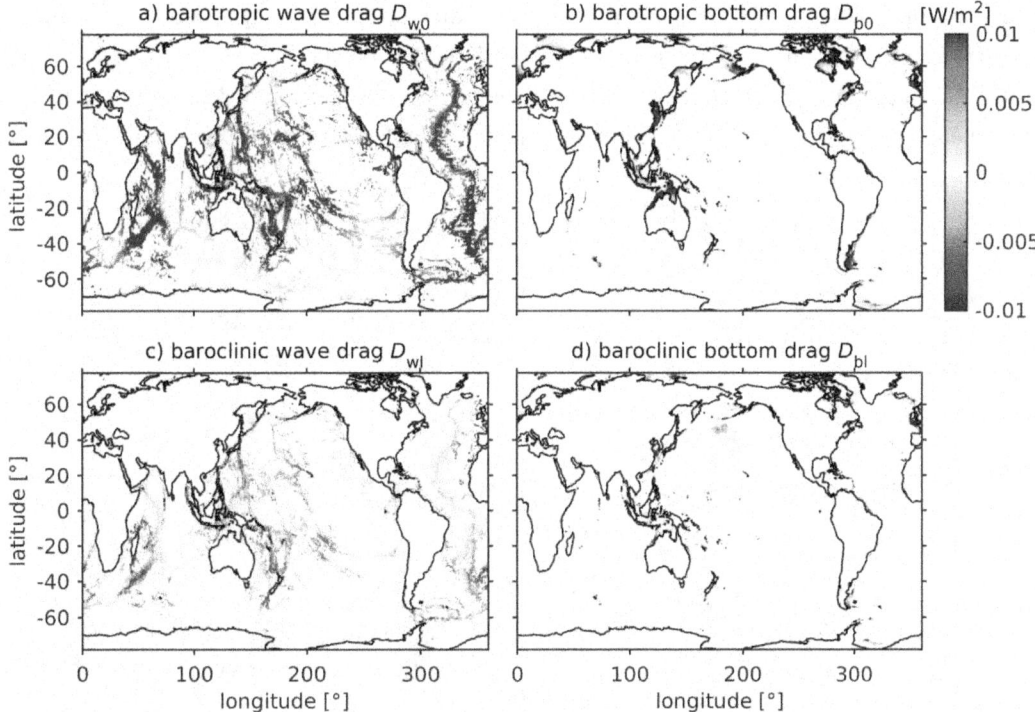

Figure 13.30. Semidiurnal barotropic dissipation due to (a) linear topographic wave drag and (b) bottom drag. Baroclinic dissipation due to (c) linear topographic wave drag and (d) bottom drag. Reproduced from Figure 2 of Buijsman et al. (2016), ©American Meteorological Society, used with permission.

The impact of the topographic wave drag on the HYCOM-altimetry internal tide comparison was comprehensively explored in Ansong et al. (2015). Fig. 13.31 displays maps of the M_2 internal tide in along-track altimetry (a), and in HYCOM simulations with varying strengths of wave drag (b-f). The HYCOM amplitudes are too large in simulations with wave drag applied only to the barotropic flow (d) and in simulations with no wave drag at all (e,f). In (f), we increased the quadratic bottom drag coefficient by a factor of about 100 along the continental shelves, as a test of the hypothesis that internal tides are dissipated primarily along shelves. However, the differences between open-ocean internal tide amplitudes in (f) and those in (e) are small. The internal tide amplitudes in simulations with wave drag acting on bottom flows (hence, on both barotropic and baroclinic motions)—the default setup in our HYCOM "wind plus tides" simulations—lie closer to altimeter observations, especially when the wave drag is strong; compare (c; stronger drag) and (b; weaker drag) to (a; observations). It should be noted that because the model contains mesoscale eddies, the internal tides in the model are scattered and become non-stationary just as internal tides are in the ocean. Thus, the model-observational discrepancies in Fig. 13.31 are not due to the lack of scattering into nonstationarity by mesoscale eddy-tide interactions in the model. Ansong et al. (2015) thus provides indirect evidence for substantial damping of low-mode internal tides in the open ocean; the damping is needed to obtain agreement with altimetry. For our current HYCOM simulations, we assume that low-mode damping is primarily due to production of high-mode internal tides over topography. The high modes break and dissipate energy in the ocean, but are not resolved in our global models. We parameterize the production and breaking of high vertical modes with topographic wave drag acting on the bottom flow. In the actual ocean, as discussed in Ansong et al. (2015) and shown schematically in Fig. 13.2, there are likely to be other damping mechanisms, which need to be better parameterized in future models in order to see continued improvements in modeled internal tides.

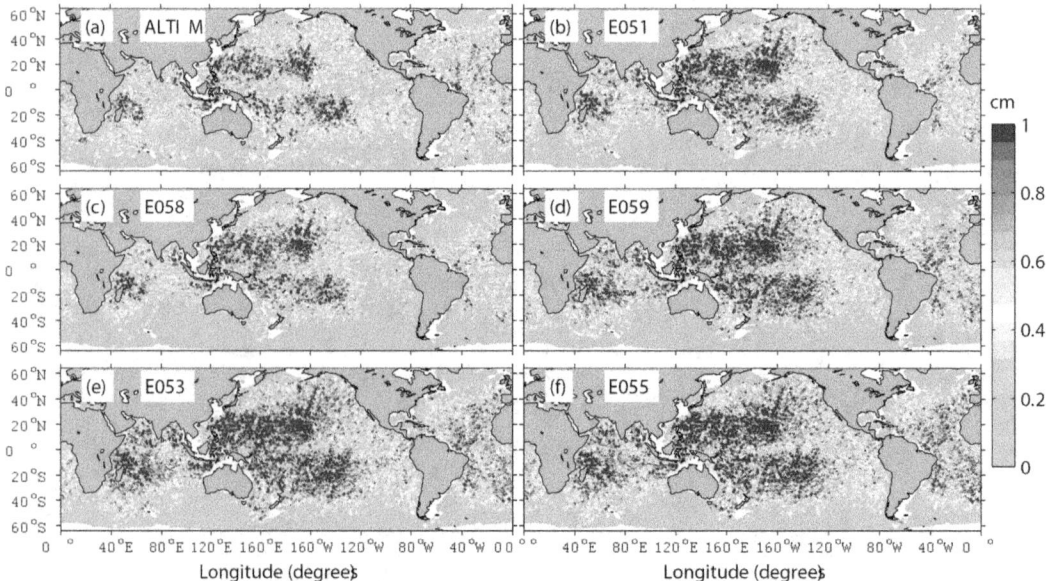

Figure 13.31. Amplitude (cm) of stationary M_2 internal tide in (a) along-track altimeter-based analyses, and in HYCOM simulations (b) E051; with wave drag (scale factor = 0.5) applied to the bottom flow, (c) E058; with wave drag (scale factor = 1.0) applied to the bottom flow, (d) E059; with wave drag (scale factor = 1.0) applied to only the barotropic flow, (e) E053; without wave drag, (f) E055; without wave drag but with the quadratic bottom drag coefficient increased by a factor of about 100 along the continental shelves. The amplitudes of the HYCOM simulations are computed from 3 months of SSH output. Reproduced from Figure 5 of Ansong et al. (2015), which can be consulted for more details of the E051-E059 simulations, ©American Geophysical Union, Wiley Online Library, used with permission.

Fig. 13.32 displays the globally averaged stationary M_2 internal tide amplitude in the HYCOM simulations with varying topographic wave drag, vs. altimetry. Again it is clear that model-altimetry agreement is better with some wave drag in the model. We point out again that HYCOM is currently the only "wave plus tides" model that employs a wave drag. Fig. 13.32 also makes clear that the record duration matters for the model-altimeter comparison. The amplitude of the modeled stationary internal tide decreases as the model record duration increases. (The altimeter analysis is based on a long—multi-year—record). The importance of record duration in the estimation of tidal quantities is seen in other studies. Nash et al. (2012), for example, finds that the amplitude of stationary tidal currents estimated from a current meter reduces as the analysis period is increased. This discussion of stationarity leads naturally into the next section, which is on non-stationary internal tides.

Internal Tide Nonstationarity

In this section, the focus turns to internal tide nonstationarity. Non-stationary internal tides are of particular interest for the SWOT mission, because it will be necessary to remove non-stationary internal tides from SWOT data in order to study other oceanic motions of interest such as mesoscale and submesoscale eddies. The HYCOM simulations discussed in this chapter have been used in a number of studies of internal tide nonstationarity. Shriver et al. (2014) computed the standard deviation of internal tide SSH signals obtained from 18 different 183-day windows, and from 60 different 30-day windows. A general finding of Shriver et al. (2014) is that the normalized internal tide variability—the standard deviation divided by the mean internal tide signal—is relatively small near internal tide generation sites and grows larger as the distance from the generation sites increases. Similar conclusions were drawn in Ansong et al. (2017)'s examination of the nonstationarity of semidiurnal internal tide energy fluxes in HYCOM vs. mooring observations. A basin-scale map, centered on the equatorial Pacific, of total and non-stationary[15] semidiurnal internal tide energy fluxes in HYCOM is shown in Fig. 13.33. While the stationary fluxes emanating from the French Polynesian Islands do not pass through the equator, the total fluxes do. Buijsman et al. (2017) shows that the internal tide nonstationarity in the equatorial Pacific is due to the scattering effects of the system of strong jets on the equator. This is consistent with the picture painted by Ponte and Klein (2015), who explored the scattering of a stationary internal tide by a strong jet in an idealized model of mid-latitude eddy-internal tide interactions. Buijsman's results suggest that the "dead zone" seen in altimeter estimates of the stationary internal tide in the equatorial Pacific may be due to internal tide nonstationarity rather than to dissipation.

The non-stationary internal tide SSH field in HYCOM has also been quantified through frequency spectra. Savage et al. (2017a) computed frequency spectra, from both steric and non-steric SSH fields, at subsampled grid points (taken every 1/4° from one year of hourly 1/25° HYCOM output, and then computed band-integrated variance over subtidal, semidiurnal, diurnal,

---

[15] Note that Buijsman et al. (2017) primarily used the terms "incoherent" and "coherent" rather than "non-stationary" and "stationary".

and supertidal bands (Table 13.2). In the semidiurnal and diurnal bands, the steric SSH variance—dominated in these bands by internal tides—is computed both before and after removal of the stationary internal tides. The variance remaining after removal of the stationary internal tides is taken to be the non-stationary internal tide signal. Fig. 13.34 displays the global map of non-stationary semidiurnal internal tide SSH variance in HYCOM simulations, determined by Savage et al. (2017a) from the method described above. The largest non-stationary internal tide signals are seen at the equator, consistent with the results in Buijsman et al. (2017). The equator is also a "hotspot" in global maps of non-stationary internal tides made from altimeter observations in Zaron (2017). Zaron's maps of non-stationary internal tides are made from a method somewhat similar to the Savage et al. (2017a) method described above, except that Zaron employed wavenumber rather than frequency spectra. It is encouraging that different analysis methods applied to different outputs (altimetry vs. model output) yield qualitatively similar findings. A comprehensive comparison of non-stationarity in HYCOM vs. altimetry is currently underway (Nelson et al., paper in preparation).

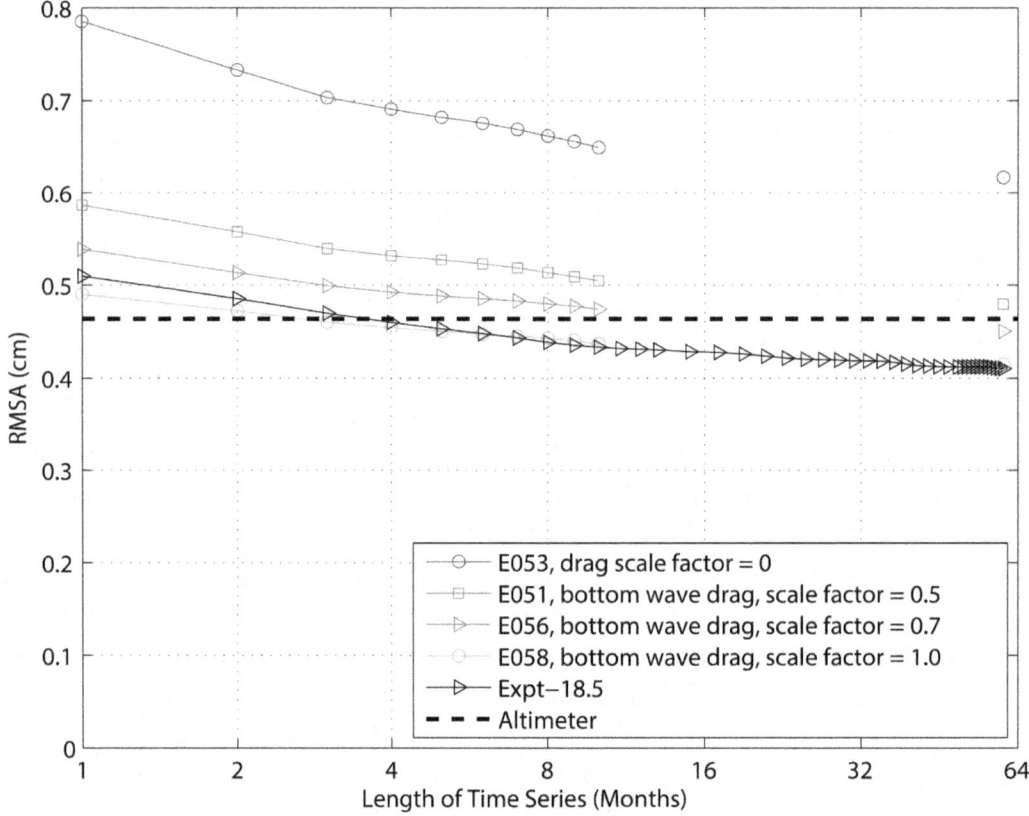

Figure 13.32. Globally averaged root-mean-square amplitude (RMSA) of stationary HYCOM baroclinic M_2 tidal elevations and along-track altimeter value. The HYCOM results are dependent on drag strength (see legend) as well as on the length of the time series used. The x-axis is a base 2 log scale. Expt-18.5 is an older HYCOM experiment about which several papers have been written (e.g., Shriver et al., 2012) and from which 5 year-long SSH output was stored. The 60-month amplitudes shown for the newer simulations are rough estimates based on the changes seen over 60 months in Expt-18.5 with record length. Reproduced from Figure 8 of Ansong et al. (2015), which can be consulted for more details of the E051-E059 simulations, ©American Geophysical Union, Wiley Online Library, used with permission.

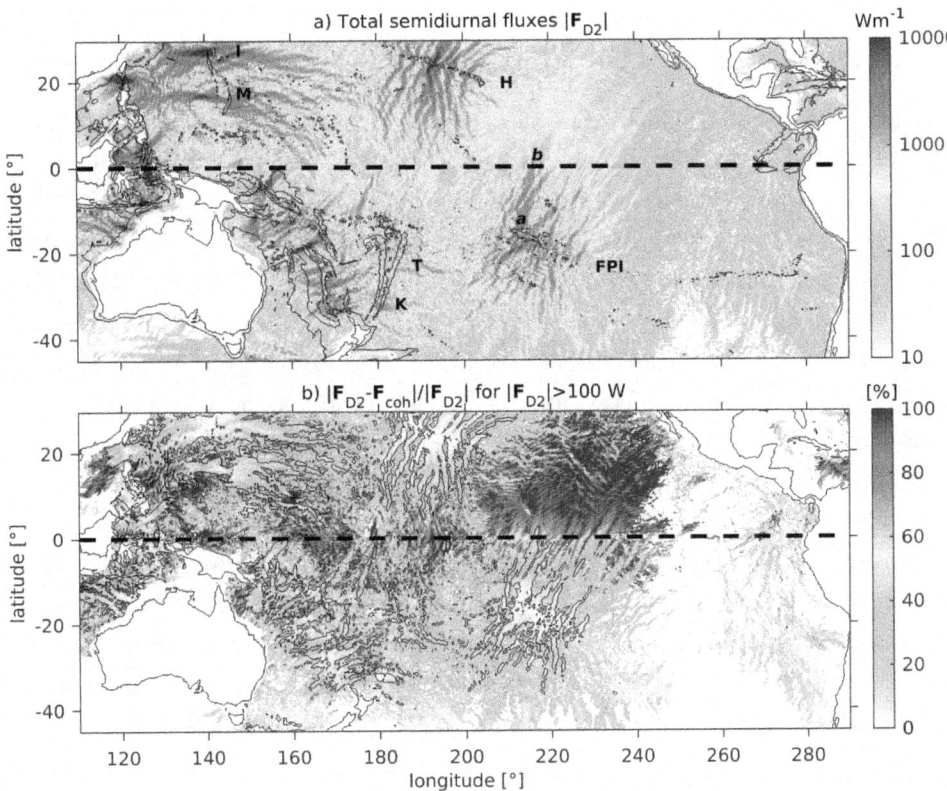

Figure 13.33. (a) The magnitude of the annual-mean semidiurnal band-passed energy fluxes $|F_{D2}|$ in the equatorial Pacific from HYCOM. Bathymetry is contoured at 0 and -2000 m. Hawai'i is abbreviated with H, the French Polynesian Islands with FPI, Tonga with T, the Kermadec Islands with K, the Izu Ogasawara Ridge with I, and the Mariana Islands with M. Spectral properties are computed in Buijsman et al. (2017) along the internal tide beam marked by a-b. (b) The percentage of the sum of the annual-mean incoherent (non-stationary) and cross-term fluxes to the band-passed fluxes. Values coinciding with $|F_{D2}| < 100$ W m^{-1} are not shown. The black contours mark 1000 W m^{-1} values in the band-passed fluxes. Bathymetry is contoured at 0 m. In both subplots, the equator is marked by the dashed black line. Reproduced from Figure 2 of Buijsman et al. (2017), ©American Geophysical Union, Wiley Online Library, used with permission.

Band	Frequencies (cpd)
Subtidal	(1/366)–0.86
Diurnal	0.87–1.05
Semidiurnal	1.86–2.05
Supertidal	2.06–12

Table 13.2. Definitions of frequency bands employed in Savage et al. (2017a).

Nonlinearities and the Internal Gravity Wave Continuum

In this section, we provide a first look at the role of nonlinearities in developing the IGW continuum. We employ a spectral kinetic energy transfer diagnostic in $K-\omega$ space, motivated by wavenumber-domain spectral transfers which have long been used to diagnose energy exchanges between length scales. For example, Scott and Wang (2005), Scott and Arbic (2007), Arbic et al. (2013), and references therein, can be consulted for discussions on using wavenumber-domain spectral transfers and fluxes (the latter being integrals of the transfers) to diagnose inverse and forward cascades of geostrophic flows in the wavenumber domain. Arbic et al. (2012b) and Arbic et al. (2014) describe extending the spectral kinetic energy transfer diagnostic to the frequency and frequency-wavenumber domains, respectively. All of the references mentioned above focus on geostrophic flows and on flows in the ocean. Shang and Hayashi (1990a,b) can be consulted for an earlier description of spectral kinetic energy transfers in frequency space, with application to flows in atmospheric models. Müller et al. (2015) applied a spectral transfer diagnostic in the wavenumber-frequency domain to total (geostrophic plus non-geostrophic) flows in the "wind plus tides" HYCOM simulations. The diagnostic is developed by multiplying Fourier transformed terms in the momentum equation by the Fourier transform of the velocity. If one writes the shallow-water momentum equation as

$$\frac{\partial \vec{u}}{\partial t} = -\vec{u} \bullet \nabla \vec{u} + OT, \tag{18}$$

where OT denotes "other terms", then multiplication by $\vec{u} \bullet$ results in a kinetic energy equation of the form

$$\frac{\partial \frac{1}{2}|\vec{u}|^2}{\partial t} = -\vec{u} \bullet [\vec{u} \bullet \nabla \vec{u}] + \vec{u} \bullet OT. \tag{19}$$

If we instead Fourier transform the terms in equation (18) and then multiply by the Fourier transform of u, we obtain

$$\frac{\partial \hat{KE}}{\partial t} = -\hat{\vec{u}} \bullet [\widehat{\vec{u} \bullet \nabla \vec{u}}] + \hat{\vec{u}} \bullet \widehat{OT}, \tag{20}$$

where KE denotes kinetic energy, $\hat{A}$ denotes the Fourier transform of A, and the spectral kinetic energy transfer T_{KE} due to nonlinear advection is given by

$$T_{KE} = -\hat{\vec{u}} \bullet [\widehat{\vec{u} \bullet \nabla \vec{u}}]. \tag{21}$$

The Fourier transform can be in the wavenumber, frequency, or wavenumber-frequency domains. Fig. 13.35 displays the results of this diagnostic applied in the wavenumber-frequency domain to the 1/12.5° and 1/25° HYCOM "wind plus tides" simulations in Müller et al. (2015). Fig. 13.36 displays the spectral kinetic energy transfers in 1/25° HYCOM as replotted in Savage et al. (2017a) with a nonlinear color bar, which makes some key features easier to see. The transfers shown in Figs. 13.35 and 13.36 reveal extraction of kinetic energy (blue colors; negative values of transfer) from near-inertial flows and semidiurnal tides, and deposition of energy (red colors; positive values of transfer) along the vertical mode-1 IGW continuum, all consistent with the classical paradigm of the formation of the IGW continuum spectrum. The spectral kinetic energy transfers become more vigorous when the model resolution increases from 1/12.5° to 1/25°, meaning that, as is also seen in the spectrum, the model has not yet achieved numerical convergence with respect to the IGW continuum. The "patchy" structure of spectra and spectral transfers was examined in Müller et al. (2015); see Figure 4 and related discussions in that paper.

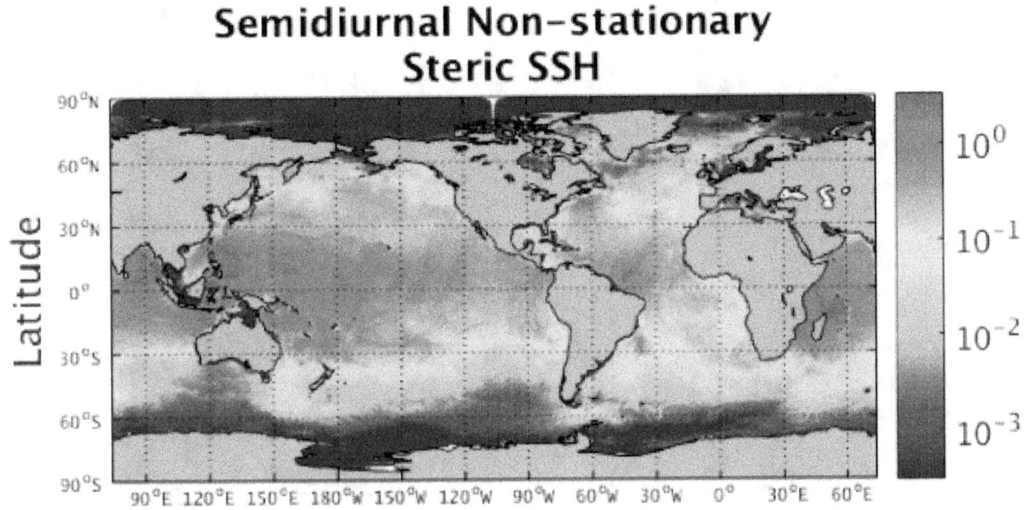

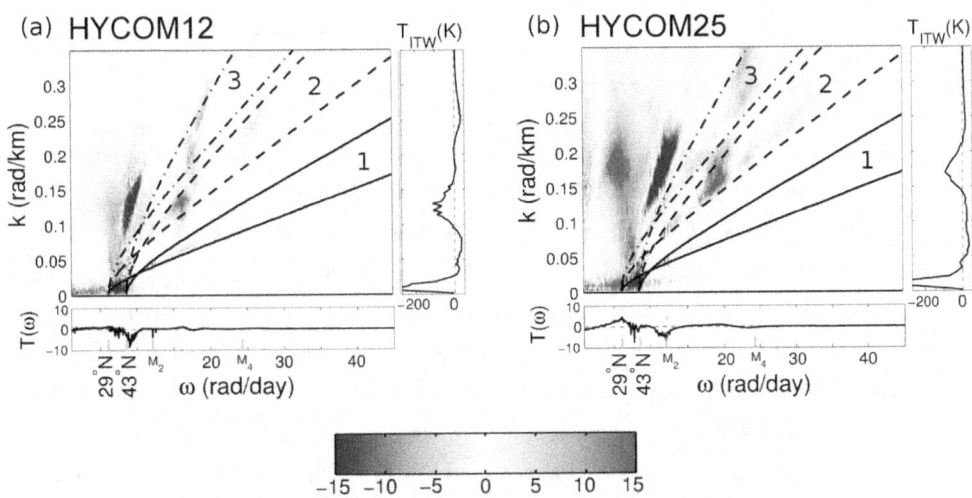

Figure 13.34. Global steric SSH variance (cm²) from HYCOM25 in the semidiurnal band (frequencies 1.86–2.05 cpd) after stationary internal tides have been removed via harmonic analysis. Reproduced from Figure 15 of Savage et al. (2017a), ©American Geophysical Union, Wiley Online Library, used with permission.

Figure 13.35. As in Fig. 13.27 but for nonlinear spectral kinetic energy transfer $T_{KE}(K, \omega)$, with units (10^{-9} W/kg)/[(rad/day)(rad/km)]. The colorbar is linear, and the dispersion curves and mode band numbers are shown in black. Energy is taken out of the blue regions (for instance, low-mode near-inertial and semidiurnal internal tidal motions) and added to the red regions (for instance, higher-frequency and higher-wave number internal waves). Reproduced from Figure 3 of Müller et al. (2015), ©American Geophysical Union, Wiley Online Library, used with permission.

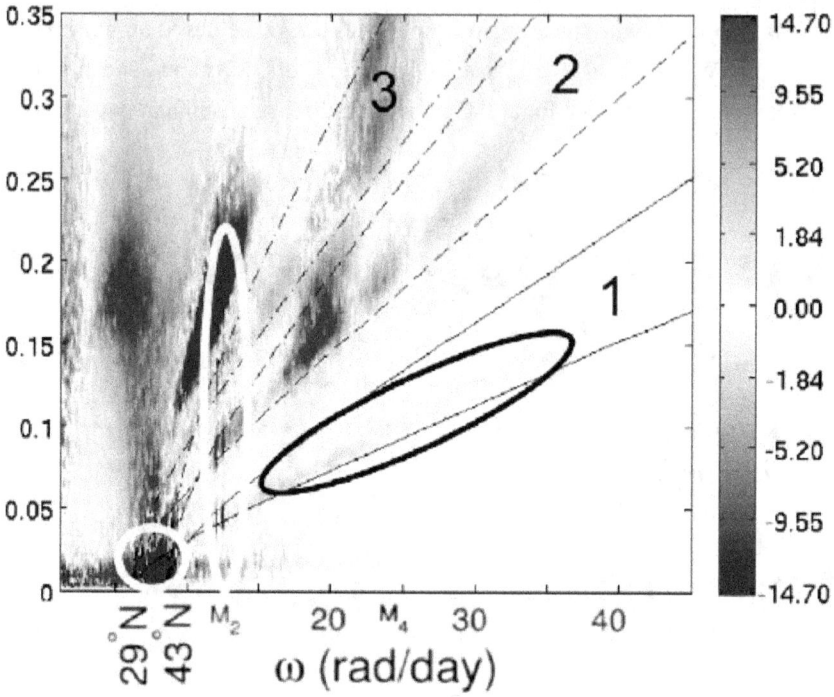

Figure 13.36. Nonlinear spectral kinetic energy transfer $T_{KE}(K, \omega)$, with units (10^{-9} W/kg)/[(rad/day)(rad/km)], computed in Müller et al. (2015) and replotted with a nonlinear color bar to emphasize some of the key features described in Fig. 13.35. Reproduced from Figure 1 of Savage et al. (2017a), ©American Geophysical Union, Wiley Online Library, used with permission.

Applications to Swath Satellite Altimetry

Next we explore more applications of global internal tide and IGW continuum models for swath satellite altimetry missions such as the SWOT mission. The SWOT swath altimeter will measure SSH in two dimensions. SWOT will measure the SSH signature of internal tides and the IGW continuum spectrum, albeit in a temporally aliased manner, and is expected to improve the mapping of these high-frequency motions. At the same time, such motions will have to be accurately subtracted out from SWOT data in order for other oceanic motions, such as mesoscale and submesoscale eddies, to be examined. Stationary internal tides, non-stationary internal tides, and the IGW continuum spectrum all contribute to SSH variance, and the latter two classes of motions will be especially challenging to predict. Savage et al. (2017a) developed global maps of the steric and non-steric SSH variance integrated over subtidal, diurnal, semidiurnal, and supertidal bands. The division of motions into steric and non-steric components, and into different frequency bands, helps to separate different dynamical regimes. For instance, the subtidal steric SSH variance is dominated by mesoscale currents and eddies. Fig. 13.37 shows the relative sizes of non-steric and steric components, and of total SSH variance, in the different frequency bands employed in Savage et al. (2017a). A few general tendencies are easily seen. The largest signals in the total and non-steric SSH are the semidiurnal and diurnal tides, followed by subtidal motions. The largest signals in the steric SSH are subtidal signals, followed by semidiurnal and diurnal tides. The fraction of semidiurnal and diurnal SSH variance that is non-stationary is significantly larger for steric signals than for non-steric signals. The supertidal steric SSH variance, dominated by the IGW continuum spectrum, is greatly enhanced with increased resolution, meaning that numerical convergence for this class of motions has not yet been achieved. This in turn implies that the steric supertidal SSH variance seen in Fig. 13.37 is likely be a lower bound on the actual value.

Fig. 13.38 displays the global map of supertidal steric SSH variance in 1/25° HYCOM, as computed in Savage et al. (2017a). To our knowledge, this is the first global map made of the IGW SSH signal. The IGW SSH signal is generally larger in the tropics than in higher latitudes, with the mid-to-high latitude North Pacific being a notable exception, and is generally larger in the Pacific and Indian Oceans than in the Atlantic. Although the models are not yet accurate enough to operationally remove such signals, we believe that simply mapping the signals as we have done here represents an important first step in understanding the challenges and opportunities that internal tides and the IGW continuum present for the SWOT mission.

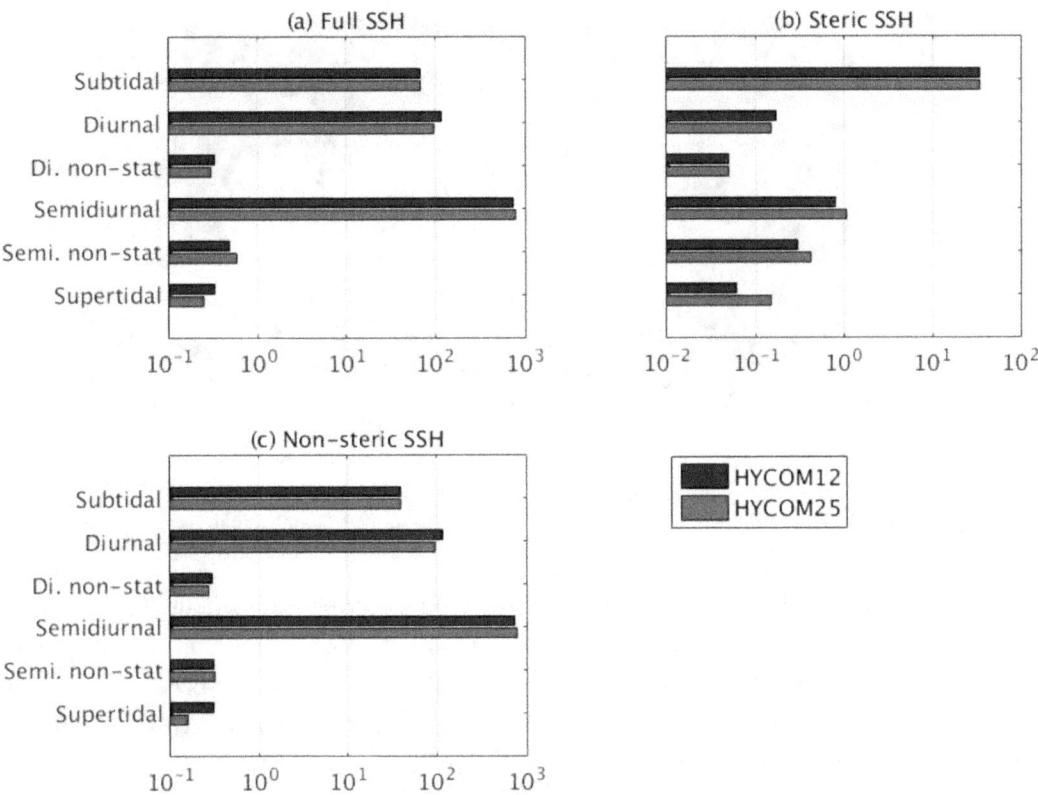

Figure 13.37. Bar graph of spatially averaged HYCOM12 and HYCOM25 variance in cm^2 in subtidal, diurnal, non-stationary diurnal, semidiurnal, non-stationary semidiurnal, and supertidal bands in (a) full, (b) steric, and (c) non-steric SSH. Variance was averaged over deep ocean grid points (seafloor depths greater than 1000 m). Axis limits differ between subplots. Reproduced from Figure 10 of Savage et al. (2017a), ©American Geophysical Union, Wiley Online Library, used with permission.

Of central interest for the SWOT mission is the SSH wavenumber spectrum. Indeed, the NASA mission requirements for SWOT are written in terms of the wavenumber spectrum (Fu et al., 2012). As discussed in Richman et al. (2012; see also references therein), the slopes of the SSH wavenumber spectrum differ according to the dominant underlying dynamics. The IGW continuum is expected to have a spectrum that falls off as K^{-2}, while "interior quasi-geostrophy" and "surface quasi-geostrophy" are expected to have spectra that are proportional to K^{-5} and $K^{-11/3}$, respectively. The slope of the SSH wavenumber spectrum therefore represents a test of whether the underlying dynamics are dominated by IGWs, "interior quasi-geostrophic" motions, or "surface quasi-geostrophic" motions. Fig. 13.39 displays the slope of the SSH wavenumber spectrum, over the 70-250 km wavelength mesoscale band defined by Xu and Fu (2012), from total and low-passed HYCOM SSH model output. The map made from total SSH, which includes high-frequency motions such as internal tides and the IGW continuum, shows flatter slopes in regions where strong IGWs can overwhelm the signals from lower-frequency motions such as mesoscale eddies.

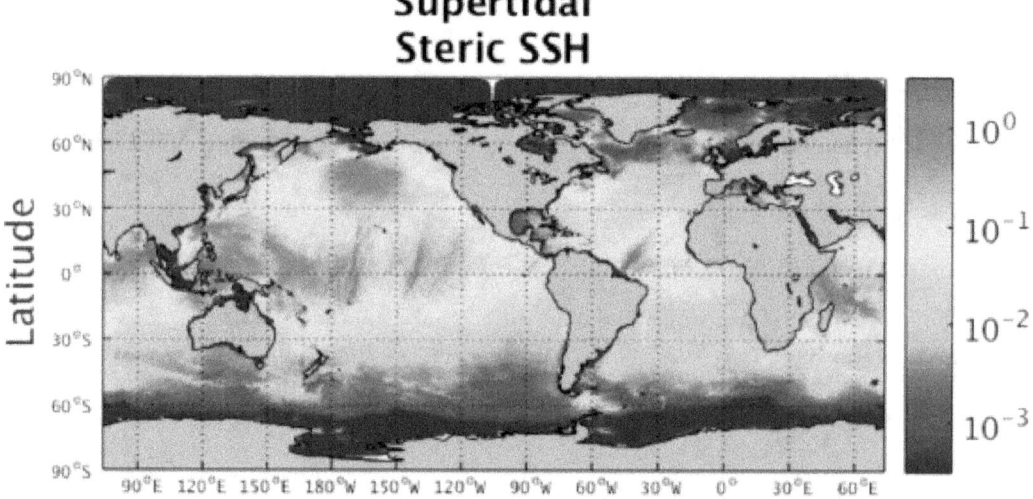

Figure 13.38. Global steric SSH variance (cm^2) from HYCOM25 in the supertidal band (frequencies 2.06–12 cpd). Reproduced from Figure 16 of Savage et al. (2017a), ©American Geophysical Union, Wiley Online Library, used with permission.

The story laid out by Richman et al. (2012) has been further corroborated by others, for instance, Callies and Ferrari (2013), who focused on kinetic energy more than on SSH, and on along-track Acoustic Doppler Current Profiler (ADCP) data rather than model output. Rocha et al. (2016a) compared along-track wavenumber spectra in the MITgcm to spectra from ADCP data, in one Southern Ocean location. Qiu et al. (2017) also analyzed along-track ADCP data, and defined a transition length scale delineating length scales over which the wavenumber spectrum is dominated by geostrophic motions from length scales over which the wavenumber spectrum is dominated by IGWs. Rocha et al. (2016b) demonstrated that the relative strengths of low- and high-frequency motions in the MITgcm output have a seasonal cycle. Qiu et al. (2018) explored seasonal changes in the transition scale using the MITgcm simulations. Wang et al. (2018) used the MITgcm output to simulate a field campaign for in-situ calibration and validation of the SWOT SSH wavenumber spectrum. Savage et al. (2017b) displayed SSH variance wavenumber spectra integrated over different frequency bands—subtidal, tidal, and supertidal. Fig. 13.40 shows the results from the MITgcm over several regions. The figure demonstrates that internal tides and the IGW continuum can dominate the spectra at high wavenumbers—the wavenumbers of interest for SWOT—at least in some regions. Because Richman et al. (2012) separated low- and high-frequency motions with a 2-day low-pass filter, rather than a frequency-wavenumber spectrum, and because we did not realize when writing Richman et al. (2012) that our models carried a partial IGW continuum spectrum, we attributed the high-frequency motions to internal tides only in that paper. The results in Fig. 13.40, and in Rocha et al. (2016a), make clear that the IGW continuum, as well as internal tides, contribute significantly to the wavenumber spectrum at higher wavenumbers.

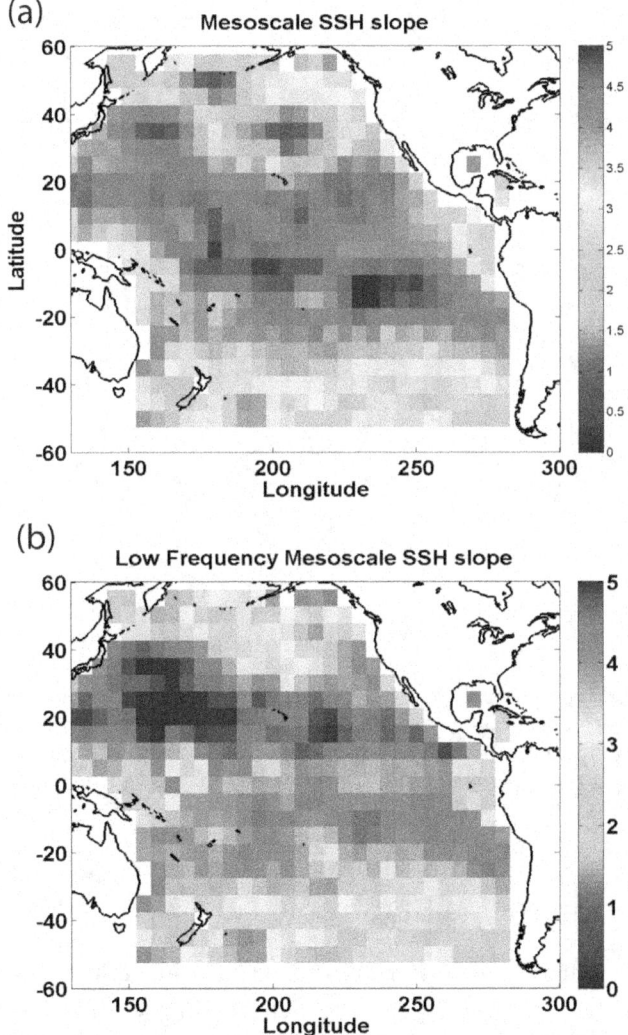

Figure 13.39. Least-squares estimate of the slope of the SSH wavenumber spectrum in the North Pacific, computed over the 70–250 km band, in HYCOM. Slopes are computed from spectra of (a) total SSH and (b) low-frequency SSH. All slopes are multiplied by -1 to make them positive. Reproduced from Figure 7 of Richman et al. (2012), ©American Geophysical Union, Wiley Online Library, used with permission.

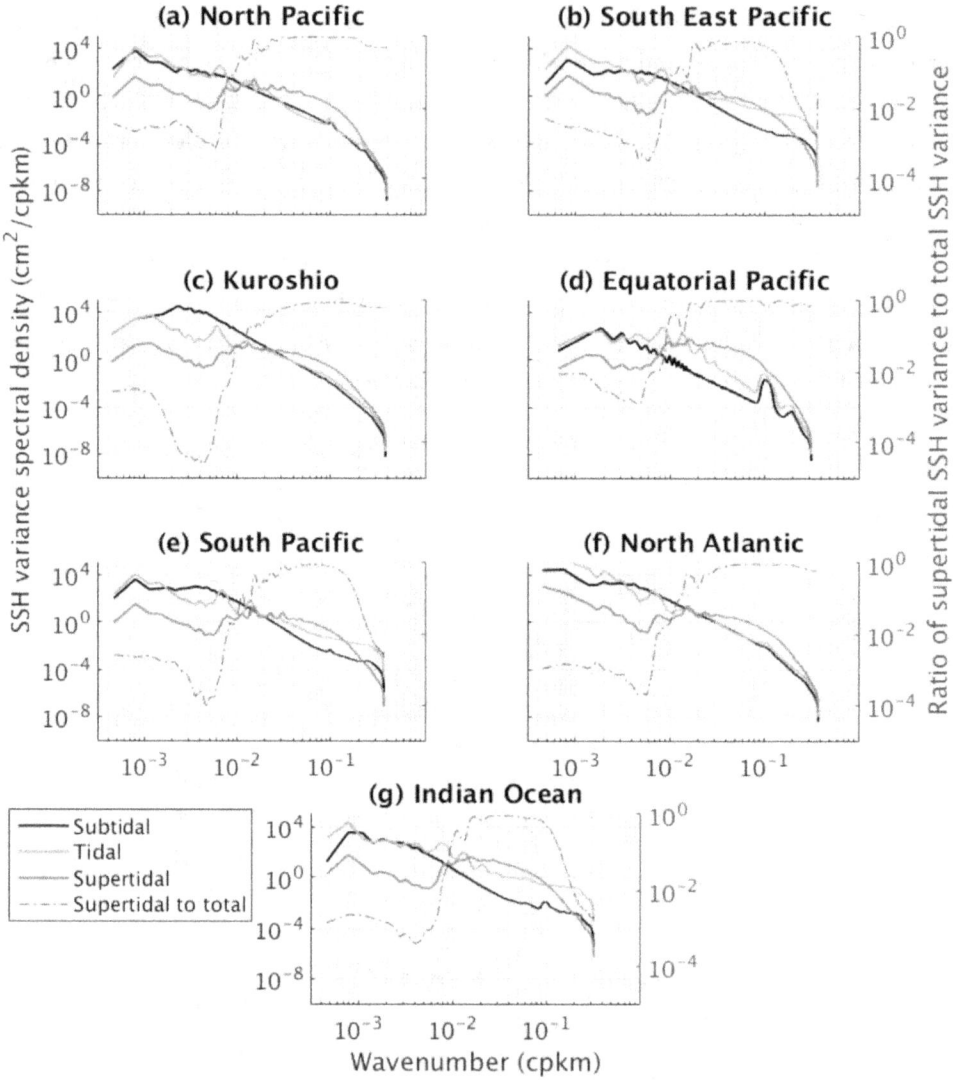

Figure 13.40. Horizontal wave number spectral density of SSH variance in seven regions in MITgcm48 [see Savage et al. (2017b) for precise locations] integrated over subtidal, tidal, and supertidal frequency bands [see Savage et al. (2017b) for definition of bands]. Right-hand axis shows ratio of supertidal to total SSH variance as a function of isotropic wave number. Reproduced from Figure 7 of Savage et al. (2017b), ©American Geophysical Union, Wiley Online Library, used with permission.

Improving Barotropic Tidal Accuracy

We continue to improve the accuracy of the barotropic tides in HYCOM. The discrepancies D of barotropic tidal elevations in the earlier HYCOM "wind plus tides" simulations are comparable to those in many other tuned forward models (e.g., Arbic et al., 2004; Egbert et al., 2004; Stammer et al., 2014; among others), but are still significantly larger than those in data-constrained tide models (e.g., Egbert et al., 1994; Ray, 1999; Lyard et al., 2006; Taguchi et al., 2014; Stammer et al., 2014; among others). Motivated by this, Ngodock et al. (2016) employed the Augmented State Ensemble

Kalman Filter (ASEnKF) technique to reduce the barotropic tidal elevation errors in HYCOM. An ensemble of perturbations to the tidal forcing was generated, with each ensemble member having horizontal length scales comparable to those in the open-ocean barotropic tides. HYCOM simulations are performed with each perturbation, and the ASEnKF machinery then selects a linear combination of the ensemble members that minimizes the barotropic tidal elevation misfits between HYCOM and TPXO. Fig. 13.41 shows the M_2 elevation errors in early HYCOM simulations, vs. HYCOM simulations with some improvements in Southern Ocean bathymetry and in iterating the SAL term, vs. HYCOM simulations that in addition employ the Ngodock et al. (2016) ASEnKF.

The ASEnKF reduces the errors throughout the globe, but large errors remain in some regions, especially in the North Atlantic. The HYCOM group is pursuing two paths for further improvements in the HYCOM barotropic tides, both based upon the fact that the North Atlantic houses three locations (the Hudson Strait, the Northwest European Shelf, and the Bay of Fundy) with large coastal tides, and the fact that such regions have been shown to have a significant back-effect upon open-ocean tides (Arbic et al., 2007, 2009; Arbic and Garrett, 2010). Because of the large back-effect, it is likely that improvement of coastal tides in key regions such as this will yield improvements in the modeled open-ocean tides. In one research path, we have developed two-way nesting in HYCOM, so that the coastal tides can be run with higher resolution, hence improving them at relatively low cost (Jeon et al., in review). In another path, we will use adjoint machinery applied to the dynamical core of TPXO to generate perturbations, for the ASEnKF machinery, that have maximum impact on the global tidal solutions. We expect that regions of large coastal tides will play a prominent role in such perturbations.

Summary of Completed Work, Ongoing Work, and Future Challenges

Completed work

To briefly summarize a large body of work, tides have been inserted into a small number of high-resolution ("eddying") global three-dimensional circulation models run by the international oceanography community, including the HYCOM simulations that are run operationally by the United States Navy, and some very-high-resolution MITgcm simulations run on NASA supercomputers. The tides in the HYCOM simulations have been compared to a number of in-situ and altimetric datasets, and a number of papers have been published. These papers have been summarized here, and the reader is referred to the original papers for more details. The HYCOM tidal simulations carry a realistic amount of internal tide nonstationarity, an important consideration for the SWOT swath altimeter mission, which will need to accurately remove both stationary and non-stationary internal tides before examination of non-tidal phenomena such as mesocale and submesoscale eddies can take place. The HYCOM simulations have also been shown to carry a partial IGW continuum spectrum, a first for global models (Müller et al., 2015). The IGW continuum spectrum is more fully developed in new simulations of the MITgcm carried out at higher horizontal resolutions (up to 1/48°) and vertical resolutions (90 z-levels) than the 1/25°, 41-hybrid level HYCOM simulations. The HYCOM and MITgcm IGW continuum spectra have been compared to dynamic height variance spectra from McLane profilers (Savage et al., 2017b) and the MITgcm spectra have been compared to kinetic energy spectra from Acoustic Doppler Current Profilers (Rocha et al., 2016a). It seems inevitable to us that in the near future many high-resolution global ocean models will include tides, and that many researchers will be exploring the applications of such models to operational oceanography, understanding satellite altimeter observations, and achieving better understanding of mixing in the global ocean.

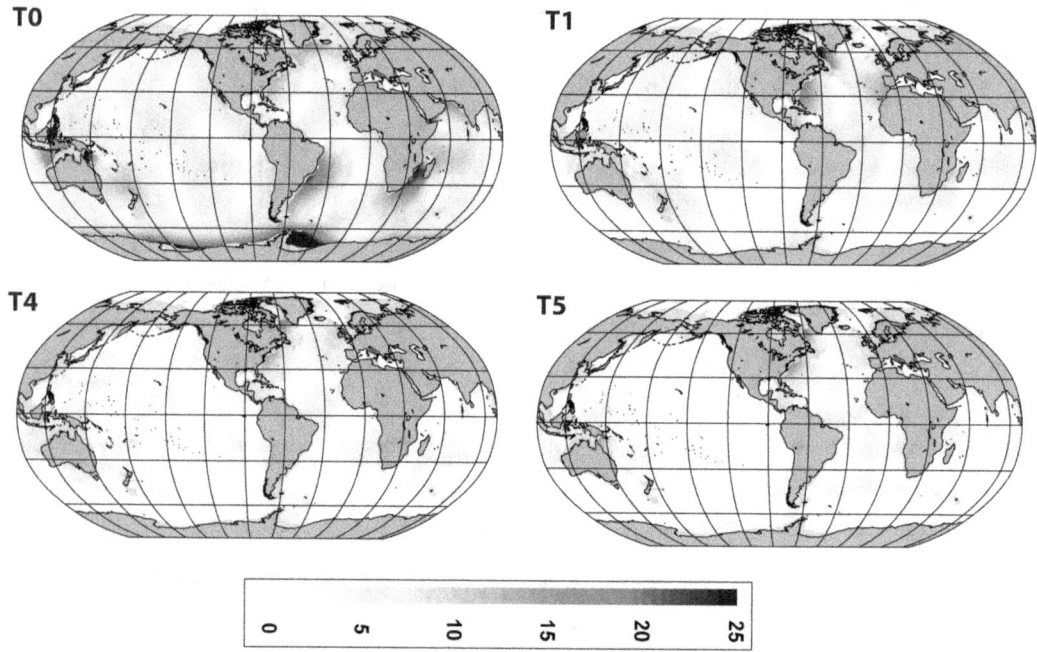

Figure 13.41. Maps of the M_2 RMS error (cm) between the HYCOM simulations and TPXO8: T0, early eight-constituent simulation with scalar SAL and Garner (2005) wave drag; T1, intermediate five-constituent simulation with bathymetry extended to include floating Antarctic ice shelves, tuned Jayne and St. Laurent (2001) wave drag, and iterated SAL; T4, ASEnKF predicted tide using a 0.5 mm constant global observation error; T5, blended ASEnKF predicted tide combining an Atlantic-only prediction with 1 cm observation error and T3 for the rest of the ocean. Reproduced from Figure 6 of Ngodock et al. (2016), which can be consulted for more details including details of the T3 simulation.

Ongoing work

In a partial list of ongoing work, we are examining tidal mixing fronts in shelves in global HYCOM (Timko et al., paper in revision), comparing non-stationary internal tides in HYCOM vs. altimetry (Nelson et al., paper in preparation), examining the IGW kinetic energy spectrum in MITgcm and HYCOM vs. McLane profilers (Ansong et al., paper in preparation), performing regional runs of the MITgcm, forced by global MITgcm at their boundaries, with even higher resolution, and comparing the vertical wavenumber spectrum of IGW kinetic energy and density variance from the regional simulations with spectra from McLane profilers (Arbic et al., paper in preparation), computing the SSH horizontal wavenumber spectra from the same regional simulations, and comparing the IGW kinetic energy and temperature variance spectra from global HYCOM and MITgcm with spectra computed from ~3000 historical moored instruments (Luecke et al., paper in preparation).

Challenges for the future

There are many remaining challenges for global internal tide and IGW continuum models. One view of the applications of such models for satellite altimetry is that the global IGW field consists of near-inertial flows, which are an important source of energy for the IGW continuum but which do not have large SSH signals, and three different classes of IGWs that do have SSH signals. Stationary internal tides will be the easiest of these to predict, although even they represent a significant challenge for global internal tide and IGW models. Quantitative feature matching between different empirical internal tide models, and between empirical models and HYCOM, has just begun (Carrère et al., paper in preparation). Non-stationary internal tides will be more difficult to predict accurately, as such a prediction would require accurate modeling of mesoscale eddies in conjunction with accurate mapping of internal tides. Our HYCOM simulations contain an ASEnKF that improves the accuracy of modeled barotropic tides, which should improve the accuracy of the modeled internal tides, at the same time that they contain data assimilation acting on mesoscale eddies. It is possible that accurate internal tide modeling in conjunction with accurate placement of mesoscale eddies will allow prediction of non-stationary internal tides. It remains to be seen whether this challenge can be met in time for the SWOT mission. The technical challenges of assimilating eddies while accurately modeling internal tides are considerable, but we note that the United States Navy has as a goal to run such a system operationally by 2018. For now, we have mapped the amount of SSH variance in non-stationary vs. stationary internal tides, as well as in the IGW continuum, in HYCOM (Savage et al., 2017a).

There are many more technical challenges for models of this size and complexity. Hourly output is standard for tidal analysis, but this makes the storage requirements 24 times more burdensome than the storage requirements of models that store daily-averaged outputs. Yet the spectra shown in Fig. 13.28 indicate that even hourly model outputs are not frequent enough to avoid aliasing of the IGW continuum in the outputs of such models. Another technical problem is that near-inertial waves, a crucial source of energy for the IGW continuum, are generated by high-frequency winds, while ocean models are often forced by atmospheric models that only update their fields every 3-6 hours. One solution to this problem would be to run high-resolution coupled ocean-atmosphere models, in which the atmosphere updates every time step, rather than high-resolution "forced ocean" models. The SAL term presents another challenge and opportunity. As noted in Hendershott (1972), the SAL term acts on all barotropic motions, not just tides, and should therefore act on the entire mass (non-steric) signal in ocean models. The computational cost and complexity of inline computations involving spherical harmonics has prevented most researchers from incorporating a "full SAL" (not just one acting on tidal motions) into global ocean general circulation models. Vinogradova et al. (2015) represents a significant recent attempt to solve this problem. The SAL term did not add greatly to the cost of running their model. However, they were running relatively coarse resolution simulations, and it is not clear whether their solution would work well in simulations with much higher horizontal resolution. We have been in frequent discussions with mathematicians who have techniques that could potentially incorporate SAL without greatly increasing the computational cost.

A technical challenge for HYCOM in particular is the presence of numerical instabilities in specific regions, noted by Buijsman et al. (2016) and others. Ongoing efforts are underway to implement the Adcroft et al. (2008) solution for this instability in HYCOM. Technical challenges for the MITgcm simulations presented here include deciding whether these simulations should continue to be developed, and implementing a plan for more extensive comparison of the simulations to observations, as has been done with the HYCOM simulations.

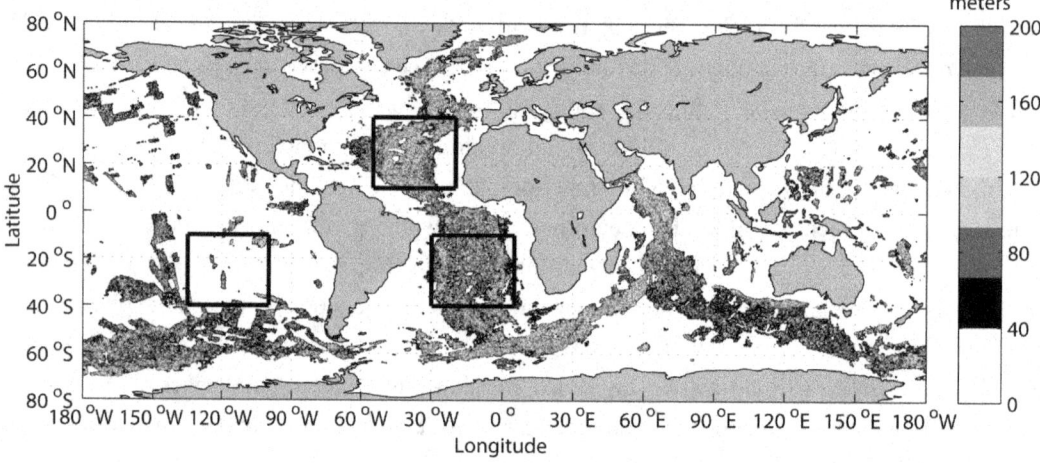

Figure 13.42. Area-weighted root mean square of the difference between model bathymetry with and without the statistical abyssal hill roughness of Goff and Arbic (2010) at 1/12.5°. The root mean square is calculated over a 1/4° × 1/4° sub-grid. Contours are drawn from 40 m to 200m at 40m intervals. The boxed regions shown in the South Pacific, North Atlantic and South Atlantic oceans are used in Timko et al. (2017) to estimate internal wave response to changes in bottom roughness. Reproduced from Figure 4 of Timko et al. (2017).

Another challenge relates to bathymetry. The Smith and Sandwell (1997) bathymetry dataset does not resolve topographic features with horizontal length scales less than about 10 km, except in limited regions where high-quality acoustic sounding data is available. As a result, abyssal hills, the most common landforms on Earth, are largely unresolved in global bathymetric datasets. It also means that the horizontal resolution of some global models, and many regional models, is quickly outstripping the actual feature resolution of the underlying topography—except, as noted above, in special regions where high-quality soundings are available throughout the domain. Goff and Arbic (2010) and Goff (2010) make note of this problem and offer, as a partial remedy, realizations, in both physical and spectral space, of topography that does include abyssal hills with the "correct" statistics. The spectral characterization of the Goff and Arbic (2010) and Goff (2010) products was used in offline computations of a parameterized topographic wave drag, that were then inserted inline into a global model of the eddying general circulation (Trossman et al., 2013, 2016). Both the spectral and physical space realizations were used by Melet et al. (2013) in a linear analysis study of tidal conversion over abyssal hills.

In Timko et al. (2017), the statistical abyssal hill roughness was employed in HYCOM simulations of the internal tides, set up under the same simplified conditions used in Arbic et al. (2004) and Simmons et al. (2004b), to examine the extra explicitly resolved internal tide activity arising from the small-scale roughness missing in global bathymetric datasets. Timko et al. (2017) found that the roughness does indeed increase internal tide generation and energy levels, especially in the higher vertical modes that are generated by smaller-scale topography (St. Laurent and Garrett, 2002; Simmons et al., 2004b). Fig. 13.42 shows the small-scale roughness that we introduce through the Goff and Arbic (2010) and Goff (2010) work, as it appears once filtered down to 1/12.5° horizontal resolution. The method employed by Timko et al. (2017) to add statistical roughness to a global internal tide model could be adapted to researchers studying other phenomena, in other models, and with different grid resolutions. In fact, in regional models with higher resolutions than the 1/12.5° and 1/25° resolutions used in Timko et al. (2017)'s global models, the effects of the statistical roughness will be greater because as resolution increases the radius of the smoother that must be applied to put the raw statistical bathymetry onto a grid will decrease.

A persistent question, discussed at length amongst the co-authors of Arbic et al. (2004) and Simmons et al. (2004b), and again amongst the co-authors of Ansong et al. (2015), is whether a parameterized topographic internal wave drag, or some other damping mechanism, should be employed in global internal tide and IGW continuum models. One point of view is that parameterizing internal wave drag while at the same time partially resolving the spectrum constitutes double counting. A point of view argued in Arbic et al. (2004, 2010), Ansong et al. (2015), and Buijsman et al. (2016) is that such models do not resolve internal wave breaking and therefore still require some damping of low-mode internal tides to make up for this deficit. Ansong et al. (2015) shows that, at least for 1/12.5° HYCOM simulations, some damping is necessary for the internal tides to agree well with altimetry. We are currently planning to test damping mechanisms other than parameterized topographic internal wave drag, such as upper-ocean wave-wave interactions, in HYCOM. Of course, at some time in the future, models will reach the point where they are resolving much of the IGW spectrum and its cascade to small-scale dissipation, at which parameterized damping will indeed have to be abandoned.

Global internal tide and IGW continuum models appear to be resolving some of the nonlinear interactions needed to produce an IGW continuum spectrum; presumably, if they were not doing so, then they would not carry a partial continuum. An important question is what such models should do to "go the rest of the way" to the even smaller scales where internal wave breaking takes place. Driving regional models which have higher horizontal and vertical resolutions is one solution; letting eddy viscosities take care of the partial IGW cascade that has taken place is another. No doubt other ideas for handling this problem will arise as such models continue to advance.

Finally, it is likely that global internal tide and IGW continuum models will help us to understand better the processes leading to the IGW continuum. The classical paradigm is that the energy source for the IGW continuum is primarily near-inertial waves and internal tides, with nonlinear interactions filling out the continuum. This view was supported by high-resolution two-dimensional (depth + one horizontal direction) numerical simulations in Sugiyama et al. (2009). However, Barkan et al. (2017) finds that a Garrett-Munk IGW continuum spectrum can develop in idealized models that contain only near-inertial waves and mesoscale eddies. Barkan's results suggest a role for mesoscale eddies in the formation of the IGW continuum spectrum. Whether there is a role for internal lee waves generated by mesoscale eddies flowing over topographic features in forming the IGW continuum, is another open question, discussed by the lead author with Stephanie Waterman and Bernard Barnier in September 2017. We conclude by stating that there are many interesting questions to explore with global internal tide and IGW continuum models.

Dedication

Brian Arbic dedicates his contributions to the work presented here to his beloved brother Joel Bernard Arbic, who passed away on December 2, 2017. Joel will forever be missed by his wife, two sons, parents, two brothers, in-laws, two nieces, and other relatives, as well as by his many friends.

Author contributions

BKA wrote this book chapter, with considerable organizational and editing help from ADN. JKA wrote the appendix. Several co-authors contributed to the editing and other presentation issues. All authors contributed to one or more of the following underpinnings of this book chapter: prior articles, comparison datasets, the development of the global oceanic internal tide and gravity wave simulations presented here, and numerous associated discussions. The first global internal tide simulations were performed at GFDL by BKA and HLS in collaboration with RWH and Stephen Garner. The first high-resolution, three-dimensional global simulations with simultaneous atmospheric and tidal forcing were performed in Naval Research Laboratory (NRL) HYCOM simulations by BKA, AJW, and EJM. Newer HYCOM simulations have been performed by JFS, JKA, and LZ, based upon coding legwork by AJW, and written about in papers by BKA, JKA, MCB, CAL, MM, HEN, JGR, ACS, JFS, and PGT. IS provided help with the Kalman filtering used in Ngodock et al. (2016). MHA, JTF, RBS, and ZZ provided observational datasets used in model-data comparisons, and helped with general interpretation of the results. The MITgcm simulations were configured and carried out by BCN, CEH, CNH, DM, and RBC. MM and RMP included the tidal forcing in the MITgcm simulations.

Acknowledgements

We thank many individuals, far too numerous to list here, for illuminating discussions on, and reviews of, our global internal tide and wave modeling research over the years. We thank Florent Lyard for a thorough review of this book chapter. We thank Richard Ray for very helpful comments on the first two sections of this chapter, and Stephen Garner and Eric Kunze for additional comments. BKA thanks Carl Wunsch for class notes on the IGW Sturm-Liouville problem, which helped greatly in formulating the appendix.

This book chapter stems from lectures at the 2017 GODAE school. BKA thanks the organizers of the GODAE school for their invitation to lecture there, and for putting together and editing a book based upon the lectures. BKA also thanks the four GODAE school students assigned to this chapter, Shaun Rigby, Alejandra Rodriguez, Pablo Rodriguez Ros, and Aurpita Saha, who provided helpful comments on early drafts. BKA participated in the GODAE school while on sabbatical. BKA thanks many French colleagues, especially Thierry Penduff, Rosemary Morrow, and Nadia Ayoub, for their help in procuring a sabbatical year in France.

The simulations in Arbic et al. (2004) and Simmons et al. (2004b) were performed on the high-performance computing cluster at the Geophysical Fluid Dynamics Laboratory (GFDL) of the National Oceanic and

Atmospheric Administration (NOAA). BKA thanks Stephen Garner, his postdoctoral advisor, for crucial collaborations on these early global internal tide papers.

Many of the papers underlying this review chapter were written by current and former members (JKA, CAL, ADN, ACS, and PGT) of BKA's research group at University of Michigan. BKA and his group members gratefully acknowledge financial support from US Naval Research Laboratory (NRL) contract N000173-06-2-C003, US Office of Naval Research (ONR) grants N00014-07-1-0392, N00014-09-1-1003, N00014-11-1-0487, N00014-15-1-2288, and N00014-17-1-2958, US National Science Foundation (NSF) grants OCE-0623159, OCE-0924481, OCE-0968783, OCE-0960820, and OCE-1351837, NASA grants NNX13AD95G, NNX16AH79G, and NNX17AH55G, a NASA Earth and Space Science Fellowship grant NNX16AO23H to ACS, and the University of Michigan Associate Professor Support Fund, derived from the Margaret and Herman Sokol Faculty Awards. Start-up and other funds, as well as local computer resources, from The University of Texas at Austin, Florida State University, and the University of Michigan are also acknowledged by BKA. MM collaborated with Arbic's group as a subcontractor and as such acknowledges support from a subcontract of ONR grant N00014-11-1-0487 to the University of Victoria. BKA's group acknowledges Michael Messina for substantial help in navigating computer resources at the University of Michigan.

Substantial analysis was also done at NRL, and most of the HYCOM simulations were performed there. All of the HYCOM simulations reported on here, except for those used in Timko et al. (2017), were supported by grants of computer time from the Department of Defense (DoD) High Performance Computing Modernization Program at the Navy DoD Supercomputing Resource Center. MCB, EJM, HEN, JGR, JFS, IS, AJW, and LZ acknowledge support from the ONR projects "Dynamics of the Indonesian Throughflow and its remote im impact", "Eddy resolving global ocean prediction including tides", "Global and remote littoral forcing in global ocean models", "Ageostrophic vorticity dynamics of the ocean", "NCOM-4DVAR, A multiscale approach for assessing predictability of ASW environment", "Extending predictability in coastal environments", "HYCOM global ocean forecast skill assessment", and the ONR and National Ocean Partnership Program (NOPP) sponsored project "Improving global surface and internal tides through two-way coupling with high resolution coastal models". MCB also acknowledges ONR grant N00014-15-1-2288.

RMP acknowledges the support of NSF grant OPP-1708308 and NASA grant NNX11AQ12G. HLS was supported by NSF (CPT) grant OCE-0968838 and ONR grant N00014-09-1-0399.

The MITgcm simulations were performed on the Pleiades supercomputer cluster at NASA Ames. We thank Ayan Chaudhuri for helping with inclusion of tidal forcing in MITgcm. PGT and BKA acknowledge the Texas Advanced Computing Center (TACC) at The University of Texas at Austin for providing HPC resources that have contributed to the research results reported in Timko et al. (2017). URL: http://www.tacc.utexas.edu. We thank the captains, crews, technical personnel and investigators who collected the in-situ observations used here over many cruises. Similarly, we thank the numerous engineers and scientists who made the satellite altimetry data used here possible.

This is NRL contribution NRL/BC/7320-18-3786 and has been approved for public release.

Appendices

A   Linear internal gravity wave dispersion relation

A.1   Development of Sturm-Liouville problem

This appendix develops the linear IGW dispersion relation, and the related eigenfunctions for the horizontal velocities and vertical displacements of density surfaces, under the simplifying assumptions of a rigid lid at the ocean surface, and plane wave solutions. The more realistic case, of a free surface, can be found in, for instance, Nugroho (2017). The orthogonality of the eigenfunctions is affected by the nature of the boundary condition (rigid lid versus free surface).

The linearized governing equations for the conservation of mass, momentum, and density, under the Boussinesq/incompressible approximation, with total density $\rho = \bar{\rho} + \rho_0(z) + \rho'(x, y, z, t)$, are given by

$$\frac{\partial u}{\partial x} + \frac{\partial v}{\partial y} + \frac{\partial w}{\partial z} = 0, \tag{22}$$

$$\frac{\partial u}{\partial t} - fv = -\frac{1}{\bar{\rho}}\frac{\partial p'}{\partial x}, \tag{23}$$

$$\frac{\partial v}{\partial t} + fu = -\frac{1}{\bar{\rho}}\frac{\partial p'}{\partial y}, \tag{24}$$

$$\bar{\rho}\frac{\partial w}{\partial t} = -\frac{\partial p'}{\partial z} - \rho'g, \tag{25}$$

$$\frac{\partial \rho'}{\partial t} + w\frac{d\rho_0}{dz} = 0, \tag{26}$$

(see Gill, 1982; Wunsch and Stammer, 1997), where (u, v, w) are the velocities in the zonal, meridional, and vertical (x, y, z) directions respectively, f is the Coriolis frequency, p is pressure, g is acceleration due to gravity, $\bar{\rho}$ is a constant density, $\rho_0(z)$ is the background density variation and $\rho'(x, y, z, t)$ is the perturbation density due to internal gravity waves. Note the slight change in notation between the appendix and the remainder of the text; in earlier sections, ρ_0 was used to denote the constant average seawater density.

Combine equations (25) and (26) to get

$$\frac{\partial^2 w}{\partial t^2} - w\frac{g}{\bar{\rho}}\frac{d\rho_0}{dz} = -\frac{1}{\bar{\rho}}\frac{\partial^2 p'}{\partial z \partial t}. \tag{27}$$

The buoyancy frequency, N, is defined as

$$N^2 = -\frac{g}{\bar{\rho}}\frac{d\rho_0}{dz}, \tag{28}$$

such that (27) becomes

$$\frac{\partial^2 w}{\partial t^2} + N^2 w = -\frac{1}{\bar{\rho}}\frac{\partial^2 p'}{\partial z \partial t}. \tag{29}$$

Combine (22)-(24) to obtain

$$\left(\frac{\partial^2}{\partial t^2} + f^2\right)\frac{\partial w}{\partial z} = \frac{1}{\bar{\rho}}\frac{\partial}{\partial t}\left(\frac{\partial^2 p'}{\partial x^2} + \frac{\partial^2 p'}{\partial y^2}\right),$$
$$\Longrightarrow \left(\frac{\partial^2}{\partial t^2} + f^2\right)\frac{\partial w}{\partial z} = \frac{1}{\bar{\rho}}\frac{\partial}{\partial t}\nabla_h^2 p'. \tag{30}$$

Introduce separation of variables (for separable solutions):

$$[u(x, y, z, t), v(x, y, z, t)] = [U(x, y, t), V(x, y, t)]F(z), \tag{31}$$

$$w(x, y, z, t) = \tilde{P}(x, y, t)G(z), \tag{32}$$

$$p'(x, y, z, t) = \bar{\rho}\tilde{P}(x, y, t)F(z). \tag{33}$$

Note that each of the equations (31)-(33) should actually be a summation over different vertical modes (e.g. Kundu, 1990, page 499); for example,

$$w(x, y, z, t) = \sum_{n=0}^{\infty} \tilde{P}_n(x, y, t)G_n(z),$$

where n is the vertical mode number. For simplicity we leave the equations as given. In addition, we set all dependent variables to be proportional to $e^{-i\sigma t}$, where σ is the wave frequency. Our goal is to determine the vertical structure functions $F(z)$ and $G(z)$. Substitute (32)-(33) into (29) to get

$$[(-i\sigma)^2 + N^2]\tilde{P} \cdot G(z) = i\sigma \tilde{P} \cdot \frac{dF}{dz},$$

$$\Longrightarrow (N^2 - \sigma^2)G(z) = i\sigma \frac{dF}{dz}, \tag{34}$$

$$\Longrightarrow \boxed{G(z) = \frac{i\sigma}{N^2 - \sigma^2}\frac{dF}{dz}}.$$

Note that because $i\sigma$ is a constant, the z dependence of the vertical displacement function is given by an expression in Wunsch and Stammer (1997), viz.

$$\boxed{G_1(z) = \frac{G(z)}{i\sigma} = \frac{1}{N^2 - \sigma^2}\frac{dF}{dz}}. \tag{35}$$

Substituting (32)-(33) into (30) yields

$$(-i\sigma)^2 \tilde{P}\frac{dG}{dz} + f^2 \tilde{P}\frac{dG}{dz} = -(i\sigma)F(z)\nabla_h^2 \tilde{P}, \tag{36}$$

$$\Longrightarrow (\sigma^2 - f^2)\tilde{P}\frac{dG}{dz} = (i\sigma)F(z)\nabla_h^2 \tilde{P},$$

$$\Longrightarrow \frac{(\sigma^2 - f^2)(dG(z)/dz)}{(i\sigma)F(z)} = \frac{\nabla_h^2 \tilde{P}}{\tilde{P}}. \tag{37}$$

Note that the left-hand-side of (37) is a function of only z and the right-hand-side is independent of z. Thus we can write

$$\frac{(\sigma^2 - f^2)(dG(z)/dz)}{(i\sigma)F(z)} = \frac{\nabla_h^2 \tilde{P}}{\tilde{P}} = -\gamma^2 \tag{38}$$

$$\Longrightarrow \boxed{(\sigma^2 - f^2)\frac{dG(z)}{dz} = -(i\sigma)\gamma^2 F(z)} \tag{39}$$

where γ^2 is a separation constant. From equations (34) and (39), we can determine equations for only the $F(z)$ and $G(z)$ functions. From these equations, we respectively obtain

$$\frac{dG}{dz} = \frac{d}{dz}\left(\frac{i\sigma}{N^2 - \sigma^2}\frac{dF}{dz}\right),$$

$$\frac{dG}{dz} = -\frac{i\sigma}{\sigma^2 - f^2}\gamma^2 F(z).$$

Equating the above equations results in

$$\boxed{\frac{d}{dz}\left(\frac{\sigma^2 - f^2}{N^2 - \sigma^2}\frac{dF}{dz}\right) + \gamma^2 F(z) = 0.} \tag{40}$$

From (39) we see that

$$(\sigma^2 - f^2)\frac{d^2 G}{dz^2} = -(i\sigma)\gamma^2 \frac{dF}{dz}.$$

From equation (34), we have

$$\frac{dF}{dz} = \frac{N^2 - \sigma^2}{i\sigma}G(z). \tag{41}$$

Therefore, we have

$$(\sigma^2 - f^2)\frac{d^2 G}{dz^2} = -\gamma^2(N^2 - \sigma^2)G(z),$$

$$\boxed{\frac{d^2 G}{dz^2} + \frac{N^2 - \sigma^2}{\sigma^2 - f^2}\gamma^2 G(z) = 0.} \qquad (42)$$

Thus the equation governing the vertical modal structure of velocity and pressure is (40) and that governing the vertical displacement is (42). The boundary conditions for these equations come from requiring that the vertical velocity vanish ($w = 0$) at the top ($z = 0$) and bottom ($z = -H$) of the domain. From equation (32), this implies that $G(z) = 0$ at $z = 0$ and $z = -H$. Equation (41) then implies that $\frac{dF}{dz} = 0$ at $z = 0$ and $z = -H$.

The main equations and their boundary conditions, with an assumed rigid lid, are rewritten below for convenience:

$$\frac{d}{dz}\left(\frac{\sigma^2 - f^2}{N^2 - \sigma^2}\frac{dF}{dz}\right) + \frac{1}{\lambda^2}F(z) = 0, \qquad (43)$$

with boundary conditions

$$\frac{dF}{dz} = 0 \quad \text{at} \quad z = 0, -H, \qquad (44)$$

where the condition at $z=0$ is the rigid lid assumption, and

$$\frac{d^2 G}{dz^2} + \left(\frac{N^2 - \sigma^2}{\sigma^2 - f^2}\right)\frac{1}{\lambda^2}G(z) = 0, \qquad (45)$$

with boundary conditions

$$G(z) = 0 \quad \text{at} \quad z = 0, -H, \qquad (46)$$

where H is the water depth and $1/\lambda^2 = \gamma^2$ represents the eigenvalues of the eigenfunctions $F(z)$ and $G(z)$. We denoted γ^2 as $1/\lambda^2$ here so that the equations resemble those solved in our MATLAB code (discussed below).

A.1.1 Dispersion relation & meaning of eigenvalues

Here we determine the meaning of the eigenvalues in the vertical mode equations, and we derive the dispersion relation for internal gravity waves. First we combine the governing equations (22)-(26) into a single equation for the vertical velocity, w. We perform $\nabla_h^2 \cdot (29) + \frac{\partial}{\partial z}(30)$ to eliminate p', and obtain, after some steps,

$$\frac{\partial^2}{\partial t^2}\left(\nabla_h^2 + \frac{\partial^2}{\partial z^2}\right)w + N^2 \nabla_h^2 w + f^2 \frac{\partial^2 w}{\partial z^2} = 0. \qquad (47)$$

We now assume a plane-wave solution:

$$w = G(z)e^{i(kx+ly-\sigma t)}. \qquad (48)$$

Note that the plane-wave assumption does not impact vertical structure; however, in the actual ocean, the horizontal structure is not always well-represented by plane waves. Substituting (48) into (47) results in

$$(-i\sigma)^2\left[G(z)(ik)^2 + G(z)(il)^2 + \frac{d^2 G}{dz^2}\right] + N^2[G(z)(ik)^2 + G(z)(il)^2] + f^2\frac{d^2 G}{dz^2} = 0,$$

$$\implies \sigma^2(k^2 + l^2)G(z) - \sigma^2\frac{d^2 G}{dz^2} - N^2(k^2 + l^2) + f^2\frac{d^2 G}{dz^2} = 0,$$

$$\implies (k^2 + l^2)(\sigma^2 - N^2)G(z) + (f^2 - \sigma^2)\frac{d^2 G}{dz^2} = 0,$$

$$\implies (k^2 + l^2)(N^2 - \sigma^2)G(z) + (\sigma^2 - f^2)\frac{d^2 G}{dz^2} = 0.$$

Thus,
$$\frac{d^2 G}{dz^2} + \left(\frac{N^2 - \sigma^2}{\sigma^2 - f^2}\right)(k^2 + l^2)G(z) = 0. \quad (49)$$

Comparison of equations (49) and (45) demonstrates that the eigenvalues, λ, are related to the horizontal wavenumber:
$$\boxed{\lambda^2 = \frac{1}{k^2 + l^2}}. \quad (50)$$

An alternative (and simpler) way to derive (50) is to employ equation (38):
$$\frac{\nabla_h^2 \tilde{P}}{\tilde{P}} = -\gamma^2 = -\frac{1}{\lambda^2}, \quad (51)$$

and assume
$$\tilde{P} = P_0 e^{i(kx + ly - \sigma t)}. \quad (52)$$

Substituting (52) into (51) gives
$$\frac{\nabla_h^2 \tilde{P}}{\tilde{P}} = -(k^2 + l^2) = -\frac{1}{\lambda^2},$$
$$\Rightarrow \lambda^2 = \frac{1}{k^2 + l^2}.$$

We rewrite equation (45) as
$$\frac{1}{N^2 - \sigma^2} \frac{d^2 G}{dz^2} + \frac{1}{(\sigma^2 - f^2)\lambda^2} G(z) = 0, \quad (53)$$

and define the eigenspeed, c_e, as
$$\frac{1}{c_e^2} = \frac{1}{(\sigma^2 - f^2)\lambda^2}, \quad (54)$$

such that
$$\boxed{c_e^2 = \frac{(\sigma^2 - f^2)}{k^2 + l^2}}, \quad (55)$$

or
$$\boxed{\sigma^2 = f^2 + c_e^2(k^2 + l^2)}. \quad (56)$$

Equation (56) is the **dispersion relation** of internal gravity waves for a general $N^2(z)$. It is analogous to the dispersion relation for Poincare waves (surface gravity waves in the absence of horizontal boundaries), with the surface gravity wave speed $\sqrt{gH}$ replaced by the eigenspeed c_e for IGWs. Equation (53) may now be written as
$$\frac{1}{N^2 - \sigma^2} \frac{d^2 G}{dz^2} + \frac{1}{c_e^2} G(z) = 0. \quad (57)$$

To see that the eigenspeed, c_e, is the speed of propagation of waves, we present two arguments: a comparison to the surface gravity wave dispersion relation, and a scaling analysis.

1) We note that equation (56) is identical to the dispersion relation for surface (external) gravity waves if c_e^2 is identified with gH. This argument is similar to that presented in Kundu (1990; page 501).
2) Since $z \sim H$, the left-hand term of (53) scales as (assuming $N \gg \sigma$) $\frac{1}{N^2 H^2} G(z) = \frac{1}{c_e^2} G(z)$. Therefore, we have $c_e \approx N H$.

To derive the dispersion relation for a constant $N(z)$, we employ equation (57) and assume $G \propto e^{imz}$, such that $\frac{d^2}{dz^2} \sim -m^2$. Thus (57) becomes

$$\frac{-m^2}{N^2 - \sigma^2} = -\frac{1}{c_e^2}, \tag{58}$$

$$\implies c_e^2 = \frac{N^2 - \sigma^2}{m^2}.$$

Substituting (58) into the dispersion relation (56) gives

$$\sigma^2 = f^2 + \frac{N^2 - \sigma^2}{m^2}(k^2 + l^2),$$

$$\implies \sigma^2 m^2 = f^2 m^2 + (N^2 - \sigma^2)(k^2 + l^2),$$

$$\implies \sigma^2(m^2 + k^2 + l^2) = f^2 m^2 + N^2(k^2 + l^2), \tag{59}$$

$$\boxed{\implies \sigma^2 = \frac{f^2 m^2 + N^2(k^2 + l^2)}{m^2 + k^2 + l^2}.}$$

We see immediately from (59) that the frequency of IGWs is bounded by f and N, as discussed previously.

A.2 Orthogonality & Orthonormality of eigenfunctions

As mentioned above, the eigenfunctions of the Sturm-Liouville equation are known to be orthogonal [see the book by Heinbockel (2003) for additional details]. We show this in the following section. Experts, or those uninterested in the details of the proof of the orthogonality of the Sturm-Liouville system, may wish to skip this section and proceed to the next section. Note once again that when the rigid lid assumption employed here is replaced by a more realistic free surface boundary condition, the orthogonality of the displacement functions becomes more involved (Nugroho, 2017). Using vertical modes computed using a rigid lid assumption to perform an energy budget in shelf regions can be problematic (Nugroho, 2017; Florent Lyard, personal communication, 2018).

A.2.1 General Orthogonal Functions

The inner product of two real functions $f(z)$ and $g(z)$ with respect to a weighting function $W(z)$ (which is never negative) over the interval $a \leq z \leq b$ is defined by

$$(f, g) = \int_a^b W(z) f(z) g(z) \mathrm{d}z. \tag{60}$$

The inner product of the function with itself is equal to the square of the norm and is written as

$$(f, f) = ||f||^2 = \int_a^b W(z) f(z)^2 \mathrm{d}z. \tag{61}$$

A set of functions $\{f_1(z), f_2(z), \cdots, f_n(z), \cdots, f_m(z), \cdots\}$ is said to be orthogonal over an interval $a \leq z \leq b$ with respect to a weight function $W(z) > 0$ if for all integer values of n and m, with $n \neq m$, the inner product of f_m with f_n satisfies

$$(f_m, f_n) = \int_a^b W(z) f_m(z) f_n(z) \mathrm{d}z = 0, \quad m \neq n. \tag{62}$$

If the sequence of functions $\{f_n(z)\}$, $n = 0, 1, 2, \cdots$ is an **orthogonal** sequence we can write for integers m and n that the inner product satisfies the relation [combining (61) and (62)]

$$(f_m, f_n) = ||f_n||^2 \delta_{mn} = \begin{cases} 0, & m \neq n \\ ||f_n||^2, & m = n \end{cases} \tag{63}$$

where δ_{mn} is the Kronecker delta. In the special case where $||f_n||^2 = 1$, for all values of n, the sequence of functions is said to be orthonormal over the interval $a \leq z \leq b$.

If a sequence $\{g_n(z)\}$ is orthogonal with respect to the weight function $W(z)$, then we can construct a new orthonormal sequence $f_n(z)$, defined by

$$f_n(z) = \frac{g_n(z)}{\|g_n(z)\|}. \tag{64}$$

We show below that equation (64) is orthogonal. Multiply (64) by $W(z)f_m(z)$ and integrate to get

$$\begin{aligned}
I &= \int W(z)f_n(z)f_m(z)\mathrm{d}z = \frac{1}{\|g_n(z)\|}\int W(z)g_n(z)f_m(z)\mathrm{d}z, \\
&= \frac{1}{\|g_n(z)\|}\int W(z)g_n(z)\frac{g_m(z)}{\|g_m(z)\|}\mathrm{d}z, \\
&= \frac{1}{\|g_n(z)\|\|g_m(z)\|}\int W(z)g_n(z)g_m(z)\mathrm{d}z, \\
&= 0, \quad \text{for} \quad n \neq m,
\end{aligned} \tag{65}$$

because the sequence $\{g_n(z)\}$ is orthogonal. Moreover, from (64)

$$\|f_n(z)\| = \frac{\|g_n(z)\|}{\|g_n(z)\|} = 1, \tag{66}$$

$$\implies \|f_n(z)\|^2 = 1,$$

and therefore, the sequence $\{f_n(z)\}$ is orthonormal.

The proof above shows that the most important property is orthogonality. Once an orthogonal sequence is obtained, we can easily construct an orthonormal sequence by dividing by the norm of the functions.

A.2.2 Orthogonality of eigenfunctions

Consider the Sturm-Liouville equation

$$\frac{\mathrm{d}}{\mathrm{d}z}\left(Q(z)\frac{\mathrm{d}F}{\mathrm{d}z}\right) + W(z)\lambda F(z) = 0, \tag{67}$$

with general boundary condition

$$\alpha F(z) + \beta \frac{\mathrm{d}F(z)}{\mathrm{d}z} = 0 \quad \text{at} \quad z = a, b, \tag{68}$$

where α and β are constants. We remark that, one could go through the following derivations using the more general boundary condition above, as is done in Nugroho (2017). However, here we use simplified boundary conditions, either

$$F(z) = 0 \quad \text{at} \quad z = a, b, \tag{69}$$

or

$$\frac{\mathrm{d}F(z)}{\mathrm{d}z} = 0 \quad \text{at} \quad z = a, b. \tag{70}$$

To determine orthogonality, we consider equation (67) with two different values of λ such that

$$\frac{\mathrm{d}}{\mathrm{d}z}\left(Q(z)\frac{\mathrm{d}F_n}{\mathrm{d}z}\right) + W(z)\lambda_n F_n(z) = 0, \tag{71}$$

$$\frac{\mathrm{d}}{\mathrm{d}z}\left(Q(z)\frac{\mathrm{d}F_m}{\mathrm{d}z}\right) + W(z)\lambda_m F_m(z) = 0. \tag{72}$$

Perform $[(72) \times F_n] - [(71) \times F_m]$ to obtain

$$F_n\frac{\mathrm{d}}{\mathrm{d}z}\left(Q(z)\frac{\mathrm{d}F_m}{\mathrm{d}z}\right) - F_m\frac{\mathrm{d}}{\mathrm{d}z}\left(Q(z)\frac{\mathrm{d}F_n}{\mathrm{d}z}\right) + (\lambda_m - \lambda_n)W(z)F_m F_n = 0. \tag{73}$$

Before returning to (73), we note that

$$\frac{d}{dz}\left(F_n Q(z)\frac{dF_m}{dz}\right) = F_n\frac{d}{dz}\left(Q(z)\frac{dF_m}{dz}\right) + Q(z)\frac{dF_m}{dz}\frac{dF_n}{dz}, \tag{74}$$

$$\frac{d}{dz}\left(F_m Q(z)\frac{dF_n}{dz}\right) = F_m\frac{d}{dz}\left(Q(z)\frac{dF_n}{dz}\right) + Q(z)\frac{dF_m}{dz}\frac{dF_n}{dz}, \tag{75}$$

so that (74)-(75) gives

$$F_n\frac{d}{dz}\left(Q(z)\frac{dF_m}{dz}\right) - F_m\frac{d}{dz}\left(Q(z)\frac{dF_n}{dz}\right) = \frac{d}{dz}\left(F_n Q(z)\frac{dF_m}{dz}\right) - \frac{d}{dz}\left(F_m Q(z)\frac{dF_n}{dz}\right). \tag{76}$$

Substituting (76) into (73) yields

$$\frac{d}{dz}\left(F_n Q(z)\frac{dF_m}{dz}\right) - \frac{d}{dz}\left(F_m Q(z)\frac{dF_n}{dz}\right) + (\lambda_m - \lambda_n)W(z)F_m F_n = 0. \tag{77}$$

Integrating equation (77) from $z = a$ to $z = b$ results in

$$\left[F_n Q(z)\frac{dF_m}{dz}\right]_a^b - \left[F_m Q(z)\frac{dF_n}{dz}\right]_a^b + (\lambda_m - \lambda_n)\int_a^b W(z)F_m F_n dz = 0.$$

The first two terms vanish due to the boundary condition (69) or (70):

$$\frac{dF_m(z)}{dz} = 0 = \frac{dF_n(z)}{dz} \quad \text{at} \quad z = a, b. \tag{78}$$

Therefore, we obtain

$$(\lambda_m - \lambda_n)\int_a^b W(z)F_m F_n dz = 0, \quad n \neq m,$$
$$\implies \int_a^b W(z)F_m F_n dz = 0, \quad n \neq m, \tag{79}$$

proving the **orthogonality** of the $F(z)$ eigenfunctions.

Note that when $m = n$, the vanishing of the left-hand side does not give us any additional information but the resulting integral is expected to be nonzero (Heinbockel, 2003). As shown above, by having an orthogonal sequence $\{F_n(z)\}$, we can construct an orthonormal sequence by dividing by the norm.

Comparing (67) to (43) shows that the weight function $W(z) = 1$ so that the sequence $\{F_n(z)\}$ may be othonormalized as

$$\frac{1}{||F_n(z)||||F_m(z)||}\int_{-H}^{0} F_n(z)F_m(z)dz = \delta_{mn}. \tag{80}$$

On the other hand, comparing (67) to (45) shows that

$$W(z) = \left(\frac{N^2 - \sigma^2}{\sigma^2 - f^2}\right),$$

so that we can orthonormalize $\{G_n(z)\}$ by

$$\frac{1}{||G_n(z)||||G_m(z)||}\int_{-H}^{0}\left(\frac{N^2 - \sigma^2}{\sigma^2 - f^2}\right)G_n(z)G_m(z)dz = \delta_{mn}, \tag{81}$$

such that

$$||G_n(z)||^2 = \int_{-H}^{0}\left(\frac{N^2 - \sigma^2}{\sigma^2 - f^2}\right)G_n^2(z)dz.$$

A.3 Computation of expansion coefficients

The vertical structure of internal gravity waves in the ocean (or from a model) can be decomposed into a linear combination of dynamical modes (Wunsch, 1975) such that

$$\mathbf{u}(z,t) = \sum_{n=0}^{M} \mathbf{u}_n(t) F_n(z), \tag{82}$$

$$\eta(z,t) = \sum_{n=1}^{M} \eta_n(t) G_n(z), \tag{83}$$

where $\mathbf{u} = (u, v)$ is the horizontal velocity and η is the isopycnal displacement, n is an index of vertical mode number, z is vertical coordinate, t is time, M is the total number of vertical modes, and $F_n(z)$ and $G_n(z)$ are the fixed vertical structures of the mth baroclinic mode respectively.

In practice, $\mathbf{u}(z, t)$ and $\eta(z, t)$ are known from either observational measurements or from a model output, and $F_n(z)$ and $G_n(z)$ can be computed from a profile of buoyancy frequency, $N(z)$. Here we discuss how to compute the expansion coefficients $\mathbf{u}_n(t)$ and $\eta_n(t)$ in (82)-(83). We first discuss a theoretical approach, and later outline a second approach via the method of least squares.

A.3.1 Theoretical solution method

Suppose that $F_n(z)$ is an eigenfunction such that

$$\frac{1}{\mathcal{H}} \int_{-H}^{0} F_n(z) F_m(z) \mathrm{d}z = \delta_{mn}, \tag{84}$$

where $\mathcal{H} = \|F_n(z)\| \|F_m(z)\|$. The F's may also be orthonormalized in a depth-averaged sense by setting

$$\mathcal{H} = \|F_n(z)\|^2 = H, \tag{85}$$

[e.g., see equation (2.3) in Flierl (1978)], and similarly for the displacement modal functions. If we expand the velocity u as (82):

$$u(z,t) = \sum u_n(t) F_n(z),$$

Then

$$\frac{1}{\mathcal{H}} \int_{-H}^{0} F_n(z) u(z,t) \mathrm{d}z = \frac{1}{\mathcal{H}} \int_{-H}^{0} F_n(z) \sum_m u_m(t) F_m(z) \mathrm{d}z$$

$$= \sum_m u_m(t) \frac{1}{\mathcal{H}} \int_{-H}^{0} F_n(z) F_m(z) \mathrm{d}z$$

$$= \sum_m u_m(t) \delta_{mn}$$

$$= u_n(t).$$

Thus,

$$\boxed{u_n(t) = \frac{1}{\mathcal{H}} \int_{-H}^{0} F_n(z) u(z,t) \mathrm{d}z.} \tag{86}$$

Following an approach similar to the one above, we can determine the expansion coefficients for the vertical displacement. For an orthogonal sequence, $\{G_n(z)\}$, we have

$$\frac{1}{\mathcal{G}} \int_{-H}^{0} W(z) G_n(z) G_m(z) \mathrm{d}z = \delta_{mn}, \tag{87}$$

where $\mathcal{G} = \|G_n(z)\| \|G_m(z)\|$ and $W(z)$ is a weighting function. Using the expansion in (83), we get

$$\frac{1}{\mathcal{G}} \int_{-H}^{0} W(z) G_n(z) \eta(z,t) \mathrm{d}z = \frac{1}{\mathcal{G}} \int_{-H}^{0} W(z) G_n(z) \sum_{m} \eta_m(t) G_m(z) \mathrm{d}z$$

$$= \sum_{m} \eta_m(t) \frac{1}{\mathcal{G}} \int_{-H}^{0} W(z) G_n(z) G_m(z) \mathrm{d}z$$

$$= \sum_{m} \eta_m(t) \delta_{mn}$$

$$= \eta_n(t).$$

Thus,

$$\eta_n(t) = \frac{1}{\mathcal{G}} \int_{-H}^{0} W(z) G_n(z) \eta(z,t) \mathrm{d}z. \tag{88}$$

A.3.2 Least-squares method

Another practical way to determine the expansion coefficients is by solving each of (82)-(83) as a least-squares problem (Dushaw et al., 1995; Nash et al., 2005). For instance, at each time, the u equation from (82) may be cast as a multivariate regression model (Emery and Thompson, 1997), such that

$$u_i = a_0 F_{i0} + a_1 F_{i1} + a_2 F_{i2} \cdots + a_M F_{iM} + \varepsilon_i \tag{89}$$

where each $i \,(= 1, \cdots, N)$ equation is for each depth and ε_i is the error to be minimized in the regression method. The set of equations in (89) can then be written as

$$\mathbf{U} = \mathbf{A} \cdot \mathbf{F} + \mathbf{E}, \tag{90}$$

where

$$\mathbf{U} = \begin{pmatrix} u_1 \\ u_2 \\ \cdots \\ \cdots \\ u_N \end{pmatrix}, \quad \mathbf{F} = \begin{pmatrix} F_{10} & F_{11} & \cdots & F_{1M} \\ F_{20} & F_{21} & \cdots & F_{2M} \\ \cdots & \cdots & \cdots & \cdots \\ \cdots & \cdots & \cdots & \cdots \\ F_{N0} & F_{N1} & \cdots & F_{NM} \end{pmatrix},$$

$$\mathbf{A} = \begin{pmatrix} a_0 \\ a_1 \\ \cdots \\ \cdots \\ a_M \end{pmatrix}, \quad \mathbf{E} = \begin{pmatrix} \varepsilon_1 \\ \varepsilon_2 \\ \cdots \\ \cdots \\ \varepsilon_N \end{pmatrix},$$

and the boldface letters indicate matrices. Solving (90) for $\mathbf{A}$, we obtain

$$\mathbf{A} = (\mathbf{F}' \cdot \mathbf{F})^{-1} \mathbf{F}' \cdot \mathbf{U} \tag{91}$$

(see Emery and Thompson, 1997, page 239-240, for more details). Note that MATLAB's *regress* and *plsregress* functions give solutions similar to (91).

A.4 Solutions to the modal equations

There are no known analytical solutions for general $N^2(z)$ (see e.g., Kundu, 1990; Gill, 1982). Equation (43) is a Sturm-Liouville type of equation for which the eigenfunctions are known to be orthogonal. Often, equation (43) with (44) is solved for $F(z)$ (sometimes the solution approach also gives dF/dz) and then equations (34)-(35) can be used to obtain $G(z)$. To be precise, the $i\sigma$ factor in (34) is a constant, and the entire function in (35) will be adjusted to be orthonormal, so we can ignore the $i\sigma$.

Also note that the λ's are unknown and must be solved for together with the eigenfunctions. Numerical solutions can be found via two main methods:

1) The matrix approach in which equation (43) is often discretized.

2) The **shooting method**, in which equation (43) is solved in two parts. This approach is described in this section and was used to compute linear IGW dispersion relations in Müller et al. (2015) and Savage et al. (2017b), based upon Matlab code that was originally developed by Glenn Flierl for solving the normal modes of the quasi-geostrophic equations (Glenn Flierl, personal communication, 1995; Flierl, 1978). See also Emery & Thomson (1997) for references to NAG Fortran shooting method routines.

   a) Integrating (43) gives

   $$\left(\frac{\sigma^2 - f^2}{N^2 - \sigma^2}\frac{dF}{dz}\right) = -\frac{1}{\lambda^2}\int_{-H}^{0} F(z)dz. \quad (92)$$

   Let

   $$d = -\frac{1}{\lambda^2}\int_{-H}^{0} F(z)dz, \quad (93)$$

   $$\Longrightarrow \frac{dd}{dz} = -\frac{1}{\lambda^2}F(z). \quad (94)$$

   b) Note that (94) is a first order differential equation which can be solved using Euler's method, for example, for a given boundary condition at one end. Because $F(z=0)$ is unknown, we set $F=1$ at the top boundary (later we will normalize the function) and make an initial guess for λ. However, d (which is proportional to the derivative of $F(z)$) is known at the first boundary from the boundary conditions. Once we have d, we can then solve (92); given in the form

   $$\frac{dF}{dz} = \left(\frac{N^2 - \sigma^2}{\sigma^2 - f^2}\right) d, \quad (95)$$

   to get $F(z)$. This gives the value of $F(z)$ at the next step, and the procedure is repeated until we get to the other boundary.

   c) We then check to see whether the solution from the step above satisfies the boundary condition at the other end, $\frac{dF}{dz}|_{z=-H} = 0$. If not, then we need to choose a different initial guess for λ.

   d) The approach used here is to provide the code with several initial guesses for λ and then determine several derivatives, $d(z)$ at $z = -H$. Because the boundary conditions require $d(z) = 0$ at $z = -H$, each pair of λ guesses that gives values of d with different signs at $z = -H$ can be used to narrow-in on the actual λ's that will give $d(-H) \approx 0$. [For example, if λ_1 results in $d = 20 > 0$ and λ_2 results in $d = -40 < 0$ at $z = -H$, then we know that the value lies between the two λ's. We may then narrow-in on the actual λ by choosing the mean of λ_1 and λ_2]. The *fzero* function in MATLAB can be used to find the zeros (i.e., the actual λ's).

   e) Now that we have the λ's that give us the correct boundary condition at the other end, we have our solution $F(z)$. The $F(z)$ functions are then normalized as described in section A.2.

   f) Determine the vertical displacement eigenfunctions $G(z)$ by dividing the d's by $\sigma^2 - f^2$ (see equation 35).

A.4.1 Example solution

Using the buoyancy frequency displayed in Fig. 13.43a, we computed the first three vertical normal modes for velocity (Fig. 13.43b) and vertical displacement (Fig. 13.43c). As is common in the ocean, the buoyancy frequency profile depicted in Fig. 13.43a is surface-intensified (i.e., has its largest values in the upper ocean). The three lowest horizontal velocity normal modes, $F(z)$, depicted in Fig. 13.43b also take on their largest values in the upper ocean. The mode-2 and mode-3 displacement normal modes, $G(z)$, take on their largest values in deeper parts of the water column. Figs. 13.43b and 13.43c demonstrate that, as is dictated by the mathematics of the Sturm-Liouville problem, the nth horizontal velocity mode has n zero crossings, while the nth displacement mode has n-1 zero crossings.

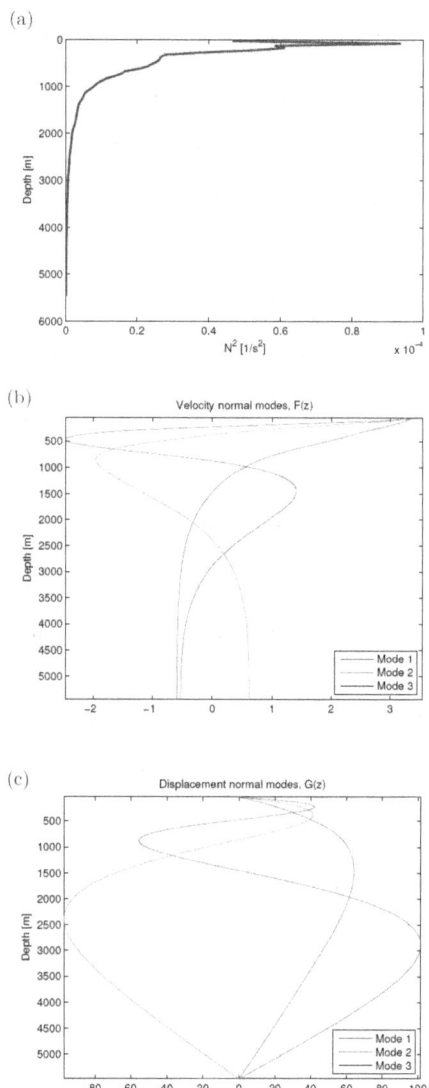

Figure 13.43. (a) Buoyancy frequency profile. Vertical mode structures for (b) horizontal velocity, and (c) vertical displacement. The modal structures are dimensionless, and satisfy orthogonality conditions as described in the text.

References

Accad, Y., and C.L. Pekeris (1978), Solution of the tidal equations for the M_2 and S_2 tides in the world oceans from a knowledge of the tidal potential alone. Philosophical Transactions of the Royal Society of London A290, 235-266, doi:10.1098/rsta.1978.0083.

Adcroft, A., R. Hallberg, and M. Harrison (2008), A finite volume discretization of the pressure gradient force using analytical integration. Ocean Modelling 22, 106-113, doi:10.1016/j.ocemod.2008.02.001.

Alford, M.H. (2003), Redistribution of energy available for ocean mixing by long-range propagation of internal waves. Nature 423, 159–162, doi:10.1038/nature01628.

Alford, M.H., J.A. McKinnon, Z. Zhao, R. Pinkel, J. Klymak, and T. Peacock (2007), Internal waves across the Pacific. Geophysical Research Letters 34, L24601, doi:10.1029/2007GL031566.

Ansong, J.K., B.K. Arbic, M.C. Buijsman, J.G. Richman, J.F. Shriver, and A.J. Wallcraft (2015), Indirect evidence for substantial damping of low-mode internal tides in the open ocean. Journal of Geophysical Research Oceans 120, 6057-6071, doi:10.1002/2015JC010998.

Ansong, J.K., B.K. Arbic, M.H. Alford, M.C. Buijsman, J.F. Shriver, Z. Zhao, J.G. Richman, H.L. Simmons, P.G. Timko, A.J. Wallcraft, and L. Zamudio (2017), Semidiurnal internal tide energy fluxes and their variability in a global ocean model and moored observations. Journal of Geophysical Research Oceans 122, 1882-1900, doi:10.1002/2016JC012184.

Arbic, B.K., S.T. Garner, R.W. Hallberg, and H.L. Simmons (2004), The accuracy of surface elevations in forward global barotropic and baroclinic tide models. Deep-Sea Research II 51, 3069-3101, doi:10.1016/j.dsr2.2004.09.014.

Arbic, B.K. (2005), Atmospheric forcing of the oceanic semidiurnal tide. Geophysical Research Letters 32, L02610, doi:10.1029/2004GL021668.

Arbic, B.K., P. St-Laurent, G. Sutherland, and C. Garrett (2007), On the resonance and influence of tides in Ungava Bay and Hudson Strait. Geophysical Research Letters 34, L17606, doi:10.1029/2007GL030845.

Arbic, B.K., R.H. Karsten, and C. Garrett (2009), On tidal resonance in the global ocean and the back-effect of coastal tides upon open-ocean tides. Atmosphere-Ocean 47, 239-266, doi:10.3137/OC311.2009.

Arbic, B.K. and C. Garrett (2010), A coupled oscillator model of shelf and ocean tides. Continental Shelf Research 30, 564-574, doi:10.1016/j.csr.2009.07.008.

Arbic, B.K., A.J. Wallcraft, and E.J. Metzger (2010), Concurrent simulation of the eddying general circulation and tides in a global ocean model. Ocean Modelling 32, 175-187, doi:10.1016/j.ocemod.2010.01.007.

Arbic, B.K., J.G. Richman, J.F. Shriver, P.G. Timko, E.J. Metzger, and A.J. Wallcraft (2012a), Global modeling of internal tides within an eddying ocean general circulation model. Oceanography 25, doi:10.5670/oceanog.2012.38.

Arbic, B.K., R.B. Scott, G.R. Flierl, A.J. Morten, J.G. Richman, and J.F. Shriver (2012b), Nonlinear cascades of surface oceanic geostrophic kinetic energy in the frequency domain. Journal of Physical Oceanography 42, 1577-1600, doi:10.1175/JPO-D-11-0151.1.

Arbic, B.K., K.L. Polzin, R.B. Scott, J.G. Richman, and J.F. Shriver (2013), On eddy viscosity, energy cascades, and the horizontal resolution of gridded satellite altimeter products. Journal of Physical Oceanography 43, 283-300, doi:10.1175/JPO-D-11-0240.1.

Arbic, B.K., M. Müller, J.G. Richman, J.F. Shriver, A.J. Morten, R.B. Scott, G. Sérazin, and T. Penduff (2014), Geostrophic turbulence in the frequency-wavenumber domain: Eddy-driven low-frequency variability. Journal of Physical Oceanography 44, 2050- 2069, doi:10.1175/JPO-D-13-054.1.

Baines, P.G. (1982), On internal tide generation models. Deep- Sea Research 29, 307–338, doi:10.1016/0198-0149(82)90098-X.

Barkan, R., K.B. Winters, and J.C. McWilliams (2017), Stimulated imbalance and the enhancement of eddy kinetic energy dissipation by internal waves. Journal of Physical Oceanography 47, 181-198, doi:10.1175/JPO-D-16-01170-1.

Bell, T.H. (1975), Lee waves in stratified flows with simple harmonic time dependence. Journal of Fluid Mechanics 67, 705–722, doi:10.1017/S0022112075000560.

Bleck, R. (2002), An oceanic general circulation model framed in hybrid isopycnic-Cartesian coordinates. Ocean Modeling 4, 55-88, doi:10.1016/S1463-5003(01)00012-9.

Buijsman, M.C., B.K. Arbic, J.A.M. Green, R.W. Helber, J.G. Richman, J.F. Shriver, P.G. Timko, and A. Wallcraft (2015), Optimizing internal wave drag in a forward barotropic model with semidiurnal tides. Ocean Modelling 85, 42-55, doi:10.1016/j.ocemod.2014.11.003.

Buijsman, M.C., J.K. Ansong, B.K. Arbic, J.G. Richman, J.F. Shriver, P.G. Timko, A.J. Wallcraft, C.B. Whalen, and Z. Zhao (2016), Impact of internal wave drag on the semidiurnal energy balance in a global ocean circulation model. Journal of Physical Oceanography 46, 1399-1419, doi:10.1175/JPO-D-15-0074.1.

Buijsman, M.C., B.K. Arbic, J.G. Richman, J.F. Shriver, A.J. Wallcraft, and L. Zamudio (2017), Semidiurnal internal tide incoherence in the equatorial Pacific. Journal of Geophysical Research Oceans 122, 5286-5305, doi:10.1002/2016JC012590.

Cairns, J.L., and G.O. Williams (1976), Internal wave observations from a midwater float, 2. Journal of Geophysical Research 81(12), 1943–1950, doi:10.1029/ JC081i012p01943.

Callies, J., and R. Ferrari (2013), Interpreting energy and tracer spectra of upper-ocean turbulence in the submesoscale range (1–200 km). Journal of Physical Oceanography 43, 2456–2474, doi:10.1175/JPO-D-13-063.1.

Capet, X., J.C. McWilliams, M.J. Molemaker, and A.F. Shchepetkin (2008), Mesoscale to submesoscale transition in the California Current System. Part I: Flow structure, eddy flux, and observational tests. Journal of Physical Oceanography 38, 29-43, doi:10.1175/2007JPO3671.1.

Carrère and Lyard (2003), Modeling the barotropic response of the global ocean to atmospheric wind and pressure forcing: Comparisons with observations. Geophysical Research Letters 30, 1275, doi:10.1029/2002GL016473.

Carter, G.S., O.B. Fringer, and E.D. Zaron (2012), Regional models of internal tides. Oceanography 25, 56–65, doi:10.5670/oceanog.2012.42.

Cartwright, D.E. (1975), A subharmonic lunar tide in the seas off Western Europe. Nature 257, 277-280, doi:10.1038/257277a0.

Cartwright, D.E. (1999), Tides: A Scientific History. Cambridge University Press, Cambridge.

Cartwright, D.E., and R.D. Ray (1990), Oceanic tides from Geosat altimetry. Journal of Geophysical Research 95, 3069–3090, doi:10.1029/JC095iC03p03069.

Cartwright, D.E., and R.J. Tayler (1971), New computations of the tide-generating potential. Geophysical Journal of the Royal Astronomical Society 23, 45–74, doi:10.1111/j.1365-246X.1971.tb01803.x.

Chapman, S., and R.S. Lindzen (1970), Atmospheric Tides. D. Reidel Press, Dordrecht.

Charette, M.A., and W.H.F. Smith (2010), The volume of the earth's ocean. Oceanography 23, 112-114, doi:10.5670/oceanog.2010.51.

Chassignet, E.P., H.E. Hurlburt, E.J. Metzger, O.M. Smedstad, J.A. Cummings, G.R. Halliwell, R. Bleck, R. Baraille, A.J. Wallcraft, C. Lozano, H.L. Tolman, A. Srinivasan, S. Hankin, P. Cornillon, R. Weisberg, A. Barth, R. He, F. Werner, and J. Wilkin (2009), Global ocean prediction with the HYbrid Coordinate Ocean Model (HYCOM). Oceanography 22, 64–76, doi:10.5670/oceanog.2009.39.

Chassignet, E.P., and X. Xu (2017), Impact of horizontal resolution (1/12° to 1/50°) on Gulf Stream separation, penetration, and variability. Journal of Physical Oceanography 47, 1999-2021, doi:10.1175/JPO-D-17-0031.1.

Clarke, A.J. (1991), The dynamics of barotropic tides over the continental shelf and slope (review). In: Tidal Hydrodynamics. B.B. Parker (Ed.). John Wiley and Sons, pp. 79–108.

Csanady, G.T. (2001), Air-sea Interaction: Laws and Mechanisms. Cambridge University Press, Cambridge.

Cummins, P.F., and L.Y. Oey (1997), Simulation of barotropic and baroclinic tides off Northern British Columbia. Journal of Physical Oceanography 27, 762-781, doi:10.1175/1520-0485(1997)027<0762:SOBABT>2.0.CO;2.

Cummins, P.F., R.H. Karsten, and B.K. Arbic (2010), The semi-diurnal tide in Hudson Strait as a resonant channel oscillation. Atmosphere-Ocean 48, 163-176, doi:10.3137/OC307.2010.

D'Asaro, E.A. (1984), Wind forced internal waves in the North Pacific and Sargasso Sea. Journal of Physical Oceanography 14, 781–794, doi:10.1175/1520-0485(1984)014<0781:WFIWIT>2.0.CO.

D'Asaro, E.A. (1985), The energy flux from the wind to near-inertial motions in the surface mixed layer. Journal of Physical Oceanography 15, 1043–1059, doi:10.1175/1520-0485(1985)015<1043:TEFFTW>2.0.CO.

Doherty, K.W., D.E. Frye, S.P. Liberatore, and J.M. Toole (1999), A moored profiling instrument. Journal of Atmospheric and Oceanic Technology 16, 1816–1829, doi:10.1175/1520-0426(1999)016<1816:AMPI>2.0.CO;2.

Doodson, A.T. (1921), Harmonic development of the tide-generating potential. Proceedings of the Royal Society of London A100, 305-329, doi:10.1098/rspa.1921.0088.

Ducousso, N., J. Le Sommer, J.-M. Molines, and M. Bell (2017), Impact of the "Symmetric Instability of the Computational Kind" at mesoscale and submesoscale-permitting resolutions. Ocean Modelling 120, 18-26, doi:10.1016/j.ocemod.2017.10.006.

Dunphy, M., and K.G. Lamb (2014), Focusing and vertical mode scattering of the first mode internal tide by mesoscale eddy interaction. Journal of Geophysical Research 119, 523-536, doi:10.1002/2013JC009293.

Dushaw, B.D., B.D. Cornuelle, P.F. Worcester, B.M. Howe, and D.S. Luther (1995), Barotropic and baroclinic tides in the central North Pacific Ocean determined from long-range reciprocal acoustic transmissions. Journal of Physical Oceanography 25, 631–647, doi:10.1175/1520-0485(1995)025<0631:BABTIT>2.0.CO.

Egbert, G.D., A.F. Bennett, and M.G.G. Foreman (1994), Topex/Poseidon tides estimated using a global inverse model. Journal of Geophysical Research 99, 24821–24852, doi:10.1029/94JC01894.

Egbert, G.D., and S.Y. Erofeeva (2002), Efficient inverse modeling of barotropic ocean tides. Journal of Atmospheric and Oceanic Technology 19, 183-204, doi:10.1175/1520-0426(2002)019<0183:EIMOBO>2.0.CO.

Egbert, G.D., and R.D. Ray (2000), Significant dissipation of tidal energy in the deep ocean inferred from satellite altimeter data. Nature 405, 775–778, doi:10.1038/35015531.

Egbert, R.D., and R.D. Ray (2001), Estimates of M_2 tidal energy dissipation from TOPEX/Poseidon altimeter data. Journal of Geophysical Research 106, 22475–22502, doi:10.1029/2000JC000699.

Egbert, R.D., and R.D. Ray (2003), Semi-diurnal and diurnal tidal dissipation from TOPEX/Poseidon altimetry. Geophysical Research Letters 30, 1907, doi:10.1029/2003GL017676.

Egbert, G.D., R.D. Ray, and B.G. Bills (2004), Numerical modeling of the global semidiurnal tide in the present day and in the last glacial maximum. Journal of Geophysical Research 109, C03003, doi:10.1029/2003JC001973.

Egbert, G.D., and R.D. Ray (2017), Tidal prediction. In: The Sea: The Science of Ocean Prediction, printed as a special issue of Journal of Marine Research 75, 189-237, doi:10.1357/002224017821836761.

Emery, W.J., and R.E. Thompson (1997), Data Analysis Methods in Physical Oceanography. Pergamon Press.

Flierl, G.R. (1978), Models of vertical structure and the calibration of two-layer models. Dynamics of Atmospheres and Oceans 2, 341-381, doi:10.1016/0377-0265(78)90002-7.

Foreman, M.G.G. (1977), Manual for tidal heights analysis and prediction. Pacific Marine Science Report 77-10, Institute of Ocean Sciences, Sidney.

Foreman, M.G.G. (2004), Manual for tidal currents analysis and prediction, Pacific Marine Science Report 78-6, Institute of Ocean Sciences, Sidney.

Foreman, M.G.G., J.Y. Cherniawsky, and V.A. Ballantyne (2009), Versatile harmonic tidal analysis: Improvements and applications. Journal of Atmospheric and Oceanic Technology 26, 806-817, doi:10.1175/2008JTECHO615.1.

Fu, L.-L., D. Alsdorf, R. Morrow, E. Rodriguez, and N. Mognard (Eds.) (2012), SWOT: The Surface Water and Ocean Topography Mission: Wide-Swath Altimetric Measurement of Water Elevation on Earth, Jet Propulsion Laboratory, Pasadena, California.

Fu, L.-L., and A. Cazenave (Eds.) (2001), Satellite Altimetry and Earth Sciences: A Handbook of Techniques and Applications, Academic Press, San Diego.

Furiuchi, N., T. Hibiya, and Y. Niwa (2008), Model predicted distribution of wind-induced internal wave energy in the world's oceans. Journal of Geophysical Research 113, C09034, doi:10.1029/2008JC004768.

Garner, S.T. (2005), A topographic drag closure built on an analytical base flux. Journal of the Atmospheric Sciences 62, 2302-2315, doi:10.1175/JAS3496.1.

Garrett, C. (1972), Tidal resonance in the Bay of Fundy and Gulf of Maine. Nature 238, 441–443, doi:10.1038/238441a0.

Garrett, C.J.R., and W.H. Munk (1975), Space-time scales of internal waves. A progress report. Journal of Geophysical Research 80, 291–297, doi:10.1029/ JC080i003p00291.

Garrett, C. and D. Greenberg (1977), Predicting changes in tidal regime: The open boundary problem. Journal of Physical Oceanography 7, 171–181, doi:10.1175/1520-0485(1977)007<0171:PCITRT>2.0.CO.

Gill, A. E. (1982), Atmosphere-Ocean Dynamics. Academic Press, New York.

Gille, S.T., K. Speer, J.R. Ledwell, and A.C. Naveira Garabato (2007), Mixing and stirring in the Southern Ocean. Eos, Transactions American Geophysical Union 88(39), 382- 383, doi:10.1029/2007EO390002.

Goff, J.A. (2010), Global prediction of abyssal hill root-mean-square heights from small-scale altimetric gravity variability. Journal of Geophysical Research 115, B12104, doi:10.1029/2010JB007867.

Goff, J.A., and B.K. Arbic (2010), Global prediction of abyssal hill roughness statistics for use in ocean models from digital maps of paleo-spreading rate, paleo-ridge orientation, and sediment thickness. Ocean Modelling 32, 36-43, doi:10.1016/j.ocemod.2009.10.001.

Gordeev, R.G., B.A. Kagan, and E.V. Polyakov (1977), The effects of loading and self-attraction on global ocean tides: The model and the results of a numerical experiment. Journal of Physical Oceanography 7, 161–170, doi:10.1175/1520-0485(1977)007<0161:TEOLAS>2.0.CO.

Gregg, M.C. (1989), Scaling turbulent dissipation in the thermocline. Journal of Geophysical Research 94, 9686-9898, doi:10.1029/JC094iC07p09686.

Griffies, S.M. (2005), Fundamentals of ocean climate models. Princeton University Press, Princeton.

Griffies, S.M., C. Böning, F.O. Bryan, E.P. Chassignet, R. Gerdes, H. Hasumi, A. Hirst, A.-M. Treguier, and D. Webb (2000), Developments in ocean climate modeling. Ocean Modelling 2, 123-192, doi:10.1016/S1463-5003(00)00014-7.

Hallberg, R. (2005), A thermobaric instability of Lagrangian vertical coordinate ocean models. Ocean Modelling 8, 279–300, doi:10.1016/j.ocemod.2004.01.001.

Heath, R.A. (1981), Estimates of the resonant period and Q in the semi-diurnal tidal band in the North Atlantic and Pacific Oceans. Deep Sea Research Part A 2, 481–493, doi:10.1016/0198-0149(81)90139-4.

Hecht, M. W., and H. Hasumi, Eds. (2008), Ocean Modeling in an Eddying Regime. Geophysical Monographs 177, American Geophysical Union.

Heinbockel, J.H. (2003), Mathematical Methods for Partial Differential Equations. Trafford Publishing, Victoria.

Hendershott, M.C. (1972), The effects of solid earth deformation on global ocean tides. Geophysical Journal of the Royal Astronomical Society 29, 389–402, doi:10.1111/j.1365-246X.1972.tb06167.x.

Hogan, T.F., M. Liu, J.A. Ridout, M.S. Peng, T.R. Whitcomb, B.C. Ruston, C.A. Reynolds, S.D. Eckermann, J.R. Moskaitis, N.L. Baker, J.P. McCormack, K.C. Viner, J.G. McLay, M.K. Flatau, L. Xu, C. Chen, and S.W. Chang (2014), The Navy Global Environmental Model. Oceanography 27, 116-125, doi:10.5670/oceanog.2014.73.

Holloway, P.E. (1996), A numerical model of internal tides with application to the Australia North West Shelf. Journal of Physical Oceanography 26, 21–37, doi:10.1175/1520-0485(1996)026<0021:ANMOIT>2.0.CO.

Jayne, S.R., and L.C. St. Laurent (2001), Parameterizing tidal dissipation over rough topography. Geophysical Research Letters 28, 811-814, doi:10.1029/2000GL012044.

Josey, S.A., S. Gulev, and L. Yu (2013), Exchanges through the ocean surface. In: Ocean Circulation and Climate, A 21st century perspective, second edition. G. Siedler, S. Griffies, J. Gould, and J. Church (Eds.) 103, Academic Press, 115-140, doi:10.1016/B978-0-12-391851-2.00005-2.

Kang, S.K., M.G.G. Foreman, W.R. Crawford, and J.Y. Cherniawsky (2000), Numerical modeling of internal tide generation along the Hawaiian Ridge. Journal of Physical Oceanography 30, 1083–1098, doi:10.1175/1520-0485(2000)030<1083:NMOITG>2.0.CO.

Kelly, S.M., N.L. Jones, J.D. Nash, and A.F. Waterhouse (2013), The geography of semidiurnal mode-1 internal tide energy loss. Geophysical Research Letters 40, 4689-4693, doi:10.1002/grl.50872.

Klymak, J.M., R. Pinkel, and L. Rainville (2008), Direct breaking of the internal tide near topography: Kaena Ridge, Hawaii. Journal of Physical Oceanography 38, 380-399, doi:10.1175/2007JPO3728.1.

Kuhlmann, J., H. Dobslaw, and M. Thomas (2011), Improved modeling of sea level patterns by incorporating self-attraction and loading. Journal of Geophysical Research 116, C11036, doi:10.1029/2011JC007399.

Kundu, P.K. (1990), Fluid Mechanics, Academic Press, New York.

Kunze, E. (2017a), Internal-wave-driven mixing: Global geography and budgets. Journal of Physical Oceanography 47, 1325-1345, doi:10.1175/JPO-D-16-0141.1.

Kunze, E. (2017b), The internal-wave-driven meridional overturning circulation. Journal of Physical Oceanography 47, 2673-2689, doi:10.1175/JPO-D-16-0142.1.

LaCasce, J.H. (2000), Floats and f/H. Journal of Marine Research 58, 61–95, doi:10.1357/002224000321511205.

Laplace, P.S. (1775), Recherches sur plusieurs points du système du monde. Mémoires de l'Académie Royale des Sciences Paris 88, 75-182. [Reprinted in Oeuvres Complètes de Laplace, Gauthier-Villars, Paris, 9 (1893)].

Laplace, P.S. (1776), Recherches sur plusieurs points du système du monde. Mémoires de l'Académie Royale des Sciences Paris 89, 177-264. [Reprinted in Oeuvres Complètes de Laplace, Gauthier-Villars, Paris, 9 (1893)].

Large, W.G., and S.G. Yeager (2004), Diurnal and decadal global forcing for ocean and sea-ice models: The data sets and climatologies. Technical Note tN-460+ST, NCAR, Boulder, Colorado, doi:10.5065/D6KK98Q6.

Le Provost, C., M.L. Genco, F. Lyard, P. Vincent, and P. Canceil (1994), Spectroscopy of the world ocean tides from a finite element hydrodynamic model. Journal of Geophysical Research 99(C12), 24,777–24,797, doi:10.1029/94JC01381.

Luecke, C.A., B.K. Arbic, S.L. Bassette, J.G. Richman, J.F. Shriver, M.H. Alford, O.M. Smedstad, P.G. Timko, D.S. Trossman, and A.J. Wallcraft (2017), The global mesoscale eddy available potential energy field in models and observations. Journal of Geophysical Research Oceans 122, 9126-9143, doi:10.1002/2017JC013136.

Lyard, F., F. Lefevre, T. Letellier, and O. Francis (2006), Modelling the global ocean tides: Modern insights from FES2004. Ocean Dynamics 56, 394–415, doi:10.1007/s10236-006-0086-x.

MacKinnon, J.A., Z. Zhao, C.B. Whalen, A.F. Waterhouse, D.S. Trossman, O.M. Sun, L.C. St. Laurent, H.L. Simmons, K. Polzin, R. Pinkel, A. Pickering, N.J. Norton, J.D. Nash, R. Musgrave, L.M. Merchant, A.V. Melet, B. Mater, S. Legg, W.G. Large, E. Kunze, J.M. Klymak, M. Jochum, S.R. Jayne, R.W. Hallberg, S.M. Griffies, S. Diggs, G. Danabasoglu, E.P. Chassignet, M.C. Buijsman, F.O. Bryan, B.P. Briegleb, A. Barna, B.K. Arbic, J.K. Ansong, and M.H. Alford (2017), Climate process team on internal-wave driven ocean mixing. Bulletin of the American Meteorological Society 98, 2429-2454, doi:10.1175/BAMS-D-16-0030.1.

Maltrud, M.E., and J.L. McClean (2005), An eddy resolving global 1/10° ocean simulation. Ocean Modelling 8, 31-54, doi:10.1016/j.ocemod.2003.12.001.

Marshall, J., A. Adcroft, C. Hill, L. Perelman, and C. Heisey (1997), A finite-volume, incompressible Navier Stokes model for studies of the ocean on parallel computers. Journal of Geophysical Research 102, 5753-5766, doi:10.1029/96JC02775.

McComas, C.H., and F.P. Bretherton (1977), Resonant interactions of oceanic internal waves. Journal of Physical Oceanography 82, 1397-1412, doi:10.1029/JC082i009p01397.

Melet, A., M. Nikurashin, C. Muller, S. Falahat, J. Nycander, P.G. Timko, B.K. Arbic, and J.A. Goff (2013), Internal tide generation by abyssal hills using analytical theory. Journal of Geophysical Research Oceans 118, 6303-6318, doi:10.1002/2013JC009212.

Merrifield, M.A., P.E. Holloway, and T.M. Shaun Johnston (2001), The generation of internal tides at the Hawaiian Ridge. Geophysical Research Letters 28, 559–562, doi:10.1029/2000GL011749.

Moum, J.N., D.M. Farmer, W.D. Smyth, L. Armi, and S. Vagle (2003), Structure and generation of turbulence at interfaces strained by internal solitary waves propagating shoreward over the continental shelf. Journal of Physical Oceanography 33, 2093-2112, doi:10.1175/1520-0485(2003)033<2093:SAGOTA>2.0.CO;2.

Müller, M. (2007), The free oscillations of the world ocean in the period range 8 to 165 hours including the full loading effect. Geophysical Research Letters 34, L05606, doi:10.1029/2006GL028870.

Müller, M., H. Haak, J. H. Jungclaus, J. Sündermann and M. Thomas (2010), The impact of ocean tides on a climate model simulation. Ocean Modelling 35, 304-313, doi:/10.1016/j.ocemod.2010.09.001.

Müller, M., J. Cherniawsky, M. Foreman, and J.-S. von Storch (2012), Global map of M_2 internal tide and its seasonal variability from high resolution ocean circulation and tide modelling. Geophysical Research Letters 39, L19607, doi:10.1029/2012GL053320.

Müller, M., B.K. Arbic, J.G. Richman, J.F. Shriver, E.L. Kunze, R.B. Scott, A.J. Wallcraft, and L. Zamudio (2015), Toward an internal gravity wave spectrum in global ocean models. Geophysical Research Letters 42, 3474-3481, doi:10.1002/2015GL063365.

Müller, P., G. Holloway, F. Henyey, and N. Pomphrey (1986), Nonlinear interactions among internal gravity waves. Reviews of Geophysics 24(3), 493–536, doi:10.1029/RG024i003p00493.

Munk, W.H., and G.J.F. MacDonald (1960), The rotation of the earth. Cambridge University Press, London.

Munk, W.H., and C. Wunsch (1998), Abyssal recipes II: energetics of tidal and wind mixing. Deep Sea Research I 45, 1977–2010, doi:10.1016/S0967-0637(98)00070-3.

Nash, J.D., M.H. Alford, and E. Kunze (2005), Estimating internal wave energy fluxes in the ocean. Journal of Atmospheric and Oceanic Technology 22, 1551-1570, doi:10.1175/JTECH1784.1.

Nash, J.D., E.L. Shroyer, S.M. Kelly, M.E. Inall, T.F. Duda, M.D. Levine, N.L. Jones, and R.C. Musgrave (2012), Are any coastal internal tides predictable? Oceanography 25(2), 80–95, doi:10.5670/oceanog.2012.44.

Newton, I. (1687), Philosophiae Naturalis Principia Mathematica.

Ngodock, H.E., I. Souopgui, A.J. Wallcraft, J.G. Richman, J.F. Shriver, and B.K. Arbic (2016), On improving the accuracy of the M_2 barotropic tides embedded in a high-resolution global ocean circulation model. Ocean Modelling 97, 16-26, doi:10.1016/j.ocemod.2015.10.011.

Nikurashin, M., and R. Ferrari (2011), Global energy conversion rate from geostrophic flows into internal lee waves in the deep ocean. Geophysical Research Letters 38, L08610, doi: 10.1029/2011GL046576.

Niwa, Y., and T. Hibiya (2001), Numerical study of the spatial distribution of the M_2 internal tide in the Pacific Ocean. Journal of Geophysical Research 106, 22229–22441, doi:10.1029/2000JC000770.

Nugroho, D. (2017), The tide in a general circulation pattern in the Indonesian Seas. PhD dissertation. Université de Toulouse.

Nugroho, D., A. Koch-Larrouy, P.Gaspar, F. Lyard, G. Reffray, and B. Tranchant (2017), Modelling explicit tides in the Indonesian Seas: An important process for sea water properties. Marine Pollution Bulletin 131, Part B, 7-18, doi:10.1016/j.marpol-bul.2017.06.033.

Nycander, J. (2005), Generation of internal waves in the deep ocean by tides. Journal of Geophysical Research 110, C10028, doi:10.1029/2004JC002487.

Paiva, A.M., J.T. Hargrove, E.P. Chassignet, and R. Bleck (1999), Turbulent behavior of a fine-mesh (1/12 degree) numerical simulation of the North Atlantic. Journal of Marine Systems 21, 307-320, doi:10.1016/S0924-7963(99)00020-2.

Parke, M.E., and M.C. Hendershott (1980), M_2, S_2, K_1 models of the global ocean tide on an elastic earth. Marine Geodesy 3, 379–408, doi:10.1080/01490418009388005.

Pawlowicz, R., B. Beardsley, S. Lentz (2002), Classical tidal harmonic analysis including error estimates in MATLAB using T-TIDE. Computers and Geosciences 28, 929–937, doi:10.1016/S0098-3004(02)00013-4.

Pedlosky, J. (1996), Ocean Circulation Theory. Springer-Verlag, Berlin.

Penduff, T., B. Barnier, J.-M. Molines, and G. Madec (2006), On the use of current meter data to assess the realism of ocean model simulations. Ocean Modelling 11, 399–416, doi:10.1016/j.ocemod.2005.02.001.

Platzman, G.W. (1991), Tidal evidence for ocean normal modes. In: Tidal Hydrodynamics. B.B.Parker (Ed.), John Wiley and Sons, pp. 13–26.

Platzman, G.W., G.A. Curtis, K.S. Hansen, and R.D. Slater (1981), Normal modes of the World Ocean. Part II: Description of modes in the period range 8 to 80 hours. Journal of Physical Oceanography 11, 579–603, doi:10.1175/1520-0485(1981)011<0579:NMOTWO>2.0.CO.

Pollard, R.T., and R.C. Millard, Jr. (1970), Comparison between observed and simulated wind-generated inertial oscillations. Deep-Sea Research 17, 813-821, doi:10.1016/0011-7471(70)90043-4.

Polzin, K. (2004), Heuristic description of internal wave dynamics. Journal of Physical Oceanography 34, 214–230, doi:10.1175/1520-0485(2004)034<0214:AHDOIW>2.0.CO.

Polzin, K.L., J.M. Toole, and R.W. Schmitt (1995), Finescale parameterization of turbulent dissipation. Journal of Physical Oceanography 25, 306-328, doi:10.1175/1520-0485(1995)025<0306:FPOTD>2.0.CO;2.

Polzin, K.L., J.M. Toole, J.R. Ledwell, and R.W. Schmitt (1997), Spatial variability of turbulent mixing in the abyssal ocean. Science 276, 93-96, doi:10.1126/science.276.5309.93.

Ponte, R. (1994), Understanding the relation between wind- and pressure-driven sea level variability. Journal of Geophysical Research 99, 8033-8039, doi:10.1029/94JC00217.

Ponte, A.L., and P. Klein (2015), Incoherent signature of internal tides on sea level in idealized numerical simulations. Geophysical Research Letters 42, 1520–1526, doi:10.1002/2014GL062583.

Price, J.F., and M. O'Neil Baringer (1994), Outflows and deep water production by marginal seas. Progress in Oceanography 33, 161-200, doi:10.1016/0079-6611(94)90027-2.

Pugh, D.T. (1987), Tides, Surges, and Mean Sea-level: A Handbook for Scientists and Engineers. Wiley, Chichester.

Qiu, B., T. Nakano, S. Chen, and P. Klein (2017), Submesoscale transition from geostrophic flows to internal waves in the northwestern Pacific upper ocean. Nature Communications 8, 14055, doi:10.1038/ncomms14055.

Qiu, B., S. Chen, P. Klein, J. Wang, H. Torres, L.-L. Fu, and D. Menemenlis (2018), Seasonality in transition scale from balanced to unbalanced motions in the world ocean. Journal of Physical Oceanography 48, 591-605, doi:10.1175/JPO-D-17-0169.1.

Ray, R.D. (1998), Ocean self-attraction and loading in numerical tidal models. Marine Geodesy 21, 181–192, doi:10.1080/01490419809388134.

Ray, R.D. (1999), A global ocean tide model from Topex/Poseidon altimetry: GOT99.2, NASA Technical Memorandum 209478, Goddard Space Flight Center, Greenbelt, MD.

Ray, R.D. (2013), Precise comparisons of bottom-pressure and altimetric ocean tides. Journal of Geophysical Research Oceans 118, 4570–4584, doi:10.1002/jgrc.20336.

Ray, R.D., and D.A. Byrne (2010), Bottom pressure tides along a line in the southeast Atlantic Ocean and comparisons with satellite altimetry. Ocean Dynamics 60, 1167–1176, doi:10.1007/s10236-010-0316-0.

Ray, R.D., and G.D. Egbert (2004), The global S_1 tide. Journal of Physical Oceanography 34, 1922-1935, doi:10.1175/1520-0485(2004)034<1922:TGST>2.0.CO.

Ray, R.D., and G.D. Egbert (2017), Tides and satellite altimetry. In: Satellite Altimetry over Oceans and Land Surfaces. D. Stammer, A. Cazenave (Eds.), Taylor and Francis, pp. 427-458.

Ray, R.D., and G.T. Mitchum (1996), Surface manifestation of internal tides generated near Hawaii. Geophysical Research Letters 23, 2101–2104, doi:10.1029/96GL02050.

Ray, R.D., and G.T. Mitchum (1997), Surface manifestation of internal tides in the deep ocean: Observations from altimetry and tide gauges. Progress in Oceanography 40, 135–162, doi:10.1016/S0079-6611(97)00025-6.

Ray, R.D., and R.M. Ponte (2003), Barometric tides from ECMWF operational analyses. Annales Geophysicae 21, 1897–1910, doi:10.5194/angeo-21-1897-2003.

Richman, J.G., B.K. Arbic, J.F. Shriver, E.J. Metzger, and A.J. Wallcraft (2012), Inferring dynamics from the wavenumber spectra of an eddying global ocean model with embedded tides. Journal of Geophysical Research 117, C12012, doi:10.1029/2012JC008364.

Rocha, C.B., T.K. Chereskin, S.T. Gille, and D. Menemenlis (2016a), Mesoscale to submesoscale wavenumber spectra in Drake Passage. Journal of Physical Oceanography 46, 601–620, doi:10.1175/JPO-D-15-0087.1.

Rocha, C.B., S.T. Gille, T.K. Chereskin, and D. Menemenlis (2016b), Seasonality of submesoscale dynamics in the Kuroshio Extension. Geophysical Research Letters 43, 11304–11311, doi:10.1002/2016GL071349.

Roosbeek, F. (1996), RATGP95: A harmonic development of the tide-generating potential using an analytical method. Geophysical Journal International 126, 197–204, doi:10.1111/j.1365-246X.1996.tb05278.x.

Rosmond, T.E., J. Teixeira, M. Peng, T.F. Hogan, and R. Pauley (2002), Navy Operational Global Atmospheric Prediction System (NOGAPS): Forcing for ocean models. Oceanography 15, 99-108, doi:10.5670/oceanog.2002.40.

Savage, A.C., B.K. Arbic, J.G. Richman, J.F. Shriver, M.H. Alford, M.C. Buijsman, J.T. Farrar, H. Sharma, G. Voet, A.J. Wallcraft, and L. Zamudio (2017a), Frequency content of sea surface height variability from internal gravity waves to mesoscale eddies. Journal of Geophysical Research Oceans 122, 2519-2538, doi:10.1002/2016JC012331.

Savage, A.C., B.K. Arbic, M.H. Alford, J.K. Ansong, J.T. Farrar, D. Menemenlis, A.K. O'Rourke, J.G. Richman, J.F. Shriver, G. Voet, A.J. Wallcraft, and L. Zamudio (2017b), Spectral decomposition of internal gravity wave sea surface height in global models. Journal of Geophysical Research Oceans 122, 7803-7821, doi:10.1002/2017JC013009.

Schiller, A., and R. Fiedler (2007), Explicit tidal forcing in an ocean general circulation model. Geophysical Research Letters 34, L03611, doi:10.1029/2006GL028363.

Schmitz, W.J., Jr. (1988), Exploration of the eddy field in the midlatitude North Pacific. Journal of Physical Oceanography 18, 459–468, doi:10.1175/1520-0485(1988)018<0459:EOTEFI>2.0.CO.

Schrama, E.J.O., and R.D. Ray (1994), A preliminary tidal analysis of TOPEX/POSEIDON altimetry. Journal of Geophysical Research 99(C12), 24799–24808, doi:10.1029/94JC01432.

Schwiderski, E.W. (1980), On charting global ocean tides. Reviews of Geophysics and Space Physics 18, 243-268, doi:10.1029/RG018i001p00243.

Scott, R.B., and B.K. Arbic (2007), Spectral energy fluxes in geostrophic turbulence: Implications for ocean energetics. Journal of Physical Oceanography 37, 673-688, doi:10.1175/JPO3027.1.

Scott, R.B., B.K. Arbic, C.L. Holland, A. Sen, and B. Qiu (2008), Zonal versus meridional velocity variance in satellite observations and realistic and idealized ocean circulation models. Ocean Modelling 23, 102–112, doi:10.1016/j.ocemod.2008.04.009.

Scott, R.B., B.K. Arbic, E.P. Chassignet, A.C. Coward, M. Maltrud, W.J. Merryfield, A. Srinivisan, and A. Varghese (2010), Total kinetic energy in four global eddying ocean circulation models and over 5000 current meter records. Ocean Modelling 32, 157-169, doi:10.1016/j.ocemod.2010.01.005.

Scott, R.B., J.A. Goff, A.C. Naveira-Garabato, and A.J.G. Nurser (2011), Global rate and spectral characteristics of internal gravity wave generation by geostrophic flow. Journal of Geophysical Research 116, C09029, doi:10.1029/2011JC007005.

Scott, R.B., and F. Wang (2005), Direct evidence of an oceanic inverse kinetic energy cascade from satellite altimetry. Journal of Physical Oceanography 35, 1650–1666, doi:10.1175/JPO2771.1.

Sheng, J., and Y. Hayashi (1990a), Observed and simulated energy cycles in the frequency domain. Journal of the Atmospheric Sciences 47, 1243–1254, doi:10.1175/1520-0469(1990)047<1243:OASECI>2.0.CO.

Sheng, J., and Y. Hayashi (1990b), Estimation of atmospheric energetics in the frequency domain during the FGGE year. Journal of the Atmospheric Sciences 47, 1255–1268, doi:10.1175/1520-0469(1990)047<1255:EOAEIT>2.0.CO.

Shriver, J.F., B.K. Arbic, J.G. Richman, R.D. Ray, E.J. Metzger, A.J. Wallcraft, and P.G. Timko (2012), An evaluation of the barotropic and internal tides in a high resolution global ocean circulation model. Journal of Geophysical Research 117, C10024, doi:10.1029/2012JC008170.

Shriver, J.F., J.G. Richman, and B.K. Arbic (2014), How stationary are the internal tides in a high resolution global ocean circulation model? Journal of Geophysical Research Oceans 119, 2769-2787, doi:10.1002/2013JC009423.

Shum, C.K., P.L. Woodworth, O.B. Andersen, G.D. Egbert, O. Francis, C. King, S.M. Klosko, C. Le Provost, X. Li, J.-M. Molines, M.E. Parke, R.D. Ray, M.G. Schlax, D. Stammer, C.C. Tierney, P. Vincent, and C.I. Wunsch (1997), Accuracy assessment of recent ocean tide models. Journal of Geophysical Research 102, 25173–25194, doi:10.1029/97JC00445.

Silverthorne, K.E., and J.M. Toole (2009), Seasonal kinetic energy variability of near-inertial motions. Journal of Physical Oceanography 39, 1035–1049, doi:10.1175/2008JPO3920.1.

Simmons, H.L., and M.H. Alford (2012), Simulating the long-range swell of internal waves generated by ocean storms. Oceanography 25, 30–41, doi:10.5670/oceanog.2012.39.

Simmons, H.L., S.R. Jayne, L.C. St. Laurent, and A.J. Weaver (2004a), Tidally driven mixing in a numerical model of the ocean general circulation. Ocean Modelling 6, 245-263, doi:10.1016/S1463-5003(03)00011-8.

Simmons, H.L., R.W. Hallberg, and B.K. Arbic (2004b), Internal wave generation in a global baroclinic tide model. Deep-Sea Research II 51, 3043-3068, doi:10.1016/j.dsr2.2004.09.015.

Skiba, A.W., L. Zeng, B.K. Arbic, M. Müller, and W.J. Godwin (2013), On the resonance and shelf/open-ocean coupling of the global diurnal tides. Journal of Physical Oceanography 43, 1301-1324, doi:10.1175/JPO-D-12-054.1.

Smith, R.D., M.E. Maltrud, F.O. Bryan, and M.W. Hecht (2000), Numerical simulation of the North Atlantic Ocean at 1/10°. Journal of Physical Oceanography 30, 1532-1561, doi:10.1175/1520-0485(2000)030<1532:NSOTNA>2.0.CO.

Smith, W.H.F., and D.T. Sandwell (1997), Global seafloor topography from satellite altimetry and ship depth soundings. Science 277, 1956-1962, doi:10.1126/science.277.5334.1956.

St. Laurent, L.C., and C. Garrett (2002), The role of internal tides in mixing the deep ocean. Journal of Physical Oceanography 32, 2882-2899, doi:10.1175/1520-0485(2002)032<2882:TROITI>2.0.CO.

St. Laurent, L.C., and H. Simmons (2006), Estimates of power consumed by mixing in the ocean interior. Journal of Climate 19, 4877-4890, doi:10.1175/JCLI3887.1.

St. Laurent, L.C, A.C. Naveira Garabato, J.R. Ledwell, A.M. Thurnherr, J.M. Toole, and A.J. Watson (2012), Turbulence and diapycnal mixing in Drake Passage. Journal of Physical Oceanography 42, 2143-2152, doi:10.1175/JPO-D-12-027.1.

Stammer, D., C. Wunsch, and R. Ponte (2000), De-aliasing of global high-frequency barotropic motions in altimeter observations. Geophysical Research Letters 27, 1175-1178, doi:10.1029/1999GL011263.

Stammer, D., R.D. Ray, O.B. Andersen, B.K. Arbic, W. Bosch, L. Carrère, Y. Cheng, D.S. Chinn, B.D. Dushaw, G.D. Egbert, S.Y. Erofeeva, H.S. Fok, J.A.M. Green, S. Griffiths, M.A. King, V. Lapin, F.G. Lemoine, S.B. Luthcke, F. Lyard, J. Morison, M. Müller, L. Padman, J.G. Richman, J.F. Shriver, C.K. Shum, E. Taguchi, and Y. Yi (2014), Accuracy assessment of global barotropic ocean tide models. Reviews of Geophysics 52, 243-282, doi:10.1002/2014RG000450.

Stepanov, V.N., and C.W. Hughes (2004), Parameterization of ocean self-attraction and loading in numerical models of the ocean circulation. Journal of Geophysical Research 109, C07004, doi:10.1029/2003JC002034.

Stewart, K.D., P. Spence, S. Waterman, J. Le Sommer, J.-M. Molines, and J.M. Lilly (2015), Anisotropy of eddy variability in the global ocean. Ocean Modelling 95, 53-65, doi:10.1016/j.ocemod.2015.09.005.

Sugiyama, Y., Y. Niwa, and T. Hibiya (2009), Numerically reproduced internal wave spectra in the deep ocean. Journal of Geophysical Research 36, L07601, doi:10.1029/2008GL036825.

Sun, S., R. Bleck, C. Rooth, J. Dukowicz, E. Chassignet, and P. Killworth (1999), Inclusion of thermobaricity in isopycnic-coordinate ocean models. Journal of Physical Oceanography 29, 2719–2729, doi:10.1175/1520-0485(1999)029<2719:IOTIIC>2.0.CO.

Taguchi, E., W. Zahel, and D. Stammer (2014), Inferring deep ocean tidal energy dissipation from the global high-resolution data-assimilative HAMTIDE model. Journal of Geophysical Research Oceans 119, 4573-4592, doi:10.1002/2013JC009766.

Taylor, G.I. (1919), Tidal friction in the Irish Sea. Philosophical Transactions of the Royal Society of London A220, 1-93, doi:10.1098/rspa.1919.0059.

Thomas, M., J. Sündermann, and E. Maier-Raimer (2001), Consideration of ocean tides in an OGCM and impacts on subseasonal to decadal polar motion excitation. Geophysical Research Letters, 28, 2457-2460, doi:10.1029/2000GL012234.

Thoppil, P.G., J.G. Richman, and P.J. Hogan (2011), Energetics of a global ocean circulation model compared to observations. Geophysical Research Letters, 38, L15607, doi:10.1029/2011GL048347.

Tierney, C., J. Wahr, F. Bryan, and V. Zlotnicki (2000a), Short-period oceanic circulation: Implications for satellite altimetry. Geophysical Research Letters 27, 1255-1258, doi:10.1029/1999GL010507.

Tierney, C.G., L.H. Kantha, and G.H. Born (2000b), Shallow and deep water global ocean tides from altimetry and numerical modeling. Journal of Geophysical Research 105, 11259-11277, doi:10.1029/1999JC900314.

Timko, P.G., B.K. Arbic, J.G. Richman, R.B. Scott, E.J. Metzger, and A.J. Wallcraft (2012), Skill tests of tidal currents in a three-dimensional ocean model: A look at the North Atlantic. Journal of Geophysical Research 117, C08014, doi:10.1029/2011JC007617.

Timko, P.G., B.K. Arbic, J.G. Richman, R.B. Scott, E.J. Metzger, and A.J. Wallcraft (2013), Skill testing a three-dimensional global tide model to historical current meter records. Journal of Geophysical Research Oceans 118, 6914-6933, doi:10.1002/2013JC009071.

Timko, P.G., B.K. Arbic, J.A. Goff, J.K. Ansong, W.H.F. Smith, A. Melet, and A.J. Wallcraft (2017), Impact of synthetic abyssal hill roughness on resolved motions in numerical global ocean tide models. Ocean Modelling 112, 1-16, doi:10.1016.j.ocemod.2017.02.005.

Trossman, D.S., B.K. Arbic, S.T. Garner, J.A. Goff, S.R. Jayne, E.J. Metzger, and A.J. Wallcraft (2013), Impact of parameterized lee wave drag on the energy budget of an eddying global ocean model. Ocean Modelling 72, 119-142, doi:10.1016/j.ocemod.2013.08.006.

Trossman, D.S., B.K. Arbic, J.G. Richman, S.T. Garner, S.R. Jayne, and A.J. Wallcraft (2016), Impact of topographic internal lee wave drag on an eddying global ocean model. Ocean Modelling 97, 109-128, doi:10.1016/j.ocemod.2015.10.013.

Vinogradova, N.T., R.M. Ponte, K.J. Quinn, M.E. Tamisiea, J.-M. Campin, and J.L. Davis (2015), Dynamic adjustment of the ocean circulation to self-attraction and loading effects. Journal of Physical Oceanography 45, 678-689, doi:10.1175/JPO-D-14-0150.1.

Wahr, J.M., and T. Sasao (1981), A diurnal resonance in the ocean tide and in the Earth's load response due to the resonant free "core nutation." Geophysical Journal of the Royal Astronomical Society 64, 747-765, doi:10.1111/j.1365-246X.1981.tb02693.x.

Wang, J., L.-L. Fu, B. Qiu, D. Menemenlis, J.T. Farrar, Y. Chao, A.F. Thompson, and M.M. Flexas (2018), An observing system simulation experiment for the calibration and validation of the surface water ocean topography sea surface height measurement using in situ platforms. Journal of Atmospheric and Oceanic Technology 35, 281-297, doi:10.1175/JTECH-D-717-0076.1.

Waterhouse, A.F., J.A. MacKinnon, J.D. Nash, M.A. Alford, E. Kunze, H.L. Simmons, K.L. Polzin, L.C. St. Laurent, O.M. Sun, R. Pinkel, L.D. Talley, C.B. Whalen, T.N. Huussen, G.S. Carter, I. Fer, S. Waterman, A. Naveira Garabato, T.B. Sanford, and C.M. Lee (2014), Global patterns of diapycnal mixing from measurements of the turbulent dissipation rate. Journal of Physical Oceanography 44, 1854–1872, doi:10.1175/ JPO-D-13-0104.1.

Weis, P., M. Thomas, and J. Sündermann (2008), Broad frequency tidal dynamics simulated by a high-resolution global ocean tide model forced by ephemerides. Journal of Geophysical Research 113, C10029, doi:10.1029/2007JC004556.

Wessel, P., and M.T. Chandler (2011), The spatial and temporal distribution of marine geophysical surveys. Acta Geophysica 59, 55-71, doi:10.2478/s-11600-010-0038-1.

Whalen, C.B., J.A. MacKinnon, L.D. Talley, and A.F. Waterhouse (2015), Estimating the mean diapycnal mixing using a finescale strain parameterization. Journal of Physical Oceanography 45, 1174-1188, doi:10.1175/JPO-D-14-0167.1.

Whalen, C.B., L.D. Talley, and J.A. MacKinnon (2012), Spatial and temporal variability of global ocean mixing inferred from Argo profiles. Geophysical Research Letters 39, L18612, doi:10.1029/2012GL053196.

Williams, J.G., and D.H. Boggs (2016), Secular tidal changes in lunar orbit and Earth rotation. Celestial Mechanics and Dynamical Astronomy 126, 89-129, doi:10.1007/s10569-016-9702-3.

Wright, C.J., R.B. Scott, P. Ailliot, and D. Furnival (2014), Lee wave generation rates in the deep ocean. Geophysical Research Letters 41, 2434-2440, doi:10.1002/2013GL059087.

Wunsch, C. (1972), Bermuda sea level in relation to tides, weather, and baroclinic fluctuations. Reviews of Geophysics 10, 1–49, doi:10.1029/RG010i001p00001.

Wunsch, C. (1975), Internal tides in the ocean. Reviews of Geophysics and Space Physics 13, 167-182, doi:10.1029/RG013i001p00167.

Wunsch, C., and D. Stammer (1997), Atmospheric loading and the oceanic "inverted barometer" effect. Reviews of Geophysics 35, 79-107, doi:10.1029/96RG03037.

Xu, Y., and L.-L. Fu (2012), The effects of altimeter instrument noise on the estimation of the wavenumber spectrum of sea surface height. Journal of Physical Oceanography 42, 2229-2233, doi:10.1175/JPO-D-12-0106.1.

Zahel, W., and M. Müller (2005), The computation of the free barotropic oscillations of a global ocean model including friction and loading effects. Ocean Dynamics 55, 137–161, doi:10.1007/s10236-005-0029-y.

Zaron, E.D. (2017), Mapping the non-stationary internal tide with satellite altimetry. Journal of Geophysical Research Oceans 122, 539–554, doi:10.1002/ 2016JC012487.

Zhao, Z., M.H. Alford, J.A. MacKinnon, and R. Pinkel (2010), Long-range propagation of the semi-diurnal internal tide from the Hawaiian ridge. Journal of Physical Oceanography 40, 713–736, doi:10.1175/2009JPO4207.1.

CHAPTER 14

Wind Waves

Fabrice Ardhuin[1] and Alejandro Orfila[2]

[1]*Univ. Brest, CNRS, IRD, Ifremer, Laboratoire d'Océanographie Physique et Spatiale (LOPS), IUEM, Brest, France* ; [2]*IMEDEA, Esporles, Spain*

Wind-generated waves dominate sea surface motions for periods shorter than 300 seconds. Waves are of interest for many applications ranging from navigation safety to ocean and coastal engineering. Waves also define air-sea fluxes and have important interactions with surface currents, upper ocean turbulence, and sea ice. Given the general focus of this book, we emphasize here the successes of wave forecasting methods, starting with a review of basic principles and how wave energy and momentum are modeled. In particular, we discuss the connection between wave modeling and remote sensing, and opportunities for joint measurements of currents and waves. A more detailed account of wave research and applications to geosciences can be found in Ardhuin (2018).

Introduction

Wind waves, which we shall simply call waves in this chapter, are surface waves that derive their energy and momentum from the wind blowing over water. A detailed knowledge of wind fields is thus necessary for the determination of wave properties, which is often sufficient in the sense that we do not need to have wave measurements to make accurate wave forecasts. Still, it is not the local wind that defines the local wave field. Indeed, most of the energy transferred from wind to waves is lost to the upper ocean turbulence by wave breaking. Like any type of wave, wind waves also carry horizontal momentum. For a sine wave, the total wave energy per unit horizontal area E_t is equal to this momentum per unit area M_w times the phase speed C, which is the speed of wave crests. This is a very general relationship that also applies to photons, internal waves, etc. Hence the wind transfers momentum from the wind to the waves, and wave breaking transfers that momentum to surface currents, as illustrated in Fig. 14.1.

Because of this near equilibrium between the wind source and the breaking sink, the wave field grows on time scales of hours to days when the wind is blowing. By definition, the waves that grow due to the local wind form the wind-sea. The degree of development, called the wave age, is measured by the ratio of the phase speed of the dominant waves C_p and the wind speed at 10 m height U_{10}. With room to propagate across ocean basins, waves can persist as swells. Swells do not gain energy from the local wind, on the contrary they lose energy to the atmosphere. Wind-sea due to the local wind and swells generated by remote winds are the two constituents of the sea state.

Ardhuin, F., and A. Orfila, 2018: Wind waves. In "*New Frontiers in Operational Oceanography*", E. Chassignet, A. Pascual, J. Tintoré, and J. Verron, Eds., GODAE OceanView, 393-422, doi:10.17125/gov2018.ch14.

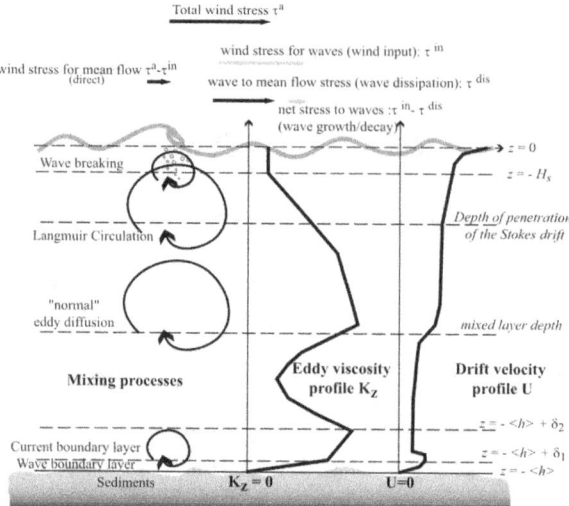

Figure 14.1. Momentum fluxes and mixing processes coupling waves and currents. Processes for horizontally uniform conditions, and possible profiles of eddy viscosity and drift velocity. (From Ardhuin et al., 2005).

A bit of history: 100 years of wave forecasting

After introducing these basic concepts of wind-sea, wave age, and swell, a short tour of wave forecasting history will provide some context. Wave forecasting is a branch of marine meteorology that started soon after the advent of operational weather forecasting in the late 19th century, with a focus on wind storms over the ocean and the associated waves. Driven by the needs of French naval operations around 1910 in Morocco, a first swell forecasting center was established in 1921 in Casablanca (Montagne, 1922) based on a method developed by Gain (1918). The center forecasts used sea level pressure measurements to track storms in the North Atlantic and swell measurements from the Azores to forecast the swells arriving in Morocco. This experience was used in the general wave forecasting method that would be adopted by the Allies for amphibious landings, from Morocco in 1942 to Normandy, and countless Pacific islands in 1944 and 1945. This work is summarized by Sverdrup and Munk (1947) in a report that was declassified after the war. The application and performance of these forecasts are well described by Bates (1949): forecasts for Normandy were good 88% of the time, corresponding to 20% error for heights around 1.5 m. Other war efforts at the UK Admiralty led to great advances on the measurement and analysis of waves, including the introduction of Fourier analysis (Barber et al., 1946). The dispersive nature of swells, with long periods propagating faster, naturally led to the adoption of spectral modeling, in which the surface elevation is represented by a sum of many sinusoidal components with different wavelengths and directions. The first numerical spectral wave model was put into operation by the Casablanca Group (Gelci et al., 1957). An important development of the 1960s was the realization that non-linear wave-wave interactions were a key process in the formation of wind wave spectra, with an inverse cascade of energy towards long period waves. The theoretical work was pioneered by Hasselmann (1962), with experimental verification in a series of experiments known as the Joint North Sea WAve Project (Hasselmann et al., 1973). Today the main physical processes leading to the evolution of wave spectra are usually well identified. In deep water without sea ice they include wave generation by the wind, non-linear interactions, wave breaking (often called whitecapping in that context), and swell dissipation presumably due to air-sea friction (Ardhuin et al., 2009). Still, the details of these processes are poorly known and involve complex and turbulent motions on both sides of the air-sea interface, such that numerical wave models rely on parameterizations that are still very empirical (e.g. Ardhuin et al., 2010; Stopa et al., 2016).

Expanding applications

Further, new applications and recent investigations have considerably broadened the scope of numerical wave models, leading to extensions towards both longer and shorter wave periods. Although they share most of their kinematic and propagation properties, it is customary to distinguish surface gravity wave categories according to their source of energy, which are indicated in green in Fig. 14.2. The wind waves covered in this chapter are taken broadly, including infragravity (IG) waves and all shorter components; we will discuss waves with periods from 1 second to 5 minutes.

The combination of coastal engineering, remote sensing, and seismology has led to the extension of regional wave models into the long IG range, showing reasonable skill that can be used to interpret future high-resolution satellite altimetry missions and microseismic records (Ardhuin et al., 2014, 2015). These same IG waves are a key component of extreme sea level and nearshore erosion (Reniers et al., 2004), and may play a role in the break-up of ice shelves (Bromirski et al., 2010).

Ocean and coastal engineering: The interest in marine renewable energies, in particular wind power to be recovered from floating platforms, has expanded the existing needs of coastal engineering. These coastal needs were behind many of the early efforts on wave research (e.g., Boussinesq, 1872; de Saint-Venant and Flamant, 1888; Miche, 1944; Iribarren and Nogales, 1949), relayed by offshore engineering when drilling for oil and gas led to many advances in wave dynamics and statistics (Cavanié et al., 1976; Tayfun, 1980).

Remote sensing: Another driver for wave research starting in the 1970s was the booming space age with the cold war in the background. The new capability to measure ocean properties from space, demonstrated in 1972-1973 with Skylab's altimeter and L-band radiometer, meant that ocean wave contributions to remote sensing data required a detailed investigation of the short ocean waves that contribute to the measured signals. Although short wave properties have long been related to wind speed alone (e.g., Cox and Munk, 1954), it is clear that long waves also contribute to the statistics of surface slope (Gourrion et al., 2002). These long waves are not fully determined by the local wind speed. Surface slope statistics can be predicted by the same numerical wave models that are used for marine weather forecasting (Ardhuin et al., 2010). These models still suffer from important errors on the directions of short waves (Peureux and Ardhuin, 2016). Remote sensing applications still largely use the wave spectrum shape of Elfouhaily et al. (1997) defined from three parameters: the wind speed, direction, and the wave age, but its directional distribution does not have the double peak (Peureux et al., 2017) that may be necessary to reproduce L-band observations (Yueh et al., 2013).

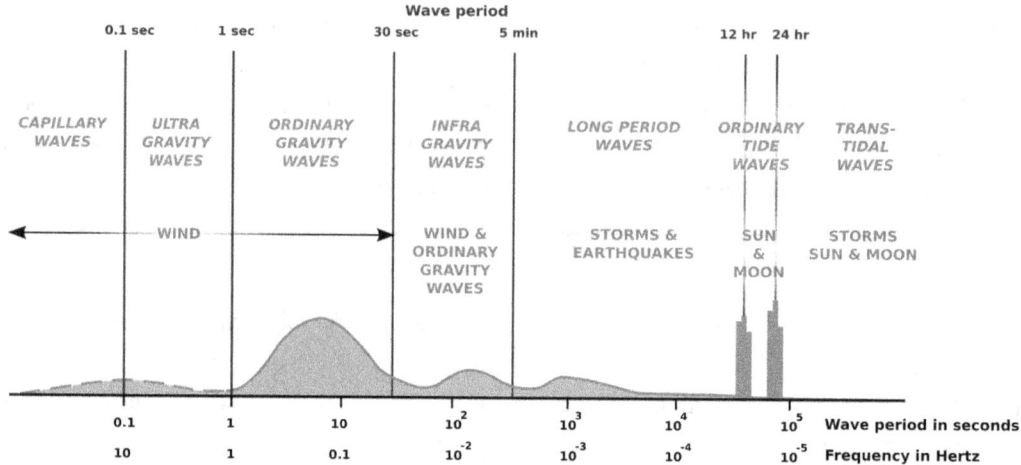

Figure 14.2. Classification of ocean waves according to wave periods. The sources of energy for the various portions of the spectrum are shown in green. The order of magnitude of the relative amplitudes is indicated by the curve (Adapted from Munk, 1950).

Weather and storm surge forecasting: These short waves are also expected to define the surface roughness for the atmospheric boundary layer (Plant, 1982; Janssen, 1989), with a large impact on storm surges (Mastenbroek et al., 1993). Hence, waves have now become an important component in weather forecasting. The ECMWF (European Centre for Medium-Range Weather Forecasts) atmospheric model is coupled to the wave model WAM, operationally since 1998 (Janssen et al., 2001), and in all ECMWF reanalyses. WAM estimates a roughness length that changes the atmospheric boundary later. The variation of wind stress with wave age was found to modify the intensification of storms and is known to have a strong effect on the life cycle of hurricanes (Jarosz et al., 2007). The actual quantitative variation of wind stress with waves is still the topic of active research with a strong impact on high wind speeds given by atmospheric models: at 30 m/s ECMWF winds and KNMI-processed ASCAT winds are both biased low by 7 m/s compared to other data sources (Pineau-Guillou et al., 2018).

Earth system science (upper ocean and sea ice): Besides atmospheric roughness and wind stress, waves define the flux of turbulent kinetic energy going into the upper ocean, dominated by wave breaking (Agrawal et al., 1992). Another source of turbulence is the stretching by the wave-induced Stokes drift, which is a source of energy for the Langmuir circulation. This circulation consists of roll vortices elongated in the wave direction that are probably the dominant source of mixing in the upper ocean (Kukulka et al., 2009; Sullivan and McWilliams, 2010; D'Asaro, 2014). Waves also contribute to the generation of air bubbles that typically double the air-sea exchange surface and facilitate gas transfer (Deane and Stokes, 2002). On the air side, spray generation is the source of salt aerosols and other atmospheric constituents.

Waves also play a dominant role at the edges of the sea ice and define a Marginal Ice Zone (MIZ) in which air-sea fluxes are enhanced by the wave motion. Ice formation is enhanced by waves via the formation of frazil and pancakes (Shen et al., 2001). Pancake ice, as illustrated on Fig. 14.3, can cover vast regions of the ocean, extending hundreds of kilometers in the Southern Ocean. Waves can also break up the pack ice, sometimes over hundreds of kilometers as well, and in the melting season it increases lateral melting. Beyond the region of broken ice, waves may propagate with low amplitudes all across the Arctic. The interaction of waves, ice, and the upper ocean is the topic of very active research.

Beyond (upper atmosphere and solid Earth): Finally, wind waves are also the main source of background seismic waves in the solid earth (these are known as microseisms) and acoustic waves in the atmosphere (microbaroms) in particular at periods around 5 s. This background signal can be used to either characterize the noise source (the wind waves) or characterize the medium through which the seismic or acoustic waves propagate. This gives access to the structure of the solid Earth, including time-variations due to volcanic activity, stress changes around faults, etc., and wind speeds in the upper atmosphere at heights of a few tens of kilometers. Fig. 14.4 shows an example of ground motion amplitude (less than 1 micrometer in this case), which is very well explained by the ocean waves activity using the model of Ardhuin et al. (2011).

Seismic measurements such as those shown in Fig. 14.4 have been routinely taken since the 19th century and they contain a climatic record of storm activity (Algué, 1900; Bernard, 1990; Grevemeyer et al., 2000). After pioneering work by Longuet-Higgins (1950) and Hasselmann (1963), we are just beginning to understand the details of the coupling of wind waves with seismo-acoustic waves (Ardhuin et al., 2015), thanks to more accurate wave models and the rapid growth of seismic observation networks. This is still a very exploratory field of research with important applications in seismology (e.g., Shapiro et al., 2005).

Figure 14.3. Picture of pancake ice in the Beaufort Sea taken during the "Sea State" cruise (Rogers et al., 2016). The yellow buoy is a SWIFT wave-measuring drifter with a diameter of 30 cm.

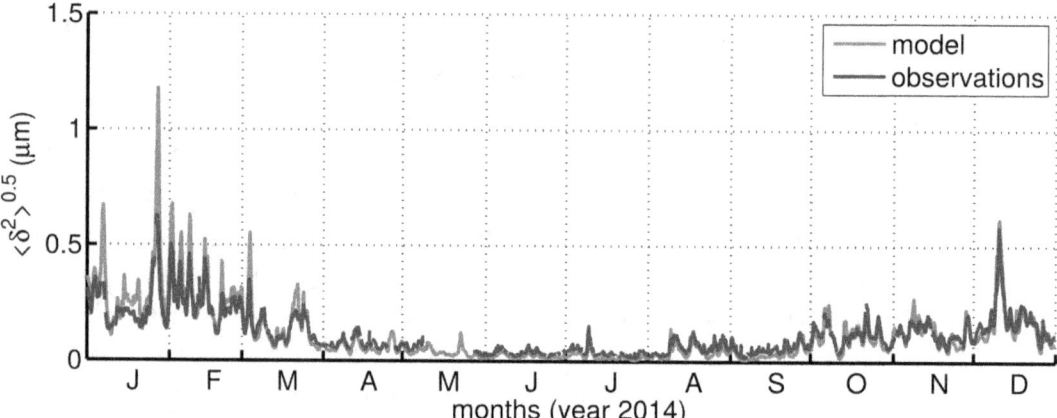

Figure 14.4. Example of measured and modeled time series of vertical ground displacement standard deviation, measured at the Royal Observatory of Uccle in Brussels, Belgium (data courtesy of T. Lecocq). Peaks are associated to storms in the North Atlantic (Ardhuin et al., 2011).

Challenges in numerical wave modeling

A key question in the development of wave models is the understanding of wave growth and dissipation processes and their parameterizations. Nonlinear wave evolution is generally thought to be well understood theoretically and relatively well treated in numerical models, provided that one is ready to pay the computational price. Still, all three aspects (growth, dissipation, and nonlinear propagation) are probably more strongly connected than is generally accepted, even in phase-resolved models. Serious theoretical challenges are also posed by the interaction of waves and currents, especially the effects of vertical current shear and waves with sea ice. New observation techniques are slowly changing our understanding of waves. These include stereo-video (Benetazzo, 2006; Leckler et al., 2015), X-band radars from ships or shore, and remote sensing from space.

This chapter is structured as follows. In the second section, we present the spectral properties of ocean waves. This section also covers a discussion about the fundamentals of numerical ocean wave models. In the third section, we discuss practical issues of phase-averaged wave modeling from the accuracy of the forcing to the choice of parameterizations. In the fourth section, we present phase-resolved modeling and nearshore applications. The fifth section discusses remote sensing of ocean waves and the impact of waves on the remote sensing of other parameters, with a particular focus on current measurements using Doppler techniques.

Ocean Waves and Their Spectral Properties

For most applications, only the statistical properties of ocean waves are of interest. In this case, the waves are referred to as 'sea state' and usually described by the power spectral density of the surface elevation. The investigation of the shape and properties of individual waves will not be covered here, but can be found in ocean engineering textbooks (e.g., Boccotti, 2000). The ocean wave spectrum is dominated by surface gravity waves with significant variations in their power over scales of a few hours, and fast modulations in the form of groups over a few wave periods due to a generally broad spectrum. Other gravity waves that take the form of short transient wave packets, such as tsunamis or ship wakes, will not be discussed here.

Linear wave properties

Waves are generally dominated by irrotational flow, and gravity is the dominant restoring force for most of the wind wave scales. We can ignore surface tension for wavelengths larger than 20 cm. These two properties give a dispersion relation that links the wavenumber k, defined from the wavelength as $k = 2\pi/L$ and radian wave frequency σ defined from the period as $\sigma = 2\pi/T$. For any water depth D, it is (de Laplace, 1776)

$$\sigma^2 = gk \tanh(kD), \qquad (1)$$

where g is the gravitational acceleration.

This dispersion relation yields the phase speed $C = \sigma/k$, which is the speed of the crests, and the group speed $C_g = \partial\sigma/\partial k$, which is the speed followed by wave groups and is also the speed at which the energy is propagated. These have two simple limits. When $kD \ll 1$, waves are in deep water, and $\sigma^2 = gk$. This gives $C = gT/(2\pi)$ and $C_g = C/2$. On the contrary, if $kD \gg 1$, waves are in shallow water and $\sigma^2 = gDk^2$. These shallow water waves are hydrostatic and not dispersive, with the same speed $\sqrt{(gD)}$ for all components. The definition of deep or shallow water is thus given by the ratio of the water depth D and wavelength $2\pi/k$ of the ocean wave.

These speeds are further modified by ocean currents, which introduce a Doppler shift. The apparent radian frequency of the waves in a frame of reference attached to the solid Earth is thus,

$$\omega = \sigma + \mathbf{k} \cdot \mathbf{U}. \qquad (2)$$

For the sake of simplicity we shall neglect the vertical variation of the current velocity and take U to be the horizontal vector of the surface current. In deep water, a 10 s ocean wave has a wavelength of 156 m, a phase speed of 15 m/s and a group speed of 7.5 m/s. Typical swells from a major hurricane have periods around 15 s, and they travel 50% faster, but it still takes them five days to cross the Atlantic from Cape Hatteras to the Bay of Biscay. Longer ocean waves propagate faster, but the upper limit is around 220 m/s for IG periods of 200 s, due to the limited water depth. Indeed, linear waves much longer than the ocean depth propagate at the so-called 'shallow water' speed $\sqrt{(gD)}$. With a water depth D = 5000 m, this speed is 220 m/s.

Typical sea states

Wind waves take all their energy from the wind, more or less directly. The direct transfer only occurs for waves slower than the wind, and the wave-wave interaction flux can push energy up to speeds 20% faster than the wind speed (Pierson and Moskowitz, 1964). As a result, ocean wave periods T are constrained by the dispersion relation (3.1) to be under 30 s for usual ocean waves. For $T = 30$ s, the phase speed C is 47 m/s in deep water, corresponding to the maximum wind speed in a category-2 tropical cyclone. However, this kind of cyclone typically does not generate wave periods larger than 16 s. This is because the wind speed is necessary, but not sufficient. Indeed, in order to reach large periods and heights, waves need time and space to develop because they grow slowly, on the time scale of days for the highest wind speeds (Hasselmann et al., 1973). The distance over which the waves develop is called fetch. Typical growth behavior of wave heights is shown in Fig. 14.5

Both give the time-limited growth for large fetches and the fetch-limited growth for large durations, with no more growth when waves reach full development at Hs 10.8 m: that value is only a function of wind speed. The dashed line is the same in both panels.

As a result, the waves with largest height and periods are not found in tropical cyclones, which move too fast, but in those extra-tropical storms that travel at the group speed of the dominant waves. For example, the largest ever sea state was measured in such an extra-tropical storm in the North Atlantic, with a maximum significant wave height Hs = 20.1 m, and a peak period of 25 s (Hanafin et al., 2012). Thus, ocean wave properties are determined by the space and time patterns of the wind speed and direction, the shoreline geometry, and the water depths. Sea ice also modifies wave propagation and dissipation by impeding the transfer of energy from wind to waves and causing an extra dissipation that is particularly strong for higher frequencies (e.g., Ardhuin et al., 2016). Swells are the waves radiated away from storms, with typically longer periods than wind sea because shorter components dissipate rapidly in the absence of wind forcing. Near the shoreline, these wind-generated waves can interact non-linearly to produce longer ocean wave periods, up to 300 s. Those long waves are called IG waves (Munk, 1950; Longuet-Higgins and Stewart, 1962). Some examples of ocean wave spectra are shown in Fig. 14.6 and discussed in the next section.

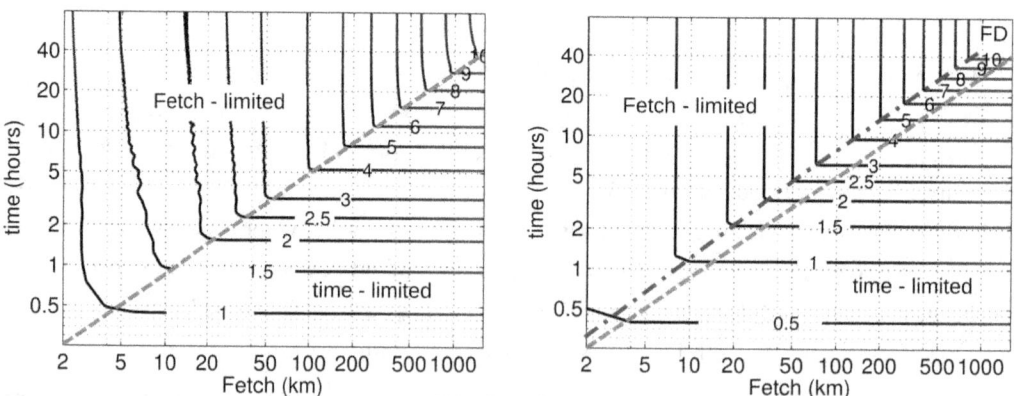

Figure 14.5. Estimation of significant wave height (H_s, contours) as a function of fetch and time for the idealized case of an infinitely long coast with a wind blowing perpendicularly offshore at a speed U10 = 20 m/s. In the left panel, the estimate is given by a numerical integration of the wave energy or action equation (3), using parametrizations for the wind-wave growth and dissipation from Rascle and Ardhuin (2013) and non-linear wave-wave interactions from Hasselmann et al. (1985) in the WAVEWATCH III model, with a third order numerical scheme Tolman (1995). The right panel combines empirical growth curves (Elfouhaily et al., 1997).

WIND WAVES • 401

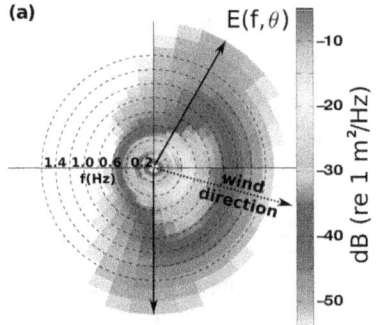
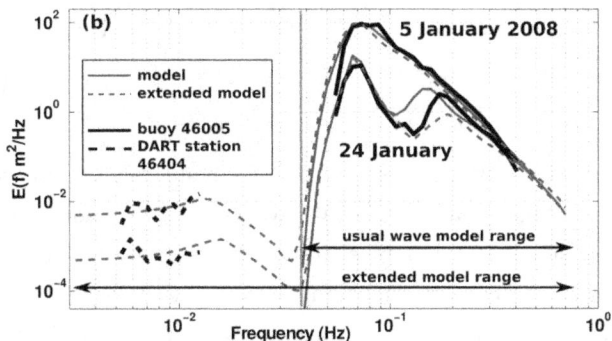

Figure 14.6. (a). Example of directional wave spectrum $E(f, \theta)$ as a function of azimuth θ and frequency f as recorded by a stereo video system in the Black Sea (Leckler et al., 2015). The high frequency of the peak, f_p 0.33 Hz, is typical of small seas or weak winds, but the directional shape of the spectrum is expected to be similar to that found in the open ocean for similar values of f/f_p. At relative high frequencies ($f/f_p > 3$) the dominant directions (solid arrows) can be 70° away from the wind direction (dashed arrow). (b) Example of modeled and measured wave spectra $E(f)$ in the North-East Pacific, on 5 (top curves) and 24 (bottom curves with 2 peaks) January 2008. Classical wave models provide wave spectra between 0.04 and 0.7 Hz (blue solid line). Ardhuin et al. (2014) have extended that range to include infragravity (IG) waves down to $f = 3 \times 10^{-3}$ Hz (blue dashed line).

Numerical wave modeling

Numerical ocean wave models were first developed for navigation safety (Gelci et al., 1957) and are generally based on an evolution equation for the wave spectrum. At each point of the ocean surface, the sea state is represented by a two-dimensional power spectral density $E(f, \theta)$ that gives the distribution of surface elevation variance across frequencies f and directions θ. Wave energy propagates in all directions θ at the speed given by the group velocity C_g that is a function of frequency f. In deep water, $C_g = g/(4\pi f)$ is half of the phase speed, hence $C_g = 8$ m/s at $f = 0.1$ Hz. For a flat ocean bottom, the evolution of the power spectral density, $E(f, \theta)$, is given by

$$\partial E(f, \theta)/\partial t + \text{propagation at speed } C_g = S(f, \theta), \qquad (3)$$

where the left side represents wave propagation, and the right side source term S represents many processes including generation by the wind (e.g., Janssen, 2004), non-linear wave evolution (Hasselmann, 1962) and dissipation by wave breaking or wave-ice interactions (Ardhuin et al., 2010, 2016).

Both E and S vary with the horizontal coordinates x and y, as well as with frequency f and azimuth θ. For each frequency and direction, the source term $S(f, \theta)$ is also a function of the spectral density $E(f, \theta)$ evaluated at all the other frequencies f and directions θ, but at the same location given by x and y. The source term S is also a function of many other parameters including wind speed and direction, water depth, ice and ocean bottom properties. In practice, there is no expression for S that is based on first principles, and empirical parameterizations are the topic of ongoing research (e.g., Ardhuin et al., 2010).

There are very few measurements of the full two-dimensional ocean wave power spectral density as a function of azimuth θ and frequency f, such as the one shown in Fig. 14.6.a. Numerical wave models have been calibrated and validated mostly in terms of dominant ocean wave frequency f_p, and significant wave height H_s, defined as

$$H_s = 4\sqrt{\int E(f)df} \quad \text{with} \quad E(f) = \int E(f,\theta)d\theta \quad (4)$$

Most wave measurements done in situ rely on time series of pressure, velocity, and/or surface elevation, so that it is usual to work with the one-dimensional power spectral density of the surface elevation $E(f)$. A typical frequency range of measurements and models goes from a minimum around 0.03 Hz to a maximum somewhat below 1 Hz. It is fairly common to have a good estimate of Hs but a not-so-good spectral distribution $E(f)$. Fig. 14.6.b shows measured spectra and results of a numerical ocean wave model (The WAVEWATCH III$^{(R)}$ Development Group, 2016) off the Oregon coast, where the water depth is 4000 m, on 5 and 24 January, 2008. The extended modeled and measured spectra are for the location of the tsunameter DART 46404 system, 200 km to the east of the wave buoy 46005. On 5 January, the model reproduces well the single peak at 0.07 Hz. On 24 January, the swell peak at 0.07 Hz is relatively well modeled, while the peak of the wind sea is at 0.15 Hz in the model and 0.18 Hz in the buoy data due to errors in the wind speed used to drive the wave model. As a result there is a large difference in the energy level at 0.15 Hz. These results depend mostly on the accuracy of winds used to force the wave model.

Operational Wave Hindcasting and Forecasting

Global to regional scales

At the scale of ocean basins, the most common models solve a wave energy equation similar to Eq. 3—usually with the addition of refraction—on regular grids in latitude and longitude (Janssen, 2008) or on multiple-grid systems that allow refinement close to shore (Tolman, 2008) where the sea state is impacted by shoreline geometry. The specific implementations depend on the sea state encountered in different regions. In particular, the frequency range over the open ocean should extend below 0.04 Hz to reproduce the rare but very strong storms (e.g., Hanafin et al., 2012). This is not necessary for a model restricted to an enclosed basin such as the Mediterranean, where long waves have no room to develop. High spatial resolution is also preferred for hurricane modeling, and moving or adaptive grids have been tested in research but, to our knowledge, are not used now in operational forecasting system.

The accuracy of wave models has benefitted from steady improvements in the accuracy of winds, in particular thanks to the assimilation of surface winds (Hersbach, 2010). For forecast ranges of a few days, the model accuracy also benefits from the assimilation of altimeter (Lionello et al. 1992) and synthetic aperture radar (SAR) wave mode data (Lefevre and Aouf, 2012; Hasselmann et al., 2012). In particular, SAR provides a measure of the wave spectrum that has a more lasting impact in the forecasts because it allows to correct initial value errors of the different wave components. The quality of the forecasts in routinely verified using moored buoys (Bidlot et al., 2007; Bidlot, 2017). The accuracy of the wind is the dominant factor for the accuracy of wave forecasts and hindcasts, followed by the effect of parameterizations, and finally a benefit from assimilation for forecasts ranges of three days or less. Given the importance of the wind forcing, it is no surprise that relatively coarse atmospheric models, such as the ERA-Interim reanalysis, produce low biases on wave heights, and generally wave models have to be tuned to specific wind forcing (Stopa, 2018).

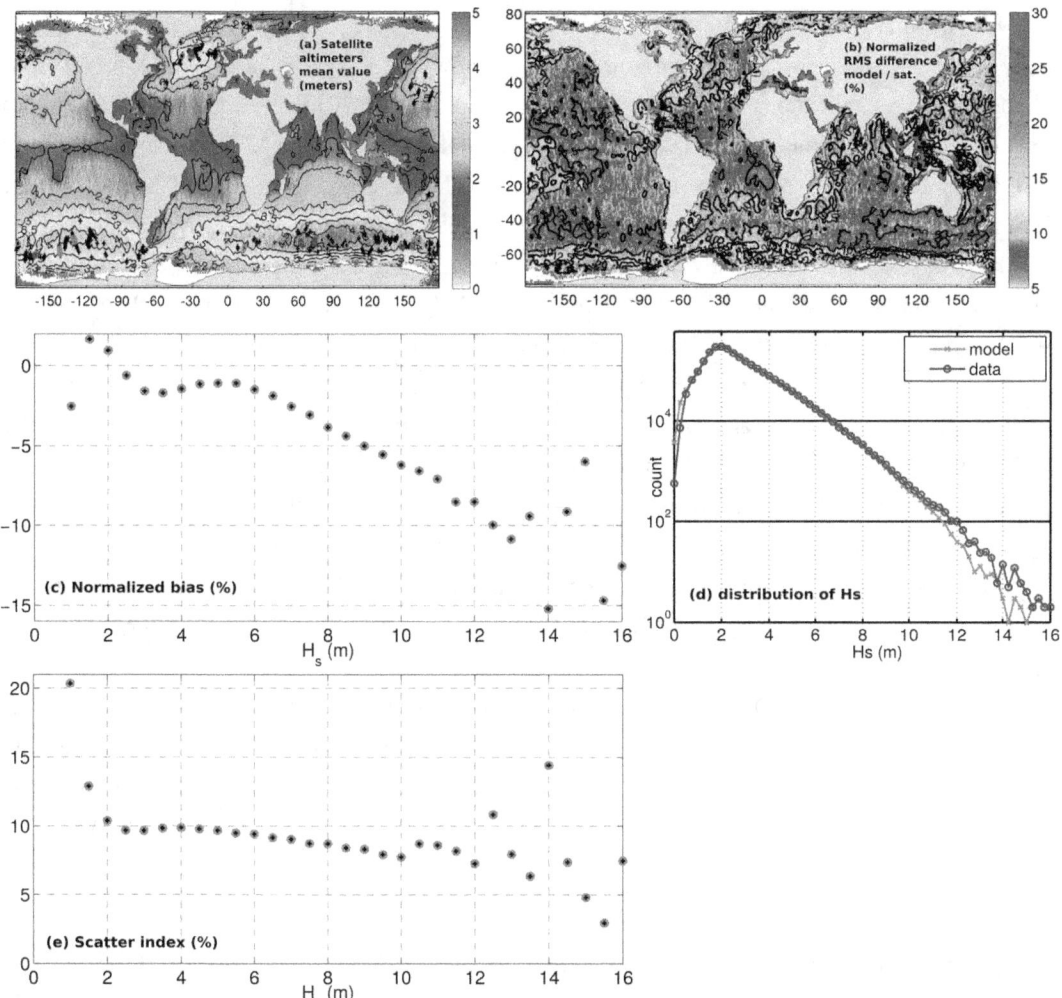

Figure 14.7. The satellite data are along-track averages over 1 degree in latitude, and statistics are based on these "super-observations", with a total of nearly 3 million super-observations and associated model match-ups.(a) Mean value of Hs for the year 2015, as recorded by Jason-2, Cryosat and SARAL-Altika. '+' marks the location of values above 12 m, red 'o' are above 14 m and blue squares are above 15 m. (b) Normalized RMS difference between model and altimeter for the year 2015, in percent. (c) Bias (model - satellite) as a function of Hs for the same year, (d) distribution of Hs in co-located model and satellite data and (e), scatter index.

Fig. 14.7.a shows the mean value of H_s in 2015, which does not vary too much from year to year. Although the largest mean values are in the Southern Ocean, the most extreme values of H_s are usually found between Iceland and Ireland. Also, it is interesting to note that extremes in the tropics are dominated by tropical storms. These give lower maximum values of H_s than the largest extra-tropical storms, because the very high winds are concentrated in the small region around the eye. Still, these extreme H_s in tropical storms are relatively high compared to the annual mean.

For all these extremes, the sampling of satellite altimeters is insufficient, even with three nadir satellite missions.

With altimeter data averaged over 120 km along-track, this average, open ocean wave heights are very accurate with typical errors under 10% of the averaged measured value. The modeled values are produced by the Laboratoire d'Océanographie Physique et Spatiale for various geophysical applications (Rascle and Ardhuin, 2013) and available at ftp://ftp.ifremer.fr/ifremer/ww3/HINDCAST/. The model results in Fig. 14.7 do not include current effects, which certainly contribute to the errors off South Africa. Other regions of large errors are in some coastal areas, especially on the west side of ocean basins, and where sea ice is important. The relative large difference between model and altimeter around Indonesia may be largely due to altimeter errors. Indeed, that region has average wave heights under 1 m, for which the altimeters with a typical vertical resolution of 0.4 m are not accurate enough (e.g., Sepulveda et al., 2015).

Predicting low to average wave heights can be important for delicate operations at sea (towing of large structures, maintenance), but very high wave heights and very large periods generally receive more attention. Fig. 14.7.e shows that the highest waves are indeed the most accurately predicted with typical random errors under 8% for Hs > 12 m, once the bias has been corrected. This bias is not completely understood but probably comes in part from an underestimation of high wind speeds in the ECMWF model (Pineau-Guillou et al., 2018). Other sources of errors are inappropriate parameterization for wind-wave interactions at very high winds where spray, bubbles, and hydrodynamic instabilities can be important (Soloviev et al., 2014).

At regional to ocean basin scales, the main differences between different wave models come first from the wind forcing, and second from parameterizations of the source term S on the right side of Eq. 3. To a lesser extent, differences can also come from numerical schemes (Ardhuin, 2018, chapter 8).

Effects of depth and currents on waves

The dispersion relation Eq. 1 is modified by surface currents **U** into Eq. 2. As a result, the phase speed is also Doppler shifted, and the phase speed of linear waves is a function of wavelength, depth, and current vector. Thus, current gradients are the cause of refraction of waves, just like changes in the water depth. A simple way to understand this effect is to consider wave rays, the trajectories followed by wave crests. Things are simpler for monochromatic waves in stationary conditions and without any dissipation nor diffraction. In the absence of current, the wave energy is conserved and the flux of energy between two rays is constant, where the rays converge with a distance l the energy flux is proportional to $1/l$ and the wave height is proportional to $1/\sqrt{(C_g\, l)}$.

In the presence of current, the wave energy is not conserved because waves and current exchange energy. This change of wave energy can be interpreted as the work of 'radiation stresses' (Phillips, 1977; Ardhuin, 2018). Still, in the absence of dissipation the wave action $\int E(f, \theta)/\sigma df\, d\theta$

is conserved. As a result, a similar ray tracing can be performed, with the ray trajectories modified by the current, and the energy flux Cg E replaced by $(Cg + \mathbf{k} \cdot \mathbf{U}/k)E/\sigma$.

In many coastal regions, the significant wave height has a strong tidal modulation, which is due to currents. This modulation is mostly caused by refraction. In the case shown in Fig. 14.8, the typical curvature radius of the rays in the current jet located south-west of the island of Ouessant is around 10 km, as materialized by the green circle. This jet peaks 1.5 hours after the low tide and deviates the waves away from the Pierres Noires buoy, located 20 km down-wave.

Similar effects explain the formation of large waves in the Agulhas current (Lavrenov, 2003), the Gulf Stream or other regions. The variability of wave heights follows that of the surface current and increases when the current gradient in the direction of the wave crests is coherent over long distances (White and Fornberg, 1998). As a result, mesoscale and submesoscale currents induce significant variations in wave heights, as shown in Fig. 14.9

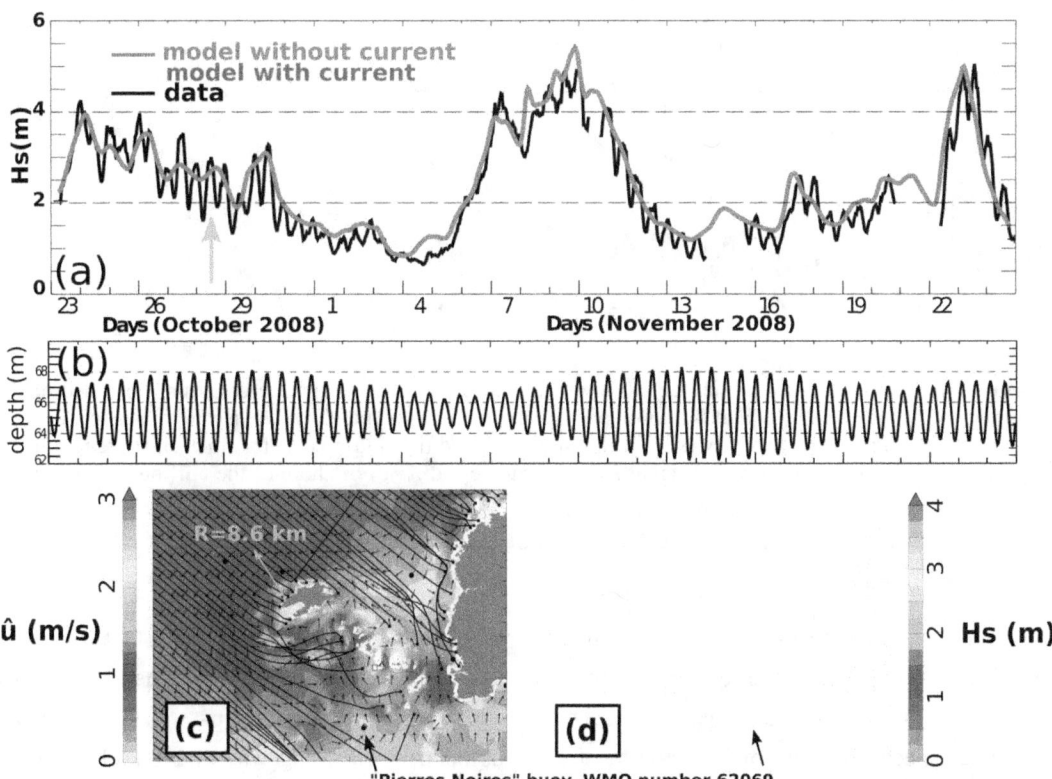

Figure 14.8. Example of strong impact of currents on wave heights due to wave refraction by currents. The top panel shows a time series of Hs recorded at the wave buoy 'Pierres Noires' (WMO number 62069) and modeled with WAVEWATCH III, while the middle panel shows the water depth at the buoy. (c) Modeled tidal current and the corresponding wave rays for $T = 10$ s (d) shows Hs and mean wave directions, both for October 28, 2008, at 11:00 AM UTC (corresponding to blue arrow in a). (Adapted from Ardhuin et al., 2012).

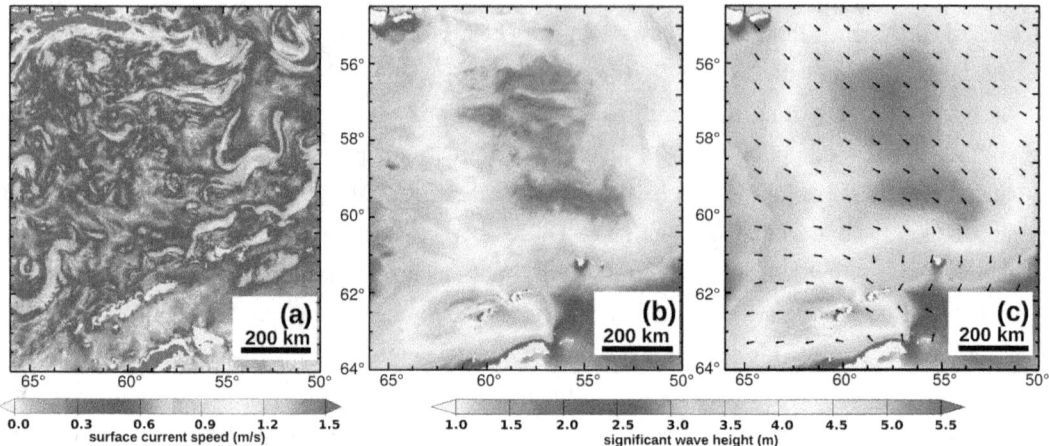

Figure 14.9. Example of (a) surface current magnitude modeled by MITgcm, courtesy of D. Menemenlis, (b) the modeled significant wave height when the current forcing is included (arrows) and (c) modeled significant wave height without effects of currents. (Adapted from Ardhuin et al., 2017b).

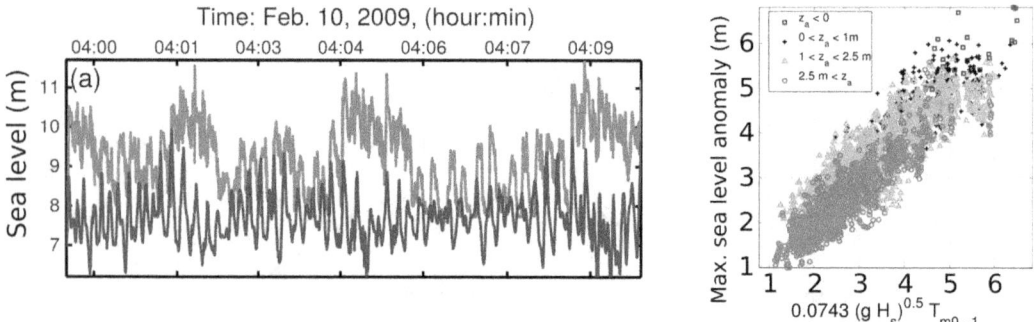

Figure 14.10. Left: example of surface elevation estimated from bottom pressure sensors, recorded on 10 February 2009 04:00–04:10 UTC at the bottom (blue) and top (red) of a cliff, on the island of Bannec, France. Right: maximum sea level anomaly (sea level minus predicted tide and inverse barometer), at the top of the cliff, as a function of modeled wave height H_s and mean period $T_{m0,-1}$ offshore of the cliff during the winter 2008-2009.

Waves at the coast

We have shown all results for wave heights, but wave periods are also very important for many applications, in particular for extreme water levels at the coast. Indeed, the amplitude of IG waves generated in shallow water is usually related to the presence of long wave groups, which occur when the wave spectrum is narrow. This is most often associated with long wave periods. Fig. 14.10 shows an example of a measured time series of sea level right at the shoreline (Sheremet et al., 2014). These contain very large IG waves with a height exceeding 2 m and a typical period of 300 s. These IG waves contribute to the maximum sea level anomalies and scale with the Hunt parameter defined as $H_H = \sqrt{(g H_s)}\, T_{m0,-1}$, in which the mean wave period $T_{m0,-1}$ is actually more important than the wave height. Mean periods, are generally defined from the spectrum as

$$T_{m0,n} = [\int f^n E(f)\, df / \int E(f) df]^{1/n}. \tag{5}$$

In the case n = −1, the mean period is also called the energy period because it corresponds to the proper frequency weighting of the energy flux. In practice, H_H and IG wave amplitudes are highly correlated (e.g., Stockdon et al., 2006), which makes possible the definition of an empirical source of IG waves at the shoreline (Fig. 14.11).

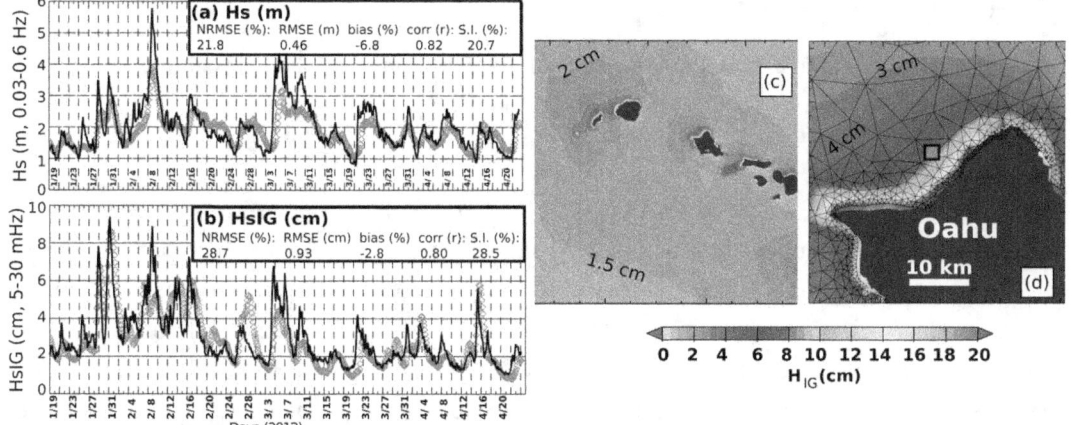

Figure 14.11. Measured (solid lines) and modelled (symbols) wave parameters for Waimea, Hawaii: (a) wave heights, (b) infragravity (IG) wave heights - the dates are written below panel (b) every four days only, starting on January 19, 2012, (c) snapshot of modelled H IG around all the islands, and (d) close-up in the Oahu north shore, where the Waimea buoy and mooring were located (red square), January 31, 2012 at 0 UTC.

Nearshore Dynamics

Given the complexity of wave evolution and the high stakes of beach or harbor management, many different approaches have been developed for dealing with the region of the ocean along the coast that is dominated by intense wave breaking, which we call the nearshore. Phase-averaged models, the same as those at larger scales such as WAVEWATCH III, are still largely used but they fail to reproduce correctly critical aspects of the nonlinear wave evolution in shallow water. Their extension with the bi-spectrum (Herbers and Burton, 1997) is an interesting alternative, but it has been only used for research, probably because users are more familiar with time series of velocities than just their statistical properties (e.g., skewness and asymmetry). Many applications use phase-resolving models, which basically solve the deterministic time dependent mass and momentum balance equations and generally can also model currents at the same time.

It is accepted that shallow water conditions correspond to $kD < 0.3$, with $k=2\pi/\lambda$ the wave number and λ the wavelength, while $kD > 3$ corresponds to deep waters (Dingemans, 1997). In intermediate waters (namely $0.3 < kD < 3$), non linearity and dispersion coexist and neither Airy theory, nor nonlinear shallow water equations can represent properly the physics of wave propagation. To overcome this problem, two main perturbation approaches are found. On the one hand, Stokes theory departs from the fully dispersive linear Airy theory to incorporate weakly nonlinear effects. On the other, Boussinesq-Type Equations (BTEs) depart from nonlinear shallow water equations and include weakly dispersive effects.

In this section, we reviewed some of the theories that have been developed for wave transformation from intermediate waters to the shore. The reader is refered for a more complete overview to Mei (1989), Dingemans (1997), and Liu and Losada (2002).

Depth integrated models

For small amplitude waves, and assuming irrotational motion and incompressibility of water, a velocity potential $\nabla \Phi = u$ exists satisfying the continuity equation, i.e.,

$$\nabla^2 \Phi = 0 \qquad -h < z < \eta \tag{6}$$

and $D = h + \eta$, the time averaged water depth. This can be solved after defining the conditions at the fixed boundaries (surface and bottom). At the surface, $z=\eta(x,y)$ two conditions have to be specified, the dynamic and the kinematic which establish continuity of stresses at the surface and that the surface is a material interface. At the bottom, $-h(x,y)$, the no-flux condition ensures the bottom as a material surface. The solution for the water wave problem represented by a linear wave $\eta(x,y,t)= a \cos(kx-\sigma t)$ with wave number k and frequency σ propagating over a flat bottom Eq. 6, subjected to the boundary conditions gives for the velocity potential,

$$\Phi = \frac{ag}{\sigma} \frac{\cosh(k(h+z))}{\cosh(kD)} \cos(kx)\sin(\sigma t) \tag{7}$$

where a is the wave amplitude and where wave frequency is related with the wave number through the dispersion relationship (Eq. 1), obtained by combining the two free boundary conditions at the surface.

As the water waves propagate to the coast, the water depth D decreases and the wave propagation becomes influenced by it. Also, nonlinear effects become important. In shallow waters, where the water depth dominates the wave propagation, the wave celerity is given by $C \approx \sqrt{gD}$, which is independent of the wave period (i.e., non-dispersive). An important physical property of shallow waters is that the horizontal velocity profile is nearly uniform in the vertical, nonlinear shallow water equations, which are vertically integrated, exploit this property and are valid for non-dispersive conditions and for arbitrary amplitudes of the wave.

Phase-resolving models need to solve the wavelength with enough resolution, which implies around 20-30 points per wavelength (10-100 time steps per wave period). By contrast, phase-averaged models only need to resolve gradients of the wave amplitude, which is generally less demanding. When a train of monochromatic waves enters a zone of slowly varying bathymetry wavelength changes and the spacing between equal phase lines also changes as the result of the adjust of the phase velocity. Phase-resolving models are still enormously expensive being still limited to specific applications or to the study of processes. The main effort in this subject has been devoted in the development of new theories that reduce the computational cost of the numerical models as well as in the development of a theory able to study the propagation of waves from deep to shallow waters including all the physics of the wave transformation phenomenon (all dispersion ranges).

Early modeling efforts for the transformation of waves from the ocean to the coast traced rays tangent to the wave number vector computing the variation of the wave envelope using the conservation of energy. However, the main problem of ray theory is that it is only valid for small amplitude waves and does not allow diffraction effects. To overcome this limitation, Eckart (1952) first and independently Berkhoff (1972) developed the mild slope equation. The main assumption is that the evanescent modes can be neglected for waves propagating over slowly varying bathymetries, i.e., $\frac{\nabla D}{kD} \ll 1$, except around obstacles. For a monochromatic wave with surface displacement η and frequency σ, Smith and Sprinks (1975) showed that to the leading order the free surface displacement satisfy,

$$\nabla \cdot (CC_g \nabla \eta) + \frac{\sigma^2}{g}\eta = 0 \qquad (8)$$

where we recall that $C=\sigma/k$ is the phase velocity and

$$C_g = \frac{\partial \sigma}{\partial k} = C\left(\frac{1}{2} + \frac{kD}{sinh(2kD)}\right) \qquad (9)$$

is the group velocity. The mild slope is suitable to propagate linear waves from deep to shallow water, reproducing correctly the velocity profile according to the linear wave theory.

The main problem of the mild slope equation is to specify the boundary at the coast. Tsay and Liu (1982) and Kirby and Dalrymple (1983) solved this problem by developing the parabolic approximation, which is suitable for waves propagating mainly in one direction. In this theory, wave energy is allowed to diffuse across wave ray including approximately the effects of diffraction. For a free surface displacement $\eta = \psi(x,y)e^{ik_0 x}$ being k_0 a reference wave number and assuming that the wave amplitude varies much slower in the wave propagation direction, and defining $K = C\,C_g$, Eq. 8 after adopting the parabolic approximation can be expressed as,

$$\frac{\partial^2 \psi}{\partial y^2} + \left(2ik_0 + \frac{1}{K}\frac{\partial K}{\partial x}\right)\frac{\partial \psi}{\partial x} + \frac{1}{K}\frac{\partial K}{\partial y}\frac{\partial \psi}{\partial y} + \left(\frac{\omega^2}{g} - k_0^2 + \frac{ik_0}{K}\frac{\partial K}{\partial x}\right)\psi = 0 \qquad (10)$$

The mild slope equation and its parabolic approximation are obtained for linear waves, and the superposition of different frequency components can be made. These models are used extensively for coastal applications since accurate propagation over two-dimensional bathymetries can be modeled with relatively low computational cost.

Boussinesq approximation

There is an intermediate zone where $H/D << 1$ and $kD << 1$. The Boussinesq Equations (BEs) were developed to represent water wave propagation in this region. BEs can be seen as an extension of the shallow water equations that includes dispersion in a perturbative way. BEs were obtained by Peregrine (1967) for weakly dispersive and weakly non-linear conditions. The extension to weakly dispersive but arbitrary (or fully) non-linear conditions are very popular nowadays (Green and Naghdi, 1976; Wei and Kirby, 1995; Madsen and Schaffer, 1998), and are usually referred to as Serre's Equations, after Serre (1953), or also as Boussinesq-type Equations (BTEs).

BEs and BTEs have ensured a good performance under weakly dispersive conditions (BEs for weakly non-linear conditions and BTEs for arbitrary non-linear conditions), by construction. In order to assess the performance under stronger dispersive conditions, BEs and BTEs are linearized and compared to linear and fully dispersive theories such as Airy and Mild Slope Equations (Dean and Dalrymple, 1984). The comparison is made in terms of wave celerity (linear dispersion) and wave shoaling over mild slopes (linear shoaling). The weakly non-linear performance is also usually compared to the second order Stokes theory for flat beds (Schaffer, 1996). Since BEs and their corresponding BTEs are identical in their linear weakly dispersive terms, the comparisons give the same results using the BEs or their corresponding fully non-linear extensions (BTEs).

The standard BEs for variable depth expressed in terms of the depth averaged velocity vector $\overline{u}$ are obtained from the conservation of momentum and mass equations by a perturbative approach as,

$$\frac{\partial \eta}{\partial t} + \nabla \cdot [(\eta + h)\overline{u}] = 0$$
$$\frac{\partial \overline{u}}{\partial t} + \frac{1}{2}\nabla \overline{u}^2 + g\nabla \eta + \left\{\frac{h^2}{6}\nabla\left(\nabla \cdot \frac{\partial \overline{u}}{\partial t}\right) - \frac{h}{2}\nabla\left(\nabla \cdot \left(h\frac{\partial \overline{u}}{\partial t}\right)\right)\right\} = 0 \quad (11)$$

which can be further coupled with the bottom boundary layer to analyze viscous damping and sediment transport processes (Liu and Orfila, 2004; Orfila et al., 2007).

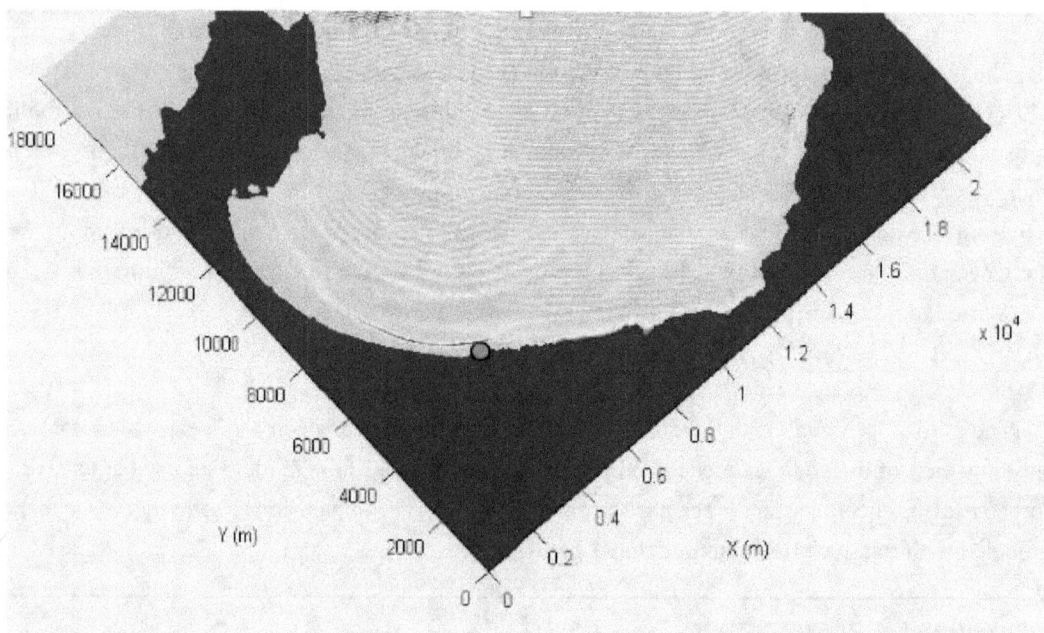

Figure 14.12. Snapshot of free surface elevation in Alcudia Bay (North of Mallorca Island). The computational domain is 20 x 18 km with a fixed grid of 1 m in order to solve the smallest wave lengths.

Although providing very accurate results compared with field and laboratory data (Liu et al., 2006), are still restricted to a small zone since it is required that both nonlinearity and frequency dispersion remain weak. Much of the research in this area during the last 20 years has been devoted to improve the linear properties of the equations. Leaving aside higher order (in dispersive terms) equations (Gobbi et al., 2000), which include spatial derivatives of order five, two main different approaches can be distinguished to this end: i) Madsen and Sorensen (1992) proposed an enhancement technique so as to introduce new terms that improve the dispersive performance, and Beji and Nadoaka (1992) proposed an alternative set of enhanced equations; and ii) Nwogu (1993) introduced a new set of BEs written for the velocity *at* $z_\alpha = \alpha h$ (instead of the depth averaged velocity), and chose $\alpha = -0.53096$ to improve the linear dispersion up to $kh\sim3$. The corresponding depth integrated and continuity equations expressed in terms of the free surface displacement η and the horizontal velocity $\boldsymbol{u}_\alpha$ at the water depth $z=z_\alpha$ are:

$$\frac{\partial \eta}{\partial t} + \nabla \cdot [(\eta + h)u_\alpha] + \nabla \cdot \left\{\left(\frac{z_\alpha^2}{2} - \frac{h^2}{6}\right)h\nabla\nabla \cdot u_\alpha\right) + \left(z_\alpha + \frac{h}{2}\right)h\nabla(\nabla \cdot hu_\alpha)\right\} = 0$$
$$\frac{\partial u_\alpha}{\partial t} + \frac{1}{2}\nabla|u_\alpha|^2 + g\nabla\eta + z_\alpha\left\{\frac{1}{2}z_\alpha\nabla\left(\nabla \cdot \frac{\partial u_\alpha}{\partial t}\right) + \nabla\left(\nabla \cdot \left(h\frac{\partial u_\alpha}{\partial t}\right)\right)\right\} = 0$$
(12)

The optimal choice of $z_\alpha=0.53h$ allows to simulate with a good accuracy wave propagation from intermediate water depth to shallow water including wave current interaction as shown in Fig. 14.12.

High order models

In the last 15 years, there has been an intense research work to further extend the range of applicability of BTEs. These models introduce higher order terms or more variables (Agnon, 1999; Gobbi et al., 2000; Lynett and Liu, 2002) and, pushed to their limits, some of them can represent highly dispersive and highly nonlinear waves. As a counterpart, since they contain either more equations or higher order derivatives in the governing equations, they are more demanding with respect to the computational resources compared to low order original models. This makes BTEs such as the above presented widely used nowadays. The highly nonlinear and weakly dispersive wave equations take the form

$$\frac{\partial \eta}{\partial t} + \nabla \cdot \left\{(h+\eta)\left[u_\alpha + \left(z_\alpha + \frac{1}{2}(h-\eta)\right)\nabla(\nabla \cdot (hu_\alpha)) + \left(\frac{1}{2}z_\alpha^2 - \frac{1}{6}(h^2 - h\eta + \eta^2)\nabla\nabla \cdot u_\alpha\right)\right]\right\} = 0 \quad (13)$$

$$\frac{\partial u_\alpha}{\partial t} + \frac{1}{2}\nabla|u_\alpha|^2 + g\nabla\eta + z_\alpha\left\{\frac{1}{2}z_\alpha\nabla\left(\nabla \cdot \frac{\partial u_\alpha}{\partial t}\right) + \nabla\left(\nabla \cdot \left(h\frac{\partial u_\alpha}{\partial t}\right)\right)\right\}$$
$$+\nabla\left\{\frac{1}{2}(z_\alpha^2 - \eta^2)(u_\alpha \cdot \nabla)(\nabla \cdot u_\alpha) + \frac{1}{2}[\nabla \cdot (hu_\alpha) + \eta\nabla \cdot u_\alpha]^2\right\} \quad (14)$$
$$+\nabla\left\{(z_\alpha - \eta)(u_\alpha \cdot \nabla)(\nabla \cdot (hu_\alpha)) - \eta\left[\frac{1}{2}\eta\nabla \cdot \frac{\partial u_\alpha}{\partial t} + \nabla \cdot \left(h\frac{\partial u_\alpha}{\partial t}\right)\right]\right\} = 0$$

which can be extended to deeper waters thereby improving the dispersive behavior (Galan et al., 2012; Simarro et al., 2015). Several models have been developed based on these equations either in finite differences, finite elements or by finite volume providing accurate results in reproducing laboratory experiments.

Fig. 14.13 shows the time history comparison between numerical results and experimental data at different reported gages. The first section has been used as the control section, allowing to synchronize model and experimental time. As seen, these models capture very well the dispersion in the approaching zone as well as the nonlinear decomposition over the bar. However, as already stated, due to the computational cost in solving the equations, these models are still far too operative for large areas of the littoral and research is devoted in code optimization.

Waves and Remote Sensing

General properties

Waves are usually the first type of motion that one can see at the sea surface, and they appear in all remotely sensed data. In some cases, the direct wave influence can be averaged out of the signal. In other cases, there is a residual bias due to the presence of waves. This is the case in range measurements with altimeters (e.g., Minster et al., 1991), velocity measurement with Doppler systems (Chapron et al., 2005; Nouguier et al., 2018), and surface brightness temperature measurements used to infer sea surface temperature or salinity (Reul and Chapron, 2003). Wave shapes and motion also introduce a variance in the measured quantity that can be useful in the case of sea level measurement with altimetry or can blur the signal beyond recognition in SAR imagery or interferometry (Peral et al., 2015).

All these effects are opportunities for measuring wave parameters or other processes thanks to their influence on waves. In the case when the source of light or radar transmitter is at the same location as the receptor (this is called monostatic geometry), radar waves reflected off from the sea surface is determined by either the variance of the surface slopes (usually called mean square slope or mss), when the source shines within about 20° of the vertical, or the wave spectral level at twice the horizontal radar wavelength for more oblique angles.

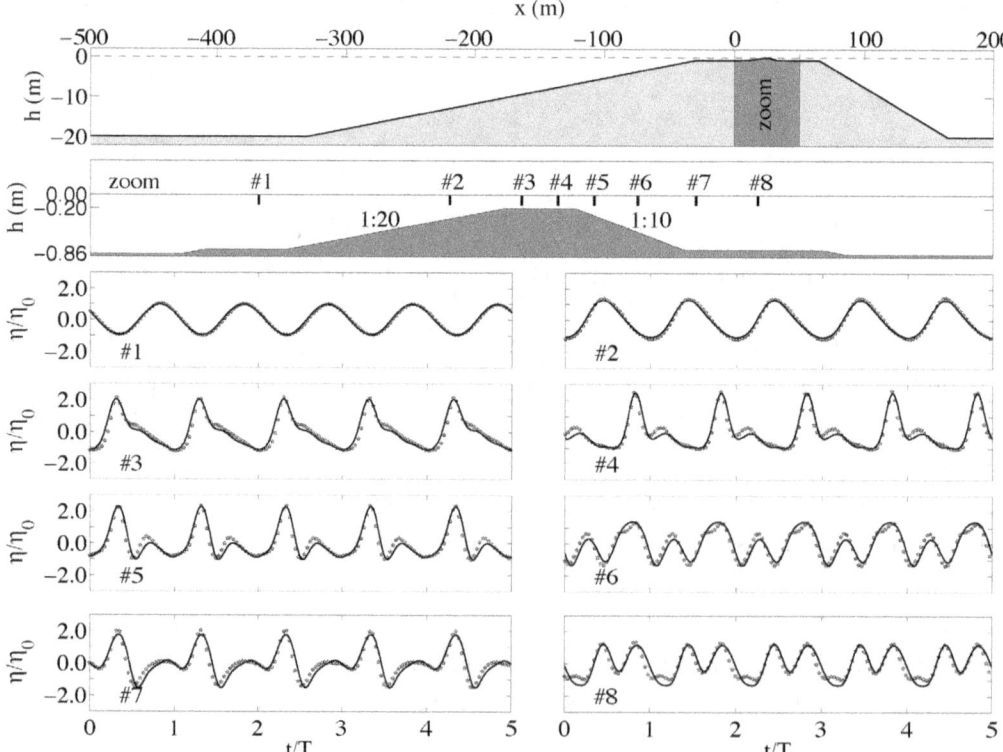

Figure 14.13. Evolution of the free surface at the different gages for the numerical results (solid lines) and experimental data (stars) from Dingemans (1994).

For near-vertical angles, the power recorded on a radar is well described by the theory for a nearly Gaussian distribution of surface slopes completely determined by the mss. In general, the mss is largely determined by the wind speed, but it is also modified by the stage of development of the wave field, increasing for more mature waves that generally correspond to higher wave heights, as shown in Fig. 14.14. A flat surface gives a strong return near zero incidence (vertical sounding) and the return power decreases as 1/mss. For incidence angles larger than about 10° in Ku-band, the return increases with the roughness. Hence, the contrast in the sun reflection in an optical or radar image depends on the incidence; near the vertical a slick flat surface will appear bright, but it will be dark at higher incidences.

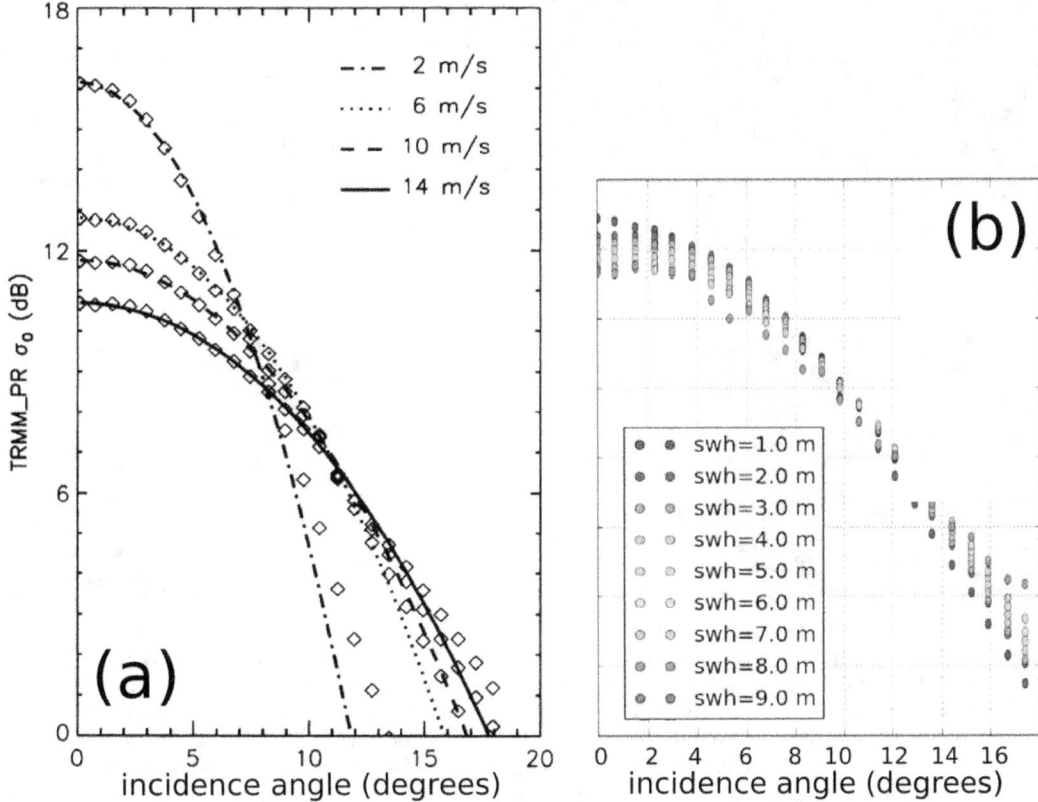

Figure 14.14. (a) Average backscatter power from the TRMM radar (from Freilich and Vanhoff, 2003). (b) Same variation for a given wind speed, as a function of wave height (from Nouguier et al., 2016).

Currents and waves in high-resolution optical imagery

The presence of currents and internal waves generally changes the amplitude of the waves, hence the mss, giving beautiful patterns in optical or SAR imagery, as illustrated in Fig. 14.15. The wind waves with wavelengths longer than a few meters also tilt the shorter waves and change the brightness. These properties can be exploited quantitatively to measure current gradients (Kudryavtsev et al., 2012; Rascle et al., 2016, 2017), or waves (Kudryavtsev et al., 2017). Also, on Sentinel-2 and other similar sensors, the different color channels are not acquired simultaneously, but with a time difference of the order of 1 s. This delay is enough to see the waves move between the different colors. From this motion, the dispersion of the waves given by Eq. 2 can be inverted to get a current velocity vector (Kudryavtsev et al., 2017), as illustrated in Fig. 14.16.

Sentinel 2, with a high resolution and large images that lead to relatively short revisit times, brings a revolution and amazing pictures that can be turned into hard numbers. Unfortunately, it is night half of the day on average and there can be many clouds. For these reasons, a routine monitoring of currents and waves is still more easily done with active radar systems.

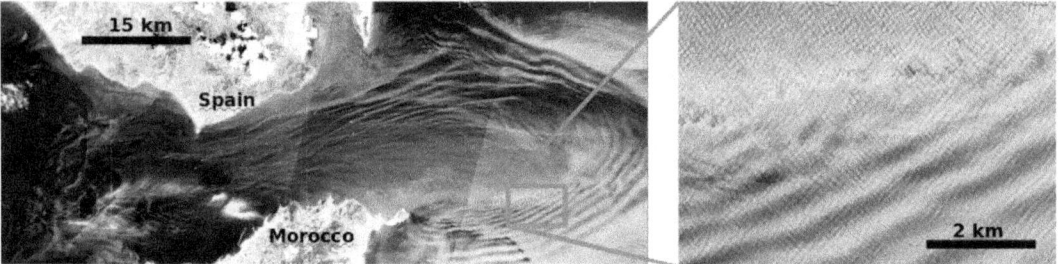

Figure 14.15. Example of a Sentinel-2 image acquired over the Straits of Gibraltar, 7 April 2016, see http://bit.ly/1SN401K for an interactive view.

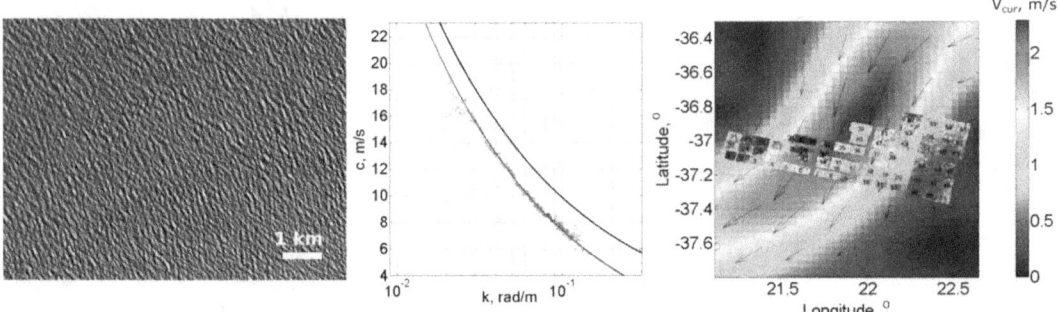

Figure 14.16. Analysis of surface current from a Sentinel-2 image over the Agulhas current (https://odl.bzh/YqP5bd6P). The left panel shows the wave pattern in the current, the middle shows the phase speed of the waves as a function of their wavenumber k with data available for wavelengths 50 to 300 m, and different symbols for different wave propagation direction. The upper curve is the dispersion without current $C = g/k$, the lower curve corresponds to $C = C + U$ with a current velocity U = 2.2 m s^{-1}. The right panel shows the result of this analysis all across the image in small windows, with a background velocity given by satellite altimetry.

Current and waves with radars

Besides the satellite altimeters that provide estimates of H_s and mss that we will not discuss here (e.g., Gower, 1979; Quartly, 2000; Queffeulou, 2004; Ardhuin et al., 2010), more information on waves can be obtained from SARs. Unfortunately, for most applications, SAR images are not like photographs; the positions of pixels in a SAR image are displaced in the azimuth direction (the direction of the satellite motion) according to the relative velocity of the target and the solid Earth along the radar line of sight. This is very useful for measuring waves under sea ice (Ardhuin et al., 2017a) where the wave-induced velocities are small, but it leads to a strong blurring of the image in the open ocean as soon as there are short energetic waves (Kerbaol et al., 1998; Stopa et al., 2015). For example, a typical vertical velocity fluctuation of 1m/s leads to a blurring on a scale of 200 m. As a result, in strong wind conditions SAR can only see waves travelling in the range direction, and very high resolution is useless for all other directions. This is illustrated with two SAR scenes in Fig. 14.17.

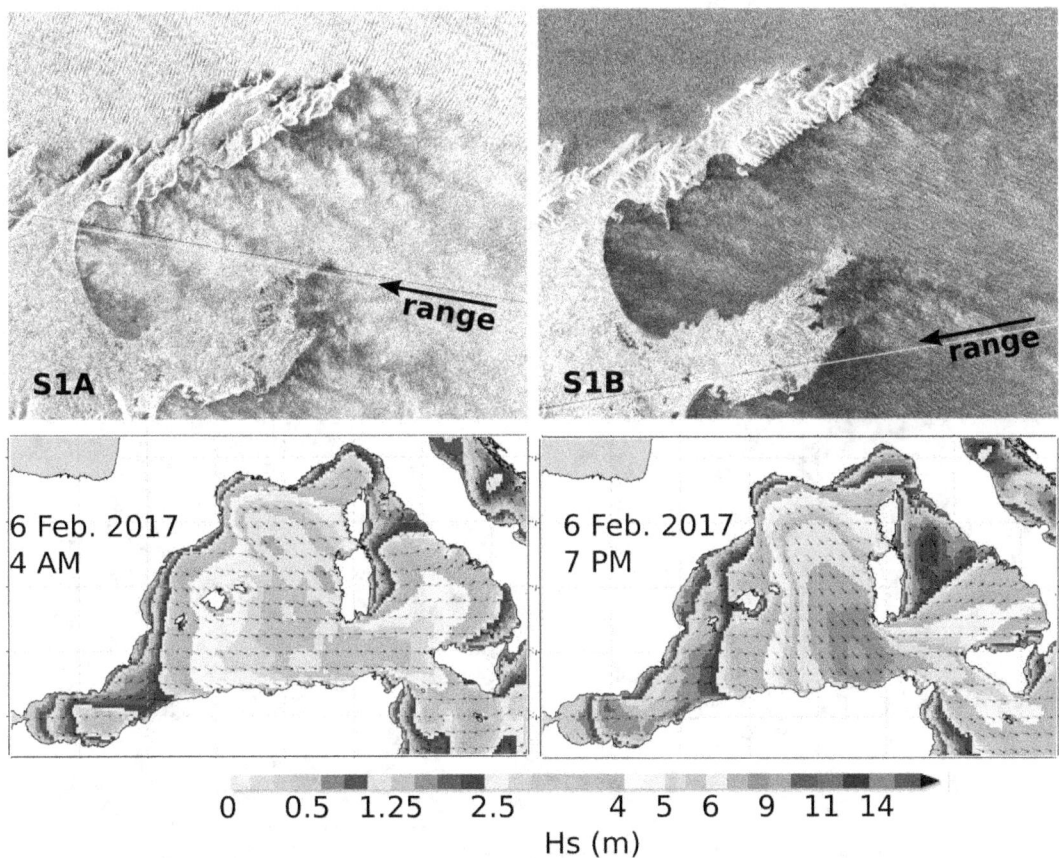

Figure 14.17. Example of waves seen by Sentinel 1A at 5:50 UTC (https://odl.bzh/z3nM6X0E) and 1B at 17:45 UTC during a severe Mediterranean storm on 6 February 2017. Top panels show the bay of Pollença, Mallorca, Spain. Range is the direction towards which the radar is pointing. The bottom panel shows model forecast wave height and directions from http://marc.ifremer.fr/.

The wind is very strong in the morning at 20 m/s and drops to 10 m/s in the evening. Strong winds, over 30 m/s, are also present off Cape Begur to the north, sending swells towards Mallorca. Waves at the northern tip of Mallorca in the morning are in the range direction of the S1A pass and thus very well-observed. However, there are also waves refracted into the Bay of Pollença (Mallorca, Spain), but they travel at an oblique direction and thus are blurred in the SAR image. In the evening, the dominant wave direction has turned to the north-west; they are invisible in the S1B pass, but we can see waves propagating into the bay as if they were coming out of nowhere. The alternative for measuring the directional wave spectrum is to use a rotating radar beam and not use synthetic aperture processing. This is the principle of the SWIM (Surface Waves Investigation and Monitoring) instrument that will be flown on CFOSAT, due for launch in 2018 (Hauser et al., 2017).

Current airborne and space-based measurements of surface velocity have been performed with across-track interferometric (ATI) SARs using two antennas (Goldstein and Zebker, 1987), giving a surface velocity projected on the satellite range direction. This has been generalized to squinted ATI SARs in order to provide the two components of the current vector (Buck, 2005; Wollstadt et al., 2016). More recently, Chapron et al. (2005) showed the potential of using the Doppler centroid of ocean backscatter received by a single antenna, as illustrated in Fig. 14.18.

Although this measurement is noisier than ATI resulting in an effective coarser resolution, the velocity given by the Doppler centroid is equivalent to an ATI measurement (Romeiser et al., 2013). Hence, the Doppler centroid method is a cost-effective solution for deriving current information from existing satellite missions such as Envisat and the Sentinel 1 constellation. This has already led to scientific application on the monitoring of intense currents (Rouault et al., 2010).

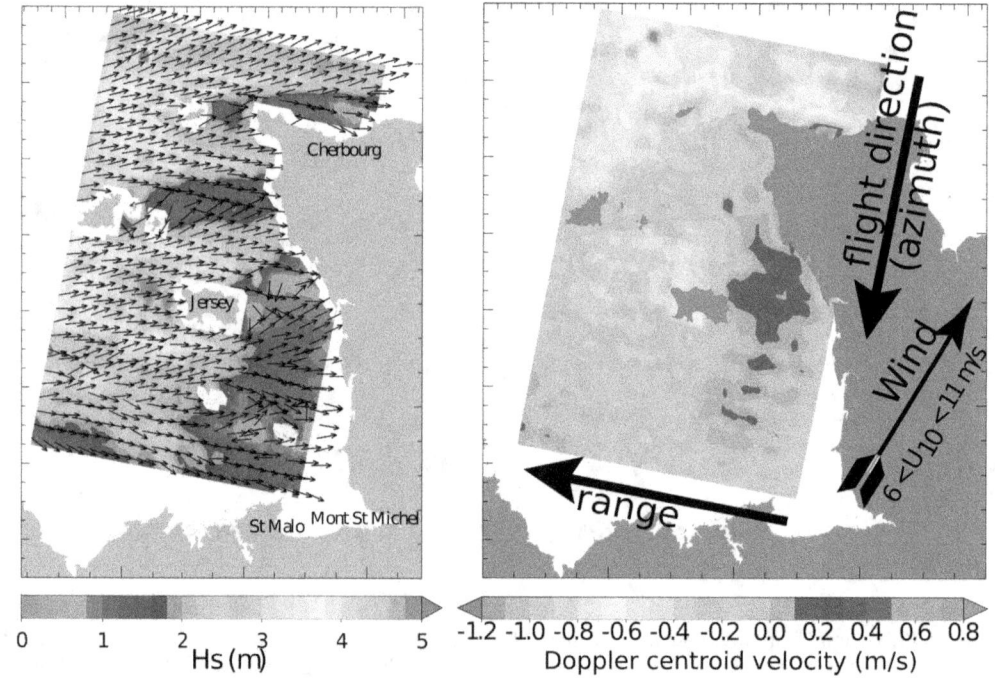

Figure 14.18. One of the first SAR images processed into a map of wave spectra giving wave heights (left) and Doppler centroid velocity (right). This was acquired by Envisat over Normandy on 10 March 2003 (Chapron et al., 2005; Ardhuin et al., 2012).

A dedicated mission that would measure surface current vector and waves based on Doppler centroid and a rotating real aperture radar has been preselected to be the ninth ESA Earth Explorer mission (Ardhuin et al., 2018). This Sea surface KInematics Multiscale monitoring (SKIM) mission would use the measurement of waves to correct for the large wave-induced bias in surface velocity that is measured in ATI or Doppler centroids. Existing European radar technology allows the use of a conical scanning radar beam over 12° of incidence, giving a typical swath width of 300 km. A preliminary error budget on this concept gives an effective resolution of surface currents with 60 km wavelength. Combined with the shorter revisit time, this is a very attractive alternative or complement to measurements based on surface height with SWOT that have a narrow swath and only measure geostrophic currents, as shown on Fig. 14.19.

The proposed instrument can reveal features on tropical ocean and marginal ice zone dynamics that are inaccessible to other measurement systems, as well as a global monitoring of the ocean mesoscale that surpasses the capability of today's nadir altimeters. Measuring ocean wave properties facilitates many applications, from wave-current interactions and air-sea fluxes to the transport and convergence of marine plastic debris and assessment of marine and coastal hazards.

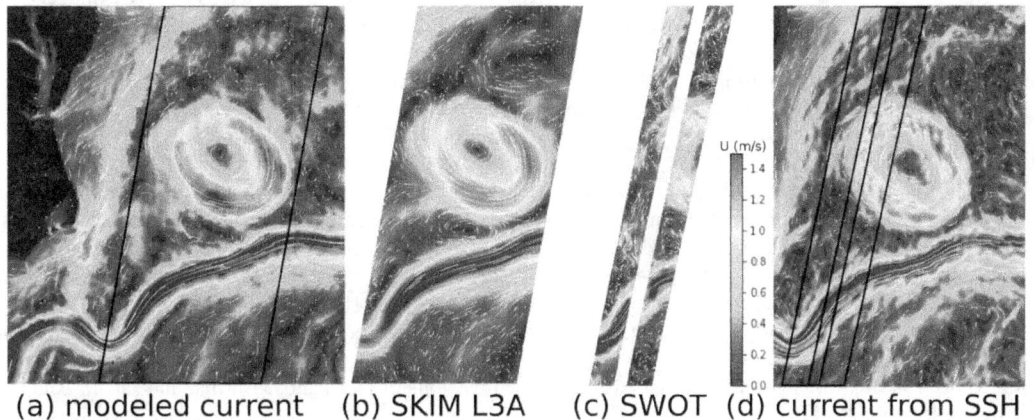

Figure 14.19. Simulated ocean currents over the Gulf Stream, and associated SKIM and SWOT simulated observations for a single satellite pass.

Acknowledgments

This work was supported by FP7-ERC grant « IOWAGA », EU-FP7 project SWARP, ONR grant number N0001416WX01117, and CNES. Visualization and analysis was greatly facilitated by the Syntool portal developed by OceanDataLab: https://swarp.oceandatalab.com. Support from LabexMer via grant ANR-10-LABX-19-01, and Copernicus Marine Environment Monitoring Service (CMEMS) as part of the Service Evolution program is gratefully acknowledged.

References

Agnon, Y. 1999. Linear and nonlinear refraction and Bragg scattering of water waves. Physical Review E, 59, R1319–R1322.

Agrawal, Y. C., Terray, E. A., Donelan, M. A., Hwang, P. A., Williams, A. J., Drennan, W., Kahma, K., and Kitaigorodskii, S. 1992. Enhanced dissipation of kinetic energy beneath breaking waves. Nature, 359, 219–220.

Algué, J. 1900. Relation entre quelques mouvements microséismiques et l'existence, la position et la distance des cyclones à Manille (Philippines). Pages 131–136 of: Congrès international de Météorlogie, Paris.

Ardhuin, F. 2018. Ocean waves in geosciences. doi:10.13140/RG.2.2.16019.78888/2

Ardhuin, F., Jenkins, A. D., Hauser, D., Reniers, A., and Chapron, B. 2005. Waves and operational oceanography: towards a coherent description of the upper ocean for applications. Eos Trans. AGU, 86(4), 37–39.

Ardhuin, F., Chapron, B., and Collard, F. 2009. Observation of swell dissipation across oceans. Geophys. Res. Lett., 36, L06607. doi:10.1029/2008GL037030.

Ardhuin, F., Rogers, E., Babanin, A., Filipot, J.-F., Magne, R., Roland, A., van der Westhuysen, A., Queffeulou, P., Lefevre, J.-M., Aouf, L., and Collard, F. 2010. Semi-empirical dissipation source functions for wind-wave models: part I, definition, calibration and validation. J. Phys. Oceanogr., 40(9), 1917–1941.

Ardhuin, F., Stutzmann, E., Schimmel, M., and Mangeney, A. 2011. Ocean wave sources of seismic noise. J. Geophys. Res., 116, C09004. doi:10.1029/2011JC006952.

Ardhuin, F., Balanche, A., Stutzmann, E., and Obrebski, M. 2012. From seismic noise to ocean wave parameters: general methods and validation. J. Geophys. Res., 117, C05002. doi:10.1029/2011JC007449.

Ardhuin, F., Rawat, A., and Aucan, J. 2014. A numerical model for free infragravity waves: Definition and validation at regional and global scales. Ocean Modelling, 77, 20–32.

Ardhuin, F., Gualtieri, L., and Stutzmann, E. 2015. How ocean waves rock the Earth: two mechanisms explain seismic noise with periods 3 to 300 s. Geophys. Res. Lett., 42, 765–772. doi:10.1002/2014GL062782.

Ardhuin, F., Sutherland, P., Doble, M., and Wadhams, P. 2016. Ocean waves across the Arctic: attenuation due to dissipation dominates over scattering for periods longer than 19 s. Geophys. Res. Lett., 43, 5775–5783. doi:10.1002/2016GL068204.

Ardhuin, F., Chapron, B., Collard, F., Smith, M., Stopa, J., Thomson, J., Doble, M., Wadhams, P., Blomquist, B., Persson, O., and Collins, III, C. O. 2017a. Measuring ocean waves in sea ice using SAR imagery: A quasi-deterministic approach evaluated with Sentinel-1 and in situ data. Remote sensing of Environment, 189, 211–222. doi:10.1016/j.rse.2016.11.024.

Ardhuin, F., Rascle, N., Chapron, B., Gula, J., Molemaker, J., Gille, S. T., Menemenlis, D., and Rocha, C. 2017b. Small scale currents have large effects on wind wave heights. J. Geophys. Res., 122(C6), 4500–4517. doi:10.1002/2016JC012413

Ardhuin, F., Aksenov, Y., Benetazzo, A., Bertino, L., Brandt, P., Caubet, E., Chapron, B., Collard, F., Cravatte, S., Delouis, J.-M., Dias, F., Dibarboure, G., Gaultier, L., Johannessen, J., Korosov, A., Manucharyan, G., Menemenlis, D., Menendez, M., Monnier, G., Mouche, A., Nouguier, F., Nurser, G., Rampal, P., Reniers, A., Rodriguez, E., Stopa, J., Tison, C., Ubelmann, C., van Sebille, E., and Xie, J. 2018. Measuring currents, ice drift, and waves from space: The sea surface kinematics multiscale monitoring concept, Ocean Sci., 14, 337-354, https://doi.org/10.5194/os-14-337-2018, 2018

Barber, N. F., Ursell, F., Darbyshire, J., and Tucker, M. J. 1946. A frequency analyser used in the study of ocean waves. Nature, 329–335.

Bates, C. C. 1949. Utilization of wave forecasting in the invasions of Normandy, Burma, and Japan. Ann. N. Y. Acad. Sci., 545–572.

Beji, S., and K. Nadaoka. 1996. A formal derivation and numerical modelling of the improved Boussinesq equations for varying depth. Ocean Engineering, 23, 691–704.

Benetazzo, A. 2006. Measurements of short water waves using stereo matched image sequences. Coastal Eng., 53, 1013–1032.

Berkhoff, J. C. W. 1972. Computation of combined refraction-diffraction. Pages 796–814 of: Proceedings of the 13th International Conference on Coastal Engineering, Vancouver. ASCE, New York, N. Y.

Bernard, P. 1990. Historical sketch of microseisms from past to future. Phys. Earth Planetary Interiors, 63, 145–150.

Bidlot et al., J.R., et al. 2007. Inter-comparison of Operational Wave Forecasting Systems, Proceedings of the 10th wave hindcasting and forecasting workshop, Oahu, Hawaii.

Bidlot, J.R 2017. Twenty-one years of wave forecast verification. ECMWF Newsletter No. 150, doi:10.21957/t4znuhb842

Boccotti, P. 2000. Wave mechanics for ocean engineering. Amsterdam: Elsevier.

Boussinesq, J. 1872. Théorie des ondes et des remous qui se propagent le long d'un canal rectangulaire horizontal, en communiquant au liquide contenu dans ce canal des vitesses sensiblement pareilles de la surface au fond. J. Math. Pures Appl., 17(2), 55–108.

Bromirski, P. D., Sergienko, O. V., and MacAyeal, D. R. 2010. Transoceanic infragravity waves impacting Antarctic ice shelves. Geophys. Res. Lett., 37, L02502. doi:10.1029/2009GL041488.

Buck, C. 2005. An extension to the wide swath ocean altimeter concept. Pages 543–5439 of: Proceedings of the IEEE International Geoscience and Remote Sensing Symposium (IGARSS), 2013., vol. 8. IEEE. doi:10.1109/IGARSS.2005.1525970.

Cavanié, A., Arhan, M., and Ezraty, R. 1976. A statistical relationship between individual heights and periods of storm waves. Pages 354–360 of: Conference on the Behaviour of Off-Shore Structures (BOSS). The Norwegian Institute of Technology.

Chapron, B., Collard, F., and Ardhuin, F. 2005. Direct measurements of ocean surface velocity from space: interpretation and validation. J. Geophys. Res., 110(C07008). doi:10.1029/2004JC002809.

Cox, C., and Munk, W. 1954. Measurement of the roughness of the sea surface from photographs of the sun's glitter. J. Opt. Soc. Am., 44(11), 838–850.

D'Asaro, E. A. 2014. Turbulence in the Upper-Ocean Mixed Layer. Annu. Rev. Mar. Sci., 6, 101–115. doi:10.1146/annurev-marine-010213-135138.

de Laplace, P. S. 1776. Suite des recherches sur plusieurs points du système du monde (XXV–XXVII). Mém. Présentés Acad. R. Sci. Inst. France, 542–552.

de Saint-Venant, A. J. C. B., and Flamant, A. 1888. De la houle et du clapotis. Annales des Ponts et Chaussées, 6, 705–773.

Dean, R. G., and Dalrymple, R. A. 1984. Water wave mechanics for engineers and scientists. Englewoods Cliffs, New Jersey 07632: Prentice-Hall, Inc.

Deane, G. B., and Stokes, M. D. 2002. Scale dependence of bubble creation mechanisms in breaking waves. Nature, 418, 839–844.

Dingemans, M. W. 1994. Comparison of computations with Boussinesq-like models and laboratory measurements. Mast-G8M technical report, Delft Hydraulics, Delft, The Netherlands.

Dingemans, M. W. 1997. Water wave propagation over uneven bottoms. Part 1 linear wave propagation. Singapore: World Scientific. 471 p.

Eckart, C. 1952. The propgation of gravity waves from deep to shallow waters. National Bureau of Standards, 20.

Elfouhaily, T., Chapron, B., Katsaros, K., and Vandemark, D. 1997. A unified directional spectrum for long and short wind-driven waves. J. Geophys. Res., 102(C7), 15781–15796.

Freilich, M. H., and Vanhoff, B. A. 2003. The Relationship between Winds, Surface Roughness, and Radar Backscatter at Low Incidence Angles from TRMM Precipitation Radar Measurements. J. Atmos. Ocean Technol., 20, 549–562.

Gain, L. 1918. La prédiction des houles au Maroc. Annales Hydrographiques, 65–75.

Galan, A., G., Orfila, A., Simarro, J., and Liu, P. L.-F. 2012. Fully Nonlinear Model for Water Wave Propagation from Deep to Shallow Waters. Journal of Waterway, Port, Coastal, and Ocean Engineering, 138(5), 362–371.

Gelci, R., Cazalé, H., and Vassal, J. 1957. Prévision de la houle. La méthode des densités spectroangulaires. Bulletin d'information du Comité d'Océanographie et d'Etude des Côtes, 9, 416–435.

Gobbi, M. F., Kirby, J. T., and Wei, G. 2000. A fully nonlinear Boussinesq model for surface waves: part II. Extension to O(kh)4. Journal of Fluid Mechanics, 405, 182–210.

Goldstein, R. M., and Zebker, H. A. 1987. Interferometric radar measurement of ocean surface current. Nature, 328, 707–709.

Gourrion, J., Vandemark, D., Bailey, S., and Chapron, B. 2002. Investigation of C-band altimeter cross section dependence on wind speed and sea state. Can. J. Remote Sensing, 28(3), 484–489.

Gower, J. F. R. 1979. The computation of ocean wave heights from GEOS-3 satellite radar altimeter. Remote sensing of Environment, 8, 97–114.

Green, A. E., and Naghdi, P. M. 1976. A derivation of equations for wave propagation in water of variable depth. Journal of Fluid Mechanics, 78, 237–246.

Grevemeyer, I., Herber, R., and Essen, H.-H. 2000. Microseismological evidence for a changing wave climate in the northeast Atlantic Ocean. Nature, 408, 349–351.

Hanafin, J., Quilfen, Y., Ardhuin, F., Sienkiewicz, J., Queffeulou, P., Obrebski, M., Chapron, B., Reul, N., Collard, F., Corman, D., de Azevedo, E. B., Vandemark, D., and Stutzmann, E. 2012. Phenomenal sea states and swell radiation: a comprehensive analysis of the 12-16 February 2011 North Atlantic storms. Bull. Amer. Meteorol. Soc., 93, 1825–1832. doi:10.1175/BAMS-D-11-00128.1.

Hasselmann, K. 1962. On the non-linear energy transfer in a gravity wave spectrum, part 1: general theory. J. Fluid Mech., 12, 481–501.

Hasselmann, K. 1963. A statistical analysis of the generation of microseisms. Rev. of Geophys., 1(2), 177–210.

Hasselmann, K., Barnett, T. P., Bouws, E., Carlson, H., Cartwright, D. E., Enke, K., Ewing, J. A., Gienapp, H., Hasselmann, D. E., Kruseman, P., Meerburg, A., Müller, P., Olbers, D. J., Richter, K., Sell, W., and Walden, H. 1973. Measurements of wind-wave growth and swell decay during the Joint North Sea Wave Project. Deut. Hydrogr. Z., 8(12), 1–95. Suppl. A.

Hasselmann, S., Hasselmann, K., Allender, J., and Barnett, T. 1985. Computation and parameterizations of the nonlinear energy transfer in a gravity-wave spectrum. Part II: Parameterizations of the nonlinear energy transfer for application in wave models. J. Phys. Oceanogr., 15, 1378–1391.

Hasselmann, K., Chapron, B., Aouf, L., Ardhuin, F., Collard, F., Engen, G., Hasselmann, S. and Heimbach, P., Janssen, P., Johnsen, H, Krogstad, H., Lehner, Susanne, Li, J.-G., Li, Xiao-Ming, Rosenthal, W. and Schulz-Stellenfleth, J. (2012) *The ERS SAR Wave Mode – A Breakthrough in global ocean wave observations.* In: ERS Missions - 20 Years of Observing Earth ESA Scientific Publications, SP-1326. ESA. pp. 1-38.

Hauser, D., Tison, C., Amiot, T., Delaye, L., Corcoral, N., and Castillan, P. 2017. SWIM: The First Spaceborne Wave Scatterometer. IEEE Trans. on Geosci. and Remote Sensing, 55(5), 3000–3014.

Herbers, T. H. C., and Burton, M. C. 1997. Nonlinear shoaling of directionally spread waves on a beach. J. Geophys. Res., 102(C9), 21,101–21,114.

Hersbach, H., 2010. Assimilation of scatterometer data as equivalent-neutral wind, Technical Memorendum 629, ECMWF.

Iribarren, R., and Nogales, C. 1949. Protection des ports. Pages 31–80 of: Proceedings XVIIth International Navigation Congress, Section II, Communication, 4, Lisbon.

Janssen, P. A. E. M. 1989. Wave-induced stress and the drag of air flow over sea waves. J. Phys. Oceanogr., 19, 745–754.

Janssen, P. 2004. The interaction of ocean waves and wind. Cambridge: Cambridge University Press.

Janssen, P., Doyle, J.-D., Bidlot, J., Hansen, B., Isaksen, L., and Viterbo, P. 2001. Impact and feedback of ocean waves on the atmosphere. Tech. rept. Memorandum 341. Research Department, ECMWF, Reading, U. K.

Janssen, P. A. E. M. 2008. Progress in ocean wave forecasting. J. Comp. Phys., 227, 3572–3594. doi:10.1016/j.jcp.2007.04.029.

Jarosz, E., Mitchell, D. A., Wang, D. W., and Teague, W. J. 2007. Bottom-up determination of air-sea momentum exchange under a major tropical cyclone. Science, 315, 1707–1709.

Kerbaol, V., Chapron, B., and Vachon, P. 1998. Analysis of ERS-1/2 synthetic aperture radar wave mode imagettes. J. Geophys. Res., 103(C4), 7833–7846.

Kirby, J. T., and Dalrymple, R. A. 1983. Propagation of obliquely incident water waves over a trench. J. Fluid Mech., 133, 47–63.

Kudryavtsev, V., Yurovskaya, M., Chapron, B., Collard, F., and Donlon, C. 2017. Sun glitter Imagery of Surface Waves. Part 1: Directional spectrum retrieval and validation. J. Geophys. Res., 122. doi:10.1002/2016JC012425.

Kudryavtsev, V., Myasoedov, A., Chapron, B., Johannessen, J. A., and Collard, F. 2012. Imaging mesoscale upper ocean dynamics using synthetic aperture radar and optical data. J. Geophys. Res., 117, C04029. doi:10.1029/2011JC007492.

Kukulka, T., Plueddemann, A. J., Trowbridge, J. H., and Sullivan, P. P. 2009. Significance of Langmuir circulation in upper ocean mixing: Comparison of observations and simulations. Geophys. Res. Lett., 36, L10603. doi:10.1029/2009GL037620.

Lavrenov, I. V. 2003. Wind-waves in oceans: dynamics and numerical simulations. Berlin: Springer.

Leckler, F., Ardhuin, F., Peureux, C., Benetazzo, A., Bergamasco, F., and Dulov, V. 2015. Analysis and interpretation of frequency-wavenumber spectra of young wind waves. J. Phys. Oceanogr., 45, 2484–2496. doi:10.1175/JPO-D-14-0237.1.

Lefèvre, J.-M., and Aouf, L. (2012) Latest developments in wave data assimilation, proceedings of the ECMWF workshop on ocean waves.

Lionello, P., Günther, H, and Janssen, PAEM, 1992. Assimilation of altimeter data in a global third generation wave model. J. Geophys. Res., 97(C9): 14,453-14,474.

Liu, P. L.-F., and Losada, I. 2002. Wave propagation modeling in coastal engineering. Journal of Hydraulic Research, 40, 229–240.

Liu, P. L.-F., and Orfila, A. 2004. Viscous effects on transient long-wave propagation. Journal of Fluid Mechanics, 520, 83–92.

Liu, P.-F., Simarro, G., Vandever, J., and Orfila, A. 2006. Experimental and numerical investigation of viscous effects on solitary wave propagation in a wave tank. Coastal Engineering, 53(2), 181–190.

Longuet-Higgins, M. S. 1950. A theory of the origin of microseisms. Phil. Trans. Roy. Soc. London A, 243, 1–35.

Longuet-Higgins, M. S., and Stewart, R. W. 1962. Radiation stresses and mass transport in surface gravity waves with application to 'surf beats'. J. Fluid Mech., 13, 481–504.

Lynett, P., and Liu, P. L.-F. 2002. A two layer approach to wave modeling. The Royal Society London, A, 46(2), 89–107.

Madsen, P. A., and Schaffer, H. A. 1998. Higher-order Boussinesq-type equations for surface gravity waves: derivation and analysis. Phil. Trans. Royal Society of London A, 356, 3123–3184.

Madsen, P. A., and Sorensen, O. R. 1992. A new form of the Boussinesq equations with improved linear dispersion characteristics. Part 2: A slowly-varying bathymetry. Coastal Engineering, 18, 183–204.

Mastenbroek, C., Burgers, G., and Janssen, P. A. E. M. 1993. The dynamical coupling of a wave model and a storm surge model through the atmospheric boundary layer. J. Phys. Oceanogr., 23, 1856–1867.

Mei, C. C. 1989. Applied dynamics of ocean surface waves. Second Edn. Singapore: World Scientific. 740 p.

Miche, A. 1944. Mouvements ondulatoires de la mer en profondeur croissante ou décroissante. Forme limite de la houle lors de son déferlement. Application aux digues maritimes. Deuxième partie. Mouvements ondulatoires périodiques en profondeur régulièrement décroissante. Annales des Ponts et Chaussées, Tome 114, 131–164,270–292.

Minster, J. F., Jourdan, D., Boissier, C., and Midol-Monnet, P. 1991. Estimation of the Sea-state bias in radar altimeter Geosat data from examination of frontal systems. J. Atmos. Ocean Technol., 9, 174–187.

Montagne, R. 1922. Le service de prédiction de la houle au Maroc. Annales Hydrographiques, 157–186.

Munk, W. H. 1950. Origin and generation of waves. Pages 1–4 of: Proceedings 1st International Conference on Coastal Engineering, Long Beach, California. ASCE.

Nouguier, F., Mouche, A., Rascle, N., Chapron, B., and Vandemark, D. 2016. Analysis of Dual-Frequency Ocean Backscatter Measurements at Ku- and Ka-Bands Using Near-Nadir Incidence GPM Radar Data. IEEE Geoscience and Remote Sensing Letters, 31, 2023–2245. doi:10.1109/LGRS.2016.2583198.

Nouguier, F., Chapron, B., Collard, F., Mouche, A., Rascle, N., Ardhuin, F., and Wu, X. 2018. Sea surface kinematics from near-nadir radar measurements. IEEE Trans. on Geosci. and Remote Sensing, in press.

Nwogu, O. 1993. Alternative form of Boussinesq equations for nearshore wave propagation. Journal of Waterway, Port, Coastal and Ocean Engineering, 119(6), 618–638.

Orfila, A., Simarro, G., and Liu, P. 2007. Bottom friction and its effects on periodic long wave propagation. Coastal Engineering, 54(11), 856–864.

Peral, E., Rodriguez, E., and Esteban-Fernandez, D. 2015. Impact of Surface Waves on SWOT's Projected Ocean Accuracy. Remote Sensing, 7(11), 14509–14529. doi:10.3390/rs71114509.

Peregrine, D. H. 1967. Long waves on a beach. J. Fluid Mech., 27, 815–827.

Peureux, C., and Ardhuin, F. 2016. Ocean bottom pressure records from the Cascadia array and short surface gravity waves. J. Geophys. Res., 121, 28622873. doi:10.1002/2015JC011580.

Peureux, C., Benetazzo, A., and Ardhuin, F. 2017. Note on the directional properties of meter-scale gravity waves. Ocean Science Discussions.

Phillips, O. M. 1977. The dynamics of the upper ocean. London: Cambridge University Press. 336 p.

Pierson, Jr, W. J., and Moskowitz, L. 1964. A proposed spectral form for fully developed wind seas based on the similarity theory of S. A. Kitaigorodskii. J. Geophys. Res., 69(24), 5,181–5,190.

Pineau-Guillou, L., Ardhuin, F., Bouin, M.-N., Redelsperger, J.-L., Chapron, B., Bidlot, J., and Quilfen, Y. 2017. Strong winds in a coupled wave-atmosphere model during a North Atlantic storm event: evaluation against observations.

Quart. Journ. Roy. Meteorol. Soc., in press.

Plant, W. J. 1982. A relationship between wind stress and wave slope. J. Geophys. Res., 87, 1961–1967.

Quartly, G. D. 2000. The Gate Dependence of Geophysical Retrievals from the TOPEX Altimeter. J. Atmos. Ocean Technol., 17, 1247–1251.

Queffeulou, P. 2004. Long term validation of wave height measurements from altimeters. Marine Geodesy, 27, 495–510. DOI: 10.1080/01490410490883478.

Rascle, N., and Ardhuin, F. 2013. A global wave parameter database for geophysical applications. Part 2: model validation with improved source term parameterization. Ocean Modelling, 70, 174–188. doi:10.1016/j.ocemod.2012.12.001.

Rascle, N., Nouguier, F., Chapron, B., Mouche, A., and Ponte, A. 2016. Surface roughness changes by fine scale current gradients: Properties at multiple Wind waves azimuth view angles. J. Phys. Oceanogr., 46, 3681–3694. doi:10.1175/JPO-D-15-0141.1.

Rascle, N., Molemaker, J., Marié, L., Nouguier, F., Chapron, B., Lund, B., and Mouche, A. 2017. Intense deformation field at oceanic front inferred from directional sea surface roughness observations. Geophys. Res. Lett., 48, 5599–5608. doi:10.1002/2017GL073473.

Reniers, A. J. H. M., Roelvink, J. A., and Thornton, E. B. 2004. Morphodynamic modeling of an embayed beach under wave group forcing. J. Geophys. Res., 109, C01030. doi:10.1029/2002JC001586.

Reul, N., and Chapron, B. 2003. A model of sea-foam thickness distribution for passive microwave remote sensing applications. J. Geophys. Res., 108(C10), 3321. doi:10.1029/2003JC001887.

Rogers, W. E., Thomson, J., Shen, H. H., Doble, M. J., Wadhams, P., and Cheng, S. 2016. Dissipation of wind waves by pancake and frazil ice in the autumn Beaufort Sea. J. Geophys. Res., 121. doi:10.1002/2016JC012251.

Romeiser, R., Runge, H., Suchandt, S., Kahle, R., Rossi, C., and Bell, P. 2013. Comparison Of Current Fields From Terrasar-X And Tandem-X Along-Track Interferometry And Doppler Centroid Analysis. In: Proceedings of the IEEE International Geoscience and Remote Sensing Symposium (IGARSS), 2013. IEEE.

Rouault, M. J., Mouche, A., Collard, F., Johannessen, J. A., and Chapron, B. 2010. Mapping the Agulhas Current from space: An assessment of ASAR surface current velocities. J. Geophys. Res., 41, C10026. doi:10.1029/2009JC006050.

Schaffer, H. A. 1996. Second order wavemaker theory for irregular waves. Ocean Engineering, 23, 47–88.

Sepulveda, H. H., Queffeulou, P., and Ardhuin, F. 2015. Assessment of SARAL AltiKa wave height measurements relative to buoy, Jason-2 and Cryosat-2 data. Marine Geodesy, 38(S1), 449–465. doi:10.1080/01490419.2014.1000470.

Serre, F. 1953. Contribution à l'étude des écoulements permanents et variables dans les canaux. Houille Blanche, 8, 830–872.

Shapiro, N. M., Campillo, M., Stehly, L., and Ritzwoller, M. H. 2005. High-Resolution Surface-Wave Tomography from Ambient Seismic Noise. Science, 307, 1615–1617. doi:10.1111/j.1365-246X.2006.03240.x.

Shen, H. H., Ackley, S. F., and Hopkins, M. A. 2001. A conceptual model for pancake ice formation in a wave field. Annales Geophysicae, 33(C1), 361–367. doi:10.3189/172756401781818239.

Sheremet, A., Staples, T., Ardhuin, F., Suanez, S., and Fichaut, B. 2014. Observations of large infragravity-wave run-up at Banneg Island, France. Geophys. Res. Lett., 41.

Simarro, G., Orfila, A., Mozos, C. M., and Pruneda, R. E. On the linear stability of one- and two-layer Boussinesq-type equations for wave propagation over uneven beds.

Smith, R., and Sprinks, T. 1975. Scattering of surface waves by a conical island. J. Fluid Mech., 72, 373.

Soloviev, A. V., Lukas, R., Donelan, M. A., Haus, B. K., and Ginis, I. 2014. The air-sea interface and surface stress under tropical cyclones. Scientific Reports, 4, 5306. doi:10.1038/srep05306.

Stockdon, H. F., Holman, R. A., Howd, P. A., and Sallenger, Jr., A. H. 2006. Empirical parameterization of setup, swash, and runup. Coastal Eng., 53, 573–588.

Stopa, J.E., 2018. Wind forcing calibration and wave hindcast comparison using multiple reanalysis and merged satellite wind datasets. Ocean Modelling, in press.

Stopa, J. E., Ardhuin, F., Chapron, B., and Collard, F. 2015. Estimating wave orbital velocity through the azimuth cutoff from space-borne satellite. J. Geophys. Res., 130, 7616–7634. doi:10.1002/2015JC011275.

Stopa, J. E., Ardhuin, F., Bababin, A., and Zieger, S. 2016. Comparison and validation of physical wave parameterizations in spectral wave models. Ocean Modelling, 103, 2–17. doi:10.1016/j.ocemod.2015.09.003.

Sullivan, P. P., and McWilliams, J. C. 2010. Dynamics of Winds and Currents Coupled to Surface Waves. Annu. Rev. Fluid Mech., 42, 19–42.

Sverdrup, H. U., and Munk, W. H. 1947. Wind, sea, and swell: theory of relations for forecasting. Tech. rept. 601. U. S. Hydrographic Office.

Tayfun, A. 1980. Narrow-band nonlinear sea waves. J. Geophys. Res., 85(C3), 1543–1552.

The WAVEWATCH III Development Group. 2016. User manual and system documentation of WAVEWATCH III version 5.16. Tech. Note 329. NOAA/NWS/NCEP/MMAB, College Park, MD, USA. 326 pp. + Appendices.

Tolman, H. L. 1995. On the selection of propagation schemes for a spectral wind wave model. Office Note 411. NWS/NCEP. 30 pp + figures.

Tolman, H. L. 2008. A mosaic approach to wind wave modeling. Ocean Modelling, 25, 35–47. doi:10.1016/j.ocemod.2008.06.005.

Tsay, T.-K., and Liu, P. L.-F. 1982. Numerical solution of water-wave refraction and diffraction problems in the parabolic approximation. Journal of Geophysical Research: Oceans, 87(C10), 7932–7940.

Wei, G., and Kirby, J. T. 1995. Time-dependent numerical code for extended Boussinesq equations. Journal of Waterway, Port, Coastal and Ocean Engineering, 121(5), 251–261.

White, B. S., and Fornberg, B. 1998. On the chance of freak waves at sea. J. Fluid Mech., 355, 113–138.

Wollstadt, S., Lopez-Dekker, P., De Zan, F., and Younis, M. 2016. Design Principles and Considerations for Spaceborne ATI SAR-Based Observations of Ocean Surface Velocity Vectors. IEEE Trans. on Geosci. and Remote Sensing, 99, 1–20. doi:10.1109/TGRS.2017.2692880.

Yueh, S. H., Tang, W., Fore, A. G., Neumann, G., Hayashi, A., Freedman, A., Chaubell, J., and Lagerloef, G. S. E. 2013. L-Band Passive and Active Microwave Geophysical Model Functions of Ocean Surface Winds and Applications to Aquarius Retrieval. IEEE Trans. on Geosci. and Remote Sensing, 51, 4619–4632.

CHAPTER 15

Sea Ice Modelling and Forecasting

Sylvain Bouillon, Pierre Rampal, and Einar Olason

Nansen Environmental and Remote Sensing Center, Bergen, Norway

Sea ice is a fascinating media, of which modelling is in its infancy compared to the ocean and atmosphere. This chapter focuses on the new frontiers in sea ice modelling and forecasting, with particular attention on sea ice dynamics. It is divided in two sections: 1) New frontiers in sea ice modelling and 2) New frontiers in sea ice forecasting. In the first section, we describe ice pack dynamics and then concentrate on the representation of sea ice dynamics in continuous models. A sub-section discusses the potential impacts on the ocean and atmosphere of explicitly resolving some features related to sea ice dynamics, in particular the opening and closing of leads, in coupled modelling systems. In the second section, we point out three important constraints on sea ice forecasting related to 1) potentially large biases in the near real-time data, 2) time-varying biases in the external forcing, and 3) far-from-equilibrium dynamical state. These points are explored by addressing the two following questions: "How can we beat ice charts persistency?" and "Can we predict sea ice fracturing and deformation days in advance?"

New Frontiers in Sea Ice Modelling

First, we describe sea ice and how the intrinsic properties of this material result in a complex dynamical behavior that is different from geophysical turbulent fluids such as the ocean and the atmosphere. We also highlight the similarities and differences between the mesoscale dynamical features present in the atmosphere (synoptic features), the ocean (eddies), and the sea ice pack (linear kinematics features), as well as some important interactions that exist between the ice cover dynamical system and the two other systems.

Then we make an attempt to define a unified modelling framework for reproducing the observed ice pack dynamics. This is achieved by combining ingredients taken from the first-generation dynamical models introduced in the 1970s (Hibler, 1979), which are still dominantly used today, with new approaches recently proposed to explicitly resolve the kinematic features (faults, ridges, and leads) present in the ice pack. This framework is in some ways the counterpart of that used in oceanography to classify models into laminar, eddy-permitting, and eddy-resolving implementations.

A unified framework, taking into account the resolved and unresolved scales, may be crucial to rethink most of the parameterizations used in sea ice modelling to date. Such a framework will likely require some adaptation of the coupling strategies that have been developed to connect the sea ice with the ocean underneath and the overlying atmosphere. Along those lines, we will wrap

---

Bouillon, S., P. Rampal, and E. Olason, 2018: Sea ice modelling and forecasting. In "*New Frontiers in Operational Oceanography*", E. Chassignet, A. Pascual, J. Tintoré, and J. Verron, Eds., GODAE OceanView, 423-444, doi:10.17125/gov2018.ch15.

up this section by looking at how sea ice lead fraction can be simulated by a next generation sea ice model and how it may impact the representation of the ocean-atmosphere interactions.

Sea ice and ice pack dynamics

The term "sea ice" is used to designate all types of ice coming from freezing sea water. Sea ice formation starts with frazil ice formation. During this process, a large amount of salt is released into the underneath water column, while some remains trapped in the ice as brine pockets. Those pockets are being emptied over time by a drainage/percolation process of the brines through the ice, allowing the ice cover to become similar to fresh ice within a couple of months. Ice growth continues through processes such as pancake formation, rafting, ridging, and consolidation, the occurrence of which depends on oceanic conditions (calm or agitated). Finally, this sometimes leads to the formation of a nearly continuous sheet of ice a few meters thick that floats on top of the ocean, extending over thousands of kilometers; this is commonly called the "sea ice pack" or "sea ice cover." A sea ice pack should not be confused with the ice sheets or ice shelves made from freshwater ice formed by a slow compaction process of snow and typically from hundreds to thousands of meters thick.

The "pack ice" (ice composing the sea ice pack) is an efficient insulator, especially when snow is present on top (e.g., Maykut et al., 1986; Andreas et al., 1979). A snow-covered, non-moving continuous ice pack drastically reduces the exchanges of heat, water, and momentum between the ocean and atmosphere, and it strongly limits long- and short-wave radiation transfers.

The simplest model for a snow-covered, non-moving continuous ice pack could be based on a 0-layer vertical heat transfer model and it would have only two prognostic variables, the snow and ice thicknesses. More complex models would consider a larger number of vertical layers in the snow and ice, with prognostic variables for the temperature and salinity in each layer, and a fine representation of snow metamorphism and upper surface processes (e.g., the formation of melt ponds).

In nature, however, the sea ice pack is not made of a continuous sheet of ice with homogeneous thickness; instead, it is highly fractured and exhibits a chaotic surface that combines pressure ridges, open leads, and plates of intact "level ice" in between. The ocean-atmosphere heat fluxes are typically much larger for open water and thin ice than for thick ice, meaning that fluxes through the leads make a large contribution to the total energy transfer even if they cover only a small percentage of the total ice pack area (see, for example, Lupkes et al., 2008a; 2008b; Marcq and Weiss, 2012; Vihma et al., 2013 for a review). In winter, this transfer corresponds to an intense and highly local injection of heat and moisture into the atmosphere, as well as formation of new ice tied to the release of large amounts of salt into the upper ocean. Ridges and leads define surface and bottom roughness, and therefore drastically affect surface and bottom stresses acting at the ice–atmosphere and ice–ocean interfaces due to the action of winds and currents.

Why does the ice pack not simply deform homogeneously as a pure linear elastic media or as a highly viscous fluid? The quick answer is because sea ice is a solid, brittle material. Under forces load, the ice pack only deforms elastically below a given critical internal stress (corresponding to the range of infinitesimal–non permanent deformation), breaks into pieces along fractures when this critical stress is reached, and then eventually continues to deform if large-scale forces are still applied (corresponding to large-scale permanent deformation), with energy dissipation at the edges of those pieces that is related to friction mechanism.

There are multiple driving forces for breaking the ice. Thermal cracking is observed when rapid and large variations of temperature occur in the vertical dimension (Weeks et al., 2010, chapter 10.4). Ice can also break due to vertical load, which is the technique used by icebreakers to progress in areas covered by high concentrations of ice (Bazant, 2000). In the marginal ice zone (i.e., the transition area between the open ocean and the pack ice), the dominant process that breaks the ice is the flexural bending due to ocean surface waves. However, the dominant cause overall for sea ice breaking, particularly in the central Arctic, is the presence of spatial gradients in the wind and ocean stresses acting on the ice.

Breaking generates blocks of ice with length scales from a few meters to hundreds of meters depending on the driving force involved (tides, waves, eddies, atmospheric features...). Based on the surrounding conditions, these blocks may be pushed against each other to form ridges that are typically tens of meters thick, further broken into smaller pieces (typically when enduring shear deformation), or moved apart to reveal open water areas between blocks. The breaking of the ice and the subsequent collisions and friction between the blocks generates large energy dissipation. Some part of the kinetic energy is also transformed into potential energy due to the build-up of ridges. Numerical simulations with discrete models (Hopkins et al., 1991) and analyses of in situ observations have yielded a solid knowledge of the ridging process at the scale of an individual ridge. As discussed in detail later in this chapter, it will be important to take those local processes into account when building a sea ice model for the ice pack.

It is only since the late 1990s, with the implementation of automated sea ice drift estimation from the synthetic-aperture radar images retrieved by RADARSAT (Kwok et al., 1990), that sea ice scientists began to get sufficient amounts of observations, both in terms of coverage and resolution in the temporal and spatial domain, to really understand the mechanisms controlling sea ice dynamics. From the sea ice drift Radarsat Geophysical Processor System (RGPS) dataset, one can derive sea ice deformation fields at a resolution of 10 km and three days over the whole Central Arctic and for an entire season. A first look at these deformation fields (see example in Fig. 15.1, first panel) reveals that most of the deformation is localized along linear features while the rest of the ice pack exhibits no or little deformation, mimicking the plate tectonic of the Earth crust characterized by fracturing processes and faulting mechanism. Those linear kinematics features can be hundreds of kilometers long and appear to be organized into systems of nearly parallel lines. New linear kinematics features are formed very quickly (much faster than three days, which is the temporal resolution of the RGPS data) and they are generally aligned with the atmospheric isobars, meaning that they are, at the time of their "activation," perpendicular to the gradient in the wind stress. These networks of fractures typically have a persistence of several weeks, during which the same set of linear kinematics features remain active. This long persistence, compared to the synoptic scale, is attributed to the fact that once sea ice is fractured and a linear feature is formed, it represents a much weaker zone in the ice pack that will likely accommodate large shear deformation. Shear deformation is also associated with large values of divergence and convergence, corresponding to leads opening and ridges building. Changes from one linear kinematics feature's system to another one can happen when the ice has recovered enough of its mechanical strength due to ice refreezing in cracks or leads or ridge consolidation, so that a next external forcing will generate another set of linear kinematics features with different directions (see discussion and images in Kwok, 2001).

Complex systems with dynamics controlled by highly localized extreme events can be analyzed quantitatively by performing scaling analysis, i.e., by looking at how statistical distributions change with the scale of observation. Fig. 15.1 shows the result of a coarse graining procedure where deformation is computed at different scales, from 7 to 200 km. It is important to note that some linear features are still visible at scales as large as 100 km, invalidating the homogenization methods on which continuous models are generally based. That is an issue we will discuss in greater detail later in this chapter.

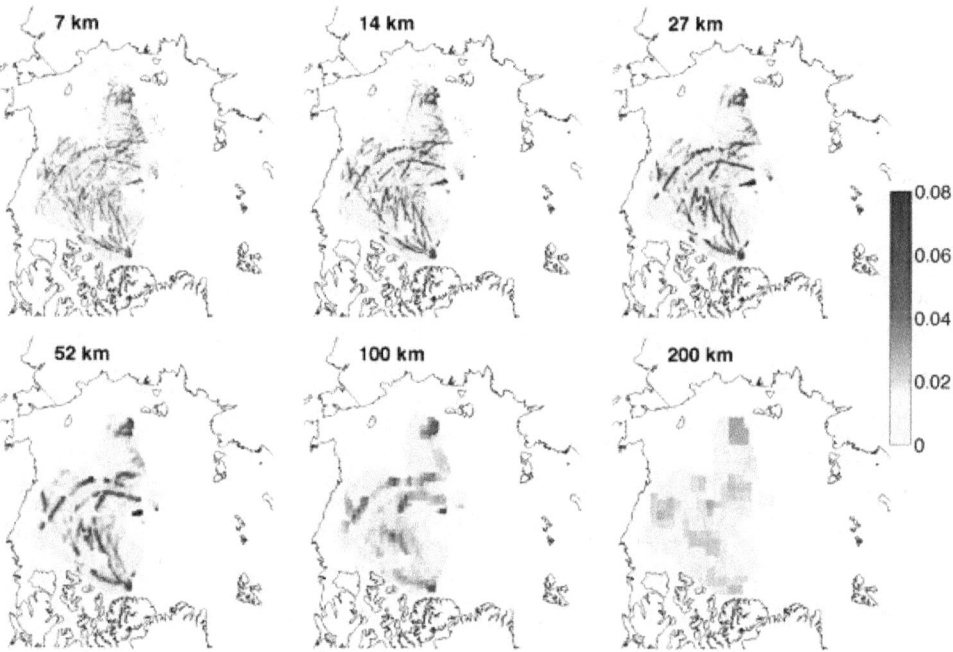

Figure 15.1. Shear rates [in one/day] computed from 3-days ice drift from the RGPS dataset at spatial scales of 7 to 200 km over the Central Arctic.

Scaling analysis of sea ice deformation can also be computed by looking at the dispersion of pairs of drifting buoys (either real buoys set in the ice or virtual buoys from a model or from the RGPS trajectories). As the same pairs of buoys can generally be followed for long periods of time, using dispersion analysis allows us to look at the temporal scaling, i.e., how the statistical distributions depend on the timescale of observation. From the RGPS trajectories dataset (see example in Fig. 15.2), we can define many pairs of buoys with initial separation $L(0)$ ranging from 10 km to hundreds of kilometers. A proxy of the deformation rate can then be defined as

$$\dot{\varepsilon} = \frac{L(t)-L(t-T)}{L(t)\,T} \qquad (1)$$

with T defining the temporal scale of observation.

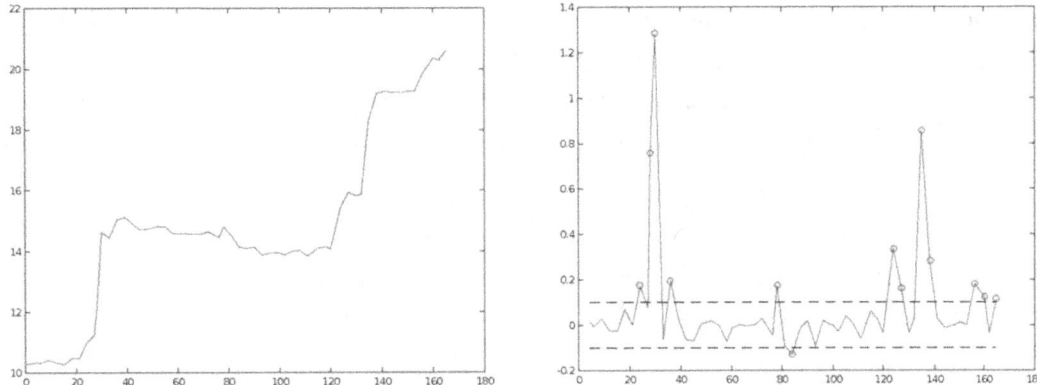

Figure 15.2. Separation [in km] and separation rate [in km/day] for a pair of RGPS trajectories followed for about 160 days in winter 2006-07. The section and circles in red indicate the deformation events when the separation rate is larger than 100 m per day.

Power-law scaling in space is observed for the moments of the distribution,

$$\langle \dot{\varepsilon}^q \rangle_{L,T} \sim L^{-\beta(q)} \tag{2}$$

with two different structure functions $\beta(q)$, one for L<=100 km and another for L>=100 km (Fig. 15.3). The analysis of different buoy trajectory datasets (Rampal et al., 2008; Hutchings and Hibler, 2008; Hutchings et al. 2011) shows that the scaling obtained from the analysis of the RGPS data holds for much smaller scales of about hundreds of meters. Some authors suggested that it may hold down to the scale of one meter (Marsan et al., 2004), which is the typical thickness of the ice pack where energy can be largely dissipated when ridges are formed or shear faulting occurs.

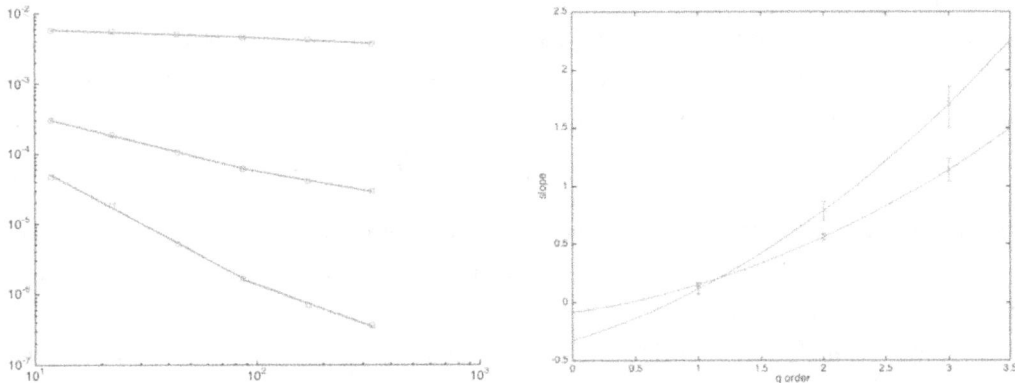

Figure 15.3. Spatial scaling analysis of the deformation for spatial scales ranging from L=10 km to L=350 km. On the left panel, the first (upper curve, q=1), second (middle curve, q=2) and third (lower curve, q=3) order moments of the distributions [in one/day] are plotted against the scale L of observation [in km]. A cut in the scaling is observed at L=100km. On the right panel the structure function $\beta(q)$ is plotted for L<=100 km and L>=100 km.

Based on the assumption that the scaling laws remain valid for smaller scales and using the information on the structure function $\beta(q)$ for L<=100 km, one can actually extrapolate the probability density functions of our proxy of the deformation to smaller scales by using the method proposed in Marsan et al. (2004). Fig. 15.4 shows that the probability density functions seamlessly transition from an exponential (straight line in the semilog axis, right panel) to a power law (straight line in the loglog axis, left panel) when going towards smaller scales of observation. Such extrapolation towards the smallest scale (1.25 km, in our example) leads to a probability density function that is very close to a power law with an exponent k close to 2 and a multiplicative factor C=0.005

$$lim_{L \to 1m} p(\dot{\epsilon}) = C\dot{\epsilon}^{-k} \tag{3}$$

Power law distributions with an exponent k close to 2 are typical of systems where the whole distribution is dominated by extreme values. For $k<=2$, even the mean becomes infinite.

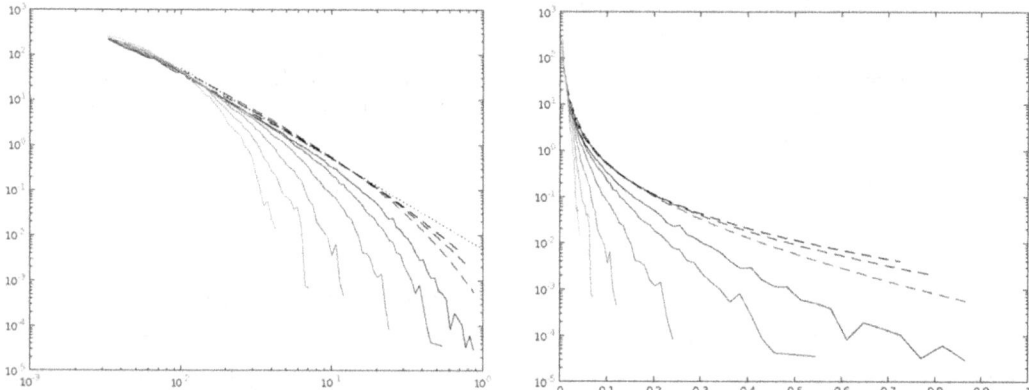

Figure 15.4. Probability density functions of the deformation rate computed at T=3 days and for L going from 360 to 10 km (solid lines from gray to dark gray) and extrapolated for L=5, 2.5 and 1.25 km (dashed line from gray to black), in a loglog plot (left) and a semi-log plot (right). The power law function $C\dot{\epsilon}^{-k}$ with C=0.005 and k=2 is plotted as a dotted line for reference.

However, in nature distributions exhibit cut-off for very large values. In the case of sea ice, one could, for example, propose an exponential cutoff so that the deformation rate probability density function at scale L would have the following shape:

$$p_L(\dot{\epsilon}) = C\dot{\epsilon}^{-k} e^{-\lambda(L)\dot{\epsilon}} \tag{4}$$

where $\lambda(L)$ is a positive monotonically increasing function of the spatial scale of observation. Preliminary test showed that $\lambda(L) = L$ is a good first estimate (not shown). Note that the numbers given here are for L in km and $\dot{\epsilon}$ in one/day.

The scaling analysis can also be performed for the temporal domain. As for the spatial domain, power law scaling is observed for the different moments (here with q=0.5 to q=3):

$$\langle \dot{\epsilon}^q \rangle_{L,T} \sim T^{-\alpha(q)} \tag{5}$$

The structure function $\alpha(q)$, defining how the moments of the distribution varies as a function of the time scale of observation, is different for scales shorter than ~ten days and longer than ~ten days (see Figure 15.5). This change in the scaling around ten days corresponds to the typical time interval between two successive synoptic atmospheric events. Probability density functions could also be extrapolated to time scales lower than 3 days (not done here) by using the structure functions $\alpha(q)$.

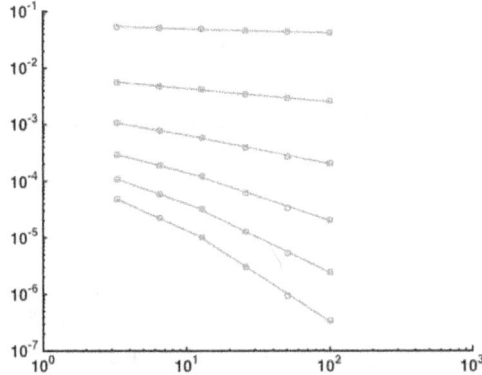
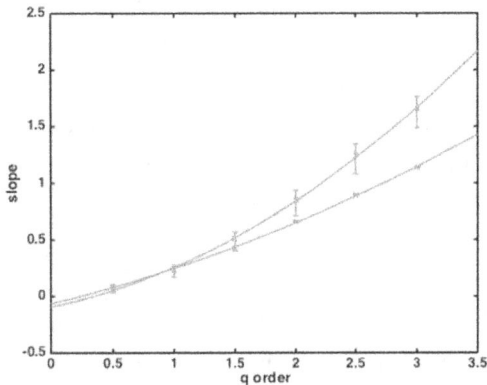

Figure 15.5. Temporal scaling analysis of the deformation rates [in 1/day] for temporal scales ranging from $T=3$ days to $T=100$ days. On the left panel, the first (upper curve), second (middle curve) and third (lower curve) order moments of the distribution are plotted against the temporal scale T of observation. A cut in the scaling is observed at $T=10$ days. On the right panel the structure function $\beta(q)$ is plotted for $T<=10$ days and $T>=10$ days.

Taking into account the observed temporal and spatial scaling of sea ice deformation is crucial to make the link between the scales at which the energy is introduced in the system (synoptic scales, ten days, hundreds of kilometers) and the scales at which it is dissipated by breaking the ice and building ridges (dissipation scale, a few hours, 1-10 m). Making this connection is at the basis of the unified modelling framework developed recently and described below.

A unified modelling framework for the ice pack

As discussed in the previous section, sea ice dynamics have many similarities with plate tectonics and earthquakes physics, but at much smaller temporal and spatial scales. The ice pack dynamics are controlled by intermittent and highly localized shearing events accommodating a large part of sea ice deformation and energy dissipation. Simulating such complex dynamics is a challenge that has still not been completely achieved. In this subsection, we briefly describe the components and results of the first generation dynamical models for sea ice, which were mainly developed to simulate the large-scale sea ice circulation. We also present a new modelling framework that takes into account the links between scales described previously.

Like the ocean and the atmosphere, the ice pack is subject to large-scale circulations, characterized in the Arctic by the Beaufort gyre and the transpolar drift, and around Antarctica by the Antarctic coastal (westward) and circumpolar (eastward) currents. For the Central Arctic, these currents have typical time- and length-scales of about 5-6 months and 400 km for winter and 2-3 months and 200 km for summer. The seasonal and interannual variability of these currents are largely controlled by the change in the sea ice extent and volume, as well as by the seasonal and interannual variability of the atmosphere and ocean. The amplified effects of climate change in polar regions also affect the trends in the large-scale circulation (Rampal et al., 2009).

The first dynamical models of sea ice were built to reproduce the large-scale circulation of sea ice. At these scales, the momentum equation basically resumes to the equilibrium between three forces: the surface stress from the wind (τ_a), the surface stress from the ocean current (τ_w), and the gradient of the internal stress integrated on the vertical. Different parameterizations for the air-ice and water-ice surface stresses are used, the simplest being that the surface stresses depend linearly on the difference between the horizontal ice velocity u and the horizontal air velocity u_a and water velocity u_w, respectively. The simplest surface stress parameterizations also neglect the turning angle and assume that ice speed is much lower than air speed, leading to the following equation for the momentum balance:

$$0 = \rho_a c_a u_a + \rho_w c_w (u_w - u) + \nabla \cdot (\sigma H) \quad (6)$$

where ρ_a and ρ_w are the air and water densities, c_a and c_w are the air and water drag coefficients for sea ice, H is the ice "thickness" (volume per unit area), and σ is the internal stress tensor.

A rheological model is needed to define the link between the deformation and internal stress. The model dominantly used is based on a plastic rheology, where the internal stress in shear and convergence depends solely on the sea ice thickness and mode of deformation. In one dimension, the plastic rheology gives a constant value $\sigma_I = -P$ when the divergence rate is negative or null, and $\sigma_I = 0$ otherwise.

The momentum equation is coupled to the volume conservation equation:

$$\frac{\partial H}{\partial t} = -\nabla \cdot (uH) + f_H \quad (7)$$

where f_H is a sink/source term related to thermodynamics. For a simple case with no thermodynamics, no ocean currents, and where the ice is pushed towards the coast with a constant wind velocity, the stationary solution corresponds to a zero velocity and a linear function for the thickness field with a slope $\frac{\partial H}{\partial x} = \frac{\tau_a}{P}$. The ice strength parameter P then defines the large scale thickness gradient. When sea ice is diverging, the internal stress is zero and the ice moves in free drift mode with a velocity set by:

$$u = \frac{\rho_a c_a}{\rho_w c_w} u_a + u_w \quad (8)$$

The ratio $\frac{\rho_a c_a}{\rho_w c_w}$ is the Nansen number that was first estimated by the Norwegian Fridtjof Nansen when comparing the drift of his boat (the Fram) trapped in the ice with the wind speed.

Such a model, with only the volume per unit area and the horizontal ice velocity as variables, is rarely used because it is important to know the sub-grid scale distribution of the ice thickness to correctly represent some processes, particularly the processes related to thermodynamics. To do so, one defines a thickness distribution $g(h)$ within each cell that has the following properties:

$$\int_0^\infty g(h)\, dh = 1 \quad (9)$$

and

$$\int_0^\infty g(h) h\, dh = H. \quad (10)$$

The governing equation for the thickness distribution is:

$$\frac{\partial g}{\partial t} = -\nabla \cdot (ug) - \frac{\partial}{\partial t}(fg) + \psi \quad (11)$$

where $f(h)$ is a sink/source term from the thermodynamics (equivalent to an advection term in the thickness space), and ψ is a redistribution term that transfers sea ice from one thickness to another. The following constraints on the redistribution term:

$$\int_0^\infty \psi \, dh = \nabla . u \qquad (12)$$

and

$$\int_0^\infty h \, \psi \, dh = 0 \qquad (13)$$

are obtained to respect equations (11) and (9). In practice all sea ice models consider a sub grid scale thickness distribution of the ice. Multi-category models usually define it using N distinct categories:

$$g(h) = \sum_{i=0}^{N-1} a_i \delta(h - h_i) \qquad (14)$$

with $h_0 = 0$ for the open water and $\delta(h)$ a Dirac delta function, the simplest being the two-category ice thickness distribution where only two prognostic variables, the ice concentration a_1 and the ice thickness h_1, and where $h_0 = 0$ and $a_0 = 1 - a_1$. Other representation of the ice thickness distribution may be defined based on observations using, for example, modal analyses that often show one mode for open water, one for refrozen leads, and one for level – non deformed ice and a negative exponential tail for the ridged ice.

Whatever the representation of the thickness distribution, one needs to define the redistribution term ψ in eq. 11. The framework to define the redistribution term was established by Thorndike (1975). The scheme is based on many assumptions that will not be listed here, but the main underlying assumption is that one needs to define a subgrid-scale distribution of the ice deformation because the deformation at the scale of one model cell will not correspond to the deformation at the scale of one lead or ridge. This is well-illustrated in Fig. 15.6, extracted from Thorndike (1975), which shows a case with no divergence at the scale of the grid cell, but with some local ridging and lead opening. The redistribution scheme proposed by Thorndike is actually equivalent to integrating a ridging model (given the evolution of the thickness at the scale of a single ridge and lead) over a subgrid-scale distribution of the ice deformation. In Thorndike (1975), the ridging model is very simple and supposes that when a ridge is formed its thickness is k times the thickness of the surrounded ice, with k set to 5. The formation of a ridge then corresponds to the redistribution process $\gamma(h_1, h_2) = \frac{1}{k} \delta(h_2 - kh_1)$.

A more complete redistribution scheme has recently been proposed to take into account the evolution of the macro-porosity (holes between ice blocks, see the presentation of A. Roberts at the Newton Institute, https://www.newton.ac.uk/seminar/20170913094510301). Such a model makes a consistent link with the ridging processes occurring at the scale of the ridges, and also estimates correctly the distributions of ridge separation, thickness, and shape.

The redistribution process is a key component in sea ice models as it makes room for open water areas through which a large amount of heat, momentum, and mass can be exchanged between the ocean and the atmosphere. It is also a way to make a consistent link between the energy dissipated and the work done during the ridging process and the energy extracted by the rheological term. In most sea ice models, however, this link is simply not made, and when it is accounted for, it is usually based on simplistic approaches where the energy dissipation is simply scaled as a constant (about 15-20) of the increase in potential energy due to ridging. The approach recently presented by Roberts tackles this issue and provides a more sound representation of the energy dissipation and work done during the ridging process.

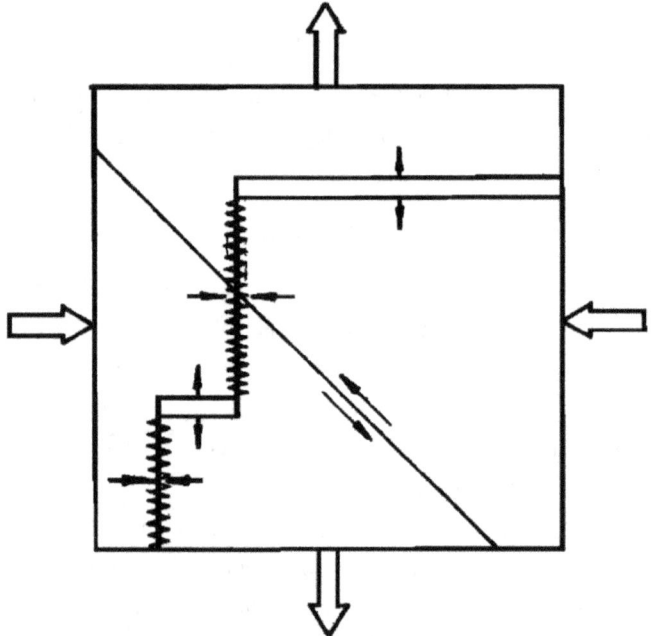

Fig. 10. Schematic diagram illustrating the formation of leads and pressure ridges during pure shearing deformation.

Figure 15.6. From Thorndike (1975; Fig. 10) illustrating the ridging scheme.

Another important assumption when defining any redistribution schemes is that the sea ice model simulates correctly the shear and divergence rate at the scale of the model cells. Knowing that below 100 km the deformation is highly localized, the rheology used for continuous models running at spatial scales smaller than 100 km ($\Delta x < 100\ km$) must then allow localization of the deformation at the scale of one model cell. A solution that has been proposed is to add a large-scale variable, referred to as the local damage of sea ice and defined at the scale of one model cell (i.e., for a 10 km x 10 km or a 100 km by 100 km cell). The ice damage is set to range between zero (for intact ice pack) and one (for totally damaged ice pack). A proposed model that includes a damage variable is the Maxwell-Elasto-Brittle (Maxwell-EB) rheology (Dansereau et al., 2016). In this model, the internal stress builds up as an elastic medium for undamaged ice. If the ice stress reaches a defined stress envelope, it fails, meaning that the damage increases and a viscous relaxation term is activated so that the accumulated stress is rapidly damped out. While the ice cell remains damaged, it can accommodate large deformation. The model assumes that the damage can be reduced via some healing processes, so that after typically several days or weeks, this cell is able to accumulate internal stress and mechanically resist external forces again. The Maxwell-EB rheology has been implemented in the finite element sea ice model neXtSIM (Bouillon and Rampal, 2015;

Rampal et al., 2016a) and used successfully to reproduce the scaling laws described previously down to the model resolution (Rampal et al, 2017). Fig. 15.7 shows an example of the sea ice drift and deformation simulated by the neXtSIM model, with localization of the deformation at the scale of the model grid cell, meaning discontinuities in the velocity field. Such a model is still continuous but it is able to generate a discrete-like behavior in which one can identify individual ice plates, moving like solid bodies and surrounded by areas of high deformation.

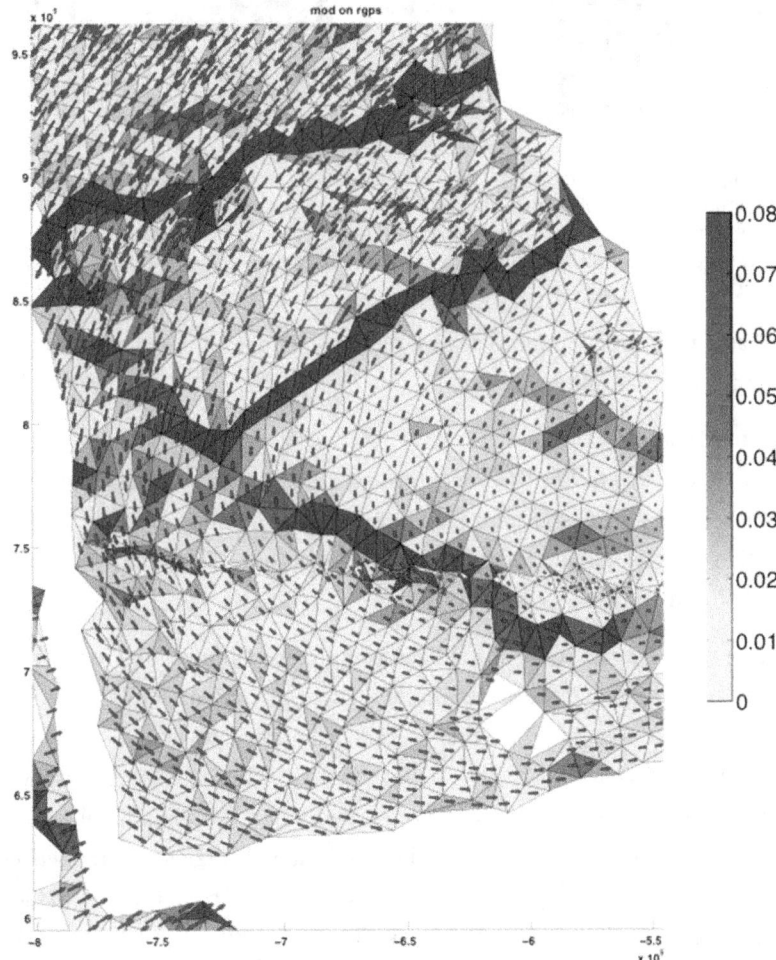

Figure 15.7. Sea ice drift and deformation simulated by the Maxwell-Elasto-Brittle model running at a resolution of 10 km and plotted as if it was computed with the RGPS system, meaning at a resolution of about 10 km and with a time interval of three days.

A model that allows strain localization will also produce sharp variations for scalar quantities such as the thickness and concentration fields. Having an advection scheme that is able to transport such sharp gradients without degrading them is a challenge for classical Eulerian advection schemes applied on a fixed grid and where advection corresponds to computing fluxes through the element's edges. An alternative is to use a Lagrangian framework where the model grid moves with the ice (see Fig. 15.8). Such an advection scheme has been implemented in the neXtSIM model (Samaké et al., 2017).

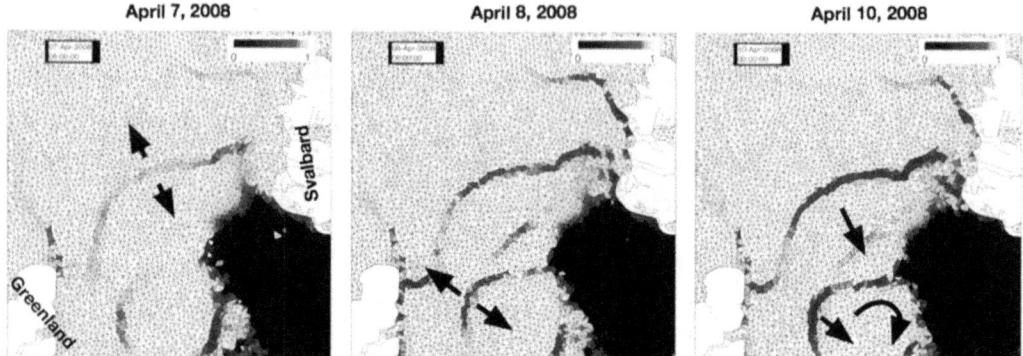

Figure 15.8. Example of sea ice concentration fields at Fram Strait and the underlying moving mesh coming from a one-year simulation using neXtSIM in a full Arctic configuration and with a resolution of about 10 km. The maps show how the localized divergence generates discontinuities in the ice cover at the scale of the mesh resolution and how these discontinuities are preserved over time by the Lagrangian advection scheme. Figure from Samaké et al. (2017).

A unified framework for continuous sea ice dynamics model running at resolutions from 0.1 to 100 km would then consists of the following:

- the momentum equation,
- a rheological model able to localise the deformation at the scale of one model cell,
- an advection scheme able to transport ice field without degrading their spatial localisation,
- a redistribution scheme making the link between the deformation at the scale of the model cells and the subgrid-scale ridging processes.

Implementing such a framework has not been completely achieved yet, because all of the above ingredients need to be present at the same time. For example, it was shown that classical viscous-plastic rheology with multi-category ice thickness distribution and energy consistent redistribution schemes perform worse than a simple two-category model with a constant ice strength parameter P (Ungermann et al., 2017). It was also recently presented that plastic-like rheologies must be run at high resolution (about 0.5-1 km) to be able to simulate some localization of the sea ice deformation at 10 km (Hutter et al, 2017). In light of the scaling laws of sea ice deformation, the redistribution schemes should also be scale-dependent and take into account the spatial and temporal scale at which the deformation is represented in the given large-scale model. Even the promising Maxwell-Elasto-Brittle approach is not complete because it lacks an energy-consistent formulation to link the energy dissipated by viscous relaxation to the ridging scheme.

However, it is now thought that a consistent modelling framework can be built to reproduce adequately sea ice deformation across a wide range of scales. At least we have now all the ingredients. The framework presented here also brings a parallel with ocean models, as we could classify the sea ice models into three categories:

1. Large-scale circulation models that correctly simulate the mean circulation at typical scales of hundreds of kilometers and hundreds of days.

2. Lead/ridge-permitting models that allow the strain localization when they run at resolutions between 0.1 and 100 km and reproduce the scaling laws from their temporal and spatial resolution up to cutting scales of 100 km and ten days.
3. Lead/ridge/floes resolving models that run at resolution of the order of 1 m and that explicitly represent individual ridges, leads, and floes.

Some may extrapolate and suggest that lead/ridge permitting continuous models could be used to simulate explicitly the deformation at the scale of one ridge/lead/floe, but this is probably unrealistic as the assumptions (i.e., two-dimensional and isostatic assumptions) made to build a continuous sea ice model are surely not valid for a resolution higher than about 100 m. This would be equivalent to suggesting that hydrostatic hydrodynamic models could explicitly reproduce non-hydrostatic flows.

Lead permitting models and impacts on the ocean-atmosphere interactions

When a lead opens in the ice during winter, relatively warm ocean waters are exposed to the cold atmosphere resulting in heat fluxes of up to 600 W/m^2 (e.g., Maykut, 1986; Andreas and Murphy, 1986). As a result, a plume of warm, moist air forms over the lead, sometimes resulting in ice fog, which significantly reduces visibility and can cause ice to accumulate on surfaces such as aircraft, power lines, and roads (e.g., Gultepe et al., 2015). This release of heat also causes convection in the predominantly stable or near-neutral Arctic atmospheric boundary layer, and as the plume rises it may penetrate the lowest levels of the capping inversion, leading to entrainment (see Lupkes et al., 2008a; 2008b; Vihma et al., 2013 for an overview). On the oceanic side, ice forming in leads removes fresh water from the ocean and releases brine (e.g., Smith et al., 1974; Kozo, 1983; Morrison et al., 1992). The brine plumes spread horizontally along the top of the halocline, reducing the depth of the mixed layer, but they do not penetrate the halocline. Nguyen et al. (2009), using subgrid-scale parameterization of brine rejection, showed that a faithful simulation of the Arctic halocline depends on the proper representation of brine release and its redistribution in the water column.

Before analyzing such processes in coupled ice-ocean-(atmosphere) systems, one should determine how well the leads are represented by the sea ice model. In this section, we recap the results of Olason et al. (2017), who performed such an analysis by comparing statistically observed lead fraction data to lead fraction simulated by the neXtSIM model described previously. We show that the model reproduces the spatial scaling of the lead fraction statistics down to its resolution, indicating that the model is "lead permitting."

Observations of lead fraction can be derived from passive microwave observations of the AMSR-E (the Advanced Microwave Scanning Radiometer for EOS). For example, the dataset produced by Ivanova et al. (2016) is available on a daily basis for the Arctic region from November to April, from 2002 to 2011. The dataset resolution is 6.25 km and the method allows leads wider than 3 km to be detected, meaning that a substantial amount of smaller leads is undetected in this product. The data show the area fraction of each grid cell covered by leads that are filled with open water and thin ice (see Fig. 15.9).

We use the latest version of the next generation sea ice model, neXtSIM, described earlier. The model uses the Maxwell-Elasto-Brittle rheology of Dansereau et al. (2016), a Lagrangian moving mesh as described in Rampal et al. (2016a), and the thermodynamic growth is modelled using the two-layer model of Winton (2000). The model has three ice categories, thick ice, open water, and newly formed thin ice. The model set-up covers the Central Arctic Ocean, with open boundaries at the Bering Strait and through the Canadian Arctic Archipelago, Greenland, Barents, and Kara Seas. The model is forced using daily mean results from the TOPAZ4 oceanic reanalysis (Sakov et al., 2012) and 6-hourly results from the CFSR and CFSv2 atmospheric reanalysis (Suranjana et al., 2010; 2012).

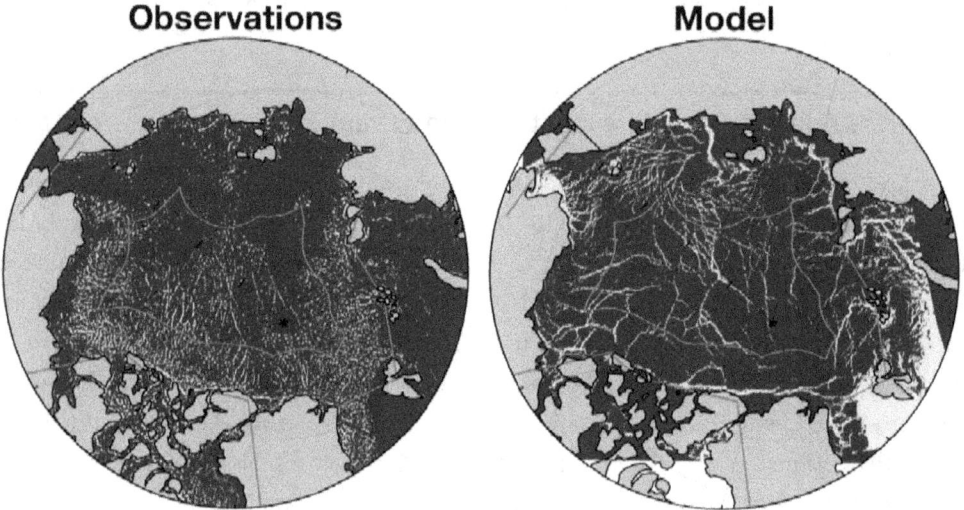

Figure 15.9. Observed and simulated leads on January 1, 2011. The figure shows the entire model domain, and the red lines indicate the boundaries of the "Arctic" (outer region) and "Central Arctic" regions used in the study. Lead fraction larger than 0.01 is indicated in yellow. Figure from Ólason et al. (2017).

As shown in Fig. 15.9, the model produces highly localized lead fraction as in the observations. The lead density is somewhat weaker than in the observations, especially in the Beaufort Sea and north of the Fram Strait; this issue is still under investigation. As for the deformation, we can explore the spatial scaling of the lead fraction simulated by the model. Due to the presence of large coastal and flaw polynyas, we confine the analysis to the Central Arctic (more than 400 km away from the coast). The results of this analysis can be seen in Fig. 15.10, which shows a clear scaling of the lead fraction statistics with a spatial scale that is similar to the one obtained from other observations (not shown here). When conducting the same analysis for simulations run at coarser resolutions, respectively 10 and 20 km, the scaling still holds down to the model resolution, meaning that the model localizes the deformation (and thus the lead fraction) down to its resolution.

There is certainly room for improvement, but the fact that the model reproduces the scaling laws of the lead fraction statistics down to its resolution means that it provides the correct estimate of the heterogeneity of that field at that scale. Subgrid-scale parameterizations are still needed to connect with the scale of the leads, and this could be accomplished based on the observed or simulated scaling laws. The best platform to investigate the effects of the better representation of the lead fraction on the heat fluxes is, of course, a coupled ice-ocean-atmosphere model that allows us to simulate the feedbacks between the localization of lead fraction, the heat fluxes, and the effects on the ocean and atmosphere.

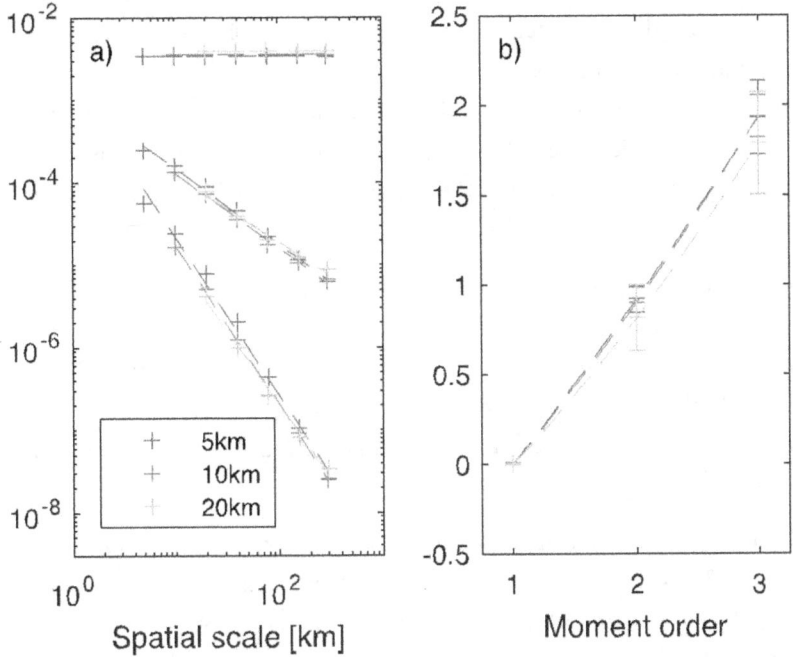

Figure 15.10. The spatial scaling of modelled lead fraction in the "Central Arctic" region over JFM, 2011. The colours denote results from runs at different model resolutions. The left panel shows the mean, variance, and skewness of the lead fraction as a function of the scale of observation. The dashed lines are linear fits for each moment. The right panel shows the slope of the scaling as a function of the moment order (q=1 for the mean, q=2 for the variance and q=3 for the skewness). Figure from Ólason et al. (2017).

New Frontiers in Sea Ice Forecasting

To operate in ice-infested water (e.g., shipping, scientific campaigns) depends heavily on local weather and sea ice conditions (see Eicken, 2013; Riska and Coche, 2013). Building forecast systems capable of reproducing and predicting local ice conditions would complement the currently available forecasting services based on ice charts, large-scale forecasts, and satellite images.

When building a forecast system, one must first determine what the users need. For example, most captains navigating close to or in the ice pack use information from daily ice charts. The ice charts define areas of different classes of ice concentration, types, and thicknesses. They also often use synthetic-aperture radar or visible satellite images where fine structures such as leads and ridges can be fairly well identified by trained sea ice experts. Therefore, systems able to provide this type of information several days in advance with enough confidence would be very valuable. Forecast systems can also provide information that does not exist in ice charts and synthetic-aperture radar images, especially the information on the sea ice drift and deformation. Such information is crucial when determining if a ship will be trapped in a converging zone, or to assist search and rescue operations in case of an incident or oil spill, for example.

The term "forecasting" is also used for seasonal and decadal predictions that aim to predicting the intra-annual and inter-annual large-scale variations of the ice pack state (total extent, volume, ...). Although not discussed in detail here, it is worth noting that this research activity has potentially large economical and societal impacts.

Operational sea ice forecasting is relatively new compared to weather and ocean operational forecasting. Initially, the sea ice forecaster community decided to apply data assimilation methods developed for the atmosphere and ocean to the sea ice. However, this strategy may not be optimal for the following reasons:

1. Data for sea ice are mainly derived from satellite products that have good coverage but potentially systematic biases as well as random noise.
2. Most sea ice forecast systems are open system forced by external forcing from ocean and/or atmosphere models, with potential time varying biases.
3. Sea ice dynamics is often far from equilibrium and characterized by large discontinuities in space and time.

In this section we present two examples where these three points are discussed. In the first, we will try to answer the question: "How can we beat ice charts persistency?" and we will discuss the use of (1) potentially biased data and (2) potentially biased forcing. Next, we will ask the question: "How can we forecast local-scale deformation and drift several days in advance?" and we will discuss data assimilation for systems that are far from dynamical equilibrium, i.e., sub-critical. In each case, we present observations, data assimilation methods, and model components that could be used in an operational system to achieve those objectives. The list of observations, assimilation methods, and model components presented here is not exhaustive and does not especially correspond to what is used today in sea ice forecasting. For such a review, please refer to the book: *Sea ice analysis and forecasting* (2017).

Can we beat ice charts persistency?

Most of the operators working in ice-infested waters rely on near real-time data acquired from satellites and distributed directly as synthetic-aperture radar images or indirectly as ice charts. One way to convince these "clients" to use model forecast outputs is to show *a posteriori* that the information we were able to provide for a given time t is in better agreement with the corresponding information observed during those days than the information available at time t. In other words, we have to prove that we are able to surpass the persistency of their preferred source of near real-time data.

Forecasting synthetic-aperture radar-like data is still a futuristic concept, as one would need to use very high-resolution systems (of the order of 10 m). On the other hand, forecasting ice charts-like maps could be achieved using existing systems. Ice charts usually cover regional or Arctic-wide domains and provide, on a daily basis, information on the ice concentration from the day before. The production of ice charts is made by sea ice analysts who combine data from various sources (synthetic-aperture radar images, passive microwave satellite data, in situ observations, forecast systems, etc.) in order to provide maps of ice classes.

Typically, operational sea ice forecasts are not based on ice charts but directly on passive microwave satellite data, providing daily estimates of sea ice concentration at resolution of about 5-20 km. However, it is important to note that these data are indirect measurements of sea ice concentration and are subject to well-known biases. They usually perform badly when ice ponds are present, for thin ice, for highly fragmented and low concentration ice, and near the ice edge. Building a system taking all these limitations into account is feasible, but has not yet been achieved. Another solution is to directly assimilate ice charts that are also provided on a daily basis and are known to have less persistent bias.

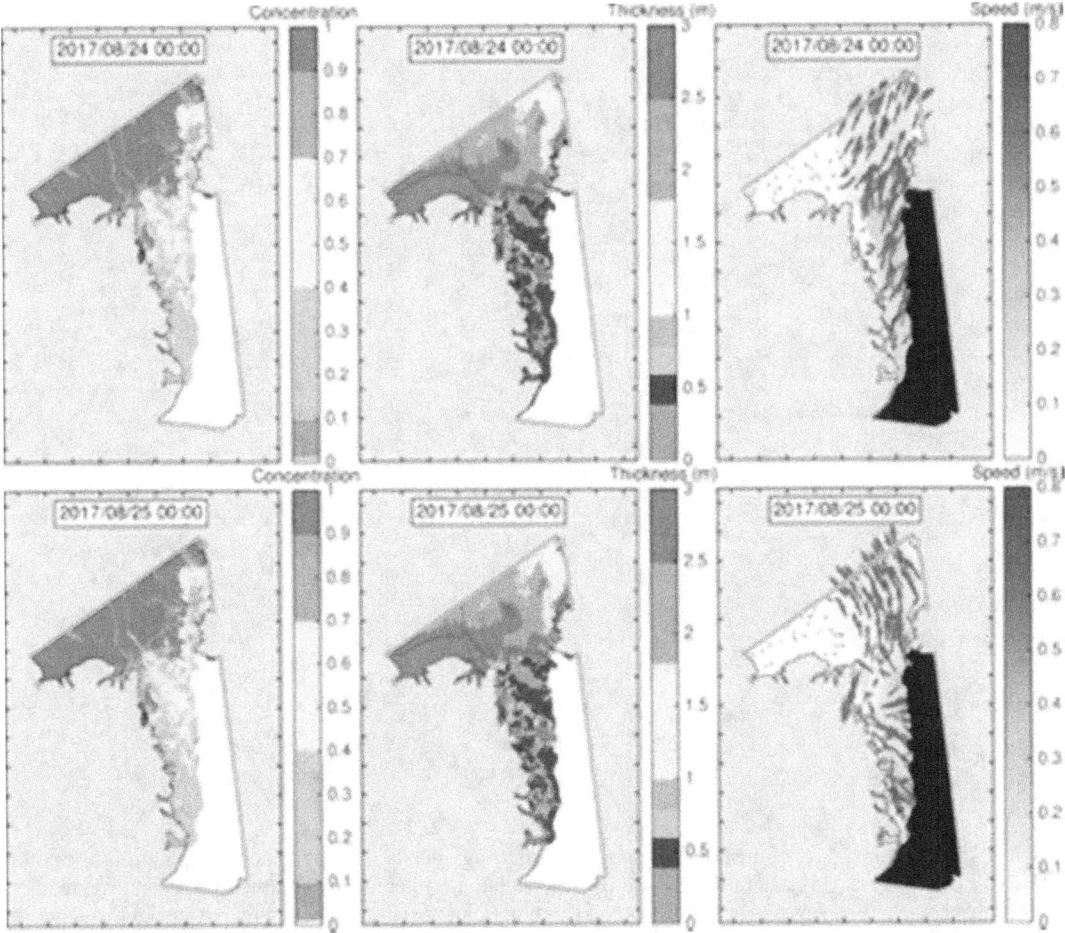

Figure 15.11. Examples of sea ice concentration, thickness and velocity forecasted by the neXtSIM-F setup for the Greenland Sea and Fram Strait for the T+24h (top panel) and T+48h (bottom panel) time horizons in August 2017. Plotted ice drift vectors are instantaneous drift at noon of that day. The concentration maps use the color code from the Norwegian ice charts and show five different ice classes. The ice thickness maps use eight different classes that matches with the ice breaker classification of Bureau Veritas.

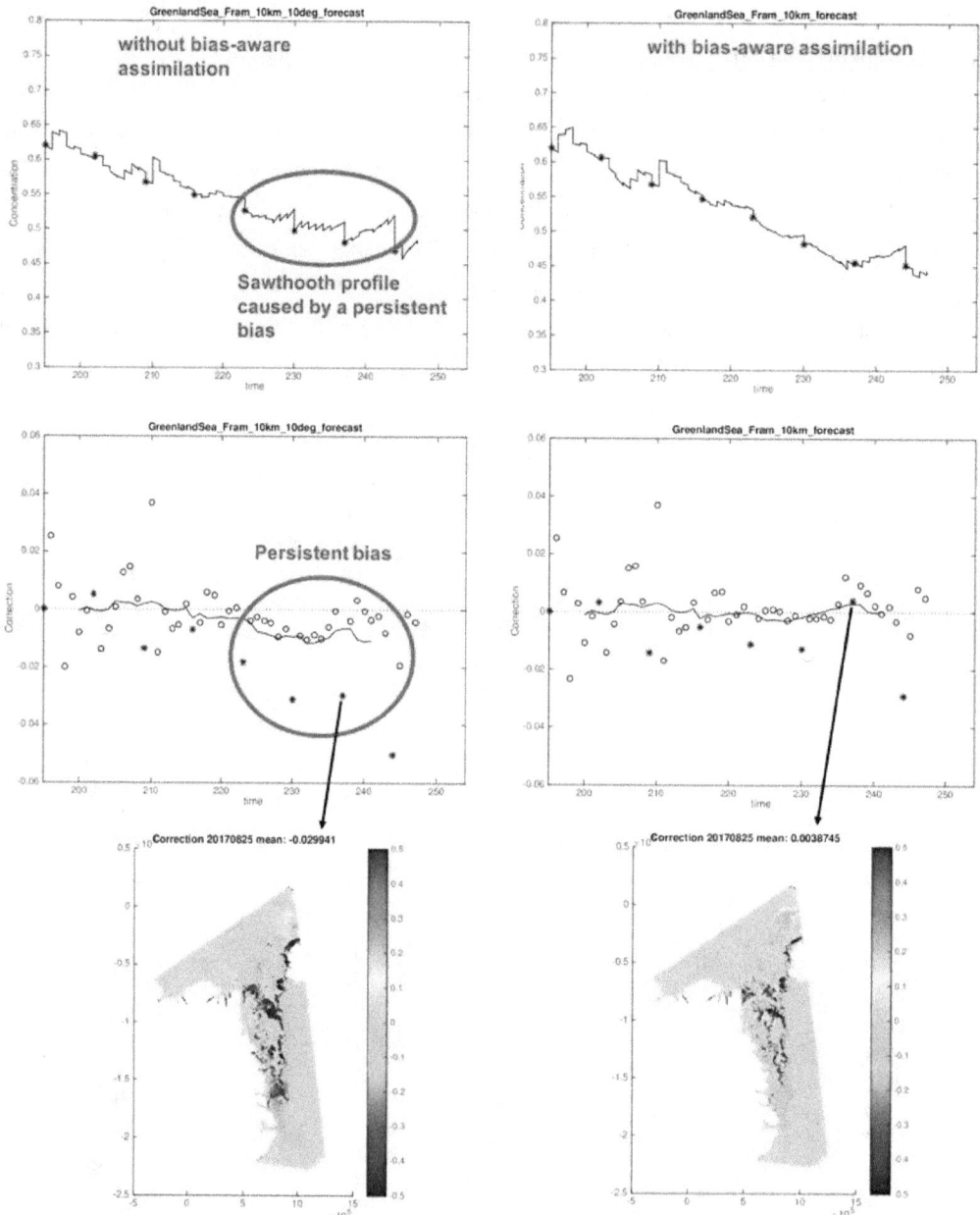

Figure 15.12. Comparison of a reforecast simulation starting on July 14, 2017 and running until September 04, 2017 (days 194 to 247), with the default data assimilation system (left) and the bias-aware assimilation system (right). The first row shows the total concentration (total area of ice divided by the area of the domain). The stars indicate when the weekly U.S. National Ice Center ice charts are assimilated (every Friday). The second row shows the mean update for the ice concentration. The third row shows an example of the concentration update field for August 25, 2017.

The neXtSIM-F is an experimental sea ice forecast platform based on the neXtSIM sea ice model (Bouillon and Rampal, 2015; Rampal et al., 2016a; 2016b; Rampal et al. 2017) and on the assimilation of multiple satellite-derived operational products and ice charts. The set-up presented here covers the Greenland Sea and Fram Strait and provides a seven-day sea ice forecast at a resolution of 10 km on a daily basis (see Fig. 15.11). The initial forecast conditions are computed by assimilating sea ice concentration from the daily and weekly ice charts produced by the U.S. National Ice Center into the model state as given by the previous forecast. The model is forced with the deterministic atmosphere forecast from the European Centre for Medium-Range Weather Forecasts and the ocean forecast is from the Arctic – Monitoring Forecasting Centre.

We implement a sequential and intermittent bias-aware assimilation scheme where the model state is modified towards observations every day. This correction step helps to start the forecast with conditions that are consistent with observations (here, the U.S. National Ice Center daily and weekly ice charts); this also helps to detect and correct biases in the model parameters or in the forcing. The daily U.S. National Ice Center ice charts contain three categories for ice concentration: below 0.1, between 0.1 and 0.8, and between 0.8 and 1 (see http://www.natice.noaa.gov/daily_graphics.htm). The weekly ice charts contain many more categories, typically [no ice, 0-0.1, 0.1-0.4, 0.4-0.6, 0.6-0.8, 0.8-1, 1].

The control parameter for the bias-aware data assimilation is defined here as a correction term applied to the near surface air temperature taken by the European Centre for Medium-Range Weather Forecasts' atmospheric forecast. In the control run, without bias-aware assimilation (Fig. 15.12, left panels), we set this correction constant and equal to -10 °C, which is a value that was fitted manually from preliminary tests. When the bias-aware assimilation is activated (Fig. 15.12, right panels), the correction term is updated every day from the analysis of two additional one-day runs using a correction equals to + or - 10 °C, respectively. In the case presented here, the correction starts from the default value (-10 °C) and varies smoothly between -15 and 0 °C.

Can we forecast local-scale deformation days in advance?

Data assimilation methods are classically ranked from the most rudimentary direct insertion to advanced Ensemble Kalman Filtering (EnKF) and 4-D variational methods. Direct insertion applies the observations directly into the corresponding model variables, but should only be used if all the model prognostic variables are observed; it then honors perfectly the assimilated observations. However, when different observation types may be inconsistent, their uncertainty levels must be taken into account; this can be done by optimal interpolation. The EnKF and 4D-Var are able to project information from observed to unobserved model variables, and are also able to assimilate observations of different types. We argue that advanced data assimilation methods are not necessary for a stand-alone sea ice forecasting model and that better forecast performance can be attained using rudimentary methods.

Sea ice deformation simulated with the neXtSIM model spontaneously localizes along linear-like faults, separating essentially undamaged ice plates/floes. Modelled faults are transient features, accommodating permanent sea ice deformation for awhile before ceasing their activity depending on the refreezing kinetics, the evolution of the wind forcing, and the internal sea ice dynamics, as observed from satellite imagery.

By initializing the model damage variable using past observations of sea ice deformation, we can transfer crucial information to the forecasting system on the history of the deformation that is likely to improve its predictive skill (see Fig. 15.13). The assimilation of information on past shear deformation (and potentially on lead fraction) allows for simulating sea ice drift and its gradient with high accuracy.

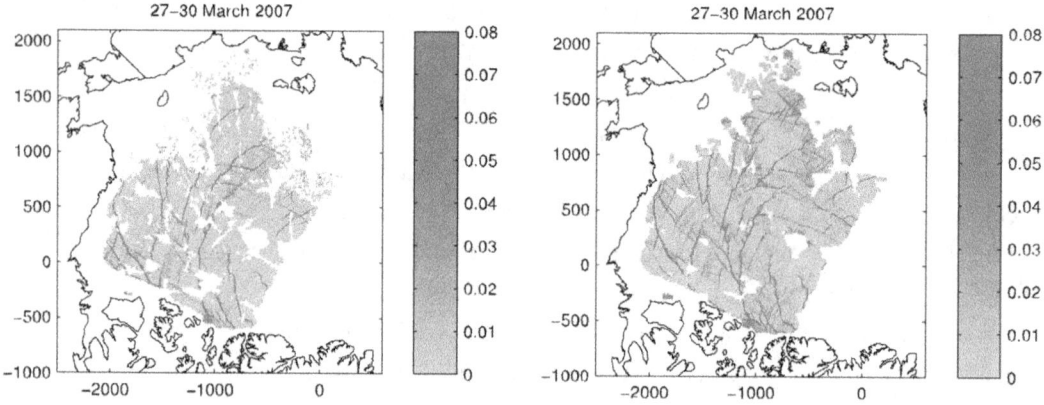

Figure 15.13. Preliminary results illustrating the potential of assimilating information on observed sea ice deformation through the damage variable in the neXtSIM model. The figure shows the observed (left) and simulated (right) shear rate for the same area and for a period centered on March 27–30, 2007. The simulation began on March 26, with a damage field initialized using the deformation rate shown on the left panel and which is computed from synthetic-aperture radar observations.

Conclusion

This chapter about new frontiers in sea ice modelling and forecasting focused on the representation of sea ice dynamics in continuous sea ice models. First, we described the complexity of ice pack dynamics and its underlying causes, defined a framework to discuss the degree of complexity of sea ice models as a function of their ability to resolve fine dynamical structures, and illustrated the potential impact of using lead-permitting continuous sea ice model in coupled systems. Next, in the section on sea ice forecasting, we presented two examples of specific data assimilation issues related to sea ice modelling. This included a solution to tackle the problem of persistent biases in the observations and external forcing and a proposition to use near real-time deformation data to better constraint sea ice forecast initial state and thus be able to forecast accurately sea ice drift and deformation.

Our overall conclusion is that sea ice remains fascinating, even for scientists having worked on the topic for years. There are still many unanswered questions, especially related to the impacts of better resolving sea ice dynamics in coupled ice-ocean-atmosphere systems. We hope that our participation in the GODAE school inspired some of the attendees to join us in investigating these issues.

References

Aagaard, Coachman, Carmack. 1981: On the halocline of the Arctic Ocean, Deep Sea Research Part A. *Oceanographic Research Papers*, 28, 529 – 545, doi:10.1016/0198- 0149(81)90115-1.

Andreas, Murphy. 1986: Bulk transfer coefficients for heat and momentum over leads and polynyas, *J. Phys. Ocean.*, 16, 1875–1883.

Andreas, Cash. 1999: Convective heat transfer over wintertime leads and polynyas, Journal of *Geophysical Research: Oceans*, 104, 25721–25734, doi:10.1029/1999JC900241, http://dx.doi.org/10.1029/1999JC900241.

Andreas, Paulson, William, Lindsay, Businger. 1979: The turbulent heat flux from arctic leads, *Boundary-Layer Meteorology*, 17, 57–91, doi:10.1007/BF00121937.

Bazant. 2000: "Scaling laws for brittle failure of sea ice." Preprints, IUTAM Symp. on Scaling Laws in Ice Mechanics (Univ. of Alaska, Fairbanks, June), J.P. Dempsey, H.H. Shen and L.H. Shapiro, eds. Paper No.3, pp. 1-23.

Bouillon, Rampal. 2015: Presentation of the dynamical core of neXtSIM, a new sea ice model, *Ocean Modelling*, 91(C), 23–37, doi:10.1016/j.ocemod.2015.04.005.

Bouillon, Rampal. 2015a: On producing sea ice deformation data sets from SAR-derived sea ice motion, *The Cryosphere*, 9(2), 663–673, doi:10.5194/tc-9-663-2015.

Dansereau, Weiss, Saramito, Lattes. 2016: A Maxwell elasto-brittle rheology for sea ice modelling, *The Cryosphere*, 10, 1339–1359, doi:10.5194/tc-10-1339-2016, https://www.the-cryosphere.net/10/1339/2016.

Eicken. 2013: Ocean science: Arctic sea ice needs better forecasts. *Nature*, 497(7450), 431–433, doi:10.1038/497431a.

Gultepe, Zhou, Milbrandt, Bott, Li, Heymsfield, Ferrier, Ware, Pavolonis, Kuhn, Gurka, Liu, Cermak. 2015: A review on ice fog measurements and modeling, Atmospheric Research, 151, 2 – 19, doi: 10.1016/j.atmosres.2014.04.014, sixth International Conference on Fog, Fog Collection and Dew.

Hebert, Allard, Metzger, Posey, Preller, Wallcraft, Phelps, Smedstad. 2015: Short-term sea ice forecasting: An assessment of ice concentration and ice drift forecasts using the U.S. Navy's Arctic Cap Nowcast/Forecast System, *J. Geophys. Res.*, 120, 8327–8345, doi:10.1002/2015JC011283.

Hibler. 1979: A dynamic thermodynamic sea ice model. *J. Phys. Oceanogr.*, 9, 815–846, doi:10.1175/1520-0485(1979)009<0815:ADTSIM>2.0.CO;2.

Hopkins, Hibler, Flato. 1991b: On the numerical simulations of the sea ice ridging process. *J. Geophysical. Res.* 96(C3): 4809-4820.

Hutchings, Hibler. 2008: Small-scale sea ice deformation in the Beaufort Sea seasonal ice zone, *J. Geophys. Res. Oceans*, 113, doi:10.1029/2006JC003971.

Hutchings, Roberts, Geiger, Richter-Menge. 2011: Spatial and temporal characterization of sea-ice deformation, *Ann. Glac.*, 2011, 52, 360–368, doi:10.3189/172756411795931769.

Hutter, Losch, Menemenlis: 2017: Scaling properties of Arctic sea ice deformation in high-resolution viscous-plastic sea ice models and satellite observations, European Geosciences Union General Assembly 2017, Vienna, 23 April 2017 - 28 April 2017.

Ivanova, Rampal, Bouillon. 2016: Error assessment of satellite-derived lead fraction in the Arctic, The Cryosphere, 10, 585–595, doi:10.5194/tc-10-585-2016, https://www.the-cryosphere.net/10/585/2016.

Kozo. 1983: Initial model results for Arctic mixed layer circulation under a refreezing lead, Journal of Geophysical Research: Oceans, 88, 2926–2934, doi:10.1029/JC088iC05p02926, http://dx.doi.org/10.1029/JC088iC05p02926.

Kwok, Curlander, McConnell, Pang. 1990: An Ice Motion Tracking System at the Alaska SAR Facility, *IEEE J. of Oceanic Engineering*, 15, 44–54.

Kwok. 2001: Deformation of the Arctic Ocean Sea Ice Cover between November 1996 and April 1997: A Qualitative Survey. In: Dempsey J.P., Shen H.H. (eds) IUTAM Symposium on Scaling Laws in Ice Mechanics and Ice Dynamics. Solid Mechanics and Its Applications, vol 94. Springer, Dordrecht.

Lüpkes, Gryanik, Witha, Gryschka, Raasch, Gollnik. 2008a: Modeling convection over arctic leads with LES and a non-eddy-resolving microscale model, *J. Geophys. Res.*, 113, C09 028, doi:10.1029/2007JC004099.

Lüpkes, Vihma, Birnbaum, Wacker. 2008b: Influence of leads in sea ice on the temperature of the atmospheric boundary layer during polar night, Geophys. Res. Lett., 35, doi:10.1029/2007GL032461.

Marcq, Weiss. 2012: Influence of sea ice lead-width distribution on turbulent heat transfer be- tween the ocean and the atmosphere, *The Cryosphere*, Volume 6, Issue 1, 2012, pp. 143-156, 6, 143–156.

Marsan, Stern, Lindsay, Weiss. 2004: Scale Dependence and Localization of the Deformation of Arctic Sea Ice, *Phys. Rev. Lett.*, 93, 178501, doi:10.1103/PhysRevLett.93.178501.

Maykut. 1986: The surface heat and mass balance, in: The Geophysics of Sea Ice, edited by Untersteiner, N., NATO ASI Series, chap. 5, pp. 395–463, Springer US.

Nguyen, Menemenlis, Kwok. 2009: Improved modeling of the Arctic halocline with a subgrid-scale brine rejection parameterization, *Journal of Geophysical Research: Oceans*, 114, n/a–n/a, doi:10.1029/2008JC005121, http://dx.doi.org/10.1029/2008JC005121, c11014.

Ólason, Rampal, Bouillon. 2017: On the statistical properties of sea ice lead fraction and heat fluxes in Arctic, submitted to *The Cryosphere*.

Rampal, Weiss, Marsan. 2009: Positive trend in the mean speed and deformation rate of Arctic sea ice, 1979–2007, *J. Geophys. Res.*, 114(C5), C05013, doi:10.1029/2008JC005066.

Rampal, Bouillon, Ólason, Morlighem. 2016: neXtSIM: a new Lagrangian sea ice model, The Cryosphere, 10, 1055–1073, doi:10.5194/tc-10-1055-2016, http://www.the-cryosphere.net/ 10/1055/2016/.

Rampal, Bouillon, Ólason, Dansereau, Williams, Samaké. 2017: On the simulation of scaling properties of sea ice deformation, submitted to *The Cryosphere*.

Riska, Coche. 2013: Station keeping in ice - challenges and possibilities, Proceedings of the 22nd International Conference on Port and Ocean Eng under Arctic Conditions (POAC), June 9-13, Espoo, Finland.

Samaké, Bouillon, Rampal, Ólason. 2017: Parallel Implementation of a Lagrangian-based Model on an Adaptive Mesh in C++: Application to Sea-Ice. *Journal of Computational Physics*, dx.doi.org/10.1016/j.jcp.2017.08.055

Schweiger, Zhang. 2015: Accuracy of short-term sea ice drift forecasts using a coupled ice-ocean model, *J. Geophys. Res.*, 120, doi:10.1002/2015JC011273.

Thorndike, Rothrock, Colony. 1975: The thickness distribution of sea ice. J. Geophys. Res.: Oceans, *80*(33), 4501–4513.

Vihma, Pirazzini, Renfrew, Sedlar, Tjernström, Nygård, Fer, Lüpkes, Notz, Weiss, Marsan, Cheng, Birnbaum, Gerland, Chechin, Gascard. 2013: Advances in understanding and parameterization of small-scale physical processes in the marine Arctic climate system: a review, Atmos. Chem. Phys. Discuss., 13, 32 703–32 816.

Weeks. 2010: On sea ice, University of Alaska Press, Fairbanks. ISBN 978-1-60223-079-8. Geol. J., 46: 657. doi:10.1002/gj.1285.

Weiss, Schulson, Stern. 2007: Sea ice rheology from in-situ, satellite and laboratory observations: Fracture and friction. Earth and Planetary Science Letters, *255*(1-2), 1–8. http://doi.org/10.1016/j.epsl.2006.11.033

Weiss. 2013: Drift, deformation and fracture of sea ice - A perspective across scales, Springer, Dordrecht, The Netherlands, doi:10.1007/978-94-007-6202-2.

Winton. 2000: A reformulated three-layer sea ice model, J. Atm. Ocean.Tech., 17, 525–531.

CHAPTER 16

Coupled Atmosphere–Ocean Modelling

Chris Harris

Met Office, Exeter, United Kingdom

The concept of "coupled modelling" is a broad one with many different meanings and understandings within the operational oceanography community and beyond. Here we focus specifically on coupled atmosphere-ocean models and how these are developing for different timescale prediction systems. After a general introduction, we briefly describe the status of coupled modelling on climate timescales (the most mature area), followed by seasonal and decadal timescales. We then consider short- and medium-range coupled timescales which are the least mature, but the area of most relevance to the future of operational oceanography (and numerical weather prediction). The third section describes new frontier applications of these systems on the different timescales. Finally, we provide some concluding remarks on coupled modelling in the fourth section.

Introduction to Coupled Modelling and the Main Challenges

The chapter is predominantly about prediction systems making use of a coupled atmosphere-ocean model. By this we would most commonly be referring to a numerical model of the atmosphere, usually with an associated land-surface model, which is 'two-way coupled' (at least daily, if not more frequently e.g. hourly) to a numerical model of the ocean, often with an associated sea-ice model. The exchange of coupling fields – variables like sea surface temperature and currents from ocean to atmosphere, and heat and momentum fluxes from atmosphere to ocean – is often accomplished by the use of a separate coupling code (like OASIS-MCT, Craig et al., 2017) or framework (like ESMF, DeLuca et al., 2012) providing a flexible way of linking component models and controlling the exchange and interpolation of coupling fields. Other systems use bespoke in-house coupling code or libraries in order to more "tightly" couple the component models. This can allow more control over when and how the models are run. Still other systems, where frequent coupling is not considered to be so essential, may write the required coupling fields as diagnostics to a file which is then used as input by another model; in the most extreme this is what might be referred to as "one-way coupling" although there is an unclear boundary between this and what would usually be considered as having separate models using each other's boundary conditions or "forcing". Many coupled models of the earth system include other physical or biogeochemical components which we mention where relevant (note that ocean-wave coupling is covered in Chapter 14 by Ardhuin and Orfila).

The advantage of genuinely coupled models is that changes in one model (e.g., evolution of sea surface temperatures in the ocean model in response to the atmospheric state) can directly and

---

Harris, C., 2018: Coupled atmosphere-ocean modelling. In "*New Frontiers in Operational Oceanography*", E. Chassignet, A. Pascual, J. Tintoré, and J. Verron, Eds., GODAE OceanView, 445-464, doi:10.17125/gov2018.ch16.

immediately influence the other model (e.g., modified heat and moisture fluxes into the atmospheric boundary layer and beyond). In ocean-only models, the absence of any feedback on the atmospheric forcing variables, e.g. winds, temperature and humidity, can cause many inaccuracies (see e.g. Griffies et al., 2009). This is true particularly if the forcing used is obtained from an atmosphere model which originally "saw" a significantly different ocean surface boundary condition; differences in areas of ice cover are especially problematic in this regard because heat and moisture fluxes over the open ocean are very different from those over sea ice.

A number of recent studies (see e.g. Renault et al., 2016) have focussed on the effect of including ocean currents in the calculation of the atmosphere-ocean momentum exchange. Although this can be included in forced ocean systems it will tend to lead to an over-damping of the eddy field if there is no attempt to parameterise the effect of the current feedback on the atmospheric winds. It should be noted that even in a coupled system there can be related subtleties regarding the exact definition of the observational wind fields which may have been assimilated in the atmospheric component (see discussion in Chapter 10 by Bourassa regarding wind stress).

In practice, on shorter timescales it may be possible to achieve some, or even the majority, of these feedbacks by something short of full "two-way coupling". However, this doesn't necessarily reduce the complexity of the system and in some cases can simply hide problems, like model drifts or incorrect fluxes, which a more fully coupled system is forced to confront.

There is much still to understand regarding air-sea coupling, particularly at short spatial and temporal scales. The impact of different choices regarding the "sequencing" of model components, (e.g., running models concurrently or sequentially, the use of averaged or instantaneous fields if not coupling every time-step) are often ignored but will have impacts for stability, conservation and accuracy. Some of the numerical issues relating to ocean-atmosphere coupling are discussed in Lemarié et al. (2015) which also demonstrates the potential of iterative methods to provide more exact solutions to overcome the problems of asynchronous coupling. There are specific numerical complexities regarding the coupling of sea ice to both atmosphere and ocean – see for example Beljaars et al. (2017) and also West et al. (2016) regarding the impact of different choices for the location of the thermodynamic atmosphere-ice interface in fully coupled models.

From what has been discussed so far, it is already becoming clear that one of the big challenges of coupled modelling is the associated complexity, and the need to link together different component models in a way which is flexible, but also scientifically sensible and computationally affordable. Additional complications arise because these models often use different horizontal grids, are developed by different communities and have different priorities, timescales and governance for their development. Decisions on the exact coupling approach will be driven by a whole variety of factors including the technical and computational infrastructure available, and the remit of a particular institution – this will affect the model outputs of most interest or relevance from each system.

Similar considerations will also affect choices about the horizontal and vertical resolution of different model components – for example, whether an ocean model in a coupled system needs to be eddy resolving, or even whether some kind of mixed layer ocean is sufficient in cases when the primary concern is feedbacks on the atmosphere over short timescales.

Many operational and research centres are now pursuing a "seamless" approach (Hurell et al., 2009; Brown et al., 2012) whereby model configurations are kept as consistent as possible, except in some cases for resolution, across a range of applications and timescales. This attempts to exploit model development efficiencies, reduce technical overheads, and facilitate increased understanding of model errors which are common across timescales. However, there are inevitably compromises due to the different requirements for different applications. For example, energy conservation and avoidance of numerical diapycnal mixing are both much more important on climate timescales.

Initialization is one of the other main challenges of coupled prediction. On longer timescales this is a particular issue for the ocean components, whereas on shorter timescales the consistent initialization of both atmosphere and ocean (and sea ice) becomes increasingly important. This is necessary to avoid spurious initialization "shocks" or adjustments (Mulholland et al., 2015) which degrade the forecast quality over the first few days. These challenges will be discussed further in the relevant sections below (the specific area of initialization using coupled data assimilation is more fully covered in Chapter 17 by Hoteit et al.).

Coupled Modelling for Different Timescales

This chapter does not attempt to provide anything close to a comprehensive history and review of coupled forecasting on these various timescales, but instead tries to highlight some of the aspects of these systems which relate most closely to the ocean, and also to emphasise the new frontiers in these areas of coupled modelling.

Climate (10 to 100+ years)

Coupled atmosphere-ocean models have been used for climate modelling since the 1960s due to the ocean's fundamental role in the global heat budget. At first these models usually required a form of "flux adjustment" in order to adequately simulate present-day climate. This was clearly undesirable and cast serious doubt on the ability of these models to make reliable projections about future climate. Ocean model improvements, including the use of the spatially varying Gent-McWilliams parameterization (Gent and McWilliams, 1990) to adiabatically release potential energy, were thought to be significant in allowing the removal of flux adjustment in the 1990s (Gordon et al., 2000). Climate models have included many new interactively coupled components over the last half century including land-surface and sea-ice models, and an increasing number of other "earth system" components as described below.

As on other timescales there is continued debate about "resolution versus complexity". Can we have confidence in climate projections from modelling systems which have a poor present-day climate mean state, or which fail to adequately simulate aspects of the present-day climate system like storm tracks, the hydrological cycle, clouds, and large-scale ocean circulation? On the other hand, can we have confidence in climate projections from models which fail to include aspects of the earth system, most notably the carbon cycle, which may become increasingly important in future? These questions motivate two of the new frontiers of climate modelling.

Firstly, there is an increasing drive to perform climate model simulations with both atmosphere and ocean resolutions comparable to, or in some cases even exceeding, those used for operational numerical weather prediction and operational ocean modelling systems. Some examples include Community Climate System Model (CCSM, Gent et al., 2011), GFDL Climate Model 2 (Delworth et al., 2006) and MIROC5 simulations using 1/10° ocean configurations; and both HadGEM3 (Hewitt et al., 2011) and EC-Earth (Hazeleger et al., 2010) simulations using a 1/12° NEMO ocean configuration. The HiresMIP initiative (Haarsma et al., 2016) and associated activities like the European Union Horizon2020 PRIMAVERA project (PRIMAVERA, 2018) are helping to drive understanding of these very high resolution models through an inter-comparison of simulations from a number of different modelling centres. These models generally show an improved present-day climate mean state compared to their lower resolution counterparts (Griffies et al., 2015; Hewitt et al., 2016) in regions including the North Atlantic and Southern Ocean. We note, however, that care should be taken when interpreting relatively short climate runs which may have been "tuned" to have an appropriate radiation balance at a particular resolution, or may still be drifting over time. Some of the improvements between eddy-resolving (e.g., 1/12° resolution) and eddy-permitting (e.g., 1/4°) ocean models may be indicative of poor parameterization choices at the lower resolutions; however, it is increasingly argued that ~1° resolution models are not appropriate even for climate studies (Bryan et al., 2010; Kirtman et al., 2012; Roberts et al., 2016). Despite running without data assimilation, high-resolution climate models can provide a useful source of information to help improve initialized coupled model predictions on shorter timescales. This includes analysis of how correlations between the sea surface temperature and the atmospheric boundary layer, and above, compare with those derived from observational products (see e.g. Bryan et al., 2010; Chelton et al., 2010; Roberts et al., 2016; Parfitt et al., 2016; Parfitt et al., 2017; many of these studies are also reviewed in Hewitt et al., 2017). Although both atmosphere and ocean components of these models are far from perfect (for example, see Chapter 2 by Fox-Kemper on the lack of a resolved sub-mesoscale) in many regards their performance does appear to be converging towards a representation of processes which is in better agreement with observations. Long, free-running high-resolution climate models can therefore inform the development of coupled models on shorter timescales, and provide confidence that these models are able to simulate features at scales the data assimilation will be trying to correct in an operational system. To some extent biases in such models can also be linked with biases on shorter timescales. However, this has to be done carefully and with a good deal of understanding of the underlying processes (see e.g. Hermanson et al., 2017).

The second significant "new frontier" of climate models is the move towards earth system models of increasing complexity, and with a whole variety of different physical and biogeochemical model components. "Earth system" is defined in various ways, but one meaning is that such models include a full representation of the carbon cycle. This allows any response of the carbon cycle to either increased carbon dioxide emissions, or a warmer climate, to be represented. The individual terms in the global carbon cycle are so large that even small changes could have a significant impact on the amount of emitted carbon dioxide which remains in the atmosphere, and therefore on the amount of future warning. In the UK, the "UKESM1" model being prepared for CMIP6 (Eyring et al., 2016) will include full atmospheric chemistry and aerosols, terrestrial carbon cycle and vegetation, marine biogeochemistry, and dynamic ice sheets, as well as "traditional" physical model components. To achieve this in a global model which can be used for century long integrations inevitably requires compromises on resolution; despite the known deficiencies a 1° ocean model will be used for most of the simulations. Both the atmospheric chemistry and ocean biogeochemistry in this model will run on a coarsened grid, and in the coming years there will most likely be further effort to ensure that the resolution in such models is only placed where it is genuinely required. This may include adoption of unstructured mesh or adaptive mesh refinement techniques (e.g., using the AGRIF package described in Debreu et al., 2008), or more sophisticated hybrid vertical coordinates (see Chapter 12 by LeSommer et al.) rather than the more traditional model grids and z-level coordinates still used for the ocean component in the majority of climate models.

Climate projections are not considered to be an "initial value problem" but that is not to say that initialization is unimportant. Projections of future climate will usually be started from a close-to-equilibrium "spun-up" initial state representative of pre-industrial (or present-day) climate forcing; this ensures that any future climate change signal is not too contaminated by model drifts. The challenge to reach such a state, which is not too far from the observed climate, remains a significant one. It is closely related to the common practice of "tuning" climate models – for example, modifying the aerosol climatology, or adjusting parameters in the cloud scheme or ocean vertical mixing scheme – to achieve a realistic net radiative balance and ocean temperature structure.

The initialization challenge becomes greater at both of the new frontiers described above. In both cases the models are very expensive to run, and additional components – particularly those poorly observed historically, or with inherently longer timescales like ice shelves – may take longer to reach a spun-up equilibrium state. The ocean carbon cycle in Earth system models will often require a spin-up of several thousand years. This would be too long to perform in a fully coupled framework and so strategies have to be devised to, for example, use a combination of coupled runs, and forced runs using atmospheric variables from those coupled simulations. This allows a suitable equilibrium state to be obtained although there is no guarantee it is representative of an actual observed realisation, particularly in a non-stationary climate.

Seasonal (3-12 months) and decadal (1-20 years)

Statistical methods have been used for seasonal forecasting for many decades, and continue to be used in combination with the dynamical coupled atmosphere-ocean models which are now also utilised. On these timescales, the atmospheric initial conditions become less important. Instead the low frequency forcing from the ocean (e.g., sea surface temperature patterns related to the El Nino Southern Oscillation, or ENSO - see Chapter 23 by McPhaden), and aspects of the land surface and sea ice, begin to provide the dominant contributions to predictability. Although the majority of this predictability originates in the tropics, the "teleconnections" to impacts elsewhere have been well known for many years. As well as direct circulation changes, Rossby wave trains can propagate from the tropics to mid and high latitudes influencing e.g. weather patterns over North America and the North Atlantic Oscillation.

Dynamical seasonal forecast models require good ocean sub-surface observational coverage for initialization – this was significantly improved with the development of the TAO/TRITON (Tropical Atmosphere Ocean/Triangle Trans-Ocean Buoy Network) moored buoy array in the 1980s. Models also need to have a good representation of ENSO variability (still problematic in some regards but improved with higher horizontal resolution in both atmosphere and ocean components; Shaffrey et al., 2010), and a good representation of the mechanisms by which the tropical predictability "leaks" to higher latitudes, as well as a correct model response to these influences. As a result of uncertainty in both initial conditions and model parameterisations, seasonal (and decadal) forecasting systems make use of ensembles of model simulations to provide probabilistic forecast information. These ensembles are generated using perturbed initial conditions and sometimes also stochastic physics.

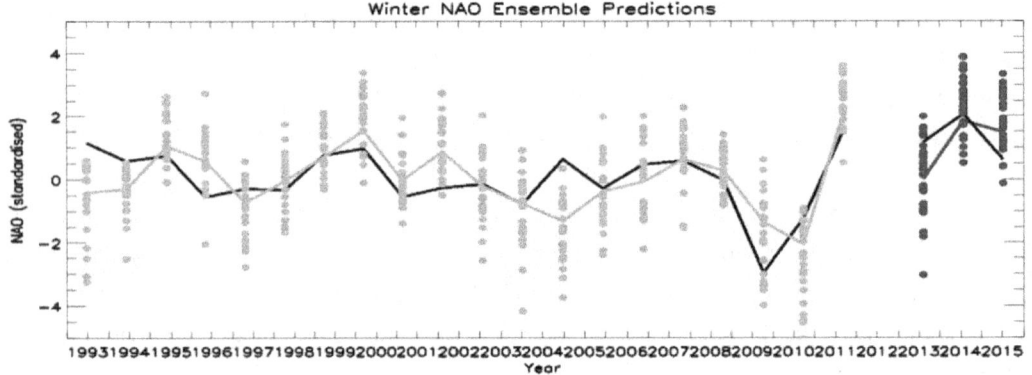

Figure 16.1. Predictability of the winter North Atlantic Oscillation, measured as the sea level pressure difference between Iceland and the Azores. The NAO in observations (black line), ensemble mean GloSea5 forecasts (orange line), and individual ensemble members (orange dots) in winter (December to February) hindcasts are shown, normalized by their respective standard deviations. Anomalies are for December to February, and forecasts were initialized from dates centred on 1 November. The correlation score of 0.62 is significant at the 99% level according to a t test and allowing for the small lagged autocorrelation in forecasts and observations. Updated from Scaife et al. (2014) with real-time forecasts (blue line and dots).

One of the notable advances in seasonal forecasting over the last few years has been the improvement in the ability to forecast the North Atlantic Oscillation (NAO). The seasonal forecast system at the UK Met Office (GloSea5) demonstrated an anomaly correlation of around 0.6 (see Fig. 16.1) for predictions of winter NAO from forecasts initialized in November (Scaife et al., 2014); similar results have subsequently been obtained in other systems (e.g., Weisheimer et al., 2017). Sources of this predictability include ENSO, upper ocean heat content anomalies in the North Atlantic which can re-emerge the following winter, Arctic sea-ice cover and the Quasi-Biennial Oscillation (QBO). Improvements in both initialization and model formulation – including horizontal and vertical resolution, as well as the inclusion of the stratosphere in the atmosphere model – are responsible for this improved correlation. However, the NAO correlation is found to increase more slowly with ensemble size than expected, indicating that the models appear to be less predictable than the real world. This has been termed the "signal-to-noise paradox" (Eade et al., 2014) and is still not fully understood. However, it is hoped that improvements in the areas of non-orographic gravity waves and stochastic physics may play a part in allowing the apparent theoretical NAO correlation, of round 0.8, to be realised without the requirement to have a very large ensemble.

It is important to recognise the role of the ocean data assimilation and operational oceanography communities in providing the initial conditions for such dynamical seasonal forecast systems. These are required both in real time to initialise the seasonal forecasts themselves, and also as reanalyses (see Chapter 19 by Haines) which are needed to initialise the seasonal hindcasts used to calibrate and "bias-correct" information from the real-time system. One of the challenges of seasonal forecasting is to ensure as much consistency as possible between these two different initializations despite a changing observational network. Improvements in mean climate in the coupled atmosphere-ocean models used within seasonal forecasting systems are important to reduce model biases, and therefore reduce the sensitivity to the bias correction and calibration procedure (which can be particularly problematic for variables like sea-ice cover). However, model biases are not likely to be completely eliminated in the foreseeable future, and so a potentially important development is the use of machine learning approaches for the calibration and bias correction of coupled ocean-atmosphere forecasts.

The computational cost of these ensemble seasonal systems, with their accompanying hindcasts, has prevented many operational centres upgrading the ocean resolution from 1° to 1/4° until the last few years. Some of the improvements in the latest UK Met Office GloSea5 system (MacLachlan et al., 2015) have been attributed to the increased ocean resolution – for example, improved Atlantic winter "blocking" due to reduced North Atlantic sea surface temperature (SST) biases at 1/4° (Scaife et al., 2011).

Decadal forecasting is a less mature field than seasonal forecasting, and in this case a key source of predictability is anthropogenic forcing from greenhouse gases and aerosols. Solar variability and volcanic forcing (once an eruption has occurred) provide additional sources of predictability, as do fluctuations of the Atlantic Meridional Overturning Circulation which are potentially predictable a few years ahead (e.g., Griffies and Bryan 1997). In common with seasonal forecasting, the QBO (predictable a couple of years ahead; Marshall and Scaife 2009) and Arctic sea ice provide further potential sources of predictability. Cassou et al. (2018) provides a comprehensive but succinct recent review on decadal variability and predictability.

Decadal forecast systems require coupled models (usually at a similar resolution to seasonal forecast systems) which can represent the ocean circulation well, including aspects such as Nordic sea overflows which are important to maintain the correct ocean water masses. Again, coupled hindcasts are required in order to allow the skill of the system to be assessed and in some cases to bias-correct the real-time forecasts. Initialization of these hindcasts is even more of a challenge than for seasonal forecasting because the timescales require a larger range of start dates to be used (e.g., back to the 1950s). This means the problem of a non-stationary ocean observing system is particularly acute, and also requires the use of a significant pre-Argo period when the deeper ocean, of greater importance to decadal predictability, was very poorly sampled. Using hindcasts for calibration of the real-time forecasts can also be more problematic because the difference between the coupled model and real-world climatology may vary over time due to climate change. This makes it harder to use hindcasts for bias correction of coupled model drift without masking what is already a smaller predictable signal than on seasonal timescales. As a result, a number of decadal prediction systems have instead used anomaly initialization (Pierce et al., 2004) where observed ocean state anomalies are added to the coupled model climatology. The relationship between model drift and model biases with alterative initialization strategies for decadal forecast systems are discussed and analysed in Sanchez-Gomez et al. (2016); this is an important area of research with the potential to reduce model biases in future. Approaches, like Empirical Orthogonal Functions, to make better use of sparse observations to initialize the historical ocean state are still developing (see Chapter 19 by Haines).

Following the progress in NAO prediction in seasonal forecasting systems, studies using UK Met Office's Decadal Prediction System showed similar correlations for the first winter (for forecasts initialized in November) and a correlation of greater than 0.4 for the second winter (Dunstone et al., 2016).

Short- to medium-range (1-2 weeks)

The one-week to two-week timescale is that on which atmosphere-ocean coupled modelling has come to the fore in the operational oceanography and numerical weather prediction communities over the last decade. In fact, for regional modelling, coupled systems can scarcely be considered as "new frontier" given models like the Hurricane Weather Research and Forecast model (Bender et al., 2007) or its predecessors have included an interactive ocean component for close to two decades. More recently, examples of operational regional coupled systems include the Coupled Ocean Atmosphere Mesoscale Prediction System (COAMPS; Chen et al., 2010; Holt et al., 2011) developed at the US Naval Research Laboratory (NRL), and the system (which also includes sea ice) run for the Gulf of St Lawrence and Great Lakes by the Canadian Centre for Meteorological and Environmental Prediction (Pellerin et al., 2004). There are an increasing number of pre-operational and research coupled systems, often using the Coupled Ocean Atmosphere Wave Sediment Transport (COAWST; Warner et al., 2010) framework and its component models: the WRF atmosphere model, the ROMS ocean model, the SWAN wave model and the Community Sediment Transport Model. NRL's COAMPS has also been used in various different domains. In the UK there is active research on a UK Environmental Prediction system which uses a 1.5 km configuration of the Met Office Unified Model coupled to configurations of both the NEMO ocean model and the WaveWatchIII wave model (also at 1.5 km) with the intention of adding additional earth system components at a later date (Lewis et al., 2018).

There are fewer global coupled configurations running at present but Table 16.1 summarises some of those which are either already operational or are planned within the next couple of years. We note here that the coupled system run at the Canadian Centre for Meteorological and Environmental Prediction (CCMEP) is currently the only operational system being used for both ocean forecasting and numerical weather prediction (NWP). The UK Met Office is at present the only provider of ocean forecasts from a coupled system using (weakly) coupled data assimilation and by 2020 this system is also expected to be used for NWP. The ECMWF operational NWP system will be (partially) coupled from 2018, and at NRL there are ambitious plans for coupled configurations with very high (1/25°) ocean resolution.

Centre	Atmosphere Model (resolution)	Ocean Model (resolution)	Wave Model (resolution)	Coupler	Data Assimilation	Year	Notes
Canadian Centre for Meteorological and Environmental Prediction (CCMEP)	GEM (25 km)	NEMO (1/4 °)	-	GOSSIP (in-house coupler)	Uncoupled	2017	Provides both NWP and ocean forecasts (since 1 Nov 2017)
	GEM (15 km)	*NEMO (1/4 °)*	-	*GOSSIP*	*Uncoupled*	*2018*	
European Centre for Medium Range Weather Forecasting (ECMWF)	*IFS (9 km)*	*NEMO (1/4 °)*	*WAM (1/8 °)*	*Single executable*	*Uncoupled*	*2018*	*Partial coupling (full coupling in tropics)*
UK Met Office	UM (40 km)	NEMO (1/4 °)	-	OASIS3	Weakly coupled	2017	Provides ocean forecasts to Copernicus Marine Service (since 11 Jul 2017)
	UM (10 km)	*NEMO (1/4 °)*	*WWIII [if used]*	*OASIS3-MCT*	*Weakly coupled*	*2020*	*Planned for NWP use; wave model may be included (not confirmed)*
US Naval Research Laboratory (NRL)	*NAVGEM (19 km)*	*HYCOM (1/25 °)*	*WAM (1/8 °)*	*ESMF / NUOPC*	*Weakly coupled*	*2019*	*This is the "Initial Operational Capability"; final capability (2022) will have a higher resolution atmosphere and strongly coupled DA*

Table 16.1. Summary of current and next planned (*in italics*) operational global coupled atmosphere-ocean systems at various national centres.

Almost all of these operational systems, or pre-operational research systems, have been able to demonstrate improvements on tropical weather and ocean forecasts. The most notable impact is on tropical cyclones where systems with an "NWP-resolution" atmosphere model have shown reductions in track errors as well as a moderation of the tendency to over-deepen storms now seen in many atmosphere-only NWP systems. This is associated with the development of a "cold wake" of sea surface temperatures which does not exist in a persisted SST analysis (and in fact may take several days to be properly represented in such an analysis due to cloud limitations in satellite retrievals; Mogensen et al., 2017). There is further work required to improve tropical storm responses in such global coupled systems because most atmosphere models – including the Unified Model configuration used in the UK Met Office NWP system, and IFS used at ECMWF – underestimate the wind strength for a given central pressure in these storms. Addressing this, while not causing excessive ocean mixing and upwelling which then weakens the storms too much, will most likely require improvements to the bulk formulae used to calculate the turbulent fluxes, and the way waves are modelled and coupled.

In a version of their deterministic operational model which was coupled to an ocean (in the tropics only), ECMWF have seen a ~10% reduction in 5-day forecast surface pressure and 500 hPa root-mean-square (rms) errors in the tropics. Smaller, but still statistically significant, improvements in winds and relative humidity from the surface up to 100 hPa were also seen (Balsamo et al., 2017). The Canadian GIOPS system has also shown similar improvements, apparently driven by improved latent heat fluxes and more realistic pumping of heat and moisture in the atmosphere (Smith et al., 2018).

Aside from resolution, one of the major differences between the operational coupled global systems shown in Table 16.1 is their approach to initialization. As on the longer timescales already discussed this is a significant challenge, but the focus is now on the avoidance of the initialization shocks mentioned earlier. Most centres are now pursuing research into coupled data assimilation. In its "weakly coupled" form, this means observations in one component can influence other components only through the background state or in some cases the outer loop (see Chapter 17 by Hoteit et al.). More "strongly coupled" data assimilation makes use of "cross-interface" covariances which allow observations in the atmosphere to directly influence the ocean and vice versa. As already mentioned, the UK Met Office coupled system is the only operational real-time system making use of (weakly) coupled data assimilation. In part this is possible because the atmosphere resolution is relatively low and it has therefore been possible to find a way of dealing with the latency of ocean observations (involving "catch-up" sub-cycles producing updated analyses) which is not prohibitively expensive. The weakly coupled data assimilation system used at the Met Office has been shown to produce ocean analyses (and short lead-time forecasts) with smaller sea surface temperature rms errors than those in the equivalent ocean-only FOAM system (see Fig. 16.2). In addition, the mean SST increments are found to be smaller than in an equivalent ocean-only analysis indicating that the ocean state is more consistent with the atmospheric forcing (Lea et al., 2015).

ECMWF and CCMEP are both running higher resolution atmosphere configurations, and have for the moment taken different approaches to reducing initialization shock without the immediate need for coupled data assimilation – instead the component models are initialized separately using their own analysis systems. The CCMEP system is careful to ensure as much consistency as possible between the atmospheric and oceanic boundary layers by using the same bulk formulae in both atmosphere and ocean, and assimilating the same SST product into the ocean analysis as is used for the atmospheric surface boundary. ECMWF instead use "full coupling" only in the tropics, and at higher latitudes apply a "partial coupling" whereby the atmosphere model sees an SST analysis combined with tendencies from the ocean model during the first week of the forecast. As well as reducing the likelihood of initialization shocks, this means the atmosphere model is not immediately exposed to sea surface temperature biases associated with an incorrect Gulf Stream position – a feature of most current ocean models, particularly at 1/4° resolution or lower.

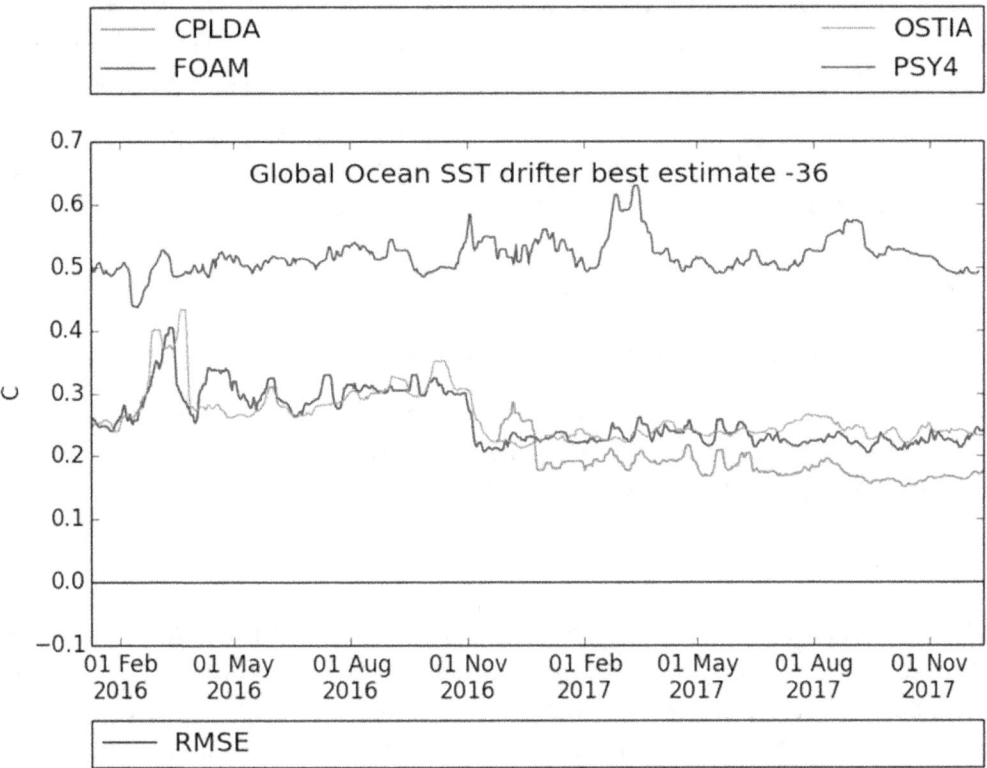

Figure 16. 2. Root-mean-square errors (RMSE) in degrees Celsius of "best estimate" sea surface temperature analyses verified using the class4 methodology (see Chapter 29 by Hernandez et al.) against in-situ observations from drifting buoys. The UK Met Office weakly coupled data assimilation system is compared (red) is compared against the operational ocean-only FOAM system (blue) and the OSTIA sea surface temperature analysis product (grey). A 10-day rolling median of globally averaged values is shown.

For ocean forecasting, there is again often a "resolution versus complexity" debate whereby a high-resolution (and eddy-resolving) ocean configuration is compared to a coupled atmosphere-ocean system with a lower (eddy-permitting) ocean resolution. In fact, the results seen in high-resolution coupled climate simulations suggest this is an artificial debate: for the benefits of coupling to be more fully seen then an eddy-resolving ocean (and comparable atmosphere resolution) is required. An obvious next step in the development of short-range coupled systems is therefore an increase of ocean resolution to match the leading eddy-resolving global systems and so allow the mesoscale and frontal coupling to the atmosphere to be better represented. This is planned at centres including the US Naval Research Laboratory and the UK Met Office within the next few years and will be accompanied by further research to allow a move beyond independent (or at best "weakly coupled") analyses for initialization. Research is also required into how to make best use of (lower resolution) coupled ensembles in order to provide flow-dependent information for improving the ocean data assimilation in a hybrid variational system.

Arguably the shortest of all short-range "forecasts" are reanalyses which only perform the data assimilation step in order to provide estimates of historical atmosphere or ocean states, but using the latest models and data assimilation codes (see Chapter 19 by Haines). Coupled data assimilation has progressed more quickly in this context, in part because there are not the real-time complications associated with late arriving ocean observations (or orbit corrections to altimetry – see Chapter 7 by Le Traon). Although ocean resolutions utilised have been relatively low – for example 1° deg for both ECDA at GFDL (Chang et al., 2013) and for CERA-20C at ECMWF (Laloyaux et al., 2016; Laloyaux et al., 2018) – coupled reanalyses have been found to improve the surface fluxes and reduce the ocean increments compared to their ocean-only counterparts (see Chapter 19 by Haines).

Applications of Coupled Modelling for Different Timescales

As with all forms of ocean and atmosphere modelling, there is only value to users, and to society in general, when the systems developed can be applied to real-world decision making. Here we briefly outline applications of coupled systems on the various timescales.

Climate

For the last three decades, the Intergovernmental Panel on Climate Change (IPCC) have reported regularly not only on the most likely changes in climate over the coming century, but also on the impacts of these future climate changes. Multi-model ensembles are used to provide some kind of estimate of the uncertainty of these projections, which are also based on assumptions derived from various possible emission pathways and potential mitigation measures. As part of CMIP6 (Eyring et al., 2016) a number of "Shared Socio-economic Pathways" have been developed covering plausible but different socio-economic, political and technological futures (O'Neill et al., 2017).

Particularly when using early climate models (with atmospheric resolution in the hundreds of km) it was difficult to make confident statements about regional climate changes and impacts (especially those related to extreme events) even if the projected large-scale climate change was assumed to be reliable. Other impacts like sea level rise are not directly predicted by the Boussinesq (i.e. volume-conserving) ocean models most frequently used and have to be inferred indirectly. However, initiatives like the "four degree map" (Met Office 2018a) produced by the UK Met Office at the time of the IPCC Fourth Assessment Report (AR4) have attempted to summarize likely impacts of a four degree rise in global mean temperatures, albeit based on a relatively low resolution climate model (HadCM3; Gordon et al., 2000).

More recently the effort devoted to assessing the socio-economic impact of potential changes in regional climate has increased significantly. In general, this still requires down-scaling techniques to obtain information on more local scales. Sophisticated statistical techniques are used, in combination with ensembles of global models, to provide probabilistic information – for example, on seasonal mean and extreme changes in temperature and precipitation, in a variety of different emissions scenarios. Recent initiatives have included UK Climate Projections 09 (Met Office 2018b) which included a marine and coastal report. These projections are being updated in 2018 to include down-scaling using a cutting-edge 2.2 km atmospheric configuration, and similar work is now being started as part of the European Union Climate Projections. Resolution increases in global climate models (as described in the previous section) bring these closer to being able to simulate realistic local "weather" in a future climate. However, the requirement to assess uncertainties in regional projections of climate change and impacts means that down-scaling is likely to have an important role for the foreseeable future. The increased complexity and additional components in the Earth system models described in in the previous section will allow more impacts to be predicted directly. They will also provide more confidence in changes caused by anthropogenic greenhouse gas emissions and, for example, melting of land ice and ice shelves.

Many of these impacts and applications are encompassed by what have become known as "climate services". This is a term with a broad meaning but is defined by the European Commission Climate Service (European Commission 2018) as "transforming climate-related data and other information into customised products such as projections, trends, economic analysis, advice on best practices, development and evaluation of solutions, and any other climate-related service liable to benefit that may be of use for the society." These services include data, information and knowledge that support adaptation, mitigation and disaster risk management on both climate and seasonal timescales.

Seasonal and decadal

The maturity of seasonal forecasting in the tropics, using both dynamical and statistical models, is such that predictions of ENSO, and impacts with teleconnections to ENSO, have been made for some time. Predictions have been made publicly available alongside other seasonal forecast information by the World Meteorological Centre Global Producing Centres (WMO 2018), and more recently as part of initiatives like the newly established Copernicus Climate Service in Europe (C3S 2018). Predictions of North Atlantic tropical cyclone activity (both number of storms, and accumulated cyclone energy over the June to November hurricane season) have also been shown to have statistically significant skill in recent years (e.g., Camp et al., 2015). This is despite the relatively low atmospheric resolution (~50 km) which is not nearly sufficient to represent realistic tropical storms. The Met Office decadal prediction system has also been able to demonstrate skill in predicting Atlantic tropical cyclone activity with lead times of several years. Probabilistic forecasts of this nature, on both one year and multi-year timescales, are of significant value, particularly for the insurance and re-insurance industries.

Recent successes in improving NAO predictability from seasonal forecast systems have opened the door for a wide of range new applications and services. At present these services largely make use of previously known, or newly identified, correlations between specific impacts and the observed winter NAO; they then use the skill in NAO predictions to show skilful predictions of these impacts. There are various reasons why the impacts themselves are often not directly predicted from the seasonal forecast model output including lack of sufficient resolution or complexity to represent the impact in the modelling system (particularly for extremes or local effects). In addition, there may be complications caused by the need to bias-correct seasonal forecast output, and difficulties due to the low signal-to-noise ratio described in x.2.2. Examples of applications where skilful winter forecasts have been shown include UK river flow (Svensson et al., 2015) and transport disruption (Palin et al., 2015), European energy demand (Clark et al., 2017) and Baltic Sea ice (Karpechko et al., 2015). In other cases, direct correlations between seasonal forecasts output and observed impacts have been identified and used to show skillful prediction of, for example, Yangtze rainfall and river flow (Li et al., 2016) and energy demand and potential renewable energy supply, from solar and wind power, in parts of China (Bett et al., 2017).

Short- to Medium-Range

The majority of the applications on short timescales are similar to those using traditional forecasting systems, albeit with some improvements in skill which may over time open up potential for new products.

Clearly more accurate forecasting of tropical storms, and particularly their tracks, is potentially of great significance. Information from deterministic models will always be of limited value and so it will be as ensembles of coupled atmosphere-ocean models start to be more routinely used that more of the benefit is realised. Some of the most beneficial impacts of coupled models for warnings and other impacts will come from the high-resolution regional coupled systems being developed. These will allow more accurate information about coupled extremes on a local basis, as well as benefitting from any improvements in lateral boundary conditions due to coupling in global, or basin-scale, models. For example, coastal flooding around the UK can be strongly affected by sea state, storm surges, and river inputs, and is therefore determined by a complex interaction of meteorology, oceanography and land-surface and hydrological processes. As more earth system components are added to these regional coupled systems then, in combination with very high resolution downscaling or nesting very close to the coast, the potential for providing improved predictions will increase. COAWST-based systems have already been used for a range of operational and research applications in different regional domains including coastal zone management, oil-spill dispersion modelling, coastal morphology changes during storms, sediment transport, egg and larvae dispersal, and hypoxic events (see e.g. Carniel et al., 2013). Other very specific applications of coupled modelling include The Balearic RIssaga Forecasting System (BRIFS) which uses a high-resolution ROMS configuration coupled to a WRF atmosphere with the aim of quantitatively predicting the occurrence of extreme sea level oscillations associated with meteo-tsunamnis in the Menorcan harbour of Ciutadella (Licer et al., 2017).

Coupled analysis (or reanalysis) and forecast systems also have the advantage a providing fully consistent surface meteorology, sea state and surface ocean products. For some users having a fully consistent coupled climatology to use in downstream statistical models may be an additional benefit. Any improvements in the ocean sub-surface which can be obtained in coupled (re)analyses are also likely to improve predictability on seasonal timescales when used for initialising seasonal hindcasts and forecasts.

Concluding Comments

This chapter has provided a selective summary of coupled atmosphere-ocean modelling on various timescales, with a particular focus on short-range global predictions. Both recent successes and some of the remaining challenges have been presented.

A common question is whether we should expect that all ocean-only systems, even for short-range forecasts, will ultimately be replaced by coupled systems. It seems very unlikely that this will happen in the foreseeable future as it will always be possible to use higher resolution in an ocean-only system than an equivalent system coupled to an atmosphere model. At least until it is possible to routinely run global coupled atmosphere-ocean models which resolve the ocean sub-mesoscale there would appear to be a clear role for ocean-only systems. However, we would expect the use of coupled models to continue to increase. Depending on application and timescale this will involve both coupled global models (e.g. climate models) being used to drive high-resolution, uncoupled down-scalers, as well as uncoupled global models being used to provide boundary conditions for high-resolution, regional coupled environmental prediction systems.

Individual operational centres will therefore make different decisions on the priority of coupling in different systems, depending on their remit and user requirements. Most users will not be concerned about whether a particular forecast output was produced from a coupled or uncoupled system; they may be more concerned about a "coupled" delivery of data from multiple model components. However, this will inevitably highlight any inconsistencies between model output which may be problematic for some users. Provision of ocean, atmosphere (as well as wave and sea-ice) data from a single coupled system is the easiest way of providing a fully consistent set of analysis and forecast data, aside from the demonstrated benefits in product quality which can be realised in a coupled system.

Acknowledgments

This chapter has been prepared by drawing on input from members of the GODAE OceanView Coupled Prediction Task Team (with particular thanks to current and previous co-chairs, Hal Ritchie and Gary Brassington, respectively). I would also like to thank all those involved in the "seamless" model development process at the UK Met Office, as well as John Siddorn and colleagues in the Ocean Forecasting Research and Development group. The chapter has also benefitted from a review by the following students at the GOV International School in Mallorca: Marcel Ricker, Sisi Qin, Safo Pineiro, Beatriz Perez Diaz, Md Jakiul Islam, Rodrigue Anicet Imbol Koungue, and Svetlana

Karimova. I am also grateful to the organisers of the GOV School for the opportunity to deliver my lectures on coupled modelling in such an enjoyable and stimulating environment.

References

Balsamo, G., K. Mogensen, S. Keeley, J.-R. Bidlot, S. Boussetta, E. Dutra, and N. Wedi, 2017: Coupling of oceans and land surfaces in the ECMWF Integrated Forecasting System: Sensitivity and impact of diurnal and synoptic variability on medium-range skill. *WGNE Blue Book 2017*, E. Astakhova, Ed., WMO, http://bluebook.meteoinfo.ru/uploads/2017/docs/09_Balsamo_Gianpaolo_CouplingOceansLandECMWF.pdf.

Beljaars, A., E. Dutra, G. Balsamo, and F. Lemarié, 2017: On the numerical stability of surface–atmosphere coupling in weather and climate models, *Geoscientific Model Development*, **10**, 977-989, doi:10.5194/gmd-10-977-2017.

Bender, M.A., I. Ginis, R. Tuleya, B. Thomas, and T. Marchok, 2007: The operational GFDL coupled hurricane-ocean prediction system and summary of its performance. *Monthly Weather Review*, **135**, 3965-3989, doi: 10.1175/2007MWR2032.1.

Bett, P., H. Thornton, J. Lockwood, A. Scaife, N. Golding, C. Hewitt, R. Zhu, P. Zhang, and C. Li, 2017: Skill and reliability of seasonal forecasts for the Chinese energy sector. *Journal of Applied Meteorology and Climatology*, doi:10.1175/jamc-d-17-0070.1.

Brown, A., S. Milton, M. Cullen, B. Golding, J. Mitchell, and A. Shelly, 2012: Unified Modeling and Prediction of Weather and Climate: A 25-Year Journey. *Bulletin of the American Meteorological Society*, **93**, 1865-1877, doi:10.1175/bams-d-12-00018.1.

Bryan, F., R. Tomas, J. Dennis, D. Chelton, N. Loeb, and J. McClean, 2010: Frontal Scale Air–Sea Interaction in High-Resolution Coupled Climate Models. *Journal of Climate*, **23**, 6277-6291, doi:10.1175/2010jcli3665.1.

C3S, 2018: Copernicus Climate Change Service. Accessed 5 January 2018, https://climate.copernicus.eu.

Camp, J., M. Roberts, C. MacLachlan, E. Wallace, L. Hermanson, A. Brookshaw, A. Arribas, and A. Scaife, 2015: Seasonal forecasting of tropical storms using the Met Office GloSea5 seasonal forecast system. *Quarterly Journal of the Royal Meteorological Society*, **141**, 2206-2219, doi:10.1002/qj.2516.

Carniel, S., and A. Russo, 2013: A review of modeling applications using ROMS model and COAWST system in the Adriatic Sea region. https://arxiv.org/abs/1309.7600.

Cassou, C., Y. Kushnir, E. Hawkins, A. Pirani, F. Kucharski, I. Kang, and N. Caltabiano, 2018: Decadal Climate Variability and Predictability: Challenges and Opportunities. *Bulletin of the American Meteorological Society*, **99**(3), 479-490, doi:10.1175/BAMS-D-16-0286.1.

Chang, Y., S. Zhang, A. Rosati, T. Delworth, and W. Stern, 2012: An assessment of oceanic variability for 1960–2010 from the GFDL ensemble coupled data assimilation. *Climate Dynamics*, **40**, 775-803, doi:10.1007/s00382-012-1412-2.

Chassignet, E., and X. Xu, 2017: Impact of Horizontal Resolution (1/12° to 1/50°) on Gulf Stream Separation, Penetration, and Variability. *Journal of Physical Oceanography*, **47**, 1999-2021, doi:10.1175/jpo-d-17-0031.1.

Chelton, D.B., and S.-P. Xie, 2010: Coupled ocean-atmosphere interaction at oceanic mesoscales. *Oceanography*, **23**(4), 52–69, doi:10.5670/oceanog.2010.05.

Chen, S., T. Campbell, H. Jin, S. Gaberšek, R. Hodur, and P. Martin, 2010: Effect of Two-Way Air–Sea Coupling in High and Low Wind Speed Regimes. *Monthly Weather Review*, **138**, 3579-3602, doi:10.1175/2009mwr3119.1.

Clark, R., P. Bett, H. Thornton, and A. Scaife, 2017: Skilful seasonal predictions for the European energy industry. *Environmental Research Letters*, **12**, 024002, doi:10.1088/1748-9326/aa57ab.

Craig A., S. Valcke, and L. Coquart, 2017: Development and performance of a new version of the OASIS coupler, OASIS3-MCT_3.0, *Geoscientific Model Development*, **10**, 3297-3308, doi:10.5194/gmd-10-3297-2017.

Debreu, L., C. Vouland, and E. Blayo, 2008: Agrif: Adaptive grid refinement in Fortran. *Computers and Geosciences*, **34**, 8–13.

DeLuca C., G. Theurich, and V. Balaji, 2012: The Earth System Modeling Framework. *Earth System Modelling - Volume 3*, SpringerBriefs in Earth System Sciences, Springer, Berlin, Heidelberg, doi: 10.1007/978-3-642-23360-9_6

Delworth, T. L., et al., 2006: GFDL's CM2 global coupled climate models. Part I: Formulation and simulation. *Journal of Climate*, **19**, 643–674, doi:10.1175/JCLI3629.1.

Dunstone, N., D. Smith, A. Scaife, L. Hermanson, R. Eade, N. Robinson, M. Andrews, and J. Knight, 2016: Skilful predictions of the winter North Atlantic Oscillation one year ahead. *Nature Geoscience*, **9**, 809-814, doi:10.1038/ngeo2824.

Eade, R., D. Smith, A. Scaife, E. Wallace, N. Dunstone, L. Hermanson, and N. Robinson, 2014: Do seasonal-to-decadal climate predictions underestimate the predictability of the real world? *Geophysical Research Letters*, **41**, 5620-5628, doi:10.1002/2014gl061146.

European Commission, 2018: Climate Services. Accessed 5 January 2018, https://ec.europa.eu/research/environment/index.cfm?pg=climate_services.

Eyring, V., S. Bony, G. A. Meehl, C. A. Senior, B. Stevens, R. J. Stouffer, and K. E. Taylor, 2016: Overview of the Coupled Model Intercomparison Project Phase 6 (CMIP6) experimental design and organization, *Geoscientific Model Development*, **9**, 1937-1958, doi:10.5194/gmd-9-1937-2016.

Gent, P., and J. McWilliams, 1990: Isopycnal Mixing in Ocean Circulation Models. *Journal of Physical Oceanography*, **20**, 150-155, doi:10.1175/1520-0485(1990)020<0150:imiocm>2.0.co;2.

Gent, P. R., G. Danabasoglu, L. J. Donner, M. M. Holland, E. C. Hunke, S. R. Jayne, D. M. Lawrence, R. B. Neale, P. J. Rasch, M. Vertenstein, P. H. Worley, Z.-L. Yang, and Z. Minghua, 2011: The Community Climate System Model Version 4, *Journal of Climate*, **24** (19), 4973-4991, doi:10.1175/2011JCLI4083.1.

Gent, P. R., S. G. Yeager, R. B. Neale, S. Levis, and D. A. Bailey, 2009: Improvements in a half degree atmosphere/land version of the CCSM. *Climate Dynamics*, **34**, 819–833,

Gordon, C., C. Cooper, C. A. Senior, H. Banks, J. M. Gregory, T. C. Johns, J. F. B. Mitchell, and R. A. Wood, 2000: The simulation of SST, sea ice extents and ocean heat transports in a version of the Hadley Centre coupled model without flux adjustments. *Climate Dynamics*, **16** (2–3), 147–168, doi:10.1007/s003820050010.

Griffies, S., and K. Bryan, 1997: A predictability study of simulated North Atlantic multidecadal variability. *Climate Dynamics*, **13**, 459-487, doi:10.1007/s003820050177.

Griffies, S. et al., 2009: Coordinated Ocean-ice Reference Experiments (COREs). *Ocean Modelling*, **26**, 1-46, doi:10.1016/j.ocemod.2008.08.007.

Griffies, S., M. Winton, W. G. Anderson, R. Benson, T. L. Delworth, C. O. Dufour, J. P. Dunne, P. Goddard, A. K. Morrison, A. Rosati, A. T. Wittenberg, J. Yin, and R. Zhang, 2015: Impacts on Ocean Heat from Transient Mesoscale Eddies in a Hierarchy of Climate Models. *Journal of Climate*, **28**, 952-977, doi:10.1175/jcli-d-14-00353.1.

Haarsma, R. et al., 2016: High Resolution Model Intercomparison Project (HighResMIP v1.0) for CMIP6. *Geoscientific Model Development*, **9**, 4185-4208, doi:10.5194/gmd-9-4185-2016.

Hazeleger W. et al., 2010: EC-Earth: a seamless earth system prediction approach in action. *Bulletin American Meteorological Society*, **91**, 1357–1363, doi:10.1175/2010BAMS2877.1.

Hermanson, L., H.-L. Ren, M. Vellinga, N. D. Dunstone, P. Hyder, S. Ineson, A. A. Scaife, D. M. Smith, V. Thompson, B. Tian, and K. D. Williams, 2017: Different types of drifts in two seasonal forecast systems and their dependence on ENSO. *Climate Dynamics*, doi:10.1007/s00382-017-3962-9.

Hewitt H., D. Copsey, I. D. Culverwell, C. M. Harris, R. S. R. Hill, A. B. Keen, A. J. McLaren, and E. C. Hunke, 2011: Design and implementation of the infrastructure of HadGEM3: the next-generation Met Office climate modelling system, *Geoscientific Model Development*, **4**, 223-253, doi:10.5194/gmd-4-223-2011.

Hewitt, H. et al., 2016: The impact of resolving the Rossby radius at mid-latitudes in the ocean: results from a high-resolution version of the Met Office GC2 coupled model. *Geoscientific Model Development*, **9**, 3655-3670, doi:10.5194/gmd-9-3655-2016.

Hewitt, H., M. J. Bell, E. P. Chassignet, A. Czaja, D. Ferreira, S. M. Griffies, P. Hyder, J. McClean, A. L. New, and M. J. Roberts, 2017: Will high-resolution global ocean models benefit coupled predictions on short-range to climate timescales? *Ocean Modelling*, doi:10.1016/j.ocemod.2017.11.002.

Holt, T., J. Cummings, C. Bishop, J. Doyle, X. Hong, S. Chen, and Y. Jin, 2011: Development and testing of a coupled ocean–atmosphere mesoscale ensemble prediction system. *Ocean Dynamics*, **61**, 1937-1954, doi:10.1007/s10236-011-0449-9.

Hurrell, J., G. Meehl, D. Bader, T. Delworth, B. Kirtman, and B. Wielicki, 2009: A Unified Modeling Approach to Climate System Prediction. *Bulletin of the American Meteorological Society*, **90**, 1819-1832, doi:10.1175/2009bams2752.1.

Kirtman, B. P., C. Bitz, F. Bryan, W. Collins, J. Dennis, N. Hearn, J. L. KinterIII, R. Loft, C. Rousset, L. Siqueira, C. Stan, R. Tomas and M. Vertenstein, 2012: Impact of ocean model resolution on CCSM climate simulations. *Climate Dynamics*, **39**, 1303–1328, doi:10.1007/s00382-012-1500-3.

Laloyaux, P., M. Balmaseda, D. Dee, K. Mogensen, and P. Janssen, 2016: A coupled data assimilation system for climate reanalysis. *Quarterly Journal of the Royal Meteorological Society*, **142**: 65–78. doi:10.1002/qj.2629

Laloyaux,, P., E. de Boisseson, M. Balmaseda, J.-R. Bidlot, S. Broennimann, R. Buizza, P. Dahlgren, D. Dee, L. Haimberger, H. Hersbach, Y. Kosaka, M. Martin, P. Poli, N. Rayner, E. Rustemeier and D. Schepers, 2018: CERA-20C: A coupled reanalysis of the Twentieth Century. *Journal of Advances in Modeling Earth Systems*, Accepted, doi:10.1029/2018MS001273.

Lea, D., I. Mirouze, M. Martin, R. King, A. Hines, D. Walters, and M. Thurlow, 2015: Assessing a New Coupled Data Assimilation System Based on the Met Office Coupled Atmosphere–Land–Ocean–Sea Ice Model. *Monthly Weather Review*, **143**, 4678-4694, doi:10.1175/mwr-d-15-0174.1.

Lemarié, F., E. Blayo, and L. Debreu, 2015: Analysis of ocean-atmosphere coupling algorithms: consistency and stability, *Procedia Computer Science*, **51**, 2066–2075, doi:10.1016/j.procs.2015.05.473.

Lewis, H. et al., 2018: The UKC2 regional coupled environmental prediction system. *Geoscientific Model Development*, **11**, 1-42, doi:10.5194/gmd-11-1-2018.

Ličer, M., B. Mourre, C. Troupin, A. Krietemeyer, A. Jansá, and J. Tintoré, 2017: Numerical study of Balearic meteotsunami generation and propagation under synthetic gravity wave forcing. *Ocean Modelling*, **111**, 38-45, doi:10.1016/j.ocemod.2017.02.001.

MacLachlan, C., A. Arribas, K. A. Peterson, A. Maidens, D. Fereday, A. A. Scaife, M. Gordon, M. Vellinga, A. Williams, R. E. Comer, J. Camp, P. Xavier, and G. Madec, 2014: Global Seasonal forecast system version 5 (GloSea5): a high-resolution seasonal forecast system. *Quarterly Journal of the Royal Meteorological Society*, **141**, 1072-1084, doi:10.1002/qj.2396.

Magnusson, L., M. Alonso-Balmaseda, and F. Molteni, 2012: On the dependence of ENSO simulation on the coupled model mean state. *Climate Dynamics*, **41**, 1509-1525, doi:10.1007/s00382-012-1574-y.

Marshall, A., and A. Scaife, 2009: Impact of the QBO on surface winter climate. *Journal of Geophysical Research*, **114**, doi:10.1029/2009jd011737.

Met Office, 2018a: The impact of four degree temperature rise. Accessed 5 January 2018, https://www.metoffice.gov.uk/climate-guide/climate-change/impacts/four-degree-rise

Met Office, 2018b: UK Climate Projections. Accessed 5 January 2018, http://ukclimateprojections.metoffice.gov.uk.

Minobe, S., A. Kuwano-Yoshida, N. Komori, S. Xie, and R. Small, 2008: Influence of the Gulf Stream on the troposphere. *Nature*, **452**, 206-209, doi:10.1038/nature06690.

Mogensen, K., L. Magnusson, and J. Bidlot, 2017: Tropical cyclone sensitivity to ocean coupling in the ECMWF coupled model. *Journal of Geophysical Research: Oceans*, **122**, 4392-4412, doi:10.1002/2017jc012753.

Mulholland, D. P., P. Laloyaux, K. Haines and M. A. Balmaseda, 2015: Origin and Impact of Initialization Shocks in Coupled Atmosphere-Ocean Forecasts. *Monthly Weather Review*, **143**, 4631-4644, doi: 10.1175/MWR-D-15-0076.1.

O'Neill, B. C., E. Kriegler, K. L. Ebi, E. Kemp-Benedict, K. Riahi, D. S. Rothman, B. J.van Ruijven, D. P.van Vuuren, J. Birkmann, K. Kok, M. Levy, and W. Solecki. 2017: The roads ahead: Narratives for shared socioeconomic pathways describing world futures in the 21st century. *Global Environmental Change*, **42**, 169-180, doi: 10.1016/j.gloenvcha.2015.01.004.

Palin, E., A. Scaife, E. Wallace, E. Pope, A. Arribas, and A. Brookshaw, 2016: Skillful Seasonal Forecasts of Winter Disruption to the U.K. Transport System. *Journal of Applied Meteorology and Climatology*, **55**, 325-344, doi:10.1175/jamc-d-15-0102.1.

Parfitt, R., A. Czaja, S. Minobe, and A. Kuwano-Yoshida, 2016: The atmospheric frontal response to SST perturbations in the Gulf Stream region. *Geophysical Research Letters*, **43**, 2299-2306, doi:10.1002/2016gl067723.

Parfitt, R., A. Czaja, and Y. Kwon, 2017: The impact of SST resolution change in the ERA-Interim reanalysis on wintertime Gulf Stream frontal air-sea interaction. *Geophysical Research Letters*, **44**, 3246-3254, doi:10.1002/2017gl073028.

Pellerin, P., H. Ritchie, F. J. Saucier, F. Roy, S. Desjardins, M. Valin, and V. Lee, 2004: Impact of a two-way coupling between an atmospheric and an ocean-ice model over the Gulf of St. Lawrence. *Monthly Weather Review*, **132**(6), 1379-1398, doi:10.1175/1520-881 0493(2004)132<1379:IOATCB>2.0.CO;2.

Pierce D. W., T. P. Barnett, R. Tokmakian, A. Semtner, M. Maltrud, J. Lysne, and A. Craig, 2004: The ACPI project, element 1: initializing a coupled climate model from observed initial conditions. *Climate Change*, **62**, 13–28, doi:10.1023/B:CLIM.0000013676.42672.23.

PRIMAVERA, 2018. Accessed 5 January 2018, https://www.primavera-h2020.eu/.

Renault, L., J. Molemaker, J. McWilliams, A. Shchepetkin, F. Lemarié, D. Chelton, S. Illig, and A. Hall, 2016: Modulation of wind-work by oceanic current interaction with the atmosphere. *Journal of Physical Oceanography*, **46**, 1685-1704, doi: 10.1175/JPO-D-15-0232.1.

Roberts, M., H. Hewitt, P. Hyder, D. Ferreira, S. Josey, M. Mizielinski, and A. Shelly, 2016: Impact of ocean resolution on coupled air-sea fluxes and large-scale climate. *Geophysical Research Letters*, **43**, 10,430-10,438, doi:10.1002/2016gl070559.

Sanchez-Gomez, E., C. Cassou, Y. Ruprich-Robert, E. Fernandez, and L. Terray, 2016: Drift dynamics in a coupled model initialized for decadal forecasts. *Climate Dynamics*, **46**, 1819-1840, doi:10.1007/s00382-015-2678-y.

Scaife, A., D. Copsey, C. Gordon, C. Harris, T. Hinton, S. Keeley, A. O'Neill, M. Roberts, and K. Williams, 2011: Improved Atlantic winter blocking in a climate model. *Geophysical Research Letters*, **38**, doi:10.1029/2011gl049573.

Scaife, A., A. Arribas, E. Blockley, A. Brookshaw, R. T. Clark, N. Dunstone, R. Eade, D. Fereday, C. K. Folland, M. Gordon, L. Hermanson, J. R. Knight, D. J. Lea, C. MacLachlan, M. Martin, A. K. Peterson, D. Smith, M. Vellinga, E. Wallace, J. Waters and A. Williams, 2014: Skillful long-range prediction of European and North American winters. *Geophysical Research Letters*, **41**, 2514-2519, doi:10.1002/2014gl059637.

Shaffrey, L. et al., 2009: U.K. HiGEM: The New U.K. High-Resolution Global Environment Model—Model Description and Basic Evaluation. *Journal of Climate*, **22**, 1861-1896, doi:10.1175/2008jcli2508.1.

Smith, D., R. Eade, and H. Pohlmann, 2013: A comparison of full-field and anomaly initialization for seasonal to decadal climate prediction. *Climate Dynamics*, **41**, 3325-3338, doi:10.1007/s00382-013-1683-2.

Smith, D., R. Eade, N. Dunstone, D. Fereday, J. Murphy, H. Pohlmann, and A. Scaife, 2010: Skilful multi-year predictions of Atlantic hurricane frequency. *Nature Geoscience*, **3**, 846-849, doi:10.1038/ngeo1004.

Smith, G. C., J.-M. Bélanger, F. Roy, P. Pellerin, H. Ritchie, K. Onu, M. Roch, A. Zadra, D. Surcel Colan, B. Winter, J.-S. Fontecilla and D. Deacu, 2018: Impact of Coupling with an Ice-Ocean Model on Global Medium-Range NWP Forecast Skill, *Monthly Weather Review*, **146**, 1157-1180, doi:10.1175/MWR-D-17-0157.1.

Svensson, C. et al., 2015: Long-range forecasts of UK winter hydrology. *Environmental Research Letters*, **10**, 064006, doi:10.1088/1748-9326/10/6/064006.

Warner, J., B. Armstrong, R. He, and J. Zambon, 2010: Development of a Coupled Ocean–Atmosphere–Wave–Sediment Transport (COAWST) Modeling System. *Ocean Modelling*, **35**, 230-244, doi:10.1016/j.ocemod.2010.07.010.

Weisheimer, A., N. Schaller, C. O'Reilly, D. A. MacLeod, and T. Palmer, 2017: Atmospheric seasonal forecasts of the twentieth century: multi-decadal variability in predictive skill of the winter North Atlantic Oscillation (NAO) and their potential value for extreme event attribution. *Quarterly Journal of the Royal Meteorological Society*, **143**, 917–926. doi:10.1002/qj.2976

West A., A. J. McLaren, H. T. Hewitt, and M. J. Best, 2016: The location of the thermodynamic atmosphere–ice interface in fully coupled models – a case study using JULES and CICE. *Geoscientific Model Development*, **9**, 1125-1141, doi:10.5194/gmd-9-1125-2016.

WMO, 2018: Global Producing Centres. Accessed 5 January 2018, http://www.wmo.int/pages/prog/wcp/wcasp/gpc/gpc.php.

Yu Karpechko, A., K. Andrew Peterson, A. Scaife, J. Vainio, and H. Gregow, 2015: Skilful seasonal predictions of Baltic Sea ice cover. *Environmental Research Letters*, **10**, 044007, doi:10.1088/1748-9326/10/4/044007.

CHAPTER 17

Data Assimilation in Oceanography: Current Status and New Directions

Ibrahim Hoteit[1], Xiaodong Luo[2], Marc Bocquet[3], Armin Köhl[4], and Boujemaa Ait-El-Fquih[1]

[1]*King Abdullah University of Science and Technology (KAUST), Thuwal, Saudi Arabia;* [2]*International Research Institute of Stavanger (IRIS), Bergen, Norway;* [3]*CEREA, Joint Laboratory of École des Ponts ParisTech and EDF R&D, Université Paris-Est, Champs-sur-Marne, France;* [4]*University of Hamburg, Hamburg, Germany*

Characterizing and forecasting the state of the ocean is essential for various scientific, management, commercial, and recreational applications. This is, however, a challenging problem due to the large, multiscale and nonlinear nature of the ocean state dynamics and the limited amount of observations. Combining all available information from numerical models describing the ocean dynamics, observations, and prior information has proven to be the most viable approach to determine the best estimates of the ocean state, a process called data assimilation (DA). DA is becoming widespread in many ocean applications; stimulated by continuous advancement in modeling, observational, and computational capabilities. This chapter offers a comprehensive presentation of the theory and methods of ocean DA, outlining its current status and recent developments, and discussing new directions and open questions. Casting DA as a Bayesian state estimation problem, the chapter will gradually advance from the basic principles of DA to its most advanced methods. Three-dimensional DA methods, 3DVAR and Optimal Interpolation, are first derived, before incorporating time and present the most popular, Gaussian-based DA approaches: 4DVAR, Kalman filters and smoothers methods, which exploit past and/or future observations. Ensemble Kalman methods are next introduced in their stochastic and deterministic formulations as a stepping-stone toward the more advanced nonlinear/non-Gaussian DA methods, Particle and Gaussian Mixture filters. Other sophisticated hybrid extensions aimed at exploiting the advantages of both ensemble and variational methods are also presented. The chapter then concludes with a discussion on the importance of properly addressing the uncertainties in the models and the data, and available approaches to achieve this through parameters estimation, model errors quantification, and coupled DA.

Introduction

Following the tremendous progress in weather monitoring and forecasting, there is increased interest in developing global and regional systems for operational oceanography in order to provide estimates and forecasts of essential ocean variables. Outputs of such systems can be used to generate data products, applications, and services through national authorities and organizations, such as metocean service providers and environmental agencies. These can include nowcasts providing the most usefully accurate description of the present state of the ocean, forecasts of the future ocean conditions as far ahead as possible (typically one to two weeks), and reanalyses (hindcasts) assembling long-term datasets to describe the history of the studied region including time series showing trends and changes. Such products can provide crucial information for a wide

Hoteit, I., et al., 2018: Data assimilation in oceanography: Current status and new directions. In "*New Frontiers in Operational Oceanography*", E. Chassignet, A. Pascual, J. Tintoré, and J. Verron, Eds., GODAE OceanView, 465-512, doi:10.17125/gov2018.ch17.

variety of marine industrial and governmental activities and societal needs, including safety of life at sea, coastal extremes, pollution and contamination management, tourism, marine conservation, fisheries and aquaculture, exploration and drilling, desalination and plant cooling operations, shipping, harbor management and national security operations, etc. Operational oceanographic programs have been recently established in several countries of the European Union, as well as the United States of America and Japan.

Operational oceanography depends on the availability of ocean observations transmitted in (near-) real-time and time-dependent numerical models to project the information gathered by the observations into the future (or the past). The models can be constructed based on the observations (e.g., Hamilton et al., 2016; Dreano et al., 2015; Lguensat et al., 2017) by exploiting their statistical properties, but their outputs could be restricted to the measured quantities, which are limited in space and time. Such models have the advantage of being computationally very efficient, but their predictive capabilities are often limited to short temporal ranges and may not be efficient for predicting extremes. More established ocean models are developed based on the physical laws that govern the oceans general circulation (Navier-Stocks equations). Despite significant computational requirements, the dynamics of these models bring more information to the otherwise underdetermined ocean state estimation problem, which enhances the accuracy and dynamical consistency of the ocean state estimates.

Whether based on statistical properties or physical laws, ocean models are not perfect and can be subject to various sources of uncertainties. Observation-based models are, for instance, constructed based on statistical assumptions which may not be always relevant. Dynamical models, on the other hand, require atmospheric forcing that is not always available at the required spatial and temporal resolutions, and are themselves distorted by uncertainties. Numerical errors, missing physics, and poorly known parameterizations coefficients (e.g., diffusivity and viscosity) are also important sources of uncertainties in such models. These errors may accumulate over time and often deviate the model from the real ocean trajectory, even when the ocean state is perfectly known at the initial time. It is now recognized that the most efficient approach to obtain reliable ocean state estimates is to routinely constrain and adjust the model outputs with incoming ocean data through a data assimilation (DA) process (Ghil and Malanotte-Rizzoli, 1991; Wunsch, 1996; Bennett, 2005). In general, DA exploits the models as spatio-temporal interpolators of the data, and the data guide the models toward the true trajectory of the system. Effective operational oceanography relies on the ability to assimilate massive amounts of data gathered by monitoring systems in real time into advanced general circulation models (GCMs) on supercomputer facilities.

Although the theoretical framework of DA methods is well established on the Bayesian estimation theory (Law et al., 2015), the applications of these with state-of-the-art ocean general circulation models (OGCMs) is still strongly hampered by the large dimensional and nonlinear characteristics of these systems. As will be further discussed later, poor knowledge of the statistical properties of the models and data uncertainties also limits the efficiency of ocean DA systems. This chapter will provide a general overview of ocean data assimilation methods, presenting the state-of-the-art methods, their origins and relations, and discussing in particular major limitations and future directions. It makes no attempt to be a comprehensive review of the extensive DA literature, which extends back to the late 1950s. This chapter will first describe the three-dimensional (3D) DA problem that only considers the current observations in the estimation of the ocean state before moving to more advanced four-dimensional (4D) DA methods, focusing on the most commonly used 4D variational (4DVAR) methods and ensemble Kalman filters. More sophisticated methods combining variational and ensemble methods and more advanced ones designed for non-Gaussian distributions and their potential use for enhancing ocean data assimilation systems will be also discussed. A summary of all the DA approaches discussed in this chapter and their main founding hypothesis is provided in Table 17.1.

DA Method	Founding Hypothesis	Derivation
3D variational assimilation (3DVar)	Gaussian model state and observation noise	Minimize a cost function that involves the model state and observation at a given time instance
4D variational assimilation (4DVar)	Gaussian model state, model errors and observation noise	Minimize a cost function that involves the model state and observation at a given time interval. Relations between model state variables are constrained by the dynamical model
Kalman filter (KF)	Linear dynamical model and observation operator; Gaussian model state, model errors and observation noise	Take the maximum a posteriori (MAP) estimate of the ocean state conditioned on previous observations
Extended Kalman filter (EKF); including reduced EKFs	Nonlinear dynamical model and/or observation operator; Gaussian model state, model errors and observation noise	Linearize nonlinear dynamical model and/or observation operator, and then take the MAP solution as in KF
Ensemble Kalman filters (EnKF), and ensemble optimal interpolation (EnOI)	As in EKF	Use an ensemble of model states to estimate the background statistics (prior mean and covariance), and corresponding KF update formulas to produce an analysis ensemble with targeted posterior mean and covariance
Particle filter (PF)	Nonlinear dynamical model and/or observation operator; non-Gaussian model state, model errors and observation noise	Use mixture of Dirac delta functions to approximate the prior and posterior distributions of the model state conditioned on previous observations
Gaussian mixture filter (GMF); including ensemble GMFs	As in PF	Use mixture of Gaussian distributions to approximate the prior and posterior distributions of the model state conditioned on previous observations, and also to approximate the distributions of the model errors and observation noise when necessary

Table 17.1. Founding hypotheses and derivations of data assimilation (DA) methods. Founding hypotheses describe assumption(s), e.g. linearity and/or Gaussianity, behind DA methods from a Bayesian perspective; whereas derivations summarize features utilized and/or actions taken to derive the DA methods. The table focuses on the filtering schemes, but the descriptions are also valid for the corresponding smoothing schemes.

Three-Dimensional Data Assimilation

The three-dimensional data assimilation (3DDA) problem refers to the space domain in which we look for the best estimate $\mathbf{x}^a$ (^{a} for analysis) of the ocean state $\mathbf{x}$ at some time given only the observation $\mathbf{y}$ of the state at that time. $\mathbf{y}$ may contain in situ measurements from cruises, profiles, gliders, and buoys, and satellite measurements of sea surface height and sea surface temperature. $\mathbf{x}$ is typically comprised of the prognostic model variables (needed to initialize the model) at every grid point of the domain, such as temperature, salinity, sea level, and velocities. We assume the observational model H, possibly nonlinear, relating the ocean state to the observation is available.

$$\mathbf{y} = H(\mathbf{x}). \tag{1}$$

Estimating $\mathbf{x}$ from $\mathbf{y}$ can be formulated as a weighted least-squares inverse problem in which we look for $\mathbf{x}$ that minimizes an objective function measuring the distance between the ocean state and the observations, of the form

$$J_{3D}(\mathbf{x}) = (\mathbf{y} - H(\mathbf{x}))^T \mathbf{W}^y (\mathbf{y} - H(\mathbf{x})) = \|\mathbf{y} - H(\mathbf{x})\|^2_{\mathbf{W}^y}, \tag{2}$$

where $\mathbf{W}^y$ is the data weight (definite positive) matrix introduced to specify the observations weights in the optimization (to assign, for example, less weights to uncertain measurements). In ocean applications, the number of observations p (i.e., dimension of $\mathbf{y}$) is typically much smaller than the number of state variables to be inferred n (i.e., dimension of $\mathbf{x}$). This makes the above problem underdetermined, and more information is needed to regularize it (Wunsch, 1996). This is commonly enforced by solving for the ocean state estimate that is not too far from a given prior state estimate $\mathbf{x}^b$ (^{b} for background), often taken as the most recent forecast (nowadays computed by the ocean model starting from the most recent state estimate). The objective function then becomes

$$J_{3D}(\mathbf{x}) = \|\mathbf{y} - H(\mathbf{x})\|^2_{\mathbf{W}^y} + \|\mathbf{x} - \mathbf{x}^b\|^2_{\mathbf{W}^b}, \tag{3}$$

where $\mathbf{W}^b$ is the background weight matrix.

This deterministic formulation of the ocean state estimation problem does not provide a framework for choosing the weight matrices or to quantify the uncertainties in the estimate. A more general approach to formulate the 3DDA problem is to encapsulate it within a Bayesian framework, which considers the ocean state $\mathbf{x}$ and the observation $\mathbf{y}$ as random variables, see for example Simon (2006) and Wikle and Berliner (2007). This naturally allows us to account for the uncertainties in the observation, which is often expressed as

$$\mathbf{y} = H(\mathbf{x}) + \varepsilon, \tag{4}$$

where ε represents the observational errors, generally assumed unbiased, and to exploit a prior knowledge of the state and its uncertainty through their probability distributions. The solution of the estimation problem is then determined as the conditional probability distribution of the state given the observation $p_{\mathbf{x}|\mathbf{y}}$, which is computed via the Bayes' rule,

$$p_{\mathbf{x}|\mathbf{y}} = \frac{p_{\mathbf{y}|\mathbf{x}}\, p_{\mathbf{x}}}{p_{\mathbf{y}}}. \tag{5}$$

$p_{\mathbf{x}|\mathbf{y}}$ is also called *posterior* and represents an update of the *prior* distribution $p_{\mathbf{x}}$, while $p_{\mathbf{y}|\mathbf{x}}$ is the likelihood function of $\mathbf{y}$ given $\mathbf{x}$, and $p_{\mathbf{y}}$ is the marginal distribution of $\mathbf{y}$ representing a normalizing constant to ensure that the final solution is a probability distribution. An ocean state estimate $\mathbf{x}^a$ can then be obtained as the maximum a posteriori (MAP) estimate maximizing $p_{\mathbf{x}|\mathbf{y}}$, or the minimum-variance (MV) estimate (or posterior mean), which are equivalent when the posterior is Gaussian.

Assuming the *prior* and observation errors follow normal (Gaussian) distributions, then the posterior is given by (Talagrand, 2010)

$$p_{\mathbf{x}|\mathbf{y}} \propto \exp\left(-\frac{1}{2}\left[\|\mathbf{y}-H(\mathbf{x})\|^2_{\mathbf{R}^{-1}} + \|\mathbf{x}-\mathbf{x}^b\|^2_{\mathbf{B}^{-1}}\right]\right), \tag{6}$$

where $\mathbf{R}$ and $\mathbf{B}$ are respectively the observation and background error covariance matrices. Maximizing $p_{\mathbf{x}|\mathbf{y}}$ is equivalent to minimizing J_{3D}, with the weight matrices as the inverse of the covariance matrices, i.e. $\mathbf{W}^y = \mathbf{R}^{-1}$ and $\mathbf{W}^b = \mathbf{B}^{-1}$.

When the observational operator is linear, and thus will be denoted by $\mathbf{H}$, the solution of the problem can be directly computed by setting the derivative of the convex objective function J_{3D} to zero, to obtain

$$\mathbf{x}^a = \mathbf{x}^b + \mathbf{K}(\mathbf{y} - \mathbf{H}\mathbf{x}^b), \tag{7}$$

and its error covariance matrix

$$\mathbf{P}^a = (\mathbf{I} - \mathbf{K}\mathbf{H})\mathbf{B}, \tag{8}$$

where $\mathbf{K}$ is the Gain matrix given by

$$\mathbf{K} = \mathbf{B}\mathbf{H}^T[\mathbf{H}\mathbf{B}\mathbf{H}^T + \mathbf{R}]^{-1}. \tag{9}$$

Note that in a non-Gaussian setting with a linear observational operator, the above solution remains the best linear unbiased estimator (BLUE; Talagrand, 2010), where "best" stands for minimum-variance.

Optimal Interpolation (OI; Bouttier and Courtier, 1999) is a popular algebraic simplification of the Gain matrix in the BLUE designed by viewing Eq. 7 as a list of scalar analysis equations, one per state variable of $\mathbf{x}$. Only observations located within a certain distance from the variable being analyzed are then used to compute the increment of that variable. This makes the OI scheme easy to parallelize and implement for efficient DA.

When the observational operator is nonlinear, the objective function is not convex and may exhibit several minima; but near the minimum one can linearize it (around the background) before computing the BLUE, exactly as above. One may also consider an iterative solution to the problem Eq. 7 by computing the linearization of the observational operator around the estimate of the last iteration (Simon, 2006).

A more straightforward approach is to apply an optimization algorithm to directly minimize the objective function J_{3D}, the most popular of which are the gradient-based optimization methods because of their fast convergence rate (Bouttier and Courtier, 1999). These methods use the gradient of the objective function to determine descent directions toward the minimum in an iterative procedure (Bouttier and Courtier, 1999). The gradient of J_{3D} with respect to $\mathbf{x}$ is

$$\nabla_{\mathbf{x}} J_{3D} = 2\mathbf{B}^{-1}(\mathbf{x} - \mathbf{x}^b) - 2\mathbf{H}_{\mathbf{x}}^T \mathbf{R}^{-1}(\mathbf{y} - H(\mathbf{x})), \quad \text{with } \mathbf{H}_{\mathbf{x}}^T = \nabla_{\mathbf{x}} H \qquad (10)$$

which only requires the computation of the product of the inverse of the background and observation error covariances by a vector. As such, this offers the possibility to accommodate more sophisticated forms of the background error covariance matrix. This framework is known as the 3D variational (3DVAR) assimilation problem and is still heavily implemented in operational weather forecast centers.

The observational and background covariance matrices $\mathbf{R}$ and $\mathbf{B}$ are very important in determining the solution of the assimilation problem. These set the extent to which the background field (forecast) will be adjusted by the data by setting the weights of the background and data terms in the inversion. In practice, however, there is insufficient information to determine these matrices and ad hoc estimates are used instead. The observation errors are often assumed to be spatially uncorrelated, so that $\mathbf{R}$ is diagonal. Imposing correlation errors for data with important spatial coverage, such as satellite and radars, is important to avoid overweighting them in the assimilation. This could be conveniently implemented through an appropriate choice of a covariance model. A simpler way is to deliberately reduce the weights (i.e., overestimates the error variances) of the "clustered" observations, whose errors are expected to be correlated. Modeling $\mathbf{B}$ is more delicate as it needs to incorporate ocean balance properties and smoothness constraints, as the analysis increment completely lies within the subspace spanned by the directions of $\mathbf{B}$. The use of such constraints helps to dynamically spread the information in the observations, which should provide an analysis that could be more conveniently assimilated by the ocean model for forecasting. This was thoroughly discussed by Weaver et al. (2003, 2005) and Blayo et al. (2014). Use of ensemble of model outputs in the modeling of $\mathbf{B}$ has also became popular (e.g., Buehner, 2005).

Examples of ocean operational systems based on 3DDA methods include the US Naval Oceanographic Office NAVOCEANO system (Smedstad et al., 2003), the Meteorological Research Institute multivariate ocean variational estimation MOVE System (Usui et al., 2006), the European ECMWF system (Balmaseda et al., 2013), the Global Ocean Forecasting System (GOFS; Cummings and Smedstad, 2013), and the UK Met Office Forecasting Ocean Assimilation System (FOAM; Blockley et al., 2014).

Four-Dimensional Variational Assimilation

Four-dimensional variational assimilation (4DVAR) is a generalization of 3DVAR to the problem of estimating the state of a dynamical system using a set of observations that are available over a time interval. This not only includes information from future and past observations to estimate the ocean state at a given time, but also enables exploitation of the dynamical information from the equations that govern the evolution of the ocean state in time (i.e., ocean model). In its most general form, the latter could be described by the Navier-Stokes equations (Temam, 1984), and we represent it here as a discrete-time dynamical operator M_k that integrates the ocean state **x** between two consecutive time steps t_{k-1} and t_k as

$$\mathbf{x}_k = M_k(\mathbf{x}_{k-1}) + \eta_k. \tag{11}$$

η_k is a stochastic term representing uncertainties in the model, referred to as model error, and is usually conveniently assumed stochastic following a Gaussian distribution of mean zero (unbiased) and covariance $\mathbf{Q}_k$.

4DVAR can be directly formulated from the least-square objective function (Eq. 3) by constraining the (sum of the) distances between the model state at the times of available data, according to some weight for each term. Here, we first derive 4DVAR as the MAP estimator of a Bayesian estimation problem before presenting the adjoint method to efficiently compute the gradient of the objective function and accordingly its optimum. We finish with a discussion on the main features and issues of this approach.

Bayesian formulation

The Bayesian estimation problem of the ocean state $\mathbf{x}_0, \ldots, \mathbf{x}_L$ over a time interval $[T_0\ T_L]$ given a set of available observations $\mathbf{y}_0, \ldots, \mathbf{y}_L$, related to the ocean state as in Eq. 1, involves the calculation of the conditional probability distribution, similar to Eq. 5.

$$p_{\mathbf{x}_0,\ldots,\mathbf{x}_l | \mathbf{y}_0,\ldots,\mathbf{y}_L} \propto p_{\mathbf{y}_0,\ldots,\mathbf{y}_L | \mathbf{x}_0,\ldots,\mathbf{x}_L}\ p_{\mathbf{x}_0,\ldots,\mathbf{x}_L}. \tag{12}$$

Assuming observation and model errors ε_k and η_k are independent in time and mutually independent, and given the Markov Chain nature of the dynamical system (Eq. 11), standard conditional probability calculations lead to. See, for example, Simon (2006) and Law et al. (2015),

$$p_{\mathbf{x}_0,\ldots,\mathbf{x}_L | \mathbf{y}_0,\ldots,\mathbf{y}_L} \propto \Pi_{k=0}^{L} p_{\mathbf{y}_k | \mathbf{x}_k} \cdot p_{\mathbf{x}_0} \cdot \Pi_{k=1}^{L} p_{\mathbf{x}_k | \mathbf{x}_{k-1}}, \tag{13}$$

and under the assumption of Gaussian probability distributions we find analogous to Eq. 6

$$p_{\mathbf{x}_0,\ldots,\mathbf{x}_L | \mathbf{y}_0,\ldots,\mathbf{y}_L} \propto \exp\left(-\frac{1}{2} J_{4D}\right), \tag{14}$$

where

$$J_{4D}(\mathbf{x}_0, \ldots, \mathbf{x}_L) = \sum_{k=0}^{L} \|\mathbf{y}_k - H_k(\mathbf{x}_k)\|^2_{\mathbf{R}_k^{-1}} + \|\mathbf{x}_0 - \mathbf{x}_0^b\|^2_{\mathbf{B}_0^{-1}} + \sum_{k=0}^{L} \|\mathbf{x}_k - M_k(\mathbf{x}_{k-1})\|^2_{\mathbf{Q}_k^{-1}}. \tag{15}$$

The MAP estimator can thus be obtained by minimizing the 4D objective function J_{4D} as defined in Eq. 15, which is the same as optimizing

$$J_{4D}(\mathbf{x}_0, \eta_1 \ldots, \eta_L) = \sum_{k=0}^{L} \|\mathbf{y}_k - H_k(\mathbf{x}_k)\|^2_{\mathbf{R}_k^{-1}} + \|\mathbf{x}_0 - \mathbf{x}_0^b\|^2_{\mathbf{B}_0^{-1}} + \sum_{k=0}^{L} \|\eta_k\|^2_{\mathbf{Q}_k^{-1}} \quad (16)$$

subject to the ocean dynamics as described by Eq. 11.

The dimension of the ocean state can be very large in realistic applications, reaching up to 10^9–10^{10} in today's applications. And given that it should be determined for every time step, the required information exceeds by far the amount of available ocean data, even for coarse resolution models. One straightforward way to reduce the dimension of the 4DVAR optimization problem and mitigate its under-determined nature is to reduce the number of parameters by allowing only certain forms of model uncertainties. The extreme case is to assume the ocean model (Eq. 11) to be perfect, i.e., $\eta_k = 0$, so the problem reduces to finding the initial condition $\mathbf{x}_0$ that best fits, within observation errors uncertainties, the model to the data by minimizing (as schematically illustrated in Fig. 17.1)

$$J_{4D}(\mathbf{x}_0) = \sum_{k=0}^{L} \|\mathbf{y}_k - H_k(\mathbf{x}_k)\|^2_{\mathbf{R}_k^{-1}} + \|\mathbf{x}_0 - \mathbf{x}_0^b\|^2_{\mathbf{B}_0^{-1}}, \quad \text{subject to} \quad \mathbf{x}_k = M_k(\mathbf{x}_{k-1}). \quad (17)$$

This is known as the strong constraint 4DVAR problem. Directly optimizing (Eq. 16) is known as the weak constraint 4DVAR problem.

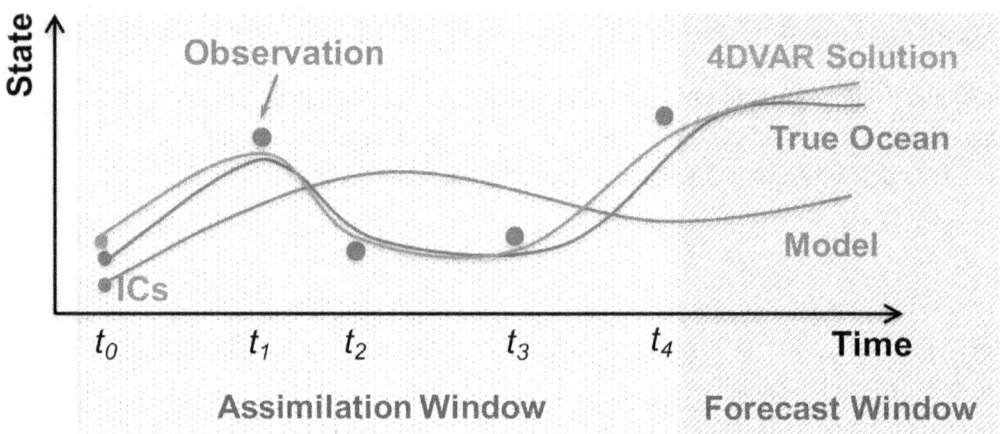

Figure 17.1. Schematic diagram of the 4DVAR assimilation procedure: fit the model to all available observations within an assimilation window to compute the analysis, from which integrate the ocean model for forecasting.

A large variety of different configurations exist between these extreme cases, for instance adjusting the ocean model parameters and inputs by including them as part of the estimation problem, i.e., as variables to be optimized in J_{4D}. This may, for example, include the external forcing fields such as atmospheric and open boundaries conditions, the ocean topography, and/or internal parameters of ocean physics such as ocean mixing parameters, as has been successfully implemented in the Estimation of the Circulation and the Climate of the Ocean (ECCO) consortium (Wunsch and Heimbach, 2007; Köhl and Stammer, 2008) and the Regional Ocean Modeling System (ROMS; Moore et al., 2011). The objective function in this case would be comprised of the standard model-data misfit term along with prior and regularization terms to constrain the adjustments to the optimized variables similar to the background term. This should be viewed as another approach to implementing a weak 4DVAR in which the model errors are non-additive but directly accounted for through appropriate dynamical parameterizations in the ocean model, which may help reduce the dimension of the optimization problem and impose dynamically balanced solutions for the adjusted variables and the estimated ocean state.

Solution of Four-Dimensional Variational (4DVAR) Assimilation

As in 3DVAR, gradient-based optimization algorithms are the standard methods to compute the 4DVAR solution. However, an important difference arises from the requirement that the solution needs to obey the model equations (Eq. 11). This leads to a so-called constrained optimization problem that is solved with the variational method. The variational principle, which is in functional space identical to setting the derivative of the objective function to zero, leads to the Euler-Lagrange equations. The latter are the adjoint equations to the tangent linear model equations, hence referred to as the adjoint method. The adjoint method provides the gradient by integrating the adjoint model backward in time, and is the most common approach to compute the gradient of the 4DVAR objective function J_{4D} (Le Dimet and Talagrand, 1986).

To understand why the backward integration of the adjoint gives the gradient, consider the strong constraint 4DVAR cost function. Using the chain rule for the derivatives of composite functions, one obtains

$$\nabla_{\mathbf{x}_0} J_{4D} = -2 \sum_{k=0}^{L} \mathbf{M}_{k:0}^T \mathbf{H}_k^T \mathbf{R}_k^{-1} (\mathbf{y}_k - H_k(\mathbf{x}_k)) + 2\mathbf{B}_0^{-1}(\mathbf{x}_0 - \mathbf{x}_0^b), \qquad (18)$$

where

$$\mathbf{M}_{k:0} = \mathbf{M}_k \cdots \mathbf{M}_1 \mathbf{M}_0, \quad \mathbf{M}_i = \nabla_{\mathbf{x}_i} M_i, \quad \text{and} \quad \mathbf{H}_i = \nabla_{\mathbf{x}_i} H_i. \qquad (19)$$

This shows that the gradient of J_{4D} can be computed as $-2\tilde{\mathbf{x}}_0$ by integrating the adjoint model backward in time

$$\begin{cases} \tilde{\mathbf{x}}_{L+1} = 0, \\ \tilde{\mathbf{x}}_k = \mathbf{M}_k^T \tilde{\mathbf{x}}_{k+1} + \mathbf{H}_k^T \mathbf{R}_k^{-1}(\mathbf{y}_k - H_k(\mathbf{x}_k)) \quad \text{for } k = L, \ldots, 0. \end{cases} \qquad (20)$$

$\tilde{\mathbf{x}}$ is the so-called adjoint variable to $\mathbf{x}$. Comparing Eq. 20 with Eq. 18 shows that $\tilde{\mathbf{x}}_0$ is the gradient of Eq. 18. Moreover, it is obvious that $\tilde{\mathbf{x}}$ does not only provide the gradient at the initial time (and any given time), but will further provide gradients to model parameters (and inputs). For linear problems, the solution can be calculated directly from the set of adjoint and forward model equations. For nonlinear cases, the solution is computed iteratively, with each optimization iteration requiring one integration of the forward model starting from the parameter changes of the most recent iteration, based on the trajectory of which another integration of the adjoint model is performed backward in time to compute the gradient of the cost function.

This same adjoint machinery is also at the basis of the weak constraint 4DVAR problem, following the dual formulation (Courtier, 1997) or the Representer method (Bennett, 2005). Both approaches transfer the inversion into the data space, which allows to drastically reduce the dimension of the weak 4DVAR optimization problem since the number of ocean data is commonly much smaller than the ocean state. The adjoint model is also used to compute the gradients of the 4DVAR objective function with respect to any model parameters, again using the chain rule (see, for example, Heimbach et al., 2002).

The weak 4DVAR methods provide very powerful tools to fit the ocean models to the available data; making the 4DVAR inversion problem highly underdetermined. Optimizing model errors at frequent model steps may enable efficient fit to the ocean observations but with a real risk of data overfitting and non-dynamical model errors adjustments. The role of the model errors covariance matrices Q_k becomes crucial, with unfortunately no established or efficient way to define these matrices (Wunsch, 1996; Hoteit et al., 2010).

Coding the adjoint model requires implementing the tangent linear model of the ocean model and its adjoint, and this can be a very demanding process. Automatic compilers have been developed to directly generate the adjoint code from the source code of the dynamical model (Giering and Kaminski, 1998). These may greatly facilitate the process of developing and maintaining the adjoint model to keep it up-to-date with forward model changes, but also impose some formats in the coding of the (forward) ocean model (Vlasenko et al., 2016). In addition to the technical challenge of generating an adjoint model, running the adjoint iteratively multiplies the cost of running a simulation by a factor of several hundreds. An additional difficulty arises in nonlinear models from the fact that the whole model trajectory needs to be known and stored at the time when the adjoint model is running. Checkpointing methods could be implemented into the adjoint code generation tools to efficiently reconstruct the trajectory (Heimbach et al., 2002).

Increasing efforts are being made to develop efficient methods that allow to either simplify the task of developing an adjoint code through reduced-order techniques, or completely by-passing the adjoint model through direct computation of the 4DVAR objective function gradients from forward model runs only. Reduced-order methods were developed around three related directions (Altaf et al., 2013a): (i) apply the optimization in a reduced space as a way to reduce the dimension of the optimization space to speed up the convergence rate (Robert et al., 2005; Hoteit and Köhl, 2006); (ii) develop a reduced-order model of the ocean model from which the adjoint model is derived (Vermeulen and Heemink, 2006; Fang et al., 2009); or (iii) directly develop a reduced-order adjoint model while still using the original forward ocean model for forward integrations (Altaf et al., 2013a; Yaremchuk et al. 2016). In this context, ensemble methods became popular as they were suggested to provide efficient tools to compute the gradients of the 4DVAR objective function, or to be used to parameterize the adjoint space in a hybrid assimilation framework (more on this in section below). Other adjoint-free optimization methods were tested, but these require dimension reduction before implementation to reduce their prohibitive computational burden (e.g., Hoteit, 2008).

The main difficulty in applying the adjoint method to ocean data assimilation problems is due to the nonlinear nature of the equations governing their dynamics. This problem is expected to become more severe as the resolution of ocean models continues to increase. In this case, the 4DVAR objective function becomes too irregular (non-convex), including multiple minima that prevent a noticeable decrease in the objective function with gradient-based optimization techniques. This is associated with rapidly-growing response to perturbations ("intrinsic variability") that ultimately become unpredictable and are thus not controllable in the system. The adjoint gradient sensitivities then grow exponentially in time and become not useful in the optimization problem because for increasing assimilation windows the parameter range of validity of the linear approximation quickly becomes smaller than the uncertainty in the control parameters, which limits the length of the assimilation windows. In other words, the nonlinearity of the system invalidates the use of the gradient for descent (Pires et al., 1996; Köhl and Willebrand, 2002). Although short assimilation windows remain feasible, this may limit the benefit of the adjoint method, particularly since ocean observations are sparse and uncertain. Therefore, larger windows are desired to properly extract the large-scale parameter information via the dynamical constraint (Köhl and Willebrand, 2002). Large windows can be also useful to reduce dependence on the background covariance matrix and to provide enough time to infer enough sensitivities to, for instance, atmospheric forcing and/or boundaries conditions and other parameters if these were also to be adjusted in the 4DVAR system.

Since large scales are associated with longer predictability timescales, a way out is to separate the small from the large scales. This could be implemented by increasing viscosity and diffusivity terms in the backward adjoint run, which becomes close to the adjoint of a coarser, more linear model without local minima (Hoteit et al., 2005a). This approach works even with the original high-resolution forward model, because secondary minima become so dense over long periods of time that they appear as stochastic perturbations (Hoteit et al., 2005a). The limitation of this approach is that it may also start filtering out large-scale features over time because of the tight coupling between the different scales in the ocean. An alternative could be based on ensemble methods (Lea et al., 2000), but since the number of required ensembles grows as the gradients increase, this method quickly becomes unfeasible. It is still not clear to what extent the smoothing nature of the reduced-order and ensemble-based approaches, whether to derive an approximate adjoint or to directly compute the gradients, could mitigate this issue and help extend the assimilation windows.

A new approach was recently borrowed from the chaos theory to tackle the issues with nonlinearities, and it was applied for ocean parameters estimation of a climate model. Noticing that a coupling leads to synchronization of similar chaotic systems over long periods of time, a parameter estimation method based on the ability of a parameter-dependent system to synchronize with observations was developed in physics (Abarbanel, 2012). The coupling to the observations is included in the model as a relaxation term that, when strong enough, ultimately will turn the system into a non-chaotic system in which parameter estimation with the adjoint method may become again feasible. A caveat to this method is that the estimation takes place in a modified system and may no longer be optimal with respect to the original system. Moreover, because synchronization turns into a damping in the adjoint model, data will have a limited effect (in time) on the estimated parameters, particularly the initial conditions.

Recently, several 4DVAR regional ocean operational systems have been successfully developed and are currently routinely providing forecasts of ocean states, including the real-time forecasting system of the Mid-Atlantic Bight (Zhang et al., 2010), the University of California Santa Cruz California Current forecasting system (Moore et al., 2011), the University of Hawaii forecasting system for the region surrounding the main Hawaiian Islands (Janeković et al., 2013), and the Navy Coastal Ocean forecasting of the Okinawa Trough (Smith et al., 2017b).

4DVAR is designed to compute the MAP as the final solution for the DA problem but not its covariance, which will be needed as a background for the next assimilation cycle. Although this could be conveniently estimated using the adjoint to compute a low-rank approximation of the Hessian matrix of the 4DVAR objective function, which is the inverse of the MAP error covariance matrix when the system is linear and the noise is Gaussian (Smith et al., 2015), strong nonlinearities mean that the Hessian will probably be computed around a local minima and may not reflect the global errors in estimation. The next section will present the Bayesian filtering approach that aims, in contrast, at directly computing the full conditional probability distribution of the ocean state given available observations.

Bayesian Filtering

The Bayesian estimation problem can be solved sequentially in time, as the observations become available. This is known as Bayesian filtering and it readily provides a suitable framework for operational oceanography where an ocean model is used for forecasting and the data are assimilated to update the model forecasts with the Bayes' rule every time they become available. Here, we are interested in estimating the ocean state at a given time t_k given all available observations up to t_k, which in a Bayesian setting involves the computation of $p_{\mathbf{x}_k | \mathbf{y}_0, \ldots, \mathbf{y}_k}$. Marginalizing Eq. 12, the filtering solution is then identical to the Bayesian estimator (Eq. 12) at the end of the assimilation window. This contrasts with a "smoother," which involves observations beyond time instant t_k, e.g., 4DVAR and ensemble Kalman smoothers (more on this in section below). This section will focus on the state estimation problem and introduce DA algorithms from a Bayesian filtering perspective. Most of these algorithms, as those presented below, aim at computing the MV estimator instead of the MAP estimator of 4DVAR.

The basis of Bayesian filtering is a state space model comprised of a dynamical model (Eq. 11) and an observation model (Eq. 1) that provides measurements of the ocean state in time. These provide $p_{\mathbf{x}_k | \mathbf{x}_{k-1}}$ and the likelihood $p_{\mathbf{y}_k | \mathbf{x}_k}$, respectively. Using standard conditional probability rules (e.g., Simon, 2006 and Law et al., 2015), one can write

$$\begin{aligned} p_{\mathbf{x}_k | \mathbf{y}_0, \ldots, \mathbf{y}_k} &\propto p_{\mathbf{y}_k | \mathbf{y}_0, \ldots, \mathbf{y}_{k-1}, \mathbf{x}_k} \cdot p_{\mathbf{x}_k | \mathbf{y}_0, \ldots, \mathbf{y}_{k-1}}, \\ &\propto p_{\mathbf{y}_k | \mathbf{x}_k} \, p_{\mathbf{x}_k | \mathbf{y}_0, \ldots, \mathbf{y}_{k-1}}. \end{aligned} \quad (21)$$

The posterior or analysis distribution is thus the product of the likelihood of the state given the new observation and the forecast (prior) distribution, which is the distribution of the state conditioned on all previous observations. This is called the update or analysis step. The forecast distribution can be computed from the analysis distribution at the previous time step by first computing the joint distribution of $(\mathbf{x}_k, \mathbf{x}_{k-1})$ conditioned on $\mathbf{y}_0, \ldots, \mathbf{y}_{k-1}$, then integrating over $\mathbf{x}_{k-1}$ to obtain the desired marginal distribution as

$$p_{\mathbf{x}_k|\mathbf{y}_0,\ldots,\mathbf{y}_{k-1}} = \int p_{\mathbf{x}_k|\mathbf{x}_{k-1}} \, p_{\mathbf{x}_{k-1}|\mathbf{y}_0,\ldots,\mathbf{y}_{k-1}} d\mathbf{x}_{k-1}. \qquad (22)$$

Therefore, if the analysis distribution is available at a given time, one can first compute the forecast distribution using Eq. 22, and then compute the analysis distribution at the next time using Eq. 21. One can then proceed recursively, starting from a prior distribution at the initial time and then alternate forecast and analysis steps to compute the analysis distribution at any given time. The filtering procedure is thus similar to a 3DDA procedure (as illustrated in Fig. 17.2), but operates on the state distribution rather than the state. This recursive framework provides a solution to the estimation problem and conceptually leads to the so-called optimal filter. In practice, however, difficulties often arise in computing the filter solution (distributions), largely due to the fact that evaluating the integrals in Eqs. 21-22 are numerically intractable in high-dimensional problems, such as the ocean. For such applications, one has to adopt certain approximations to derive some sub-optimal filters that provide accurate enough results at reasonable computational requirements. Reviewing different approaches for sub-optimal filtering is the focus of this section.

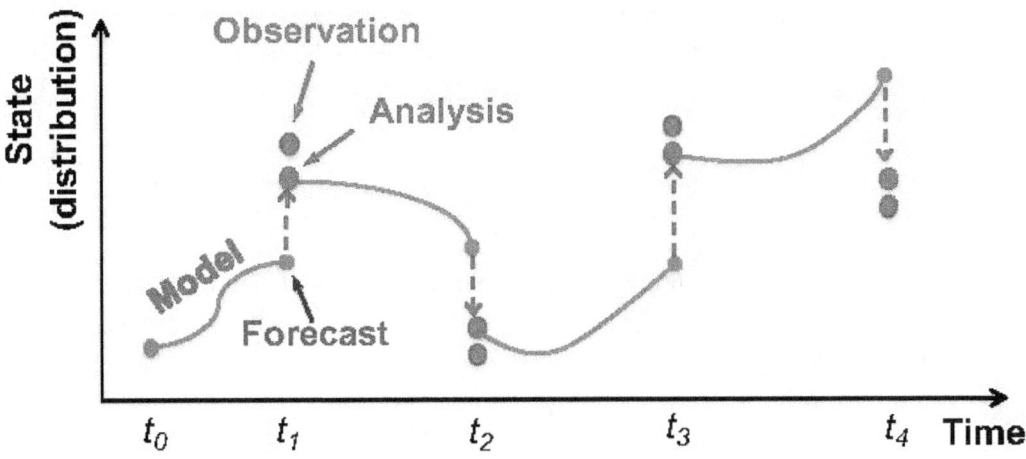

Figure 17.2. Schematic diagram of the sequential (3D and filtering) assimilation procedures, including 3DVAR, OI, KF, EnKFs, and PF. During the forecast step, the ocean model is integrated to the time of the next available observation starting from the most recent analysis. During the analysis step, the forecast is updated using the incoming observation to compute the analysis. Four assimilation cycles are shown.

Starting with the celebrated Kalman filter (KF; Kalman, 1960), which is designed for linear systems (dynamics and observations) and Gaussian errors, we present variants of the KF that enable its implementation for DA into large-scale nonlinear models. This includes the low-rank extended Kalman filters (LR-EKFs) and ensemble Kalman filters (EnKFs). The KF linear update step does not hold with the nonlinear ocean dynamics and for some of the ocean observations (e.g., acoustics). We will also present two nonlinear/non-Gaussian filters that are currently being investigated for potential use with realistic ocean data assimilation problems, the particle filter (PF) and the

The Kalman Filter (KF)

In the KF (Kalman, 1960), the dynamical and observation operators in (Eqs. 11 and 1 are linear, and thus denoted respectively by $\mathbf{M}$ and $\mathbf{H}$ in this section; and the associated noise ε_k and η_k are Gaussian (i.e., the likelihood function $p_{\mathbf{y}_k|\mathbf{x}_k}$ and the transition distribution $p_{\mathbf{x}_k|\mathbf{x}_{k-1}}$ in (21) and (22) are Gaussian), independent in time and mutually independent. Starting from some initial Gaussian distribution, the forecast and posterior distributions, $p_{\mathbf{x}_k|\mathbf{y}_0,\ldots,\mathbf{y}_{k-1}}$ and $p_{\mathbf{x}_k|\mathbf{y}_0,\ldots,\mathbf{y}_k}$ remain Gaussian at subsequent time instants. The algorithm of the KF therefore reduces to recursively computing the means and covariance matrices of $p_{\mathbf{x}_k|\mathbf{y}_0,\ldots,\mathbf{y}_{k-1}}$ and $p_{\mathbf{x}_k|\mathbf{y}_0,\ldots,\mathbf{y}_k}$, which fully characterize their distributions. These represent the forecast and analysis MAPs (and MVs, which are identical in this case) and their errors covariance matrices, and are computed by recursive cycles of the following forecast and analysis steps.

<u>Forecast step</u>: Integrate the posterior mean, i.e., analysis state, $\mathbf{x}^a_{k-1}$ at time instant t_{k-1} and the associated error covariance $\mathbf{P}^a_{k-1}$ forward with the dynamical model (Eq. 11) to compute the forecast state $\mathbf{x}^f_k$ and the associated error covariance $\mathbf{P}^f_k$ at the time of the next available observation t_k, as

$$\mathbf{x}^f_k = \mathbf{M}_k \mathbf{x}^a_{k-1}, \tag{23a}$$

$$\mathbf{P}^f_k = \mathbf{M}_k \mathbf{P}^a_{k-1} \mathbf{M}^T_k + \mathbf{Q}_k. \tag{23b}$$

<u>Analysis step</u>: Once the new observation $\mathbf{y}_k$ is available, update the forecast statistics $\mathbf{x}^f_k$ and $\mathbf{P}^f_k$ to their analysis counterparts, $\mathbf{x}^a_k$ and $\mathbf{P}^a_k$, using the BLUE (which is also the MAP here) in Eq. 7 with $\mathbf{B} = \mathbf{P}^f_k$,

$$\mathbf{x}^a_k = \mathbf{x}^f_k + \mathbf{K}_k(\mathbf{y}_k - \mathbf{H}_k \mathbf{x}^f_k), \tag{24a}$$

$$\mathbf{P}^a_k = (\mathbf{I} - \mathbf{K}_k \mathbf{H}_k)\mathbf{P}^f_k, \tag{24b}$$

$$\mathbf{K}_k = \mathbf{P}^f_k \mathbf{H}^T_k (\mathbf{H}_k \mathbf{P}^f_k \mathbf{H}^T_k + \mathbf{R}_k)^{-1}. \tag{24c}$$

Therefore, the only difference from a 3D assimilation setting is in the use of a time- (or flow-) dependent forecast (background) error covariance matrix that is updated in the analysis step as in Eq. 24b to account for the reduction in the estimation error (uncertainties) after the assimilation of an observation. The resulting analysis error covariance is then integrated by the model forward as in Eq. 23b to reflect an increase, or eventual decrease, in the initial analysis error during forecasting, depending on the ocean dynamics during that period (Pham et al., 1997), plus the contribution of the model errors.

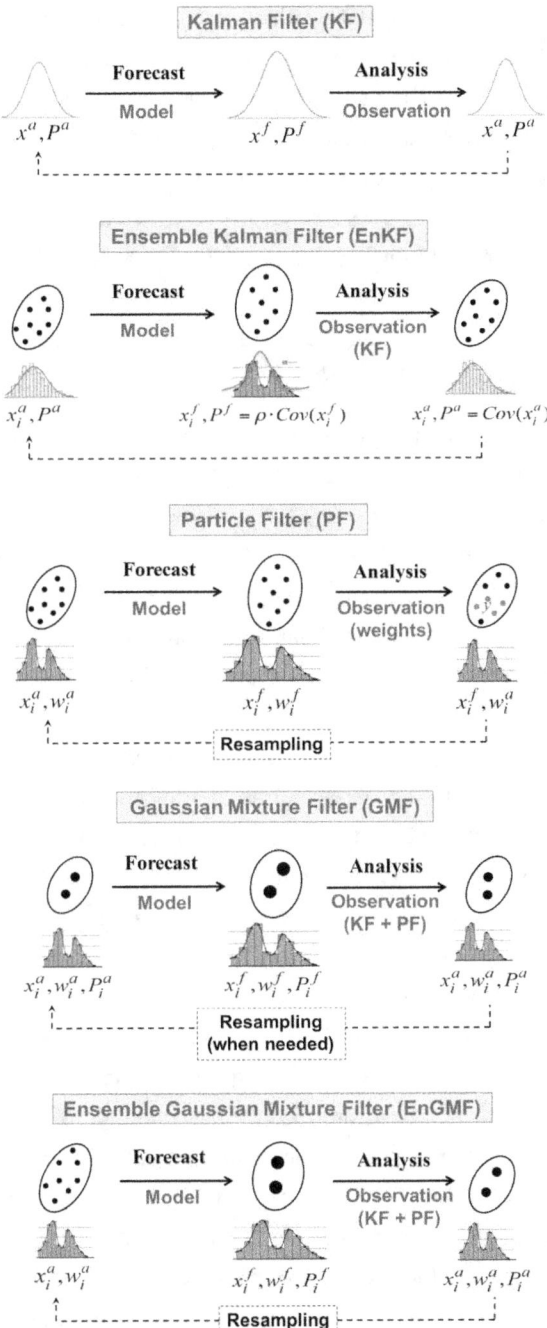

Figure 17.3. Schematic illustration of the different filtering strategies presented in Section 4. The Kalman filter (KF) only involves Gaussian distributions, characterized by their mean and covariances. Both the ensemble Kalman filter (EnKF) and the Particle filter (PF) integrate a set of ocean states sampled from the present distribution forward during the forecast step. While the EnKF assumes Gaussian background at the analysis step so that the forecast ensemble is updated with the incoming observation as in the KF, the PF applies a non-Guassian update to the samples weights only. The Gaussian mixture filter (GMF) maintains Gaussian mixture distributions at the forecast and analysis steps, in which the mixture weights are updated as in the PF and the mixture covariances as in the KF. As the EnKF, the ensemble GMF (EnGMF) integrates an ensemble of ocean states at the forecast step, but assumes Gaussian mixture background so that it applies a GMF analysis.

The application of the KF for ocean DA is hampered by (i) the nonlinear nature of the ocean dynamics (and eventually of some ocean observations), and (ii) the large dimension of the OGCMs (which is reaching up to $n \sim O(10^{10})$ in today's numerical ocean models). The first means that the KF cannot be directly implemented, and the second implies prohibitive computational burden (in term of storage and computation) in order to manipulate the KF error covariance matrices, of dimensions $n \times n$. Different simplified variants of the KF have been therefore proposed for ocean data assimilation, which can be split into two main categories, reduced extended Kalman filters and ensemble Kalman filters.

Reduced Extended Kalman Filters (REKFs)

To apply the KF to ocean assimilation problems, one can compute the forecast state as in Eq. 23a by just integrating the analysis state forward with the nonlinear model. Implementing the KF forecast error covariance calculation step (Eq. 23b) is, however, not as straightforward. A popular approach is to linearize the model (and eventually the observation operator) using, for instance, a first-order Taylor expansion, and then apply the KF to the linearized system. This leads to the popular, but no longer optimal, extended Kalman filter (EKF) (Jazwinski 1970), and eventually its higher-order variants depending on the retained order of the Taylor expansion (Anderson and Moore, 1979; Simon, 2006).

To avoid the prohibitive computational requirements of the EKF due to the large numerical dimension of realistic ocean models, different forms of reduced-state space or reduced-error space (i.e., low-rank error covariance matrix) approximations have been proposed (e.g., Fukumori and Malanotte-Rizzoli, 1995; Cane et al., 1996); Cohn and Todling, 1996; Verlaan and Heemink, 1997; Pham et al., 1997; Lermusiaux and Robinson, 1999; Farrell and Ioannou, 2001; Hoteit et al., 2002). A common feature of these REKFs is that they exploit information from a representative subspace of the full ocean state, or error subspace, and ignore information from the less influential complement subspace. This is supported by the dissipative and driven nature of the ocean dynamics, which concentrates energy at large scales, imposing a red spectrum of variability (Daley, 1991; Pham et al., 1997; Lermusiaux and Robinson, 1999). Consequently, EKF calculations are conducted on the retained subspace only, dramatically reducing the computational cost. The reduced state/error spaces, denoted by **L**, can be set invariant in time, as in the reduced order EKF (ROEKF), or left to evolve with the model dynamics as in the singular evolutive extended Kalman (SEEK) filter, with the latter leading to more robust state estimates during periods of strong ocean variability (Hoteit and Pham, 2003). Both schemes operate with low-rank r EKF error covariance matrices, with r the dimension of the reduced space, while keeping the rest of the EKF algorithm mostly unchanged. The error covariance matrices are only evaluated through their low-rank counterparts, **L** and **U**, where $\mathbf{P} = \mathbf{LUL}^T$ and **U** a $r \times r$ matrix representing the error variance in the reduced space, which avoids the storage of **P** and drastically reduces the EKF computational burden. More details can be found in Cane et al. (1996) and Pham et al. (1997). One caveat of this approach is in the treatment of the model error covariance matrix **Q** in Eq. 23b, as the rank of the forecast error covariance matrix **P** cannot indeed be preserved after adding **Q** unless the model error is projected on (and therefore only treated in) the reduced space **L**, or simply neglected assuming perfect model (**Q** = 0) (Hoteit et al., 2007). This implies another approximation in the final EKF algorithm and would inevitably lead to an underestimation of the forecast error covariance.

REKFs have been applied to ocean DA in the global ocean as part of the Estimating the Circulation and Climate of the Global Ocean Data Assimilation Experiment (ECCO-GODAE) system (Kim et al., 2006), in the Pacific Ocean (Cane et al., 1996; Verron et al., 1999; Hoteit et al., 2002), and regionally in a nested implementation of the Ligurian Sea (Barth et al. 2007). They are also used operationally in, for instance, Monterey Bay (Haley et al., 2009), the Greek national POSEIDON-II system for the Mediterranean (Korres et al., 2010), and the European MERCATOR system (Lellouche et al., 2013). Given the complexity of the linearization step and its limitation with strongly nonlinear models (Evensen, 1994), as well as the difficulty of specifying and evolving a reduced subspace, these methods have dramatically lost popularity in recent years owing to the advances in ensemble Kalman filtering methods.

Ensemble Kalman Filters (EnKFs)

The main idea behind the EnKFs is to apply a Monte Carlo-like forecast step to integrate the KF analysis state and its error covariance forward through a set, or ensemble, of ocean states sampled from these two statistical moments (Evensen, 2003). The sampled analysis ensemble is then integrated forward with the (nonlinear) model to obtain the forecast ensemble, from which the forecast state and error covariance are taken as the sample mean and covariance of the ensemble. A KF analysis step is then applied to update the forecast ensemble every time a new observation is available. The ensemble formulation allows to avoid the manipulation of the KF error covariance matrices by performing the calculations on the ensemble members, which enables the implementation of the filter on large-scale ocean applications. Generally speaking, the Monte-Carlo forecast step requires N (= ensemble-size) ocean model integrations to compute the forecast ensemble, and the KF update step is applied in the low-rank ensemble subspace, typically of a dimension $N-1$ (Pham, 2001). Another important advantage of the Monte Carlo forecast step is the possibility of implicitly accounting for the model errors through perturbations sampled from their distributions and then carried with the ensemble model runs (Evensen, 2003; Hoteit et al., 2007). This further allows avoiding the additive model error assumption, which is otherwise less general and difficult to account for in the REKFs (Pham et al., 1997; Hoteit et al., 2005b).

Because of their non-intrusive formulation and ease of implementation, remarkable robustness and effectiveness, and reasonable computational requirements, EnKF methods have become very popular in the geosciences. Many variants of the EnKF have been proposed in the literature, but a full review is beyond the scope of this chapter. They all operate as cycles of Monte Carlo forecast and KF update steps involving only the first two moments of the ocean state posterior, basically only differing in the sampling scheme of their analysis ensembles. Depending on whether the observations are perturbed before assimilation or not, the EnKFs are customarily classified as one of two types (Tippett et al., 2003): stochastic EnKFs (Burgers et al., 1998; Houtekamer et al., 2005; Hoteit et al., 2015) and deterministic EnKFs (Anderson, 2001; Bishop et al., 2001; Whitaker and

Hamill, 2002; Hoteit et al., 2012; Luo and Hoteit, 2014c). A stochastic EnKF essentially updates each forecast ensemble member with perturbed observations during the KF correction step. By contrast, a deterministic EnKF updates the ensemble mean and a specific (square-root) form of the sample (ensemble) error covariance matrix exactly as in the KF, without perturbing the observations. An analysis ensemble is then produced from the updated mean and covariance prior to the forecast step. The most popular deterministic EnKFs with publicly available codes are the singular evolutive interpolated KF–SEIK (Pham, 2001; Hoteit et al., 2002), the ensemble transform KF–ETKF (Bishop et al., 2001; Wang et al., 2004; Hunt et al., 2007), and the ensemble adjustment KF–EAKF (Anderson, 2001, 2009). With the continuous advances in computing capabilities, EnKF methods are becoming increasingly popular in the development of ocean operational systems, e.g. Toye et al. (2017). An EnKF is, for instance, already used operationally in the Norwegian North Atlantic and Arctic forecasting system, TOPAZ (Sakov et al., 2012). Below we focus on presenting the algorithms of the two "basic" forms of ensemble Kalman filtering; the original stochastic EnKF (Burgers et al. 1998; Houtekamer and Mitchell 1998) and two standard, closely-related deterministic EnKFs.

1) STOCHASTIC EnKF (SEnKF)

Assume an N-member analysis ensemble $\mathbf{X}_{k-1}^a = [\mathbf{x}_{k-1}^{a,i}, i = 1, 2 \ldots, N\}$ is available at the end of the $(k-1)^{th}$ assimilation cycle. The forecast ensemble at the next time t_k is obtained by integrating $\mathbf{x}_{k-1}^{a,i}$ with the dynamical model (11), i.e.

$$\mathbf{X}_k^f = \left\{ \mathbf{x}_k^{f,i} = M_k(\mathbf{x}_{k-1}^{a,i}) + \eta_k^i, \ i = 1, \ldots, N \right\}, \tag{25}$$

where η_k^i are sample dynamical noise drawn from the distribution of the model error term. The ensemble sample mean and covariance are taken as the forecast state and its error covariance matrix, respectively as

$$\mathbf{x}_k^f = \frac{1}{N} \sum_{i=1}^N \mathbf{x}_k^{f,i}, \qquad \mathbf{P}_k^f = \frac{1}{N-1} \sum_{i=1}^N \left(\mathbf{x}_k^{f,i} - \mathbf{x}_k^f \right) \left(\mathbf{x}_k^{f,i} - \mathbf{x}_k^f \right)^T. \tag{26}$$

In practice, $\mathbf{P}_k^f$ needs not be calculated. Instead, it is customary to approximate

$$\mathbf{P}_{xh}^k = \frac{1}{N-1} \sum_{i=1}^N \left(\mathbf{x}_k^{f,i} - \mathbf{x}_k^f \right) \left(\mathbf{y}_k^{f,i} - \mathbf{y}_k^f \right)^T, \tag{27}$$

$$\mathbf{P}_{hh}^k = \frac{1}{N-1} \sum_{i=1}^N \left(\mathbf{y}_k^{f,i} - \mathbf{y}_k^f \right) \left(\mathbf{y}_k^{f,i} - \mathbf{y}_k^f \right)^T,$$

where

$$\mathbf{y}_k^{f,i} = H_k(\mathbf{x}_k^{f,i}) \quad \text{and} \quad \mathbf{y}_k^f = \frac{1}{N} \sum_{i=1}^N \mathbf{y}_k^{f,i}, \tag{28}$$

which enables to avoid the linearization of the nonlinear observation operator. The Kalman gain $\mathbf{K}_k$ is then approximated as

$$\mathbf{K}_k \approx \mathbf{P}_{xh}^k \left(\mathbf{P}_{hh}^k + \mathbf{R}_k \right)^{-1}. \tag{29}$$

When a new observation $\mathbf{y}_k$ is available, one computes the analysis ensemble from the forecast ensemble using the KF update step, as

$$\mathbf{x}_k^{a,i} = \mathbf{x}_k^{f,i} + \mathbf{K}_k \left(\mathbf{y}_k^{\varepsilon,i} - H_k(\mathbf{x}_k^{f,i}) \right), \quad \text{for} \quad i = 1, \cdots, N, \qquad (30)$$

where $\mathbf{y}_k^{\varepsilon,i}$ are perturbed observations generated by adding to the observation $\mathbf{y}_k$ random perturbations sampled from the distribution of the observational error. Accordingly, the sample mean and covariance of the analysis ensemble $\mathbf{X}_k^a = \{\mathbf{x}_k^{a,i} : i = 1, \cdots, N\}$ are obtained in the spirit of Eq. 26, and the observations perturbations guarantee that they converge to the KF analysis and its error covariance with increasing ensemble size N. Integrating $\mathbf{X}_k^a$ forward to the time of the next available observation, one starts a new assimilation cycle, and so on.

The first two moments of the SEnKF analysis may only asymptotically match those of the KF (Evensen, 2003). In this sense, the SEnKF update step always introduces noise during the analysis (Nerger et al., 2005). The noise may become pronounced in typical oceanic data assimilation applications where the rank of the observational error covariance matrix $\mathbf{R}_k$ is much larger than the ensemble size, meaning that $\mathbf{R}_k$ will be greatly under-sampled (Altaf et al., 2014). Spurious correlations between the observation perturbations and the forecast perturbations could also lead to errors in the EnKF sample analysis error covariance matrices (Pham, 2001; Bowler et al., 2013). To mitigate this issue, one can either introduce a certain correction scheme as in Hoteit et al. (2015), or simply avoid perturbing the observations following a deterministic EnKF formulation, which will be discussed next. In the opposite case, that is when the ensemble size is larger than the number of observations, the SEnKF was shown to perform better than other EnKFs without perturbations in many situations (Anderson, 2010; Hoteit et al., 2012). Lawson and Hansen (2004) argued that the observation perturbations in the SEnKF tend to re-Gaussianize the ensemble distribution to explain the improved stability. Lei et al. (2010) also demonstrated that the SEnKF is generally more stable in certain circumstances, especially in the presence of wild outliers in the data. An important advantage of the SEnKF update step is that it readily provides an analysis ensemble for forecasting (that is randomly sampled from the assumingly Gaussian analysis distribution), avoiding the deterministic updating step that may distort some of the features of the forecast ensemble distribution as in the other ensemble KFs without perturbations (Lei et al., 2010; Hoteit et al., 2015). This allows more straightforward implementation of some auxiliary techniques, such as covariance localization and hybrid schemes, as will be further discussed below.

2) DETERMINISTIC EnKFS (DEnKFS)

The DEnKFs analysis ensemble is deterministically generated in order to perfectly match the KF estimate, and thus avoid the random perturbations of the SEnKF. There are infinite ways to match a mean and a covariance by an ensemble and accordingly various DEnKFs have been proposed, many of which are based on the square-root formulation of the KF, which was introduced as an approach to improve the stability of the KF by working on a certain square-root of the filter covariance matrix. Anderson (2001), Bishop et al. (2001), and Whitaker and Hamill (2002) exploited the readily square-root form of the ensemble-based covariances to propose deterministic EnKFs assimilating the data serially, one at a time, assuming uncorrelated observational errors (i.e., diagonal $\mathbf{R}_k$)[1]. This enables efficient (parallel) assimilation of very large number of observations (Houtekamer and Zhang, 2016). In contrast, the ensemble transform Kalman filter (ETKF; Bishop et al., 2001) and the singular evolutive interpolated Kalman (SEIK) filter (Pham, 2001; Hoteit et al., 2002) can directly handle any form of $\mathbf{R}_k$ by computing the analysis increment in the ensemble subspace. To assimilate large numbers of observations, one may apply local analysis steps using only neighbor observations (as in optimal interpolation), which needs to be used anyway to deal with the low-rank nature of ensemble sampled covariances, as will be further discussed below.

Here, we present the ETKF for illustration and discuss its similarities with SEIK (Nerger et al., 2012). Following the same forecast step as the SEnKF, a forecast ensemble $\mathbf{X}_k^f = \{\mathbf{x}_k^{f,i}, i = 1, \cdots, N\}$ is available at time t_k as in Eq. 25, with the forecast state as the sample mean $\mathbf{x}_k^f$ and its error covariance as $\mathbf{P}_k^f$ given by Eq. 26. Instead of directly working on $\mathbf{P}_k^f$, one first constructs a square-root $\mathbf{S}_k^f$ of $\mathbf{P}_k^f$ from the forecast ensemble perturbations,

$$\mathbf{S}_k^f = \frac{1}{\sqrt{N-1}} \left[\mathbf{x}_k^{f,1} - \mathbf{x}_k^f, \cdots, \mathbf{x}_k^{f,N} - \mathbf{x}_k^f \right]; \tag{31}$$

and its equivalent in the observation space,

$$\mathbf{x}_k^{a,i} = \mathbf{x}_k^a + \sqrt{N-1}\, (\mathbf{S}_k^a)_i, \quad \text{for } i = 1, \cdots, N. \tag{32}$$

ETKF updates the ensemble forecast and error covariance exactly as in the KF, using Eqs. 24a and 24c,

$$\mathbf{x}_k^a = \mathbf{x}_k^f + \mathbf{K}_k \left(\mathbf{y}_k - \mathbf{y}_k^f \right), \tag{33a}$$

$$\mathbf{K}_k = \mathbf{S}_k^f (\mathbf{S}_k^h)^T \left(\mathbf{S}_k^h \left(\mathbf{S}_k^h \right)^T + \mathbf{R}_k \right)^{-1}, \tag{33b}$$

whereas the covariance update formula (Eq. 24b) can be computed as

$$\mathbf{P}_k^a = \mathbf{S}_k^f \mathbf{V}_k \left(\mathbf{S}_k^f \right)^T, \tag{34a}$$

$$\mathbf{V}_k = \left((\mathbf{S}_k^h)^T \mathbf{R}_k^{-1} \mathbf{S}_k^h + \mathbf{I}_N \right)^{-1}, \tag{34b}$$

where $\mathbf{I}_N$ is the N-dimensional identity matrix. A spectral decomposition is then applied, so that

$$(\mathbf{S}_k^h)^T \mathbf{R}_k^{-1} \mathbf{S}_k^h = \mathbf{E}_k \mathbf{D}_k \mathbf{E}_k^T,$$

where $\mathbf{E}_k$ consists of eigenvectors of $(\hat{\mathbf{S}}_k^h)^T \mathbf{R}_k^{-1} \hat{\mathbf{S}}_k^h$, and $\mathbf{D}_k$ is a diagonal matrix whose diagonal elements are the corresponding eigenvalues. One can then express $\mathbf{V}_k$ as

$$\mathbf{V}_k = \mathbf{T}_k (\mathbf{T}_k)^T, \quad \text{with} \quad \mathbf{T}_k = \mathbf{E}_k (\mathbf{D}_k + \mathbf{I}_N)^{-1/2}. \tag{35}$$

so that

$$\mathbf{P}_k^a = \mathbf{S}_k^a (\mathbf{S}_k^a)^T, \quad \text{where} \quad \mathbf{S}_k^a = \mathbf{S}_k^f \mathbf{T}_k. \tag{36}$$

---

[1] For correlated observational errors, one may transform the observations by the inverse of the square-root of the observation error covariance matrix $\mathbf{R}_k$ to obtain a new set of uncorrelated observations that could be serially assimilated.

Accordingly, the analysis ensemble $\mathbf{X}_k^a = \{\mathbf{x}_k^{a,i}, i = 1, \cdots, N\}$ can then be generated using

$$\mathbf{x}_k^{a,i} = \mathbf{x}_k^a + \sqrt{N-1}\,(\mathbf{S}_k^a)_i, \quad \text{for } i = 1, \cdots, N, \tag{37}$$

where $(\mathbf{S}_k^a)$ denotes the ith column of $\mathbf{S}_k^a$, so that the sample covariance of $\mathbf{X}_k^a$ is exactly $\mathbf{P}_k^a$. This, however, does not guarantee that its sample mean is $\mathbf{x}_k^a$ in Eq. 33a unless

$$\sum_{i=1}^{N} (\mathbf{S}_k^a)_i = 0. \tag{38}$$

To avoid this bias, Wang et al. (2004) latter followed SEIK formulation and proposed to take $\mathbf{S}_k^a = \mathbf{S}_k^f \mathbf{T}_k \mathbf{Z}_k$, where

$$\mathbf{Z}_k \mathbf{1} = \mathbf{0} \quad \text{and} \quad \mathbf{Z}_k \mathbf{Z}_k^T = \mathbf{I}_N, \tag{39}$$

with $\mathbf{1}$ being a vector whose elements are all 1s. Another feature of the SEIK filter is the use of a "random" matrix $\mathbf{Z}_k$, aiming at "redistributing" the error variance among the ensemble members to help in the mitigation of ensemble degeneracy (Sakov and Oke, 2008).

Many other variants of the KF are implemented in a similar way to the DEnKFs, i.e. ensemble forward propagation for forecasting and Kalman-based update with the observations, with the only differences in the formulation of the analysis step to update the forecast ensemble to the analysis ensemble. We cite here, for example, the unscented Kalman filter Julier et al. (2000); Julier and Uhlmann (2004); Luo and Moroz (2009), the divided difference filter (Ito and Xiong, 2000; Luo et al., 2012). However, these generally require an ensemble size that is larger than the dimension of ocean state, which makes them prohibitive for large-scale applications. In contrast, the EnKF formulation is more efficient and has found numerous applications in many ocean data assimilation problems.

3) ENSEMBLE OPTIMAL INTERPOLATION (EnOI)

Integrating large ensembles with an OGCM is computationally demanding. Following the optimal interpolation formulation of the DA problem, which uses a static pre-selected background covariance in the update step, ensemble optimal interpolation (EnOI) methods were proposed (Evensen, 2003; Hoteit et al., 2002; Oke et al., 2007). EnOI is a very cost-effective alternative to an EnKF, in which the static background covariance is estimated as the sample covariance matrix of an adequately pre-selected ensemble, generally describing the error growing modes or representing the variability of the studied ocean. Ocean large-scale dynamics evolve slowly within relatively short time windows, which justifies keeping the ensemble members static in time or over certain periods (Hoteit et al., 2002). In doing so, the EnKF reduces to an OI scheme in which only the analysis mean is integrated forward with the model for forecasting. The update step could be implemented based on a stochastic (e.g., SEnKF) update scheme (Counillon and Bertino, 2009) or a deterministic (e.g. SEIK) update scheme (Hoteit et al., 2002). Dropping the model integration of the ensemble members from the EnKF algorithm not only allows to drastically reduce the computational burden, but also to avoid the degeneracy of its members (more on this below). The method was found to be quite competitive compared to an EnKF at a fraction of the computational cost (Hoteit et al., 2002; Oke et al., 2007; Sakov and Sandery, 2015; Toye et al., 2017). However, its performance may be limited during periods of rapidly evolving dynamics, which are generally not well captured by a static background covariance (Hoteit et al., 2002; Hoteit and Pham, 2004). To account for the seasonal and intra-seasonal variability of the ocean flow, Xie and Zhu (2010) proposed to implement EnOI with an ensemble selected at every assimilation cycle from monthly climatology ocean states with a three-month moving window centered at the assimilation time. Currently, EnOI is used operationally in the Australian Bluelink system (Oke et al., 2008).

4) AUXILIARY TECHNIQUES TO ENHANCE THE PERFORMANCE OF ENKFS

Realistic EnKF ocean data assimilation problems are typically implemented with small ensembles of the order of 100 members or less (Aanonsen et al., 2009; Hoteit et al., 2015; Houtekamer and Zhang, 2016) to restrict the computational cost. The downside of using small ensembles is, however, twofold: (i) rank-deficient (low rank or degrees-of-freedom) forecast/background ensemble compared to the ocean state and observations dimensions (Hamill et al., 2009; Houtekamer and Zhang, 2016), and (ii) important Monte Carlo sampling errors (Anderson, 2012; Hamill et al., 2001; Luo et al., 2018). Together with the ubiquitous nonlinearity of the ocean dynamics, the implementation of EnKFs with small ensembles for OGCM DA requires explicit compensation for the effects of a finite ensemble. For instance, Bocquet et al. (2015) derived a prior probability density function conditional on the background ensemble to account for the sampling errors due to a small ensemble. Iterative methods were also introduced in the framework of the EnKFs to deal with strong nonlinearities. Many other auxiliary techniques have been proposed in the literature, including the most popular ones presented here: *covariance inflation*, *localization*, and *hybrid covariance*.

(i) Covariance inflation: As suggested by its name, covariance inflation "inflates" the covariance of an EnKF forecast, or analysis, ensemble by some positive factor at each assimilation cycle. The rationale behind covariance inflation can be explained from different points of views. It is, for instance, often justified based on the observation that the EnKF ensemble covariances are systematically underestimated due to the effect of finite ensembles (Whitaker and Hamill, 2002) and/or to account for neglected model errors (Pham et al., 1997; Hoteit and Pham, 2004). This helps mitigateg the ensemble collapse due to a lack of spread. In such cases, covariance inflation may be directly applied to the background ensemble through either an additive (Houtekamer and Mitchell, 2005; Lee et al., 2017; Yang et al., 2015) or multiplicative (Anderson and Anderson, 1999; Anderson, 2007a, 2009; Miyoshi, 2011) factor, or to the analysis ensemble through a certain relaxation term (Whitaker and Hamill, 2012; Zhang et al., 2004). An alternative point of view is to relate covariance inflation to robust filtering (Luo and Hoteit, 2011) in the context of an ensemble implementation of the H_∞ filter (Simon, 2006). This leads to different forms of inflation, including the conventional additive, multiplicative, or relaxation methods, and also encompasses the less conventional inflation methods such as those modifying the eigenvalues of the estimation error covariance in the ensemble space (Altaf et al., 2013b; Ott et al., 2004; Bai et al., 2016).

In practice, the value of the inflation factor is often set by trial and error, but adaptive inflation methods, spatially and in time, have gained popularity recently (e.g., Hoteit et al., 2002; Anderson, 2007a; Anderson, 2009; Li et al., 2009a; Bocquet, 2011; Luo and Hoteit, 2013; Miyoshi, 2011; and

Lee et al., 2017). In Anderson (2009), the inflation factor is treated as a random variable, and is then updated at each assimilation cycle. Similar ideas have been applied by Li et al. (2009a), Miyoshi (2011), and Gharamti (2018). Other approaches to "estimating" the value of the inflation factor have been also proposed, including the use of the forecast error statistics to guide the choice of the inflation factor to stabilize the EnKF and prevent filter divergence (Hoteit et al., 2005b; Luo and Hoteit, 2013, 2014c; Lee et al., 2017).

(ii) Localization: A small ensemble not only introduces spurious correlations between physically uncorrelated model variables in the ensemble covariance, but also provides limited degrees-of-freedom (rank of the ensemble) to fit the observations (Whitaker and Hamill, 2002). A straightforward way shown to be very efficient in many applications for dealing with such problems is to taper the long-range correlations in the EnKFs ensemble covariance matrices (Hamill et al., 2001; Houtekamer and Mitchell, 1998), a technique known as "localization." Most localization schemes are based on the distances between the physical locations of model variables and/or observations. For instance, Houtekamer and Mitchell (1998) introduced a local analysis scheme that updates an ocean state variable using only the observations located in its neighborhood. Hamill et al. (2001) adopted a covariance localization scheme in which one replaces the background covariance matrix with a Schur product between the background covariance matrix and a tapering matrix, whereas each element of the tapering matrix is computed using the Gaspari-Cohn function that depends on the physical distance between a pair of model variable and an observation (Gaspari and Cohn, 1999). For large-scale problems, directly manipulating the background covariance may become quite demanding in terms of computer memory. To alleviate this problem, other localization schemes have been proposed in which the Schur product is conducted between a certain tapering matrix and other quantities, such as the cross-covariance matrix between the forecast state and observation ensembles, the covariance matrix of the forecast observations ensemble (Houtekamer and Mitchell, 2001), and the Kalman gain matrix (Anderson, 2007b; Zhang and Oliver, 2010), or even the ensemble of forecast perturbations (Sakov and Bertino, 2011). Alternatively, to reduce the size of the involved matrices, localization has been also implemented in the observation space (Fertig et al., 2007) and in the context of a local EnKF (Bishop and Hodyss, 2007; Hunt et al., 2007; Ott et al., 2004).

The distance-based localization methods discussed above require that both the model variables and the observations have associated physical locations. In certain applications, however, some ocean variables/observations may not be associated with a physical location, e.g. non-local or spatial-temporal (or 4D) observations such as acoustics. In these situations, it becomes difficult to "localize" such observations with a distance-based localization method (Bocquet, 2016; Fertig et al., 2007; Luo et al., 2018). Therefore, adaptive localization methods have been proposed, some of them not based on physical distances. For instance, Bishop and Hodyss (2007) conducted a Schur product between the background error covariance matrix and a tapering matrix, whereas the latter was conducted by raising each element of a sample correlation matrix of model variables to a certain power. Anderson (2007b) and Zhang and Oliver (2010) used multiple background ensembles to compute a set of Kalman gain matrices, and then constructed the tapering matrices based on the sample statistics of the Kalman gain matrices. Anderson (2012, 2016) and De La Chevrotière and Harlim (2017) also proposed localization methods correcting for sampling errors of the correlation coefficients between the pairs of model variables and observations. Sample correlation coefficients were also used in adaptive localization schemes (Evensen, 2009; Rasmussen et al., 2015a). More recently, Luo et al. (2018) elaborated on how to "localize" non-local and/or 4D observations through detections of causal relations between model variables and simulated observations.

(iii) Hybrid covariance: The use of small ensembles generally means that a significant part of the state (error) space is not represented by the ensemble. This implies that the ensemble subspace will not offer enough degrees-of-freedom to fit a large number of observations, and produces unrealistic confidence in the filter forecast (Song et al., 2010). The hybrid EnKF-OI/3DVAR method (Hamill and Snyder, 2000) is another approach that one could consider to enhance the performance of the EnKF without significantly increasing its computational cost. In this method, and at every assimilation cycle, the filter forecast covariance is estimated as a linear combination of a flow-dependent ensemble covariance sampled by an EnKF and a (pre-selected) static background covariance, i.e.,

$$\mathbf{P}_k^{Hybrid} = \alpha \mathbf{P}_k^{EnKF} + \beta \mathbf{B}^{Static}, \tag{40}$$

as a way to compensate for the complement of the ensemble's subspace α and β are tuning parameters that are usually set by trial and error, but could also be optimized adaptively as in Gharamti et al. (2014a). This technique has been successfully applied in several ocean applications (see e.g. Counillon and Bertino, 2009; and Tsiaras et al., 2017) and was shown to be quite efficient at improving the EnKFs robustness and performances.

Other forms of hybrid methods have been also proposed, for example, the semi-evolutive SEIK filter (Hoteit et al., 2001) in which only a selected part of the ensemble is updated by the model, and the adaptive EnKF, which selects new members from a static ensemble to enrich the EnKF ensemble based on the analysis error (Song et al., 2010).

Computing the ensemble SEnKF update based on a hybrid covariance is rather straightforward; obtaining it from a square-root EnKF is not (Bocquet et al., 2015; Auligné et al., 2016). Currently, many DA systems, including most operational ones, make use of hybrid covariances.

(iv) Iterative EnKFs: To handle nonlinear observations, one may adopt an iterative optimization scheme for the update step, similar to the variational DA methods (Courtier et al., 1994). In the context of ensemble DA, one may interpret iterative methods through a Bayesian perspective (Emerick and Reynolds, 2012). Alternatively, one can recast ensemble DA as a stochastic optimization problem (Oliver et al., 1996), and solve the problem using different optimization algorithms. Various iterative methods were introduced in the context of the EnKF (Zupanski, 2005; Gu and Oliver, 2007; Lorentzen and Nævdal, 2011; Sakov et al., 2012; Bocquet and Sakov, 2012; Luo and Hoteit, 2014c; Gharamti et al., 2015a), the ensemble smoother (EnS) (Emerick and Reynolds, 2012; Chen and Oliver, 2013; Luo et al., 2015), and the iterative ensemble Kalman smoother (IEnKS) (Bocquet and Sakov, 2014).

Non-Gaussian Filtering

The different Kalman filter options presented above are all based, in some way or another, on Gaussian distributions for the background/forecast and the noise (and linear observation operator). Given the nonlinear nature of the ocean dynamics, the forecast distribution will not be Gaussian even when the analysis distribution at the previous step is Gaussian. As such, all Kalman-type filters are sub-optimal in the context of nonlinear Bayesian filtering. Relaxing the assumption of Gaussian distributions is an active area of research in ocean DA. This field is very well developed in the mathematics and electric engineering communities, and mathematically sound non-Gaussian Bayesian filters have been already developed, the most famous of which is the PF. The excessively large dimension of the ocean models means a prohibitive number of realizations to sample the ocean state distribution, which precludes any brute force implementation of these techniques for ocean DA. Given this hard constraint on the number of samples that could be considered in a realistic ocean application, our goal here should be more practical to derive approximate nonlinear/non-Gaussian Bayesian filtering schemes that are robust and efficient, in terms of computational cost and performance, at least competitive with EnKFs, for potential application on realistic ocean data assimilation problems.

Nonlinear/non-Gaussian Bayesian filtering recently became an active area of research in the ocean community, which has mainly focused on two types of filters, namely the PF (Gordon et al., 1993) and the GMF (Sorenson and Alspach, 1971). Both approaches resort to some (truncated) statistical mixture models to describe the forecast and analysis distributions of the Bayesian filter, so that an approximate numerical solution can be computed. The two filtering strategies are summarized below with appropriate references.

1) PARTICLE FILTERING (PF)

The PF uses mixture models of Dirac delta densities (or a random set of ocean states) to approximate/discretize the prior (forecast) and posterior (analysis) state distributions. More specifically, suppose that at the $(k-1)^{th}$ assimilation cycle, the posterior is approximated by

$$p_{\mathbf{x}_{k-1}|\mathbf{y}_0,\ldots,\mathbf{y}_{k-1}} \approx \sum_{i=1}^{N} w_{k-1}^{i} \delta(\mathbf{x}_{k-1} - \mathbf{x}_{k-1}^{a,i}), \qquad (41)$$

where δ denotes the Dirac delta function, $\mathbf{x}_{k-1}^{a,i}$ are the particles at the analysis step (similar to the ensemble members in the EnKF), w_{k-1}^{i} are the associated weights, and N is the total number of particles. The parameters of this mixture are then updated recursively based on the Bayesian filter steps (Gordon et al., 1993; Doucet et al., 2001; Van Leeuwen, 2009; Bocquet et al., 2010) as follows.

<ins>Forecast step</ins>: As in the EnKF, the analysis particles $\mathbf{x}_{k-1}^{a,i}$ are integrated forward with the dynamical model to obtain the forecast particles $\mathbf{x}_{k}^{f,i}$ at the next time t_k. The associated weights w_{k-1}^{i} remain unchanged.

Analysis step: The incoming observation y_k is used to update the weights only, while the particles themselves are kept unchanged, i.e. $\mathbf{x}_k^{a,i} = \mathbf{x}_k^{f,i}$. Roughly speaking, the particles will see their weights increase if they are close to y_k and decrease otherwise, according to

$$w_k^i = \frac{1}{c_k} w_{k-1}^i \, p(\mathbf{y}_k | \mathbf{x}_k^{f,i}), \tag{42}$$

where c_k is a constant that normalizes the weights of the posterior distribution, but does not need to be computed in practice. In the case of Gaussian observational error, the likelihood $p(y_k|\mathbf{x}_k^{f,i})$ becomes the Gaussian distribution of mean $H_k(\mathbf{x}_k^{f,i})$ and covariance $\mathbf{R}_k$.

After an analysis (or forecast) step, the MV estimate of the ocean state is then obtained as the weighted-average of the particles, i.e. $\sum_{i=1}^{N} w_k^i \mathbf{x}_k^{a(f),i}$.

The theory of the PF is well established and the convergence of the PF state distribution toward the Bayesian filter distribution has been proven given an infinite number of particles (Doucet et al., 2001). In practice, however, one is restricted to a finite number of particles and the PF will suffer from the degeneracy of its particles, a phenomenon in which most of the weights concentrate on very few particles after only a few assimilation cycles. The effective number of the particles then decreases (Snyder et al., 2008) and the filter often collapses. This happens because the particles drift away from the true state, with the observations exerting no feedback on the particles. To overcome this, a resampling technique is needed. The basic idea of resampling is to draw new particles according to their estimated weights, and then assign them uniform weights (Gordon et al., 1993).

Many forms of PFs have been suggested, mainly differing in their resampling strategies. There is a rich literature on this topic and readers are referred to, for example, Arulampalam et al. (2002), Doucet et al. (2001), Doucet and Johansen (2011), Duan et al. (2010) and the references therein, for more information. However, even with resampling, the PF still requires a large number of particles to achieve an accurate solution (Doucet et al., 2000b). This makes the brute force implementation of the PF for DA with computationally demanding, realistic OGCMs a challenging problem (Anderson, 2003; Snyder et al., 2008; Van Leeuwen, 2009). Several strategies are currently being investigated to enable the implementation of the PF for large-scale ocean DA problems, many of which try to somehow exploit the future observations for efficient sampling of a limited number of particles (e.g., Van Leeuwen, 2010; Chorin et al., 2010); split the ocean state into smaller vectors to reduce the dimension of the problem, a form of localization (e.g., Ait-El-Fquih and Hoteit, 2016; Penny and Miyoshi, 2016; Poterjoy, 2016); or apply some transformation to move the particles toward high-probability regions so that a single particle does not dominate the total weight (e.g., Luo and Hoteit, 2014a; Reich, 2013; El-Sheikh et al., 2014). A more recent school of thought is investigating combinations of the EnKF and the PF (Frei and Künsch, 2013; Shen and Tang, 2015; Ait-El-Fquih and Hoteit, 2017), exploiting the robustness of the EnKF and (asymptotic) optimality of the PF.

2) GAUSSIAN MIXTURE FILTERING (GMF)

The idea here is to use mixture models of Gaussian densities to approximate the state distributions, i.e., of the form

$$\sum_{i=1}^{N} w^i \Phi(\mathbf{x} : \mathbf{x}^i, \mathbf{P}^i), \tag{43}$$

where w^i are the weights of each Gaussian component Φ in the mixture of mean $\mathbf{x}^i$ and covariance $\mathbf{P}^i$. The $\mathbf{x}^i$s are also called particles in analogy to the PF. This is based on Alspach and Sorenson (1972) who demonstrated that when the likelihood is Gaussian and the observation operator is linear, a Bayesian update of a Gaussian mixture (GM) prior leads to a GM posterior for which the parameters of the Gaussian components are updated as in the KF, and their weights are updated as in the PF. Two GMF strategies can be then distinguished for ocean DA, depending on whether the forecast step is initiated from GM or Dirac mixture (DM) posteriors. If initiated from a GM, applying a local linearization around the centers of the Gaussian components of the mixture allows to carry the forecast step as a set of EKFs operating in parallel. More precisely, the GM posterior is integrated by an ocean model into a GM prior to the next assimilation step with the same weights, and centers and covariances computed from their prior counterparts using the EKF (Sorenson and Alspach, 1971; Chen and Liu, 2000; Bengtsson et al., 2003; Hoteit et al., 2008). In doing so, the prior and posterior distributions always remain as GMs, with the weights updated as in the PF, and the parameters of the Gaussian components as in the EKF, or any other nonlinear Kalman filter, such as an EnKF, (Hoteit et al., 2008; Luo et al., 2010; Hoteit et al., 2012; Sondergaard and Lermusiaux, 2013). This is generally referred to as Gaussian mixture filtering (GMF). If initiated from a DM, the forecast step is identical to that of the PF/EnKF, but the GM posterior needs to be resampled into a DM before forecasting (after every update step). We refer to this type of filtering the ensemble GMF (EnGMF; Anderson and Anderson, 1999; Liu et al., 2015, 2016). Resampling is then part of the EnGMF, but could be conducted only when needed (Hoteit et al., 2008, 2012). Even though the GMFs are less prone to the degeneracy of their particles, because the PF update of its weights is normalized by the covariance of the innovation vector (or observation prediction error) instead of the observation error covariance as in the standard PF, particles might still collapse and resampling a new GM with uniform weights may improve the filter robustness and performance (Hoteit et al., 2008). Another approach to mitigate the weight's collapse is to somehow follow an approach combining EnKF and PF, as suggested by Bengtsson et al. (2003).

GMFs are basically implemented as an ensemble of nonlinear KFs running in parallel, and as such can be computationally very demanding. Hoteit et al. (2008) suggested using a uniform low-rank covariance for all GM components to reduce the computational burden. Hoteit et al. (2012) later investigated a GMF in which each component of the forecast/analysis GM is represented by a different ensemble as a way to use a different covariance matrix for each Gaussian component. Sondergaard and Lermusiaux (2013) constructed the GM (from the forecast particles) in a reduced state space using an expectation-maximization (EM) algorithm. This enabled them to consider a variable number of Gaussian components in the GM, which was determined based on the Bayesian information criterion. The implementation of an EnGMF can be more straightforward, as it can be conveniently done via an EnKF code since they share the same forecast step, and a Kalman update step to each particle (or ensemble member). It is important to realize, however, that the EnGMF does not need to perturb the observations and may eventually use a different covariance matrix to update each particle depending on how the GM prior is estimated from the forecast particles (Liu et al., 2016). Of course one needs to add a resampling step for the EnGMF; a deterministic resampling step by analogy to the DEnKF was found beneficial when dealing with a small number of particles (Liu et al., 2015), but defining an optimal strategy for resampling from a GM remains an open problem.

Smoothing

Filters condition the ocean state with past (including current) observations, and are thus naturally designed for operational/forecasting settings. In contrast, smoothers condition the ocean state with past and future observations, exactly as 4DVAR. As such, smoothing estimates are expected to be more accurate than filtering estimates because of the information gain from future observations, and are thus more relevant to performing re-analyses. Smoothing is also useful to estimate model parameters whose influence on the state may be spread in time. Unlike 4DVAR, which is designed to compute the MAP estimator only, smoothers are usually implemented through additional update steps of the filters' estimates, with the future data, involving the associated distributions.

The ensemble smoothing update presented in this section is based on a linear analysis, which is an approximation for nonlinear models. More advanced but significantly more costly schemes that could go beyond the Gaussian approximation of the EnKF are available, such as the iterative ensemble Kalman smoother (Bocquet and Sakov, 2014) with multi-pass iterative updates. All these smoothers should be equivalent with linear model and Gaussian statistics (Cosme et al., 2012), and sometimes even with a nonlinear model (Raanes, 2016).

We are, therefore, interested here in estimating the ocean state $\mathbf{x}_k$ (in the past) at t_k, given all the observations up to T_L ($\geq t_k$). Let $\mathbf{E}_k^f$ be the matrix of the forecast ensemble members at t_k, column-wise, and consider the collection of these matrices within the assimilation window $[T_0\ T_L]$. This collection represents an ensemble of trajectories of the model, which can be seen as a prior within $[T_0\ T_L]$. Let $\mathbf{E}_k^a$ be the matrix of the EnKF-updated ensemble members at $t_k \in [T_0\ T_L]$ after assimilating the observation vector $\mathbf{y}_k$. Whatever the EnKF's choice, the ensemble member updates are linear combinations of the prior members, so that one can actually write,

$$\mathbf{E}_k^a = \mathbf{E}_k^f \mathbf{T}_k, \tag{44}$$

where $\mathbf{T}_k$ is an update transform matrix that depends on the observations, the prior and the EnKF flavor (Evensen, 2003). In the case of the stochastic and the deterministic EnKFs, these could be obtained from Eqs. 29 and 36, respectively.

The advantage of expressing the ensemble members, instead of the perturbations, update as in Eq. 44 is to include the analysis as part of the transformation. Then, the ensemble update at t_l ($t_k \leq t_l \leq T_L$) given this observation vector at t_k is

$$\mathbf{E}_l^a = \mathbf{M}_{k:l}\mathbf{E}_k^a = \mathbf{M}_{k:l}(\mathbf{E}_k^f \mathbf{T}_k) = (\mathbf{M}_{k:l}\mathbf{E}_k^f)\mathbf{T}_k = \mathbf{E}_l^f \mathbf{T}_k, \quad (45)$$

where $\mathbf{M}_{k:l}$ is the tangent linear model from t_k to t_l. As a consequence, to perform an update of the ensemble at any time within the assimilation window, one just needs to apply the transform $\mathbf{T}_k$ on the right to the ensemble. This generalizes to the case where $T_0 \leq t_l \leq t_k \leq T_L$ by setting $\mathbf{M}_{k:l} = \mathbf{M}_{k:l}^{-1}$.

From this translational invariance of the update, which was put forward by Evensen (2003) and Hunt et al. (2004), one can build the following ensemble Kalman smoothing (EnKS) algorithm. First, run the EnKF throughout the assimilation window up to T_L and store the ensemble $\mathbf{E}_l^a$ and the ensemble transform matrices $\mathbf{T}_l$ at any time t_l where a smoothing analysis is needed. Second, to compute the smoothing updated ensemble $\mathbf{E}_k^s$ at t_k within the assimilation window, one simply needs to apply the translational invariance principle and use the transform matrices from t_l to T_L:

$$\mathbf{E}_k^s = \mathbf{E}_k^a \prod_{l=k}^{L} \mathbf{T}_l. \quad (46)$$

Hence, this smoother consists of two passes: an EnKF forward run, followed by a retrospective update of the ensembles (that need updating). This makes this EnKS a very elegant procedure, but with the drawback of requiring important storage.

Domain localization of the EnKS can be implemented as in the EnKF. A common mistake is to enforce inflation, often required by an EnKF, in the forward EnKF pass as well as in the backward smoothing pass. Instead, inflation should only be applied in the forward EnKF step, since sampling errors counteracted by inflation have already been accounted for in the forward pass. Cosme et al. (2010) and Nerger et al. (2014) offer examples of application of the EnKS in ocean data assimilation.

Hybrid Ensemble-Variational Methods

As discussed in the previous sections, there are benefits and drawbacks in using an EnKF or a 4DVAR. The EnKF involves a flow-dependent representation of the errors, via the ensemble. In applications involving high-dimensional models, this representation is nevertheless rank-deficient and relies on localization. The computational cost of the ensemble propagation is demanding, but this can be mitigated by trivial parallelization. On the downside, the EnKF is generally efficient for moderate model nonlinearity because of its second-order moments approximation of the error statistics. 4DVAR, as a nonlinear variational method, can handle nonlinearities. This, however, requires the adjoint of the observation and propagation models, which is a strong technical drawback as it is time-consuming to derive and maintain the adjoint code. Moreover, the traditional 4DVAR does not propagate the flow-dependent error statistics, but only the state analysis. This is however mitigated by the possibility of using a full-rank background error covariance matrix, which is not possible in a standard EnKF. Finally, 4DVAR does not lend itself easily to parallelization. A full discussion can be found in Lorenc (2003), Kalnay et al. (2007), and Chapter 7 of Asch et al. (2016).

There have been various attempts to merge these two types of methods in order to combine their strengths while avoiding some of their drawbacks, which we refer to as hybrid or hybrid ensemble-variational methods.

(i) Ensemble of Data Assimilations: To implement flow-dependence of the error representation in a DA system, one could consider an ensemble of such systems. The outcomes of the analyses from the DA ensemble could then be combined to form a flow-dependent background matrix, as in the EnKF. The implementation of such an approach is straightforward, requiring minimal changes to an existing operational DA system. It is called ensemble of data assimilations, and has mostly been used in National Weather Prediction centers that operate 4DVAR assimilation and, in particular, Météo-France (Raynaud et al., 2009; Berre et al., 2015) and the European Centre for Medium-Range Weather Forecasts (ECMWF; Bonavita et al., 2011, 2012). In such a setting, each 4DVAR analysis, indexed by i, uses a different first guess $\mathbf{x}_0^i$ in the optimization procedure, and perturbed observation vectors $\{\mathbf{y}_k^i\}_{k=0,\ldots,L}$, as in the SEnKF, to maintain statistical consistency. The background covariance is typically a hybrid one as in Eq. 40, comprised of the static covariances of the traditional 4DVAR and incorporating the sample covariances from the prior ensemble. The resulting ensemble of analyses is then propagated to the next analysis. The sample covariances may need to be regularized by localization. In this context, covariance localization has been implemented via the so-called α control variable trick (Lorenc, 2003; Buehner, 2005; Wang et al., 2007), or using wavelet truncations (Berre et al., 2015).

(ii) Iterative Ensemble Kalman Filter and Smoother: Most ensemble variational and hybrid methods have been designed empirically. By contrast, the iterative ensemble Kalman smoother (IEnKS) has been derived from Bayes' rule (Bocquet and Sakov, 2013, 2014) with well-identified approximations. The analysis step of the IEnKS consists in a nonlinear minimization of the 4DVAR cost function over an assimilation window $[T_0, T_L]$, with the goal of estimating the initial state at T_0, assuming a perfect model. It uses a previously forecast ensemble at T_0 to generate second-order background statistics in exactly the same way as the ETKF. Different from 4DVAR, the optimization is carried out in the space spanned by the ensemble rather than the full state-space. This restricts the search of the analysis state, but enables the use of efficient minimization techniques such as Gauss-Newton, Levenberg-Marquardt, and trust-region methods. Analogous to the strong 4DVAR cost function in Eq. 17, the IEnKS cost function is typically of the form

$$J_{IEnKS}(\mathbf{w}) = \|\mathbf{w}\|^2 + \sum_{k=L-S+1}^{L} \left\| \mathbf{y}_k - H_k\left(M_{0:k}(\mathbf{x}_0^b + \mathbf{S}_0^b \mathbf{w})\right) \right\|_{\mathbf{R}_k^{-1}}^2, \qquad (47)$$

where $\mathbf{w}$ is the vector of unknown coordinates in the ensemble space, $\mathbf{x}_0^b$ and $\mathbf{S}_0^b$ are respectively the prior ensemble mean and the matrix of the prior ensemble anomalies at T_0, and $M_{0:k}$ is the model resolvent integrating the ocean state from T_0 to t_k. The sensitivities of the observations to $\mathbf{w}$ required for the calculation of the gradient of the cost function can be computed by finite-difference methods, or using finite-spread representation as in the EnKF. This avoids the need for the tangent-linear and adjoint models. Once a minimum is found, an approximate Hessian can be obtained at the corresponding minimizer in the ensemble space. Just like with the ETKF, this allows to generate an updated ensemble of anomalies. In the forecast step, the newly generated ensemble is integrated with the model from t_L to t_{L+S}, to be used as the prior of the next assimilation cycle. The only approximations in the scheme are the Gaussian modeling of the prior and posterior ensemble, and, to a lesser extent, the finite-size of the ensemble.

The "shift" parameter S can obviously be chosen to be L or, more generally, any value between 1 and $L + 1$, if all observations are to be assimilated. This degree of freedom, barely exploited in 4DVAR, is interesting because it allows flexibility in the transfer of information from one assimilation window to the next through the ensemble. The case $S = 1$, $L = 0$ with a linear observation model, coincides with the ETKF (Hunt et al., 2007). The case $S = 1$, $L = 0$ and a general (nonlinear) observation model coincides with the maximum likelihood ensemble filter (MLEF) (Zupanski, 2005). The case $S = 1$, $L = 1$ is known as the iterative ensemble KF (IEnKF, Sakov et al., 2012; Bocquet and Sakov, 2012). The case $S = L$ with a single iteration in the analysis and further mild restrictions corresponds to the 4D-ETKF (Hunt et al., 2004).

The name smoother comes from Bell (1994), who first proposed an iterative (full-rank) Kalman smoother. The IEnKS can be used as a smoother as well as a filter. As opposed to the EnKS, and with a nonlinear model, each iteration should improve the filtering solution, i.e., the state estimate at the present time. With low-order chaotic models, the IEnKS has been shown to outperform any scalable known method such as the EnKF, the EnKS, and 4DVAR, for both filtering and smoothing.

As any ensemble method, the IEnKF and IEnKS require localization. This is not as straightforward to implement as in the EnKF since it should be applied within the full assimilation window. Because the dynamics of the error do not generally commute with localization (Fairbairn et al., 2014; Bocquet and Sakov, 2014), suboptimalities can appear for long windows. A solution—the so-called dynamically covariant localization—has been proposed in Bocquet (2016) and yields good results with simple models. If the adjoint model is available, then this issue could be dealt with in a more efficient way (Bocquet, 2016).

The IEnKS is a mathematically justified Hybrid Ensemble-Variational method that can also be also useful in understanding and rationalizing the so-called Four-Dimensional Ensemble Variational (4DEnVAR) methods.

(iii) Four-Dimensional Ensemble Variational (4DEnVAR): This class of methods was developed by the National Weather Prediction centers within a 4DVAR framework. The primary goal was to avoid maintaining the adjoint of the forecast model. Like the IEnKS and similar to the reduced 4DVAR put forward early in oceanography by Robert et al. (2005) and Hoteit and Köhl (2006), the analysis is performed within the subspace spanned by the ensemble (Liu et al., 2008). Observation perturbations are usually generated stochastically using, for instance, a stochastic EnKF (Liu et al., 2009; Buehner et al., 2010a). Hence, in addition to avoiding the need for an adjoint model, flow-dependent error estimation is introduced. The 4DEnVAR implementations usually come with a hybrid background. Just like the IEnKS, localization is necessary and theoretically more challenging than with an EnKF (Desroziers et al., 2016). Many 4DEnVAR variants have been

suggested, depending on the availability of the adjoint models and how the perturbations are generated (Buehner et al., 2010a,b; Zhang and Zhang, 2012; Clayton et al., 2013; Poterjoy and Zhang, 2015). Several 4DEnVar weather systems are now operational or on the verge of being so (Buehner et al., 2013; Gustafsson et al., 2014; Desroziers et al., 2014; Lorenc et al., 2015; Kleist and Ide, 2015; Buehner et al., 2015; Bannister, 2017). A recent further sophistication is to construct an ensemble of data assimilations of 4DEnVAR in order to generate the perturbations (Bowler et al., 2017; Arbogast et al., 2017).

Discussion and Future Developments

The theoretical framework of ocean DA methods is now well established around the Bayesian estimation theory, and many robust methods have been developed for efficient assimilation of available ocean data into state-of-the-art general circulation ocean models. These provide various tools to predict the past and/or future state of the ocean, conditioned on available data, and also to quantify its uncertainties. Evaluating the uncertainties is important for decision making and proper weighting of the most recent estimate against incoming observations for computing the next estimate. At present, 4DVAR and EnKFs are the state-of-the-art techniques for ocean data assimilation, and most ocean centers are currently developing their operational systems around these approaches. Both have been extensively studied and their characteristics are now well understood. They were further shown to provide viable solutions in many ocean DA applications.

4DVAR and EnKF have their own advantages and weaknesses, and this lead to the development of a new class of Hybrid Ensemble-Variational methods, aiming at combining the strength of these two approaches. The hybrid techniques are currently being actively investigated by the atmospheric community, but nothing should preclude their applications for ocean DA. Other advances in the developments of auxiliary techniques that were proven important for enhancing the performances of 4DVAR and EnKF are also expected to improve the accuracy of the ocean DA products, which may include working on more sophisticated parameterizations of (ensemble) covariance matrices, enhanced resampling techniques, enforcing dynamical balances and constraints, etc.

In addition, recent advances in the development of non-Gaussian/nonlinear DA methods for large-scale DA problems hold the promise of edging us closer toward the optimal Bayesian estimate. These techniques were indeed demonstrated to be superior to EnKFs and 4DVAR in many idealized DA problems. However, many challenges still require more investigations into, for example, how best to parameterize the involved probability distributions or to resample a reasonable-size ensemble/sample from the estimated distributions. These should be further complemented by further studies to better understand the relevance of more advanced assimilation for the estimation of the various ocean spatial and temporal scales (Subramanian et al., 2012), particularly in relation to the associated computational cost and the spatial and temporal coverage of the assimilated observations.

All of these developments focused on introducing new DA techniques or enhancing existing ones. Another very important aspect of any assimilation system is the treatment of the underlying uncertainties in the system, which basically characterize the final assimilation solution. These, however, have not received the attention they deserve by the ocean data assimilation community, with the exception of some uncoordinated attempts to address some of the challenges, and the problem remains open. Uncertainties can be inherited from the incomplete ocean models dynamics, measurements sensors, and various parameters and inputs. These could also be introduced by poor knowledge of the statistics of the modeled noise, which are required by the assimilation algorithms. There is a variety of approaches to quantify, reduce, and account for the uncertainties in an assimilation system that are yet to be fully exploited by the ocean community. We end this chapter with a general discussion and a summary of available promising tools to efficiently handle the uncertainties in an ocean DA system.

(i) State-Parameters Estimation: Various parameterizations are used in the general circulation ocean models (e.g., mixing parameters) and, as such, the ocean state depends on a set of (physical) parameters that may not be perfectly known. Estimating these parameters along with the state should reduce the uncertainties in the system and eventually lead to an improved final assimilation solution.

Estimating the parameters of an ocean model is technically an inverse problem that can be addressed by a deterministic (least-squares or variational-like) approach aiming at determining the parameters that best fit the model to available data, or as a Bayesian inverse problem computing the posterior distribution of the parameters given a prior and the likelihood (e.g., Aster et al., 2005; Tarantola, 2005). The first approach is basically an optimization problem, very similar to 4DVAR (Wunsch, 1996; Elbern et al., 2007). Markov Chain Monte Carlo (MCMC) is the reference numerical method to compute the posterior solution of the Bayesian inverse problem, but it can require a prohibitive number of ocean model runs to converge (Metropolis et al., 1953; Malinverno, 2002; Sraj et al., 2016). Surrogate models have been used to enable the application of such methods for large-scale problems, see for example Li et al. (2015), Sraj et al. (2016) and Sripitana et al. (2017) for ocean applications. An advantage of DA, even though its methods are often suboptimal, is that its framework may allow for joint state-parameters estimations. The 4DVAR framework readily enables for ocean parameters estimation (Wunsch and Heimbach, 2007; Liu et al., 2012) within a strong or weak constraint formulation, but it is an offline approach. The recursive nature of filters and smoothers allows for simultaneous online estimation of the state and parameters, and accordingly straightforward updates with new observations.

The technical framework for online estimation of the state and the parameters of a dynamical system based on observations is well established (Harvey and Phillips, 1979; Pagan, 1980; Hamilton, 1986; Aksoy et al., 2006; Rasmussen et al., 2015b; Ait-El-Fquih et al., 2016) and is already being heavily exploited in hydrology and subsurface flow applications (e.g., Oliver and Chen, 2011; Moradkhani et al., 2005; Gharamti et al., 2015b). It has been recently applied in the context of a storm surge ocean model by Sripitana et al. (2018). Within the Bayesian estimation framework, and similarly to the state, the parameters are treated as random variables with a given prior distribution. The state-parameters estimation problem then consists of computing the distribution of the augmented (or joint) state-parameters vector, say $[\mathbf{x}, \theta]$, conditioned on available observations. The assimilation methods presented in this chapter can be then directly used to address this problem by applying them to [x, θ] instead of x, assuming constant dynamics for the time-evolution of the parameters. This is the joint state-parameters estimation approach. Another method that became popular more recently is the so-called dual approach, which consists of separately updating the state and the parameters using two interactive filters, one acting on the parameters and the other on the state given the parameters. This approach is commonly referred to as Rao-blackwelisation in the PF community where it has been introduced to reduce the variance of the joint state and parameters estimation (Doucet et al., 2000a). It was applied early on in an ensemble context in hydrology (Moradkhani et al., 2005) to mitigate for noticeable inconsistencies issues in the joint approach (Chen and Zhang, 2006), but only recently was used in a Bayesian consistent framework (Ait-El-Fquih et al., 2016). The dual estimation approach has been proven to be especially beneficial in strongly nonlinear applications (Gharamti et al., 2014b; Ait-El-Fquih et al., 2016).

Another state-parameters estimation problem that is important to address considers the statistical parameters instead of (or together with) the physical parameters. This mainly involves the parameters of the dynamical model errors and observations noise distributions (Särkkä and Hartikainen, 2013; Ardeshiri et al., 2015), and could be particularly useful in the context of stochastic parameterizations. It should help sampling relevant stochastic terms based on the available data and the dynamics in hand.

The most common approach to simultaneously estimate these so-called hyper-parameters along with the state consists of maximizing the likelihood of the whole set of available observations given these parameters (Hamilton, 1986). However, because of the complexity of the likelihood function, an analytical evaluation of the maximum likelihood estimate of the hyper-parameters is generally not feasible. A number of approximate numerical solutions have therefore been proposed, mostly relying on gradient-based iterative optimization methods (Harvey and Phillips, 1979; Pagan, 1980). To overcome the common limitations of gradient-based methods, the popular iterative Expectation-Maximization (EM) algorithm has been first introduced to linear systems (Shumway and Stoffer, 1982), and later extended to nonlinear systems involving PF- and EnKF-like assimilation algorithms (e.g., Cappé et al., 2005; Stroud and Bengtsson, 2007; Frei and Künsch, 2012; Ueno et al., 2010; and Dreano et al., 2017). The EM iterations alternate between an expectation (E) step, which constructs an expectation-type cost function of the log-likelihood evaluated at the parameters current estimate, and a maximization (M) step, which computes new parameters maximizing the expected log-likelihood created in the E step. The resulting parameters' estimates are then used to determine the distribution of the hyper-parameters in the following E step.

More recently a Bayesian approach has been proposed, considering the hyper-parameters as random variables with a prior distribution, commonly a Wishart distribution. This leads to a posterior that is also a Wishart distribution and for which sufficient statistics can be obtained in closed forms (Robert, 2007). The Bayesian approach is an online procedure that computes a new estimate at each observation time step, while the maximum likelihood approach is viewed as an off-line (or smoother) approach. Furthermore, the maximum likelihood approach computes only a point estimate of the hyper-parameters, while the Bayesian approach provides a full distribution from which any (point) estimate can be obtained.

(ii) Accounting for Model Errors: It is important to properly quantify the statistics of the forecast (background) in the assimilation (Dee, 1995; Li et al., 2009b). In the description of the state-space model (dynamical and observation models) of the DA problem, model errors were traditionally represented as white noise, Gaussian of mean zero and a given covariance. This assumption is at the basis of the KF- and 4DVAR- like assimilation algorithms, even though the contribution of this additive error term, through its covariance, is often treated crudely or even neglected. The model bias, or mean of model errors, can be conveniently treated as a state-parameters estimation problem (Zupanski and Zupanski, 2006) or closely related techniques (Dee and da Silva, 1998; Chepurin et al., 2005). This can be further extended to the case of time-dependent bias, for which a computationally efficient scheme expressing the errors in terms of very few degrees-of-freedom has been proposed by Danforth and Kalnay (2008). The dimension of the model error covariance can be prohibitively large; accurately estimating its coefficients would require a large amount of data that is simply not available (Mehra, 1970).

In 4DVAR, the model error is accounted for as a control term in the objective cost function (Trémolet, 2007). Strong constraint 4DVAR completely ignores it to simplify its optimization problem, projecting all errors on the initial conditions or further on other model parameters and inputs in slightly modified variants. Weak constraint 4DVAR methods, such as the representer method, provide elegant ways to directly tackle the problem but the end result heavily depends on the specification of appropriate model error covariance matrices to properly weight these terms in the cost function and spread the information to non-observed locations (Di Lorenzo et al., 2007). Imposing dynamical constraints is also important, as for the background covariance (Ngodock and Carrier, 2014).

EnKFs (smoothers) algorithms can directly handle the model error covariance matrix as in the KF (smoother), but this is not as straightforward for the large dimensional ocean DA problem. In this context, it is more efficient to exploit the Monte Carlo framework of the ensemble methods and account for the model errors through perturbations sampled from given statistics (covariance) (Mitchell and Houtekamer, 2000; Hamill and Whitaker, 2005; Hoteit et al., 2007), as in a fully nonlinear DA scheme. Likewise, this could be also applied in the case of non-additive noise, e.g. as a stochastic parameterization scheme (Buizza et al., 1999; Wu et al., 2008). These methods, or a combination of them, may provide an efficient approach, computationally and dynamically, to account for ocean models errors in an ensemble (nonlinear) setting. Additive or not, the parameters and statistics of such terms could be quantified using the state-parameters and hyper-parameters methods discussed in the previous section (Ardeshiri et al., 2015; Dreano et al., 2017). One may also opt for an offline approach to quantify the statistics of the model errors or of their parameterizations based on the available data (Daley, 1992) or a set of assimilated increments (analysis minus model forecast), which became popular recently (van Leeuwen, 2015). One may further consider identifying missing physics in the model following a similar approach (van Leeuwen, 2015).

(iii) Coupled Data Assimilation: Building more accurate ocean models with better predictive skills can be achieved through parameters estimation and quantification of modeling errors, but also by implementing more complete dynamics. Ocean models are designed to be forced by prescribed atmospheric fields, a framework that does not allow to fully account for the ocean feedback to the atmosphere. This is also true for the atmospheric models, which require ocean surface fields on the ocean-atmosphere boundary layer. Following the continuous progress in computing resources and the desire to solve the most complete dynamics and extend the predictability of the atmospheric and oceanic forecasting systems, fully coupled ocean-atmosphere models have been under development for many years (e.g., Delworth et al., 1993; Stockdale et al., 1998). Another important benefit from models coupling is that it provides a dynamical framework to exploit the observations of the state of one system in the assimilation of the other model, e.g. use atmospheric observations to update the ocean model state, and vice-versa. This is referred to as the coupled ocean-atmosphere data assimilation (CDA) problem.

CDA allows to exchange information between the models through the coupling dynamics and the assimilated observations. This should further provide state estimates that are more consistent with the dynamics of both systems. Nowadays, most operational centers agree that CDA is the goal for analysis and prediction of the climate system, particularly on subseasonal-to-seasonal and longer timescales (Penny and Hamill, 2017). The most straightforward CDA method would be to follow an augmented state approach in which the states and observations of the coupled models are concatenated into one state vector and one observation vector, basically considering the coupled system as a single state-space model. One could then directly apply any of the DA methods presented in this chapter for simultaneous assimilation of all available observations.

Because of the complexity of coding and maintaining the adjoint of a coupled ocean-atmosphere system, the non-intrusive and portable nature of the EnKF algorithms have made them more popular for CDA (Tardif et al., 2015; Sluka et al., 2016). These methods may also offer more flexibility in implementation, such as using different filters and ensemble sizes for each model (Luo and Hoteit, 2014b). However, the multi-scale nature, in space and time, of the dynamics of the coupled system requires revisiting the calculations of the ensembles cross-correlations (Luo and Hoteit, 2014b; Lu et al., 2015) and accordingly adapting the necessary auxiliary techniques, such as the ensemble localization and inflation (Frolov et al., 2016). Another important complexity in CDA is related to the turbulent dynamics of the atmospheric component together with the longer than synoptic timescales that make coupled problems interesting, but difficult to handle with a linear ensemble assimilation technique. Nonlinear/non-Gaussian assimilation methods may thus prove to be useful for such systems. However, the algorithmic and computational complexity of the coupled system, involving two or more different general circulation models, remains an important factor in limiting the development of CDA systems. To simplify the CDA problem, the notion of "weak" CDA has been introduced (e.g. Lu et al., 2015), in which each model of the coupled system assimilates its own observations. The coupling is achieved only via the forecast step, as opposed to performing coupled analysis steps, which is referred to as "strong" CDA. Weak CDA relies solely on the coupling dynamics to spread the observations information between the models and, thus, may miss opportunities to exploit some useful information from the observations of the other models during assimilation. Nevertheless, the development of CDA systems is underway and the characteristics of cross-covariances between the errors in the atmosphere and ocean model forecasts are being explored (Smith et al., 2017a), demonstrating the additional potential of the "strong" CDA.

Despite these challenges, the promises of delivering more skillful assimilative and predictive models, and particularly long-term (e.g., subseasonal-to-seasonal and longer) forecasts, will make CDA an active area of research in the years to come. Ocean models are also often coupled with many other components of the earth system, such as wave models to better describe wave-ocean interactions and properly resolve the surface roughness that is important for the atmosphere, various transport models for tracking, and biogeochemical models to simulate the ocean ecosystem variability. This presents endless possibilities for developing multi-assimilative models that combine different models, observations, and assimilation methods.

Acknowledgements

Xiaodong Luo was partly supported by the project "Advancing permafrost carbon climate feedback-improvements and evaluations of the Norwegian Earth System Model with observations" funded by the Research Council of Norway (ID 250740), and the project (ID 230303) funded by the Research Council of Norway and the industry partners – ConocoPhillips Skandinavia AS, Aker BP ASA, Eni Norge AS, Maersk Oil Norway AS, DONG Energy A/S, Denmark, Statoil Petroleum AS, ENGIE E&P NORGE AS, Lundin Norway AS, Halliburton AS, Schlumberger Norge AS, Wintershall Norge AS – through The National IOR Centre of Norway. CEREA is a member of Institute Pierre-Simon Laplace (IPSL). The authors would like to thank Naila Raboudi for her help and comments.

References

Aanonsen, S., G. Nævdal, D. Oliver, A. Reynolds, and B. Valls, 2009: The ensemble Kalman filter in reservoir engineering: a review. SPE Journal, 14, 393–412, SPE-117274-PA.

Abarbanel, H., 2012: Analysis of observed chaotic data. Springer Science & Business Media.

Ait-El-Fquih, B., M. E. Gharamti, and I. Hoteitl, 2016: A Bayesian consistent dual ensemble Kalman filter for state-parameter estimation in subsurface hydrology. Hydrology and Earth System Sciences, 20, 3289–3307.

Ait-El-Fquih, B., and I. Hoteit, 2016: A variational Bayesian multiple particle filtering scheme for large-dimensional systems. IEEE Transactions on Signal Processing, 64 (20), 5409–22.

Ait-El-Fquih, B., and I. Hoteit, 2018: An efficient state-parameter filtering scheme combining ensemble Kalman and particle filters. Mon. Wea. Rev., https://doi.org/10.1175/MWR-D-16-0485.1.

Aksoy, A., F. Zhang, and J. Nielsen-Gammon, 2006: Ensemble-based simultaneous state and parameter estimation with MM5. Geophysical Research Letters, 33, L12 801.

Alspach, D., and H. Sorenson, 1972: Nonlinear Bayesian estimation using Gaussian sum approximations. IEEE transactions on automatic control, 17 (4), 439–448.

Altaf, M. U., M. E. Gharamti, A. W. Heemink, and I. Hoteit, 2013a: A reduced adjoint approach to variational data assimilation. Computer Methods in Applied Mechanics and Engineering, 254, 1–13.

Altaf, U., T. Buttler, T. Mayo, C. Dawson, A. Heemink, and I. Hoteit, 2014: A comparison of ensemble Kalman filters for short range storm surge assimilation. Mon. Wea. Rev., 142, 2899–2914.

Altaf, U. M., T. Butler, X. Luo, C. Dawson, T. Mayo, and H. Hoteit, 2013b: Improving short range ensemble Kalman storm surge forecasting using robust adaptive inflation. Mon. Wea. Rev., 141, 2705–2720.

Anderson, B. D. O., and J. B. Moore, 1979: Optimal filtering. Englewood Cliffs, 21, 22–95.

Anderson, J. L., 2001: An ensemble adjustment Kalman filter for data assimilation. Mon. Wea. Rev., 129, 2884–2903.

Anderson, J. L., 2003: A local least squares framework for ensemble filtering. Mon. Wea. Rev., 131 (4), 634–642.

Anderson, J. L., 2007a: An adaptive covariance inflation error correction algorithm for ensemble filters. Tellus, 59A (2), 210–224.

Anderson, J. L., 2007b: Exploring the need for localization in ensemble data assimilation using a hierarchical ensemble filter. Physica D: Nonlinear Phenomena, 230 (1), 99–111.

Anderson, J. L., 2009: Spatially and temporally varying adaptive covariance inflation for ensemble filters. Tellus, 61A, 72–83.

Anderson, J. L., 2010: A non-Gaussian ensemble filter update for data assimilation. Mon. Wea. Rev., 138, 4186–4198.

Anderson, J. L., 2012: Localization and sampling error correction in ensemble Kalman filter data assimilation. Mon. Wea. Rev., 140 (7), 2359–2371.

Anderson, J. L., 2016: Reducing correlation sampling error in ensemble Kalman filter data assimilation. Mon. Wea. Rev., 144 (3), 913–925.

Anderson, J. L., and S. L. Anderson, 1999: A Monte Carlo implementation of the nonlinear filtering problem to produce ensemble assimilations and forecasts. Mon. Wea. Rev., 127, 2741–2758.

Arbogast, E., G. Desroziers, and L. Berre, 2017: A parallel implementation of a 4DEnVar ensemble. Q. J. R. Meteor. Soc., doi:10.1002/qj.3061.

Ardeshiri, T., E. Özkan, U. Orguner, and F. Gustafsson, 2015: Approximate Bayesian smoothing with unknown process and measurement noise covariances. IEEE Signal Processing Letters, 22 (12), 2450–2454.

Arulampalam, M. S., S. Maskell, N. Gordon, and T. Clapp, 2002: A tutorial on particle filters for online nonlinear/non-Gaussian Bayesian tracking. IEEE Transactions on Signal Processing, 50, 174–188.

Asch, M., M. Bocquet, and M. Nodet, 2016: Data assimilation: Methods, algorithms, and applications. SIAM.

Aster, R. C., B. Borchers, and C. H. Thurber, 2005: Parameter Estimation and Inverse Problems. Elsevier, New York, 301 pp.

Auligné, T., B. Ménétrier, A. C. Lorenc, and M. Buehner, 2016: Ensemble–variational integrated localized data assimilation. Mon. Wea. Rev., 144 (10), 3677–3696.

Bai, Y., Z. Zhang, Y. Zhang, and L. Wang, 2016: Inflating transform matrices to mitigate assimilation errors with robust filtering based ensemble Kalman filters. Atmospheric Science Letters, 17, 470–478.

Balmaseda, M. A., K. Mogensen, and A. T. Weaver, 2013: Evaluation of the ECMWF ocean reanalysis system ORAS4. Q. J. R. Meteor. Soc., 139 (674), 1132–1161.

Bannister, R. N., 2017: A review of operational methods of variational and ensemble-variational data assimilation. Q. J. R. Meteor. Soc., 143 (703), 607–633.

Barth, A., A. Alvera-Azcárate, J.-M. Beckers, M. Rixen, and L. Vandenbulcke, 2007: Multigrid state vector for data assimilation in a two-way nested model of the ligurian sea. J. Mar. Sys., 65 (1), 41–59.

Bell, B. M., 1994: The iterated Kalman smoother as a Gauss–Newton method. SIAM Journal on Optimization, 4 (3), 626–636.

Bengtsson, T., C. Snyder, and D. Nychka, 2003: Toward a nonlinear ensemble filter for high-dimensional systems. Journal of Geophysical Research: Atmospheres, 108 (D24).

Bennett, A. F., 2005: Inverse modeling of the ocean and atmosphere. Cambridge University Press.

Berre, L., H. Varella, and G. Desroziers, 2015: Modelling of flow-dependent ensemble-based background-error correlations using a wavelet formulation in 4DVar at Météo-France. Q. J. R. Meteor. Soc., 141 (692), 2803–2812.

Bishop, C. H., B. J. Etherton, and S. J. Majumdar, 2001: Adaptive sampling with ensemble transform Kalman filter. Part I: Theoretical aspects. Mon. Wea. Rev., 129, 420–436.

Bishop, C. H., and D. Hodyss, 2007: Flow-adaptive moderation of spurious ensemble correlations and its use in ensemble-based data assimilation. Q. J. R. Meteor. Soc., 133, 2029–2044.

Blayo, E., M. Bocquet, E. Cosme, and L. F. Cugliandolo, 2014: Advanced data assimilation for geosciences. Oxford University Press, lecture Notes of the Les Houches School of Physics: Special Issue, 608 pp.

Blockley, E. W., M. J. Martin, A. J. McLaren, A. G. Ryan, J. Waters, D. J. Lea, I. Mirouze, K. A. Peterson, A. Sellar, and D. Storkey, 2014. Recent development of the Met Office operational ocean forecasting system: an overview and assessment of the new Global FOAM forecasts. Geosci. Model. Dev., 7, 2613–2638.

Bocquet, M., 2011: Ensemble Kalman filtering without the intrinsic need for inflation. Nonlin. Proc. Geophys., 18, 735–750.

Bocquet, M., 2016: Localization and the iterative ensemble Kalman smoother. Q. J. R. Meteor. Soc., 142, 1075–1089.

Bocquet, M., C. Pires, and L. Wu, 2010: Beyond Gaussian statistical modeling in geophysical data assimilation. Mon. Wea. Rev., 138 (8), 2997–3023.

Bocquet, M., P. N. Raanes, and A. Hannart, 2015: Expanding the validity of the ensemble Kalman filter without the intrinsic need for inflation. Nonlin. Processes Geophys., 22, 645.

Bocquet, M., and P. Sakov, 2012: Combining inflation-free and iterative ensemble Kalman filters for strongly nonlinear systems. Nonlin. Processes Geophys., 19 (3), 383–399.

Bocquet, M., and P. Sakov, 2013: Joint state and parameter estimation with an iterative ensemble Kalman smoother. Nonlin. Processes Geophys., 20 (5), 803–818.

Bocquet, M., and P. Sakov, 2014: An iterative ensemble Kalman smoother. Q. J. R. Meteor. Soc., 140 (682), 1521–1535.

Bonavita, M., L. Isaksen, and E. Hólm, 2012: On the use of EDA background error variances in the ECMWF 4DVar. Q. J. R. Meteor. Soc., 138 (667), 1540–1559.

Bonavita, M., L. Raynaud, and L. Isaksen, 2011: Estimating background-error variances with the ECMWF ensemble of data assimilations system: Some effects of ensemble size and day-to-day variability. Q. J. R. Meteor. Soc., 137 (655), 423–434.

Bouttier, F., and P. Courtier, 1999: Date assimilation concepts and methods. meteorological training lecture notes, ECMWF, shinfield park. Reading.

Bowler, N. E., J. Flowerdew, and S. R. Pring, 2013: Tests of different flavors of EnKF on a simple model. Q. J. R. Meteor. Soc., 139, 1505–1519.

Bowler, N. E., and Coauthors, 2017: Inflation and localization tests in the development of an ensemble of 4D-ensemble variational assimilations. Q. J. R. Meteor. Soc., 143 (704), 1280–1302.

Buehner, M., 2005: Ensemble-derived stationary and flow-dependent background-error covariances: Evaluation in a quasi-operational NWP setting. Q. J. R. Meteor. Soc., 131 (607), 1013–1043.

Buehner, M., P. L. Houtekamer, C. Charette, H. L. Mitchell, and B. He, 2010a: Intercomparison of variational data assimilation and the ensemble Kalman filter for global deterministic NWP. Part-I: Description and single-observation experiments. Mon. Wea. Rev., 138 (5), 1550–1566.

Buehner, M., P. L. Houtekamer, C. Charette, H. L. Mitchell, and B. He, 2010b: Intercomparison of variational data assimilation and the ensemble Kalman filter for global deterministic NWP. Part II: One-month experiments with real observations. Mon. Wea. Rev., 138 (5), 1567–1586.

Buehner, M., J. Morneau, and C. Charette, 2013: Four-dimensional ensemble-variational data assimilation for global deterministic weather prediction. Nonlin. Processes Geophys., 20 (5), 669–682.

Buehner, M., and Coauthors, 2015: Implementation of deterministic weather forecasting systems based on ensemble–variational data assimilation at environment Canada. Part I: The global system. Mon. Wea. Rev., 143 (7), 2532–2559.

Buizza, R., M. Milleer, and T. N. Palmer, 1999: Stochastic representation of model uncertainties in the ECMWF ensemble prediction system. Q. J. R. Meteorol. Soc., 125 (560), 2887–2908.

Burgers, G., P. J. van Leeuwen, and G. Evensen, 1998: On the analysis scheme in the ensemble Kalman filter. Mon. Wea. Rev., 126, 1719–1724.

Cane, M. A., A. Kaplan, R. N. Miller, B. Tang, E. C. Hackert, and A. J. Busalacchi, 1996: Mapping tropical pacific sea level: Data assimilation via a reduced state space Kalman filter. Journal of Geophysical Research: Oceans, 101 (C10), 22 599–22 617.

Cappé, O., E. Moulines, and T. Rydén, 2005: Inference in Hidden Markov Models. Springer-Verlag.

Chassignet, E. P., H. E. Hurlburt, O. M. Smedstad, G. R. Halliwell, A. J. Hogan, P. J. Wallcraft, R. Baraille, and R. Bleck, 2007: The HYCOM (Hybrid Coordinate Ocean Model) data assimilative system. J. Mar. Sys., 65 (1), 60–83.

Chen, R., and J. Liu, 2000: Mixture Kalman filters. Journal of the Royal Statistical Society: Series B (Statistical Methodology), 62 (3), 493–508.

Chen, Y., and D. Oliver, 2013: Levenberg-Marquardt forms of the iterative ensemble smoother for efficient history matching and uncertainty quantification. Computational Geosciences, 17, 689–703.

Chen, Y., and D. Zhang, 2006: Data assimilation for transient flow in geologic formations via ensemble Kalman filter. Advances in Water Resources, 29, 1107–1122.

Chepurin, G. A., J. A. Carton, and D. P. Dee, 2005: Forecast model bias correction in ocean data assimilation. Mon. Wea. Rev., 133, 1328–1342.

Chorin, A., M. Morzfeld, and X. Tu, 2010: Implicit particle filters for data assimilation. Commun. Appl. Math. Comput. Sci., 5, 221–240.

Clayton, A. M., A. C. Lorenc, and D. M. Barker, 2013: Operational implementation of a hybrid ensemble 4DVar global data assimilation system at the met office. Q. J. R. Meteor. Soc., 139 (675), 1445–1461.

Cohn, S., and R. Todling, 1996: Approximate data assimilation schemes for stable and unstable dynamics. J. Meteor. Soc. Japan, 74, 63–75.

Cosme, E., J.-M. Brankart, J. Verron, P. Brasseur, and M. Krysta, 2010: Implementation of a reduced rank square-root smoother for high resolution ocean data assimilation. Ocean Modelling, 33 (1), 87–100.

Cosme, E., J. Verron, P. Brasseur, J. Blum, and D. Auroux, 2012: Smoothing problems in a Bayesian framework and their linear Gaussian solutions. Mon. Wea. Rev., 140 (2), 683–695.

Counillon, F., and L. Bertino, 2009: Ensemble optimal interpolation: multivariate properties in the Gulf of Mexico. Tellus, 61A, 296–308.

Courtier, P., 1997: Dual formulation of four-dimensional variational assimilation. Q. J. R. Meteor. Soc., 123 (544), 2449–2461.

Courtier, P., J.-N. The´paut, and A. Hollingsworth, 1994: A strategy for operational implementation of 4DVar using an incremental approach. Q. J. R. Meteor. Soc., 120, 1367–1387.

Cummings, J.A., O.M. Smedstad. 2013. Variational data assimilation for the global ocean. In: Park S., Xu L. (eds) Data Assimilation for Atmospheric, Oceanic and Hydrologic Applications (Vol II). Springer, Berlin, Heidelberg.

Daley, R., 1991: Atmospheric data analysis. Cambridge Atmospheric and Space Science Series, Cambridge University Press, 6966, 25.

Daley, R., 1992: Estimating model-error covariances for application to atmospheric data assimilation. Mon. Wea. Rev., 120, 1735–1746.

Danforth, C. M., and E. Kalnay, 2008: Using singular value decomposition to parameterize state-dependent model errors. J. Atmos. Sci., 65, 1467–1478.

De La Chevrotie`re, M., and J. Harlim, 2017: A data-driven method for improving the correlation estimation in serial ensemble Kalman filters. Mon. Wea. Rev., 145, 985–1001.

Dee, D. P., 1995: Online estimation of error covariance parameters for atmospheric data assimilation. Mon. Wea. Rev., 123, 1128–1145.

Dee, D. P., and A. M. da Silva, 1998: Data assimilation in the presence of forecast bias. Q. J. R. Meteorol. Soc., 124, 269–295.

Delworth, T., S. Manabe, and R. J. Stouffer, 1993: Interdecadal variations of the thermohaline circulation in a coupled ocean-atmosphere model. Journal of Climate, 6, 1993–2011.

Desroziers, G., E. Arbogast, and L. Berre, 2016: Improving spatial localization in 4DEnVar. Q. J. R. Meteor. Soc., 142 (701), 3171–3185.

Desroziers, G., J.-T. Camino, and L. Berre, 2014: 4DEnVar: Link with 4D state formulation of variational assimilation and different possible implementations. Q. J. R. Meteor. Soc., 140 (684), 2097–2110.

Di Lorenzo, E., A. M. Moore, H. G. Arango, B. D. Cornuelle, A. J. Miller, B. Powell, B. S. Chua, and A. F. Bennett, 2007: Weak and strong constraint data assimilation in the inverse regional ocean modeling system (ROMS): Development and application for a baroclinic coastal upwelling system. Ocean Modelling, 16, 160–187.

Doucet, A., N. De Freitas, and N. Gordon, Eds., 2001: Sequential Monte Carlo methods in practice. Springer Verlag.

Doucet, A., N. De Freitas, K. P. Murphy, and S. J. Russell, 2000a: Rao-blackwellised particle filtering for dynamic Bayesian networks. Proceedings of the 16th world Conference on UAI, Stanford, California, USA, 176–83.

Doucet, A., S. Godsill, and C. Andrieu, 2000b: On sequential Monte Carlo sampling methods for Bayesian filtering. Statistics and Computing, 10 (3), 197–208.

Doucet, A., and A. M. Johansen, 2011: A tutorial on particle filtering and smoothing: fifteen years later. Oxford Handbook of Nonlinear Filtering.

Dreano, D., B. Mallick, and I. Hoteit, 2015: Filtering remotely sensed chlorophyll concentrations in the Red Sea using a space–time covariance model and a Kalman filter. Spatial Statistics, 13, 1–20.

Dreano, D., P. Tandeo, M. Pulido, B. Ait-El-Fquih, T. Chonavel, and I. Hoteit, 2017: Estimating model-error covariances in nonlinear state-space models using Kalman smoothing and the expectation-maximization algorithm. Q. J. R. Meteorol. Soc., 143, 1877–1885.

Duan, L., C. L. Farmer, and I. M. Moroz, 2010: Regularized particle filter with Langevin resampling step. AIP Conference Proceedings, 1281 (1), 1080–1083.

El-Sheikh, A., I. Hoteit, and M. Wheeler, 2014: Nested sampling particle filter for nonlinear data assimilation. Q. J. R. Meteorol. Soc., 140, 1640–1653.

Elbern, H., A. Strunk, H. Schmidt, and O. Talagrand, 2007: Emission rate and chemical state estimation by four-dimensional variational inversion. Atmos. Chem. Phys., 7, 3749–3769.

Emerick, A. A., and A. C. Reynolds, 2012: Ensemble smoother with multiple data assimilation. Computers & Geosciences, 55, 3–15.

Evensen, G., 1994: Sequential data assimilation with a nonlinear quasi-geostrophic model using Monte Carlo methods to forecast error statistics. Journal of Geophysical Research: Oceans, 99 (C5), 10 143– 10 162.

Evensen, G., 2003: The ensemble Kalman filter: Theoretical formulation and practical implementation. Ocean Dynamics, 53, 343–367.

Evensen, G., 2009: The ensemble Kalman filter for combined state and parameter estimation. IEEE Control Syst., 29 (3), 83–104.

Fairbairn, D., S. Pring, A. Lorenc, and I. Roulstone, 2014: A comparison of 4DVar with ensemble data assimilation methods. Q. J. R. Meteor. Soc., 140 (678), 281–294.

Fang, F., C. C. Pain, I. M. Navon, M. D. Piggott, G. J. Gorman, P. E. Farrell, P. A. Allison, and A. J. H. Goddard, 2009: A POD reduced-order 4DVar adaptive mesh ocean modelling approach. International Journal for Numerical Methods in Fluids, 60, 709–732.

Farrell, B. F., and P. J. Ioannou, 2001: State estimation using a reduced-order Kalman filter. Journal of the Atmospheric Sciences, 58 (23), 3666–3680.

Fertig, E. J., B. R. Hunt, E. Ott, and I. Szunyogh, 2007: Assimilating non-local observations with a local ensemble Kalman filter. Tellus A, 59, 719–730.

Frei, M., and H. Künsch, 2012: Sequential state and observation noise covariance estimation using combined ensemble Kalman and particle filters. Mon. Wea. Rev., 140, 1476–95.

Frei, M., and H. R. Künsch, 2013: Bridging the ensemble Kalman and particle filters. Biometrika, 100, 781–800.

Frolov, S., C. H. Bishop, T. Holt, J. Cummings, and D. Kuhl, 2016: Facilitating strongly coupled ocean-atmosphere data assimilation with an interface solver. Mon. Wea. Rev., 144, 3–20.

Fukumori, I., and P. Malanotte-Rizzoli, 1995: An approximate kaiman filter for ocean data assimilation: An example with an idealized gulf stream model. J. Geophys. Res.: Oceans, 100 (C4), 6777–6793.

Gaspari, G., and S. E. Cohn, 1999: Construction of correlation functions in two and three dimensions. Q. J. R. Meteor. Soc., 125, 723–757.

Gharamti, M., J. Valstar, and I. Hoteit, 2014a: An adaptive hybrid EnKF -OI scheme for efficient state-parameter estimation of reactive contaminant transport models. Advances in Water Resources, 71, 1–15.

Gharamti, M. E., 2018: Enhanced adaptive inflation algorithm for ensemble filters. Mon. Wea. Rev., 146, 623–640.

Gharamti, M. E., B. Ait-El-Fquih, and I. Hoteit, 2015a: An iterative ensemble Kalman filter with one-step-ahead smoothing for state-parameters estimation of contaminant transport models. Journal of Hydrology, 527, 442–457.

Gharamti, M. E., B. Ait-El-Fquih, and I. Hoteit, 2015b: An iterative ensemble Kalman filter with one-step-ahead smoothing for state-parameters estimation of contaminant transport models. Journal of Hydrology, 527, 442–457.

Gharamti, M. E., A. Kadoura, S. Sun, J. Valstar, and I. Hoteit, 2014b: Constraining a compositional flow model with flow-chemical data using an ensemble Kalman filter. Water Resources Research, 50, 2444–2467.

Ghil, M., and P. Malanotte-Rizzoli, 1991: Data assimilation in meteorology and oceanography. Advances in Geophysics, 33, 141–266.

Giering, R., and T. Kaminski, 1998: Recipes for adjoint code construction. ACM Transactions on Mathematical Software (TOMS), 24 (4), 437–474.

Gordon, N. J., D. J. Salmond, and A. F. M. Smith, 1993: Novel approach to nonlinear and non-Gaussian Bayesian state estimation. IEE Proceedings F in Radar and Signal Processing, 140, 107–113.

Gu, Y., and D. Oliver, 2007: An iterative ensemble Kalman filter for multiphase fluid flow data assimilation. SPE Journal, 12, 438–446.

Gustafsson, N., J. Bojarova, and O. Vignes, 2014: A hybrid variational ensemble data assimilation for the high resolution limited area model (HIRLAM). Nonlin. Processes Geophys., 21 (1), 303–323.

Haley, P. J., and Coauthors, 2009: Forecasting and reanalysis in the Monterey Bay/California current region for the autonomous ocean sampling network-ii experiment. Deep Sea Research Part II: Topical Studies in Oceanography, 56 (3), 127–148.

Hamill, T. M., and C. Snyder, 2000: A hybrid ensemble Kalman filter-3D variational analysis scheme. Mon. Wea. Rev., 128 (8), 2905–2919.

Hamill, T. M., and J. S. Whitaker, 2005: Accounting for the error due to unresolved scales in ensemble data assimilation: A comparison of different approaches. Mon. Wea. Rev., 133, 3132–3147.

Hamill, T. M., J. S. Whitaker, J. L. Anderson, and C. Snyder, 2009: Comments on "Sigma-point Kalman filter data assimilation methods for strongly nonlinear systems". J. Atmos. Sci., 66, 3498–3500.

Hamill, T. M., J. S. Whitaker, and C. Snyder, 2001: Distance-dependent filtering of background error covariance estimates in an ensemble Kalman filter. Mon. Wea. Rev., 129, 2776–2790.

Hamilton, F., T. Berry, and T. Sauer, 2016: Ensemble Kalman filtering without a model. Physical Review X, 6 (1), 011 021.

Hamilton, J. D., 1986: State-space models. R.F. Engle & D. McFadden, H. of Economics, Ed., Vol. 4, 1st ed., Elsevier, chap. 50, 3039–80.

Harvey, A. C., and G. D. A. Phillips, 1979: Maximum likelihood estimation of regression models with autoregressive-moving average disturbance. Biometrika, 66, 49–58.

Heimbach, P., C. Hill, and R. Giering, 2002: Automatic generation of efficient adjoint code for a parallel Navier-Stokes solver. Computational Science ICCS 2002, 1019–1028.

Hoteit, I., D.-T. Pham, M. E. Gharamti, and X. Luo, 2015: Mitigating observation perturbation sampling errors in the stochastic EnKF. Mon. Wea. Rev., 143 (7).

Hoteit, I., X. Luo, and D. T. Pham, 2012: Particle Kalman filtering: A nonlinear Bayesian framework for ensemble Kalman filters. Mon. Wea. Rev., 140, 528–542.

Hoteit, I., B. Cornuelle, and P. Heimbach, 2010: An eddy permitting variational data assimilation system for estimating the of the tropical pacific. J. Geophys. Res., 115, C03001, doi:10.1029/2009JC005 347.

Hoteit, I., 2008: A reduced-order simulated annealing approach for four-dimensional variational data assimilation in meteorology and oceanography. Int. J. Numer. Meth. Fluids, 58 (11), 1181–1199.

Hoteit, I., D. T. Pham, G. Triantafyllou, and G. Korres, 2008: A new approximate solution of the optimal nonlinear filter for data assimilation in meteorology and oceanography. Mon. Wea. Rev., 136, 317–334.

Hoteit, I., G. Triantafyllou, and G. Korres, 2007: Using low-rank ensemble Kalman filters for data assimilation with high dimensional imperfect models. J. Num. Ana. Ind. Appl. Math., 2 (1-2), 67–78.

Hoteit, I., and A. Köhl, 2006: Efficiency of reduced-order, time-dependent adjoint data assimilation approaches. Journal of Oceanography, 62 (4), 539–550.

Hoteit, I., B. Cornuelle, A. Köhl, and D. Stammer, 2005a: Treating strong adjoint sensitivities in tropical eddy-permitting variational data assimilation. Q. J. R. Meteorol. Soc., 131, 3659–3682.

Hoteit, I., G. Korres, and G. Triantafyllou, 2005b: Comparison of extended and ensemble based Kalman filters with low and high-resolution primitive equations ocean models. Nonlin. Processes Geophys., 12, 755–765.

Hoteit, I., and D.-T. Pham, 2004: An adaptively reduced-order extended Kalman filter for data assimilation in the tropical pacific. J. Mar. Sys., 45 (3), 173–188.

Hoteit, I., and D. T. Pham, 2003: Evolution of the reduced state space and data assimilation schemes based on the Kalman filter. Journal of the Meteorological Society of Japan, 81, 21–39.

Hoteit, I., D. T. Pham, and J. Blum, 2002: A simplified reduced order Kalman filtering and application to altimetric data assimilation in tropical pacific. J. Mar. Sys., 36, 101–127.

Hoteit, I., D. T. Pham, and J. Blum, 2001: A semi-evolutive partially local filer for data assimilation. Marine Pollution Bulletin, 43, 164–174.

Houtekamer, P., and F. Zhang, 2016: Review of the ensemble Kalman filter for atmospheric data assimilation. Mon. Wea. Rev., 144, 4489–4532.

Houtekamer, P. L., and H. L. Mitchell, 1998: Data assimilation using an ensemble Kalman filter technique. Mon. Wea. Rev., 126, 796–811.

Houtekamer, P. L., and H. L. Mitchell, 2001: A sequential ensemble filter for atmospheric data assimilation. Mon. Wea. Rev., 129, 123–137.

Houtekamer, P. L., and H. L. Mitchell, 2005: Ensemble Kalman filtering. Q. J. R. Meteor. Soc., 131, 3269–3289.

Houtekamer, P. L., H. L. Mitchell, G. Pellerin, M. Buehner, M. Charron, L. Spacek, and B. Hansen, 2005: Atmospheric data assimilation with an ensemble Kalman filter: Results with real observations. Mon. Wea. Rev., 133 (3), 604–620.

Hunt, B. R., E. J. Kostelich, and I. Szunyogh, 2007: Efficient data assimilation for spatiotemporal chaos: A local ensemble transform Kalman filter. Physica D, 230 (1), 112–126.

Hunt, B. R., and Coauthors, 2004: Four-dimensional ensemble Kalman filtering. Tellus A, 56 (4), 273–277.

Ito, K., and K. Xiong, 2000: Gaussian filters for nonlinear filtering problems. IEEE Transactions on Automatic Control, 45, 910–927.

Janekovic, I., B. Powell, D. Matthews, M. McManus, and J. Sevadjian, 2013: 4DVar data assimilation in a nested, coastal ocean model: A hawaiian case study. Journal of Geophysical Research: Oceans, 118 (10), 5022–5035.

Jazwinski, A. H., 1970: Stochastic Processes and Filtering Theory. Academic Press.

Julier, S., J. Uhlmann, and H. Durrant-Whyte, 2000: A new method for the nonlinear transformation of means and covariances in filters and estimators. IEEE Transactions on Automatic Control, 45, 477–482.

Julier, S., and J. K. Uhlmann, 2004: Unscented filtering and nonlinear estimation. Proc. IEEE, 92, 401–422.

Kalman, R., 1960: A new approach to linear filtering and prediction problems. J. Basic Eng., 82 (1), 35–45.

Kalnay, E., H. Li, T. Miyoshi, S.-C. Yang, and J. Ballabera-Poy, 2007: 4DVar or ensemble Kalman filter? Tellus A, 59 (5), 758–773.

Kim, S.-B., I. Fukumori, and T. Lee, 2006: The closure of the ocean mixed layer temperature budget using level-coordinate model fields. Journal of Atmospheric and Oceanic Technology, 23 (6), 840–853.

Kleist, D. T., and K. Ide, 2015: An OSSE-based evaluation of hybrid variational–ensemble data assimilation for the ncep gfs. Part I: System description and 3D-hybrid results. Mon. Wea. Rev., 143 (2), 433–451.

Köhl, A., and D. Stammer, 2008: Variability of the meridional overturning in the North Atlantic from the 50 years GECCO state estimation. J. Phys. Oceanogr., 38, 1913–1930.

Köhl, A., and J. Willebrand, 2002: An adjoint method for the assimilation of statistical characteristics into eddy-resolving ocean models. Tellus, 54, 406–425.

Korres, G., K. Nitti, L. Perivoliotis, K. Tsiaras, A. Papadopoulos, G. Triantafyllou, I. Hoteit, and K. Abdullah, 2010: Forecasting the Aegean Sea hydrodynamics within the Poseidon-ii operational system. Journal of Operational Oceanography, 3 (1), 37–49.

Law, K., A. Stuart, and K. Zygalakis, 2015: Data assimilation: a mathematical introduction, Vol. 62. Springer.

Lawson, W. G., and J. A. Hansen, 2004: Implications of stochastic and deterministic filters as ensemble-based data assimilation methods in varying regimes of error growth. Mon. Wea. Rev., 132 (8), 1966–1981.

Le Dimet, F. X., and O. Talagrand, 1986: Variational algorithms for analysis and assimilation of meteorological observations: theoretical aspects. Tellus A, 38A (2), 97–110.

Lea, D., M. Allen, and T. Haine, 2000: Sensitivity analysis of the climate of a chaotic system. Tellus A, 52, 523–532.

Lee, Y., A. J. Majda, and D. Qi, 2017: Preventing catastrophic filter divergence using adaptive additive inflation for baroclinic turbulence. Mon. Wea. Rev., 145, 669–682.

Lei, J., P. Bickel, and C. Snyder, 2010: Comparison of ensemble Kalman filters under non-Gaussianity. Mon. Wea. Rev., 138, 1293–1306.

Lellouche J.-M., O. Le Galloudec, M. Drévillon, C. Régnier, E. Greiner, G. Garric, N. Ferry, C. Desportes, C.-E. Testut, C. Bricaud, R. Bourdallé-Badie, B. Tranchant, M. Benkiran, Y. Drillet, A. Daudin, and C. De Nicola, 2013. Evaluation of global monitoring and forecasting systems at Mercator Océan. Ocean Science, 9, 57–81,

Lermusiaux, P., and A. Robinson, 1999: Data assimilation via error subspace statistical estimation. Part I: Theory and schemes. Mon. Wea. Rev., 127, 1385–1407.

Lguensat, R., P. Tandeo, P. Ailliot, M. Pulido, and R. Fablet, 2017: The analog data assimilation. Mon. Wea. Rev., 145 (10), 4093–4107.

Li, G., M. Iskandarani, M. Le Henaff, J. Winokur, O. Le Maitre, and O. Knio, 2015: Quantifying initial and wind forcing uncertainties in the Gulf of Mexico. Comput. Geosci., 20, 1133–1153.

Li, H., E. Kalnay, and T. Miyoshi, 2009a: Simultaneous estimation of covariance inflation and observation errors within an ensemble Kalman filter. Q. J. R. Meteor. Soc., 135, 523–533.

Li, H., E. Kalnay, T. Miyoshi, and C. M. Danforth, 2009b: Accounting for model errors in ensemble data assimilation. Mon. Wea. Rev., 137, 3407–3419.

Liu, B., B. Ait-El-Fquih, and I. Hoteit, 2016: Efficient kernel-based ensemble Gaussian mixture filtering. Mon. Wea. Rev., 144 (2), 781–800.

Liu, C., A. Köhl, and D. Stammer, 2012: Adjoint-based estimation of eddy-induced tracer mixing parameters in the global ocean. Journal of Physical Oceanography, 42 (7), 1186–1206.

Liu, C., Q. Xiao, and B. Wang, 2008: An ensemble-based four-dimensional variational data assimilation scheme. Part I: Technical formulation and preliminary test. Mon. Wea. Rev., 136 (9), 3363–3373.

Liu, C., Q. Xiao, and B. Wang, 2009: An ensemble-based four-dimensional variational data assimilation scheme. Part II: Observing system simulation experiments with advanced research WRF (ARW). Mon. Wea. Rev., 137 (5), 1687–1704.

Liu, L., H. Ji, and Z. Fan, 2015: Improved iterated-corrector-PHD with Gaussian mixture implementation. Signal Processing, 114, 89–99.

Lorenc, A. C., 2003: The potential of the ensemble Kalman filter for NWP a comparison with 4DVar. Q. J. R. Meteor. Soc., 129 (595), 3183–3203.

Lorenc, A. C., N. E. Bowler, A. M. Clayton, S. R. Pring, and D. Fairbairn, 2015: Comparison of hybrid-4DEnVar and hybrid-4DVar data assimilation methods for global NWP. Mon. Wea. Rev., 143 (1), 212–229.

Lorentzen, R., and G. Nævdal, 2011: An iterative ensemble Kalman filter. IEEE Transactions on Automatic Control, 56, 1990–1995.

Lu, F., Z. Liu, S. Zhang, and Y. Liu, 2015: Strongly coupled data assimilation using leading averaged coupled covariance (lacc). Part I: Simple model study. Mon. Wea. Rev., 143, 3823–3837.

Luo, A., L. M. Moroz, and I. Hoteit, 2010: Scaled unscented transform Gaussian sum filter: Theory and application. Physica-D, 239, 684–701.

Luo, X., T. Bhakta, and G. Nædal, 2018: Correlation-based adaptive localization with applications to ensemble-based 4D seismic history matching. SPE Journal, in press, SPE-185936-PA.

Luo, X., and I. Hoteit, 2011: Robust ensemble filtering and its relation to covariance inflation in the ensemble Kalman filter. Mon. Wea. Rev., 139, 3938–3953.

Luo, X., and I. Hoteit, 2013: Covariance inflation in the ensemble Kalman filter: a residual nudging perspective and some implications. Mon. Wea. Rev., 141, 3360–3368.

Luo, X., and I. Hoteit, 2014a: Efficient particle filtering through residual nudging. Q. J. R. Meteor. Soc., 140, 557–572.

Luo, X., and I. Hoteit, 2014b: Ensemble Kalman filtering with a divided state-space strategy for coupled data assimilation problems. Mon. Wea. Rev., 142, 4542–4558.

Luo, X., and I. Hoteit, 2014c: Ensemble Kalman filtering with residual nudging: an extension to the state estimation problems with nonlinear observations. Mon. Wea. Rev., 142, 3696–3712.

Luo, X., I. Hoteit, and I. M. Moroz, 2012: On a nonlinear Kalman filter with simplified divided difference approximation. Physica D, 241, 671–680.

Luo, X., and I. M. Moroz, 2009: Ensemble Kalman filter with the unscented transform. Physica D, 238, 549–562.

Luo, X., A. Stordal, R. Lorentzen, and G. Nævdal, 2015: Iterative ensemble smoother as an approximate solution to a regularized minimum-average-cost problem: theory and applications. SPE Journal, 20, 962–982.

Malinverno, A., 2002: Parsimonious Bayesian Markov chain Monte Carlo inversion in a nonlinear geophysical problem. Geophysical Journal International, 151, 675–688.

Mehra, R., 1970: On the identification of variances and adaptive Kalman filtering. IEEE Transactions on Automatic Control, 15, 175–184.

Metropolis, N., A. W. Rosenbluth, M. N. Rosenbluth, A. H. Teller, and E. Teller, 1953: Equation of state calculations by fast computing machines. J. Chem. Phys., 21, 1087–1092.

Mitchell, H. L., and P. L. Houtekamer, 2000: An adaptive ensemble Kalman filter. Mon. Wea. Rev., 128, 416–433.

Miyoshi, T., 2011: The Gaussian approach to adaptive covariance inflation and its implementation with the local ensemble transform Kalman filter. Mon. Wea. Rev., 139, 1519–1535.

Moore, A. M., and Coauthors, 2011: The regional ocean modeling system (roms) four-dimensional variational data assimilation systems: Part II–performance and application to the California Current system. Progress in Oceanography, 91 (1), 50–73.

Moradkhani, H., S. Sorooshian, H. V. Gupta, and P. R. Houser, 2005: Dual state–parameter estimation of hydrological models using ensemble Kalman filter. Advances in Water Resources, 28 (2), 135–147.

Nerger, L., W. Hiller, and J. Schröter, 2005: A comparison of error subspace Kalman filters. Tellus A, 57 (5), 715–735.

Nerger, L., T. Janjic, J. Schrter, and H. W., 2012: A unification of ensemble square root Kalman filters. Mon. Wea. Rev., 140 (7), 2335–2345.

Nerger, L., S. Schulte, and A. Bunse-Gerstner, 2014: On the influence of model nonlinearity and localization on ensemble Kalman smoothing. Q. J. R. Meteor. Soc., 140 (684), 2249–2259.

Ngodock, H., and M. Carrier, 2014: A 4D-Var system for the Navy coastal ocean model. Part II: Strong and weak constraint assimilation experiments with real observations in Monterey Bay. Mon. Wea. Rev., 142 (6), 2108–2117.

Oke, P. R., G. B. Brassington, and A. Griffin, and D. A. Schiller, 2008: The Bluelink ocean data assimilation system (bodas). Ocean Modelling, 21 (1), 46–70.

Oke, P. R., P. Sakov, and S. P. Corney, 2007: Impacts of localisation in the EnKF and EnOI: experiments with a small model. Ocean Dynamics, 57 (1), 32–45.

Oliver, D., and Y. Chen, 2011: Recent progress on reservoir history matching: A review. Computational Geosciences, 15, 185–221.

Oliver, D. S., N. He, and A. C. Reynolds, 1996: Conditioning permeability fields to pressure data. ECMOR V-5th European Conference on the Mathematics of Oil Recovery.

Ott, E., and Coauthors, 2004: A local ensemble Kalman filter for atmospheric data assimilation. Tellus, 56A, 415–428.

Pagan, A., 1980: Some identification and estimation results for regression models with stochastically varying coefficients. Journal of Econometrics, 13, 341–63.

Penny, S. G., and T. M. Hamill, 2017: Coupled data assimilation for integrated earth system analysis and prediction. Bulletin of the American Meteorological Society, 97 (7), ES169–ES172.

Penny, S. G., and T. Miyoshi, 2016: A local particle filter for high-dimensional geophysical systems. Nonlin. Processes Geophys, 23, 391–405.

Pham, D. T., 2001: Stochastic methods for sequential data assimilation in strongly nonlinear systems. Mon. Wea. Rev., 129, 1194–1207.

Pham, D. T., J. Verron, and C. Roubaud, 1997: Singular evolutive Kalman filter with EOF initialization for data assimilation in oceanography. J. Mar. Syst., 16, 323–340.

Pires, C., R. Vautard, and O. Talagrand, 1996: On extending the limits of variational assimilation in nonlinear chaotic systems. Tellus, 48, 96–121.

Poterjoy, J., 2016: A localized particle filter for high-dimensional nonlinear systems. Mon. Wea. Rev., 144, 59–76.

Poterjoy, J., and F. Zhang, 2015: Systematic comparison of four-dimensional data assimilation methods with and without the tangent linear model using hybrid background error covariance: E4DVar versus 4DEnVar. Mon. Wea. Rev., 143 (5), 1601–1621.

Raanes, P. N., 2016: On the ensemble Rauch-Tung-Striebel smoother and its equivalence to the ensemble Kalman smoother. Q. J. R. Meteor. Soc., 142 (696), 1259–1264.

Rasmussen, J., H. Madsen, K. H. Jensen, and J. C. Refsgaard, 2015a: Data assimilation in integrated hydrological modeling using ensemble Kalman filtering: evaluating the effect of ensemble size and localization on filter performance. Hydrology and Earth System Sciences, 19, 2999–3013.

Rasmussen, J., H. Madsen, K. H. Jensen, and J. C. Refsgaard, 2015b: Data assimilation in integrated hydrological modeling using ensemble Kalman filtering: Evaluating the effect of ensemble size and localization on filter performance. Hydrology Earth System Science, 19, 2999–3013.

Raynaud, L., L. Berre, and G. Desroziers, 2009: Objective filtering of ensemble-based background-error variances. Q. J. R. Meteor. Soc., 135 (642), 1177–1199.

Reich, S., 2013: A nonparametric ensemble transform method for Bayesian inference. SIAM Journal on Scientific Computing, 35, A2013–A2024.

Robert, C., 2007: The Bayesian choice: From decision-theoretic foundations to computational implementation. Springer Science & Business Media, New York.

Robert, C., S. Durbiano, E. Blayo, J. Verron, J. Blum, and F.-X. Le Dimet, 2005: A reduced-order strategy for 4DVar data assimilation. J. Mar. Sys., 57 (1), 70–82.

Sakov, P., and L. Bertino, 2011: Relation between two common localisation methods for the EnKF. Computational Geosciences, 15 (2), 225–237.

Sakov, P., and P. R. Oke, 2008: Implications of the form of the ensemble transformation in the ensemble square root filters. Mon. Wea. Rev., 136 (3), 1042–1053.

Sakov, P., D. S. Oliver, and L. Bertino, 2012: An iterative EnKF for strongly nonlinear systems. Mon. Wea. Rev., 140 (6), 1988–2004.

Sakov P., F. Counillon, L. Bertino, K.A. Lisæter, P. Oke, A. Korablev. 2012. TOPAZ4: an ocean-sea ice data assimilation system for the North Atlantic and Arctic. Ocean Science, 8, 633–656.

Sakov, P., and P. A. Sandery, 2015: Comparison of EnOI and EnKF regional ocean reanalysis systems. Ocean Modelling, 89, 45–60.

Särkkä, S., and J. Hartikainen, 2013: Non-linear noise adaptive Kalman filtering via variational Bayes. Proceedings of the IEEE 2013 International Workshop on Machine Learning for Signal Processing.

Shen, Z., and Y. Tang, 2015: A modified ensemble Kalman particle filter for non-Gaussian systems with nonlinear measurement functions. Journal of Advances in Modeling Earth Systems, 7, 50–66.

Shumway, R. H., and D. S. Stoffer, 1982: An approach to time series smoothing and forecasting using the EM algorithm. Journal of Time Series Analysis, 3 (4), 253–264.

Simon, D., 2006: Optimal State Estimation: Kalman, H-Infinity, and Nonlinear Approaches. Wiley-Interscience, 552 pp.

Sluka, T. C., S. G. Penny, E. Kalnay, and T. Miyoshi, 2016: Assimilating atmospheric observations into the ocean using strongly coupled ensemble data assimilation. Geophysical Research Letters, 43, 752–759.

Smedstad, O. M., H. E. Hurlburt, E. J. Metzger, R. C. Rhodes, J. F. Shriver, A. J. Wallcraft, and A. B. Kara, 2003: An operational eddy resolving 1/16 global ocean nowcast/forecast system. J. Mar. Sys., 40, 341–361.

Smith, K. D., A. M. Moore, and H. G. Arango, 2015: Estimates of ocean forecast error covariance derived from hessian singular vectors. Ocean Modelling, 89, 104–121.

Smith, P. J., A. S. Lawless, and N. K. Nichols, 2017a: Estimating forecast error covariances for strongly coupled atmosphere–ocean 4D-Var data assimilation. Mon. Wea. Rev., 145 (10), 4011–4035.

Smith, S., H. Ngodock, M. Carrier, J. Shriver, P. Muscarella, and I. Souopgui, 2017b: Validation and operational implementation of the Navy Coastal Ocean Model Four-Dimensional Variational Data Assimilation system (NCOM 4DVar) in the Okinawa Trough. Data Assimilation for Atmospheric, Oceanic and Hydro- logic Applications (Vol. III), Springer, 405–427.

Snyder, C., T. Bengtsson, P. Bickel, and J. Anderson, 2008: Obstacles to high-dimensional particle filtering. Mon. Wea. Rev., 136, 4629–4640.

Sondergaard, T., and P. F. Lermusiaux, 2013: Data assimilation with Gaussian mixture models using the dynamically orthogonal field equations. Part I: Theory and scheme. Mon. Wea. Rev., 141 (6), 1737–1760.

Song, H., I. Hoteit, B. Cornuelle, and A. C. Subramanian, 2010: An adaptive approach to mitigate background covariance limitations in the ensemble Kalman filter. Mon. Wea. Rev., 138 (7), 2825–2845.

Sorenson, H. W., and D. L. Alspach, 1971: Recursive Bayesian estimation using Gaussian sums. Automatica, 7, 465–479.

Sraj, I., S. Zedler, C. Jackson, O. Knio, and I. Hoteit, 2016: Polynomial chaos-based Bayesian inference of K-profile parameterization in a general circulation model of the Tropical Pacific. Mon. Wea. Rev., 144 (12), 4621–4640.

Sripitana, A., T. Mayo, I. Sraj, O. Knio, C. Dawson, O. Le Maitre, and I. Hoteit, 2017: Assessing an ensemble Kalman filter inference of manning's n coefficient of a storm surge model against a polynomial chaos-based MCMC. Ocean Dynamics, 67, 1067–1094.

Stockdale, T. N., D. L. T. Anderson, J. O. S. Alves, and M. A. Balmaseda, 1998: Global seasonal rainfall forecasts using a coupled ocean-atmosphere model. Nature, 392, 370–373.

Stroud, J. R., and T. Bengtsson, 2007: Sequential state and variance estimation within the ensemble Kalman filter. Mon. Wea. Rev., 135, 3194–3208.

Subramanian, A. C., I. Hoteit, I. Cornuelle, A. J. Miller, and H. Song, 2012: Linear versus nonlinear filtering with scale-selective corrections for balanced dynamics in a simple atmospheric model. Journal of the Atmospheric Sciences, 69, 3405–3419.

Talagrand, O., 2010: Variational assimilation. Data assimilation, Springer, 41–67.

Tarantola, A., 2005: Inverse problem theory and methods for model parameter estimation. Society for Industrial and Applied Mathematics (SIAM), 339 pp.

Tardif, R., G. J. Hakim, and C. Snyder, 2015: Coupled atmosphere–ocean data assimilation experiments with a low-order model and CMIP5 model data. Climate Dynamics, 45, 1415–1427.

Temam, R., 1984: Navier-Stokes equations: Theory and Numerical Analysis, Vol. 343. AMS Chelsea Publishing.

Tippett, M. K., J. L. Anderson, C. H. Bishop, T. M. Hamill, and J. S. Whitaker, 2003: Ensemble square-root filters. Mon. Wea. Rev., 131, 1485–1490.

Toye, H., P. Zhan, G. Gopalakrishnan, A. Kartadikaria, H. Huang, O. Knio, and I. Hoteit, 2017: Ensemble data assimilation in the Red Sea: sensitivity to ensemble selection and atmospheric forcing. Ocean Dynamics, 67, 915–933.

Trémolet, Y., 2007: Model-error estimation in 4D-Var. Q. J. R. Meteorol. Soc., 133, 1267–1280.

Tsiaras, K. P., I. Hoteit, S. Kalaroni, G. Petihakis, and G. Triantafyllou, 2017: A hybrid ensemble-OI Kalman filter for efficient data assimilation into a 3-D biogeochemical model of the Mediterranean. Ocean Dynamics, 67, 673–690.

Ueno, G., T. Higuchi, T. Kagimoto, and N. Hirose, 2010: Maximum likelihood estimation of error covariances in ensemble-based filters and its application to a coupled atmosphere–ocean model. Q. J. R. Meteorol. Soc., 136, 1316–1343.

Usui, N., S. Ishizaki, Y. Fujii, H. Tsujino, T. Yasuda, and M. Kamachi, 2006: Meteorological research institute multivariate ocean variational estimation (MOVE) system: Some early results. Advances in Space Research, 37 (4), 806–822.

Van Leeuwen, P. J., 2009: Particle filtering in geophysical systems. Mon. Wea. Rev., 137, 4089–4114.

Van Leeuwen, P. J., 2010: Nonlinear data assimilation in geosciences: An extremely efficient particle filter. Q. J. R. Meteor. Soc., 136 (653), 1991–1999.

van Leeuwen, P. J., 2015: Representation errors and retrievals in linear and nonlinear data assimilation. Q. J. R. Meteorol. Soc., 141, 1612–1623.

Verlaan, M., and A. W. Heemink, 1997: Tidal flow forecasting using reduced rank square-root filters. Stochastic hydrology and Hydraulics, 11 (5), 349–368.

Vermeulen, P. T., and A. W. Heemink, 2006: Model-reduced variational data assimilation. Mon. Wea. Rev., 134, 2888–2899.

Verron, J., L. Gourdeau, D. T. Pham, R. Murtugudde, and A. Busalacchi, 1999: An extended Kalman filter to assimilate satellite altimeter data into a nonlinear numerical model of the tropical Pacific Ocean: Method and validation. Journal of Geophysical Research: Oceans, 104 (C3), 5441–5458.

Vlasenko, A. V., A. Köhl, and D. Stammer, 2016: The efficiency of geophysical adjoint codes generated by automatic differentiation tools. Computer Physics Communications, 199, 22–28.

Wang, X., C. H. Bishop, and S. J. Julier, 2004: Which is better, an ensemble of positive-negative pairs or a centered simplex ensemble. Mon. Wea. Rev., 132, 1590–1605.

Wang, X., C. Snyder, and T. M. Hamill, 2007: On the theoretical equivalence of differently proposed ensemble–3DVAR hybrid analysis schemes. Mon. Wea. Rev., 135 (1), 222–227.

Weaver, A., J. Vialard, and D. L. T. Anderson, 2003: Three-and four-dimensional variational assimilation with a general circulation model of the tropical Pacific Ocean. Part I: Formulation, internal diagnostics, and consistency checks. Mon. Wea. Rev., 131 (7), 1360–1378.

Weaver, A. T., C. Deltel, É. Machu, S. Ricci, and N. Daget, 2005: A multivariate balance operator for variational ocean data assimilation. Q. J. R. Meteor. Soc., 131 (613), 3605–3625.

Whitaker, J. S., and T. M. Hamill, 2002: Ensemble data assimilation without perturbed observations. Mon. Wea. Rev., 130 (7), 1913–1924.

Whitaker, J. S., and T. M. Hamill, 2012: Evaluating methods to account for system errors in ensemble data assimilation. Mon. Wea. Rev., 140, 3078–3089.

Wikle, C. K., and L. M. Berliner, 2007: A Bayesian tutorial for data assimilation. Physica D: Nonlinear Phenomena, 230 (1), 1–16.

Wu, L., V. Mallet, M. Bocquet, and B. Sportisse, 2008: A comparison study of data assimilation algorithms for ozone forecasts. Journal of Geophysical Research: Atmospheres, 113 (D20), doi:10.1029/2008JD009991.

Wunsch, C., 1996: The ocean circulation inverse problem. Cambridge University Press, Cambridge, UK.

Wunsch, C., and P. Heimbach, 2007: Practical global oceanic state estimation. Physica D, 230, 197–208.

Xie, J., and J. Zhu, 2010: Ensemble optimal interpolation schemes for assimilating Argo profiles into a hybrid coordinate ocean model. Ocean Modelling, 33 (3), 283–298.

Yang, S.-C., E. Kalnay, and T. Enomoto, 2015: Ensemble singular vectors and their use as additive inflation in EnKF. Tellus A: Dynamic Meteorology and Oceanography, 67, 26 536.

Yaremchuk, M., P. Martin, A. Koch, and C. Beattie, 2016: Comparison of the adjoint and adjoint-free 4DVar assimilation of the hydrographic and velocity observations in the Adriatic Sea. Ocean Modelling, 97, 129–140.

Zhang, F., C. Snyder, and J. Sun, 2004: Impacts of initial estimate and observation availability on convective-scale data assimilation with an ensemble Kalman filter. Mon. Wea. Rev., 132, 1238–1253.

Zhang, M., and F. Zhang, 2012: E4DVar: Coupling an ensemble Kalman filter with four-dimensional variational data assimilation in a limited-area weather prediction model. Mon. Wea. Rev., 140 (2), 587–600.

Zhang, W. G., J. L. Wilkin, and H. G. Arango, 2010: Towards an integrated observation and modeling system in the New York Bight using variational methods. Part I: 4D-Var data assimilation. Ocean Modelling, 35 (3), 119–133.

Zhang, Y., and D. S. Oliver, 2010: Improving the ensemble estimate of the Kalman gain by bootstrap sampling. Mathematical Geosciences, 42, 327–345.

Zupanski, D., and M. Zupanski, 2006: Model error estimation employing an ensemble data assimilation approach. Mon. Wea. Rev., 134, 1337–1354.

Zupanski, M., 2005: Maximum likelihood ensemble filter: theoretical aspects. Mon. Wea. Rev., 133, 1710–1726.

CHAPTER 18

Operational Ocean Data Assimilation

Gregg A. Jacobs[1], Charlie N. Barron[1], Cheryl A. Blain[1], Matthew J. Carrier[1], Joseph M. D'Addezio[2], Robert W. Helber[1], Jackie C. May[1], Hans E. Ngodock[1], John J. Osborne[4], Mark D. Orzech[1], Clark D. Rowley[1], Innocent Souopgui[3], Scott R. Smith[1], Jay Veeramony[1], and Max Yaremchuk[1]

[1]*Oceanography Division, Naval Research Laboratory, Stennis Space Center, Mississippi, USA;* [2]*University of Southern Mississippi, Stennis Space Center, Mississippi, USA;* [3]*University of New Orleans, Stennis Space Center, Mississippi, USA;* [4]*ASEE Postdoc, Naval Research Laboratory, Stennis Space Center, Mississippi, USA*

Operational ocean data assimilation is necessary to continually correct and maintain accurate ocean forecasts. The primary GODAE objective was prediction of mesoscale eddies, and the science community has successfully addressed this issue. Here, we examine a data assimilation process for this problem, beginning with a generalized solution of 4D variation assimilation (4DVar) so that assumptions will be clear as we reduce to a 3DVar that is often used operationally. The primary difficulty lies in specifying the covariances that relate variables at different locations in space and time. Simplifications are applied to provide covariances that sufficiently describe the relations and are computationally feasible. Some deficiencies are introduced through the assumptions leading to the 3DVar and within the covariances, and this points to areas of future research. Prior assumptions were predicated on the expected observing systems and numerical model capabilities, which were all consistent with prediction of mesoscale features. We believe that numerical models and observations will surpass present capability, and there is strong motivation to move data assimilation forward to achieve prediction at scales not now feasible.

Introduction

The goal for operational ocean prediction is to enable decisions that address societal problems across time scales from hours to years. For example, flooding and other impacts due to storms require accurate forecasts to determine evacuations, coastal ocean conditions off Louisiana and Texas leading to hypoxia determine the quantity of fertilizer that farmers in Iowa may use, and growth in aquaculture requires consideration of material transport over years as well as its impact on the ocean environment. For all these problems and more, the end measure of success is the accuracy of the ocean prediction.

We must consider the physics governing ocean evolution to understand the application of the assimilation process. Incoming solar radiation drives the global physical environment. The relatively higher energy input per unit of Earth surface at the equator versus the poles results in spatial density and pressure gradients that drive atmosphere flows, ocean flows, surface waves, ice, and a hydrological cycle of rainfall and river runoff. Energy flowing within and between these connected systems converts into a myriad of forms that we see as storms, eddies, and many other individual events. For the ocean, nonlinear instability processes develop into mesoscale eddies,

---

Jacobs, G.A., et al., 2018: Operational ocean data assimilation. In "*New Frontiers in Operational Oceanography*", E. Chassignet, A. Pascual, J. Tintoré, and J. Verron, Eds., GODAE OceanView, 513-544, doi:10.17125/gov2018.ch18.

which is a result of energy exchange from the large-scale potential and kinetic energy fields into the growth of mesoscale features. These eddy features have been observed in situ and remotely to have timescales on the order of ten days and spatial scales from hundreds of km in the tropics to tens of km in higher latitudes (IMAWAKI, 1981; Richman et al., 1977; Le Traon, 1991). Mesoscale perturbations grow exponentially in time (Charney, 1947), and thus errors in a model initial condition grow exponentially through the forecast. The ocean eddy physics limit the time period over which accurate predictions may be made, and these features are described as nondeterministic. The time and location of specific features are of primary interest in operational decisions. This mandates a process to continually correct the model evolution for mesoscale prediction. Operational assimilation provides this process.

This chapter begins with a general formulation of the problem. There are many excellent sources that examine the historical evolution of the data assimilation and forecasting problem (Daley, 1993; Kalnay, 2003). The history provides insight into how considerations have continually extended the data assimilation problem leading from initial simple solutions to the present day complex systems. In this chapter, rather than examine the historical development, we begin with general formulation of a four-dimensional variational (4DVar) data assimilation that incorporates many of the past considerations (Bennett, 2002).

Following the general formulation, we outline simplifications of the 4DVar for the ocean mesoscale prediction problem. In past years, computational resources in the operational environment were insufficient to apply a 4DVar approach. At present, the 4DVar approach may be applied to the ocean in limited regions (Moore et al., 2011; Ngodock and Carrier, 2014; Powell et al., 2008), though the global ocean 4DVar assimilation problem is not presently computationally feasible. Explicit assumptions are made to reduce the 4DVar to a three-dimensional variational solution (3DVar), optimal interpolation, or Kalman Filter. The general assimilation problem contains statistical covariance information relating error of one variable at one space and time location to another variable at a different location, and the relations between all variables are contained in the covariance matrix. The assumptions simplifying the covariance matrix reduce the 4DVar to a 3DVar solution.

The covariance relations are extremely important for ocean predictions. Relative to the scales of eddies, present ocean observations are somewhat sparse. Recent observations are the input information to the data assimilation process. For GODAE mesoscale forecasts, the primary data sources have been satellite observations of sea surface height (SSH), sea surface temperature (SST), and profile observations of temperature and salinity from ARGO amongst others (Le Traon, Rienecker et al. 1999). It has been recognized that even with the range of satellite and in situ global systems, the spatial and temporal coverage is relatively sparse, and efforts to augment the regular observations have been proposed (Testor, Meyers et al. 2010). In particular, for ocean prediction, the data assimilation process must extend the influence of observations from individual points or small area averages (such as satellite footprints) of the ocean state. The covariances provide the relations of errors in variables at observation locations to other variables at other locations in space and time. For example, the covariances must extend observations such as SSH to the ocean interior temperature and salinity.

When computing the corrections to a model state at a particular time, the covariance relations depend on the present conditions such as the positions of mesoscale features. The covariances between variables across an eddy front will be very different from those within an eddy. The covariances also change over time. Before observations correct the model, there is a background covariance, and once observations correct the model the covariance magnitudes decrease in the analysis covariance. During the time period from one correction to the next, the forecast covariance magnitudes increase until the forecast covariance becomes the background for the next assimilation cycle. There are analytic solutions in a linear system for the covariance decrease in the analysis and increase in the forecast period. However, the high degree of freedom nonlinear ocean system makes the problem intractable.

Present operational systems utilize many simplifying assumptions. Because covariance relations provide the critical function of extending sparse observations, covariance simplifications must be carefully considered due to effects on reduced forecast skill. The covariance matrix in the reduced 3DVar remains daunting, and specifying the statistical information has required further simplifying assumptions. One example we examine assumes an analytic representation of the covariances without considering dependence on feature location, and a simple representation of the time variation is incorporated. Because there are simplifying assumptions to the covariance matrix, these restricts predictive skill.

We must recognize such assumptions because these imply where the frontiers of future research lie. The simplifications reducing the problem from 4DVar to 3DVar also remove time correlated information. Ocean features have long time correlations. The 4DVar problem retains these correlations, and additional considerations of the covariances are required. Approaches such as ensemble methods are intended to incorporate temporally and spatially changing information within the covariances based on the dynamical evolution (Evensen, 2003). All these methods are at the leading edge of research as the next generation of operational systems is developed, and we examine an example of considerations within the 4DVar system.

Operational data assimilation requires considerations beyond just the data assimilation process. We must keep in mind other parts of the overall prediction system. Predicting features in an operational environment must consider a balance of 1) physical understanding and representation in numerical models, 2) computational capability that limits resolution of features in numerical models, 3) systems that are typically insufficient to observe the features high resolution numerical models may represent, and 4) assimilation capability to specify covariances and correct numerical forecasts with the given observations. In all of these areas, significant limitations exist. Understanding physics within just the ocean is incomplete. Computational capacity limits model resolution and, therefore, physics. Observations typically are sparse. It is not possible to fully and accurately specify the full time-evolving covariance. All of these limitations motivate advancing the data assimilation problem.

The measure of success can be traced from perspective of the operational decision. Are decisions made correctly with greater frequency? Is the environmental information on which decisions are based more accurate? Are the environmental features that affect the decisions information more accurately forecast? Following the chain from operational decisions to accuracy in forecasts is a challenging problem. We know that GODAE systems have skill in forecasting mesoscale eddies (Dombrowsky et al., 2009). For many applications, such as search and rescue and oil spill response, the trajectory of objects or material is of high interest. For these problems, trajectory accuracy and forecast placement of ocean features on the order of 10 km is required. We provide a recent example showing that present systems do not meet this challenging requirement, which implies there is much work to be done in the future.

In this chapter, we examine the operational ocean data assimilation problem to understand the considerations. The second section provides a brief general formulation of the problem, and the third section provides simplifying assumptions leading from the general 4DVar solution to a 3DVar solution used operationally. Within any assimilation approach are covariance relations between variables, space, and time. The fourth section provides one example of a covariance implementation in 3DVar that considers maintaining consistency with ocean physics. One simplifying assumption in reaching the 3DVar reduces time correlations, and the considerations and impact when incorporating time correlations are examined in the fifth section. Finally, looking to the future, in the final section, we gauge the performance of a present system prediction to independent observations in a difficult problem of providing drift predictions.

Formulation

Data assimilation corrects operational forecast systems on a regular basis. The frequency varies across GODAE systems from every six hours (Saha et al., 2010b) to daily (Cummings et al., 2009) to weekly (Brasseur et al., 2005). The typical process uses a prior forecast as a background, and the background is subtracted from observations over a time period leading up the present to form a set of innovations. The data assimilation is often a 3DVar that uses the innovations to compute an increment that corrects the model at the present time, and the corrected model state is integrated forward over a forecast interval. We first formulate the problem in a more generalized manner to understand what assumptions lead to this particular procedure and therefore the inherent limitations.

A well-posed forecast problem consists of initial and boundary conditions (IBCs) and dynamical specifications. If we consider the global ocean system with a given initial condition, the boundary conditions are surface momentum flux, surface heat flux, surface turbulence injection, and incoming solar radiation. From the IBCs, an initial forecast is made. For a regional ocean application, lateral boundary conditions of the dynamical system augment the IBCs. Conceptually, specific model implementation information also augments the IBCs. This includes bathymetry, tidal boundary conditions, bottom friction parameterization, lateral river transports, and other coefficients within horizontal and vertical mixing. While typical operational applications do not make regular corrections of model implementation information, many data assimilation problems address these as optimal parameter estimations. Including these as IBCs in the general formulation allows extension to many problems.

Define the state vector x as the concatenation of all IBCs and model variables over the time period of the observations and over the forecast period. In addition to the ocean temperature, salinity, velocity, and surface elevation, every intermediary variable calculation in the model solution can be considered as part of the state. This includes intermediary fluxes of momentum and heat, horizontal and vertical diffusivity, and others. For convenience, we assume that the state is ordered in the same order as variables are computed in the model solution. A subset of the state is provided by the IBCs, and the dynamics of a numerical model provides the complementing set. The IBCs can be written in a matrix form as $A_{IBC} x = b_{IBC}$, where each row of the matrix A_{IBC} corresponds to setting one state value corresponding to b_{IBC}. The dynamics are nonlinear, and these can be written in a similar matrix form as $D = b_D$, where coefficients of the ith row of D may depend on state variables with indices less than i. Thus D is lower left triangular. As the problem is well posed, A_{IBC}, D, b_{IBC}, and b_D define a unique solution, and we refer to this solution as the background state x^b. The background state is traditionally the forecast conducted after the prior data assimilation cycle. Together, the IBCs and dynamics compose the model, so that the full model can be cast as:

$$\begin{bmatrix} A_{IBC} \\ D \end{bmatrix} x^b = \begin{bmatrix} b_{IBC} \\ b_D \end{bmatrix}, \text{ or } Mx^b = b_M \qquad (1)$$

The observations are y^o, and the operator H acts on the model state to simulate the observing systems. The observation innovation vector is $d = y^o - Hx^b$. In some instances, the observation value may depend on the state, so that H is nonlinear. This is typically not the case in the ocean, but it does occur in examples such as acoustic or optical data. We seek a state that satisfies all the IBCs, dynamics, and matches the observations. However, the problem with only IBCs and dynamics is well-posed. Adding the additional requirement to match the observations over-determines the problem. Thus, our objective is to compute an analysis state that minimizes the squared errors to the background IBCs, the dynamical equations, and the observations:

$$\begin{bmatrix} M \\ H \end{bmatrix} x^a = \begin{bmatrix} b_M \\ y^o \end{bmatrix} + \begin{bmatrix} \eta \\ \varepsilon \end{bmatrix} \qquad (2)$$

where the errors to IBCs and dynamics are contained in the model errors η, the observations are y^o, and observations errors are ε. The problem is cast to find a correction δx to the background x^b so that $x^a = x^b + \delta x$. In this context, the dynamical operator is linearized around the background state so that the tangent linear model is M'. The assumption is that the solution time interval is short enough that a linearized dynamical operator is sufficiently accurate. This can be tested by comparing the tangent linear solution to a full nonlinear solution, both with a small perturbation from the background. The optimal increment should provide a minimum squared error to

$$\begin{bmatrix} M' \\ H \end{bmatrix} \delta x = \begin{bmatrix} 0 \\ d \end{bmatrix} + \begin{bmatrix} \eta \\ \varepsilon \end{bmatrix}, \text{ or simply } A\delta x = b + e. \qquad (3)$$

The weighted minimum error variance solution to Eq. 3 is given by $\delta x = \left(A^T W A\right)^{-1} A^T W b$, where the weighting $W = B^{-1}$, or the (pseudo) inverse of the covariance matrix B. The covariance matrix is composed of the values

$$B = \left\langle \begin{bmatrix} \eta \\ \varepsilon \end{bmatrix}, \begin{bmatrix} \eta^T & \varepsilon^T \end{bmatrix} \right\rangle \qquad (4)$$

where $\langle , \rangle$ indicates an expected value that is an average over many realizations of the forecast under similar conditions, and we will see that many simplifications are required to estimate this covariance matrix. Assumptions underlying this solution are that the corrections δx are sought in the range of B and are Gaussian-distributed random variables with zero means. For some variables this is appropriate, but there are exceptions to keep in mind. Ice concentration and salinity, for example, cannot have negative values. Therefore, the corrections are not simply Gaussian-distributed with zero mean. Methods to contend with non-Gaussian errors are an active area of research (Bocquet et al., 2010), but for operational applications, we continue with this assumption.

The covariance matrix describes how errors in one IBC, dynamical equation, or observation are related to others. For example, an initial condition temperature error in one location will be correlated to an error in temperature nearby horizontally and vertically. Likewise, an error in the dynamical equation for diffusivity at one location and time will have a spatial and a temporal correlation. In this solution process, we must specify the statistical relation of the errors in every state variable to every other state variable at every location and at every time over the entire forecast trajectory.

When this process is applied to only one system, such as only the ocean or only ice or only the atmosphere, we must realize an assumption has already been made. The state can contain all variables of all the Earth systems together such as $x^T = \begin{bmatrix} x^T_{ocean} & x^T_{atmosphere} & x^T_{ice} & x^T_{waves} \cdots \end{bmatrix}$, and we should recognize that a correction to ocean surface temperature should lead to a correction of near surface atmospheric properties. By applying assimilation to individual systems, we are assuming that the covariances between the systems are zero. If we assume that errors in other dynamical systems are not correlated to errors in the ocean, the covariance matrix becomes block diagonal with vast areas assumed to be zero. Each block may be inverted separately. This assumption allows us to conduct ocean data assimilation separately from other systems. There are coupled fluxes between all these dynamical systems, and the justification for separating them is mainly to reduce the problem to a tractable size rather than considering a scale analysis. The decoupling certainly simplifies the problem, and we should be aware that addressing these covariances within coupled systems remains an area of research (Zhang et al., 2007). There are also intermediary steps toward fully-coupled data assimilation. For example, the computation of fluxes between the ocean and atmosphere may be considered a separate model, and observations are made that allow estimation of the fluxes. If we include the fluxes as a separate model in the state, we can again assume the error covariance matrix to be block diagonal so that data assimilation may be performed separately on the ocean, atmosphere, and on fluxes between the two.

The optimal correction given by (3) is the solution of a weighted least squares problem, however, the covariance matrix poses two challenges. The first challenge is the size of the matrix that prohibits direct solution by the usual weighted least squares approach. The direct inversion of the large matrices is problematic. If we consider a 1/25°L40 global ocean model with a five-day period prior to the present containing observations and a five-day forecast with a one-minute time step, the state size for just temperature, salinity, velocity, and surface height is about 10^{14} variables. The covariance matrix then contains 10^{28} variables. No computer can store the state in memory, and thus storing the error covariance matrix is orders of magnitude beyond present capacity. The problem is formulated to reduce the size. Here, we start by casting it through a 4DVar approach:

$$\delta x = M'B_4 M'^T H^T \left(HM'B_4 M'^T H^T + R \right)^{-1} d \qquad (5)$$

The covariance matrix in Eq. 4 is separated into two components under the assumption that observation errors ε are not correlated to model errors η. The matrix R is the error covariance of the observations $R = \langle \varepsilon, \varepsilon^T \rangle$, and the covariance for the 4DVar contains the model error covariances $B_4 = \langle \eta, \eta^T \rangle$ of dynamics, IBCs and other model input parameters. Both R and the matrix $HM'B_4 M'^T H^T$ are the size of the number of observations rather than the size of the full state. So the problem is reduced. The inversion is usually through a conjugate gradient with an appropriate preconditioner so that the actual inverse is never computed. It is only the action of $\left(HM'B_4 M'^T H^T + R \right)$ on a vector that is required in the conjugate gradient process. Even when cast in this form, the problem remains very large and must be reduced further.

The second covariance matrix challenge is that there is insufficient observational data on which to compute the statistics represented in the matrix. We must make further simplifying assumptions to enable progress on this problem. We consider further assumptions to address the ocean mesoscale problem in the next section, and further simplifying assumptions are required for the covariance matrix in the subsequent section.

Simplifications for Mesoscale Assimilation

The initial focus of the Global Ocean Data Assimilation Experiment (GODAE) has been the prediction of the mesoscale eddy field. Internal ocean dynamics lead to instabilities that convert potential energy stored in the displacement of isopycnals into kinetic energy. The western boundary currents, such as the Gulf Stream, continually generate very strong eddies, and eddies are pervasive in the interior oceans.

The process outlined in the prior section makes regular corrections to the ocean forecast. The prior forecast is used as the background state x^b, observation differences to the background provide the observation innovations d, and the data assimilation computes an analysis increment δx. In the formulation leading up to the 4DVar in Eq. 5, the analysis increment is computed for every variable over all space and time, as well as the model inputs. This leads to the 4DVar covariance matrix being very large, which requires us to make simplifying assumptions. The temporal representation of corrections can be reduced by assuming a functional form over time, which may be as simple as linear interpolation between correction estimates at times spaced much greater than the model time step. The 3DVar takes this simplification several steps further. The formal assumptions leading from the 4DVar to the 3DVar should be stated explicitly so that we are aware.

The 3DVar first assumes that errors exist only at the times when a new forecast is made, which are the analysis times and that all observations exist only at the analysis times. In addition, assume that the errors are not correlated between analysis times. Finally, assume the dynamical errors are negligible so that $\eta = 0$, which is the strong constraint problem.

The effect of these assumptions is to set areas of the 4DVar error covariance matrix B_4 to zero. Because we have ordered the state vector to correspond to time, the assumption that the errors are not correlated at different analysis times leads to B_4 being zero except at diagonal blocks that represent the covariances of errors at the analysis times. The solution process at each analysis time is now independent, which greatly reduces the problem. The assumption that the dynamical errors are negligible allows reformulating the problem in terms of the Lagrangian multipliers so that the solution from one analysis time to the next exactly satisfies the model dynamics. The assumption that observations are only at the analysis time leads to having to compute the analysis increment only at the analysis time.

At each analysis time, the 3DVar solution is computed:

$$\delta x_3 = B_3 H^T \left(H B_3 H^T + R \right)^{-1} d \qquad (6)$$

The analysis increment δx_3 is computed just at the analysis time, and the covariance matrix B_3 is the error covariance at the one analysis time. The dynamical operator in time no longer enters into the assimilation problem. The error covariance matrix B_3 now represents covariances of errors in the background at the analysis time. Because the background field from a prior forecast is typically the initial condition, B_3 is the background error covariance in the 3DVar approach. As in the 4DVar, the matrix $\left(H B_3 H^T + R \right)$ has dimension of the number of observations rather than the state size. At the initial time of GODAE, these assumptions were necessary to reduce the problem to a tractable size.

We still must specify B_3, and this requires some dynamical insight. How are errors of one variable at one (x, y, z) related to another at a different position? This is the most critical part of the problem. If there were full observations of all variables over all (x, y, z), the form of B_3 would not be so important as it would serve to only reduce the observation error from one point to another. However, the ocean is under-sampled. It is rare that even one aspect of the ocean in one area is sufficiently sampled, let alone over-sampled. This has significant implications. The analysis that serves as the initial condition for the forecast is critically dependent on our ability to specify the error covariances because this matrix greatly extends the influence of observations. This one matrix is the point at which the disciplines of oceanography and data assimilation meet.

As a first example, consider one primary source of observational information for operational ocean prediction, satellite altimeter data. Assume one observation of sea surface height (SSH) occurs at a model grid location. In this case, the observation operator H is a vector of zeros except one element that has a value of one, and the position of that element corresponds to the appropriate observation location. In this case, the matrix $\left(HB_3H^T + R\right)^{-1}$ is a scalar that provides the relative weight of the background error in SSH and the observation error. This scalar multiplies one column of the covariance provided by B_3H^T. The observation operator is selecting one column of B_3 that specifies the relation between SSH errors to all other variables over space. Thus, the correction due to one observation is proportional to one column of the covariance matrix.

How should the background state be changed given SSH observations? Suppose the column of B_3 under consideration specifies that SSH is not correlated to other model variables. The resulting analysis increment would change just the SSH. Prior experiments show that changing only model SSH dissipates quickly as a barotropic wave. The oceanography science community experience is that the primary errors in initial conditions are the positions of mesoscale eddies. The error in eddy feature position leads to errors of internal thermohaline structure that results in SSH errors. Therefore, B_3 must provide the covariances between SSH and the underlying temperature and salinity. Additionally, observations have shown that the dynamical balance between the mesoscale pressure field and the velocity field is primarily geostrophic at low Rossby number (large spatial scale and long time periods). Corrections to temperature, salinity, and SSH that lead to pressure must relate to corresponding changes in the velocity field. The background covariance matrix must specify these properties.

There is no entirely satisfactory method to determine B_3, and there are many proposed. All of these methods have advantages and disadvantages. Two considerations are important when evaluating methods. The first is physical representation and the second is a large sample size to confidently compute statistics. Ideally, we would compute the error covariances based on observations (Helber et al., 2013). This removes the question of any bias of the physical system being used since we would be using the true ocean. However, we require many realizations to compute each element of B_3. There are not sufficient data to compute statistics of errors in all variables and the relation to errors in all other variables at other points in space. A numerical model run can provide many events over a very long period of time to build confident statistics such as those by Oke et al. (2005). However, biases in models can bias the statistics. Relations should be expected to change depending local conditions such as if two variables inside a cyclone, inside an anticyclone, or on opposite sides of a front. The relations also depend on the season since thermocline structure changes from summer to winter These considerations result in conditional statistics, which are covariance relations given certain conditions such as time of year or information on the local background features. Covariance estimates from observations or historical model runs typically do not include substantial conditional information. Estimates from ensemble-based systems allow the conditional information to be included more naturally (Evensen, 2003). However, the number of ensemble members is typically small, which leads to errors in the estimated covariances and techniques such as localization (Anderson, 2012). At this point, we must make the reader aware that one particular formulation of the covariance will be examined here, but it is an important area of active work with many possible approaches given the references above.

The approach of Brandt and Zaslavsky (1997) initially separates B_3 into variance S_b and a correlation function C_b as $B_3 = S_b^{1/2} C_b S_b^{1/2}$. A decomposition of C_b into separable functions is then made so that the correlation between two field variables v and v' is given by

$$C_b(x,y,z,t,x',y',z',t') = C_{vv'}^H(x,y,x',y') C_{vv'}^V(z,z') C_{vv'}^{FDB}(x,y,x',y') \qquad (7)$$

The correlation between horizontal locations x, y and x', y' and vertical positions z and z' is the product of the separable components of horizontal $C_{vv'}^H$, vertical $C_{vv'}^V$ and flow-dependent $C_{vv'}^{FDB}$ correlation functions. This separation of variables is relatively typical since there is not sufficient a priori information to provide the full correlations between the values of $v(x,y,z,t)$ and $v'(x',y',z',t')$. When separated in this manner, there begins to be sufficient observational data for certain aspects as examined in the next section. Such a decomposition is applied in general circulation models (Derber and Rosati, 1989), the Harvard Ocean Prediction System (HOPS; Lozano et al., 1996), MERCATOR (Brasseur et al., 2005), the Forecasting Ocean Assimilation Model (FOAM; Martin et al., 2007) and the Climate Forecast System (CFS) reanalysis (Saha et al., 2010b).

A basic assumption in specifying B_3 is that the covariances between errors should have characteristics similar to the ocean features themselves. Vertical structures of corrections should be similar to the vertical structures of eddies; this vertical structure being critical for operational ocean prediction. The basic dynamics of mesoscale eddies are results of thermocline displacement, which change across the globe and seasonally. With historical expendable bathythermograph, CTD, and ARGO data, sufficient observational information to specify the vertical covariance structure of temperature and salinity from observations is becoming available.

Specifying the Background Covariance

An example of the vertical structure within C_{vb} of Eq. 7 computed from historical observations at one location (275°E, 24°N, which is in the Gulf of Mexico Loop Current just southwest of Florida) during February is shown in Fig. 18.1. The diagonal submatrices from lower-left to upper-right show the correlations over depth for temperature, salinity, and geopotential respectively. These correlations are computed globally on a 1/2° grid monthly at 72 vertical levels. The data size is reduced by retaining only the 20 leading eigenmodes of the correlation matrices, which is a further simplifying assumption. The results in Fig. 18.1 show diagonal correlations that are slightly less than 1 due to this. The temperature and salinity correlations show distinct layers. Temperatures in depths less than 50 m are correlated to one another more strongly that to those greater than 50 m, while temperatures greater 50 m are correlated to one another more strongly than those less than 50 m. Salinity exhibits a similar pattern with the separation depth at about 150 m. We must relate the thermohaline variations to the pressure field given by the geopotential. Define the vector of temperature and salinity anomalies (deviations from the mean vertical structure during the month) as $X = \begin{bmatrix} T_1' \cdots T_N' S_1' \cdots S_N' \end{bmatrix}^T$, where N is the number of vertical depths. The geopotential is a linear transformation of the temperature and salinity given by $\varphi = G\kappa X$. The operator κ provides specific volume anomaly (the operator may be linearized about the monthly mean), and the operator G provides an integral over pressure. The cross correlation of temperature, salinity, and geopotential is then computed as

$$C_{vb} = \begin{bmatrix} \langle XX^T \rangle & \langle XX^T \kappa^T G^T \rangle \\ \langle G\kappa X^T X \rangle & \langle G\kappa XX^T \kappa^T G^T \rangle \end{bmatrix} \tag{8}$$

One interesting aspect of the vertical structure in Fig. 18.1 can be seen by the correlation of surface T, S, or G to the values deeper in the water column. The correlation of G at the surface to the underlying T and S is much stronger than the correlation of surface T or S to the subsurface. This is a result of geopotential at the surface (which is SSH multiplied by the gravitational acceleration) being an integral of the underlying properties as well as the large vertical correlation scales of T and S over depth. The SSH is a vertical integral of the effects and is strongly related to T and S. This leads to satellite altimeter data being a very influential observation for thermocline variability.

As in the vertical structure, we commonly assume horizontal structures of the estimated corrections are similar to the ocean features themselves. The horizontal correlations $C_{vv'}^H$ in equation (7) are computed using a second order auto-regressive function (Gaspari and Cohn, 1999) with a length scale specified as a fraction of the Rossby radius of deformation (Chelton et al., 1998). An example of the temperature analysis increments at 190 m depth is shown in Fig. 18.2 (top figure). The lines of features are due to the satellite SSH observations, and the vertical covariance provides the relation to the 190 m depth. The horizontal correlation relates increments along the ground tracks to values away from the ground tracks. The length scale varies spatially and has an average value of 21 km in the example, which is smaller than the Rossby radius.

The flow-dependent correlation for each variable is specified as $C^{FDB} = (1+s_f)e^{-s_f}$, where $s_f = \delta p / dh$, δp is the difference in model forecast pressure between two observation locations and dh is the specified flow-dependence scale factor. This increases the correlations between points that have little pressure difference and decreases correlation across pressure gradients. The pressure field is used in the flow-dependent correlation under the assumption that the flow is in geostrophic balance and directed along pressure surfaces due to mesoscale features. Obviously, for some processes such as tides, this assumption does not hold. However, spatial scales for tides are typically well-separated from the mesoscale in the open ocean. Thus, it is not expected to be a significant problem. The flow field itself could be used in the formulation, though some influence of the tidal signal would still contribute. An example of strong flow-dependence influence is provided in Fig. 18.2 (bottom figure). The flow-dependence remains a minor influence on the accuracy of ocean prediction as the major covariance deficiencies lie mainly in the amplitudes as will be shown (Jacobs et al., 2014a).

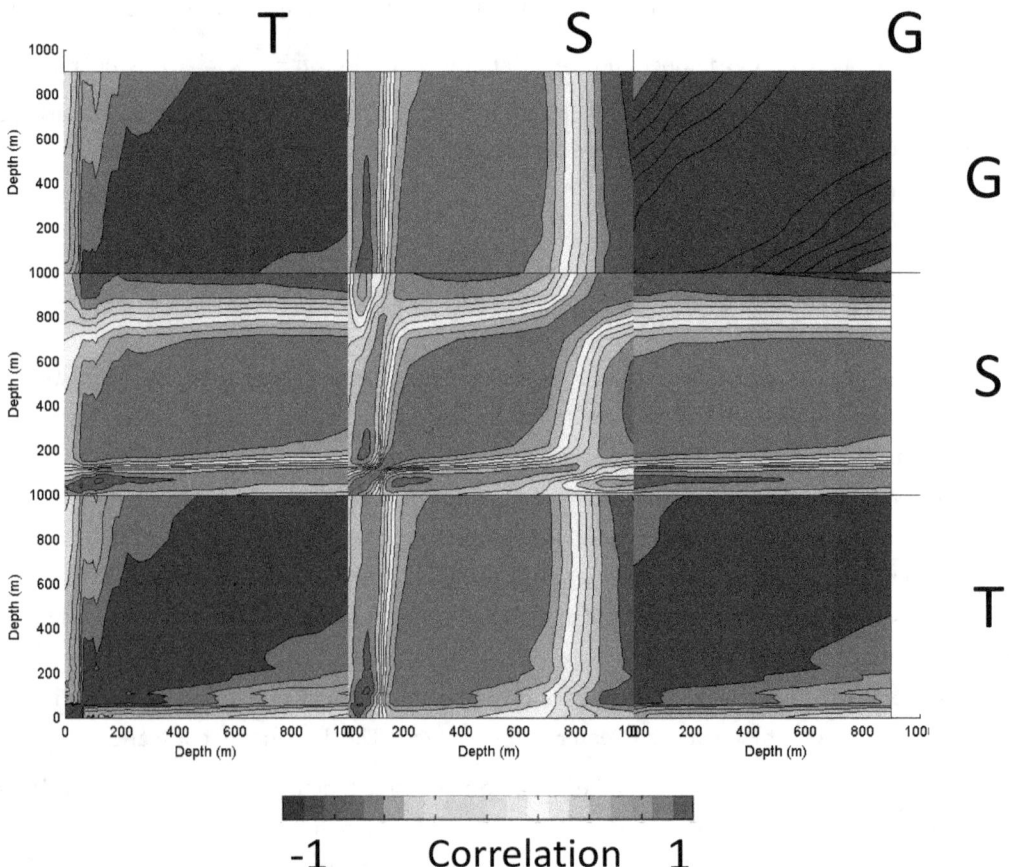

Figure 18.1. Cross correlations between temperature, salinity, and geopotential from the ocean surface to 1000m depth. This is based on historical data in February around the point 275°E, 24°N, which is in the Gulf of Mexico Loop Current just northeast of Cuba. The 20 most significant eigenmodes of the corelation matrix are retained to reduce data storage requirements of the global database, so the diagonal values (particularly in S-S correlation) are slightly less than 1. Distinct layers of correlation are apparent. The layers are different in temperature and salinity.

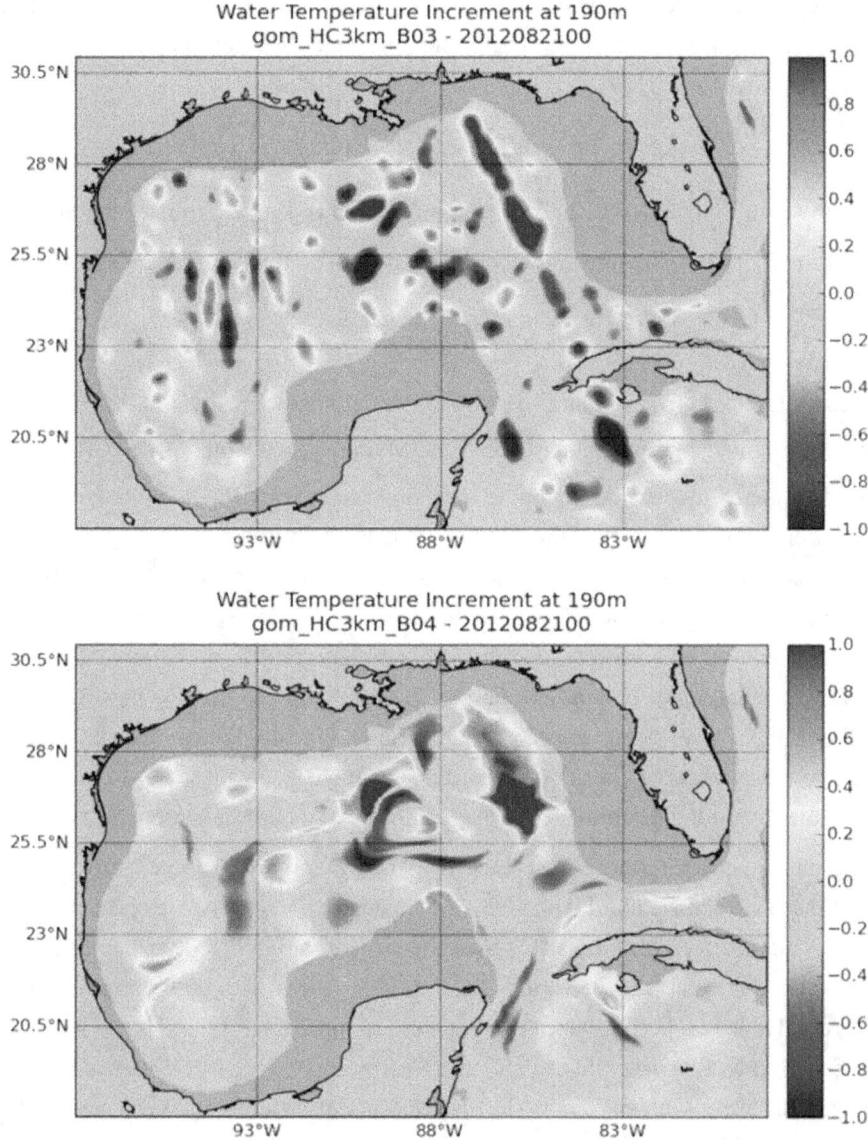

Figure 18.2. The analysis increments during one assimilation cycle of a 3DVar using (top) no flow dependence and (bottom) high flow dependence.

The background variance S_b is a diagonal matrix providing the amplitudes of the error variances, but is very difficult to compute. As discussed in the introduction, the 3DVar analysis should use the observations with background variance to compute the analysis covariance. The observations reduce the errors, and the amplitude of elements in the analysis covariance are smaller than those in background. The errors then increase in amplitude to provide the forecast error covariance that becomes the background error covariance for the next assimilation. For small dimensional problems to which Kalman filters are applied, propagating the analysis error covariance forward in time by applying the system dynamical operator and its adjoint is straight forward. However, an ocean prediction system dimension is far too large. Ensemble methods provide a mechanism to accomplish this (Evensen, 2003). The example here uses a simplified approach. Given a first estimate of the background variance and the observations, the expected analysis error variance S_a is:

$$S_a = \left(I - BH^T \left(HB_3 H^T + R\right)^{-1} H\right) S_b \qquad (9)$$

where I is the identity matrix, B_3 and R are as defined in Eq. 6, and only the diagonal values are computed in S_a with off-diagonal values being zero. The forecast variance may be estimated by

$$S_f = \left(1 - 1/\gamma^c\right) S_a + 1/\gamma^c \left(S_c - S_a\right) \qquad (10)$$

which increases forecast error variance S_f toward the climatological variability S_c provided by the Generalized Digital Environmental Model (GDEM) database (Carnes et al., 2010) with $\gamma^c = 20$ days. This value generally provides an upper limit on variance amplitude in the thermocline in the absence of observations. In practice, the forecast variance amplitude shows the desired behaviors: observations reduce the background errors, and the errors increase to the upper limit of climatologically observed variance if no observations occur over several weeks. One example of the background error variance is provided in Fig. 18.3. The results are the forecast error standard deviation after the analysis of each day from Aug 21 to Aug 24 of 2012. As new data are received each day, the analysis error decreases according to Eq. 9. On subsequent days, the forecast errors increase in previously observed locations according to Eq. 10.

It is possible to test the performance of the estimated standard deviation $S_b^{1/2}$ by comparing the observation minus background difference over many assimilation cycles to build statistical estimates of the observed standard deviation. Two methods are tested with results shown in Fig. 18.4. The first method (referred to as R1) uses the system variability from one analysis cycle to another to estimate the standard deviation. This provides areas with high variability with higher expected errors. However, because ocean timescales are long, the estimate is too low. The second method (referred to as R2) uses the estimation given by Eqs. 9 and 10. The results over a three-month experiment in the Gulf of Mexico in Fig. 18.4 indicate the forecast error estimate of R2 is much more consistent with the root mean square (RMS) of observation minus background. In addition, the RMS of observation minus background is lower for R2 than for R1 indicating that the background field has lower errors relative to observations. As the background error covariance is a critical part of the data assimilation and thus prediction system, verification of the accuracy as in Fig. 18.4 is a very important part of the process of overall system validation for operational application.

Finally, we must specify how velocities are related to other variables. This is based on the correlation between geopotential and velocity given the horizontal $C_{vv'}^H$ spatial structure as determined by a second order auto-regressive function. The derivative of this spatial function provides a geostrophic velocity balancing geopotential height (Daley, 1993).

All these details in the background error covariance are typical from the beginning of mesoscale ocean prediction. Over time, a range of new assimilation considerations and observations have made advancements incrementally. However, the results are still constrained by assumptions and limitations that reduced the 4DVar of Eq. 5 to the 3DVar of Eq. 6.

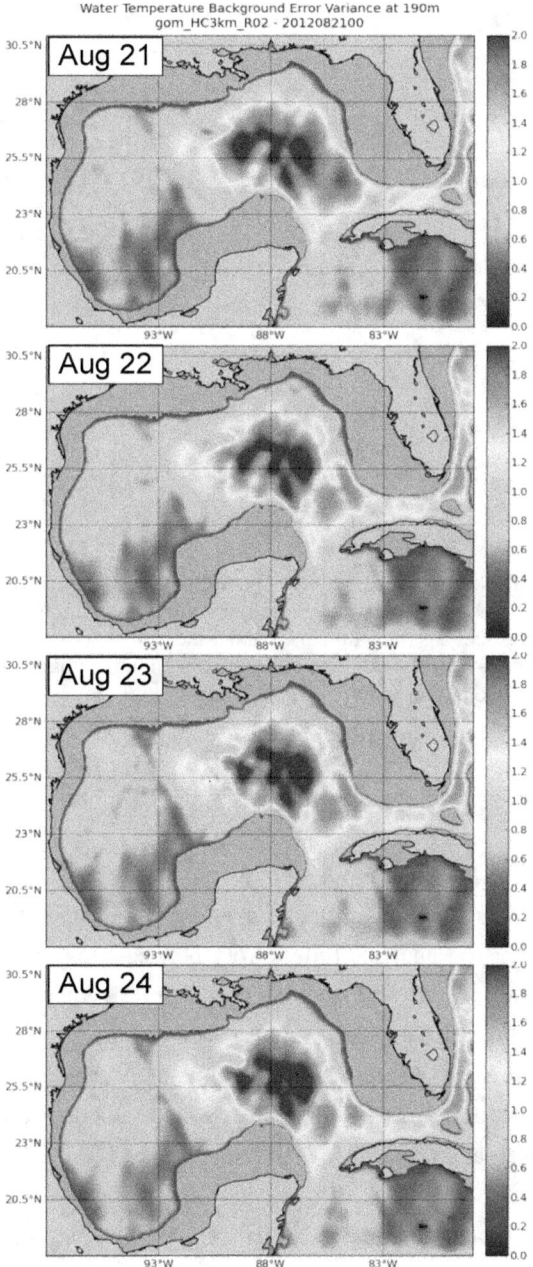

Figure 18.3. Temperature error standard deviation at 190 m temperature on August 21-24, 2012 (top to bottom rows).

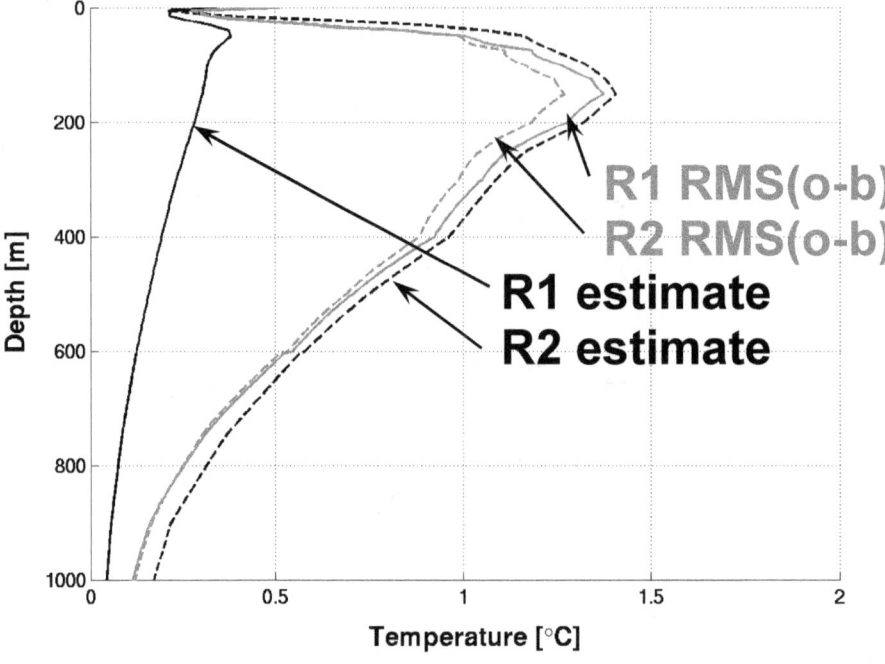

Figure 18.4. The estimated (black lines) temperature standard deviation error over depth from two methods (solid versus dashed lines) is compared to differences in the observations minus background (red lines). The background error estimate using analysis reduction in variance and growth to climatological variance (R2 black line) indicates a closer agreement with the RMS of observations minus the background state (red lines) than the methodology using only the temporal variability of the background state (R1 black line). In addition, the RMS difference between observations and background is smaller for R1 than R2 (red lines) indicating the background is more accurate.

Time Correlation

The temporal correlation of errors is a significant issue. In simplifying the 4DVar to a 3DVar, we assumed errors exist only at the analysis time, errors between analysis times are uncorrelated, observations exist only at the analysis time, and that the dynamical equations have negligible errors. The 3DVar analysis in Eq. 6 computes analysis increments at sequential analysis times independently. Errors in the assumption that observations be at the analysis time are reduced by using the First Guess at Appropriate Time (FGAT) in which the observation operator acts on the background field at the observation time [J A Cummings, 2005]. If the model has skill at predicting deterministic high frequency variability such as wind-driven events or internal tides, the FGAT approach reduces the high frequency signal in the data and the 3DVar corrects the slowly evolving field. However, the temporal influence of the data remains limited.

From the 4DVar perspective in Eq. 5, assuming errors between different analysis times are uncorrelated implies the temporal correlation is a set of Dirac delta functions. Impulsive forcing is not physically realistic. If the initial condition is reset at the analysis time, transients such as inertial oscillations develop even though the analysis may be geostrophically balanced. The source is the ageostrophic flow present in a model system as well as nonlinear advective processes that are not geostrophic. Often, rather than resetting the initial condition at the analysis time, the analysis increment is inserted over a time interval. The total increment is divided by the time interval, and at every time step the model state is adjusted. This significantly reduces initial transients. The insertion also treats the corrections as though there is a long decorrelation time rather than an impulsive forcing at a single time. The time correlation function when inserting the increment is a boxcar function. The insertion approach recognizes some time correlations in errors, though errors from one analysis time to another remain uncorrelated.

Timescales in the deep ocean are long. Motivated by the assumption that errors have characteristics similar to the ocean mesoscale processes, the errors and subsequent analysis increments should also have long timescales. The improved forecast skill when moving from a time correlation of a delta function to a boxcar function is consistent with errors having long timescales. The assumptions leading to the 3DVar preclude observations from making corrections beyond the system assimilation interval. If the assimilation interval is one day, observations cannot influence subsequent days.

Attempts to take the time correlation into account mainly consist of including observations over a data window T_{obs} that is long. This consideration appears in many GODAE systems. MERCATOR assimilates altimeter data covering the prior seven days in an assimilation cycle that occurs every seven days (Brasseur et al., 2005); FOAM has a daily assimilation cycle with data used over multiple cycles and an error variance increasing linearly with data age (Martin et al., 2007); the Bluelink Ocean Data Assimilation System (BODAS) uses an 11-day T_{obs} to assimilate observations (Oke et al., 2008); and CFS conducts a six-hour assimilation cycle using data in the prior ten days with a weighting based on data age (Saha et al., 2010a). The Global Ocean Forecast System uses a daily assimilation cycle with observations covering the past several days so that data are used more than once (Cummings et al., 2009). Formally, observations should be assimilated only once. Otherwise, observation error variances are not accurate. There are competing considerations when determining the time period between analysis cycles to correct the model. A long time between assimilation cycles allows more data to influence the initial condition, however this implies that data soon after the assimilation could improve the forecast. A one-day assimilation cycle does use newly acquired observations, but it does not allow a long time period correlation. These problems arise because of the assumptions simplifying 4DVar to 3DVar. Early in GODAE, the development of 3DVar was more mature, and infrastructure supported it. There has been strong motivation to enable observations to influence corrections over a long time period, and this motivates a 4DVar approach.

The infrastructure and capability within ocean data assimilation has advanced to the point where 4DVar assimilation is feasible at least in regional nested areas (Moore et al., 2011; Ngodock and Carrier, 2014; Powell et al., 2008). The 4DVar solution in Eq. 5 contains the ocean tangent linear operator M' and its adjoint M'^T. The adjoint provides a way to compute the derivative of the observation innovations (difference between the observation and background field) with respect to all variables over all time and space. We immediately know how to change the observation errors by a change to the state at any time, boundary condition, or any dynamical equation throughout the solution period. Moving to 4DVar overcomes the inherent limitations of 3DVar. Observations can influence the analysis over long time periods.

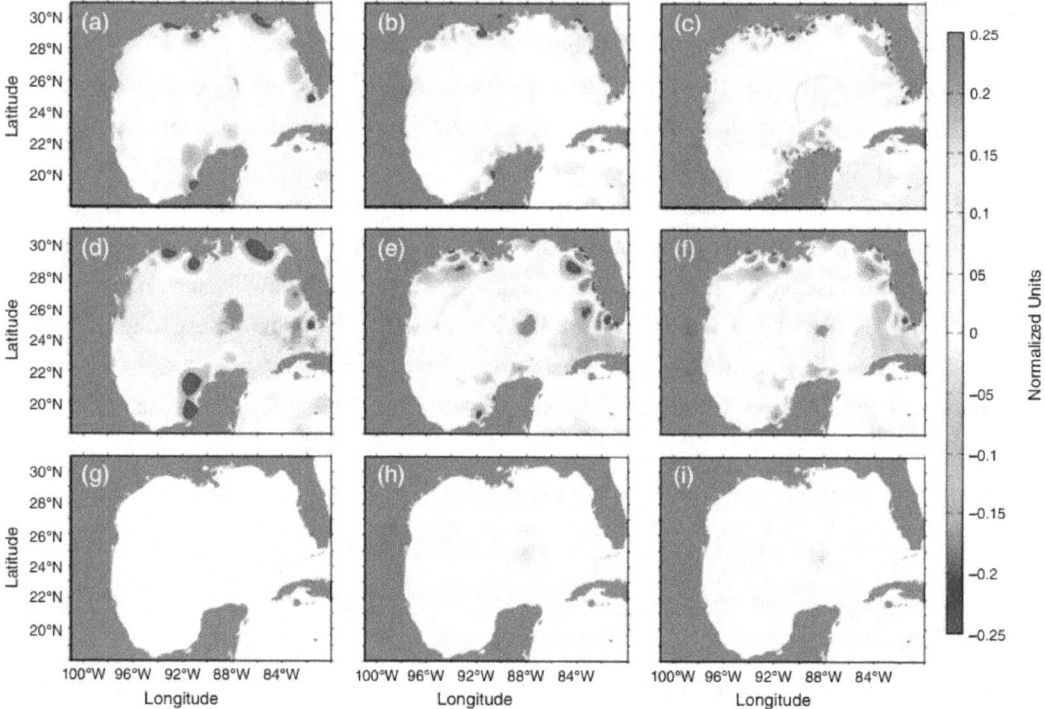

Figure 18.5. An example of gravity waves in the adjoint free-surface solution 96-, 48- and 3-h into the adjoint integration respectively (a–c), and the corresponding free surface for the forward solutions at the corresponding adjoint times, one forced with the adjoint solution (d–f) and the other without adjoint forcing of the free surface (g–i).

4DVar does not alleviate the need for dynamical consideration of the covariance matrix. The covariance matrix in the 4DVar of Eq. 5 contains not only the error covariance of the state at a specific time within 3DVar in Eq. 6 but also the error covariances between all the dynamical equations of the dynamical operator M'. The problem is larger, and the science communities must address the problem more closely. The consideration of errors in the dynamical operator requires further insight. One example is provided by considering the dynamical errors associated with SSH observations in the 4DVar (Ngodock et al., 2016). Consider the difference between observed and background SSH. A number of physical mechanisms may result in the difference. A mesoscale structure could produce a temperature and salinity anomaly deep within the thermocline that changes the specific volume and thus SSH. Alternatively, a barotropic wave generated at a particular location and time (e.g., a tsunami) may arrive at exactly the observation location to produce the observed SSH discrepancy. A sudden wind stress or any ocean process that may change SSH is an equally feasible source of error. The dynamic operator and its adjoint in the 4DVar do not differentiate between these solutions.

An example of this problem is shown in Fig. 18.5. The result of the operator $M'B_4 M'^T H^T$ for one observation of SSH in the central Gulf of Mexico is computed by first applying the adjoint

M'^T to the observation operator and integrating backward in time (Fig. 18.5, top row). The error covariance and tangent linear model operators are then applied sequentially. The result (Fig. 18.5, middle row) shows the impact that one SSH observation would have on the analysis at the observation time under the assumptions contained in the covariance matrix B_4. The resulting analysis appears outside our normal expectations. The corrections 96 hours prior to the observed SSH (middle row left) indicate features around the coastlines, and if we intend to correct mesoscale features, these are not consistent. However, the analysis is physically correct. Stating that the result is outside our expectations implies that it is statistically anomalous, which implies a problem in the covariance that embodies our expectations of errors.

The covariances in Eq. 5 include the error in the barotropic equations used in the numerical model, which express the time rate of change of the vertically integrated velocities and SSH. The horizontal divergence integrated over the water column is the time rate of change of SSH. The results above included errors to the vertical integration of continuity, even though there is little error expected in this equation. This equation is a result of applying the conservation of mass property. Ocean models are typically constructed in a flux conservative form so that properties are conserved. The experience of physicists and oceanographers is that there is no error to the conservation of mass equation. Thus, the covariance matrix components describing these errors should be set to zero, and the minimization should have Lagrangian multipliers applied. When this is done, SSH observation stimulation of barotropic waves is greatly reduced (Fig. 18.8, bottom row). The data assimilation community has led the development of the 4DVar optimization approach. The expectation of where errors lie must be included in the error covariances of the 4DVar, and this requires the insight and experience of oceanographers. Expanding application of 4DVar to alleviate the time correlation issues inherent in 3DVar remains a challenge. Experiments conducted with 4DVar indicate good advancement beyond the 3DVar, and some of these are examined later on.

A Recent Example and Implications

The Lagrangian Submesoscale Experiment (LASER) as part of the Consortium on Advanced Research for Transport of Hydrocarbons in the Environment (CARTHE) released over 1,000 surface drifters starting January 2016 in the northeastern Gulf of Mexico. The objective was understanding submesoscale features and sharp horizontal buoyancy gradients. The persistence of drifters allowed observation of a wide range of additional features including the Loop Current Eddy (LCE) and associated cyclonic eddies. To highlight the types of features that are not well-predicted, we examine the details of when the assimilative forecast system diverges significantly from the observed drift trajectories. These are important in helping to understand the limits of all the components of the prediction system including the assimilation and its simplifying assumptions. The assimilation system is a 3DVar described by Eq. 6 with error covariances specified by the standard deviation, vertical correlations, and horizontal correlations described in the previous sections.

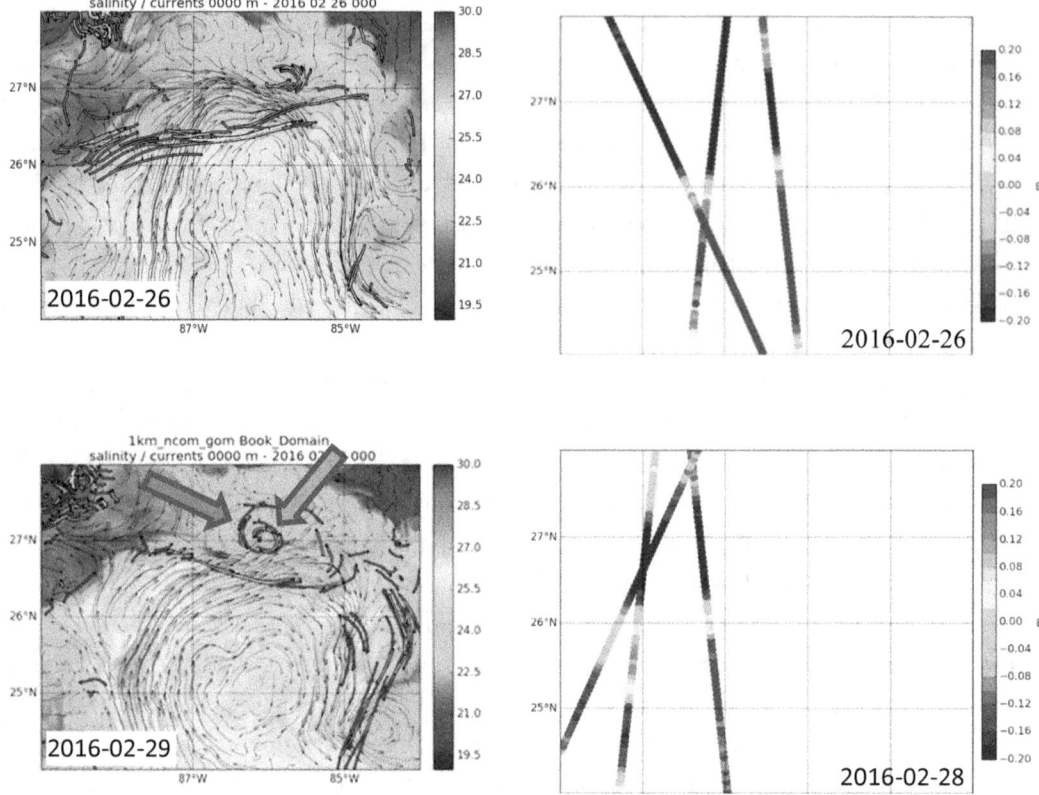

Figure 18.6. On the left are assimilative model results showing surface currents (vectors) and salinity (color) with observed drifter trajectories (white lines). On the right are satellite SSH anomaly observations assimilated into the model. The model shows a Loop Current Eddy (LCE) flow far to the north up to 27°N, which is not observed by the drifters. After assimilating the satellite observations, the position of the front is more correctly located. Also note the cyclonic eddy on the northern edge of the eddy at 27°N on 2016-02-29. The red arrow notes the position in the model, and the blue arrow notes the position observed by drifters.

Figs. 18.6 to 18.8 display three events. In each, the LCE covers the lower portion of the domain. The size of this feature is about 300 km in diameter, which is in the targeted features of prediction for GODAE. The figures show surface currents and salinity from the system assimilating all the regular observations but not the LASER drifters. The general location of the LCE is consistent with the satellite SSH anomaly observations shown in the figures as well as the LASER drifters. The main discrepancies are in the position of the eddy front and cyclonic eddies on the periphery of the LCE.

Fig. 18.6 shows an event starting February 26, 2016 when the model LCE extends up to 27°N to the north. The LASER drifters indicate the LCE front is around 26.3°N. The SSH anomaly observed by the altimeter data in subsequent days shows the LCE front much further to the south as well. After the satellite observations have been assimilated to correct the model, on February 29, 2016 the LCE front is much closer to the observed location February 29, 2016. This is an error in frontal position on the order of 50 to 100 km. At the end time of this event, also note the cyclonic eddy to the north of the LCE front at about 27°N 86°W. After the assimilation has corrected the frontal position, the forecast model has a cyclonic feature near the cyclonic circulation observed by the LASER drifters. This particular cyclonic feature is the focus of the subsequent two events.

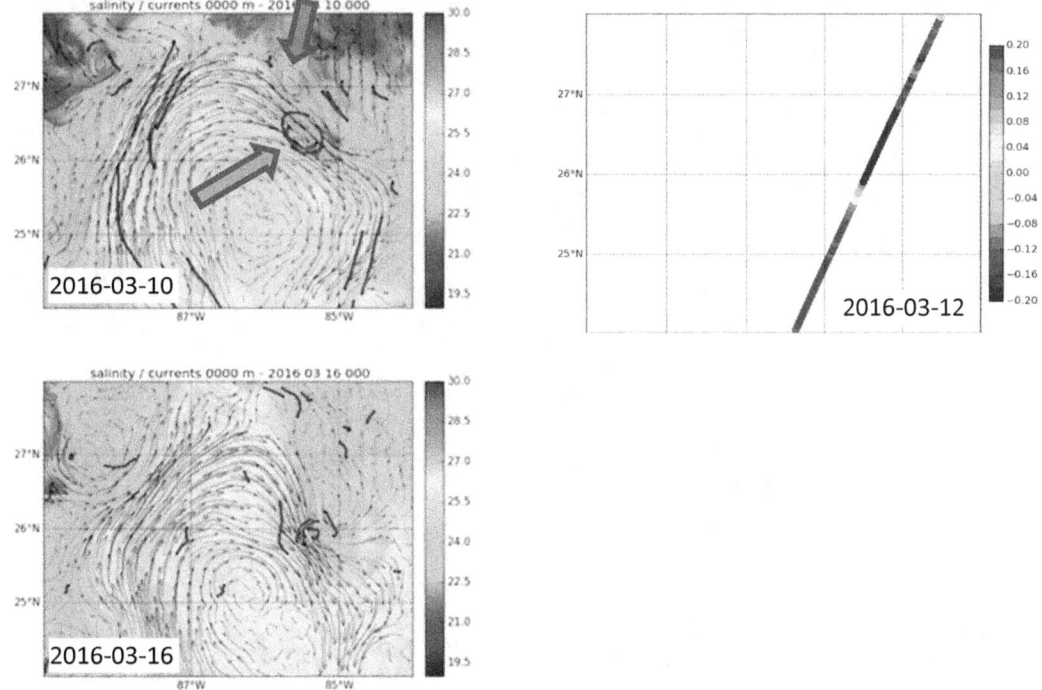

Figure 18.7. On the left are assimilative model results showing surface currents (vectors) and salinity (color) with observed drifter trajectories (white lines). On the right are satellite SSH anomaly observations assimilated into the model. At 2016-03-10, the drifters are entrained in the cyclonic eddy at 26.3°N 85.5°W, though the model indicates cyclonic circulation either at 27°N or at 85°W. The red arrow notes the position in the model, and the blue arrow notes the position observed by drifters. After assimilating the satellite data shown on the right, the model cyclonic circulation is similar to the drifter observed cyclone at 26°N on 2016-03-16.

The second event begins on March 10, 2016 in Fig. 18.7. Note the error in positioning of the cyclonic feature has grown significantly compared to the end time of the first event in Fig. 18.6. The model cyclonic feature is at 27°N while the drifter-observed feature is at 26.3°N. By the end time of the event on March 16, 2016, the model cyclonic feature is much closer to the observed. However, the shape is quite different. The most significant satellite SSH anomaly data available during this interval are shown in Fig. 18.7. This one satellite pass transects the cyclone, and these are the main data affecting the model cyclone position. On March 20, 2016, the model cyclone is at 26°N while the observed cyclone is at 25.3°N (Fig. 18 8). There are several satellite passes observing the area, and by March 27, 2016 the model cyclone position is much better aligned with the observations.

These errors are typical and to be expected. The main features on which GODAE systems have focused have been the larger mesoscale eddies. The time and spatial scales of the features are roughly consistent with the observation systems, from satellite to in situ. The data assimilation assumptions of horizontal and vertical correlation structures are consistent as well. The results from the LASER experiment highlight the edge of technical capability within present prediction systems.

The forecast skill is limited by the observation density as well as the prescribed assumptions within the data assimilation systems that have been built to be consistent with the expected observation density. Technology continues to move forward, and we must consider how the observing systems will evolve, how numerical models will evolve, and therefore what changes are necessary to the data assimilation to remain consistent with expected future development.

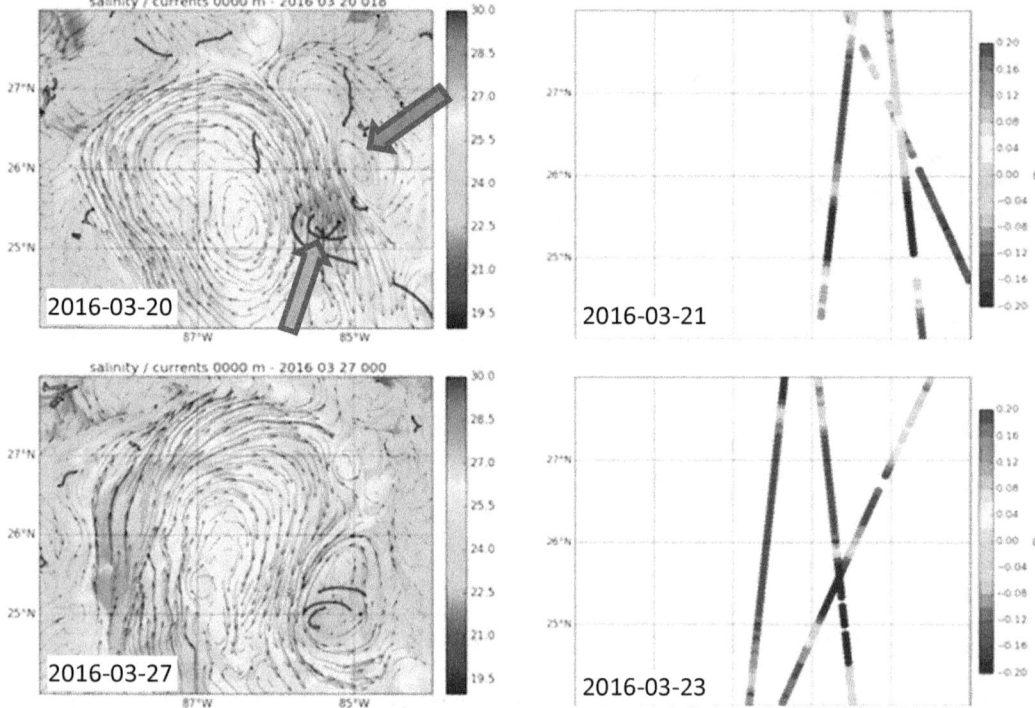

Figure 18.8. On the left are assimilative model results showing surface currents (vectors) and salinity (color) with observed drifter trajectories (white lines). On the right are satellite SSH anomaly observations assimilated into the model. At 2016-03-20, the drifters are entrained in the cyclonic eddy at 25.3°N 85.5°W, though the model indicates cyclonic circulation is at 26°N 85°W. The red arrow notes the position in the model, and the blue arrow notes the position observed by drifters. After assimilation of the satellite data shown on the right, the model cyclonic circulation is similar to the drifter observed cyclone at 25°N on 2016-03-27.

Future Advancements

From the 3DVar perspective of forecasting mesoscale features, the primary model state errors to be corrected are at the analysis time. The 4DVar approach is demonstrated to be superior to the 3DVar (Powell et al., 2008; Smith et al., 2017; A Weaver et al., 2003), and the errors are corrected in both the initial conditions and the state trajectory over time. In addition, the ability to advance The largest impediment to advancing operational ocean prediction is the density of observations. It is important to identify the metric by which forecasts are evaluated as we move forward. Many examinations quantify the error in the SSH of a forecast system. While this was very useful for initially gauging skill, it is not a strongly sensitive measure of accuracy. An example in Fig. 18.9 shows the skill of several ocean variables as a function of the number of altimeter data streams (Jacobs et al., 2014b).

The data were constructed from a series of observation system experiments in which all data were assimilated into one system, which is referred to as the nature run. This is the best solution possible. Subsequent experiments use the same boundary conditions and forcing but varying numbers of altimeter satellite data streams from TOPEX/Poseidon, Jason-1, GFO, and ENVISAT over a 1.5-year period. The initial condition at the beginning of the period for the observation system experiments is different from the nature run to ensure nondeterministic features deviate from the nature run. The spatial anomaly correlation between the observation system experiments and the nature run is averaged over the 1.5 years to produce the results plotted in Fig. 18.9. Anomaly correlations must be 0.6 or greater to achieve skillful prediction. The correlation of steric height relative to 1000 m (a proxy for SSH) with one altimeter is about 0.85 and asymptotically moves toward 1.0 with additional data. This reflects that the observation system experiments have the main mesoscale features in roughly the correct position. The steric height correlation is not greatly sensitive to small errors in mesoscale feature position. Other variables such as the surface divergence or the frontogenesis forcing, Q_1 (Hoskins, 1982; Pollard and Regier, 1992), linearly increase from about 0.25 with one data stream to just under 0.6 with four data streams. The frontogenesis is driven by confluence of water masses, which occurs along the fronts of eddies, and the surface divergence is controlled by this frontal process. Thus, these two variables measure the ability to accurately predict frontal positions. Typical frontal position errors are on the order of 50 -100 km at present, as we have seen in the LASER results (Figs. 18.6 to 18.8).

As observation density increases, the processes associated with fronts become predictable. The present regular observing systems such as altimeter satellites, ARGO, and others are not sufficient to resolve the positions of fronts. There are two approaches to increase the observation density, either by new in situ dense observations or new remote sensing instruments. An example of in situ dense observations is provided by the Grand Lagrangian Deployment (GLAD) during which 200 CODE-like drifters were deployed in the northeastern Gulf of Mexico in July 2012 (Ozgokmen et al., 2013). The forecast of Lagrangian trajectories is one of the most difficult problems we face. Errors grow exponentially in time. Small displacements of ocean features cause trajectory forecasts to rapidly diverge from observations. A pair of assimilation experiments is conducted, in which the first experiment uses the standard 3DVar approach for all regular observations and the second experiment uses the 4DVar approach and infers the velocity from drifter trajectories as observations in the assimilation. Forecast trajectories are then compared to the observed trajectories (Fig. 18.10). The general area of the LCE (around 90°-88°W, 25°-27°N) is similar in both results, but the shape and, therefore, the frontal location is quite different. The 4DVar experiment assimilating the inferred velocity shows forecast trajectories that are much more accurate than the 3DVar that is not assimilating the trajectories

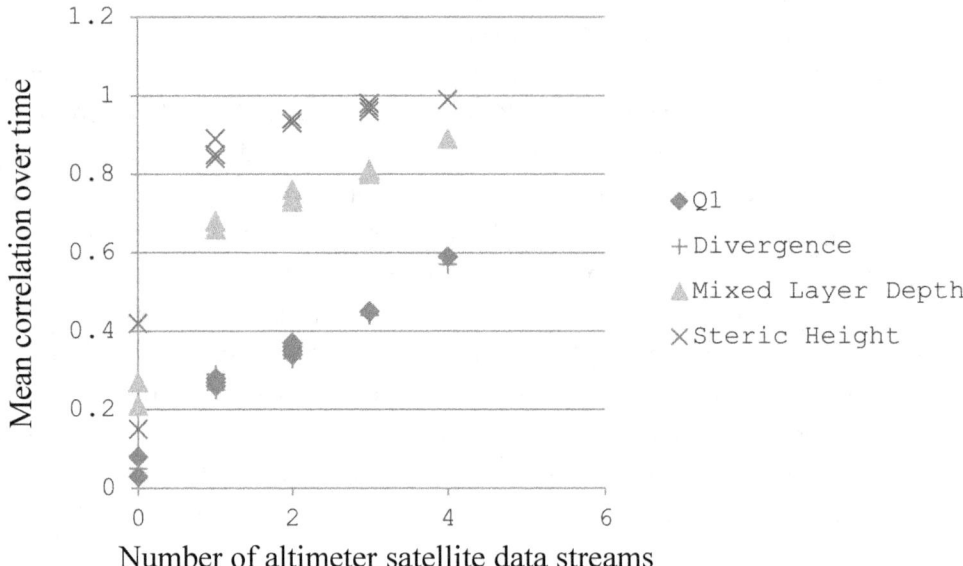

Figure 18.9. The time-average correlation as a function of number of satellite altimeters assimilated. The steric height correlation increases rapidly with just one data stream, and the marginal improvement of additional altimeters is small and decreases with increasing altimeters. The mixed layer depth marginal improvement is relatively constant from 1 to 4 altimeters. Frontogenesis Q_1 and surface divergence (vertical velocity) show a nearly linear increase in correlation as the number of altimeters assimilated increases.

The drifters prove to be a powerful observing system. The sensors are relatively simple and inexpensive. Large quantities can be deployed. In mesoscale features, the drifters persist for weeks while providing data continually, which is contrasted to satellite observations which typically require more than a week to return to the same area. If we assume the drifters are affected by geostrophic currents, an analogy is that the drifters are observing SSH horizontal gradients. Ageostrophic effects should be accounted for as representativeness errors, and these may be primarily a function of wind speed. Thus, the drifters provide information similar to altimeters in a selected area for a persistent period. This makes drifters an ideal observing system at the mesoscale for urgent events during which accurate forecasts become necessary, such as oil spills or search and rescue.

These high-density persistent observations affect considerations from the data assimilation perspective. The spatial scales require relating the observations across time as drifters move along fronts. Therefore, we must retain the time-correlated errors that are contained within the 4DVar in Eq. 5. The spatial scales presently used in the 3DVar are not consistent with the small scales of observations. Much work is under way to construct analysis increments that successively correct the larger-scale and the smaller-scale features ((Brandt and Zaslavsky, 1997; Choi et al., 2008; Haley and Lermusiaux, 2010; Lermusiaux, 2002; (Li, McWilliams et al. 2015)). These approaches become necessary as the dynamical processes of smaller scales change from the larger scales. The dynamical change across scales implies that a change in error covariances is required as well. As we consider new observing systems, the data assimilation of operational prediction systems must advance.

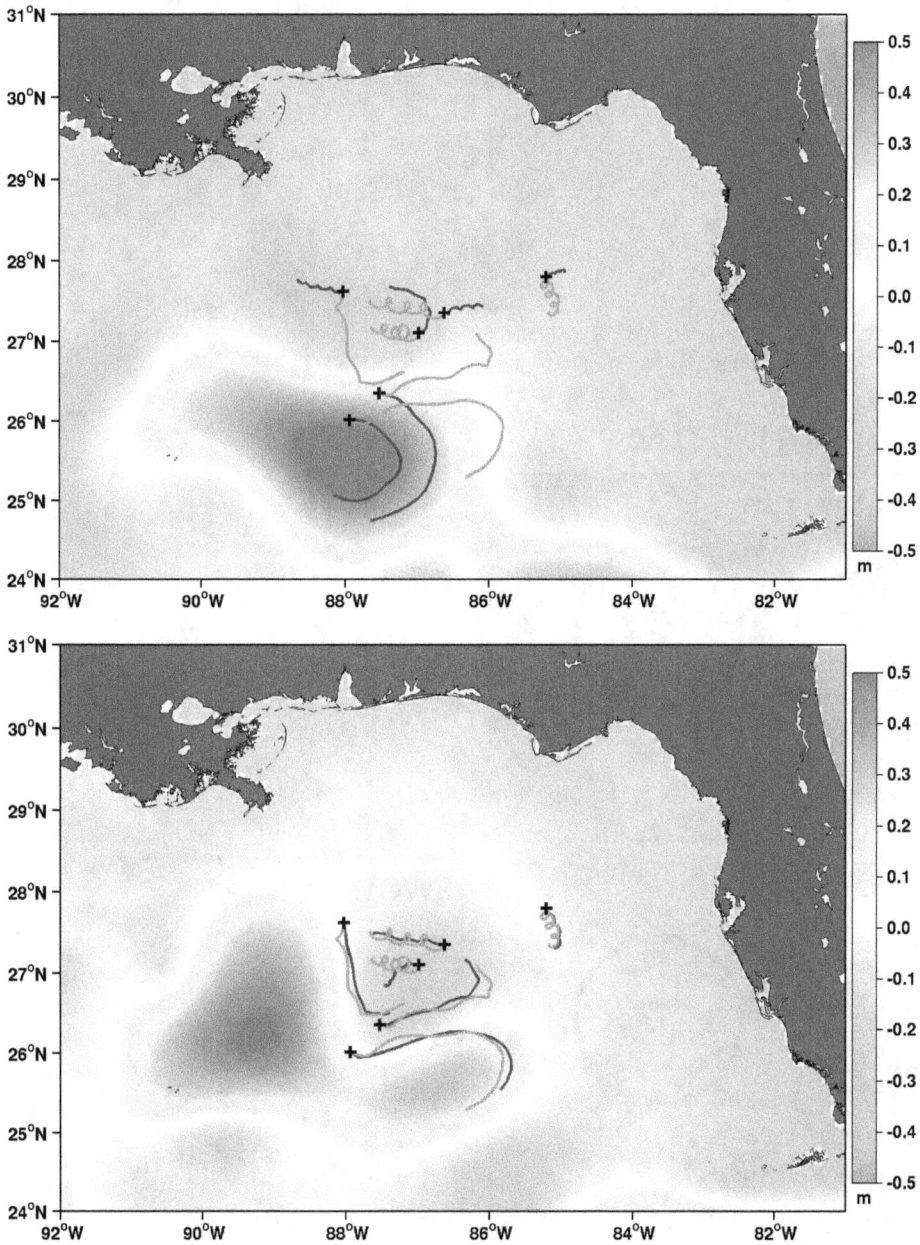

Figure 18.10. Two experiments are conducted during the July 2012 time period when 200 drifters were released in the northeast Gulf of Mexico. The 3DVar experiment (top) does not assimilate the drifters while the 4DVar experiment (bottom) does assimilate the drifters. The examples are shown one month after the assimilation experiments begin. Model forecast trajectories (purple lines) are compared to the observed trajectories (green lines). The background color is the model SSH.

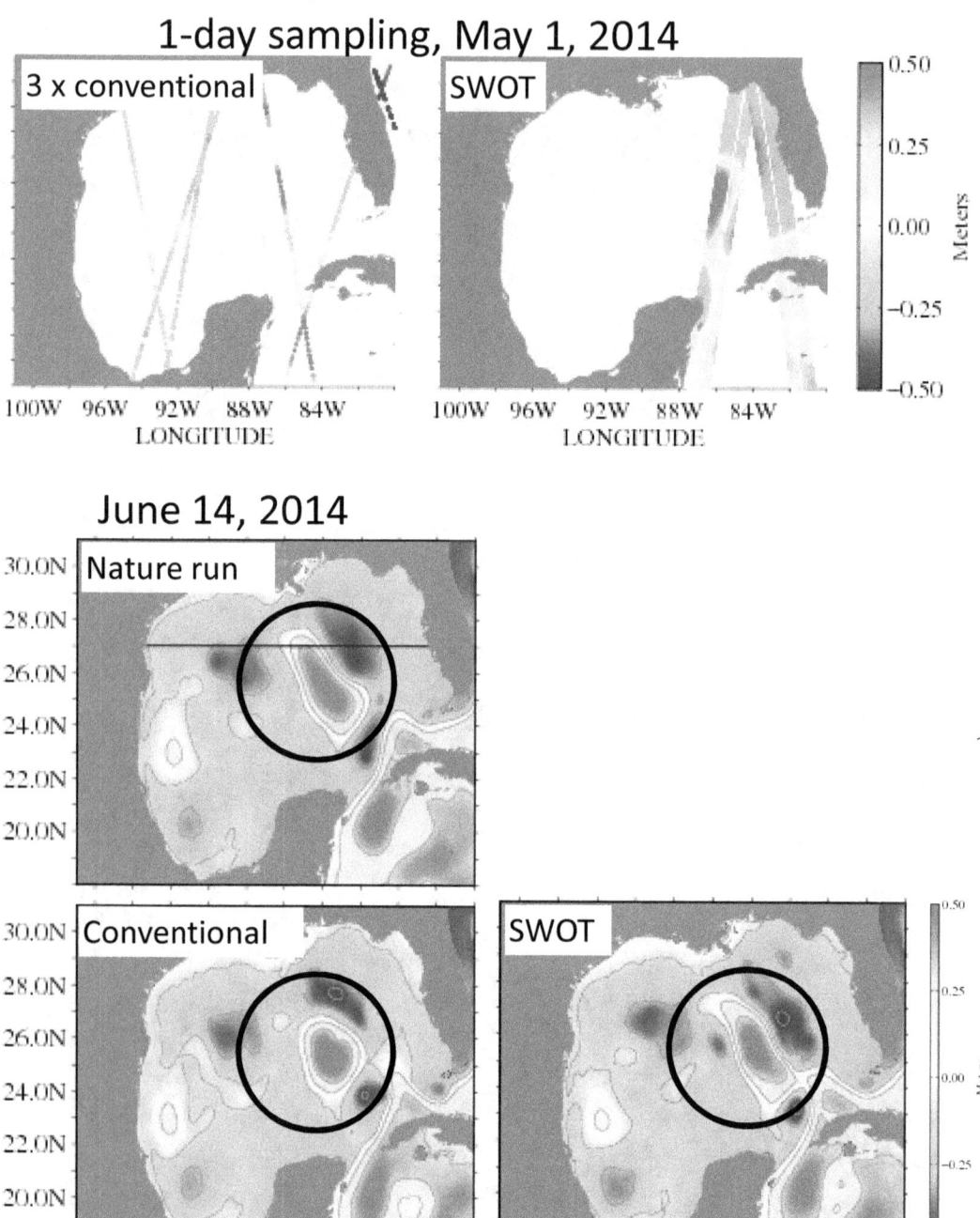

Figure 18.11. Observation System Simulation Experiments (OSSEs) simulate three conventional altimeter satellites and one Surface Water Ocean Topography (SWOT) satellite in terms of predicing the SSH of the LCE. The top row shows one day of sampling from the two OSSEs near the beginning of the experiment. The middle row shows the nature run SSH on June 14 with the LCE circled. The bottom row is the SSH from the conventional (left) and SWOT (right) simulations with the circle in the same location.

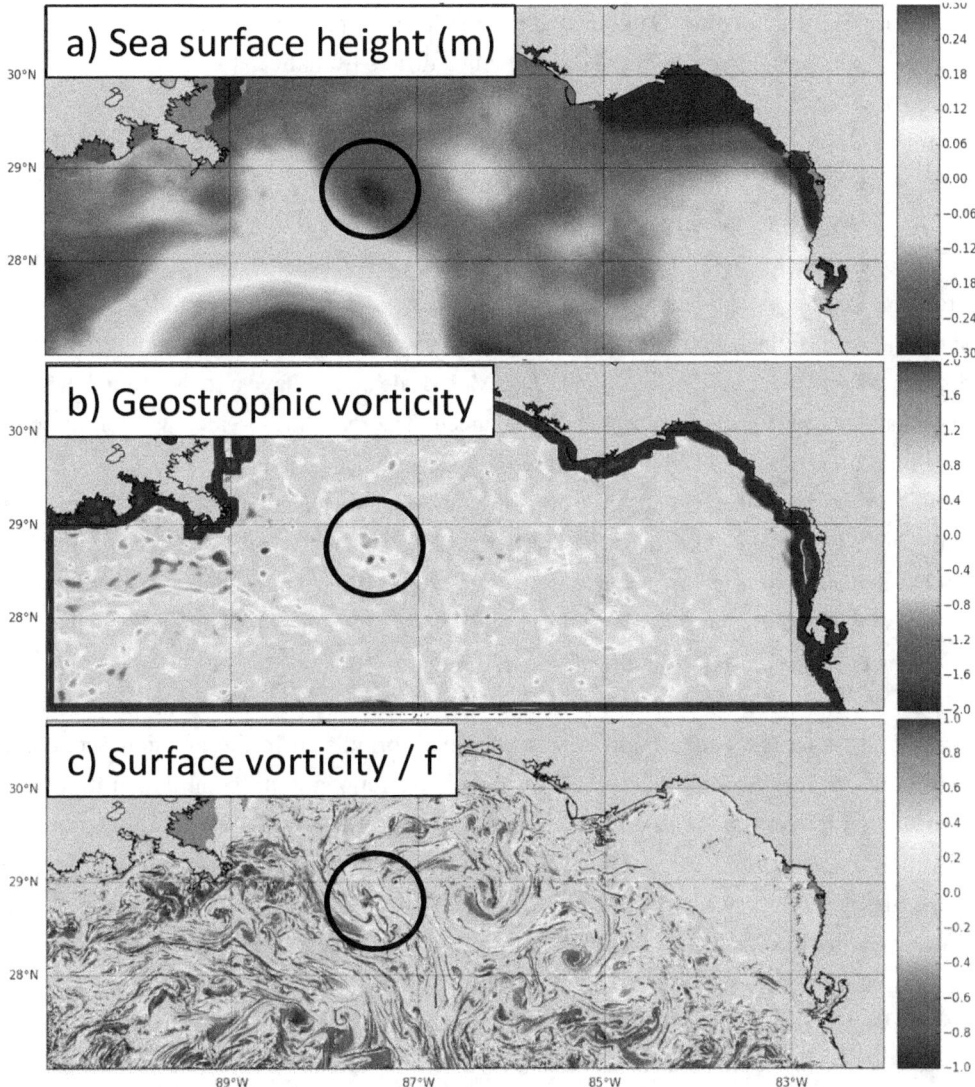

Figure 18.12. A numerical model shows the submesoscale difference from mesoscale. The SSH (top) is dominated by large amplitdue large horizontal scale features. The geostrophic vorticity computed from the SSH and normalized by the Coriollis parameter (middle) indicates small scale features with order one vorticity. The vorticity from the model normalized by the Coriollis parameter (bottom) contains these features, as well as the latteral shear associated with fronts.

A significant advancement in satellite observations will be provided by the Surface Water Ocean Topography (SWOT) mission (Durand et al., 2010). This instrument is an interferometric synthetic aperture radar (INSAR) that measures SSH across a 120 km wide swath (with a gap at the nadir). The ocean data are expected to be about 1 km resolution in the directions both along and across the ground track. Two Observation System Simulation Experiments (OSSEs) were conducted to gauge the influence of SWOT on assimilation and model skill. One experiment was run without any assimilation to provide the nature run that was sampled for the two OSSEs. One sampling scheme is provided by using the ground tracks of three conventional altimeters (Jason-1, AltiKa, Jason-2 in the interleaved orbit). The second sampling scheme uses the sampling of SWOT by linking the nature run to the SWOT simulator (Gaultier et al., 2015). The sampled data are assimilated into two separate model runs starting from an initial condition that differs from the nature run, and the results are examined one month later (Fig. 18.11). The shape and frontal positions of the LCE in the SWOT OSSE are much closer to the nature run than the features in the conventional altimeter OSSE. We expect three conventional altimeters to be operating during the SWOT time period. Together, we can expect these systems to advance ocean predictive skill significantly.

SWOT also presents a possibility of predicting much smaller scales throughout the globe, which moves from mesoscale to submesoscale prediction. The dynamics change significantly and it requires reevaluation of the assimilation approach. We can examine the Rossby number based on the ratio of vorticity to the local Coriolis parameter. Mesoscale eddies have a Rossby number much less than one and are primarily geostrophically balanced. The main mesoscale variation in ocean structure is due to vertical movement of the thermocline in balance with the SSH (Fig. 18.12a). Spatial scales are on the order of 200 km in this area. The submesoscale has order one or larger Rossby number. The main vertical structure variability is in the mixed layer depth. The submesoscale deviates significantly from the observed historical vertical structure in Fig. 18.1, where the ocean acts mainly as a two-layer fluid with the interface at the thermocline depth. The submesoscale acts with an interface at the mixed layer depth.

Small scale features (Fig. 18.12b,c) are dominated by submesoscale eddies with spatial scales on the order of 10 km. The background error covariance formulated in the previous sections is targeted specifically at mesoscale features. The present operational data assimilation process is oriented to this underlying assumption. The vertical structure, horizontal structure, geostrophic balance, cycling frequency, and application of data all have underlying assumptions that the primary features are mesoscale. As we move to forecast submesoscale features, the dynamics change dramatically, and the features themselves are strongly advected by the mesoscale flow.

Addressing these requires a multiscale approach in which the analysis of the two dynamical regimes is conducted sequentially (Li, McWilliams et al. 2015). A correction is first computed for the mesoscale using appropriate error covariance, and this correction is applied to ensure the mesoscale is as accurate as possible as it will advect the submesoscale field. A second correction is computed for the submesoscale. The error covariances for the submesoscale must be different from the mesoscale, but it would be convenient if the infrastructure for both analyses were as reusable as possible. This leads to a generalization of the assimilation problem.

As discussed earlier, the analysis increment is not constructed by explicitly computing a matrix inverse. Iterative techniques, such as conjugate gradient descent with preconditioners, are often used. In this case, all we require is the action of matrices on vectors. The effects of the covariance operating on state vector can also be applied through means polynomials of the diffusion operators have been routinely used to provide the action (Carrier and Ngodock, 2010; Derber and Rosati, 1989; Weaver and Mirouze, 2013; Yaremchuk et al., 2013). Dynamical balance operators may be generalized similarly. There is an operation that provides density from temperature and salinity by equation of state. Integration of the continuity equation relates density to SSH. Hydrostatic balance provides pressure from density and SSH, and finally the geostrophic relation provides velocity from the pressure field. The action of these operators maps from one variable to another by the specified dynamics, and the covariance matrix may be specified in terms of these operators (Yaremchuk et al., 2017; Yaremchuk and Martin, 2016). For example, the sequence of computations described above relates temperature and salinity to SSH and velocity by a mesoscale operator:

$$\begin{bmatrix} \zeta \\ \vec{u} \end{bmatrix} = \mathbf{L_M} \begin{bmatrix} T \\ S \end{bmatrix} \tag{11}$$

The full covariance matrix may be provided by:

$$\left\langle \begin{bmatrix} T \\ S \\ \zeta \\ \vec{u} \end{bmatrix} \begin{bmatrix} T \\ S \\ \zeta \\ \vec{u} \end{bmatrix}^T \right\rangle = \begin{bmatrix} B_{TS} & B_{TS} L_\mathbf{M}^\mathbf{T} \\ L_\mathbf{M} B_{TS} & L_\mathbf{M} B_{TS} L_\mathbf{M}^\mathbf{T} \end{bmatrix} \tag{12}$$

where B_{TS} is the cross-covariance of temperature and salinity as provided in Fig. 18.1. Note the similarity between Eqs. 8 and 12. This generalizes the assimilation problem so that it may be applied to the submesoscale. The only required change is replacing the mesoscale dynamical operator $\mathbf{L_M}$ with an appropriate dynamical operator for the submesoscale $\mathbf{L_S}$. The challenge becomes to provide an operator that appropriately relates ocean properties for the submesoscale. The same solution process may then be used in a sequential multiscale analysis to predict ocean submesoscale features.

Conclusions

GODAE has led to operational data assimilation systems that produce regular forecasts in many centers across the globe. The implementations have primarily been based on 3DVar analysis, and by deriving this from a 4DVar analysis we can explicitly understand the inherent assumptions and impact on the forecast skill. Many of the assumptions were appropriate for the observing systems and numerical model predictions. New observing systems can provide high density observations, both in local settings and globally. GODAE systems predict the general locations of mesoscale eddy features. However, the errors in the frontal positions are on the order of 50 to 100 km. At these scales and smaller, dynamics become more ageostrophic. The continued development of data assimilation approaches promises to extend existing observations and exploit new ones. We expect the present errors in frontal positions to decrease dramatically. The movements toward multiscale analysis and relieving the 3DVar assumptions are required to advance. The frontal positions are critical aspects to many operational problems for fisheries, search and rescue, oil drilling, and aquaculture. With the progression of computational capability in operational centers along with numerical model representation of physics and observing systems, we have the opportunity to extend operational data assimilation to meet many new challenges and applications.

References

Anderson, J. L. (2012), Localization and sampling error correction in ensemble Kalman filter data assimilation, Mon Weather Rev, 140(7), 2359-2371.

Bennett, A. F. (2002), Inverse modeling of the ocean and atmosphere, Cambridge University Press.

Bocquet, M., C. A. Pires, and L. Wu (2010), Beyond Gaussian statistical modeling in geophysical data assimilation, Mon Weather Rev, 138(8), 2997-3023.

Brandt, A., and L. Y. Zaslavsky (1997), Multiscale algorithm for atmospheric data assimilation, Siam J Sci Comput, 18(3), 949-956.

Brasseur, P., et al. (2005), Data assimilation for marine monitoring and prediction: The MERCATOR operational assimilation systems and the MERSEA developments, Q J Roy Meteor Soc, 131(613), 3561-3582.

Carnes, M. R., R. W. Helber, C. N. Barron, and J. M. Dastugue (2010), Validation test report for GDEM4Rep., DTIC Document.

Carrier, M. J., and H. Ngodock (2010), Background-error correlation model based on the implicit solution of a diffusion equation, Ocean Model, 35(1-2), 45-53.

Charney, J. G. (1947), The dynamics of long waves in a baroclinic westerly current, Journal of Meteorology, 4(5), 136-162.

Chelton, D. B., R. A. Deszoeke, M. G. Schlax, K. El Naggar, and N. Siwertz (1998), Geographical variability of the first baroclinic Rossby radius of deformation, J Phys Oceanogr, 28(3), 433-460.

Choi, M. J., V. Chandrasekaran, D. M. Malioutov, J. K. Johnson, and A. S. Willsky (2008), Multiscale stochastic modeling for tractable inference and data assimilation, Comput Method Appl M, 197(43-44), 3492-3515.

Cummings, J., et al. (2009), Ocean Data Assimilation Systems for Godae, Oceanography, 22(3), 96-109.

Cummings, J. A. (2005), Operational multivariate ocean data assimilation, Q J Roy Meteor Soc, 131(613), 3583-3604.

Daley, R. (1993), Atmospheric data analysis, Cambridge university press.

Derber, J., and A. Rosati (1989), A global oceanic data assimilation system, J Phys Oceanogr, 19(9), 1333-1347.

Dombrowsky, E., L. Bertino, G. B. Brassington, E. P. Chassignet, F. Davidson, H. E. Hurlburt, M. Kamachi, T. Lee, M. J. Martin, and S. Mei (2009), GODAE systems in operation, Oceanography, 22(3), 80-95.

Durand, M., L.-L. Fu, D. P. Lettenmaier, D. E. Alsdorf, E. Rodriguez, and D. Esteban-Fernandez (2010), The surface water and ocean topography mission: Observing terrestrial surface water and oceanic submesoscale eddies, Proceedings of the IEEE, 98(5), 766-779.

Evensen, G. (2003), The ensemble Kalman filter: Theoretical formulation and practical implementation, Ocean dynamics, 53(4), 343-367.

Gaspari, G., and S. E. Cohn (1999), Construction of correlation functions in two and three dimensions, Q J Roy Meteor Soc, 125(554), 723-757.

Gaultier, L., C. Ubelmann, and L.-L. Fu (2015), SWOT Simulator DocumentationRep., Tech. Rep. 1.0. 0, Jet 422 Propulsion Laboratory, California Institute of Technology. 423.

Haley, P. J., and P. F. J. Lermusiaux (2010), Multiscale two-way embedding schemes for free-surface primitive equations in the "Multidisciplinary Simulation, Estimation and Assimilation System", Ocean Dynamics, 60(6), 1497-1537.

Helber, R. W., et al. (2013), Validation test report for the Improved Synthetic Ocean Profile (ISOP) system, Part I: Synthetic profile methods and algorithm, Naval Research Lab Stennis Detachment, Stennis Space Center, MS, Oceanography Div.

Hoskins, B. J. (1982), The Mathematical Theory of Frontogenesis, Annu Rev Fluid Mech, 14, 131-151.

IMAWAKI, S. (1981), Vertical Structure and Horizontal Scales of the Mesoscale Baroclinic Variability in the Western North Pacific.

Jacobs, G. A., B. P. Bartels, D. J. Bogucki, F. J. Beron-Vera, S. S. Chen, E. F. Coelho, M. Curcic, A. Griffa, M. Gough, and B. K. Haus (2014a), Data assimilation considerations for improved ocean predictability during the Gulf of Mexico Grand Lagrangian Deployment (GLAD), Ocean Model, 83, 98-117.

Jacobs, G. A., J. G. Richman, J. D. Doyle, P. L. Spence, B. P. Bartels, C. N. Barron, R. W. Helber, and F. L. Bub (2014b), Simulating conditional deterministic predictability within ocean frontogenesis, Ocean Model, 78, 1-16.

Kalnay, E. (2003), Atmospheric modeling, data assimilation, and predictability, Cambridge university press.

Lermusiaux, P. (2002), On the mapping of multivariate geophysical fields: sensitivities to size, scales, and dynamics, J Atmos Ocean Tech, 19(10), 1602-1637.

Le Traon, P.-Y., (1991), Time scales of mesoscale variability and their relationship with space scales in the North Atlantic, J Mar Res, 49(3), 467-492.

Le Traon, P.-Y., et al. (1999). Operational oceanography and prediction-a GODAE perspective, OceanObs' 09.

Li, Z., J. C. McWilliams, K. Ide, and J. D. Farrara (2015), A multiscale variational data assimilation scheme: formulation and illustration, Mon Weather Rev, 143(9), 3804-3822.

Lozano, C. J., A. R. Robinson, H. G. Arango, A. Gangopadhyay, Q. Sloan, P. J. Haley, L. Anderson, and W. Leslie (1996), An interdisciplinary ocean prediction system: Assimilation strategies ana structured data models, Elsevier Oceanography Series, 61, 413-452.

Martin, A. J., A. Hines, and M. J. Bell (2007), Data assimilation in the FOAM operational short-range ocean forecasting system: A description of the scheme and its impact, Q J Roy Meteor Soc, 133(625), 981-995.

Moore, A. M., H. G. Arango, G. Broquet, B. S. Powell, A. T. Weaver, and J. Zavala-Garay (2011), The Regional Ocean Modeling System (ROMS) 4-dimensional variational data assimilation systems Part I - System overview and formulation, Prog Oceanogr, 91(1), 34-49.

Ngodock, H., and M. Carrier (2014), A 4DVAR system for the Navy Coastal Ocean Model. Part I: System description and assimilation of synthetic observations in Monterey Bay, Mon Weather Rev, 142(6), 2085-2107.

Ngodock, H., M. Carrier, I. Souopgui, S. Smith, P. Martin, P. Muscarella, and G. Jacobs (2016), On the direct assimilation of along-track sea-surface height observations into a free-surface ocean model using a weak constraint four-dimensional variational (4D-Var) method, Q J Roy Meteor Soc, 142(695), 1160-1170.

Oke, P., A. Schiller, D. Griffin, and G. Brassington (2005), Ensemble data assimilation for an eddy-resolving ocean model of the Australian region, Q J Roy Meteor Soc, 131(613), 3301-3311.

Oke, P. R., G. B. Brassington, D. A. Griffin, and A. Schiller (2008), The Bluelink ocean data assimilation system (BODAS), Ocean Model, 21(1-2), 46-70.

Ozgokmen, T., A. Poje, B. Lipphardt Jr, A. Haza, B. Haus, G. Jacobs, A. Reniers, J. Olascoaga, E. Ryan, and G. Novelli (2013), Grand LAgrangian Deployment (GLAD): Surface Dispersion Characteristics Near the Deepwater Horizon Oil Spill Site, paper presented at EGU General Assembly Conference Abstracts.

Pollard, R., and L. Regier (1992), Vorticity and vertical circulation at an ocean front, J Phys Oceanogr, 22(6), 609-625.

Powell, B., H. Arango, A. Moore, E. Di Lorenzo, R. Milliff, and D. Foley (2008), 4DVAR data assimilation in the intra-Americas sea with the Regional Ocean Modeling System (ROMS), Ocean Model, 25(3), 173-188.

Richman, J. G., C. Wunsch, and N. G. Hogg (1977), Space and time scales of mesoscale motion in the western North Atlantic, Rev Geophys, 15(4), 385-420.

Saha, S., S. Moorthi, H.-L. Pan, X. Wu, J. Wang, S. Nadiga, P. Tripp, R. Kistler, J. Woollen, and D. Behringer (2010a), The NCEP climate forecast system reanalysis, B Am Meteorol Soc, 91(8), 1015-1057.

Saha, S., et al. (2010b), The Ncep Climate Forecast System Reanalysis, B Am Meteorol Soc, 91(8), 1015-1057.

Smith, S., H. Ngodock, M. Carrier, J. Shriver, P. Muscarella, and I. Souopgui (2017), Validation and Operational Implementation of the Navy Coastal Ocean Model Four Dimensional Variational Data Assimilation System (NCOM 4DVAR) in the Okinawa Trough, in Data Assimilation for Atmospheric, Oceanic and Hydrologic Applications (Vol. III), edited, pp. 405-427, Springer.

Testor, P., et al. (2010). Gliders as a component of future observing systems, OceanObs' 09.

Weaver, A., J. Vialard, and D. Anderson (2003), Three-and four-dimensional variational assimilation with a general circulation model of the tropical Pacific Ocean. Part I: Formulation, internal diagnostics, and consistency checks, Mon Weather Rev, 131(7), 1360-1378.

Weaver, A. T., and I. Mirouze (2013), On the diffusion equation and its application to isotropic and anisotropic correlation modelling in variational assimilation, Q J Roy Meteor Soc, 139(670), 242-260.

Yaremchuk, M., M. Carrier, S. Smith, and G. Jacobs (2013), Background error correlation modeling with diffusion operators, in Data Assimilation for Atmospheric, Oceanic and Hydrologic Applications (Vol. II), edited, pp. 177-203, Springer.

Yaremchuk, M., P. Martin, G. Panteleev, C. Beattie, and A. Koch (2017), Adjoint-Free 4D Variational Data Assimilation into Regional Models, in Data Assimilation for Atmospheric, Oceanic and Hydrologic Applications (Vol. III), edited, pp. 83-114, Springer.

Yaremchuk, M., and P. J. Martin (2016), Implementation of a balance operator in NCOMRep., Naval Research Lab Stennis Detachment, Stennis Space Center, MS, Ocean Dynamics and Prediction Branch.

Zhang, S., M. Harrison, A. Rosati, and A. Wittenberg (2007), System design and evaluation of coupled ensemble data assimilation for global oceanic climate studies, Mon Weather Rev, 135(10), 3541-3564.

CHAPTER 19

Ocean Reanalyses

Keith Haines

Meteorology Dept., University of Reading, Earley Gate, Reading, UK

Ocean reanalyses are becoming increasingly available and useful, and may eventually attract a similar applications base as atmospheric reanalyses. Here we look at how they are being evaluated against both assimilated and independent data, and emphasise that circulation and transport estimates are critical. The Ocean Reanalysis Intercomparison project, ORA-IP, has been comparing many products for consistency on a regional and global basis, including ocean heat content, air-sea fluxes, and recently polar properties including sea ice. The Atlantic meridional overturning circulation as measured by the RAPID array at 26N, is now a challenging new target for simulation. This chapter shows that reanalyses may represent interior ocean basin circulations well (better than free-running models) but they still fail to consistently constrain boundary currents, where most meridional heat transport takes place. There is new work ongoing to try to physically interpret observation increments in reanalysis products, and to look at how to best develop long period reanalysis in earlier years when ocean observations were scarce. Finally, we look at new coupled ocean-atmosphere reanalysis that, by always maintaining a coupled ocean-atmospheric boundary layer, may lead to reduced assimilation increments and air-sea fluxes across domains.

Introduction

This chapter will focus on reanalyses in the oceans and will not look at any of the assimilation methods involved in any detail. Although atmospheric reanalyses have become ever more widely used in the past two decades, with much of the pioneering work done at the European Centre for Medium-Range Weather Forecasts (ECMWF), the corresponding use of ocean data to produce ocean reanalyses is still rather less established. First some terms:

Analysis: It is a complete discretised description of the atmosphere or ocean based on assimilating data against model background data that contains information about past data brought forward to the current time thorough model prediction. Analysis is performed at operational weather forecasting centres every day or more frequently, in order to initialise numerical forecasts. It always uses the most up-to-date version of the data, the numerical weather prediction products, or the operational oceanography, model and data assimilation system. Because the model and data assimilation systems are constantly evolving and new ideas are being incorporated, the analyses on which the forecasts are based would not be suitable to interpret through time.

Reanalysis: The same procedure is used as in making an analysis, but the model and assimilation system are kept constant through time and the reanalysis uses this fixed system to produce the best estimate it can of the whole history through the observational record. It is a very expensive procedure because modern models and assimilation systems are expensive to run, having expanded to run on the biggest computers for daily data. But now we want to run multiple decades

---

Haines, K., 2018: Ocean reanalyses. In "*New Frontiers in Operational Oceanography*", E. Chassignet, A. Pascual, J. Tintoré, and J. Verron, Eds., GODAE OceanView, 545-562, doi:10.17125/gov2018.ch19.

of data through covering past years. Therefore, reanalyses are generally run with lower spatial resolution than current operational suites, at least for the atmosphere.

State estimation: In the ocean, timescales of water mass changes below the surface are long, therefore data on water mass properties may be useful for long timescales, e.g. decades. Long-window data assimilation methods have been pioneered by the Estimating Climate and Circulation of the Oceans consortium to fit ocean data over long time periods. Essentially, this is tuning of an ocean model by modifying surface forcing (winds, fluxes) or interior mixing to fit observations, using 4Dvar. The strength of this approach is that unphysical data increments are avoided, however the method is hard to apply if the system is very non-linear, as is the case in high-resolution models, and limited convergence to the observed data can be obtained.

Both reanalyses and state estimates are starting to be compared and used for similar purposes, so it remains to be seen whether one methodology is any better than the other. Reanalyses have the advantage of being available as an, albeit expensive, by-product of forecasting operations and thus there are many ocean reanalyses available to the community.

Assimilated Datasets

Ocean datasets available to be assimilated into reanalysis products vary hugely through time. Operational oceanography took off in 1992 with the arrival of satellite altimetry giving 10-day views of the global sea surface height. This is the mainstay for providing eddy information in the upper ocean, as well as for monitoring global sea level change. It is also enormously helpful that this dataset has had consistent data coverage since 1992, with the TOPEX and Jason satellite series (and future Sentinel 6), supplemented by one to three additional altimeters for shorter periods, allowing long-term comparisons to be made throughout the reanalyses. Sea surface temperatures (SSTs) have been measured globally using satellite infrared radiation since 1979, but atmospheric correction is challenging in the early years and cloud cover often prevents infrared radiation measurements of the surface. Improving the quality of older satellite SST observations is an active research topic, e.g., through the European Space Agency Climate Change Initiative, (http://www.esa-sst-cci.org/). Ship-based observations go back a century or more (e.g., ICOADS), but the quality and coverage is very variable. Reanalyses before 1990 or so tend to rely on gridded SST products based on statistical modelling from the available observations, e.g. HadISST, Rayner et al. (2003). Sea ice extent also has a good satellite observational record and has become an important target for ocean and climate reanalyses in recent years because of the rapid decline in Arctic sea ice, and operational centres have made increasing efforts to use these data in the last few years. The upper 400 m of the equatorial ($\pm 10°$) oceans began to be monitored from the early 1990s with moored arrays, starting with TAO/TRITON in the Pacific, and extended to include PIRATA in the Atlantic and RAMA in the Indian Ocean. These were established to enable El Niño–Southern Oscillation operational forecasting and are the basis upon which ocean reanalyses at many national weather forecast centres have been built.

Since around 2004, global subsurface observations have been dominated by Argo profiling floats, measuring temperature and salinity down to 2000 m or so. This still leaves the deep ocean below 2000 m, the polar seas with ice cover, and critical flow regions such as western boundary currents, under-observed. Prior to this, Argo subsurface observations were very northern hemisphere and summer dominated with many more temperature-only measurements (salinity is harder to measure accurately), and they were confined increasingly nearer to the surface going back in time. The World Ocean Database maintained by NOAA (www.nodc.noaa.gov/OC5/WOD/pr_wod.html), the ENSEMBLES database (e.g., EN4, www.metoffice.gov.uk/hadobs/en4) maintained by the Hadley Centre, or the CORA product maintained by IFREMER (www.coriolis.eu.org/Data-Products/Products/CORA), all provide historical in situ ocean profile data available for ocean reanalyses. Some centres also provide gridded in situ products using statistical data melding methods, e.g., the CMEMS ARMOR product. Older mechanical bathythermograph (MBT) and expendable bathythermograph (XBT) observations require careful calibration before assimilation into long period reanalyses, as is underway in the International Quality Controlled Ocean Database (IQUOD) project, http://www.iquod.org. More recent observational datasets such as sea surface salinity from SMOS and Aqua, and surface water (mass) distributions from GRACE will provide new targets for inclusion in reanalyses, but currently these data are not accurate or long enough to have been used much in reanalyses.

Ocean Reanalyses and Metrics: Examples

Table 19.1 shows a selection of currently available ocean reanalyses. In Europe the operational community has selected the NEMO (Nucleus for European Modelling of the Ocean) as a community model for development, and several reanalyses use this with fairly similar modelling although different assimilation configurations. U.S. reanalyses have greater diversity, using models such as the Modular Ocean Model (MOM), the MIT General Circulation Model (MITGCM), or the hybrid isopycnal model HYCOM (Hybrid Coordinate Ocean Model). Global ocean reanalysis products are also available from Japan and Australia. Many of these products from operational centres focus on the period after 1992 when operational oceanography started to take off, but there is increasing interest in longer term products, e.g. from 1950s onwards, when current atmospheric reanalyses are now available for ocean forcing. There is a strong interest in ocean initial conditions for decadal prediction purposes from these earlier periods. With the recent arrival of century-long atmospheric reanalyses (e.g. CFSR, ERA-20C, the 20th Century Reanalysis Project), such ocean or coupled reanalyses are now also being attempted (see later in the chapter).

Two kinds of metric might generally be considered to benchmark ocean reanalyses: the fits to the assimilated data and the fits to independent, non-assimilated data. Fig. 19.1 shows data from the Copernicus quality control document (see legend) on the fit of global subsurface temperature and salinity data to observations in the GLORYS reanalysis product. The rapid improvement as Argo data arrives is clear for both temperature and salinity, although seasonal temperature root mean square (RMS) errors up to 1.5 °C can still be seen in the top 150 m. A surface to subsurface mean salinity bias dipole can be seen in the same depth range, perhaps reflecting problems with vertical mixing in the upper ocean? More challenging is to assess the circulation in these reanalysis systems. Currents and transports are generally not directly observed or assimilated and yet such information is critical for both climate and forecasting applications of reanalysis data. Near-surface currents are monitored independently by drifters, and deeper currents at 1000 m by the drifts in Argo float positions. Major current systems are reproduced but near-surface currents are more confined than in the drifter data, and the deep currents are more dispersed than the Argo drifts reveal. The combination of hydrography data (from Argo floats or CTDs) and altimetry are both needed to reproduce upper ocean current structures (Fig. 19.2). Transports of mass heat and freshwater have been estimated at World Ocean Circulation Experiment sections for example, and are being monitored continuously at a few sections such as by the RAPID array at 26°N in the Atlantic. We will look at this transport section in more detail later.

	Name	Model Resolution	DA Scheme	Atmospheric forcing
1	CGLORS025v5	NEMO3.2-LIM2; ¼ x ¼ 50z-levels	3DVar (Sea-ice conc+relaxed ice-thickness)	ERA-Interim
2	ECDA3	MOM4-SIS, 1 x 1 50z-levels	EnKF coupled DA	Coupled O-A model
3	GECCO2	MITgcm, 1x1 (1/3) 50z-levels	4DVar	NCEP
4	GLORYS2v4	NEMO3.1-LIM2 ¼ x ¼ 75z-levels	Reduced-Kalman SEEK filter	ERA-Interim
5	Glosea5-GO5	NEMO3.4-CICE, ¼ x ¼ 75z-levels	NEMOVAR 3D FGAT	ERA-Interim
6	MOVE-G2i	MRI.COM3-CICE4, ½ x 1 52z-levels	Multivariate 3DVar	JRA55
7	ORAP5	NEMO3.4-LIM2, ¼ x ¼ 75z-levels	NEMOVAR 3D FGAT	ERA-Interim
8	SODA3.3.1	GFDL-MOM5-SIS, ¼ x ¼ 50z-levels	OI	NASA MERRA2
9	TOPAZ4	HYCOM-EVP-SI, 12-16km 28hybrid levels	EnKF	ERA-Interim
10	UR025.4	NEMO3.2=LIM2, ¼ x ¼ 75z-levels	OI	ERA-Interim

Table 19.1. A selection of ocean reanalysis products using different ocean models and configurations, all global except for TOPAZ4 (Arctic), different assimilation schemes and atmospheric forcings. This particular set of products has been used for the ORA-IP Polar oceans intercomparison, and more details of the reanalyses can be seen in Table 1 of Uotila et al. (2018).

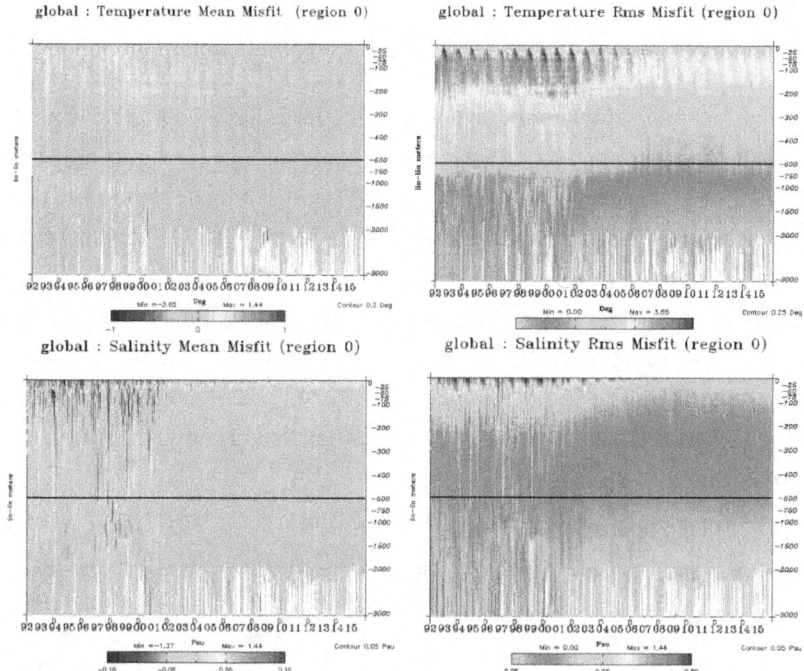

Figure 19.1. Temperature (top) and salinity (bottom) misfits (Obs-model) to global ocean data through time from the Copernicus services global model (GLORYS), reproduced from the online quality assurance document (http://marine.copernicus.eu/documents/QUID/CMEMS-GLO-QUID-001-025.pdf). Left figures show the global mean biases as function of depth and right figures show the RMS errors. Red color, +ve, in the bias shows a cold or fresh bias. The reduction in errors is very clear after the introduction of Argo data, 2005-07.

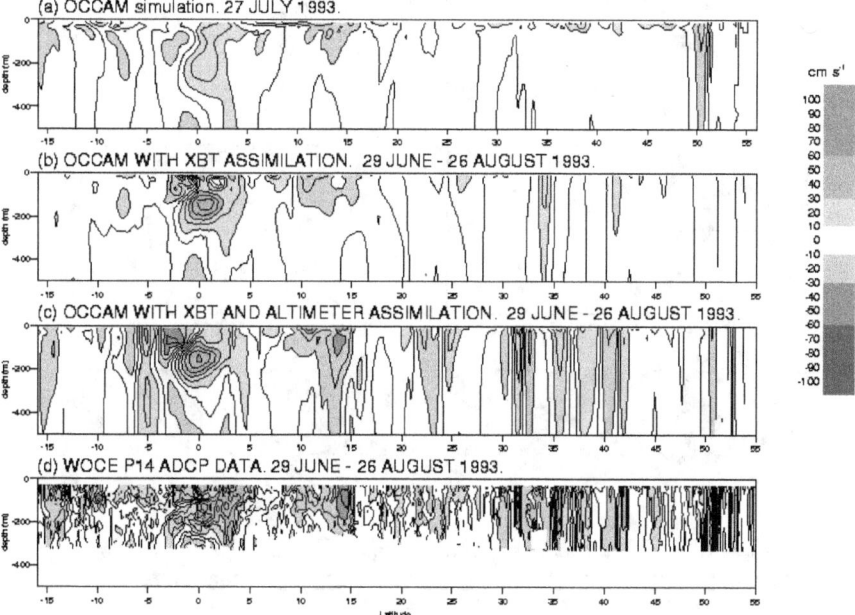

Figure 19.2. Zonal currents in simulation and assimilation experiments with a ¼ global ocean model (OCCAM, Webb et al., 1998) shown as a N-S section through the Pacific in 1993 following the World Ocean Circulation Experiment P14 section occupation. The bottom plot shows independent current data from the on-board acoustic Doppler current profiler. From Fox et al. (2000).

Ocean Reanalysis Intercomparison

The Ocean Reanalysis Intercomparison Project (ORA-IP) was initiated in 2011 with the backing of both the GODAE-Oceanview and CLIVAR-GSOP communities, recognising that high-quality ocean reanalysis products are of value to both forecasting and climate variability communities. A special issue of Climate dynamics, and an introduction to the ORA-IP (Balmaseda et al., 2017) covers aspects of many of the ocean reanalysis products as they were five or so years ago. The project saw the population of an ocean reanalysis data server at the Integrated Climate Data Centre in Hamburg (http://icdc.cen.uni-hamburg.de/1/daten/reanalysis-ocean/oraip.html), where data can be found for further studies. The ORA-IP project recently completed a polar oceans intercomparison paper (Uotila et al., 2018) and a North Atlantic intercomparison is now underway (Laura Jackson, personal communication).

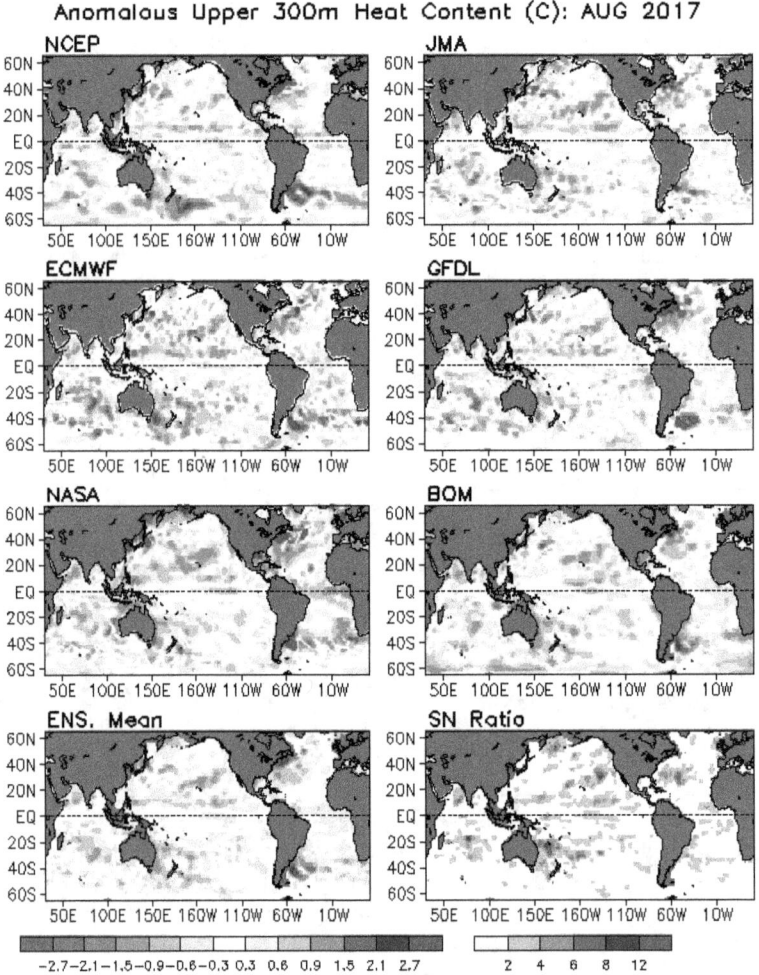

Figure 19.3. Upper ocean (300 m) heat content anomalies for August 2017 from several operational ocean assimilation systems, presented as deviations from each product's 1980-2010 climatology. These plots are available in near real-time from the Real-time Ocean Reanalysis Intercomparison (NOAA Climate Prediction Center).

To further promote and disseminate data, a website has been set up between the operational seasonal forecasting groups from the NOAA Climate Prediction Center (CPC; http://www.cpc.ncep.noaa.gov/products/GODAS/multiora_body.html) to distribute real-time ocean reanalysis intercomparison data. These data are needed for initialising seasonal forecasts, and making relevant comparisons between the groups will allow better interpretation of forecast differences while presenting a continuously evolving record of the state of the oceans. Fig. 19.3 shows an example of the upper 300 m ocean heat content anomalies from six different products in August 2017, along with an ensemble mean and a signal/noise ratio map, allowing regions of agreement between the products to be easily identified.

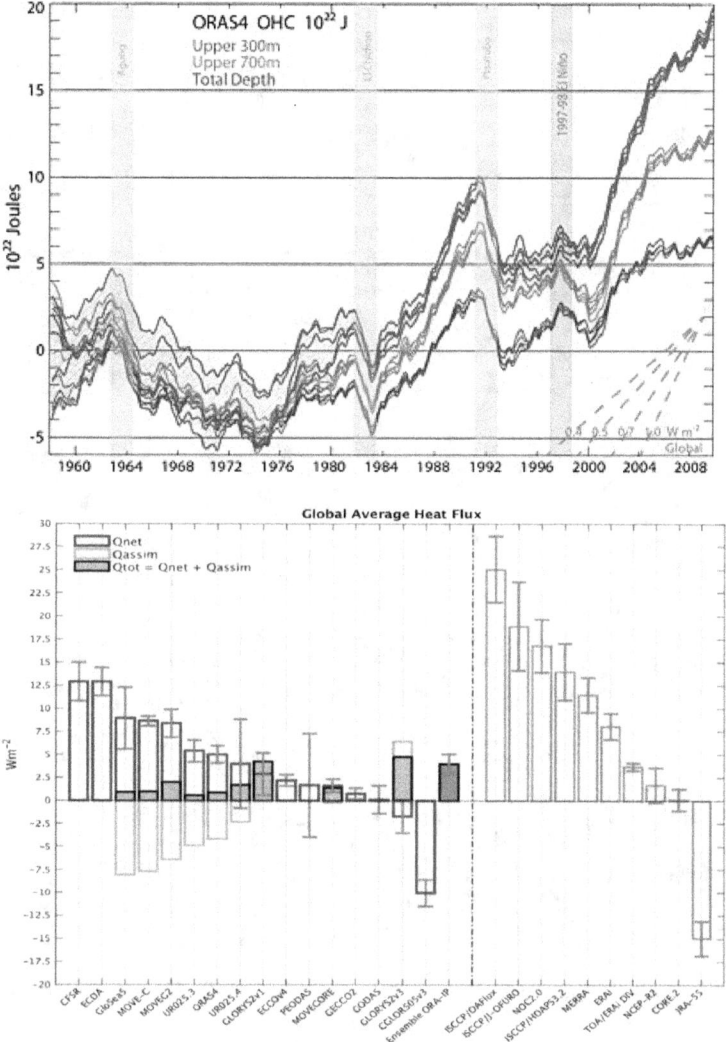

Figure 19.4. Top: Global ocean heat content in the ORAS4 reanalysis for different depth layers through time. The reanalysis is run as an ensemble of five members, which are shown by the different lines. From Balmaseda et al. (2013). Bottom: Global ocean surface heat fluxes from various ocean and atmospheric reanalysis and independent observation based products. From Valdivieso et al. (2017).

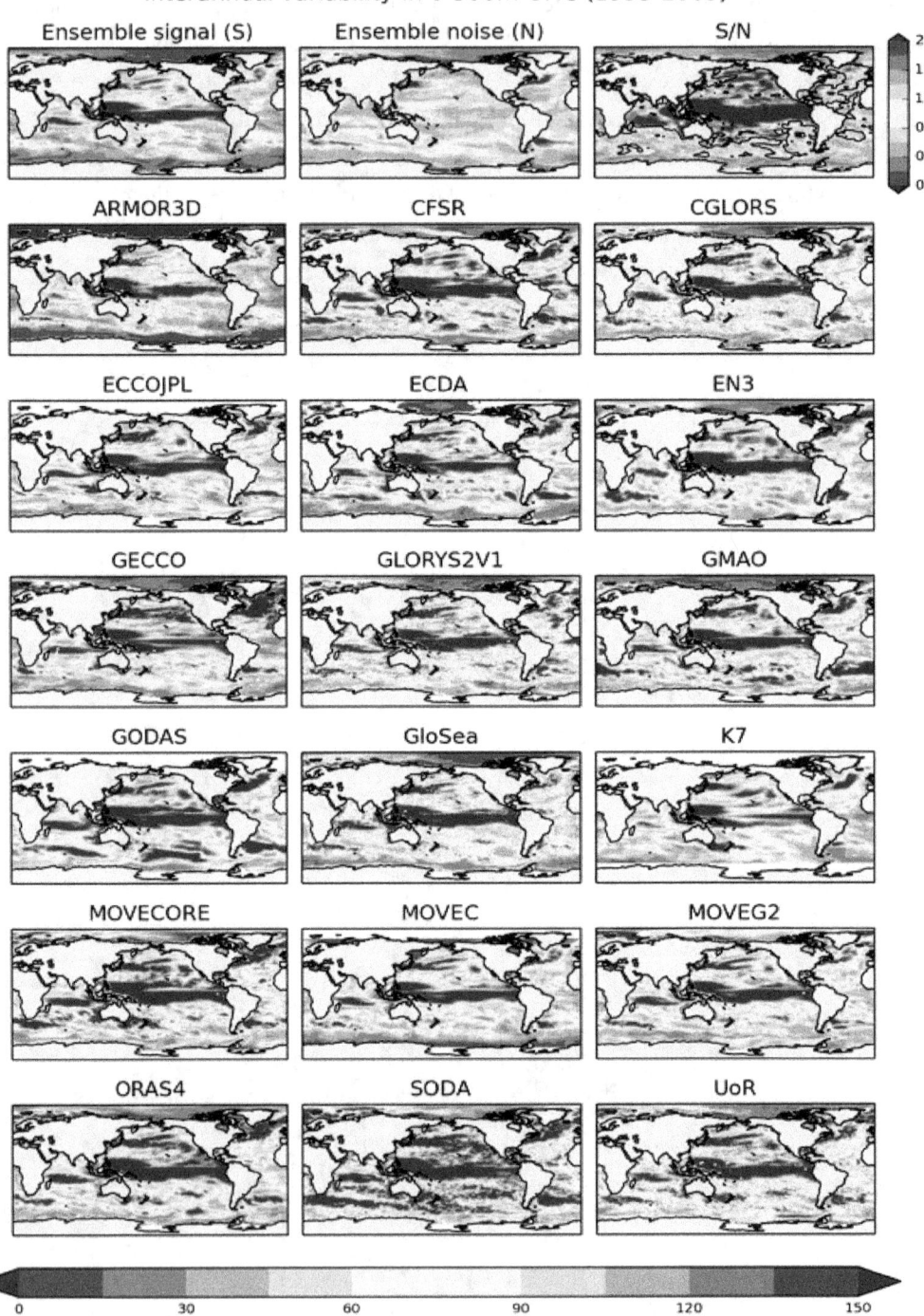

Figure 19.5. Maps of interannual variability in top 300 m ocean heat content for 1993-2010.

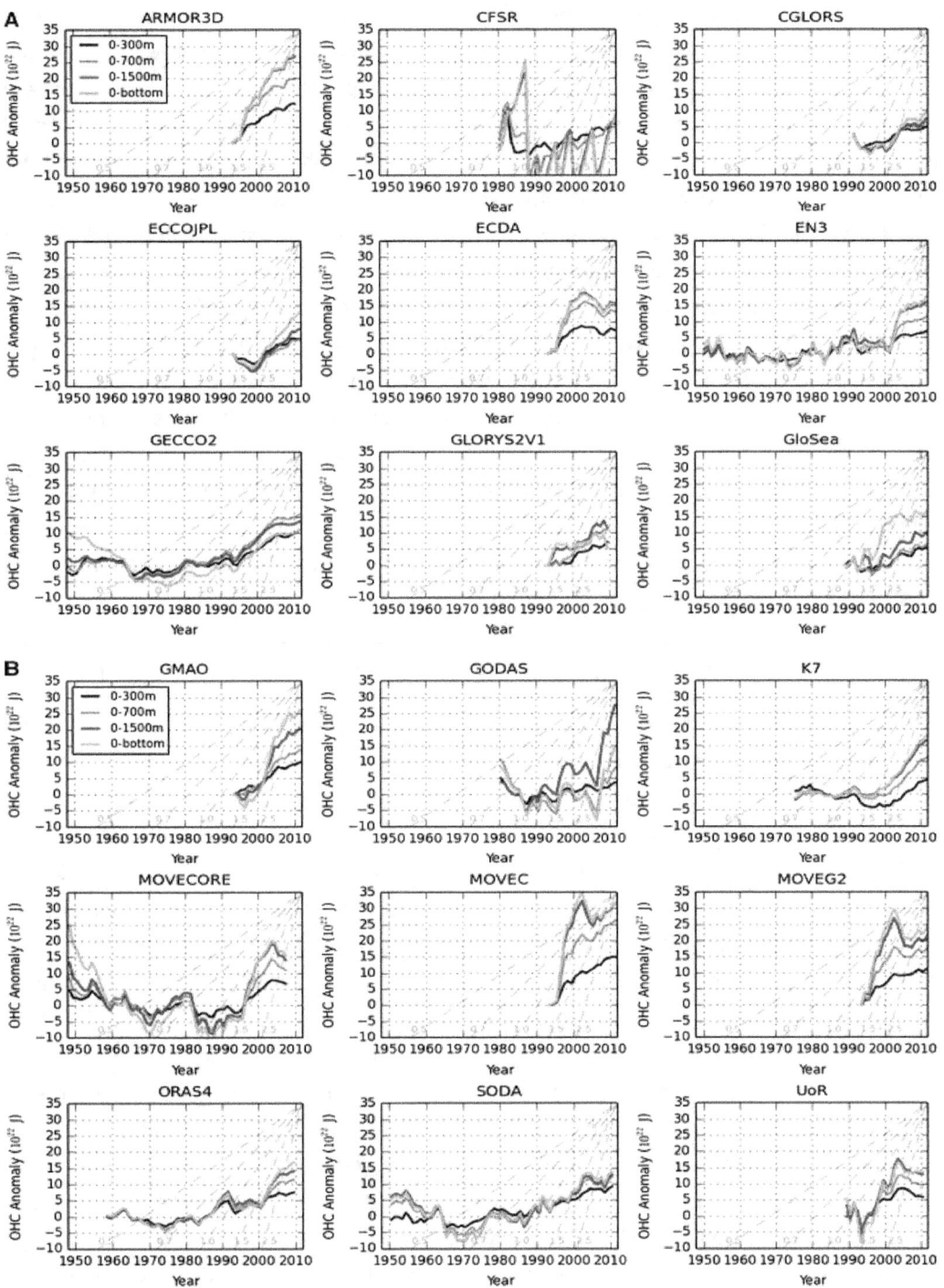

Figure 19.5 (continued). Time evolution of ocean heat content (OHC) in different layers from the different products. Most products only have data back to 1993, but some are longer. From Palmer et al. (2017).

The ocean heat content's depth distribution and trends are particularly important from a climate perspective as they can shed light on global warming of the oceans and the temporal and spatial variability. Balmaseda et al. (2013) used the European Centre for Medium-Range Weather Forecasts ORAS4 ocean reanalysis to look at the ocean's response to volcanic eruptions and the warming trends at different depth (Fig. 19.4a). While the cooling from volcanic eruptions is all found in the upper ocean, the strong warming since the mid-1990s shows a robust contribution from below 700 m depth. Palmer et al. (2017) used the ORA-IP project to look at ocean heat content in a wider range of reanalysis products (Fig 19.5). Most but not all show a similar signal of warming below 700 m depth, although the majority of these products were not long enough to study the signals from volcanic eruptions since the last big eruption event, which was Pinatubo in 1991. Valdivieso et al. (2017) looked at surface heat fluxes within the ORA-IP project (Fig. 19.4b). Observations, e.g., Roemmich et al. (2016), suggest the Argo-monitored oceans (above 2000 m) are warming at ~1 W/m^2. Although most ocean reanalysis products show slightly larger warming rates, these rates are comparable (if not lower and hence more consistent) than the fluxes from many atmospheric reanalysis products. Most ocean reanalyses show compensation between surface fluxes adding heat at the ocean surface, and assimilation increments removing heat. This may be due to excess mixing of heat downwards in the models, although GLORYS2v3 is an exception, as also suggested by the weak mixing signature in Fig. 19.1.

Intercomparison of Atlantic Meridional Overturning Circulation Transports

Karspeck et al. (2017) was the only ORA-IP paper to look at transports, focussing on the Atlantic meridional overturning circulation at 26°N and using only six long reanalysis products stretching back to the 1950s but stopping in 2007 before the RAPID monitoring time series became available. These results were not very encouraging, with more disagreement in interannual and interdecadal variability between products being shown than for model simulations without assimilation. However, Jackson et al. (2016) showed that the current UK Met Office reanalysis system appears to capture well the recent geostrophic variations in the Atlantic meridional overturning circulation at 26°N. Encouraged by this, further studies have been conducted of the North Atlantic interannual heat budgets in the subtropical and subpolar gyres, see Fig. 19.6.

Different contributors to ocean heat content change in the Met Office Glosea5 reanalysis are followed from 1997-2013 in the Atlantic subtropical (25°-45°N) and subpolar (45°-65°N) gyres. In many cases the ocean assimilation increments make a large contribution to the changes.

However, the subtropical cooling in 2009-2011 is strongly driven by both surface cooling and reduced heat transport across 25°N (see also Cunningham et al., 2013); and the slower subpolar cooling after 2005 (see Robson et al., 2016) is contributed to by reduced heat transport across 45°N into the subpolar gyre. Clearly further work is needed, however these kinds of studies are proving very useful in highlighting problems with the current generation of reanalyses. So how reliable are these poleward heat transports in current reanalysis products in general?

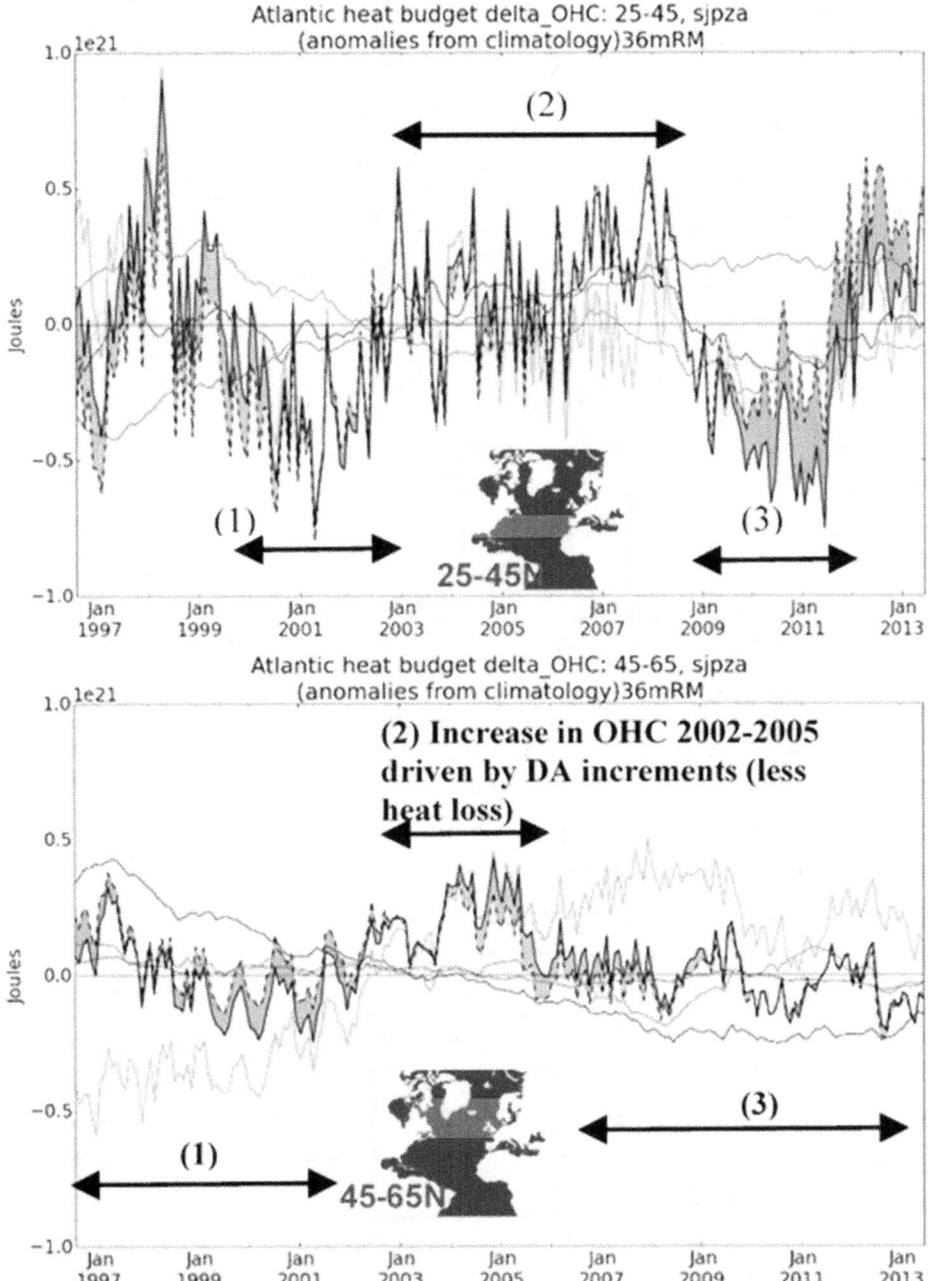

Figure 19.6. North Atlantic regional heat budgets from the Met Office Glosea5 reanalysis. Monthly integrated heat fluxes and the month-to-month changes in ocean heat content (OHC) are shown, so that positive/negative values correspond to positive/negative trends in OHC (Lines are black dashed = monthly change in OHC; green = surface fluxes; blue = ocean heat transport across southern boundary; red = ocean heat transport across northern boundary; cyan = heat input by data assimilation increments; black solid = sum of heat budget components) (Lesley Allison personal communication). Red shading indicates where heat budget contributions (black solid) appear greater than the OHC changes (black dashed), and blue areas show the reverse. Generally, residuals are small. The numbers are periods with gyre warming (2) or cooling (1,3).

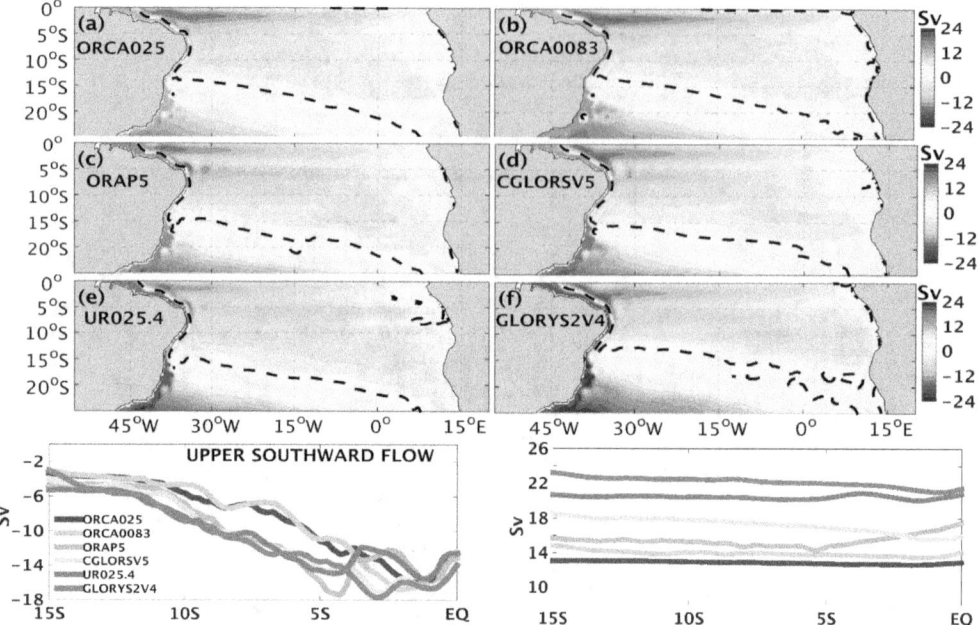

Figure 19.7. (a)–(f) Northward volume transport in the top 1000m in the South Atlantic, integrated westwards from the eastern boundary and averages from 1997-2010, dashed lines zero contour. The top two products ORCA025 and ORCA0083 are model simulations at ¼° and 1/12°, respectively. The other four products are ocean reanalyses all at ¼°. (g) Shows the peak southward flow (-ve) before the E-W integrated flow turns northward near the western boundary. This represents the gyre component of the top 1000 m ocean circulation. (h) Shows the total northward flow continuing the integration to the western boundary, and is thus a measure of the overturning component of the circulation. From Mignac et al. (2017).

Mignac et al. (2017) showed results for South Atlantic transports in four reanalyses from Table 19.1, along with two free-running models in the more recent 1997-2010 period. They demonstrated that the mean ocean interior circulation in the reanalyses are in fact quite consistent with each other and in better agreement with independent estimates of known ocean currents (for example, Fig. 19.7 shows south tropical current systems reproduced better than in the free-running models). However, they also showed the reanalysis products disagree near the western boundary in both the upper and deep circulations, as illustrated by the schematic in Fig. 19.8. This message is more encouraging than that of Karspeck et al. (2017) as it suggests the current ocean observing system may allow reproduction of many aspects of circulation, but that more work is required to understand how to better represent and constrain the critical western ocean boundary currents where a lot of the climatically important heat and freshwater transport occurs.

Despite these problems with ocean reanalysis transports, they sometimes still provide useful climatic transport information. When we look at ocean reanalysis heat transport convergences over different ocean basins and compare these with atmospheric reanalysis heat transport convergences for the same regions we often find more agreement between the ocean reanalysis products. This is particularly true for smaller ocean basins and regions where inflow-outflow is particularly restricted in the oceans, such as the Mediterranean and the Arctic.

Analysis of atmospheric heat transports show that by far the most energy transport in the atmosphere is in the form of latent heat in water vapour. It is, therefore, important to follow freshwater transports in the oceans that compensate these within the global water cycle. For the first time, Argo data should start to make this possible. Valdivieso et al. (2014) sought to compare the freshwater transports in one ocean reanalysis with what has been published from a hydrographic section data analysis (their Table 4, Table 19.2 below). The considerable disagreement between many of the observational estimates clearly shows that better measurement and monitoring methods are needed, but for most sections the reanalysis freshwater transports are in reasonable consistency. In the next (6th) Intergovernmental Panel on Climate Change assessment, there will be a stronger focus on the global heat and water cycles because climate models are known to also strongly disagree and there is an important role for using ocean (and atmospheric) reanalysis transports to develop more reliable observation-based estimates.

Freshwater Convergence	Hydrographic Section-Based		UR025.4	
	Wijffels [2001]	*Talley* [2008]	Mean	Total (Including Eddies)
Arctic/Atlantic				
45°–47°N to Bering	−0.25	−0.32	−0.39	−0.45 ± 0.04
35°N to Bering	−0.37	−0.33	−0.44	−0.49 ± 0.06
24°–26°N to Bering	+0.12	−0.29	−0.16	−0.13 ± 0.03
16°–19°S to Bering	−0.10	−0.18	−0.07	−0.04 ± 0.03
North of 30°–32°S	+0.24	+0.28	+0.23	+0.33 ± 0.04
Indian				
North of 20°S	−0.03	+0.07	+0.10	+0.20 ± 0.06
North of 32°S	+0.31	+0.38	+0.49	+0.65 ± 0.07
Pacific				
47°N to Bering	−0.27	−0.11	−0.16	−0.18 ± 0.02
35°N to Bering	−0.56	−0.15	−0.32	−0.44 ± 0.03
24°N to Bering	−0.21	−0.19	−0.20	−0.28 ± 0.04
17°S to Bering	−0.30		−0.34	−0.36 ± 0.06
32°S to Bering	+0.06	−0.04	+0.03	+0.09 ± 0.04

[a]The total transports (last column) include eddies and the ±values represent annual standard deviations over the 14 year period. Positive is implied net evaporation; negative is implied net precipitation/runoff.

Table 19.2. Comparison of freshwater convergences (Sv) from the UR025.4 reanalysis over the period 1997–2010, with direct estimates from ocean hydrographic sections compiled by Wijffels (2001, Table 6.2.2), and Talley's (2008) hydrographic section-based analysis[a] (Valdivieso et al., 2014).

Interpreting assimilation increments?

A number of groups have investigated using assimilation increments from ocean reanalyses to interpret physical process errors in models. The suggestion of mixing errors from the persistent dipole biases in Fig. 19.1 was noted earlier. Fig. 19.9 shows a demonstration experiment assimilating the same ocean data into the same ocean model but with the incoming shortwave flux coming from two alternative sources, ERAInterim and the Drakkar forcing (DFS4) based on ERA40. The difference in the assimilation ocean heat content increments closely compensates for the shortwave flux differences, although evidence of the location of some of the assimilated data can still be clearly seen. If other sources of differences (errors) could be discounted it would suggest that surface flux errors may be diagnosable from the assimilation increments. However, this attribution relies mainly on having confidence in the veracity of advective components of the ocean heat content, and we have seen that results presented above suggest this is not yet generally possible with ocean reanalysis products. Nevertheless, as models improve it remains a possibility for the future.

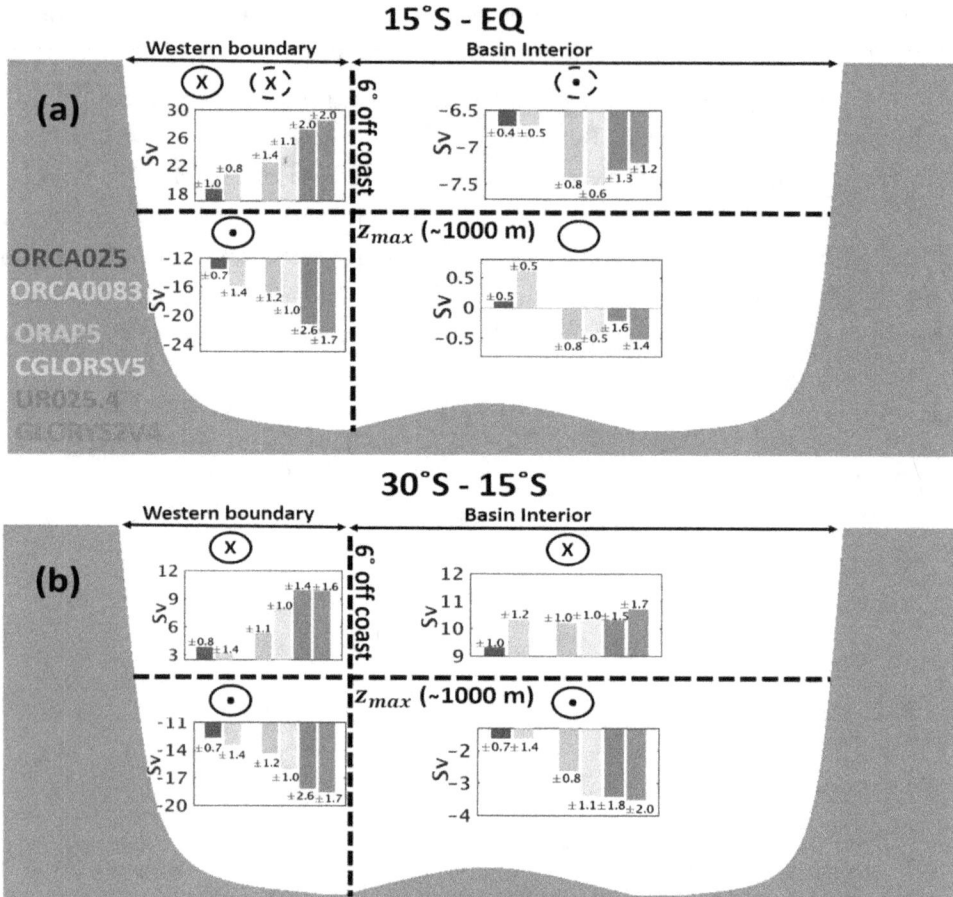

Figure 19.8. Schematic of the South Atlantic 1997-2010 mean circulation averaged over different latitude ranges, and shown for the two model simulation products and the four ocean reanalysis products. The ocean interior and the western boundary flows are separated. From Mignac et al. (2017).

Ocean reanalyses for initialising decadal predictions and the need for reanalyses with sparse data from the pre-Argo period

One major driver for ocean reanalysis has been to initialise decadal (or at least interannual) climate predictions, which will contribute a great deal to CMIP6. On these timescales, long periods of reanalysis are needed in order to develop sufficient hindcasts of past periods so as to gain statistical confidence. The UK Met Office has been using the DePreSys system to initialise such hindcasts since the pioneering work of Smith et al. (2007). Initialising the upper ocean temperatures in the 1950s or 1960s is very challenging given the sparsity of any subsurface data. That this can be done relies on two factors: (1) the ability to model the upper ocean based on forcing with atmospheric reanalysis data from the period, which is generally much more extensive; and (2) the use of modelling-derived large-scale error covariance patterns based on ocean property variability from within the more extensively observed recent periods.

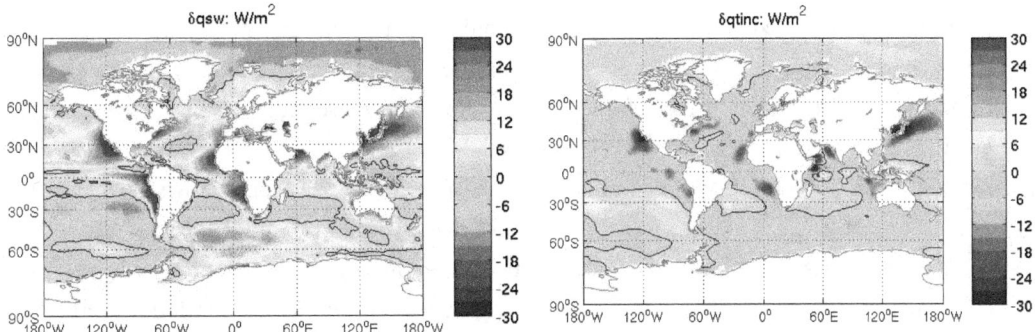

Figure 19.9. Left: Difference in downward shortwave radiation over the oceans at the surface between ERA-Interim and the Drakkar forcing set 4 (DFS4, Brodeau et al., 2010). Right: Difference in ocean heat content increments (expressed as surface flux) from two data assimilation experiments differing only in the shortwave radiation forcing, shown left. The model is NEMO ORCA1 with data assimilation described in Smith et al. (2009) (Valdivieso, personal communication).

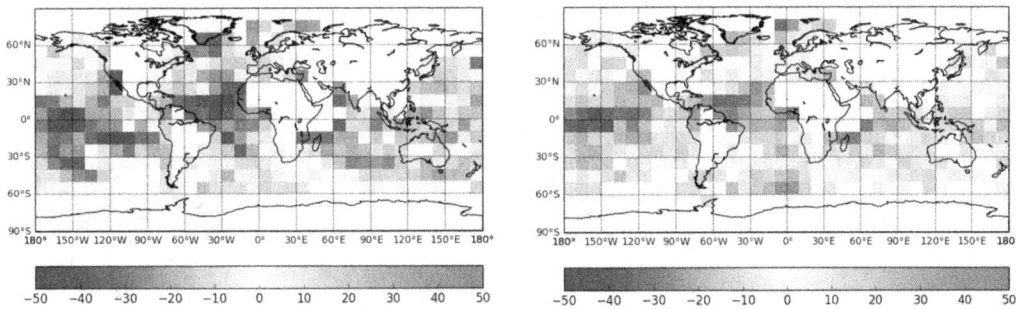

Figure 19.10. Percentage change in RMS SST error when compared to unassimilated observation data (those data removed by subsampling) when assimilating 2010 SST and subsurface profile data subsampled to 1953 densities, into climatological data. Left: standard assimilation method. Right: EOF assimilation method. (Dan Lea, personal communication).

The DePreSys system develops initial ocean conditions from sparse observations using model-based covariances developed through iterating the reanalysis for the last 50-year period, and recycling the large-scale state covariances for the whole period for the next 50-year iteration. This has the effect of developing large-scale covariance patterns from the recent well-observed period and using them at earlier times. This can also be done more directly and explicitly as in the ongoing work at the Met Office (Dan Lea, personal communication). Seasonal Empirical Orthogonal Functions (EOFs) are developed for SST and subsurface temperature and salinity using 20 years of reanalysis data (1989-2010). The current observations are then subsampled to 1953 densities and a climatology used as background data onto which the subsampled data are assimilated using the EOFs, the current operational covariance suite, or a hybrid of the two approaches. The hybrid has been found to perform best with the large-scale EOFs, allowing better reconstruction based on sparse data away from the observation locations, see Fig. 19.10.

Coupled and 20th Century Reanalysis

All seasonal and decadal reanalyses have always used coupled models, and medium- to long-range weather forecasting from 10 days to monthly timescales are also starting to use coupled models. A new generation of coupled atmosphere-ocean reanalyses are currently under development at operational centres. Long period "20th Century" reanalyses make heavy use of surface observations so it is important that the near surface boundary layers are well reproduced. Fig. 19.11 is based on the new the European Centre for Medium-Range Weather Forecasts 20th Century Coupled Reanalysis (CERA-20C) using coupled data assimilation, which essentially shows how the near surface heat budget is balanced in a long ocean-only reanalysis and a coupled reanalysis, which uses the same ocean assimilation data. As the data volume increases in the ocean-only system a strong cancellation between surface heat flux cooling the ocean surface, and assimilation increments warming the ocean develops. This is avoided in the coupled system because the ocean and atmospheric marine boundary layers are in better balance. The CERA-20C reanalysis, despite using variational approach (with a 4DVar atmosphere and 3DVar ocean), is also an ensemble product with 10 members, which gives some measure of variable uncertainty in the product through the century. This varies with data volume and quality, as discussed in Feng et al. (2017).

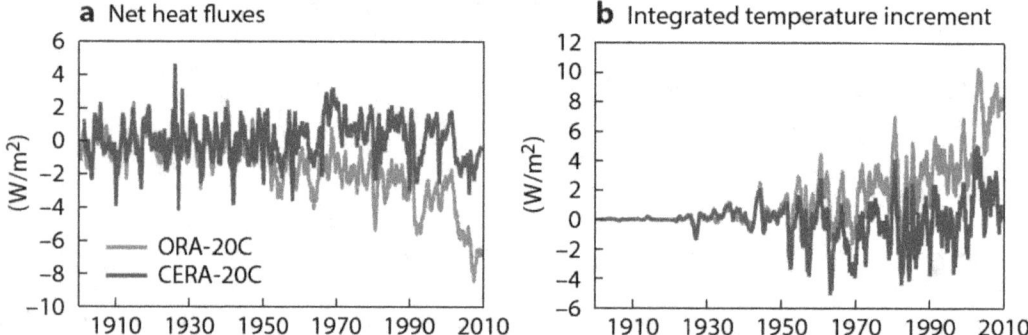

Figure 19.11. Net global surface heat fluxes and integrated temperature increments from 2 the European Centre for Medium-Range Weather Forecasts 20th Century Reanalyses. ORA-20C is an ocean only reanalysis, while CERA-20C is a coupled atmosphere-ocean reanalysis. From Laloyaux et al. (2017).

Discussion and Conclusions

This chapter has presented a range of ocean reanalysis results looking at multiple products and how well they agree, for example from the ORA-IP, and on applications of the longer data records that they provide. We specifically looked at circulation and transports, as these cannot be observed directly and yet they are critical to both climate change and prediction studies. The current generation of ocean reanalyses are improving, especially for currents and transports in the ocean interior, due to good Argo and altimeter data, but the strong boundary currents are still hard to confidently correct through assimilating ocean current data. Studies of the AMOC from observations such as the RAPID program at 26°N are providing useful test cases for assessing ocean reanalysis performance.

We have not covered higher resolution reanalyses, e.g., at 1/12° resolution or higher, with which assimilation is currently performed for marine forecasting applications, both for regional ocean modelling, and for at least two global products: the US Navy HYCOM project and the CMEMS NEMO-based global product. Details of the operational advantages of these models are discussed in other chapters of this book.

On longer timescales ocean and coupled reanalyses are starting to be performed for the whole 20th century, and these early reanalysis products are also needed to initialise decadal prediction systems, for example. Improved quality control of some of the early ocean observations, as well as better coupling between the atmosphere and the upper ocean in coupled reanalysis systems, and new ways of using the older sparse observational datasets are all important areas of ongoing development that will have a big impact on ocean reanalyses over the coming years.

Acknowledgements

I would particularly like to thank contributions from Met Office colleagues, Dan Lea and Lesley Allison, for allowing me to use some of their current research material.

References

Balmaseda, M. A., K. E. Trenberth, and E. Källén (2013), Distinctive climate signals in reanalysis of global ocean heat content. *Geophys. Res. Lett.,* 40, 1754–1759, doi:10.1002/grl.50382.

Balmaseda, M.A. et al. (2017) The Ocean Reanalysis Intercomparison project (ORA-IP). *J. Op. Oceanogr.,* 8, S1, s80–s97, http://dx.doi.org/ 10.1080/1755876X.2015.1022329

Brodeau et al. (2010) An ERA40-based atmospheric forcing for global ocean circulation models. *Ocean Modelling,* 31, 88-104, ISSN 1463-5003.

Cunningham, S. A., C. D. Roberts, E. Frajka-Williams, W. E. Johns, W. Hobbs, M. D. Palmer, D. Rayner, D. A. Smeed, and G. McCarthy (2013), Atlantic Meridional Overturning Circulation slowdown cooled the subtropical ocean. *Geophys. Res. Lett.,* 40, 6202–6207, doi:10.1002/2013GL058464.

Feng, X., K. Haines, and E. de Boissason (2017) Coupling of surface air and sea surface temperatures in the CERA-20C reanalysis, *Quart. J. Roy. Met. Soc.,* doi:10.1002/qj.3194/full

Fox, A.D., K. Haines, B. De Cuevas and D.J. Webb (2000) Altimeter assimilation in the OCCAM global model, Part II: TOPEX/POSEIDON and ERS1 data, *J. Marine Sys.,* 26, 323-347.

Jackson, L.C., K. Andrew Peterson, C.D. Roberts & R.A. Wood (2016) Recent slowing of Atlantic overturning circulation as a recovery from earlier strengthening, Nature Geoscience 9, 518–522 (2016) doi:10.1038/ngeo2715

Karspeck et al. (2017) Comparison of the Atlantic meridional overturning circulation between 1960 and 2007 in six ocean reanalysis products. *Climate Dynamics,* doi:10.1007/s00382-015-2787-7

Laloyaux et al. (2017) CERA-20C: An earth system approach to climate reanalysis, *ECMWF Newsletter,* 150, 25-30. doi:10.21957/ffs36birj2

Mignac D., D Ferreira, and K. Haines (2017) South Atlantic meridional transports from NEMO-based model simulations and reanalyses. *Ocean Science,* 14(1), 53-68. ISSN 1812-0784, doi: https://doi.org/10.5194/os-14-53-2018

Palmer M.D et al. Ocean heat content variability and change in an ensemble of ocean reanalyses. *Climate. Dynamics,* doi:10.1007/s00382-015-2801-0

Rayner, N. A.; Parker, D. E.; Horton, E. B.; Folland, C. K.; Alexander, L. V.; Rowell, D. P.; Kent, E. C.; Kaplan, A. (2003) Global analyses of sea surface temperature, sea ice, and night marine air temperature since the late nineteenth century *J. Geophys. Res,* 108, No. D14, 4407 doi:10.1029/2002JD002670

Robson et al. (2016) A reversal of climatic trends in the North Atlantic since 2005. *Nature Geoscience,* 9; 513-51,7 doi:10.1038/ngeo2727

Roemmich et al. (2015) Unabated planetary warming and its ocean structure since 2006. *Nature Climate Change, 5; 240-245,* doi:10.1038/nclimate2513

Smith et al. (2007) Improved Surface Temperature Prediction for the Coming Decade from a Global Climate Model. *Science,* 10 Aug 2007, Vol. 317, Issue 5839, pp. 796-799, doi: 10.1126/science.1139540

Uotila et al. (2018) An assessment of ten ocean reanalyses in the Polar regions. Climate Dynamics, doi:10.1007/s00382-018-4242-z, https://rdcu.be/PkKi.

Valdivieso, M., K. Haines, H. Zuo and D. Lea (2014) Freshwater and heat transports from global ocean syntheses, *J. Geophys. Res. Oceans,* 119, doi:10.1002/2013JC009357.

Valdivieso, M., K. Haines, et al. (2017)Surface heat fluxes in ocean and coupled reanalyses. Climate Dynamics, doi:10.1007/s00382-015-2843-3.

Webb et al. (1998) The first main run of the OCCAM global ocean model. Southampton Oceanography Centre Internal Document No. 34. ftp://ftp.soc.soton.ac.uk/pub/occam/papers/occam_tech1.pdf

CHAPTER 20

The Mercator Ocean Global High-Resolution Monitoring and Forecasting System

Jean-Michel Lellouche[1], Eric Greiner[2], Olivier Le Galloudec[1], Charly Régnier[1], Mounir Benkiran[1], Charles-Emmanuel Testut[1], Romain Bourdallé-Badie[1], Marie Drévillon[1], Gilles Garric[1], and Yann Drillet[1]

[1]*Mercator Océan, Ramonville Saint Agne, France;* [2]*CLS, Ramonville Saint Agne, France*

Mercator Ocean monitoring and forecasting systems are routinely operated in real time since early 2001. They have been regularly upgraded through several systems of increasing complexity, expanding the geographical coverage from regional to global, improving models and assimilation schemes. In this chapter we give a description of the current Mercator Ocean real-time, global high-resolution system. The ocean model, the observations, and the data assimilation scheme are detailed with a particular focus to the specifics of the Mercator Ocean system. Technical details about the real-time operation of the system are given. The system is then examined through a scientific evaluation, highlighting the level of performance and the reliability of the system. User needs and evolutions of the system are finally drawn.

Introduction

Over the past 20 years, the use of data assimilation methods for operational ocean forecasting systems has been extensively developed. A number of operational systems have emerged at the national and international scale within the international GODAE (Global Ocean Data Assimilation Experiment) project, and then within the GODAE OceanView (Schiller et al., chapter 2 of this book). Mercator Ocean is the French contribution to this initiative that aims to describe and forecast changing ocean conditions by running predictive models that work on a principle used successfully by meteorological models to forecast atmospheric conditions.

Mercator Ocean monitoring and forecasting systems have been routinely operated in real-time in Toulouse since early 2001 and regularly upgraded through four prototypes of increasing complexity (PSY1, PSY2, PSY3, and PSY4). These upgrades expanded the geographical coverage from regional to global and improved models and assimilation schemes (Brasseur et al., 2006; Lellouche et al., 2013). After having successfully coordinated the European MyOcean and MyOcean2 projects (http://www.myocean.eu), Mercator Ocean was officially entrusted by the European Commission on November 11, 2014 to implement and operate the Copernicus Marine Environment Monitoring Service (CMEMS), as part of the European Earth observation program

Lellouche, J.-M.., et al., 2018: The Mercator Ocean global high-resolution monitoring and forecasting system. In *"New Frontiers in Operational Oceanography"*, E. Chassignet, A. Pascual, J. Tintoré, and J. Verron, Eds., GODAE OceanView, 563-592, doi:10.17125/gov2018.ch20.

Copernicus (http://marine.copernicus.eu). Mercator Ocean opened the CMEMS in May 2015 and is in charge of the global high resolution ocean analyses and forecasts. Since then, research and development activities have been conducted to improve the real-time 1/12° high-resolution (eddy-resolving) global analysis and forecasting system. The main ingredients of an analysis and forecasting system are 1) an ocean numerical model, 2) available observations in the past, and 3) data assimilation techniques based on mathematical methods. The goal of a data assimilation method is to force the ocean model to be as close as possible to the observations available in the past in order to obtain the best forecasts in the future, taking into account observations and model errors, as illustrated in Fig. 20.1. Since October 19, 2016, Mercator Ocean has delivered real-time daily services (weekly analyses and daily 10-day forecasts) with an updated global 1/12° system. In the latter, the ocean/sea-ice model and the assimilation scheme benefit from the following main updates: atmospheric forcing fields are corrected at large-scale with satellite data; freshwater runoff from ice sheets melting is added to river runoffs; a time varying global average steric effect is added to the model sea level; the last version of Gravity field and steady-state Ocean Circulation Explorer (GOCE) geoid observations are taken into account in the mean dynamic topography used for sea level anomalies assimilation; adaptive tuning is used on some of the observational errors; a dynamic height criteria is added to the quality control of the assimilated temperature and salinity vertical profiles; satellite sea-ice concentrations are assimilated; and climatological temperature and salinity in the deep ocean (below 2000 m) are assimilated to prevent drifts in those sparsely observed depths.

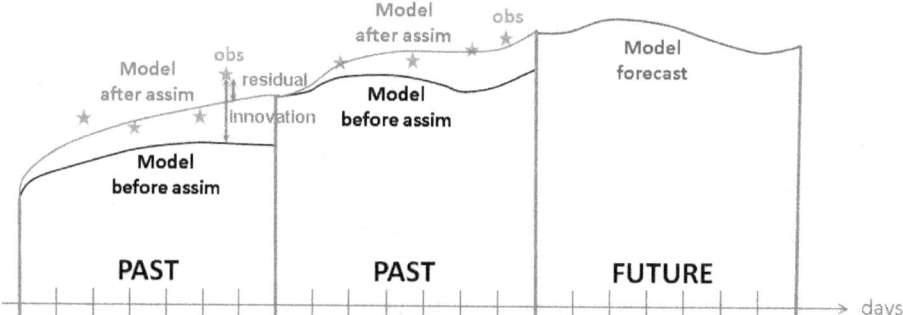

Figure 20.1. General principle of an analysis and forecasting system. The model is run a first time and the gap between the observation (green stars) and the model (black line) is called "innovation". An analysis (data assimilation scheme) is done and the model is run a second time, taking into account the correction given by the analysis. The gap between the observations and the model (red line) is then called "residual". Several data assimilation cycles are done in the past to perform the model forecast (purple line) in the future.

This chapter will concentrate on the way the three components of the Mercator Ocean operational system (observation, model, and data assimilation) are integrated to improve the behavior of the system, while maintaining a reasonable elapsed time between the data load and the service delivered to the users. The chapter is organized as follows. The ingredients of the system, with a particular focus on the specifics of the Mercator Ocean, are described first. Technical details about the real-time operation of the system are then given. The next section gives the validation methodology adopted by Mercator Ocean. The system is then examined through a scientific evaluation, highlighting its level of performance and reliability. User needs and evolutions of the system are discussed in the last section.

Description of the Monitoring and Forecasting System

Physical model

The high-resolution global analysis and forecasting Mercator Ocean system uses version 3.1 of the NEMO ocean model (Madec et al., 2008). The physical configuration is based on the tripolar ORCA12 grid type (Madec and Imbard, 1996) with a horizontal resolution of 9 km at the equator, 7 km at Cape Hatteras (mid-latitudes) and 2 km toward the Ross and Weddell seas. The 50-level vertical discretization retained for this system has a decreasing resolution from 1m at the surface to 450 m at the bottom, and 22 levels within the upper 100 m. A "partial cells" parameterization (Adcroft et al., 1997) is chosen for a better representation of the topographic floor (Barnier et al., 2006) and the momentum advection term is computed with the energy- and enstrophy-conserving scheme proposed by Arakawa and Lamb (1981). The advection of the tracers (temperature and salinity) is computed with a total variance diminishing advection scheme (Lévy et al., 2001; Cravatte et al., 2007). We use a free surface formulation. External gravity waves are filtered out using the Roullet and Madec (2000) approach. A laplacian lateral isopycnal diffusion on tracers and a horizontal biharmonic viscosity for momentum are used. In addition, the vertical mixing is parameterized according to a turbulent closure model (order 1.5 and mixing length of 30 m) adapted by Blanke and Delecluse (1993). The lateral friction condition is a partial-slip condition with a regionalisation of a no-slip condition (over the Mediterranean Sea) and the elastic-viscous-plastic rheology formulation for the LIM2 ice model (Fichefet and Maqueda, 1997) has been activated (Hunke and Dukowicz, 1997). Instead of being constant, the depth of light extinction is separated in red-green-blue bands depending on the chlorophyll data distribution from mean monthly SeaWiFS (Sea-viewing Wide Field-of-view Sensor) climatology. The bathymetry used in the system is a combination of interpolated ETOPO1 (Amante and Eakins, 2009) and GEBCO8 (Becker et al., 2009) databases. ETOPO1 datasets are used in regions deeper than 300 m and GEBCO8 is used in regions shallower than 200 m, with a linear interpolation in the 200 m – 300 m layer. Moreover, the bathymetry benefits from a specific correction in the Indonesian Sea inherited from the INDESO system (Tranchant et al., 2016).

Internal tide-driven mixing is parameterized following Koch-Larrouy et al. (2008) for tidal mixing in the Indonesian Seas. The atmospheric fields forcing the ocean model are taken from the European Centre for Medium-Range Weather Forecasts' Integrated Forecast System. A three-hour sampling is used to reproduce the diurnal cycle. Momentum and heat turbulent surface fluxes are computed from the Large and Yeager (2009) bulk formulae using the following set of atmospheric variables: surface air temperature and surface humidity at a height of 2 m, mean sea level pressure and wind at a height of 10 m. Downward longwave and shortwave radiative fluxes and rainfall (solid + liquid) fluxes are also used in the surface heat and freshwater budgets. Due to large known biases in precipitations, a satellite-based, large-scale correction of precipitations has been performed, except at high latitudes (poleward of 65°N and 60°S). Moreover, in order to avoid any mean SSH drift due to the large uncertainties in the water budget closure, the following two treatments were applied. First, a trend of 2.2 mm/year has been added to the runoffs in order to somewhat represent the recent estimate of the global mass addition to the ocean (from glaciers, land water storage changes, Greenland and Antarctica ice sheets mass loss) (Chambers et al., 2017). Second, the surface freshwater budget has been set to zero at each time step with a superimposed seasonal cycle (Chen et al., 2005). The monthly runoff climatology is built with data on coastal runoffs and 100 major rivers from the Dai et al. (2009) database. This database uses data from recent years, streamflow simulated by the Community Land Model version 3 (CLM3) to fill the gaps, in all lands areas except Antarctica and Greenland. In addition, we built the runoff fluxes coming from Greenland and Antarctica ice sheets and glaciers melting using the Altiberg icebergs database project (Tournadre et al., 2013). This complements the estimate of Silva et al. (2006) for Antarctica. Lastly, as the Boussinesq approximation is applied to the model equations, conserving the ocean volume and varying its mass, the simulations do not properly directly represent the steric effect on the sea level (Greatbatch, 1994). For improved consistency with satellite observations of sea level anomalies, which are unfiltered from the steric component, a time-evolving global average steric effect is added to the sea level in the simulation. This global average steric effect has been computed as the time derivative between two successive daily global mean dynamic heights (vertical integration, from the surface to the bottom, of the specific volume anomaly).

Observations and data assimilation scheme

Data assimilation scheme

The data assimilation method SAM (Système d'Assimilation Mercator) relies on a reduced-order Kalman filter based on the Singular Evolutive Extended Kalman Filter (SEEK) formulation introduced by Pham et al. (1998). In the Mercator Ocean system, the forecast error covariance is based on the statistics of a collection of three-dimensional ocean state anomalies, typically a few hundred. This approach is based on the concept of statistical ensembles in which an ensemble of anomalies is representative of the error covariances. In our case, the anomalies are computed from a long numerical experiment with respect to a running mean in order to estimate the seven-day scale error on the ocean state at a given period of the year for temperature (T), salinity (S), zonal velocity (U), meridional velocity (V), sea surface height (SSH) and sea ice concentration (SIC). More precisely, each temporal anomaly corresponds to the difference between the model state M and a running mean $\langle M \rangle_{-\tau}^{+\tau}$ over a fixed time period window ranging from $-\tau$ to τ (Fig. 20.2a).

Moreover, the signal at a few horizontal grid "Δx" intervals in the model outputs on the native full grid is not physical, only numerical (Grasso, 2000). This signal should not be taken into account when updating an analysis. This is why several passes of a Shapiro filter are applied in order to remove the very short scales that, in practice, correspond to numerical noise. Consequently, a little subsampling of the model state is applied without aliasing error and the anomalies are thus calculated on a reduced horizontal grid (one out of every two points in both horizontal directions and all the points along the coast) to limit the storage and the load cost during the analysis stage.

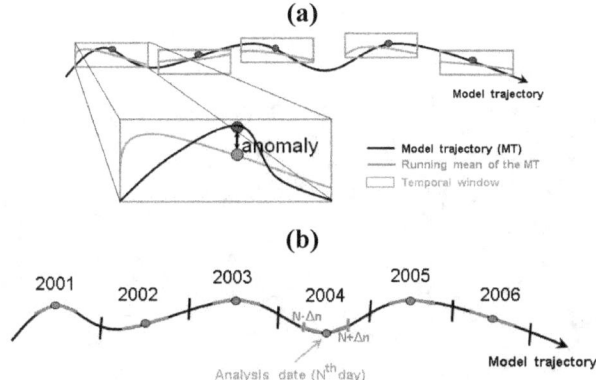

Figure 20.2. (a) Schematic representation of the anomalies calculation along a model trajectory and (b) of the use of these anomalies to build the model forecast covariance.

To create the running mean $\langle M \rangle_{-\tau}^{+\tau}$, a Hanning low-pass filter is used:

$$Ha(v) = \begin{cases} 0.5 + 0.5 \cos\left(\dfrac{\pi.v}{v_{max}}\right) & for \quad |v| \leq v_{max} \\ 0 & for \quad |v| > v_{max} \end{cases} \qquad (1)$$

where v is the temporal frequency of the model state and v_{max} is the cut-off frequency (equal to 1/36 days^{-1} in our case). The main characteristic of the anomaly calculation is to filter out temporal scales at low frequencies in order to keep high frequencies for which the period is shorter than two or three assimilation cycles. For an assimilation cycle centered on the Nth day of a given year, ocean state anomalies falling in the window [N-Δn; N+Δn] of each year of the simulation are gathered and define the covariance of the model forecast error (see Fig. 20.2b). In our case, Δn is equal to 45 days, which means that anomalies are selected over 90-day windows centered on the Nth day of each year of the simulation. So, in SAM, the forecast error covariances rely on a fixed-basis, seasonally-variable ensemble of anomalies. This method implies that at each analysis step a subset of anomalies is used that improves the dynamic dependency. A significant number of anomalies are kept from one analysis to the other, thus ensuring error covariance continuity. It should also be noted that the analysis increment is a linear combination of these anomalies and depends on the innovation (observation minus model forecast equivalent) and on the specified observation errors. A particular feature of the SEEK formulation is that the error covariance only gives the direction of the model error and not its intensity. An adaptive scheme for the model error variance calculates an optimal variance of the model error based on a statistical test formulated by Talagrand (1998). The last feature of the model forecast covariance employed is a localization technique that sets the covariances to zero beyond a distance defined as twice the local spatial correlation scale. Because a finite number of ocean state anomalies have been used to build the model forecast covariance, the latter is not significant when further away than this particular distance from the analysis point. This is why it is preferable to not use this information and to set the covariance to zero. Spatial (zonal and meridional directions) and temporal correlation scales (Fig. 20.3) are then used to define an "influence bubble" around the analysis point in which data are also selected.

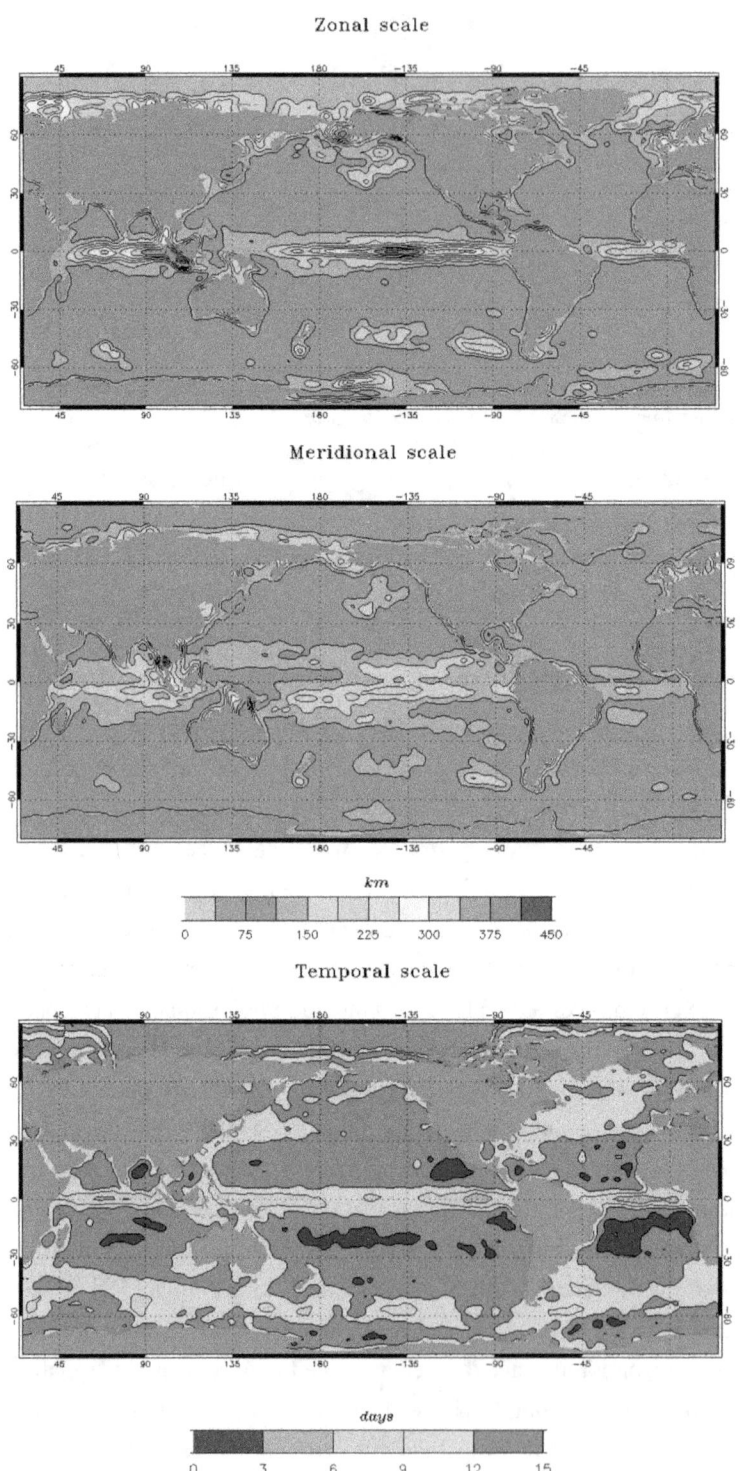

Figure 20.3. Zonal, meridional (km), and temporal (days) correlation scales (from top to bottom) used by the Mercator Ocean system.

To save computing time, the analysis is performed on a reduced grid (one out of every four points in both horizontal directions, all the points along the coast and one out of every two points in the first 150 km from the coast). An important distinction of the Mercator Ocean system with respect to more classical forecasting systems is that the analysis is not performed at the end of the assimilation window, but instead at the middle of the seven-day assimilation cycle. The objective is to take into account both past and future information and to provide the best estimate of the ocean centered in time. With such an approach, the analysis, to some extent, acts like a Smoother algorithm.

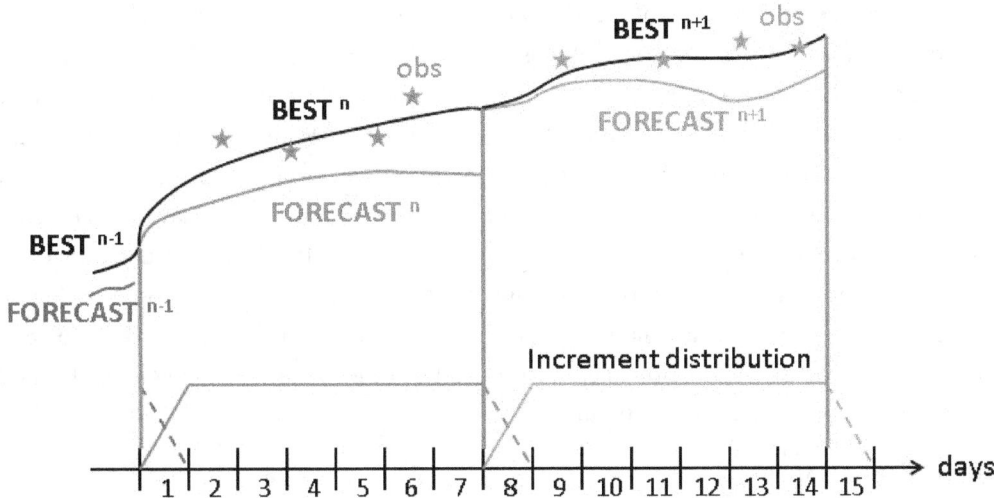

Figure 20.4. Schematic representation of the incremental analysis update procedure for three consecutive cycles n-1, n and n+1. Following the analysis performed at the end of the forecast (or background) model trajectory (referred to as "FORECAST" first trajectory, with analysis time at the fourth day of the cycle), the incremental analysis update scheme rewinds the model and starts again from the beginning of the assimilation cycle, integrating the seven-day run (referred to as "BEST" second trajectory) with a tendency term added in the model prognostics equations and modulated by an increment distribution function. The time integral of this function equals one over the cycle length.

After each analysis, the data assimilation produces increments of SSH, temperature, salinity, zonal velocity, meridional velocity, and sea ice concentration. All these increments are applied progressively using the incremental analysis update method (Bloom et al., 1996; Benkiran and Greiner, 2008), which makes it possible to avoid model shock every week due to the imbalance between the analysis increments and the model physics. In this way, the incremental analysis update reduces spin-up effects. Following the analysis performed at the end of the forecast (or background) model trajectory (referred to as "FORECAST" first trajectory, with analysis time at the fourth day of the cycle), a classical forward scheme would continue straight on from this analysis, integrating from day seven until day 14. Instead, the incremental analysis update scheme rewinds the model and starts again from the beginning of the assimilation cycle, integrating the model for seven days (referred to as "BEST" second trajectory) with a tendency term added in the model prognostics equations for temperature, salinity, sea surface height, sea ice concentration, and horizontal velocities. The tendency term (which is equal to the increment divided by the length of the cycle) is modulated by an increment distribution function shown in Fig. 20.4. The time integral of this function equals one over the cycle length. In practice, the incremental analysis update scheme is costlier than the "classical" model correction (increment applied on one time step) because of the additional model integration ("BEST" trajectory) over the assimilation window.

In addition to this assimilation scheme, a method of bias correction has been developed. This method is based on a 3D-VAR approach, which takes into account cumulative three-dimensional temperature and salinity innovations over the last month in order to estimate large-scale temperature and salinity biases when enough observations are available. The aim of the bias correction is to correct the large scale, slowly-evolving error of the model whereas the SAM assimilation scheme is used to correct the smaller scales of the model forecast error. The bias correction involves several steps. First, temperature and salinity innovations over the last three months are binned and averaged on a coarse resolution (1° x 1°) grid. The two variables are treated separately because temperature and salinity biases are not necessarily correlated. Then, the 3D-VAR method is used to analyze the bias. The bias covariance is constrained by the structures of density fronts in the ocean (the bias can be large on the sides of a front). There is little bias correction in the mixed layer if the vertical gradient of the thermocline is sharp. The bias correction is fully effective under the thermocline, away from density gradients. The correlations are modelled by means of an anisotropic Gaussian recursive filter. Bias correction of temperature, salinity, and dynamic height are then computed and interpolated on the model grid. Lastly, these bias corrections are applied as tendencies in the model prognostic equations, with a one-month timescale.

Assimilated observations

Altimeter data, in situ temperature and salinity vertical profiles, satellite sea surface temperature (SST), and sea ice concentration are assimilated to estimate the initial conditions for numerical ocean forecasting. These observations come from CMEMS Thematic Data Assembly Centers, which provide real-time and reprocessed (historic years) satellite and in situ products to be assimilated by the monitoring and forecasting centers.

Altimeter data consist of along-track sea level anomalies (SLA). Along each track of SLA, only one point in two is conserved to avoid redundant information. Moreover, observations along-tracks are smoothed by several altimetric corrections (Le Traon et al., 2001). A mean dynamic topography (MDT) is also used as a reference for SLA assimilation. This MDT is based on the "CNES-CLS13" MDT (Rio et al., 2014) with adjustments made using high-resolution analyses, the last version of the GOCE geoid, and an improved post-glacial rebound (also called a glacial isostatic adjustment). The accuracy of the MDT is very important because it constrains the mean circulation of the model indirectly. For instance, the mean positions of the Gulf Stream, the Kuroshio, or the North Atlantic Current are dictated by the fronts in the MDT. The mesoscale activity given by the SLA is superimposed, but the mean advection is greatly dependent on the MDT.

Temperature and salinity in situ vertical profiles from the CORA 4.1 database (Cabanes et al., 2013) has been assimilated for the hindcast calibration run of the system. This database includes temperature and salinity vertical profiles from the sea mammal database (Roquet et al., 2011) to compensate for the lack of such data at high latitudes. Near real-time satellite operational sea surface temperature and sea ice analysis (OSTIA) SST and the EUMETSAT Ocean and Sea Ice Satellite

Application Facility (OSI-SAF) sea ice concentration are assimilated. For the sea-ice concentration, a separate monovariate/monodata analysis is carried out for the ice variables, in parallel to that for the ocean. The two analyses are completely independent.

Due to unresolved processes and inaccurate forcing, the model may drift at depth. Unfortunately, there are very few temperature and salinity profiles below 2,000 m to constrain the model drift. Hence, the climatology is currently the only source of information at depth to prevent the model from drifting. Virtual vertical profiles of temperature and salinity below 2000m are built from monthly World Ocean Atlas 2013 climatology. These virtual observations are geographically positioned on the model horizontal grid with a coarse resolution (1° x 1°) and on the model vertical levels from 2,200 m to the bottom. A non-Gaussian error is used to impose a weak constraint on the model at depth. This allows the system to capture a potential climate drift at depth.

The concept of "pseudo-observations" or "observed-no change" (innovation equal to zero) is also used to overcome the deficiencies of the background errors, in particular for extrapolated and/or poorly observed variables. We apply this approach to the barotropic height and the three-dimensional coastal salinity at river mouths and all along the coasts (run off rivers). Pseudo-observations are also used for the three-dimensional variables temperature, salinity, zonal velocity, meridional velocity under the ice and between 6°S and 6°N below a depth of 200 m. These observations are geographically positioned on the analysis grid points rather than on a coarser grid in order to avoid generating aliasing on the horizontal. The time of these observations is the same as for the analysis, namely the fourth day of a seven-day assimilation cycle. Given ongoing concern about the need to reduce costs in an operational context, the three-dimensional variables mentioned above were sampled on the vertical in order to keep only about ten model levels.

Observation operators

In data assimilation schemes, the computation of innovation requires defining an observation operator to represent the model equivalent of the observation. In the Mercator Ocean system, prognostic variables are interpolated on a quadrilateral grid, i.e., on the four canvas grid points surrounding the observation. The four weights are calculated with a bilinear remapping interpolation.

For in situ vertical profiles, a mapping from the profiles onto the model levels is used. Whenever possible, the nearest data is associated to those levels. In order to prevent large errors near the pycnocline, this procedure is only applied if the data depth is close to the model depth (not further than half the model level thickness). When more than three observations are contained in the in situ profile, a spline algorithm interpolates the profile on the model levels which are in between the observations, only if the distance between the two data depths is less than the model level thickness. No extrapolation is performed at the top or at the bottom of the profile.

For the SST and the sea ice concentration, only the daily data corresponding to the fifth day of the seven-day window is assimilated. For sea ice concentration, the model equivalent corresponds to a simple daily average of the model variable. As OSTIA provides the foundation SST (considered nominally at 10 m depth), the SST model equivalent is performed by calculating the nighttime average of the first level of the model temperature. For SLA, the problem is much more complex.

In order to obtain a consistent model equivalent for SLA, different space/time filters are applied. These time filters act both on the sea level height and the barotropic height computed by the ocean model in order to remove high frequency barotropic signals in a closer manner than what is done in the altimetric data processing (Carrère at al., 2003; Dibarboure et al., 2011). First, the SSH is averaged over a one-day window, and two running means of barotropic height spatially averaged (over a box of 5°) are removed from it.

The model equivalent for SLA is:

$$SLA_{equivalent} = \overline{SSH}_{1day} - \text{LargeScale}\,(\overline{HBAR}_{1day} - \overline{HBAR}_{21days}) - MDT \qquad (2)$$

The running mean over a one-day window eliminates barotropic high frequencies (periods less than one day) and over a 21-day window eliminates barotropic signals with periods less than 21 days, which correspond to the low-pass filter applied to along-track altimetric data.

Lastly, the First Guess at Appropriate Time (FGAT) method (Huang et al., 2002) is used, which means that the forecast model equivalent of the observation for the innovation computation is taken at the time for which the data are available, even if the analysis is delayed.

Quality control on in situ observations and feedback to input data providers

To minimize the risk of erroneous observations being assimilated in the model, the system carries out two successive quality controls (QC1 and QC2) on the assimilated temperature and salinity vertical profiles. These are done in addition to the quality control procedures performed by the data producers.

Quality Control (QC1)

QC1 allows the detection of spikes and large biases. It can be summarized as follows. An observation is considered suspicious if the following two conditions are satisfied:

$$\begin{cases} |innovation| > threshold \\ |observation - climatology| > 0.5 * |innovation| \end{cases} \qquad (3)$$

where the spatially and seasonally varying $threshold$ value comes from statistics (mean, standard deviation) computed with the very large number of temperature and salinity innovations collected in GLORYS2V1 Mercator Ocean reanalysis (1993-2009). The first condition is a test on the innovation. It determines whether the innovation is abnormally large, which would most likely be due to an erroneous observation. The second condition avoids rejecting "good" observations (i.e., an observation close to the climatology) even if the innovation is high due to the model background being biased. Fig. 20.5 shows an example of a wrong temperature profile detected by the quality control in the GLORYS2V1 simulation. Below 400 m in depth, innovations are no longer valid. The two conditions described previously are satisfied and the profile is rejected (Fig. 20.5a). When this profile is assimilated, an abnormal salinity value appears at its temporal and geographical positions (Fig. 20.5b). This is due to the fact that the assimilation algorithm used is multivariate, meaning that an observation of temperature leads to corrections of all of the model variables and especially, in this case, the surface salinity.

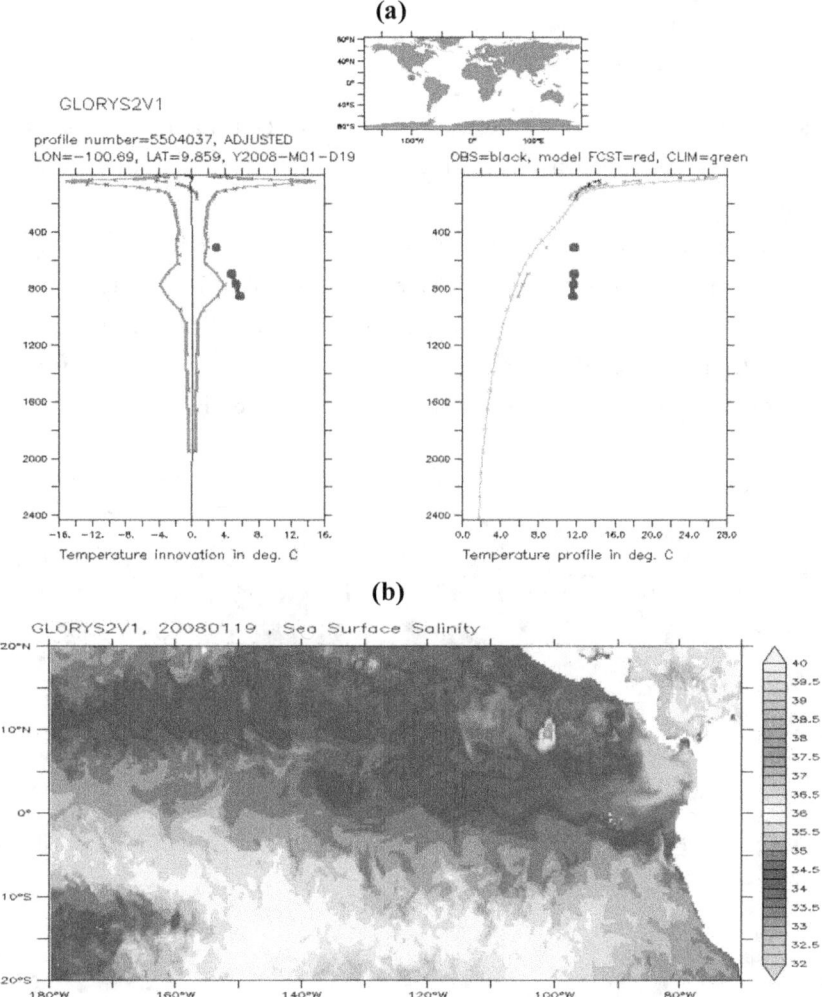

Figure 20.5. Example of a suspicious temperature vertical profile at 100.69° W - 9.86° N, highlighted by the quality control on the CORA3.1 dataset. **(a)** Left panel shows temperature innovation profile in blue and temperature innovation threshold in red. Right panel shows the absolute vertical temperature profile (observation in black, climatology in green and model in red). Large blue dots correspond to "bad" innovations or "bad" observations. **(b)** When this profile is assimilated, an abnormal value of salinity appears at the position of this profile.

Quality Control (QC2)

QC2 is based on dynamic height innovation statistics and allows detection of small biases, which are present in the whole water column and thus can induce large errors. It basically says that the thermosteric or halosteric information cannot exceed some threshold in height. It can be summarized as follows.

An observation is considered to be suspect if:

$$\begin{cases} For\ temperature: \ \frac{|C*hdyn(innov_T)|}{\sum dz_T} > threshold_T \\ For\ salinity: \ \frac{|C*hdyn(innov_S)|}{\sum dz_S} > threshold_S \end{cases} \quad (4)$$

574 • JEAN-MICHEL LELLOUCHE ET AL.

where
$$\begin{cases} C = 200/\sum dz & \text{if } 0 < \sum dz \leq 200 \\ C = 500/\sum dz & \text{if } 200 < \sum dz \leq 500 \\ C = \sum dz & \text{if } \sum dz > 500 \end{cases} \quad (5)$$

and dz_T is the model layer thickness corresponding to the temperature observation (same for dz_S and salinity). These last conditions (Eq. 5) prevent the $threshold$ from being reached too quickly in shallow areas.

The parameters used by QC2 (average and standard deviation of the dynamical height innovations, and therefore threshold value) have been calculated from a global simulation at 1/4°, which is a twin simulation of the high-resolution one. Note that the simulation at 1/4° also assimilates the CORA 4.1 CMEMS in situ database. The threshold two-dimensional field is then computed as the average plus six times the standard deviation of the dynamical height innovations (Fig. 20.6).

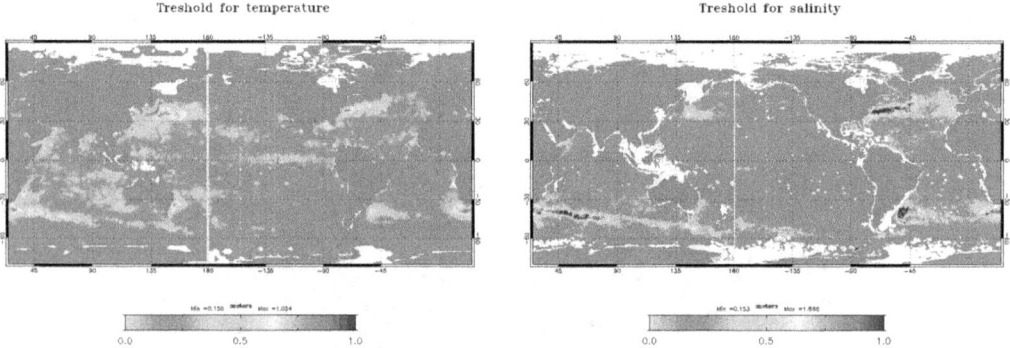

Figure 20.6. Threshold used for QC2 for temperature (left panel: $threshold_T$) and salinity (right panel: $threshold_S$).

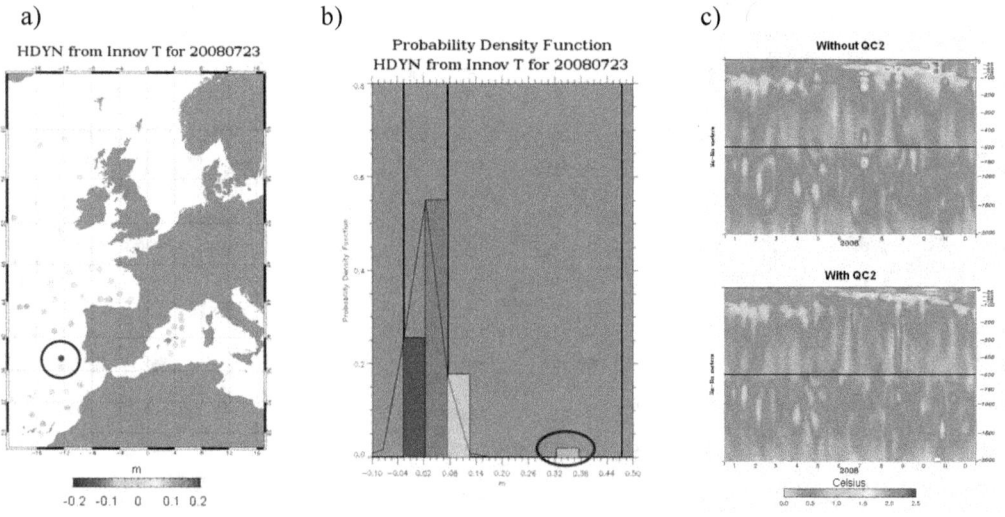

Figure 20.7. Statistics in the Azores region: a) absolute value of dynamical height innovations from temperature innovations for the 7-day assimilation cycle from 16 July 2008 to 23 July 2008, b) Probability density function (PDF) of theses dynamical height innovations (the value 0.3 appears in the tail of the PDF), c) Root mean square (RMS) innovation with respect to the vertical temperature profiles over the year 2008 for two "twins" simulations (without and with QC2). These last scores are averaged over all seven days of the data assimilation window, with a lead time equal to 3.5 days.

Fig. 20.7a shows an example of a "wrong" temperature profile detected by the QC2 at the end of July 2008. In this case, $threshold_T = 0.3$ (Fig. 20.7b). The first condition of Eq. 4 is satisfied and the profile is rejected. When this profile is assimilated (simulation without QC2), abnormal temperature root mean square (RMS) innovation values appear at the temporal position (July 2008) of this profile in the Azores region (Fig. 20.7c). Using QC2 quality control allows for solving the problem.

Adaptive tuning of observation error and implementation

In order to refine the prescription of observation errors, adaptive tuning of observation errors for the SLA and SST has been implemented. The method has not been used for temperature and salinity vertical profiles because of the lack of in situ data. Then, three-dimensional fixed observation errors are used for the assimilation of in situ temperature and salinity vertical profiles. These three-dimensional errors have been computed from the last MyOcean global 1/4° Mercator system using an offline version of the adaptive tuning method mentioned above. Fig. 20.8 shows these observation errors at surface and at 100 m depth.

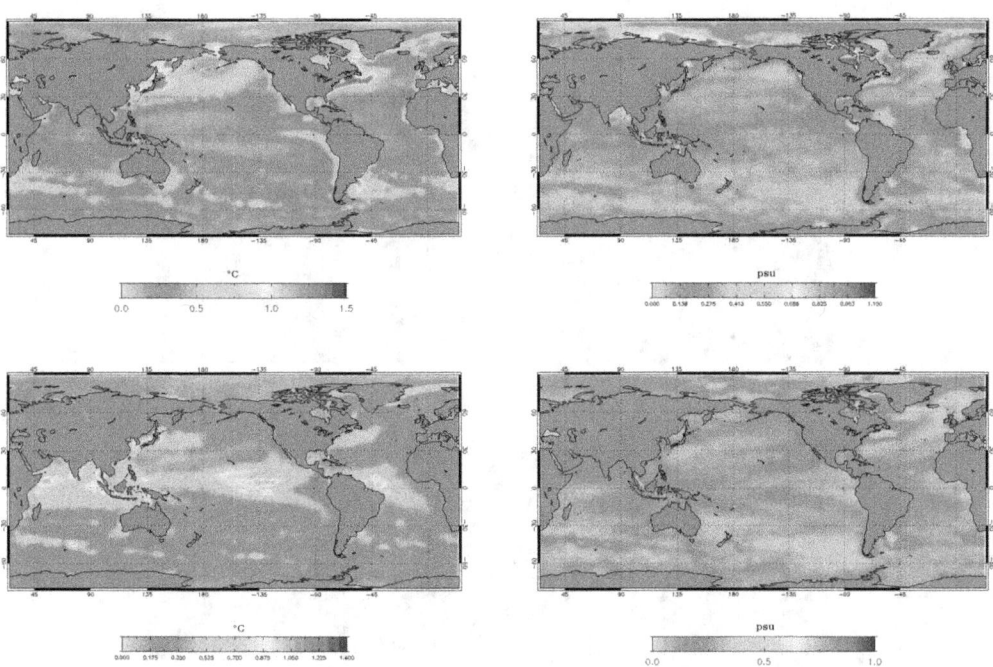

Figure 20.8. Observation errors used in SAM for in situ temperature (left panels) and salinity (right panels) profiles, at surface (top) and at 100 m depth (bottom).

The adaptive tuning method consists in the computation of a ratio, which is a function of observation errors, innovations, and residuals. It helps correct inconsistencies on the specified observation errors. Following Desroziers et al. (2005), this ratio can be expressed as:

$$ratio = \frac{residual\ (innovation)^T}{observation\ error} \qquad (6)$$

Ideally, *ratio* is equal to one. When the ratio is less (respectively larger) than one, it means that the observation error is overestimated (respectively underestimated). The objective of this diagnostic is to improve the error specification by tuning an adaptive weight coefficient acting on the error of each assimilated observation. As a first guess of the method, the initial prescribed observation error matches the one used in the previous system (Lellouche at al., 2013), where the observation error variance was increased near the coast and on the shelves for the assimilation of SLA, and increased only near the coast (within 50 km of the coast) for the assimilation of SST. Only an illustration of the method on the SLA is shown here. Similar results are obtained for SST. Fig. 20.9 represents the temporal evolution of the ratio for Envisat satellite. At the beginning of the simulation, the observation error is overestimated (ratio less than one). The ratio tends to one after only a few weeks of simulation.

For SLA (Fig. 20.10), the a priori prescribed observation error is globally significantly reduced. The median value of the error changed from 5 cm to 2.5 cm in a few assimilation cycles and allows for better results. This method allows us to have more realistic and evolutive observation error maps which can provide valuable information to space agencies.

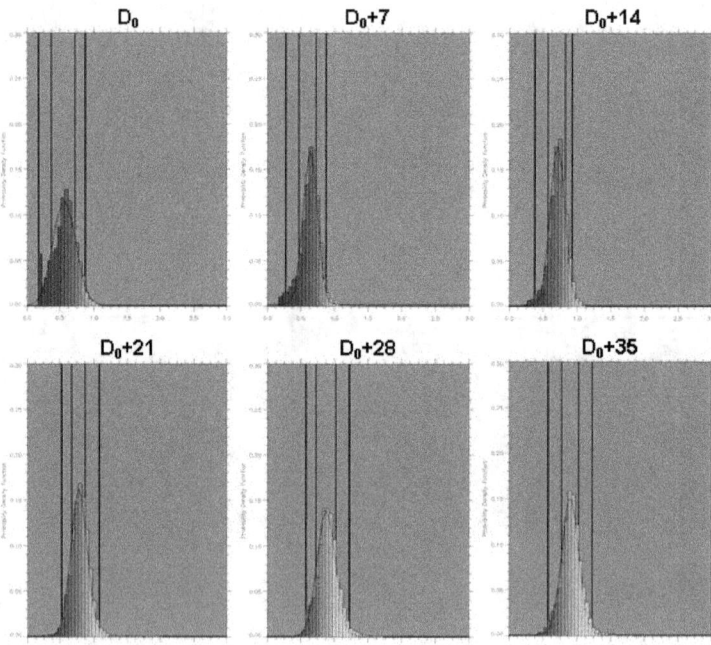

Figure 20.9. Evolution of the PDF of the ratio for Envisat satellite. D_0 corresponds to the first day where Envisat is assimilated by the system.

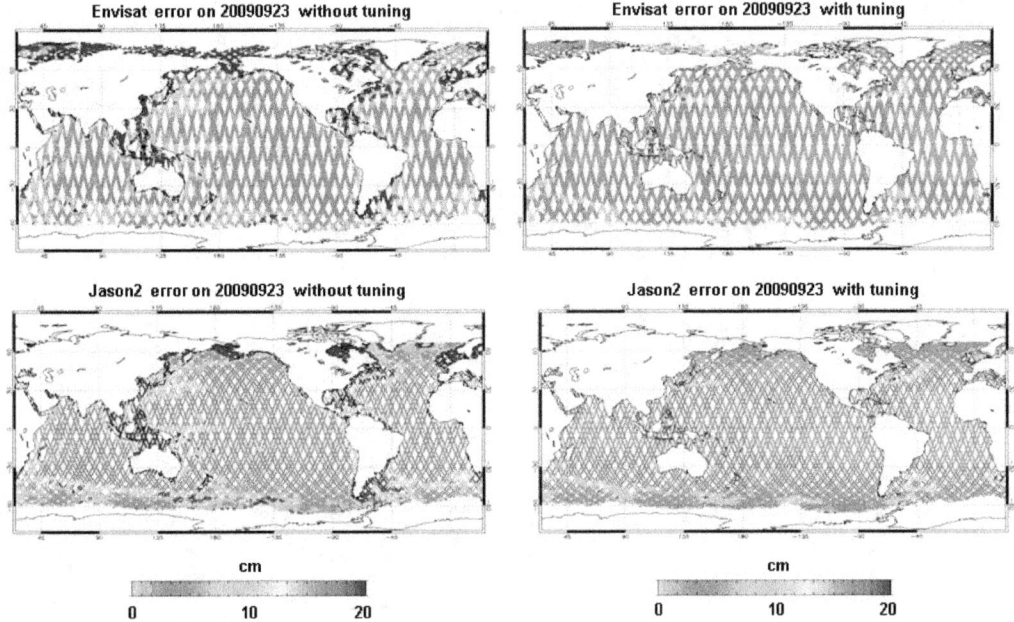

Figure 20.10. Envisat (top panels) and Jason2 (law panels) observation errors used on seven-day assimilation cycle ending September 23, 2009 without tuning (left panels) and with tuning (right panels) method.

Real-time Operation of the System, Computing Facility, Dissemination Capacity and Service Delivery

The system is run weekly on Wednesdays and provides two weekly analyses and two weeks of forecast (Fig. 20.11). The first analysis (hindcast) is the better one and is performed 14 days behind the real-time (from D-14 to D-8 included). The second analysis (nowcast) is performed seven days behind the real-time (from D-7 to D-1 included). Moreover, each day, the system is run without assimilation for days D-1 to D+9, providing ten days of forecast by updating the atmospheric forcing of the first day. The daily runs are initialized with the previous day's run, except on Thursdays, when they start from the weekly analysis run.

The system runs on 54 nodes (1296 processors) of Météo France BULL machine. The ocean model alone takes one hour (elapsed time) for a seven-day simulation and the analyses (SEEK and 3D-VAR) take 0.6 h. This means a total of about eight hours elapsed (two analyses, 42 model days, and computation of diagnostics) every week to deliver the 14-day forecast. This illustrates the fact that computing resources remains a key limiting factor to the development of global high resolution systems to be able to deliver the service to users within a reasonable time.

Moreover, the system generates a huge amount of data (0.5Tb per week) that have to be physically stored to provide the services. The storage capacity is important, but even more important is the question of efficient access to the data stored by the users in a timely manner. To address this issue, the web server "Motu" is used, which allows for the distribution of met/ocean gridded data files through the web. Motu is a highly efficient and robust software tool that fills the gap between heterogeneous data providers to end users and allows for handling, extracting, and transforming oceanographic huge volumes of data without performance collapse. Lastly, a service desk has been put in place to help users with the products, register the requests, and make sure that the service is well-delivered.

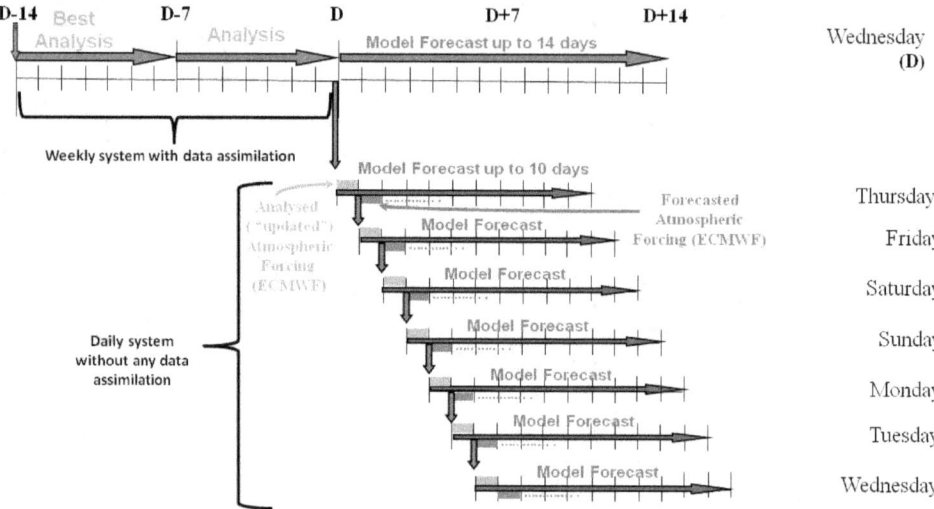

Figure 20.11. Schematic representation of the operational schedule for operation of the system. Real-time corresponds to the base date D. Vertical arrows in red represent the writing of files allowing to the model to restart.

Validation Methodology

Metrics and calibration period

Scientific quality is one of the key criteria for the continuous improvement of CMEMS products. Moreover, metadata on scientific quality helps users understand the content of products and their usefulness. Therefore, it is very important for monitoring and forecasting centers to agree on quality standards and to produce homogeneous and accessible information on the scientific quality of their analyses and forecasts. The scientific assessment procedure applied for CMEMS consists of two phases. During the first "calibration phase," new products or developments are checked with a series of metrics before being commissioned. Once the product has been commissioned it then undergoes an "operational validation phase" during which the products are checked against the reference calibration results. Standards and metrics were defined during the MERSEA (Marine Environment and Security for the European Area) integrated project and in the context of GODAE. These standardized diagnostics have enabled comparative exercises between European operational oceanography monitoring and forecasting centers (Crosnier and Le Provost, 2007) and others outside of the EU (Hernandez et al., 2009).

Some of these metrics were proposed for calibration and validation purposes. An efficient validation procedure has already been defined by Mercator Ocean for its model, including scientific assessment (calibration) and quarterly control bulletins (validation). These documents give a general picture of the normal behavior of the system in terms of accuracy and realism of the ocean physics. The accuracy is measured by the differences between simulations and observations and the realism by studying particular oceanic processes. In addition, more than a thousand diagnostic checks are routinely performed every day at Mercator Ocean.

The metrics that can be divided into four main categories derived from Crosnier and Le Provost (2007). The consistency between two-system solutions or between a system and observations can be checked by "eyeball" verification. This consists in comparing subjectively two instantaneous or time mean spatial maps of a given parameter. Coherent spatial structures or oceanic processes such as main currents, fronts, and eddies are evaluated. This process is referred to as CLASS1 metrics. The consistency over time is checked using CLASS2 metrics, which include comparisons of moorings time series, and statistics between time series. Space and/or time integrated values such as volume and heat transports, heat content, and eddy kinetic energy are referred to as CLASS3. Their values are generally compared with literature values or values obtained with past time observations such as climatologies or reanalyses. Finally, CLASS4 metrics give a measure of the real-time accuracy of systems by calculating various statistics of the differences between all available oceanic observations (in situ or satellite) and their model equivalent at the time and location of the observation.

The scientific assessment or calibration procedure thus involves all classes of metrics. It checks improvements between versions of a system, and ensures that a version is robust and its performance stable over time. The assessment must be conducted through a one-year numerical experiment at least, in order to obtain representative results. It is currently very difficult to run real-time systems over many years in the past, for computational reasons, but also due to the recent (and ongoing) evolution of the ocean observational network. Different data densities imply different tunings of the data assimilation system.

Quality check of real-time analyses and forecasts

Once the scientific assessment has been done and the system's nominal accuracy values and consistent behavior have been described, it is possible to apply a regular quality check to the real-time analyses and forecasts. Due to the very large amount of information produced by a global system, a real-time quality check is based on a reduced number of metrics, and comparisons with observations are constrained by their availability and timeliness. However, more than a thousand graphics are checked each week (weekly monitoring of the analysis) and each day (consistency check of the daily forecast) by Mercator Ocean. The major part of this procedure is currently being automated with indicators based on distribution (percentiles) thresholds computed from the scientific assessment stage. To record the strengths and weaknesses of forecasting systems, Mercator Ocean has published the Quarterly Ocean Validation Bulletin *"QuO Va Dis?"* since July 2010. It is available at http://www.mercator-ocean.fr/eng/science/qualification. There, one can find observation minus analysis (called "residual") and observation minus forecast (called "innovation") statistics for temperature and salinity vertical profiles, SST, SLA, and sea ice concentration observations that are assimilated. Comparisons are also made with independent observations, such as currents at 15 m derived from drifting buoys, sea-ice drift, or tide gauges (the low frequency component of the tide gauges' elevation signal). Integrated parameters such as sea ice extent and global mean SST are monitored. Process studies focusing on one process or region, or short research and development validation studies complement the bulletins.

In the next section, we illustrate the scientific assessment results mostly with the same metrics.

Assessment of the System

This section describes the Mercator Ocean system's quality assessment with diagnostics over particular years, assorted with time series over multiyear periods. To give an idea of the quality of the system, the distance to the assimilated observations (SST, SLA, temperature and salinity vertical profiles, and sea ice concentration) is measured. Moreover, the analyses are also compared with observations that have not been assimilated by the system.

SST

Assimilation diagnostics

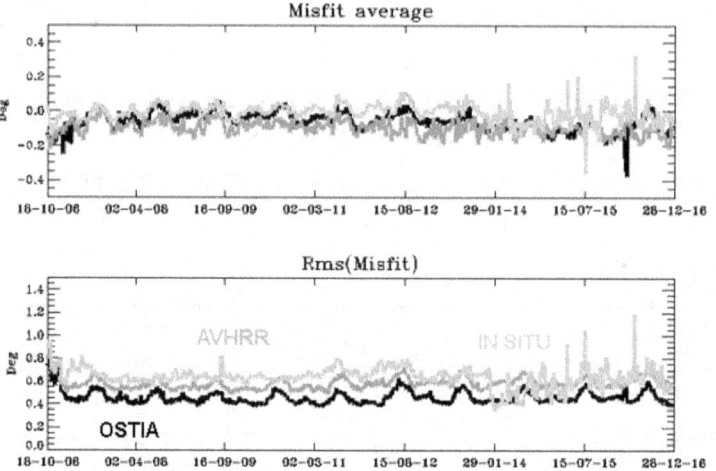

Figure 20.12. SST (°C) global misfit average (top) and RMS (bottom) for OSTIA observations (black line, assimilated), NOAA AVHRR observations (blue line, not assimilated), and in situ observations (orange line, assimilated), for the (October 2006 - December 2016) hindcast period.

Here we checked temporal series of the mean and the RMS of the misfit between the observed SST and the model. For OSTIA SST, we obtain a mean warm bias of -0.1°C and a mean error of 0.45°C (Fig. 20.12). Seasonal fluctuations of the SST biases on a global average can be seen as a lack of stratification in the model, which causes stronger mid-latitude cold biases during (boreal) summer (and a warm bias between 50 m and 100 m). For in situ SST, the bias is smaller, suggesting that OSTIA might be colder than in situ near surface observations on global average. We can notice a drop in the RMS of in situ surface data at the end of the period, which is due to the use of near real-time observations, where most of the surface observations do not have sufficient quality flag (drop in the number of data).

SST mode decomposition

An empirical orthogonal function (EOF) analysis is used to assess the variability of the SST over the time period 2010-2014. An EOF decomposition for the system (hindcast period) and the OSTIA SST has been performed.

Fig. 20.13 shows the mode 3 corresponding to the ENSO signal, which is very faithfully reproduced by the system. The SST EOFs are very robust and there are very few differences between the system and the observations up to mode 10. The only differences concern the coastal structures that are sharper with the system.

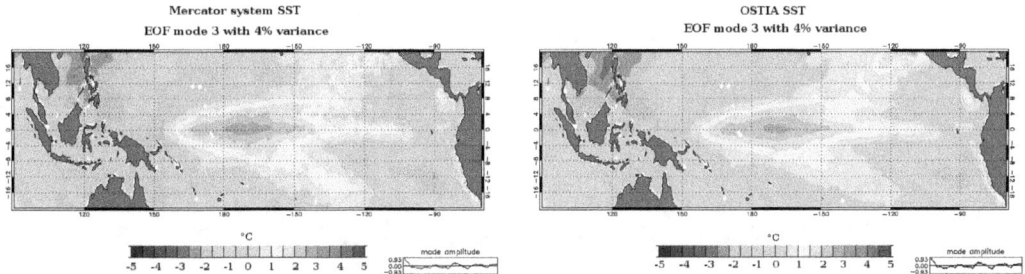

Figure 20.13. 3th EOF (bottom panel) of SST (°C) over the time period 2010-2014 (left panel: model SST, right panel: OSTIA SST). The time series at the bottom of each panel correspond to the mode amplitude.

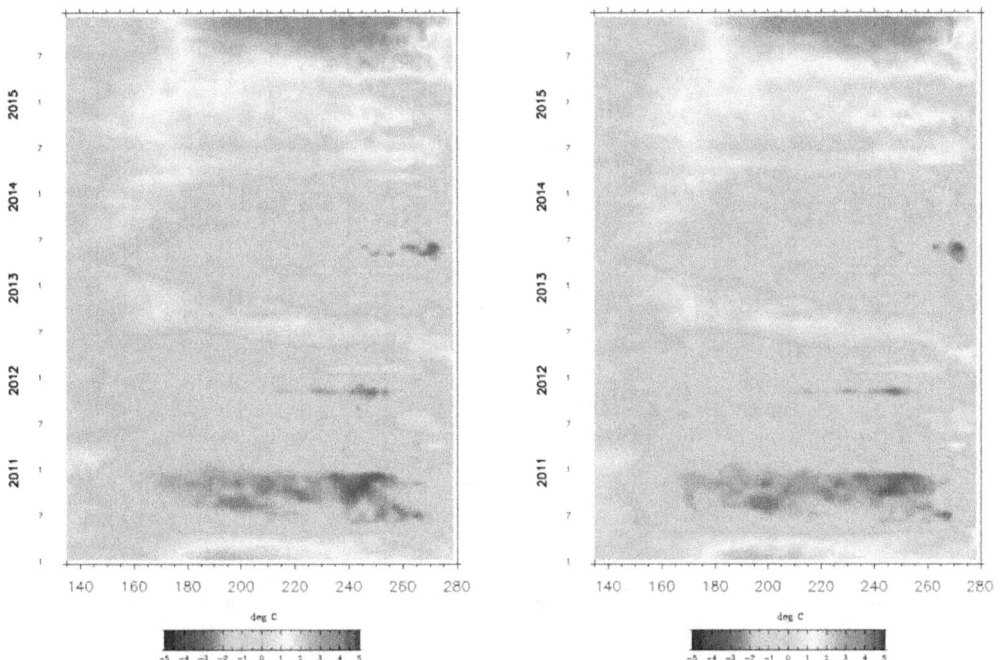

Figure 20.14. SST longitude time diagrams within 2°N-2°S over the Pacific (left panel: model; right panel: OSTIA observation). Monthly anomalies relatively to the 2010-2015 seasonal signal.

Equatorial propagation

The 2010-2015 monthly anomalies averaged within 2°N-2°S are used to validate the equatorial propagation with the SST.

Fig. 20.14 shows the coherence between the system and the OSTIA SST, both in terms of amplitude and phase. There is a little westward shift in longitude of the system structures, but the equatorial propagation can be seen in the system and the observations. Following the 2010-2011 La Niña, the westward propagation of the warm anomaly is clear in the system and the observation (second half of 2012). The equatorial slope returns to normal in 2013 and warm waters accumulate in the west. The El Niño that did not quite happen is visible in winter 2013-2014 in the west. The eastward Kelvin wave in May-June 2014 warms the eastern side and reflects into a westward Rossby wave. This reinforces the 2015 El Niño warming east of the dateline.

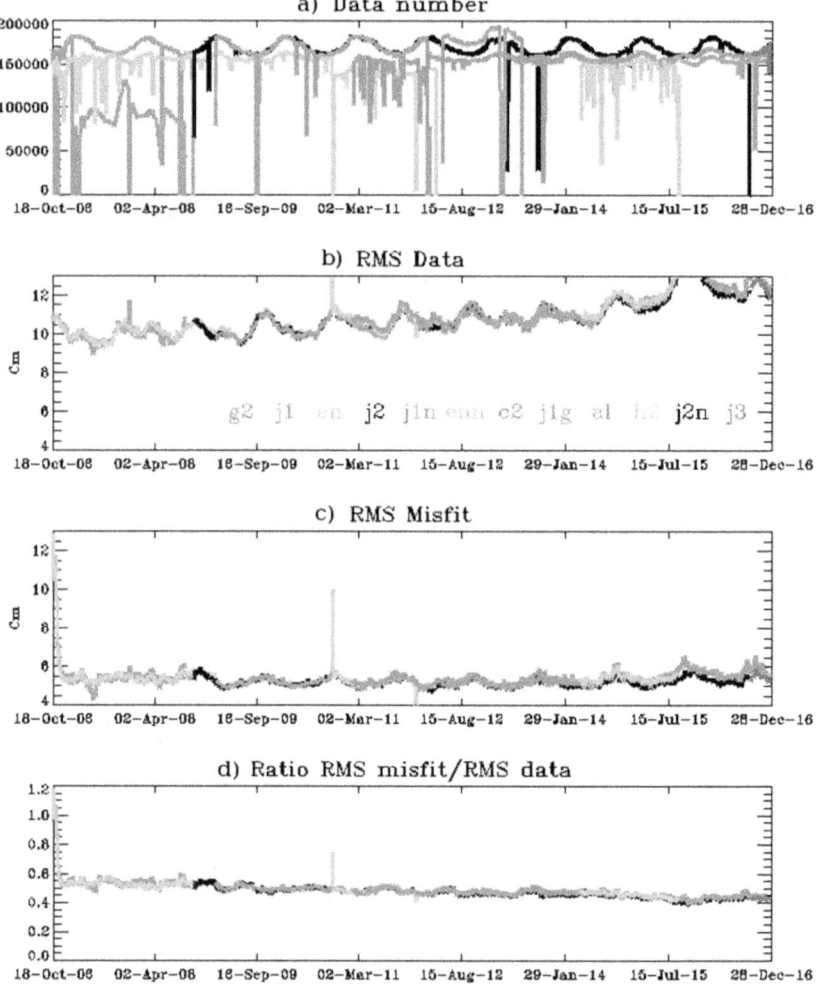

Figure 20.15. Time evolution of SLA (cm) data assimilation statistics averaged over the whole domain: a) data number, b) quadratic mean of the SLA data, c) RMS of innovations, d) RMS of innovations divided by quadratic mean of SLA observations. The scores are averaged over all seven days of the data assimilation window, with a lead time equal to 3.5 days.

Sea level anomalies (SLAs)

Assimilation diagnostics

Fig. 20.15 shows time series of assimilation statistics. The system is close to altimetric observations with a forecast RMS difference of the order of 5.5 cm. It does better than the previous system, which had a forecast RMS difference of the order of 6.5 cm. This is mainly due to the use of the "Desroziers" method to adapt the observations errors online, which yields to more information from the observations being used. Moreover, the model is able to explain the observed signal as shown by the ratio of RMS innovation to RMS data, which decreases with time and converges towards a value less than one.

Fig. 20.16 illustrates the skill related to the different changes of versions of the global system. For this figure, only statistics with Jason1 (J1), Jason2 (J2), and Jason3 (J3) altimeters are shown. The improvement between successive versions can be seen thanks to the increase of horizontal resolution (from 1/4° to 1/12°) and to refinements or adjustments to the system (model and data assimilation updates).

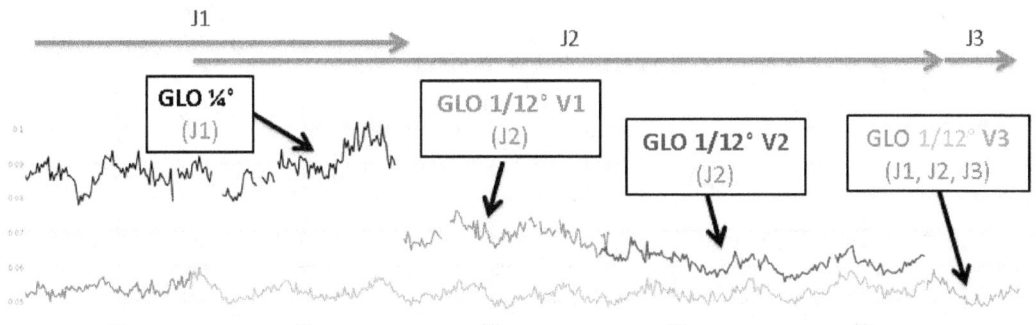

Figure 20.16. Time evolution of RMS of SLA innovations (m) related to the different changes of versions of global systems at ¼° or 1/12° resolution. Only statistics with Jason1 (J1), Jason2 (J2), and Jason3 (J3) altimeters are shown. Early performances were about 10 cm whereas they are now close to 5 cm.

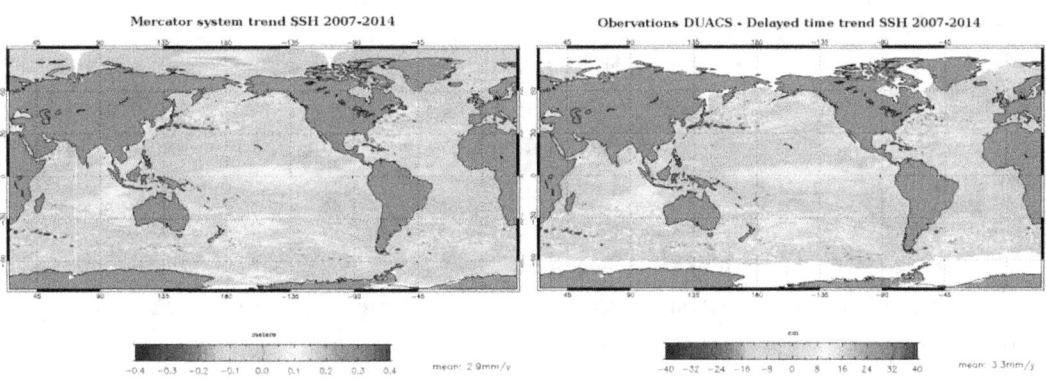

Figure 20.17. Model (on the left) and observations (on the right) SSH trends over the time period 2007-2014.

SLA trends

Next, we compare the sea level trends from the system and the Data Unification and Altimeter Combination System (DUACS) delayed time SLA for the time period 2007-2014 in order to avoid the signature of the strong 2015 El Niño. Fig. 20.17 shows good spatial agreement between the system and the data. The shift in global mean sea level (2.9 mm/year for the system and 3.3 mm/year for the data) is due to the glacial isostatic adjustment (0.3 mm/year), which must be added to the ocean model.

SLA mode decomposition

Next, we compare the EOF decomposition of the SLA from the system and data for the time period 2000-2014. Again, the EOFs are very similar in terms of spatial patterns and time amplitudes up to mode 10. Fig. 20.18 shows this property for mode 4. It corresponds to the extinction of the 2010-2011 La Niña (the amplitude is negative at the beginning). The joined effect on the equatorial SST can be seen in Fig. 20.14.

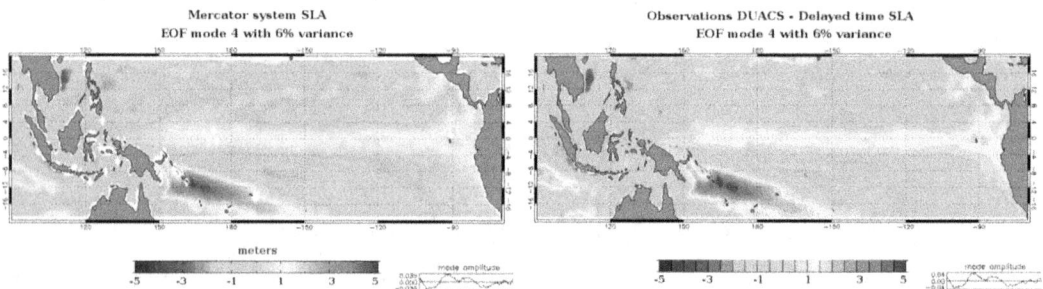

Figure 20.18. Fourth EOF (bottom panel) of SLA (m) over the time period 2010-2014 (left panel: model SLA, right panel: DUACS SLA). The time series at the bottom of each panel correspond to the mode amplitude.

Temperature and salinity vertical profiles

Next, we check time series of the RMS of the difference between the model analysis and the observations, for temperature on the left and for salinity on the right (Fig. 20.19) in the whole water column. We compare observation and climatology (red line), the previous system (blue line), and the current system (black line).

On global average, the current system slightly degrades the temperature statistics (-0.03°C) but greatly improves the salinity statistics (+0.1 psu). This enables us to get a more accurate description of the water masses and better balance arises from the new in situ errors that give more weight to the salinity data. Note that the systems are always better than the climatology.

The current system experiences a slight warm bias (negative observation – forecast difference) in subsurface (25-500 m) on global average (not shown). For the year 2015, part of this signal comes from the strong interannual ENSO signals in the Tropical Pacific, where the near-surface bias is also warm, as well as in the Antarctic Circumpolar Current and the Gulf Stream. Seasonal cold surface biases appear in the mid latitudes, linked with a lack of stratification during summer. Summer warming is injected too deep which results in subsurface spurious warming and a mixed layer that is too shallow. However, these biases remain small on global average.

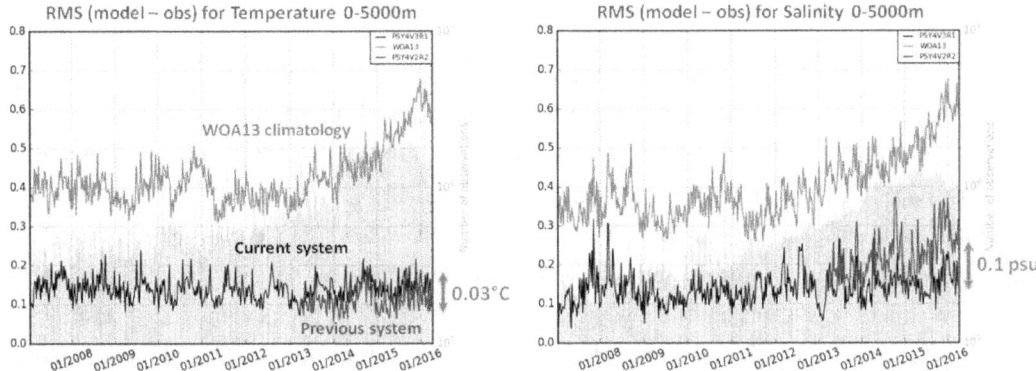

Figure 20.19. Time series of the RMS of the difference between the model analysis and the in situ observations (in the whole water) column for previous and current systems and the World Ocean Atlas 13 climatology. Left panel: temperature (°C), right panel: salinity (psu). Time series of the number of available observations appear in grey.

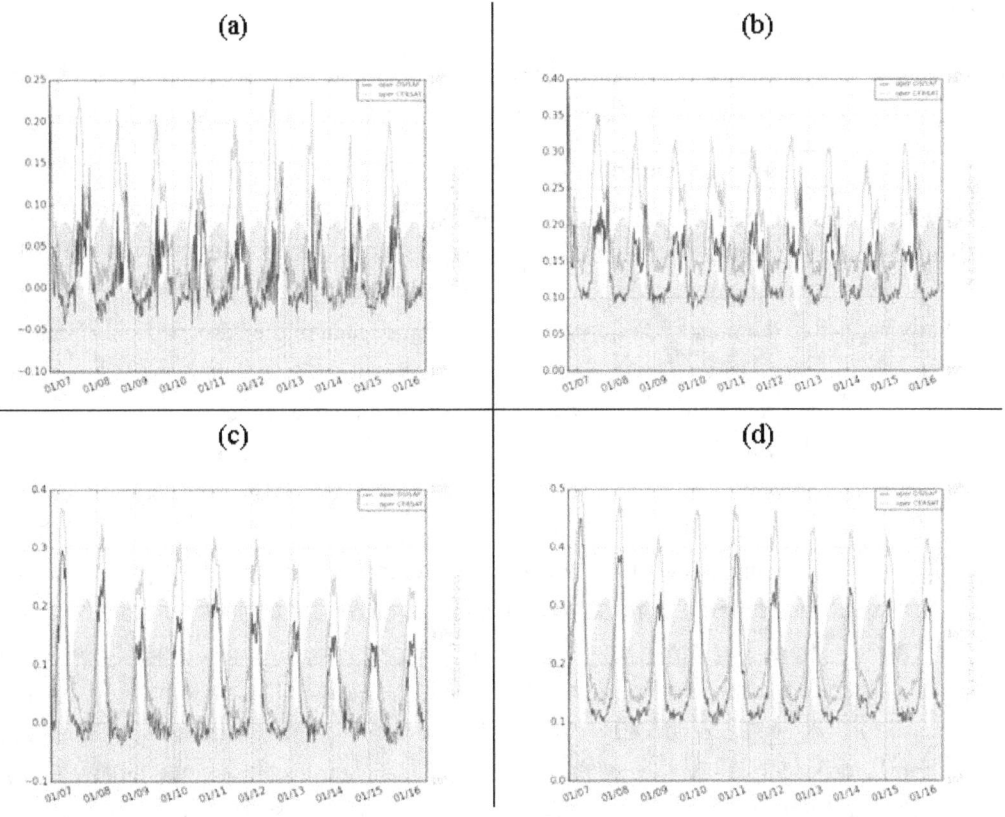

Figure 20.20. (observation-forecast) mean (a and c) and RMS (b and d) differences of sea ice concentration (0 means no ice, 1 means 100% ice cover) in the Arctic Ocean (a and b) and Antarctic Ocean (c and d) over the time period 2007-2016. Dark blue is the mean or RMS error of the system with respect to assimilated OSI Thematic Data Assembly Centers' sea ice concentration, light blue is the mean or RMS of the system with respect to CERSAT sea ice concentration observations. Time series of the number of available observations appear in grey.

Sea ice concentration

As shown in Fig. 20.20, the system sees slightly too much ice during the beginning of the melting season in summer (up to 3% overestimation on average in both the Arctic and Antarctic basins), while the mean error is stronger on average during winter (10-20% underestimation, depending on the year). RMS errors are also larger during the summer melt season (up to 20% in the Arctic and up 30% in the Antarctic with respect to OSI Thematic Data Assembly Centers' observations), and they drop to less than 10% in winter. These RMS errors quantify the capacity of the system to capture weekly time changes in the ice cover. In the Arctic, the error peaks without data assimilation occur at the sea ice cover minimum in September, while the error peaks with data assimilation occur earlier in July-August when the sea ice melts. This may be linked to the fact that the weekly assimilation of sea ice concentration is not fitted to constrain rapid changes in the sea ice cover in the marginal seas.

Currents

In this section, we use velocity observations that were not assimilated in the system or a case study to assess the level of performance of the current system compared to the previous ones.

Drifter velocities: near surface validation

The 15 m currents are checked by comparing the model to velocity observations coming from Argo surface floats and in situ Atlantic Oceanographic and Meteorological Laboratory drifters. Grodsky et al. (2011) revealed that an anomaly in the drogue loss detection system of the Surface Velocity Program buoy had led to the presence of undetected undrogued data in the "drogued-only" dataset distributed by the Surface Drifter Data Assembly Center. Rio (2012) applied a procedure using altimeter and wind data to produce an updated dataset, including a wind slippage correction. Therefore, we use this new in situ dataset coming from CMEMS Thematic Data Assembly Centers to check mean model currents. The "YoMaHa" real-time surface ARGO drifts are utilized, as well (Lebedev, 2007).

Fig. 20.21 shows a comparison of the 2007-2015 averaged drifts from the system and the observations over the Indonesian region. Currents in this region are very difficult to resolve because of the many narrow straits and strong tidal mixing. The retroflection of the westward South and North Equatorial Currents (along Papua and near 12°N) into the eastward North Equatorial Counter Current (near 4°N) are well-reproduced structures in the Pacific. The system South Equatorial Current is a little too strong at the edge of the warm pool. The complex flow in the Sulawesi Sea, the Makassar Strait and the South China Sea is also well-reproduced by the system. The correlation is 0.70 (respectively 0.64) for the zonal (respectively meridional) velocity.

THE MERCATOR OCEAN GLOBAL HIGH-RESOLUTION MONITORING AND FORECASTING SYSTEM • 587

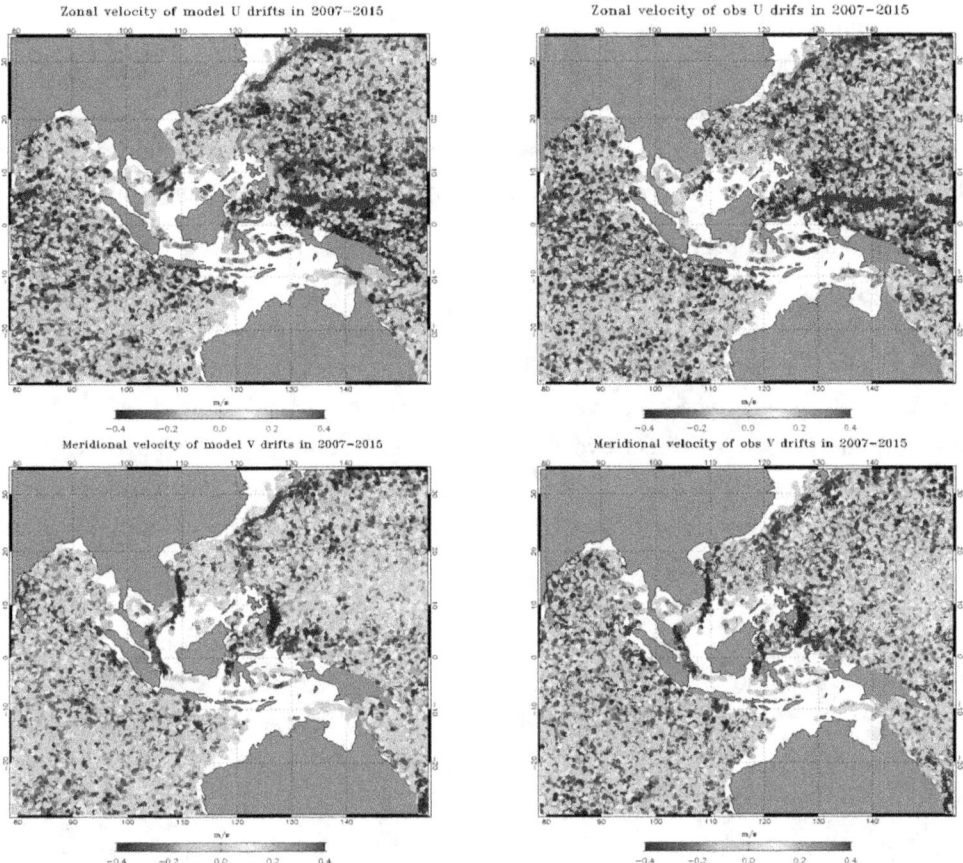

Figure 20.21. System 15 m current (left panels) and observation near surface drifts (right panels) from Argo surface floats and a Surface Velocity Program-corrected dataset (Rio, 2012). The zonal (top panels) and meridional (bottom panels) velocity information is indicated by a colored dot. Velocity information is averaged over the time period 2007-2015

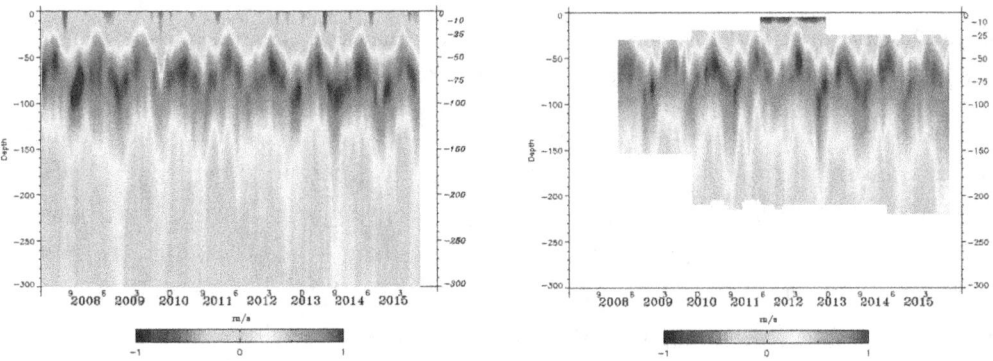

Figure 20.22. Vertical profiles of zonal velocity in the equatorial Atlantic at 0°N-23°W: system (on the left) and mooring data (on the right). Note that the strong westward data near the surface is questionable.

Mooring validation

Next, we compare the Equatorial Undercurrent in the Atlantic at 0°N-23°W, as seen by the system and the PIRATA (Prediction and Research Moored Array in the Atlantic) mooring. Even if this velocity data was not assimilated into the system, the phase and intensity of the undercurrent are well-reproduced. The only concern is the core, which seems slightly too strong (Fig. 20.22).

ALBOREX multi-platform experiment

This multi-platform experiment was conducted from 25 to 31 May 2014. The week-long experiment was designed with the objective of sampling the intense front where Atlantic and Mediterranean waters meet in the Eastern Alboran Sea and capture the intense, but transient, vertical motion associated with mesoscale and sub-mesoscale features such as ocean eddies, filaments, and fronts. Toward this goal, 24 drifters, two gliders, and three Argo floats were deployed. Drifters from the ALBOREX campaign are used here to assess the current system and the previous one. Fig. 20.23 shows SSH and velocity field for previous and current systems. Magenta dots correspond to ALBOREX drifters' positions from May 25-31 2014.

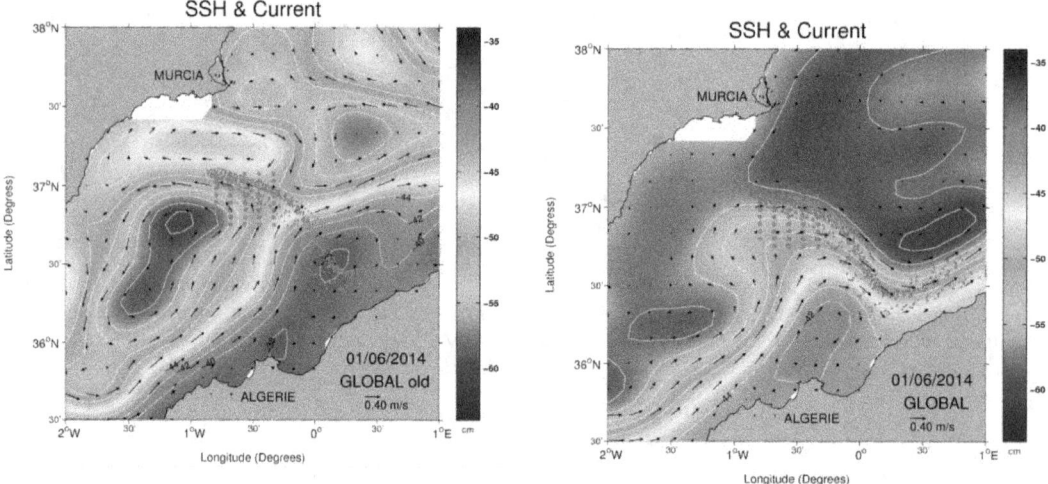

Figure 20.23. SSH and velocity field for previous (on the left) and current (on the right) systems. Magenta dots correspond to ALBOREX drifters' positions from May 25-31, 2014.

The previous system forecasts the intense signal of the Algerian Current flowing to the east. However, it doesn't reproduce the anticyclonic eddy and its associated front sampled during the field ALBOREX experiment. The current system significantly improves the local circulation. SSH and velocity field are in very good agreement with observations, in terms of position of the frontal zone. This improvement in the representation of the eddy and frontal zone can be attributed to the new MDT and the better-adapted prescribed SLA observation error in this region.

Position of the Air France AF447 wreckage

On the evening of June 1, 2009, a Rio-Paris Air France flight (AF447) disappeared in a highly variable and poorly observed part of the western tropical Atlantic Ocean. The first debris from the aircraft was found five days after the accident. Several reverse drift computations were conducted in order to define the likely position of the wreckage. Tailored high-resolution atmosphere reanalysis at Météo France and several ocean reanalyses at Mercator Ocean were produced but failed to locate the wreckage (Drévillon et al., 2013). So, in order to measure the impact of the current system updates, a new reverse drift computation was produced. Fig. 20.24 illustrates these reverse drifts using the best reanalysis from a previous Mercator system (left panel) and the current system (right panel). Circles (respectively squares) represent the model (respectively the observations) positions; the color indicates the day. The position of the wreckage is very close to the last known position (per the aircraft communications addressing and reporting system or ACARS, indicated by a black or white dot). PolSar indicates pollution observed by SAR. The best reanalysis from the previous Mercator system gave a position too far north with an estimated uncertainty of 32 km, while the new system (shown at right) locates the crash with an estimated uncertainty of only 17 km.

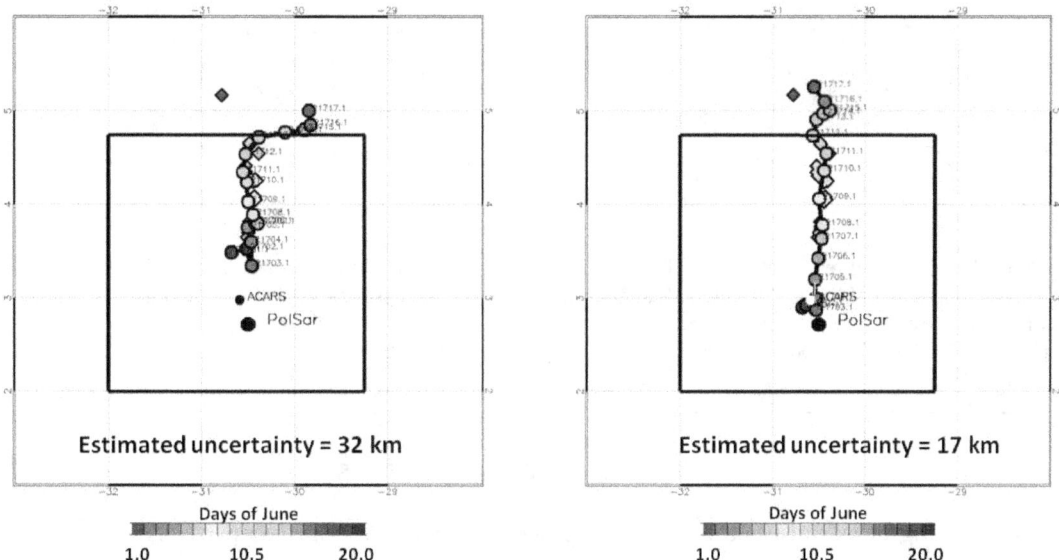

Figure 20.24. Illustration of reverse drifts using the best reanalysis from a previous Mercator system (left panel) and the current system (right panel). Circles (respectively squares) represent the model (respectively the observations) positions; the color indicates the day of June. The position of the wreckage is very close to the last known position (aircraft communications addressing and reporting system or ACARS, black or white dot). PolSar indicates pollution observed by SAR.

Daily forecast validation

The performance of the daily ten-day forecasts has been checked. Fig. 20.25 represents temperature RMS differences (model minus observation) for best analysis and for one-day, three-day, five-day, seven-day, and nine-day forecasts. As expected, the best analysis has the lower RMS and this RMS increases with the forecast length. Similar results are obtained for salinity, SLA, and SST.

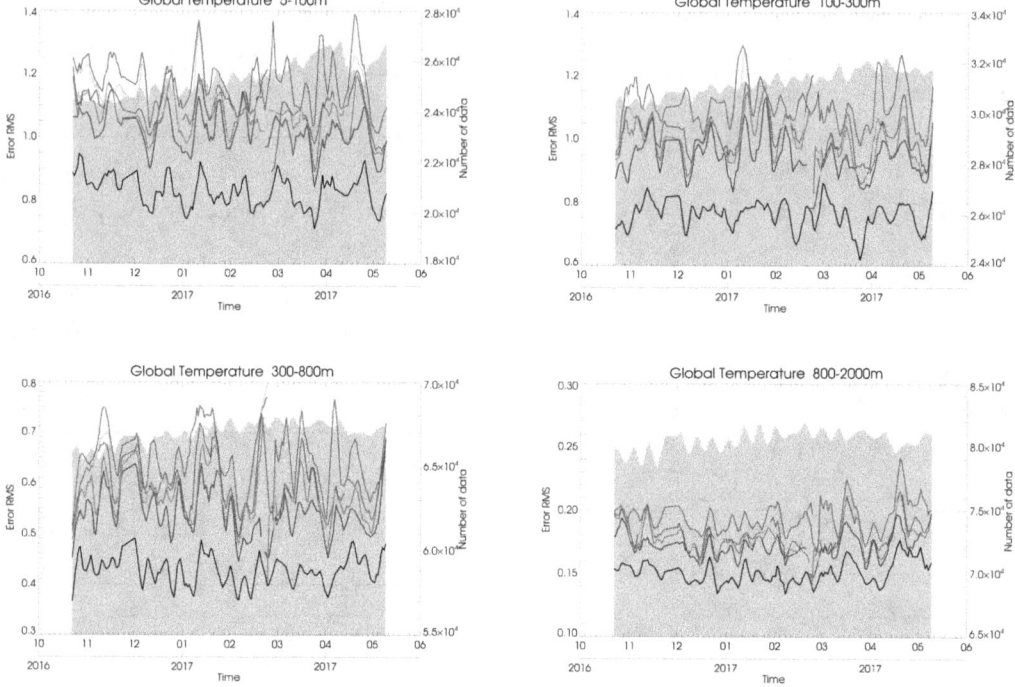

Figure 20.25. Global temperature (°C) RMS differences (model minus observation) in the 5-100 m, 100-300 m, 300-800 m and 800-2000 m layers. Statistics are displayed for best analysis (black line) and for one-day (blue line), three-day (red line), five-day (green), seven-day (orange line), and nine-day (brown line) forecasts. The number of available observations appears in grey in the background.

Users and Developments

In the coming decades, strong growth of the ocean economy is expected particularly in marine aquaculture, renewable marine energies, and port activities. Scientific and technological advances will play a crucial role in addressing many of the environmental challenges of oceans and in the development of ocean-related economic activities. The ability of operational oceanography to provide relevant responses to current or future applications remains highly dependent on research and development and supply improvements. Several thematic research and development projects have been recently defined at Mercator Ocean. These projects are associated with major scientific challenges and aim to better meet the needs of users and the different applications for products delivered in real and delayed time. Products must meet the specific needs of the applications classified in the four CMEMS themes: maritime safety; marine resources management; coastal and marine environment; and weather, climate and seasonal forecasting.

The main challenges for the design of the post-2020 global Mercator Ocean system corresponds to evolutions in progress:
- Four-dimensional and ensemble-based methods to improve the analysis and provide forecast uncertainty.

- Higher 1/36° resolution consistent with high-resolution future observations (SWOT mission).
- Interaction and retroaction between ocean, waves, and atmosphere, as well as improvement of the surface layer, high-frequency phenomena including tides.
- Data assimilation of satellite sea surface salinity, ocean color, and bio-ARGO observations.

Conclusion

The current global Mercator high-resolution system has a quite good statistical behavior with an accurate representation of the water masses, the surface fields and the mesoscale activity. Most of the components of the system have been improved compared to the previous system: global mass balance, three-dimensional T/S, sea level, sea-ice, and currents. Major variables, such as sea level and surface temperature, are hard to distinguish from the data. A 1992-to-present reanalysis is ongoing using this system, with seasonal errors for in situ vertical profiles. Future updates will deal with the assimilation of satellite sea surface salinity and the development of a four-dimensional analysis in the assimilation scheme. The remaining issues are being addressed within national and international scientific collaborations (Météo France, CNRS, CNES, ESA).

Acknowledgments

This study has been conducted using E.U. CMEMS information. The figure related to drifters from the ALBOREX campaign was carried out as part of the CMEMS MedSUB project (PI: Simon Ruiz, CSIC). We would like to thank Hanna Kauko, Jean-Baptiste, and Christina Eunjin Kong for their participation in the summer school, as well as for their review of the manuscript and for their remarks.

References

Amante C. and Eakins, B. W., 2009: ETOPO1 1 Arc-minute global relief model: procedures, data sources and analysis, *NOAA Technical Memorandum NESDIS NGDC-24*, 25 pp.

Arakawa, A. and Lamb, V. R., 1981: A potential enstrophy and energy conserving scheme for the shallow water equations, *Mon. Weather. Rev.*, 109, 18–36.

Barnier, B., Madec, G., Penduff, T., Molines, J. M., Treguier, A. M., Le Sommer, J., Beckmann, A., Biastoch, A., Böning, C., Dengg, J., Derval, C., Durand, E., Gulev, S., Remy, E., Talandier, C., Theetten, S., Maltrud, M., McClean, J., and De Cuevas, B., 2006: Impact of partial steps and momentum advection schemes in a global circulation model at eddy permitting resolution, *Ocean Dynam.*, 56, 543-567.

Becker, J. J., Sandwell, D. T., Smith, W. H. F., Braud, J., Binder, B., Depner, J., Fabre, D., Factor, J., Ingalls, S., Kim, S.H., Ladner, R., Marks, K., Nelson, S., Pharaoh, A., Trimmer, R., Von Rosenberg, J., Wallace, G., and Weatherall, P., 2009: Global Bathymetry and Elevation Data at 30 Arc Seconds Resolution: SRTM30_PLUS, *Mar. Geod.*, 32, 355-371, doi: 10.1080/01490410903297766.

Blanke, B. and Delecluse, P., 1993: Variability of the tropical Atlantic-Ocean simulated by a general-circulation model with 2 different mixed-layer physics, *J. Phys. Oceanogr.*, 23, 1363-1388.

Cabanes, C., Grouazel, A., Von Schuckmann, K., Hamon, M., Turpin, V., Coatanoan, C., Paris, F., Guinehut, S., Boone, C., Ferry, N., de Boyer Montégut, C., Carval, T., Reverdin, G., Pouliquen, S., and Le Traon, P.-Y., 2013: The CORA dataset: validation and diagnostics of in-situ ocean temperature and salinity measurements, *Ocean Sci.*, 9, 1-18, https://doi.org/10.5194/os-9-1-2013.

Carrère, L. and Lyard, F., 2003: Modelling the barotropic response of the global ocean to atmospheric wind and pressure forcing - comparisons with observations, *Geophys. Res. Let.*, 30(6), pp 1275.

Chambers, D.P., A. Cazenave, N. Champollion, H. Dieng, W. Llovel, R. Forsberg, K. von Schuckmann, and Y. Wada, 2017: Evaluation of the global mean sea level budget between 1993 and 2014, *Surv. Geophys.*, 38, no. 1, 309-327, doi:10.1007/s10712-016-9381-3.

Chen, J. L., Wilson, C. R., Tapley, B. D., Famiglietti, J. S., and Rodell, M., 2005: Seasonal global mean sea level change from satellite altimeter, GRACE, and geophysical models, *J. Geodesy*, 79, 532-539, doi:10.1007/s00190-005-0005-9.

Cravatte, S., Madec, G., Izumo, T., Menkes, C., and Bozec, A., 2007: Progress in the 3-D circulation of the eastern equatorial Pacific in a climate, *Ocean Model.*, 17, 28-48.

Dai A., Qian, T., Trenberth, K., and Milliman, J.D., 2009: Changes in Continental Freshwater Discharge from 1948 to 2004, *J. Climate*, vol. 22, 2773-2792.

Desroziers, G., Berre, L., Chapnik, B., and Polli, P., 2005: Diagnosis of observation, background and analysis-error statistics in observation space, *Q. J. R. Meteorol. Soc.*, 131, pp. 3385–3396, doi: 10.1256/qj.05.108.

Dibarboure, G., M.I. Pujol, F. Briol, P. Y. Le Traon, G. Larnicol, N. Picot, F. Mertz, M. Ablain, 2011: Jason-2 in DUACS: Updated System Description, First Tandem Results and Impact on Processing and Products, *Marine Geodesy*, Vol. 34, 214-241.

Drévillon, M., Greiner, E., Paradis, D., Payan, C., Lellouche, J.M., Reffray, G., Durand, E., Law-Chune, S., and Cailleau, S., 2013: A strategy for producing refined currents in the Equatorial Atlantic in the context of the search of the AF447 wreckage, *Ocean Dynamics*, 63, 63-82, DOI 10.1007/s10236-012-0580-2.

Fichefet, T. and Maqueda, M. A., 1997: Sensitivity of a global sea ice model to the treatment of ice thermodynamics and dynamics, *J. Geophys. Res.*, 102, 12609-12646.

Grasso, L. D., 2000: The differentiation between grid spacing and resolution and their application to numerical modelling, *B. Am. Meteor. Soc.*, 81, 579-580.

Grodsky, S. A., Lumpkin, R., and Carton, J. A., 2011: Spurious trends in global surface drifter currents, *Geophys. Res. Lett.*, 38, L10606, doi: 10.1029/2011GL047393.

Koch-Larrouy, A., Madec, G., Blanke, B. and Molcard, R., 2008: Water mass transformation along the Indonesian throughflow in an OGCM, *Ocean Dynam.*, 58, 289-309, doi: 10.1007/s10236-008-0155-4.

Large, W. G. and Yeager, S. G., 2009: The global climatology of an interannually varying air–sea flux data set, *Clim. Dynam.*, 33, 341-364, doi:10.1007/s00382-008-0441-3.

Lebedev, K.V., Yoshinari, H., Maximenko, N.A. and Hacker, P.W., 2007: YoMaHa'07: Velocity data assessed from trajectories of Argo floats at parking level and at the sea surface, *IPRC Technical Note*, No. 4(2)..

Lellouche, J.-M., Le Galloudec, O., Drévillon, M., Régnier, C., Greiner, E., Garric, G., Ferry, N., Desportes, C., Testut, C.-E., Bricaud, C., Bourdallé-Badie, R., Tranchant, B., Benkiran, M., Drillet, Y., Daudin, A., and De Nicola, C., 2013: Evaluation of global monitoring and forecasting systems at Mercator Océan, *Ocean Sci.*, 9, 57-81, doi:10.5194/os-9-57-2013.

Lévy, M., Estublier, A., and Madec, G., 2001: Choice of an advection scheme for biogeochemical models, *Geophys. Res. Lett.*, 28, 3725–3728, doi: 10.1029/2001GL012947.

Madec, G. and Imbard M., 1996: A global ocean mesh to overcome the North Pole singularity, *Clim. Dynam.*, 12, 381-388.

Madec, G. and the NEMO team, 2008: NEMO ocean engine. *Note du Pôle de modélisation, Institut Pierre-Simon Laplace (IPSL), France*, No. 27 ISSN, 1288-1619.

Rio, M.H., 2012: Use of altimeter and wind data to detect the anomalous loss of SVP-type drifter's drogue, *Journal of Atmospheric and Oceanic Technology*, DOI:10.1175/JTECH-D-12-00008.1.

Rio, M.-H., Mulet, S. and Picot, N., 2014: Beyond GOCE for the ocean circulation estimate: Synergetic use of altimetry, gravimetry, and in situ data provides new insight into geostrophic and Ekman currents, *Geophys. Res. Lett.*, 41, doi: 10.1002/2014GL061773.

Roquet, F., Charrassin, J. B., Marchand, S., Boehme, L., Fedak, M., Reverdin, G., and Guinet, C., 2011: Delayed-mode calibration of hydrographic data obtained from animal-borne satellite relay data loggers, *J. Atmos. Ocean. Tech.*, 28, 787–801.

Roullet, G., and Madec, G., 2000: Salt conservation, free surface, and varying levels: a new formulation for ocean general circulation models, *J. Geophys. Res.*, 105, 23927–23942.

Silva, T. A. M., Bigg, G. R., and Nicholls, K. W., 2006: Contribution of giant icebergs to the Southern Ocean freshwater flux, *J. Geophys. Res.*, 111, C03004, doi: 10.1029/2004JC002843.

Tranchant, B., Reffray, G., Greiner, E., Nugroho, D., Koch-Larrouy, A., and Gaspar, P., 2016: Evaluation of an operational ocean model configuration at 1/12° spatial resolution for the Indonesian seas (NEMO2.3/INDO12) – Part 1: Ocean physics, *Geosci. Model Dev.*, 9, 1037-1064.

CHAPTER 21

A Coastal Ocean Forecast System for the U.S. Mid-Atlantic Bight and Gulf of Maine

John Wilkin, Julia Levin, Alexander Lopez, Elias Hunter,
Javier Zavala-Garay, and Hernan Arango

*Department of Marine and Coastal Sciences, Rutgers, The State University of New Jersey,
New Brunswick, NJ, USA*

Coastal ocean models that downscale global operational models are widely used to study regional circulation at enhanced resolutions. When operated as nowcast/forecast systems, these models offer predictions that can provide actionable guidance for maritime applications. A nowcast/forecast system for the northeast U.S. coastal ocean is described in this chapter to illustrate, by example, the many practical issues to be considered when configuring such a model for operational oceanography applications. The system uses the Regional Ocean Modeling System (ROMS) and four-dimensional variational data assimilation of observations from a comprehensive network of in situ platforms, coastal radars, and satellites. The emergence of open access web data services that adhere to community conventions for metadata descriptions for coordinate systems and geo-scientific data types, and support geospatial search and sub-setting, are shown to foster inter-operability of data and model usage, accelerate the test, validate and acceptance cycle for modeling system enhancements, streamline the addition of new data streams, facilitate operational monitoring of the system, and enable novice users to view and download model outputs to underpin the generation of higher level ocean information products.

Introduction

Hydrodynamic models are used in coastal oceanography to simulate the circulation of limited-area domains for studies of regional ocean dynamics, biogeochemistry, geomorphology, and ecosystem processes; for example, to deduce transport pathways for nutrients, sediments, pollutants, or larvae. When operated as real-time nowcast or forecast systems, these models offer predictions that assist in decision-making related to water quality and public health, coastal flooding, shipping, maritime safety, and other applications.

Here we describe the configuration and operation of an ocean modeling nowcast and forecast system, with data assimilation, for the northeast North American coast extending from Cape Hatteras, North Carolina, northward to near Halifax on the Scotian Shelf of Canada. This domain encompasses two very different dynamical regimes in the Mid-Atlantic Bight (MAB) and the Gulf of Maine (GOM).

In the MAB (Cape Hatteras to Cape Cod, Massachusetts; Fig. 21.1), a permanent front at the shelf-break separates relatively fresh and cool waters on the broad (~100 km-wide) shelf from saltier, warmer Slope Sea water (Mountain, 2003). This shelf-break front is prone to instabilities with wavelengths on the of order 40 km that evolve on timescales of a few days (Fratantoni and Pickart, 2003; Gawarkiewicz et al., 2004; Linder and Gawarkiewicz, 1998), and sustain along-shelf currents that reach the seafloor driving significant flow-bathymetry interactions. Eddy shelf interactions tied to Gulf Stream-induced warm core rings (Zhang and Gawarkiewicz, 2015) also lead to cross-shelf exchange with surface and sub-surface structure at scales of 10-30 km and days to weeks. Across-shelf fluxes of heat, freshwater, nutrients, and carbon control water mass characteristics and impact ecosystem processes in the MAB.

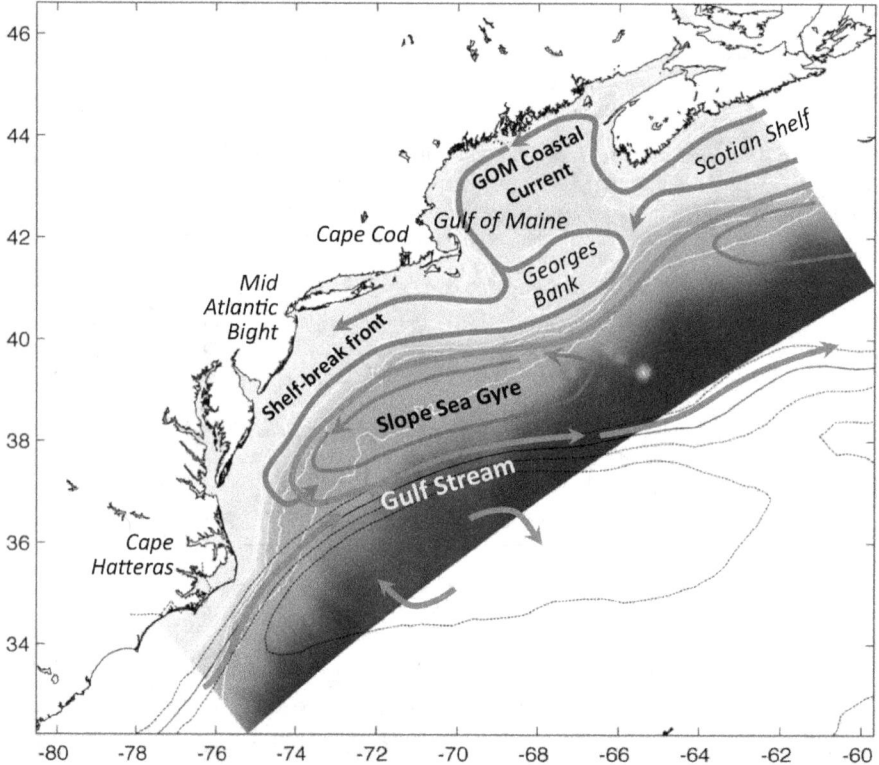

Figure 21.1. Bathymetry (colors) in the MARACOOS ROMS Doppio ocean model domain. Heavy lines and arrows show, schematically, the mean circulation of cool (blue) and warmer (red) currents. White lines show major bathymetric contours. Dotted black lines show major contours of the mean dynamic topography; solid black line is the north wall of the Gulf Stream.

The GOM is a relatively shallow, semi-enclosed marginal sea with local depth maxima of ~400 m in three distinct sub-basins. This basin topography exerts an influence on the overlying pattern of geostrophic flow. Water mass characteristics in the Gulf are determined by significant river inflows and multi-layer exchange flows through two narrow passages that flank Georges Bank. The region has famously strong tides that cause vigorous vertical mixing.

Therefore, all of the following processes influence circulation in this region: winds, tides, buoyancy input from rivers, air-sea heat fluxes, strong inflows from boundary currents at both the southern and northern extremities of the domain, and mesoscale Gulf Stream eddies that impinge upon the shelf edge. A large-scale, along-shelf pressure gradient is also significant in the along-shelf momentum budget (Csanady, 1976; Lentz, 2008).

This spectrum of forcing mechanisms, and the dynamic shelf-edge frontal zone, make the region a challenging laboratory for testing the skill of coastal ocean models and data assimilation methodologies. However, a great advantage to studying and modeling this region is that the MAB and GOM are quite densely observed compared to coastal oceans globally, with much of the local data acquisition coordinated by the Mid-Atlantic Regional Association Coastal Ocean Observing System (MARACOOS; maracoos.org) and the Northeastern Regional Association for Coastal Ocean Observing Systems (NERACOOS; neracoos.org), both members of the U.S. Integrated Ocean Observing System (IOOS) network of coastal regional associations, which are ad hoc consortia of federal, state, academic and commercial partners. MARACOOS operations include an extensive CODAR (Coastal Ocean Dynamics Applications Radar) HF-radar network that observes surface currents from the coast to the shelf edge, and deployments of autonomous underwater glider vehicles to acquire subsurface temperature, salinity and biogeochemical data along transects throughout the MAB. NERACOOS operations emphasize a network of several telemetering buoys that observe atmospheric conditions and surface and subsurface ocean conditions, and prototype and sustained programs observing biogeochemical properties of the Gulf. The MAB is also home to the National Science Foundation Ocean Observatories Initiative's (http://oceanobservatories.org) coastal Pioneer Array of seven profiling moorings and multiple deployments of autonomous vehicles, which provide very high-resolution observations in the vicinity of the MAB shelf-break front.

Traditionally, federal agencies have been the primary organizations implementing operational models in the U.S., while academic institutions have concentrated on process studies and model development and experimentation. A global example, operated by NOAA National Centers for Environmental Prediction (NCEP), is the Real-Time Ocean Forecasting System (RTOFS; http://polar.ncep.noaa.gov/global), a hybrid coordinate, 1/12° global ocean model that runs once a day and serves 3-hour interval forecasts of surface values and daily interval full-volume forecasts from the initial time out to 144 hours (six days). The NOAA Center for Operational Oceanographic Products and Services (CO-OPS) also operates forecast systems for several critical ports, harbors, bays, and estuaries on the U.S. coastline (https://tidesandcurrents.noaa.gov/models.html) with robust and automated model runs, quality checking, and forecast product generation.

But increasingly, non-federal regional association (RA) partners in IOOS are running ocean forecast systems. These systems might not meet federal requirements for operational robustness and reliability; nevertheless, many user communities find the immediate environmental information served by regional association models to be valuable (Wilkin et al., 2017). This may be because: the systems are superior in intrinsic skill having been carefully configured with local knowledge; they assimilate local datasets not adopted for operations by global centers; and they offer regional products, higher resolution, or local interpretive expertise that are not matched by larger domain systems.

Regional forecasting systems require open boundary condition information at the domain perimeter, which is typically drawn from global- or basin-scale model-based analysis systems, such as the products delivered by the Mercator-Océan system of the Copernicus Marine Environmental Monitoring Service (CMEMS) or the HYCOM system operated by the U.S. Naval Research Laboratory and NOAA. Both Mercator-Océan and HYCOM are elements in the international GODAE OceanView (Bell et al., 2015) program, developing and evaluating global ocean forecast systems. Regional modeling is thus a downscaling exercise, and a measure of success is whether the resulting ocean state analyses and forecasts achieve skill that exceeds that of the driving model in the same region. Other performance metrics will be based on the utility of valued-added products designed to meet specific stakeholder needs such as actionable guidance for marine operations, hazards, water quality, and regional ecosystems and fisheries.

The Rutgers University Ocean Modeling Group has operated a real-time forecasting system for the MARACOOS region since September 2009, assimilating in situ temperature and salinity data, CODAR velocities, satellite sea surface height (SSH) and satellite surface temperature (SST) for all available platforms. The system uses the Regional Ocean Modeling System (ROMS) model and its 4D-Var (four-dimensional variational) data assimilation system.

This article documents the set-up and operation of the MARACOOS ROMS forecast system, and is intended as a tutorial, by example, for others who might wish to emulate our effort to establish a similar downscaling model system in other coastal regions. The principal features of ROMS are described in the second section of this chapter. The third section outlines the particular configuration of ROMS for the MAB and GOM domains, including the information used for meteorological, river, and boundary forcing. The fourth section describes our choices in configuring 4D-Var for this application, the datasets assimilated, steps taken in the preparation and pre-processing of the observations, a discussion of practical issues regarding data access and monitoring the real-time system, and preliminary results of the performance of the prototype system still under development. We conclude with summary remarks on general lessons and challenges for the development and operation of models that enhance the value of observations through model-based synthesis and data assimilation to provide robust and reliable past, present, and forecasted ocean conditions.

Regional Ocean Modeling System – ROMS

Dynamical and numerical core

The ROMS computational kernel is described in detail elsewhere (Shchepetkin and McWilliams, 2005; Shchepetkin and McWilliams, 2009) and need not be reiterated here, but several aspects of the kernel are notable for the advantages they bring to coastal ocean and shelf sea simulation.

ROMS solves the hydrostatic, Boussinesq, Reynolds-averaged Navier-Stokes equations in terrain-following vertical coordinates. It employs a split-explicit formulation wherein the two-dimensional depth-integrated continuity and momentum equations are advanced using a much smaller time step than the three-dimensional baroclinic momentum and tracer equations. Time-weighted averaging of the barotropic mode prevents aliasing of unresolved signals into the slow baroclinic mode while accurately representing barotropic motions resolved by the baroclinic time step (e.g., tides and coastal-trapped waves). A formulation of the Equation of State and the density Jacobian is implemented that minimizes the pressure gradient truncation error that can be problematic in other terrain-following coordinate models. The terrain-following coordinate can be stretched vertically to better resolve surface and bottom boundary layers. Collectively, these features enhance the representation of friction, baroclinicity, and the vortex stretching of flow adjacent to steep bathymetry that are fundamental to steering sub-inertial frequency circulation in the coastal ocean and continental shelf-break region.

A finite-volume, finite-time-step discretization for the tracer equations improves integral conservation and constancy preservation properties associated with the variable free surface, which is important in coastal applications where the free surface displacement represents a significant fraction of the water depth.

By virtue of these many features, ROMS is a particularly attractive choice as a hydrodynamic modeling platform for achieving accurate, efficient, high-resolution ocean simulations of mesoscale processes in the coastal ocean and adjacent deep sea.

Vertical turbulent mixing closure

ROMS presents users with several options for how vertical eddy viscosity for momentum and eddy diffusivity for tracers are parameterized. The many coastal applications of the model have used them all: (i) a k-profile parameterization (KPP) in surface- and bottom-boundary layers (Large et al., 1994; Durski et al., 2004), (ii) Mellor-Yamada level 2.5 (MY25) (Mellor and Yamada, 1982), or (iii) the generic length-scale (GLS) method (Umlauf and Burchard, 2003), which includes several sub-options for closure and stability function.

The KPP scheme specifies turbulent mixing coefficients in the boundary layers based on Monin-Obukhov similarity theory, and in the interior principally as a function of the local gradient Richardson number (Large et al., 1994; Wijesekera et al., 2003). The KPP method is diagnostic in the sense it does not solve a time-evolving (prognostic) equation for any of the elements of the turbulent closure, whereas the MY25 and GLS schemes are of the general class of closures where two prognostic equations are solved – one for turbulent kinetic energy and the other related to turbulence length scale.

The implementation of GLS in ROMS is described by Warner et al. (2005), who also contrasted the performance of the various GLS sub-options and the historically widely used MY25 scheme. While the differing schemes lead to differences in the vertical eddy mixing profiles, the net impact on profiles of model state variables (velocities and tracers) is relatively minor. Similar conclusions were reached by Wijesekera et al. (2003). In the model set-up described here we use the GLS k-kl closure option, which is essentially an implementation of MY25 within the GLS conceptual framework. Our experience is that the model solutions are not particularly sensitive to the choice of sub-options within GLS, but the k-kl closure appears to be somewhat more stable in routine operations.

Two-way nesting

The ROMS code allows for one-way and two-way nesting for refinement grids (https://www.myroms.org/wiki/Nested_Grids) to provide increased resolution in specific regions, with future code developments planned to extend these capabilities to composite grids that only partially overlap, or are not aligned in their local grid coordinates.

The methodology for two-way nesting follows the same paradigm as the information exchange methodology used by ROMS to evaluate horizontal advection and diffusion operators across periodic boundaries or parallel subdomain partitions (in the MPI coarse-grained parallel execution option). In refinement nesting, a so-called "child receiver" grid obtains the information it needs to complete the high-order spatial stencils for the advection and diffusion operators surrounding the grid perimeter by having this information interpolated from the "parent donor" grid into a "contact region. This exchange is made on every model time-step, in both the predictor and corrector partial time steps. Lateral open boundary conditions are not applied since the extension of the numerical stencil into the contact region evaluates the full primitive equations horizontal operators at the perimeter of the receiver grid. This approach is preferable to providing donor grid information to the receiver strictly at the perimeter via open boundary conditions because open boundary conditions formulations are inevitably lacking in aspects of the ocean dynamics. Indeed, if the donor and receiver grids are identical (i.e., there is no refinement and all grid points coincide) then this approach is exact because it emulates the regular ROMS MPI parallel tiled domain decomposition (Warner et al., 2010).

This configuration can be strictly one-way (downscaling) with information flowing only from parent to child, or two-way. To achieve two-way nesting, i.e., including upscaling of information from the fine to coarse grid, the roles of donor and receiver are reversed in the complimentary step. The child grid becomes the donor; its solution is averaged to the parent grid resolution and replaces the parent solution where they overlap. Thus, the added physical realism achieved by increased resolution in the refined child grid is communicated back to the larger domain.

Variational data assimilation

ROMS supports a suite of differing implementations of 4D-Var data assimilation that are complemented by post-processing and analysis tools for a variety of applications. The system is described in detail by Moore et al. (2011a; 2011b; 2011c), and only a brief review of the important features will be presented here.

If we denote by $\mathbf{x}$ the state vector of the ocean (i.e., T, S, u, v, and SSH), the goal of 4D-Var is to identify the model initial conditions, surface forcing, and open boundary conditions (collectively referred to as the control vector $\mathbf{z}$) that yield the "best" ocean circulation estimate, $\mathbf{x}_a$. The "best" circulation estimate is that associated with the $\mathbf{z}$ that minimizes a cost function given by $J_{NL} = (\mathbf{z} - \mathbf{z}_b)^T \mathbf{B}^{-1} (\mathbf{z} - \mathbf{z}_b) + (\mathbf{y} - H(\mathbf{z}))^T \mathbf{R}^{-1} (\mathbf{y} - H(\mathbf{z}))$ where $\mathbf{z}_b$ is a *prior* or background estimate of the control vector, $\mathbf{y}$ is the vector of observations, $\mathbf{B}$ and $\mathbf{R}$ are respectively the background error and observation error covariance matrices, and H is the observation operator that maps **z** to the space-time observation locations. In the case of 4D-Var, the operator H includes the ROMS model and information is dynamically interpolated in time via the model dynamics. According to Bayes' theorem, the cost function J_{NL} corresponds to the logarithm of the *posterior* probability distribution of **z**, so finding the minimum of J_{NL} is equivalent to identifying the most likely ocean circulation state, described by z_a, given the prior circulation estimate z_b and the observations **y**. The topology of J_{NL} may be very complicated so in general the minimum of J_{NL} is found using an iterative truncated Gauss-Newton method, which takes the form of a sequence of linear minimization problems. During each sequence of linear minimizations, the cost function that is actually minimized is given by:

$$J_k = \delta \mathbf{z}_k^T \mathbf{B}^{-1} \delta \mathbf{z}_k + (\mathbf{H}_{k-1} \delta \mathbf{z}_k - \mathbf{d})^T \mathbf{R}^{-1} (\mathbf{H}_{k-1} \delta \mathbf{z}_k - \mathbf{d}) \qquad (1)$$

where $\delta \mathbf{z}_k = \sum_{i=1}^{k-1} \delta \mathbf{z}_i = \mathbf{z}_k - \mathbf{z}_b$ represents the departure of the kth iterate from the background, $\mathbf{d} = (\mathbf{y} - H(\mathbf{z}_b))$ is referred to as the innovation vector, and $\mathbf{H}_{k-1}$ is the tangent linearization of the observation operator H linearized about $\mathbf{z}_{k-1}$. Linearization of the minimization problem in this way is referred to as the "incremental" approach (Courtier et al., 1994) and is used in ROMS 4D-Var. The primary workhorse algorithm that will be used here is the incremental, strong constraint, dual formulation of ROMS 4D-Var in which J is minimized directly in the space spanned by the observations.

The minimization of (1) proceeds via a Lanczos formulation of the restricted B-preconditioned conjugate gradient (CG) method (Gürol et al., 2014), with each iteration in the sequence of so-called inner loops identifying a new search direction in the control vector space that is orthogonal to prior search directions. When the $\delta \mathbf{z}_k$ that minimizes J_k has been identified, the estimate $\mathbf{z}_k$ about which **H** is linearized is updated (a so-called outer loop), and minimization of (1) proceeds again. The inner and outer loops are continued until J_{NL} is reduced to an acceptable level or until the iterative procedure has converged to the point where further reductions in are negligible.

Details of specific algorithmic choices regarding background error and observation error for our application are detailed in the fourth section.

ROMS configuration for MAB and GOM – Doppio

Doppio – A double espresso

The model configuration used here builds upon experience with an established model of the MAB region termed ESPreSSO (Experimental System for Predicting Shelf and Slope Optics; Zavala-Garay et al., 2014) that has underpinned numerous regional studies related to ecosystems (Hu et al., 2012; Xu et al., 2013), biogeochemical cycles (Mannino et al., 2016), sediment transport (Dalyander et al., 2013; Miles et al., 2015), storm-driven circulation (Miles et al., 2017; Seroka et al., 2017), and underwater acoustics (Lin et al., 2017), as examples. In a comparison of seven real-time models encompassing the MAB region (Wilkin and Hunter, 2013), no model was more skillful than data assimilative ESPreSSO in capturing MAB circulation.

Model resolution

The present domain, depicted in Fig. 21.1, is twice the size of the ESPreSSO grid – hence the moniker "Doppio" that we use to refer to the model. The grid has uniform horizontal resolution of 7 km. Finer resolution (at ~2 km, or less) would be desirable from the perspective of allowing the emergence of submesoscale variability, but the modest resolution greatly facilitates experimentation with 4D-Var assimilation. Given the effective resolution of CODAR data, along-track altimeter data, other satellite data, and the relatively sparse in situ observations, this resolution is adequate for the length scales that data assimilation might reasonably be expected to constrain.

Two-way nesting grid refinement offers the capability to refine model forecast resolution in selected regions using initial conditions drawn from the data assimilative analysis, through this is not currently an approach we implement. We are experimenting with two-way nested 4D-Var (which must apply nesting also in the adjoint and tangent linear models as they are iterated by the inner and outer loops of 4D-Var) as a rigorous approach to propagating the information content of local high-resolution observations through the grid hierarchy, such as the closely spaced moorings and multiple concurrent underwater vehicles operating as part of the Ocean Observatories Initiative Pioneer Coastal Array.

The model vertical resolution is 40 terrain-following levels, with stretching chosen such that throughout the continental shelf ocean the vertical resolution is finer than 1.5 m at the sea surface and better than 3 m in the bottom boundary layer.

Forcing

Air-sea fluxes of momentum and heat are computed from atmospheric conditions in the marine boundary layer (air pressure, temperature, relative humidity, and 10 m winds) using standard bulk formulae (Fairall et al., 2003). The SST predicted by ROMS is used in the calculation of out-going long-wave radiation, sensible heat exchange, and the transfer coefficients for boundary layer turbulence. In the forecast system, meteorological conditions are specified using 3-hour interval fields from the NCEP North American Mesoscale (NAM) weather forecast model. In longer retrospective simulations we use fields from the North American Regional Reanalysis (Mesinger et al. 2006). The net shortwave radiation is known to have a significant positive bias in these models (Kennedy et al., 2011) so is decreased by a factor of 25%. The diurnal cycle of net shortwave is poorly resolved by 3-hour interval data, so these are converted to daily average values that ROMS modulates internally with an idealized local diurnal cycle that is a function of longitude, latitude and year-day.

The direct influence of sea level atmospheric pressure gradient enters the momentum equations via the sea surface boundary condition to the hydrostatic balance, i.e. the model simulates its own dynamical response to the inverse barometer effect.

The NAM forecast we use is the CONUS (Continental U.S.) nest computed at 12 km resolution in a Lambert-conic projection, though the data we access via NOMADS (NOAA Operational Model Archive and Distribution System https://nomads.ncep.noaa.gov) is re-mapped to uniform 0.11°

longitude/latitude rectangular grid. The NAM forecast runs every six hours, but our ocean forecast only runs daily so we choose to access only the 00:00 UTC forecast cycle since this is reliably loaded on NOMADS by late evening local time.

The calculation of air-sea momentum flux (i.e., stress) in ROMS can be configured to account for the relative speed of the air and water by subtracting the ocean surface vector current from the 10 m boundary layer wind prior to applying the bulk formulae following Bye et al. (1979), but with no explicit account taken of the influence of Stokes drift associated with a wave field. This option was not active in the Doppio configuration for which we show results below, but following a systematic analysis showing it adds skill, the option is now standard in the latest prototype system.

Air-sea freshwater flux is set by the precipitation from NAM or North American Regional Reanalysis subtracted from the evaporation associated with the latent heat loss computed by the bulk formulae.

Open boundary conditions are always a vexing issue for regional ocean models. We draw information on the 3-D ocean state at the model open boundary from the Mercator-Océan system (Drévillon et al., 2008) using daily mean fields from the CMEMS PSY4QV3R1 Global Ocean Physics Analysis and Forecast. In the forecast window these fields are refreshed daily with each new Mercator-Océan cycle. An appealing feature of CMEMS is the provision of a consistent product suite spanning several years back into the past. This allowed us to compute a long-term mean of the PSY4QV3R1 results, which we adjust to remove moderate biases (most noticeable in the salinity of the inflow from the Labrador Sea that enters our model northern boundary) by replacing the Mercator-Océan mean with our own Mid-Atlantic region Ocean Climatological Hydrographic Analysis (MOCHA) annual mean (Fleming and Wilkin, 2010; Fleming, 2016). The MOCHA climatology is based on hydrographic observations from the World Ocean Database (WOD) (Boyer et al., 2009) augmented by CTD data from the NOAA Northeast Fisheries Science Center (NEFSC) and inner shelf CTD observations acquired by MARACOOS institutions, mapped to an equal angle 0.05° grid (~5 km) on 57 standard depths by Fleming (2016) using an adaptation of the weighted least squares method of Ridgway et al. (2002). Bias adjustments to mean velocity and mean dynamic topography (MDT) are made using dynamically balanced velocity and sea level consistent with MOCHA computed by the same climatological mean 4D-Var approach used by Levin et al. (2018) for ESPreSSO. These adjustments to Mercator-Océan conditions affect the long-term mean only; mesoscale variability is retained. The boundary data are used in ROMS via a combination of active/passive perimeter radiation and nudging (Marchesiello et al., 2001) and flow relaxation (Blayo and Debreu, 2006) in a "nudging zone" some 70 km wide around the perimeter.

The Mercator-Océan model does not include tides, so harmonic tidal sea level and current variability derived from a regional model (Mukai et al., 2002) is added to the depth-averaged velocity and sea level boundary data, and imposed on the dynamics using methods adapted from Flather (1976) and Chapman (1985) following Mason et al. (2010).

River inflows are important sources of buoyancy that contribute to coastal current dynamics and the salt budget throughout the GOM and MAB. While the continental U.S. and Canada are relatively well-instrumented with real-time monitoring stream gauges, still an appreciable portion of the watershed is un-gauged and it is necessary to account for this to capture the full measure of freshwater inflow. We have taken a statistical approach to rescaling real-time gauge data using results from an analysis of 11 years (2000-2010) of river flow and precipitation data, coupled with water balance and water transport models to infer the daily discharge from land to sea at 3 arc minute latitude and longitude resolution (Stewart et al., 2013). At this resolution there are 403 sources along the coast from Nova Scotia to Cape Hatteras that we aggregated into 22 major discharge sites coinciding with significant named rivers. Maximum covariance analysis was used to correlate the net river discharge with observations at stream gauges operated by the USGS and Water Service of Canada that reliably report in near real-time. This gave us a statistical model to infer the full discharge from the watershed based on the partial direct observations.

Free-running model performance

The set-up described above is our prior configuration for Doppio running freely as a "forward" model without data assimilation. Using the data sets to be described in the next section, and that we ultimately will use for data assimilation, we next present some results on the skill of this forward model configuration.

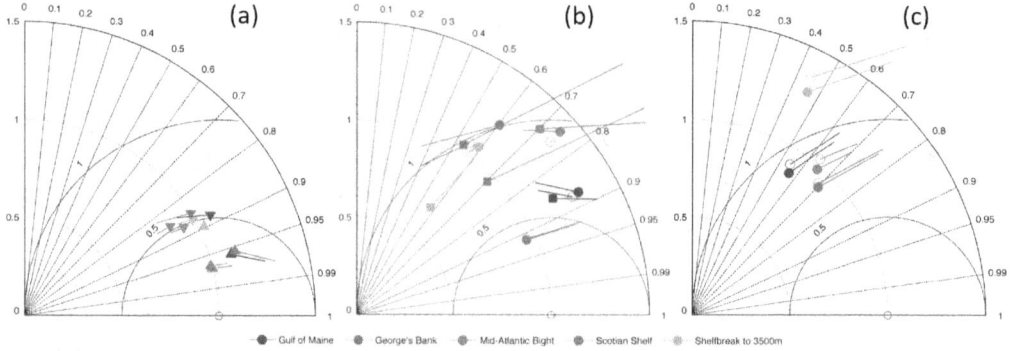

Figure 21.2. Taylor diagrams for model skill. Colors indicate sub-region of domain as shown in key. (a) Temperature at surface ▲ and sub-surface ▼ for control case. (b) Salinity for satellite precipitation forcing ●, HYCOM open boundary conditions ☐☐, and control case (open circles). (c) Surface current skill for wind minus current surface stress calculation ●, and control case (open circles).

The skill of the free-running forward model, which we term the Doppio *control* case, is shown in a set of Taylor diagrams (Taylor, 2001) in Fig. 21.2. The radial distance to the symbols is the model standard deviation normalized by observation standard deviation, the azimuth is the arc cosine of the correlation, and the distance to the point (1,0) on the x-axis is therefore the normalized centered RMS error. Less conventionally for Taylor diagrams, we add sticks whose length is the normalized mean bias of the model; then the distance from the end of the stick to (1,0) is the overall normalized RMS error including bias. The closer to (1,0), the better the performance.

The number of satellite versus in situ observations of temperature differs by orders of magnitude, so surface and sub-surface skill is shown separately in Fig. 21.2a. Symbols are color-coded by sub-regions within the model domain. Skill is consistently higher for surface than sub-surface temperature, but we make little of this. Generally speaking, SST is a poor metric of model skill because it is so strongly constrained by the sea surface boundary condition imposed by air temperature – it's relatively easy to get SST right. Sub-surface temperature, on the other hand, is the consequence of vertical mixing and lateral transport processes acting over quite some time and distance and is a more demanding skill metric. We see that the sub-surface temperature skill of the control simulation is consistently high in all geographic sub-regions, and almost without bias.

Our model development and assessment framework makes use of the ROMS observation file format for 4D-Var assimilation (described more fully later on) through a convenient model option (VERIFICATION; https://www.myroms.org/blog/archives/128) that samples the model at the 4-D (space and time) locations of each datum in the unified observation file using the same observation operator, H, used by 4D-Var. When enabled, this option causes the model to compute, as it runs, a full suite of model minus observation statistics ready for analysis. This accelerates the experimentation cycle of model change, quantitative evaluation, and acceptance/rejection of configuration modifications.

To illustrate, Fig. 21.2b shows skill for in situ salinity for runs with different precipitation forcing data sets: the control case with rainfall from the NAM analysis (unfilled circles) and a run using rainfall derived from satellite (Huffman et al., 2007; filled circles). The two cases are almost indistinguishable, so we elected to retain the model-based precipitation data rather than further pursue use of satellite derived products.

Also shown in Fig. 21.2b (filled squares) is the salinity skill of a run using output from the GODAE HYCOM model for open boundary conditions. Centered RMS error is comparable in the two cases with the exception of a decrease in correlation on the Scotian Shelf. But when the bias is considered (the end of the sticks) there is a more noticeable decrease in skill for the case using HYCOM, hence our choice to retain Mercator-Océan for the control configuration.

Fig. 21.2c illustrates another configuration test, which is to take account of the relative speed of water and air in the bulk formula for surface stress. When this modification is applied (filled circles) there is a modest but consistent shift toward higher skill, and therefore this change is being adopted in future model simulations.

As a final comment on model performance we compare the along-isobath component of velocity between the model and two NERACOOS moored current meters in the GOM (Figure 21.3). The blue line is the squared coherency in the time series, with faint red lines indicating 95% confidence limits. Where the lower limit falls to zero, the coherence is not statistically significant and the blue line is erased. At the western site, within the GOM coastal current, the time series are coherent across all timescales. At the eastern site in the Northeast Channel, variability in the mesoscale band (20–50 days period) is not coherent, which encourages us that the assimilation of observations (which are ample to capture this timescale) from current meters and coastal altimetry has the potential to bring model variability into mesoscale event-wise agreement with the data. The coherence in variability at high frequencies at both mooring sites is likely testament that the local meteorological forcing, which dominates ocean surface current response at these time scales, is sufficiently accurate.

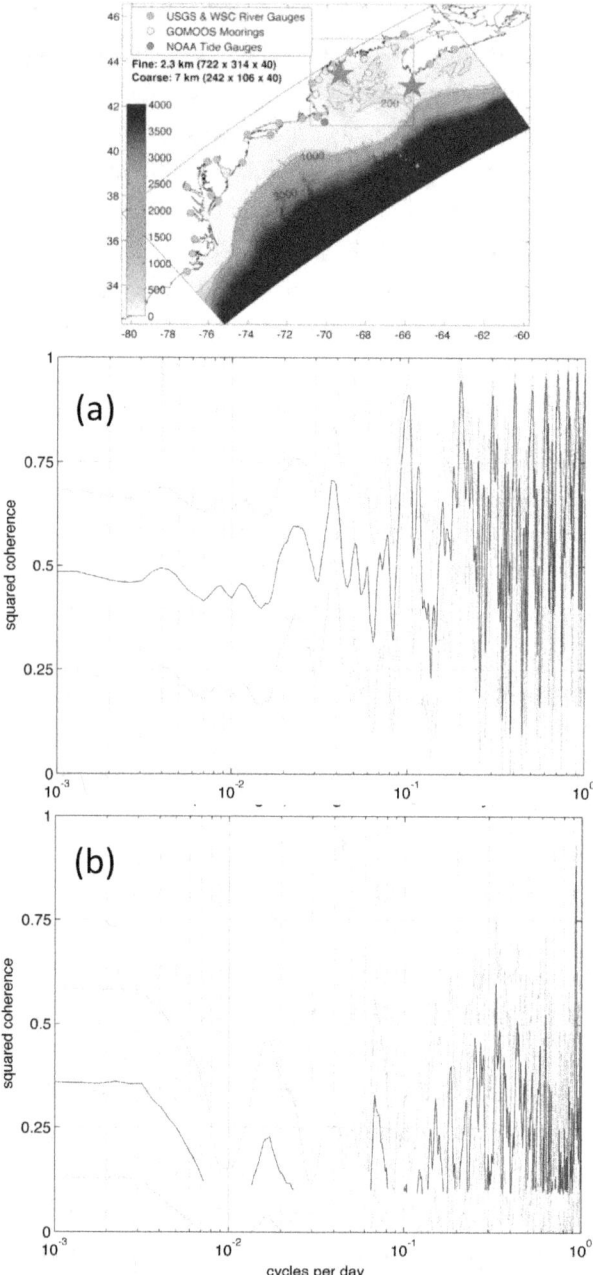

Figure 21.3. Squared coherency between along-isobath component of near-surface velocity and NERACOOS mooring data for (a) western site in GOM Coastal Current, and (b) eastern site in Northeast Channel entrance to GOM. Red stars in map show mooring locations.

Doppio Data Assimilation

4D-Var error hypothesis

4D-Var requires a prior estimate of the model background error covariance for initial, boundary and air-sea forcing conditions. ROMS specifies these as a univariate correlation matrix scaled by the square of standard deviations provided by the user. We computed background standard deviations from the mesoscale variability in a multi-year forward run (i.e., without data assimilation). The user also provides initial condition and boundary condition error de-correlation scales that determine the univariate correlation. In Doppio, we use 50 km in the horizontal, and 50 m in the vertical; surface forcing de-correlation scale is 100 km.

Observation errors are estimated from accepted practice with respect to the various observation platforms. For example, we would anticipate infrared satellite SST imagery to have the widely accepted expected error of 0.4 °C. Likewise, CODAR currents errors are nominally 0.1 m s^{-1}, but we inflate this by the mapping error returned by the optimal interpolation step that combines radial components into velocity vectors in CODAR data processing. CTD and current-meter errors for temperature and velocity, respectively, are expected to be substantially less than these. In practice, we find convergence of 4D-Var is aided by scaling these observation errors by the corresponding background error variance. This is an acknowledgment that our error model is imperfect, and most likely due to an incomplete account of representation error, which is the inability of the model to represent processes that are not captured by the model resolution or possibly aliased by the data sampling.

Observations for assimilation

Temperature and salinity

We have gone to extensive lengths to assemble the most comprehensive set possible of all observations of ROMS state variables – temperature, salinity, velocity and sea level – in the MAB and GOM. For in situ temperature and salinity, MARACOOS and NERACOOS are key providers on the MAB continental shelf shoreward of the shelf-break, and within the GOM. We augment these data streams with further in situ observations from the National Data Buoy Center, National Marine Fisheries Ecosystem Monitoring voyages, and data reported by in situ platforms of opportunity (Argo profiling floats, Expendable Bathythermographs [XBTs], drifting buoys, vessel underway thermo-salinograph). Contrasting the distribution of data accessible to us in near real time with the comparable data set assimilated in Mercator-Océan (Fig. 21.4) is readily apparent that our data assembly activity has very effectively harvested a wealth of information with the potential to significantly improve state estimation by data assimilation.

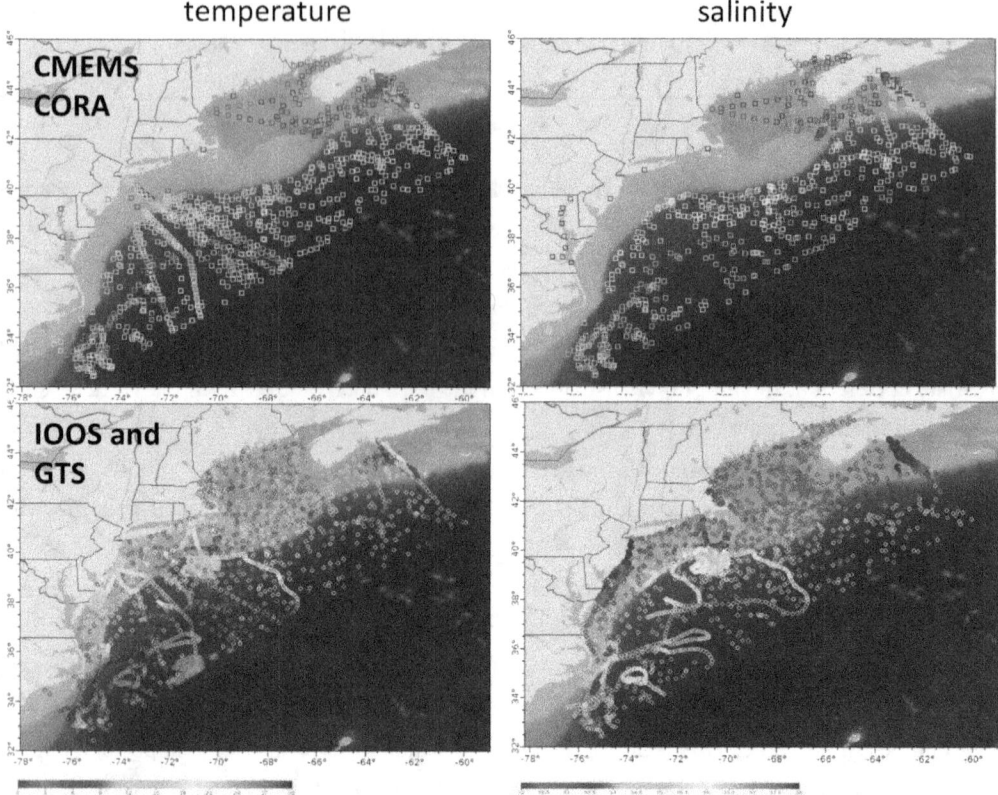

Figure 21.4. All sub-surface observations of (left) temperature and (right) salinity during 2015. Top row: Distribution of data in the CMEMS CORA database assimilated in Mercator-Océan. Bottom row: Distribution of data assembled for Doppio from U.S. IOOS data sources and the WMO Global Telecommunication System (GTS).

Though not presently available as near real time data streams, we have also expanded the data available in delayed mode (for retrospective reanalyses or skill assessment) by incorporating data from the eMOLT (Environmental Monitoring on Lobster Traps) project that returns bottom temperature data from sensors on lobster traps, water column and bottom temperature from the Northeast Cooperative Research Study Fleet Program derived from sensors mounted on fishing trawl doors, and water column temperatures (surface to 200 m depth) from sensor tags on loggerhead turtles that migrate throughout the estuaries of the MAB and across the continental shelf ocean. The wide geographic spread of these data is shown in Fig. 21.5.

The various observational assets we access, their approximate resolution, their latency, and the near real time sources from which we acquire the data operationally, are summarized in Table 21.1. From experience, we cannot overemphasize the value of more and varied sources of data in a coastal ocean analysis system.

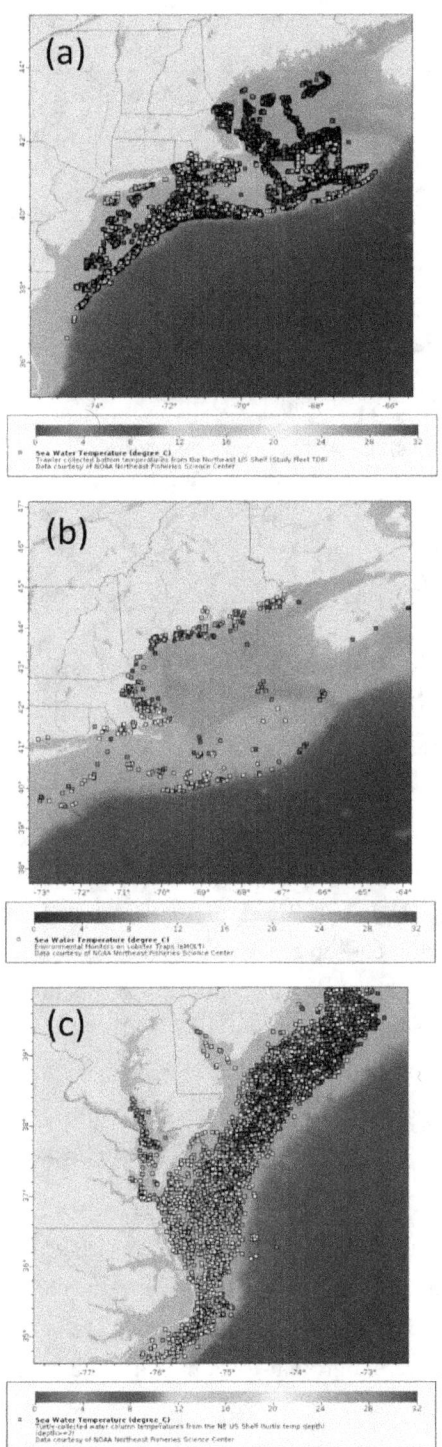

Figure 21.5. Observations of sub-surface temperature from sensors on novel observing platforms. (a) Northeast trawl fishing fleet. (b) Lobster traps. (c) Sea turtles. At publication these data are not accessible to the Doppio forecast system, but will soon become accessible in near real time through the NOAA Ocean Technology Transfer program for operational oceanography applications.

Observation type and platform	Source	Sampling frequency and resolution	Latency
AVHRR infrared SST	MARACOOS.org and NOAA CoastWatch	4 passes per day, 1 km	2 hr
GOES infrared SST	NOAA CoastWatch	hourly, 6 km	12 hr
AMSR2 and WindSat microwave SST	NASA JPL PO-DAAC	daily, 15 km	24 hr
SSH: 4 satellite altimeters: Jason, AltiKa, CryoSat and Sentinel-3 with coastal corrections	Radar Altimeter Database System at TU Delft	~1 pass daily in domain, ~4 km	4 hr
in situ T, S on GTS from National Data Buoy Center buoys, Argo floats, shipboard XBT, surface drifters	NOAA Observing System Monitoring Center (OSMC)	varies with platform	~12 hr
Surface currents from CODAR HF-radar	MARACOOS.org	hourly, 1km	4 hr
Glider T, S ~1-2 deployments per month in domain by MARACOOS	IOOS Glider Data Assembly Center	dense along trajectory	2 hr

Table 21.1. A summary of the observational data streams accessed for the ROMS Doppio near real time data assimilation system.

Where the observation sampling in space and/or time is higher than the model spatial resolution and model time step, observations should be combined to form "super-observations," a standard practice in data assimilation (Daley, 1991). Super-observations are data averages, weighted by inverse observation error, within chosen space and time bins. The formation of super-observations reduces data redundancy and poor conditioning of the cost function with respect to minimization.

The preparation of temperature and salinity data for ROMS 4D-Var from satellite and in situ platforms is relatively straightforward, as is their merger into super-observations. There are some challenges and subtleties, however, to the use of satellite altimeter sea level and HF-radar currents due to high frequency motions – principally tides. We detail these next.

Sea level and velocity

The Jason series of radar altimeter satellites measure sea surface height (SSH) along ~12 ground-tracks that traverse the Doppio domain with a 10-day repeat cycle. Adding in the other satellites in the altimeter constellation (presently CryoSat, AltiKa and Sentinel-3A), this coverage is complemented by a mix of different ground-track patterns and repeat cycles that combine to form a very comprehensive data set.

Historically, a great deal has been inferred about coastal ocean dynamics from the analysis of coastal sea level data from tide gauges, yet relatively little similar analysis has been conducted using altimetry. This is due in large measure to assertions that errors in altimeter data near the coast render the data unusable, yet significant progress has been made over the past decade in extending the validity of altimeter data to within a few kilometers of the coast by the appropriate application of altimeter radar range corrections and re-tracking of radar waveforms proximate to land (Cipollini et al., 2017; Vignudelli et al., 2011). This opens up to coastal oceanographers the opportunity to exploit the dynamical information content of so-called "coastal corrected altimeter" data.

In the MAB and GOM, altimeter data that would ordinarily be rejected by conventional quality control in coastal regimes can be reclaimed by judicious application of the data error flags and a revised wet tropospheric radar range correction (Feng and Vandemark, 2011). We extract 1 Hz along-track (approximately 6 km interval) Jason data from the Radar Altimeter Database System (rads.tudelft.nl) (Scharroo et al., 2013), making coastal corrections that retain data close to land (up to the 25-m isobath). These are to (i) use the European Centre for Medium-Range Weather Forecasts Wet Troposphere radar range correction in place of the onboard microwave radiometer correction that is contaminated by land within 50 km of the coast, and (ii) to ignore entirely the rain error flag that rejects numerous valid observations in the GOM. Data from CryoSat, AltiKa and Sentinel-3A are similarly downloaded and coastal-corrected for assimilation.

Our ROMS configuration simulates its own response to atmospheric pressure at the sea surface, so we do not make the dynamic atmosphere correction to altimeter range because that would be dynamically inconsistent and would inflate the model-data error.

In the coastal ocean where steep and variable bathymetry exacerbates uncertainty in the geoid and mean sea surface at short length scales (several tens of kilometers) there is an acute need to improve the precision of the mean dynamic topography (MDT) that is summed with altimeter sea level anomaly data to give an absolute dynamic topography (sea level above geoid) for assimilation that corresponds to the ROMS sea surface height prognostic variable.

Unfortunately, global MDT products such as the CNES-CLS13 MDT (Rio et al., 2014) (also generically referred to as "AVISO MDT" – Archiving, Validation and Interpretation of Satellite Oceanographic data – www.altimetry.fr) exhibit features in the MAB and GOM that oceanographers familiar with the locale recognize as unrealistic. These include contours of MDT strongly orthogonal to the coast that indicate landward geostrophic flow, some closed contours that imply isolated recirculation in shelf waters, and an intense boundary current adjacent to the coast of northern Virginia. Instead of AVISO MDT, we use a mean sea surface height computed by the climatological mean 4D-Var analysis (Levin et al., 2018) mentioned earlier in the context of bias correction of the open boundary condition data from Mercator-Océan. This approach has the added advantage that the mean sea level in the assimilated altimeter data is consistent with the mean dynamical balance of a free-running Doppio model.

MARACOOS CODAR systems began observing surface currents in the MAB in 2001, with the network reaching near complete coverage of the region by 2009. Radial component data from more than a dozen sites are gridded by optimal interpolation into a 6-km resolution vector velocity product with mapping error depending on the number, extent of overlap, and relative direction of the individual radial current observations (Roarty et al., 2010). In preparation for data assimilation, these data were further binned to 15 km resolution, but with velocities with large normalized optimal interpolation mapping errors ignored. The size of the bins was chosen to provide independent super-observations within the background error de-correlation scale.

Concern that small phase errors in the barotropic tide might dominate model-data misfit for sea level and velocity prompted us to apply pre-processing steps to reduce this possibility. Using harmonic analysis of long time series of CODAR data we de-tide those observations and replace the tidal variability with a signal computed from tidal analysis of a long free run of Doppio. Through this action it is our conjecture that the model-data misfit will not be dominated by tidal energy but instead will be due principally to dynamical responses that are within the scope to be modified by adjustment of the 4D-Var control variables. We adopt a similar approach in the pre-processing of altimeter sea level for assimilation, but where de-tiding of the observations is by application of the GOT4.10 harmonic tide altimeter range correction (Ray, 2013) in the extraction of data from the Radar Altimeter Database System.

This step is admittedly ad hoc, and has not been rigorously evaluated for its impact. There is the possibility that it carries little advantage because the tidal response in the free-running model is quite skillful.

A final point of detail in the handling of altimeter data is that early on in developing the prototype ESPreSSO system, when experimenting with assimilating satellite data only (SST and altimetry), we encountered a tendency for 4D-Var increments to excite unnaturally energetic surface gravity waves. Though unphysical to an oceanographer, these waves are nevertheless valid solutions to the ROMS governing equations. They can arise if not explicitly penalized in the cost function minimization. In the absence of subsurface temperature and salinity observations that would be inconsistent with simple gravity wave dynamics and discourage their excitation, and without the implementation of a thermal wind balance constraint (Weaver et al., 2005) on the multivariate component of the error covariances, we adopted a simple approach aimed at suppressing the generation of these waves. This was to repeat the satellite SSH observations one hour prior to, and one hour following, the actual observation time. For these "pseudo-observations" the ROMS harmonic tidal sea level at the appropriate phase is added to the de-tided satellite value. Augmenting the observational data set in this way has the effect of encouraging 4D-Var to choose a solution more in accord with slowly varying sub-tidal sea level dynamics than with gravity waves excited by an impulse in the model-data misfit.

Practical operational cycle

Daily forcing and observation data ingest

Our near real time 4D-Var analysis system runs daily. Each evening (U.S. local time) a set of automated scripts runs to acquire forecast meteorology and open boundary data, and conduct the various pre-processing steps noted above. The stream gauge data are also acquired nightly, inflated by the maximum covariance analysis, and extrapolated into the forecast interval by persisting the last observation. The scripts that execute these tasks employ a mix of software tools including Matlab, Python, NCO toolbox, and Perl. The daily timeline of data gathering, processing, analysis, and output, is illustrated in Fig. 21.6.

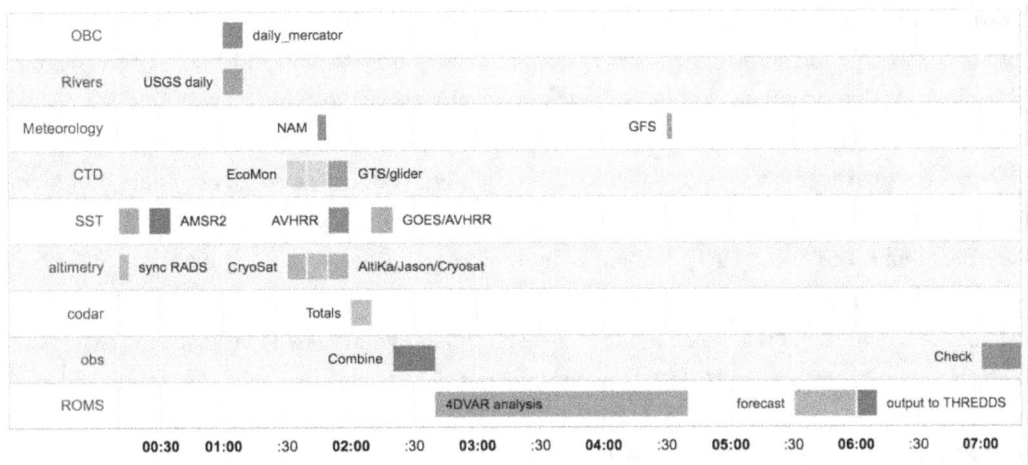

Figure 21.6. Schedule of data assembly steps that proceed daily via automated scripts to acquire open boundary conditions from Mercator-Océan, river discharge data from USGS and Water Service of Canada, metrological forcing data from NOAA NOMADS, in situ CTD data from numerous sources, satellite SST from infrared and microwave platforms, altimeter SSH (following local synchronization with the Radar Altimeter Database System database), and CODAR surface currents. The combine step forms super-observations. 4DVAR analysis takes approximately 155 minutes. All times are local U.S. Eastern Standard Time.

For some data, the latency from observation time to availability is predictable (e.g. polar orbiting satellites and HF-radar), while for others the schedule is more erratic. For these, remote data servers are polled a relatively short time in advance of the merger of common observation types into super-observations – effectively operating a "last chance" for inclusion strategy.

The data assimilation analysis step runs daily, but it uses three days of data. Data that are delayed additions to the database (more than 24 hours latency, but less than 60 hours) will therefore miss out on inclusion in the analysis at first, but could still subsequently enter the analysis on a following day. Therefore, all data acquisition queries are configured to request data for a full three days prior to analysis time.

Open data access and web services

The trend in the ocean science community toward providing access to data in a manner that follows the so-called FAIR data principles (findable, accessible, interoperable, reusable) (Wilkinson et al., 2016) has proven a great help to implementing the Doppio real-time forecast system. The vast majority of data we use are available via openly accessible web services such as THREDDS (Thematic Real-time Environmental Distributed Data Services) (Unidata 2018b) and ERDDAP (Simons 2018) and are formatted according to agreed conventions for metadata descriptions, e.g. the Climate-Forecast (CF) Conventions (http://cf-conventions.org) (Gregory, 2003). This convergence of standards makes it possible in many instances to re-use code with only minor modification to bring a new data set into the real-time data stream. Spatial and temporal sub-setting facilities in THREDDS and ERDDAP make software tools for data acquisition easily re-useable for new downscaling applications, requiring little more than the redefinition of the bounding box that encompasses the model domain.

Our experience is that searchable catalogs at data assembly centers such as NOAA CoastWatch (https://coastwatch.pfeg.noaa.gov/erddap/griddap) and the NASA Physical Oceanography Distributed Active Archive (PO.DAAC; https://podaac.jpl.nasa.gov) have made satellite data sets and services readily findable and accessible.

Near real time access to in situ data sets is less straightforward, especially coastal observations; applied coastal ocean modelers would enjoy greater access to existing data if there were more widespread embrace of the FAIR principles by the coastal ocean observing community. There is, however, a comprehensive near real time global aggregation of open-ocean in situ data that also encompasses many shelf regions, boundary currents and marginal seas. This is via the WMO Global Telecommunication System (GTS) that coordinates timely delivery of atmosphere and ocean observations to numerical weather prediction centers. By international convention (Resolution 40 of the Twelfth World Meteorological Congress, 1995), ocean physical observations (sea level, temperature, salinity and velocity) are considered to be meteorological data that may be acquired and exchanged without restriction due to maritime borders. Ocean observations in the GTS data stream are principally acquired beyond the continental shelf break using surface drifters, Argo profiling floats, or instrumentation mounted on or deployed by vessels participating in volunteer observing networks (XBT, XCTD, and underway thermo-salinograph). The data from shallow coastal waters that reach the GTS are predominantly from fixed moorings such as U.S. territorial sea observations from the NOAA data buoy network.

The CMEMS CORA (Cabanes et al., 2013) global data set of in situ ocean observations (CMEMS Product Identifier INSITU_GLO_TS_REP_OBSERVATIONS_013_001_b at https://marine.copernicus.eu) depicted in Fig. 21.4 that are assimilated in the Mercator-Océan model that Doppio uses for open boundary conditions comprises essentially the same data are as available in near real time via GTS, but with post-processing quality control and checking to reject bad profiles and apply delayed mode corrections.

Prior to launching the ROMS data assimilation step, the forcing and observation data assembly process (Fig. 21.6) concludes with the merger of observations of the same state variable into super-observations, for reasons noted previously. For satellite SST and SSH, super-observations are formed at the model grid resolution (here ~7 km) for each satellite pass. For in situ data, super-observations are grouped into 14 km spatial (2 model grid cells) and 12-minute time interval (2 model time steps) bins. These are then written to a single ROMS "obsfile" for 4D-Var analysis and assessment of the subsequent forecast, or to be used as verification data in freely running retrospective simulations for model experimentation such as presented earlier in the chapter. There are typically of order 200,000 independent super-observations in a three-day analysis interval.

Monitoring Doppio operational inputs

The daily process of assembling boundary condition and forcing inputs for Doppio, and aggregating observations for assimilation for skill assessment, requires continual monitoring to ensure continuity and integrity of the data streams. This is somewhat automated, since batch scripts and the ROMS model itself will register errors when web services are unavailable or file creation fails, but other failure modes are more subtle and require the vigilance of an operator to detect.

The ROMS forcing information (meteorology, river sources, etc.) netCDF file formats (Unidata, 2018c) are easily aggregated over time using THREDDS and made accessible in a graphical browse format using an ERDDAP Slide Sorter interface. Fig. 21.7 shows an example of a Slide Sorter web page configured to display the last day of NAM meteorology data, and the last three days of river discharge data. The date stamp on the NAM plot quickly reveals whether forecast data are present – which may not be so if NCEP services were unavailable when this job executed in the schedule (Fig. 21.6); the river discharge plot would display a warning that the query produced no matching results if the data acquisition failed because, for example, a gauge malfunctioned. The plots will reveal numerically valid but geophysical unreasonable data to an operator who takes time to browse the display. Any such adverse outcome would alert an operator that the Doppio system lacked valid inputs required to deliver a complete forecast.

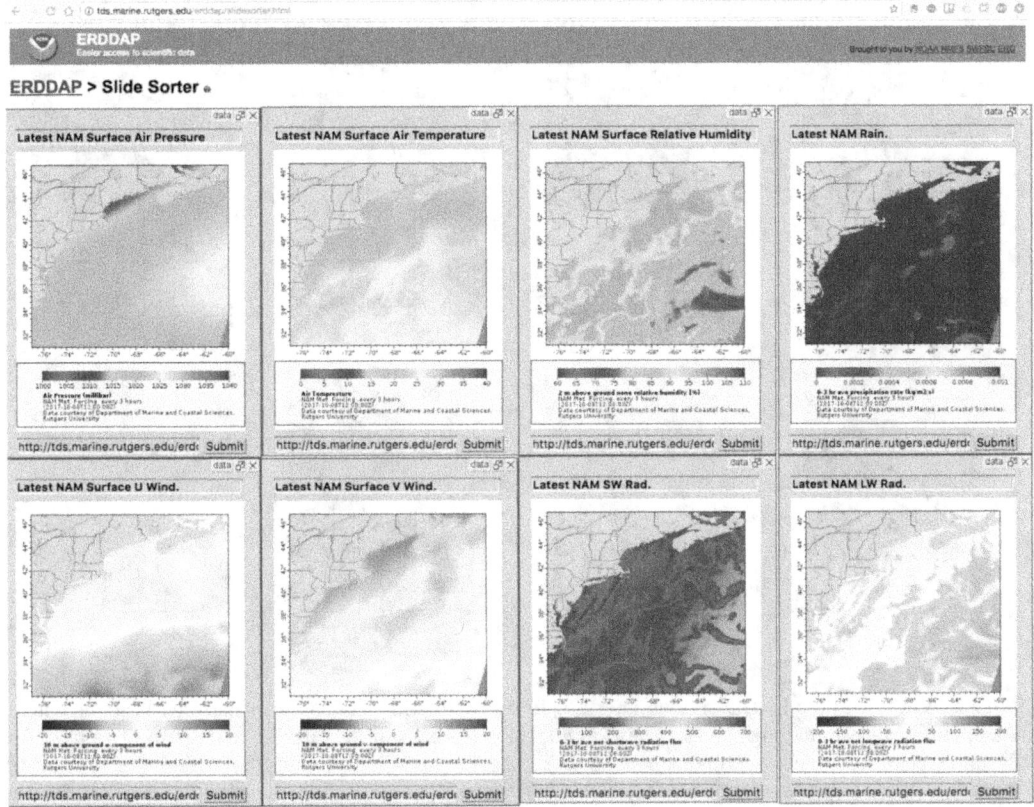

Figure 21.7. Partial view of the ERDDAP web interface that serves as an operator control panel to visualize inputs (meteorology and rivers) for the Doppio forecast system.

The ROMS netCDF *obsfile* format holds the observation geo-location and acquisition time, and also tracks the provenance of the observing platform and data provider. Making an aggregation of these files accessible with ERDDAP provides a convenient service to monitor the data entering the DA analysis. Fig. 21.8 shows an ERDDAP Slide Sorter web page configured to display subsets of the data for an example three-day analysis interval. In separate panels, user-defined optional constraints restrict views into the *obsfile* to highlight only subsurface observations, or certain provenance codes, to quickly reveal the presence or absence of anticipated data. Typical checks of the data stream we might use this system for include: If an autonomous underwater glider vehicle is known to have been recently deployed in the region it can be checked whether those data are entering the assimilation data stream via GTS; distinct provenance codes identify each altimeter satellite in the constellation, so it can be checked whether all anticipated data are flowing through the ground segment of each mission to the Radar Altimeter Database System; SST from different sensors and satellites can be visually browsed for consistency or noise that may be indicative of incomplete cloud-clearing.

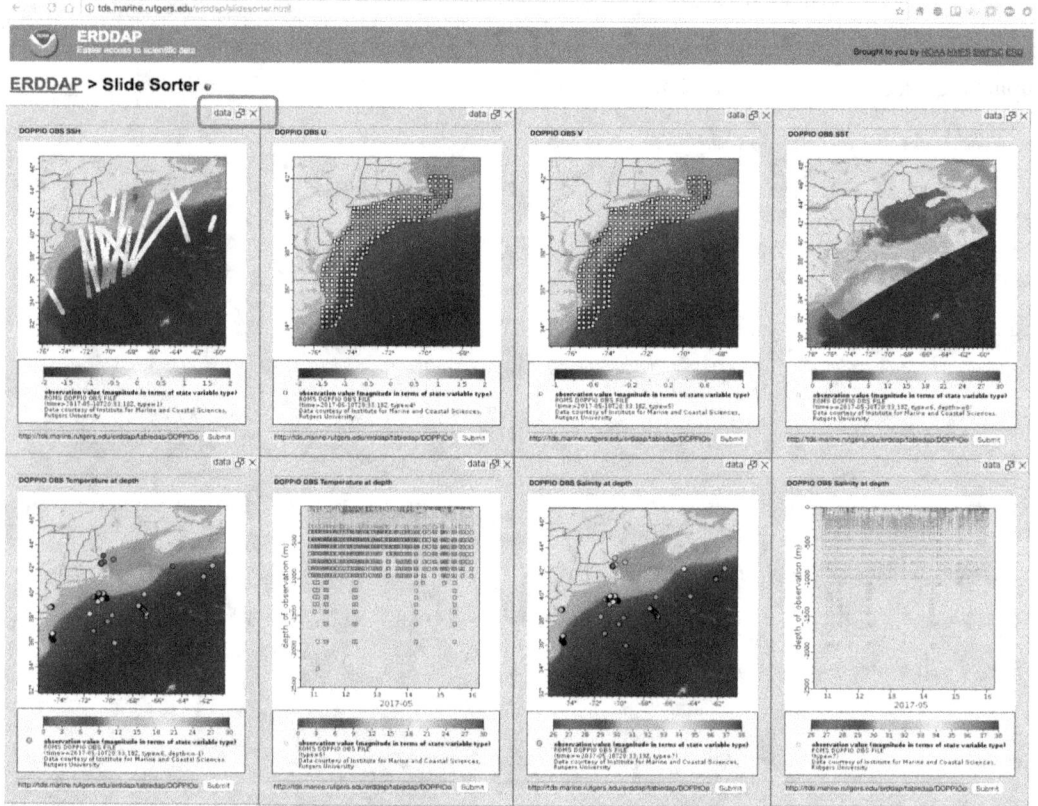

Figure 21.8. View of the ERDDAP web interface for browsing the data base of super-observations entering the Doppio data assimilation analysis. The example shows 6 days of data during May 2015. Top row, from left: altimeter SSH, u- and v-component of CODAR currents, and all satellite SST. Bottom row, from left: Sub-surface temperature as geo-locations and as a function of time and depth, all salinity as geo-locations and as a function of time and depth. The *data* button highlighted in the top left panel launches a data set browse page similar to Figure 21.9.

The Slide Sorter views are not previously created static images, but rather are generated anew from data in the *obsfile* when the page is loaded. An operator can modify the content displayed by selecting an individual browse plot to launch the underlying ERDDAP page in a new window. This allows customization of the geospatial search (i.e., zooming, or depth range constraints), constraining the display by data provenance, or modification of the image appearance (symbols, colors, etc.). This Slide Sorter view was easily configured to monitor near real time operational data streams for Doppio because ERDDAP accepts "now" as time query search, as in requesting "time>now-3days".

A further useful feature of ERDDAP is the ability to immediately download the displayed data in formats readable by a wide range of scientific, GIS and spreadsheet software via a RESTful (Representational State Transfer) interface. This facilitates further analysis, such as the calculation of data statistics, and comparisons to independent observations or other models.

In an ERDDAP data view or download request the dataset identifier, variable name, search constraints and plot commands are fully described in the browser URL, so it can be bookmarked, shared or scripted via the UNIX "curl" command. These and many other features of ERDDAP are described by Simons (2018) and documented in web links included on every ERDDAP web page.

Analysis/forecast cycle – Doppio 4D-Var

Once the data ingest and preparation steps are complete, the ROMS 4D-Var sequence of inner and outer loops iterates toward an optimal solution. Upon convergence, the time varying 3-D ROMS solution through the three-day analysis window represents the maximum likelihood estimate of the ocean state during that interval. The conditions at the end of interval, notionally a "nowcast", become the initial conditions for a single 72-hour forecast. The forecast horizon of 72 hours is set by the scope of the NAM meteorological forecast.

The 4D-Var analysis uses two outer and eight inner loops. Our experience with Doppio is that further iterations typically accomplish little in reducing the cost function. For this model grid of 240 by 104 horizontal points and 40 vertical levels, with 720 time steps in the three-day analysis interval, execution of the iterative 4D-Var analysis typically takes 155 minutes of walk clock time on 12 cores of a modest UNIX computer (3.5 GHz Intel Xeon processor). The subsequent forecast takes 15 minutes to complete on the same machine.

Operational system outputs

ROMS output files conform to CF-Conventions and the Common Data Model (CDM) API (Unidata, 2018a) that describes semantic layers for coordinate systems and scientific data types common in geophysics. Libraries for Python, Matlab, and many other scientific analysis software tools support the CF and CDM data models, and this standardization in output format greatly facilitates the inter-operability of the Doppio modeling system with partners in MARACOOS, IOOS and the broader user community when model outputs are made available via open access web services such as THREDDS.

We serve Doppio model output on the full ROMS 3-D grid at 1-hourly intervals of simulated time using the Forecast Model Run Collection (FMRC) facility in THREDDS. Every daily run generates three days of analysis and three days of forecast output, so there are multiple realizations of any given date. We choose to serve a FMRC "best time series" aggregation formulated as the concatenation of the central 24 hours of each three-day analysis, appended with the final day of the latest analysis and the forecast. With a data URL end-point that does not change, this FMRC "best time series" therefore provides users with a continuous, hourly, monotonic 3-D ocean state retrospective estimate, plus the forecast of the day. Users need only specify date and time in a data request, and FMRC will return an unambiguous result.

The "best time series" is the mostly widely used FMRC product, but FMRC preserves each individual cycle of analysis and forecast in its entirety and these are equally easily accessed as part of the THREDDS collection. Forecast skill can therefore be evaluated in posterior analyses targeted at appraising performance with respect to user-specific metrics, or against newly acquired delayed mode data that were not available to the data assimilation.

For the model grid dimensions noted above, 1-hourly interval 3-D model state snapshots and 24-hour average files add 7.2 Gb of output to the THREDDS collection each day. In addition, we add 6.1 Gb each day to an offline archive comprising the observation files and saved boundary conditions and forcing files for future re-runs and re-evaluations of the system.

The Doppio forecast is harvested by MARACOOS for ingest to the U.S. Coast Guard Environmental Data Server (EDS) that provides guidance to the USCG Search and Rescue Optimal Planning System (SAROPS), and by the National Marine Fisheries Service of NOAA for guidance on bottom temperatures in the MAB that strongly influence regional fisheries. MARACOOS also make views of the surface current and bottom temperature available through their *OceansMap* graphical browse service (http://oceansmap.maracoos.org), which allows easy qualitative comparison to many other observations and model-based products.

Supplementary special output products are three-day predictions of the drifting trajectories of mobile observing assets starting from their last reported position. These are computed for skill assessment versus the observed path of passive drifters, and to have advice ready in real time should an autonomous underwater glider vehicle become disabled and float passively at the surface; the prediction aids the mobilization of a vessel to rendezvous with a disabled vehicle for recovery. In addition, each day we predict drift trajectories originating at the locations of fixed moorings in the Ocean Observatories Initiative Pioneer Coastal Array in anticipation of the break out of a mooring; again, to provide guidance in mobilizing a recovery response.

Skill assessment – Prototype near real time Doppio output

The Doppio system described here was presented to students and lecturers at the GODAE International School on "New Frontiers in Operational Oceanography" in October 2017. At that time, the system operated in near real time on a "best effort" daily basis, generating ocean state analyses and 72-hour forecasts on most days, but accepting that occasional failures of either the data assimilation or forecast run are inevitable while in prototype.

Forecast failures typically stem from incomplete download of the meteorological forcing or river flow data – problems that can be identified but not remedied by the ERDDAP monitoring described earlier. Failures of the data assimilation analysis through instability of the 4D-Var iterations are rare, with most difficulties arising due to insufficiently quality-controlled data. Incomplete forecasts can also occur due to loss of power or network given that the system runs on a university computing infrastructure designed for research and education. Whatever the cause, once remedied the analysis and/or "forecast" (now actually a hindcast) are restarted so that the data can be added to the THREDDS FMRC catalog to complete best time series aggregation for later evaluation or applications.

The Doppio system could be hardened to forcing data interruptions by instituting redundant systems that enable a fail-over to alternative data streams or a fall back to a statistical model that combines climatology and persistence of prior valid data. The Mid-Atlantic Regional Association Coastal Observing Systems (MARACOOS) ROMS group may pursue such enhancements in the process of transitioning the Doppio system to near real-time sustained operation.

However, whatever steps are taken, vulnerabilities in the system will remain as long as there are dependencies on computing and cyberinfrastructure environments that have the potential to go down, and data streams that are operating on open research networks. Given these constraints, a system such as Doppio is unlikely to ever achieve "operational" status as would be defined by a national meteorological agency such as the National Centers for Environmental Protection (NCEP). Nevertheless, user communities exist for near real-time coastal ocean information products from research operators, whether they are model-based analyses or the underlying data stream themselves.

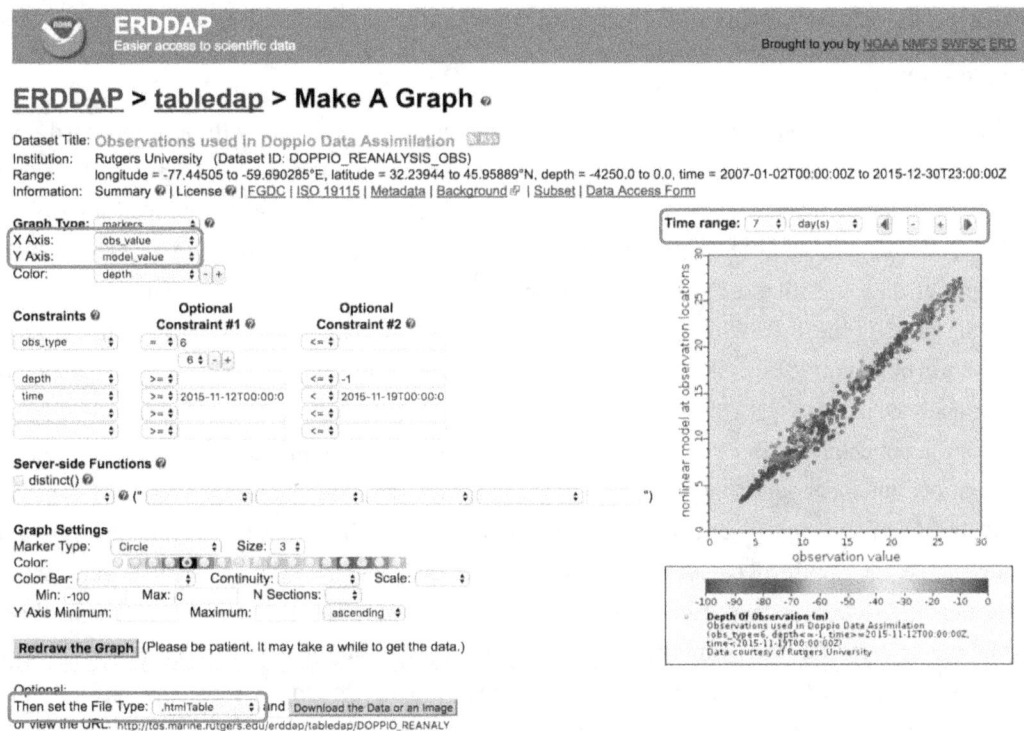

Figure 21.9. View of the ERDDAP web interface for browsing the data base of Doppio super-observations, but here set so as to display the ROMS 4DVAR analysis versus observation value as a scatter plot via the x-axis/y-axis controls (highlighted in the upper left red box). Also highlighted are the controls to select file type (lower left red box), where the RESTful interface offers netCDF, CSV, Matlab etc. download options, and the time range slider (upper right red box) which enables quick browse forward and backward through weekly time intervals.

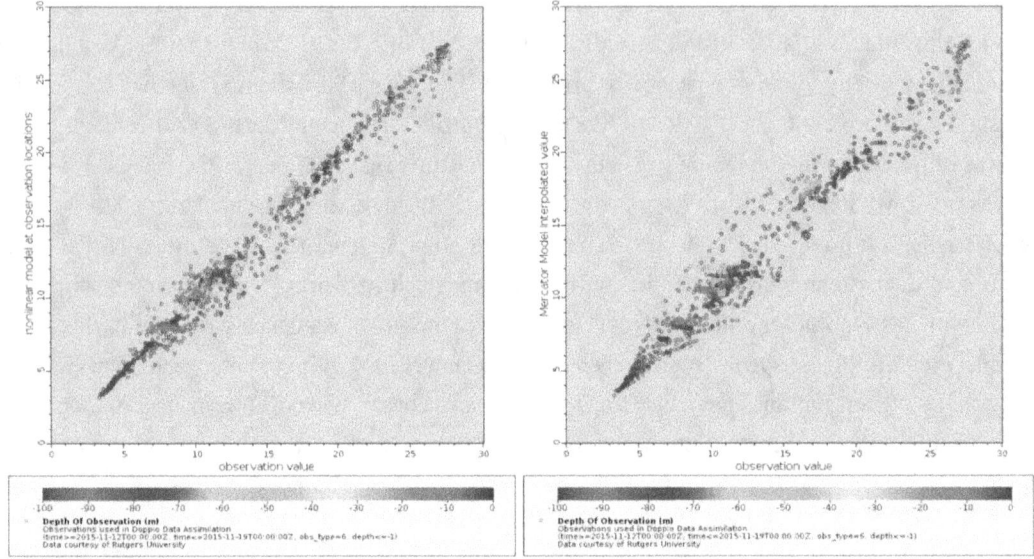

Figure 21.10. Model versus observed salinity comparison visualized with ERDDAP (from the FMRC best time series aggregation) for two 3-day analysis cycles in December 2015. Left: for Doppio ROMS. Right: for Mercator-Océan.

In a previous section, we showed that a free running Doppio model with bias-corrected open boundary condition data from Mercator-Océan retained useful skill throughout the model domain. Next we evaluate whether assimilation of the satellite and in situ data sets assembled for Doppio improves ocean state analyses compared to the free model. While this is to be expected – it is the principle of data assimilation to bring the analysis into agreement with data – what we wish to illustrate most is the ease with which the ROMS output formats and ERDDAP facilitate rapid browse and quantitative assessment of models and data for appraisal of the system in the context of operational oceanography.

It is fundamental to 4D-Var to minimize the innovation, this being the difference between the observations and the model state interpolated to the positions and times where the observations were made. In ROMS, this one-to-one match-up of model state to each observation is retained as output in a separate file. With ERDDAP, we virtually aggregate these two data sets (observations and corresponding ROMS estimate) and augment them with the independent Mercator-Océan analyses (without bias correction) also interpolated to the same observation locations. The merger of these products is accomplished in the back-end to ERDDAP; it does not require significant reformatting or re-writing of any of the ROMS output files.

It is then straightforward within ERDDAP to create scatter plots of observation value versus model, and to further focus the comparison using geospatial or time constraints or according to other metadata such as observing platform provenance. The RESTful interface enables download of the match-up data sets for further analysis. Some of these features are highlighted in Figure 21.9, which shows the ERDDAP interface one of our users would see.

Figure 21.10 shows examples of using this interface to quickly compare ROMS to observations and encode the plotted values according to observation depth or provenance. Comparison is also made to Mercator-Océan, from which it is evident that Doppio is much closer to the observations, though it must be recalled Mercator-Océan does not assimilate the vast majority of these data (Figure 21.4).

In closing, we note that the Doppio system as described here is still in prototype. Aspects of the system configuration may change before it enters sustained near real time operation, such as the background and observation error hypothesizes, data quality control and pre-processing practices, and the meteorological and river discharge inputs. The model and data browse interfaces described above are valuable for the way they accelerate the model prototype and update cycle by facilitating rapid qualitative and quantitative assessment of configuration changes and system performance within an operational oceanography environment.

Summary

We have described the configuration and operation of a modeling system that downscales output from a global GODAE forecast model in order to provide skillful estimates of ocean circulation in coastal, shelf and adjacent deep ocean waters of the northeast U.S. The Doppio modeling system is designed to assist the MARACOOS and NERACOOS Regional Associations of U.S. IOOS in the near real time delivery of coastal ocean information products in support of maritime safety, the marine economy, and the health and sustainable use of the coastal environment and coastal marine living resources.

Key elements of the system are (i) a regional ROMS model encompassing major estuaries and shallow coastal waters and extending far enough offshore to enable representation of oceanic mesoscale variability that drives coastal circulation, (ii) accurate surface meteorological forcing from the best available operational forecast of the national meteorological agency, (iii) attention to fully representing the buoyancy input from coastal river inflows, (iv) pre-processing to decrease biases in open boundary conditions by reference to a local, high resolution ocean hydrographic climatology, (v) assembly of a comprehensive suite of remote and in situ regional observations from all available platforms, and (vi) assimilation of these observations by 4D-Var.

Specifically, we have noted the essentials of our choices in configuring ROMS for the dynamical regime of the MAB and GOM region, and provided a brief overview of ROMS 4D-Var and how we have configured it for Doppio. At some length we have described the many observing platforms we access in near real time for assimilation prior to forecasting, and in delayed mode for further model evaluation. We have taken pains to detail many of the pre-processing steps required to adapt data streams typically utilized in mesoscale operational oceanography to the coastal environment. These details are documented for users who might wish to emulate our efforts and develop similar GODAE model downscaling system for other coastal regions globally.

To efficiently implement the many data ingest and model output steps that are part of Doppio, we make extensive use of web services that embrace community conventions for metadata descriptions and enable open access with geospatial searching and sub-setting capabilities, principally THREDDS and ERDDAP. When the providers of observations follow FAIR principles for serving their data they can be quickly incorporated into our near real time operations with minimal effort.

We have illustrated a number of instances of ERDDAP interfaces to observations and model outputs that enable rapid browse and data access and are readily customized for particular user purposes. We consider ERDDAP a transformational technology that through the versatility of the RESTful interface for data queries and download makes themed collections of observations and full model output data sets readily accessible to novice users with a minimum of training and little ongoing intervention on the part of the data providers and model operators.

The data distribution plots in Fig. 21.4 reveal that, on this coastline at least, there exists a substantial data stream of observations that are available in near real time yet do not reach the global GODAE systems. It remains an open question the extent to which the skill of analyses from Mercator-Océan and similar systems might rival Doppio were they able to access this more extensive suite of observations. But even if this were so, there remains a strong case to undertake region specific downscaling with data assimilation endeavors such as Doppio in order to better link local coastal ocean knowledge to end users through communities such as the IOOS Regional Associations that can be more responsive to local stakeholder needs. A challenge for the existing Task Teams of GODAE, and a potential new frontier for a future GODAE School, is how outputs from the many near real time research systems like Doppio that are emerging globally might be communicated back to global GODAE and used to augment their skill locally through merged multi-model ensembles.

Acknowledgements

We thank Olmo Zavala, Peng Zhan, Xueming Zue and Misha Zujev for helpful comments on the manuscript. Grants NA11NOS0120038 and NA16NOS0120020 from the National Oceanic and Atmospheric Administration fund the implementation of ROMS for MARACOOS. Other grants from the Office of Naval Research, NASA, NSF and NOAA have contributed to elements of the Doppio system development, and to the ROMS model enterprise in general. The ROMS project benefits from the collaboration of the entire international ROMS user community at the www.myroms.org User Portal.

References

Bell, M., A. Schiller, P. Le Traon, N. SMith, E. Dombrowsky, and K. Wilmer-Becker, 2015: An introduction to GODAE OceanView. Journal of Operational Oceanography. Journal of Operational Oceanography, 8 (sup1), s2-11, doi:10.1080/1755876X.2015.1022041.

Blayo, E., and L. Debreu, 2006: Nesting Ocean Models. Ocean Weather Forecasting: An integrated view of oceanography, E. P. Chassignet, and J. Verron, Eds., Springer, 127-146.

Boyer, T. P., J. I. Antonov, O. K. Baranova, H. E. Garcia, D. R. Johnson, R. A. Locarnini, A. V. Mishonov, T. D. O'Brien, D. Seidov, I. V. Smolyar, and M. M. Zweng, 2009: World Ocean Database 2009, S. Levitus, Ed., NOAA Atlas NESDIS 66, U.S. Government Printing Office, Washington, D.C., 216 pp., DVDs.

Bye, J. A. T., R. A. Heath, and T. W. Sag, 1979: A Numerical Model of the Oceanic Circulation around New Zealand. Journal of Physical Oceanography, 9, 892-899.

Cabanes, C., A. Grouazel, K. von Schuckmann, M. Hamon, V. Turpin, C. Coatanoan, F. Paris, S. Guinehut, C. Boone, N. Ferry, C. de Boyer Montégut, T. Carval, G. Reverdin, S. Pouliquen, and P. Le Traon, 2013: The CORA dataset: validation and diagnostics of in-situ ocean temperature and salinity measurements. Ocean Science, 9, 1-18, doi:10.5194/os-9-1-2013.

Chapman, D., 1985: Numerical treatment of cross-shelf open boundaries in a barotropic coastal ocean model. Journal of Physical Oceanography, 15, 1060-1075.

Cipollini, P., J. Benveniste, F. Birol, J. Fernandes, E. Obligis, M. Passaro, P. T. Strub, G. Valladeau, S. Vignudelli, and J. Wilkin, 2017: Satellite Altimetry in Coastal Regions. Satellite altimetry over oceans and land surfaces, D. Stammer, and A. Cazenave, Eds., CRC Press, 343-380.

Courtier, P., J.-N. Thépaut, and A. Hollingsworth, 1994: A strategy for operational implementation of 4DVAR using an incremental approach. Quarterly Journal of the Royal Meteorological Society, 120, 1367-1388.

Csanady, G., 1976: Mean circulation in shallow seas. Journal of Geophysical Research, 81, 5389-5399.

Daley, R., 1991: Atmospheric Data Analysis. Cambridge University Press, 472 pp.

Dalyander, P. S., B. Butman, C. R. Sherwood, R. P. Signell, and J. Wilkin, 2013: Characterizing wave- and current--induced bottom shear stress: U.S. Middle Atlantic continental shelf. Continental Shelf Research, 52, 73-86, doi:10.1016/j.bbr.2011.03.031.

Drévillon, M., R. Bourdallé-Badie, C. Derval, J. M. Lellouche, E. Rémy, B. Tranchant, M. Benkiran, E. Greiner, S. Guinehut, N. Verbrugge, G. Garric, C. E. Testut, M. Laborie, L. Nouel, P. Bahurel, C. Bricaud, L. Crosnier, E. Dombrowsky, E. Durand, N. Ferry, F. Hernandez, O. Le Galloudec, F. Messal, and L. Parent, 2008: The GODAE/Mercator-Ocean global ocean forecasting system: results, applications and prospects. Journal of Operational Oceanography, 1, 51-57, doi:10.1080/1755876X.2008.11020095.

Fairall, C., E. Bradley, J. Hare, A. Grachev, and J. Edson, 2003: Bulk parameterization of air–sea fluxes: Updates and verification for the COARE algorithm. Journal of Climate, 16, 571-591.

Feng, H., and D. Vandemark, 2011: Altimeter Data Evaluation in the Coastal Gulf of Maine and Mid-Atlantic Bight Regions. Marine Geodesy, 34, 340-363, doi:10.1080/01490419.2011.584828.

Flather, R. A., 1976: A tidal model of the northwest European continental shelf. Mem. Soc. Roy. Sci. Liege, Ser. 6, 10, 141-164.

Fleming, N., and J. Wilkin, 2010: MOCHA: A 3-D climatology of the temperature and salinity of the Middle Atlantic Bight. Eos, Trans. Amer. Geophys. Union, 91, Abstract PO35G-08.

Fleming, N. E., 2016: Seasonal and spatial variability in temperature, salinity and circulation of the Middle Atlantic Bight. PhD, 336 pp, Rutgers University, New Brunswick, NJ, doi:10.7282/T3XW4N4M.

Fratantoni, P., and R. Pickart, 2003: Variability of the shelf break jet in the Middle Atlantic Bight: Internally or externally forced. J. Geophys. Res, 108, 3166.

Gawarkiewicz, G., K. Brink, F. Bahr, R. Beardsley, M. Caruso, J. Lynch, and C. Chiu, 2004: A large-amplitude meander of the shelfbreak front during summer south of New England: observations from the Shelfbreak PRIMER experiment. Journal of Geophysical Research Oceans, 109, C03006, doi:10.1029/2002JC001468.

Gregory, J., 2003: The CF Metadata standard. CLIVAR Exchanges, 8, 4.

Gürol, S., A. T. Weaver, A. M. Moore, A. Piacentini, H. Arango, and S. Gratton, 2014: Preconditioned minimization algorithms for variational data assimilation with the dual formulation. Quarterly Journal of the Royal Meteorological Society, 140, 539-556, doi:10.1002/qj.2150.

Hu, J., K. Fennel, J. Mattern, and J. Wilkin, 2012: Data assimilation with a local Ensemble Kalman Filter applied to a three-dimensional biological model of the Middle Atlantic Bight. Journal of Marine Systems, 94, 145-156, doi:10.1016/j.jmarsys.2011.11.016.

Huffman, G. J., D. T. Bolvin, E. J. Nelkin, D. B. Wolff, R. F. Adler, G. Gu, Y. Hong, K. P. Bowman, and E. F. Stocker, 2007: The TRMM multisatellite precipitation analysis (TMPA): Quasi-global, multiyear, combined-sensor precipitation estimates at fine scales. Journal of Hydrometeorology, 8, 38-55.

Kennedy, A. D., X. Dong, B. Xi, S. Xie, Y. Zhang, and J. Chen, 2011: A Comparison of MERRA and NARR Reanalyses with the DOE ARM SGP Data. Journal of Climate, 24, 4541-4557.

Large, W. G., J. C. McWilliams, and S. C. Doney, 1994: Oceanic vertical mixing: A review and a model with a nonlocal boundary layer parameterization. Reviews of Geophysics, 32, 363-403.

Lentz, S., 2008: Observations and a Model of the Mean Circulation over the Middle Atlantic Bight Continental Shelf. Journal of Physical Oceanography, 38, 1203-1221, doi:10.1175/2007JPO3768.1.

Levin, J., J. Wilkin, N. Fleming, and J. Zavala-Garay, 2018: Mean circulation of the Mid-Atlantic Bight from a climatological data assimilative model. Ocean Modelling, 128, 1-14, doi:10.1016/j.ocemod.2018.05.003.

Lin, Y.-T., A. Newhall, G. R. Potty, and J. H. Miller, 2017: A preliminary numerical model of three-dimensional underwater sound propagation in the Block Island Wind Farm area. Journal of the Acoustical Society of America, 141, 3933, doi:10.1121/1.4989146.

Linder, C., and G. Gawarkiewicz, 1998: A climatology of the shelfbreak front in the Middle Atlantic Bight. Journal of Geophysical Research, 103, 18,405-418,423.

Mannino, A., S. Signorini, M. Novak, J. Wilkin, M. Friedrichs, and R. Najjar, 2016: Dissolved organic carbon fluxes in the Middle Atlantic Bight: An integrated approach based on satellite data and ocean model products. Journal of Geophysical Research Biogeosciences, 121, doi:10.1002/2015JG003031.

Marchesiello, P., J. McWilliams, and A. Shchepetkin, 2001: Open boundary conditions for long-term integration of regional oceanic models. Ocean Modelling, 3, 1-20.

Mason, E., J. Molemaker, A. Shchepetkin, F. Colas, J. McWilliams, and P. Sangrà, 2010: Procedures for offline grid nesting in regional ocean models. Ocean Modelling, 35, 1-15.

Mellor, G. L., and T. Yamada, 1982: Development of a Turbulence Closure Model for Geophysical Fluid Problems. Reviews of Geophysics and Space Physics, 20, 851-875.

Mesinger, F., G. DiMego, E. Kalnay, K. Mitchell, P. E. Shafran, W, D. Jovic, J. Woollen, E. Rogers, E. E. Berbery, M, Y. Fan, R. Grumbine, W. Higgins, H. Li, Y. Lin, G. Manikin, D. Parrish, and W. Shi, 2006: North American Regional Reanalysis. Bulletin of the American Meteorological Society, 87, 343-360.

Miles, T., G. Seroka, and S. Glenn, 2017: Coastal ocean circulation during Hurricane Sandy. Journal of Geophysical Research, 122, 7095-7114, doi:10.1002/2017JC013031.

Miles, T., G. Seroka, J. Kohut, O. Schofield, and S. Glenn, 2015: Glider observations and modeling of sediment transport in Hurricane Sandy. Journal of Geophysical Research, 120, 1771-1791, doi:10.1002/2014JC010474.

Moore, A. M., H. Arango, G. Broquet, B. Powell, A. T. Weaver, and J. Zavala-Garay, 2011a: The Regional Ocean Modeling System (ROMS) 4-dimensional variational data assimilations systems, Part I - System overview and formulation. Progress in Oceanography, 91, 34-49.

Moore, A. M., H. Arango, G. Broquet, C. Edwards, M. Veneziani, B. Powell, D. Foley, J. Doyle, D. Costa, and P. Robinson, 2011b: The Regional Ocean Modeling System (ROMS) 4-dimensional variational data assimilations systems, Part II – Performance and application to the California Current System. Progress in Oceanography, 91, 50-73.

——, 2011c: The Regional Ocean Modeling System (ROMS) 4-dimensional variational data assimilations systems, Part III - Observation impact and observation sensitivity in the California Current. Progress in Oceanography, 91, 74-94.

Mountain, D., 2003: Variability in the properties of Shelf Water in the Middle Atlantic Bight, 1977–1999. Journal of Geophysical Research, 108, 3014.

Mukai, A. Y., J. J. Westerink, R. A. Luettich, and D. Mark, 2002: Eastcoast 2001, A tidal constituent database for the western North Atlantic, Gulf of Mexico and Caribbean Sea. Tech. Rep. ERDC/CHL TR-02-24, 196 pp.

Ray, R., 2013: Precise comparisons of bottom-pressure and altimetric ocean tides. Journal of Geophysical Research, 118, 4570-4584, doi:10.1002/jgrc.20336.

Ridgway, K. R., J. R. Dunn, and J. L. Wilkin, 2002: Ocean interpolation by 4-dimensional weighted least squares: Application to the waters around Australasia. Journal of Atmospheric and Oceanic Technology, 19, 1357-1375.

Rio, M.-H., S. Mulet, and N. Picot, 2014: Beyond GOCE for the ocean circulation estimate: Synergetic use of altimetry, gravimetry, and in situ data provides new insight into geostrophic and Ekman currents. Geophysical Research Letters, 41, 8918-8925, doi:10.1002/2014GL061773.

Roarty, H., S. Glenn, J. Kohut, D. Gong, E. Handel, E. Rivera Lemus, T. Garner, L. Atkinson, C. Jakubiak, W. Brown, M. Muglia, S. Haines, and H. Seim, 2010: Operation and application of a regional high frequency radar network in the Mid Atlantic Bight. Marine Technology Society Journal, 44, 133-145.

Scharroo, R., E. Leuliette, J. Lillibridge, D. Byrne, M. Naeije, and G. Mitchum, 2013: RADS: Consistent multi-mission products. in Proc. of the Symposium on 20 Years of Progress in Radar Altimetry, Venice, 20-28 September 2012, Eur. Space Agency Spec. Publ., ESA SP-710, 4 pp.

Seroka, G., T. Miles, Y. Xu, J. Kohut, O. Schofield, and S. Glenn, 2017: Rapid shelf-wide cooling response of a stratified coastal ocean to hurricanes. Journal of Geophysical Research, 122, 4845-4867, doi:10.1002/2017JC012756.

Shchepetkin, A., and J. McWilliams, 2005: The regional oceanic modeling system (ROMS): a split-explicit, free-surface, topography-following-coordinate oceanic model. Ocean Modelling, 9, 347-404.

Shchepetkin, A. F., and J. C. McWilliams, 2009: Computational Kernel Algorithms for Fine-Scale, Multiprocess, Longtime Oceanic Simulations. Handbook of Numerical Analysis, Computational Methods for the Atmosphere and the Oceans, 14, 121-183, doi:10.1016/S1570-8659(08)01202-0.

Simons, R. A.: ERDDAP, NOAA/NMFS/SWFSC/ERD, Monterey, CA. [Available online at https://coastwatch.pfeg.noaa.gov/erddap/information.html.]

Stewart, R., W. Wollheim, A. Miara, C. Vorosmarty, B. Fekete, R. Lammers, and B. Rosenzweig, 2013: Horizontal cooling towers: riverine ecosystem services and the fate of thermoelectric heat in the contemporary Northeast US. Environmental Research Letters, 8, 025010, doi:10.1088/1748-9326/8/2/025010.

Taylor, K., 2001: Summarizing multiple aspects of model performance in a single diagram. Journal of Geophysical Research, 106, 7183-7192.

Umlauf, L., and H. Burchard, 2003: A generic length-scale equation for geophysical turbulence models. Journal of Marine Research, 61, 235-265.

Unidata, 2018a: Common Data Model (CDM) [software], UCAR/Unidata, Boulder, CO, https://www.unidata.ucar.edu/software/thredds/current/netcdf-java/CDM.

——, 2018b: Thematic Real-time Environmental Distributed Data Services (THREDDS) Data Server [software], UCAR/Unidata, Boulder, CO, doi:10.5065/D6N014KG, https://www.unidata.ucar.edu/software/thredds/current/tds/.

——, 2018c: Network Common Data Form (NetCDF) [software], UCAR/Unidata, Boulder, CO, doi:10.5065/D6H70CW6, https://www.unidata.ucar.edu/software/netcdf.

Vignudelli, S., A. Kostianoy, P. Cipollini, and J. Benveniste, 2011: Coastal Altimetry. Springer, 565 pp.

Warner, J., C. Sherwood, H. Arango, and R. Signell, 2005: Performance of four turbulence closure models implemented using a generic length scale method. Ocean Modelling, 8, 81-113.

Warner, J. C., B. Armstrong, R. He, and J. B. Zambon, 2010: Development of a Coupled Ocean–Atmosphere–Wave–Sediment Transport (COAWST) Modeling System. Ocean Modelling, 35, 230-244, doi:210.1016/j.ocemod.2010.1007.1010.

Weaver, A., C. Deltel, E. Machu, S. Ricci, and N. Daget, 2005: A multivariate balance operator for variational ocean data assimilation. Quarterly Journal of the Royal Meteorological Society, 131, 3605-3625.

Wijesekera, H., J. Allen, and P. Newberger, 2003: Modeling study of turbulent mixing over the continental shelf: Comparison of turbulent closure schemes. Journal of Geophysical Research, 108, 3103.

Wilkin, J., and E. Hunter, 2013: An assessment of the skill of real-time models of Mid-Atlantic Bight continental shelf circulation. Journal of Geophysical Research, 118, 2919-2933, doi: 2910.1002/jgrc.20223.

Wilkin, J., L. Rosenfeld, A. Allen, R. Baltes, A. Baptista, R. He, P. Hogan, A. Kurapov, A. Mehra, J. Quintrell, D. Schwab, R. Signell, and J. Smith, 2017: Advancing coastal ocean modeling, analysis, and prediction for the U.S. Integrated Ocean Observing System. Journal of Operational Oceanography, doi:10.1080/1755876X.2017.1322026

Wilkinson, M., M. Dumontier, I. Aalbersberg, G. Appleton, M. Axton, A. Baak, N. Blomberg, J. Boiten, L. da Silva Santos, P. Bourne, J. Bouwman, A. Brookes, T. Clarke, and a. others, 2016: The FAIR Guiding Principles for scientific data management and stewardship. Scientific Data, 3, 160018, doi:10.1038/sdata.2016.18.

Xu, Y., B. Cahill, J. Wilkin, and O. Schofield, 2013: Role of wind in regulating phytoplankton blooms on the Mid-Atlantic Bight. Continental Shelf Research, 63, S26-S35, doi:10.1016/j.csr.2012.09.011.

Zavala-Garay, J., J. Wilkin, and J. Levin, 2014: Data assimilation in coastal oceanography: IS4DVAR in the Regional Ocean Modeling System (ROMS). Advanced Data Assimilation for Geosciences: Lecture Notes of the Les Houches School of Physics: Special issue June 2012, E. Blayo, M. Bocquet, E. Cosme, and L. Cugliandolo, Eds., Oxford University Press, 555-576, doi:10.1093/acprof:oso/9780198723844.003.0024.

Zhang, W., and G. Gawarkiewicz, 2015: Dynamics of the direct intrusion of Gulf Stream ring water onto the Mid-Atlantic Bight shelf. Geophysical Research Letters, 42, 7687-7695, doi:10.1002/2015GL065530.

CHAPTER 22

Marine Biogeochemical Modelling and Data Assimilation for Operational Forecasting, Reanalysis, and Climate Research

David Ford[1], Susan Kay[1,2], Robert McEwan[1], Ian Totterdell[1], and Marion Gehlen[3]

[1]*Met Office, FitzRoy Road, Exeter, EX1 3PB, UK;* [2]*Plymouth Marine Laboratory, Prospect Place, The Hoe, Plymouth, PL1 3DH, UK;* [3]*Laboratoire des Sciences du Climat et de l'Environnement/Institut Pierre-Simon Laplace (LSCE/IPSL), Orme des Merisiers, 91191 Gif-sur-Yvette, France*

Predictions of marine biogeochemistry are of importance for a range of applications, from operational forecasting of harmful algal blooms, to seasonal prediction of primary production, to understanding the influence of the marine carbon cycle on future climate change. Reanalyses, which include data assimilation in model hindcasts, are also required for the assessment of long-term environmental change. The inclusion of marine biogeochemistry in ocean forecasting and reanalysis systems is still in its early stages, but is already providing valuable insights. This chapter begins by giving an overview of biogeochemical modelling and data assimilation, and discussing challenges around physical-biogeochemical coupling and the use of observations. A summary of current applications to operational forecasting, reanalysis and climate studies is then given, before a vision is presented for a fully integrated prediction framework, linking five-day regional forecasting to global climate research.

Introduction

Marine biogeochemistry is the study of chemical elements in the ocean, and their interactions with marine life. Chief amongst these elements is carbon, the building block of life and a key influence on Earth's climate. Others of importance include nitrogen, oxygen, phosphorus, silicon, and iron. Biogeochemical cycling happens through physical transport, chemical reactions, and uptake and processing by plankton, which are organisms unable to swim against ocean currents. Phytoplankton, microscopic photosynthesising algae, form the base of the ocean food web and contribute about half of Earth's primary production. The zooplankton that consume them also process a significant quantity of carbon and nutrients. In addition, many plankton produce shells or skeletons mostly made of calcium carbonate or silicate (also referred to as biogenic opal). Higher trophic levels such as fish and marine mammals play a lesser role in elemental cycling, and so are generally considered separately.

Ford, D., et al., 2018: Marine biogeochemical modelling and data assimilation for operational forecasting, reanalysis and climate research. In *"New Frontiers in Operational Oceanography"*, E. Chassignet, A. Pascual, J. Tintoré, and J. Verron, Eds., GODAE OceanView, 625-652, doi:10.17125/gov2018.ch22.

Broadly speaking, there are two main motivations for the modelling and prediction of marine biogeochemistry. The first is to study variability and trends in the uptake of carbon dioxide by the ocean, and the influence of this on global climate, as well as the process of ocean acidification. The second is to understand the impact of the physical-chemical marine environment on ecosystems and human activities. For instance:

- Algal blooms can be harmful to human health, either directly or via commercial fisheries.
- Visibility and turbidity is important for seafloor life and high biodiversity ecosystems such as tropical reefs, as well as for recreational, commercial, and naval diving activities.
- River runoff with increased nutrient contents, often with altered elemental ratios, can lead to excessive nutrient levels and increased algal growth, a process called eutrophication. This can be the result of fertiliser use or the discharge of domestic and agricultural waste, as well as natural processes. When these organisms die and decay the water can become depleted of oxygen, leading to conditions which are detrimental to many organisms, and can even result in widespread fish kills (Diaz et al., 2008).
- Different plankton species prefer different environmental conditions, and changes in temperature and nutrients can alter the composition of the whole marine ecosystem. The impacts of this can range from effects on commercial fish stocks, to large jellyfish blooms clogging the cooling vents of nuclear power stations.

Water quality and marine biodiversity are increasingly regulated by international law, so governments must monitor, anticipate and respond to environmental changes.

Operational oceanography primarily focuses on short-term impacts on the marine environment, and climate research largely focuses on long-term impacts on ecosystems, the carbon cycle and ocean acidification. The main focus of this chapter is on the applications and challenges of operational biogeochemical forecasting, but for completeness and context, a brief overview of reanalysis and climate research is also presented, focussing on how these relate to operational applications. In most cases, operational biogeochemical forecasting systems have developed by extending existing physical forecasting systems, as described elsewhere in this book, by combining them with existing biogeochemical models developed either for climate research or for ecological modelling. As such, operational biogeochemistry is less mature than operational physical forecasting, and often fundamental model development remains driven by research priorities, with reanalyses and hindcasts providing a natural link between different applications.

Numerous textbooks have already been dedicated to the field of marine biogeochemistry (e.g. Sarmiento and Gruber, 2006), as well as reviews of modelling (Heinze and Gehlen, 2013), and operational forecasting (Gehlen et al., 2015), and this chapter will not try to replicate them. A high-level overview is given in order to introduce physical oceanographers to the concepts and methods of biogeochemical modelling and data assimilation. The focus is on how biogeochemistry interacts with the various systems described in the other chapters of this book, and the motivations for its inclusion. A summary of existing applications is presented, and future challenges and ambitions are explored, with a vision given of a fully integrated physical-biogeochemical prediction framework. Examples are drawn primarily from work performed at the Met Office, but also more broadly from international initiatives such as the Copernicus Marine Environment Monitoring Service (CMEMS), the GODAE OceanView Marine Ecosystem Analysis and Prediction Task Team (GOV MEAP-TT), and the Climate Model Intercomparison Projects (CMIP) which inform the reports of the Intergovernmental Panel on Climate Change (IPCC).

Biogeochemical Models

General formulation

Most marine biogeochemical models are based on an "NPZD" approach (Fasham et al., 1990), standing for nutrient-phytoplankton-zooplankton-detritus (see Fig. 22.1). One or more state variables are used to represent each of these compartments, which together are used to calculate the evolution of the lower trophic levels of the pelagic (water column) ecosystem. Additionally, models may contain variables describing the carbon and oxygen cycles, sediments, benthic (seafloor) ecosystem, bacteria, and viruses. The general principles of ocean modelling are described elsewhere in this book (Chapter 2 by Fox-Kemper), and the same considerations of numerical methods apply to biogeochemistry, so are not repeated here. Biogeochemical models are reliant on physical models, and either the two are coupled and run together online, or physical model output is used to force the biogeochemistry offline (Heinze and Gehlen, 2013). The state variables are advected and diffused by the physical model, in the same way as temperature and salinity.

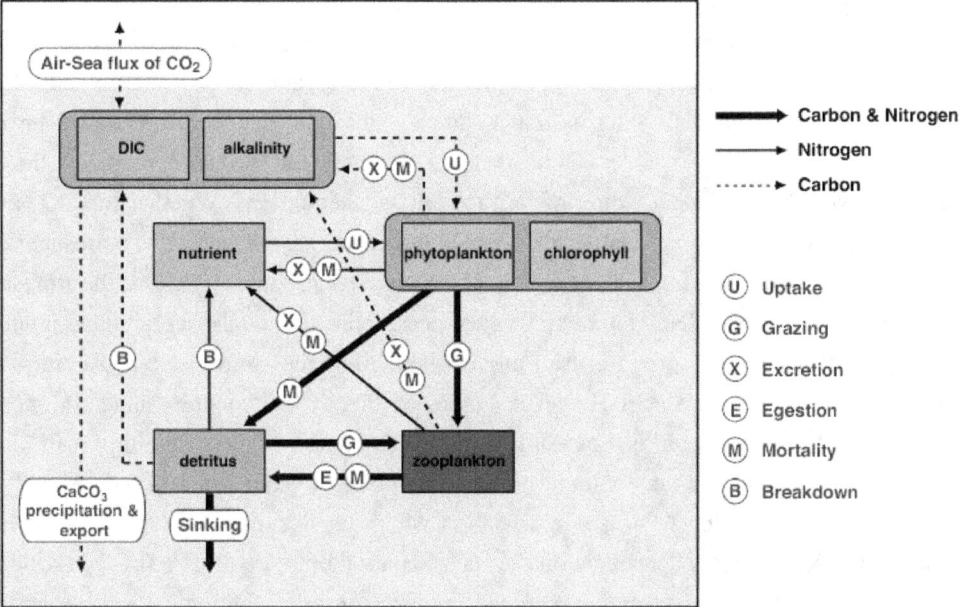

Figure 22.1. Schematic of the Hadley Centre Ocean Carbon Cycle model (HadOCC): an NPZD model with a coupled carbon cycle (Palmer and Totterdell, 2001). The NPZD component is represented by the four boxes labelled nutrient, phytoplankton, zooplankton, and detritus, with the carbon cycle components to the left and top of the diagram. Processes linking the components are shown in circles, and are represented in the model by equations.

Each of the NPZD variables are modelled as tracers, and are formulated as concentrations of (typically) nitrogen or carbon, rather than individual organisms. The nutrient compartment usually includes a variable for dissolved inorganic nitrogen (DIN), sometimes separated into nitrate and ammonium, as DIN is the main limiting nutrient throughout much of the ocean, particularly coastal regions. Phosphate, silicate, and iron may also be included. Plankton are split into phytoplankton, organisms which obtain their energy from sunlight through photosynthesis (autotrophy), and zooplankton, which obtain their energy by consuming other organisms (heterotrophy). Classical NPZD models just have a single variable for each of these, making no consideration of differences between species. A more complex approach is to split each up into two or more plankton functional types (PFTs), which group together species based on their function within the ecosystem. Phytoplankton grow by synthesising organic matter through photosynthesis, a process referred to as primary production. This may be limited by availability of light or nutrients, and is further regulated by temperature. Phytoplankton losses occur through grazing by zooplankton, natural mortality, and respiration. Zooplankton experience similar losses. Some of the organic matter lost from the plankton variables returns directly to the nutrient pool, and the remainder forms detritus. This is either grazed on by zooplankton, or broken down by bacterial and other processes and converted back to dissolved nutrients as it sinks to the deep ocean, in a process called remineralisation. Nutrient concentrations are typically high at depth, but become depleted at the surface through biological utilisation.

Models which represent the carbon cycle typically include additional state variables for dissolved inorganic carbon (DIC) and alkalinity. DIC accounts for the vast majority of carbon in the ocean, with the remainder stored as either dissolved or particulate organic carbon, including the carbon in plankton and detritus. Alkalinity is a measure of the capacity of seawater to neutralise an acid. This is distinct from pH, which is not a state variable, but can be diagnosed from other biogeochemical and physical variables (Orr et al., 2015). At the ocean surface, carbon dioxide (CO_2) is exchanged with the atmosphere, altering surface DIC concentrations. Whether the CO_2 flux is into or out of the ocean varies regionally. In the pre-industrial period the ocean and atmosphere CO_2 concentrations were in approximate balance, with the ocean a weak source of CO_2 to the atmosphere to balance riverine inputs of DIC (although in open ocean climate models riverine DIC inputs are not usually considered, so in pre-industrial model simulations the atmosphere and ocean are in equilibrium through the sea surface). However, in the present day, due to increasing anthropogenic CO_2 emissions, the global ocean is a net sink for CO_2. This helps mitigate the impact of climate change. However, CO_2 is a weak acid, and when it dissolves in water it reacts with it, and in so doing lowers the pH, leading to ocean acidification which impacts marine chemistry and living organisms. The air-sea flux is dependent on the surface partial pressure of CO_2 (pCO_2), which is a function of DIC, alkalinity, temperature, salinity, and pressure (with biological activity acting to modulate DIC and alkalinity). As with nutrients, the vertical distribution of DIC is such that the surface concentration is much lower than at depth. Physical processes typically act to bring carbon to the surface, whilst two mechanisms allow carbon to be transported to the deep ocean: the solubility pump and the biological pump. The solubility pump acts in regions of deep water formation such as the North Atlantic and Southern Ocean, where a strong uptake of atmospheric CO_2 is mixed to the deep ocean by the physically-driven thermohaline circulation. The biological pump acts through the gravitational sinking and recycling of detritus. The deeper the remineralisation takes place, the longer the carbon is shielded from exchange with the atmosphere. A small fraction of detritus will escape remineralisation and be buried into sediments, where it will be stored on geological timescales. Systems with a strong seasonal variability, such as high latitude bloom systems, are believed to be particularly efficient in exporting particulate carbon to depth. The carbon brought to the deep ocean by these mechanisms may be stored for centuries or longer, so studying their strength and variability is important for understanding climate change.

How much complexity?

Biogeochemical models can be very simple, with just a single variable for each of the NPZD compartments (or even omitting detritus) (Edwards, 2001), or much more complex, with dozens of state variables representing different species and processes (Le Quéré et al, 2005; Butenschön et al., 2016). The desirable level of complexity is a matter of great contention within the scientific community (Anderson, 2005; Flynn, 2005), and fundamentally boils down to a simple trade-off: more complex models include processes which are known to be important but are not fully understood, whereas simpler models only include processes which are much better understood, but neglect or amalgamate key aspects of the real world. It is vital to choose a model that is fit for answering the question under consideration.

A common difference between biogeochemical models is the number of PFTs included. Simple models such as HadOCC (Palmer and Totterdell, 2001) include a single phytoplankton and a single zooplankton functional type, as shown schematically in Fig. 22.1. Other models commonly used for global-scale forecasting and climate research, such as MEDUSA (Yool et al., 2013) and PISCES (Aumont et al., 2015), extend this to two phytoplankton and two zooplankton. Zooplankton are split into microzooplankton and mesozooplankton, which have differing sizes and diets, with mesozooplankton preferentially grazing on larger phytoplankton and microzooplankton. Meanwhile, phytoplankton are split between diatoms, which are relatively large (2-200 μm) plankton that form silicate shells and play a particularly important role in the sinking of carbon, and the remaining non-diatom species. The more complex model ERSEM (Baretta et al., 1995; Butenschön et al., 2016) introduces further PFTs, splitting non-diatoms into picophytoplankton (< 2 μm), nanophytoplankton (2-20 μm), and microphytoplankton (> 20 μm), and adding heterotrophic nanoflagellates to zooplankton. The PFTs are distinguished in the model by differing parameters for traits such as growth rates, grazing, and nutrient affinity. PFTs are typically grouped according to cell size, as different-sized organisms play different biogeochemical roles, and their relative distributions can define the entire food chain (Finkel et al., 2010). An alternative approach, taken by the DARWIN model (Follows et al., 2007), is to include tens or hundreds of PFTs with randomly prescribed parameters, allowing the fittest to emerge in the resulting ecosystem. Despite the relative complexity of some models, they still do not generally consider processes such as day-night cycles, buoyancy adjustment, or diurnal migration of zooplankton.

As well as the number of PFTs, the number of independently varying elements is a key difference between models. In 1934, Alfred Redfield reported that the ratios of carbon (C), nitrogen (N), and phosphorus (P) within phytoplankton and the deep oceans were remarkably constant at C:N:P = 106:16:1, which has become known as the Redfield ratio (Redfield, 1934). This ratio allows models to be formulated with fixed stoichiometry (elemental ratios), meaning PFTs can be represented in terms of a single element, with the quantity of other elements derived using the Redfield ratio. However, it has become clear that while the Redfield ratio may hold on average, stoichiometry varies between species and environmental conditions, and plays an important role in phytoplankton growth and diversity (Finkel et al., 2010). Therefore, some models include variable stoichiometry (e.g. Vichi et al., 2007), necessitating the inclusion of multiple state variables for each PFT, and additional nutrients: for instance, the nitrogen biomass and silicon biomass of diatoms would be separate state variables, with both DIN and silicate required. Chlorophyll also needs to be represented, either as a fixed or variable ratio to the biomass, as it is essential for photosynthesis. This can be in addition to the explicit inclusion of bacteria, sediments, the benthic ecosystem, and more. For instance, ERSEM 15.06 (Butenschön et al., 2016), a model with fully flexible stoichiometry, includes up to 59 pelagic and 36 benthic state variables.

An issue with including additional complexity is being able to validate each component. Biogeochemical observations are sparse at the best of times, and become all the more so the more specific the variable. Therefore, additional processes are often only validated in terms of their contribution to quantities such as total chlorophyll concentration. Of studies which have compared models of different complexities, it has generally been found that adding complexity does not necessarily improve model skill (Friedrichs et al., 2007; Kriest et al., 2010; Ward et al., 2013; Kwiatkowski et al., 2014; Xiao and Friedrichs, 2014).

Even the most complex models obviously remain a simplification, and the traditional split between phytoplankton and zooplankton may itself want revisiting. It is increasingly clear that the majority of plankton are not in fact exclusively autotrophs or heterotrophs, but mixotrophs: individual organisms that gain energy by both photosynthesis and the consumption of others, analogous to the Venus flytrap on land. This is not yet commonly represented in models, and may prove to be of great importance (Flynn et al., 2013; Mitra et al., 2014).

In some cases, choice of model complexity will be limited by computational cost. Each extra state variable has to be physically transported at every time step, and once coupled with state-of-the-art physics models, biogeochemical models can become extremely expensive to run. When operational forecasts must be generated within a limited timeframe, or climate simulations run for thousands of years, this often restricts the choice of model complexity (as well as physical resolution) to the simpler end of the spectrum.

In practice, the optimal complexity will depend on the available computing resources and the task at hand: models are tools, and different jobs require different tools. For instance, if the aim is to simulate decadal change in basin-scale primary production and air-sea CO_2 flux, many processes can be safely neglected or parameterised, and a simpler model may be the preferred choice. Whereas simulating small-scale variations in phytoplankton community structure and nutrient ratios will require more variables and processes by necessity. Furthermore, accurate modelling of a coastal ecosystem may require different components than for an open ocean ecosystem.

Biogeochemical Data Assimilation

The theory of data assimilation and its application to physical ocean forecasting systems is introduced elsewhere in this book (Chapter 17 by Hoteit et al., Chapter 18 by Jacobs et al.). The same principles and techniques apply to biogeochemical data assimilation, so are not repeated, but particular considerations for biogeochemical variables are discussed here. A number of reviews and discussions have also been published in recent years (e.g., Gregg et al., 2009; Dowd et al., 2014; Gehlen et al., 2015; Ford and Barciela, 2015).

As with biogeochemical modelling, the lack of first principles equations (such as the Navier-Stokes equations for physics) can potentially lead to a greater flexibility of approaches. In particular, there is scope for data assimilation methods to inform the building of models, and the appropriate variables and processes to include (Ward et al., 2013). In theory, this approach could be developed as a way to construct models based on observations.

More so than in physical data assimilation, the primary application of data assimilation to biogeochemistry has been for parameter estimation [see Schartau et al. (2017) for a recent and comprehensive review]. This aims to adjust model parameters so that a resulting model run better fits a given set of observations, and is usually applied *a priori* as a tuning exercise. As such, while a valuable part of model development, parameter estimation is not routinely applied during the production process of operational forecasts and reanalyses. But there is no reason why state and parameter estimation cannot be combined, so that model parameters are also adjusted during a model run, based on the assimilation of observations. Time-dependent parameters have previously been estimated this way (Losa et al., 2003; Mattern et al., 2012; Doron et al, 2013; Simon et al., 2015), but combined state/parameter estimation has yet to make the transition to operational forecasting and reanalysis systems (Matear and Jones, 2011). The first challenge for most models will be to select appropriate parameters (growth and grazing rates are likely candidates), convert these single numbers to fields that can vary spatially and evolve in time, and define sensible ranges for parameters that are mostly poorly constrained by experimental studies. The bigger challenge is then to vary these realistically, either directly during the data assimilation process, or as a balance relationship, perhaps extending the approach of Hemmings et al. (2008).

As in physical forecasting and reanalysis systems, biogeochemical data assimilation is increasingly used for state estimation, updating the model state variables to produce an analysis, the best estimate of the ocean state at a given time. The most common set of data assimilated is satellite ocean colour (McClain, 2009), as this is the only source of routine global observations of marine biogeochemistry. A full description of the ocean colour processing chain is provided elsewhere in this book (Chapter 9 by Volpe et al.), but the basic procedure is as follows. What ocean colour sensors directly measure is the radiation at different wavelengths leaving the top of the atmosphere. An atmospheric correction model is then used to calculate the water-leaving radiances, converted to remote sensing reflectance, from which optical properties and chlorophyll concentration can be estimated. Algorithms have also been developed to estimate primary production (Carr et al., 2006), and to split chlorophyll up into contributions from different PFTs (Brewin et al., 2011, 2017). The further up the processing chain the more biologically descriptive the product, but the greater the uncertainties.

The majority of ocean colour assimilation studies assimilate total chlorophyll concentration. This has some special considerations compared with physical variables such as sea surface temperature (SST). In particular, chlorophyll distributions are highly non-Gaussian, which poses issues for traditional assimilation methods, which typically assume a Gaussian error distribution. Often a logarithmic (Campbell, 1995) or anamorphic (Brankart et al., 2012) transformation is applied in order to normalise the error distribution, but some studies are exploring assimilation methods such as particle filters (van Leeuwen, 2009), which do not require a Gaussian distribution. Non-Gaussianity is a fundamental issue for data assimilation systems (Bocquet et al., 2010), and assimilation with non-Gaussian observations and models may lead to undesirable effects if not properly treated, potentially including issues with physical data assimilation, as discussed in the fourth section.

Furthermore, chlorophyll products are generated from satellite observations at different wavelengths using an empirical model, and can have relatively large observation errors. Meanwhile, chlorophyll is not the most straightforward variable to accurately model, so model errors are often large, too. Perhaps the most demanding challenge though is in using these observations of near-surface chlorophyll to also improve the sub-surface and non-observed biogeochemical model variables such as nutrients and carbon.

In part, this last point drives the choice of assimilation method. Many centres choose to use a scheme which is inherently multivariate, with the most common being the ensemble Kalman filter (EnKF; Evensen, 2003) and singular evolutive extended Kalman (SEEK) filter (Pham et al., 1998). These generate increments to multiple state variables based on relationships in the background error covariances. This brings obvious advantages in terms of updating the full model state, but relies on accurate evolution of the background error covariances, which is not straightforward. Furthermore, such methods can be computationally expensive, particularly the EnKF which often requires an ensemble of around 100 members. An alternative is to use a univariate assimilation method such as optimal interpolation or 3D-Var which just updates the model chlorophyll, but with the option of using a set of balance relationships to update other state variables. The most sophisticated such scheme in use is the nitrogen balancing scheme of Hemmings et al. (2008). This seeks to determine whether phytoplankton growth or loss errors dominate at each grid point, and partitions increments to the other state variables accordingly, constrained by the principle of conservation of mass.

Because of the reduced uncertainties, there is a great attraction in assimilating optical properties rather than chlorophyll. This is particularly the case in a shelf seas environment with large concentrations of suspended particulate matter (SPM) and coloured dissolved organic matter (CDOM, also known as gelbstoff), where ocean colour chlorophyll algorithms struggle, and where light directly affects more components of the ecosystem than chlorophyll does. Optical properties have been successfully assimilated by Shulman et al. (2013), Ciavatta et al. (2014), Jones et al. (2016), and Gregg and Rousseaux (2017). This has been shown to improve the representation of the ecosystem as a whole, and even of chlorophyll in particular, over assimilating chlorophyll, and is an approach likely to be increasingly explored.

However, it does rely on the biogeochemical model having a sufficiently sophisticated optical model. This remains lacking from most models, and such treatment of optics is of secondary importance for simpler models designed for the open ocean, and when computational restrictions limit model complexity. In these cases, it may be of benefit to move the other way up the ocean colour processing chain, and develop the assimilation of chlorophyll separated by plankton functional type. This has been investigated by Ciavatta et al. (2018) and Skákala et al. (2018), and results suggest that it improves the internal ecosystem dynamics over assimilating total chlorophyll, increasing reanalysis and forecasting skill. Further investigation is required, but it could be that, whilst undoubtedly beneficial, total chlorophyll is often the least appropriate ocean colour product to assimilate into a biogeochemical model.

On the other hand, there is a largely untapped potential to directly assimilate chlorophyll imagery into physical models (Gaultier et al., 2013; Titaud et al., 2010). High-resolution chlorophyll images contain a lot of information about eddies and surface currents, which could be used to update sea surface height fields, complementing existing altimetry products. This could either be done directly, or as part of a fully coupled physical-biogeochemical data assimilation scheme, which would help maintain consistency between physics and biogeochemistry.

With the increasing availability of in situ biogeochemical observations, more studies will incorporate assimilation of data other than ocean colour, which will allow more accurate updating of the multivariate biogeochemical state. However, observational coverage is still likely to remain relatively sparse for years to come, especially in shelf seas, so assimilating these data will provide its own challenges. In situ biogeochemical observations are even sparser than physical ones, and very rarely available in near-real-time, limiting the potential to assimilate them into operational models. Use of in situ biogeochemistry for state estimation has so far been largely limited to research studies related to specific cruises (e.g. Anderson et al., 2000), with observations instead providing valuable independent data sets for validation. An exception is the pCO_2 assimilation scheme of While et al. (2012).

But a revolution (of sorts) is promised over the coming years. The Argo programme has transformed physical oceanography, and is now being extended to include biogeochemistry under the flag of Biogeochemical-Argo (Johnson and Claustre, 2016; Claustre et al., 2010). The prospect of regular global profiles of chlorophyll, nitrate, pH, and oxygen means researchers are gearing up to assimilate these. Further observations will be provided through the increasing deployment of gliders. Advances in biogeochemical in situ observing systems are detailed elsewhere in this book (Chapter 6 by Telszewski et al.). Development of schemes to assimilate Biogeochemical-Argo data is underway at a number of centres. In common with some other areas of data assimilation, there are challenges around how to make the fullest use of sparse observations. More specific to marine biogeochemistry is the challenge of optimally integrating the assimilation of different in situ variables, and ocean colour. This will be a key focus in the development of these schemes.

Physical-biogeochemical Coupling

Impact of model physics on biogeochemistry

As biogeochemical state variables are modelled as tracers, model physics is of fundamental importance. Physical and biogeochemical models may be coupled online, in which case they are integrated together, with physical fields used directly by the biogeochemical model at each time step. Alternatively, the biogeochemical model may be forced offline using previously-generated physical model output, usually at a lower temporal resolution. Online coupling is in theory most accurate, but offline coupling allows more flexibility and spreading of computational cost.

Whether the coupling is online or offline, the equations are the same. Currents transport plankton and nutrients, and vertical velocities drive mixing in the water column. The same advection and diffusion schemes as used by the physical model for temperature and salinity are often used for the biogeochemical state variables too, but this need not be the case. A particular consideration is that tracer concentrations should never be negative, meaning a positive definite advection scheme should be used.

Despite the similarity in approach to ocean circulation, the biogeochemical modeller often has different priorities for physical model accuracy than the physical modeller does. In particular, tracers are very sensitive to mixed layer depth (MLD) and vertical velocities, which are notoriously difficult to model with accuracy. The MLD is critical in determining the mixing of nutrients and carbon into the surface ocean, as well as the depth from the surface to which phytoplankton are mixed. Small changes in the concentration of nutrients and plankton in the euphotic zone, where there is enough light available for photosynthesis, can have a large impact on primary production, and the maximum MLD over the integration period is more important than the mean. Furthermore, stratified regimes with a shallow MLD but deeper euphotic zone can lead to a deep chlorophyll maximum developing at the base of the mixed layer, where nutrient concentrations are greater than at the surface, and there remains sufficient light for growth. Similarly, vertical velocities and other contributors to vertical mixing control the amount of nutrients and carbon brought from the deep to the surface ocean. Biogeochemical quantities tend to be much more sensitive to this than physical ones such as SST or sea level.

Spatial scales and model resolution play an important role too. On longer timescales the thermohaline circulation and carbon pumps (Volk and Hoffert, 1985) control the transport of carbon and nutrients in the deep ocean, as well as patterns of ocean productivity. On shorter timescales, the mesoscale and submesoscale are a major control on primary production and air-sea fluxes, and these are not always well represented in models. Upwelling of nutrients along fronts can result in phytoplankton blooms, whilst eddies can either supply nutrients and fuel blooms (Lévy et al., 2001, 2012a), or aid the sequestering of carbon to the deep ocean.

A good discussion of these and other challenges is given by Holt et al. (2014). Given their sensitivity, tracers and biogeochemical models can also act as a valuable diagnostic for issues with modelled ocean physics.

As well as controlling the physical transport of tracers, temperature and salinity have direct effects on biogeochemistry. Many organisms grow at different rates depending on the temperature, as commonly represented by the parameterisation of Eppley (1972). The carbon cycle is particularly sensitive, with pCO_2 being a function of temperature and salinity as well as DIC and alkalinity. Decreasing temperature or increasing salinity increases the solubility of CO_2, with impacts on air-sea CO_2 flux and the solubility pump. Alkalinity is a charge balance and as such is closely tied to seawater constituents and hence salinity (Zeebe and Wolf-Gladrow, 2001).

Sometimes, the impact of physics on biology in the real ocean itself is unclear, an example of which is the mechanism behind triggering the spring bloom. In 1953, Harald Sverdrup published the critical depth hypothesis (Sverdrup, 1953). The critical depth is defined as the depth at which phytoplankton losses integrated over the water column equal phytoplankton growth, so that net primary production is zero. In winter, light levels are low and so the critical depth is shallow, whilst the MLD is deep, mixing phytoplankton away from the surface, prohibiting growth. During spring, the MLD shallows and the critical depth deepens. The critical depth hypothesis states that the spring bloom is initiated at the point where the MLD becomes shallower than the critical depth, because nutrients and phytoplankton are being kept in the euphotic zone and so growth must exceed loss. This hypothesis has been widely accepted and been a key component of biological oceanography for decades.

Recently though, studies have started to question this paradigm (Behrenfeld and Boss, 2014), arguing that the physical events described above are merely correlated with the spring bloom, and are not the cause. Many observational studies, both in situ and based on ocean colour, suggest that net primary production often becomes positive during winter, earlier than would be explained by the critical depth hypothesis. Furthermore, phytoplankton loss rates are not, as assumed by Sverdup (1953), constant. It has also been shown that bloom initiation coincides with the spring switch in net heat flux from out of the sea to downwards into the ocean (Smyth et al., 2014). As a replacement theory, Behrenfeld (2010) proposed the dilution-recoupling hypothesis. This states that in winter, deep mixing causes a decoupling between growth rates and losses from grazing ("dilution phase"), by reducing predator-prey interactions. This allows phytoplankton biomass to increase (bloom) until environmental conditions lead to a recoupling of growth and loss rates.

The debate over the exact mechanisms at play continues, and the study by Kuhn et al. (2015) provides a fine example of the opportunity for biogeochemical models to inform the understanding of the fundamentals of biological oceanography.

Impact of physical data assimilation on biogeochemistry

As discussed in other chapters of this book, assimilation of physical data, such as temperature, salinity, and altimetry, is a fundamental component of operational forecasting and reanalysis systems. It brings major improvements to analysis and forecast skill, and few operational centres would be without it. Given the reliance of biogeochemistry on model physics, physical data assimilation should also improve biogeochemical simulations.

Counter-intuitively though, the opposite is usually found, and physical data assimilation can significantly degrade biogeochemical fields. This is most commonly the case around the equator in global models (While et al., 2010; El Moussaoui et al., 2011; Park et al., 2018), but has also been observed in regional models (Raghukumar et al., 2015).

The reason for this is an impact of physical data assimilation on vertical mixing processes. The assimilation can result in spuriously large and noisy vertical velocities, bringing excessive concentrations of nutrients and carbon to the surface, as shown in Fig. 22.2. In the nutrient-limited tropics this leads to massively increased primary production and chlorophyll concentrations, and anomalous outgassing of CO_2.

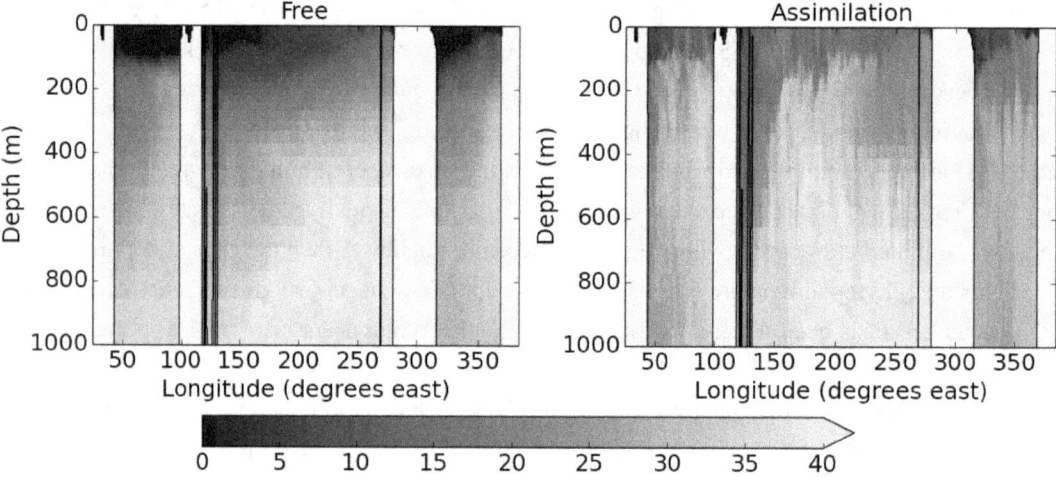

Figure 22.2. Equatorial cross-section of DIN (mmol m^{-3}) from equivalent runs of the NEMO-HadOCC physical-biogeochemical model with no data assimilation (left) and assimilation of physical data (right).

The exact mechanisms remain unclear, and addressing the issue has proven far from trivial. The blame is often laid on vertical velocities, but experiments have shown that filtering the vertical velocities alone does not prevent spurious tracer transport. Other fields relevant to vertical mixing are also affected, including vertical eddy diffusivity and density gradients. The impact on density may be particularly key. Major issues are typically only found when the assimilation updates the 3D temperature and salinity fields, either by directly assimilating temperature and salinity profiles, or as a balance when assimilating sea level anomaly from satellite altimetry. It is desirable for the assimilation to introduce finer-scale eddy structures than in the original model, but doing so without introducing undesirable levels of mixing (to quote Holt et al. (2014): "The last thing many ocean models need is more vertical mixing") will be a challenge.

Preventing the issue at source is the long-term goal, though a number of methods to alleviate the issue in the short-term have been proposed. These have had limited success however. Not assimilating physics data in the affected regions does generally work (Park et al., 2018), but is not entirely satisfactory. Divergence damping (Dobricic et al., 2007) and incremental pressure correction (Waters et al., 2017) each help reduce vertical velocities, but may need to be used in conjunction with other techniques. The sensitivity of the biogeochemistry to elevated nutrients in nutrient-limited regions means that any truly effective solution needs to be comprehensive, as even a small increase in surface nutrients can significantly increase primary productivity. Biogeochemical model parameters could potentially be tuned to compensate, but probably not within sensible ranges. Furthermore, tuning biogeochemical parameters to compensate for biases in model physics is neither satisfying nor recommended, and may prevent the model from accurately responding to long-term changes. One method that may work in some regions is the nutrient balance proposed by While et al. (2010). This applies direct adjustments to nutrient concentrations based on the physical assimilation increments, to ensure the nutricline and pycnocline remain aligned. This addresses the problem found in some regions of the assimilation deepening the MLD, instantly putting increased nutrients into the mixed layer. However, it does not address the increased vertical mixing seen in other regions, particularly the tropics.

The issue does not hinder physics forecasting in the way it does biogeochemistry for two main reasons. First, accurate simulation of vertical mixing is not as critical to the variables that other users of physics output are primarily interested in, such as SST. Second, assimilating physical data will continually correct these variables, masking any problems. Nonetheless, it will still negatively affect physical as well as biogeochemical forecasts, and a robust solution is likely to lead to improvements to all systems. A long-term approach will include the development of strongly coupled physical-biogeochemical data assimilation schemes.

Impact of biogeochemistry on physical fields

Whilst biogeochemical models cannot be run without physical forcing, it is less common for ocean models to contain any feedback from the biogeochemistry to the physics, particularly on the timescales of operational forecasting and reanalysis. However, a number of such feedbacks exist in the real world, and will increasingly be represented as models improve.

Most obviously, as included in Earth system models (ESMs) used for climate projections, is the marine carbon cycle. By altering air-sea CO_2 flux and global ocean carbon uptake, this helps determine atmospheric CO_2 concentrations and therefore global climate. These processes give rise to complex feedback loops linking ocean physics, chemistry and biology (Heinze and Gehlen, 2013; Gehlen et al., 2014; Riebesell et al., 2009).

Another feedback beginning to be included in state-of-the-art ESMs is the cycling of the gas dimethyl sulphide (DMS). Some species of phytoplankton, notably coccolithophores, naturally synthesise dimethyl sulphoniopropionate (DMSP). This breaks down in the water to become DMS, which is outgassed to the atmosphere once it reaches the sea surface. The DMS then reacts with oxygen in the atmosphere to form sulphur dioxide, leading to the creation of sulphate aerosols, which act as cloud condensation nuclei. This process has been widely studied in the context of the CLAW hypothesis (Charlson et al., 1987), which proposes a negative feedback loop in the climate system. It suggests that global temperature rises will lead to increased phytoplankton growth through physiological effects, and so increased DMS production. In turn, this will lead to more clouds being formed, reflecting solar radiation and so cooling the planet. It is unclear how big a role this feedback actually plays, and there is also evidence that global warming will instead decrease phytoplankton growth through increased stratification and restricted nutrient supply, thereby turning it into a positive climate feedback. DMS production may also decrease as a result of ocean acidification (Six et al., 2013; Schwinger et al., 2017). No consensus has yet been reached in the scientific community (Gunson et al., 2006), but recent studies have suggested that the overall impact of DMS cycling on long-term climate change may in fact be minimal (Quinn and Bates, 2011). It is likely to contribute to natural variability though, and its inclusion in operational forecasting models and reanalyses is a challenge for the future.

Organic matter from phytoplankton and bacteria can also form cloud condensation nuclei directly (Wilson et al., 2015). Bubbles form on the surface of the ocean, and organic matter can become incorporated. The bubbles burst, releasing salt and organic matter into the atmosphere, known as sea spray aerosol, acting as cloud condensation nuclei.

Phytoplankton directly impact atmospheric physics further through their contribution to ocean albedo. The presence of chlorophyll colours the surface ocean green, altering the wavelengths of light which are absorbed and reflected by the ocean, and so the amount of shortwave radiation heating the lower atmosphere. Whilst a smaller contribution than factors such as sea state, sea surface chlorophyll is included in the widely-used albedo parameterisation of Jin et al. (2004), and a climatological representation of this is now included in weather forecasts produced at the Met Office. In future, a direct feedback from a marine biogeochemical model may be included in coupled weather and climate models.

A larger and direct biophysical feedback exists within the ocean, operating on similar principles (Morel, 1988). The depth over which shortwave solar radiation is absorbed is controlled by the presence of chlorophyll in the open ocean, as well as SPM and CDOM in coastal waters. When chlorophyll is present, more radiation is absorbed in the surface layers, and so the associated warming is concentrated nearer the surface. This raises SST, cools the sub-surface, and shallows the MLD. It has also been found to amplify seasonal cycles of temperature and sea ice cover (Manizza et al., 2005), and enhance rainfall during the South Asian monsoon (Turner et al., 2012). It is beginning to be included in ocean models, with chlorophyll taken either from a satellite climatology or directly from a biogeochemical model, and is expected to have impacts on all time scales.

A more controversial suggestion of a significant biophysical feedback in the ocean is that creatures such as copepods and jellyfish make a major contribution to physical mixing (Dewar et al., 2006). One study (Katija and Dabiri, 2009) measured the amount of turbulent mixing generated by motile plankton, extrapolated the results to the global ocean, and concluded that their contribution to ocean mixing could be greater even than that provided by wind, which would be a huge omission from current ocean models and generally accepted theory. The suggestion has obviously been treated with caution, but the contribution of different processes to ocean mixing remains uncertain (Dewar et al., 2006), and so the role of zooplankton is not implausible. Even if only partially the case, it is a phenomenon warranting further study and potential inclusion in state-of-the-art ocean models.

Finally, (perhaps), sea ice can sustain large populations of algae, the presence of which can potentially modify the ice albedo and speed up seasonal melting, accelerated further by their impact on light penetration once in the water (Taskjelle et al., 2017). Whilst the overall significance of this is unclear, a number of models of the ice ecosystem have now been developed (Vancoppenolle and Tedesco, 2017).

The impact of biophysical feedbacks on forecasting skill has yet to be investigated, but any process found to make a relevant contribution will progressively be incorporated into operational systems.

Overview of Current Applications

Operational forecasting

Most operational biogeochemical forecasting systems operate at regional scales, as these are the scales where there is most human interest in short-term change of the marine environment. Shelf seas provide 90% of fisheries and almost all leisure use of the sea. Short-range forecasts of algal blooms, water clarity and environmental indicators are of interest to fisheries, regulators, navies and others. At the global scale, forecasts of primary production can be used as inputs to fisheries models (Cheung et al., 2009; Lehodey et al., 2015) or as boundary conditions for regional models. In addition, applications for monitoring the global carbon cycle are emerging (Gehlen et al., 2015).

In Europe, operational forecasts for a number of regions are produced and made publicly available as part of CMEMS (http://marine.copernicus.eu/). These cover the global ocean, Arctic Ocean, Baltic Sea, Black Sea, Mediterranean Sea, Iberian-Biscay-Irish (IBI) Seas, and Northwest European Shelf Seas. Regional forecasts are also produced in other countries around the world, including Australia, Canada, Indonesia, USA, and others. Many of the centres involved in producing operational biogeochemical forecasts collaborate as part of the GOV MEAP-TT. A related, emerging field is the operational forecasting of marine ecology (Payne et al., 2017).

Biogeochemical forecasting systems vary considerably in terms of model complexity, data assimilation, and coupling techniques (Brasseur et al., 2009; Gehlen et al., 2015). This depends on available resources, customer requirements, and how closely tied a biogeochemical system is to a physical counterpart. Systems generally run on either a daily or a weekly basis, either coupled online to an equivalent physics forecasting system, or forced offline by forecast physical fields. Data assimilation, normally of ocean colour, is increasingly employed, but this is not ubiquitous. Forecast accuracy and validation strategies are covered elsewhere in this book (Chapter 19 by Hernandez et al.).

At the Met Office, a daily analysis and six-day forecast is produced for the Northwest European Shelf Seas (NWS), and made available through CMEMS. This is run operationally as part of the FOAM suite (Blockley et al., 2014), which provides global and regional ocean forecasts. The physical model used is NEMO (Madec, 2008), in a series of one-way nested configurations. An analysis and forecast is initially produced for the 1/4° global configuration. This then provides boundary conditions for a 1/12° model of the North Atlantic, which in turn provides the boundary conditions for a 7 km resolution model of the NWS (O'Dea et al., 2012, 2017). Each uses atmospheric forcing from the Met Office numerical weather prediction (NWP) model. The NWS configuration of NEMO is coupled online with the ERSEM biogeochemical model (Edwards et al., 2012), providing coupled physical-biogeochemical forecasts for CMEMS and other customers. Physics data are assimilated using the 3D-Var NEMOVAR scheme (Waters et al., 2015). Assimilation of ocean colour data has recently been developed, and will be implemented operationally in the near future. The operational system is accompanied by a comprehensive monitoring system, as shown in Fig. 22.3.

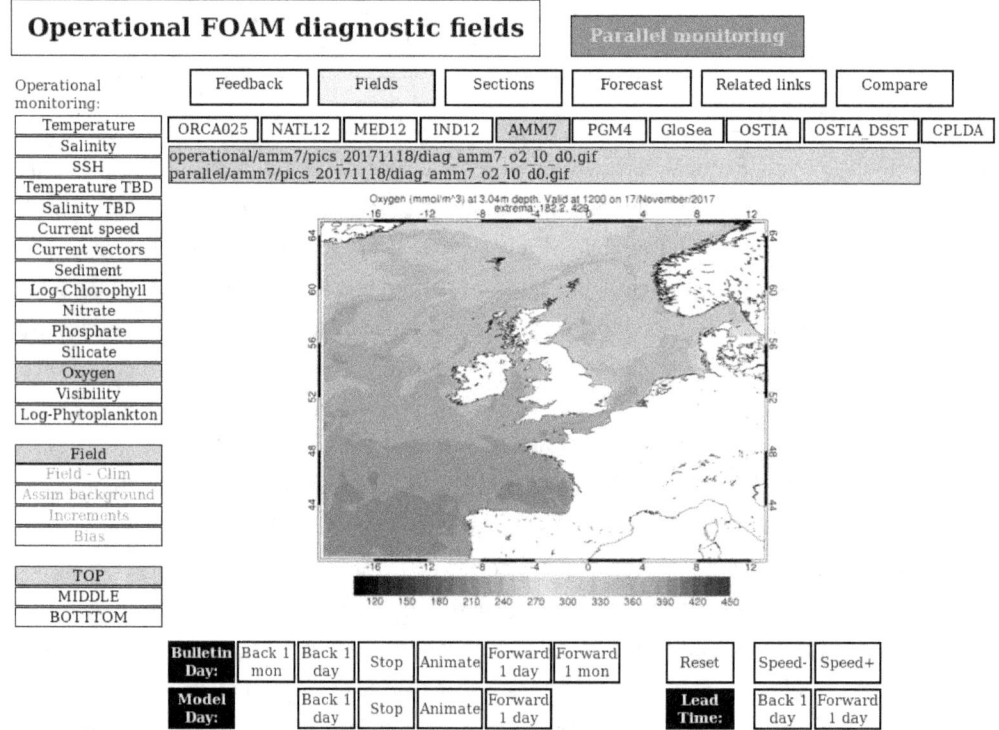

Figure 22.3. Monitoring of the Met Office operational forecasting system.

A similar approach is taken by other centres contributing to CMEMS, although forcing strategies and update schedules differ. For instance, the global ocean product from Mercator Ocean is updated weekly, and run using offline physical forcing downscaled from daily physics forecasts. The Mediterranean Sea product run by OGS is updated biweekly, also using offline forcing, delivering seven days of analysis/hindcast and ten days of forecast, and incorporating ocean colour data assimilation (Teruzzi et al., 2014). Despite some differences in methodology, in general a relatively standardised approach is taken to the generation and delivery of products within CMEMS.

Outside of CMEMS, there are currently few examples of short-range operational biogeochemical forecasts being produced, with any operational applications typically more focussed on ecology and fisheries (Payne et al., 2017), and often based on statistical relationships with physical fields and satellite ocean colour data, rather than coupled physical-biogeochemical modelling. Such applications can include forecasting distributions of pathogens such as bacteria of the genus Vibrio (Jacobs et al., 2014), risk of coral bleaching (Liu et al., 2018), and harmful algal blooms (Davidson et al., 2016). In India, potential fishing zone advisories are issued by INCOIS based on satellite SST and ocean colour data (Deshpande et al., 2011), with plans to develop a model forecasting system (Thushara and Vinayachandran, 2016; Vijith et al., 2016). In Australia, an operational physical-biogeochemical forecasting system for the Great Barrier Reef is being assembled as part of the eReefs project (Schiller et al., 2014; Baird et al., 2016; Jones et al., 2016), for applications including sediment transport and water quality. In Indonesia, weekly regional biogeochemical forecasts are produced operationally using the NEMO-PISCES model (Gutknecht et al., 2016), as part of the INDESO project which monitors and forecasts variables from physics to fish.

Aside from short-range forecasts, there is an increasing demand for seasonal and decadal forecasts of marine biogeochemistry (Rousseaux and Gregg, 2017; Siedlecki et al., 2016; Hobday et al., 2018; Séférian et al., 2014). This is of importance for fisheries and aquaculture, as well as informing environmental commitments, for instance relating to the commitment to Good Environmental Status under the European Union's Marine Strategy Framework Directive (MSFD).

Climate studies

For many biogeochemical models, the primary motivation for development is inclusion in ESMs, for use in climate projections. This allows full consideration of the marine carbon cycle, how ocean carbon uptake will vary into the future, and the resulting feedback on global climate (Friedlingstein et al., 2006). Ocean acidification and ecosystem change can also be studied. The regular reports of the IPCC draw on results submitted as part of the international CMIP projects, most recently CMIP5 (Taylor et al., 2012).

As climate models potentially need to be run for thousands of years, computational expense is a critical factor. Biogeochemical models run as part of ESMs will therefore normally be relatively simple with only a small number of PFTs, and most are still only run at 1° or lower resolution, where eddies must be parameterised. Results from multiple ESMs are often combined in a multi-model ensemble for the analysis of climate change and ocean acidification impacts on marine biogeochemistry and ecosystems (Orr et al., 2005; Bopp et al., 2013; Gehlen et al., 2014). This approach gives insights into regional differences in ecosystem stressors that may occur as a result of climate change. Current models suggest a seesaw pattern with increased primary productivity at high latitudes and decreased productivity over the tropics and mid-latitudes. These changes are at first order explained by increased stratification in response to warming. A warmer ocean leads to enhanced stratification and fewer nutrients upwelled to the euphotic zone. At low- to mid-latitudes this increases the nutrient limitation and results in lower primary productivity. At high latitudes, however, enhanced stratification reduces light limitation as phytoplankton remain nearer the surface, which along with warming and an extended growing season, results in increased primary productivity (Bopp et al., 2013). However, this broad picture hides significant regional and inter-model variability (Laufkötter et al., 2015). More complex ecosystem models, with multiple and flexible plankton types, can also give insight into how changes in temperature and nutrient supply may affect ecosystem productivity and the makeup of plankton communities in different regions (Dutkiewicz et al., 2013; Barton et al., 2016).

Regional studies use global climate models as drivers to produce downscaled projections at higher resolution. For example, biogeochemical modelling has been used to investigate the impact of climate change and other anthropogenic stressors on the marine ecosystem of the NWS (Wakelin et al., 2015). Regional model outputs can be combined with fish models to investigate how climate change may affect food supply and promote sustainable fisheries (Barange et al., 2014; Fernandes et al., 2015).

Reanalysis

Reanalyses, in which models and data assimilation are used to recreate the past ocean state, provide a link between operational forecasting and climate studies. They can be used to assess long-term environmental change, such as eutrophication or alterations to plankton community structure and phenology (e.g. bloom timings). They can also be used to monitor the global ocean carbon sink, and study its variability. Reanalyses could also provide validation of climate models, by seeing how well they reproduce the observed ocean when given realistic forcing and initial conditions. The requirements for physical reanalyses, such as consistent and stable inputs, are discussed elsewhere in this book (Chapter 19 by Haines), and these considerations apply equally to biogeochemistry.

The same NEMO-ERSEM configuration used for operational forecasting of the NWS at the Met Office is also used to produce reanalyses of the region covering recent decades, again available through CMEMS. The version currently available (at CMEMS v3) just assimilates SST, but the latest reanalysis, being produced for CMEMS v4, introduces the assimilation of ocean colour data. This will allow greater exploration of long-term variability and change, and can be compared to other regional reanalyses (Ciavatta et al., 2016). Reanalyses are also available for the other regions considered as part of CMEMS.

The Met Office also produces reanalyses of the global ocean (Ford and Barciela, 2017), using NEMO coupled with the HadOCC biogeochemical model (Fig. 22.1). Ocean colour is assimilated using 3D-Var plus the nitrogen balancing scheme of Hemmings et al. (2008), and there is the option to assimilate pCO_2 (While et al., 2012) and physics (Waters et al., 2015) data. A particular motivation for the work is to explore variability in the global ocean carbon cycle. An example is shown in Fig. 22.4, which shows the monthly mean air-sea CO_2 flux in the Tropical Pacific from a NEMO-HadOCC reanalysis, and how this compares to climatology (Takahashi et al., 2009) and variability of the multivariate El Niño Southern Oscillation (ENSO) index (Wolter and Timlin, 1993, 1998). The Tropical Pacific is a major upwelling region, which brings nutrient- and carbon-rich waters to the surface, resulting in strong outgassing of CO_2 to the atmosphere. In El Niño periods, such as 1997/1998 and 2009/2010, upwelling is weaker, reducing CO_2 outgassing. In La Niña periods, such as 1999 and 2007/2008, upwelling is enhanced, increasing CO_2 outgassing. This strong inter-annual variability is captured well by the reanalysis. NEMO-HadOCC reanalyses are also being used to learn about biases in climate model projections submitted to CMIP5.

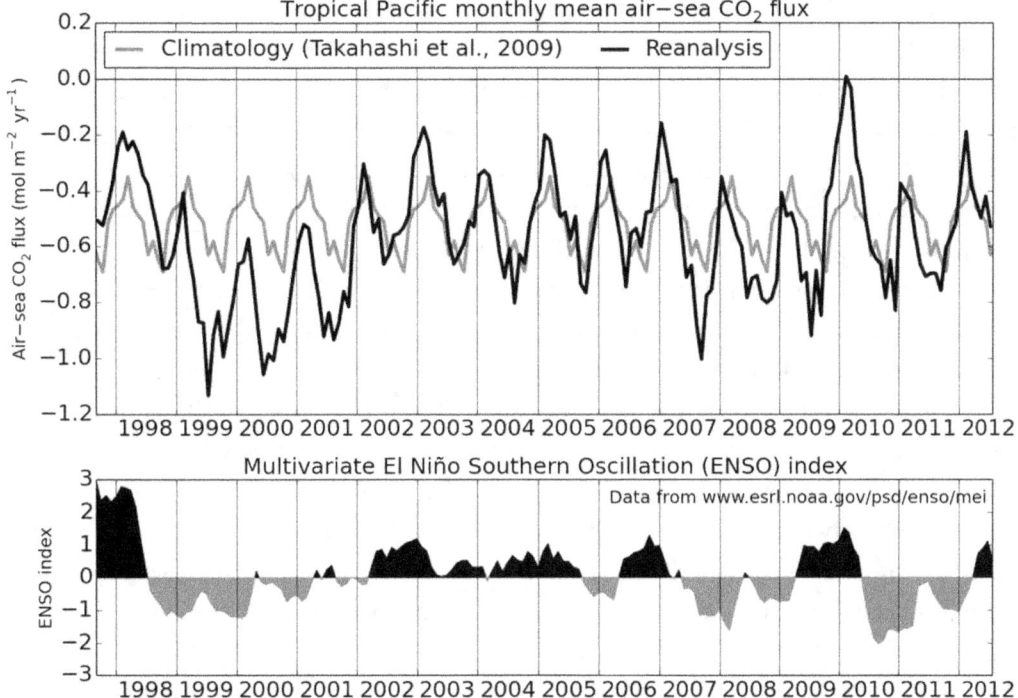

Figure 22.4. Top: Tropical Pacific monthly mean air-sea CO_2 flux (negative means outgassing) from a repeating climatology (Takahashi et al., 2009) and a reanalysis assimilating ocean colour data (Ford and Barciela, 2017). Bottom: multivariate ENSO index (Wolter and Timlin, 1993, 1998).

Towards a Fully Integrated Prediction Framework

As scientific understanding and computing resources continue to advance, so do the modelling systems used to predict the weather, ocean, and climate. As discussed elsewhere in this book (Chapter 16 by Harris), coupled ocean-atmosphere forecasting models are becoming increasingly mature, and starting to be used operationally. On climate time scales, the next generation of ESMs aim to include ever more components of the Earth system. There is also a strong desire to ensure consistency and traceability between regional and global, and weather and climate, simulations. Marine biogeochemistry should be actively included, and there is presently an opportunity to increase the value of biogeochemical predictions to society.

For modelling and forecasting biogeochemistry, there are a number of particular considerations, which are already facing the current generation of systems. Physical ocean forecasting systems are often run at the highest spatial resolution that can be afforded, meaning that coupling a biogeochemical model at the same resolution is not an option. This may necessitate offline coupling using downscaled physical fields, or the online coarsening approach of Lévy et al. (2012b), in which physics and biogeochemistry are coupled online, but the biogeochemical transport is calculated at lower resolution. Alternatively, given the issues around biogeochemical model complexity discussed in the second section, a possible approach might be a "plug-and-play" framework, in which the same model could be run with fewer PFTs, or different parameterisations, depending on the application. This could be achieved using a dedicated coupler such as FABM (Bruggeman and Bolding, 2014). Even if high-resolution physical fields can be used though, exploiting these to benefit biogeochemistry is not necessarily straightforward. The (sub-)mesoscale processes represented, and amount of extra energy present in high-resolution simulations, can potentially have large impacts on biogeochemical fields (Lévy et al., 2001, 2012a), which may not be beneficial if the biogeochemical model parameterisations have been developed and tuned at coarser resolution. Furthermore, currently neglected biogeochemical processes may need to be included.

A further challenge is understanding the limits of biogeochemical predictability, and properly characterising and presenting uncertainty information. Errors can arise and propagate from a variety of sources, both physical and biogeochemical, and may either add or cancel. The ability to quantify uncertainty through comparison with observations is limited due to the sparsity of data, the high level of uncertainty present in both satellite and in situ observations themselves, and the fact that observed and modelled quantities are often not a like-for-like match. Furthermore, the spatial and temporal scales at which models have skill must be considered. Models may be able to capture large-scale patterns but not small-scale features (Saux Picart et al., 2012), and understanding and quantifying these limits of predictability is important for communicating with, and gaining the confidence of, end users (Hyder et al., 2015).

When taken to its logical conclusion, the current direction of travel in weather and climate modelling leads to centres developing fully integrated prediction frameworks, allowing two-way coupling between all relevant components of the Earth system, which can be used for modelling and assimilation on all time and space scales. These would have the flexibility to allow components and processes to be switched on and off, parameterisations to be changed, and resolutions to vary. In the Met Office, for example, the aim is to maintain a seamless approach to atmosphere and ocean modelling across time scales (Martin et al., 2010). The UM (Cullen, 1993) atmosphere and NEMO (Madec, 2008) ocean models are used for regional and global, weather and climate modelling, with the aim of traceability between applications. These are increasingly used for coupled modelling and data assimilation (Lea et al., 2015). The United Kingdom Environmental Prediction (UKEP) project (Lewis et al., 2018) aims to extend this by also including the land surface, waves, and marine biogeochemistry, for regional predictions covering the UK. The components and processes included will develop further as these projects mature.

Such an approach has various potential benefits. It would allow the different biophysical feedbacks detailed in the fourth section to be included, benefitting physics simulations and in turn allowing the biogeochemistry to benefit from improved physics. Physical and biogeochemical forecasts would be fully consistent, providing clarity to end users. Conclusions and issues from simulations on one time scale could directly inform others, for instance using reanalyses to learn about climate models. This potentially makes the model much more robust, particularly when predicting future change, having been tested and validated in a wider range of situations. It would also make it quicker and easier to extend to new regions and novel applications, as developments made for one purpose would be compatible elsewhere. This is further the case due to the shared development effort of a large community working on the same system.

A fully integrated system would present a number of challenges though. Different applications may desire specialist approaches, and compromises may have to be made that otherwise would not. For marine biogeochemistry in particular, where there remains considerable model uncertainty, there is much to be said for maintaining a diverse set of different models, and this should be balanced with the advantages that shared developments towards a community model would bring. Furthermore, with every extra component and process added, particularly with two-way feedbacks, it becomes increasingly difficult to diagnose issues, and validate the impact of any changes. It is likely that many developments would improve some components and degrade others, and marine biogeochemical forecasts will always be of lower priority than weather forecasts when deciding what changes are acceptable. The computational expense of a fully integrated high-resolution system would also be rather large.

As well as the potential for better including marine biogeochemistry in physical systems, improved coupling with the wider marine ecosystem needs to be considered. An integrated approach to this is already taken by end-to-end ecosystem models (Fulton, 2010), which aim to include marine physics, biogeochemistry, fish, marine mammals, seabirds, and humans in a single model. Opinions differ as to whether it is better to use a single end-to-end model or a series of more specialist coupled models. Again, there is a benefit to maintaining a diversity of approaches, particularly whilst the subject is still maturing, and coupled physical-biogeochemical models can feed into these developments. This can help address the challenges of increasing the use of biogeochemical and ecosystem models by policy makers, as discussed by Hyder et al. (2015).

Summary

Marine biogeochemistry is increasingly being included in operational forecasting and reanalysis systems, using progressively more sophisticated models and data assimilation schemes. These systems are less mature than their physical counterparts, but are starting to provide valuable information to a range of users. Future development of biogeochemical systems promises many exciting challenges, through more accurate modelling, exploiting new data sources, increased integration with physical ocean and atmosphere forecasting systems, and better communication of products to end users (Hyder et al., 2015). This will help address important societal needs, such as the impact of toxic algae on human health, the maintenance of diverse and sustainable ecosystems, and understanding the role of the ocean in a changing climate.

Acknowledgements

The authors would like to thank Rabitah Daud, Dianne Deauna, Matías Dinapoli, Jenny Machinford, and an anonymous reviewer for their comments on the draft manuscript, which helped to improve the quality of the final version.

References

Anderson, L. A., A. R. Robinson, and C. J. Lozano, 2000: Physical and biological modeling in the Gulf Stream region: I. Data assimilation methodology. *Deep Sea Research Part I: Oceanographic Research Papers*, **47**, 1787-1827, https://doi.org/10.1016/S0967-0637(00)00019-4.

Anderson, T. R., 2005: Plankton functional type modelling: running before we can walk? *Journal of Plankton Research*, **27**, 1073–1081, https://doi.org/10.1093/plankt/fbi076.

Aumont, O., C. Ethé, A. Tagliabue, L. Bopp, and M. Gehlen, 2015: PISCES-v2: an ocean biogeochemical model for carbon and ecosystem studies. *Geoscientific Model Development*, **8**, 2465-2513, https://doi.org/10.5194/gmd-8-2465-2015.

Baird, M. E., and Coauthors, 2016: Remote-sensing reflectance and true colour produced by a coupled hydrodynamic, optical, sediment, biogeochemical model of the Great Barrier Reef, Australia: comparison with satellite data. *Environmental Modelling & Software*, **78**, 79-96, https://doi.org/10.1016/j.envsoft.2015.11.025.

Barange, M., and Coauthors, 2014: Impacts of climate change on marine ecosystem production in societies dependent on fisheries. *Nature Climate Change*, **4**, 211-216, https://doi.org/10.1038/nclimate2119.

Baretta, J. W., W. Ebenhöh, and P. Ruardij, 1995: The European regional seas ecosystem model, a complex marine ecosystem model. *Netherlands Journal of Sea Research*, **33**, 233-246, https://doi.org/10.1016/0077-7579(95)90047-0.

Barton, A. D., A. J. Irwin, Z. V. Finkel, and C. A. Stock, 2016: Anthropogenic climate change drives shift and shuffle in North Atlantic phytoplankton communities. *Proceedings of the National Academy of Sciences*, **113**, 2964-2969, https://doi.org/10.1073/pnas.1519080113.

Behrenfeld, M. J., 2010: Abandoning Sverdrup's Critical Depth Hypothesis on phytoplankton blooms. *Ecology*, **91**, https://doi.org/10.1890/09-1207.1.

Behrenfeld, M. J., and E. S. Boss, 2014: Resurrecting the ecological underpinnings of ocean plankton blooms. *Annual Review of Marine Science*, **6**, 167-194, https://doi.org/10.1146/annurev-marine-052913-021325.

Blockley, E. W., and Coauthors, 2014: Recent development of the Met Office operational ocean forecasting system: an overview and assessment of the new Global FOAM forecasts. *Geoscientific Model Development*, **7**, 2613-2638, https://doi.org/10.5194/gmd-7-2613-2014.

Bocquet, M., C. A. Pires, and L. Wu, 2010: Beyond Gaussian statistical modeling in geophysical data assimilation. *Monthly Weather Review*, **138**, 8, 2997-3023, https://doi.org/10.1175/2010MWR3164.1.

Bopp, L., and Coauthors, 2013: Multiple stressors of ocean ecosystems in the 21st century: projections with CMIP5 models. *Biogeosciences*, **10**, 6225-6245, https://doi.org/10.5194/bg-10-6225-2013.

Brankart, J-M., C-E. Testut, D. Béal, M. Doron, C. Fontana, M. Meinvielle, P. Brasseur, and J. Verron, 2012: Towards an improved description of ocean uncertainties: effect of local anamorphic transformations on spatial correlations. *Ocean Science*, **8**, 121-142, https://doi.org/10.5194/os-8-121-2012.

Brasseur, P., and Coauthors, 2009: Integrating biogeochemistry and ecology into ocean data assimilation systems. *Oceanography*, **22**, 206-215.

Brewin, R. J. W., and Coauthors, 2011: An intercomparison of bio-optical techniques for detecting dominant phytoplankton size class from satellite remote sensing. *Remote Sensing of Environment*, **115**, 325-339, https://doi.org/10.1016/j.rse.2010.09.004.

Brewin, R. J., and Coauthors, 2017: Uncertainty in Ocean-Color Estimates of Chlorophyll for Phytoplankton Groups. *Frontiers in Marine Science*, **4**, 104, https://doi.org/10.3389/fmars.2017.00104.

Bruggeman, J., and K. Bolding, 2014: A general framework for aquatic biogeochemical models. *Environmental Modelling & Software*, **61**, 249-265, https://doi.org/10.1016/j.envsoft.2014.04.002.

Butenschön, M., and Coauthors, 2016: ERSEM 15.06: a generic model for marine biogeochemistry and the ecosystem dynamics of the lower trophic levels. *Geoscientific Model Development*, **9**, 1293-1339, https://doi.org/10.5194/gmd-9-1293-2016.

Campbell, J. W., 1995: The lognormal distribution as a model for bio-optical variability in the sea. *Journal of Geophysical Research: Oceans*, **100**, 13237–13254, https://doi.org/10.1029/95JC00458.

Carr, M. E., and Coauthors, 2006: A comparison of global estimates of marine primary production from ocean color. *Deep Sea Research Part II: Topical Studies in Oceanography*, **53**, 741-770, https://doi.org/10.1016/j.dsr2.2006.01.028.

Charlson, R. J., J. E. Lovelock, M. O. Andreaei, and S. G. Warren, 1987: Oceanic phytoplankton, atmospheric sulphur, cloud. *Nature*, **326**, 655-661.

Cheung, W. W. L., V. W. Y. Lam, J. L. Sarmiento, K. Kearney, R. Watson, and D. Pauly, 2009: Projecting global marine biodiversity impacts under climate change scenarios. *Fish and Fisheries*, **10**, 235–251, https://doi.org/10.1111/j.1467-2979.2008.00315.x.

Ciavatta, S., R. Torres, V. Martinez-Vicente, T. Smyth, G. Dall'Olmo, L. Polimene, and J. I. Allen, 2014: Assimilation of remotely-sensed optical properties to improve marine biogeochemistry modelling. *Progress in Oceanography*, **127**, 74-95, https://doi.org/10.1016/j.pocean.2014.06.002.

Ciavatta, S., S. Kay, S. Saux-Picart, M. Butenschön, and J. I. Allen, 2016: Decadal reanalysis of biogeochemical indicators and fluxes in the North West European shelf-sea ecosystem. *Journal of Geophysical Research: Oceans*, **121**, 1824–1845, https://doi.org/10.1002/2015JC011496.

Ciavatta, S., R. J. W. Brewin, J. Skákala, L. Polimene, L. de Mora, Y. Artioli, and J. I. Allen, 2018: Assimilation of Ocean-Color Plankton Functional Types to Improve Marine Ecosystem Simulations." *Journal of Geophysical Research: Oceans*, **123**, no. 2, 834-854, https://doi.org/10.1002/2017JC013490.

Claustre, H., and Coauthors, 2010: Bio-optical profiling floats as new observational tools for biogeochemical and ecosystem studies. *Proceedings of OceanObs'09: Sustained Ocean Observations and Information for Society (Vol. 2)*, Venice, Italy, 21-25 September 2009, Hall, J., D. E. Harrison, and D. Stammer, Ed., ESA Publication WPP-306, https://doi.org/10.5270/OceanObs09.cwp.17.

Cullen, M. J. P., 1993: The unified forecast/climate model. *Meteorological Magazine*, **122**, 81-94.

Davidson, K., D. M. Anderson, M. Mateus, B. Reguera, J. Silke, M. Sourisseau, and J. Maguire, 2016: Forecasting the risk of harmful algal blooms. *Harmful Algae*, **53**, 1-7, https://doi.org/10.1016/j.hal.2015.11.005.

Deshpande, S. P., K. V. Radhakrishnan, and U. G. Bhat, 2011: Direct and indirect validation of potential fishing zone advisory off the coast of Uttara Kannada, Karnataka. *Journal of the Indian Society of Remote Sensing*, **39 (4)**, 547-554, https://doi.org/10.1007/s12524-011-0104-4.

Dewar, W. K., R. J. Bingham, R. L. Iverson, D. P. Nowacek, L. C. St Laurent, and P. H. Wiebe, 2006: Does the marine biosphere mix the ocean? *Journal of Marine Research*, **64**, 541-561, https://doi.org/10.1357/002224006778715720.

Diaz, R. J., and R. Rosenberg, 2008: Spreading dead zones and consequences for marine ecosystems. *Science*, **321**, 926-929, https://doi.org/10.1126/science.1156401.

Dobricic, S., N. Pinardi, M. Adani, M. Tonani, C. Fratianni, A. Bonazzi, and V. Fernandez, 2007: Daily oceanographic analyses by Mediterranean Forecasting System at the basin scale. *Ocean Science*, **3**, 149-157, https://doi.org/10.5194/os-3-149-2007.

Doron M., Brasseur P., Brankart J.-M., Losa S. and Melet A., 2013: Stochastic estimation of biogeochemical parameters from Globcolour ocean colour satellite data in a North Atlantic 3D coupled physical-biogeochemical model, *Journal of Marine Systems.*, **117–118**, 81–95, https://doi.org/10.1016/j.jmarsys.2013.02.007.

Dowd, M., E. Jones, and J. Parslow, 2014: A statistical overview and perspectives on data assimilation for marine biogeochemical models. *Environmetrics*, **25**, 203–213, https://doi.org/10.1002/env.2264.

Dutkiewicz, S., J. R. Scott, and M. J. Follows, 2013: Winners and losers: Ecological and biogeochemical changes in a warming ocean. *Global Biogeochemical Cycles*, **27**, 463–477, https://doi.org/10.1002/gbc.20042.

Edwards, A. M., 2001: Adding Detritus to a Nutrient–Phytoplankton–Zooplankton Model: A Dynamical-Systems Approach. *Journal of Plankton Research*, **23**, 389–413, https://doi.org/10.1093/plankt/23.4.389.

Edwards, K. P., R. Barciela, and M. Butenschön, 2012: Validation of the NEMO-ERSEM operational ecosystem model for the North West European Continental Shelf. *Ocean Science*, **8**, 983-1000, https://doi.org/10.5194/os-8-983-2012.

El Moussaoui, A., C. Perruche, E. Greiner, C. Ethé, and M. Gehlen, 2011: Integration of biogeochemistry into Mercator Ocean systems. *Mercator Océan Newsletter*, **40**, 3-14.

Eppley, R. W., 1972: Temperature and phytoplankton growth in the sea. *Fishery Bulletin*, **70**, 1063-1085.

Evensen, G., 2003: The ensemble Kalman filter: Theoretical formulation and practical implementation. *Ocean Dynamics*, **53**, 343-367, https://doi.org/10.1007/s10236-003-0036-9.

Fasham, M. J. R., H. W. Ducklow, and S. M. McKelvie, 1990: A nitrogen based model of plankton dynamics in the oceanic mixed layer. *Journal of Marine Research*, **48**, 591–639.

Fernandes, J. A., S. Kay, M. A. R. Hossain, M. Ahmed, W. W. L. Cheung, A. N. Lazar, and M. Barange, 2015: Projecting marine fish production and catch potential in Bangladesh in the 21st century under long-term environmental change and management scenarios. *ICES Journal of Marine Science*, **73**, 1357–1369, https://doi.org/10.1093/icesjms/fsv217.

Finkel, Z. V., J. Beardall, K. J. Flynn, A. Quigg, T. A. V. Rees, and J. A. Raven, 2010: Phytoplankton in a changing world: cell size and elemental stoichiometry. *Journal of Plankton Research*, **32**, 119–137, https://doi.org/10.1093/plankt/fbp098.

Flynn, K. J., 2005: Castles built on sand: dysfunctionality in plankton models and the inadequacy of dialogue between biologists and modellers. *Journal of Plankton Research*, **27**, 1205–1210, https://doi.org/10.1093/plankt/fbi099.

Flynn, K. J., D. K. Stoecker, A. Mitra, J. A. Raven, P. M. Glibert, P. J. Hansen, E. Granéli, J. M. Burkholder, 2013: Misuse of the phytoplankton–zooplankton dichotomy: the need to assign organisms as mixotrophs within plankton functional types. *Journal of Plankton Research*, **35**, 3–11, https://doi.org/10.1093/plankt/fbs062.

Follows, M. J., S. Dutkiewicz, S. Grant, and S. W. Chisholm, 2007: Emergent biogeography of microbial communities in a model ocean. *Science*, **315**, 1843-1846, https://doi.org/10.1126/science.1138544.

Ford, D., and R. Barciela, 2015: Marine biogeochemical data assimilation - literature review and scoping report. Met Office Forecasting Research Technical Report 609, 46 pp, https://www.metoffice.gov.uk/learning/library/publications/science/weather-science-technical-reports.

Ford, D., and R. Barciela, 2017: Global marine biogeochemical reanalyses assimilating two different sets of merged ocean colour products. *Remote Sensing of Environment*, **203**, 40-54, https://doi.org/10.1016/j.rse.2017.03.040.

Fox-Kemper, B., 2018: Chapter 2: Notions for the motions of the oceans. In "New Frontiers in Operational Oceanography", E. Chassignet, A. Pascual, J. Tintoré, and J. Verron, Eds., GODAE OceanView.

Friedlingstein, and Coauthors, 2006: Climate-carbon cycle feedback analysis: results from the C^{4}MIP model intercomparison. *Journal of Climate*, **19**, 3337-3353, https://doi.org/10.1175/JCLI3800.1.

Friedrichs, M. A. M., and Coauthors, 2007: Assessment of skill and portability in regional marine biogeochemical models: Role of multiple planktonic groups. *Journal of Geophysical Research: Oceans*, **112**, https://doi.org/10.1029/2006JC003852.

Fulton, E. A., 2010: Approaches to end-to-end ecosystem models. *Journal of Marine Systems*, **81**, 171-183, https://doi.org/10.1016/j.jmarsys.2009.12.012.

Gaultier, L., J. Verron, J. M. Brankart, O. Titaud, and P. Brasseur, 2013: On the inversion of submesoscale tracer fields to estimate the surface ocean circulation. *Journal of Marine Systems*, **126**, 33-42, https://doi.org/10.1016/j.jmarsys.2012.02.014.

Gehlen, M., and Coauthors, 2014: Projected pH reductions by 2100 might put deep North Atlantic biodiversity at risk. *Biogeosciences*, **11**, 6955-6967, https://doi.org/10.5194/bg-11-6955-2014.

Gehlen, M., and Coauthors, 2015: Building the capacity for forecasting marine biogeochemistry and ecosystems: recent advances and future developments. *Journal of Operational Oceanography*, **8**, sup1, s168-s187, https://doi.org/10.1080/1755876X.2015.1022350.

Gregg, W. W., M. A. M. Friedrichs, A. R. Robinson, K. A. Rose, R. Schlitzer, K. R. Thompson, and S. C. Doney, 2009: Skill assessment in ocean biological data assimilation. *Journal of Marine Systems*, **76**, 16-33, https://doi.org/10.1016/j.jmarsys.2008.05.006.

Gregg, W. W., and C. S. Rousseaux, 2017: Simulating PACE Global Ocean Radiances. *Frontiers in Marine Science*, **4**, 60, https://doi.org/10.3389/fmars.2017.00060.

Gunson, J. R., S. A. Spall, T. R. Anderson, A. Jones, I. J. Totterdell, and M. J. Woodage, 2006: Climate sensitivity to ocean dimethylsulphide emissions. *Geophysical Research Letters*, **33**, L07701, https://doi.org/10.1029/2005GL024982.

Gutknecht, E., G. Reffray, M. Gehlen, I. Triyulianti, D. Berlianty, and P. Gaspar, 2016: Evaluation of an operational ocean model configuration at 1/12° spatial resolution for the Indonesian seas (NEMO2.3/INDO12)–Part 2: Biogeochemistry. *Geoscientific Model Development*, **9 (4)**, 1523-1543, https://doi.org/10.5194/gmd-9-1523-2016.

Haines, K., 2018: Chapter 19: Ocean reanalysis. In "New Frontiers in Operational Oceanography", E. Chassignet, A. Pascual, J. Tintoré, and J. Verron, Eds., GODAE OceanView.

Harris, C., 2018: Chapter 16: Coupled atmosphere-ocean modelling. In "New Frontiers in Operational Oceanography", E. Chassignet, A. Pascual, J. Tintoré, and J. Verron, Eds., GODAE OceanView.

Heinze, C., and M. Gehlen, 2013: Chapter 26: Modelling ocean biogeochemical processes and resulting tracer distributions. *Ocean Circulation and Climate*, G. Siedler, S. Griffies, J. Gould, and J. Church, Ed., Academic Press, 667-694.

Hemmings, J. C. P, R. M. Barciela, and M. J. Bell, 2008: Ocean color data assimilation with material conservation for improving model estimates of air-sea CO_2 flux. *Journal of Marine Research*, **66**, 87-126, https://doi.org/10.1357/002224008784815739.

Hernandez, F., and Coauthors, 2018: Chapter 29: Measuring performances, skill and accuracy in operational oceanography: New challenges and approaches. In "New Frontiers in Operational Oceanography", E. Chassignet, A. Pascual, J. Tintoré, and J. Verron, Eds., GODAE OceanView.

Hobday, A. J., C. M. Spillman, P. Eveson, J. Hartog, X. Zhang, and S. Brodie, 2018: A framework for combining seasonal forecasts and climate projections to aid risk management for fisheries and aquaculture. *Frontiers in Marine Science*, **5**, 137, https://doi.org/10.3389/fmars.2018.00137.

Holt, J., and Coauthors, 2014: Challenges in integrative approaches to modelling the marine ecosystems of the North Atlantic: Physics to fish and coasts to ocean. *Progress in Oceanography*, **129**, 285-313, https://doi.org/10.1016/j.pocean.2014.04.024.

Hoteit, I., 2018: Chapter 17: Data assimilation in oceanography: Status and new directions. In "New Frontiers in Operational Oceanography", E. Chassignet, A. Pascual, J. Tintoré, and J. Verron, Eds., GODAE OceanView.

Hyder, K., and Coauthors, 2015: Making modelling count - increasing the contribution of shelf-seas community and ecosystem models to policy development and management. *Marine Policy*, **61**, 291-302, https://doi.org/10.1016/j.marpol.2015.07.015.

Jacobs, G., and Coauthors, 2018: Chapter 18: Operational ocean data assimilation. In "New Frontiers in Operational Oceanography", E. Chassignet, A. Pascual, J. Tintoré, and J. Verron, Eds., GODAE OceanView.

Jacobs, J. M., M. Rhodes, C. W. Brown, R. R. Hood, A. Leight, W. Long, and R. Wood, 2014: Modeling and forecasting the distribution of Vibrio vulnificus in Chesapeake Bay. *Journal of Applied Microbiology*, **117** (5), 1312-1327, https://doi.org/10.1111/jam.12624.

Jin, Z., T. P. Charlock, W. L. Smith Jr., and K. Rutledge, 2004: A parameterization of ocean surface albedo. *Geophysical Research Letters*, **31**, L22301, doi:10.1029/2004GL021180.

Johnson, K., and H. Claustre, 2016: The scientific rationale, design, and implementation plan for a Biogeochemical-Argo float array. *Biogeochemical-Argo Planning Group*, 58, https://doi.org/10.13155/46601.

Jones, E. M., and Coauthors, 2016: Use of remote-sensing reflectance to constrain a data assimilating marine biogeochemical model of the Great Barrier Reef. *Biogeosciences*, **13**, 6441-6469, https://doi.org/10.5194/bg-13-6441-2016.

Katija, K., and J. O. Dabiri, 2009: A viscosity-enhanced mechanism for biogenic ocean mixing. *Nature*, **460**, 624-626, https://doi.org/10.1038/nature08207.

Kriest, I., S. Khatiwala, and A. Oschlies, 2010: Towards an assessment of simple global marine biogeochemical models of different complexity. *Progress in Oceanography*, **86**, 337-360, https://doi.org/10.1016/j.pocean.2010.05.002.

Kuhn, A. M., K. Fennel, and J. P. Mattern, 2015: Model investigations of the North Atlantic spring bloom initiation. *Progress in Oceanography*, **138**, 176-193, https://doi.org/10.1016/j.pocean.2015.07.004.

Kwiatkowski, L., and Coauthors, 2014: iMarNet: an ocean biogeochemistry model intercomparison project within a common physical ocean modelling framework. *Biogeosciences*, **11**, 7291-7304, https://doi.org/10.5194/bg-11-7291-2014.

Laufkötter, C., and Coauthors, 2015: Drivers and uncertainties of future global marine primary production in marine ecosystem models. *Biogeosciences*, **12**, 6955-6984, https://doi.org/10.5194/bg-12-6955-2015.

Le Quéré, C., and Coauthors, 2005: Ecosystem dynamics based on plankton functional types for global ocean biogeochemistry models. *Global Change Biology*, **11**, 2016–2040, https://doi.org/10.1111/j.1365-2486.2005.1004.x.

Lea, D. J., I. Mirouze, M. J. Martin, R. R. King, A. Hines, D. Walters, and M. Thurlow, 2015: Assessing a new coupled data assimilation system based on the Met Office coupled atmosphere–land–ocean–sea ice model. *Monthly Weather Review*, **143**, 4678-4694, https://doi.org/10.1175/MWR-D-15-0174.1.

Lehodey, P., I. Senina, S. Nicol, and J. Hampton, 2015: Modelling the impact of climate change on South Pacific albacore tuna. *Deep Sea Research Part II: Topical Studies in Oceanography*, **113**, 246-259, https://doi.org/10.1016/j.dsr2.2014.10.028.

Lévy, M., P. Klein, and A. M. Treguier, 2001: Impact of sub-mesoscale physics on production and subduction of phytoplankton in an oligotrophic regime. *Journal of Marine Research*, **59**, 535-565, https://doi.org/10.1357/002224001762842181.

Lévy, M., D. Iovino, L. Resplandy, P. Klein, G. Madec, A. M. Tréguier, S. Masson, and K. Takahashi, 2012a: Large-scale impacts of submesoscale dynamics on phytoplankton: Local and remote effects. *Ocean Modelling*, **43**, 77-93, https://doi.org/10.1016/j.ocemod.2011.12.003.

Lévy, M., L. Resplandy, P. Klein, X. Capet, D. Iovino, and C. Éthé, 2012b: Grid degradation of submesoscale resolving ocean models: Benefits for offline passive tracer transport. *Ocean Modelling*, **48**, 1-9, https://doi.org/10.1016/j.ocemod.2012.02.004.

Lewis, H.W., and Coauthors, 2018: The UKC2 regional coupled environmental prediction system. *Geoscientific Model Development*, **11**, 1–42, https://doi.org/10.5194/gmd-11-1-2018.

Liu, G., and Coauthors, 2018: Predicting Heat Stress to Inform Reef Management: NOAA Coral Reef Watch's 4-Month Coral Bleaching Outlook. *Frontiers in Marine Science*, **5**, 57, https://doi.org/10.3389/fmars.2018.00057.

Losa, S. N., G. A. Kivman, J. Schröter, and M. Wenzel, 2003: Sequential weak constraint parameter estimation in an ecosystem model. *Journal of Marine Systems*, **43**, 31-49, https://doi.org/10.1016/j.jmarsys.2003.06.001.

Madec, G., 2008: NEMO ocean engine. Note du Pôle de modélisation, Institut Pierre-Simon Laplace (IPSL), France, No 27, ISSN No 1288-1619.

Manizza, M., C. Le Quéré, A. J. Watson, and E. T. Buitenhuis, 2005: Bio-optical feedbacks among phytoplankton, upper ocean physics and sea-ice in a global model. *Geophysical Research Letters*, **32**, L05603, https://doi.org/10.1029/2004GL020778.

Martin, G. M., S. F. Milton, C. A. Senior, M. E. Brooks, S. Ineson, T. Reichler, and J. Kim, 2010: Analysis and reduction of systematic errors through a seamless approach to modeling weather and climate. *Journal of Climate*, **23**, 5933-5957, https://doi.org/10.1175/2010JCLI3541.1.

Matear, R. J., and E. Jones, 2011: Marine biogeochemical modelling and data assimilation. *Operational Oceanography in the 21st Century*, A. Schiller, and G. B. Brassington, Ed., Springer, 295-317.

Mattern, J. P., K. Fennel, and M. Dowd, 2012: Estimating time-dependent parameters for a biological ocean model using an emulator approach. *Journal of Marine Systems*, **96**, 32-47, https://doi.org/10.1016/j.jmarsys.2012.01.015.

McClain, C. R., 2009: A decade of satellite ocean color observations. *Annual Review of Marine Science*, **1**, 19-42, https://doi.org/10.1146/annurev.marine.010908.163650.

Mitra, A., and Coauthors, 2014: The role of mixotrophic protists in the biological carbon pump. *Biogeosciences*, **11**, 995-1005, https://doi.org/10.5194/bg-11-995-2014.

Morel, A., 1988: Optical modeling of the upper ocean in relation to its biogenous matter content (case I waters). *Journal of Geophysical Research: Oceans*, **93**, 10749–10768, https://doi.org/10.1029/JC093iC09p10749.

O'Dea, E. J., and Coauthors, 2012: An operational ocean forecast system incorporating NEMO and SST data assimilation for the tidally driven European North-West shelf. *Journal of Operational Oceanography*, **5**, 3-17, https://doi.org/10.1080/1755876X.2012.11020128.

O'Dea, E., and Coauthors, 2017: The CO5 configuration of the 7 km Atlantic Margin Model: large-scale biases and sensitivity to forcing, physics options and vertical resolution. *Geoscientific Model Development*, **10**, 2947-2969, https://doi.org/10.5194/gmd-10-2947-2017.

Orr, J. C., and Coauthors, 2005: Anthropogenic ocean acidification over the twenty-first century and its impact on calcifying organisms. *Nature*, **437**, 681-686, https://doi.org/10.1038/nature04095.

Orr, J. C., J.-M. Epitalon, and J.-P. Gattuso, 2015: Comparison of ten packages that compute ocean carbonate chemistry. *Biogeosciences*, **12**, 1483-1510, https://doi.org/10.5194/bg-12-1483-2015.

Palmer, J. R., and I. J. Totterdell, 2001: Production and export in a global ocean ecosystem model. *Deep Sea Research Part I: Oceanographic Research Papers*, **48**, 1169-1198, https://doi.org/10.1016/S0967-0637(00)00080-7.

Park, J.-Y., C. A. Stock, X. Yang, J. P. Dunne, A. Rosati, J. John, and S. Zhang, 2018: Modeling global ocean biogeochemistry with physical data assimilation: A pragmatic solution to the equatorial instability. *Journal of Advances in Modeling Earth Systems*, **10 (3)**, 891-906, https://doi.org/10.1002/2017MS001223.

Payne, M. R., and Coauthors, 2017: Lessons from the First Generation of Marine Ecological Forecast Products. *Frontiers in Marine Science*, **4**, 289, https://doi.org/10.3389/fmars.2017.00289.

Pham, D. T., J. Verron, and M. C. Roubaud, 1998: A singular evolutive extended Kalman filter for data assimilation in oceanography. *Journal of Marine Systems*, **16**, 323-340, https://doi.org/10.1016/S0924-7963(97)00109-7.

Quinn, P. K., and T. S. Bates, 2011: The case against climate regulation via oceanic phytoplankton sulphur emissions. *Nature*, **480**, 51-56, https://doi.org/10.1038/nature10580.

Raghukumar, K., C. A. Edwards, N. L. Goebel, G. Broquet, M. Veneziani, A. M. Moore, and J. P. Zehr, 2015: Impact of assimilating physical oceanographic data on modeled ecosystem dynamics in the California Current System. *Progress in Oceanography*, **138**, 546-558, https://doi.org/10.1016/j.pocean.2015.01.004.

Redfield, A. C., 1934: On the proportions of organic derivatives in sea water and their relation to the composition of plankton. *James Johnson Memorial Volume*, R. J. Daniel, Ed., University Press of Liverpool, 177-192.

Riebesell, U., A. Körtzinger, and A. Oschlies, 2009: Sensitivities of marine carbon fluxes to ocean change. *Proceedings of the National Academy of Sciences*, **106**, 20602-20609, https://doi.org/10.1073/pnas.0813291106.

Rousseaux, C.S., and W. W. Gregg, 2017: Forecasting Ocean Chlorophyll in the Equatorial Pacific. *Frontiers in Marine Science*, **4**, 236, https://doi.org/10.3389/fmars.2017.00236.

Sarmiento, J.L., and N. Gruber, 2006: *Ocean Biogeochemical Dynamics*. Princeton University Press, 528 pp.

Saux Picart, S., M. Butenschön, and J. D. Shutler, 2012: Wavelet-based spatial comparison technique for analysing and evaluating two-dimensional geophysical model fields. *Geoscientific Model Development*, **5 (1)**, 223-230, https://doi.org/10.5194/gmd-5-223-2012.

Schartau, M., and Coauthors, 2017: Reviews and syntheses: parameter identification in marine planktonic ecosystem modelling. *Biogeosciences*, **14**, 1647-1701, https://doi.org/10.5194/bg-14-1647-2017.

Schiller, A., M. Herzfeld, R. Brinkman, and G. Stuart, 2014: Monitoring, predicting, and managing one of the seven natural wonders of the world. *Bulletin of the American Meteorological Society*, **95 (1)**, 23-30, https://doi.org/10.1175/BAMS-D-12-00202.1.

Schwinger, J., J. Tjiputra, N. Goris, K. D. Six, A. Kirkevåg, Ø. Seland, C. Heinze, and T. Ilyina, 2017: Amplification of global warming through pH dependence of DMS production simulated with a fully coupled Earth system model. *Biogeosciences*, **14**, 3633-3648, https://doi.org/10.5194/bg-14-3633-2017.

Séférian, R., L. Bopp, M. Gehlen, D. Swingedouw, J. Mignot, E. Guilyardi, and J. Servonnat, 2014: Multi-year prediction of Tropical Pacific Marine Productivity. *Proceedings of the National Academy of Sciences*, **111 (32)**, 11646–11651, https://doi.org/10.1073/pnas.1315855111.

Shulman, I., S. Frolov, S. Anderson, B. Penta, R. Gould, P. Sakalaukus, and S. Ladner, 2013: Impact of bio-optical data assimilation on short-term coupled physical, bio-optical model predictions. *Journal of Geophysical Research: Oceans*, **118**, 2215–2230, https://doi.org/10.1002/jgrc.20177.

Siedlecki, S. A., and Coauthors, 2016: Experiments with Seasonal Forecasts of ocean conditions for the Northern region of the California Current upwelling system. *Scientific Reports*, **6**, 27203, https://doi.org/10.1038/srep27203.

Simon, E., A. Samuelsen, L. Bertino, and S. Mouysset, 2015: Experiences in multiyear combined state–parameter estimation with an ecosystem model of the North Atlantic and Arctic Oceans using the Ensemble Kalman Filter. *Journal of Marine Systems*, **152**, 1-17, https://doi.org/10.1016/j.jmarsys.2015.07.004.

Six, K. D., S. Kloster, T. Ilyina, S. D. Archer, K. Zhang, and E. Maier-Reimer, 2013: Global warming amplified by reduced sulphur fluxes as a result of ocean acidification. *Nature Climate Change*, **3**, 975-978, https://doi.org/10.1038/nclimate1981.

Skákala, J., D. A. Ford, R. J. Brewin, R. McEwan, S. Kay, B. H. Taylor, L. de Mora, and S. Ciavatta, 2018: The assimilation of phytoplankton functional types for operational forecasting in the North-West European Shelf. Journal of Geophysical Research: Oceans, https://doi.org/10.1029/2018JC014153.

Smyth TJ, I. Allen, A. Atkinson, J. T. Bruun, R. A. Harmer, R. D. Pingree, C. E. Widdicombe, and P. J. Somerfield, 2014: Ocean Net Heat Flux Influences Seasonal to Interannual Patterns of Plankton Abundance. *PLoS ONE*, **9**, e98709, https://doi.org/10.1371/journal.pone.0098709.

Sverdrup, H. U., 1953: On Conditions for the Vernal Blooming of Phytoplankton. *ICES Journal of Marine Science*, **18**, 287–295, https://doi.org/10.1093/icesjms/18.3.287.

Takahashi, T., and Coauthors, 2009: Climatological mean and decadal change in surface ocean pCO_2, and net sea–air CO_2 flux over the global oceans. *Deep Sea Research Part II: Topical Studies in Oceanography*, **56**, 554-577, https://doi.org/10.1016/j.dsr2.2008.12.009.

Taskjelle, T., M. A. Granskog, A. K. Pavlov, S. R. Hudson, and B. Hamre, 2017: Effects of an Arctic under-ice bloom on solar radiant heating of the water column. *Journal of Geophysical Research: Oceans*, **122**, 126–138, https://doi.org/10.1002/2016JC012187.

Taylor, K. E., R. J. Stouffer, and G. A. Meehl, 2012: An overview of CMIP5 and the experiment design. *Bulletin of the American Meteorological Society*, **93**, 485-498, https://doi.org/10.1175/BAMS-D-11-00094.1.

Telszewski, M., A. Palacz, and A. Fischer, 2018: Chapter 6: Biogeochemical In-situ Observations – Motivation, Status, and New Frontiers. In "New Frontiers in Operational Oceanography", E. Chassignet, A. Pascual, J. Tintoré, and J. Verron, Eds., GODAE OceanView.

Teruzzi, A., S. Dobricic, C. Solidoro, and G. Cossarini, 2014: A 3-D variational assimilation scheme in coupled transport-biogeochemical models: Forecast of Mediterranean biogeochemical properties. *Journal of Geophysical Research: Oceans,* **119 (1)**, 200-217, https://doi.org/10.1002/2013JC009277.

Thushara, V., and P. N. Vinayachandran, 2016: Formation of summer phytoplankton bloom in the northwestern Bay of Bengal in a coupled physical-ecosystem model. *Journal of Geophysical Research: Oceans,* **121 (12)**, 8535-8550, https://doi.org/10.1002/2016JC011987.

Titaud, O., A. Vidard, I. Souopgui, and F. X. Le Dimet, 2010: Assimilation of image sequences in numerical models. *Tellus A*, **62**, 30–47, https://doi.org/10.1111/j.1600-0870.2009.00416.x.

Turner, A. G., M. Joshi, E. S. Robertson, and S. J. Woolnough, 2012: The effect of Arabian Sea optical properties on SST biases and the South Asian summer monsoon in a coupled GCM. *Climate Dynamics*, **39**, 811-826, https://doi.org/10.1007/s00382-011-1254-3.

Van Leeuwen, P. J., 2009: Particle filtering in geophysical systems. *Monthly Weather Review*, **137**, 4089-4114, https://doi.org/10.1175/2009MWR2835.1.

Vancoppenolle, M., and L. Tedesco, 2015: Numerical models of sea ice biogeochemistry. *Sea Ice*, D. N. Thomas, Ed., John Wiley & Sons, Ltd, https://doi.org/10.1002/9781118778371.ch20.

Vichi, M., N. Pinardi, and S. Masina, 2007: A generalized model of pelagic biogeochemistry for the global ocean ecosystem. Part I: Theory. *Journal of Marine Systems*, **64**, 89-109, https://doi.org/10.1016/j.jmarsys.2006.03.006.

Vijith, V., P. N. Vinayachandran, V. Thushara, P. Amol, D. Shankar, and A. C. Anil, 2016: Consequences of inhibition of mixed-layer deepening by the West India Coastal Current for winter phytoplankton bloom in the northeastern Arabian Sea. *Journal of Geophysical Research: Oceans*, **121**, 6583–6603, doi:10.1002/2016JC012004.

Volk, T. and M. I. Hoffert, 1985: Ocean carbon pumps: Analysis of relative strengths and efficiencies in ocean-driven atmospheric CO2 changes. *The carbon cycle and atmospheric CO2: Natural variations Archean to present*, E.T. Sundquist, and W. S. Broecker, Ed., American Geophysical Union, 99–110.

Volpe, G., B. Buongiorno Nardelli, S. Colella, A. Pisano, and R. Santoleri, 2018: Chapter 9: An operational interpolated ocean colour product in the Mediterranean Sea. In "New Frontiers in Operational Oceanography", E. Chassignet, A. Pascual, J. Tintoré, and J. Verron, Eds., GODAE OceanView.

Wakelin, S. L., Y. Artioli, M. Butenschön, J. I. Allen, and J. T. Holt, 2015: Modelling the combined impacts of climate change and direct anthropogenic drivers on the ecosystem of the northwest European continental shelf. *Journal of Marine Systems*, **152**, 51-63, https://doi.org/10.1016/j.jmarsys.2015.07.006.

Ward, B. A., M. Schartau, A. Oschlies, A. P. Martin, M. J. Follows, and T. R. Anderson, 2013: When is a biogeochemical model too complex? Objective model reduction and selection for North Atlantic time-series sites. *Progress in Oceanography*, **116**, 49-65, https://doi.org/10.1016/j.pocean.2013.06.002.

Waters, J., D. J. Lea, M. J. Martin, I. Mirouze, A. Weaver, and J. While, 2015: Implementing a variational data assimilation system in an operational 1/4 degree global ocean model. *Quarterly Journal of the Royal Meteorological Society*, **141**, 333–349. https://doi.org/10.1002/qj.2388.

Waters, J., M. J. Bell, M. J. Martin, and D. J. Lea, 2017: Reducing ocean model imbalances in the equatorial region caused by data assimilation. *Quarterly Journal of the Royal Meteorological Society*, **143**, 195–208, https://doi.org/10.1002/qj.2912.

While, J., K. Haines, and G. Smith, 2010: A nutrient increment method for reducing bias in global biogeochemical models. *Journal of Geophysical Research: Oceans*, **115**, C10036, https://doi.org/10.1029/2010JC006142.

While, J., I. Totterdell, and M. Martin, 2012: Assimilation of pCO_2 data into a global coupled physical-biogeochemical ocean model. *Journal of Geophysical Research: Oceans*, **117**, C03037, https://doi.org/10.1029/2010JC006815.

Wilson, T.W., and Coauthors, 2015: A marine biogenic source of atmospheric ice-nucleating particles. *Nature*, **525**, 234-238, https://doi.org/10.1038/nature14986.

Wolter, K., and M.S. Timlin, 1993: Monitoring ENSO in COADS with a seasonally adjusted principal component index. Proc. of the 17th Climate Diagnostics Workshop, Norman, OK, NOAA/NMC/CAC, NSSL, Oklahoma Clim. Survey, CIMMS and the School of Meteor., Univ. of Oklahoma, 52-57, https://www.esrl.noaa.gov/psd/enso/mei/WT1.pdf.

Wolter, K., and M. S. Timlin, 1998: Measuring the strength of ENSO events - how does 1997/98 rank? *Weather*, **53**, 315-324, https://doi.org/10.1002/j.1477-8696.1998.tb06408.x.

Xiao, Y., and M. A. M. Friedrichs, 2014: Using biogeochemical data assimilation to assess the relative skill of multiple ecosystem models in the Mid-Atlantic Bight: effects of increasing the complexity of the planktonic food web. *Biogeosciences*, **11**, 3015-3030, https://doi.org/10.5194/bg-11-3015-2014.

Yool, A., E. E. Popova, and T. R. Anderson, 2013: MEDUSA-2.0: an intermediate complexity biogeochemical model of the marine carbon cycle for climate change and ocean acidification studies. *Geoscientific Model Development*, **6**, 1767-1811, https://doi.org/10.5194/gmd-6-1767-2013.

Zeebe, R.E., and D. Wolf-Gladrow, 2001: *CO_2 in seawater: equilibrium, kinetics, isotopes*. Elsevier, 360 pp.

CHAPTER 23

Understanding and Predicting El Niño and the Southern Oscillation

Michael J. McPhaden

NOAA/Pacific Marine Environmental Laboratory, Seattle, WA, USA

This chapter reviews basic concepts about the El Niño/Southern Oscillation (ENSO) cycle and its global climatic impacts. It also highlights progress in understanding, observing, and predicting ENSO timescale variations, focusing on the 2015–16 El Niño as a case study. This El Niño was one of the strongest on record; its evolution and many of its far-field impacts were remarkably well predicted at lead times of 6–9 months. Despite progress to date, however, there are many outstanding issues that need to be addressed to improve our understanding and ability to predict ENSO.

Introduction

ENSO is the most prominent year-to-year climate fluctuation affecting the globe. It originates in the tropical Pacific through coupled ocean-atmosphere interactions mediated by wind and sea surface temperature (SST) feedbacks (Fig. 23.1). We refer to the warm phase of ENSO as El Niño and the cold phase as La Niña; individual events recur roughly every 2–7 years. Its influence extends worldwide through atmospheric teleconnections that shift the probabilities for drought, flood, heat waves, and other extreme weather events (Fig. 23.2).

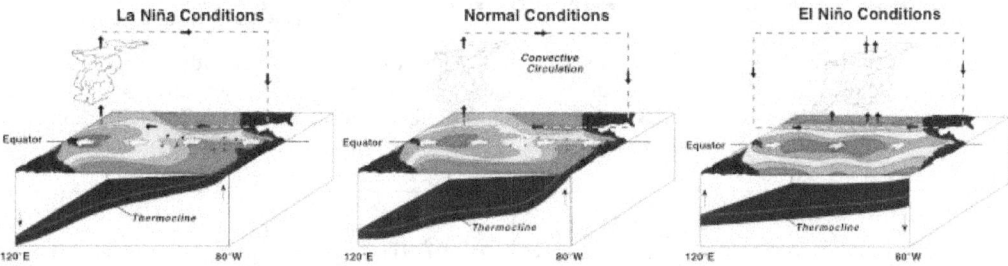

Figure 23.1. A schematic of La Niña, normal and El Niño conditions in the tropical Pacific.

ENSO is the largest source of predictability in the climate system on seasonal timescales, other than the normal march of the seasons. Over the past 30 years, there have been major advances in our understanding of ENSO variations and their impacts, in the development of ocean observing systems to support ENSO prediction, and in the development of seasonal forecast models capable of skillful forecasts with lead times of 6–9 months. Satellites, moored buoys (the so-called TAO-

---

McPhaden, M.J., 2018: Understanding and predicting El Niño and the Southern Oscillation. In "*New Frontiers in Operational Oceanography*", E. Chassignet, A. Pascual, J. Tintoré, and J. Verron, Eds., GODAE OceanView, 653-662, doi:10.17125/gov2018.ch23.

TRITON array in the Pacific), Argo floats, and other measurement systems (Fig. 23.3) provide critical data for developing improved forecast models and for initializing and validating those models.

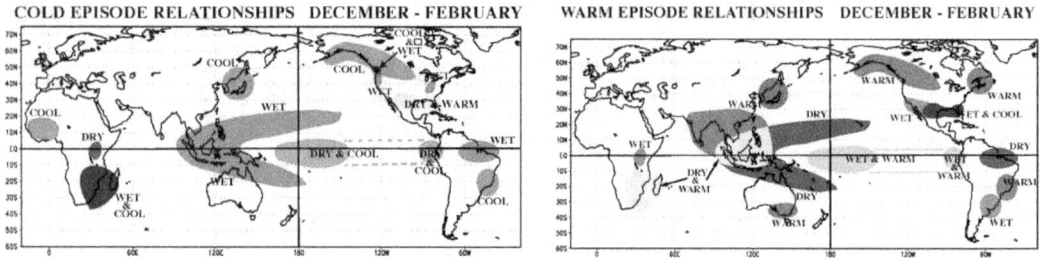

Figure 23.2. Summary of La Niña (left) and El Niño (right) climate impacts in terms of temperature and precipitation for December–February, which is the season of ENSO event peak development. Figure courtesy of NOAA Climate Prediction Center.

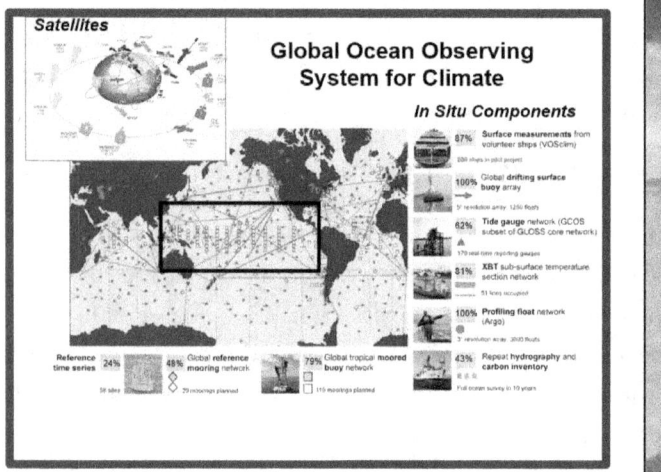
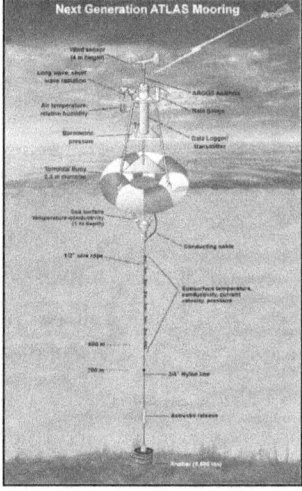

Figure 23.3. Schematic of the Global Ocean Observing System for Climate (left) with the tropical Pacific outlined. The TAO/TRITON array (yellow squares in the tropical Pacific) was implemented specifically for ENSO research and forecasting. ATLAS moorings (right) or their equivalent, which provide oceanic and atmospheric data in real-time via satellite relay, make up the majority of TAO/TRITON moorings.

The 2015–16 El Niño

The 2015–16 El Niño was of comparable magnitude to the other two major El Niños in the historical record (1982–83 and 1997–98) and one of the strongest on record (McPhaden, 2015; L'Heureux et al., 2017). Thus, it provides a timely case study for highlighting advances in understanding, observing, and predicting El Niño and its impacts. Typical of most El Niños, the event developed early in the calendar year, reached its peak development in the boreal winter, and terminated in the following spring (Fig. 23.4). With the large-scale collapse of the trade winds in 2015, the western Pacific warm pool (surface water >28–29 °C) migrated eastward along the equator, the thermocline flattened out, and the normally cold upwelled water that forms the equatorial "cold tongue" in the eastern Pacific was only weakly evident at the height of the event in December 2015 (Fig. 23.5).

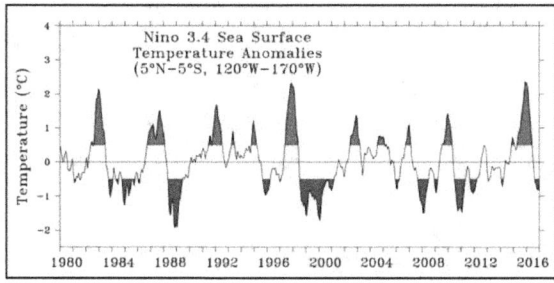 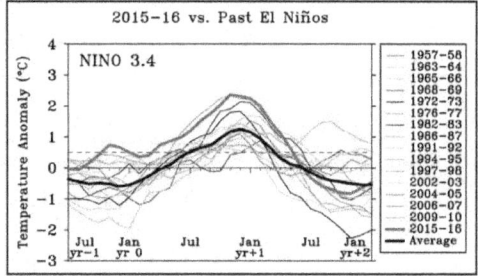

Figure 23.4. Monthly mean values of the Nino3.4 SST index since 1980 (left) for El Niño events (red peaks above 0.5 °C threshold) and La Niña events (blue troughs below 0.5 °C threshold). Comparative amplitudes of El Niños in the Nino3.4 region since the late 1950s (right) based on 2.5-year segments of the Nino3.4 record centered on the season of peak development in boreal winter. The thick black line is the average of 14 El Niños during this time, excluding 2015–16, which is shown as a thick red line. The dashed horizontal line is the 0.5 °C threshold for El Niño. Nino3.4 SST is one of the most commonly used ENSO indices because of strong ocean-atmosphere coupling on seasonal timescales in this region (shown in Fig. 23.5) and because of the robust relationship between Nino3.4 SST variations and global climate impacts. The Nino3.4 time series were not detrended in this figure since trends are weak in this particular region. Base period for computing anomalies is 1981-2010.

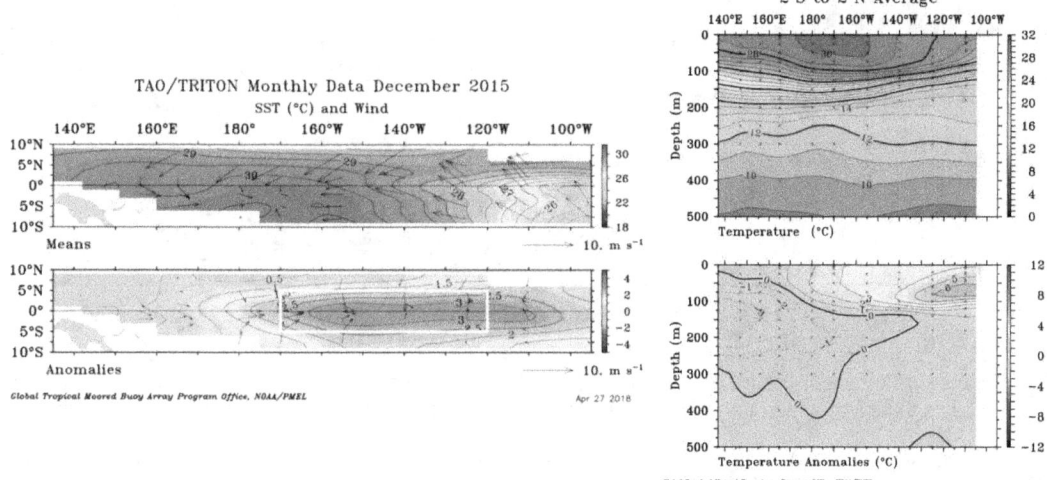

Figure 23.5. Monthly means and anomalies of surface winds and SST (left) and subsurface temperature along the equator (right) from the TAO/TRITON array for December 2015 at the height of the 2015-16 El Niño. The Nino3.4 region is outlined in the anomaly plot of the left panel. More information about gridding procedures and base periods for computing anomalies can be found at www.pmel.noaa.gov/gtmba.

A comparison of SST anomalies for December 1997, 2009, and 2015 (Fig. 23.6) illustrates the concept of ENSO diversity (Capotondi et al., 2015). All El Niños are characterized by warmer than normal SSTs and weakened trade winds along the equator but the location of maximum SST anomalies can vary significantly from event to event. For example, compared to the previous major El Niño of 1997–98 (which has been referred to as an "eastern Pacific" or EP El Niño), SST anomalies were shifted westward along the equator in 2015–16 and reached historical highs west of the Nino3.4 region (Xue and Kumar, 2017). However, compared to the 1997–98 El Niño, SST anomalies were weaker east of about 140°W in December 2015. The 2009–10 El Niño, a "central

Pacific" or CP El Niño, was different from either of these other two El Niños, with the largest warm anomalies confined to the central Pacific and very little warming in the eastern Pacific cold tongue. Compared to 1997–98 and 2009–10, the 2015–16 event appears to be a hybrid between these two types of El Niño (L'Heureux et al., 2017; Paek et al., 2017). The details of these SST patterns are important because they can affect atmospheric teleconnections and climate impacts associated with different El Niño events (Capotondi et al., 2015).

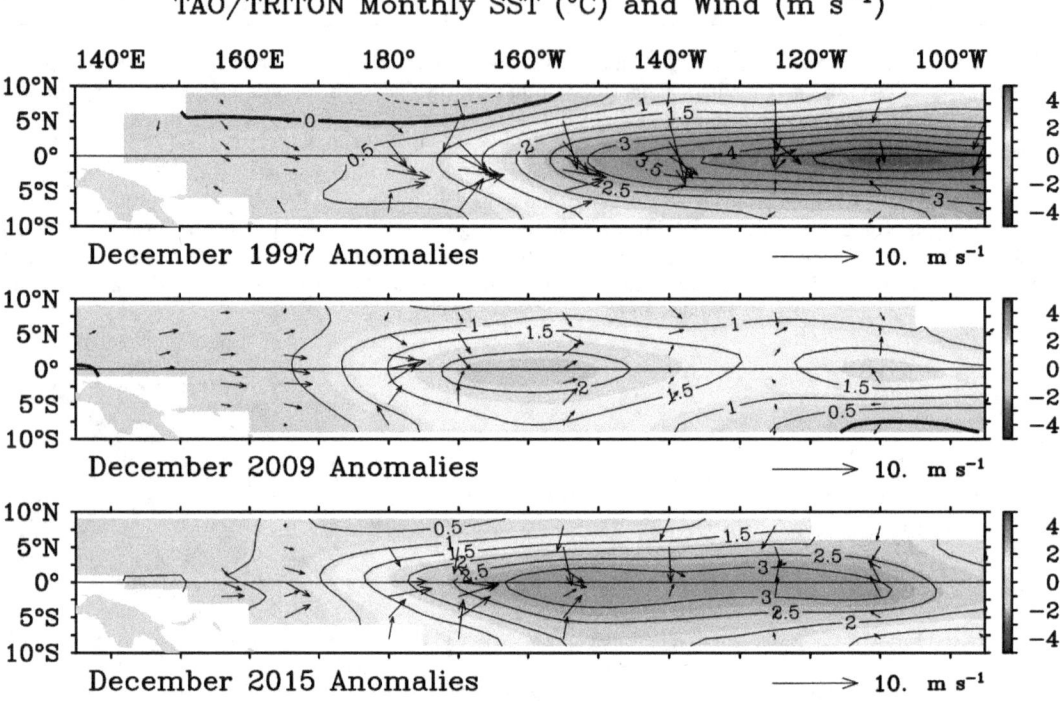

Figure 23.6. Comparison of surface wind and SST anomalies for December 1997, 2009, and 2015. The 1997–98 El Niño was an eastern Pacific (EP) event while 2009–10 was a central Pacific (CP) event. The 2014–15 El Niño appears to be a hybrid of these two types of El Niño.

The 2015–16 El Niño also illustrates the interplay between large-scale deterministic seasonal timescale dynamics associated with theories such as the recharge oscillator (Jin, 1997) and delayed oscillator (e.g., Battisti, 1988; Schopf and Suarez, 1988) and higher frequency stochastic processes. Collapse of the trade winds during El Niño in particular is very episodic, punctuated by a series of westerly wind bursts (Fig. 23.7a) that lead to warming along the equator in the central and eastern Pacific. Warming in the central Pacific in 2015 (Fig. 23.7b) was caused by very strong eastward wind-driven flows that advected the western Pacific warm pool eastward. Episodic westerly wind forcing also generated downwelling equatorial Kelvin waves that crossed the basin in about 45 days (Fig. 23.7c), leaving in their wake a thermocline depressed by up to 40 m so as to reduce the efficiency of equatorial upwelling to cool the surface. The net effect of these processes was to shift the locus of warm surface water eastward and with it deep atmospheric convection. Positive feedbacks between the ocean and the atmosphere reinforced the surface warming and the weakening trade winds, which allowed the El Niño to grow to a large amplitude. Delayed negative feedbacks in the form of an upwelling Kelvin wave along the equator in early 2016 began to lift the thermocline, eventually initiating surface cooling on the onset of a weak La Niña. This Kelvin wave may have emanated from the western boundary, consistent with delayed oscillator theory, or have been forced by easterly wind anomalies in the far western Pacific, or some combination of both these effects.

It is also noteworthy that the westerly wind bursts in 2015 appear to be mostly confined to the region west of the 29 °C isotherm and disappear as the western Pacific cools in 2016. Thus, the stochastic forcing of ENSO is not completely random but depends on the state of ENSO itself. This is referred to as state-dependent noise forcing and it is an important element of ENSO dynamics (Eisenmann et al., 2015; Levine and McPhaden, 2016).

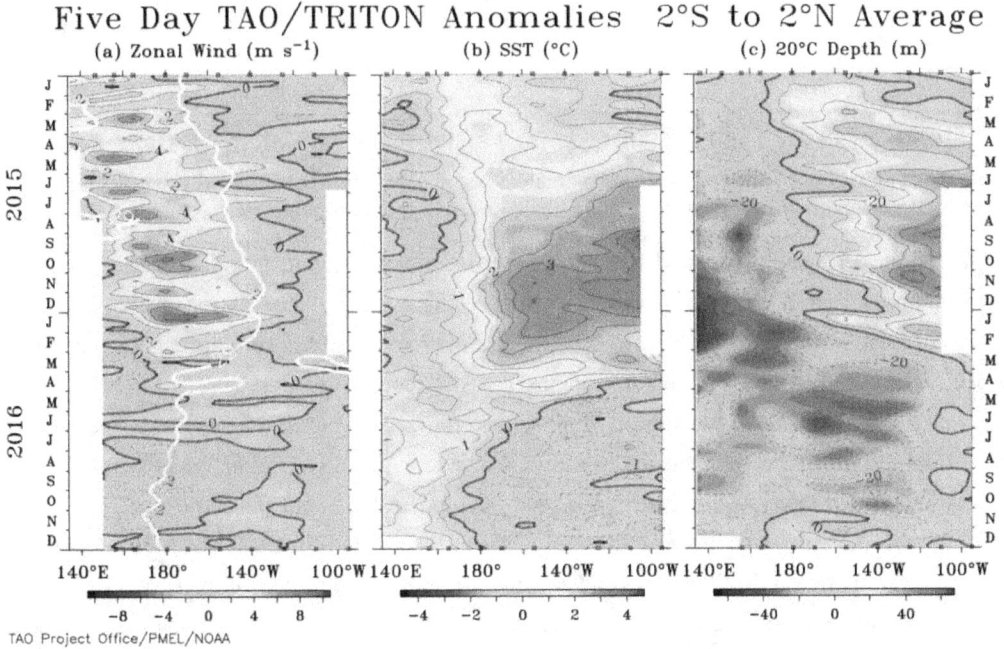

Figure 23.7. Five-day analyses of (a) zonal wind, (b) SST, and (c) 20 °C depth anomalies averaged 2°N–2°S based on TAO/TRITON moored time series data for 2015–16. The depth of the 20 °C isotherm is a measure of the depth of the thermocline. The white line in the left panel shows the longitudinal position of the 29 °C SST isotherm, which is an indication of the eastern edge of the western Pacific warm pool. Dots on the upper and lower axes show longitudes where data are available. More information about gridding procedures and base periods for computing anomalies can be found at www.pmel.noaa.gov/gtmba.

It has long been known that a build-up of excess heat content along the equator preconditions the tropical Pacific to the development of El Niño and that heat content is a useful predictor of ENSO development (Jin, 1997; Meinen and McPhaden, 2000). This concept is illustrated by comparing Nino3.4 SST with upper ocean heat content as measured by the depth averaged temperature anomaly in the upper 300 m integrated from the coast of western South America to the coast of New Guinea between 5°N and 5°S. Heat content leads Nino3.4 SST by typically 6–9 months except for during the first decade of the 21st century (McPhaden, 2012). Moreover, the largest build-up of heat content since 1980, other than that observed in 1997, occurred in 2015 prior to the full development of the 2015–16 El Niño. With such a strong build-up of heat content, one might have anticipated the development of a strong El Niño.

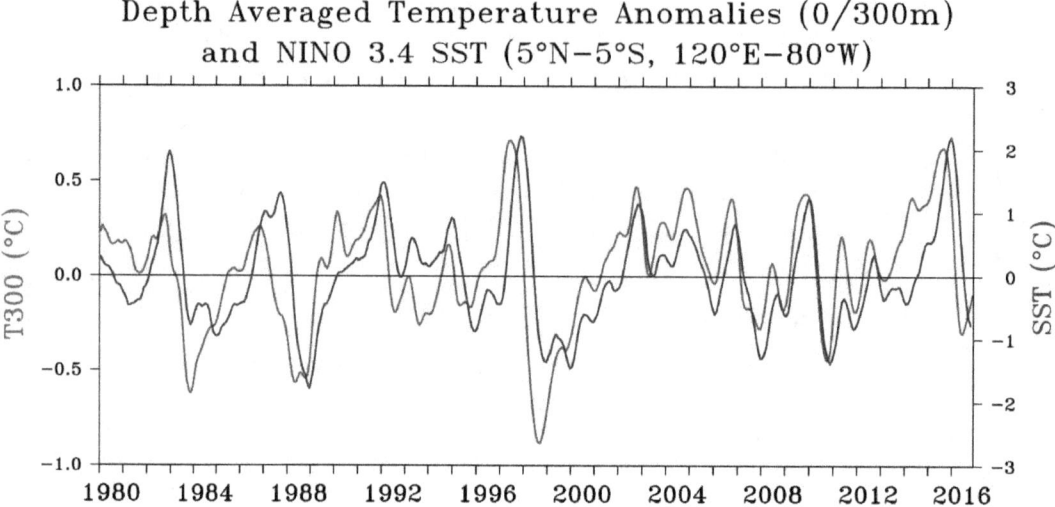

Figure 23.8. Time series of Nino3.4 SST and anomalous ocean heat content along the equator (T300) between 1980 and 2016. Monthly means have been smoothed with a five-month running mean.

Forecast models used to predict ENSO range include purely statistical models, hybrid statistical-dynamical models, and coupled global ocean-atmosphere general circulation models. Seasonal forecasts of Nino3.4 SSTs from these models, beginning in mid-2015, were more accurate than for any event since 2002 when systematic tracking of skill scores for multi-model ensembles began (L'Heureux et al., 2017). As an example, forecasts initialized in July 2015 verified extremely well as an ensemble against the observations over the next year (Fig. 23.9). Global precipitation forecasts also verified well in many regions of the globe at the peak of the event (December 2015–February 2016), with the notable exception of the west coast of the United States (Fig. 23.10). Overall, El Niño forcing from the tropical Pacific could account for about 25% of the variance in seasonal mean precipitation anomalies during the peak of the event (L'Heureux et al., 2017). On the other hand, the highly anticipated rains that El Niño was supposed to bring to California after several years of severe drought failed to materialize. Kumar and Chen (2017) have suggested this failed forecast may simply have been due to chance, given the high degree of random variability in seasonal rainfall totals along the west coast of the U.S. even in the face of strong El Niño forcing from the tropical Pacific.

Summary

Advances in understanding, observing system development, and forecast model development have made skillful ENSO forecasts routinely possible 2–3 seasons in advance. Successful prediction of the 2015–16 El Niño and its impacts demonstrate these advances. Even so, many challenges remain. For example, while seasonal forecasts for 2015–16 were very successful, this is not always the case. Predictive skill was low in the first decade of the 21st century (Barnston et al., 2012) and notably low in 2014, when a highly anticipated strong El Niño failed to develop (McPhaden, 2015). Understanding what factors limit ENSO predictability is thus a major outstanding issue.

A related question involves the processes that give rise to ENSO diversity. Various hypotheses have been proposed, such as variations in stochastic forcing (Levine et al., 2016), decadal changes in the background state of the tropical Pacific (Choi et al., 2012), and forcing from regions outside the tropical Pacific (Paek et al., 2017). How these and other factors may combine to influence the evolution of individual events is a subject of great interest given that diversity in ENSO characteristics can result in a diversity of climatic impacts.

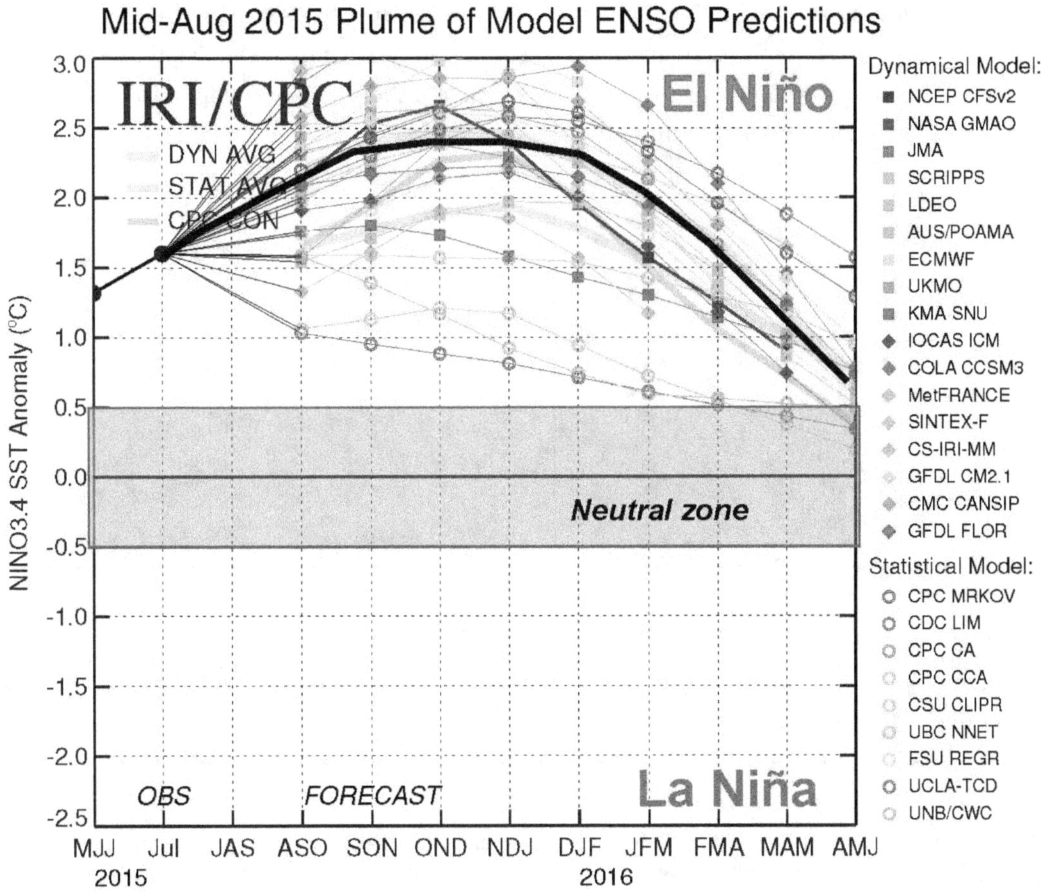

Figure 23.9. Forecasts of seasonal mean Nino3.4 SST anomalies using different forecast models for July 2015 initial conditions. Thick solid black line is the verification. Adapted from the International Research Institute for Climate and Society.

There is also the question of whether global warming has already affected the ENSO cycle and how it will affect it in the future. This is an area of lively debate but much uncertainty. Analysis of historical data, paleoclimate records, and numerical model simulations often lead to conflicting results. Perhaps the most robust conclusion so far is that in a warmer world the frequency of extreme El Niños like those observed in 1982–83, 1997–98, and 2015–16 will increase in the future (Cai et al., 2014). But much more work needs to be done on this topic, which is a rich and fertile ground for further research.

Finally, the TAO/TRITON array, which was designed in the 1980s and implemented over the 10-year period 1985–94, has served as the cornerstone of the tropical Pacific Ocean observing system for the past 30 years (McPhaden et al., 2010). Data from the moored buoy array are distributed via the Global Telecommunications System to operational centers around the globe for routine ocean, weather, and climate forecasting. They also serve as a primary dataset for many oceanic and atmospheric databases and for virtually all oceanic and atmospheric reanalysis products. Since the 1980s, there have been advances in our understanding of the processes involved in ENSO dynamics and new technologies such as Argo profiling floats have become available. As a result of these developments, an international committee is currently reviewing the design of the tropical Pacific Ocean Observing System, with recommendations for how to optimize it for the 21st century still pending (Cravatte et al., 2016).

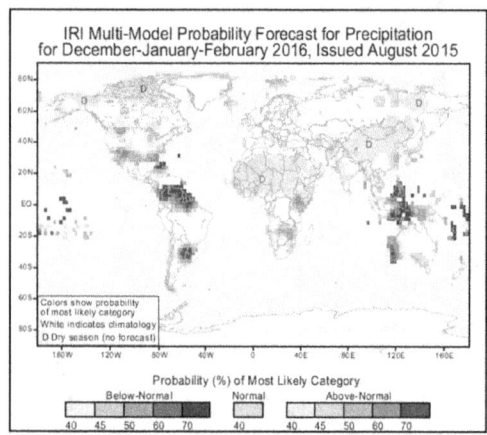
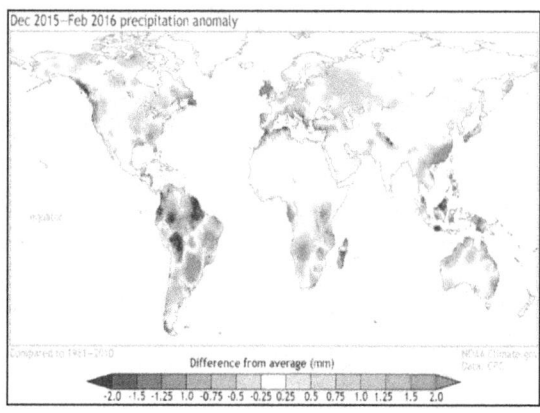

Figure 23.10. Precipitation forecasts for December 2015–February 2016 based on August 2015 initial conditions (left). Actual rainfall anomalies for December 2015–February 2016 (right). Forecasts and observed anomalies are courtesy of the International Research Institute for Climate and Society (https://iri.columbia.edu).

References

Barnston, A. G., M. K. Tippett, M. L. L'Heureux, S. Li, and D. G. DeWitt, 2012: Skill of real-time seasonal ENSO model predictions during 2002–11: Is our capability increasing? Bull. Amer. Meteor. Soc., 93, 631–651.

Battisti, D. S., 1988: Dynamics and thermodynamics of a warming event in a coupled atmosphere-ocean model, J. Atmos. Sci., 45, 2889–2919.

Cai, W., S. Borlace, M. Lengaigne, P. van Rensch, M. Collins, G. Vecchi, A. Timmermann, Santoso, M. J. McPhaden, L. Wu, M. England, E. Guilyardi, and F.-F. Jin, 2014: Increasing frequency of extreme El Niño events due to greenhouse warming. Nature Climate Change, 4, 111–116.

Capotondi et al., 2015: Understanding ENSO diversity. Bull. Am. Metoerol. Soc., 96, 921-938.

Choi, J., S.-I. An, and S. W. Yeh, 2012: Decadal amplitude modulation of two types of ENSO and its relationship with the mean state. Clim. Dyn., doi:10.1007/s00382-011- 1186-y.

Cravatte, S., W. S. Kessler, N. Smith, S. E. Wijffels, and Contributing Authors, 2016: First Report of TPOS 2020. GOOS-215, 200 pp. (http://tpos2020.org/first-report/)

Eisenman, I., L. Yu, and E. Tziperman, 2005: Westerly wind bursts: ENSO's tail rather than the dog? J. Clim., 18, 5224–5238.

Jin, F.F., 1997: An equatorial recharge paradigm for ENSO. Part I: Conceptual model. J. Atmos. Sci., 54, 811-829.

Kumar, A. and M. Chen, 2017: What is the variability in US west coast winter precipitation during strong El Niño events? Clim. Dyn., 49, 2789–2802.

Levine, A. F. Z. and M. J. McPhaden, 2016: How the July 2014 easterly wind burst gave the 2015–2016 El Niño a head start. Geophys. Res. Lett., 43, 6503–6510.

Levine, A.F.Z., F.F. Jin, M.J. McPhaden, 2016: Extreme noise-extreme El Niño: How state-dependent noise forcing creates El Niño-La Niña asymmetry. J. Climate, 29, 5483-5499.

L'Heureux, M. et al., 2017: Observing and predicting the 2015/16 El Niño. Bull. Am. Metoerol. Soc., 98, 1363-1382.

McPhaden, M. J., A. J. Busalacchi, and D. L. T. Anderson, 2010a: A TOGA retrospective. Oceanography, 23, 86-103.

McPhaden, M. J., 2012: A 21st Century Shift in the Relationship between ENSO SST and Warm Water Volume Anomalies. Geophys. Res. Lett., 39, L09706, doi:10.1029/2012GL051826.

McPhaden M J. 2015: Playing hide and seek with El Niño. Nat Clim Change, 5, 791–795

Meinen, C.S. and M.J. McPhaden, 2000: Observations of warm water volume changes in the equatorial Pacific and their relationship to El Niño and La Niña. J. Climate, 13, 3551–3559.

Paek, H., J.-Y. Yu, and C. Qian, 2017: Why were the 2015/16 and 1997/98 Extreme El Niños different? Geophys. Res. Lett., 44, 1848–1856.

Suarez, M. J., and P. S. Schopf, 1988: A delayed action oscillator for ENSO, J. Atmos. Sci., 45, 3283-3287.

Xue, Y. and A. Kumar, 2017: Evolution of the 2015/16 El Niño and historical perspective since 1979. Science China, 60, 1572–1588.

CHAPTER 24

Assessment of High-Resolution Regional Ocean Prediction Systems Using Multi-Platform Observations: Illustrations in the Western Mediterranean Sea

Baptiste Mourre[1], Eva Aguiar[1], Mélanie Juza[1], Jaime Hernandez-Lasheras[1], Emma Reyes[1], Emma Heslop[3], Romain Escudier[4], Eugenio Cutolo[1], Simon Ruiz[2], Evan Mason[2], Ananda Pascual[2], and Joaquin Tintoré[1,2]

[1]SOCIB, Balearic Islands Coastal Observing and Forecasting System, Palma, Spain; [2]IMEDEA, UIB-CSIC, Esporles, Spain; [3]IOC-UNESCO, Paris, France; [4]Department of Environmental Sciences, Rutgers, The State University of New Jersey, USA

High-resolution regional models of the ocean circulation are now operated on a routine basis using realistic setups in many regions of the world, with the aim to be used for both scientific purposes and practical applications involving decision-making processes. While the evaluation of these simulations is essential for the provision of reliable information to users and allows the identification of areas of model improvement, it also highlights several challenges. Observations are limited and the real state of the ocean is, to a large extent, unknown at the short spatiotemporal scales resolved in these models. The skill of the model also generally varies with the region, variable, depth and the spatiotemporal scale under consideration. Moreover, the increased spatial resolution might require ad hoc metrics to properly reflect the model performance and reduce the impact of so-called "double-penalty" effects occurring when using point-to-point comparisons with features present in the model but misplaced with respect to the observations. Multi-platform observations currently collected through regional and coastal ocean observatories constitute very valuable databases to evaluate the simulations. Gliders, high frequency radars, moorings, Lagrangian surface drifters, and profiling floats all provide, with their own specific sampling capability, partial but accurate information about the ocean and its variability at different scales. This is complementary to the global measurements collected from satellites. Using a case study in the Western Mediterranean Sea, this chapter illustrates the opportunities offered by multi-platform measurements to assess the realism of high-resolution regional model simulations.

Introduction

The development of operational oceanography has enabled the production of high-resolution regional models and predictions. On the one hand, large-scale models have gained sufficient maturity over the last decades to provide robust and reliable initial and boundary conditions for regional simulations with an increased resolution (the so-called downscaling approach). On the other hand, oceanographic centers have benefitted from enhanced computing capabilities through the development of high-performance computing technology. Today, short-

---

Mourre, B., et al., 2018: Assessment of high-resolution regional ocean prediction systems using multi-platform observations: Illustrations in the Western Mediterranean Sea. In "*New Frontiers in Operational Oceanography*", E. Chassignet, A. Pascual, J. Tintoré, and J. Verron, Eds., GODAE OceanView, 663-694, doi:10.17125/gov2018.ch24.

term (a few days ahead) predictions over extended areas $\mathcal{O}(1000$ km) with spatial resolutions of $\mathcal{O}(1$ km) and $\mathcal{O}(50)$ vertical levels can be produced sufficiently quickly to be delivered on an operational basis. In parallel, developments in ocean modelling science (e.g., treatment of boundary conditions and sub-grid scale parameterizations) have allowed us to address some of the challenges associated with the increased resolution and its application to coastal areas. These regional implementations are now operated on a routine basis using realistic setups in many regions of the world, often covering open ocean to coastal areas (e.g., Nittis et al., 2001; Wilkin et al., 2005; Siddorn et al., 2007; Chao et al., 2009; Stanev et al., 2011; Kerry et al., 2016; Sakamoto et al., 2016; Kurapov et al. 2017; Kourafalou et al., 2015; and references therein). They aim to reproduce a wide range of spatiotemporal scales, ranging from the variability of the main currents to mesoscale eddies and small-scale features of the coastal circulation such as filaments, coastal eddies or river plumes.

Models provide approximations of reality that are inevitably affected by different types of errors related to inaccuracies in initial and boundary conditions and in the sub-grid scale parameterizations. However, and in spite of this, there is a high demand for products generated from these high-resolution regional models due to their capacity to 1) provide four-dimensional estimates of multiple oceanic variables which are not accessible through observations, and 2) reach coastal zones, raising interest for a diversity of applications. The scientific community uses models to study processes, analyze ocean variability, evaluate energy budgets, or investigate relationships between ocean circulation and ecosystems, for instance. The desire to make informed decisions also motivates the use of models for practical applications. This is the case for search-and-rescue operations, analysis of pollutant drift, science-based management of coastal areas and fisheries, support to ship routing, and forensic science as examples.

Model users from the scientific community and the public have diverse requirements. They might, for instance, ultimately be interested in the strength and direction of the currents over a specific area, their associated variability, the water temperature at the surface and along the vertical, the position and strength of density fronts and eddies, the existence and variability of accumulation zones, the time of residence of water particles in specific areas, or the connectivity between oceanic regions. As a consequence, evaluation of operational model outputs is strongly needed, covering a wide range of different properties (see Hernandez et al., 2015, for a recent review).

Several challenges are associated with the model assessment exercise. First, it is important to keep in mind that evaluation is fundamentally partial since observations are limited in space, time, and variables with respect to the full four-dimensional and multi-parametric extension of the model (Oreskes et al., 1996). This means that only specific properties of the model can be properly evaluated. In particular, high-resolution models produce energetic small-scale oceanic features, such as small eddies and filaments, which are of critical importance for the circulation and energy transfers in the ocean. These structures affect users, in particular through their significant impact on surface currents, but by and large they are not accurately monitored by present observations. Another difficulty is that the model performance generally depends on the region, variable, depth, and spatial and temporal scales under consideration, so that the model assessment presents multiple facets. Moreover, the traditional point-to-point evaluation might become problematic with high-resolution models where oceanic structures like eddies are present, driven by non-deterministic generation mechanisms, and therefore generally out of sync with the reality. For the structures that can be observed, a small misplacement in time or space in the model generally leads to a large error when using traditional point-to-point evaluation metrics (Ziegeler et al., 2012; Sandvik et al., 2016). Alternative model-data comparison approaches could potentially provide complementary information leading to a fairer evaluation of the simulations. This aspect is illustrated in the second section of this chapter.

From the observational point of view, the developments of operational oceanography and marine technologies has allowed a transition from ship-based observations to multi-platform, integrated observing systems based on moorings, tide gauges, gliders, high frequency radars, drifters, Argo floats, and satellites among other platforms, all providing openly-accessible measurements available in real time. Ocean observatories combine a variety of sampling technologies, which provide access to a diversity of measurements, all with a specific spatial coverage, spatiotemporal resolution, and accuracy. This multi-platform observation paradigm (Tintoré et al., 2013) provides new opportunities for the assessment of numerical simulations based on multiple insights into the skill of the models at different scales. This chapter illustrates such multi-platform model assessment opportunities in the Western Mediterranean Sea, taking as an example the Balearic Islands Coastal Observing and Forecasting System (SOCIB).

The chapter is organized as follows: it first addresses the problem of model performance quantification, highlighting the importance of the choice of the metrics. Then, it introduces SOCIB observatory and model simulations in the Western Mediterranean Sea, before presenting multiple facets of the multi-platform assessment of the model from a qualitative perspective. This chapter is complementary to Chapter 29 by Hernandez et al. in this book, presenting a higher level view of coordinated GODAE and CMEMS (Copernicus Marine Service) model validation activities.

Quantifying Model Performance

In this section, before getting into the details of the Western Mediterranean study case, we first introduce the standard statistical metrics used in model error quantification as well as the potential of alternative approaches.

Standard statistical metrics

Estimating the performance of a model to represent the real state of the ocean commonly goes through the construction of a model equivalent to the available observations. This model equivalent might be a simple interpolation in space and time onto the position of the observations. It may also involve some filtering or grid-cell averaging to mimic the observation process as much as possible. The differences between observations and model equivalents can be quantified using different statistical measures. While the term "error" is used in the following, it is worth mentioning that this "error" represents model-observation differences, which have contributions from both model and observation errors.

Given vectors of N observation and model values o_i and m_i, with respective standard deviations σ_o and σ_m, we briefly summarize here the most usual statistical quantities computed to assess model performance:

- The Mean Error (ME)

$$ME = \frac{1}{N}\sum_{i=1}^{N}(m_i - o_i)$$

- The Mean Absolute Error (MAE)

$$MAE = \frac{1}{N}\sum_{i=1}^{N}|m_i - o_i|$$

- The Root-Mean-Square-Error (RMSE)

$$RMSE = \sqrt{\frac{1}{N}\sum_{i=1}^{N}(m_i - o_i)^2}$$

- The standard deviation error (SDE)

$$SDE = \sigma_m - \sigma_o$$

- The cross-correlation coefficient (CC)

$$CC = \frac{\frac{1}{N}\sum_{1}^{N}(m_i - \bar{m}_i)(o_i - \bar{o}_i)}{\sigma_m \sigma_o}$$

- From which the cross-correlation error (CCE) can be computed:

$$CCE = \sqrt{2\sigma_m \sigma_o (1 - CC)}$$

There is an important relationship between these quantities allowing to decompose the overall RMSE into different contributions (e.g. Murphy, 1995; Oke et al., 2002):

$$RMSE^2 = ME^2 + SDE^2 + CCE^2$$

The unbiased or centered Root-Mean-Square Error (CRMSE), defined as follows, quantifies the agreement between the fluctuating parts of observations and model values:

$$CRMSE = \sqrt{\frac{1}{N}\sum_{i=1}^{N}((m_i - \bar{m}_i) - (o_i - \bar{o}_i))^2}$$

It is linked to other quantities through the following expressions:

$$CRMSE^2 = SDE^2 + CCE^2$$

$$RMSE^2 = ME^2 + CRMSE^2$$

While ME accounts for the mean difference between model and observations, SDE compares the standard deviations of the two fields and CCE, through the cross-correlation coefficient, provides a measure of their general correspondence in phase. Each one of these terms affects the total error provided by the RMSE. A low RMSE is obtained when 1) the mean error is low, 2) models and observations have similar standard deviations, and 3) the cross-correlation coefficient is close to unity.

The geometric relationships between these quantities allow to represent these multiple statistical measurements of model performance in single summary diagrams such as Taylor (Taylor, 2001) or Target (Jolliff et al., 2009) diagrams. Moreover, skill scores (Murphy, 1995; Willmott, 1981) can also be defined to determine the improvement with respect to specific references such as a climatology, the persistence field, or any other reference simulation. An example of skill score is:

$$SS = 1 - \frac{RMSE^2}{RMSE_{ref}^2}$$

A positive skill score indicates a better performance with respect to the reference (the closer to unity, the better the match with the observations).

Alternative approaches: neighborhood methods

Given a set of observations, these statistical measures of model accuracy using point-wise comparisons are very practical in that they allow provision of single numbers for the model performance. However, the use of point-wise statistics might become problematic when oceanic features (typically an ocean eddy) are represented in the model but with a mismatch in space and/or time with respect to the observations. In this case, the RMSE will be affected by the so-called "double penalty" error, the first penalty being the non-representation of the observed feature and the second the representation of a non-existing pattern. Alternative evaluation methods that have been developed in meteorology (e.g., Casati et al., 2008; Mittermaier and Csima, 2017) are destined to be more widely used in oceanography in the future to better address these issues. As an illustration of these possibilities, we propose here an alternative evaluation of the performance of different simulations in representing a mesoscale eddy. Fig. 24.1 illustrates a situation where an eddy was observed by altimetry south of Mallorca Island in the Western Mediterranean Sea. Five simulations generated at SOCIB varying model parameters and boundary conditions provide five different representations of sea level anomalies for this date, some of them with an eddy in the neighborhood of the observed structure.

The simulations representing an eddy with similar characteristics to that observed in altimetry, but in a slightly different position, are penalized in terms of RMSE with respect to simulations without any marked eddy. Therefore, an evaluation based on this RMSE might not be a completely fair assessment of the performance of the model.

An alternative framework to evaluate these simulations is provided by neighborhood methods (Ebert, 2009), consisting of characterizing model-data correspondences within space-time neighborhoods not limited to the observation point. In the example shown in Fig. 24.1, the Okubo-Weiss parameter (a particular metric allowing the identification of two-dimensional vortices based on the separation of strain-dominated and vorticity-dominated regions (Okubo, 1970; Weiss, 1991)) has been calculated from the sea level anomalies altimeter map and from the different simulations to focus on the characterization of the eddy. Values of the Okubo-Weiss parameter lower than the specific threshold of minus two standard deviations (Isern et al., 2003) determines the presence of the eddy in the data and in the simulations. Given a neighborhood distance and a particular grid point where an eddy is detected in the observations, the skill of a simulation for this grid point is here set to one if an eddy is found (from the value of the Okubo-Weiss parameter) in that neighborhood, zero otherwise. The overall skill of a simulation for this specific neighborhood distance is then computed as the spatial average of all local skills computed at every grid point where an eddy is detected in the observations. The overall skill of the different simulations for different neighborhood scales are presented in Fig. 24.2.

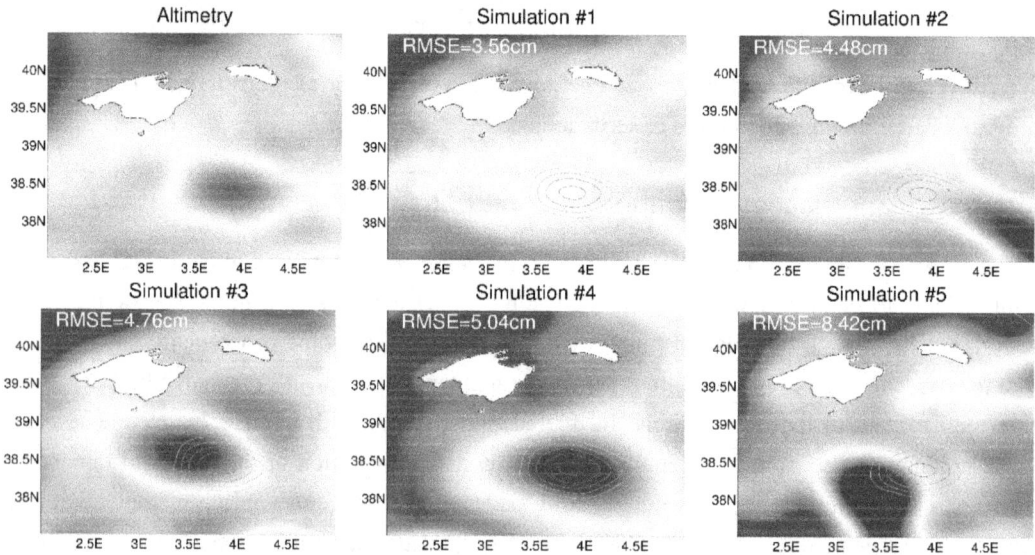

Figure 24.1. Sea level anomalies (cm) from altimetry and five different simulations for 26 September 2014. The point-wise RMSE is specified in the upper-left corner of the simulation panels. The contours of the observed eddy are also plotted in these panels.

Fig. 24.2 shows how simulations 1 and 2, which were the best ranked according to the RMSE, have the lowest skill when using this metric focusing on the representation of the eddy. Simulations 3 and 5 have a maximum skill from 24 km and 44 km distances, respectively. Simulation 4, which represents an eddy centered at the right place but with an overestimation of the radius, exhibits the best skill of all the simulations.

This example illustrates how two different metrics (the standard RMSE and a more sophisticated distance-dependent skill measure focusing on a parameter identifying a specific ocean structure) lead to very different results in terms of model performance. The particular purpose of the use of the model might require the definition of very specific metrics to quantify the model performance.

Other alternatives to point-wise comparisons in deterministic simulations include probabilistic verification approaches, which also aim at dealing with the uncertainty in the location and timing of ocean features. In that framework, an ensemble of model simulations is generated, which allows us to quantify the probability of occurrence of specific oceanic events (e.g., Candille and Talagrand, 2005). Lagrangian methods identifying Lagrangian coherent structures associated with currents, eddies and filaments also provide interesting alternative approaches for the evaluation of high-resolution models. Lagrangian coherent structure diagnostics allow to detect material surfaces hidden in the ocean circulation (Peacock and Haller, 2013), but with a crucial role on the transport, dispersion and mixing properties of the flow.

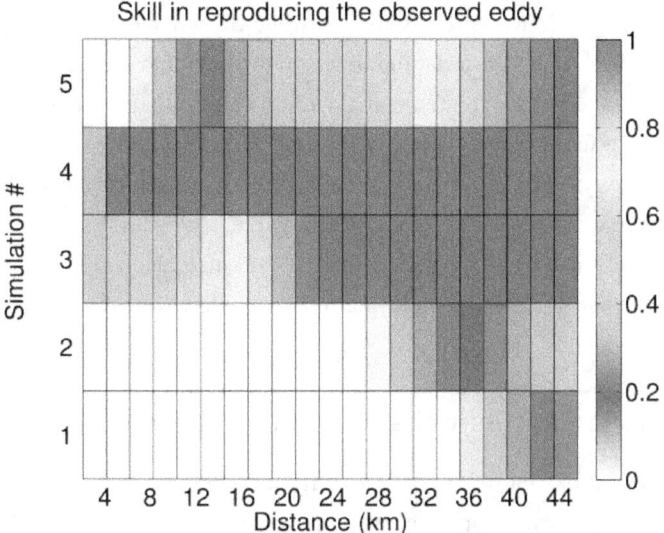

Figure 24.2: Overall skill of the different simulations in reproducing the observed eddy, based on Okubo-Weiss parameter computations and application of a neighborhood approach.

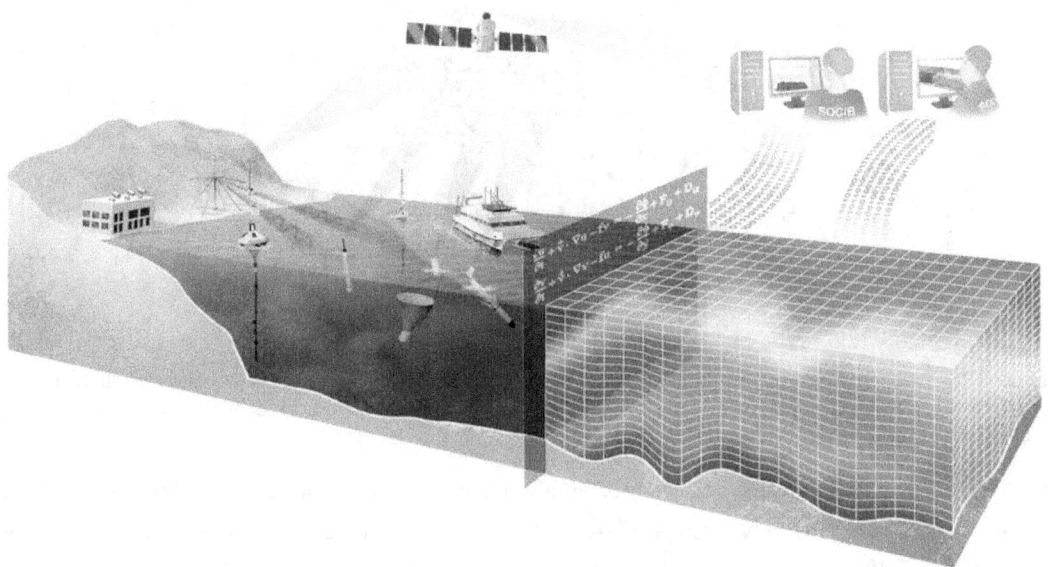

Figure 24.3. Illustration of SOCIB components, with observing facilities (ARGO floats, gliders, research vessel, surface drifters, moorings, beach monitoring, satellites, radars), a modelling and forecasting facility and a data centre for the management and distribution of observations and simulations.

SOCIB Observatory, Study Area, and Modelling System

SOCIB coastal observatory

SOCIB, the Balearic Islands Coastal Observing and Forecasting System (Tintoré et al., 2013), is a coastal observatory located in Mallorca, Spain, with the objectives to collect, quality-control, and distribute multi-platform ocean observations from both fixed and Lagrangian platforms. SOCIB capacities extend from the coastal to the open ocean. In close collaboration with researchers at the Mediterranean Institute for Advanced Studies (IMEDEA, UIB-CSIC), SOCIB aims at supporting research and technology and has a strong orientation towards applications for society. To complement the observations and to provide added-value products, it also produces forecast and hindcast simulations of the ocean circulation, waves, and meteotsunamis affecting the harbour of Ciutadella in Menorca Island. SOCIB multi-platform and modelling components are illustrated in Fig. 24.3.

Circulation and water masses in the Western Mediterranean Sea

SOCIB activities are mainly centred in the Western Mediterranean Sea, which we briefly describe here in terms of its main oceanographic characteristics (see Fig. 24.4 for a scheme of the surface circulation).

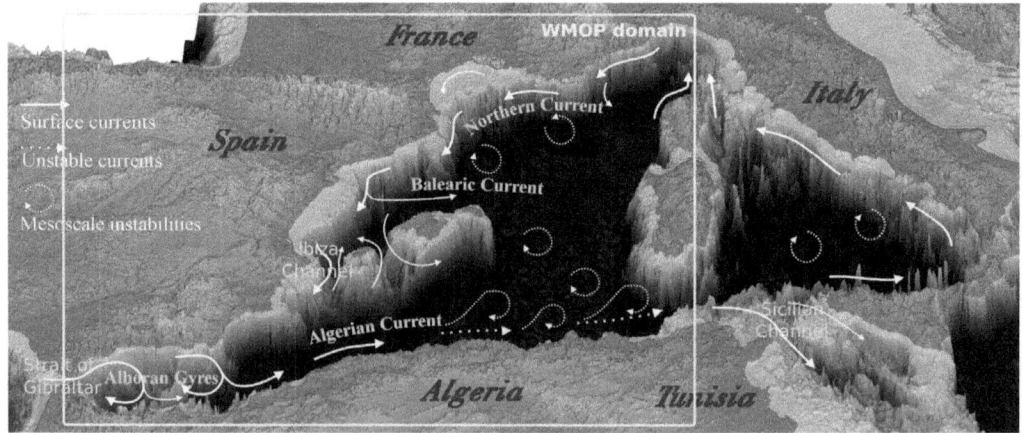

Figure 24.4. Schematic representation of the surface circulation in the Western Mediterranean Sea. The Western Mediterranean Operational Model (WMOP) model domain is indicated in yellow. Adapted from Escudier et al. (2016a) and Millot (1999).

The Western Mediterranean Sea sub-basin is connected to the Atlantic Ocean through the Strait of Gibraltar, where relatively fresh Atlantic Water (AW) is injected with a ~1Sv transport, giving rise to the so-called Atlantic Jet. The Atlantic Jet is an intense meandering and unstable current that interacts with two large anticyclonic gyres in the Alboran Sea until reaching the Almería-Oran front, which separates AW from saltier waters having already recirculated in the Mediterranean Sea. The surface AW then flows along the northern coast of Africa forming the Algerian Current. This current is highly unstable and associated with an intense mesoscale activity. In particular, baroclinic instabilities generate meanders and eddies that detach from the main flow and propagate along two preferred paths in the southern part of the Western Mediterranean basin (Escudier et al., 2016b). The surface water splits into two branches when reaching Sicily. The first branch crosses the Sicilian Channel and flows cyclonically towards the Eastern Mediterranean Basin where it becomes more saline due to intense evaporation. The second branch flows to the north along the western coast of Italy through the Tyrrhenian Sea. This northward current converges north of Corsica with the Western Corsica Current to give rise to the Northern Current, which flows westwards along the shelf break until reaching the Balearic Sea. It then splits into two branches: the Balearic Current flows north-eastwards along the northern shelf of the Mallorca and Menorca islands, also fed by AW inflows through the Ibiza and Mallorca channels, and the remaining part flows southwards through the Ibiza Channel into the Algerian Basin. Further details can be found in Millot (1999), Millot and Taupier-Letage (2005), or Robinson et al. (2001).

Different water masses are found in the Western Mediterranean Sea. At the surface, the relatively fresh AW becomes progressively more saline along its path in the Mediterranean basin under the effects of evaporation. In particular, two types of AW can be distinguished in the Balearic sub-basin according to their salinity: 1) AW of recent Atlantic origin that flows from the south (AWr, salinity<37.5), and 2) more saline AW having circulated in the Western Mediterranean Sea and flowing from the north (AWo, salinity>37.5). At the intermediate levels, a warm and salty Levantine Intermediate Water, which originates from convection processes in the Levantine sub-basin of the Eastern Mediterranean Sea, enters the Western basin through the Sicilian Channel at depths between 200 and 500 m. A second intermediate water mass, relatively cold and fresh and known as the Western Intermediate Water, is formed during winter cooling events in the Gulf of Lion, Ligurian Sea, or Ebro river area. This regional winter water mass flows southwards between AW and Levantine Intermediate Water, intermittently reaching the Ibiza Channel. The Western Mediterranean Deep Water is found at deeper levels (>1000 m), formed during extreme winter weather events leading to deep convection in the Gulf of Lion.

The Mediterranean Sea is often considered as a reduced ocean laboratory for ocean studies, due to the presence of ocean processes of global relevance such as mesoscale and submesoscale activity, eddy propagation, water mass formation and spreading, and the smaller Rossby radius (~10-15 km) compared to the large oceans. In particular, the narrow Ibiza Channel (850 m deep and 90 km wide), and to a lesser extent the Mallorca Channel (650 m deep and 80 km wide), represent key "choke points" for the north/south exchanges of the different water masses, and concentrate signals from different processes including surface circulation, mesoscale activity, and intermediate water mass formation and propagation. This is the reason why they have been the focus of specific regional sampling programs, including a glider endurance line operated by SOCIB since 2011.

Modelling system

The Balearic Sea is a particularly challenging region for ocean modelling due to the complexity of the topography and the interaction of multiple ocean processes, in which salinity gradients in particular play a key role. To simulate the ocean circulation and mesoscale variability in the Balearic Sea and adjacent sub-basins, SOCIB runs the Western Mediterranean Operational Model (WMOP, Juza et al., 2016). WMOP provides both daily predictions and hindcast simulations over the recent years. The model has a spatial resolution of 2 km, and is nested in the larger-scale 1/16° Mediterranean model from the Copernicus Marine Service (CMEMS-MED, Clementi et al., 2017; Simoncelli et al., 2014) through a dynamical downscaling approach. WMOP uses a regional configuration of the ROMS (Shchepetkin and McWilliams, 2005) model covering an area from the Strait of Gibraltar to the Sardinia/Corsica Islands. The vertical grid is made up of 32 stretched sigma levels. The high resolution HIRLAM atmospheric fields (Undén et al., 2002) from the Spanish Meteorological Agency are used to compute surface fluxes through bulk formulae. They have a resolution of 5 km in space and 1 hour in time. Runoffs from the six major rivers of the domain are also specified as point sources of low saline water with their corresponding volume transports. The vertical mixing is determined in the model using the generic length-scale method described in Umlauf and Burchard (2003). At the boundaries, mixed active-passive conditions (Marchesiello et al., 2001) are imposed using daily forcing data from the CMEMS-MED simulations. A particular treatment, including the alignment of bathymetries between the external model and WMOP and a correction of interpolated velocities, is applied at the Strait of Gibraltar to ensure that the boundary forcing field properly represents the original inflow and outflow transports of the large-scale model.

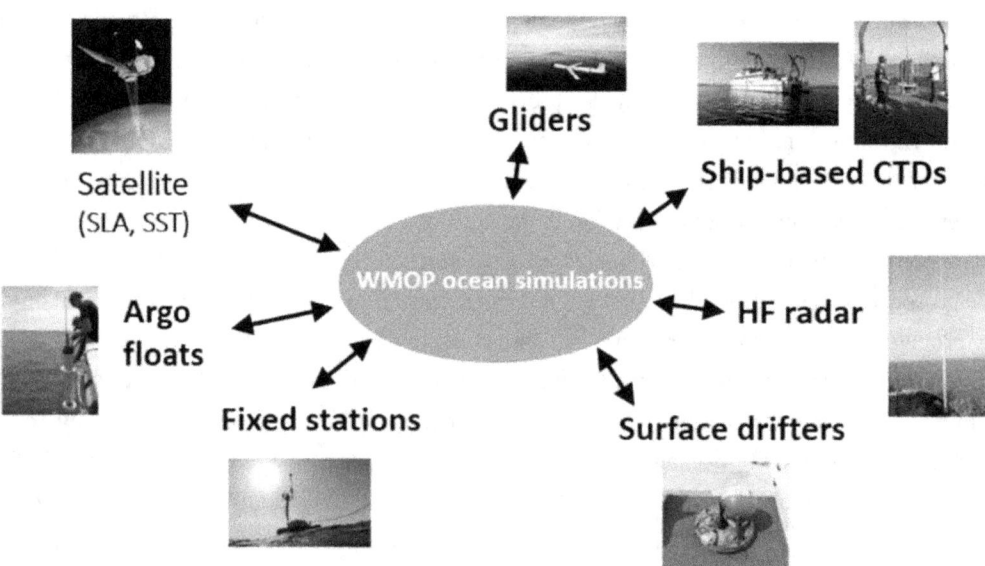

Figure 24.5. Multi-platform observations used for WMOP model evaluation in near real-time and delayed modes.

The model is run in both hindcast and forecast modes. None of them include data assimilation at the moment. On the one hand, free run hindcast simulations (hereafter the *WMOP_hindcast*) are performed over the period 2009-2015 using initial and boundary fields from the CMEMS

Mediterranean Sea physical reanalysis (CMEMS-MED-REAN, Simoncelli et al., 2014), with an initialization on 15 September 2008. Observed daily river discharge values are prescribed in the *WMOP_hindcast*. These free-running simulations are mainly used for detailed studies of dynamical ocean processes.

On the other hand, WMOP forecasts use the CMEMS Mediterranean Sea Analysis and Forecast products (CMEMS-MED-AN-FC, Clementi et al., 2017) and climatological values of river discharges to provide daily operational 72-hour predictions (hereafter the *WMOP_forecast*). *WMOP_forecast* is reinitialized every week from the output of a three-week spin-up simulation initialized from the CMEMS-MED-AN-FC model fields. This regular restart aims to avoid significant drifts of the model in the absence of data assimilation. The *WMOP_forecast* uses climatological values of river discharges. Both the *WMOP_hindcast* and *WMOP_forecast* use HIRLAM outputs as atmospheric forcing.

Multi-platform Model Assessment in the Western Mediterranean Sea

WMOP ocean simulations are systematically compared to multi-platform observations from satellite, SOCIB platforms, and other national and international ocean monitoring systems, both in real-time and delayed modes (Fig. 24.5). In particular, the *WMOP_forecast* model-data comparisons are updated daily on SOCIB website (www.socib.es). In the following section, we illustrate the multiple possibilities of model assessment provided by these different platforms, remaining intentionally qualitative in the evaluation. Yet, all types of model-data comparisons illustrated in this section can be quantified through the standard statistical measures introduced in Section 2 or through any other specific metrics.

Satellite altimetry and Mean Dynamic Topography

Satellite altimeters are unambiguously essential ocean observing platforms that enable repetitive and global measurements of sea level anomalies with respect to a long-term mean. These sea level anomalies provide estimates of the variability of surface geostrophic currents. Furthermore, advanced methods combining altimeter time series, hydrographic profiles, and long-term model averages allow to estimate the Mean Dynamic Topography (MDT), i.e. the long-term mean sea level elevation with respect to the marine geoid. This, in turn, allows for provision of accurate estimates of the mean surface geostrophic circulation over the period of a model simulation. Note that the MDT is not directly measured by the altimeter, but instead estimated from a combination of observation and model data sources. The scarcity of measurements in specific areas might still lead to uncertainties in this estimate of the mean circulation. The comparison of the mean model sea surface with the MDT allows us to evaluate the large-scale mean circulation in the model, which is an essential aspect to be verified before considering further details. Also note that the MDT provides mean sea level elevations over a reference period. When the period of the model simulation differs from this reference period, this mean state needs to be corrected by the mean altimeter anomalies over the simulation period to represent the mean dynamic topography over the same period as the numerical simulation.

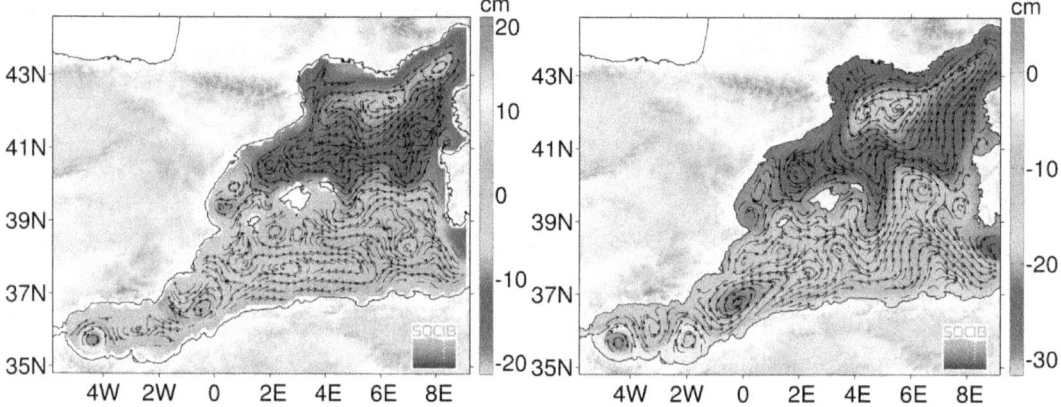

Figure 24.6. Mean absolute dynamic topography and associated surface geostrophic currents over the period 2009–2015. Left: from the MDT (Rio et al., 2014) corrected by mean altimeter anomalies over the period 2009–2015; Right: from the *WMOP_hindcast* simulation.

The comparison displayed in Fig. 24.6 shows that the regional WMOP model is able to properly represent the mean surface circulation, including the Atlantic Jet, the western and eastern Alboran gyres (with a stronger signal of the eastern Alboran gyre with respect to the MDT), the Algerian Current, the cyclonic circulation in the Ligurian Basin and Gulf of Lion, and the Northern and Balearic currents. Details of eddies and meanders differ between the two estimates, especially in the Algerian Basin. Similar analyses with other Western Mediterranean Sea models can be found in Pascual et al. (2014) and Escudier et al. (2016a). Note that the color bars in both panels have the same range but are not centered on the same values. Indeed, here the model mean sea level is mainly determined by initial and boundary conditions coming from the larger scale CMEMS-MED-REAN model and does not necessarily match the mean sea level provided by the MDT. The relevant aspects of the comparison concern the spatial variability and the associated gradients of the mean sea surface height, which determine the mean ocean circulation. The mismatch between the spatial mean model sea surface and the MDT will, however, require special care when assimilating data in the model.

The fundamental strength of altimetry is its ability to measure the variability of the sea level around this mean state. In particular, sea surface height variance or eddy kinetic energy (EKE) computed from the altimeter-derived geostrophic currents allows us to quantify the energy associated with the mesoscale activity. Fig. 24.7 shows several estimates of the EKE from a) altimeter-derived geostrophic velocities (calculated from the daily $1/8°$-resolution, delayed-time, all-sat-merged Mediterranean Sea gridded product distributed by CMEMS); b) total model velocities; c) model surface geostrophic velocities; and d) spatially- and temporally-filtered model surface geostrophic velocities to remove the effects of the small scales unresolved by altimetry.

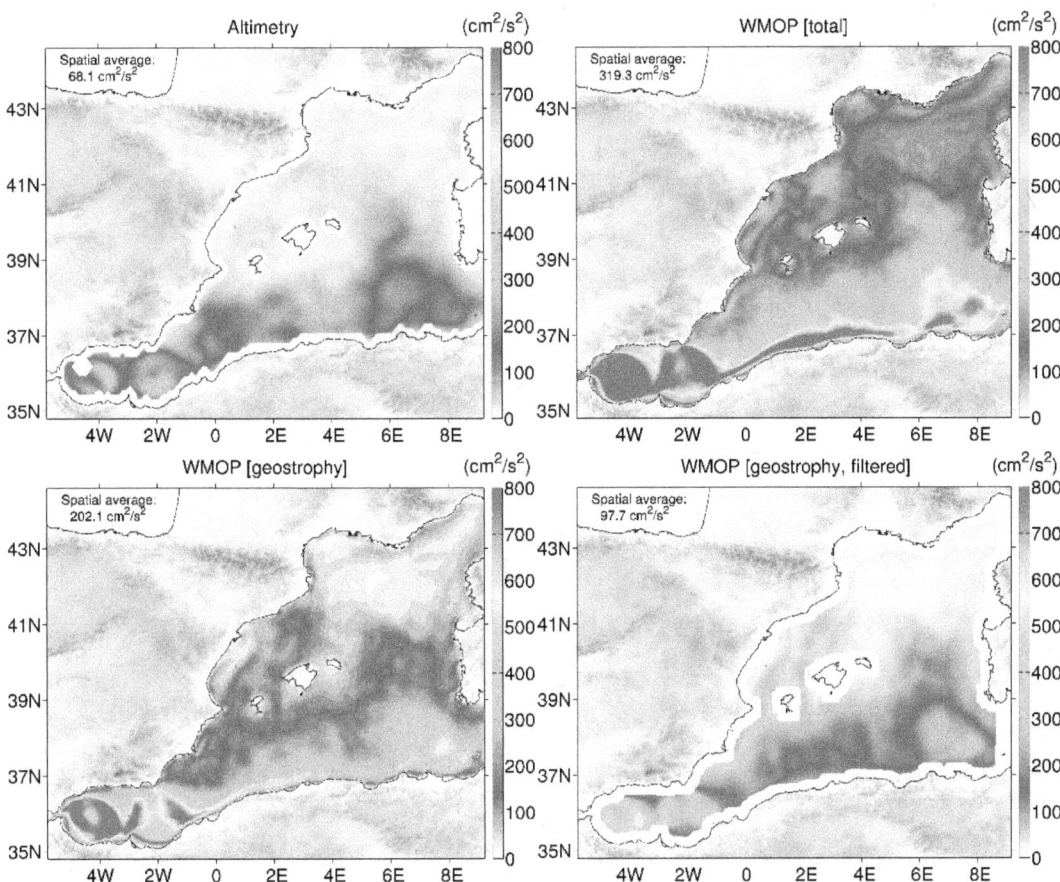

Figure 24.7. Mean eddy kinetic energy (EKE) over the period 2009–2015, from: a) gridded altimetry; b) *WMOP_hindcast* total surface currents; c) *WMOP_hindcast* geostrophic surface currents; d) *WMOP_hindcast* filtered (44 km and 10-day moving average) geostrophic surface currents.

Comparisons of the different panels reveal that the eddy activity is larger in the southern part of the model (Alboran Sea and Algerian Basin), both in the model and altimeter estimates. However, the mean EKE is found to be more than two times larger in the total model velocities than in the altimeter-derived geostrophic currents. Indeed, the high-resolution model represents 1) ageostrophic processes and 2) energetic small eddies and filaments associated with the mesoscale structures, which are not present in the altimeter estimates. Gridded altimetry is only able to represent the geostrophic mesoscale activity associated with oceanic structures with a minimum radius around 40-50 km (Chelton et al., 2011), only allowing us to evaluate the realism of the model at those scales. Comparisons of panels b) and c) illustrate the importance of the ageostrophic component in the total model surface currents. The EKE is reduced by up to 40% in the energetic areas of the Alboran Sea and Algerian Basin. When the model surface geostrophic velocities are filtered to remove the impact of small spatiotemporal scales that are not resolved by altimetry (panel d), the EKE gets much closer to that provided by altimeter estimates, indicating a realistic amount of mesoscale variability of the model at these scales. In the model assessment exercise, the representativity of model estimates with respect to specific observations needs to be properly taken into account to interpret model-data differences. Here, only complementary data from surface drifters and high frequency (HF) radar would allow us to assess the realism of the model EKE at the smaller scales.

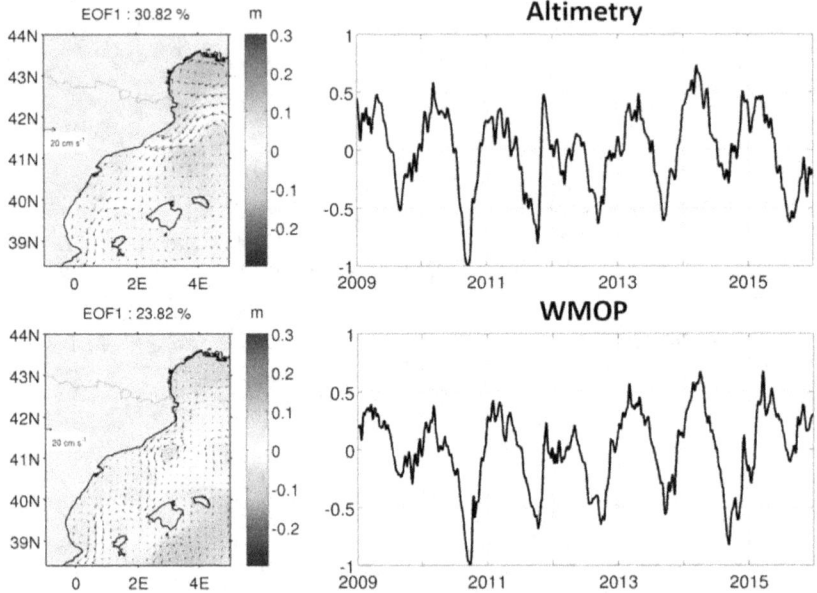

Figure 24.8. First empirical orthogonal function (EOF) mode of sea level anomaly variability from altimetry (upper panels) and *WMOP_hindcasts* (lower panels). Left: spatial pattern and explained variance; Right: associated time series.

Sea level map time series available through altimetry are also frequently analyzed in terms of empirical orthogonal functions (EOFs), which consist of decomposing the signal in terms of orthogonal basis functions (e.g., Emery and Thompson, 2001). As a result of this analysis, a set of dominant spatial patterns of variability can be identified together with their corresponding time series. Fig. 24.8 illustrates one such analysis, comparing the first EOF mode of sea level anomalies variability from altimetry and from the model, in terms of spatial pattern and corresponding temporal coefficients. In both cases, this mode is associated with an acceleration (when the associated coefficient is positive) or deceleration (when the coefficient is negative) of the slope current.

This comparison shows that the main mode is associated with the velocity of the slope current in both cases, but with some differences such as a larger variability in the Gulf of Lion in the observations and the significant signature of an eddy north of Mallorca in the model. Both time series have a strong seasonal component with a good correspondence in phase, modulated by a significant interannual variability. The deceleration of the Northern Current in 2010, characterized by a significant negative value of the EOF coefficient is well represented in the model. EOF patterns and associated time series are useful diagnostics to detect whether the model properly captures the main dynamical properties in the area of study.

Other sophisticated analyses of sea level signals can be carried out to evaluate the model performance. For instance, along-track altimeter and model data can be analyzed in terms of sea level wavenumber power spectra to examine the energy distribution across different scales (e.g. Le Traon et al., 2008). Finite size Lyapunov exponents computations in the model and the data can also be applied to evaluate the representation of Lagrangian coherent structures associated with eddies and fronts in the model. Here, we illustrate another advanced analysis consisting of characterizing eddies using an automatic detection method.

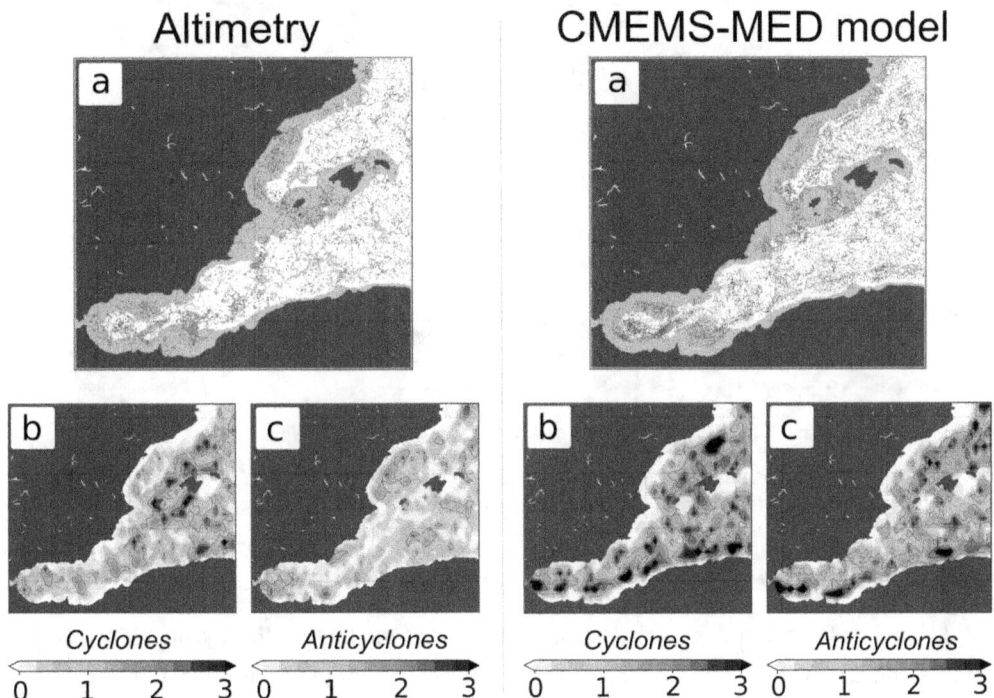

Figure 24.9. Selection of eddy properties in the western Mediterranean from altimetry and the CMEMS-MED-AN-FC model between 2013 and 2016. (a) Cyclone (blue) and anticyclone (red) eddy tracks detected in the altimetry and the model. Light blue and orange markers indicate eddies with lifetimes greater than six months (cyclones and anticyclones, respectively). Bottom: eddy counts per 0.2°x0.2° bins per year for altimetry and CMEMS-MED-AN-FC model for (b) cyclones and (c) anticyclones.

Indeed, the availability of sea level altimetry maps has motivated the development of these automatic eddy detection methods based on either closed contours of sea level anomalies, geometry of surface velocities, or local deformation properties of the flow (Chelton et al., 2011; Nencioli et al., 2010). They allow quantification of the number, position, polarity, amplitude, size, and lifetime of eddies, which provides an alternative and novel view of the mesoscale variability that can also be evaluated in model outputs (Halo et al., 2014; Escudier et al., 2016a).

The examples given here use the automated eddy tracker *py-eddy-tracker* (v3; Mason et al., 2014), which analyzes sea surface height fields, searching for closed contours that are associated with the surface signature of eddies. The method is applied here to altimetry and to the CMEMS-MED-AN-FC model, which is used as the "parent" model of the *WMOP_forecast* simulations. Fig. 24.9 shows eddy locations for cyclones (blue) and anticyclones (red) from altimetry and for the CMEMS-MED-AN-FC model over the period 2013–2016. Eddies are present over the whole domain, with higher densities in the Algerian basin, especially around longitude 0°, where instabilities of the Algerian current occur (Millot, 1985, Escudier et al., 2016b).

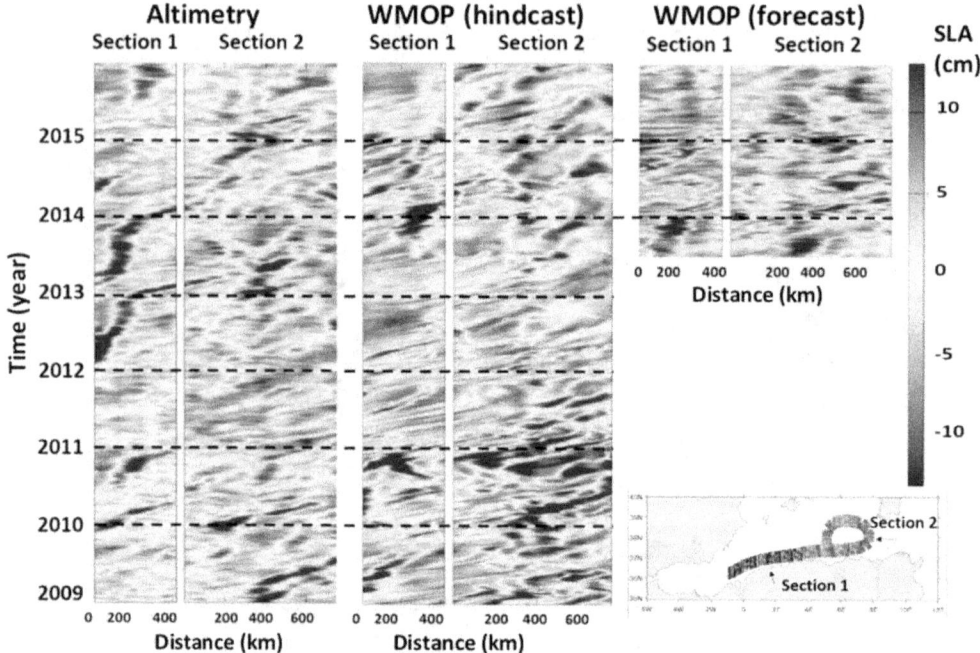

Figure 24.10. Hovmöller diagrams of sea level anomalies along sections 1 and 2 of Algerian eddy propagation paths for altimetry, *WMOP_hindcast* and *WMOP_forecast* simulations.

Eddy counts from the eddy tracker are higher from the model than those from altimetry, for both cyclones and anticyclones. This result is to be expected given the higher resolution of the model compared to altimetry. Furthermore, the presence of eddies in gridded altimetry products was shown to be affected by the particular position of satellite tracks (Escudier et al., 2016a), which does not happen in the model. These advanced automatic detection methods have been successfully applied with gridded altimeter data in many regions of the world oceans, while also highlighting different outcomes between distinct detection methods in some cases (Escudier et al., 2016a, 2016b). Their application to high-resolution models representing smaller and highly dynamic eddies is relatively recent and probably still requires a special attention in the interpretation of the results.

Eddy statistics derived from the application of multiple eddy trackers to altimeter map time series have permitted the identification of eddy propagation paths in the Algerian Basin (Escudier et al., 2016b). Algerian eddies preferentially propagate around two gyres with two marked detachment points from the African coast. Once these propagation paths have been identified, Hovmöller diagrams of sea surface anomalies along these paths allow us to analyze the presence and propagation of eddies and evaluate these particular aspects in model simulations. Fig. 24.10 presents such an analysis for altimetry maps, *WMOP_hindcast* and *WMOP_forecast* models. Even if the number of eddies generated in the model along section 1 (see map in lower right corner of the figure) is quite consistent with the observations (one large eddy per year with a significant interannual variability), as well as the propagation speed of these eddies, their timing is not properly represented in the model. This comparison indicates that instabilities of the Algerian current occur in the model, but the proper representation of particular eddies and meanders needs to be constrained through data assimilation.

Finally, it is worth mentioning that the advent of high-resolution altimetry, including delay-Doppler/SAR (synthetic aperture radar) technology, Ka-band missions, 20Hz high-sampling rate signals, or the forthcoming SWOT wide-swath instrument, all providing an enhanced capacity to approach closer to the coast, also offers new capabilities to evaluate the smaller-scale features represented in high-resolution ocean models.

Satellite sea surface temperature

Sea surface temperature (SST) is a key ocean variable routinely observed by satellites with a global coverage and high resolution. By measuring the radiation emitted by the ocean surface, thermal infrared and passive microwave radiometers provide SST images with different resolution, accuracy and spatiotemporal coverage due to the specific influence of clouds and other atmospheric effects. Advanced optimal interpolation and blending allow to combine the measurements from different satellites and provide improved gap-free SST products (e.g., GODAE High-Resolution SST GHRSST project) that are useful for model validation. Model-data SST comparisons allow us to evaluate the representation of air-sea interactions and vertical mixing in the model and provide indications on the validity of both model parameterizations and external forcing fields. High-resolution SST is also useful to monitor mesoscale structures like eddies, fronts, or filaments since these structures often have significant signatures in temperature.

The use of combined products provides access to gap-free SST time series allowing us to precisely assess the seasonal cycle in the model, the main modes of variability through EOF analysis, or the regional distribution of model surface temperature errors. In cloud-free areas, the use of images from infrared radiometers might be more adapted when focusing on the evaluation of specific mesoscale and submesoscale features due to their high resolution (~1km) and accuracy, which allows for the precise identification of surface temperature gradients associated with fronts and filaments. Advanced detection algorithms (e.g., Belkin and O'Reilly, 2009) can be applied to both satellite imagery and model outputs to evaluate the representation of oceanic fronts in the model. One aspect to be considered when comparing model and satellite SSTs is that these estimates do not generally represent the temperature at the same effective depth. On the one hand, infrared and microwave radiometers measure the so-called skin and sub-skin SST, which correspond to depths around 10 μm and 1 mm, respectively. The foundation SST, which corresponds to night-time SST free of any diurnal variability at a depth around 10 m, is often provided in blended products. On the other hand, the effective depth of the model SST depends on the thickness of the uppermost level. Fig. 24.11 illustrates the richness of satellite SST data, showing an example of comparisons of SST with the *WMOP_forecast* and Level-4 optimally interpolated product from CMEMS (Buongiorno-Nardelli et al., 2012) for a specific day on 19 September 2017.

It shows good overall agreement between the two fields, with differences in the small-scale features associated with the frontal areas, in the coastal zones of the Gulf of Lion under the influence of upwellings and the Rhône river discharge, and in the Alboran Sea where the absence of tidal mixing in the model tends to generate an overestimation of the SST. Notice the representation of many details in the observed SST field, such as the long filament of relatively colder water with respect to the surroundings extending south of Ibiza Island towards the African coast around 2°E.

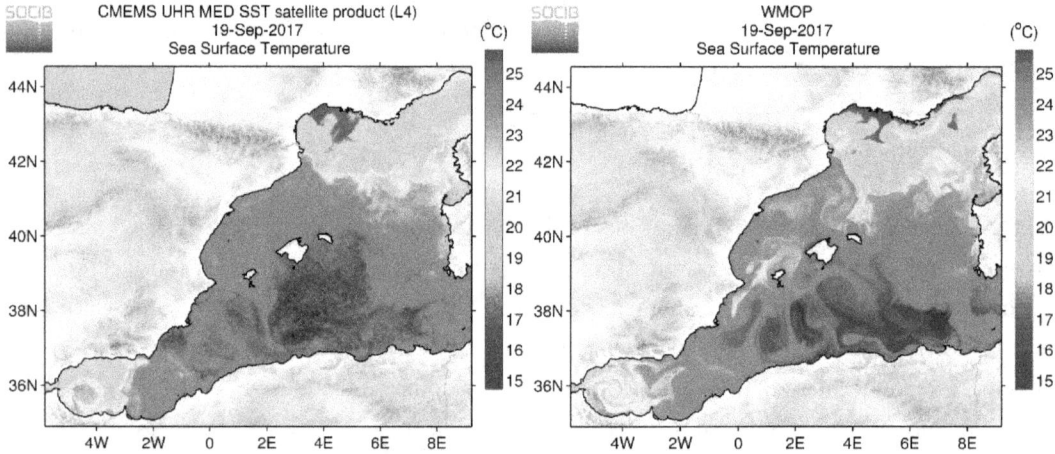

Figure 24.11. Sea surface temperature (SST) on 19 September 2017 from a) a satellite-derived optimally interpolated product distributed through CMEMS, and b) the *WMOP_forecast* model.

Argo floats

Argo floats are Lagrangian platforms that provide regular profiles of the upper thermohaline structure in the ocean, allowing a vertical characterization that is complementary to surface satellite measurements. In the Mediterranean Sea, Argo floats are programmed to execute five-day cycles. SOCIB participates in the Argo program, deploying several floats each year. At the time of the writing of this chapter, there were 63 floats distributed over the whole Mediterranean Sea, 25 of them located in the Western Mediterranean sub-basin. This system provides an average of five profiles per day with a random distribution over the domain. The distance between two profiles is on the order of 200-300 km. While Argo profiles do not allow us to characterize small-scale variability, they are very useful in assessing the seasonal and interannual variations of water mass properties over the basin. They provide estimates of vertical stratification and mixed layer depth and so they can be used to evaluate the vertical mixing properties in the model. They might also be used to test the performance of the model in representing intermediate and deep water masses after convection events that are parameterized in the model. Argo floats provide systematic salinity data, which is an essential variable affecting sea water density and ocean circulation and of particular importance in the Mediterranean Sea. This is highly valuable for the assessment of ocean models, which are generally found to be affected by significant errors in the representation of ocean salinity. Fig. 24.12 illustrates Argo-model comparisons available for a specific day in September 2017, allowing us to evaluate the realism of the model mixed layer and thermocline in particular. In the recent years, the Argo program has been extended to include biogeochemical observations. More than 100 Biogeochemical-Argo profiling floats were deployed in the world oceans, providing measurements of oxygen, nitrates, chlorophyll fluorescence, or particulate backscattering. The Mediterranean Sea is one of the regions with the densest network of Biogeochemical-Argo floats.

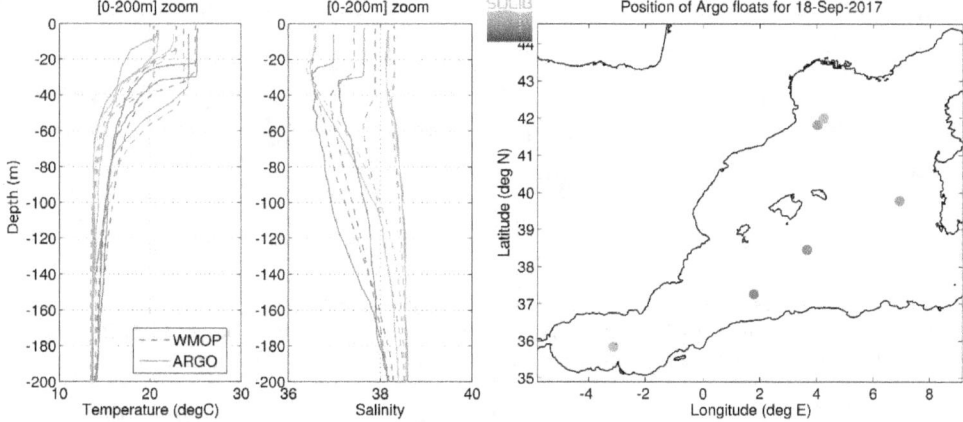

Figure 24.12. Temperature and salinity vertical profiles from Argo and from the *WMOP_forecast* on 18 September 2017.

Fixed moorings

Moorings offer the capability to measure multiple ocean variables at a fixed location over long time periods. They allow us to characterize both high-frequency variability, sporadic events, and longer period signals at a wide range of timescales on the shelf and in the coastal area. Fixed moorings can be equipped with CTDs, acoustic Doppler current profilers (ADCPs) or thermistor chains to provide measurements of temperature, salinity, and velocities at different depths. They also frequently host meteorological instruments. SOCIB maintains two fixed moorings, in Ibiza Channel and in the Bay of Palma, that are complementary to other stations operated by Puertos del Estado along the continental slope of the Iberian Peninsula. Measurements from fixed moorings can be used to verify whether the model properly represents the frequency content of the variations of oceanic properties. They can also be used to identify and analyze specific events in the time series. When equipped with meteorological sensors, they might provide insights into the realism of air-sea fluxes in the model. Importantly, moorings are also platforms used to collect biogeochemical measurements.

The left panel in Fig. 24.13 illustrates the evolution of SST over a one-month period. It shows an overall decrease in the SST associated with the seasonal cycle, combined with 1) SST variations over the atmospheric weather scale of a few days and 2) the daily warming cycle due to solar radiation. On the longer timescale (right panel), the seasonal cycle and interannual variations are also precisely measured by the instrument and can be evaluated in the model, illustrating how model-data comparisons can be used to evaluate the model performance at these different scales. Moorings equipped with temperature and salinity sensors along the vertical also allow to monitor the variability of the mixed layer depth and to precisely evaluate vertical mixing properties in the model.

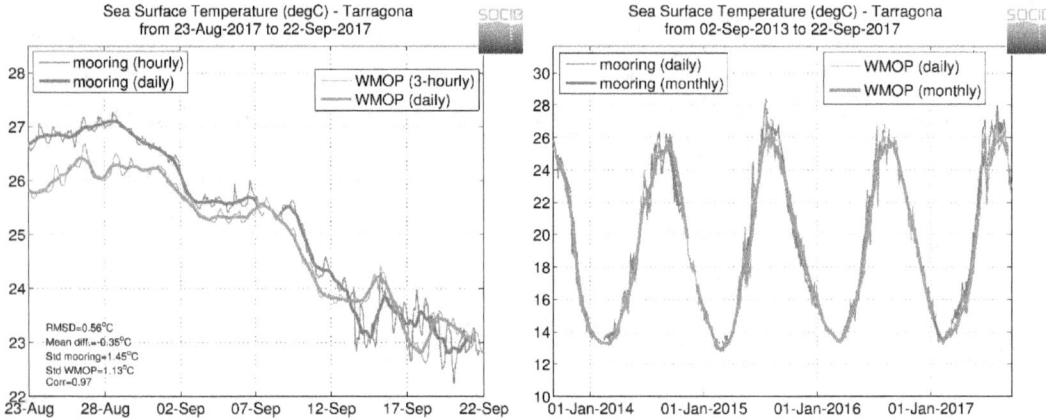

Figure 24.13. Time series of sea surface temperature from the *WMOP_forecast* model and mooring data at Tarragona station (northernmost mooring in Fig. 24.15): a) over a one-month period from 23 August to 22 September 2017; b) over a four-year period from 02 September 2013 to 22 September 2017.

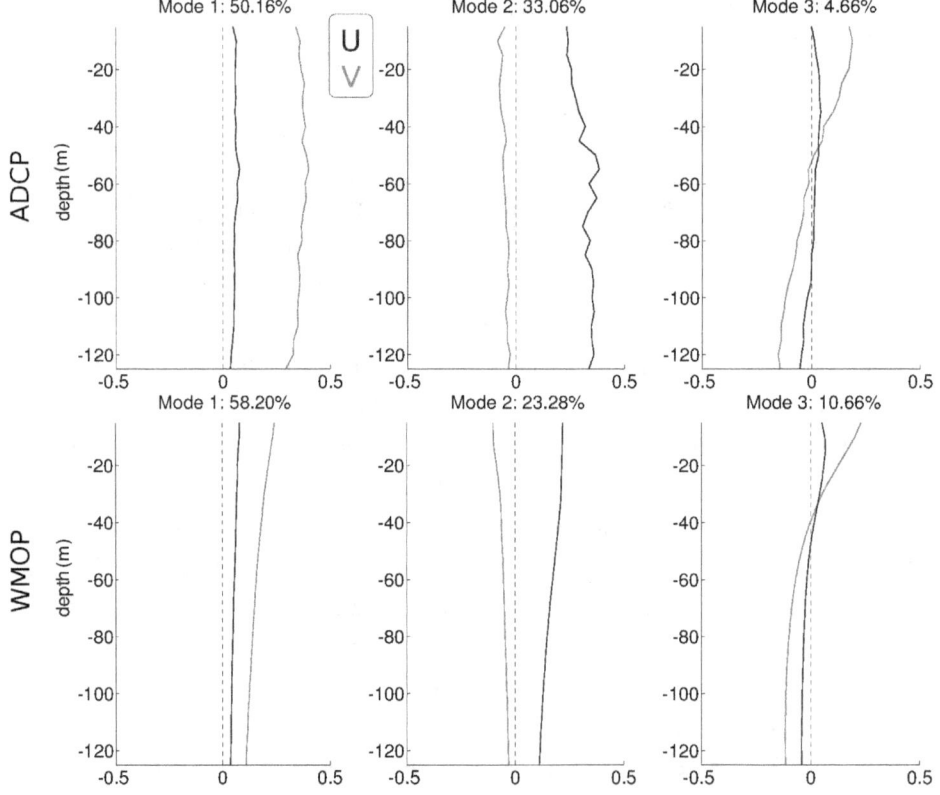

Figure 24.14. Three main modes of vertical EOF patterns for ADCP observations and the *WMOP_hindcast* at Ibiza Channel mooring. The meridional component of the velocity is plotted in red, the zonal component in blue.

Moorings often host current meters and ADCPs, providing time series of ocean velocities at different depths. These time series along the vertical can be analyzed in terms of EOFs to determine the vertical modes of variability. The vertical EOFs shown in Fig. 24.14 are qualitatively close to each other between model and observations: a first mode mainly representing the meridional variability with a small zonal component, a second mode representing the zonal variability, and a third mode with a marked baroclinic structure and a zero-crossing around 40 m depth. The modes have the same order of magnitude in the model and observations. A noticeable difference is the decay of the variability of velocities with depth in modes 1 and 2 in the model, which is not depicted in ADCP observations. This comparison provides unique information about the capability of the model to represent the variability patterns along the vertical. Cross-correlations between different depths can also be estimated from the ADCP data, providing an assessment of vertical model correlations used in data assimilation schemes, for instance (Oke et al., 2002).

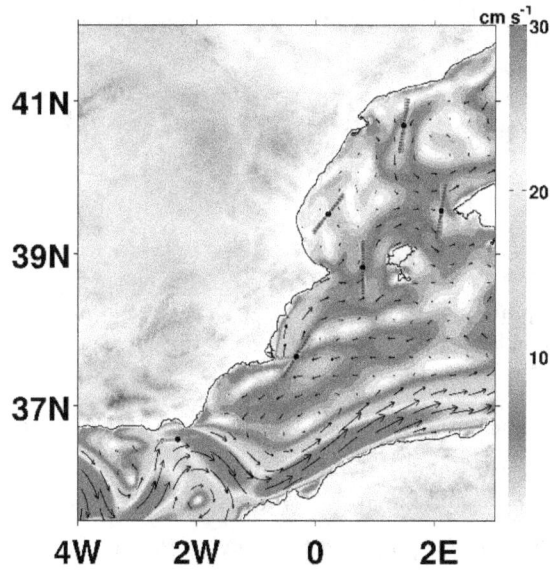

Figure 24.15. Main axes of variability computed at different mooring locations along the Iberian shelf break (moorings operated by Puertos del Estado in Tarragona, Valencia, Dragonera, Cabo de Palos and Cabo de Gata, and SOCIB in the Ibiza Channel), plotted over the mean velocities of the *WMOP_hindcast* simulation. Blue: mooring, Green: altimetry, Red: the *WMOP_hindcast*.

The computation of the main axis of variability is another example of how we can exploit fixed location measurements to evaluate the model currents. The availability of current time series from moorings allows us to determine the axis along which the variance in the observed velocity fluctuations is maximum. The angle of maximum variance can also be computed from the model and compared to the observational estimates. In the comparison presented in Fig. 24.15, model and data are found to be in general good agreement. Discrepancies between model and mooring data of about 20° in terms of mean axis orientation are found in Ibiza Channel and Cabo de Palos moorings, highlighting errors in the variability of the surface circulation in the model. Altimeter estimates coincide with mooring data except at Tarragona station where the model slightly better matches the mooring than altimetry.

Surface drifters

Hourly surface ocean velocities can be obtained from surface drifters, providing essential information for model validation. Surface drifters are driven by the total currents including the contributions of winds, waves, mesoscale and submesoscale structures, tides, or inertial oscillations. They provide a complementary source of information on surface currents with respect to altimetry, which is limited to the geostrophic component and larger scales. When deployed in coastal areas they provide details of the circulation that are generally not accessible from other observations except in areas covered by HF radars. Surface drifters often have a drogue so that they drift according to the currents at a specific depth (a few meters below the surface); this needs to be carefully taken into account before comparing with model currents. Eulerian model-data validation using drifters is possible if a sufficient number of floats is available to map the currents over a specific area. An illustration is provided in Fig. 24.16, computing mean values of the velocities at 15 m depth from a set of 33 drifter trajectories over spatial bins of $0.1° \times 0.1°$.

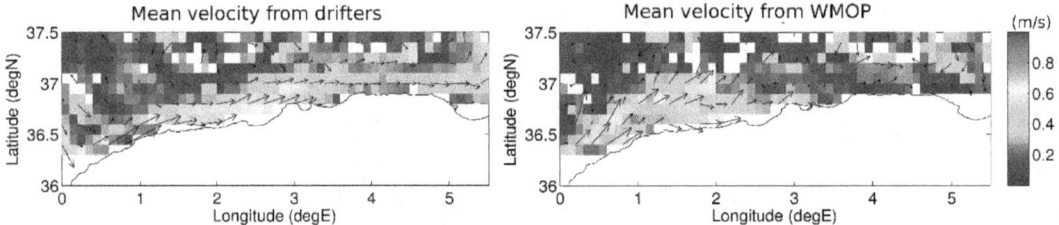

Figure 24.16. Left: mean Eulerian velocity fields computed from 33 drifter trajectories with a drogue at 15 m depth in the Algerian Basin during the period 2011–2015. Right: mean velocity fields at 15 m depth from *WMOP_forecast* simulations sampled at the time and locations of the drifters.

Fig. 24.16 depicts values of drifter velocities larger than 0.6 m/s along the Algerian coast west of 3°E. Such values are also found in the *WMOP_forecast* but the current has a wider extension with respect to the data. Moreover, while the current continues to drive the drifters east of 3°E with a velocity around 0.3-0.4 m/s, it appears to fluctuate more in the model. One important limitation here is that the reliability of Eulerian statistics inferred from surface drifters is limited by the amount of available drifters. Also, the distribution of real drifters might be biased by a convergence effect attracting the drifting floats towards the areas of larger velocities (Davis, 1985).

The most common way to compare drifter data with models is to use Lagrangian diagnostics based on the comparison of observed and modeled trajectories. Virtual drifters can be launched in the model from the real position of the platforms and advected by the model velocities, including a diffusive component to represent the effect of unresolved processes. The evolution of separation distances with time between model and real trajectories then provides an evaluation of the model surface velocity errors (e.g. Liu and Weisberg, 2011).

Fig. 24.17 illustrates how drifters can be used to evaluate two different coastal processes in the model. First, the presence of a coastal gyre close to the northern cape of Mallorca Island is revealed by the trajectory of two drifters, with a good correspondence with model surface velocities. In the second example, the model is found to properly represent the mean current direction and magnitude over a three-day period, in particular also including cross-shore oscillations associated to sea breeze effects in the Bay of Palma. This evaluation of coastal processes is very valuable for applications based on surface drifts such as search-and-rescue or responses to local emergencies on short timescales. Notice that the SST is also generally recorded along the drifter trajectory and can be used for a complementary model validation.

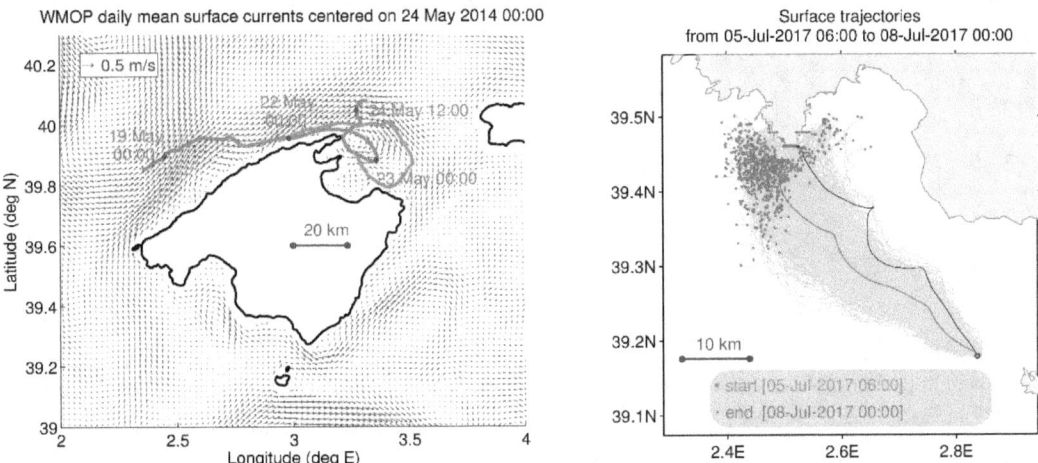

Figure 24.17. Left: An illustration of the trajectories of two surface drifters (in red and green) in May 2014 on the northern coast of Mallorca Island. *WMOP_forecast* velocities are represented with blue arrows. Right: trajectories (in grey) of a cluster of 1000 particles advected by the *WMOP_forecast* model surface velocities (the mean trajectory is represented in red, the final positions with red dots), and trajectory of a drifter (in black) during the same t-day period in July 2017, highlighting the importance of daily sea-breezes on the drifter and model-derived trajectories.

High frequency (HF) radar

Using radio waves emitted from fixed antennas on shore, HF radars provide continuous (hourly) real-time coastal surface current mapping, with a limited geographical coverage but a high spatiotemporal resolution. They provide insights into the small-scale variability of coastal ocean currents, knowledge of which is in high demand for practical applications but extremely challenging for operational ocean models. SOCIB operates a 2-CODAR-Sea-Sonde-antenna system, transmitting at 13.5 MHz and monitoring the eastern side of the Ibiza Channel. The spatial resolution is 3 km. Velocity data are delivered hourly and are available in real-time. The area covered by the HF radar extends from the coast to 60 km offshore.

The comparison of the mean velocity fields between model and measurements allows us to detect circulation biases in the model at the coastal scale, which is not properly resolved by altimetry. In the illustration in Fig. 24.18, the model represents an energetic northward vein in the mean circulation in the area covered by the HF radar, overestimating velocities compared to the measurements during this period. This vein is associated with inflows of water of Atlantic origin through the Ibiza Channel. The change of direction of the surface velocities from northwards to southwards observed by the HF radar around 0.9°E is found to be misplaced here in the model with a spatial error around 40 km towards the north-west.

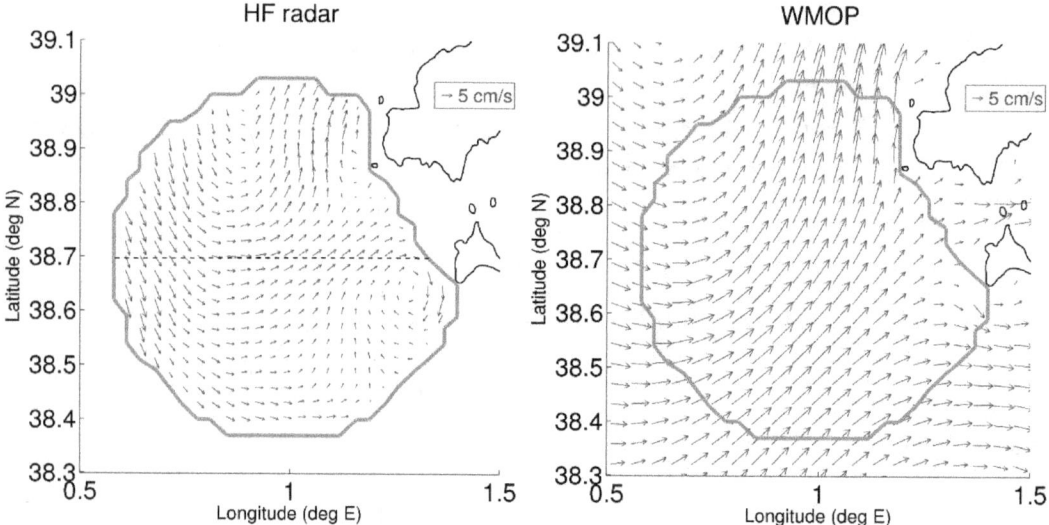

Figure 24.18. Mean surface velocity field from the radar and the *WMOP_hindcast* simulation in the Ibiza Channel over the period 01-June-2012 to 30-Sept-2014. The dashed line at 38.7°N on the left panel shows the position of the section illustrated in Fig. 24.20.

HF radar measurements also provide EKE estimates of the surface ocean at high resolution, which can be compared to models and complement the comparison based on altimetry. Figure 24.19 shows that while EKE have similar magnitudes in the model and in the data in the northern part of the area covered by the HF radar, the model is found to overestimate the eddy activity in the southern part of the region.

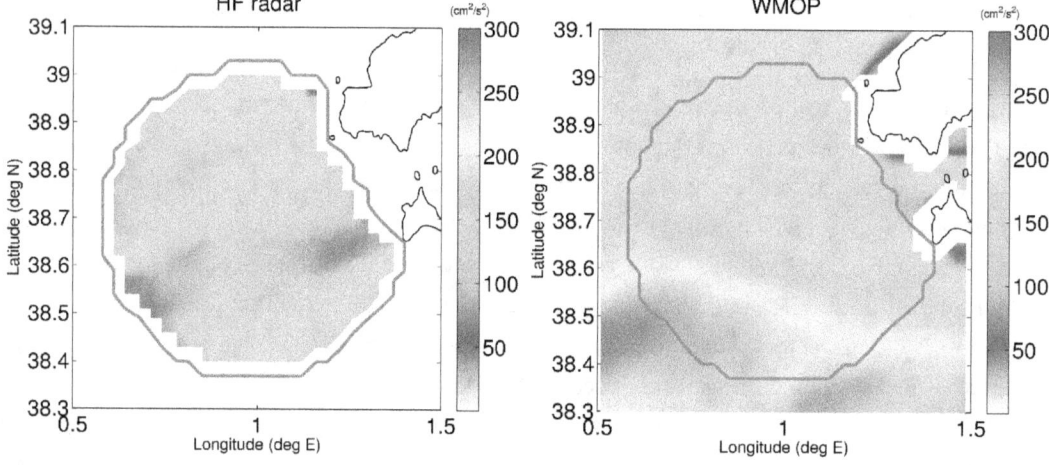

Figure 24.19. Mean EKE estimated from daily mean HF radar data and the *WMOP_hindcast* simulation over the period 01-June-2012 to 30-Sept-2014.

The availability of time series also allows for the assessment of the model capacity to represent sporadic events. Fig. 24.20 shows such a HF radar-model data comparison in the form of a Hovmöller diagram along a zonal section at 38.7°N, considering the *WMOP_hindcast*, the *WMOP_forecast*, the CMEMS-MED model (REAN until 31-Dec-2015, AN-FC afterwards), as well as mapped altimetry. This representation is useful to examine the representation of northwards and southwards flows through the Ibiza Channel. It reveals the HF variability in the radar and in the different models, which is larger than that provided by interpolated altimetry. Larger velocity standard deviations are found in the model when compared to the radar, especially in the case of the *WMOP_forecast*. The overall pattern correspondence between models and radar is relatively poor, except the strong northward event in July 2016, which is represented in the radar, the *WMOP_forecast*, CMEMS-MED, and also in altimeter data.

Finally, it is worth noting that spatiotemporal HF radar data also allow for the evaluation of the model in terms of EOFs modes, complex correlations, or variability ellipses. Advanced gap-filling methods can be implemented to fill short spatial data gaps in the HF radar data time series (e.g., Lekien et al., 2004; Erick et al., 2016).

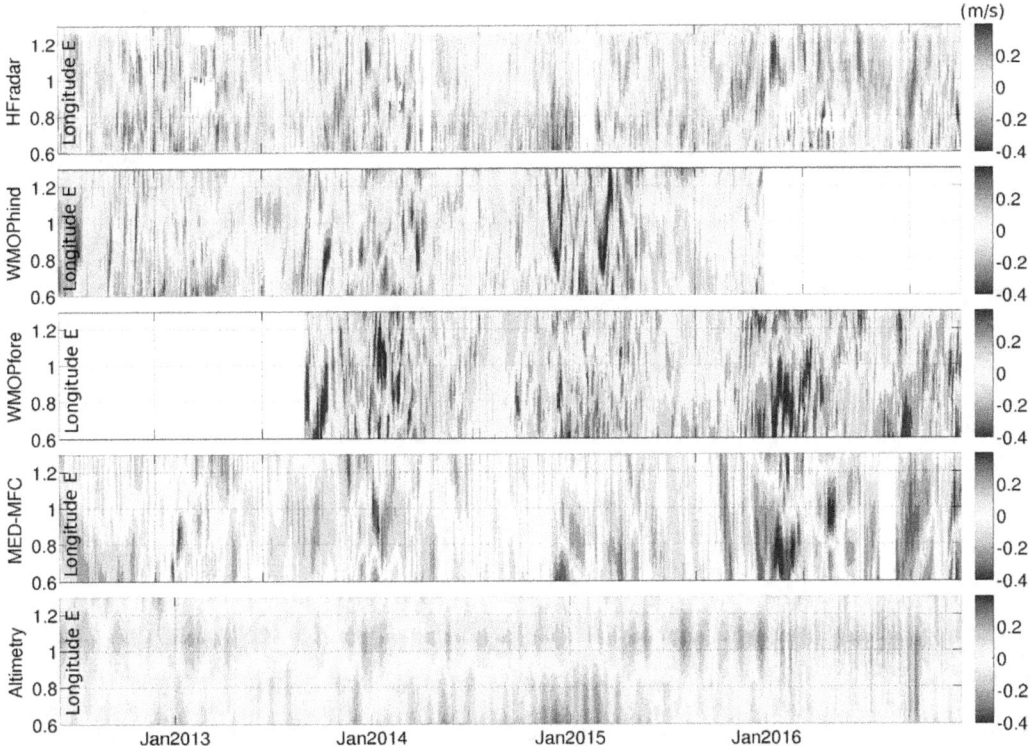

Figure 24.20. Hovmöller diagram of meridional surface velocities (m/s) across the zonal section at 38.7°N shown in Fig. 24.18, from the radar, *WMOP_hindcast*, *WMOP_forecast*, CMEMS-MED model (REAN until 31-Dec-2015, AN-FC afterwards) and altimetry. Red (respectively blue) represents northward (resp. southwards) velocities.

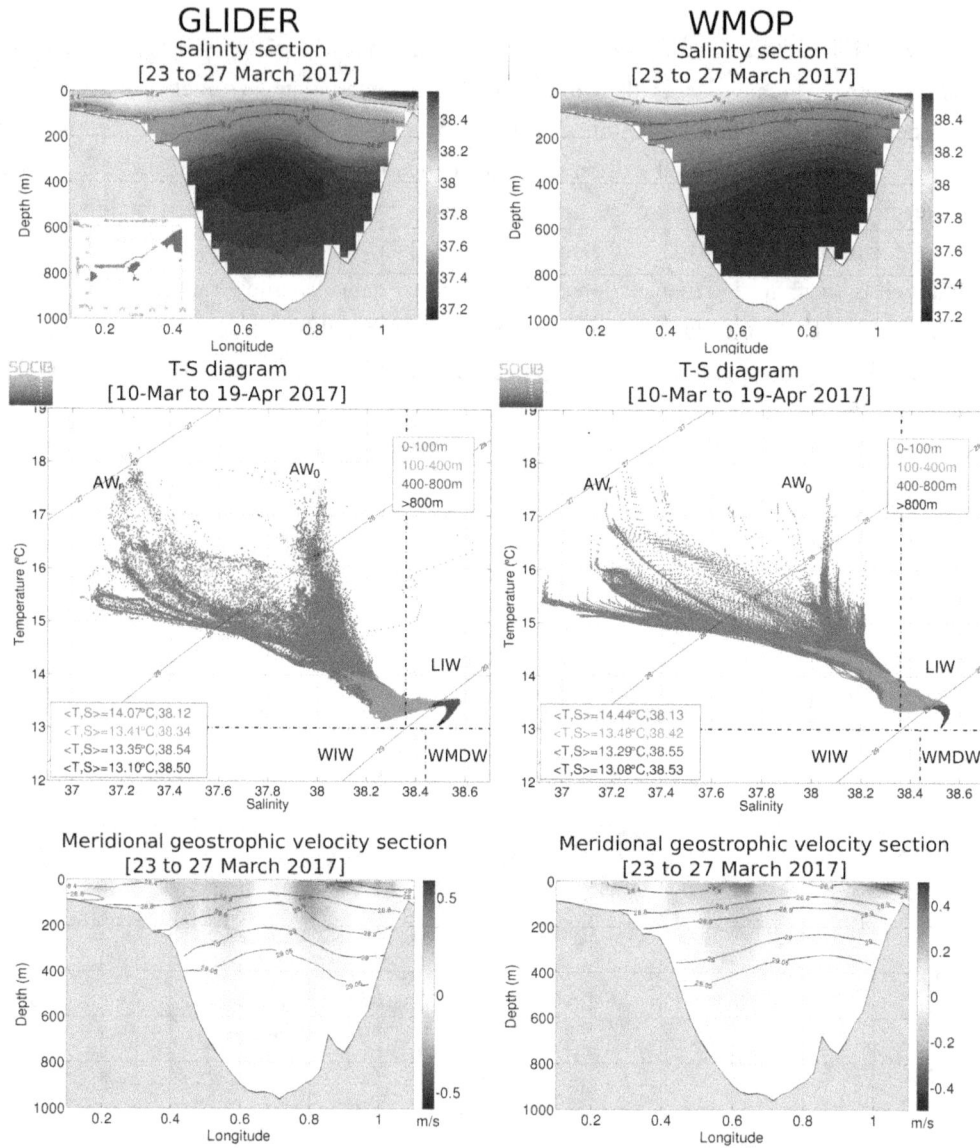

Figure 24.21. Top: vertical salinity section from the glider (left) and the *WMOP_forecast* model (right) during the Canales mission in March 2017 (glider tracks are plotted in the insert). Middle: associated T-S diagrams. Bottom: cross-section geostrophic velocities estimated from the glider and model hydrographic sections.

Gliders

Gliders are steerable autonomous vehicles that have the capacity to monitor oceanic sections down to 1000 m depth with high resolution at a forward speed around 20 km per day and transmit data in real time. Typical glider sensors collect data of conductivity, temperature, chlorophyll, and oxygen. The data are unique to detect fine-scale structures in the ocean and to relate the variability of biochemical and physical variables. Gliders are low energy consumers and may be used for missions of several weeks at sea. They are often deployed repetitively along endurance lines, thus providing time series of hydrographic sections and associated cross-section geostrophic transport variability. Glider data also allow us to distinguish different water masses present along a specific section and then compute the across-section transports of the different water masses. Thanks to their controllability, fleets of gliders can be deployed in a coordinated manner to provide an adaptive sampling of specific targeted ocean circulation features (e.g., Leonard et al., 2010, Alvarez and Mourre, 2014).

SOCIB operates an endurance glider line in the Ibiza Channel in order to monitor meridional water mass exchanges in the Western Mediterranean Sea. The data from the Ibiza Channel glider section (shown in Fig. 24.21) reveals the presence of different water masses: AWo and AWr with significant salinity differences at the surface, and more saline Levantine Intermediate Water at intermediate levels. The model properly represents the presence of AWr on the eastern side of the Ibiza Channel, less accurately on the western side. The subsurface salinity maximum associated with Levantine Intermediate Water is not as marked in the model as it is in the observations, highlighting some model deficiencies in representing this intermediate water mass.

Since 2011, the repetitivity of the glider sections has revealed the high temporal variability of the meridional water mass exchanges through the Ibiza Channel (Heslop et al., 2012). After a few years, the time series has also allowed for the identification of the main modes of variability of temperature, salinity, and cross-section geostrophic velocities, and for the analysis of seasonal and interannual signals in the water mass exchanges, as well as changes in the water mass properties. All of these aspects can be evaluated in the model. Fig. 24.22 illustrates the variability of the transports per water mass for all available Ibiza Channel transects between January 2011 and December 2015. It shows the high transport variability in both the model and the observations. The model generally fails in exactly representing the real transport fluctuations. However, large transports in 2012 are properly described, as well as southward events in 2014 and 2015. The model is also able to form Western Intermediate Water and propagate it until the Ibiza Channel in 2011, 2012, and 2013, as also detected by the glider. No Western Intermediate Water is present in the Ibiza Channel in 2014 and 2015 in model or in observations.

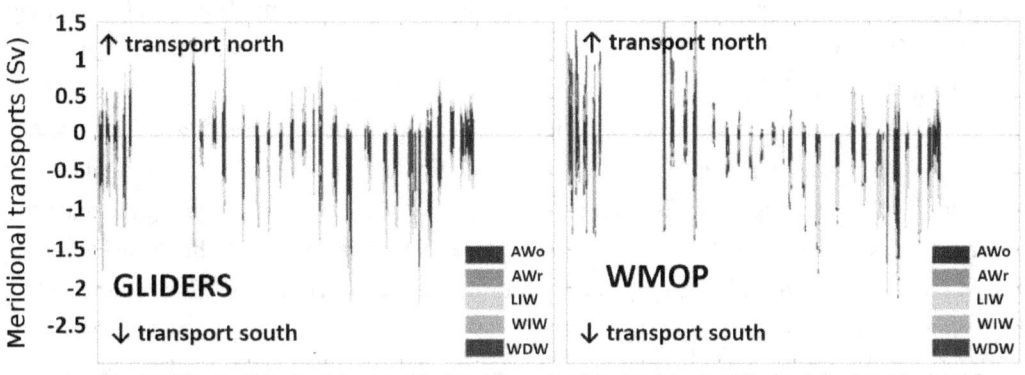

Figure 24.22. Meridional geostrophic transport (Sv) of the different water masses across the Ibiza Channel endurance line as seen by the gliders (left) and the *WMOP_hindcast* simulation (right).

Conclusions

Ocean model assessment is a multi-variable, region-dependent, multi-scale problem constrained by available observations and specific user needs requiring specific measures of the model performance. No single metrics can fairly represent the whole performance of a model. In addition, the standard point-wise statistics might penalize high-resolution simulations properly representing ocean structures but misplaced with respect to observations, making necessary the definition of advanced metrics for the quantification of model-data differences.

Multiple platforms collecting measurements of different variables at different scales provide various insights into the model performance. Satellites, Argo floats, fixed moorings, surface drifters, HF radar, and gliders all provide specific and complementary assessments of the models. They allow to evaluate a wide range of different properties, from the large-scale to the fine-scale and coastal variability represented in high-resolution models. The following oceanographic features have been illustrated in this chapter:

- the mean large-scale surface circulation
- the associated EKE and modes of variability
- the presence and propagation of eddies
- the spatial and temporal variability of the SST
- the vertical stratification and the depth of the mixed layer
- the vertical modes of variability of ocean currents at a fixed location
- the representation of specific coastal circulation processes (coastal gyre, sea breeze effects)
- the coastal surface circulation, EKE, and representation of inflow events in the Ibiza Channel
- the representation of surface and intermediate water masses in the Ibiza Channel and their corresponding meridional transport

These multi-platform data allow us to evaluate the realism of the model both in terms of statistical behaviour and representation of sporadic events. The following references are highly recommended for the reader interested in more detailed quantitative multi-platform ocean regional model assessment exercises (Oke et al., 2002; Warner et al., 2005; Penduff et al., 2006; Chiggiato and Oddo, 2008; Chao et al., 2009; Liu et al., 2009; Wilkin and Hunter, 2013; Lorente et al., 2014; Juza et al., 2015; Capó et al., 2016).

The integration of this multiple information into synthetic metrics addressing specific user needs now constitutes an important challenge in operational oceanography. The application of neighborhood validation methods, probabilistic approaches, advanced Lagrangian diagnostics, front detection algorithms from high-resolution satellite imagery, and the analysis of high-resolution altimetry certainly also represent important future directions for the evaluation of high-resolution ocean simulations.

Acknowledgements

We are very grateful for the helpful comments on the manuscript by the students of the GODAE OceanView Summer School, Hanna Kauko, Christina Kong and Jean-Baptiste Kassi, as well as the anonymous reviewer. The WMOP hindcast simulation and corresponding assessment has been developed in the framework of the MEDCLIC project funded by La Caixa Fundation. Validation using coastal platforms is part of the H2020 JERICO-NEXT project. Eddy tracker outputs have been obtained during the MEDSUB project funded by the CMEMS Service Evolution Programme. We gratefully acknowledge Puertos del Estado for making data available from their mooring network, AEMET for providing HIRLAM atmospheric fields, and CMEMS for the large-scale Mediterranean model products.

References

Alvarez A. and B. Mourre (2014). Cooperation or coordination of underwater glider networks? An assessment from Observing System Simulation Experiments in the Ligurian Sea. J. Atmos. And Ocean. Technology, 31, 2268-2277.

Belkin I.M. and J.E. O´Reilly (2009). An algorithm for oceanic front detection in chlorophyll and SST satellite imagery. Journal of Marine Systems, 78, 319-326.

Buongiorno Nardelli B., C. Tronconi, A. Pisano, R. Santoleri, (2013). High and Ultra-High resolution processing of satellite Sea Surface Temperature data over Southern European Seas in the framework of MyOcean project, Rem. Sens. Env., 129, 1-16, doi:10.1016/j.rse.2012.10.012

Capó E., A.Orfila, Sayol J.-M., Conti D., Juza, M., Ruiz S., Sotillo M.G., García-Ladona E., Simarro G., Mourre B. and Tintoré J. (2016). Assessment of operational models in the Balearic Sea during the MEDESS experiment. Deep Sea Research, 133, 118-131, doi:10.1016/j.dsr2.2016.03.009.

Chao Y, Li Z, Farrara J, McWilliams C, Bellingham J, Capet X, Chavez F, Choi J-K, Davis R, Doyle J, et al. (2009). Development, implementation and evaluation of a data-assimilative ocean forecasting system off the central California coast. Deep-Sea Res II. 56: 100–126.

Candille G. and O. Talagrand (2005). Evaluation of probabilistic prediction systems for a scalar variable. Q.J.R. Meteorol. Soc. , 131, pp. 2131-2150.

Casati, B. et al., (2008). Forecast verification: Current status and future directions. Meteor. Appl., 15, 3–18.

Chelton, D. B., M. G. Schlax, and R. M. Samelson (2011). Global observations of nonlinear mesoscale eddies, Prog. Oceanogr., 91, 167–216, doi:10.1016/j.pocean.2011.01.002.

Chiggiato J, Oddo P. (2008). Operational ocean models in the Adriatic Sea: a skill assessment. Ocean Sci. 4:61–71.

Clementi E., Pistoia J., Fratianni C., Delrosso D., Grandi A., Drudi M., Coppini G., Lecci R., Pinardi N. (2017). "Mediterranean Sea Analysis and Forecast (CMEMS MED-Currents 2013-2017)". [Data set]. Copernicus Monitoring Environment Marine Service (CMEMS). doi:10.25423/MEDSEA_ANALYSIS_FORECAST_PHYS_006_001

Davis, R. E. (1985). Drifter observations of coastal surface currents during CODE: The method and descriptive view, J. Geophys. Res., 90, 4741-4755, doi:10.1029/JC090iC03p04741.

Ebert, E.E., (2009).Neighborhood Verification: A Strategy for Rewarding Close Forecasts. Wea. Forecasting, 24, 1498–1510, doi:10.1175/2009WAF2222251.1

Emery, W. J. and Thompson, R. E. (2001). Data analysis methods in physical oceanography. Elsevier, Oxford, UK

Erick, F., Hugh, R., Josh, K., Michael, S., Scott, G. (2016). Gap Filling of the Coastal Ocean Surface Currents from HFR Data: Application to the Mid-Atlantic Bight HFR Network. Journal of Atmospheric and Oceanic Technology 33, 1097-1111.

Escudier, R., L. Renault, A. Pascual, P. Brasseur, D. Chelton, and J. Beuvier (2016a). Eddy properties in the Western Mediterranean Sea from satellite altimetry and a numerical simulation, J. Geophys. Res. Oceans, 121, 3990–4006, doi:10.1002/2015JC011371.

Escudier, R., B. Mourre, M. Juza, and J. Tintore (2016b). Subsurface circulation and mesoscale variability in the Algerian subbasin from altimeterderived eddy trajectories, J. Geophys. Res. Oceans, 121, 6310–6322, doi:10.1002/2016JC011760.

Halo, I., B. Backeberg, P. Penven, I. Ansorge, C. Reason, and J. Ullgren (2013). Eddy properties in the Mozambique Channel: A comparison between observations and two numerical ocean circulation models, Deep Sea Res., Part II, 100, 38–53.

Hernandez F., E. Blockley, G. B. Brassington, F. Davidson, P. Divakaran, M. Drévillon, S. Ishizaki, M. Garcia-Sotillo, P. J. Hogan, P. Lagemaa, B. Levier, M. Martin, A. Mehra, C. Mooers, N. Ferry, A. Ryan, C. Regnier, A. Sellar, G. C. Smith, S. Sofianos, T. Spindler, G. Volpe, J. Wilkin, E. D. Zaron and A. Zhang (2015). Recent progress in performance evaluations and near real-time assessment of operational ocean products, Journal of Operational Oceanography, 8:2, s221-s238, doi: 10.1080/1755876X.2015.1050282.

Heslop, E.E., Ruiz, S., Allen, J., López-Jurado, J.L., Renault, L., Tintoré, J. (2012). Autonomous underwater gliders monitoring variability at "choke points" in our ocean system: A case study in the Western Mediterranean Sea. Geophysical Research Letters, 39.

Isern-Fontanet, J., García-Ladona, E., Font, J., (2003). Identification of marine eddies from altimetry. Journal of Atmospheric and Oceanic Technology 20, 772-778.

Jolliff JK, Kindle JC, Shulman I, Penta B, Friedrichs MAM, Helber R, Arnone R (2009). Summary diagrams for coupled hydrodynamic-ecosystems model skill assessment. J Mar Syst, 76, 64–82

Juza, M., Mourre, B., Lellouche, J.-M., Tonani, M., and Tintoré, J. (2015). From basin to sub-basin scale assessment and intercomparison of numerical simulations in the western Mediterranean Sea. *Journal of Marine System*, 149, 36-49, doi;10.1016/j.jmarsys.2015.04.010.

Juza, M., Mourre, B., Renault, L., Gómara, S., Sebastián, K., Lora, S., Beltran, J.P., Frontera, B., Garau, B., Troupin, C., Torner, M., Heslop, E., Casas, B., Escudier, R., Vizoso, G., Tintoré, J. (2016). SOCIB operational ocean forecasting system and multi-platform validation in the Western Mediterranean Sea. Journal of Operational Oceanography, 9, s155-s166.

Kerry C., B. Powell, M. Roughan and P. Oke (2016). Development and evaluation of a high-resolution reanalysis of the East Australia Current region using the Regional Ocean Modelling System (ROMS3.4) and Incremental Strong-Constraint 4-Dimensional Variational (IS4D-Var) data assimilation. Geoscientific Model Development, 9(10), pp. 3779-3801.

Kourafalou VH, De Mey P, Le Hénaff M, Charria G, Edwards CA, He R, Herzfeld M, Pasqual A, Stanev E, Tintoré J, Usui N, Van Der Westhuysen A, Wilkin J, Zhu X. (2015). Coastal Ocean Forecasting: system integration and validation. J Oper. Oceanogr. 8(1).

Kurapov. A., S. Erofeeva and E. Myers (2017). Coastal sea level variability in the US West Coast Ocean Forecast System (WCOFS). *Ocean Dynamics*, 67, 23-36.

Lekien, F., Coulliette, C., Bank, R., Marsden, J. (2004). Open-boundary modal analysis: Interpolation, extrapolation, and filtering. Journal of Geophysical Research: Oceans 109, C12004.

Leonard N.E., D. Paley, R. Davis, D. Fratantoni, F. Lekien and F. Zhang (2010). Coordinated control of an underwater glider fleet in an adaptive sampling filed experiment in Monterey Bay. J. Field Robotics, 27, 718-740.

Liu, Y., and R. H. Weisberg (2011). Evaluation of trajectory modeling in different dynamic regions using normalized cumulative Lagrangian separation, J. Geophys. Res., 116, C09013, doi:10.1029/2010JC006837.

Le Traon, P. Y., P. Klein, B. L. Hua, and G. Dibarboure (2008). Do altimeter wavenumber spectra agree with the interior or surface quasigeostrophic theory?, J. Phys. Oceanogr., 38, 1137-1142, doi:10.1175/2007JPO3806.1.

Liu, Y., P. MacCready, B. M. Hickey, E. P. Dever, P. M. Kosro, and N. S. Banas (2009). Evaluation of a coastal ocean circulation model for the Columbia River plume in summer 2004, J. Geophys. Res., 114, C00B04, doi:10.1029/2008JC004929.

Lorente P., S. Piedracoba, M. García Sotillo, R. Aznar, A. Amo-Baladrón, A. Pascual, J. Soto-Navarro and E. Álvarez-Fanjul (2016). Ocean model skill assessment in the NW Mediterranean using multi-sensor data, Journal of Operational Oceanography, 9:2, 75-92, doi:10.1080/1755876X.2016.1215224

Marchesiello P, McWilliams JC, Shchepetkin A. (2001). Open boundary conditions for long-term integration of regional oceanic models. Ocean Modell. 3:1–20.

Mason, E., A. Pascual, and J.C. McWilliams (2014). A New Sea Surface Height–Based Code for Oceanic Mesoscale Eddy Tracking. J. Atmos. Oceanic Technol., 31, 1181–1188, doi;10.1175/JTECH-D-14-00019.1.

Millot, C. (1985). Some features of the Algerian current. Journal of Geophysical Research 90 (C4), 7169–7176.

Millot, C. (1999). Circulation in the western Mediterranean Sea, J. Mar. Syst., 20(1-4), 423–442, doi:10.1016/S0924-7963(98)00078-5.

Millot C. and I. Taupier-Letage (2005). Circulation in the Mediterranean Sea. Handbook of Environmental Chemistry, Vol. 5, Part K: 29–66, doi:10.1007/b107143

Mittermaier and Csima (2017). Ensemble versus Deterministic Performance at the Kilometer scale. Weather and Forecasting, 32, 1697-1709.

Murphy AH (1995), The coefficients of correlation and determination as measures of performance in forecast verification. Wea. Forecasting, 10, 681-688.

Nencioli, F., C. Dong, T. Dickey, L. Washburn, and J. C. McWilliams (2010). A vector geometry-based eddy detection algorithm and its application to a high-resolution numerical model product and high-frequency radar surface velocities in the southern California bight, J. Atmos. Oceanic Technol., 27(3), 564–579.

Nittis K, Zervakis V, Perivoliotis L, Papadopoulos A, Chronis G. (2001). Operational monitoring and forecasting in the Aegean Sea: system limitations and forecasting skill evaluation. Marine Pollut Bull. 43: 154–163.

Oke, P. R., J. S. Allen, R. N. Miller, G. D. Egbert, J. A. Austin, J. A. Barth, T. J. Boyd, P. M. Kosro, and M. D. Levine (2002). A modeling study of the three-dimensional continental shelf circulation off Oregon. Part I: Model-data comparisons, *J. Phys. Oceanogr.*, 32, 1360–1382.

Okubo, A. (1970). Horizontal dispersion of floatable particles in the vicinity of velocity singularities such as convergences. Deep-Sea Res., 17, 445–454.

Oreskes, N., Shrader-Frechette, K., Belitz, K. (1994). Verification, validation and confir- mation of numerical models in the earth sciences. Science 263 (5147), 641–646.

Pascual, A., E. Vidal-Vijande, S. Ruiz, S. Somot and V. Papadopoulos (2014). Spatiotemporal variability of the surface circulation in the Western Mediterranean: a comparative study using altimetry and modeling, In: The Mediterranean Sea: Temporal variability and spatial patterns, John Wiley & Sons, Inc., doi:10.1002/9781118847572.ch2

Peacock T. and G. Haller (2013). Lagrangian coherent structures: the hidden skeleton of luid flows. Phys. Today, 66(2), 41.

Penduff, T., B. Barnier, J.-M. Molines, and G. Madec. (2006). On the use of current meter data to assess the realism of ocean model simulations, Ocean Modelling, 11, 3-4, 399-416.

Rio, M.-H., A. Pascual, P.-M. Poulain, M. Menna, B. Barcelo, and J. Tintore (2014). Computation of a new mean dynamic topography for the Mediterranean Sea from model outputs, altimeter measurements and oceanographic in-situ data, Ocean Sci. Discuss., 11(1), 655–692.

Robinson, A. R., W. G. Leslie, A. Theocharis, and A. Lascaratos (2001). Mediterranean Sea circulation, Ocean Currents: A Derivative of the Encyclopedia of Ocean Sciences, Academic Press, pp. 1689–1705.

Sakamoto K., G. Yamanaka, H. Tsujino, H. Nakano, S. Urakawa, N. Usui, M. Hirabara and K. Ogawa (2016). Development of an operational coastal model of the Seta Inland Sea, Japan, Ocean Dynamics, 66 (1), pp. 77-97.

Sandvik A.D., Ø. Skagseth, M. D. Skogen (2016). Model validation: Issues regarding comparisons of point measurements and high-resolution modeling results, Ocean Modelling, 106, 68-73.

Shchepetkin, A.F., McWilliams, J.C. (2005). The regional oceanic modeling system (ROMS): a split-explicit, free-surface, topography-following-coordinate oceanic model. Ocean Modelling, 9, 347-404.

Siddorn JR, Allen JI, Blackford JC, Gilbert FJ, Holt JT, Holt MW, Osborne JP, Proctor R, Mills DK. (2007). Modelling the hydro-dynamics and ecosystem of the North-west European Continental Shelf for operational oceanography. J Mar Syst 65: 417–429.

Simoncelli, S., Fratianni, C., Pinardi, N., Grandi, A., Drudi, M., Oddo, P., and Dobricic, S. (2014). "Mediterranean Sea physical reanalysis (MEDREA 1987-2015)". [Data set]. EU Copernicus Marine Service Information. DOI: https://doi.org/10.25423/medsea_reanalysis_phys_006_004

Stanev E, Schulz-Stellenfleth J, Staneva J, Grayek S, Seemann J, Petersen W. (2011). Coastal observing and forecasting system for the German Bight. Estimates of hydro-physical states. Ocean Sci. 7: 1–15.

Taylor KE (2001). Summarizing multiple aspects of model performance in a single diagram. J Geophys Res, 106, 7183–7192

Tintoré, J., et al. (2013). SOCIB: The Balearic Islands Coastal Ocean Observing and Forecasting System Responding to Science, Technology and Society Needs. *Marine Technology Society Journal*, 47, 101-117.

Undén, P., et al. (2002). The HIRLAM version 5.0 model. *HIRLAM documentation manual (HIRLAM Scientific Documentation)*. Available at http://www.knmi.nl/hirlam/SciDocDec2002.pdf or from Hirlam-5 Project, SMHI, S-60176, Norrkoping, Sweden.

Umlauf, L., Burchard, H. (2003). A generic length-scale equation for geophysical turbulence models. J. Marine Res. 61,235–265.

Warner, J. C., W. R. Geyer, and J. A. Lerczak (2005). Numerical modeling of an estuary: A comprehensive skill assessment, J. Geophys. Res., 110, C05001, doi:10.1029/2004JC002691.

Weiss, J. (1991). The dynamics of enstrophy transfer in two-dimensional hydrodynamics. Physica D, 48, 273–294.

Wilkin J. L., H.G. Arango, D.B. Haidvogel, C.S. Lichtenwalner, S.M. Glenn and K. S. Hedström (2005). A regional ocean modelling system for the Long-term Ecosystem Observatory, J. Geophys. Res., 110, C06S91, doi:10.1029/2003JC002218.

Wilkin, J. L., and E. J. Hunter (2013). An assessment of the skill of real-time models of Mid-Atlantic Bight continental shelf circulation, J. Geophys. Res. Oceans, 118, 2919–2933, doi:10.1002/jgrc.20223.

Willmott, C. J. (1981). On the validation of models, Phys. Geogr., 2, 184–194.

Ziegeler S., Dykes J. and J. Shriver (2012). Spatial error metrics for oceanographic model verification. J. Atmos. Oceanic Technol., 29, 260-266.

CHAPTER 25

Learning about Copernicus Marine Environment Monitoring Service "CMEMS": A Practical Introduction to the Use of the European Operational Oceanography Service

Marie Drévillon, Pierre Bahurel, David Bazin, Mounir Benkiran, Jonathan Beuvier, Laurence Crosnier, Yann Drillet, Edmée Durand, Michèle Fabardines, Isabel Garcia Hermosa, Cédric Giordan, Elodie Gutknecht, Fabrice Hernandez, Stéphane Law Chune, Pierre-Yves Le Traon, Jean-Michel Lellouche, Bruno Levier, Angelique Melet, Dominique Obaton, Julien Paul, Mathieu Peltier, Diane Peyrot, Elizabeth Rémy, Karina von Schuckmann, and Cécile Thomas-Courcoux

MetOcean Mercator Océan, Ramonville St Agne, France

The Copernicus Marine Environment Monitoring Service (CMEMS; http://marine.copernicus.eu) is one of the six services of the European Copernicus Programme for Earth Observation (http://www.copernicus.eu). CMEMS was implemented by Mercator Ocean beginning in 2014, under a delegation agreement from the European Commission. The operational services of CMEMS were set up gradually as part of a series of European projects, starting with MERSEA (2004-2008), and followed by MyOcean (2009-2012) under FP7, and MyOcean2 (and its follow-on) from 2012 through 2015.

The development of the Copernicus Marine Environment Monitoring Service (CMEMS) has required collaboration and innovation across research and technology in observations, modelling, data assimilation, and product and service delivery (Le Traon et al., 2017). There is a growing need for accurate and timely oceanographic information for defense, weather and seasonal forecasts, maritime transports security and routing; and coastal management. Since 2008, the European Union's (EU's) Marine Strategy Framework Directive aims to achieve good environmental status of the EU's marine waters by 2020 and to protect the resource base upon which marine-related economic and social activities depend. This directive ensures that member states put in place the assessment of the marine environment of the European Seas. Cooperation has extended across Europe and into the international community to achieve this aim, and also in support of the sustainable development of downstream economic activity based on the exploitation of marine resources (energy, food, oil and minerals, health, and tourism), often called "Blue Growth" or the "Blue Economy."

Drévillon, M., et al., 2018: Learning about Copernicus Marine Environment Monitoring Service "CMEMS": A practical introduction to the use of the European operational oceanography service. In "*New Frontiers in Operational Oceanography*", E. Chassignet, A. Pascual, J. Tintoré, and J. Verron, Eds., GODAE OceanView, 695-712, doi:10.17125/gov2018.ch25.

In order to stimulate Blue Economy innovation and progress, CMEMS provides open, free, regular, and systematic reference information on the physical state, variability, and dynamics of the ocean, sea ice, and marine ecosystems for the global ocean and the European regional seas. This capacity encompasses the description of the current ocean state (analysis), the variability at different spatial and temporal scales, the prediction of the ocean state 10 days ahead (forecast), and the provision of consistent retrospective data records for recent years (reprocessing and reanalysis).

After a short description of CMEMS services, we will illustrate the benefit of CMEMS operational oceanography products through a selection of use cases. Scientific collaboration is an important asset of CMEMS, in particular through validation and Ocean State Reporting activities, which we will explain in the third section of this chapter. In the fourth section, we will briefly address how the calls for tenders entitled service evolution and user uptake guarantee the connection of CMEMS with the research community and with the future needs in operational oceanography. We will reference training materials currently available from CMEMS in the fifth section and we will describe the "hands-on CMEMS" tutorial given during the GODAE international school "New Frontier in Operational Oceanography," with a focus on the main take-home-messages to be derived from this tutorial.

Open, Free, and Easy Access to a European Operational Oceanography Service

A vast portfolio of physical and biogeochemical ocean variables is available for download from marine.copernicus.eu, as summarized in Table 25.1. Fourteen different parameters are estimated by both observations and, except for winds, ocean models (most of them assimilating these observations). They are produced, quality controlled, updated, and delivered daily on several platforms (ftp, subsetter, and direct getfile).

As shown in Fig. 25.1, CMEMS relies on a central product management system (the Central Information System) as well as on production centres for observations (Thematic Assembly Centres – TACs) and modelling/assimilation (Monitoring and Forecasting Centres – MFCs). The TACs gather observational data and generate elaborate products (e.g., multi-sensor and gridded observational products) derived from these observations. The TACs are fed data by the operators of the space and in situ observational infrastructure. The global MFC and six regional MFCs generate model-based analyses, reanalyses, and forecasts of the ocean physical state and biogeochemical characteristics. The six regional MFCs take advantage of regional modelling advances for the European seas (i.e., the better description of the physical and biogeochemical local processes, higher resolution). The Central Information System is in charge of the management and organization of CMEMS information and products, as well as a unique user interface. Under Mercator Ocean's coordination, the TACs and MFCs meet at least once a year during CMEMS operation reviews, and their scientific experts collaborate on a regular basis around the definition of quality control procedures and around the development of ocean monitoring activities. Additionally, TACs and MFCs share knowledge to support the development of models and data assimilation techniques (in particular for biogeochemistry) and for the assimilation of new types of observations.

PARAMETER	MODEL			SATELLITE		IN SITU	
	20 years in the past	Today	10-day forecast	20 years in the past	Today	20 years in the past	Today
Sea Surface Height	X	X	X	X	X	X	X
Temperature	X	X	X	X	X	X	X
Salinity	X	X	X			X	X
Waves	X	X	X				
Currents/velocity	X	X	X	X	X	X	X
Mixed Layer Depth	X	X	X			X	X
Sea ice	X	X	X	X	X		
Turbidity/Transparency				X	X		
Reflectance				X	X		
Nutrients	X	X	X			X	
Primary Production	X	X	X			X	
Oxygen	X	X	X			X	
Plankton	X	X	X			X	
Wind				X	X		

Table 25.1. Summary of (global or regional) ocean parameters available from the CMEMS portfolio. The online catalogue available at marine.copernicus.eu provides information on the contents and scientific qualification of each product.

CMEMS evolves continuously using a rigorous change management process in order to maintain state-of-the-art services and to answer requirements from its users. Short-term (< 1 year) R&D and part of the mid-term R&D (1 to 3 years) are carried out by CMEMS production centres. Longer-term R&D is fostered by service evolution calls for tenders, and user uptake calls for tenders build dedicated collaborations with users and efficient feedback on future needs (see section below).

CMEMS is distributed across Europe. Each TAC and MFC is led by a different institute, and most of them rely on consortiums of pan-European companies, including oceanographic research laboratories, marine environment monitoring institutes, meteorological agencies, or IT companies.

Some of them provide R&D while others produce observational products, analyses or forecasts, and others provide IT services. All companies involved in CMEMS are listed in the Appendix.

CMEMS' operational oceanography community continually faces scientific and technical challenges in order to improve the products for users, starting with the increase of spatial and temporal resolution of the products. Other challenging evolutions of the catalogue in the coming years are the dissemination of reliable ocean monitoring indicators close to real time such as regionally averaged heat content time series or pH time series, new observational products for surface currents (satellite, high frequency radars), and in situ biogeochemical measurements. Big data technologies and new visualization tools are also being utilized for the future versions of viewing and downloading capabilities.

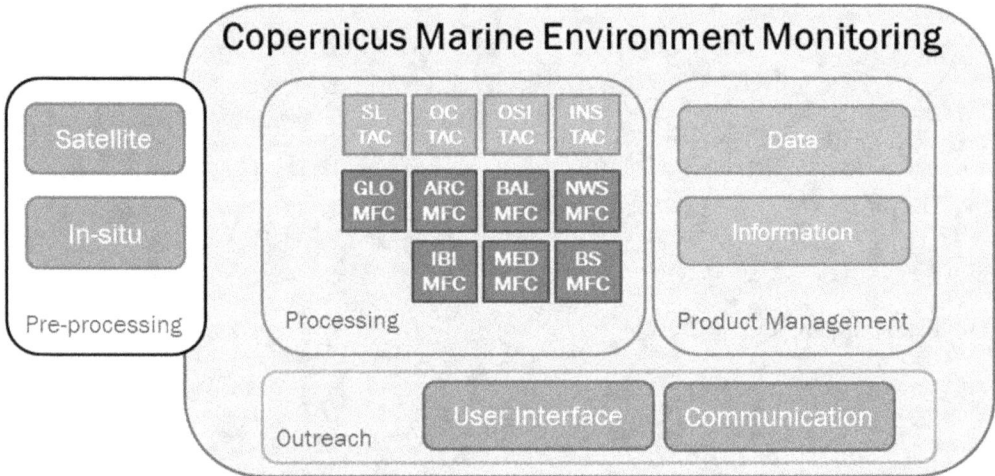

Figure 25.1. Schematic of the CMEMS organization in 2017. The products use satellite and in situ upstream data, they are produced and disseminated by four TACs (satellite sea level –SL TAC-, In situ observations – INS TAC-, satellite Ocean Colour –OC TAC-, and satellite SST, Sea Ice, and Winds -OSI TAC-) and seven MFCs (Global Ocean –GLO MFC-, Arctic Ocean –ARC MFC-, Baltic Sea –BAL MFC-, North Western Shelves –NWS MFC-, Iberian Biscay Ireland –IBI MFC-, Mediterranean Sea –MED MFC-, Black Sea –BS MFC-). The CMEMS is managed (administration, product management) and coordinated (technical and scientific coordination, outreach) by Mercator Ocean.

Use Cases

Copernicus services are designed to stimulate and facilitate the development of innovative downstream applications that produce effective economical value and societal benefit. Use cases are available online and are a good communication tool to demonstrate how CMEMS open data is used within the CMEMS community of users. CMEMS data serve many marine applications that can be broken up into four categories: Coastal & Marine Environment; Marine Resources; Maritime Safety; and Weather, Climate & Seasonal Forecasting.

Use cases highlight the use of CMEMS data by a large panel of users including scientific institutions, governments, European agencies and business. The CMEMS use cases web page[1]

---

[1] http://marine.copernicus.eu/markets

advertises many applications developed in various countries. Use cases are developed into factsheets that advertise how users transform CMEMS data to create the Blue Economy. Use case factsheets highlight the work of users and their organization.

All users can visit the CMEMS use case website, see examples, and learn more about the domains where CMEMS open data can be applied. Users can also submit their own use cases and fill out a short form with details of how they have used CMEMS data. A PDF will automatically be generated (after a validation process) that can then be downloaded by the user to share. CMEMS also employs these use cases for promotion during its various events and among its stakeholders, including the European Commission.

Moreover, CMEMS invites its users to many events where they can testify about how CMEMS data are useful for their applications and where they can express their requirements and provide feedback to drive CMEMS service evolution. Many users' feedback focuses on the added value of an open and free service for observations and model estimates available from a single website. The accuracy of CMEMS ocean model products in general, and the reliability and timeliness of their delivery, is also very important for many users of near real-time products.

Figure 25.2. Example of a use case summary that can be downloaded as a PDF from CMEMS webpage, here for offshore wind farms in the Mediterranean Sea.

Among the many use cases displayed on the CMEMS webpage, we briefly highlight three examples:

- OCEAN ENERGY: Several technologies have emerged to harness the energy of the seas (see Fig. 25.2 the example of the floating wind farm). The CMEMS ocean models provide

key input to estimate the ocean energy resources, minimize the risks and help with the mandatory environmental monitoring of offshore sites[2].
- SHIP ROUTING: Ship routing allows maritime shipping companies to reduce fleet navigation risks, save fuel and reduce CO2 emissions. The CMEMS products provide input conditions for ship routing software for safer navigation routes including ice covered areas[3].
- SUSTAINABLE FISHERY MANAGEMENT: A new sustainable strategy for fisheries is becoming possible thanks to numerical modelling of the marine ecosystem and its food chain from small organisms to top predators. The CMEMS products are used in modelling of fish habitats[4].

A Community of Scientists

Products quality monitoring

One of the major objectives of CMEMS is to deliver useful scientific quality information, such as reliability of the forecasts, accuracy of the analyses, quality control of the observations, etc. Moreover, this quality information has to be scientifically sound and as consistent as possible across the variety of products that CMEMS delivers. Validation is usually two-fold: 1) new products have to be validated prior to their entry into service (operational delivery to users) and their quality communicated effectively to users; and 2) the operational products' quality has to be checked routinely in order to ensure that the quality standards are met. In CMEMS, the TACs and MFCs validate all their products following a well-defined protocol. The validation is mostly based on the verification of the distribution, mean, and mean square of the differences between CMEMS products and reference datasets. Deviations between climatologies and observations are used for the control of observations, while differences between observations (or observational products) and their model counterparts are used for the control of model outputs. The scales and processes represented by either models or observational products must be taken into account when producing this verification, and observation operators (as in data assimilation) usually have to be applied to model outputs, for instance daily averages of the models sea surface temperature are performed before comparing to OSTIA (Operational Sea Surface Temperature and Sea Ice Analysis) daily sea surface temperature products. Estimated accuracy numbers are derived from these verifications. Currently, for model products these numbers are based on root-mean-square (RMS) departures between models and observations—and will probably evolve towards scatter index RMS/mean in the future—while for observations, those accuracy numbers rely on comparisons with climatologies or on upstream observations availability. The estimated accuracy numbers provide indicators of the

---

[2] http://marine.copernicus.eu/wp-content/uploads/use-cases/environmental-monitoring-offshore-wind-farm-offshore-leucate-mediterranean-sea.pdf
[3] http://marine.copernicus.eu/wp-content/uploads/use-cases/ship-routing-save-fuel-reduce-co2-emissions.pdf
[4] http://marine.copernicus.eu/wp-content/uploads/use-cases/supporting-jrc-eu-common-fisheries-policy-cfp-1.0.pdf

average quality level expected over basin scale areas. These statistics and their history are published on a central webpage, which is updated quarterly. Reference values computed on a long period are available in the Product Quality Information Documents (QuIDs) associated with each product, together with a summary of the quality of the product.

The evolution of the CMEMS validation protocol is coordinated by Mercator Ocean and relies on a product quality working group involving experts from all production centres (TACs and MFCs). This group aims at developing validation metrics and associated validation capabilities, and is also making the link with state-of-the-art validation practices and international standards and metrics, such as those defined by MERSEA (Crosnier and Le Provost, 1997) and GODAE OceanView (Hernandez et al., 2015, 2018 this book). In particular, the CLASS4 approach for the computation of the analysis error and of the forecast error in the observation space (at the time and spatial location of the observation), which allows deriving skill scores, was adopted as a standard by the MERSEA project. The product quality working group gathers scientists with various backgrounds (observations or models, in situ or satellite, global or regional) and each participant has the opportunity to share his/her own expertise by proposing dedicated metrics for specific variable and/or region of interest.

Following results obtained by this group of experts, categorical and site-specific metrics, as well as specific biogeochemistry metrics (Maksymczuk, 2016) will be implemented in the future. User feedback also provides indirect quality measurements—external qualification—through the evaluation of the CMEMS products for specific applications. The latter approach will be encouraged and developed in the future.

Ocean state monitoring and reporting

One of the main requirements from CMEMS users is to have long time series of data that can be used to produce a statistical and quality reference framework for their applications. CMEMS ensures the collection of "best quality" input data and maximal use of multiple observation systems and, on the long term, aims at a fully consistent approach across global and regional reanalyses, organizing their interoperability, their inter-dependencies, and joint operations closer to real time (a few months only) with a systematic yearly update. CMEMS reprocessing aims at an optimal use of high-resolution input data and at a seamless connection with CMEMS real-time observations. CMEMS reanalyses aim at a seamless connection with CMEMS real-time analyses and forecast, thus CMEMS produces regional reanalyses that benefit from both high-resolution and specific regional tunings. Specific efforts are made on the processing of sea ice and biogeochemistry components for all CMEMS reanalyses, global and regional. To reach those ambitious objectives, the coordination of the production of multi-year products was set up at the beginning of CMEMS and a working group of reanalysis and reprocessing experts from each production centre has been created. The other responsibility of this group is to coordinate the ocean state reporting activities. In the context of the Marine Strategy Framework Directive, environmental agencies require ocean state and marine environment monitoring. This is achieved through the annual release of the CMEMS Ocean State Report to monitor and describe ocean variability and change from the past to present, and through the development of operational ocean monitoring indicators (OMIs), and related error bars. The OMIs are an ensemble of average or integrated quantities describing the state and evolution of the oceanic environment, such as heat content, sea level rise, sea ice extent, and pH. These developments must rely on continuous and high-quality time series from reanalyses and reprocessed observations, which go up to real time and which ensure high-resolution coverage of the European regional seas (e.g., those implemented by CMEMS). The OMIs can be downloaded from the CMEMS website, together with a short scientific context description and a dedicated QuID.

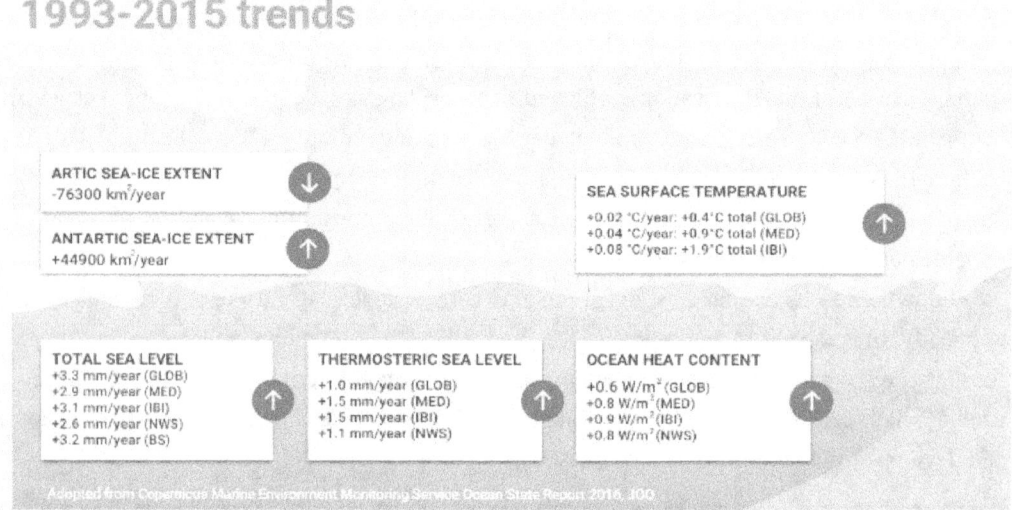

Figure 25.3. Summary of major ocean trends reported in the Ocean State Report #1 von Schuckmann et al. (2016). See also http://marine.copernicus.eu/wp-content/uploads/2017/03/Ocean-State-Report-Summary.pdf

The CMEMS Ocean State Report provides a comprehensive and advanced assessment of the state of the global ocean and European regional seas for the ocean scientific community as well as for policy and decision makers. It will contribute to the reporting tasks and activities of European Environmental Agencies (EEA) and international organizations (e.g., the Intergovernmental Panel on Climate Change, United Nations Sustainable Development Goal 14). In addition, the report aims at increasing general public awareness about the status of, and changes in, the marine environment. The Ocean State Report draws on expert analyses and provides a four-dimensional view (reanalysis systems) from above (through remote sensing data) and directly from the interior (through in situ measurements) of the blue (e.g., hydrography, currents), white (e.g., sea ice) and green (e.g., chlorophyll) global ocean and European regional seas. The first issue was prepared in collaboration with ~80 scientists involved in CMEMS. It provides information on the physical ocean state and change over the period 1993–2015 and has been published in the *Journal of Operational Oceanography*. The first issue reports on a number of trends (Fig. 25.3), including decreasing Arctic and increasing Antarctic sea ice extent, global and regional sea level rise, sea surface temperature rise, and the warming of the global and European regional seas[5].

---

[5] http://marine.copernicus.eu/science-learning/ocean-state-report/ocean-state-report-2016-1st-issue/

The second issue was accepted for publication in April 2018 and highlights changes in the marine environment for the period 1993–2016 as well as specific remarkable events during the year 2016. Most diagnostics from this second issue will be implemented as OMIs on CMEMS website, and subsequent issues of the Ocean State Report will serve as a peer-reviewed platform for the definition and acceptance of new OMIs. Future issues will include additional essential variables, not yet published in previous issues, describing changes in ocean climate at global and regional scale. Strong efforts are put on monitoring biogeochemical changes, improving progressively the accuracy and uncertainty assessment for biogeochemical OMIs, with a specific focus on European regional seas.

New Frontiers in CMEMS

Mid-term and long-term R&D activities are mainly addressed through calls for tenders for CMEMS service evolution and through external programmes (e.g., Horizon 2020 and national R&D programmes). Long-term R&D activities, although implemented in those external programmes, are as crucial for the sustainable evolution of the service as short- and mid-term activities led at the CMEMS production centre level. High level strategy documents for CMEMS service evolution[6], as well as for product quality and multi-year products coordination, were prepared by Mercator Ocean and the Scientific and Technical Advisory Committee of CMEMS. The service evolution strategy document describes four key areas of innovation and research: ocean circulation, ocean-wave and ocean-ice coupling; biogeochemistry and ecosystems in the marine environment; coastal environment; and ocean, atmosphere and climate. Regarding the last key area "ocean, atmosphere and climate," high-level interactions are established with the Copernicus Climate Change Service (C3S), which is in charge of providing global climate change scenarios and developing earth system models including all components of the climate system. The CMEMS focuses on complementary high-resolution ocean models and observations, including scenarios of the regional/coastal impacts of climate change on the marine environment. At the mid- to long-term planning scales, making innovations in the four key areas will require overcoming the following challenges: very high-resolution (kilometric) modelling for ocean and sea ice, including tides, coupling of wave and ocean circulation, with data assimilation; improvement of biogeochemical products and their assessment, including the data assimilation of ocean colour observations and bio-Argo profiles, and in the longer term modelling of higher trophic levels; ocean, waves and atmosphere coupling for improved ocean analyses and forecasts; improved satellite products [higher resolution, new sensors such as SMOS (Soil Moisture Ocean Salinity) or Sentinel missions], and new in situ data for coastal areas, river inputs, stronger links with downstream coastal systems; longer reanalyses time series, seasonal/decadal/climate projections and on the longer-term downscaled scenarios including impacts on the ecosystems and coastal environment. There will also be challenges to improving the CMEMS system such as improving the access to Sentinel data and all Copernicus Services through technological advances such as DIAS (Data and Information Access Services) platforms. A more

---

[6] http://marine.copernicus.eu/science-learning/service-evolution/service-evolution-strategy/

detailed description of each of these research topics and how they will be addressed with mid-term or long-term objectives can be found in the "CMEMS Service Evolution strategy: R&D priorities" document, available online.[7]

CMEMS will have to keep up with technological advances and to implement the service upgrades or changes expected by users, while also keeping them explicitly involved, to evolve together with the downstream sector to provide the level of service expected by EU member states and other national users, for instance. The user uptake calls for tenders intend to build privileged relationships with a series of users in order to help them evolve towards the use of CMEMS core products, but also to answer to the growing need for cooperation with downstream users in the mid- to long-term.

A Series of Practical Tools to Introduce CMEMS

As part of CMEMS, expert information on products and their quality, as well as outreach documentation, is continuously improved upon thanks to dedicated communication activities, exchanges in between TACs and MFCs production centers, and user feedback. Focused tutorials[8] are available on the CMEMS website, which provide scientific or technical assistance to both beginner and advanced users on the CMEMS portfolio and access to service.

In-person training sessions and user workshops are also regularly organized by CMEMS. Practical trainings were developed using Jupyter notebooks, which allows interactive navigation into Python scripts developed for downloading and handling data files[9].

For instance, the "Global Ocean Week," which took place in October 2016, included a training session co-sponsored by Mercator Ocean as leader of CMEMS Global Monitoring and Forecasting Centre and the COST Action "Evaluation of Ocean Syntheses"[10]. Keynote lectures on ocean reanalyses evaluation were filmed[11]. In this context, several practical exercises were also prepared, introducing the core scientific activities of global ocean forecasters, from the design of the analysis system to its validation. Specific attention was paid to ocean variability monitoring capacities thanks to ocean reanalyses. Based on this material, a tutorial called "Hands on CMEMS" was proposed during the 2017 GODAE-OceanView international school, "New frontiers in operational oceanography." This tutorial from the Global MFC uses Jupyter notebooks and will be available for download on the marine.copernicus.eu website together with Jupyter notebooks developed by other TACs and MFCs. It allows interactive evaluation of several CMEMS reanalyses, with Python and Ferret routines, as used by ocean reanalysis producers to explore their experiments. It provides practical illustrations of the main strengths and weaknesses of ocean reanalyses. In the following section, we will explore the three different themes developed in the hands-on CMEMS tutorial: the

---

[7] http://marine.copernicus.eu/wp-content/uploads/2017/06/CMEMS-Service_evolution_strategy_RD_priorities_V3-final.pdf
[8] http://marine.copernicus.eu/training/online-tutorials/
[9] See for instance the In-situ TAC training (http://marineinsitu.eu/material)
[10] http://eos-cost.eu
[11] available on YouTube (search COST-EOS training or CMEMS training)

balance between statistical and physical processes analysis when evaluating reanalyses; the importance of the atmospheric boundary, forced or coupled, and its resolution; and the impact of changes in the observations network onto the quality of the reanalysis.

A view on the ocean: the statistical view or the physical view?

CMEMS reanalyses are global as well as regional. Their aim is providing reference data over the last decades using optimal resolution and observations coverage, and using an analysis system consistent with the one producing real-time analyses and forecasts. Regional reanalyses usually benefit either from better physics, from a model configuration specifically tuned for the region of interest, and/or from higher resolution. The IBIRYS reanalysis at 1/12° for the Iberian Biscay Ireland area is developed at Mercator Ocean, as well as the GLORYS global reanalysis at ¼°. Both reanalyses are based on the NEMO model and the SAM2 data assimilation system (Lellouche et al., 2013) and are forced with ERA-Interim atmospheric reanalysis, but their physics and resolution are different. For instance, IBIRYS explicitly resolves tides and benefits from variable volume-free surface and a state-of-the-art mixing scheme consistent with the near real-time Iberian-Biscay-Ireland (IBI) monitoring and forecasting system (Maraldi et al., 2013).

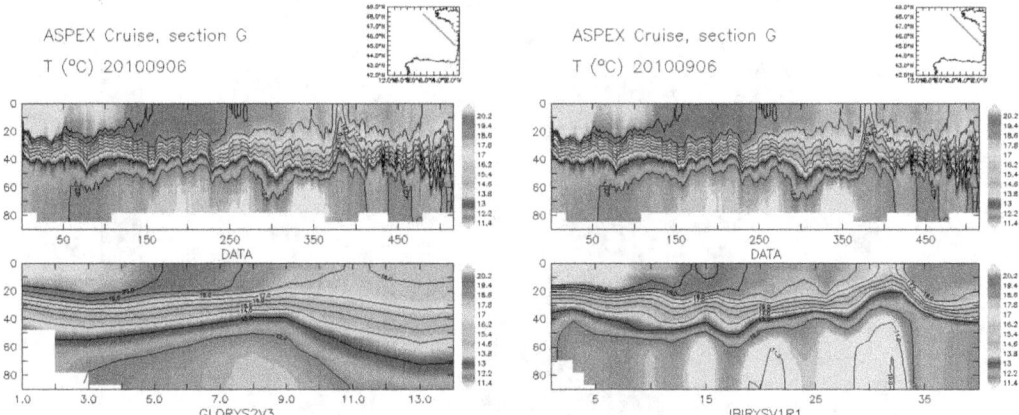

Figure 25.4. Illustration from Tutorial Hands on CMEMS #1 "A view on the ocean: the statistical view or the physical view?" The regional high-resolution model (IBIRYS) produces a better stratification compared to the global model with a low resolution (GLORYS) and using data of the Aspex campaign as a reference (courtesy of L. Marie; Ifremer, from ASPEX3 cruise).

In the first part of the tutorial, we compare GLORYS and IBIRYS with a series of standard GODAE metrics, at the time and location of available in situ observations. Statistical comparisons (CLASS4 type) and analysis of the physical processes (CLASS1 eyeball validation, for instance Fig. 25.4.) are complementary techniques used to explore the differences between the two experiments and highlight the added value of IBIRYS with respect to GLORYS. It is particularly important to consider scales and representativity of both observations and models when evaluating high-resolution model products in respect to coarser ones. The high-resolution models produce small-scale features that may be shifted in space or time with respect to observations, inducing higher RMS errors. These small scales are not present in coarser models, which will give smoother statistics but will not be as realistic in terms of dynamics. For this reason, standard deviations from observations may not be significantly improved in higher-resolution models with respect to coarser ones. A complementary look at physical processes and scales shows the interest of dynamical solution of the high-resolution model. The benefit of high-resolution biogeochemistry models is shown in the second part of the tutorial, comparing model outputs with CMEMS Ocean Colour and sea surface temperature observations.

Finally, this tutorial stresses the need for higher-resolution observations in order to constrain small scales in high-resolution models, as well as the improvement expected on the short term from operational assimilation of biogeochemistry observations. Another mid-term challenge, for the IBI analysis system in particular, is to constrain tidal signal with data assimilation.

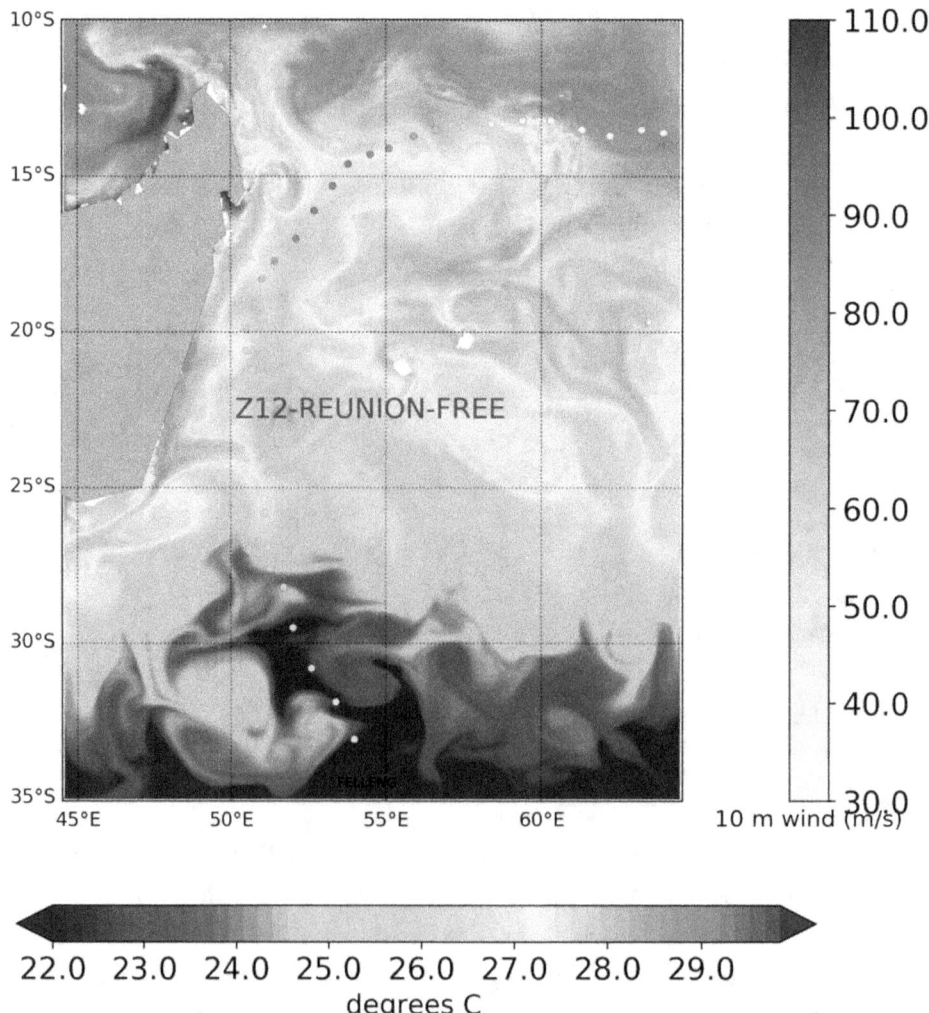

Figure 25.5. Illustration from Tutorial Hands on CMEMS #2 « Sail the global ocean: at the interface with the atmosphere", showing the signature of the Felleng cyclone (2013/02/04) on the sea surface temperature (°C) of a model zoom at 1/12° embedded in the GLORYS reanalysis. The observed wind velocity (m/s) of the cyclone is reported inside color dots showing the track of the cyclone from north to south.

Sail the global ocean: at the interface with the atmosphere

At the surface of the ocean, the accuracy of currents, temperature, and salinity strongly depends on the accuracy of the atmospheric forcing. Ocean reanalyses are often forced with atmospheric reanalyses. Most CMEMS reanalyses use ECMWF ERAInterim forcing (Dee et al., 2011). One known limitation of this forcing is the underestimation of cyclonic winds. In this tutorial, we look at the oceanic impact of the Felleng cyclone, which ran along the coasts of La Reunion Island in January-February 2013 (Fig. 25.5). We compare the global reanalysis GLORYS at ¼° (forced with ERA Interim) with the near real-time CMEMS global ocean analyses at 1/12°, which are forced with near real-time atmospheric forcings from ECMWF (with realistic cyclonic winds amplitude). We also look at the outputs of a nested NEMO configuration at higher resolution (1/12°), with updated physics but no data assimilation, embedded into GLORYS and forced with near real-time atmospheric forcings from ECMWF. This tutorial shows how downscaling is not only about improving the resolution, but also allows users to add the missing physical parameterizations in order to correctly capture oceanic phenomena. This tutorial also demonstrates that ocean atmosphere coupling at the global scale or using a downscaling approach is a major research topic for cyclone forecasting activities.

In the second part of this tutorial, we run the ARIANE software (Blanke and Raynaud, 1997, see also Beuvier et al., 2012 for a use case example) to compute water particles trajectories using CMEMS surface currents in the Indonesian Throughflow area. The spread of Lagrangian trajectories after a few days illustrates the importance of improving the quality of small-scale representation. Currently, the forecast skill is low after one or two days. Improvement is expected in the coming years thanks to data assimilation of current observations and high-resolution observations.

Dive into a 3D virtual ocean: in between observations and model solution

Three-dimensional (3D) ocean analyses produced in near real-time, thanks to models and data assimilation of ocean observations, provide gridded and dynamically consistent 3D estimates of the ocean variables. Ocean *re*-analyses use comparable model plus data assimilation systems (although often with coarser horizontal resolution), but they use re-analysed atmospheric forcings, better quality-controlled input ocean observations, and aim at providing four-dimensional views of the state of the ocean over the last decades, as homogeneous as possible in space and time. These tools are powerful and very useful for a variety of applications, especially because they provide estimates when and where no observations are available. By construction, the quality of ocean analyses and re-analyses should be lower when and where there is a lack of observations; but also, because of this lack of observations, there are often no means to quantify properly the uncertainty.

Interannual variability in the Leeuwin Current along the western coast of Australia is strongly linked with the El Nino Southern Oscillation (Feng et al., 2013), and the Leeuwin Current experiences a strong seasonal cycle that is well captured by altimetry (Ridgway and Godfrey, 2015). In the first part of this tutorial, we quantify with a few simple metrics how data assimilation modifies the 3D solution, first looking at the variability in the ocean (interannual and seasonal variability) and then comparing that to independent results from the literature.

In the second part of the tutorial, we inter-compare a series of sensitivity experiments where only the altimetry assimilation changes. These Observing System Experiments (OSEs; Oke et al., 2015a, 2015b, Lea et al., 2014, Turpin et al., 2016) were performed with an eddy-permitting-resolution analysis system for the Atlantic Ocean and Mediterranean Sea (20°S-70°N) assimilating sea surface temperature, in-situ temperature and salinity profiles, and sea level anomalies. The number of altimeters considered for data assimilation varies from zero to three, and the user observes how the increase in assimilated data improves the overall solution (Fig. 25.6). This is confirmed by Observing System Simulation Experiments (OSSE; Verrier et al., 2017), which allow testing observing network configurations using synthetic observations from a model as assimilated observations. We explore some of the limitations of data assimilation and stress the need for a sustainable observation network filling the gaps of unexplored oceanic areas or phenomena (higher-resolution observations such as SWOT, observations of the deep ocean such as Deep Argo, biogeochemical observations such as Bio Argo).

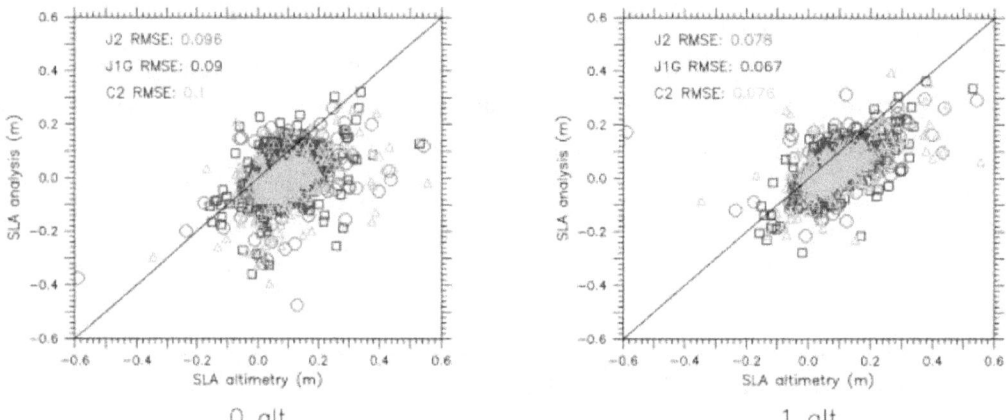

Figure 25.6. Illustration from Tutorial Hands on CMEMS #3 "Dive into a three-dimensional virtual ocean: in between observations and model solution" showing how the assimilation of observations from one altimeter improve the statistics with respect to all altimeters.

Conclusion

The Copernicus Marine Service reaches the end of its first phase in 2017, and it is an achievement of European Operational Oceanography. It currently provides observations, analyses, and forecasts of the ocean, reanalyses, and reprocessing of observations, as well as regular reports on the state of the ocean. Last, but not least, it is a *service* and it is organized around its users. CMEMS is about helping the users of operational oceanography and training them, but also about knowing the users and noting their needs and collecting feedback. The Copernicus Marine Service is also an investment of the European Commission to stimulate the Blue Economy and foster innovative downstream activities. The main challenges for the future will be to improve the uptake of the products and the interaction with coastal users, and to utilize big data capabilities to improve the service. The scientific challenges will be to increase the resolution of the products (observations and models), while also improving the representation of the interactions between the ocean, the sea ice, the waves, and the atmosphere in the models. Documentation and tutorials about CMEMS are referenced that can provide a practical view on the scientific content of operational oceanography products, their strengths, and current limitations. Useful and practical tutorials to ease the download and handling of CMEMS data are also referenced. The hands-on CMEMS tutorial helps users understand the strengths and weaknesses of ocean analyses and reanalyses, and what will be the major sources of improvement in the future (higher-resolution observations, sustainable ocean observing network, higher-resolution models with better physics, improvements of the biogeochemical models, …). Now you can become an actor of the Copernicus Marine service!

Acknowledgments

We thank the authors of Jupyter notebook tutorials from the CMEMS scientific community who kindly agreed to share their notebooks in addition to the ones described in this article: Paz Rotllan (SOCIB, for INS TAC), Pierre Prandi (CLS, for SL TAC), Ben Loveday (PML, for OC TAC), Patrick Raanes (NERSC, for ARC MFC). We also thank the GODAE international school organizers for this very nice opportunity to improve our tutorials. Finally, we warmly thank the students, Jessica Anderson, Barbara Barcelo-Llull, and Theo Baracchini, who reviewed this article, and the anonymous reviewer for their careful reading and very pertinent comments.

References

Beuvier, J., K. Beranger, C. Lebeaupin Brossier, S. Somot, F. Sevault, Y. Drillet, R. Bourdallé-Badie, N. Ferry, and F. Lyard (2012): Spreading of the Western Mediterranean Deep Water after winter 2005: Time scales and deep cyclone transport. *J. Geophys. Res.*, **117**, C07022, doi:10.1029/2011JC007679.

Blanke, B., and S. Raynaud (1997), Kinematics of the Pacific Equatorial Undercurrent: An Eulerian and Lagrangian approach from GCM results, *J. Phys. Oceanogr.*, **27(6)**, 1038–1053.

Crosnier, L., & Le Provost, C. (2007): Inter-comparing five forecast operational systems in the North Atlantic and Mediterranean basins: The MERSEA-strand1 Methodology. *Journal of Marine Systems*, **65(1)**, 354-375.

Dee, D. P., Uppala, S. M., Simmons, A. J., Berrisford, P., Poli, P., Kobayashi, S., Andrae, U., Balmaseda, M. A., Balsamo, G., Bauer, P., Bechtold, P., Beljaars, A. C. M., van de Berg, L., Bidlot, J., Bormann, N., Delsol, C., Dragani, R., Fuentes, M., Geer, A. J., Haimberger, L., Healy, S. B., Hersbach, H., Hólm, E. V., Isaksen, L., Kållberg, P., Köhler, M., Matricardi, M., McNally, A. P., Monge-Sanz, B. M., Morcrette, J.-J., Park, B.-K., Peubey, C., de Rosnay, P., Tavolato, C., Thépaut, J.-N., and Vitart, F. (2011): The ERA-Interim reanalysis: configuration and performance of the data assimilation system. *Q.J.R. Meteorol. Soc.*, **137**, 553–597, doi:10.1002/qj.828

Feng, M., McPhaden, M.J., Xie, S. & Hafner, J. (2013) La Nina forces unprecedented Leeuwin Current warming in 2011. *Sci. Rep. 3*, 1277, doi:10.1038/srep01277.

Hernandez F., E. Blockley, G. B. Brassington, F. Davidson, P. Divakaran, M. Drévillon, S. Ishizaki, M. Garcia-Sotillo, P. J. Hogan, P. Lagemaa, B. Levier, M. Martin, A. Mehra, C. Mooers, N. Ferry, A. Ryan, C. Regnier, A. Sellar, G. C. Smith, S. Sofianos, T. Spindler, G. Volpe, J. Wilkin, E. D. Zaron & A. Zhang (2015). Recent progress in performance evaluations and near real-time assessment of operational ocean products. *Journal of Operational Oceanography*, **8(sup2)**, s221-s238.

Lea, D. J., Martin, M. J. and Oke, P. R. (2014): Demonstrating the complementarity of observations in an operational ocean forecasting system. *Q.J.R. Meteorol. Soc.*, **140**: 2037–2049. doi:10.1002/qj.2281

Lellouche J.-M., O. Le Galloudec, M. Drevillon, C. Regnier, E. Greiner, G. Garric, N. Ferry, C. Desportes, C.-E. Testut, C. Bricaud, R. Bourdalle-Badie, B. Tranchant, M. Benkiran, Y. Drillet, A. Daudin, and C. De Nicola (2013): Evaluation of global monitoring and forecasting systems at Mercator Océan. *Ocean Science*, **9(1)**, 57.

Le Traon P.Y., A. Ali, E. Alvarez Fanjul, L. Aouf, L. Axell, M. Ballarotta, M. Benkiran, A. Bentamy, L. Bertino, L.A. Breivik, S. Cailleau, N. Mc Connell, G. Coppini, E. O'Dea, M. L. Grégoire, S. Guinehut, C.

Harris, V. Huess, S. Kay, J. de Kloe, G. Korres, F. Dinessen, M. Drevillon, Y. Drillet, Y. Faugère, I. Garcia Hermosa, G. Garric, I. Golbeck, J. Gourrion, J. Johannessen, F. Hernandez, R. King, P. Lagemaa, J.F. Legeais, M. Martin, A. Melet, J. Murawski, E. Özsoy, A. Palazov, E. Peneva, D. Peterson, L. Petit de la Villeon, N. Pinardi, S. Pouliquen, M.I. Pujol, P. Rampal , A. Ryan, A. Samuelsen, A. Saulter, J. She, M. Sotillo, A. Storto, T. Szekely, G. Taburet, M. Tonani, L. Tuomi, D. van Zanten, K. von Schuckmann, E. Stanev, P. Sykes, A. Stoffelen , T. Williams, H. Zuo , J. Xie (2017). The Copernicus Marine Environmental Monitoring Service: Main Scientific Achievements and Future Prospects. *Mercator Ocean Journal,* **56**, https://www.mercator-ocean.fr/en/science-publications/mercator-ocean-journal/mercator-ocean-journal-56-special-issue-cmems/

Maksymczuk J., F. Hernandez, A. Sellar, K. Baetens, M. Drevillon, R. Mahdon, B. Levier, C. Regnier, A. Ryan, (2016), Product Quality Achievements Within MyOcean, *Mercator Ocean Journal* **#54**

Maraldi C., J. Chanut, Bruno Levier, Nadia Ayoub, Pierre De Mey, G. Reffray, Florent Lyard, Sylvain Cailleau, Marie Drévillon, E. Fanjul, M. G. Sotillo, Patrick Marsaleix, and the Mercator Team (2013): NEMO on the shelf: assessment of the Iberia-Biscay-Ireland configuration. *Ocean Science, European Geosciences Union,* 2013, **9**, 745-771

Marie L. (2016) ASPEX8 cruise, RV Thalia, http://dx.doi.org/10.17600/16006500

Oke, P.R., G. Larnicol, Y. Fujii, G.C. Smith, D.J. Lea, S. Guinehut, E. Remy, M. Alonso Balmaseda, T. Rykova, D. Surcel-Colan, M.J. Martin, A.A. Sellar, S. Mulet, V. Turpin. (2015): Assessing the impact of observations on ocean forecasts and reanalyses: Part 1, Global studies. *Journal of Operational Oceanography,* **8**:sup1, s49-s62.

Oke, P.R., G. Larnicol, E.M. Jones, V. Kourafalou, A.K. Sperrevik, F. Carse, C.A.S. Tanajura, B. Mourre, M. Tonani, G.B. Brassington, M. Le Henaff, G.R. Halliwell Jr., R. Atlas, A.M. Moore, C.A. Edwards, M.J. Martin, A.A. Sellar, A. Alvarez, P. De Mey, M. Iskandarani. (2015): Assessing the impact of observations on ocean forecasts and reanalyses: Part 2, Regional applications. *Journal of Operational Oceanography* **8**:sup1, s63-s79.

Verrier, S., Le Traon, P.-Y., and Remy, E. (2017): Assessing the impact of multiple altimeter missions and Argo in a global eddy permitting data assimilation system, *Ocean Sci. Discuss.,* doi:10.5194/os-2016-104, in review.

von Schuckmann K., Le Traon P.Y., Alvarez-Fanjul E., Axell L, Balmaseda M., Breivik L.-A., Brewin R.J.W., Bricaud C., Drevillon M., Drillet Y., Dubois C., Embury O., Etienne H., García Sotillo M., Garric G., Gasparin F., Gutknecht E., Guinehut S., Hernandez F., Juza M., Karlson B., Korres G., Legeais J.-F., Levier B., Lien V. S., Morrow R., Notarstefano G., Parent L., Pascual A., Pérez-Gómez B., Perruche C., Pinardi N., Pisano A., Poulain P.-M., Pujol I. M., Raj R.P., Raudsepp U., Roquet H., Samuelsen A., Sathyendranath S., She J., Simoncelli S., Solidoro C., Tinker J., Tintoré J., Viktorsson L., Ablain M., Almroth-Rosell E., Bonaduce A., Clementi E., Cossarini G., Dagneaux Q., Desportes C., Dye S., Fratianni C., Good S., Greiner E., Gourrion J., Hamon M., Holt J., Hyder P., Kennedy J., Manzano-Muñoz F., Melet A., Meyssignac B., Mulet S., Buongiorno Nardelli B., O'Dea E., Olason E., Paulmier A., Pérez-González I., Reid R., Racault M.-F., Raitsos D.E., Ramos A., Sykes P., Szekely T. & Verbrugge N. (2016): The Copernicus Marine Environment Monitoring Service Ocean State Report, *Journal of Operational Oceanography,* **9** , sup2

Ridgway, K. R., and J. S. Godfrey (2015): The source of the Leeuwin Current seasonality, *J. Geophys. Res. Oceans,* **120**, 6843–6864, doi:10.1002/2015JC011049.

Turpin, V., Remy, E., and Le Traon, P. Y. (2016): How essential are Argo observations to constrain a global ocean data assimilation system?, *Ocean Sci.,* **12**, 257-274, doi:10.5194/os-12-257-2016.

Appendix

The Copernicus Marine Thematic Assembly Centres - TACs -

The Copernicus Marine Monitoring and Forecasting Centres - MFCs -

CHAPTER 26

Operational Oceanography and the Management of Marine Living Resources: The Mediterranean Sea as a Case Study

Patricia Reglero[1], Diego Alvarez-Berastegui[2], Francisco Javier Alemany[1], Vincent Rossi[3], Asvin P. Torres[1], Rosa Balbin[1], and Manuel Hidalgo[1,4]

[1]*Instituto Español de Oceanografía, Centre Oceanogràfic de les Balears, Palma de Mallorca, Spain;*
[2]*Balearic Islands Coastal Observing and Forecasting System, Palma de Mallorca, Balearic Islands, Spain;*
[3]*Mediterranean Institute of Oceanography, Marseille, France;* [4]*Instituto Español de Oceanografía, Centro Oceanográfico de Málaga, Malaga, Spain*

This chapter provides examples demonstrating the relevance of integrating operational oceanography data from ocean observing systems and ecological and fisheries assessment models to improve the management of marine living resources. Commercial fisheries exploit coastal, demersal, and pelagic marine resources. Many marine organisms merge in the water column and most of them occupy the pelagic habitat as planktonic organisms (egg and larval stages), during which time they are subjected to the highest mortalities of their life cycle as they are transported by oceanic currents. Hence, it is important to determine how environmental processes control survival rates and dispersal patterns of the early life stages of the species. We offer a general view of a multidisciplinary research field that aims at the protection and exploitation/management of marine living resources by documenting some current strategies and recent advances in the Mediterranean Sea. We include a short introduction of the current strategies for the protection and exploitation of living resources and the recent advances of the field and present four practical examples, which show how the integration of operational oceanography into the management of living resources has improved our knowledge of: 1) the spatial distribution of adult fish, 2) the connection among management areas, 3) the redefinition of management areas, and 4) the use of marine protected areas for the conservation of coastal ecosystems.

Marine Living Resources

Since life first appeared on the Earth, the oceans have been a first home for living organisms. A wide variety of creatures, from primitive bacteria to complex forms of life, inhabit the sea. Their survival depends very much upon their relationship with the surrounding environment, their capacity to respond and adapt to environmental variability, and their relationship with other players in the ocean ecosystem. Feeding, settling in their preferential habitats, using currents for dispersal and migrations, limiting their spatial distribution, and timing their migrations and reproductive events are only some of the biological processes that can be influenced by the abiotic and biotic environment in which marine organisms live.

Reglero, P., et al., 2018: Operational oceanography and the management of marine living resources: The Mediterranean Sea as a case study. In "*New Frontiers in Operational Oceanography*", E. Chassignet, A. Pascual, J. Tintoré, and J. Verron, Eds., GODAE OceanView, 713-728, doi:10.17125/gov2018.ch26.

Humans have been using the sea for goods (living and mineral resources) and services, especially as a source of food (fishing and shellfish), since immemorial time. Living marine resources are renewable, but not unlimited, and their exploitation must ensure the preservation of habitats, with fishing activities subjected to management systems aimed at safeguarding their sustainability. Therefore, effective management strategies that lead to sustainable development must be implemented.

Practically all marine habitats are currently being exploited, from intertidal areas to deep-banked bottoms and the open ocean. We can distinguish three main types of commercial fisheries (Fig. 26.1, note that recreational fisheries are becoming more and more important in some areas):

- The artisanal fisheries that exploit the **coastal resources** with a great diversity of gears, fixed nets, or hand lines. This type of fisheries is also the most diverse in the number of species captured, targeting a huge variety of taxa, from shellfishes (clams...) or crustaceans (crabs, lobsters...) to shore fishes (sea-breams, mullets, soles, groupers...).
- The industrial or semi-industrial fisheries, as trawl fleets, that exploit **demersal resources** - the organisms that preferentially inhabit or are in close association with the seabed (cod, hake, flatfish ...) - in shelf or slope areas.
- The fleets exploiting the **pelagic resources** – those that inhabit the water column including both species in neritic (sardines, anchovies ...) and oceanic areas (tunas, swordfish ...) which are captured mainly by purse seines, trawl lines or longlines.

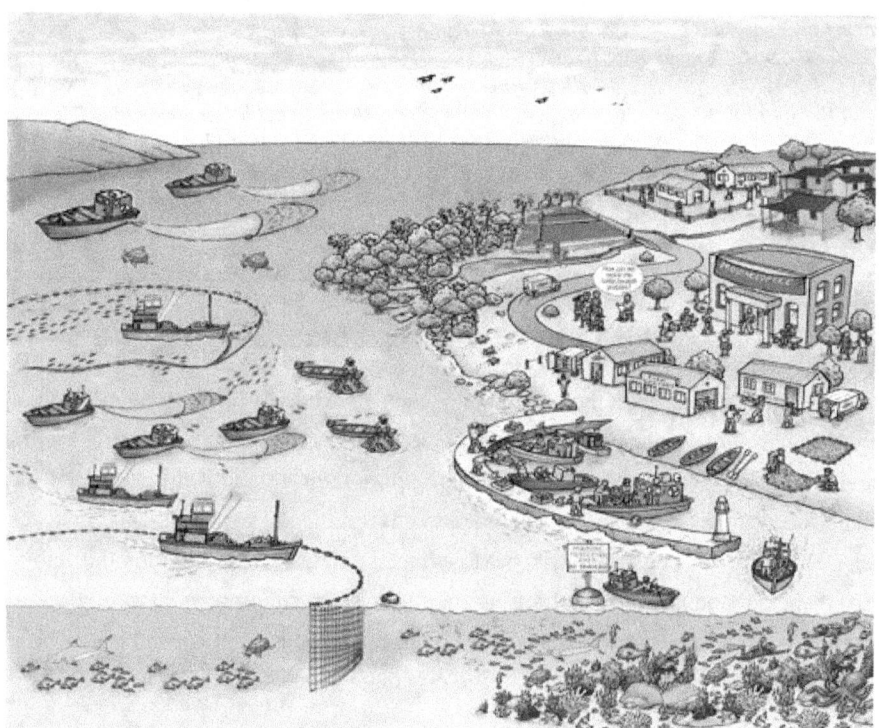

Figure 26.1. Illustration showing how living resources are being exploited in most marine habitats. Source: image adapted from FAO EAF Nansen Project in Staples et al. (2014).

All living resources, regardless of their habitat—coastal or offshore, pelagic or demersal—merge in the water column at some stage of their life cycle. This is because most marine organisms pass through a planktonic egg and larval stage, typically lasting from weeks to few months. Pelagic species live their entire life cycle in the water column. For example, Atlantic bluefin tuna, a pelagic top predator, release their 1 mm size eggs at the surface, which hatch into larvae (around 2-3 mm long) after few hours and continue growing several days to attain juvenile size. All of these developmental stages take place in the upper water column (around 20 m depth in the Mediterranean, Fig. 26.2). Even at the adult stage, this species will conduct large migrations within their habitat range and will perform extensive movements throughout the water column.

In contrast to pelagic species, demersal or sessile benthic species (those living fixed to the substrate) only spend the early stages of their life cycle in the water column. For many of these species most of the mid- or large-scale displacements will occur while they occupy the pelagic habitat as planktonic organisms (during the egg and larval stages), at which time they are transported by oceanic currents from their natal place to distant habitats where they may settle and continue living as juveniles and adults. Larvae from most exploited species inhabit surface waters, where light and food resources are abundant. Nevertheless, some larvae from demersal species such as deep lobsters can also take advantage of resources at deeper layers where there is less light (Fig. 26.3). Among marine invertebrates, decapod crustaceans have relatively long larval stages during which they can be especially sensitive to any environmental factors (Giménez, 2010).

In general, the survival rates during the planktonic stages of different exploited resources (fish or invertebrates) are very low and highly variable depending on environmental conditions. this means that planktonic stages are not only important for the spatial dynamics of many marine living resources serving as the dispersive phase of the life cycle, but also constitute a critical period in which the success of the recruitment process is determined.

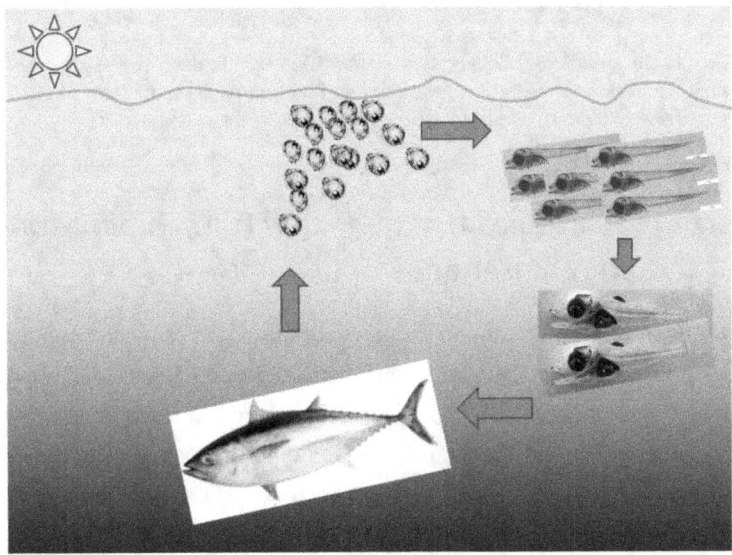

Figure 26.2. The life cycle of a pelagic species. Example of Atlantic bluefin tuna (*Thunnus thynnus*) from eggs, then larvae (urostile pre-flexion and post-flexion), juvenile, and finally pelagic adult phases. All of the developmental stages of this species take place in the water column.

In the context of fisheries, recruitment is defined as the number of juveniles that survive and grow long enough that they can be caught by a specific fishery. Recruitment is the result of complex physical and ecological processes that act at various spatial and temporal scales and throughout the life stages (both larvae and juvenile) leading up to the recruitment itself. As a result, it is important to determine the impacts of environmental processes on survival rates and movements during the early life stages of a species. Therefore, knowledge of the pelagic environment and the characteristics and dynamics of water masses, is crucial for the understanding the processes driving marine population dynamics, and hence for evaluation and planning of management measures aimed at ensuring the sustainability of pelagic and demersal living resources.

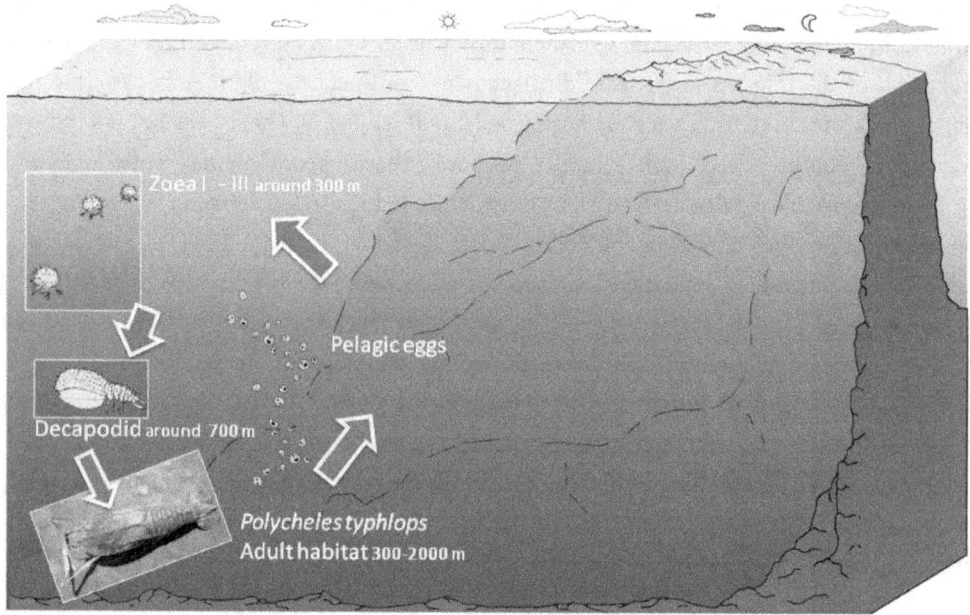

Figure 26.3. Life cycle for a demersal decapod as exemplified by the deep sea lobster *Polycheles typhlops*. Larval stages (from zoea to decapodid) are represented as function of depth (Torres et al., 2014), with adults living around 300 to 2000 m depth in Mediterranean Sea.

Current Strategies for the Protection, Exploitation, and Maintenance of Living Resources

Methodologies for the formal evaluation of the "state of the stocks" (i.e., the population of a certain species occupying a certain geographic area considered as management unit) was initially applied in the 1950s. Within this framework, fisheries are managed as populations isolated from the surrounding environment in order to maximize production of the targeted species. This is based on models that allow determining the biomass of the stock from fishery data alone, such as catches and effort (production models), or using a combination of fishery data (fishing mortality) and biological/ecological parameters (growth rates, natural mortality, survival to the adult population or recruitment). In addition, these models provide projections of how the biomass of the stock will change in relation to fishing efforts in order to estimate the suitable harvest levels for sustainable exploration. Following this approach, stocks have been managed through the control of fishing effort, quotas of capture, and technical measures such as minimum sizes, based mostly on fishery-dependent data and with a focus on maximizing production of the targeted species irrespective of the type of fisheries, either coastal, benthic, or pelagic (Fig. 26.4). In this context, they are still considered isolated from the environment, without taking into account environmental condition variations or spatially explicit issues.

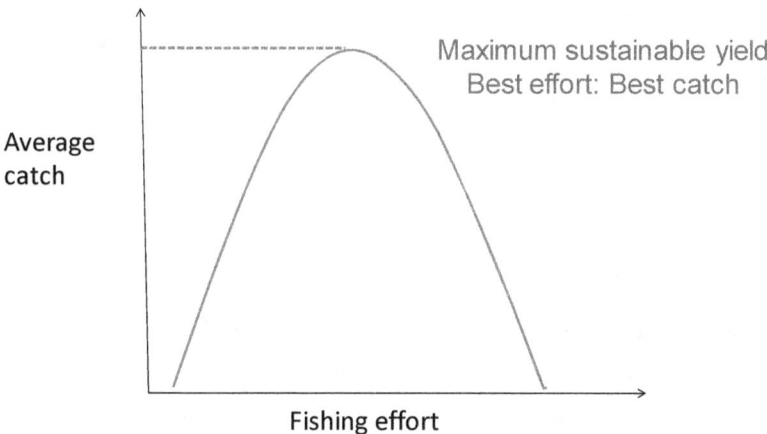

Figure 26.4. The maximization of production of a targeted species. The maximum sustainable yield is estimated from the best relationship between fishing effort and average catch.

Let us take, as an example, the Atlantic bluefin tuna in the Eastern Atlantic and the Mediterranean Sea. After single population models suggested an important decrease in the biomass of the stock, the European Commission established a 15-year recovery plan for Atlantic bluefin tuna starting in 2007 that included, as management measures, decreasing the total allowable catch limit, restricting the fishing seasons for certain types of fishing vessels, prohibiting the use of airplanes and helicopters to search for Atlantic bluefin tuna, and establishing a minimum size of 115 cm and minimum weight of 30 kg (except for juveniles of 8 kg in the Adriatic Sea). Using this approach, the assessment of the eastern Atlantic bluefin tuna does not take into consideration natural fluctuations in abundance due to environmental factors' impact the survival of the early life stages, although inter-annual variability is expected (Fig. 26.5).

Over time and after some failures by fisheries management systems based on reductionist approaches—perhaps because of their intrinsic shortcomings or failures in governance or enforcement issues or as a result of the growing social awareness on environmental protection issues—a new paradigm, the *ecosystem approach,* has been proposed. This approach, which emerged in the early 1980s and was accepted internationally at the Rio Earth Summit in 1992, began to be incorporated into fisheries management at the 24th session of the Food and Agriculture Organization of the United Nations (FAO) Fisheries Committee and at the World Conference on Responsible Fisheries in Reykjavik (2001). Its implementation is now an obligation in some areas, such as the European seas, having been formally adopted by the European Commission's Common Fisheries Policy.

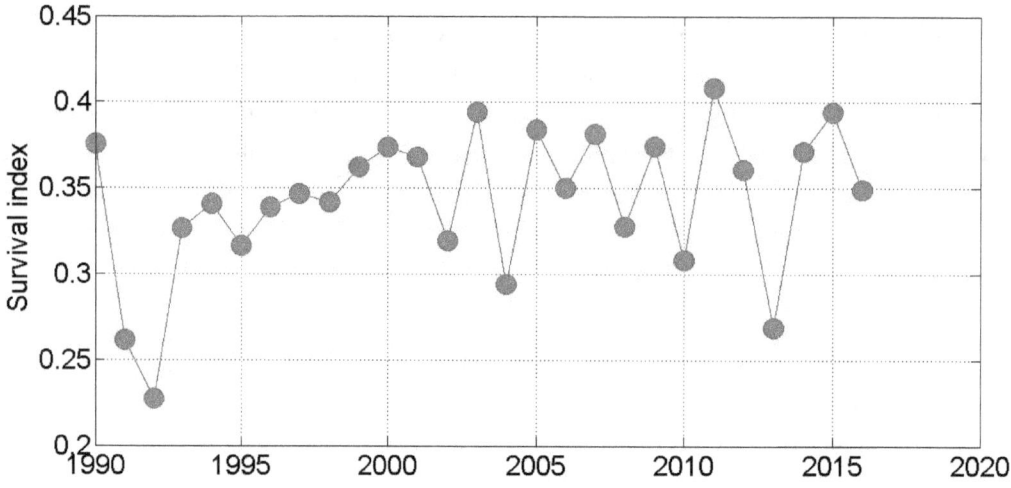

Figure 26.5. Natural fluctuations in larval survival probability due to the effect of temperature on growth.

Applying the ecosystem approach involves promoting multidisciplinary scientific research that provides quality scientific information and advice on fisheries, associated ecosystems taken as a whole and their relevant environmental factors, as well as their interactions. Besides fishing activity, variations in the Earth's climate, such as warming, heat waves or increasing sea level, are expected to affect marine ecosystems. While it is difficult to know exactly the systems' responses to climate warming, a broad range of observations during the past decades as well as modeling exercises predict that species can change their distribution, migration routes, or their reproductive timing and location. One example is the "tropicalization" of ecosystems, or the immigration of exotic species from warmer neighboring areas, when species change their distribution area along with changes of ecological niches (usually northward migration). Improved medium- to long-term projections based on understanding the physiology of organisms and the ecosystem functioning is important if climate change drives the environment out of the scenarios for which historical datasets are available (Fox and Aldridge, 2008).

The application of the ecosystem approach to the evaluation and management of living marine resources inevitably requires information on the characteristics and dynamics of the bodies of water provided by studies of physical oceanography. Also, in order to assess the vulnerability of fisheries to climate change, it is necessary to properly forecast future scenarios of environmental conditions. The incorporation of environmental information into models for the assessment of fishery resources is an emerging field of research, but there are already relevant examples of products based on operational oceanography data directly applicable to fishery management. Some of these applications are described below to demonstrate how they relate to each other to improve the management of living resources in a dynamic context. One of the current and future challenges is the need to work toward integration of oceanography into the assessment of marine living resources, allowing designing management actions to take into account environmental variability and climatic trends and, thus, ensure the sustainability of resources.

From a practical perspective there is a need to better link the knowledge of physical oceanography to fisheries assessments. Future research and development of data tools by oceanographers will have a higher impact on the sustainability of living resources if they are designed in multidisciplinary working groups. In fact, different international bodies are starting to establish collaborative networks with this in mind (e.g., http://www.imber.info/science/regional-programmes/cliotop/task-teams). To facilitate a better understanding of how operational oceanography is being linked in practice with pelagic, demersal, and coastal fisheries assessments, below are a number of study cases.

Current Advances and Challenges Integrating Operational Oceanography in the Management of Living Resources: Applied Examples.

The spatial distribution of adult fish

The fisheries of tuna species in the Mediterranean Sea mainly target adults during their reproductive season. Mature fish are captured in areas where they go to mate and reproduce. The spatial location for tuna species to reproduce can be linked to a certain spatially-fixed recurrent feature in a specific region (geographically-driven spawning) or it can be strongly influenced by oceanographic features that vary in their position from year to year, such as fronts (environmentally-driven spawning) (Reglero et al., 2012; Ciannelli et al., 2014). Thus, tuna species may be able to change their spawning distribution between years in relation to environmental changes within the ecological limits for the offspring success imposed by larval survival (Ciannelli et al., 2014).

Worldwide, tuna spawning grounds are usually found in areas with mesoscale oceanographic features such as fronts and eddies (Bakun, 2006; Reglero et al., 2014). There are several tuna species that reproduce in the Western Mediterranean (Fig. 26.6). The bullet tuna (*Auxis rochei*) is a small-size tuna that inhabits the coastal areas year-round. The albacore tuna (*Thunnus alalunga*) is a medium-size tuna that inhabits the open waters, and the Atlantic bluefin tuna (*Thunnus thynnus*) is a large-size tuna that undertakes far-reaching seasonal migrations in spring-summer from their feeding grounds in the Atlantic Sea into the Mediterranean Sea to reproduce. These three tuna species select their reproductive habitats based on different environmental signals (Reglero et al., 2012).

The location of reproduction habitats for Atlantic bluefin tuna shows two distribution patterns depending on the position of the salinity front that recurrently occurs and persists during the summer off Balearic Islands (Fig. 26.7a-b). Albacore reproductive locations are geographically fixed and linked to a quasi-permanent eddy (Fig. 26.7c), whereas bullet tuna reproduce mainly in the coastal areas (Fig. 26.7d).

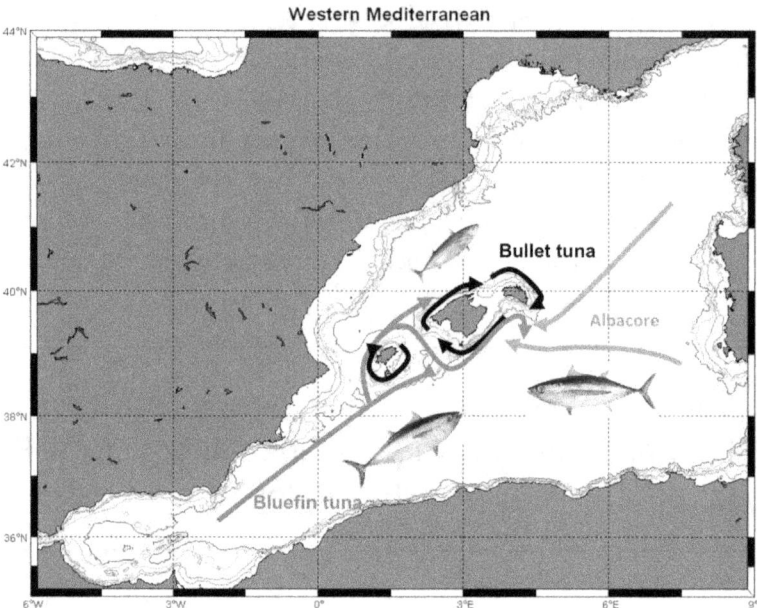

Figure 26.6. Conceptual figure summarizing the spawning geography of the three species of tuna—the large migratory Atlantic bluefin tuna (red), the Mediterranean albacore (green), and the coastal bullet tuna (dark blue)—and their plausible migration pattern around the Balearic Islands (Mediterranean Sea) from Reglero et al. (2012). Tuna pictures from S. P. Iglésias.

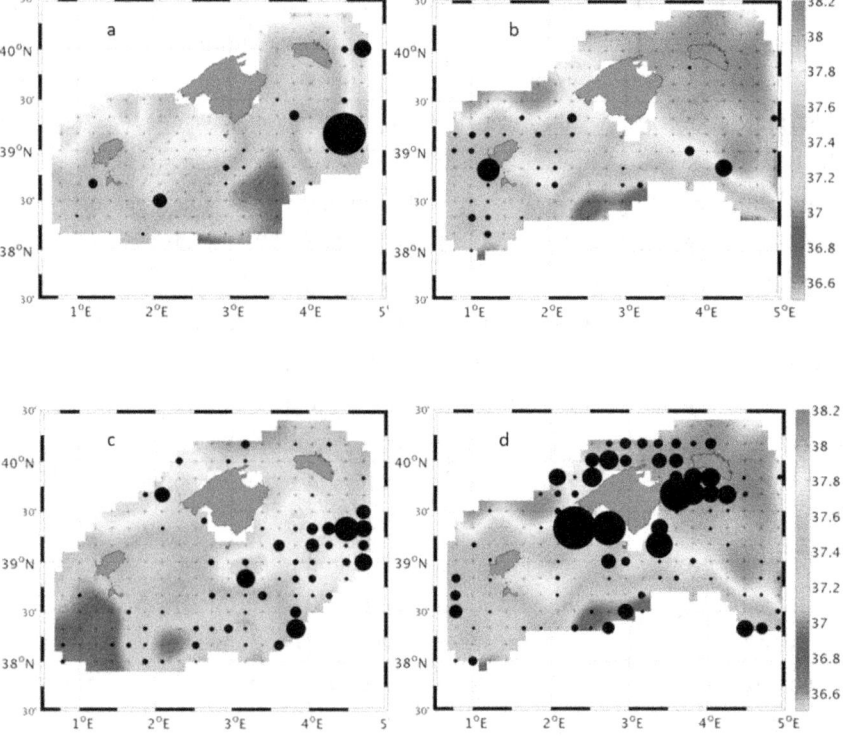

Figure 26.7. Spatial distribution of tuna larvae (black dots size represent larval density in larvae m^{-3}) in relation to salinity. a-b) Atlantic bluefin tuna spatial distribution varies with the position of the salinity front. c) Albacore spatial distribution is associated to a fixed mesoscale structure south of Menorca. d) Bullet tuna spawning in mostly in coastal areas. From Reglero et al. (2012).

Spawning locations can be estimated from models developed from data on spatial distributions of egg and larval abundances in relation to oceanographic operational data. Forecasting oceanographic summer situations, based on a long-term data series of environmental data from periodic research vessels surveys and new oceanography products based on the knowledge of environmental variability and ecological process, can help in the development of decision support tools for the spatial conservation and management of tuna species (Alvarez-Berastegui et al., 2016).

Management areas in the NW Mediterranean: Connectivity among subpopulations

One the most illuminating examples of the paramount need for oceanographic modeling (both hindcast and forecast) in fisheries research is the current challenge of embracing the spatial pattern of commercially-exploited populations to their temporal assessment. Indeed, better understanding the complex dynamics of large marine fish stocks is a critical need in fisheries science, as it is essential to know how many fish are in a given area and at a given time. Marine populations are often structured as a set of sub-populations connected through the exchange of individuals and whose spatial boundaries rarely coincide with management units. This is the case in the Mediterranean Sea (Fig. 26.8) where more than 90% of the stocks are considered over-exploited, with the European hake (*Merluccius merluccius*) being the most representative example.

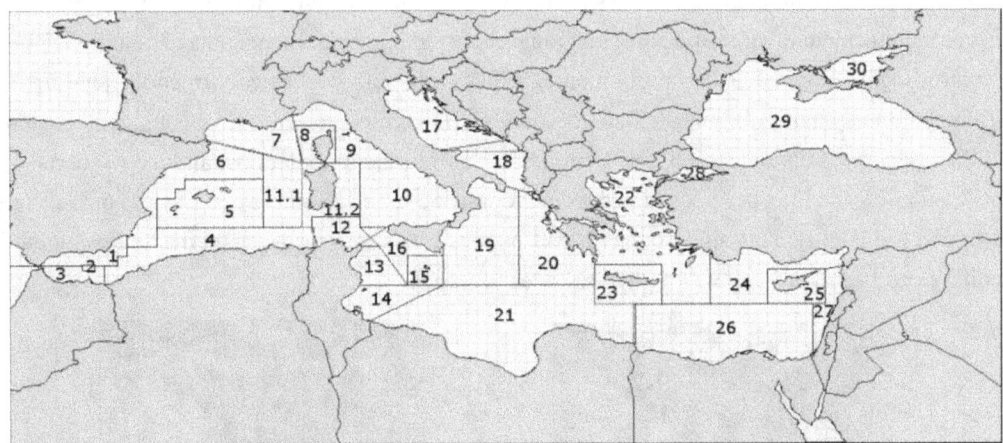

Figure 26.8. Geographical subareas used in the Mediterranean Sea as data collection and management frameworks (FAO).

Although fisheries models are increasingly considering adults' movement estimated from tagging experiments (i.e., marked-recaptured individuals), they still overlook the dynamics of early-life stages (eggs and larvae) potentially connecting subpopulations. Connectivity during early life stages is largely driven by physical dispersion due to multi-scale oceanic currents (Fig. 26.9) and has a profound impact on fish recruitment (e.g., the smallest fish or juveniles being captured by fisheries). Estimates of connectivity among subpopulations in the northwestern Mediterranean can be compared using ocean circulations models, with the recruitment estimates (number of young fish at sea per year) from the commercial fishery.

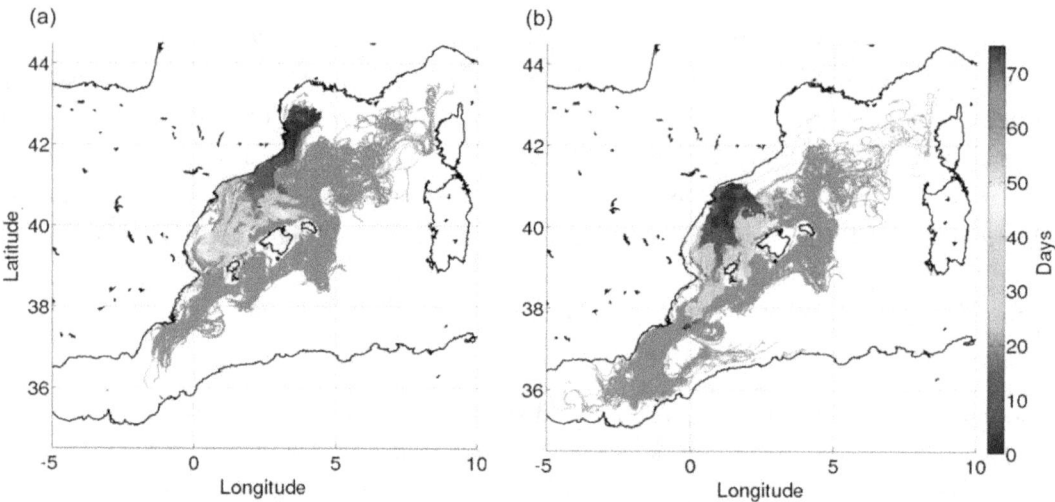

Figure 26.9. Lagrangian trajectories of Winter Intermediate Water (WIW) particles from the Gulf of Lion (a) and (b) the Ebro Delta regions. Color indicates time for particles with WIW that could simulate ELS in this habitat. Trajectories indicated in gray lost WIW properties (adapted from Juza et al., 2013)

A recent study demonstrated for the first time that the inter-annual variability of recruitment of large fish stocks can be modeled in relation to physical pattern of currents, calculating spatio-temporal connectivity estimates derived from high-resolution circulation models (Fig. 26.10). Fisheries science would benefit from efficient ways to include physics and fisheries assessment to improve management of large and complex populations, integrating the analysis of a limited number of controlling ecological and environmental processes that are critical to understanding and reproducing the dynamics of marine fish populations. This research provides a considerable step in that direction by acknowledging the complexity of marine populations and ecosystems in a relatively simple manner, as opposed to the development of overly complex ecosystem models (e.g., end-to-end modeling). This study opens broad opportunities to improve fisheries management by including short-term projections of physical oceanography.

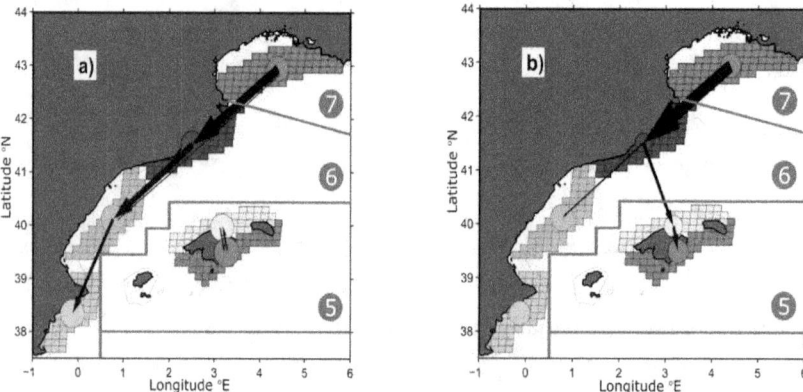

Figure 26.10. Spatial variability of the connectivity processes (export as arrows and retention as bubbles) between hake (*Merluccius merluccius*) subpopulations for two contrasting years. (a) Year 1989 shows the main south-westward transport pattern along the mainland with almost no import into the Balearic archipelago. (b) Year 2005 exhibits a reduced southwestward transport and stronger connections with the Balearic Islands (adapted from Rossi et al. in review).

Characterizing the dispersion potential of the seascape for a better informed marine spatial planning

As dispersion and connectivity are expected to affect both genetic structures and demographic rates of local populations, a careful evaluation of these bio-physical processes is crucial to understanding and modelling population dynamics. Thus, characterizing the connectivity regimes of all location of the ocean provides useful information for ecosystem management. Through the direct incorporation of population genetic concepts into a basin-scale biophysical model (i.e., Lagrangian module coupled with state-of-the-art operational ocean model), Dubois et al. (2016) proposed a common platform for geneticists, marine ecologists, and oceanographers to evaluate connectivity for management applications. By manipulating connectivity matrices with network theory tools, Dubois et al. (2016) analyzed complementary connectivity metrics: local retention (the ratio of locally produced settlement to local larval release), self-recruitment (the ratio of locally produced settlement to the overall settlement), and the source/sink proxy (relative importance of larval export versus import) (Botsford et al., 2009; Bode et al., 2006). The model predicts that retention processes are favored along certain coastlines due to specific oceanographic features, such as a sluggish circulation and extended continental shelves (Fig. 26.11b). Moreover, wind-driven divergent (convergent, respectively) oceanic regions are systematically characterized by larval sources (sinks, respectively) (Fig. 26.11a).

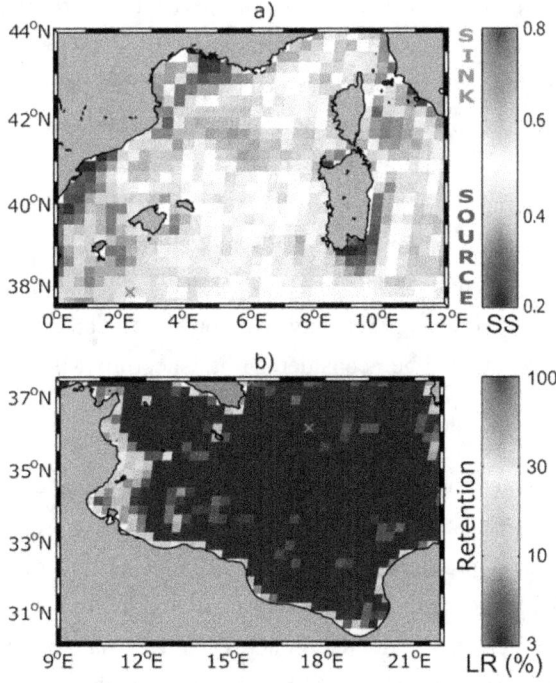

Figure 26.11. Connectivity studies conducted using a biophysical model simulating the dispersal of passive larvae in the upper water column for a pelagic dispersal during 30 days. a) Seasonal mean sink and source areas (relative importance of larval export versus import) in winter. b) Seasonal mean local retention (%) in summer. Adapted from Dubois et al. (2016).

Using biophysical modelling together with an integrated interpretation of retention and exchange connectivity indices give insight into how subpopulations are connected through larval transport; and, as such, helps to predict the effects of management measures or disturbances on both local and surrounding subpopulations. An accurate depiction of both local and broad-scale connectivity, as is allowed by connectivity proxies, is necessary to appropriately implement effective, spatially explicit management measures, such as marine reserves or fishery restrictions, by providing relevant information for managers and scientists to discuss implementation guidelines "case-by-case," in accord with sustainable objectives.

Marine protected areas for the conservation of coastal ecosystems

Marine protected areas (MPAs) are widely applied for the conservation of coastal ecosystems. Moreover, extensive research has made the case for promoting their application as fisheries management tool (Pérez-Ruzafa et al., 2008). Reducing fish mortality in the MPAs allows recovery of fish populations, which can result in spillover of larvae and adults to adjacent waters to the benefit of local fisheries (Goñi et al., 2008). In most cases, fisheries MPAs have been directed to enhance small-scale (artisanal) fisheries, operating in coastal waters and targeting highly resident species, which are the ones most benefitting from this conservation approach. MPAs are now a priority in the political environmental roadmap at a global scale, and member countries of the Convention on Biological Diversity agreed to reach a 10% of the marine real to be protected in 2020.

MPAs should meet specific conditions, such as incorporating the heterogeneity of the seascape and matching the biogeography of organisms, to ensure their functionality (Rees et al., 2017). Of particular relevance to the oceanography of regions where the protected areas are located, the MPAs should be connected to each other in order to build networks. The International Union for Conservation of Nature defines a MPA network as "a collection of individual MPAs operating cooperatively and synergistically at various spatial scales and with a range of protection levels that are designed to meet objectives that a single reserve cannot achieve." The Convention on Biological Diversity also specifies that currents, gyres, physical bottlenecks, and species dispersal processes, among other parameters, must be considered at specific MPA sites (see Annex II, https://www.cbd.int/decision/cop/?id=11663). The objective is to be able to design MPA networks were the size of each MPA and spacing between MPAs are based on scientific data informing by the patterns of adult movement and larval dispersal of protected species (see International Union for Conservation of Nature, https://www.iucn.org). For instance, Rossi et al. (2014) introduced the concepts of hydrodynamical provinces, which allow for a systematic characterization of the seascape connectivity. This information is useful when attempting to better define the relevant size and spacing guidelines of a given oceanic region in accord with management objectives.

The Mediterranean Sea is characterized by having a large number of small protected areas (Fig. 26.12). Most of them have been designed with the objective of increasing the abundance of living resources targeted by the fisheries or to act as areas that export biomass to neighboring areas.

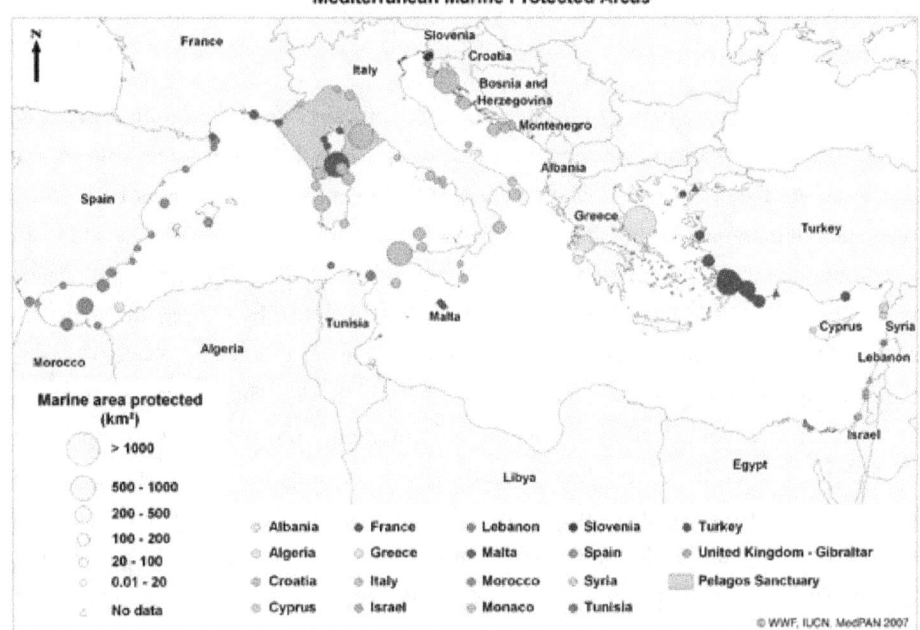

Figure 26.12. Marine protected areas in the Mediterranean Sea. Source: WWF, International Union for Conservation of Nature, Medpan (2007).

There are two main mechanisms through which marine protected areas act to improve fisheries: 1) larval exportation and 2) adult spillover from the protected to surrounding areas. Operational oceanography can expand our understanding of the distribution of species and life stages in order to address the biomass of targeted fish. One example is from Coll, et al. (2013) who looked at the biomass of fish as it was related to the coastal exposure, the benthic slope, rock rugosity, and depth. Continuous monitoring of these variables provides useful insight into the spatiotemporal dynamics of the species. On the other hand, life cycles develop in different habitats where the coastal and pelagic interphase plays a major role in connecting different life stages of the same species. Thus, there is a need for hydrodynamic and drifting models coupling the different spatiotemporal scales. Finally, operational oceanography can be used to develop tools for spatially dynamic marine protected areas since they enable the improvement of habitat prediction and management lines based on the spatial distribution of the species in relation to environmental variables.

The capability of the countries and international organizations to promote advances towards the implementation of effective MPA networks highly depends on our capability to assess the connectivity processes among specific locations in the sea (e.g., Rossi et al., 2014, Dubois et al., 2016). Toward this objective, we should consider which species are the object of protection and what the link is between their live cycles and the local oceanography. To accomplish this, the following biological characteristics must be considered: timing of the reproduction, early life biology, and developmental time to settlement. Operational oceanography should be encouraged to consider these issues in the development of hydrodynamic and dispersal models, considering that the life cycle occurs within different habitats (coastal/pelagic), and advocating for the development of complex hydrodynamic models coupling various spatiotemporal scales.

Challenges for the future of operational oceanography and marine living resources

There are only few examples of a systematic application of operational oceanography in the management of Mediterranean fisheries. Current assessments have not been able to ensure the sustainability of the fisheries, and nowadays 90% of the resources are overexploited (GFCM-FAO 2016). They do not integrate, or they include very poorly, environmental variability and its influence on the distribution and abundance of fish. The successful integration of operational oceanography into fisheries in the Mediterranean Sea has been recently discussed by experts in the framework of the Mediterranean Operational Network for the Global Ocean Observing System, identifying a number of gaps and challenges (Alvarez-Berastegui et al., 2017). These authors have proposed a roadmap based on the identification of specific oceanographic processes driving species ecology and the design of operational products specifically informing on those processes. The data accessibility and the data quality should be improved and new software should be created that enables data handling and end-to-end user communication. The usefulness of the operational oceanographic products can be proved by evaluating the impact of introducing environmental variability into a specific assessment model. This is one objective that can be tackled by the scientific community through European funding initiatives (e.g., EU Horizon 2020 funded project no. 773713, acronym PANDORA). Biochemical and physical oceanographic models are well developed, but extending them further up the food web to include plankton and fish has long been a major challenge for (de Young et al., 2004). As we have revealed in this chapter, most fish species pass through different developmental stages during their life cycle with changing habitats, thus varying their relationship with the environment. Such variability should be taken into account when linking biochemical models to stock assessment models.

References

Alvarez-Berastegui D., Hidalgo M., Tugores M.P., Reglero P., Aparicio-González A., Ciannelli L., Juza M., Mourre B., Pascual A., López-Jurado J.L., García A., Rodríguez J.M., Tintoré J., Alemany F. (2016) Pelagic seascape ecology for operational fisheries oceanography: modeling and predicting spawning distribution of Atlantic bluefin tuna in Western Mediterranean. *ICES Journal of Marine* Science fsw041

Alvarez-Berastegui D., Reglero P., Hidalgo M., Balbín R., Mourre B., Coll J., Alemany F., Tintoré J. (2017) Towards Operational Fisheries Oceanography in the Mediterranean, 6th Mediterranean Oceanography Network for the Global Ocean Observing System (MonGOOS), (Athens, Greece. 14,15, 16 November 2017). Article for the Scientific plan 2017-2020 of the Mediterranean Operational Network for the Global Ocean Observing System, doi:10.13140/RG.2.2.30708.86400

Bakun A. (2006) Fronts and eddies as key structures in the habitat of marine fish larvae: opportunity, adaptive response and competitive advantage. *Scientia Marina*, 70: 105–122

Bode M., Bode L., Armsworth P.R. (2006) Larval dispersal reveals regional sources and sinks in Great Barrier Reef. *Marine Ecology Progress Series*, 308, 17–25

Botsford L.W., Brumbaugh D.R., Grimes C., Kellner J.B., Largier J., O'Farrell M.R., Ralston S., Soulanille E., Wespestad V. (2009) Connectivity, sustainability, and yield: bridging the gap between conventional fisheries management and marine protected areas. *Reviews in Fish Biology and Fisheries*, 19, 69–95.

Ciannelli L., Bailey K., Olsen E.M. (2014) Evolutionary and ecological constraints of fish spawning habitats. *ICES Journal of Marine Science*, 72: 285-296

deYoung B., Heath M., Werner F., Chai F., Megrey B., Monfray P. (2004). Challenges of modeling ocean basin ecosytems. *Science*, 304(5676): 1463-1466

Dubois M., Rossi V., Ser-Giacomi E., Arnaud-Haond S., López C., Hernández-García E. (2016) Linking basin-scale connectivity, oceanography and population dynamics for the conservation and management of marine ecosystems, *Global Ecology and Biogeography*, 25, 503–515, doi:10.1111/geb.12431

Fox CJ, Aldridge JN (2008) Simulating the marine environment and its use in fisheries research. In Advances in Fisheries Science, 50 years on from Beverton and Holt. Blackwell Publishing.

Giménez L. (2010) Relationships between habitat conditions, larval traits, and juvenile performance in a marine invertebrate. *Ecology,* 91(5):1401-1413.

Goñi R., Adlerstein S., Alvarez-Berastegui D., Forcada A., Reñones O., Criquet G., and Bonhomme P. (2008). Spillover from six western Mediterranean marine protected areas: evidence from artisanal fisheries. *Marine Ecology Progress Series*, 366, 159-174.

Juza M., Renault L., Ruiz S., Tintoré J. (2013) Origin and pathways of Winter Intermediate Water in the Northwestern Mediterranean Sea using observations and numerical simulation. *Journal of Geophysical Research: Oceans* 118 (12), 6621-6633

Pérez-Ruzafa A., Marcos C., García-Charton J. A., Salas F. (2008) European marine protected areas (MPAs) as tools for fisheries management and conservation.

Rees S. E., Foster N. L., Langmead O., Pittman S., Johnson D. E. (2017) Defining the qualitative elements of Aichi Biodiversity Target 11 with regard to the marine and coastal environment in order to strengthen global efforts for marine biodiversity conservation outlined in the United Nations Sustainable Development Goal 14. *Marine Policy.*

Reglero P, Ciannelli L, Alvarez-Berastegui D, Balbín R, López-Jurado JL, Alemany F (2012) Geographically and environmentally driven spawning distributions of tuna species in the western Mediterranean Sea. *Mar. Ecol. Prog. Ser.*, 463:273–284

Reglero P., Tittensor D. P., Álvarez-Berastegui D., Aparicio-González A., Worm B. (2014) Worldwide distributions of tuna larvae: revisiting hypotheses on environmental requirements for spawning habitats. *MEPS*, 501:207-224.

Rossi V., Ser-Giacomi E., Lopez C., Hernandez-Garcia E. (2014) Hydrodynamic provinces and oceanic connectivity from a transport network help designing marine reserves. *Geophysics Research Letters,* 41, 2883–2891

Staples D., Brainard R., Capezzuoli S., Gunge-Smith S., Grose C., Heenan A., Hermes R., Maurin P., Moews M., O'Brien C., Pomeroy R. (2014) Essential EAFM. Ecosystem Approach to Fisheries Mangament Training Course. Volume 1-For Trainees. FAO Regional Office for Asia and the Pacific, Bangkok, Thailand, RAP Publication 2014/13, 318 pp.

Torres A. P., Palero F., Dos Santos A., Abelló P., Blanco E., Boné A., Guerao, G (2014) Larval stages of the deep-sea lobster Polycheles typhlops (Decapoda, Polychelida) identified by DNA analysis: morphology, systematic, distribution and ecology. *Helgoland Marine Research*, 68(3), 379.

CHAPTER 27

Operational Oceanography at the Service of the Ports

Enrique Alvarez Fanjul[1], Marcos García Sotillo[1], Begoña Pérez Gómez[1], José María García Valdecasas[1], Susana Pérez Rubio[1], Pablo Lorente[1], Álvaro Rodríguez Dapena[1], Isabel Martínez Marco[2], Yolanda Luna[2], Elena Padorno[2], Inés Santos Atienza[2], Gabriel Díaz Hernandez[3], Javier López Lara[3], Raúl Medina[3], Manel Grifoll[4], Manuel Espino[4], Marc Mestres[4], Pablo Cerralbo[4], and Agustín Sánchez Arcilla[4]

[1]*Puertos el Estado, Madrid, Spain;* [2]*Agencia Estatal de Meteorología, Madrid, Spain;* [3]*Instituto de Hidráulica de Cantabria, Santander, Spain;* [4]*Laboratorio de Ingenieria Marítima-Universidad Politécnica de Cataluña, Barcelona, Spain*

The Spanish ports demand operational oceanography products for their operation. In recent years, this demand has been fulfilled by SAMOA project. SAMOA (Sistema de Apoyo Meteorológico y Oceanográfico a las Autoridades portuarias - System of Meteorological and Oceanographic Support for Port Authorities) is revolutionary in the way solutions are provided to the operational oceanography needs of port authorities. An integrated system, ultimately based on Copernicus Marine Environment Monitoring Service (CMEMS) data, has been developed. A total of 10 new high-resolution atmospheric models (1 km resolution, based on Harmonie), 10 wave models (5 m, mild slope), and nine circulation models (70 m, ROMS) have been developed and operationally implemented. In terms of instrumentation, SAMOA has improved the preexisting large network of Puertos del Estado by means of 13 new meteorological stations and three global navigation satellite systems associated with the tide gauges. Twenty-five ports from 18 port authorities will benefit from these new modeling and monitoring advances.

Introduction

Approximately 85% of total imports and 60% of Spanish exports are channeled through ports, a fact that speaks for itself of the vital role they play in Spain's national economy. The ports suffer the extreme events of essential physical variables, especially wind, waves, and sea level. This affects the installations during all phases of the harbor life, from design to operation. To respond to these complex needs, the SAMOA initiative (Sistema de Apoyo Meteorológico y Oceanográfico a las Autoridades portuarias - System of Meteorological and Oceanographic Support for Port Authorities) was born, co-financed by Puertos del Estado and the Port Authorities.

All previously existing (models at the regional scale and measuring networks: buoys, tide gauges, and high-frequency radars) and new products (the models and instrumentation described in this paper) have been fully integrated in a specific SAMOA visualization tool that is being managed by administrators in the harbors. In addition, there is a new alert system via e-mail and SMS (short

---

Alvarez Fanjul, E., et al., 2018: Operational oceanography at the service of the ports. In "*New Frontiers in Operational Oceanography*", E. Chassignet, A. Pascual, J. Tintoré, and J. Verron, Eds., GODAE OceanView, 729-736, doi:10.17125/gov2018.ch27.

message service), that is fully configurable by the users in the port community. Finally, an extended set of applications, including oil spill models and air pollution monitoring tools, have been fully integrated into the system.

Uses of the system, among others, include: (1) provision of information for knowledge-based operation of infrastructures (i.e., to forecast port closings due to extreme events); (2) aid for pilot operations; (3) safer and more efficient port operations (i.e., crane operations affected by winds, planning of roll-on/roll-off operations); (4) fighting against oil spills in the interior of the harbours, and (5) control of water and air quality.

SAMOA is based on activities of the SAMPA project (Sistema Autónomo de Monitorización y Previsión en Algeciras – Algeciras autonomous monitoring and forecasting System), developed with the Algeciras Port Authority, which in the present framework functioned as a laboratory for the Spanish Port System. SAMOA has been co-financed by the Spanish Ports (75%) and by Puertos del Estado (25%).

The Monitoring Component

SAMOA relies heavily on previously existing Puertos del Estado monitoring networks (25 buoys, eight high-frequency radars, and 40 tide gauges). As such, SAMOA has been used to fill detected knowledge gaps, improving the coverage of meteorological stations on the ports (13 new stations) and the control of the tide gauges by means of continuous global navigation satellite systems (three new stations).

The Atmospheric Component

SAMOA's meteorological component provides a high-resolution forecast for surface weather variables, especially winds, over harbour influence areas. This is done using a new configuration and integration of the HARMONIE-AROME model running semi-operationally at 1 km resolution over the ports of Almería, Avilés, Baleares, Barcelona, Gijón, Las Palmas, Málaga, Melilla, Tenerife, and Tarragona.

HARMONIE (Hirlam-Aladin Research on Mesoscale Operational NWP in Europe) is a spectral bi-Fourier limited area numerical weather prediction (NWP) model. The main features of the model's deterministic system are thorough data assimilation (Brousseau et al., 2011) and surface treatment, a non-hydrostatic dynamics core with semi-Lagrangian semi-implicit discretization over the horizontal grid, and hybrid coordinate on the vertical. The AROME implementation (Seity et al., 2011) is designed for high resolution, 2.5 km by default, including new parametrizations like convection or solar radiation, but it can be forced to reach higher horizontal resolutions providing even better results. Running HARMONIE-AROME at the very high resolution of 1.0 km includes several challenges. Numerical time step is adjusted to 30 seconds to keep a continuous air flow without overloading computational resources. Topography is based on Global Multi-resolution Terrain Elevation Data (GMTED2010; Danielson and Gesch, 2011), a global digital elevation model with 250 m resolution, appropriate for generating a smooth and realistic field through upscaling without introducing orographic noise. The boundary conditions are obtained from the 0.1 degree resolution integrated forecasting system model from the European Centre for Medium-Range Weather Forecasts, as testing proved (not shown) that it provides better results than using an intermediate Agencia Estatal de Meteorología (AEMET; the Spanish meteorological agency) HARMONIE-AROME model.

The model runs twice per day on the European Centre for Medium-Range Weather Forecasts' supercomputer, at 00 UTC and 12 UTC, with a forecast length of 48 hours over the areas of interest. The results obtained reveal much richer dynamics than the AEMET official 2.5 km forecast (Fig. 27.1), capturing smaller eddies and more local behaviors. The quantitative verification, performed with the MONITOR tool developed by HIRLAM, shows an improvement of the statistical scores for the highest resolution model wind forecast both for deep convection episodes and long-term periods over the four main areas. However, several issues remain unresolved for very high-resolution mesoscale models, such as the use of a more suitable verification method, adequate data assimilation, or the continuous improvement of forecast accuracy through model improvements.

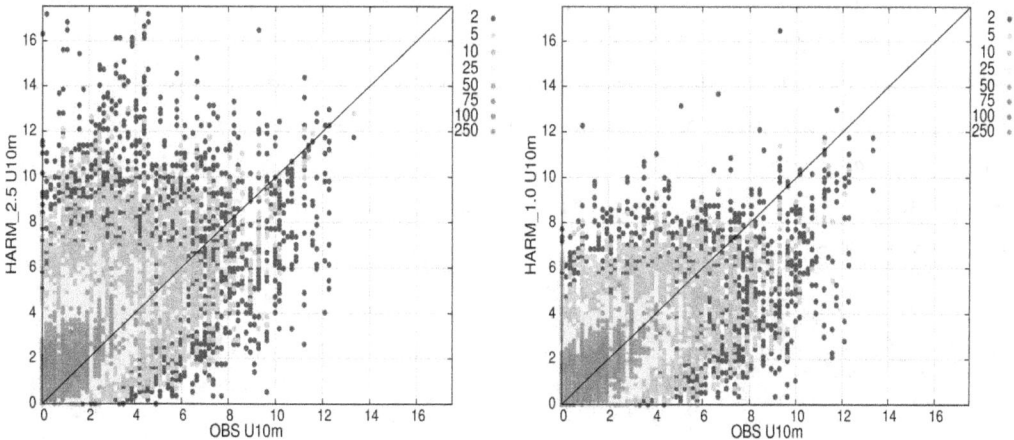

Figure 27.1. Validation over the Gulf of Biscay area during March 2017. Left: Scatter plot of 10 m wind observations against operational 2.5 km HARMONIE-AROME; Right: Against new 1.0 km HARMONIE-AROME. The different point colours indicate the range of cases that occurred.

The Circulation Component

The SAMOA circulation component produces, on a daily basis, a short-term (three-day) forecast of three-dimensional currents and other oceanographic variables, such as temperature, salinity, and sea level for nine Spanish ports in the Mediterranean (Barcelona, Tarragona, Almería), the Iberian Atlantic (Bilbao, Ferrol), and the Canary Islands (Las Palmas, Tenerife, La Gomera, and Santa Cruz de la Palma).

The three-dimensional hydrodynamic model used in the SAMOA circulation systems is the Regional Ocean Modeling System (ROMS; Shchepetkin and McWilliams, 2005).

The SAMOA model applications consist of two nested regular grids with spatial resolutions of ~350 m and ~70 m for the coastal and harbour domains, respectively. The chosen vertical discretization consists of 20 sigma levels for the coastal domains (except for the Canary Island implementations where, due to the deepest bathymetry, 30 levels are used) and 15 levels for the port domains. Bathymetry of the SAMOA coastal systems is built using a combination of bathymetric data from the General Bathymetric Chart of the Oceans (GEBCO; Becker et al., 2009) and from specific local high-resolution sources provided by local port authorities.

At the surface, models are forced by daily updated high frequency (hourly) winds and heat and water fluxes from AEMET forecast services. The SAMOA forecast systems are Copernicus Marine Environment Monitoring Service (CMEMS) downstream services, being the coastal models nested into the regional CMEMS Iberia Biscay Irish forecast solution. In those ports where river freshwater discharges may be relevant (Ferrol, Bilbao, Barcelona, and Tarragona) it is included in terms of climatological data, but developments are underway to include a hydrological model.

Prior to the operational launch of the SAMOA systems, the quality of a one-year of SAMOA coastal and port products was assessed by comparingn model solutions with observations, both from in situ moorings and remotely-sensed products. A validation tool has been implemented to evaluate the downscaled SAMOA local solutions, using available operational observational sources (both remotely-sensed, including high-frequency radar and in situ). Apart from validating the local solution, the objective of this tool is to evaluate the effectiveness of the dynamical downscaling performed, providing an objective measure of potential value-added with respect to the regional Copernicus solutions (in which local models are nested). An example of this potential value-added evaluation is shown in Fig. 27.2.

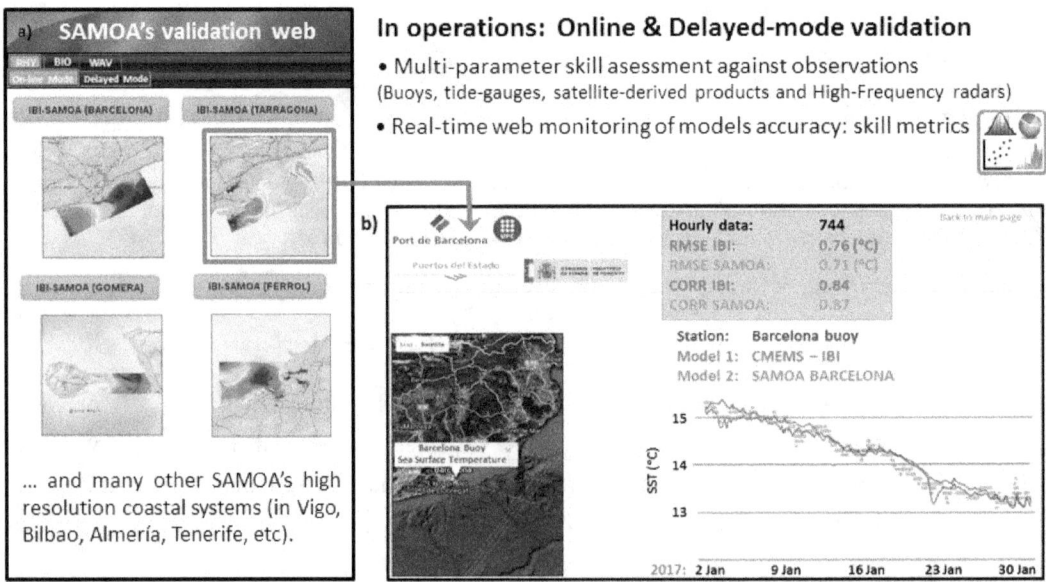

Figure 27.2. Snapshot of the multi-parametric ocean model skill assessment tool, implemented by Puertos del Estado to validate operationally SAMOA downscaled local solutions and to compare them with the "parent" regional solution in which they are nested (the CMEMS IBI MFC forecast product).

The Wave Component

The SAMOA wave component has been designed to provide a three-day forecast of agitation (significant wave height in the interior of the port) inside 10 Spanish ports of special interest: Almería (two ports), Gijón, Las Palmas (three ports), Málaga and Santa Cruz de Tenerife (three ports). Prior to SAMOA, Puertos del Estado was running an operational wave forecast able to provide, using the SWAN model, wave forecasts at the harbours mouth. Thanks to the SAMOA developments, this forecast has now been downscaled to the interior of the ports at extremely high resolution (2 m).

The new system is based on the spectral reconstruction technique of sea states, using a monochromatic wave catalogue previously computed in advance by means of a model based on the elliptical approximation of the mild slope equation. The process includes some ad hoc innovative improvements:

- Updated numerical solver to speed up the simulations (reaching 10x times faster than similar models), which has enabled changes in the meshing tool to reach the above-mentioned resolution.
- Introduction of the reflection response algorithm in the numerical contours as a function of the incident wave period and based on the functional response of each type of pier and breakwaters analysed.

Results of the system have been validated and calibrated using agitation data measured by the Puertos del Estado tide gauges (Fig. 27.3), which are able to measure sea level at a 2 Hz sampling rate and, therefore, are capable of measuring this variable. The model output is validated on a daily basis with this instrumentation.

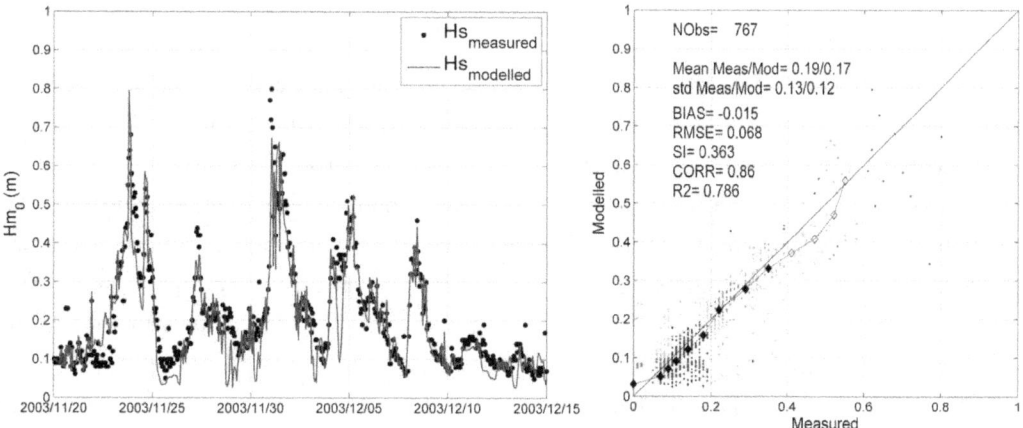

Figure 27.3. Example of validation of the operational predictions in the ports of Barcelona, using agitation data provided by the tide gauge. Measured and simulated time series comparison (left). Scatter plot of modelled and measured time series (right).

Integration and Downstream Tools

The SAMOA model outputs are freely accessible through the Puertos del Estado's THREDDS catalogue (the THREDDS data server is a web server that provides metadata and data access for scientific datasets using a variety of remote data access protocols.). Likewise, free access to some products is granted via the Puertos del Estado's Portus web interface (http://portus.puertos.es). Additionally, a specific tool for the port authorities—the CMA (Cuadro de Mando Ambiental - see Fig. 27.4)—has been developed to properly exploit all SAMOA products, and has been implemented in 25 ports, to date. The CMA is based on a web interface (http://cma.puertos.es) and provides easy access to all information generated by the SAMOA systems, both in real time and in forecast mode. Users can define thresholds for all spatial points inside the application (model points and measuring stations) that are used to trigger alerts. The CMA is also capable of creating customised PDF reports for each forecast point.

Additionally, a user-friendly oil spill model and an atmospheric dispersion model have been developed and incorporated into the CMA.

Port managers granted the access to the tool can define the level of permission that users have. For example, some users can get permission for visualization, but might not have access to the oil spill model. Algeciras harbour is a very good example of how the CMA tool should be used, wherein a community of 250 port users utilize the CMA tool (including the companies located at the facilities).

The CMA is also is utilized to configure personalised alert systems, defining the points and the alerts to be triggered, as well as the reception method (e-mail or SMS). The alerts can be defined as a combination of parameters, conditions, and/or thresholds as complex as desired by the user.

Future Plans

Building on the success of SAMOA, a SAMOA 2 project is being launched. This second phase will include new components, such as a wave overtopping forecast and extremely high-resolution wind prediction (2 m). In 2020, by the end of SAMOA 2, the system will have the following components: 44 CMA implementations in different ports (of a total of 46 ports in the national system), 20 1 km resolution atmospheric forecasts, 21 agitation systems, 31 circulation systems, 19 new meteorological stations, eight global navigation satellite systems, 15 very high wind forecast systems, plus other additional modules. There are plans to use the new models to explore, for example, the wave current interactions.

With the new system in place, the most important challenge for the Spanish Port System will be to implement the methodologies necessary to make proper use of all the new available tools. While some ports are very active and are already making good use of the new information, other are still not able to fully exploit it. Several initiatives will be launched to reduce this gap. Making operational oceanography a core part of the port management business is probably the most significant result of SAMOA.

Figure 27.4. The SAMOA visualization tool, showing the circulation at Barcelona Port (upper panel) and the real-time measurements at Algeciras Harbour (lower panel).

References

Becker, J.J., D.T. Sandwell, W.H.F. Smith, J. Braud, B. Binder, J. Depner, D. Fabre, J. Factor, S. Ingalls, S-H. Kim, R. Ladner, K. Marks, S. Nelson, A. Pharaoh, R. Trimmer, J. Von Rosenberg, G. Wallace, and P. Weatherall (2009). Global Bathymetry and Elevation Data at 30 Arc Seconds Resolution: SRTM30_PLUS, *Marine Geodesy*, **32**:4, 355-371.

Brink, H., and J. Allen (1978). On the effect of bottom friction on barotropic motion over the continental shelf. *Journal of Physical Oceanography*, **8**, 919-922.

Brousseau P., L. Berre, F. Bouttier, and G. Desroziers (2011). Background-error covariances for a convective-scale data-assimilation system: AROME-France 3D-Var. *Quarterly Journal of the Royal Meteorological Society*, **137**, 409-422.

Danielson J., and D. Gesch (2011). Global Multi-resolution Terrain Elevation Data 2010 (GMTED2010). U.S. Geological Survey Technical Report, Open-File Report 2011-1073, 26 pp.

Seity Y., P. Brousseau, S. Malarden, G. Hello, P. Bernard, F. Bouttier, C. Lac, and V. Masson (2011). The AROME-France convective-scale operational model. *Monthly Weather Review*, **139**, 976.

Shchepetkin, A F., and J.C. McWilliams (2005). The regional oceanic modeling system (ROMS): a split-explicit, free-surface, topography-following-coordinate oceanic model. *Ocean Modelling*, **9**(4), 347-40.

Sotillo, M.G., P. Cerralbo, P. Lorente, M. Grifoll, M. Espino, A. Sanchez-Arcilla, and E. Álvarez-Fanjul (2018). High Resolution Coastal Ocean Forecasting in Spanish Ports: The SAMOA Operational Service. Submmited to *Journal of Operational Oceanography*

CHAPTER 28

Diagnosis, Prognosis, and Management of Jellyfish Swarms

Laura Prieto

Ecosystem Oceanography Group, Departamento de Ecología y Gestión Costera, Instituto de Ciencias Marinas de Andalucía (ICMAN), Consejo Superior de Investigaciones Científicas (CSIC), Cádiz, Spain

Jellyfish includes creatures that are mostly constituted by water and have a gelatinous consistency. In this chapter, after providing a biological description of these organisms, the scales of variability associated to their life cycle and framing their dynamics in the context of the climate change, I review the diverse initiatives and management of coastal jellyfish swarms. Jellyfish swarms have relevant social and economic implications; however, systematic and periodic data of jellyfish occurrences along beaches is sparse. This data would help us to understand the inter-annual variability of the episodes of high jellyfish abundances and its potential relation to variable environmental conditions. Joint strategies with tools available to scientist, administration, policymakers, and stakeholders can optimize the cost of gathering these in situ data and maximize the benefit obtained from its scientific analysis. Three case studies of jellyfish blooms are presented, from which we can infer the importance of co-creation with stakeholders emerges as a key issue to allow for a solid understanding of the episodes and the implementation of appropriate knowledge-based future mitigation actions.

Jellyfish: Biological Description and Scales of Variability Associated to Their Life Cycle

Jellyfish, as a general term, includes creatures that are mostly constituted by water and have a gelatinous consistency. As gelatinous zooplankton they include a broad taxonomy, including numerous groups such as Ctenophores, Cnidarians, salps, larvaceans, molluscs, and worms. This chapter focuses on two phyla, namely Ctenophora and Cnidaria, which are the most relevant due to their great impact on the ecosystem and economy.

Biological description of jellyfish

Among the gelatinous zooplankton, there is a distinct group of carnivores in the pelagic environment, the Ctenophores (commonly known as comb jellies). Recently, a new hypothesis of animal evolution emerged indicating that all animal phyla originated from this group (Dunn et al., 2015; Jager and Manuel, 2016; Presnell et al., 2016; Shen et al., 2017), having regenerative capacities as adults (Martindale, 2016). They have been found to be important predators in the surface waters of the open sea in the North Atlantic and Indian Ocean (Harbison et al., 1978), in the

---

Prieto, L., 2018: Diagnosis, prognosis, and management of jellyfish swarms. In "*New Frontiers in Operational Oceanography*", E. Chassignet, A. Pascual, J. Tintoré, and J. Verron, Eds., GODAE OceanView, 737-758, doi:10.17125/gov2018.ch28.

Arctic Ocean (Purcell et al., 2010), in coastal areas of North and South America, Mediterranean, Black, Azov, Caspian, and Marmara seas (reviewed in Shiganova et al., 2001; Purcell et al., 2001, 2007) and recently in the North and Baltic seas (Hamer et al., 2011; Javidpour et al., 2009).

The other important group of gelatinous zooplankton is the phylum Cnidaria, which has an immense diversity and consist of five classes: Staurozoa, Hydrozoa, Anthozoa, Cubozoa, and Scyphozoa (Daly et al., 2007). Hydrozoa includes the family Physaliidae or the Portuguese Man-O-War (*Physalia physalis*), well-known due to its impact on human activities on the shore. Anthozoa includes all the corals species. Cubozoa has includes the deadliest species of jellyfish, which cause the Irukandji syndrome in tropical waters, while Scyphozoa includes species that the general public consider "true jellyfish" (Purcell and Arai, 2001).

A common feature of the phylum Cnidaria is the presence of stinging organelles, named cnidocysts or nematocysts, which are triggered by mechanical or chemical stimuli and are predominantly used for prey and defense (Beckman and Özbek, 2012). These cnidocysts are formed by a cnidoblast in which nematocyst are rolled in and prepared to be discharged in a harpoon-like fashion at high velocity (2 m/s) in order to inject the venom inside the target (Holstein and Tardent, 1984). The phylum Cnidaria englobes thousands of species that are present in all oceans, from the tropics to the polar areas, from the surface to the bottom. There are also a few freshwater species. Jellyfish belong to the mid-trophic level within the food web and are predatory zooplankton (Lehodey et al., 2010).

In general, the family Physaliidae are pleustonic, live in the surface of the water, and are open ocean organisms. They are common in tropical and subtropical regions in the oceans, ranging from 55°N to 40°S. The Portuguese Man-of-War is a colony of organisms constituted by groups of distinct forms and functional individuals: a polyp named pneumatophore is the sailing bag-form that provides the rest of the colony of the floating device; under the pneumatophore there are three different types of polypoids (gastrozooids for feeding, dactylozooids for defence and gonozooids for reproduction) and three types of medusoids (gonophores, siphosomal nectophores and vestigial siphosomal nectophores). Reproduction is sexual and gametes are spawned in the water, being the fertilization external. Portuguese Man-of-War possesses a singularly potent toxin (Burnett, 2000; Edwards and Hessinger, 2000), being physalitoxin the major protein of the venom, an hemolysin (Tamkun and Hessinger, 1981), which is contained in the nematocysts. The nematocyst venom of *P. physalis* can be lethal to animals and humans (Edwards and Hessinger, 2000), and the envenomation syndromes in humans are extensive (Burnett, 2000).

Regarding Scyphozoa, their size can vary widely from 12 mm to 2 m. They do not have specialized organs for respiration or excretion and generally, during their life cycle, alternate between pelagic and benthic phases. A typical life cycle of a scyphozoan is as follows: They are pelagic and reproduce sexually forming a ciliate planktonic larva named planulae. These larvae search for a surface to attach to and transform into polyps, starting the benthic phase of their life cycle. If the environmental conditions are favorable, polyps can reproduce asexually by different modes: budding, partial fission, and pedal cysts. Nevertheless, under adverse environmental conditions, some species can form podocyst (resting stages). The transition of jellyfish from the benthic to pelagic stages occurs during strobilation. Through this process, individual polyps metamorphose to form the juvenile pelagic stage (free-swimming ephyra) that mature into adult medusae. Depending on the species, strobilation is triggered by warming or cooling of the water (Sugiura, 1965; Calder, 1973; Hofmann et al., 1978; Prieto et al., 2010).

Scales of variability associated to the life cycle of jellyfish

Due to their complex and diverse life cycle, the scales of variability at which jellyfish are exposed to differ: from the ones affecting the benthic ecosystem to the ones acting on the pelagic system, both at coastal and open ocean scales. All those spatial scales also are subject to different temporal scales of variability: climate, seasonal, and meteorological. The variability of the ocean, both temporal and spatial, is described in the first chapter of this book.

The benthic ecosystem has some peculiarities from the point of view of a jellyfish population. For the polyps it has a "biological-spatial-limited" constraint associated to competition and predation. Also, if the environmental conditions (temperature, salinity, food availability) are not favorable, they cannot move to better settlement surfaces. Thus, survival during this life stage is closely linked to the local environment. Therefore, relative short periods of time with very cold temperatures (Prieto et al., 2010), drops in salinity associated to rivers outflows, or long periods of food scarcity (Prieto et al., 2010) would result in a population decrement.

During the pelagic vital phases, jellyfish are exposed both to open ocean and coastal scales of variability. Some species are rarely seen in open waters during their pelagic phase (e.g., *Cotylorhiza tuberculata*). Other species are common in the open ocean (e.g. *Physalia physalis* and *Pelagia noctiluca*) and are only observed in coastal areas when the currents and the meteorological conditions transport them to the shores (Licandro et al., 2010; Prieto et al., 2015).

Some live on the surface of the water (e.g., *Physalia physalis*). If they have a symbiotic relationship with some species of algae in, then these Cnidaria species need light for the algae to perform the photosynthesis and they live in the first few meters of the water column (e.g, *Cotylorhiza tuberculata*). Therefore, all these species are more exposed to surface currents and winds. Other species perform vertical migrations (e.g., *Pelagia noctiluca*), being at the surface at night and at considerable depth by day (Franqueville, 1971; Larson et al., 1991; Mariottini et al., 2008) in response to zooplankton, on which they prey (Giorgi et al., 1991; Zavodnik, 1991; Malej et al., 1993).

Jellyfish Dynamics and Climate Change

The Intergovernmental Panel on Climate Change (IPCC) Fifth Assessment Report confirms that warming of the climate system is unequivocal (IPCC, 2013). More than 60% of the net energy increase in the climate system is stored in the upper ocean (0–700 m) during the 40-year period from 1971 to 2010 (IPCC, 2013). But also the ocean is an important sink of atmospheric carbon dioxide (CO_2), as it has absorbed close to 30% of the CO_2 emitted since the industrial revolution (Khatiwala et al., 2013; Le Quéré et al., 2015). This sink of CO_2, even though it has been shown beneficial at a global scale by mitigating the greenhouse effect, has been demonstrated to carry out deleterious effects on the oceanic environment. This is due to the fact that it provokes a gradual decrease in the seawater pH, a phenomenon known as oceanic acidification (Caldeira and Wickett, 2003).

Independently of future predictions scenarios, there are several general hypotheses of the marine ecosystem change associated with global warming. First, the increase in temperature in the surface ocean provokes an increase in stratification and therefore a decrease in winter mixing with the concomitant decrease in nutrient supply to the surface waters. This change in the nutrient concentration results in a decrease in phytoplankton biomass and, at the same time, a shift in the dominant phytoplankton group, from large diatoms to small cocolithophorids (Watanabe et al., 2003). Because of these differences in the phytoplankton, a decrease in the zooplankton biomass and a shift in the dominant zooplankton group occurs, being that jellyfish are the dominant group.

There are more changes in the ecosystem associated with global warming that affect jellyfish population dynamics, such as changes in fish resources (due to their predator-prey interaction), the increase in temperature, and ocean acidification. At the community level, it has been suggested that within the same trophic level, the most tolerant species to the increment of the level of dissolved CO_2 could replace more vulnerable species (Kroeker et al., 2013). This change in community structure towards the dominance of few generalist species would benefit non-calcifying and opportunistic species (e.g., jellyfish and anemonies; Winans and Purcell, 2010).

Purcell et al. (2007) performed a revision of the response to global warming for eight long-term datasets of gelatinous organisms at a global scale, confirming that in five of those time series an increase in frequency has been registered. More recently, Condon et al. (2012) carried out a study to question how much of the perception of the increment on gelatinous organisms was real or just a vision of change from a human perspective. Nevertheless, in diverse sub-basins of the Mediterranean Sea a change in the frequency of the blooms of *Pelagia noctiluca* has been confirmed (Daly Yahia et al., 2010; Kogovsek et al., 2010). This species also proliferates in the North Atlantic (Licandro et al., 2010). Additionally, a greater abundance of *Cotylorhiza tuberculata* and *Rhizostoma pulmo* have been observed in the last 100 years in the Mediterranean Basin (Kogovsek et al., 2010). In the case of *C. tuberculate,* it has been verified that mild winters and an earlier onset of spring will cause a proliferation of *Cotylorhiza tuberculata* during the summer season (discussed later in the chapter).

Laboratory analyses have shown that the increment of water temperature provokes an increase in the asexual reproduction rates of the polyps in Scyphozoa (Widmer, 2005; Wilcox et al., 2007; Purcell et al., 2009). Regarding the capacity of these organisms to tolerate the increase in dissolved CO_2, it has been demonstrated that the associated decrease of seawater pH decreases the size of the statoliths since these are formed by calcic carbonate or calcic sulphate (Winans and Purcell, 2010). The statoliths are sensor organs for orientation and they are essential for survival in the jellyfish life stage (the pelagic one). The interactive effect of warming and acidification on the formation of the statoliths has been evaluated also in the cubozoan *Alatina nr mordens* (Klein et al., 2014). The authors concluded that the number of polyps decreased, suggesting that the asexual reproduction would be slower in an ocean more acid and warmer. Although the polyps could tolerate future conditions of climate change scenarios, it was less probable that they would proliferate in the long-term. Nevertheless, it was pointed out that if the acidification occurred gradually, *A. mordens* could expand to colder waters in short periods of time, as has been evidenced previously in this and other tropical species (Richardson et al., 2009). These results agree with observations that point out that also non-calcifying organisms are not immune to the ocean acidification, although the future evolution of their communities in an increasingly acid ocean is not yet known (Doney et al., 2009; Kroeker et al., 2013).

Initiatives and Management of Coastal Jellyfish Swarms

There are several environmental conditions that favor jellyfish blooms. First, jellyfish can survive environmental conditions that are deleterious to most other organism, such as contaminated water or water with very low oxygen concentrations also known as "dead zones" (Richardson et al., 2009). Second, overfishing eliminates their predators and at the same time their competitors on resources since fish and jellyfish both prey on zooplankton (Purcell et al., 2007).

Figure 28.1. Example of a jellyfish swarm of *Pelagia noctiluca* in the Mediterranean Sea.

When a jellyfish swarm occurs, it has different implications at the social, economic, and ecologic levels (Fig. 28.1). Compared to 100 years ago, the impact of this phenomenon on the recreational uses of the shore is higher due to the simple fact that now there are "more people in the water" (Brinkman and Burnell, 2009). Additionally, jellyfish swarms can interfere with aquaculture (Bosch-Belmar et al., 2017), fisheries activities (reviewed by Purcell et al., 2007), and clog the cooling systems of power plants (Angel et al., 2016).

At the community level, the proliferation of jellyfish has implications for the flux of carbon along the food web, as has been noted by Condon et al. (2011) in their hypothesis of the "jelly carbon shunt." According to the authors, when a jellyfish swarm occurs, less carbon reaches the economically important upper trophic levels such as fish, and more carbon is retained in jellyfish biomass.

An invasion of non-native jellyfish can occur through the ballast-water of commercial ships. This occurred in the late 1980s in the Black Sea where the ctenophore *Mnemiopsis leidy* was introduced (Kideys, 2002). The species provoked a decrease in non-gelatinous zooplankton biomass, which increased the biomass of phytoplankton causing an increase in turbidity. All these changes in the ecosystem were reflected in a marked decrease in the Turkish anchovy landings (Kideys, 2002).

There are several international jellyfish database initiatives, most based on citizen science (public participation in scientific research) activities, that appeared recently due to the socio-economic impact of jellyfish swarms and the public interest that emerged. One is "JellyWatch" (http://www.jellywatch.org/), which covers all oceans and seas. For the Mediterranean Basin there is a similar platform that originated in 2013 known as the "Jellyfish Spotting Campaign" of the European project PERSEUS (http://www.ciesm.org/marine/programs/jellywatch.htm). Another initiative at the Mediterranean scale within the framework of the Mediterranean Science Commission (CIESM) is the "CIESM JellyWatch Program". The philosophy of this initiative is different from the other two, in that the information is obtained from focal points of diverse operational areas (http://www.ciesm.org/marine/programs/jellywatch.htm).

At the national level, there are several citizen science initiatives including: the app "MeteoMeduse" in Italy (https://www.focus.it/ambiente/natura/meteomeduse), "Seawatchers" in Spain (http://www.observadoresdelmar.es/projecte-3-que-pots-fer-tu.php), or "Spot the jellyfish" in Malta (http://oceania.research.um.edu.mt/jellyfish/ReportForm.html). Some campaigns are even focused solely in one species (e.g., the dangerous box-jellyfish) as seen in this one for Alatina alata (https://www.surveymonkey.com/r/XCFQTJ9).

The National Center for Ecological Analysis and Synthesis houses a compilation of scientific data regarding gelatinous zooplankton. The Jellyfish Database Initiative (JeDI; http://people.uncw.edu/condonr/JeDI/JeDI.html) is a scientifically-coordinated global jellyfish database spanning the past two centuries.

Regarding management, in areas where the jellyfish sting is potentially dangerous, there are some web pages that provide warnings or alerts, as is the case in Hawaii in the United States (http://beatofhawaii.com/hawaii-jellyfish-stings-2013-caution-dates-and-new-treatment/) or in Australia (http://www.outback-australia-travel-secrets.com/box-jellyfish.html#box-jellyfish-season). In other areas, where the jellyfish stings are not dangerous but still the use of the shore is intense and the general public needs to know where the jellyfish are located, there are web pages such as that in the France Mediterranean coast (http://meduse.acri.fr/carte/carte.php) or the Red Cross (http://jav.cruzroja.es/appjv/consPlayas/fichaPlaya.do) in the Spanish coasts.

Applied Cases and the Importance of Co-creation with Stakeholders

Jellyfish swarms have relevant social and economic implications; however, systematic and periodic data of jellyfish occurrences along beaches is sparse. The availability of this data would help us understand the inter-annual variability of the episodes of high jellyfish abundances and its potential relation to variable environmental conditions. Joint strategies with tools available to scientists, administration, policy makers, and stakeholders can optimize the cost of obtaining these in situ data and the benefits achieved from its scientific analysis. An international workshop dedicated to improving the monitoring and research of jellyfish was held in Cadiz (Spain) in 2015 (Prieto et al., 2016). It brought together scientists and stakeholders interested in research and management of jellyfish bloom phenomena. Workshop participants concluded that jellyfish should be monitored on a regular basis, taking advantage of both citizen science and a systematic, robust monitoring program, and that ultimately the monitoring of jellyfish should become mandatory. For both approaches, one has to:

- define the purpose of monitoring;
- establish a sampling scheme (where, when, and how often);
- standardize methodologies and set up monitoring protocols; and
- establish training programs.

Additionally, the various citizen science initiatives need to coordinate. Moreover, scientists could benefit from including data collected from fishermen or other sectors, such as offshore aquaculture operators, dive clubs, marine protected area supervisors, lifeguards, sailors, or enthusiast naturalists. Including these stakeholders in observation programs would increase the spatial coverage of observational data and generally increase the number of observations (both coastal and non-coastal observations).

The lack of standardized approaches and methodologies for jellyfish monitoring was also recognized as an important issue with regards to systematic monitoring programs by the United Nations Educational, Scientific and Cultural Organization (UNESCO; see https://en.unesco.org/themes/monitoring-ocean) and previous programs such as the Global Ocean Ecosystem Dynamics program or the U.S. Joint Global Ocean Flux Study (Knap et al., 1996). Although there is still a relative lack of methodological papers concerning jellyfish monitoring, the United Nations Environment Programme released such a report 25 years ago that can be built upon (UNEP, 1991). The creation of a jellyfish occurrence database in the Ocean Biogeographic Information System and the formulation of data-sharing policy were also mentioned as important (Prieto et al., 2016).

The Balearic observation system

A joint stakeholder-scientist pilot strategy was designed and tested in the Balearic Islands during Summer 2014, in the framework of the Balearic Islands Coastal Observing and Forecasting System (SOCIB) activities of the Strategic Issues and Applications Division (Tintoré et al., 2013). In a pioneering effort, several governmental services from the Balearic Islands worked together with scientists to upload jellyfish observations in real time, establishing a new database generated under scientific standards that will help us gain a solid understanding of the episodes and the implementation of appropriate knowledge-based future mitigation actions.

The system is still operative and now involves the regional fisheries, environmental and emergency administrations, charter associations, as well as Consejo Superior de Investigaciones Científicas (CSIC) institutes and SOCIB. For the first time, a systematic program of routine jellyfish observations was established with qualified and trained personnel, monitoring at high spatial and temporal resolution in three different types of coastal areas: marine reserves, coastal waters around one mile offshore, and beaches. The system includes a web platform and an associated database that compiles the daily sightings in five marine protected areas, with several observation sites, each manned by the General Directorate of Fisheries and Aquaculture personnel from 33 routes (with 66 sites) of the coastal area boat cleaning services from the General Directorate of Water Quality, and at 120 beaches where monitoring is carried out by lifeguards from the General Directorate of Emergencies (Fig. 28.2).

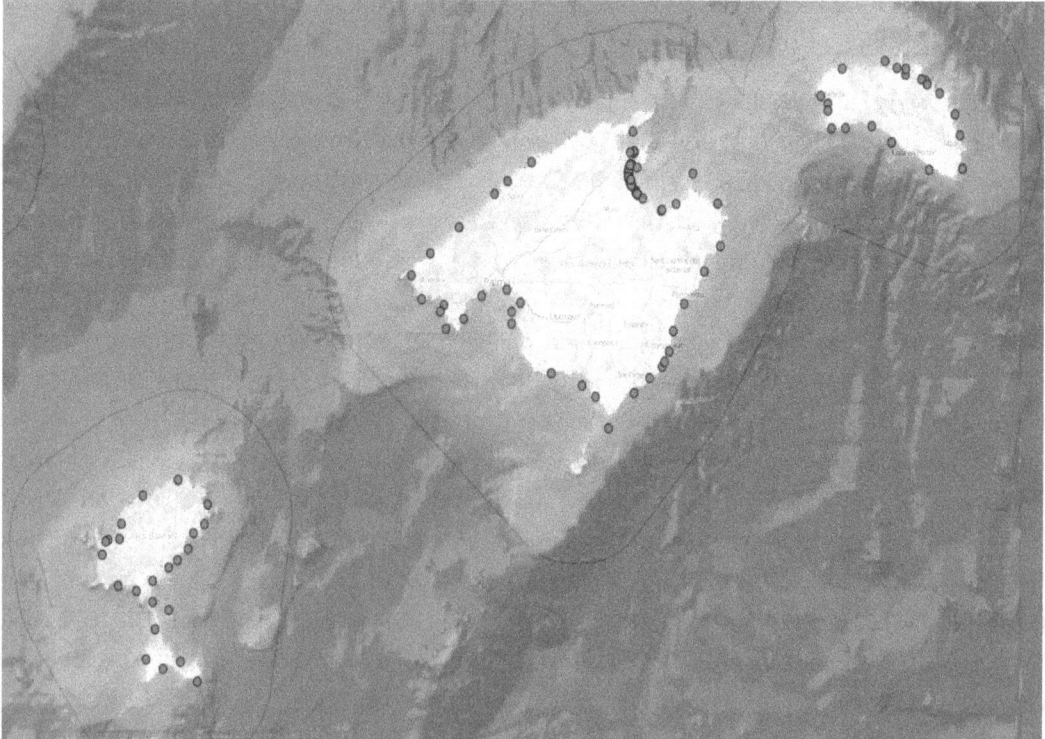

Figure 28.2. Observations points of the Balearic Jellyfish Observation System.

All observations are performed following established protocols to obtain a systematic, periodic, routine monitoring, collecting information on the location and time each species observed. The application allows the filtering per species, location, or period. The web platform provides "heat maps" of jellyfish observation abundance (Fig. 28.3) and enables users to download the entire dataset. Currently, access is restricted to participating institutions.

Since 2014, more than 92,200 observations have been registered on the website. These include observations of the absence of jellyfish, and from the total number of observations only in 6,039 cases were jellyfish observed. Over the years, the most abundant species was *Pelagia noctiluca* (4097), followed by *Cotylorhiza tuberculata* (827), *Rhizostoma pulmo* (545) and *Velella velella* (240). The remaining jellyfish species were one order of magnitude lower, such as *Physalia physalis* (10) and *Aurelia aurita* (9). The most active users of the system are the beach emergency staff (with more than 70,200 observations), followed by employees of the boat cleaning services (20,700+ observations) and from the marine protected areas (1,250+ observations). In the future, the database will be expanded to incorporate information from meteorology and oceanography, in order to advance our understanding of the links among the jellyfish swarms in the shore and the different scales of variability at which they are affected.

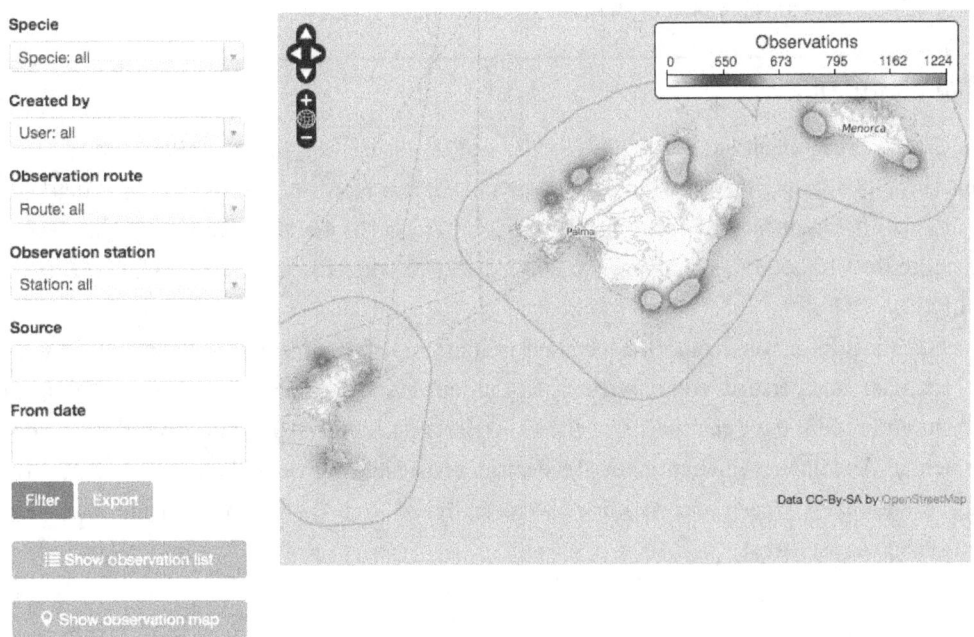

Figure 28.3. Example of an observation heat map from the GRUMERS web-platform for jellyfish observations.

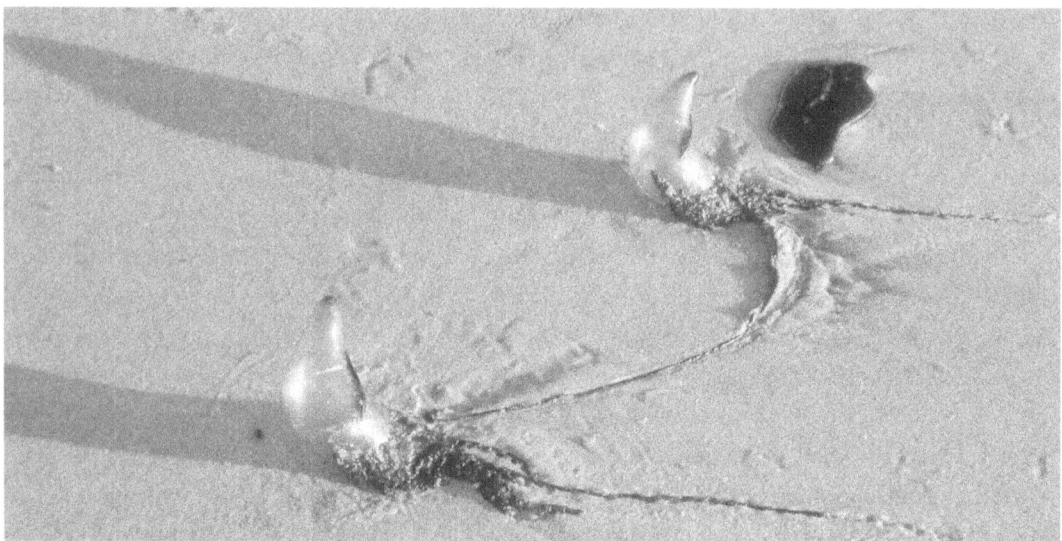

Figure 28.4. Two stranded colonies of *Physalia physalis* on the Camposoto beach (Gulf of Cadiz).

The effectiveness of this new jellyfish observation system is clearly based on the quantity and quality of the data obtained being up to date. Using jellyfish database, it is possible to create a tool of operational oceanography in order to predict the occurrence of jellyfish swarms in the four Balearic Islands. To achieve this goal, the diverse coastal orientation, different currents, and variability of coastal winds on each of the four islands will be taken into account.

Physalia bloom of 2010

The year 2010 registered an unusually high record of Portuguese Man-of-War (*Physalia physalis*) sightings (Fig. 28.4) along the Mediterranean Sea and eastern North Atlantic. It was a remarkable year in terms of the frequency of occurrences, but also in terms of the total number of colonies that arrived (more than 100,000), as compared to 2009 and 2012 when there were less than 60 colonies (Fig. 28.5).

P. physalis sightings were compiled for eight years from different sources: media, national and regional agencies, and personal communications. A unique event between February-April 2010 was carefully monitored by the Technicians of the Consejeria de Medio Ambiente from the Regional Government of Andalucia, which monitored the entire coast and counted and measured all stranded colonies. Additionally, *P. physalis* sightings were analyzed from the database of the Jellywatch Program (Prieto et al., 2015).

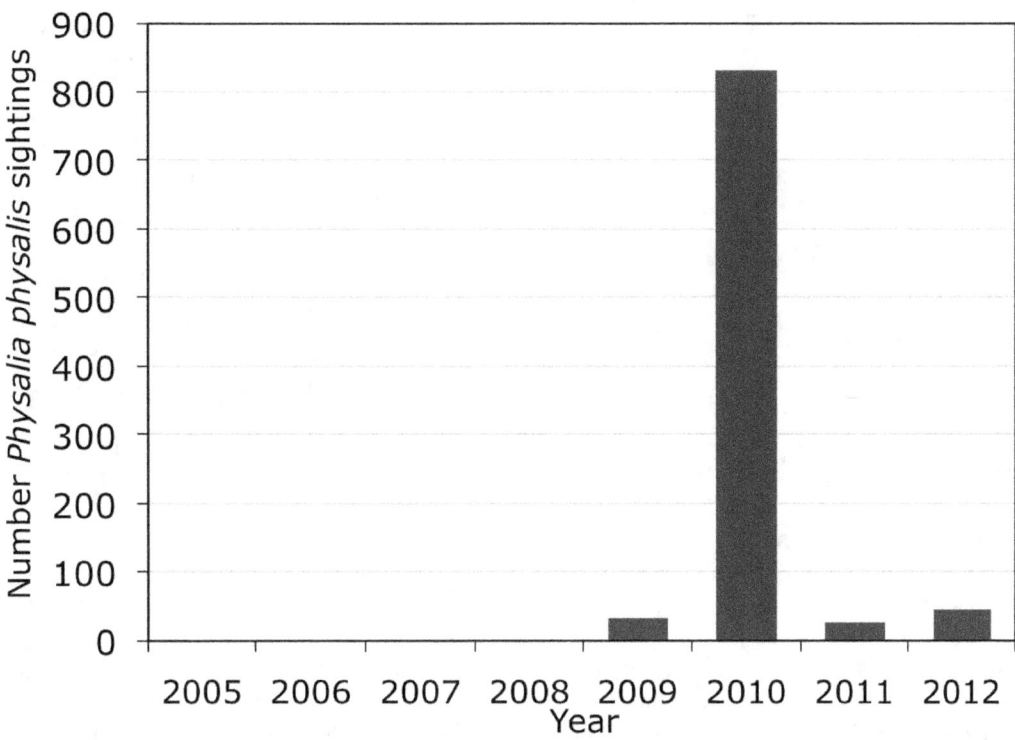

Figure 28.5. Total number of *Physalia physalis* sightings on the Mediterranean Sea and the Spanish and Portuguese Atlantic coasts during eight consecutive years (2005-2012).

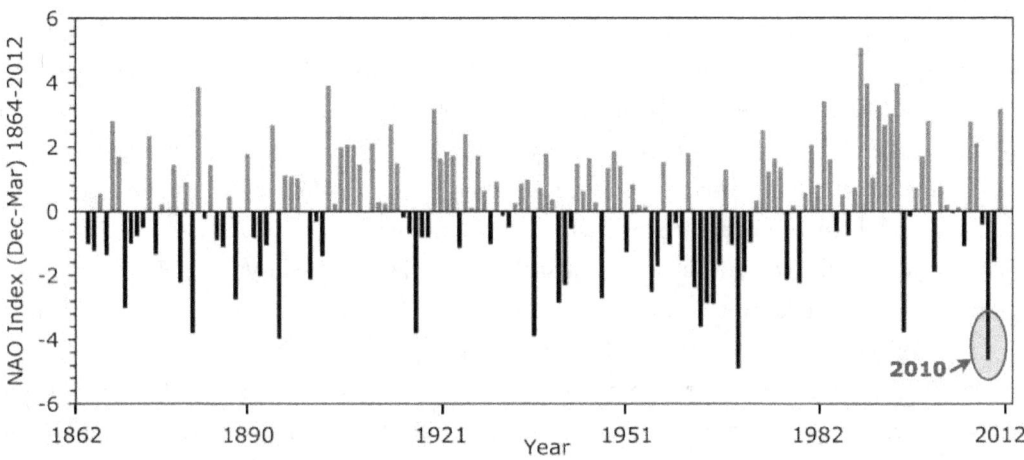

Figure 28.6. Winter index of North Atlantic Oscillation (NAO) based on the difference of the normalized sea level pressure between Lisbon, Portugal and Stykkisholmur/Reykjavik, Iceland, since 1864. The station index value for year N refers to an average of December for year N-1 and January, February, and March for year N. The sea level pressure anomalies at each station were normalized by the division of each seasonal mean pressure by the long-term mean (1864–1983) standard deviation. Normalization is used to avoid the series being dominated by the greater variability of the northern station.

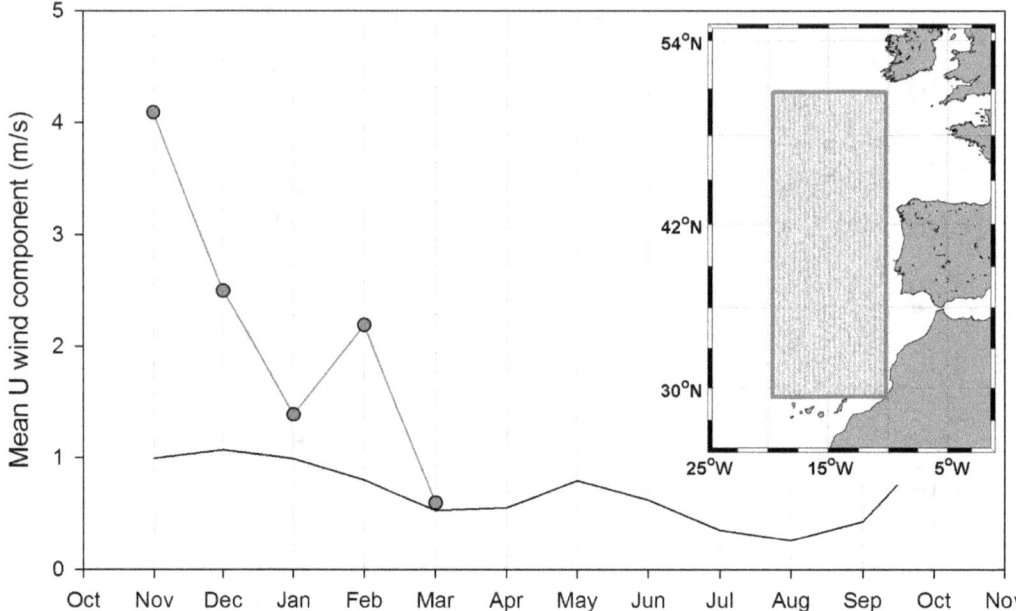

Figure 28.7. Winds components in the eastern North Atlantic from 50°N to 28°N and from -20°W to -10°W (red square in the map). The black line is the monthly climatology from 1979 to 2012 of the wind (U component). The red line shows the data from the 2009–2010 winter when westerlies (positive values) were much stronger in the entire basin compared to the climatology. The data are ERA-Interim analysis daily products (http://www.ecmwf.int/research/era/do/get/era-interim).

A reasonable explanation for the massive occurrence of the Portuguese Man-of-War within the Mediterranean Basin in summer 2010 is that specific climatic and oceanographic conditions during the previous winter in the North Atlantic favored the transport of these colonies into the Mediterranean Sea. The 2009–2010 winter had one of the most negative North Atlantic Oscillation (NAO) indices (− 4.64) measured during the nearly 150-year record (Fig. 28.6; Hurrel, 2012).

This climatic condition caused a stormy mid-latitude Atlantic, with increased storm activity and rainfall in southern Europe, the western Mediterranean, and North Africa. Thus, the climatic/oceanographic conditions have been analyzed for that particularly year, which turned out to be one year of stronger westerlies winds in the Northeast Atlantic basin compared to the time series from 1979 (Fig. 28.7).

A virtual experiment of the drifting of the individuals was performed using a hydrodynamical model. This consisted of using a Regional Ocean Modeling System (ROMS)-based numerical simulation forced with realistic winds (Advanced Scatterometer, ASCAT) and heat fluxes from ERA-Iterim, together with individual based model simulations (Prieto et al., 2015). The oceanic population of *Physalia physalis* started to appear stranded on the beach on February 22, 2010, and observations on the coast occurred from west to east advancing towards the Mediterranean, passing the Strait of Gibraltar and to the far east of the Alboran Sea (Fig. 28.8). The beaching timing observed was highly correlated to the simulations ($r=0.81$, $p<0.001$, $n=18$). The results showed small differences in the overall estimated arrival of *Physalia* between the model experiment and the real observations.

Such high abundances of *P. physalis* were not observed during subsequent years (Fig. 28.5). According to the literature and observations, this species has been observed in the Mediterranean with Malta being the easternmost site (Deidun, 2010). Therefore, it is widely accepted that the presence of *P. physalis* along Mediterranean beaches is not likely to become a continuous problem (Prieto et al., 2015).

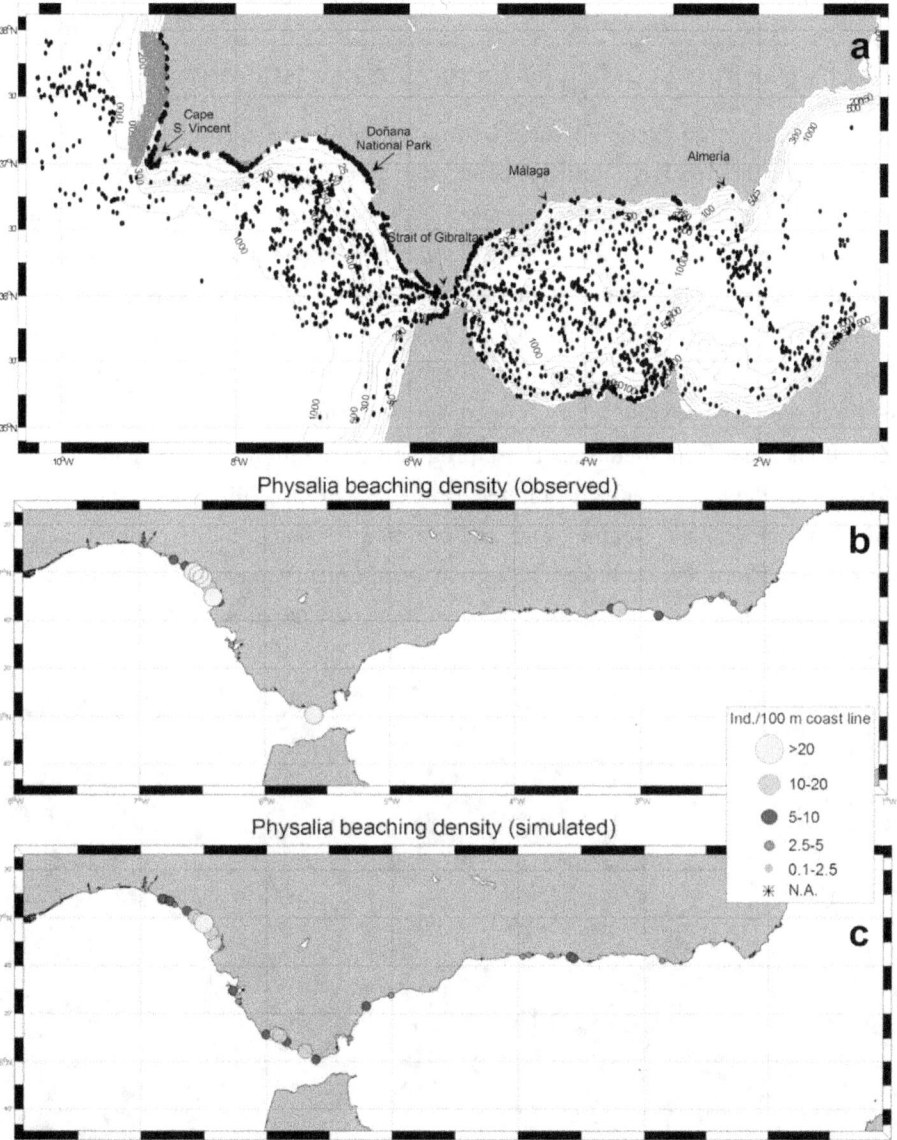

Figure 28.8. Simulation of Portuguese Man-of-War jellyfish drifting and real beachings during January–March 2010. (a) The virtual position of *P. physalis* on 26 January 2010 (beginning of the simulation) indicated by red dots and on 30 March 2010 (end of the simulation) in black dots. Cumulative density, colonies per 100 m of coastline and observed (b) and simulated (c) *P. physalis* arrivals to the Atlantic and Mediterranean coasts of the South Iberian Peninsula. The arrivals of *P. physalis* occurred from west to east both in the observations and in the simulation from 22 February to 30 March (from Prieto et al., 2015).

This applied case study was the scientific community's response to a policymaker's petition to better understand 2010's extraordinary event. Based in the knowledge obtained from this study, the next step will be to integrate the findings into a future operational oceanography tool. This will make it possible to forecast the potential occurrence of *P. physalis* in the Mediterranean Sea each year by evaluating the winter situation at the other side of the Strait of Gibraltar based on climatological, meteorological, and *P. physalis* population abundance data.

Decades of social impact of jellyfish blooms in a coastal lagoon

Cotylorhiza tuberculata, an exotic Scyphozoan in the Mar Menor, was selected as a case study for parameterization of an invasive jellyfish species. Mar Menor is the largest coastal lagoon exposed to intensive tourism in the western Mediterranean with a surface area of about 135 km^2 and a mean depth of about 3.5 m. *C. tuberculata* was not found in the lagoon prior to the lagoon's connection with the Mediterranean, when it was made deeper and wider in the 1970s to facilitate navigation. Since the early 1990s, high abundances of *C. tuberculata* during the summer have become an increasing problem for the recreational use of the lagoon. Local authorities have implemented programs for the removal of adult jellyfish by means of fishing vessels and they have installed nets to protect bathing areas from medusae. These programs have removed more than 5,000 tons of *C. tuberculata* during the most abundant summers (Fig. 28.9; ECOS, 2004).

Planulae appear in late summer and early autumn when adult females reach maturity. Consequently, their survival is linked to lagoon conditions in that season, when the physical environment can change abruptly in association with the passage of low pressure weather systems across eastern Spain.

Figure 28.9. Removal of *C. tuberculata* from Mar Menor using fishing nets (from La Opinión de Murcia, June 15, 2016).

Once polyps fix to the substrate during late summer and early autumn, they must survive until the following spring when strobilation occurs. Although medusae are more visible, polyps are the main stage of *C. tuberculata* in the lagoon in terms of residence time. Pelagic stages of *C. tuberculata* happen from June to September, whereas the benthic phase lasts for the rest of the year. Polyps must survive winter conditions, when light and food are lower than in spring and summer. In addition, although polyp sensitivity to salinity is very low, temperatures can drop dramatically in a shallow lagoon such as Mar Menor. One example of this control of polyp survival by temperature occurred in 2005 when, aside from being one of the warmest months of June in 20 years (mean water temperatures of 23 °C), no outbreak of *C. tuberculata* was detected within the lagoon. To clarify this exception, it is necessary to consider the effect of temperature not only on strobilation but also on polyp survival. Fig. 28.10a shows the air temperature at the meteorological station of San Javier airport (in Mar Menor) for the winters between 1986 and 2005. During the winter of 2005 (from Dec 2004 to Feb 2005) a strong cold event with temperatures close to freezing during several days was apparent. As can be observed such a severe and persistent cold event was not present in other years (Fig. 28.10a). In addition, this event was accompanied by strong winds (Fig. 28.10b), which increased heat loss from the shallow lagoon in a period of very cold air temperature.

Although there are no in situ data for the winter of 2005, the water temperature during that time is likely to have dropped below 10 °C, which poses severe stress on polyps as shown from laboratory experiments (Fig. 28.11). This may have caused the reduction of *C. tuberculata*'s polyp population in the lagoon. Thus, the high mortality of polyps at low winter temperatures seems to be a critical factor that controls polyp population in Mar Menor.

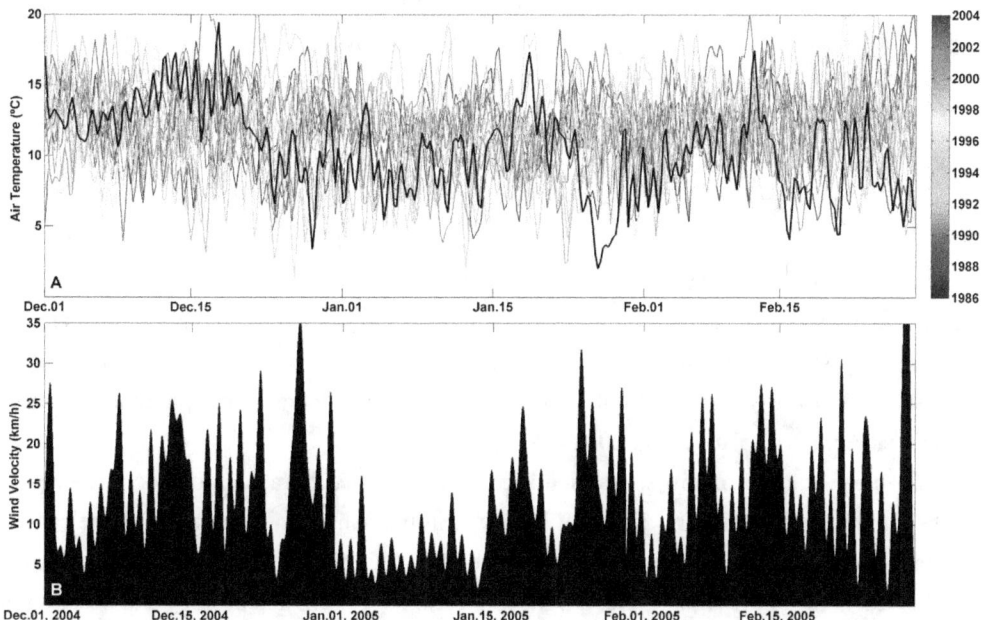

Figure 28.10. Temperature and wind in situ during winter. a) winter air temperature (from December to February, in °C) at the meteorological station in the airport of San Javier (Mar Menor) during 20 years (each line, one winter). Thick black line is the winter from December 2004 to February 2005. b) wind velocity (km/h) at the airport of San Javier (Mar Menor) during winter of 2004-2005 (from December 2004 to February 2005).

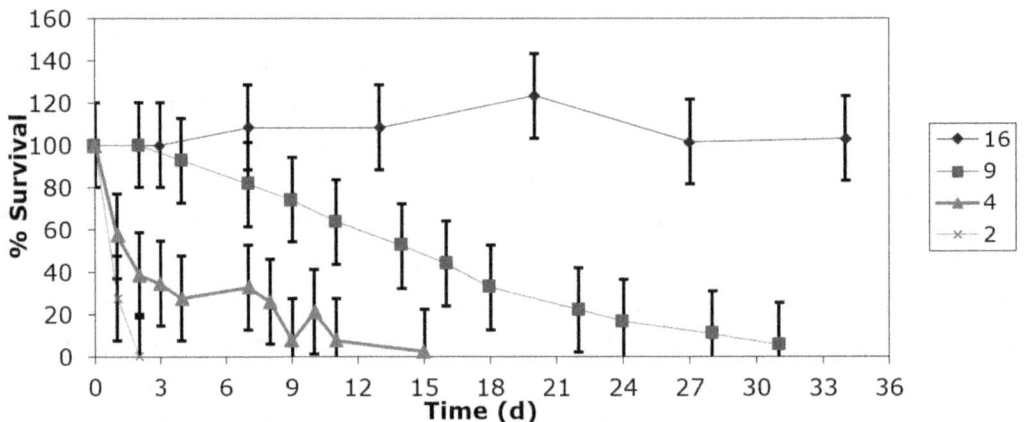

Figure 28.11. Polyp survival at low temperatures. Blue rhomboid, green square, red triangle and purple circle stand for 16, 9, 4 and 2 °C respectively. Error bars are the standard deviation of the three replicates'. Survival is greater than 100% at 16°C because of budding (asexual reproduction).

An increase in temperature results in higher strobilation rates compared to stable temperatures. Laboratory experiments confirmed that the scale of variability of the environment temperature to trigger the strobilation process needs to be seasonal (Prieto et al., 2010). No strobilation occurred in any of the treatments checked at synoptic scales (i.e., meteorological events with a time scale between days to weeks such as a storm or a drop of temperatures associated to a low pressure situation).

Early life stages of *Cotylorhiza tuberculata* are not sensitive to salinity variations or the availability of light or nutrients. However, temperature critically controls polyp survival and strobilation. Low temperatures imply reduced polyp survival during the winter. Abrupt water warming during spring triggers strobilation and, therefore, the start of the medusa phase of the life cycle (Fig. 28.12).

In coherence with laboratory results, these thermal controls determine the inter-annual presence/absence of outbreaks of this jellyfish in the Mar Menor lagoon (Prieto et al., 2010). Therefore, *C. tuberculata* populations fluctuate under the simple rule of "the warmer the better", with collapses after polyp mortality in severe winters and peaks in years with mild winters and long summers. A Bayesian model was developed to implement this simple rule in a tool with the potential to forecast the probability of medusa outbursts to help the management of the lagoon by the different stakeholders (Fig. 28.13; Ruiz et al., 2012).

This is a clear example of how solid scientific data can be incorporated into a social management tool to help marine policymakers. This tool is easy to implement, as demonstrated in the case of *C. tuberculata* populations in Mar Menor, as it only needs the air temperature from the nearby airport. The shallowest areas of the lagoon infer a tight connection between meteorology and oceanography in this case study. However, it can be adapted for analyzing the outburst risk of other jellyfish species thriving in other coastal confined areas. This can also help to explain past and future fluctuations of abundance in a thermally-changing ocean.

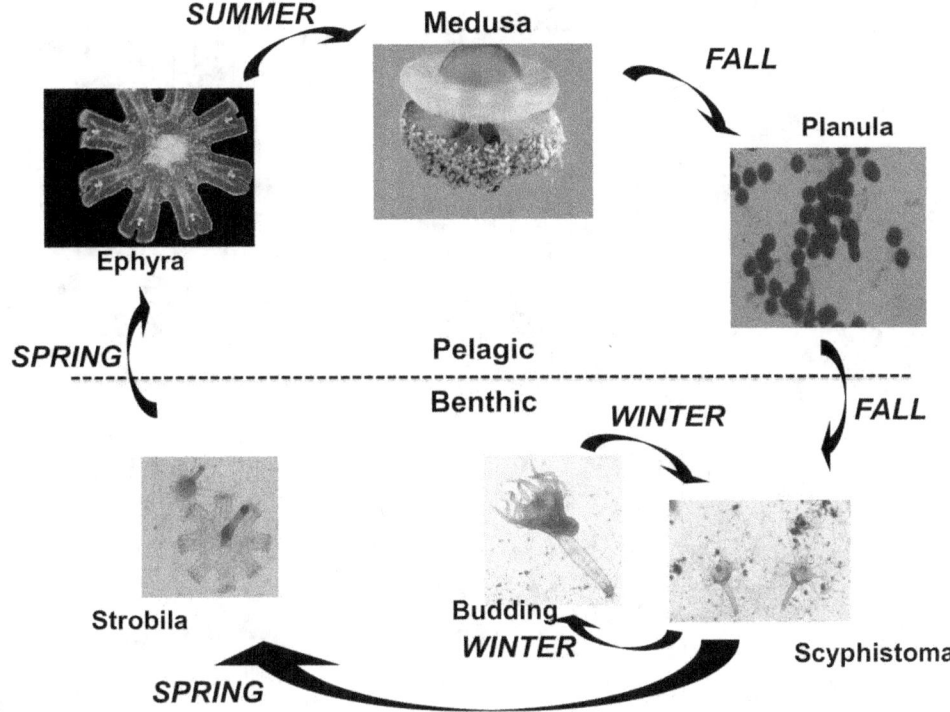

Figure 28.12. Illustrating thermal control and seasonal occurrence of *C. tuberculata* in Mar Menor.

Summary and conclusions of applications of operational oceanography on jellyfish swarms: Present and future

Due to their diversity and the complexity of their life cycle, jellyfish can be used as a proxy to applied operational oceanography as applications of predictions systems. In the three case studies presented in this chapter, operational oceanography was applied in different ways. The first case study, the Balearic Jellyfish Observing System, is an example of taking a first step towards pre-operational oceanography applied directly to jellyfish swarm coastal management. The second applied case study, the Portuguese Man-O-War in the Mediterranean Basin, shows how operational oceanography has been already applied to effectively diagnose the past, and efforts are underway that will help operational oceanography be able to predict future events. In the last case study, management of jellyfish swarms in a marine lagoon, the tool obtained is already available for policymakers and stakeholders. Of course, we are now facing a moment where the oceanography technology for observation is advancing rapidly, together with models and data assimilation. We should include jellyfish research and monitoring as one of the targets to link operational oceanography to both coastal and open ocean environments.

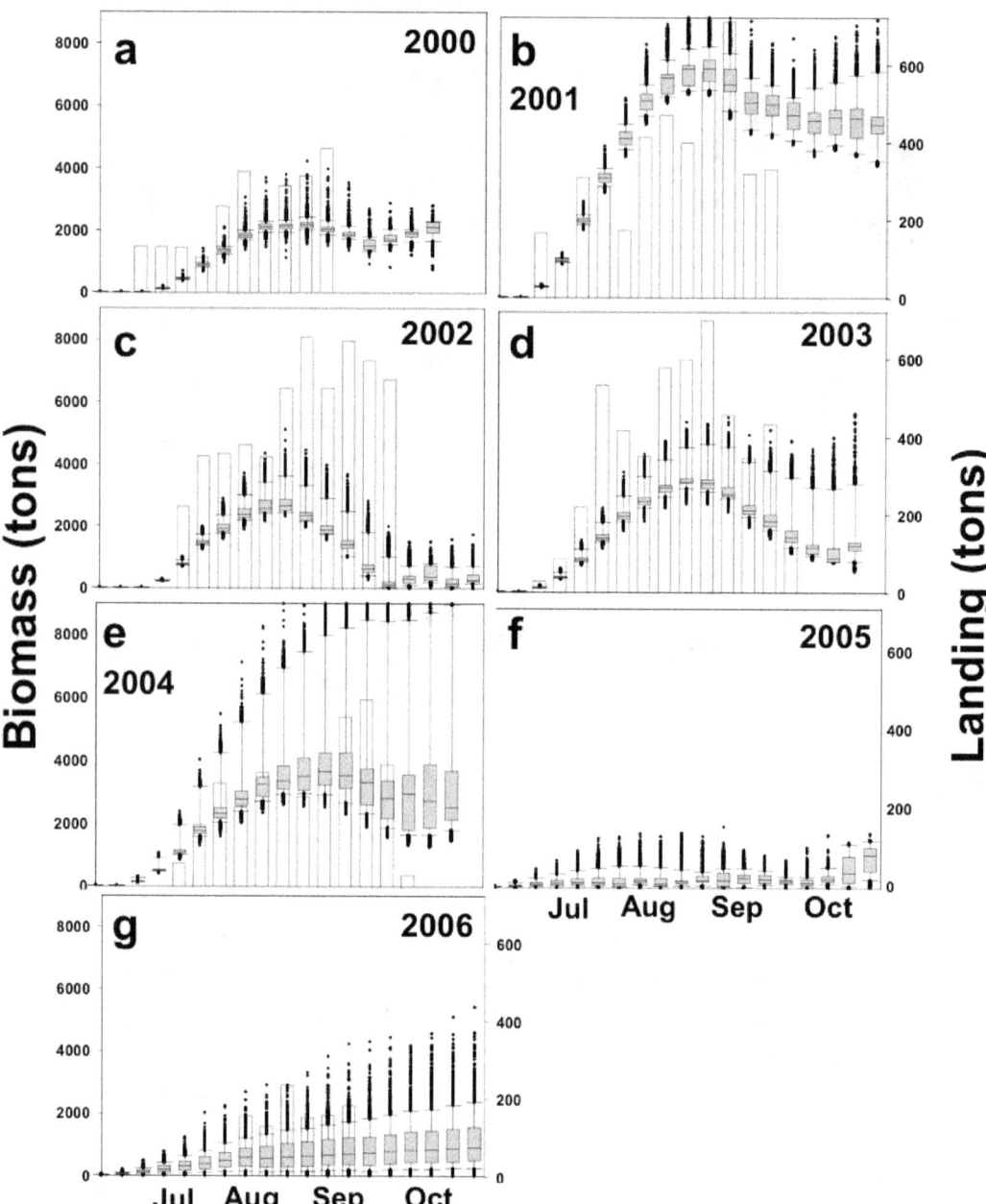

Figure 28.13. Medusa biomass and landings. Box plots for the posteriors of medusa biomass of the weekly-resolved model during years 2000 to 2006. Box limits and whiskers indicate respectively the 25–75 and the 10–90 percentile limits, dots are outliers. Vertical bars (right axis) are the weekly landing data from Consejería de Agricultura y Pesca (Región de Murcia; from Ruiz et al., 2012).

Acknowledgments

Thanks to Dr. Javier Ruiz and Dr. Ananda Pascual for encouraging me to write this chapter, to the research group of Ecosystem Oceanography (ICMAN-CSIC) for their assistance, to Dr. Joaquín Tintore, and to Antonio M. Grau for his support of starting the endeavor of the jellyfish observation system in the Balearic Sea (reflected in case study 1). Thanks to the Consejeria de Medio Ambiente of Junta de Andalucia (for *P. physalis* stranding data for case study 2). Also acknowledged is the Consejeria de Agricultura y Pesca of Region de Murcia for the jellyfish fishing data and the Agencia Estatal del Meteorologia of Ministerio de Medio Ambiente for wind data (both for case study 3). This work was support financially by the following projects: "Respuesta de medusas mediterráneas al efecto interactivo de motores climáticos de impacto: supervivencia en un mediterráneo más cálido y ácido" Med2CA (CTM2016-75487-R), "Forzamientos fisicos en la proliferacion costera de organismos gelatinosos (2017301072) and "Deteccion de medusas en el mar Balear y su relacion con las conduciones ambientales: hacia el desarrollo de un sistema de prediccion pre-operacional" (Govern de les Illes Balears, SOCIB and CSIC).

References

Angel, D. L., D. Edelist, and S. Freeman, 2016: Local perspectives on regional challenges: jellyfish proliferation and fish stock management along the Israeli Mediterranean coast. *Reg. Environ. Change*, **16(2)**, 315–323.

Beckman, A., and S. Özbek, 2012: The nematocyst: a molecular map of the cnidarian stinging organelle. *Int. J. Dev. Biol.*, **56**, 577–582.

Bosch-Belmar, M., E. Azzurro, K. Pulis, and Coauthors, 2017: Jellyfish blooms perception in Mediterranean finfish aquaculture. *Marine Policy*, 76, 1–7.

Brinkman, D. L., and J. N. Burnell, 2009: Biochemical and molecular characterization of cubozoan protein toxins. *Toxicon*, **54(8)**, 1162–1173.

Burnett, J., 2000: Medical aspects of jellyfish envenomation: pathogenesis, case reporting and therapy. *Hydrobiology*, **155**, 1–9.

Caldeira, K, and M. E. Wickett, 2003: Oceanography: anthropogenic carbon and ocean pH. *Nature*, **425**, 365.

Calder, D. R.,1973: Laboratory observations on the life history of Rhopilema verrilli (Scyphozoa: Rhizostomae). *Mar. Biol.*, **21**, 109–114.

Condon, R. H., D. K. Steinberg, P. A. del Giorgio, and Coauthors, 2011: Jellyfish blooms result in a major microbial respiratory sink of carbon in marine systems. *P. Natl. Acad. Sci.*, **108**, 10225–10230.

Condon, R.H., W. M. Graham, C. M. Duarte, and Coauthors, 2012: Questioning the Rise of Gelatinous Zooplankton in the World's Oceans. *BioScience*, **62 (2)**, 160–169.

Daly, M., M. R. Brugler, P. Cartwright, A. G. Collins, M. D. Dawson, D. G. Fautin, S. C. France, C. S. McFadden, D. M. Opresko, E. Rodriguez, and S. L. Romano, 2007: The phylum Cnidaria: A review of phylogenetic patterns and diversity 300 years after Linnaeus. *Zootaxa*, **1668**, 127–182.

Daly Yahia, M. N., M. Batistic, I. D. Lu, , and Coauthors, 2010. Are outbreaks of *Pelagia noctiluca* (Forskål, 1771) more frequent in the Mediterranean basin? *ICES Co-operative Res. Rep.*, **300**, 8–14.

Deidun. A., 2010: Notes on the recent occurrence of uncommon pelagic "jellyfish" species in Maltese coastal waters. *Naturalista sicil*, **S. IV, XXXIV (3-4)**, 375–384.

Doney, S. C., V. J. Fabry, R. A. Feely, and J. A. Kleypas, 2009: Ocean Acidification: The Other CO_2 Problem. *Annu. Rev. Mar. Sci.*, **1**, 169–192.

Dunn, C. W., S. P. Leys, and S. H. D. Haddock, 2015: The hidden biology of sponges and ctenophores, *Trends Ecol. Evol.*, **30**, 282–291.

ECOS, 2004: Poblaciones de medusas en el Mar Menor. *Murcia: CARM.* 12 p.

Edwards, L., and D. A. Hessinger, 2000: Portuguese Man-of-war (*Physalia physalis*) venom induces calcium influx into cells by permeabilizing plasma membranes. *Toxicon*, **38**, 1015–1028.

Franqueville, C., 1971: Macroplancton profond (invertébrés) de la Méditerranée nord-occidentale. *Tethys*, **3**, 11–56.

Giorgi, R., M. Avian, S. De Olazabal, S., and Coauthors, 1991: Feeding of *Pelagia noctiluca* in open sea. In *Jellyfish blooms in the Mediterranean: proceedings of the II Workshop on Jellyfish in the Mediterranean Sea, Trieste, 2–5 September 1987.* United Nations Environment Programme (eds.), Trieste, Italy, pp. 102–111.

Hamer, H. H., A. M. Malzahn, and M. Boersma, 2011: The invasive ctenophore *Mnemiopsis leidyi*: a threat to fish recruitment in the North Sea? *J. Plankton Res.*, **33**,137-144.

Harbison, G. R., L. P. Madin, and N. R. Swanberg, 1978: On the natural history and distribution of oceanic ctenophores. *Deep-Sea Res.*, **25**, 233–256.

Hofmann, D. K., R. Neumann, and K. Henne, 1978: Strobilation, budding and initiation of scyphistoma morphogenesis in the rhizostome *Cassiopea andromeda*. *Mar. Biol.*, **47**, 161–179.

Holstein, T. and P. Tardent, 1984: An ultrahigh-speed analysis of exocytosis: nematocyst discharge. *Science*, **223**(4638), 830–833.

Hurrel, J. and National Center for Atmospheric Research Staff (Eds), 2012: The Climate Data Guide: Hurrell North Atlantic Oscillation (NAO) Index (station-based), https://climatedataguide.ucar.edu/guidance/hurrell-north-atlantic-oscillation-nao-index-station-based. Accessed on 21/03/2013.

IPCC, 2013: Synthesis Reports available in http://www.ipcc.ch.

Jager, M., and M. Manuel, 2016: Ctenophores: and evolutionary-developmental perspective. *Curr. Opin. Genetics Dev.*, **39**, 85–92.

Javidpour, J., J. C. Molinero, A. Lehmann, and Coauthors, 2009: Annual assessment of the predation of *Mnemiopsis leidyi* in a new invaded environment, the Kiel Fjord (Western Baltic Sea): a matter of concern. *J. Plankton Res.*, **31**, 729–738.

Khatiwala, S., Tanhua, T., M. Fletcher, and Coauthors, 2013: Global ocean storage of anthropogenic carbon. *Biogeosciences*, **10**, 2169–2191, doi:10.5194/bg-10-2169-2013.

Kideys, A. E., 2002: Fall and rise of the Blask Sea ecosystem. *Science*, **297(5586)**, 1482–1484.

Klein, S. G, K. A. Pitt, K. A. Rathjen, and J. E. Seimour, 2014: Irukandji jellyfish polyps exhibit tolerance to interacting climate change stressors. *Global Change Biol.*, **20**, 28–37.

Knap, A., A. Michaels, A. Close, H. Ducklow, and A. Dickson (eds.), 1996: Protocols for the Joint Global Ocean Flux Study (JGOFS) Core Measurements. *JGOFS Report Nr. 19*, 170 pp.

Kogovšek, T., B. Bogunović, and A. Malej, 2010: Recurrence of bloom forming scyphomedusae: wavelet analysis of a 200-year time series. *Hydrobiologia*, **645**, 81–96.

Kroeker, K. J., R. L. Kordas, R. Crim, and Coauthors, 2013: Impacts of ocean acidification on marine organisms: quantifying sensitivities and interaction with warming. *Global Change Biol.*, **19**, 1884–1896.

Larson, R., C. Mills, and G. Harbison, G. 1991: Western Atlantic midwater hydrozoan and scyphozoan medusae: in situ studies using manned submersibles. *Hydrobiologia*, **216**, 311–317.

Le Quéré, C., R. Moriarty, R. M. Andrew, and Coauthors, 2015: Global Carbon Budget 2015. *Earth Syst. Sci. Data*, **7**, 349–396.

Lehodey, P., R. Murtugudde, and I. Senina, 2010: Bridging the gap from ocean models to population dynamics of large marine predators: A model of mid-trophic functional groups. *Prog. Ocn.*, **84**, 69–84.

Licandro, P., D. V. P. Conway, M. N. Daly Yahia, and Coauthors, 2010: A blooming jellyfish in the northeast Atlantic and Mediterranean. *Biol. Lett.*, **6(5)**, 688–691. doi 10.1098/rsbl.2010.0150.

Malej, A., J. Faganeli, and J. Pezdic, 1993: Stable isotope and biochemical fractionation in the marine pelagic food chain: the jellyfish *Pelagia noctiluca* and net zooplankton. *Mar. Biol.*, **116**, 565–570.

Mariottini, G. L., E. Giacco, and L. Pane, 2008: The mauve stinger *Pelagia noctiluca* (Forsskal, 1775). Distribution, ecology, toxicity and epidemiology of stings. A review. *Mar. Drugs*, **6**, 496–513.

Martindale, M. Q., 2016: The onset of regenerative properties in ctenophores. *Curr. Opin. Genetics Dev.*, **40**, 113–119.

Perez-Ruzafa, A., J. Gilabert, J. M. Gutiérrez, and Coauthors, 2002: Evidence of a planktonic food web response to changes in nutrient input dynamics in the Mar Menor coastal lagoon, Spain. *Hydrobiologia*, **475–476**, 359–369.

Presnell, J. S., L. E. Vandepas, K. J. Warren, B. J. Swalla, C. T. Anemiya, and W. E. Browne, 2016: The presence of a functionally tripartite through-gut in Ctenophora has implications for Metazoan character trait evolution. *Current. Bio.*, **26**, 2814–2820.

Prieto, L., A. Deidun, A. Malej, T. Shiganova, and V. Tirelli, 2016. Coming to grips with the jellyfish phenomenon in the Southern European and other Seas: research to the rescue of coastal managers. ISBN 978-960-9798-23-5, 43 pp.

Prieto, L., D. Astorga, G. Navarro, and J. Ruiz, 2010: Environmental control of phase transition and polyp survival of a massive-outbreaker jellyfish. *PLoS One*, **5**, e13793.

Prieto, L., D. Macías, A. Péliz, and J. Ruiz, 2015: Portuguese Man-of-War (*Physalia physalis*) in the Mediterranean: A killer jellyfish invasion? *Sci. Rep.-UK*, **5**, 11545. doi 10.1038/srep11545.

Purcell, J. E., R. A. Hoover, and N. T. Schwarck, 2009: Interannual variation of strobilation by the scyphozoan *Aurelia labiata* in relation to polyp density, temperature, salinity, and light conditions in situ. *Mar. Ecol. Prog. Ser.*, **375**, 139–149.

Purcell, J. E., S.-I. Uye, and W.-T. Lo, 2007: Anthropogenic causes of jellyfish blooms and direct consequences for humans: a review. *Mar. Ecol. Prog. Ser.*, **350**, 153–174.

Purcell, J. E., and M. N. Arai, 200: Interactions of pelagic cnidarians and ctenophores with fish: a review. *Hydrobiologia*, **451**, 27–44.

Purcell, J. E., T. A. Shiganova, M. B. Decker, and Coauthors, 2001: The ctenophore *Mnemiopsis* in native and exotic habitats: US estuaries versus the Black Sea basin. *Hydrobiologia*, **451**, 145–176.

Purcell, J. E., V. L. Fuentes, D. Atienza, and Coauthors, 2010: Use of respiration rates of scyphozoan jellyfish to estimate their effects on the food web. *Hydrobiologia*, **645**, 135–152.

Richardson, A. J., A. Bakun, G. C. Hays, and M. J. Gibbons, 2009: The jellyfish joyride: causes, consequences and management responses to a more gelatinous future. *Trends Ecol. Evol.*, **24**, 312–322.

Ruiz, J., L. Prieto, and D. A. Astorga, 2012: Model for temperature control of jellyfish (*Cotylorhiza tuberculata*) outbreaks: a causal analysis in a Mediterranean coastal lagoon. *Ecol. Model.*, **23**, 59–69.

Shen, X.-X., C. T. Hittinger, and A. Rokas, 2017: Contentious relationships in phylogenomic studies can be driven by a handful of genes. *Nature Ecol. Evol.*, **1**, 0126.

Shiganova, T. A., and Coauthors, 2001: Population development of the invader ctenophore *Mnemiopsis leidyi* in the Black Sea and other seas of the Mediterranean basin. *Mar. Biol.*, *139*, 431–445.

Sugiura, Y., 1965: On the life-history of rhizostome medusae. III. On the effect of temperature on the strobilation of Mastigias papua. *Biol. Bull.*, **128**, 493–496.

Tamkun, M. M., and D. A. Hessinger, 1981: Isolation and partial characterization of a hemolytic and toxic protein from the nematocyst venom of the Portuguese Man-of-war, *Physalia physalis. Biochim. Biophys. Acta*, **667**, 87–98.

Tintoré, J., G. Vizoso, B. Casas, and Coauthors, 2013: SOCIB: the Balearic Islands Observing and Forecasting System responding to science, technology and society needs. *Mar. Tech. Soc. J.*, **47(1)**, 17 pp. http://doi.org/10.4031/MTSJ.47.1.10

UNEP 1991: Jellyfish blooms in the Mediterranean. *MAP Technical Reports Series*, **47**, 338 pp.

Watanabe, O. J. Jouzel, S. Johnsen, and Coauthors, 2003: Homogeneous climate variability across East Antarctica over the past three glacial cycles. *Nature*, **422**, 509–512.

Widmer, C. L., 2005: Effects of temperature on growth of north-east Pacific moon jellyfish ephyrae, *Aurelia labiata* (Cnidaria: Scyphozoa). *J. Mar. Biol. Ass. U.K.*, **85(3)**, 569–573.

Willcox, S., N. A. Moltschaniwskyj, and C. Crawhord, 2007: Asexual reproduction in scyphistomae of *Aurelia sp.*: Effects of temperature and salinity in an experimental study. *J. Exp. Mar. Bio. Ecol.*, **353**, 107–114.

Winans, A. K., and J. E. Purcell, 2010: Effects of pH on asexual reproduction and statolith formation of the scyphozoan, *Aurelia labiata. Hydrobiologia*, **645**, 39–52.

Zavodnik, D., 1991: On the food and feeding in the North Adriatic of *Pelagia noctiluca* (Scyphozoa). In *Jellyfish Blooms in the Mediterranean: Proceedings of the II Workshop on Jellyfish in the Mediterranean Sea, Trieste, 2–5 September 1987*. United Nations Environment Programme (eds.), Trieste, Italy, pp. 212–216.

CHAPTER 29

Measuring Performances, Skill and Accuracy in Operational Oceanography: New Challenges and Approaches

Fabrice Hernandez[1], Greg Smith[2], Katrijn Baetens[3], Gianpiero Cossarini[4], Isabel Garcia-Hermosa[1], Marie Drévillon[1], Jan Maksymczuk[5], Angélique Melet[1], Charly Régnier[1], and Karina von Schuckmann[1]

[1]*MetOcean Mercator Océan, Ramonville St Agne, France;* [2]*Environment Canada, Montréal, Canada;* [3]*Royal Belgian Institute of Natural Sciences, Brussels, Belgium;* [4]*National Institute of Oceanography and Experimental Geophysics, Sgonico, Italy;* [5]*Met Office, Exeter, UK*

Operational oceanography is now established in many countries, focusing on global, regional, or coastal areas, and targeting different aspects of the « blue », « white » or « green » ocean processes in order to provide reliable information to users. There are nowadays a large variety of interests and users, with different disciplines and levels of expertise. Validation and verification of operational products and systems are evolving in order to anticipate user's needs, and better quantify the level of confidence on all these variety of ocean products. Operational oceanography evaluation development is in front of key issues: Ocean models are reaching the submesoscale description, which is currently not adequately observed; many products are available now for a given ocean variable, and often discrepancies are larger than similarities; real time forecasting systems are also challenged by reanalyses or reprocessed time series; operational systems are getting more complex, with coupled modelling, where errors from the different compartment need to be carefully addressed in order to measure their performance and provide further improvements. In parallel, the global ocean observing system is continuously completed with additional satellites in the constellation, with innovative sensors on new satellite missions, with efforts to better integrate the global, regional and coastal in-situ observing capabilities, and the design of new instrument, like the BGC-Argo that should bring an enhanced description of the ocean biogeochemical variability. This book chapter provides an overview of the existing, mature, validation and verification science in operational oceanography; discusses the ongoing efforts and new strategies; presents some of the structured groups and outcomes; and lists a series of challenges on the field.

Introduction

Operational oceanography has reached a mature stage. Now established in many countries, operational centres began by providing ocean products to a small, select group of experts. However, over the past several years, operational oceanography has expanded and now provides services and ocean monitoring to a wide community of users. Operational Ocean Forecasting and Monitoring Systems (OOFMS) can focus on global, regional, or coastal areas, and target different aspects of the « blue » physical ocean processes, « green » biological/biogeochemical low trophic level processes, or « white » sea ice and cryosphere

---

Hernandez, F., et al., 2018: Measuring performances, skill and accuracy in operational oceanography: New challenges and approaches. In "*New Frontiers in Operational Oceanography*", E. Chassignet, A. Pascual, J. Tintoré, and J. Verron, Eds., GODAE OceanView, 759-796, doi:10.17125/gov2018.ch29.

processes over ocean, in order to inform a large variety of interests and marine users within different disciplines and with varying levels of expertise (see, for example, Fig. 1 of Schiller et al., 2016).

Operational ocean global and regional initiatives expanded a great deal over the past two decades, working to overcome major issues under the auspices of community of experts such as those involved in the Global Ocean Data Assimilation Experiment (GODAE), which was followed by GODAE OceanView (Bell et al., 2015; Tonani et al., 2015). Operational oceanography is based on three pillars: 1) ocean observing systems; 2) modelling tools; and 3) data assimilation or other estimation and control techniques. These are structured to provide descriptions and predictions in the marine environment and offer dedicated services to marine stakeholders. The 2017 GODAE OceanView summer school addresses many of the advances and challenges that arise in these three areas, in particular the main limitations and weaknesses that directly impact contemporary performances of OOFMS. The reader is invited to look at all of these GODAE OceanView (GOV) summer school contributions where observing system limitations, model errors, and data assimilation performances are discussed as a way to begin to understand assessment and evaluation approaches designed and implemented by the operational oceanography community.

Although initially, operational oceanography developments were heavily science-driven (Schiller et al., 2015), they evolved and are now more user-driven (Schiller et al., 2016). This has broadened the prediction and monitoring capabilities in a seamless way closer to the coast (Kourafalou et al., 2015) in order to fulfil the UN sustainable development goals for the marine ecosystems and taking into account the huge needs of the permanently growing "blue economy."

From very local applied operational systems to global monitoring and forecasting ocean centres, the main goal of operational oceanography is to *provide timely and accurate information, including prediction and projections, about the marine environment*. Consequently, validation and verification of ocean numerical simulations and estimations are core activities in operational oceanography in order to anticipate user needs and better quantify the level of confidence on a variety of ocean products (Hernandez et al., 2015).

This chapter provides an overview of the validation and verification framework in operational oceanography. It also presents the standards and methods adopted by the GOV community since the beginning of GODAE. It then introduces different evaluation approaches and key issues based on the evaluation framework raised by the European Union (EU) Copernicus Marine Environment Monitoring Service (CMEMS) program. And finally, it presents some recent validation approaches and new metrics.

General Validation and Verification Background

Introductory considerations on evaluating performance of operational systems and quality of products

The ocean operational community is challenged in the way assessment is performed. Evaluation tools are now widely implemented in operational centres. But consider the overview proposed by

Hernandez et al. (2015), as a companion or introductory work to this chapter. Hernandez et al. (2015) reviewed the principles and main concepts driving evaluation approaches in operational oceanography, specifically the methodology raised by GODAE and GOV, with standardized methods such as Class 1, 2, 3, and 4 metrics, which are largely implemented now (e.g., Ryan et al., 2015). Hernandez et al. (2015) also detailed a series of recent metrics designed to assess specific variables (e.g., sea ice, chlorophyll, water masses) or focus on particular OOFMS assessment aspects (e.g., long-term forecast, assimilation performance, regional systems efficiency, upstream quality control, and/or ensemble assessment).

Most operational centres have implemented an assessment framework dedicated to:

- Evaluating and monitoring the performance of operational systems, considering the:
    - impact of the observing system,
    - model errors, and
    - data assimilation efficiency.
- Evaluating the accuracy of products such as the:
    - products derived from observation (real time or reprocessed),
    - routine hindcast and forecast (and their predictive skill), and
    - reanalyses.
- Measuring the strengths and weaknesses of the system operated in order to make further improvements
- Assessing each product's reliability in light of user needs

It is important to keep in mind the way that these errors and existing observations that represent the "ocean truth" are part of the "three pillars" of the operational oceanography structure mentioned above. Fig. 29.1 describes the main errors associated with each product and notes some of the difficulties in using observations for evaluation of an ocean product quality. It becomes evident that, for the complete OOFMS, many factors and errors limit a product's quality. In particular, observation distribution in space and time as well as sparseness strongly impact our capacity to assess how efficient operational systems are in representing ocean processes. An effectual validation and verification framework should take into account these factors and characterize the contribution of each error type.

Finally, considering user needs is a new aspect in the operational oceanography evaluation framework that has required consideration of several "internal" and "external" metrics. These include:

***Internal metrics*: verification that the systems satisfy initial requirements**. In other words, verify that a system reacts and behaves as expected with regard to its own representativity. For example, let's look at evaluation of the eddy-permitting system representation of the Gulf Stream path and the associated gradient. It would not be appropriate to evaluate the capability of this system in producing correctly tidal fronts, since tidal dynamics are not part of the "ocean model engine" used for this system. But it would be appropriate to measure the Gulf Stream frontal position against near real time satellite sea surface temperature images which indicates its surface thermal signature.

Other appropriate examples include position and seasonal variability of the subtropical gyres, which should be verified in a 1,000-year coupled climate simulation, while M2 harmonic phase and amplitude should be assessed in a barotropic tidal model.

***External metrics*: reliability of the product based on user requirements.** In many cases, external metrics look to measure the departures of the products against real ocean processes, whatever the representativity of the ocean model. So, for instance, if a user is looking at the extension of harmful algae blooms, proper assessment of the hourly rate of nutrients, oxygen, turbidity, mixed layer dynamics, fresh water run-off on the shelf is needed. Obviously, a basin scale eddy-permitting biogeochemical system would not be suitable for this purpose; a high-resolution regional model with high-frequency forcings and the relevant local ecosystem dynamics would offer a user better predictions in a case such as this.

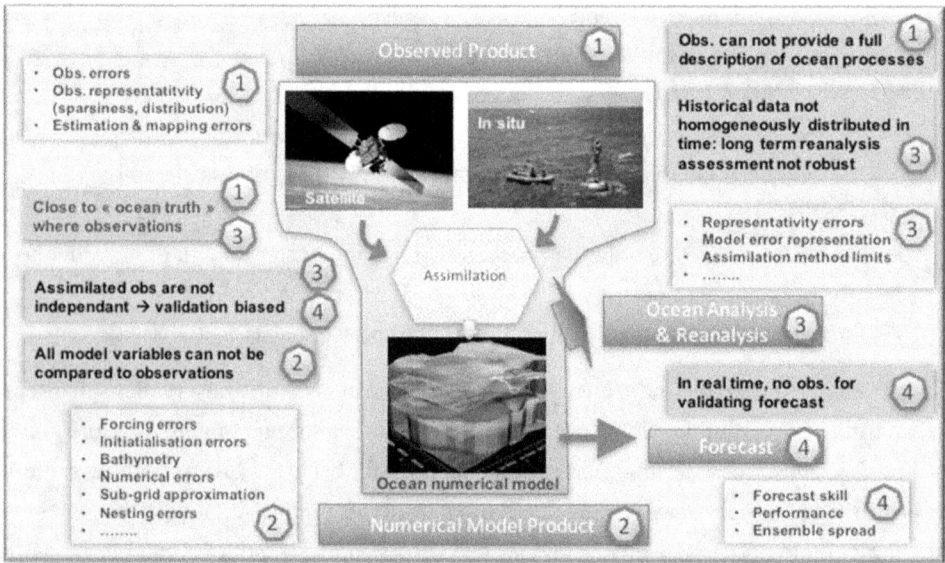

Figure 29.1. Different operational oceanography products (blue), their associated errors (yellow), positive aspects (light blue), and some drawbacks in using observations for the evaluation (pink).

The operational ocean assessment framework is essentially built upon comparison to observations. It is thus worth to note that a given set of observations might be used both for internal or external assessment. For instance, let's consider a metrics aiming at measuring sea level differences at high-frequency (e.g., a few minutes) between a tide gauge along the coast and model solutions. The low-pass filtering of the sea level time series would allow an "internal" assessment of any eddy-permitting system. This is what is carried out when GOV global systems are compared to the international tide gauge network (e.g., see Fig. 7 of Zuo et al., 2015). Now, let's consider the same tide gauge time series non-filtered on an hourly basis; comparing it to the same GOV global system would demonstrate the inability of this GOV system to represent coastal tidal harmonics. This would be an "external" evaluation of this global product for a user interested in high-frequency, on-shelf circulation. Meanwhile, if the tested system was a high-frequency, high-resolution regional shelf model that included tidal dynamics, then the comparison between hourly tide gauge data and the model output would again be considered an "internal" evaluation, where one tries just to assess what such a model should represent (e.g., tidal assessment in the CMEMS IBI configuration by Maraldi et al., 2013). Finally, considering the complete non-filtered tide gauge time series, e.g., at five-minute frequency, and comparing the extreme sea level events with this regional high-resolution regional system would show the capability of this system in predicting and phasing extreme storm surges. This would be a valuable "external" assessment for any decision-maker in charge of coastal management and warnings.

Operational oceanography community effort on evaluation and verification

Through GOV, some in the international community work to address common OOFMS assessment challenges. The GOV Intercomparison and Validation Task Team, together with other task teams, is focusing on some of these issues and experimenting with several Class 1 and Class 4 metric approaches (Divakaran et al., 2015; Hernandez et al., 2015; Ryan et al., 2015). The GOV Coastal and Shelf Seas Task Team has proposed new approaches to the evaluation of regional operational systems (De Mey et al., 2017; Kourafalou et al., 2015), and highlighted coastal modelling, coastal observations, and nesting assessment issues. As part of the GOV 2017 summer school lectures, Mourre et al. (2018) for the Balearic Sea OOFMS and Roughan et al. (2018) for the New South Wales Australian coast integrated observing system, offered a comprehensive overview of regional assessment objectives, methods, and challenges of the operational ocean framework.

Complementary other assessment challenges have been tackled by GOV community efforts. Recently, the GOV Marine Ecosystem Analysis and Prediction Task Team began to address validation issues of the biogeochemical component of global or regional operational systems (Gehlen et al., 2015). In the GOV framework, task teams such as the Observing System Evaluation Task Team and the Coupled Prediction Task Team have also shared and rely upon evaluation approaches proposed by the other task teams (Brassington et al., 2015; Oke et al., 2015a; 2015b).

Moreover, some international initiatives such as the CLIVAR Global Synthesis and Observations Panel (GSOP) Ocean Reanalyses Intercomparison Project (http://www.clivar.org/panels-and-working-groups/gsop/gsop.php), along with EU COST Action Evaluation of Ocean Syntheses (EOS) project (http://www.eos-cost.eu/the-action/about-eos), allow for exploration of evaluation approaches for ocean reanalyses (Balmaseda et al., 2015) at both global and regional scales. The GOV 2017 summer school lecture by Haines (2018) proposed an updated overview of reanalyses development and assessment.

In parallel, the EU CMEMS is pioneering many aspects of global and operational oceanography development and issues, supported by the expertise and organization of most "marine" nations in Europe (Drévillon et al., 2018, this book, as part of GOV summer school contribution). The CMEMS developed a dedicated, cross-cutting activity on validation and verification activities, with a long-term strategy and plans to address the most recent issues on the validation, verification, and performance of OOFMS (Hernandez and Melet, 2016; Le Traon et al., 2017).

Evaluation in operational oceanography: some identified challenges

Above we provide some examples of the operational oceanography community's ongoing efforts to organize and develop adequate evaluation tools. Their main achievements are reviewed in Hernandez et al. (2015), and some challenging issues are listed in Schiller et al. (2015). However, the community is facing new emerging questions:

First, operational oceanography is continuously evolving toward more complex systems:

- Global and regional open ocean models are reaching the submesoscale description, typically less than 10 km (see, for example, discussion in the lecture by Jacobs et al., 2018), and the ocean dynamics at these fine scales are not adequately observed;
- Through nesting strategies, increasingly, local, coastal modelling tools are exchanging poorly controlled information with larger-scale systems at their boundaries;
- The diversity of products available for a given ocean interest/variable increases among operational ocean catalogues (e.g., gridded observed products from in situ or satellite measurements, or products merging both in situ and remote sensing data; model forecasts, model reanalyses from different kind of models; global or regional products etc.). Sometimes these discrepancies are larger than their similarities; assessment is needed to evaluate their quality and to inform users about their usefulness for a given application.
- Real-time forecasting systems are also challenged by reanalyses or reprocessed time series. All of these products need to be evaluated and their respective quality communicated to users. Then, real-time system performance must be revisited with regard to better quality reanalyses, with the goal being to motivate improvements.
- Operational systems are getting more complex with coupled modelling (e.g., Brassington et al., 2015; Harris, 2018; Tonani et al., 2015), where errors from the different compartments need to be carefully addressed in order to measure performance and provide further improvements, but also in order design ways to limit impacts of these errors from one compartment to another.

Second, operational oceanography, originally a product of academia, must now communicate more efficiently with a wide range of user communities and practices:

- Traditionally, for a given ocean domain, evaluation in research mode tried to assess the performance of the system as a whole. Now, in the same domain, different types of users and different types of applications (economic, security, health, biodiversity, regulation etc.) can have entirely different and dedicated assessment procedures carried out in parallel.
- Representativity in a given OOFMS needs to be taken into account and quantified in terms of "representativity errors" (also called "representation" or "representativeness" errors) when communicating a product's reliability: through a comprehensive assessment based on external metrics, one can inform users expecting the full

discrepancies and errors from what really happens at sea, and not providing confidence levels through few metrics related to the specific capabilities of the system.
- Informing users about product reliability also requires adopting new approaches: communicating error and accuracy levels must take into account the user's expertise, awareness, and their effective use of the products.

This chapter, associated with Hernandez et al. (2015), addresses validation, verification, and assessment concepts by structuring most of the evaluation methods and tools implemented in operational oceanography centres. For additional practical examples, the reader is invited to read other chapters in this book that describe specific OOFMS and their identified errors from different components (e.g., chapters by Bouillon et al., 2018; Ford et al., 2018; Harris, 2018; Jacobs et al., 2018; Le Sommer et al., 2018) as well as their validation framework (e.g., chapters by Lellouche et al., 2018; Mourre et al., 2018; Roughan et al., 2018; Wilkin et al., 2018).

The GODAE OceanView Common Evaluation Framework

Any ocean operational centre can design, implement, and perform evaluation of its forecasting tools on its own if worldwide ocean observations are easily accessible. However, when GODAE began, it was clear that the operational oceanography community could follow in the footsteps of the weather forecast and climate communities in the way that they work together under the patronage of the World Meteorological Organisation (WMO). For validation and verification activities, this allows for the 1) sharing of best practices and innovations; 2) inter-comparing performances of operational systems and evaluating the GODAE system against other systems; 3) using other estimates to improve the operational service; 4) responding consistently, particularly when requesting information, tools, and support from other parties (e.g., observations from space agencies, national marine institutions, other expert groups etc.); and 5) understanding common requests from users and applications. The ability to act in unison has allowed GODAE to establish and engage in legitimate dialogs with other communities. Over the past several years, the GOV Intercomparison and Validation Task Team has begun to exchange with expert groups, such as the Working Group on Numerical Experimentation from the WMO and World Climate Research Program, and the WMO Joint Working Group on Forecast Verification Research (https://www.wmo.int/pages/prog/arep/wwrp/new/Forecast_Verification.html), whose expertise in weather forecast verification offers numerous valuable approaches (e.g., Casati et al., 2008; Ebert et al., 2013; Gilleland et al., 2009).

With regard to understanding common user requests and applications, one of the unfortunate illustrations are airplane accidents at sea. As discussed in Hernandez et al. (2015), the AF447 Air France Rio-Paris airplane crash in June 2009 led to a series of published improvements in ensemble estimation and assessment from OOFMS in order to better specify rescue and search activities at sea (e.g., Drévillon et al., 2013). This permitted several organizations to initiate their search activities after the crash of the Malaysia Airline MH370 plane in 2014 and to refine their techniques using model simulations and observations while debris and evidences were appearing (Griffin and

Oke, 2017; Griffin et al., 2016). Despite the fact that the crash location was not yet known, this work resulted in new insights into ways to deal with the reliability of ocean products and give confidence to the relevant users. Specific events such as the Malaysia Airline crash are valuable as case studies for evaluating the performance of incoming OOFMS. Similarly, the AF447 accident was used by Mercator Océan to evaluate the surface dispersion provided by the new high-resolution global system, and showed large improvements and skill in positioning search areas (see the chapter in this book by Lellouche et al., 2018).

Clearly, the performance of common model and forecast evaluations requires the adoption of common objectives and principles. Thus, from the start GODAE validation experts adopted approaches in line with those of the weather forecast community. First, assessment is sought and expected in terms of "consistency," then "accuracy," and then "performance" evaluation (see Murphy, 1993, and Hernandez et al. [2015] for details on these evaluation principles). Finally, the "fit-for-purpose" principle associated with "external" metrics discussed previously is applied with the goal being to provide a more user-oriented evaluation.

Additionally, a common framework and inter-comparison implies the use of a common vocabulary and standardized tools. To accomplish this, Class 1, 2, 3, and 4 metrics categories were defined in order to implement a technical framework for model-to-model and model-to-observation comparisons (see details in Hernandez et al., 2009). Initially, this framework was defined for open ocean, physical eddy-permitting modelling assessments. Subsequently, this framework has been used for regional, higher-frequency, higher-scale forecasting systems as well as for reanalyses intercomparison. And most recently, it has been extended to sea ice or biogeochemical variables.

As part of the GOV Intercomparison and Validation Task Team activities, intercomparison tasks based on Class 1, 2, and 4 metrics have been conducted, resulting in a number of global and regional published results (Hernandez et al., 2015; Divakaran et al., 2015; Oke et al., 2012; Ryan et al., 2015). A particular effort on Class 4 metrics is ongoing, based on model-to-observation comparisons for routine monitoring of hindcasts and forecasts. This approach compares water column temperature and salinity assessment to profilers, sea level data to along-track satellite altimetry data, and sea surface temperature (SST) to drifter measurements. The initiative has been expanded to include sea ice concentrations using satellite data, and efforts are ongoing to extend it to comparisons of surface velocity to drifter trajectories. The goal is to develop global and regional OOFMS with eddy-permitting to eddy-resolving capability. However, the scales (observability) provided by these data do not allow assessment of shorter scales (see further discussion below).

Centres that have adopted this international intercomparison framework on a routine basis are now using it for their internal system evaluation (e.g., Blockley et al., 2014; Lellouche et al., 2013). As described in Hernandez and Melet (2016), this framework has also been adopted for the CMEMS product quality assessment framework (see below).

Additionally, the GOV Intercomparison and Validation Task Team intercomparisons initiatives are providing experiences and feedback with regards to community best practices, in particular how to structure the activity among international partners and what to ask, in a context of where an operational centre's expertise, tools, and goals are evolving rapidly. The Class 4 intercomparison projects called attention to the global OOFMS involved and strengthened their comparisons with regional OOFMS at their national levels (e.g., Australia and the US east coast; chapters by Roughan et al., 2018; Wilkin et al., 2018). That said, the GOV intercomparison framework has also generated a great deal of interest among regional operational centres worldwide that are working to enhance certain comparison partnerships between global OOFMS and their regional system (e.g., the China Sea; Zhu et al., 2016).

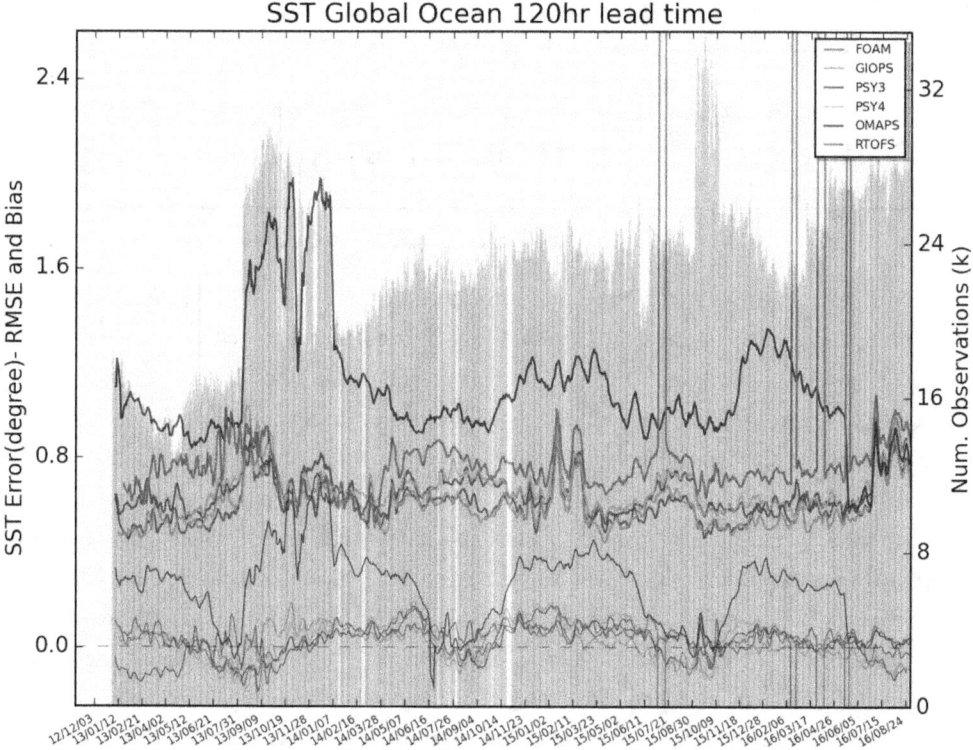

Figure 29.2. Time series of SST Class 4 global statistics from the 2013-2016 period. Six operational systems' five-day forecast real-time evaluation comparing SST to drifters. FOAM (¼°, UK-Met), GIOPS (¼° coupled, ECCC, Canada), PSY3 & PSY4 (1/4° & 1/12° Mercator Océan, France), OMAPS (1/4°-1/10°, Bureau of Met. Australia), RTOFS (1/12°, NOAA/NCEP USA). Statistics of differences between model forecasts and drifter measured temperature: global root mean square differences (thick lines) and global biases (or mean differences, thin lines). Shaded gray bars corresponds to the number of drifter data. This synthetic assessment in the frame of GOV Intercomparison and Validation Task Team was computed by the Department of Fisheries and Ocean and the Environment and Climate Change Canada.

The GOV Intercomparison and Validation Task Team Class 4 intercomparison involves monitoring the overall performance of the participating global system. In recent years, an annual review was presented to GOV scientists, addressing the strengths and weaknesses of these systems and revealing problems with the evaluation approach. Fig. 29.2 shows that most of the systems involved (with the exception of the Australian system, OMAPS) offer similar skill scores meaning resolution was a strong penalty compared to SST from drifters. Notably, the OMAPS system was upgraded in 2016 and Class 4 monitoring shows that it now offers the same overall quality as the others. Also noteworthy are peaks associated with the RTOFS time series: here, the computation of Class 4 differences was performed without quality control of the drifter SST data and outliers were not removed, thus greatly increases the measure of misfit. Based on this, the evaluation procedure at NOAA/NCEP has been upgraded as well. This example demonstrates another key objective of the GOV intercomparison tasks: strengthening and maturing the operational oceanography community toward organized, standardized, and shared practices. Later on, this evaluation framework could also be adopted by international organizations such as the WMO and the Intergovernmental Oceanographic Commission's Joint Technical Commission for Oceanography and Marine Meteorology (e.g., Schiller et al., 2016).

Overview of Existing Evaluation Approaches in Operational Oceanography with the CMEMS Product Quality Policy

General evaluation concepts in CMEMS operational centres

The CMEMS aims to provide regional and global products of the "blue" (physical variables), "green" (biological/biogeochemical low trophic level variables), and "white" (sea ice variables) ocean (Drévillon et al., 2018, this issue, as part of the GOV summer school contribution). The product scales cover from global mesoscale (20-50 km) to regional submesoscale (2-10 km) in the European seas, at frequencies of hourly to monthly. The CMEMS delivers real-time: hindcasts, short-term forecasts (three to 15 days). The service also delivers three-dimensional ocean temperature, salinity, currents, chlorophyll, nutrient content, dissolved oxygen, as well as other parameters at the surface such as sea level, waves, and sea ice variables. And, in delayed mode, the CMEMS provides ocean reanalyses and reprocessed observed products for the same parameters.

The CMEMS put a tremendous amount of effort into the assessment of the OOFMS' performance and skill as well as the evaluation of the product's accuracy (Hernandez and Melet, 2016; Le Traon et al., 2017). The CMEMS evaluation approach and the tasks performed can be split in several distinct categories:

Calibration of ocean models and estimation tools. This task is carried out when models or estimation tools are revisited, and when their algorithms need to be adjusted. Most often, comparison to the "ocean truth" based on observations or reference data is performed. Also very often, these situations (in time and space) are chosen in the most favorable way. For instance, a model will be compared to a comprehensive dataset from various oceanographic campaigns, where a large amount data has been recorded. This permits testing of new algorithms under different sea conditions and accommodates a large range of ocean process behaviors of interest.

Pre-operational qualification of the OOFMS. This task is performed in the CMEMS when the existing OOFMS is going to be replaced by a new and improved one. In such cases, the new system is tested over a given period in pre-operational conditions and compared to the existing system in order to measure improvements and potential benefits of the upgrade. Observations are most often used to provide an "ocean truth," where the existing and new OOFMS are compared. Of course, "non-regression" is an important criteria and it is expected that a new OOFMS will beat the performance of an existing one. Typically, this qualification is completed over one year or more of simulations in order to test the OOFMS for various seasons and to take into account the ocean variability. This type of evaluation is "internal."

Routine validation of OOFMS. This task is carried out in real-time or near real-time. The goal is to monitor performance of the system on a daily basis in order to alert operational teams to major mismatches of the system against the "ocean truth." Every observation available in real-time conditions is usually taken into account, and most of the time the same observation is used by the assimilation procedure. When observations are not available, reference information such as climatologies can be used. Additionally, dedicated metrics can be applied to provide a specific user assessment in order to characterize the reliability of ocean products against "ocean truth" or the fit-for-purpose when considering specific applications (see earlier discussion on external metrics).

Off-line validation of reprocessed products and ocean reanalyses. For all real-time products, the CMEMS associates provision of reprocessed, observation-based products or reanalyses. The objective of this task is to offer an accurate description of past ocean conditions by providing reprocessed information rather than an accumulation of real time hindcasts. Dedicated validation tasks are performed with two main goals in mind: 1) characterizing the accuracy level of these products in order to communicate their reliability, and 2) measuring their quality as compared to real-time products. For comparison to the "ocean truth" off-line validation usually benefits from using a reprocessed, more accurate and more comprehensive dataset. Many types of observations not available in real-time, and specifically quality-controlled and calibrated, offer a more complete description of the three-dimensional ocean variability for many physical and biogeochemical variables such as datasets of vessel-mounted acoustic Doppler current profilers, tide gauge data not transmitted in real-time, CTDs from sea experiments, fluorescence measurements, etc. Moreover, as discussed later in this chapter, several products from different origins can provide past period estimations for the same ocean variables. In such cases, through intercomparison and by taking into account their relative strengths and weaknesses, one can infer their relative accuracy or, at least, propose an overall accuracy level.

Quality control of upstream information used by the OOFMS. Both for real-time forecasts and for production of reanalyses, errors caused by external information used to run the systems are considered, tracked, and sometimes reduced with adapted corrections. Several types of information are currently subject to dedicated quality control: forcing fields, observations used in the assimilation, and boundary conditions from other systems.

Internal control of coupled component of the OOFMS. Coupled systems are used in the most recent versions of the CMEMS, initially for the biogeochemical forecast and coupled with the physical ocean model, but also coupling atmospheric, waves and ocean models; or coupling optical and biogeochemical models. The newest approach consists of monitoring the exchanged variables at the interface or some key parameters (e.g., the mixed-layer depth variations, whose errors may exaggerate the vertical fluxes of nutrients and impact the quality of the primary production by the biogeochemical model).

For all categories, assessment is always sought and expected in terms of "consistency," but the majority of efforts are dedicated to evaluating "accuracy" and "performance" (see Murphy, 1993 and Hernandez et al. 2015 for details on these evaluation criteria). The "fit-for-purpose" criterion is associated with "external" metrics evaluation and aims, through routine and off-line validation tasks, to provide a more user-oriented assessment of the CMEMS products.

Whenever possible, the CMEMS metrics are based on comparisons to observations and upstream information is sought internally during production through Thematic Assembly Centre deliveries. This guarantees that 1) all observations used are quality-controlled and sometimes reprocessed (for off-line validation); 2) producers of observation datasets are known internally by all CMEMS; and 3) in cases where spurious observations are detected when used for validation, a blacklisting procedure can inform Thematic Assembly Centre experts and trigger a corrective task.

These observations may be used through assimilation by the CMEMS OOFMS. However, the CMEMS product quality strategy does not rely on assimilation statistics (i.e., statistics on misfits and increments) for evaluating accuracy and performance. Even if observations are used through assimilation, dedicated Class 2 and Class 4 metrics comparing model values and observations, are carried out (see Hernandez et al., 2015 for an introduction to these metrics). In such cases, the evaluation is not fully independent. However, these approaches limit the filtering effects of the observation operator used by the assimilation scheme that take into account the model's representativity and compute misfits.

Providing quality information to CMEMS users

The CMEMS product quality strategy is primarily dedicated to informing users about the reliability of the products available for download. A series of difficulties have been identified in this communication effort:

- Most calibration, qualification, off-line, or routine validation tasks are based on "internal" metrics that characterize the accuracy and performance of the OOFMS with respect to its representativity. Following the example above: a sea level assessment will filter out tidal signals in the observation dataset if the operating system does not contain tidal dynamics. Consequently, it is important that users be informed of that limitation.
- As reviewed in Hernandez et al. (2015), sometimes there is a failure of a model to accurately represent a specific ocean process. However, sometimes this is a failure of the evaluation approach, in that it is not able to assess the effective skill of the OOFMS because it is not using the appropriate metrics. This issue may arise when the "geography" of an assessed area contains various scales and processes (e.g., open ocean and coastal zones), with a metrics design to characterize some of these scales and processes. For instance, the use of global SST-mapped products is not adequate to identify coastal fronts when assessing both the open Celtic Sea area and the Bay of Biscay in some of the regional CMEMS production centres.
- One aspect of the above issue is the "traditional" use of metrics based on Gaussian statistics of model-to-observation comparisons. That is, mean differences (also called biases) and

root mean square differences (also called root-mean-square errors) usually tend to hide the real nature of the discrepancies. For instance, a model field may contain well-shaped fronts and eddies that are not properly phased in time, which may create very large errors. Users would need to know that the forecast provides appropriate scales and features, but at the wrong time. In a case such as this, a metrics that provides some uncertainty on the phase lag of these predicted features would be valuable. Conversely, low bias or root-mean-square difference statistics over large areas may not reflect some high and localized errors, unidentified due to their relative weight in the overall metric's computation.

- Alternatively, "tradition" can also be a drawback to user expectations. For example, users expecting an evaluation based on comparison to observations consider it a paradigm to evidence departure from the "ocean truth." But given the scarcity of ocean observations, evaluation sometimes must rely on other approaches such as ensemble assessment. In such cases, there is no "ocean truth" but rather the comparison of several estimates of a given ocean process with the assumption that some of the estimate's errors are not correlated. Therefore, a probability level of accuracy is provided, but users need to be properly informed on how to manage it.

Over time, the CMEMS has adopted various methods for communicating product quality to users. First, it provides a quality information document (QuID for every product in the CMEMS' product catalogue (http://marine.copernicus.eu/services-portfolio/access-to-products/). The QuID offers a comprehensive description of the OOFMS itself, the way the system's performance and the product's accuracy are assessed, and it discusses results of this validation. In order to offer a quick understanding to non-expert users, QuIDs include an executive summary that provides tables of estimated accuracy numbers. The goal of these estimations is to communicate some overall error level for a given product. The QuIDs are produced each time the OOFMS is upgraded or when the quality or reliability of the associated products is changed (typically every one to two years).

For the most recent accuracy numbers, users can visit the CMEMS website (http://marine.copernicus.eu/services-portfolio/scientific-quality/). There, using Class 4 metrics, every production centre provides model-to-observation statistics for their region of interest on a quarterly basis. Users can learn about recent changes, time series, quality improvements over time, and successive upgrades of the operational systems (going back to 2013, when this service was established). However, this monitoring is standardized for all types of products across all areas, so it is complemented by specific product quality information available from the different CMEMS production centres.

Soon, links to regional websites will be added in order that will connect site visitors to related regional overviews and expertise. These websites are designed to emphasize local aspects of the CMEMS product accuracy or highlight specific aspects of the quality, reliability, or added value of the products in light of a particular downstream user's activities. For instance, the Baltic Operational Oceanography System website (www.boos.org) provides a comprehensive overview of operational oceanography tools around the Baltic Sea (e.g., the community and tools, real-time observation information, forecasts, etc.). In particular, it provides a daily multi-model assessment that informs the Baltic stakeholders about the reliability of Baltic ocean products with a real-time warning system. A similar initiative has been started by the Baltic Monitoring and Forecasting Centre (MFC) for the North Sea (http://noos.eurogoos.eu/model-results/), with plans to build a dedicated regional quality monitoring website. Forecasting centre regional verification websites have also been developed for the Arctic (http://cmems.met.no/ARC-MFC/V2Validation/index.html) and the Mediterranean Sea (http://medforecast.bo.ingv.it/mfs-copernicus-evaluation/ [illustrated in Fig. 29.3] and http://medeaf.inogs.it/nrt-validation). At the same time, the CMEMS Thematic Assembly Centres are developing ways to provide users with online information about their observed products, e.g., for sea level (https://duacs.cls.fr) or for ocean colour (http://octac.acri.fr). Plans call for these websites to be integrated into the CMEMS framework.

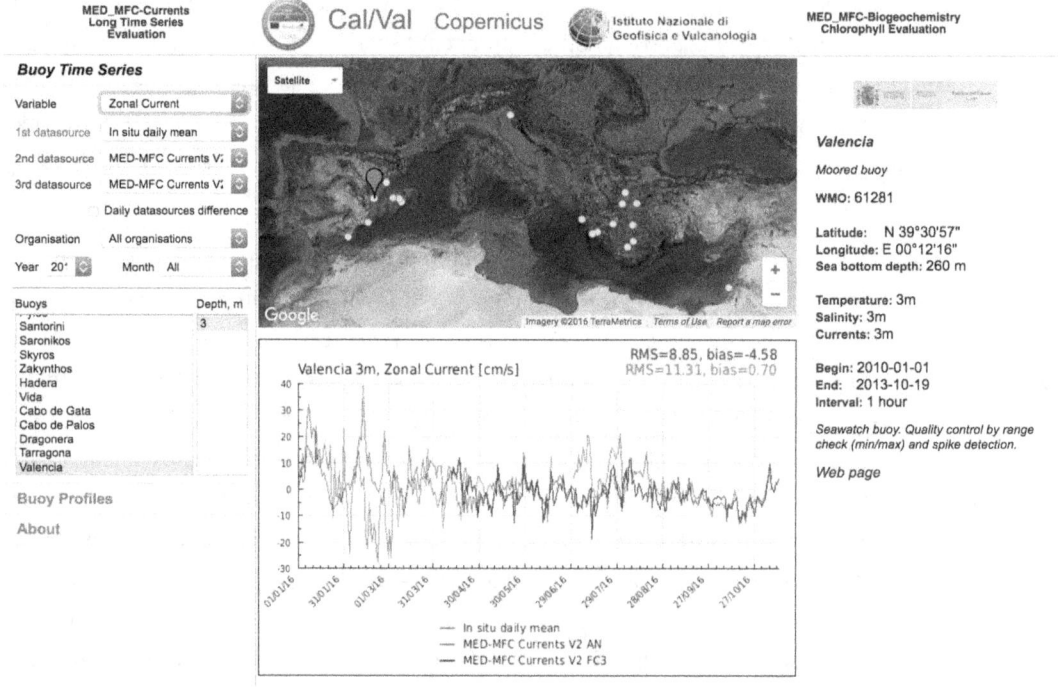

Figure 29.3. Screen copy of the monitoring website developed for the Mediterranean Sea MFC assessment. Against fixed moored platform observations, the time series of measured zonal currents are compared to analysis and three-day forecast. Courtesy of G. Coppini, CMCC and N. Pinardi, INGV, Italy.

On an annual basis, the CMEMS has started to provide regular description of the ocean climate changes and variability dedicated to the scientific community, the decision makers or the general public. As shown with the first initiative: the annual Ocean State Report 2015 (von Schuckmann et al., 2016), the accuracy levels of the CMEMS product used are detailed. For this ocean climate assessment annual reporting it was decided to: 1) use only verified and quality-controlled CMEMS products for inferring and discussing the ocean state; 2) provide a level of confidence, whenever possible (i.e., error bars for every ocean indicator illustrated in the report); and 3) move toward reliance on an ensemble assessment in order to be more confident of error levels.

Continuing to improve the way that product quality information is shared with users is critical. While users are certainly interested in product quality and reliability, on a dedicated user forum set up by the CMEMS they appear to prioritize timeliness and continuous delivery as their most immediate concerns (D. Obaton, CMEMS Service Manager, personal communication, 2017). This makes clear how important it is that users understand and be aware of any potential areas for product anomalies.

Estimated accuracy numbers, while informative, should be considered by users as the most basic level of product quality information They speak to the overall quality of a product based on first and second order Gaussian statistics. However, these statistics do not allow for characterizing specific anomalous behaviours of the OOFMS in specific areas or events. For example, estimated accuracy numbers for satellite altimetry reflect aggregate information provided by satellites (Table 29.1) deduced from cross-over difference statistics but they should be complemented with a specific error analysis for along-track noise or large wavelength errors (e.g., Le Traon, 2013) as illustrated in Fig. 29.4. For model products, estimated accuracy numbers are typically obtained through comparisons to observations or climatology over one-year periods or longer (as illustrated for the North West Shelf Regional MFC in Table 29.2). However, nothing in the estimated accuracy numbers would alert a user to potential product error(s). In Table 29.2, values are given with two-digit precision, which is not necessary for an overall estimation of errors; however, this level of accuracy does allow experts to identify further improvements when these numbers are compared in time over successive versions of the operational systems.

More process-oriented metrics are now being tested to increase the value of estimated accuracy numbers. The CMEMS Arctic MFC is pioneering a new approach for sea ice extent, sea ice concentration, and sea ice type assessment contingency table metrics that characterizes the number of good forecasts or occurrences compared to observations, and discussed against persistence score, as it is also done worldwide by other operational centres (e.g., in Canada, see Smith et al., 2016).

Altimeter	NRT errors (cm rms)	DT errors (cm rms)
OSTM/Jason-2	< 4	< 3
AltiKa	< 4	<3
Cryosat2	< 6	<5
HY-2A	< 6	< 5

Table 29.1. Estimated accuracy numbers provided for satellite altimetry in the CMEMS Sea Level QuID, for Near Real Time (left) and Delayed Time (right) products.

In weather forecasting, verification tools are used to characterize the occurrence of specific events such as rain with a confidence level. The WMO Joint Working Group on Forecast Verification Research puts forth diagnostic metrics, but many of these meteorological assessment methods are based on ensemble forecasts (Ebert et al., 2013). Currently, the CMEMS products considered to be "core products" of the marine environment may not reach the scales necessary to described many of the synoptic events of interest, in particular extreme events that may develop on short timescales near the coast.

Variable	Location	Supporting observations	RMS error	Mean error
M2 tidal harmonic (amplitude)	Whole region	Tide gauge data	12 cm	0.59 cm
M2 tidal harmonic (phase)	Whole region	Tide gauge data	12.3°	1.9°
SST	Full domain	In situ observations	0.51°C	-0.025°C
	Continental shelf	In situ observations	0.52°C	-0.032°C
T profiles	Full domain	In situ observations	0.65°C	-0.076°C
S profiles	Full domain	In situ observations	0.15	-0.0
Bottom temperature	Continental shelf	Climatology	1.5°C	0.95°C
Mixed layer depth	Full domain	In situ observations	121.77 m	21.7m

Table 29.2. Estimated accuracy numbers available for the CMEMS North West Shelf Regional MFC, for different variables, from QuID.

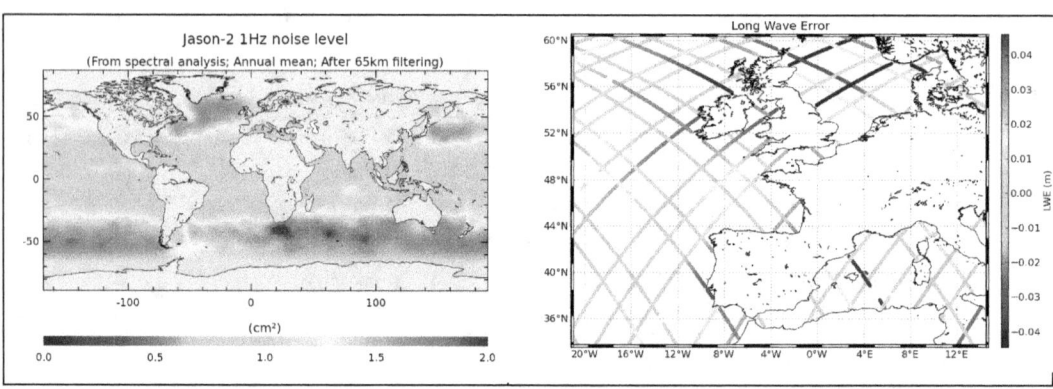

Figure 29.4. At left: Jason-2 along-track sea level anomaly noise as provided by the CMEMS Sea Level Thematic Assembly Centre (I. Pujol, CLS, as part of CMEMS, 2017). At right: Example of along-track sea level anomaly long wavelength errors for Jason-1 (C. Dufau, CLS, as part of CMEMS, 2017).

Finally, when it comes to informing users about product reliability, the CMEMS does this mainly through static documents such as the QuIDs. But ideally, these should be complemented with near real-time information published online. That said, dedicated user-oriented applications and downstream services that use the CMEMS products and provide services on websites, tablets, and smartphones are now available (e.g., www.sea-condition.com; personal communication, Coppini, 2017). Users must be informed of the potential errors and reliability of these services based on the CMEMS product quality information.

Identified path of improvements in the CMEMS product quality assessment framework

Within the CMEMS framework, areas for improvement have been identified and described by Hernandez and Melet (2016).

Reference information used for evaluation.

Due to some national, institutional or private sector data policy, ocean observations are not freely available to all operational centres or the oceanographic community. Some production centres may access and use them for validation while others do not. Sharing data through CMEMS Thematic Assembly Centres would ensure free access to all (particularly to historical datasets), if data owner give access to them. If not, and in cases difficulties arise, the growing influence of the CMEMS at the European level may help to push owners to share them.

Ocean observations used in the CMEMS validation framework are quality controlled. Furthermore, the implementation of black and grey listing internal mechanisms through assimilation procedures from MFCs toward Thematic Assembly Centres is also a way to evidence observation anomalies. Both procedures allow Thematic Assembly Centres to better manage outliers or instrumental problems.

Issues related to fixed measurement data also need to be addressed. Historical data from moorings, tide gauges, other platforms and instruments usually located near the coast is often difficult to access. Many countries and groups do not submit their historical or real-time observations to a common database or a Global Data Assembly Centre. When subjected to the standardization of a Global Data Assembly Centre, fixed instrument measurements can be quality-controlled, which allows for multi-model assessments. Such is the case with the BOOS framework in the Baltic Sea, which makes it possible to compare forecasts from the Danish, Swedish, Finish, and Norwegian regional operational systems. A global system could be handled similarly; this would allow us to measure the benefit and added value of these regional 1-3 km resolution systems in contrast to the CMEMS global 1/12° eddy-resolving system, forced by global European Centre for Medium-Range Weather Forecasts atmospheric fluxes and not considering the tides. Fig. 29.3 illustrates this type of OOFMS monitoring based on fixed platforms. Zonal current time series are compared to a three-day forecast and hindcast from the Mediterranean Sea MFC. The metrics used are biases and root-mean-square differences. Other CMEMS velocity products could also be compared and their skill characterized in a Taylor diagram (Taylor, 2001), with other metrics evidencing specific weaknesses of the OOFMS (e.g., rotary spectrum, threshold and contingency table metrics, etc.). Because many of these platforms are located near the coast where many applications can be anticipated, they could support the design of user-oriented external metrics.

Strengthen the validation and verification activities on most-used variables.

Surface variables (such SST, mixed layer depth, surface current, sea level, chlorophyll) are among the most used because they have numerous applications. Therefore, it makes sense to focus the quality assessment of these surface variables on high-frequency variability and assessment of the diurnal cycle. Until recently, the CMEMS typically provided product delivery on a daily basis whereas today, it provides hourly products for surface parameters.

The CMEMS is a multidata framework. This means that for a parameter of interest (e.g., SST), several products are available in a given area: from the regional and global MFCs (for model products), to Thematic Assembly Centres (for observed products), as well as, of course, from real-time or delayed-mode reanalysis/reprocessing. Hence, comparison between these different estimates of a given parameter will allow for a better characterization of the accuracy and weaknesses of the products and ultimately evaluation of the reliability of a given product for a given application. The CMEMS multi-data framework also encourages the sharing of best practices. For instance, the SST satellite community may share their metrics with MFCs to the benefit of modellers, who sometimes lack the expertise that SST producers have when considering SST assessment.

Better monitoring of the information used in the OOFMS.

Data for assimilation should be systematically quality-controlled against predefined criteria or using a departure from the model forecast. If rejected, grey or black listing has to be considered, taking into account model forecast errors and representativity. External forcing functions (atmospheric, bathymetry, river run-off....) also need to be monitored and possibly corrected, in cases of anomalies.

To improve nesting OOFMS, boundary conditions should be monitored. Errors from the large-scale model need to be identified in order to infer their impact in the nested system. Moreover, inconsistencies between large-scale and regional-scale dynamics will first appear at the boundary, and also need to be characterized.

Furthermore, coupling strategies are becoming widespread in operational centres and seamless atmosphere-wave-ocean modelling approaches are being developed. Likewise, all CMEMS MFCs now offer biogeochemical modelling, coupling the physical components on a daily basis. But biogeochemical models need dedicated assessments based on biogeochemical observations, which presents a challenge because of the sparseness of the measurements and the fact most variables in the biogeochemical models are not observed. Also, the performance of a biogeochemical model can be modelled with consideration for the accuracy of the physical forcing model. Some physical variables, such as vertical fluxes, have a large influence on the biogeochemical model. Therefore, a number of key physical parameters have been identified for dedicated monitoring including: bottom temperature, stratification and length of the stratification period, mixed layer depth variability, vertical velocities and diffusivity, euphotic layer depth, solar radiation and its penetration into the sea water, and wind stress. Dedicated monitoring of these parameters would allow for better characterizations of the behaviour of a biogeochemical model in direct response to the physical forcing, which has the potential to be erroneous.

It is also important to remember that the CMEMS is a distributed network of centres and experts who sometimes collaborate on validation/verification development in other frameworks. Therefore, when innovations are transferred to the CMEMS processing chains, it directly benefits the overall product quality of the CMEMS organisation. For example, the CMEMS global systems are benefitting from developments carried on in the GOV intercomparison framework. A similar benefit is seen when the CMEMS products in the Arctic are able to be part of the evaluations performed in the frame of the Year of Polar Prediction project (YOPP) and in the CMEMS ocean reanalyses being part of the international intercomparison framework lead by the CLIVAR/GSOP community.

Novel Evaluation Approaches in Operational Oceanography

Benefiting from existing observations

As discussed above, observations need to be considered first with regard to what part of the "ocean truth" they represent, taking into account the type of sampling they offer (their observation representativity). Then, their reliability (i.e., accuracy and precision) at different scales of time and space should be considered.

Table 29.3 summarizes the use of existing observation datasets in operational oceanography for validation/verification purposes. The first and second columns list the instruments and measured parameters, the third column details the characteristics of each type of measurement, and the fourth column indicates the type of assessment that is performed by each instrument or measured parameter.

When validating OOFMS, it is important to remember that most raw measurements require dedicated expertise and permanent quality control monitoring in order to be used for validation purposes. Most of these measurements depend on a lot of additional information in order to be derived into the final observations. This is particularly true for satellite measurements, which need to account for the satellite platform behaviour (e.g., orbit, rolling), noise of the instrument, modification of the measured information through the atmosphere and at the surface, and other ancillary information.

Even though real-time data quality is not fully satisfactory given there is no time to obtain the complementary information needed to correct raw data, provision of measurements in real-time is a key element in operational oceanography. For most users, observations collected in real-time provide some evidence of the reliability of estimates and forecasts. Of course, corrected datasets are more precise, thus allowing us to calibrate operational systems, evaluate ocean reanalyses in greater detail, and evaluate the performance of the operational system in delayed-mode. But all measurements cannot be collected in real-time, which means that real-time assessment suffers from less accurate reference data and consists of only some of the observations that describe the ocean processes.

Note also that scales and representativity of observations need always be considered before OOFMS validation. As mentioned earlier in this chapter, editing, filtering, or averaging might be necessary to compare model values and measurements at the same time and space scales. However, for external metrics, the full measurements are preferred. Observations might also be rejected. For instance, coastal data that describe very local processes that cannot be filtered out before comparisons with larger-scale model.

Observation type	Parameters	Measurement characteristics	Use for OOFMS evaluation
CTD + additional sensor on rosette (ADCP…)	• T/S • Additional parameters (U/V)	• Vertical profile • Non frequent • Not systematic real time transmission • High resolution, high quality (high precision instrument, then quality control on profiles and possible corrected values)	• Real time (RT) –sometimes– validation of water masses and stratification • Delayed mode (DM) precise assessment of water masses and stratification • RT/DM validation of additional parameter • Unless for a dense section, where synoptic scales can be evaluated, used for large scale assessment
Water samples from experiments	• T/S • Chemical properties of sea water • Biogeochemical properties of sea water • Biological analysis	• Non frequent, sparse • Time and space sampling depending on experiments • Mostly processes off-line in labs and available with substantial delay • Top quality measures	• DM Validation or dedicated calibration of physical or biogeochemical parameters • Unless for a dense section, where synoptic scales can be evaluated, used for large scale assessment
XBT / XCTD	• T • T/S	• Vertical profile • Repeat sections / sea experiments • Mostly real time transmission • Low quality temperature and salinity profiles	• RT/DM validation of temperature/MLD/thermocline • RT/DM validation of water masses and stratification • For frequent and dense section, used for synoptic scales assessment
Argo profiler	• T/S	• Vertical profile • 5-10 day cycling • real time transmission	• RT/DM validation of water masses and stratification deep to 2000m • Global • Depending on density, can be used for synoptic or large scales assessment
BGC Argo profiler	• T/S • NO_3, O_2, Chl-a (fluorescence), downward irradiances and derived optical properties	• Vertical profile • 5-10 day cycling real time transmission • Bio sensors quality still under improvements	• RT/DM validation of low trophic levels / Dissolved oxygen • At present, used for specific assessment

Observation type	Parameters	Measurement characteristics	Use for OOFMS evaluation
Gliders	• T/S • Additional parameters	• High frequency vertical profiles • Real time transmission	• RT/DM validation of water masses and stratification deep to 1000m • RT/DM validation using additional sensors • At specific locations of interest • Used for synoptic assessment
On board TSG or FerryBox	• T/S • Fluorescence • Turbidity • pH • Oxygen • Phyto/zoo	• Along the route of ship (merchant or oceanographic vessels) • Need careful calibration and processing using water samples, in particular for S and biogeochemical measurements	• DM (sometimes RT) validation of surface T/S properties • DM validation of biogeochemical properties • Used both for synoptic and large scale assessment
Miscellanous opportunistic T sensors (Recopesca, net sensor)	• T/S • Turbidity	• Fishermen nets : follow their route (mostly tested over continental shelves) • Low quality T/S sensors attached to the net • Real time transmission	• Need dedicated QC • DM validation of T/S, dense measurements • Tested on coastal waters
Sensors on sea mammals	• T/S • Additional parameters	• Depending on sea mammal, vertical profiles at various depth and frequencies • Instruments may be biased • Transmission not always real time	• Need dedicated QC • RT/DM validation of water masses and additional sensors • Useful to cover high latitudes and near sea-ice areas. • Used like Argo profilers
Ice tethered - profiler	• Ice temperature • Near-surface water temperature	• Over the sea-ice: may move with the ice, or in water if melting • Very few profilers • Profile down to 800m • Real time transmission	• DM/RT validation of water mass below the ice
Drifters	• Trajectories • Temperature • Air pressure • Salinity (specific) • Wind (specific) • Rain (specific) • Wave (specific)	• Real time transmission • Global array • Different type of buoys, different depth of drogues • Post-processing of trajectories to infer U/V (drogue loss estimation)	• RT/DM validation of U/V at given depth, Eulerian and Lagrangian assessment • RT/DM validation of T • RT/DM validation of using specific sensors • Unless dense coverage, used for large scale assessment
VM-ADCP	• U/V	• Ship route measurement below the hull • Not real time • Need post-processing	• DM validation of U/V for surface layers

Observation type	Parameters	Measurement characteristics	Use for OOFMS evaluation
Fixed/moored ADCP	• U/V	• Fixed location measurement • Often not real time • High frequency measures stored on memory, often transmission of time-averages • Need post-processing	• DM validation of U/V for specific layer • Local measurement
Tide gauges	• Sea level	• Fixed location measurement • On the coast (most of the time) • Real-time • High frequency measures stored on memory, often transmission of time-averages • Need post-processing for exact positioning (e.g. land vertical displacement)	• RT/DM sea level validation • Local measurement, but used for large scale assessment of sea level
Bottom pressure gauges	• Bottom pressure	• Fixed location measurement on sea floor • Often not real time (unless cable transmission) • High frequency measures stored on memory • Need post-processing	• Bottom pressure or sea level assessment • DM validation
Wave buoys	• SWH • Dominant part of the wave spectrum (peak, period and direction)	• Fixed location measurement • Real-time • High frequency measures stored on memory, often transmission of time-averages	• RT/DM validation of wave parameters
High frequency (HF) radar	• Waves • U/V	• Fixed location measurement • Real-time • High frequency measures • Coastal measurements from few km to few hundred km • Need post-processing • Near-real time	• U/V DM validation on very specific locations along the coast

Observation type	Parameters	Measurement characteristics	Use for OOFMS evaluation
Satellite nadir radar altimeter & laser altimeter (ICEsat)	• Sea level • SWH • Wind intensity • Ice thickness	• Along-track measurements (e.g., Earth revolution in 100 minutes) • Global coverage • Different repeat or non-repeat satellites and track distances • Different high latitude coverage • Real-time transmission • Specific post processing • Sea surface height scale resolved: 50 km at best	• RT/DM validation of sea level, SWH and wind • DM validation of ice thickness
Satellite IR radiometer	• Surface brightness temperature	• Along swath (variable length) or geostationary measurements • Different orbits • Global coverage • Real time transmission • Specific post processing • Cloudiness problem • Transform in skin, bulk or foundation SST values	• RT/DM SST validation • Used for meso- and large-scale assessment
Satellite microwave radiometer & spectrometer	• Surface brightness temperature • Surface salinity • Surface roughness • Sea Ice Thickness (thin)	• Along swath (variable length) measurements • Less precise than IR radiometers for SST • Different orbits • Global coverage • Real time transmission • Specific post processing • Transform in skin, bulk or foundation SST values • Large scale SSS • Thin Ice thickness	• RT/DM SST validation • DM SSS validation (large scale) • DM validation for ice thickness
Satellite scatterometer	• Surface roughness (wind)	• Along swath (variable length) measurements • Different orbits • Global coverage • Real time transmission • Specific post processing	• RT/DM validation of wind
Satellite Synthetic Aperture Radar (SAR)/InSAR	• Sea Ice concentration • Wave spectrum • U/V	• Along swath (variable length) measurements • Different orbits • Global coverage • Real time transmission • Specific post processing	• RT validation of sea ice • DM validation of sea-ice • DM validation for U/V

Observation type	Parameters	Measurement characteristics	Use for OOFMS evaluation
Satellite Imager	• Reflectances (ocean colour) for different visible and near-IR spectral bands	• Along swath (variable length) measurements • Different orbits • Global coverage • Real time transmission • Specific post processing for Chl content • Strong dependence on type of waters and algorithms • Depending on euphotic depth and turbidity, different colours at different layers	• RT/DM validation of reflectances, and chlorophyl content • Used for meso- and large-scale assessment

Table 29.3. List of observations, the parameters measured, and uses for OOFMS validation and verification.

Since the 1970s, satellite instruments and constellations have been constantly improving the observability of increasingly more parameters from space. That said, there has been the strong limitation of measuring only the ocean surface. Although some measurements, such as sea surface height from altimetry or earth potential/geoid from gravimetry, offer integrated estimates of the full water column.

The in situ observing system offers the full three-dimensional description of ocean parameters, although data is more sparse. Nevertheless, the Argo program provided a noticeable improvement of the observability of water masses. The Biogeochemical (BGC)-Argo initiative is expected to make equally significant contributions to the in situ system and to bring new, potentially revolutionary insight by allowing for a synoptic assessment of some key parameters of the biogeochemical processes. The other "new player" will be the coastal high-frequency (HF) radar network. Effort to allow a continuous monitoring of coastal waters is difficult but ongoing, and in the future we might expect HF radar networks in some areas (the US/Canada east and west coasts, European coasts, Australia, China Sea) that will provide surface velocity and sea-state measurement to monitor the reliability of coastal operational systems.

Along-track satellite altimetry: finer scales to be observed

Since the launch of Geosat in 1985 (the earlier SeaSat mission was more a pioneering effort), the sea level and surface geostrophic flow assessment is traditionally performed using along-track satellite radar altimeters with the so-called "conventional" nadir low resolution mode. The recent availability of Sentinel-3A (S-3A) data using an along-track Synthetic Aperture Radar (SAR) mode that reduces noise level of along track sea surface height retrievals (see details in the chapter by Morrow et al., 2018) changes the sea level spatial scales that can be observed and compared to model products. Hence, instrumental noise of the 1 Hz "conventional" data since Geosat needed to be filtered out and was giving access to wavelength larger than 70-80 km. Due to changes of frequency and a finer footprint, the AltiKA Saral altimeter provides access to 40-50 km closer to the coast, and the S3 SAR mode allows us to reach the 30-50 km along-track resolution (see also the Mesoscale Capability Determination estimation by Dufau et al., 2016).

Satellite altimetry repetitivity and time sampling have always been the main limitations for describing the ocean two-dimensional turbulence. However, the constellation of satellites has increased and the combination of several satellite passages allows for better two-dimensional descriptions through mapping techniques. However, two-dimensional reconstructed maps of sea level cannot be considered as reference observations for exact validation. Mapping techniques offer gridded fields, such as ocean models, but also present weaknesses and erroneous extrapolation features, even if new optimal interpolation dynamical techniques are now proposed (Rogé et al., 2017; Ubelmann et al., 2015). This is the reason why the Class 4 approach (see GOV intercomparison above) is based on along-track comparison that provides the evaluation only on one direction, but that allows assessing in a rigorous way specific aspects of the mesoscale dynamics (fronts, size of meanders and eddies, strength of currents etc...).

Recently, in order to assess dynamical turbulence in model fields, comparisons between along-track satellite altimetry spectrum of sea level and similar quantities from the model have been conducted. For a given area, sea surface height wavenumber spectra can be computed and spectral slopes can be determined and mapped globally (Xu and Fu, 2012) or by season (Dufau et al., 2016). The power law of the spectrum can be compared to the turbulence theory, as well as compared between model and satellite data (see the chapter by Morrow et al., 2018). Fig. 29.5 illustrates an evaluation performed between Jason-1 along-track data and best estimates of the CMEMS 1/12° global forecasting system. In this example, when compared to Fig. 1 of Dufau et al. (2016), the model sea surface height presents sharper slopes in the equatorial band.

The incoming Surface Water and Ocean Topography (SWOT) mission's new technology will offer a new paradigm for sea level and ocean turbulence assessments. The interferometric SAR (InSAR) will measure 70 km on both sides of the satellite course, with a 10 km gap at the nadir that will be partially compensated by measurements from a classical altimeter. The swath resolution will be 1 km, but in practice, SWOT is expected to provide a 15 km-scale two-dimensional resolution. This will allow us to better infer vertical movements associated with quasi-geostrophic turbulence at mesoscale (see chapter by Morrow et al., 2018). It will also allow us to monitor errors on coastal models and forecasts, and small mesoscale behaviors of eddy-resolving operational systems.

Sea ice parameters assessment

Sea ice concentration, thickness, drift, and type are operational products of great interest for navigation and off-shore industry. To a lesser extent, these parameters are also needed to close the budget of the Earth climate system, the water budget, and its monitoring on short-to-medium timescales in terms of seasonal forecasts. Ice modelling also needs to progress (see chapter by Bouillon et al., 2018). Sea ice concentrations in the Arctic and Antarctic basins are typically assessed using satellite microwave measurements from instruments such as the Special Sensor Microwave Imager (SSM/I) or, more recently, the Advanced Microwave Scanning Radiometer 2 (AMSR2). The latter is used to perform Class 4 metrics in the GOV Intercomparison and Validation

Task Team's intercomparison framework. However, in real time, marine navigation users prefer to rely on ice charts deduced by ice centres from direct observations or advanced very-high-resolution radiometer (AVHRR), Moderate Resolution Imaging Spectroradiometer (MODIS), Radarsat, or Sentinel-1 SAR imagery. Users look at ice model based products for prediction of ice edge, extent, and thickness. Let's look at why ice operational centres pay particular attention to the validation of these last parameters.

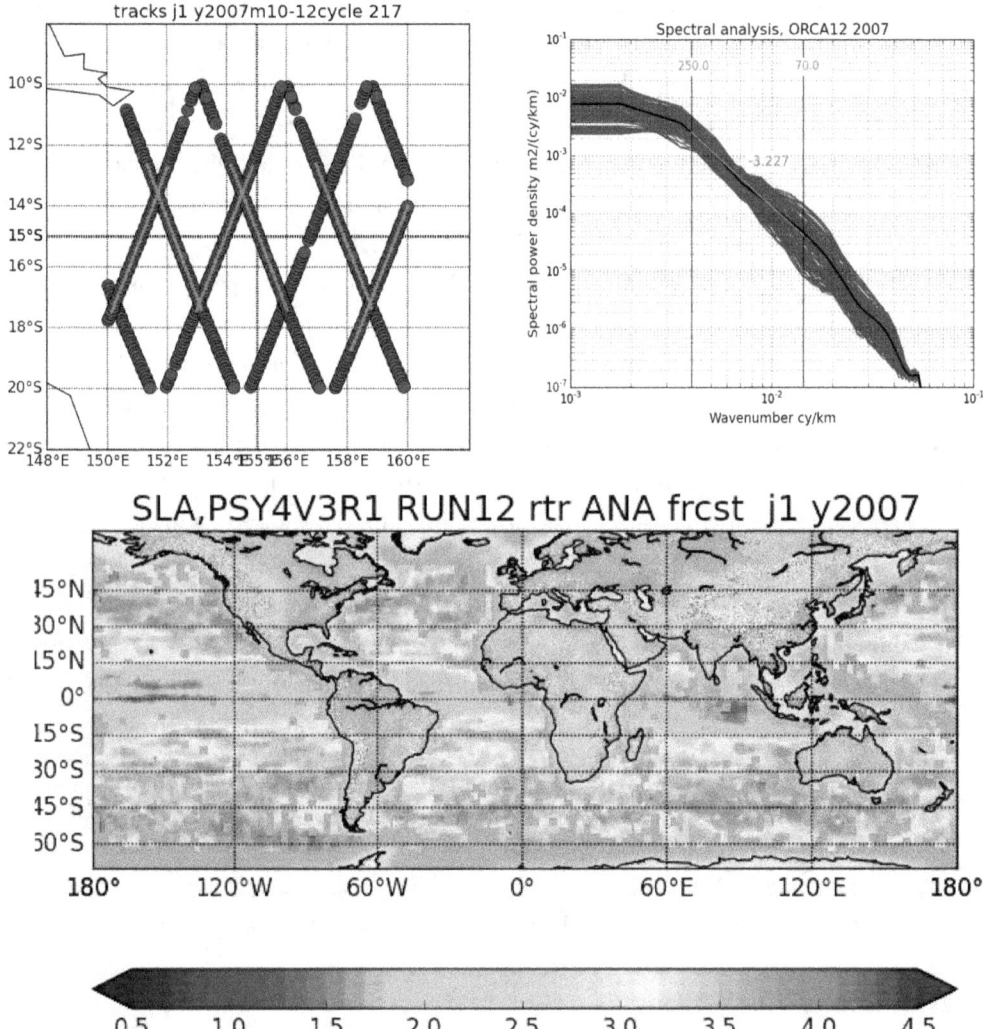

Figure 29.5. Top: Wave number spectra computation over 10°x10° boxes (only tracks longer than 560 km). Top left: example of satellite tracks from Jason-1. Top right: Corresponding spectrum on the along track interpolated SSH fields from the CMEMS 1/12° global model in 2007, with the estimated slope (in red). Bottom: global map of the wave number spectra slope from the CMEMS 1/12° global system.

Hernandez et al. (2015) presented contingency metrics designed to measure skill prediction for correct water or correct ice areas, looking specifically at scores on marginal ice zone. Sea ice edge position is subject to particular assessment due to high user interest in these zones. For instance, the Arctic forecasting system of the CMEMS provides near real-time validation bulletins (Bertino and

Melsom, pers. comm., 2017) in which sea ice edge forecasts are compared to persistence. Moreover, sea ice concentration is assessed using different categories of ice fraction cover and then compared to satellite data (see http://cmems.met.no/ARC-MFC/V2Validation/index.html). Fig. 29.6 compares a Class 4 metrics assessment against AMSR2 data for the global CMEMS operational system. Scores, in terms of mean biases and root-mean-square differences, are plotted for different operational products (these are best estimates that gives the maximum performance of the system). Then, forecast and persistence scores are compared for one to 10 days lead time. In both the Arctic and Antarctic basins, forecast beat persistence in winter, which is not the case in summer (this demonstrates that the system still need to be improved). Next, nominal and former CMEMS global systems are compared, looking at the proportion of correct hits versus the total (PCT, sum of the correct ice and correct water, or CP and CN in the table of Fig. 29.7 below, divided by the total number of samples, i.e., [a+d]/[a+b+c+d]). Improvements are observed in summer. This is due to an enhancement of the full PSY4V3R1 system as well as the implementation of the assimilation of sea ice concentration (see chapter by Lellouche et al., 2018).

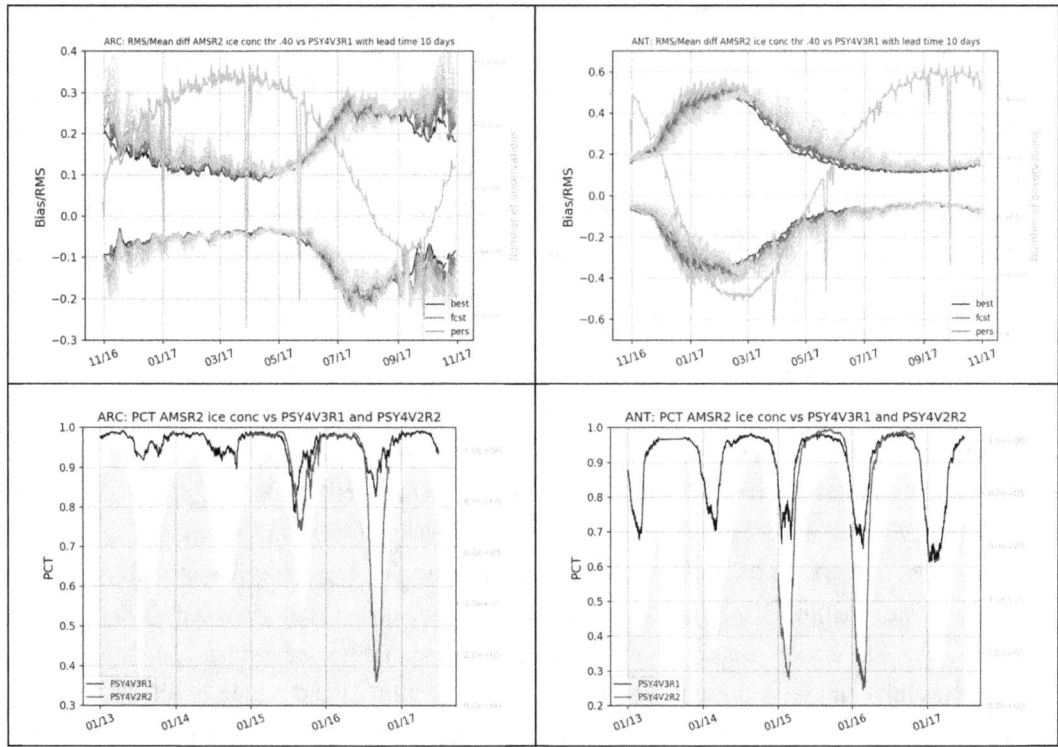

Figure 29.6. Sea ice concentration assessments based on contingency table for the Arctic (left column) and the Antarctic (right column) basins. Top row: Time series of Class 4 metrics for the global operational 1/12° CMEMS system PSY4V3R1 against AMSR2 data. Grey curve: number of non-zero observations. Lower curves: biases. Upper curves: root mean square differences. The black curve for the best estimate. The blue/cyan curves the forecast from 1- to-10 day lead time. The associated persistence score from 1- to 10-day are plotted in red/magenta. Bottom row: time series comparison of PCT (see text) for the nominal CMEMS global system PSY4V3R1 (black) and the former PSY4V2R2 operational system (blue), with a sea ice concentration threshold of 0.4.

New approaches propose to combine ice concentration and ice type or ice thickness information in order to evaluate for which category of ice the operational system is skillful. This is the way Smith et al. (2016) performed their recent evaluation of the Canadian Global Ice Prediction System using a contingency table on sea ice concentration per sea ice thickness categories, and characterizing performances for the different seasons. With the availability of altimeter Cryosat-2 data together with the SMOS satellite retrievals, which provide reliable estimates for ice thicknesses smaller than 50 cm (Tian-Kunze et al., 2014), other new ice thickness assessments are emerging. Dedicated assessment focuses on first-year ice and multi-year ice categories, mostly due to larger ice melting in the Arctic during recent summers, in order to measure the capability of models to correctly estimate multi-year ice location and drifts (Melsom et al., 2017).

Assessment of biogeochemical parameters and models

As discussed in Hernandez et al. (2015), biogeochemical operational products that provide information about marine ecosystems and marine life state and dynamics are increasingly in demand by a wide range of users and policy makers, particularly in the context of global change (e.g., Gehlen et al., 2015). Hernandez et al. (2015) also points out that until now available observations did not allow for a detailed assessment of biogeochemical and ecosystem models. Moreover, these models, while complex, with many variables, and specifically tailored for the different domains, still require significant improvement (see chapter by Ford et al., 2018). In practice, biogeochemical modelling key parameters, such as nutrients (e.g. nitrogen, phosphorus, silicon and iron), oxygen, and carbon biomass of plankton compartments, are very sparsely measured across the global ocean. Meanwhile, in situ historical datasets, which are not always distributed, cannot provide comprehensive and reliable climatology (e.g., the main physical parameters).

An alternative has been to rely on satellite measurements of ocean colour in order to assess the concentration of chlorophyll that can be linked to the carbon biomass of phytoplankton. Ocean colour reflectance measurements can be combined with sea water optical property parametrisations to provide estimates of chlorophyll concentration. In real-time, ancillary data (e.g., meteorological data) are not available and ocean colour estimates are less accurate (for more discussion, see chapter by Volpe et al., 2018). Furthermore, up until now, in situ data were not available for real-time validation of the satellite reflectance measurements, as can be done in delayed-mode. The CMEMS Ocean Colour Production Centres have developed a method for real-time comparison to a chlorophyll climatology, as presented in Hernandez et al. (2015) and Volpe et al. (2018). This method makes it possible, through consistency assessments, to evaluate if ocean colour estimates suffer from peculiar peaks and outliers. Moreover, a multi-sensor comparison (e.g., between VIIRS and MODIS or between VIIRS and OLCI reflectances) performed between the satellite swath make it possible to evidence biases and apply corrections (Garnesson and Mangin, as part of the CMEMS validation, pers. comm. 2017).

Comparison of model chlorophyll content against satellite ocean colour retrieval is widely used. However, due to the very specific aspect of distribution in time and space of phytoplankton blooms, the classical statistics can be complemented by image oriented assessment. Size and intensity of the observed bloom can be compared with time lag or space shift through windowing techniques, and then contingency metrics can be applied in order to measure if and when the model could totally or partially reproduce the bloom. Imagery techniques are being used for process-oriented metrics in weather forecast evaluation (e.g., Gilleland et al., 2009). These new approaches have been tested in the CMEMS framework (see http://marine.copernicus.eu/services-portfolio/scientific-quality/#novelmetrics|novelmetrics|biogeochemistry). Fig. 29.7 illustrates the daily performance through a contingency table, also called the ROC (Relative Operational Characteristics) index, for a given threshold value of the chlorophyll field of 2 mg/m3 in the North Sea. Performance time series can then be drawn using the Hanssen-Kuipers Discriminant, also known as the Peirce Skill Score (e.g., Gordon, 1982), where the False Alarm Rate (FAR, the incorrect negative divided by the sum of the incorrect negative and the correct negative, IN/[IN+CN]) is subtracted from the Hit Rate (HR, the correct positive divided by the sum of the correct positive and the incorrect positive, CP/[CP+IP]). Both the HR and the FAR range from 0 to 1. If higher than 0.5, the rates are meaningful; thus, when the Hanssen-Kuipers Discriminant is higher than 0.5, it implies that a meaningful hit rate (HR>0.5) is not neutralized by a high and meaningful false alarm rate. Since the value of the threshold is arbitrary, the Hanssen-Kuipers Discriminant needs to be calculated for a range of thresholds between the minimum and maximum chlorophyll values.

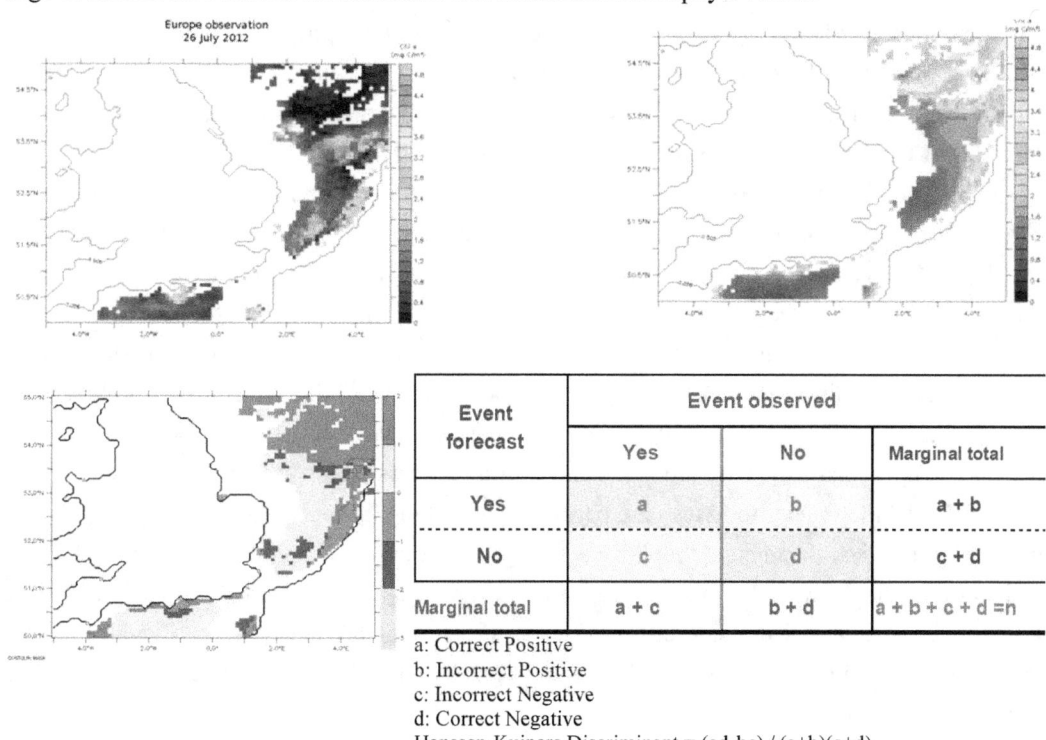

a: Correct Positive
b: Incorrect Positive
c: Incorrect Negative
d: Correct Negative
Hanssen-Kuipers Discriminant = (ad-bc) / (a+b)(c+d)

Figure 29.7. Top left: Chlorophyll concentration (mg/m3) in the North Sea from the 1 km MyOcean satellite ocean colour product merging several sensors the 26/07/2012. Top left: chlorophyll content for the same day from the 7 km FOAM ERSEM (UK-Met Office) operational system. Bottom left: Contingency metrics map, with a chlorophyll threshold of 2 mg/m3, comparing the model and observations. Green (CP): hit, observed and forecast. Yellow (CN): correct non-event, observed and forecast. Blue (IP): false alarm, forecast but not observed. Red (IN): miss, observed but not forecast.

The development of Argo floats equipped with biogeochemical sensors, the so-called BGC-Argo floats, provides access to vertical profiles of downward irradiances, photosynthetically available radiation, turbidity, coloured dissolved organic matter, chlorophyll, dissolved oxygen, and nitrate (NO_3). For more details, see Xing et al. (2011). These data allow us to analyse a model's performance in reproducing key biogeochemical parameters (e.g., oxygen, nitrate, chlorophyll) and the vertical profile shapes that are the result of the interaction between physical and biogeochemical processes. Further, BGC-Argo data can be used to intercalibrate ocean colour variables such as chlorophyll, diffuse attenuation coefficient for a given downwelling irradiance wavelength, and turbidity, as proposed by the CMEMS Ocean Colour experts (http://octac.acri.fr/ and http://seasiderendezvous.fr/matchup.php). Fig. 29.8 shows an observed (i.e., a BGC-Argo float in the Balearic Sea) and forecasted (i.e. CMEMS MED-MFC model) vertical time series of chlorophyll concentration. Interestingly, we can see that the chlorophyll maximum occurs below 50 m most of the time, except during winter when mixed layer depth deepens. Obviously, such a subsurface maximum could not be observed through remote sensing, which means that ocean colour data are not (at least for some periods of the year) able to provide adequate information for assessing key biogeochemical processes. BGC-Argo profiles, once validated, would allow us to design more efficient metrics: the surface chlorophyll and timing of the surface blooms, as well as the total chlorophyll content in the top layers (i.e., photic layer or 0-200m), the deep chlorophyll maximum in the summer, and the depth of the vertically mixed bloom in the winter. Then, since vertical biogeochemical profile shapes are tightly linked to physical vertical processes, these novel metrics might help to identify possible mismatches of physical processes. Fig. 29.9 shows the nitrate comparison for the same model system with another BGC-Argo float in the western Mediterranean Sea. One can see, the shallowing of the nitracline is captured by the BGC-Argo float during vertical mixing winter events, but it stays in the 50-100 m depth layers the rest of the time. Here, metrics are used to compare nitrate concentration at the surface, nitracline depth, and the vertical integrated content, which is a measure of the potential fertilization occurring during winter mixing. Similar comparisons using the global CMEMS system BIOMER and BGC-Argo floats in the Labrador Sea (not shown here) allow measurements of dissolved oxygen, characterizing the winter ventilation, the oxygen uptake, and then the loss due to respiration processes along the water column.

The CMEMS strategy to assess the performance of biogeochemical systems focuses also on monitoring physical parameters that have a strong impact on the coupled biogeochemical model, and with "physical forcing" errors that may strongly alter the biogeochemical results. In particular, the vertical mixing near the surface and the mixed-layer depth changes used to evidence when erroneous chlorophyll concentration forecasts, might be caused by unrealistic surface dynamics behaviour.

Another emerging aspect is the uncertainty assessment of important essential climate variables such as pH and pCO_2. The CMEMS product accuracy can be indirectly estimated by evaluating prognostic model carbonate system variables (e.g., dissolved inorganic carbon and alkalinity) with historical datasets and climatology, or directly by using BGC-Argo floats data planned to measure also the pH.

MEASURING PERFORMANCES, SKILL AND ACCURACY IN OPERATIONAL
OCEANOGRAPHY: NEW CHALLENGES AND APPROACHES • 789

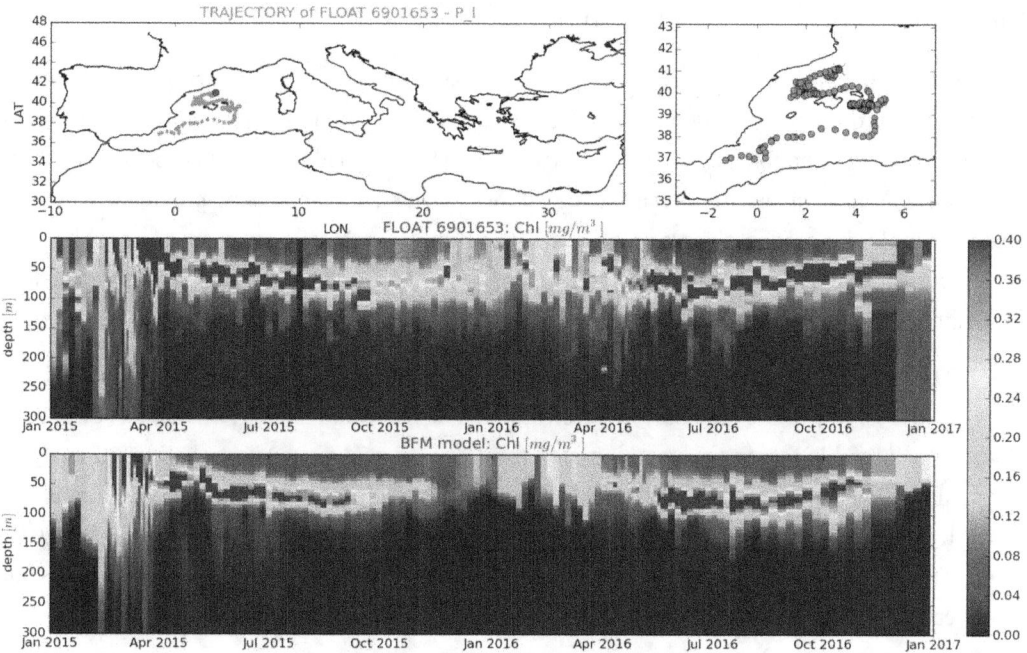

Figure 29.8. Time series of vertical chlorophyll profiles: comparison between the CMEMS Mediterranean biogeochemical system BFM (bottom) and a BGC-Argo float (middle panel), in mg/m3, in the Balearic Sea (float location at the top figures).

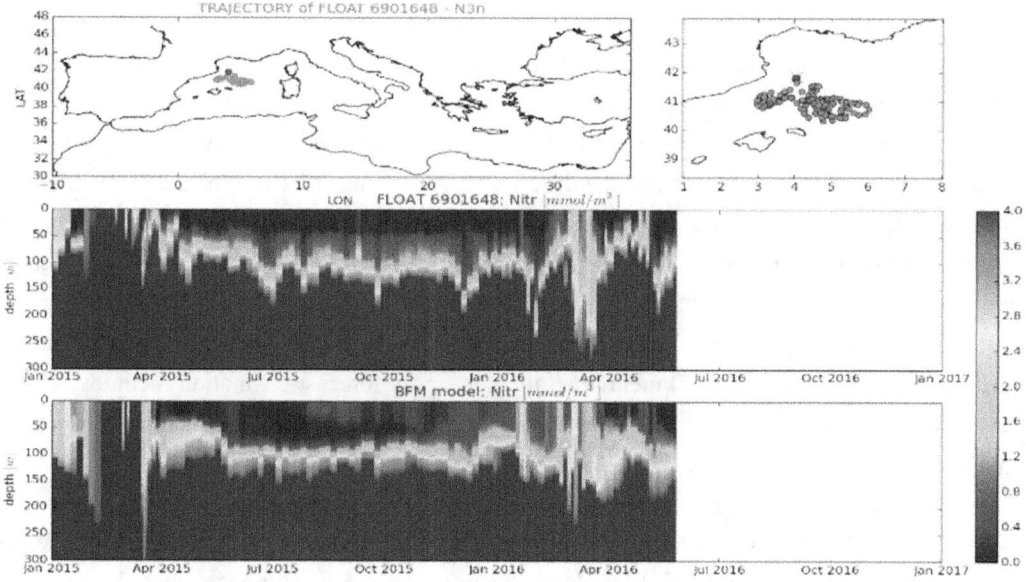

Figure 29.9. Time series of vertical profiles of nitrate: comparison between the CMEMS Mediterranean biogeochemical system BFM (bottom) and a BGC-Argo float (middle panel), in mmol/m3, in the Western Mediterranean Sea (float locations in the top figure).

Multi-model and ensemble assessment

For a given ocean parameter, OOFMS may provide a variety of estimates: observed values, gridded products based on observed values, gridded products from models corrected by assimilation (hindcast, nowcast), and forecasts. The OOFMS can also be designed to perform ensemble forecasts and to provide an estimate of forecast uncertainties through the spread of predicted values. In addition to real-time products, there are also offline products such as ocean reanalyses.

Consequently, the following ensemble approaches provide different ways to evaluate uncertainties and performance of systems:

- The operational systems based on ensemble assimilation allows us, through an ensemble of analyses, to explore the analyses error patterns in space and time (Hernandez et al., 2015).
- The ensemble forecast system, initiated from several analyses or perturbations of analyses, also allows us to explore statistically the forecast errors patterns and their variations over lead time (short- to medium-range) and over different seasons, etc. (Hernandez et al., 2015).
- The collection of estimates (hindcast, nowcast, or forecast) from several models and estimating techniques can be compared in order to characterise their uncertainties.

With the first two approaches, a preliminary work is typically carried out in order to provide users with the best options and the most relevant products from the ensemble, associated with some ways to understand uncertainty level deduced from the spread.

The third approach is becoming more relevant. In practice, in a given area, users are now in front of many ocean products for a requested parameter. Which implies that evaluation strategies need to 1/ characterise the accuracy of each product and estimate how close these products represent ocean "truth" and 2/ how relevant is a given product for a given application. Evaluating as a given ensemble several products allows to carry out the same metrics for all of them, compute an average of this ensemble, and characterise the departure of each product from this average. If product's errors are non-correlated, the spread and the distance to this average offers an objective estimate of the relative quality of each product. If errors are correlated, the mean biases should be estimated separately, ideally with independent reference data. All these products may also offer different spatial and temporal spectral content. So it is important in the evaluation mentioned above to perform metrics that separate or filter out scales that are non-shared by the all products under evaluation. Hernandez et al. (2015) mentioned the consensus forecast estimation methods, where an ensemble average computed in an adapted way (e.g., clustering) offers increased accuracy compared to every individual member.

Operational centres are preparing to share forecasts and collaborate on ensemble assessments at global and regional scales. A good example of this already underway are the Baltic and North West Shelf CMEMS multi-system projects (http://www.boos.org, http://noos.eurogoos.eu/model-results/), which allow every contributing system to make its own evaluations of real-time departures from others and from the average.

Several ongoing real-time ocean monitoring initiatives are also using ensemble evaluations. For instance, on a monthly basis, the NOAA/NCEP ocean monitoring group gathers updates of temperature from different operational global systems and uses spreading and signal-to-noise ratios to perform ocean analysis (Xue et al., 2017). This monitoring is a direct outcome of the Ocean Reanalyses Intercomparison Project and its focus is on thermal content (Balmaseda et al., 2015; Xue et al., 2012). Following this same strategy, the CMEMS is now providing several eddy-permitting ocean reanalyses as well as its suite of global ocean reanalysis multi-model ensemble products (GREP-V1; see http://marine.copernicus.eu/documents/QUID/CMEMS-GLO-QUID-001-026.pdf) average, similar to the SST-gridded products which are averaged into the Global Ocean Sea Surface Temperature Multi Product Ensemble (GMPE, Martin et al., 2012). Fig. 29.10 illustrates the evaluation of each reanalyses product as compared to observed values, averaged monthly. This example shows global salt content and the score of the GREP-V1 ensemble average as compared to the climatology. Clearly the GREP-V1 offers better data than the individual estimates that fall below the ensemble spread. It is noteworthy that in 2002, with addition of Argo data, discrepancies between each reanalyses salt content estimates were reduced. But the significant increase of salinity data (gray shading on right figure) does not appear to improve noticeably after 2008. A similar assessment can be performed for other model variables and can be carried out for derived quantities whenever reference quantities are available.

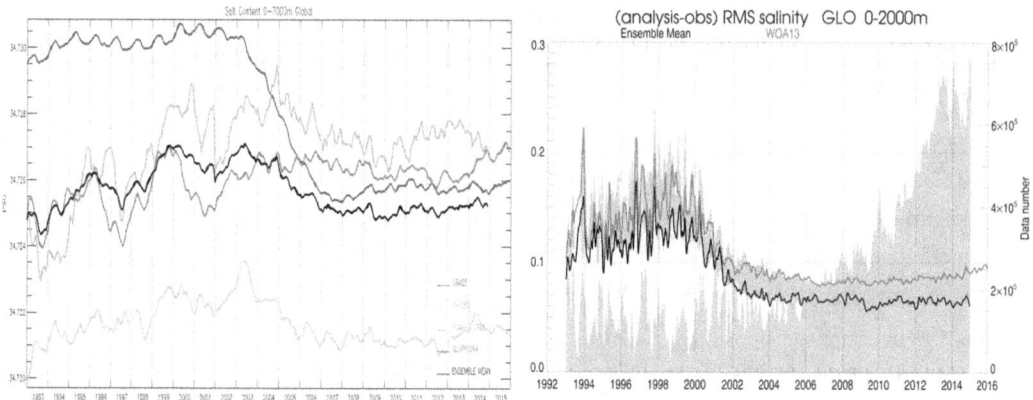

Figure 29.10. Left: Time evolution of the salt content [0-bottom] global average (psu) for the ensemble mean (black), GLORYS2V4 (red), ORAS5 (blue), CGLORS (cyan), FOAM (green) reanalyses. Right: time evolution of the root-mean-square differences with respect to observations from the Coriolis database, computed in the observations' space, between monthly ocean reanalyses estimates and daily observations for salinity in psu: for the ensemble mean GREP-V1 (black) and using the WOA 13 monthly climatology for the 2005-2012 decade (red). The green shading indicates the spread of the RMS of the four members of GREP-V1.

Concluding remarks

In support of wider user interest, particularly in the marine industry and among policy makers, operational oceanography is now able to deliver products on longer timescales and at finer space- and timescales, covering the "blue", the "green" and the "white" ocean. Validation and verification are core functions of operational centres, measuring the system performance, accuracy of ocean products, and the reliability of these products for targeted applications.

A review of validation and verification approaches designed and implemented by operational oceanography was proposed in Hernandez et al. (2015). Here, we update this review and discuss the primary principles and strategies followed in developing these evaluation tools.

Ocean observations are key elements of validation and verification approaches, although the sparseness of in situ data limits the robustness of this method. Another challenge to using observations is timeliness, which limits the capability of real-time validation in operational centres. Furthermore, operational products are now provided at higher resolution (i.e., full eddy-resolving and submesoscale representations at hourly frequencies). Observation representativity is thus an aspect that needs to be taken into account with regards to validation and verification practices. The good news is that the global observing system continue to become more effective: coastal networks of high-frequency radar integrated into coastal observing strategies are emerging; the use of gliders on regular routes is becoming more common; BGC-Argo floats are now equipped with more robust and reliable biogeochemical sensors; and opportunity measurements using sea mammals or other animals are taking place. At the same time, remote sensing capabilities are expanding: constellations are larger, and new sensors are coming online (for the future SWOT or SKIM satellite missions) that should generate new insights into surface circulation. Of course, these new datasets will not impact our ability to evaluate ocean reanalyses for the past. For that, we will need to rely on new techniques, such as multi-model assessments, to help us better understand errors and accuracy of historical ocean estimations.

Operational systems are also becoming more complex, coupling various models along the full range of causalities and mechanisms: atmosphere, wave and ocean dynamics, physical and biogeochemical processes, optical, biological, sea floor sedimentology, chemistry, etc. As such, the performance and robustness of these more complex systems must be able to pay attention to and monitor the efficiency of interfaces. Verification techniques are now systematically taking into account the information shared by the different compartments of an overall full system.

Evaluation of the accuracy of operational ocean products is also evolving, focusing more on user requests and areas of specific interest. Here, we introduced the idea of "external" assessment and metrics in contrast to the more classical "internal" test-bed academic evaluation of an ocean model's performance. This new "external" or "user-oriented" metrics strategy need to be complemented by new and expanded ways of communicating the reliability of products to users given new technological options (e.g., smart phones) and the fact that the user community is more diverse, less academic, and more oriented toward societal decision-making.

The validation and verification activities in operational oceanography are maturing, first in terms of more structuration. Initiatives such as the CMEMS or GOV groups, which are associated with standardization mechanisms and the endorsement of best practices by the Joint Technical Commission for Oceanography and Marine Meteorology, demonstrate that operational centres are organizing and developing networks in order to leverage data and expertise, and to associate their efforts through multi-model assessment. These groups allow for the sharing of experiences with user interactions and expectations. They also function as a bridge between the oceanographic and other scientific communities, such as the atmospheric and weather prediction verification groups, often resulting in the adoption if innovative approaches and metrics that are more user- or process-oriented.

Finally, this chapter provides an overview of the validation and verification of operational system performance evaluation. As discussed, it is important to keep in mind that validation techniques and metrics must be continually revisited in order to become more robust in characterizing products accuracy and reliability.

Acknowledgements

The authors want to thank the GODAE OceanView Intercomparison and Validation Task Team members as well as the CMEMS Product Quality Working Group experts for the valuable exchanges and discussions on metrics development and validation/verification approaches over the years. Work on this review is partly funded by Mercator Océan, as part of the CMEMS activities.

References

Balmaseda, M. A., and Coauthors, 2015: The Ocean Reanalyses Intercomparison Project (ORA-IP). Journal of Operational Oceanography, 8:sup1, s80-s97. doi:10.1080/1755876X.2015.1022329

Bell, M. J., A. Schiller, P.-Y. Le Traon, N. R. Smith, E. Dombrowsky, and K. Wilmer-Becker, 2015: An introduction to GODAE OceanView. Journal of Operational Oceanography, 8, s2-s11, doi: 10.1080/1755876X.2015.1022041.

Blockley, E. W., and Coauthors, 2014: Recent development of the Met Office operational ocean forecasting system: an overview and assessment of the new Global FOAM forecasts. Geosci. Model Dev., 7, 2613-2638, doi:10.5194/gmd-7-2613-2014.

Bouillon, S., P. Rampal, and E. Olason, 2018: Sea ice modelling and forecasting. GODAE Oceanview International School in "New Frontiers in Operational Oceanography", E. P. Chassignet, A. Pascual, J. Tintore, and J. Verron, Eds.

Brassington, G. B., and Coauthors, 2015: Progress and challenges in short- to medium-range coupled prediction. Journal of Operational Oceanography, 8, s239-s258. doi:10.1080/1755876X.2015.1049875.

Casati, B., and Coauthors, 2008: Forecast verification: current status and future directions. Meteorological Applications, 15, 3-18. doi:10.1002/met.52.

Coppini, G., 2017: New operational tools at sea: www.sea-conditions.com. Personal Communication.

De Mey, P., E. Stanev, and V. H. Kourafalou, 2017: Science in support of coastal ocean forecasting—part 1. Ocean Dynamics, 67, 665-668, doi:10.1007/s10236-017-1048-1

Divakaran, P., and Coauthors, 2015: GODAE OceanView Class 4 inter-comparison for the Australian Region. Journal of Operational Oceanography, 8:sup1, s112-s126, doi:10.1080/1755876X.2015.1022333

Drévillon, M., and Coauthors, 2013: A strategy for producing refined currents in the Equatorial Atlantic in the context of the search of the AF447 wreckage. Ocean Dynamics, 63, 63-82, doi:10.1007/s10236-012-0580-2

Drévillon, M., and Coauthors, 2018: Learning about Copernicus Marine Environment Monitoring Service "CMEMS": a practical introduction to the use of the European operational oceanography service. GODAE Oceanview International School in "New Frontiers in Operational Oceanography", E. P. Chassignet, A. Pascual, J. Tintore, and J. Verron, Eds.

Dufau, C., M. Orsztynowicz, G. Dibarboure, R. Morrow, and P.-Y. Le Traon, 2016: Mesoscale resolution capability of altimetry: Present and future. Journal of Geophysical Research: Oceans, 121, 4910-4927, doi: 10.1002/2015JC010904.

Ebert, E., and Coauthors, 2013: Progress and challenges in forecast verification. Meteorological Applications, 20, 130-139, doi:10.1002/met.1392.

Ford, D., S. Kay, R. McEwan, I. Totterdell, and M. Gehlen, 2018: Marine biogeochemical modelling and data assimilation for operational forecasting, reanalysis and climate research. GODAE Oceanview International School in "New Frontiers in Operational Oceanography", E. P. Chassignet, A. Pascual, J. Tintore, and J. Verron, Eds.

Gehlen, M., and Coauthors, 2015: Building the capacity for forecasting marine biogeochemistry and ecosystems: recent advances and future developments. Journal of Operational Oceanography, 8, s168-s187, doi:10.1080/1755876X.2015.1022350.

Gilleland, E., D. Ahijevych, B. G. Brown, B. Casati, and E. E. Ebert, 2009: Intercomparison of Spatial Forecast Verification Methods. Weather Forecast., 24, 1416-1430, doi:10.1175/2009waf2222269.1.

Gordon, N. D., 1982: Evaluating the Skill of Categorical Forecasts. Monthly Weather Review, 110, 657-661, doi:10.1175/1520-0493(1982)110<0657:etsocf>2.0.co;2.

Griffin, D. A., and P. R. Oke, 2017: The search for MH370 and ocean surface drift – Part III. (Report number EP174155), ed. by CSIRO Oceans and Atmosphere, Australia. [Available at https://www.atsb.gov.au/media/5773371/mh370_csiro-ocean-drift-iiil.pdf.]

Griffin, D. A., P. R. Oke, and E. Jones, M., 2016: The search for MH370 and ocean surface drift – Part II. (Report number EP167888), ed. by CSIRO Oceans and Atmosphere, Australia. [Available at https://www.atsb.gov.au/media/5772119/mh370_ocean_driftv29.pdf.]

Haines, K., 2018: Ocean Reanalysis. GODAE Oceanview International School in "New Frontiers in Operational Oceanography", E. P. Chassignet, A. Pascual, J. Tintore, and J. Verron, Eds.

Harris, C. M., 2018: Coupled atmosphere-ocean modelling. GODAE Oceanview International School in "New Frontiers in Operational Oceanography", E. P. Chassignet, A. Pascual, J. Tintore, and J. Verron, Eds.

Hernandez, F., and A. Melet, 2016: Product Quality Strategic Plan. in CMEMS, (CMEMS-PQ-StrategicPlan), ed. by Mercator Océan, Toulouse.

Hernandez, F., and Coauthors, 2009: Validation and intercomparison studies within GODAE. Oceanography Magazine, 22, 128-143, doi:10.5670/oceanog.2009.71.

Hernandez, F., and Coauthors, 2015: Recent progress in performance evaluations and near real-time assessment of operational ocean products. Journal of Operational Oceanography, 8, s221-s238, doi:10.1080/1755876X.2015.1050282.

Jacobs, G. A., and Coauthors, 2018: Operational Ocean Data Assimilation. GODAE Oceanview International School in "New Frontiers in Operational Oceanography", E. P. Chassignet, A. Pascual, J. Tintore, and J. Verron, Eds.

Kourafalou, V. H., and Coauthors, 2015: Coastal Ocean Forecasting: system integration and evaluation. Journal of Operational Oceanography, 8, s127-s146, doi:10.1080/1755876X.2015.1022336.

Le Sommer, J., E. Chassignet, and A. Wallcraft, 2018: Ocean circulation modelling for operational oceanography: current status and future challenges. GODAE Oceanview International School in "New Frontiers in Operational Oceanography", E. P. Chassignet, A. Pascual, J. Tintore, and J. Verron, Eds.

Le Traon, P.-Y., 2013: From satellite altimetry to Argo and operational oceanography: three revolutions in oceanography. Ocean Sci., 9, 901-915, doi:10.5194/os-9-901-2013.

Le Traon, P.-Y., and Coauthors, 2017: The Copernicus Marine Environmental Monitoring Service: Main Scientific Achievements and Future Prospects. Mercator Ocean Journal, 56, 100.

Lellouche, Jean-Michel, and E. Greiner, 2018: The Mercator Ocean Global High Resolution Monitoring and Forecasting System. GODAE Oceanview International School in "New Frontiers in Operational Oceanography", E. P. Chassignet, A. Pascual, J. Tintore, and J. Verron, Eds.

Lellouche, J.-M., and Coauthors, 2013: Evaluation of global monitoring and forecasting systems at Mercator Océan. Ocean Sci., 9, 57-81, doi:10.5194/os-9-57-2013.

Maraldi, C., and Coauthors, 2013: NEMO on the shelf: assessment of the Iberia-Biscay-Ireland configuration. Ocean Sci., 9, 745-771, doi:10.5194/os-9-745-2013

Martin, M., and Coauthors, 2012: Group for High Resolution Sea Surface temperature (GHRSST) analysis fields inter-comparisons. Part 1: A GHRSST multi-product ensemble (GMPE). Deep Sea Research Part II: Topical Studies in Oceanography, 77-80, 21-30, doi:10.1016/j.dsr2.2012.04.013.

Melsom, A., S. Eastwood, J. Xie, S. Aaboe, and L. Bertino, 2017: Challenges in validating model results for first year ice. EGU 2017 General Assembly, EGU.

Morrow, R. A., D. Blumstein, and G. Dibarboure, 2018: Fine-scale altimetry and the future SWOT mission. GODAE Oceanview International School in "New Frontiers in Operational Oceanography", E. P. Chassignet, A. Pascual, J. Tintore, and J. Verron, Eds.

Mourre, B., and Coauthors, 2018: Assessment of high-resolution regional ocean prediction systems using multi-platform observations: illustrations in the Western Mediterranean Sea. GODAE Oceanview International School in "New Frontiers in Operational Oceanography", E. P. Chassignet, A. Pascual, J. Tintore, and J. Verron, Eds.

Murphy, A. H., 1993: What is a Good Forecast - An essay on the nature of goodness in weather forecasting. Weather Forecast, 8, 281-293, doi:10.1175/1520-0434(1993)008<0281:wiagfa>2.0.co;2.

Oke, P. R., G. B. Brassington, J. A. Cummings, M. J. Martin, and F. Hernandez, 2012: GODAE inter-comparisons in the Tasman and Coral Seas. Journal of Operational Oceanography, 5, 11-24.

Oke, P. R., and Coauthors, 2015a: Assessing the impact of observations on ocean forecasts and reanalyses: Part 1, Global studies. Journal of Operational Oceanography, 8, s49-s62, doi:10.1080/1755876X.2015.1022067.

Oke, P. R., and Coauthors, 2015b: Assessing the impact of observations on ocean forecasts and reanalyses: Part 2, Regional applications. Journal of Operational Oceanography, 8, s63-s79, doi: 10.1080/1755876X.2015.1022080.

Rogé, M., R. Morrow, C. Ubelmann, and G. Dibarboure, 2017: Using a dynamical advection to reconstruct a part of the SSH evolution in the context of SWOT, application to the Mediterranean Sea. Ocean Dynamics, 67, 1047-1066, doi:10.1007/s10236-017-1073-0.

Roughan, M., C. Kerry, and P. McComb, 2018: Shelf and Coastal Ocean Observing and Modelling Systems – A New Frontier for Operational Oceanography. GODAE Oceanview International School in "New Frontiers in Operational Oceanography", E. P. Chassignet, A. Pascual, J. Tintore, and J. Verron, Eds.

Ryan, A. G., and Coauthors, 2015: GODAE OceanView Class 4 forecast verification framework: Global ocean inter-comparison. Journal of Operational Oceanography, 8:sup1, s98-s111, doi: 10.1080/1755876X.2015.1022330

Schiller, A., F. Davidson, P. M. DiGiacomo, and K. Wilmer-Becker, 2016: Better Informed Marine Operations and Management: Multidisciplinary Efforts in Ocean Forecasting Research for Socioeconomic Benefit. Bul. Amer. Met. Soc., 97, 1553-1559, doi:10.1175/bams-d-15-00102.1.

Schiller, A., and Coauthors, 2015: Synthesis of new scientific challenges for GODAE OceanView. Journal of Operational Oceanography, 8, s259-s271, doi:10.1080/1755876X.2015.1049901.

Smith, G. C., and Coauthors, 2016: Sea ice forecast verification in the Canadian Global Ice Ocean Prediction System. Quarterly Journal of the Royal Meteorological Society, 142, 659-671. doi: 10.1002/qj.2555.

Taylor, K. E., 2001: Summarizing multiple aspects of model performance in a single diagram. Journal of Geophysical Research: Atmospheres, 106, 7183-7192, doi:10.1029/2000JD900719.

Tian-Kunze, X., L. Kaleschke, N. Maaß, M. Mäkynen, N. Serra, M. Drusch, and T. Krumpen, 2014: SMOS-derived thin sea ice thickness: algorithm baseline, product specifications and initial verification. The Cryosphere, 8, 997-1018, doi:10.5194/tc-8-997-2014.

Tonani, M., and Coauthors, 2015: Status and future of global and regional ocean prediction systems. Journal of Operational Oceanography, 8, s201-s220, doi:10.1080/1755876X.2015.1049892

Ubelmann, C., P. Klein, and L.-L. Fu, 2015: Dynamic Interpolation of Sea Surface Height and Potential Applications for Future High-Resolution Altimetry Mapping. Journal of Atmospheric & Oceanic Technology, 32, 177, doi:10.1175/JTECH-D-14-00152.1.

Volpe, G., B. Buongiorno Nardelli, S. Colella, and R. Santoleri, 2018: An Operational Interpolated Ocean Colour Product in the Mediterranean Sea. GODAE Oceanview International School in "New Frontiers in Operational Oceanography", E. P. Chassignet, A. Pascual, J. Tintore, and J. Verron, Eds.

von Schuckmann, K., and Coauthors, 2016: The Copernicus Marine Environment Monitoring Service Ocean State Report. Journal of Operational Oceanography, 9, s235-s320, doi: 10.1080/1755876X.2016.1273446.

Wilkin, J. L., J. Levin, A. Lopez, H. Arango, E. J. Hunter, and J. Zavala-Garay, 2018: A coastal ocean forecast system for U.S. Mid-Atlantic Bight and Gulf of Maine. GODAE Oceanview International School in "New Frontiers in Operational Oceanography", E. P. Chassignet, A. Pascual, J. Tintore, and J. Verron, Eds.

Xing, X., A. Morel, H. Claustre, D. Antoine, F. D'Ortenzio, A. Poteau, and A. Mignot, 2011: Combined processing and mutual interpretation of radiometry and fluorimetry from autonomous profiling Bio-Argo floats: Chlorophyll a retrieval. Journal of Geophysical Research: Oceans, 116, doi: 10.1029/2010JC006899.

Xu, Y., and L.-L. Fu, 2012: The Effects of Altimeter Instrument Noise on the Estimation of the Wavenumber Spectrum of Sea Surface Height. J. Phys. Oceanogr., 42, 2229, doi:10.1175/JPO-D-12-0106.1.

Xue, Y., and Coauthors, 2012: A Comparative Analysis of Upper-Ocean Heat Content Variability from an Ensemble of Operational Ocean Reanalyses. J. Climate, 25, 6905-6929. doi:10.1175/jcli-d-11-00542.1.

Xue, Y., and Coauthors, 2017: A real-time ocean reanalyses intercomparison project in the context of tropical pacific observing system and ENSO monitoring. Climate Dynamics, 49, 3647-3672, doi:10.1007/s00382-017-3535-y.

Zhu, X., and Coauthors, 2016: Comparison and validation of global and regional ocean forecasting systems for the South China Sea. Nat. Hazards Earth Syst. Sci., 16, 1639-1655. doi: 10.5194/nhess-16-1639-2016.

Zuo, H., M. Balmaseda, and K. Mogensen, 2015: The new eddy-permitting ORAP5 ocean reanalysis: description, evaluation and uncertainties in climate signals. Climate Dynamics, 1-21, doi:10.1007/s00382-015-2675-1.

INDEX

A

Advection 46, 47, 49, 54, 57, 62, 64, 66, 110, 204, 218, 220, 221, 265, 277, 358, 431, 433, 434, 565, 570, 598, 634
Air-sea fluxes 79, 95, 131, 134, 137-139, 141, 150, 154, 220, 245, 246, 248, 249, 263, 266, 294, 393, 396, 417, 545, 600, 628, 634, 681
Air-sea interaction 79, 93, 95, 172, 271, 272, 679
Altimetry 4, 5, 11, 12, 117, 121, 126, 127, 161, 162, 164, 166, 168-172, 182-186, 191-193, 195, 196, 200, 201, 203-205, 207, 211, 212, 214, 218, 221, 222, 279, 307, 308, 312, 313, 323, 337-339, 341, 353-356, 360, 366, 367, 369, 371, 395, 412, 414, 456, 546, 548, 603, 608, 610, 633, 635, 636, 667, 668, 673-679, 683-687, 690, 707, 708, 766, 773, 782, 783

B

Baroclinic instability 44, 53, 56, 59, 61, 671
Biogeochemical modelling 625, 626, 631, 642, 776
Bluelink 9, 10, 92, 100, 101, 486, 529
Boundary conditions 11, 12, 28, 38, 41, 42, 49, 58, 101, 107, 183, 293, 374, 381, 408, 445, 458, 459, 516, 535, 598, 601-603, 612, 619, 639, 640, 664, 667, 674, 731, 769, 776
Bulk formulae 41, 260, 261, 300, 323, 454, 565, 600, 601, 603, 672

C

Climatology 27, 100, 109, 110, 165, 170, 174, 221, 230-238, 240-243, 282, 449, 451, 452, 458, 486, 550, 559, 565, 566, 571-573, 584, 585, 601, 617, 619, 638, 642, 643, 667, 748, 773, 774, 786, 788, 791
CMEMS 12-14, 128, 183, 184, 227-234, 236-238, 240, 241, 243, 298, 417, 547, 561, 563, 564, 570, 574, 578, 586, 590, 591, 596, 601, 606, 612, 627, 639, 640, 642, 665, 673, 674, 680, 691, 695-709, 729, 732, 760, 763, 766, 768-777, 783-793
Convection 36, 59, 118, 258, 265, 271, 435, 656, 671, 680, 730, 731
Coordinates 38, 209, 321, 344, 401, 449, 494, 596, 598, 612

D

Diagnostics 445, 577, 578, 580, 583, 669, 676, 684, 690, 703
Diffusion 34-36, 38, 39, 42, 47, 54, 256, 297, 540, 565, 598, 634
Drift 6, 16, 18, 49, 56-58, 62, 63, 65, 66, 180, 183, 217, 254, 282, 344, 394, 396, 425, 426, 429, 430, 433, 438, 439, 442, 451, 452, 490, 516, 531, 565, 571, 580, 588, 589, 601, 616, 664, 684, 783

E

ECCO 279, 283, 472
ECMWF 323, 396, 404, 452-454, 456, 470, 494, 545, 707
Ecosystem models 175, 176, 179, 230, 641, 645, 722, 786

F

Flux 11, 37-40, 54, 61, 76, 79, 134, 139, 140, 144, 148, 149, 155, 246, 250-261, 265, 271, 272, 277, 279, 294, 396, 399, 405, 407, 447, 516, 531, 557, 559, 560, 601, 628, 630, 635, 637, 642, 643, 742, 743
FOAM 454, 455, 470, 522, 529, 639, 767, 787, 791

H

HYCOM 9, 100, 206, 216, 308, 316, 317, 319-322, 328-331, 334-342, 344-350, 353-358, 361-371, 453, 547, 561, 596, 602, 603

J

Jellyfish 626, 638, 737-746, 749, 750, 752, 753, 755

K

Kalman filter 8, 11, 18, 100, 307, 321, 365, 467, 477, 479, 480, 483, 485, 489, 491, 494, 514, 566, 632

M

MERSEA 578, 695, 701
Mesoscale eddies 27, 53-56, 60, 66, 119, 127, 212, 213, 294, 299, 310-315, 319, 321, 323, 330, 337, 341, 343, 344, 354, 362, 367, 370, 513, 516, 519, 521, 522, 533, 540, 541, 664, 667
Mesoscale processes 119, 529, 597, 644
Meteorology 2, 79, 394, 458, 610, 613, 667, 745, 752

N

Nesting 19, 365, 458, 598, 600, 763, 764, 776
Numerical models 59, 119, 125, 128, 149, 292, 293, 308, 398, 408, 465, 466, 513, 515, 534
NWP 80, 184, 452-454, 640, 730

O

Oil spill 10, 21, 438, 458, 516, 536, 730, 734

P

Parameterization 46, 50, 57, 59, 60, 62, 66, 107, 229, 253, 299, 300, 404, 435, 447, 448, 499, 516, 565, 597, 750
Precipitation 27, 30, 32, 38, 39, 82, 256, 277, 294, 322, 456, 601-603, 654, 658, 660

R

Radiation 11, 29, 31, 39, 47, 63, 82, 149, 165-168, 174-177, 179, 245, 247, 250, 256-258, 261, 264, 405, 424, 448, 513, 516, 546, 559, 600, 601, 631, 637, 638, 679, 681, 730, 776, 788
Reanalyses 1, 12, 16, 19, 170, 396, 451, 456, 465, 545-548, 554, 556-558, 560, 561, 579, 589, 606, 625, 626, 631, 638, 642, 644, 696, 701-705, 707-709, 759, 761, 763, 764, 766, 768, 769, 777, 790-792
ROMS 10, 92, 101, 452, 458, 472, 593, 594, 596-601, 603, 605, 608-610, 612, 613, 615, 617-620, 672, 729, 731, 748
Runoff 27, 141, 176, 277, 513, 564, 566, 626

S

Salt 8, 37, 38, 40, 41, 93, 117, 396, 424, 602, 638, 791
Satellite oceanography 161, 162, 165, 183, 274
Search and rescue 10, 12, 15, 16, 21, 183, 438, 516, 536, 541, 616, 664, 685
SEEK filter 8, 548
SSH 4, 102-106, 168-170, 175, 183, 191, 193, 194, 200-206, 209, 210, 212-222, 228, 229, 243, 312-314, 316, 320, 334, 335, 337, 341, 349, 350, 354-356, 359-364, 367, 514, 515, 521, 523, 530-541, 565, 566, 569, 572, 583, 588, 596, 598, 608, 610-612, 614, 784
SST 2, 4, 6, 10, 11, 13, 15, 80, 94, 102-104, 106, 109, 119, 120, 122, 126, 163, 164, 172-175, 181, 184, 185, 220, 222, 228, 229, 243, 253, 261-266, 271-275, 278-284, 451, 453, 454, 514, 546, 559, 570, 571, 575, 576, 579-582, 584, 589, 596, 600, 603, 605, 608, 610-612, 614, 632, 634, 637, 638, 641, 642, 653, 655-659, 679-681, 685, 690, 698, 766-768, 774, 776, 781
Submesoscale processes 57, 100, 117, 118

T

Topography 12, 57, 59, 111, 120, 128, 169, 170, 184, 191, 192, 195, 206, 212, 309, 310, 312, 314-316, 323, 354, 368, 369, 472, 538, 539, 564, 570, 594, 601, 609, 672-674, 730, 783
Tracers 38, 45, 49, 54, 62, 79, 117, 124, 139, 148, 150, 151, 156, 175, 206, 213, 221, 222, 317, 565, 596, 597, 628, 634-636
Transport processes 144, 213, 407, 410, 603

U

U.S. Navy 100, 191, 312, 322, 365, 367, 371, 476, 561
Users 4, 5, 9, 12, 15, 18, 20, 22, 76, 79, 80, 86, 111-113, 132, 170, 178, 179, 199, 203, 230, 231, 237, 246, 292, 302, 407, 437, 456, 458, 459, 564, 577, 578, 590, 593, 597, 615, 616, 618-620, 637, 644, 645, 663, 664, 697-701, 704, 707-709, 730, 734, 745, 759, 760, 764-766, 770-775, 777, 784, 786, 790, 792

W

Wind stress 11, 248, 250, 260, 262, 263, 294, 322, 396, 425, 446, 530, 776

www.ingramcontent.com/pod-product-compliance
Lightning Source LLC
Chambersburg PA
CBHW062346220526
45472CB00008B/1717